S0-ATI-199

Perspectives on

Human Biology

Perspectives on

Human Biology

Stuart I. Fox
Pierce College

Wm. C. Brown Publishers

Book Team

Editor *Kevin Kane*
Developmental Editor *Carol Mills*
Production Editor *Michelle M. Campbell*
Designer *David C. Lansdon*
Art Editor *Jess Schaal*
Photo Editor *Carrie Burger*
Permissions Editor *Karen L. Storlie*
Visuals Processor *Andrêa Lopez-Meyer*

Wm. C. Brown Publishers

President *G. Franklin Lewis*
Vice President, Publisher *George Wm. Bergquist*
Vice President, Publisher *Thomas E. Doran*
Vice President, Operations and Production *Beverly Kolz*
National Sales Manager *Virginia S. Moffat*
Advertising Manager *Ann M. Knepper*
Marketing Manager *Craig S. Marty*
Editor in Chief *Edward G. Jaffe*
Managing Editor, Production *Colleen A. Yonda*
Production Editorial Manager *Julie A. Kennedy*
Production Editorial Manager *Ann Fuerste*
Publishing Services Manager *Karen J. Slaght*
Manager of Visuals and Design *Faye M. Schilling*

Cover photo © Russ Kinne/Comstock

The credits section for this book begins on page 513, and is considered an extension of the copyright page.

Library of Congress Catalog Card Number: 90-80895

ISBN 0-697-10785-X

Printed in the United States of America by Wm. C. Brown Publishers, 2460 Kerper Boulevard, Dubuque, IA 52001

10 9 8 7 6 5 4 3 2 1

To my daughter Laura and others of her generation, in whose care spaceship earth will be entrusted at the dawn of the second millennium

Brief Table of Contents

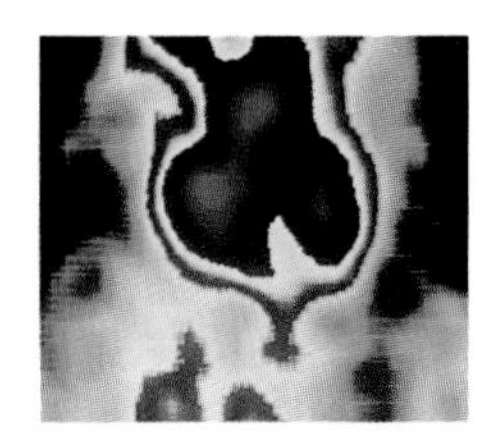 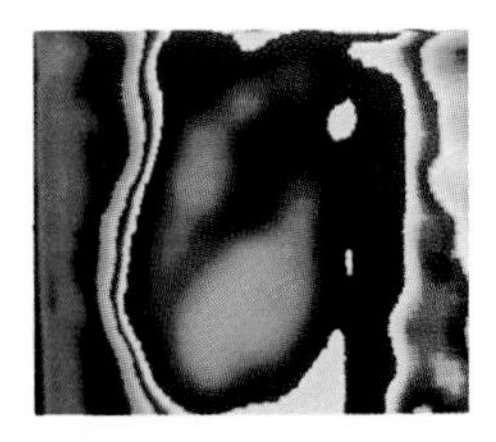 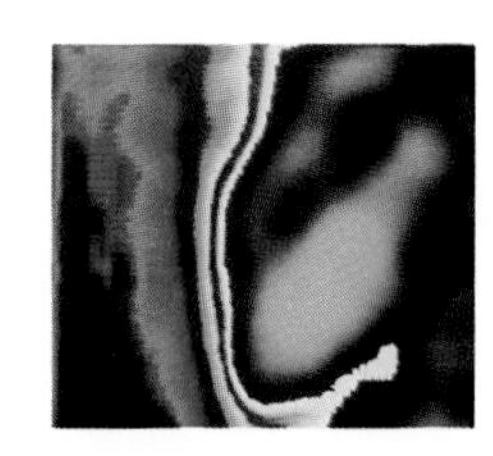

Expanded Table of Contents

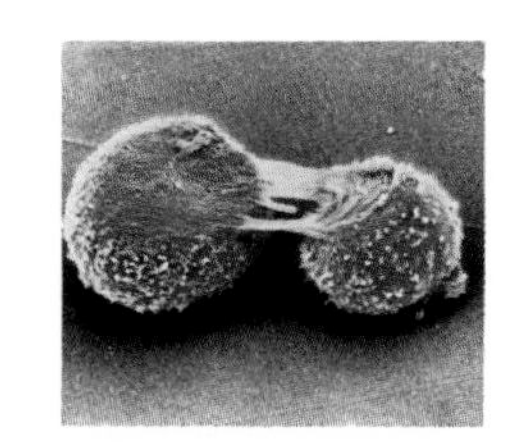

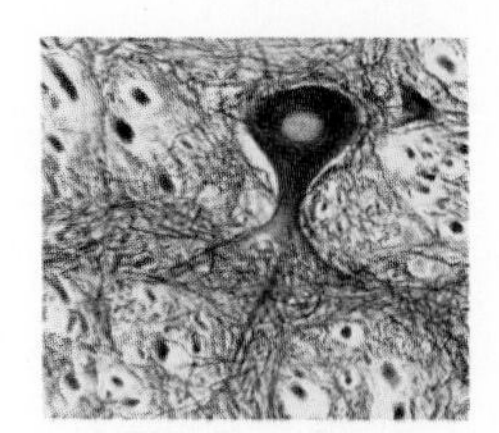

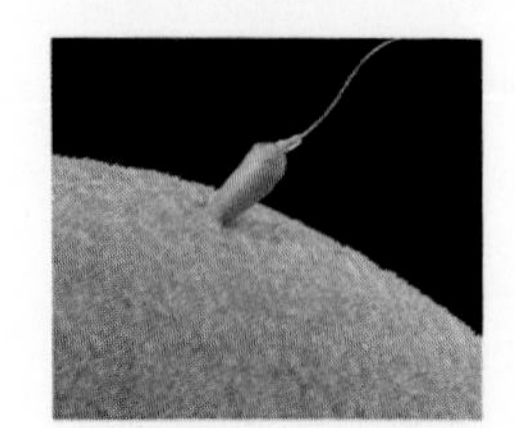

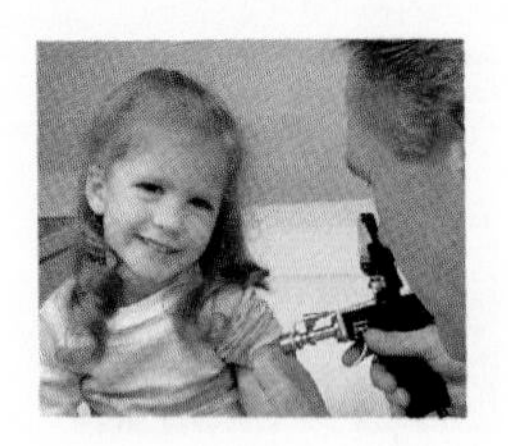
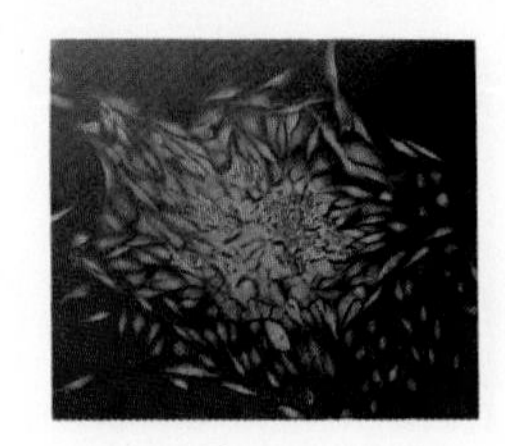
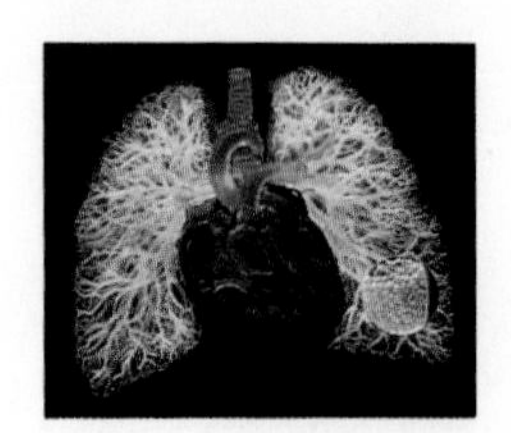

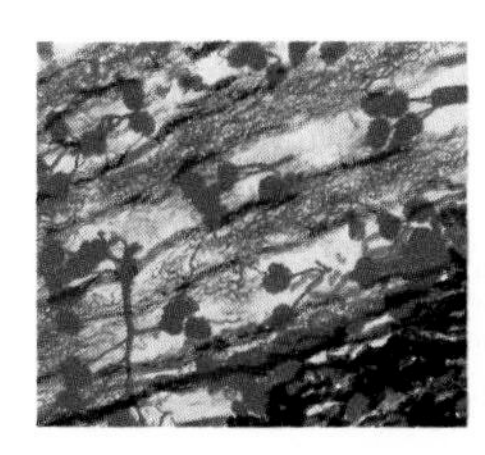
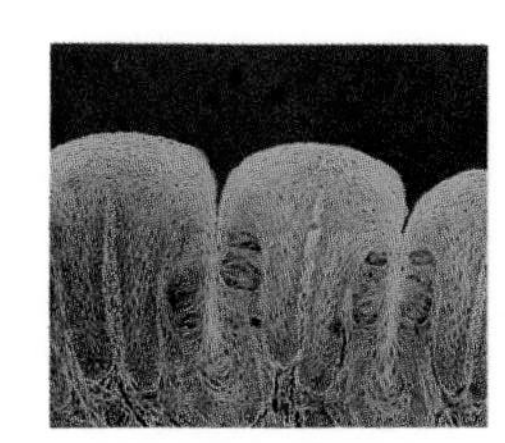

16

The Urinary System 355

17

The Digestive System 373

18

The Metabolism of the Body 399

19

Nutrition 421

20

Human Perception of the Environment: The Sensory System 437

21

Human Movement through the Environment: The Musculoskeletal System 455

22

Human Interaction with the Environment: Ecology 477

Preface

Human biology is an extraordinarily interesting subject to study and a challenging one to teach. Its interest to the student is derived from its root in personal concerns: "This is *my* body, *my* sex life, and *my* environment." The challenge of teaching human biology derives from its seemingly all-encompassing nature; instructors must choose which topics to include and emphasize, and which to exclude or de-emphasize. This unusual flexibility provides a unique opportunity for instructors to tailor their human biology courses to the specific desires and needs of their students.

Most human biology courses share certain common features. Basic concepts in biology and other sciences are introduced and explained succinctly so that students can advance to the more interesting applications of these concepts. Some knowledge of systematics, evolutionary mechanisms, and ecology is needed, for example, to understand man's place in the natural world. Similarly, a basic understanding of cells, tissues, organs, systems, and the principle of homeostasis is required before biomedical advances can be appreciated. Ideally, all of these basic concepts are explained clearly but relatively quickly; they serve as a necessary preamble to the topics of major interest to the students.

This text presents all of the background information needed to understand the subjects most often emphasized in a human biology course, and presents these subjects in the most current and interesting fashion possible. Human reproduction, development, and aging, for example, are emphasized by two separate chapters following a chapter on the endocrine system, which provides the basic background information. Genetic engineering is a common thread throughout the text, and is directly addressed following a thorough and modern introduction to gene action. Basic concepts and exciting discoveries about the immune system, as another example, precede a separate and current chapter on cancer. An entire chapter on human nutrition logically follows separate chapters on metabolism and the digestive system.

The ability of students to use this text and understand the concepts presented is aided by the logical organization of topics and by extensive use of beautifully rendered, full-color figures. In addition, there are numerous pedagogical devices that can help students to better understand the fascinating subject of human biology.

Student Aids

The following information about the organization and pedagogical devices in the text will help you to derive maximum benefit from this book.

Chapter Openers

Each chapter begins with three aids to learning: (1) an *Outline,* which lists the headings within the chapter and their page numbers for easy reference; (2) a list of *Objectives,* which tells you what you can expect to learn from the chapter; and (3) *Keys to Pronunciation,* which helps you to pronounce many new words in the chapter. These materials should be quickly read for familiarity before beginning the chapter, and then should be used for reference as you proceed through the chapter.

Perspectives

These are the paragraphs at the beginning of each major heading that are set off in different type and color from the main body of the text. They are summaries of the major concepts to be presented in that section, and provide a bird's-eye view of that section. Read these carefully, because they will help you to identify the organizing concepts of the section and prevent you from becoming distracted by the details. The details provided later breathe life into these concepts, but should not be allowed to obscure the major themes covered in the section.

Boxed Information

Following a discussion of a basic concept in the text, you may find a colored box of text. These contain short discussions of clinical or practical applications of the information preceding the boxes. You will find it enjoyable, as well as instructive, to see how your newly acquired basic knowledge can be applied to practical problems.

Social Issues

Within most of the chapters of this book are larger boxed essays labeled *Social Issues.* These are devoted to current ethical and political concerns raised by a topic in human biology covered in that chapter. These issues are hotly debated by many segments of society, and should be familiar to all educated citizens. The points of view expressed are those of the author, and are written in such a way as to stimulate debate in the classroom. If you do not agree with some aspect of the essay, speak up! These important issues can be resolved only when each person voices their opinion and is honestly open to the opinion of others.

Footnotes

The derivations of many of the new words introduced in the chapter are provided in footnotes. These can help you to understand why a particular word is used, and this understanding makes it easier for you to remember the word.

Study Activities

Each major heading in the chapters ends with a list of study activities: pictures and flowcharts to draw, essays to answer, and other activities. The purpose of these activities is to help you to interact with the information presented, and provide a "reality check" to see if you really did understand the information. These activities will be more useful to you if you actually write them out, rather than just think about them.

Chapter Summaries

At the end of each chapter, the material is summarized for you in outline form. This summary is organized by major headings followed by the major points of information. Read the summary after studying the chapter to be sure that you have not missed any points, and use the summaries to help you review for examinations.

Review Activities

The Review Activities follow each chapter summary, and include objective and essay questions. The answers to the objective questions are provided in the Appendix at the back of the book. The first essay question in each chapter is answered in the Student Study Guide. Be sure to take these self-quizzes in a "closed-book" fashion before looking up the answers.

Appendix

The Appendix contains the answers to the objective questions in the Review Activities at the end of each chapter.

Glossary

The Glossary provides definitions of the more important terms used in the text. Whenever you encounter an unfamiliar term or would like additional information about a term, look it up in the Glossary.

Supplementary Materials

Student Study Guide

Written by Dr. Lawrence G. Thouin, Jr., this is an optional book that can help you to derive more benefit from the text. The answer to the first question in the Review Activities at the end of each chapter is provided here, together with helpful hints about how to answer essay questions on human biology. The study guide also provides additional objective questions (with answers), fill-in-the-blank questions, crossword puzzles, and other learning devices.

Instructor's Manual-Test Item File

The Instructor's Manual-Test Item File was written by the author to assist instructors in preparing for their classes. Each chapter includes a chapter outline and objectives, a list of suggested discussion topics, objective questions with answers, essay questions with answers, and a list of suggested films relating to the chapter. Addresses of film suppliers are provided in appendix I, and a list of transparencies that accompany the text is provided in appendix II. The Test Item File contains additional objective questions with answers for each textbook chapter. These can be used to construct examinations.

wcb TestPak

A computerized testing service, provides instructors with either a mail-in/call-in testing program or the complete test item file on diskette for use with the Apple and IBM PC computers. wcb TestPak requires no programming experience.

Transparencies

This text is accompanied by 100 transparencies in two and full color. The transparencies feature text illustrations with oversized labels, facilitating their use in large lecture rooms. The transparencies are free to adopters.

Acknowledgments

I am indebted to the entire book staff at Wm. C. Brown Publishers, but would particularly like to thank Bea Sussman, Jess Schaal, Michelle M. Campbell, and Carol Mills, for their contributions. Their skill and perseverance are evident throughout this book. I am in awe at the talents of the many artists who were able to take my chicken scratchings and convert them into respectable, even beautiful, figures. *Perspectives in Human Biology* could not have been written without the aid of dedicated reviewers, who provided expert suggestions and needed encouragement. These reviewers are:

Gary Bradley
Loma Linda University

David E. Fulford
Edinboro University of Pennsylvania

Edwin J. Spicka
State University of New York–Geneseo

Leigh Auleb
San Francisco State University

Ellie Skokan
Wichita State University

Judith P. Downing
Bloomsburg University

John Zavodni
Pennsylvania State University–McKeesport

Perspectives on

Human Biology

1

Introduction to Human Biology

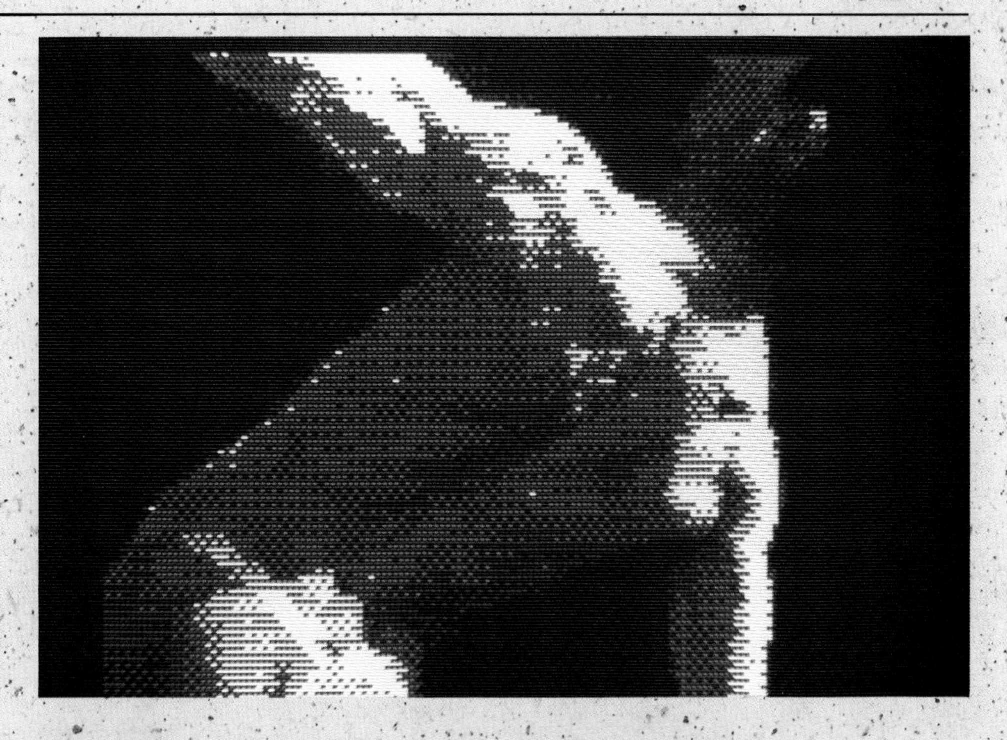

Outline

Objectives

By studying this chapter, you should be able to

1. list the specialties of general biology that pertain to the study of human biology
2. explain the characteristics of the scientific method
3. list the five kingdoms, and the characteristics of chordates, mammals, and primates
4. define the term *species,* and explain why all living humans are considered to be members of the same species
5. list the different genera and species included in the hominid family
6. describe the anatomical characteristics of humans

Keys to Pronunciation

Hominidae: *ho-min′ĭ-de*
pharyngeal: *fah-rin′je-al*
pharynx: *far′inks*
phylum: *fi′lum*
physiology: *fiz″e-ol′o-je*

Photo: The human body. Modern medical imaging techniques allow us to see ourselves in new ways.

Introduction to Human Biology

The study of human biology involves many specialized areas common to other fields of biology, but is unique in its practical applications and cultural implications. Human biology is a scientific study of man, and as such utilizes the scientific method to enable us to know ourselves better and to function more effectively within our society and global environment.

What is the subject of human biology, and why study it? Human biology is actually difficult to define. It is less than "the biology of humans," because this could, directly and indirectly, include all of the topics in general biology. Human biology is also more encompassing than its name implies, because human biology and culture interact to a degree not seen in other organisms studied by biologists. The nature of human biology as a subject might best be grasped indirectly, by answering the second part of the question—"Why study it?" One of the best answers was provided almost three hundred years ago:

> Know then thyself, presume not God to scan,
> The proper study of mankind is man.
> Alexander Pope (1688–1744)

At the time these words were written, the scientific method of studying living organisms was in its infancy. It had been applied to the study of human **anatomy**[1] by Andreas Vesalius (1514–1564), whose revolutionary book *De Humani Corporis Fabrica* (fig. 1.1) has earned him the title of "father of human anatomy." In 1628, William Harvey published *On the Movement of the Heart and Blood in Animals* (fig. 1.2) in which he proved by means of experimentation that the blood circulates and does not flow back and forth through the same vessels. William Harvey is thus considered to be the father of modern **physiology.**[2]

We don't know if human anatomy and physiology is what Alexander Pope had in mind when he urged people to "know thyself," but certainly some knowledge of anatomy and physiology is required to follow this admonition. Further, such knowledge has a practical benefit that would have been inconceivable to a person from the seventeenth century. Modern medicine is founded on the sciences of human anatomy and physiology, and thus a basic understanding of these sciences is required if a person is to be an informed consumer of modern medical technology.

The science of physiology and almost all of modern medicine is based on animal experimentation. This has reaped enormous benefits for mankind, but has raised ethical questions that relate to the nature of mankind and the relationship of humans to other animals. People from Aesop to Walt Disney have endowed other animals with humanlike characteristics. On the other hand, poets have viewed humans as more like gods than like other animals:

> What a piece of work is man! how noble in reason! how infinite in faculty! in form and moving how express and admirable! in action how like an angel! in apprehension how like a god! the beauty of the world! the paragon of animals!
> *Hamlet,* by William Shakespeare (1564–1616)

It would seem, therefore, that part of the requirement for understanding ourselves is to learn the relationship that humans have to other aspects of the biological world. This involves a knowlege of the way that scientists group living organisms according to their relationships—a study known as *taxonomy*—and the scientific study of how these relationships were established. A basic knowledge of the processes of **evolution,** one of the major organizing concepts of biology, is thus required for the scientific study of man. Also, a knowledge of the interactions between humans and their environment—which is part of the science of **ecology**—is necessary to properly understand the relationship between humans and other members of the biological world.

In the 1920s through the 1940s, the explosion in knowledge about the physical sciences—atomic structure and energy, electronics, relativity and quantum mechanics—transformed the world. Our reality today, regardless of our occupation or interests, is very different from the reality of our grandparents' youth. A similar revolution, perhaps of even greater magnitude, is occurring today in our knowledge of **molecular genetics** and **immune function.** All educated people must thus have a basic understanding of these fields in order to make informed decisions that will affect the future of ourselves and our descendants.

Human culture cannot be divorced from a study of human biology. Culture has developed along the lines constrained by biology and geography. Agriculture, technology, politics, economics, and cultural history, in turn, influence the interaction between people and their environment. This has always been true, but—because of modern technology and the bloated size of the human population—the implications today are graver than ever. Religion and ethics are part of human culture, and influence the way that people interact with each other and with their environment. Thus, strictly biological topics—such as ecology, sexual reproduction, and embryonic development—have ethical and political correlates (as in the issues of pollution, birth control, and abortion).

In this text, the study of human biology will encompass the topics outlined in this introduction and will attempt to correlate the scientific topics with their practical, social, and political applications. Biologists use the scientific method as a means for studying mankind, as opposed to the nonscientific method employed by the poets previously quoted. The results may not be as lyrical, but the picture that emerges from the scientific method of studying man is awesome, at least as beautiful, and far more practical.

[1]anatomy: Gk. *ana,* up; *tome,* a cutting
[2]physiology: Gk. *physis,* nature; *logos,* study

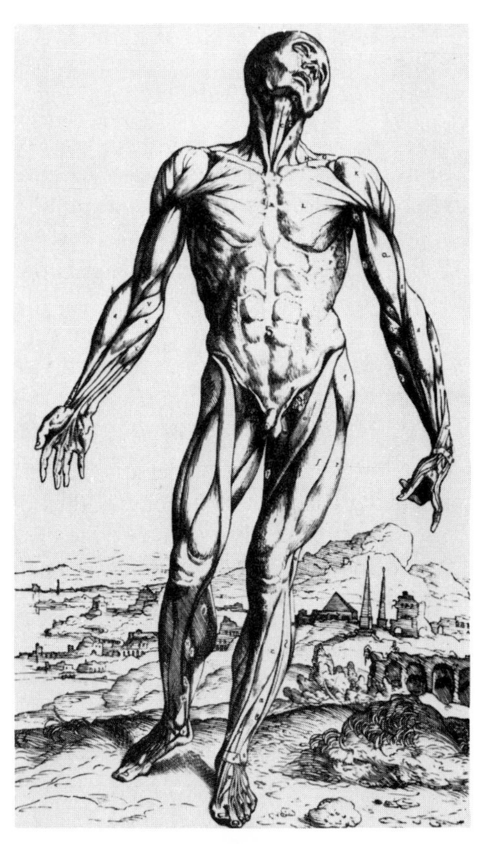

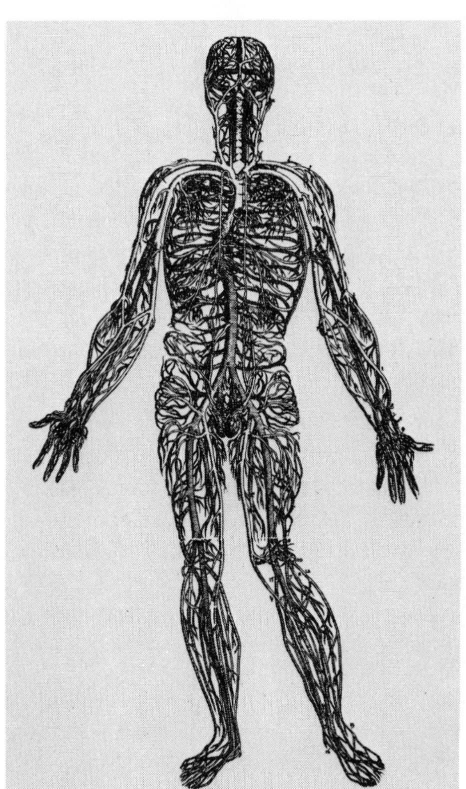

Figure 1.1. *Plates from* De Humani Corporis Fabrica, *which Vesalius completed at the age of twenty-eight. This book, published in 1543, revolutionized the sciences of anatomy and physiology.*

(Courtesy of The New York Academy of Medicine Library).

Figure 1.2. *The English physician William Harvey demonstrated with experiments in 1628 that blood circulates and does not flow back and forth through the same vessels.*

Scientific Method

All of the information in this text has been gained by the application of the scientific method. Although many different techniques are involved in the scientific method, all share three attributes: (1) confidence that the natural world, including ourselves, is ultimately explainable in terms we can understand; (2) descriptions and explanations of the natural world that are honestly based on observations and that could be modified or refuted by other observations; and (3) humility, that is, the willingness to accept the fact that we could be wrong. If further study should yield conclusions that refute all or part of an idea, the idea must be accordingly modified. In short, the scientific method is based on a confidence in our rational ability, honesty, and humility. Practicing scientists may not always display these attributes, but the validity of the large body of scientific knowledge that has been accumulated—as shown by the technological applications and the predictive value of scientific hypotheses—are ample testimony to the fact that the scientific method works.

The scientific method involves specific steps. In the first step, a **hypothesis** is formulated. In order for this hypothesis to be scientific, it must be capable of being refuted by experiments or other observations of the natural world. An example of a scientific hypothesis might be "people who perform endurance exercise regularly have a lower resting pulse rate than other people." Experiments or other observations are performed and the results analyzed. Conclusions are then drawn that these observations either refute or support the hypothesis. If the hypothesis is proven to be true, it might be incorporated into a more general **theory.** Scientific theories are statements about the natural world that incorporate a number of proven hypotheses, serve as a logical framework by which these hypotheses can be interrelated, and provide the basis for predications that can be tested by future research.

The hypothesis in the preceding example is scientific because it is *testable;* the pulse rates of 100 endurance-trained athletes and 100 sedentary people could be measured, for example, to see if statistically significant differences were obtained. If they were, the statement that athletes, on the average, have lower resting pulse rates than sedentary people would be justified *based on this data.* One must still be open to the fact that this conclusion could be wrong (the measurement techniques may have been biased, or the sample of people tested may not have been representative of the general population). Before the discovery could become generally accepted and included in textbooks, other scientists must consistently replicate the results—scientific theories are based on *reproducible* data.

It is quite possible that when others attempt to replicate the experiment their results will be slightly different. They may then construct scientific hypotheses that the differences in resting pulse rate also depend on factors such as the nature of the exercise performed, or on other variables. When scientists attempt to test these hypotheses, they will likely encounter new problems, requiring new explanatory hypotheses, which must be tested by additional experiments.

In this way, a large body of highly specialized information is gradually accumulated and a more generalized explanation (a scientific theory) can be formulated. This explanation will almost always be different from preconceived notions. People who follow the scientific method will then appropriately modify their concepts, realizing that their new ideas will probably have to be changed again in the future as additional experiments are performed.

---

1. Name four fields within the general biological sciences that are applicable to the study of humans.
2. Describe three characteristics of the scientific method.
3. Is an explanation of the natural world based on biblical interpretation scientific? Use the characteristics of the scientific method in your analysis.
4. "The theory of evolution should not be taught in schools, because, after all, it's just a theory." Analyze this statement in the light of the definition of a scientific theory.

Classification of Humans

Humans are in the kingdom Animalia, but can be distinguished from other animals by lower classification categories. Humans are in the phylum Chordata (along with other vertebrates), the class Mammalia, the order Primates, and the family known as the Hominidae. All modern humans are in the same genus and species: *Homo sapiens.* The most distinguishing anatomical features of humans are their large, well-developed brain, upright posture and bipedal locomotion, and the presence of opposable thumbs.

The classification, or taxonomic, scheme has been established by biologists to organize the structural and evolutionary relationships of organisms. Each category of classification is called a *taxon.* The highest and most general taxon is the **kingdom,** and the most specific is the **species.**

Five kingdoms are now generally recognized: *Animalia* (animal); *Plantae* (plant); *Fungi* (such as mushrooms and yeast); *Protista* (single-celled organisms with a true nucleus, such as an amoeba); and *Monera* (single-celled organisms that lack a true nucleus, such as bacteria). Since these five kingdoms include the entire living world, it should be obvious that humans are members of the animal kingdom.

There are several levels of classification below kingdom. Of these, only the species level has a biological definition. A species is *the largest population of organisms capable of interbreeding and producing fertile offspring.* From this definition, and even a rudimentary knowledge of human history, it is clear that all humans are members of the same species. Horses and donkeys are separate species, but mules—which are hybrids of horses and donkeys—are not, because mules are sterile. Sometimes the capacity for interbreeding cannot be known, as when the fossil remains of extinct organisms are examined. In these cases the assignment of species is based only on morphological[3] evidence. The taxa above the species level are arranged only by shared characteristics, and thus are somewhat arbitrary and subject to frequent reinterpretation.

Similar species are grouped together into a **genus.** When you refer to a particular species, you usually include the genus name as well. The genus name is always capitalized, while the species name is in lowercase—as in *Homo sapiens* (modern humans). Similar genera (plural of genus) are grouped in a **family.** Family names end in the suffix *-idae;* the family of humans is thus called *Hominidae.* Similar families form an **order,** similar orders a **class,** similar classes a **phylum,** and similar phyla (the plural of phylum) a kingdom. These taxa can also be subdivided into sub- and super-categories, as in subphylum, subspecies, and superfamilies. The classification of humans is summarized in table 1.1.

Phylum Chordata

Humans belong to the phylum **Chordata,** along with fish, amphibians, reptiles, birds, and other mammals. All chordates have three structures in common: a *notochord,* a *dorsal hollow nerve cord,* and *pharyngeal pouches* (fig. 1.3). A notochord is a flexible tube located on the dorsal (back) side of the embryo which eventually disappears (parts become incorporated into the intervertebral discs of the vertebral column, shown in fig. 1.4). The dorsal hollow nerve cord becomes the brain and spinal

---

[3]morphological: Gk. *morphe,* form; *logy,* study

Taxon	Designated Grouping	Characteristics
Kingdom	Animalia	Eucaryotic cells that lack walls, plastids, and photosynthetic pigments
Phylum	Chordata	Dorsal hollow nerve cord; notochord; pharyngeal pouches
Subphylum	Vertebrata	Vertebral column
Class	Mammalia	Mammary glands; hair
Order	Primates	Well-developed brain; prehensile hands
Family	Hominidae	Large cerebrum; bipedal locomotion
Genus	*Homo*	Flattened face; prominent chin and nose with inferiorly positioned nostrils
Species	*sapiens*	Largest cerebrum

Table 1.1 Classification scheme of human beings

From Kent M. Van De Graaff, *Human Anatomy,* 2d ed.

cord—together known as the central nervous system—in the adult. Pharyngeal pouches (referring to pouches in the throat area) form gill openings in fish and some amphibians. In other chordates, including humans, only one of the embryonic pharyngeal pouches persists, becoming the eustachian (auditory) canal connecting the middle ear cavity with the pharynx.

Probably all the chordates with which you are familiar are grouped into the **subphylum Vertebrata.** These are animals that have vertebral columns, or backbones. This subphylum designation is required because there are other chordates, such as tunicates, or sea squirts, that do not develop backbones.

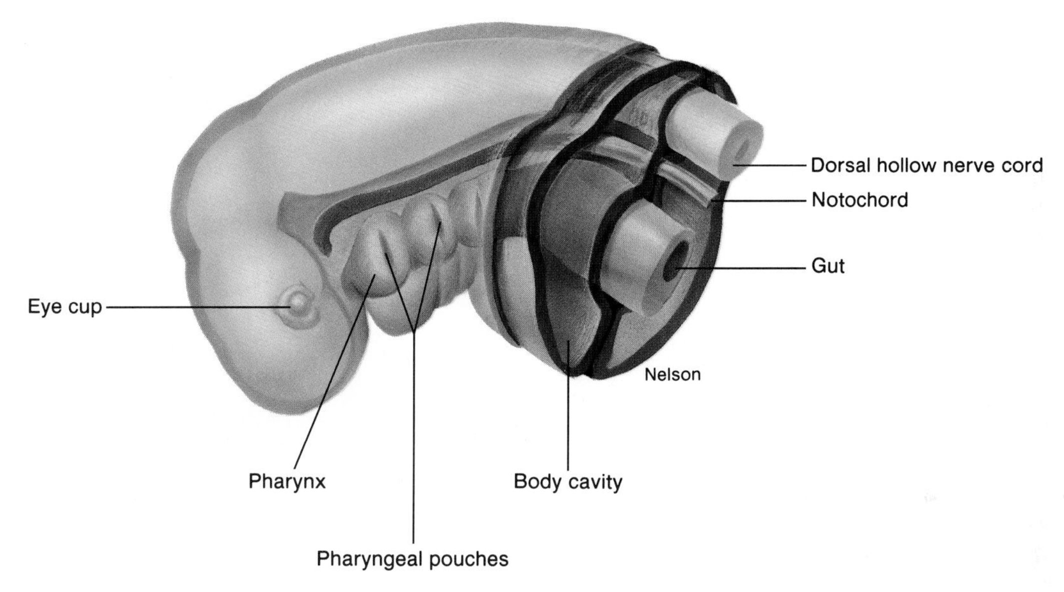

Figure 1.3. *A schematic diagram of the front part of the embryo of a chordate.*

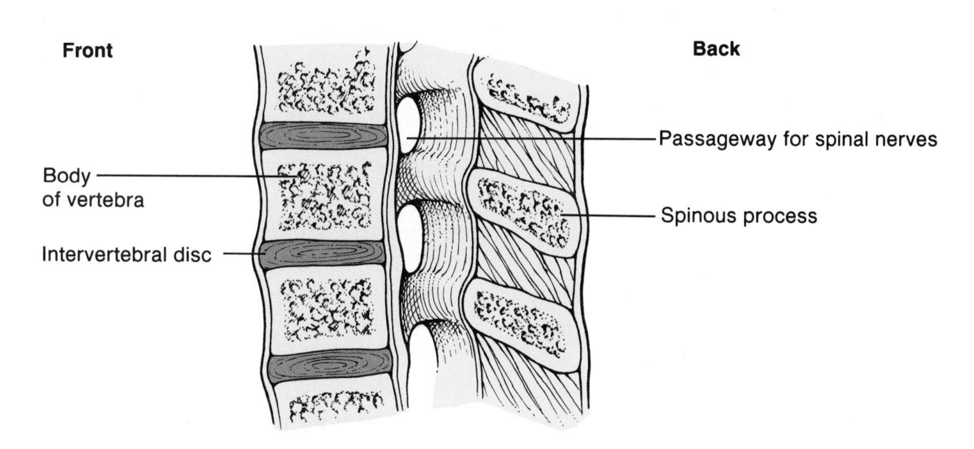

Figure 1.4. *The vertebral column, as seen in midsagittal section (lengthwise cut along the midline). The vertebrae are composed of bone and the intervertebral discs are composed of cartilage.*

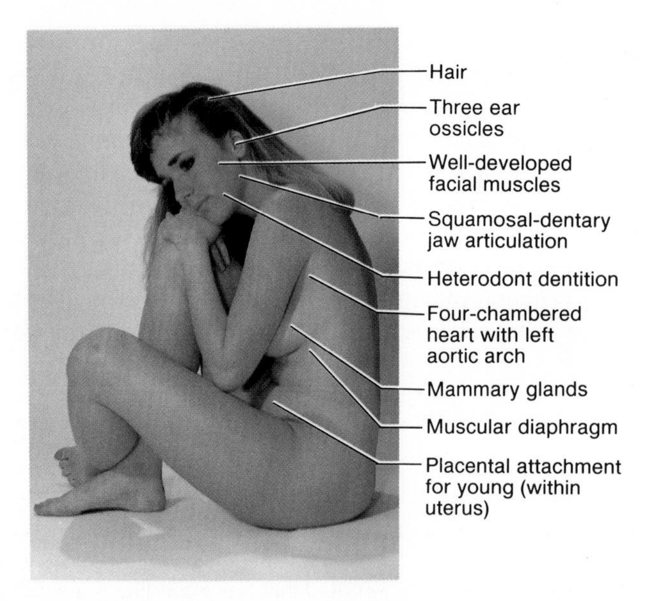

Figure 1.5. Characteristics of mammals—the class to which humans belong.

Class Mammalia

Mammals are vertebrates with hair and mammary glands (fig. 1.5). Other characteristics of mammals include three middle ear ossicles (small bones), fleshy external ear, heterodont dentition (differently shaped teeth), a joint between the lower jaw and skull, usually seven cervical (neck) vertebrae, an attached placenta, well-developed facial muscles, a muscular diaphragm (a sheet of muscle separating the thoracic from the abdominal cavities), and a four-chambered heart with a left aortic arch (described in chapter 12).

Order Primates

There are almost 200 species of primates,[4] an order that includes humans, the *great apes* (chimpanzees, gorillas, and orangutans), *lesser apes* (gibbons and siamangs), *Old World monkeys* (such as the rhesus monkey, long-tailed macaque, and baboon), *New World monkeys* (such as the squirrel monkey), and the *prosimians*. Most members of this order have prehensile (grasping) tails, digits modified for grasping, and relatively large, well-developed brains (fig. 1.6).

Family Hominidae

The family Hominidae[5] includes only one living genus and species—*Homo sapiens* ("thinking man"). The fossil record, however, shows that at different periods of geological time the family Hominidae included two genera—*Australopithecus* and *Homo*, and that each of these genera included a number of species. According to most current classifications, *Australopithecus* was a genus that contained four recognized species: *A. afarensis, A. africanus, A. robustus,* and *A. boisei*. All are extinct. There are currently three recognized species within the genus *Homo:* the extinct species *H. habilis* and *H. erectus*, and us—*H. sapiens*.

It should be noted that the classification scheme for the hominids is based on limited fossil evidence, and is, as a result, subject to differing interpretations and frequent revision as new fossils are discovered. The evolutionary relationships between the different members of the hominid family are discussed in chapter 2.

[4]primates: L. *primas*, first

[5]Hominidae: L. *homo*, man; Gk. *eidos*, resemblance

(a)

(b)

(c)

(d)

(e)

Figure 1.6. Examples of different primates. (a) A prosimian known as a tarsier; (b) a spider monkey; (c) a baboon; (d) macaque monkey; (e) a chimpanzee.

Primates in Research

The DNA of humans and chimpanzees is 98%–99% similar. Old World monkeys have 92%, and New World monkeys 85%, similarity of DNA structure with humans. These molecular observations reinforce the anatomical and behavioral observations that have led layman as well as scientists to conclude that humans, apes, and monkeys are closely related. Indeed, a controversial DNA matching technique has led some scientists to believe that chimpanzees are more closely related to humans than they are to gorillas. This has important implications for the evolution of humans (discussed in chapter 2) and for the use of "lower" primates in biomedical research.

Approximately twenty million animals are used annually in biomedical research. Of these, about 90% are rodents (mainly rats and mice). Still, about 60,000 are primates. The three most commonly used primates in research are rhesus monkeys, long-tailed macaques, and squirrel monkeys. Most of the experiments that involve the use of these animals utilize noninvasive techniques, so that the animals can be reused in other experiments. The reasons for the limited use of primates for research include high cost, limited numbers, and ethical considerations that limit the use of these animals to just those projects for which they would be the most valuable. Table 1.2, for example, presents a checklist used by the Yerkes Regional Primate Research Center for evaluating requests for use of primates in specific research proposals.

Investigations of the poliovirus and the development of vaccines against it required the use of primates, because this virus only attacked the nervous system of primates. The development of a vaccine against hepatitis B also required the use of primates. The virus that causes acquired immune deficiency syndrome (AIDS) attacks only humans and chimpanzees, and thus investigations into the nature of this virus at present necessitates the use of chimpanzees.

Though their use may at present be necessary for specific medical research projects, most scientists would agree that the use of chimpanzees in research should be as limited and humane as possible. Jane Goodall, the director of the Gombe Stream Research Center in Tanzania and a noted expert in chimpanzee behavior in the wild, however, believes that we do not devote sufficient attention to the emotional and psychological health of chimpanzees used for laboratory research. Caretakers, she argues, need to be thoroughly trained to become trusted friends of the chimpanzees, the animals should be housed together or allowed to interact socially on frequent occasions, and cages should be large and provide a rich, stimulating environment with bedding and toys. Dr. Goodall has noted that these conditions are met by some laboratories, but not by most.

Table 1.2 Checklist of criteria used to evaluate research proposals at Yerkes Regional Primate Research Center

1. Are primates necessary for the proposed study, or can the work be as well conducted with another species or an alternative, nonanimal method?
2. Is the particular primate species selected appropriate biologically or behaviorally for the proposed investigation?
3. Is the study likely to contribute significantly to scientific knowledge or to human or animal health?
4. Is the investigator scientifically and technically qualified to conduct the study?
5. Will the study be conducted in a humane fashion, with proper consideration for the welfare of the animal, and in compliance with existing regulations?
6. If invasive procedures or others likely to produce pain or discomfort are proposed, are they essential to the study?
7. In proposals involving potentially painful procedures or surgery, has provision been made for elimination or minimization of pain or discomfort including proper anesthesia, analgesia, and round-the-clock postoperative care and surveillance?
8. If the research is replication of previous or other ongoing studies, is it justified and needed?
9. Is the number of animals to be used and the research design adequate to produce clearly interpretable results, but not excessive?
10. Will the study limit reproductive capacity in a way that will be injurious to breeding in the particular primate colony or to the species itself?

From King, et al., *Science*, 240(1988):1475. Copyright 1988 by the AAAS.

Characteristics of Humans

Humans have a few anatomical characteristics that, taken together, are diagnostic in separating them from closely related mammals. These characteristics are:

1. **Size and development of the brain.** The average human brain weighs 1,350–1,400 grams (g). This gives humans a large brain-to-body weight ratio. Of greater importance, however, is the development of certain portions of the brain. Specialized regions of the brain process thought and reason, emotions, memory, coordinated movements, and speech and language ability.
2. **Style of locomotion.** Because humans stand and walk on two legs, their style of locomotion is said to be bipedal.[6] The upright posture results in other diagnostic features, such as sigmoid (S-shaped) curvature of the spine, the anatomy of the hips and thighs, and arched feet.
3. **Opposable thumb.** The human thumb joint is structurally adapted for tremendous versatility in grasping objects. Most primates have opposable thumbs.

[6]bipedal: L. *bi,* two; L. *pes,* foot

4. **Vocal structures.** Humans, unlike any other animal, have developed articulated speech. This is made possible by the anatomy of the vocal organs and the high development of the brain.
5. **Stereoscopic vision.** Although this characteristic is well developed in many other animals, it is also keen in humans. Human eyes are directed forward, so that when they focus on an object the object is viewed from two angles. Stereoscopic vision gives depth perception, or a three-dimensional image.

Humans also differ from other animals in the number and arrangement of their vertebrae (called their vertebral formula), the kind and number of their teeth (tooth formula), their well-developed facial muscles, and the specializations of various body organs.

---

1. List the taxonomic categories that describe modern humans. Where possible, give an example of one other animal that shares each taxon with humans.
2. Explain why there is a genus and family level of classification of humans, even though all living humans are members of the same genus and species.
3. Discuss the relationships between humans and other living primates, and the issue of the use of primates for medical research.
4. Name some of the anatomical features that distinguish humans. Speculate on the interrelationships between these features.

Summary

Introduction to Human Biology p. 4

I. The study of human biology involves elements of the sciences of anatomy, physiology, taxonomy, evolutionary biology, and ecology. Human biology is also concerned with the practical, ethical, and political implications of the scientific study of man.

II. The study of human biology uses the scientific method to gain information about the nature of humans.
   A. The scientific method is based on honest observations of the natural world, testable hypotheses, and the humility to change preconceived notions in the face of new objective evidence.
   B. General scientific concepts and explanations are derived from a large body of facts accumulated by use of the scientific method.

Classification of Humans p. 6

I. Humans are members of the kingdom Animalia.
   A. Humans are in the phylum Chordata, which includes all of the animals that have vertebral columns (backbones).
   B. Humans are in the class Mammalia, which is characterized by the presence of hair and mammary glands.
   C. Humans are in the order Primates, which includes humans, apes, and monkeys.
   D. Humans are members of the family known as the Hominidae, which includes modern humans together with extinct members of the hominid family.
   E. All living humans are members of a single genus and species, *Homo sapiens*.

II. Anatomical features that help to distinguish humans are a large brain, adaptations to an upright posture, bipedal locomotion, and the presence of an opposable thumb.

Review Activities

Objective Questions

Match the correct taxonomic categories of humans:

1. Class	(a) Hominidae
2. Phylum	(b) *Homo*
3. Order	(c) Mammalia
4. Family	(d) Chordata
5. Genus	(e) Primates

6. The father of physiology was
   (a) Alexander Pope
   (b) Andreas Vesalius
   (c) Charles Darwin
   (d) William Harvey
7. Which of the following is *not* a characteristic of the scientific method?
   (a) honest observation
   (b) reference to an authority
   (c) testable hypotheses
   (d) willingness to change a hypothesis because of new information
   (e) reliance on reproducible experimental results
8. What percentage of the DNA is similar in humans and chimpanzees?
   (a) 10%–15%
   (b) 25%–28%
   (c) 50%–55%
   (d) 90%–92%
   (e) 98%–99%
9. About 90% of the animals used in biomedical research are
   (a) primates
   (b) dogs
   (c) rodents
   (d) cats
   (e) pigs
10. Which of the following is *not* a characteristic of humans?
   (a) large brain
   (b) opposable thumb
   (c) prehensile digits
   (d) upright posture
   (e) quadrupedal locomotion

Essay Questions

1. Define the term *species,* and explain why all humans on earth are considered to be members of the same species.
2. Is astrology (the belief that human activities and fates are determined by the stars) a science? Explain your answer.

2

Evolution

Outline

Objectives

By studying this chapter, you should be able to
1. describe the influence of Charles Darwin's contemporaries on the development of the theory of evolution
2. describe the observations and deductions that led Darwin to develop the theory of evolution
3. explain what survival of the fittest means in terms of evolutionary biology
4. explain how natural selection operates to either maintain or change the characteristics of a species
5. describe how evolution of new species occurs
6. define the terms *ecological niche, adaptive radiation, geographic speciation,* and *phylogeny*
7. contrast gradualism and punctuated equilibrium
8. explain how human activities have produced extinction of many species
9. explain how radioisotope dating is accomplished
10. explain the mechanisms that caused industrial melanism in moths and resistance to antibiotics in bacteria
11. describe the sources of evidence for the evolution of species over long periods of time
12. describe the hominid ancestors of modern *Homo sapiens,* and explain the relationships among different species of the hominid family
13. distinguish between modern *Homo sapiens* and other hominids in terms of anatomy and culture

Keys to Pronunciation

Cretaceous: *cre-ta'ceous*
extinction: *eks-ting'shun*
heterozygous: *het''er-o-zi'gus*
homologous: *ho-mol'o-gus*
homozygous: *ho''mo-zi'gus*
Neanderthal: *ni-an'der-tal*
niche: *nich*
ontogeny: *on-toj'e-ne*
phalanges: *fah-lan'jez*
phylogeny: *fi-loj'e-ne*
Tertiary: *ter'she-er-e*

Photo: Visage of Java man, or Homo sapiens soloensis, *as he may have appeared 50,000 to 250,000 years ago.*

The Theory of Evolution

The theory of evolution was developed by Charles Darwin from numerous observations of the natural world. He observed that there is variation among individuals of a species, and that some of these variants are able to survive and reproduce better than others can within their environment. These individuals are described as being more fit, and they contribute proportionately more to the genetic makeup of the next generation. The variation found within a species is produced randomly rather than in a directed fashion, and nature provides a selective advantage to some of these variants.

Evolution is one of the most important theories in the history of science, and serves as one of the major organizing principles in the study of biology. Although there are still some people who refuse to accept the fact that species evolve, every shred of evidence derived from the use of the scientific method points to the fact that species are not immutable (unchangeable). The mountain of evidence in support of evolution is indeed so great, it is difficult for an educated person today to imagine a time when this fundamental process was unknown.

Yet prior to the latter part of the eighteenth century the process of organic evolution would have been impossible to accept, because people believed that the earth was only a few thousand years old. Fossils were known, but their age could not be judged. Comte de Buffon (1707–1788), a French naturalist, proposed that fossils closer to the surface in a layer of rock must be more recent than those more deeply buried. He further suggested that species could change over time, and that humans and apes shared a common ancestor. This, however, was just speculation; he had little evidence to support his theory and could propose no mechanism that would explain how evolution occurred.

In 1795, James Hutton argued that fossils located in sedimentary rock were buried by the same gradual accumulation of sediment carried by rivers, streams, and ocean tides that is observable today. Since the buildup of sedimentary rock by this process is extremely slow, he reasoned that the earth must be far older than was generally believed. These proposals were extended and integrated into a consistent framework by Charles Lyell (1797–1875), who is considered to be the father of the science of geology. In the first volume of his famous *Principles of Geology,* he presented multiple sources of evidence that the earth was millions of years old. In a move that was perhaps pivotal to his career, the young Charles Darwin decided to take this recently published volume with him on his voyages around the world.

Charles Darwin and the Theory of Evolution

> These are the voyages of the starship *Enterprise.* . . . Its five-year mission: to seek out new life . . . to boldly go where no man has gone before.
>
> Gene Roddenberry, *Star Trek*

Charles Darwin, age twenty-two, also embarked on a five-year voyage of discovery (1831–1836). Analogous to the mythical

Figure 2.1. *The* H. M. S. Beagle, *which carried Charles Darwin on his five-year voyage of discovery.*

Mr. Spock, science officer on the starship *Enterprise,* Darwin was the assistant naturalist on board the H.M.S. *Beagle* (fig. 2.1). Though this was not a five-year voyage "where no man has gone before" in a physical sense, it was analogous to that intellectually; it seems probable that the theory of evolution was taking shape in Darwin's mind by the end of that time.

Charles Darwin (1809–1882, fig. 2.2) was born to a prominent and wealthy English family, and so had the time and resources to devote to his passion for natural history (the study of animal and plant life). His father, a physician, wanted him to study medicine; he balked at this, and instead resolved to prepare for the clergy. This was done, however, only out of a sense of duty; when the position of assistant naturalist on board the H.M.S. *Beagle* was offered to him he eagerly seized the opportunity.

The influence of Charles Lyell's *Principles of Geology,* and the many observations of fossil and living organisms he encountered on his voyage, provided the evidence Darwin was later to use in support of his theory of evolution. Twenty-three years after his voyage, at the age of fifty, Darwin published *On the Origin of Species by Means of Natural Selection, or the Preservation of Favoured Races in the Struggle for Life* (1859). It was an instant best-seller and has become one of the most influential books ever written.

Darwin's theory of evolution is often said to be based on four observations and two deductions. The four observations are:

1. **Great numbers of individuals are born.** This is obvious for organisms such as insects, fish, and frogs, but it is also true of other, more slowly reproducing species.
2. **The number of individuals in a population remains relatively constant.** Although great numbers are born, the population size does not increase geometrically. There may be cyclic fluctuations in population size, but over an extended period of time the size of a population is relatively stable.
3. **There is individual variation within a species.** Humans, clearly, differ from each other in ways that allow us to tell people apart. This is also true in other animals and

Figure 2.2. A portrait of Charles Darwin at the age of thirty-one.

plants, although it may not always be as obvious. Individual variation, however, was thoroughly documented in many species by Darwin.

4. **The variation in a species is heritable.** Individual variation among members of a population is passed to the next generation.

From these four observations Darwin drew two deductions:

1. **There is a constant struggle for existence.** Given observations number 1 and 2, this must be true. In making this deduction, Darwin was influenced by Thomas Malthus (1766–1834), an English economist who proposed in his *Essay on the Principles of Population* that human populations were held in check by famine, disease, and war.
2. **The survival of the fittest occurs by natural selection.** Given the struggle for existence, and the variation that occurs within a species, it is reasonable to assume that some individuals will be better able to survive and reproduce (will be more *fit*) than others. Since the physical variations in members of a species are based on genetic differences, the individuals who are more fit will have more influence on the genetic constitution of the next generation than those who are less fit.

Survival of the Fittest

Darwin knew less than you probably do about genetics. He was unaware of the basic principles of inheritance, which were discovered by Gregor Mendel (1822–1884) and published in 1866. Darwin did not know that genes are made of DNA and that they are located on the chromosomes within the cell nucleus. The explosion in knowledge of molecular genetics (chapter 10) has added immensely to our understanding of the processes of evolution. Nevertheless, it is quite possible to understand, as Darwin did, how evolution occurs without knowledge of modern molecular biology.

Darwin confirmed and extended the observation made by the governor of the Galápagos Islands that the tortoises found on one island could be distinguished from those on another island. These were members of the same species but were quite distinctive, much like the different breeds of pigeons or dogs that are produced artificially by breeders (fig. 2.3). The natural selection that produced the different types of tortoises seen in nature must thus be analogous to the *artificial selection* for particular traits made by animal and plant breeders. In the case of artificial selection, people make the decision about what trait should be strengthened by allowing those individuals which possess that trait to interbreed. The different tortoises that were characteristic of each island could have been produced in an analogous fashion, with nature as the selecting agent.

Survivial of the fittest refers to reproductive success. In practice, the term "fit" is defined in a somewhat circular fashion. Those individuals who are most fit reproduce more, and so most influence the gene pool (all of the genes of a population considered together) of the next generation. Consequently, those who contribute most to the gene pool of a population are, by definition, the fittest. Notice that the term "fit" does not necessarily refer to the biggest, strongest, or most aggressive. In certain cases these qualities may contribute to fitness, but it is possible that an individual that is too large may not get enough to eat, or may not be able to hide as well, and an individual that is too aggressive may not be able to attract a mate or successfully raise its young.

Natural Selection

Darwin's achievement was amazing; in the absence of modern knowledge of molecular biology, he realized that *the variation among members of a species is random.* In more modern terms, genes mutate (change) randomly; the specific genetic change that occurs during a mutation is not directed or predictable. Nature, in turn, selects the genetic variants offered up by the random mutation process by allowing the most fit to contribute proportionately more genes to the next generation. This is a most important concept: nature does not cause a particular variant to appear. It simply selects from what is available.

This is simple but it is not obvious. Indeed, a great biologist, Jean Baptiste Lamarck (1744–1829), made a reasonable error that led many biologists astray well into the twentieth century. Examination of this error is instructive in understanding the principles of evolution. Lamarck believed that simple, primitive

Figure 2.3. *Artificial selection was used to generate the different breeds of pigeons. Darwin reasoned that a similar process may occur in nature.*

species evolve into more complex, more perfect species through the *inheritance of acquired characteristics.* The classic example of this concept is his explanation of why giraffes have long necks: it is because, he argued, the ancestors of giraffes had to stretch their necks to reach the high leaves. By the same argument, parents who study mathematics will produce children who are good in math, and people who lift weights will produce muscular children. It is an attractive concept, but, unfortunately, it is wrong. Modern molecular biology has confirmed Darwin's concepts: (1) mutation and thus variation is random, not directed; (2) natural selection allows the most fit to contribute proportionately more genes, so that (3) the species becomes better adapted to its environment. Modern species are thus not more perfect than older, now extinct species—they are simply better adapted to their present environment.

Lamarck was wrong about the mechanism of biological evolution, but the inheritance of acquired characteristics does provide a correct explanation for the evolution of human culture. The culture acquired by the creativity and hard work of parents *is* "inherited"—through education—by the next generation. This has resulted in extremely rapid changes in human society. In a span of a mere 40,000 years, which was too short to produce biological changes in *Homo sapiens,* the Lamarckian inheritance of culture has transformed human society from the Stone Age to the age of personal computers, space shuttles, and genetic engineering.

1. Explain how the development of the modern science of geology influenced the development of the theory of evolution.
2. Describe the observations and deductions made by Charles Darwin in the development of the theory of evolution.
3. Define *natural selection* and *fitness* as they are used in evolutionary theory.
4. Explain what is meant by "Lamarckian inheritance" and why this concept is incorrect as an explanation of biological evolution.

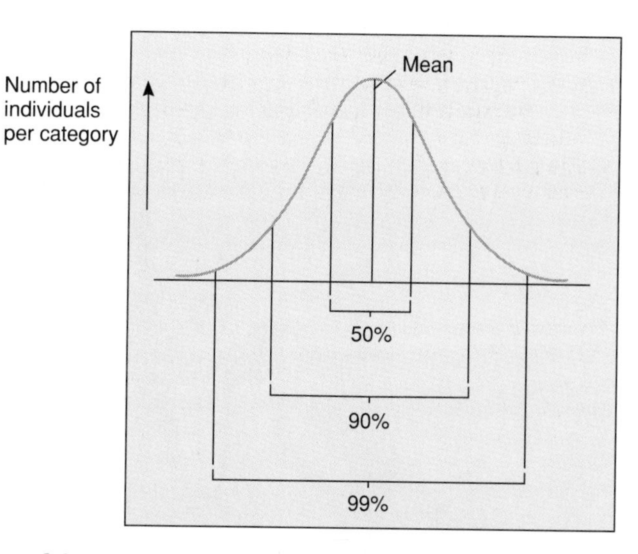

Figure 2.4. *An idealized normal distribution. Percentages indicate the percentage of the population that would fall within a bracketed area of the graph. For example, 90% of the population would fall within the brackets indicated for 90%, while 10% of the population would be outside the brackets.*

In a stable environment, natural selection acts to maintain the characteristics of a species by eliminating less fit variants. In a changed environment, individuals who were previously less fit may now be more fit. Natural selection in this case causes the species to change in the direction of this previously less common variant. This is directional selection, and can result in the evolution of new species.

Natural selection in a stable environment helps to prevent significant changes from occurring in a species. This is known as **stabilizing selection.** If a particular characteristic of a species is measured among many individuals—for example, the weight of newborn babies—it will be found to vary somewhat. Most newborns will have weights within a certain range (such as six to eight pounds); fewer will have weights slightly above or slightly below these more typical values. The reasons for this distribution are obvious; a fetus that is too large could not get through the birth canal, and a newborn that is too small would be too weak and underdeveloped to survive. A graph that plots the value of the characteristic being measured (such as birth weight) against the number of individuals in each category (such as the number of four-pound babies, five-pound babies, etc.) will approach being a bell-shaped curve (fig. 2.4). A graph of this type is known as a **normal distribution.** Stabilizing selection is thus also known as *normalizing selection.*

Directional Selection and the Origin of Species

When a change occurs in the environmental requirements for survival, an individual who was less fit in the old environment may be more fit in the changed environment. Since this individual would therefore contribute proportionately more genes to the gene pool of the population, over many generations the population will come to more closely resemble this previously less common variant (fig. 2.5). This process is known as **directional selection,** and is largely responsible for the origin of new species.

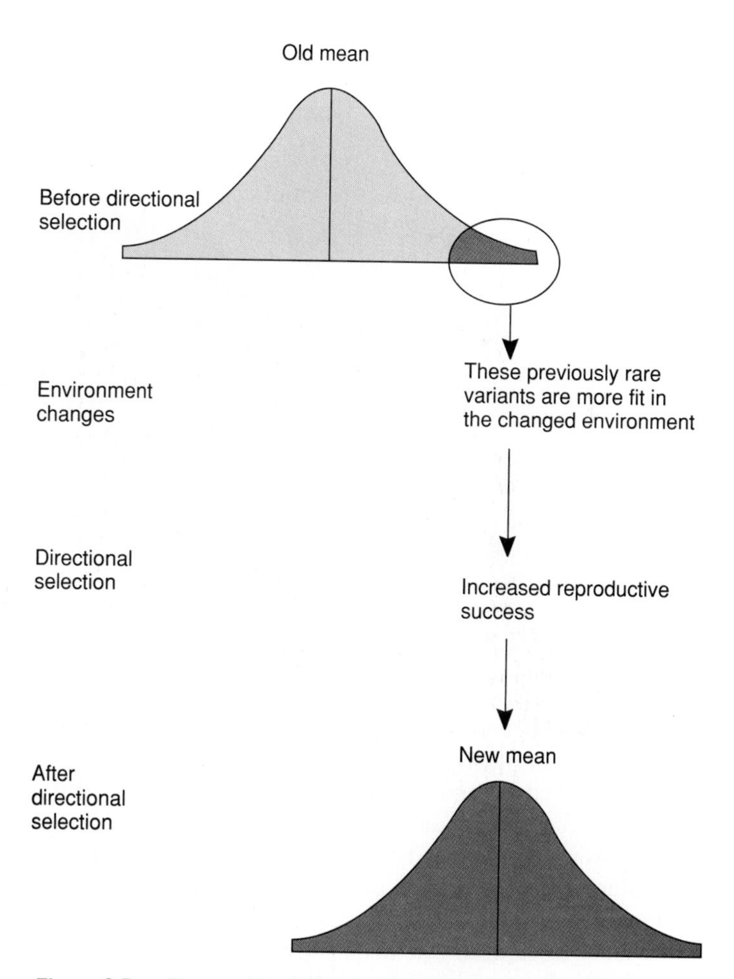

Figure 2.5. *The results of directional selection. The frequency of a particular trait in a population increases if this trait makes individuals more fit in a changed environment.*

Figure 2.6. *Darwin's finches. Only a few of the fourteen species of finches found on the Galápagos Islands are shown here. Darwin believed, as later scientists have confirmed, that these species originated by adaptive radiation from an ancestral species on the mainland.*

The process of directional selection was inferred by Charles Darwin from observations of the finches on the Galápagos Islands. These finches—and indeed all life on the islands—must have originated on the mainland, which is located 950 kilometers (km) from the islands. Yet each of the Galápagos Islands has a different species of finch (fig. 2.6).

Darwin reasoned that, because of the rarity with which finches could travel from the mainland to the islands and between islands, the finches on each island were *reproductively isolated* from each other as a result of their geographic separation. Since there were few other species of birds on the islands to compete with the finches, the finches on each island could become specialized to exploit different ways of life. For example, the finches on the mainland are primarily seed eaters, but some of the Galápagos finches became insect eaters. This required changes in the length and shape of the beak, behavioral changes, and other modifications.

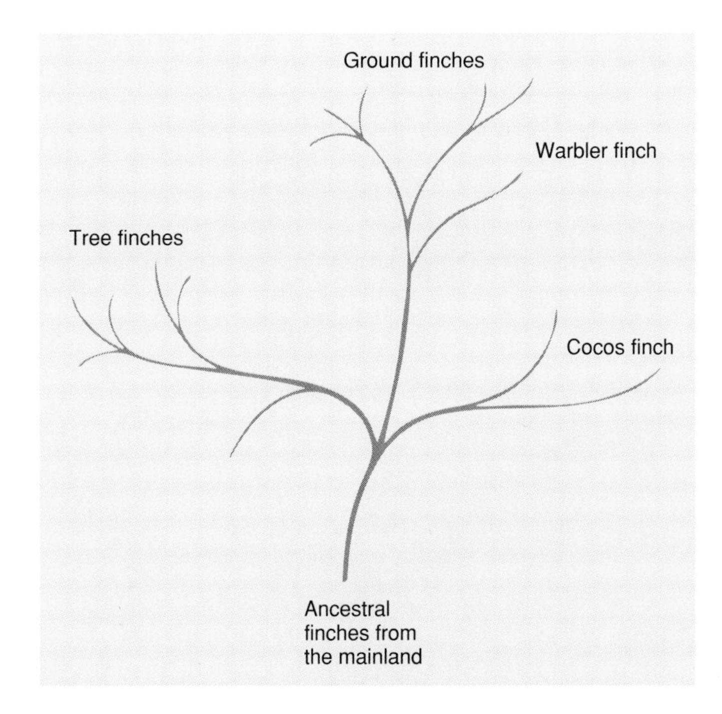

Figure 2.7. *A single ancestral species can give rise to a number of new species in the process of adaptive radiation.*

Eventually, the changes became sufficiently marked that, even when finches from different islands are artificially brought together, they are incapable of mating. They have become separate species. One species—the one on the mainland—thus gave rise to a number of different species in a process known as **adaptive radiation** (fig. 2.7). The formation of new species as a result of geographic separation of different populations of a species is known as **geographic,** or **allopatric, speciation.** This and other processes that result in the production of new species from other, older species is termed **evolution.**

Ecologists use the term *ecological niche* to refer to the environmental requirements of a species, and to all of the interactions between a species and its environment. In the case of Darwin's finches, differences in directional selection on each island favored different characteristics. A characteristic that adapted the finches to one ecological niche and thus enhanced fitness on one island was different from that which enhanced fitness on another island. The finches on each island, by this means, became adapted to different ecological niches.

Gradualism versus Punctuated Equilibrium

Darwin believed that evolutionary changes occurred gradually and very slowly, with small changes accumulating over long periods of time. This interpretation of the course of evolution is termed **gradualism,** and is still accepted as the way that many, and perhaps most, evolutionary changes occur. In certain cases, however, there is evidence that relatively rapid spurts of changes occur between long periods of stability. This concept of a jumpy rate of evolution is called **punctuated equilibrium.** It should be noted, however, that "rapid" is a relative term. Over the geologic time periods in which evolution occurs, 100,000 years could be considered rapid. For a species that matures and reproduces in a year, this amounts to 100,000 generations. Many species mature and reproduce in less time, so there is ample opportunity for great changes in the gene pool to occur.

Extinction

As new species arise due to the forces of natural selection, other species may become extinct. **Extinction** refers to the irretrievable loss of a species. The fossilized remains of extinct species provide scientists with clues as to the origin of their living descendants. The stages in the evolution of the horse (fig. 2.8), for example, can be clearly followed by examining the fossilized remains of the extinct ancestors of the modern horse.

The ages of fossils can be determined by a number of methods. Fossils that are buried in a particular layer of rock are assumed to be at least as old as the age of the rock in which they are encased. The rock, the fossils, and remains that are not old enough to have become fossilized, can be dated by means of **radioisotope dating.** This method utilizes the fact that a radioactive atom, by emitting radiation, decays to a different, stable, atom at a known rate. Radioactive carbon (^{14}C) decays to a nonradioactive, stable isotope at such a rate that half of a given amount of ^{14}C will decay in 5,730 years. This method of measuring the rate of decay of a radioactive atom is termed its *half-life.* Since ^{14}C has a half-life of 5,730 years, there is very little of it left after about 40,000 years. Objects older than this, therefore, must be dated using other radioactive atoms. Those that are used include uranium (that decays to lead) and potassium (that decays to argon), which have half-lives in the billions of years.

Through examination of the ages of fossils and the layers of rocks in which they are associated, the course of the evolution of life on earth is fairly well understood in its broad outlines (fig. 2.9). At least it is well understood by some people. A survey, published by the National Science Foundation in 1989, revealed that nearly two-thirds of adult Americans believe that people inhabited the earth at the same time as dinosaurs! This erroneous idea probably originates from exposure to such T.V. shows as *The Flintstones* and movies such as *One Million Years BC.* The facts, however, are that dinosaurs became extinct about sixty-six million years ago, whereas the first members of the family *Hominidae* did not appear until five million years ago (and these early australopithecines were a far cry from modern *Homo sapiens*). At the time the dinosaurs became extinct, the only mammals on earth were small, rodentlike creatures.

Figure 2.8. *The evolution of the modern horse as deduced from fossil evidence.*

Not only dinosaurs, but a great many other types of organisms became extinct by about sixty-six million years ago. The time of these mass extinctions corresponds to the boundary between the *Cretaceous* and the *Tertiary* periods (see fig. 2.9), as can be seen in rock layers formed at this time. A group of scientists led by Luis Alvarez (a Nobel prize winner in physics) proposed that the mass extinction that occurred at this time was due to the impact of a very large meteorite. This was suggested by the fact that the rocks at the Cretaceous-Tertiary boundary were enriched in the element *iridium,* which is rare on earth but common in meteorites.

Although some scientists believe that the extinctions were caused more gradually by extreme volcanic activity rather than more abruptly by a meteorite, the basic concept is similar. Whether caused by vulcanism or a meteorite, huge amounts of debris were thrown high into the air. This blocked out the sun for years, causing the death of many plants. The extinction of particular plants produced a drastically altered ecology that in turn resulted in the extinction of many interdependent species. Frighteningly, a number of eminent scientists have shown that nuclear war could have basically the same effect on the species in existence today. Nuclear explosions would directly throw debris into the air, and—of far greater consequence—would cause planetwide fires that have the same effect. The resulting changes in climate that could occur have been labeled **nuclear winter.**

Era	Period	Epoch	Millions of years from present	Animal life
Cenozoic	Quarternary	Recent	.01	Age of human civilization
		Pleistocene	2.5	Many mammals became extinct. First human culture.
	Tertiary	Pliocene Miocene Oligocene Eocene Paleocene	7 25 38 54 65	Dominance of land mammals, birds, insects Mammalian radiation First hominids
Mesozoic	Cretaceous		136	Dinosaurs reach peak, then become extinct Second great radiation of insects First primates
	Jurassic		195	Dinosaurs large, specialized, more abundant First mammals appear First birds appear
	Triassic		225	First dinosaurs appear Mammallike reptiles evolve
Paleozoic	Permian		280	Expansion of reptiles Decline of amphibians
	Carboniferous	Pennsylvanian Mississippian	321 345	"Age of Amphibians" First great radiation of insects First reptiles appear
	Devonian		395	"Age of Fishes" First land vertebrates, the amphibians, appear
	Silurian		435	First air-breathing land animals, such as land scorpion, appear Rise of fishes
	Ordovician		500	Many marine invertebrates First vertebrates appear as fish
	Cambrian		570	Diverse primitive marine invertebrates, trilobites common Animals with skeletons appear
Proterozoic	Precambrian		1500	Eukaryotes evolve, multicellular organisms evolve
			2500	Age of prokaryotes (single-celled organisms without a true nucleus)

Figure 2.9. *The history of life on earth, as reconstructed from the fossil evidence.*

Extinctions Caused by Human Activity

The fact that species can become extinct should be obvious to anyone who has heard about our currently endangered species, such as the California condor (fig. 2.10). In fact, the selective pressures imposed by humans and our associated animals (including dogs, cats, pigs, and rats) have caused the extinction of many species. This can occur either as a result of excessive hunting, the destruction of the environmental requirements of a species by human activity (such as destruction of forests, creation of dams, etc.), or the human introduction of foreign species that outcompete the indigenous species for an ecological niche.

Figure 2.10. *The California condor. This very large vulture is almost extinct. Scientists are attempting to save this species through a captive breeding program.*

Figure 2.11. *The white dodo.*

The California condor is endangered primarily as a result of human destruction of its habitat. The *dodo,* a flightless bird (fig. 2.11) that was once found on the Mascarene Islands, was brought to extinction in the seventeenth century through excessive hunting. (Actually, the term "hunting" is inappropriate; the dodo was so stupid a person could walk right up and kill it—its name is derived from the Portuguese *doudo,* which means simpleton.) A bird twelve feet tall (known as a *moa*) once lived in New Zealand; it was hunted to extinction by the Maori people who landed in New Zealand in A.D. 1300.

In North America in 1672, an observer wrote: "I have seen a flight of pigeons that to my thinking had neither beginning nor ending, length nor breadth, and so thick I could not see the sun." These were the now-extinct *passenger pigeons.* A later observer calculated a single flock at 240 miles long, containing 2,300,000,000 birds. John James Audubon (1785–1851), a famous American naturalist, described what happened to one flight of pigeons over Kentucky in 1813:

> The people were all in arms, and the banks of the Ohio were crowded with men and boys incessantly shooting at the pilgrims. . . . Multitudes were thus destroyed. For a week or more, the population fed on no other flesh than that of Pigeons, and talked of nothing other but Pigeons. . . . No one dared venture near the devastation. The picking up of the dead and wounded birds was put off till morning. The Pigeons were constantly coming, and it was passed midnight before I noticed any decrease in the number of those arriving. . . . Then the authors of all of this devastation began to move among the dead, the dying, and the mangled, picking up the Pigeons and piling them in heaps. When each man had as many as he could possibly dispose of, the hogs were let loose to feed on the remainder.

In the span of about one hundred years, the population of passenger pigeons went from billions to zero. A bill to protect the birds was proposed in 1857, but was defeated because most people could not imagine that a species as prolific as the passenger pigeon could be in danger. The last of the great pigeon hunts occurred in 1878. Hunters killed three hundred tons of birds in a month, enough to load five railroad freight cars per day for thirty days. The last passenger pigeon in the wild was shot in 1907; the last one in captivity died on September 1, 1914.

1. Explain why the characteristics of a species remain stable as long as the environment remains stable.
2. Explain how a drastic change in the environment can produce changes in a species over time.
3. Describe how adaptive radiation occurs in the process of geographic speciation.
4. Did cavemen eat dinosaur meat? Explain your answer.

Evidence for Evolution

Evolutionary processes on a small scale have been directly observed and documented. Such direct evidence of evolutionary processes include changes in the English pepper moth and the development of antibiotic resistance in bacteria and of pesticide resistance in insects. Larger evolutionary changes over long geological time periods are documented by indirect evidence. These sources of evidence include the fossil record;

anatomical and molecular similarities between related species; repetition of evolutionary changes during embryonic development; and the existence of vestigial structures.

Techniques in radioisotope dating, and evidences from many of the physical sciences (geology, astronomy, physics, and chemistry), provide overwhelming support for the idea that the earth is about 4.6 billion years old. There thus appears to be ample time for the environment to have caused changes in species. The sources of evidence that this evolution of species has occurred can be grouped into two categories: changes in species over relatively short periods of time, which is documented by human observation; and changes in species over longer time periods, which could not have been directly observed by humans but which can only be inferred from indirect evidence.

Direct Observations of Evolution

There is no question that evolution can occur, because people have observed it. These observed evolutionary changes are necessarily limited in scope—great evolutionary changes could not have occurred in the short time that the science of biology has been in existence. Nevertheless, the changes that have occurred due to natural selection acting over a short time show that larger changes over long geologic time periods could have been produced by similar processes.

Industrial Melanism

One of the best documented examples of evolutionary change is the case of **industrial melanism.** Prior to the mid-nineteenth century, the *peppered moth* in England had wings that were light colored with some speckles. The darker speckles are areas rich in the pigment *melanin.*[1] There were occasional variants of this species that had completely dark wings (with more melanin), but they were so rare that they were prized by collectors of the time. In the major cities of England during the industrial revolution, when soot filled the air and covered the surfaces of buildings and trees, the dark-colored variants were observed to become much more common. In time they became the dominant type of the peppered moth.

Evolutionary theory easily explains this change. Due to random mutation, some dark moths were produced prior to the industrial revolution. These variants were less fit than the lighter moths, however, because they could not be as well camouflaged against the light background of the trees (fig. 2.12). Birds that ate the moths provided the selective pressure that made the darker variant less fit. After the industrial revolution, when soot darkened the trees, the darker variants were better camouflaged and thus more fit than the lighter moths. This theory was validated by experiments in the 1950s, in which both light and dark moths were grown and released within given environments. The lighter ones became predominant in unpolluted areas, whereas the darker variants became predominant in areas that were blackened by soot.

[1]melanin: Gk. *melas,* black

Figure 2.12. *The light-colored pepper moth (a) would be well camouflaged against its normal unpolluted background (b), but is starkly revealed against the polluted background. The dark moth (b), however, is well adapted to be hidden from predators in the polluted environment (a).*

Resistance to Antibiotics and Pesticides

It is unfortunately a common observation that certain strains of bacteria that are resistant to particular antibiotics develop in hospitals. The development of this antibiotic resistance is easy to explain and demonstrate in a laboratory. First, a culture of bacteria that is susceptible to a particular antibiotic is layered on a nutrient medium (*agar*) which contains this antibiotic (fig. 2.13). Millions of bacteria are present on the agar; of these millions, perhaps one or two contain a mutation that makes them resistant to the antibiotic. The antibiotic is the selective agent. Within a few days, all of the susceptible bacteria have died, whereas the antibiotic-resistant bacteria have thrived and reproduced. Reproduction is by cell division and is thus extremely rapid. In a short period of time the dish will contain millions of these antibiotic-resistant bacteria.

Through similar mechanisms, certain insect pests become resistant to pesticides sprayed on crop plants. It is important to point out that an individual insect does not develop the resistance. If it is susceptible to the poison, it dies. Some mutants that are resistant to the pesticide, however, will be present out of the millions of insects in the population. These mutants will have a selective advantage and thus will survive and reproduce. Within a few seasons, the same pesticide may not provide protection against that species of insect. This poses a continuing problem for farmers, and has led to the greater use of biological mechanisms for the control of insect pests.

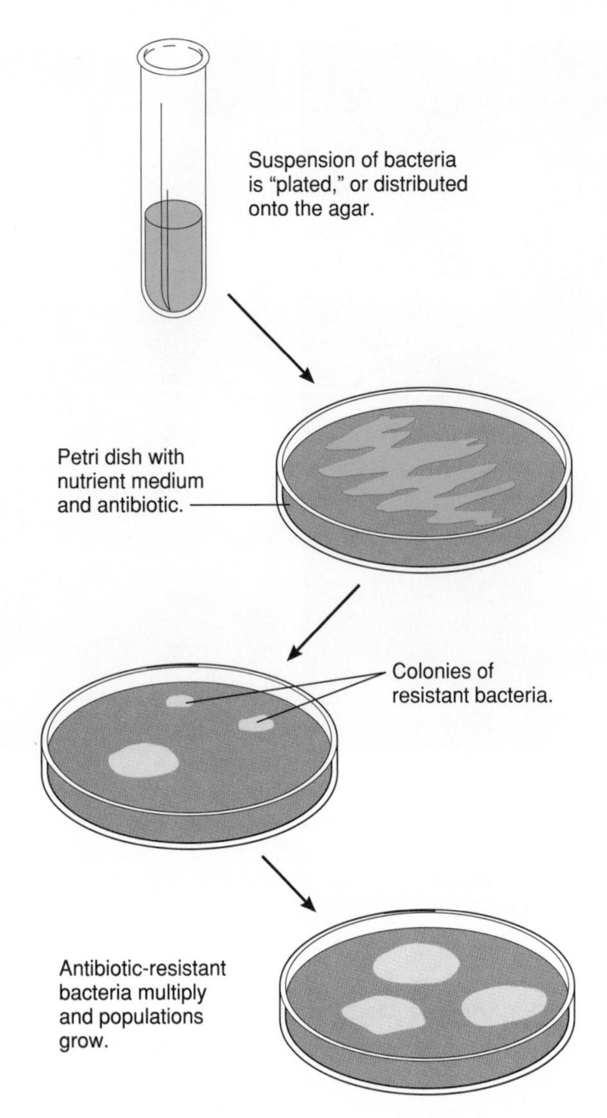

Figure 2.13. *An antibiotic can serve as an agent of natural selection for bacteria. Those bacteria that are resistant to the antibiotic have a selective advantage and can grow and multiply to produce new populations that have this characteristic.*

Indirect Evidence for Evolution

Evolutionary change that we can directly observe or produce artificially is sometimes called *microevolution.* Larger evolutionary change that occurs over geologic time periods is known as *macroevolution.* In the absence of a time machine, it is as impossible to observe macroevolution as it is to observe the signing of the Declaration of Independence. Like the signing of the Declaration of Independence, or the existence of the ancient Roman Empire, the proof of macroevolutionary changes must be indirect. Actually, this is true of most of the subjects studied by scientists. Nobody has ever seen an atom, much less an electron; nobody has ever directly seen energy or a photon; and nobody has ever directly observed the earth orbiting the sun. Indirect evidence may not be as emotionally satisfying as direct observation, but it is all we can have for the study of most natural processes.

Figure 2.14. *Cast of an egg and hatchling of a duck-billed dinosaur. The adults were more than 30 feet long.*

Fossil Evidence

We know that there was such a thing as the Declaration of Independence because we have a copy of it; we know that there was such a thing as cavepeople because, among other reasons, we can see their paintings (see fig. 2.24). Similarly, we know that there were such things as dinosaurs and other creatures no longer among us because we have their fossilized remains (fig. 2.14).

In many cases, such as the evolution of the modern horse previously discussed (fig. 2.8), the fossil record provides us with a succession of similar species that lead to their modern representative. Radioisotope dating tells us their approximate age, and so we can determine the evolutionary development, or **phylogeny,**[2] of the modern species. The principles of evolution are thus the most logical and simple explanation for the existence of the extinct species that provided the fossils, and for the relationship between the extinct species and their modern descendants.

Evidence from Anatomical Homology

Similar anatomical structures in related organisms that appear to have been derived from a common ancestor are called **homologous**[3] **structures.** The forelimbs of the tetrapod (four-limbed) vertebrates, for example, all contain the same bones (fig. 2.15). There is no functional reason why the flipper of a whale, the wing of a bat, and the arm of a human should have the same basic structure, but they do. All have one arm bone leading from the shoulder (the *humerus*), two bones in the forearm (*radius* and *ulna*), wrist bones (*carpals*), hand bones (*metacarpals*), and finger bones (*phalanges*). Again, the most logical explanation is that they are all derived from the forelimb of a common ancestor.

[2]phylogeny: Gk. *phylon,* tribe; *genesis,* generation
[3]homologous: Gk. *homologos,* correspondent

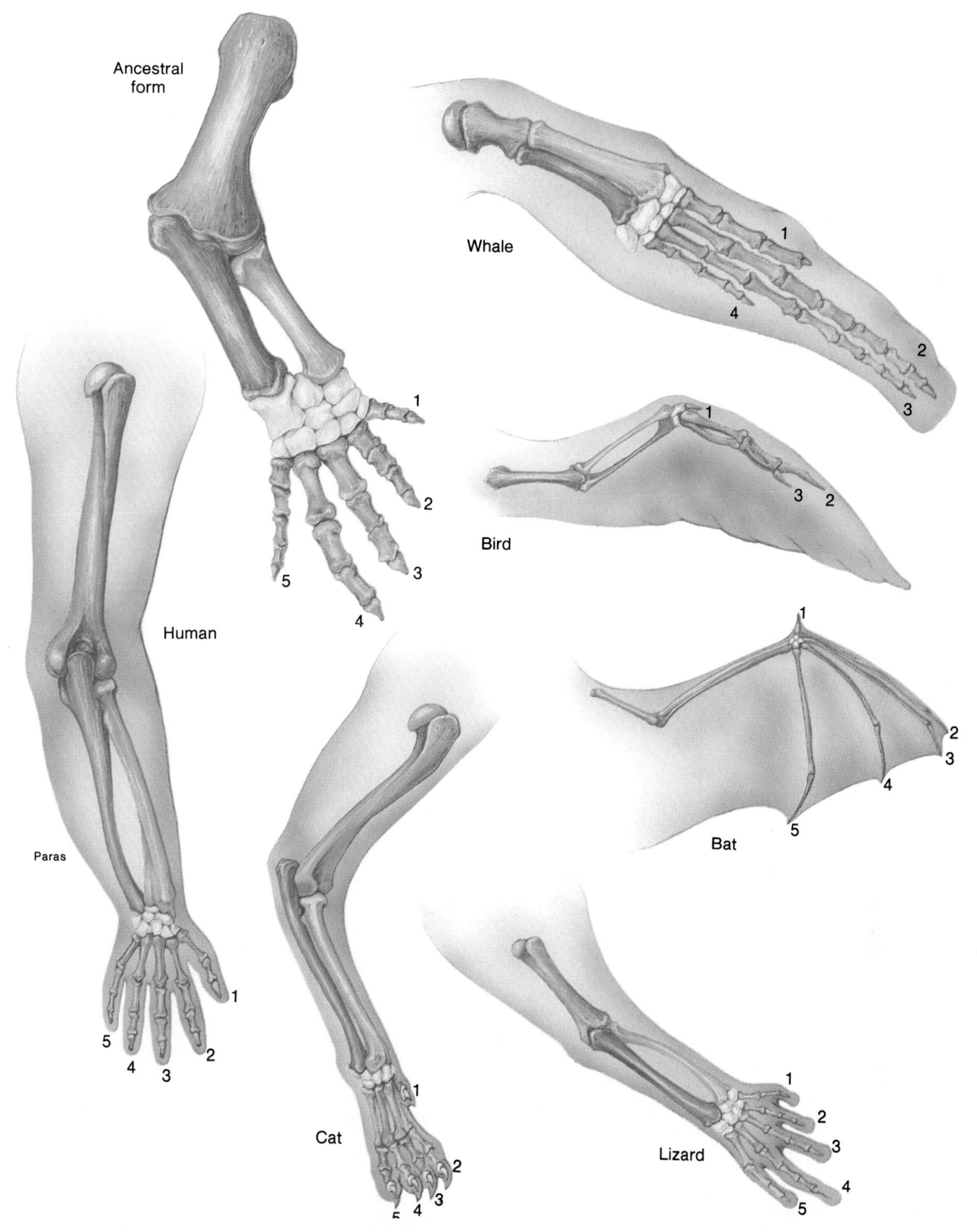

Figure 2.15. *The bones of the forelimbs of all vertebrates are similar, and appear to be derivatives of a common ancestral form. These bones are thus said to be* homologous.

Molecular Evidence of Homology

Not only are anatomical structures similarly constructed in related species, the fundamental molecules within their bodies are likewise similar. In recent years, scientists have been able to map the structure of a number of proteins. Proteins are very big molecules composed of subunits known as *amino acids* (chapter 3). There are over twenty different types of amino acids, and they are put together to make proteins according to the instructions contained in the genes (chapter 10). If there is a mutation at a single point in the genetic code (DNA), a different amino acid

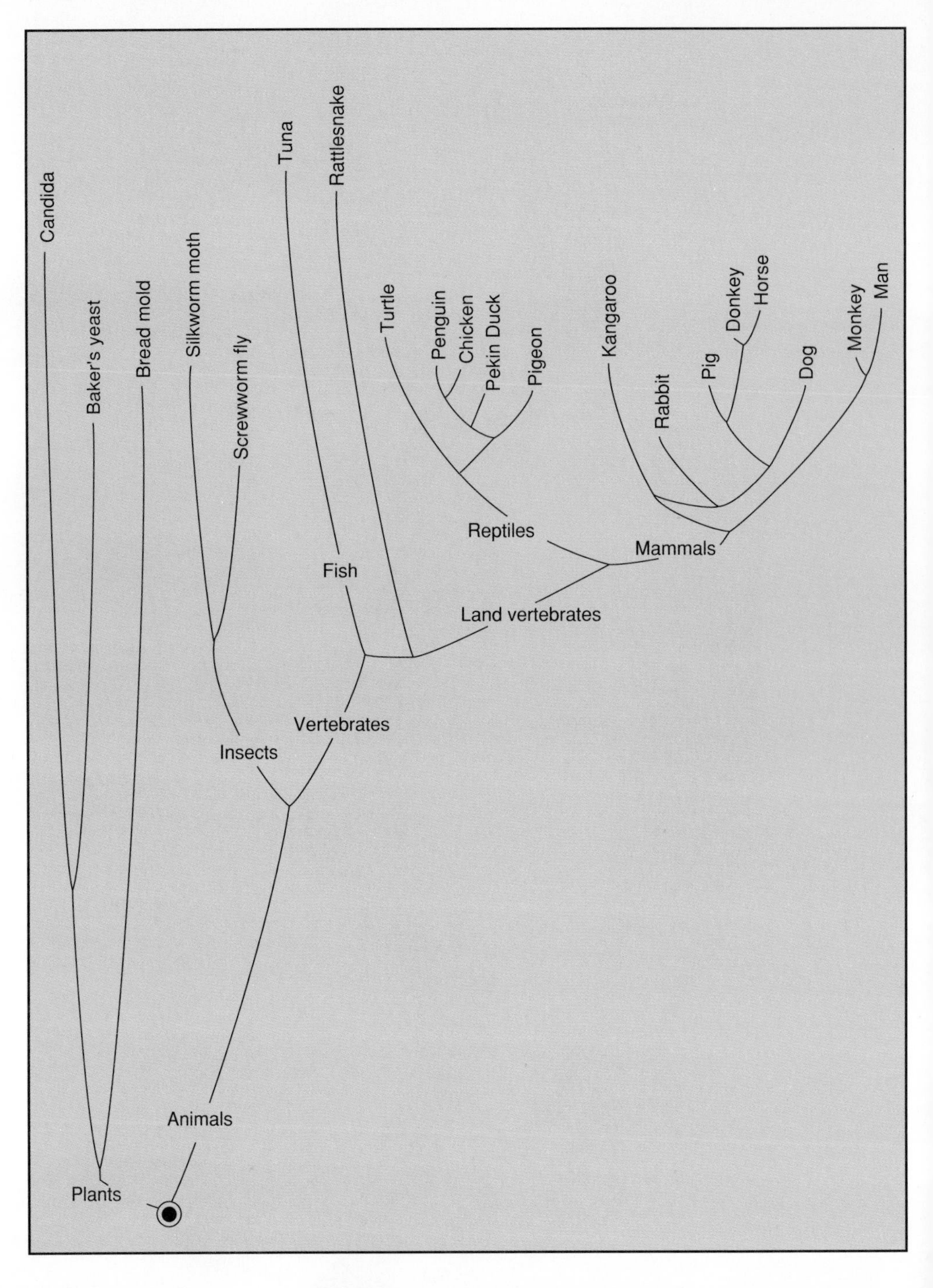

Figure 2.16. *Phylogenetic tree based entirely on the differences in the amino acid composition of cytochrome c of different organisms. The lengths of the vertical lines are correlated to the degree of differences in the amino acid sequence of cytochrome c of the organisms.*

will be placed into a particular position in a specific type of protein. Changes in the amino acid sequence of a specific protein, therefore, indicate that such mutations in the DNA have occurred.

One of the proteins that has been studied intensively for amino acid homologies is *cytochrome c*. It appears that the gene that codes for this protein mutates at a fairly constant rate. Therefore, two species that are closely related will be more similar in the amino acid composition of their cytochrome c than will be two species that are more distantly related. Study of the amino acid composition of cytochrome c and other proteins has produced phylogenetic trees (fig. 2.16) of evolutionary relationships that are similar to those produced by anatomical homologies. These similarities confirm the validity of using homologies to determine the evolution of species.

Evidence from Development

During the embryonic and fetal development of all chordates, including humans, the embryo passes through common stages. Fish, frogs, ducks, and humans all have a rudimentary tail, a notochord (chapter 1), and pharyngeal pouches (fig. 2.17) at some time during their development. The pharyngeal pouches become gills in fish; they disappear or become partially incorporated into other structures in land vertebrates.

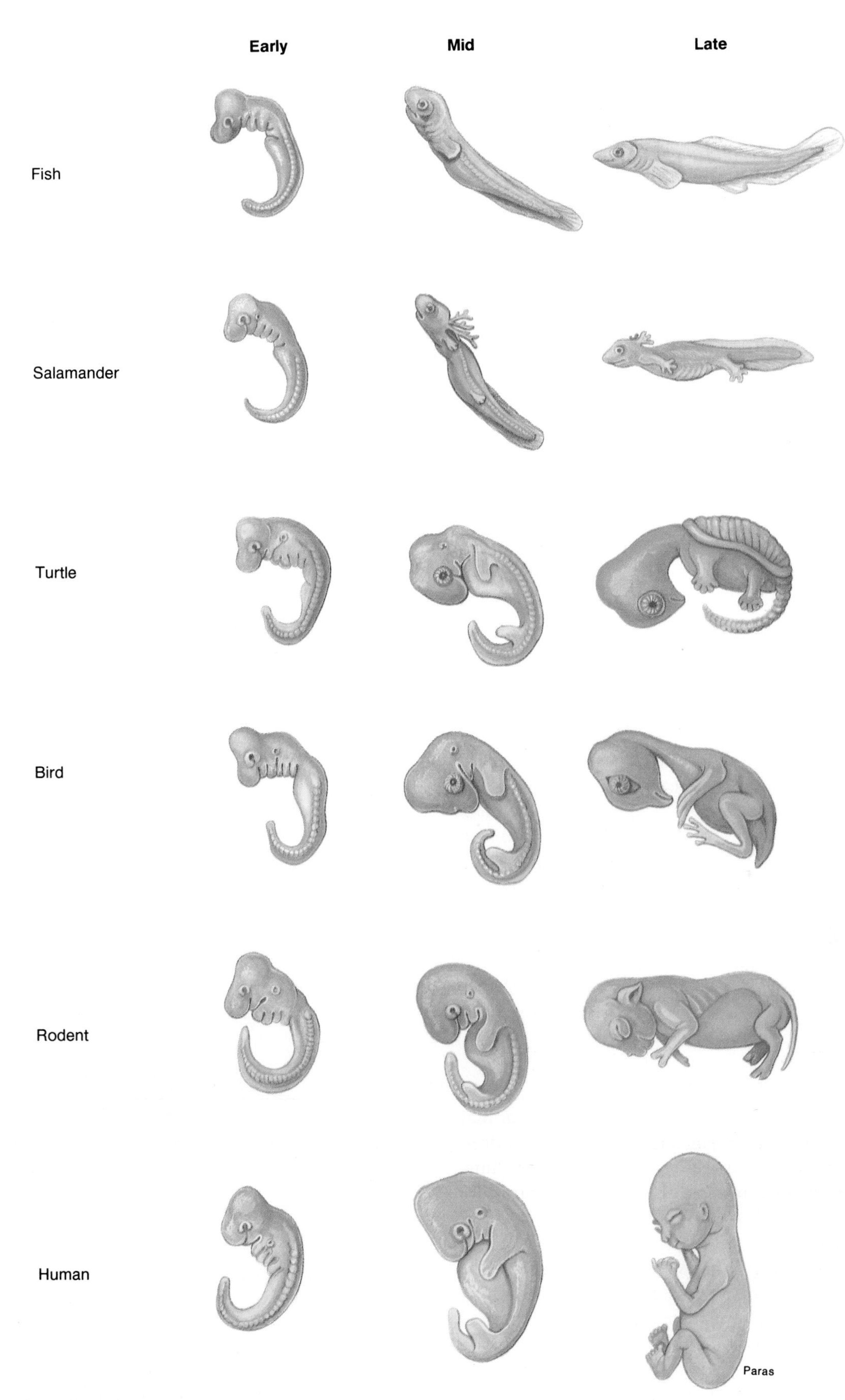

Figure 2.17. *Stages in the embryonic development of a number of vertebrates. Note the similarities of structure.*

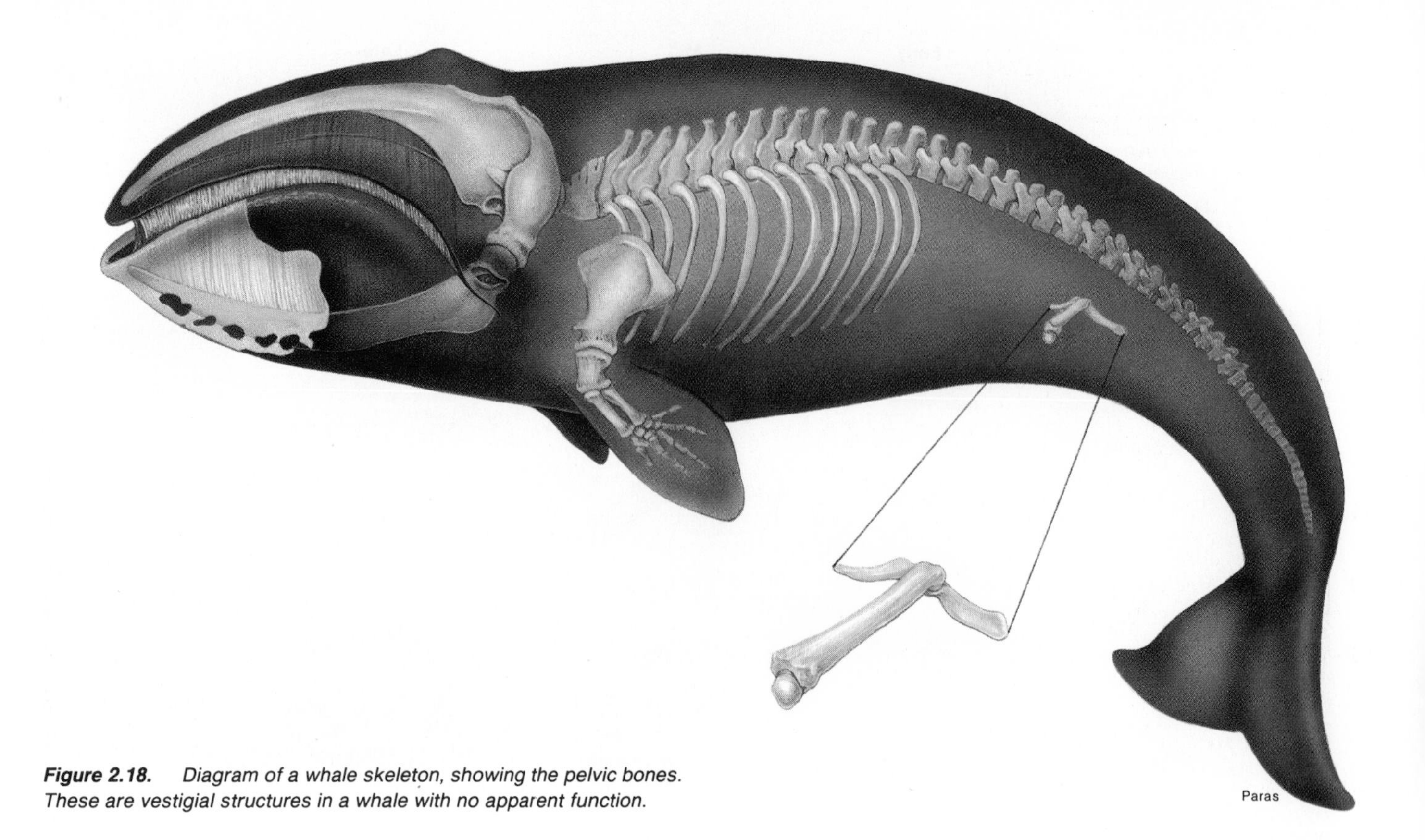

Figure 2.18. *Diagram of a whale skeleton, showing the pelvic bones. These are vestigial structures in a whale with no apparent function.*

The most logical explanation for these similarities in development is that the genetic changes that were added later in evolution are superimposed on the earlier genetic plan. The course of the evolution of the species is seen as the individual develops. This is described by the phrase *ontogeny recapitulates phylogeny.* The term *ontogeny*[4] refers to the changes that occur during the embryonic and fetal development of an individual; the term *phylogeny* refers to the changes that occur during the evolutionary development of a species. The phrase is not entirely accurate—humans do not become fish, frogs, and so on during their embryonic development. Still, the appearance of human embryos at different stages of development support the basic concept of a recapitulation of phylogeny.

Evidence from Vestigial Structures

Vestigial[5] structures are anatomical features of a species that have no apparent functional purpose, but appear to be derived from stuctures in an ancestral species that were at one time functional. One fascinating example of vestigial structures are the pelvic bones of a whale (fig. 2.18). The pelvic bones of four-legged vertebrates articulate with the lower limbs and help to support the body in walking. Whales have no lower limbs, and do not walk, yet they still have a pelvis. It is a useless structure, only present because it is the evolutionary "baggage" inherited from their terrestrial ancestors.

[4]ontogeny: Gk. *on,* existing; *gennan,* to produce
[5]vestigial: L. *vestigium,* a trace

Humans have a famous vestigial structure—the *coccyx.* This is a short and functionless tailbone composed of three to five rudimentary vertebrae at the end of our vertebral column (backbone). The *vermiform appendix,* a wormlike projection from the beginning of our large intestine, may also be a vestigial organ. This is controversial, because the appendix may have some immune function in young children. In adults, it serves absolutely no worthwhile function, and in fact is a handicap. Inflammation of the appendix—*appendicitis*—can be a dangerous condition. The presence of the coccyx, and perhaps the appendix, is not an adaptation; its presence is best explained as a vestigial structure inherited from a more primitive ancestor.

1. What is the evidence that species are capable of changing in the natural world? Give at least three examples of such changes that have been directly observed.
2. Define the term *homologous structures,* and provide examples that suggest that all vertebrates descended from a common ancestor.
3. Define the term *vestigial structures* and give examples of such structures.
4. Explain the phrase *ontogeny recapitulates phylogeny.*

Human Evolution

The australopithecines were the first hominids. They arose nearly five million years ago, and appear to be the ancestors of *Homo habilis.* This was the first member of the genus *Homo,*

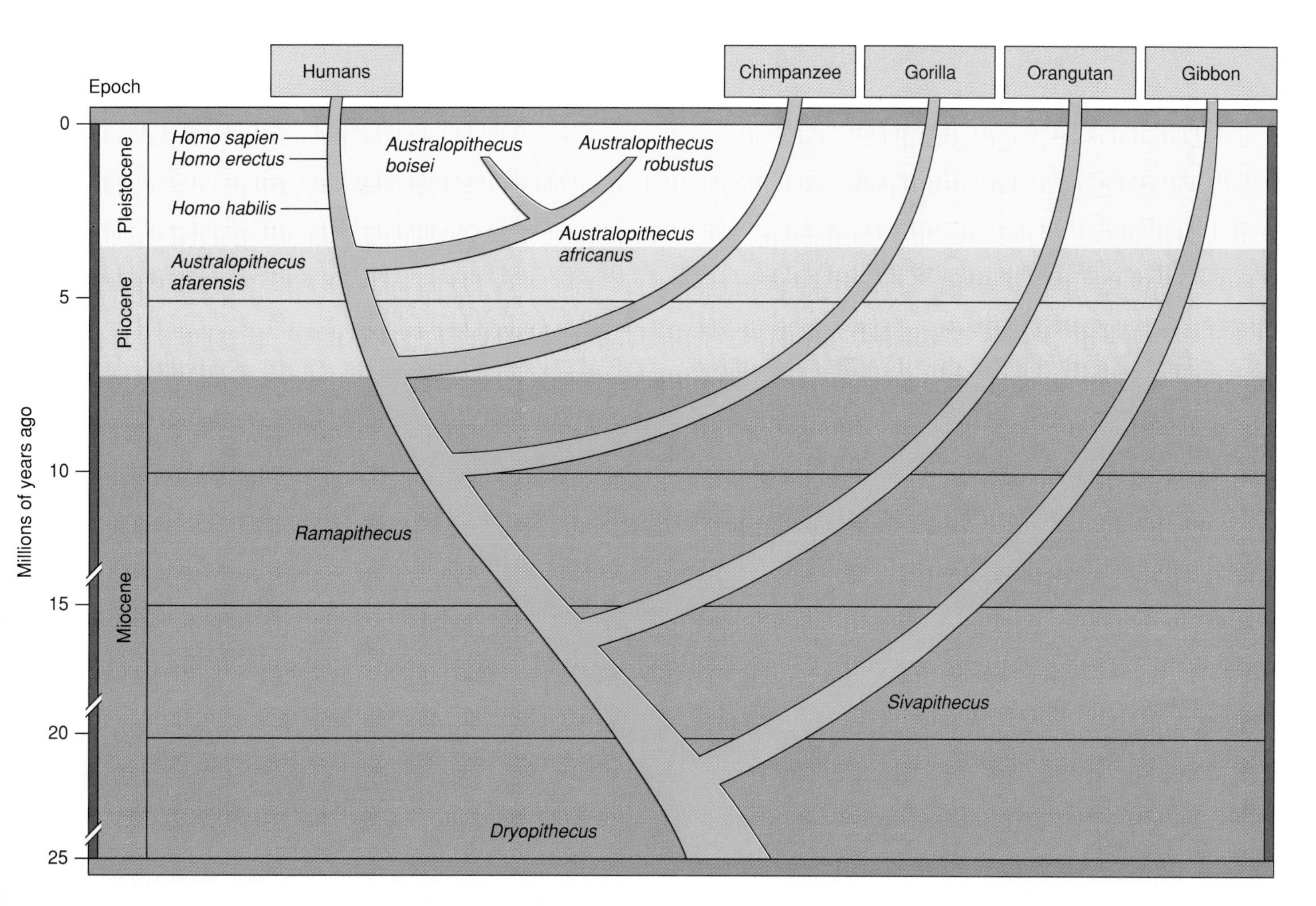

Figure 2.19. *The phylogeny of humans and apes. It should be noted that opinions differ in regard to the specific relationships of the hominid lineage.*

and appeared about two million years ago. *Homo habilis* may have given rise to *Homo erectus,* which spread out of Africa. *Homo erectus,* in turn, probably gave rise to *Homo sapiens.*

The hominid family has been in existence for about five million years (fig. 2.19). If we had a time machine that went back that far, however, we would only find these ancestors of modern man in a limited area of Africa. Indeed, that would be the case up to about 1.5 million years ago. Afterward, we would also find hominids in Asia and Europe, but we would have to search intensively for these uncommon animals. As late as 10,000 years ago, when completely modern *Homo sapiens* was dominant and starting the agricultural revolution, there were only ten million people on the entire planet! (This is one-third the current population of California alone.) Considering the scarcity of human ancestors, it is no wonder that scientists have to labor so hard to find ancient hominid fossils. As a result of these labors we now have a fairly complete picture of the evolution of humans. As in all scientific endeavors, however, there are areas of intense controversy and our current concepts of human evolution will almost certainly be modified as new fossils are discovered.

Australopithecines

The term *australopithecines* refers to members of the genus *Australopithecus.* As described in chapter 1, members of this genus first appeared about five million years ago and became extinct about 1.3 million years ago. The first evidence of *Australopithecus* was the skull of a child from Taung in South Africa (fig. 2.20). Later, a complete skeleton (named "Lucy," after a Beatles song that was playing at the time of its discovery) was found and dated at three million years old (fig. 2.21).

An adult australopithecine weighed up to 20 kilograms (kg), was up to 1.3 meters (m) tall, and had a brain size of 350–450 cubic centimeters (cc). This is approximately the size of the brain of a modern gorilla. Though the skeleton suggests that the australopithecines looked more like apes than humans, they had humanlike teeth and walked upright (were bipedal). The upright posture and bipedal locomotion were dramatically revealed by fossilized footprints (fig. 2.22) discovered at Laetoli in Africa by Mary Leakey. These footprints were created by a female *A. afarensis* walking across an area of damp volcanic

Figure 2.20. Raymond Dart, its discoverer, holds a cast of the Taung child skull. This was the first australopithecine found.

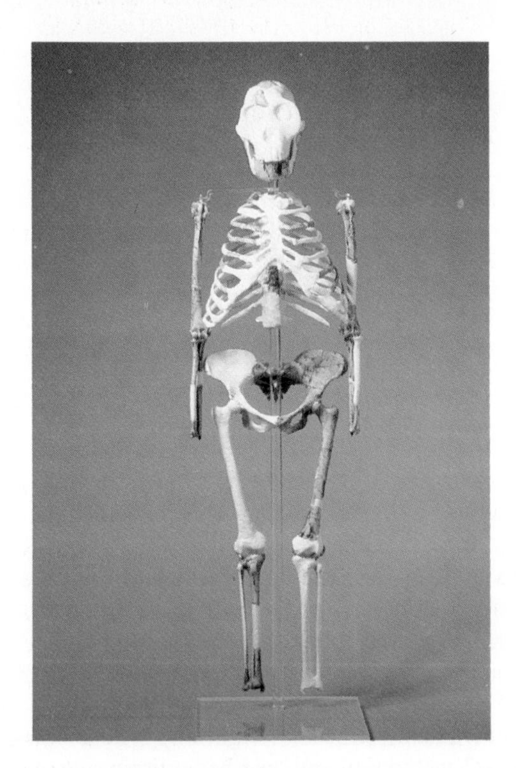

Figure 2.21. The skeleton of "Lucy." This is the most complete australopithecine skeleton yet discovered.

Figure 2.22. Footprints at Laetoli in Tanzania, made by an australopithecine 3.6 million years ago.

ash. The footprints were subsequently covered by additional volcanic ash and other deposits, to be revealed ages later by the forces of erosion.

Homo habilis and *Homo erectus*

Homo habilis was the first member of the genus *Homo* to appear approximately two million years ago. These hominids may have been more apelike than was previously believed. Whereas the humerus (arm bone) of modern humans is only about 70% as long as the femur (thighbone), the humerus of *H. habilis* was 95% as long as its femur. This is similar to the relationship seen in modern chimpanzees. Nevertheless, *H. habilis* had a larger brain (650–800 cc) than that of modern apes, and was the first hominid to flake stone to produce tools for cutting and scraping. They ate meat, although it is not known if they were hunters or scavengers.

Homo habilis probably became extinct about 1.6 million years ago. Most scientists believe that it was the ancestor of *Homo erectus,* which appeared about 1.7 million years ago. *H. erectus* had a brain volume of 800–1,000 cc, fashioned stone tools and could use fire, and appear to have developed cooperative methods of hunting. *H. erectus* was the first hominid to spread out of Africa, as skeletons of this species have been found in Asia and Europe. Since *H. erectus* may have lived until as

Social Issues

The Teaching of Evolution

Ever since the *Origin of Species* was first published, the theory of evolution has been attacked by those who insist that the biblical account of genesis be accepted literally. Of course, a literal interpretation of biblical genesis and the theory of evolution cannot both be true. People who follow the scientific method regard the biblical account as allegory. People who want to believe that the biblical description is literally true must also believe that the evidence for evolution is somehow false. This conflict has resulted in political repercussions that continue even today.

This conflict is presented very forcefully and dramatically in the book and movie *Inherit the Wind.* The story is based on a trial that actually took place in the United States in the 1920s. One of the states passed a law outlawing the teaching of evolution, and a high-school teacher was tried for breaking this law. Clarence Darrow was a famous lawyer who defended the teacher. William Jennings Bryan—a nationally renowned figure and three-time presidential candidate—was the attorney for the prosecution. Because of the fame of the central protagonists, the seeming backwardness of the law to most of the country, and the acerbic commentary by one of America's foremost newpaper columnists, H. L. Menken, the trial received national attention. The teacher was found guilty and received a negligible fine. More importantly, the state law banning the teaching of evolution was subsequently declared unconstitutional.

In 1988, a different state law requiring that evolution and "creation science" be taught as equally possible scientific theories was declared unconstitutional. This is because the Supreme Court found that "creation science" was not science at all, but religion. It is important for people to understand the reasoning behind this assessment.

As described in chapter 1, a scientific hypothesis must be (1) based upon observation; and (2) testable—it must be able to be proven false by observations of the natural world. "Creation science" fails on both accounts. It is based, not upon observations of the natural world, but on the belief in the literal truth of biblical genesis. Thus, evidence supporting the concept of evolution must be made to fit this interpretation. For example, some creationists explain the fossil record as due to successive God-induced catastrophes and subsequent creations of new species. That may be true. It also may be false; the scientific method has no way of distinguishing between these two possibilities. And that is precisely the point. The creationists' interpretation of the fossil evidence is not derived from scientific observations, nor could it ever be falsified by additional observations of the natural world.

Despite claims to the contrary by creationists, nearly all biological scientists accept the fact of evolution (one scientist cited by creationists as rejecting evolution was, in fact, playing devil's advocate at a scientific meeting and was quoted out of context). Also, despite claims to the contrary by creationists, scientists have found numerous intermediate forms that show descent between extinct and living species (as in the evolution of the horse, shown in fig. 2.8). The evidence in support of evolution is overwhelming; and evolution, in turn, is a fact that serves as the central organizing principle of all the life sciences, from molecular genetics to ecology and animal behavior. The substitution of religious dogma for evolution would make it impossible for a student to understand modern biology.

Religion and science are certainly compatible. Religion deals with spiritual matters and the belief in a Supreme Being. Science deals with the physical world and natural laws that humans are capable of understanding. Neither may be complete without the other, but neither is well served by erroneously mixing the two together.

late as 200,000 years ago, they were on earth for more than one million years. This is considerably longer than can yet be said about our own species.

Homo sapiens

Homo sapiens appeared about 500,000 years ago. These older versions of our own species appeared to more closely resemble *H. erectus* than modern *H. sapiens,* however, and so they have been termed **archaic** *H. sapiens* to distinguish them from **modern** *H. sapiens.* Approximately 200,000 years ago, a group of these archaic *H. sapiens* in Europe and western Asia became a distinct group known as **Neanderthals** (fig. 2.23) The Neanderthals were stocky, powerful individuals with a heavy ridge of bone over their eyebrows. Because of their brutish appearance and the fact that they lived in caves, the Neanderthals are the prototype of the "cavemen" caricatures. It should be realized, however, that the Neanderthals had a brain as large as modern *H. sapiens* (1,400 cc), and developed a distinctly human culture. There is evidence that Neanderthals and modern *H. sapiens* coinhabited areas of the world for many tens of thousands of years, and they may have interbred. Because of interbreeding or for reasons not known, the Neanderthals disappeared between 100,000 and 32,000 years ago (depending on the part of the world).

Modern *Homo sapiens* first appeared 100,000 to 200,000 years ago. Anatomically and intellectually, they were the equals of humans today. Those modern humans who displaced the Neanderthals and occupied the caves of Europe are often referred to as **Cro-Magnons.** They had language and an intricate culture. Indeed, some of the cave paintings found in France (fig. 2.24) that were produced by the Cro-Magnons are awesome in their beauty. It is fascinating to realize that all of the changes that modern *H. sapiens* have undergone from the time we occupied caves to the present is due to cultural rather than biological evolution.

Figure 2.23. *(a) The skull of a Neanderthal (top) and a modern human (bottom). The Neanderthal had larger brow ridges and a more sloping forehead than modern humans. (b) An artist's conception of the appearance of a Neanderthal which is consistent with the skeletal evidence.*

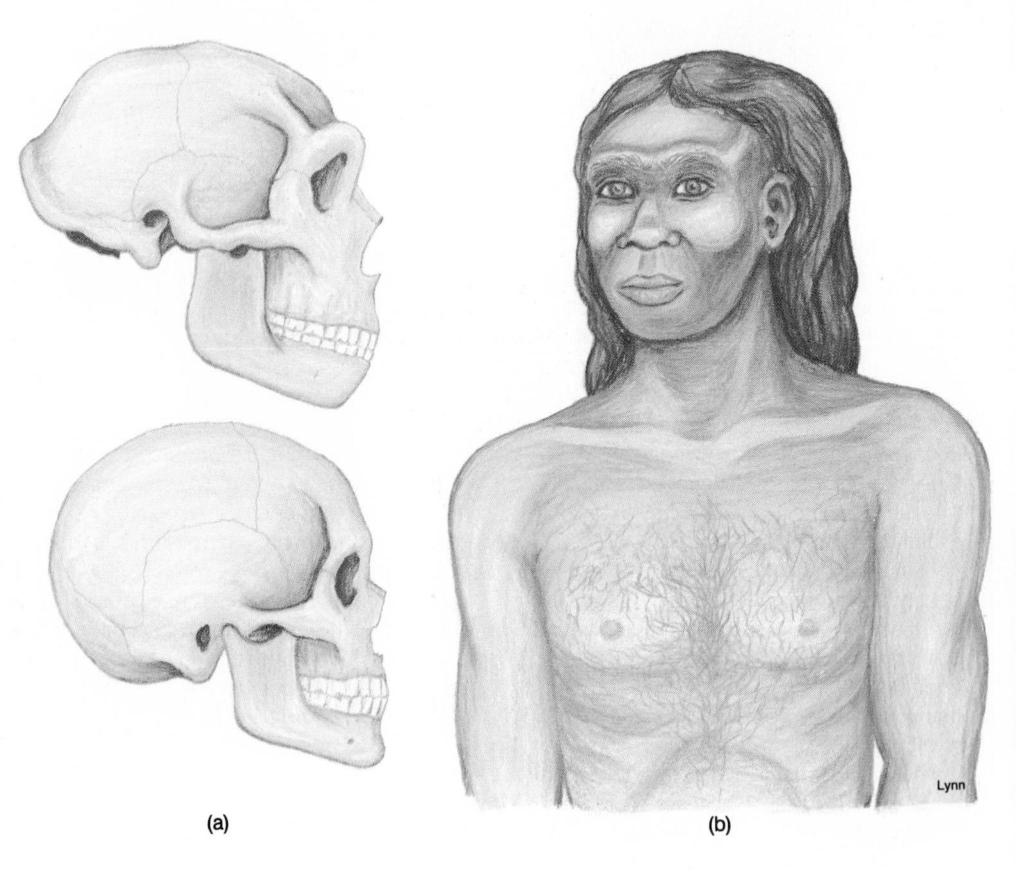

Figure 2.24. *Paintings of wild horses and aurochs at Lascaux cave in France. These paintings were made by modern Homo sapiens about 17,000 years ago.*

Headlines in 1988 proclaimed that scientists had discovered the existence of Eve, or more accurately, a **mitochondrial Eve.** The adjective refers to *mitochondria,* which are small, sausage-shaped structures within our cells that produce most of the energy used by the cell (chapter 4). Mitochondria have their own DNA (genes), which is a simpler form of DNA than that found within the cell nucleus. Since ova (egg cells) contain mitochondria but the head of sperm do not, all of the mitochondria within a fertilized ovum are derived from the mother. The only way this DNA can change is through mutations, which occur at a known rate.

Scientists studied the mitochondrial DNA of people from diverse racial and ethnic groups and discovered a region of this DNA that all people on earth share in common. Using information about the rate of mutation of mitochondrial DNA and the size of this shared region, they determined that all humans alive today may be derived from only one, or at most a few, individual females. This is the mitochondrial Eve. She lived in Africa about 200,000 years ago, and so probably was a member of archaic *H. sapiens.* This conclusion—certainly controversial—suggests that modern *H. sapiens* originated only once, from a small group of archaic *H. sapiens* in Africa.

1. Using a flowchart (with arrows), trace the course of hominid evolution from *Australopithecus* to modern *Homo sapiens.*
2. Compare the size of the brain of modern people to those of our hominid ancestors.
3. Were Neanderthals and Cro-Magnons members of separate species? Explain.
4. Explain the meaning of the term "mitochondrial Eve."

Summary

The Theory of Evolution p. 12

I. Darwin's development of the theory of evolution was influenced by Charles Lyell's work in geology and Thomas Malthus's theories in economics.

II. Darwin observed that there is variation in each species.
   A. He also observed that the number of individuals within a population remains relatively stable, even though great numbers are born.
   B. Darwin thus deduced that there is a struggle for existence among the variants of a population, and that those variants that are most fit survive and reproduce to a greater extent than the other variants.

III. Natural selection allows those individuals that are most fit to contribute proportionately more genes to the gene pool of the population.
   A. The term "fitness" is only an indication of the ability of an individual to reproduce; it does not refer to any other specific attribute or trait.
   B. Natural selection favors particular traits that are available in the population due to random mutation; it does not produce these traits.
   C. Only those traits that a particular individual inherits can be passed to the next generation; acquired characteristics cannot be inherited.

The Causes of Evolution p. 15

I. Natural selection may be described as stabilizing or directional.
   A. In a stable environment, natural selection reduces the fitness of individuals that differ too markedly from the average of the population.
   B. In a changed environment, individuals that have particular traits that made them less fit in the old environment may become more fit in the changed environment.
   C. Individuals that have the traits that make them more fit in the changed environment will contribute proportionately more genes to the gene pool of the population; over many generations, this directional selection will result in a change, or evolution, of the species.

II. When different populations of a species are separated geographically, they become reproductively isolated.
   A. The force of directional selection in one population may help that population to adapt to a new ecological niche.
   B. Over generations, the two populations that were originally members of the same species may become two new and different species, which are no longer capable of interbreeding.
   C. The derivation of new species in this way is called adaptive radiation; the evolutionary history of a species is called its phylogeny.
   D. New species may evolve very slowly through the gradual accumulation of genetic changes, or evolution may occur at varying rates; this latter process is called punctuated equilibrium.

III. Species can become irretrievably lost, or extinct; human activity has hastened the extinction of many species.

Evidence for Evolution p. 20

I. Small evolutionary changes have been directly observed by scientists.
   A. In the case of industrial melanism, the peppered moth of England was observed to evolve to a dark form as an adaptation to an environment darkened by pollution.
   B. Bacteria and insects have evolved resistances to antibiotics and pesticides, respectively.

II. Large evolutionary changes that have occurred over long geological time periods are documented by indirect evidence.
   A. The fossil record shows different stages in the evolution of modern species from their ancestors, whose fossils can be dated and organized to show the progressive stages in the phylogeny of the modern species.
   B. Anatomical structures, such as the forelimbs of vertebrates, are built in a homologous fashion, suggesting that the homologous structures were derived from a common ancestor.
   C. Homologous structures are also observed on a molecular level.
   D. Structures that are characteristic of the lower vertebrates are seen during the embryonic development of the higher vertebrates, suggesting that some of the changes occurring during phylogeny are repeating during ontogeny.
   E. Vestigial structures indicate descent from differently formed ancestral species.

Human Evolution p. 26

I. The first members of the hominid family were the australopithecines, which first appeared about five million years ago.

II. *Homo habilis* first appeared about two million years ago, and is believed to have given rise to *Homo erectus.*
   A. The australopithecines and *Homo habilis* walked upright, but resembled apes more than modern humans.
   B. *Homo erectus* appeared about 1.7 million years ago and may not have disappeared until as late as 200,000 years ago.
   C. *Homo erectus* could use stone tools; they are the first hominids to spread out of Africa to Asia and Europe.

III. An archaic form of *Homo sapiens* may have appeared as early as 500,000 years ago.
   A. Neanderthals were one group of archaic *Homo sapiens* that occupied caves and lived from as early as 200,000 years ago to as late as 32,000 years ago.
   B. Neanderthals had the same brain size as modern humans (1,400 cc), and had a distinctly human culture.
   C. The Neanderthals were eventually displaced by modern *Homo sapiens,* which first appeared between 100,000–200,000 years ago.

Review Activities

Objective Questions

Match the following people to their contributions or publications:

1. Gregor Mendel
2. Charles Lyell
3. Jean Baptiste Lamarck
4. Thomas Malthus

(a) inheritance of acquired characteristics
(b) father of modern geology
(c) wrote *Essay on the Principles of Population*
(d) father of genetics

5. Which of the following is a deduction, rather than an observation, made by Darwin?
   (a) Great numbers of individuals are born.
   (b) The survival of the fittest occurs by natural selection.
   (c) The number of individuals in a population remains relatively constant.
   (d) There is individual variation within a species.
6. Which of the following individuals is the fittest, as the term is used in evolutionary biology?
   (a) the largest individual
   (b) the most aggressive individual
   (c) the smartest individual
   (d) the individual who contributes the most genes to the future population
7. Which of the following statements is *true?*
   (a) Genetic variation is caused by natural selection.
   (b) Genetic variation can be influenced by the activity of an individual.
   (c) Genetic variation is random.
   (d) None of these.
8. In a stable environment, natural selection
   (a) maintains the genetic stability of a population
   (b) causes genetic change in a population
   (c) causes evolution of new species
   (d) none of these

9. The term "adaptive radiation" refers to
   (a) beneficial mutations caused by radioactive substances
   (b) the formation of new ecological niches
   (c) the adaptation of an individual to an ecological niche
   (d) the formation of a number of new species from a single-parent species
10. If a radioactive material has a half-life of one million years, how much of the original material will be left after five million years?
   (a) 1/2
   (b) 1/4
   (c) 1/8
   (d) 1/16
   (e) 1/32

Match the observation with its category as a source of evidence for evolution:

11. Human embryos have pharyngeal pouches
12. The human hand and the wing of a bat each contain five carpal bones
13. The existence of a tailbone in humans

(a) vestigial structure
(b) ontogeny recapitulates phylogeny
(c) molecular homology
(d) anatomical homology

Match the hominid with its description:

14. The first to use stone tools
15. The first to use bipedal locomotion
16. Produced the cave paintings in France
17. The first to leave Africa

(a) *Australopithecus*
(b) *Homo erectus*
(c) *Homo habilis*
(d) Neanderthals
(e) Cro-Magnons

Essay Questions

1. Do giraffes have long necks because their ancestors strained to reach leaves on tall trees? Explain your answer.
2. Are modern species more fit than ancestors that lived millions of years ago? Explain your answer.
3. Arrange these hominid characteristics in the order in which they first evolved: large brains, bipedal locomotion, use of stone tools. Speculate as to why these characteristics may have evolved in this order.
4. How do creationists explain the existence of a fossil record? Explain why this explanation is not regarded as being scientific, and distinguish between the terms *untrue* and *unscientific.*

3

The Chemical Basis of Human Biology

Outline

Objectives

By studying this chapter, you should be able to
1. describe the structure of an atom
2. explain how covalent bonds are formed
3. explain how ions and ionic bonds are formed
4. define the terms *acid* and *base* and explain the meaning of the pH scale
5. describe the different types of carbohydrates and give examples of each type
6. explain the mechanisms and give examples of dehydration synthesis and hydrolysis reactions
7. explain the common characteristic of lipids and describe the different categories of lipids
8. describe how peptide bonds are formed and the different orders of protein structure
9. list some of the functions of proteins and explain why proteins can have a wide variety of specific functions
10. describe the principles of catalysis and explain how enzymes function as catalysts
11. describe how enzymes are named
12. describe the effects of pH and temperature on the rate of enzyme-catalyzed reactions
13. explain how enzymes cooperate to produce a metabolic pathway
14. describe how ATP is produced and explain the significance of this molecule as the universal energy carrier

Keys to Pronunciation

amino acid: *ah-me'no as'id*
anatomy: *an-nat'o-me*
disaccharide: *di-sak'ah-rid*
carboxyl: *kar-bok'sil*
catalyst: *kat'ah-list*
glycerol: *glis'er-ol*
hydrolysis: *hi-drol'i-sis*
hydrophobic: *hi''dro-fo-bik*
hydroxyl: *hi-drok'sil*
ion: *i'on*
lecithin: *les'i-thin*
metabolism: *me-tab'o-lizm*
monosaccharide: *mon''o-sak'ah-rid*
myosin: *mi'o-sin*
phenylalanine: *fen''il-al'ah-nin*
phospholipid: *fos''fo-lipid*
physiology: *fiz-e-ol'o-je*
protein: *pro'te-in*

Photo: Crystals of vitamin B_{12}.

Atoms, Ions, and Chemical Bonds

The study of the functions of the human body requires some familiarity with the concepts and terminology of basic chemistry. A knowledge of atomic and molecular structure, the nature of chemical bonds, and the nature of pH and associated concepts provide the foundation for much of physiology.

Physiology is the study of biological function—of how the body works, from cell to tissue, tissue to organ, organ to system, and of how the organism as a whole accomplishes particular tasks essential for life. In the study of physiology, the emphasis is on *mechanisms*—with questions that begin with the word *how* and answers that involve cause-and-effect sequences. These sequences can be woven into larger and larger stories that include descriptions of the structures involved (anatomy) and that overlap with the sciences of chemistry and physics.

The anatomical structures and physiological processes of the body are based, to a large degree, on the properties and interactions of atoms, ions, and molecules. Water is the major solvent in the body and contributes 65% to 75% of the total weight of an average adult. Dissolved in this water are many organic molecules (carbon-containing molecules such as carbohydrates, lipids, proteins, and nucleic acids) and inorganic molecules and ions (atoms with a net charge). Before describing the structure and function of organic molecules within the body, some basic chemical concepts, terminology, and symbols will be introduced.

Atoms

Atoms are much too small to be seen individually, even with the most powerful electron microscope. Through the efforts of many generations of scientists, however, the structure of an atom is now well understood. At the center of an atom is its *nucleus.* The nucleus contains two types of particles—*protons,* which have a positive charge, and *neutrons,* which are noncharged. The mass of a proton is approximately equal to the mass of a neutron, and the sum of the protons and neutrons in an atom is equal to the **atomic mass** of the atom. For example, an atom of carbon, which contains six protons and six neutrons, has an atomic mass of 12 (table 3.1).

The number of protons in an atom is given as its **atomic number**. Carbon has six protons and thus has an atomic number of six. Outside the positively charged nucleus are negatively charged *electrons.* Since the number of electrons in an atom is equal to the number of protons, atoms have a net charge of zero.

Although it is often convenient to think of electrons as orbiting the nucleus like planets orbiting the sun, this older view of atomic structure is no longer believed to be correct. A given electron can occupy any position in a certain volume of space called the *orbital* of the electron. The orbital is like an energy "shell," or barrier, beyond which the electron usually does not pass.

There are potentially several such orbitals around a nucleus, with each successive orbital being farther from the nucleus. The first orbital, closest to the nucleus, can contain only two electrons. If an atom has more than two electrons (as do all atoms except hydrogen and helium), the additional electrons must occupy orbitals that are farther removed from the nucleus. The second orbital can contain a maximum of eight electrons; the third can contain a maximum of eighteen. Thus, the orbitals are filled from the innermost outward. Carbon, with six electrons, has two electrons in its first orbital and four electrons in its second orbital (fig. 3.1).

It is always the electrons in the outermost orbital, if this orbital is incomplete, that participate in chemical reactions and form chemical bonds. These outermost electrons are known as the *valence electrons* of the atom.

Isotopes

A particular atom with a given number of protons in its nucleus may exist in several forms that differ from each other in their number of neutrons. The atomic number of these forms is thus the same, but their atomic mass is different. These different forms are called **isotopes**. All of the isotopic forms of a given atom are included in the term **chemical element**. The element hydrogen, for example, has three isotopes. The most common of these has a nucleus consisting of only one proton. Another isotope of hydrogen (called *deuterium*) has one proton and one neutron in the nucleus, whereas the third isotope (*tritium*) has one proton and two neutrons. Tritium is a radioactive isotope that is commonly used in physiological research and in many clinical laboratory procedures.

Table 3.1 Atoms commonly present in organic molecules

Atom	Symbol	Atomic Number	Atomic Mass	Orbital 1	Orbital 2	Orbital 3	Number of Chemical Bonds
Hydrogen	H	1	1	1	0	0	1
Carbon	C	6	12	2	4	0	4
Nitrogen	N	7	14	2	5	0	3
Oxygen	O	8	16	2	6	0	2
Phosphorus	P	15	31	2	8	5	5
Sulfur	S	16	32	2	8	6	2

Chemical Bonds, Molecules, and Ionic Compounds

Molecules are formed when two or more atoms are joined together by sharing or some other interaction of the electrons in their outer orbitals. Such sharing of electrons or other attractive forces produces *chemical bonds* (fig. 3.2). The number of bonds that each atom can have is determined by the number of electrons in its outer orbital. Hydrogen, for example, must obtain only one more electron—and can thus form one chemical bond—to complete the first orbital of two electrons. Carbon, in contrast, must obtain four more electrons—and can thus form four chemical bonds—to complete the second orbital of eight electrons (fig. 3.3).

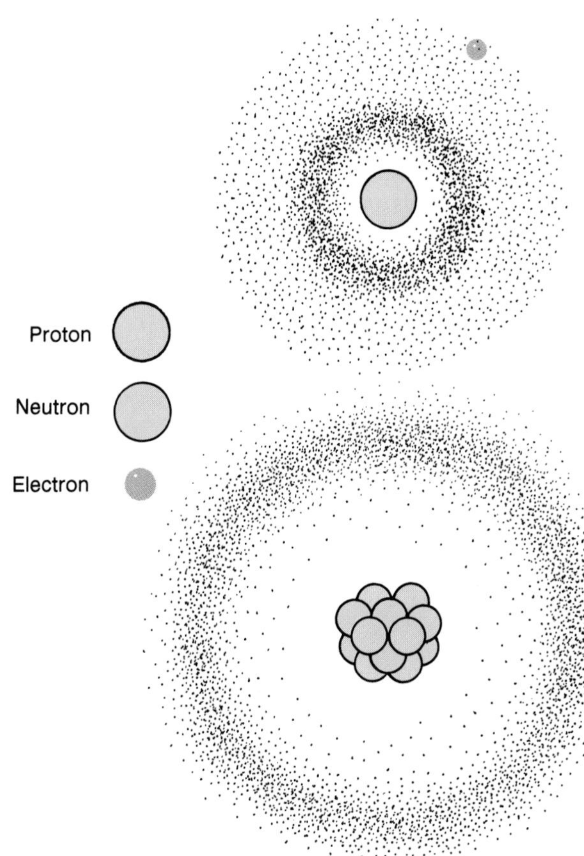

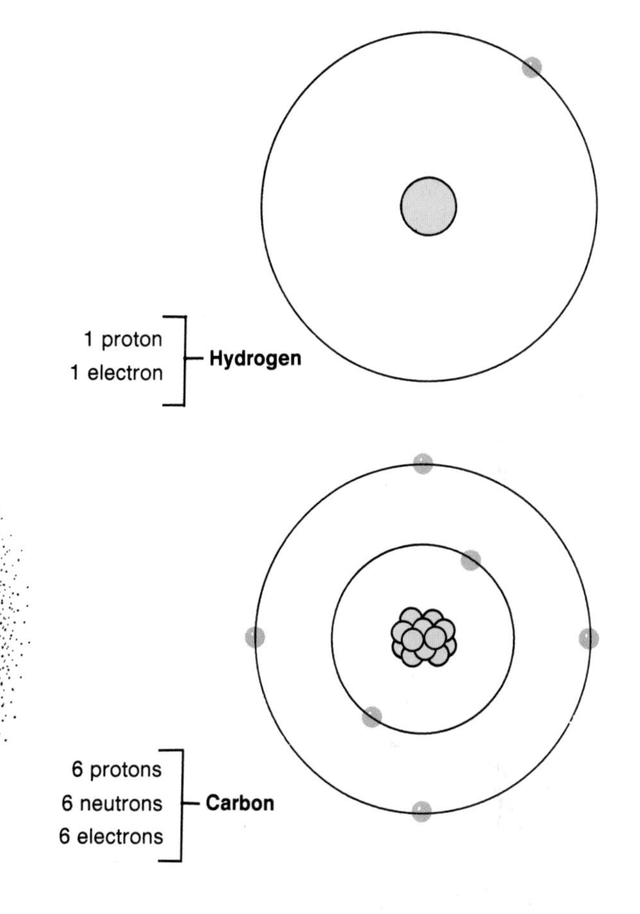

Figure 3.1. *Diagrams of the hydrogen and carbon atoms. The electron orbitals on the left are represented by dots indicating probable positions of the electrons. The orbitals on the right are represented by concentric circles.*

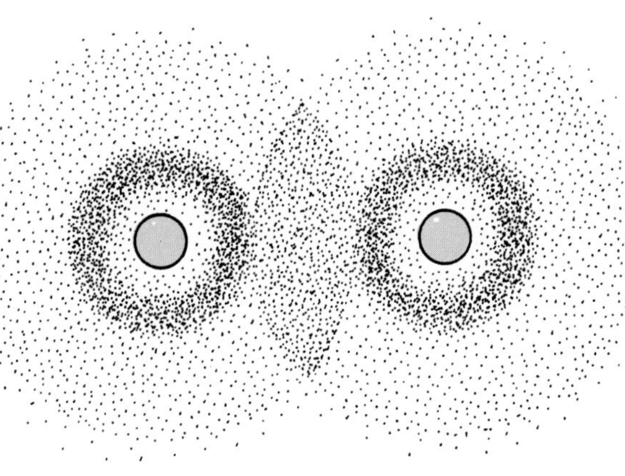

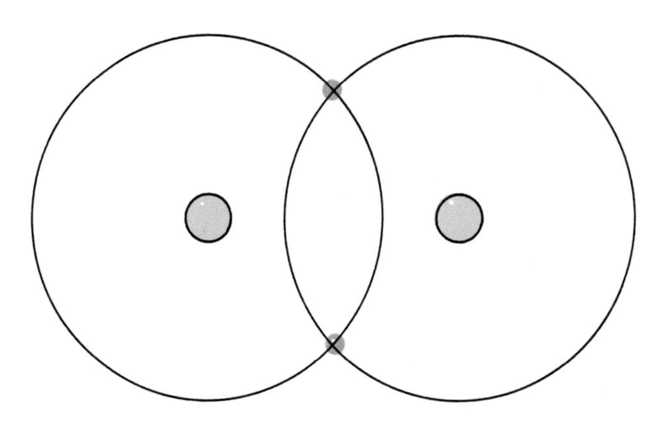

Figure 3.2. *The hydrogen molecule, showing the covalent bonds between hydrogen atoms formed by the equal sharing of electrons.*

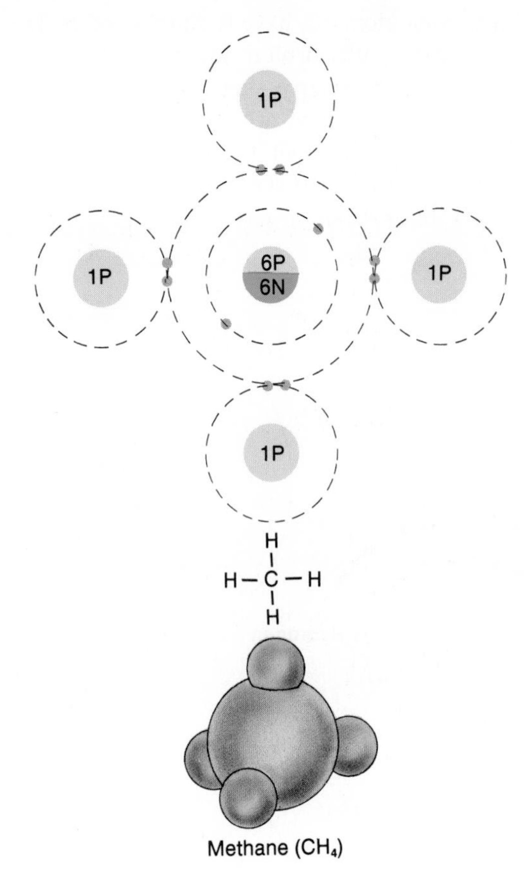

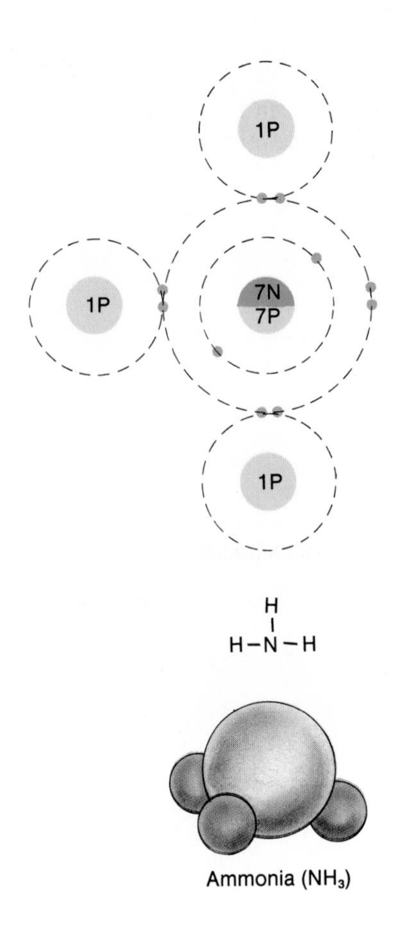

Figure 3.3. *The molecules methane and ammonia represented in three different ways. Notice that a bond between two atoms consists of a pair of shared electrons (the electrons from the outer orbital of each atom).*

Covalent Bonds

In **covalent bonds,** atoms are joined together by the sharing of electrons. Covalent bonds that are formed between identical atoms—as in oxygen gas (O_2) and hydrogen gas (H_2)—are the strongest because their electrons are equally shared. Since the electrons are equally distributed between the two atoms, these molecules are said to be *nonpolar.* When covalent bonds are formed between two different atoms, however, the electrons may be pulled more toward one atom than the other. The side of the molecule toward which the electrons are pulled is electrically negative in comparison to the other end. Such a molecule is said to be *polar* (has a positive and negative "pole"). Atoms of oxygen, nitrogen, and phosphorus have a particularly strong tendency to pull electrons toward themselves when they bond with other atoms.

Water is the most abundant molecule in the body and serves as the solvent of body fluids. Water is a good solvent because it is polar; the oxygen atom pulls electrons from the two hydrogens toward its side of the water molecule, so that the oxygen side is negatively charged compared to the hydrogen side of the molecule (fig. 3.4). The significance of the polar nature of water in its function as a solvent is discussed in the next section.

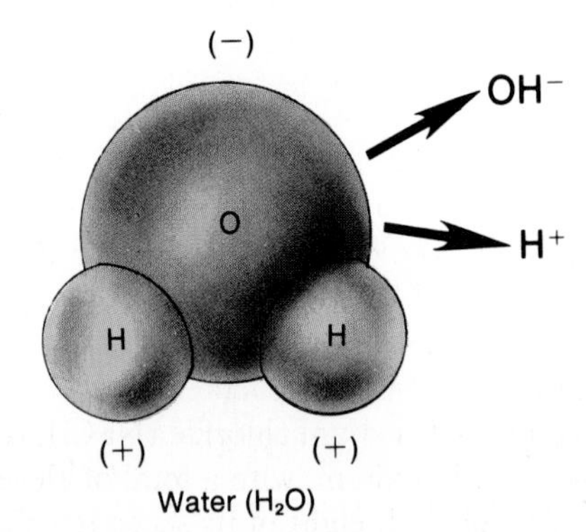

Figure 3.4. *A model of a water molecule showing its polar nature. Notice that the oxygen side of the molecule is negative whereas the hydrogen side is positive. Polar covalent bonds are weaker than nonpolar covalent bonds. As a result, some water molecules ionize to form OH^- (hydroxyl ion) and H^+ (hydrogen ion).*

Ionic Bonds

In **ionic bonds** the electrons are not shared at all. Instead, one or more valence electrons from one atom are completely transferred to a second atom. The first atom thus loses electrons, so that its number of electrons becomes less than its number of

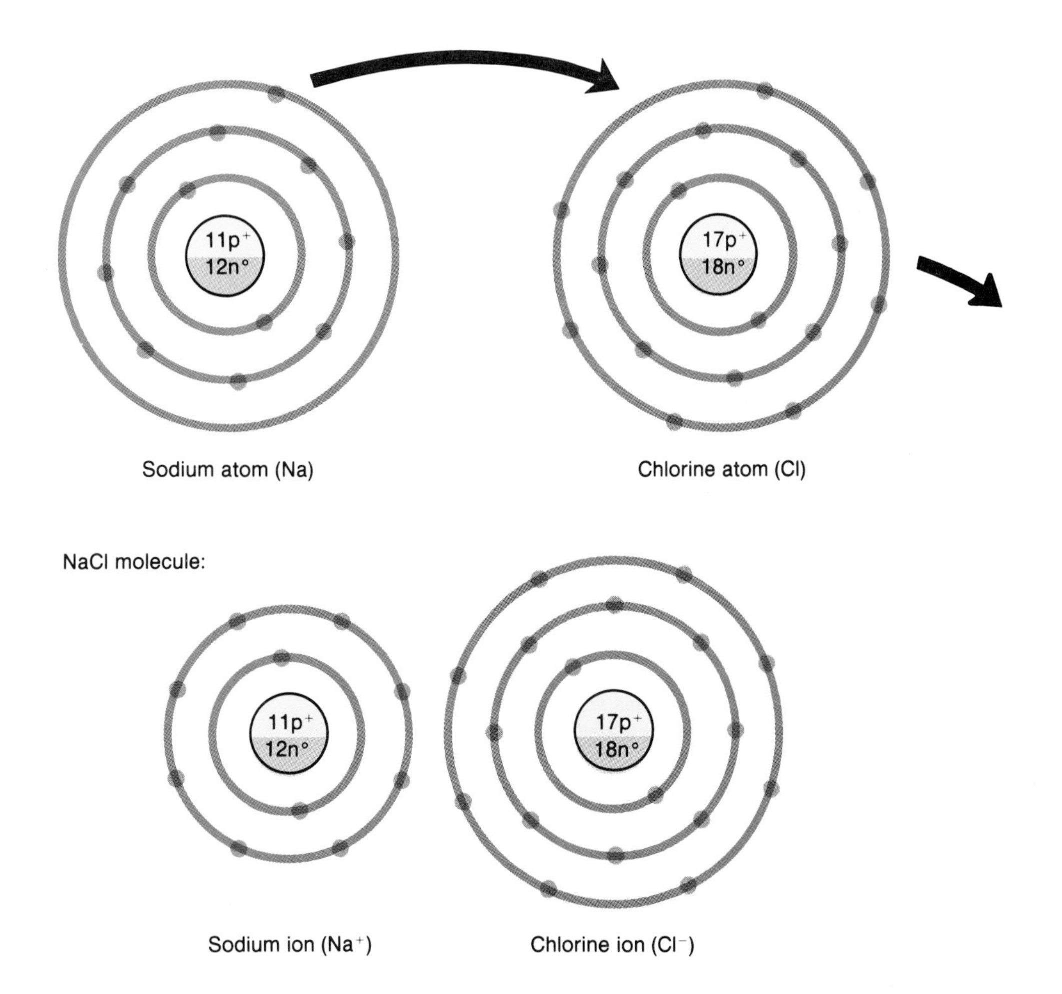

Figure 3.5. *The ionization of sodium chloride to produce sodium and chloride ions. The positive sodium and the negative chloride ions attract each other to produce the ionic compound sodium chloride (NaCl).*

protons; it becomes a positively charged **ion**. Positively charged ions are called *cations* because they move toward the negative pole, or cathode, in an electric field. The second atom gains more electrons than it has protons and becomes a negatively charged ion, or *anion* (so-called because it moves toward the positive pole, or anode, in an electric field). The cation and anion attract each other to form an **ionic compound**.

Common table salt, sodium chloride (NaCl), is an example of an ionic compound. Sodium, with a total of eleven electrons, has two in its first orbital, eight in its second orbital, and only one in its third orbital. Chlorine, conversely, is one electron short of completing its outer orbital of eight electrons. The lone electron in sodium's outer orbital is attracted to chlorine's outer orbital. This creates a chloride ion (represented as Cl^-) and a sodium ion (Na^+). Although table salt is shown as NaCl, it is actually composed of Na^+Cl^- (fig. 3.5).

Ionic bonds are weaker than polar covalent bonds, and therefore ionic compounds easily dissociate when dissolved in water to yield their separate ions. Dissociation of NaCl, for example, yields Na^+ and Cl^-. Each of these ions attracts polar water molecules; the negative ends of water molecules are attracted to the Na^+, and the positive ends of water molecules are attracted to the Cl^- (fig. 3.6). The water molecules that surround these ions in turn attract other molecules of water to form *hydration spheres* around each ion.

The formation of hydration spheres makes an ion or a molecule soluble in water. Glucose, amino acids, and many other organic molecules are water-soluble because hydration spheres can form around atoms of oxygen, nitrogen, and phosphorus, which are joined by polar covalent bonds to other atoms. Such molecules are said to be **hydrophilic.**[1] In contrast, molecules composed primarily of nonpolar covalent bonds, such as the hydrocarbon chains of fat molecules, have few charges and, thus, cannot form hydration spheres. They are insoluble in water and, in fact, actually avoid water—they are **hydrophobic.**[2]

[1]hydrophilic: Gk. *hydor*, water; *philos*, fond
[2]hydrophobic: Gk. *hydor*, water; *phobos*, fear

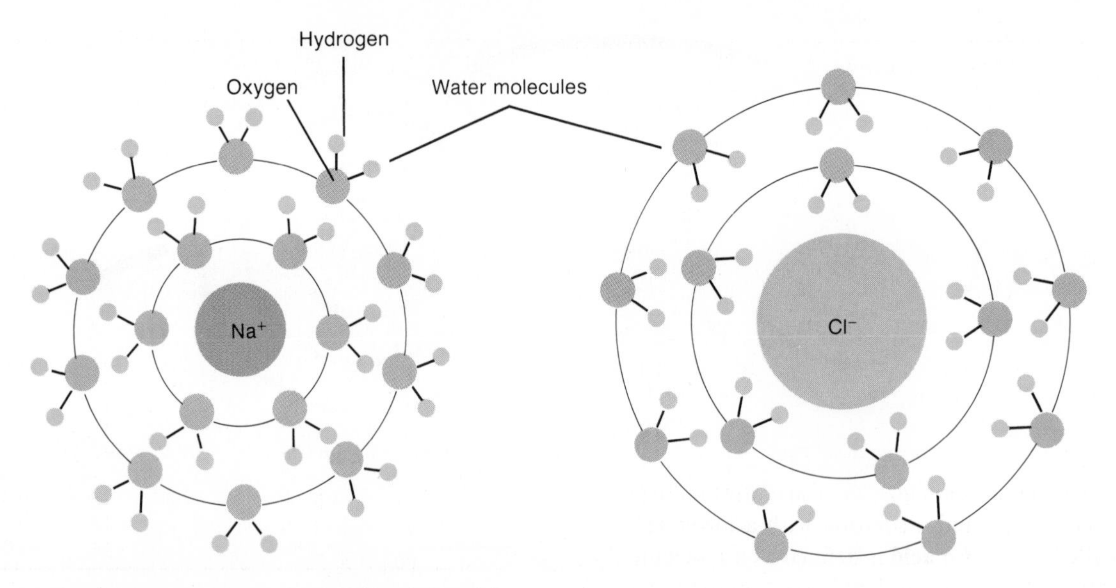

Figure 3.6. *The negatively charged oxygen-ends of water molecules are attracted to the positively charged* Na^+*, whereas the positively charged hydrogen-ends of water molecules are attracted to the negatively charged* Cl^-*. Other water molecules are attracted to this first concentric layer of water, forming hydration spheres around the sodium and chloride ions.*

Hydrogen Bonds

Hydrogen bonds are very weak bonds that help to stabilize the delicate folding and bending of long organic molecules such as proteins. When hydrogen forms a polar covalent bond with an atom of oxygen or nitrogen, the hydrogen gains a slight positive charge as its electron is pulled toward the other atom. This also imparts a slight negative charge to the other atom, which is said to be electronegative. Since the hydrogen has a slight positive charge, it will have a weak attraction for a second electronegative atom (oxygen or nitrogen) that may be located nearby. This weak attraction is called a **hydrogen bond**.

Hydrogen bonds are usually shown with dotted lines (fig. 3.7) to distinguish them from strong covalent bonds, which are shown with solid lines. As shown in figure 3.7, hydrogen bonds form between adjacent water molecules. Hydrogen bonds also help to stabilize the structure of large molecules of protein, as will be described in a later section.

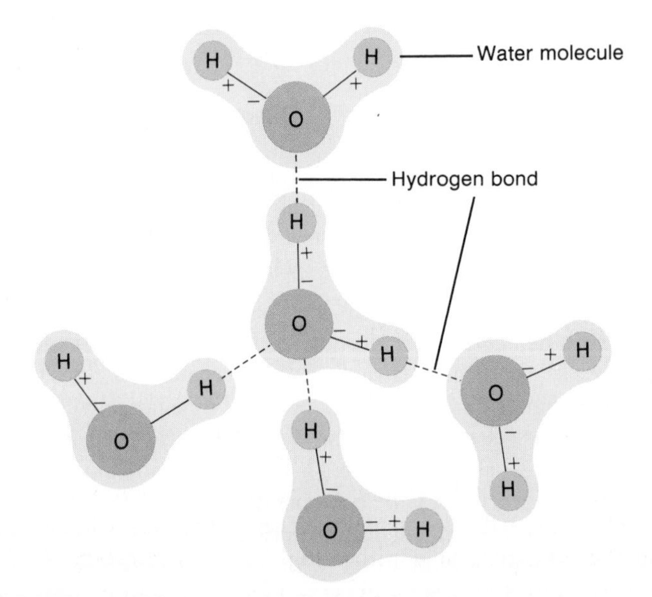

Figure 3.7. *The oxygen atoms of water molecules are weakly joined together by the attraction of the electronegative oxygen for the positively charged hydrogen. These weak bonds are called* hydrogen bonds.

Acids, Bases, and the pH Scale

The bonds in water molecules joining hydrogen and oxygen atoms together are, as previously discussed, polar covalent bonds. Although these bonds are strong, a proportion of them break as the electron from the hydrogen atom is completely transferred to oxygen. When this occurs, the water molecule ionizes to form a *hydroxyl* ion (OH^-) and a hydrogen ion (H^+), which is simply a free proton.

Ionization of water molecules produces equal amounts of OH^- and H^+. Since only a small proportion of water molecules ionize, the concentration of H^+ is only equal to 10^{-7} molar (the term *molar* is a unit of concentration; for hydrogen, one molar

Table 3.2 Common acids and bases

Acid	Symbol	Base	Symbol
Hydrochloric acid	HCl	Sodium hydroxide	$NaOH$
Phosphoric acid	H_3PO_4	Potassium hydroxide	KOH
Nitric acid	HNO_3	Calcium hydroxide	$Ca(OH)_2$
Sulfuric acid	H_2SO_4	Ammonium hydroxide	NH_4OH
Carbonic acid	H_2CO_3		

equals one gram per liter). A solution with 10^{-7} molar hydrogen ion concentration, which is produced by the ionization of water molecules in which the H^- and OH^- concentrations are equal, is said to be **neutral**.

A solution that contains a higher H^+ concentration than that of water is called *acidic*, and a solution with a lower H^+ concentration is called *basic*. An **acid** is defined as a molecule that can release protons (H^+) to a solution; it is a "proton donor." A **base** is a molecule or a negatively charged ion (anion) which can combine with H^+ and thus remove the H^+ from solution; it is a "proton acceptor." Most strong bases release OH^- into a solution, which combines with H^+ to form water and which thus lowers the H^+ concentration. Examples of common acids and bases are shown in table 3.2.

pH

The H^+ concentration of a solution is usually indicated in pH units on a pH scale that runs from 0 to 14. The pH number is equal to the logarithm of one over the H^+ concentration:

$$pH = \log \frac{1}{[H^+]} \quad \text{where } [H^+] = \text{molar } H^+ \text{ concentration.}$$

Pure water has a H^+ concentration of 10^{-7} molar and, thus, has a pH of 7 (neutral). Because of the logarithmic relationship, a solution with ten times the hydrogen ion concentration (10^{-6} M) has a pH of 6, whereas a solution with one-tenth the H^+ concentration (10^{-8} M) has a pH of 8. The pH number is easier to write than the molar H^+ concentration, but it is admittedly confusing because it is *inversely related* to the H^+ concentration: a solution with a higher H^+ concentration has a lower pH number; one with a lower H^+ concentration has a higher pH number. A strong acid with a high H^+ concentration of 10^{-2} molar, for example, has a pH of 2, whereas a solution with only 10^{-10} molar H^+ has a pH of 10. **Acidic solutions,** therefore, have a pH of less than 7 (that of pure water), whereas **basic solutions** have a pH of between 7 and 14 (table 3.3).

Organic Molecules

Organic molecules are those that contain the atom *carbon.* Since the carbon atom has four electrons in its outer orbital, it must share four additional electrons by covalent bonding with other

Table 3.3 The pH scale

	H^+ Concentration (molar)	pH	OH^- Concentration (molar)
Acids	1.0	0	10^{-14}
	0.1	1	10^{-13}
	0.01	2	10^{-12}
	0.001	3	10^{-11}
	0.0001	4	10^{-10}
	10^{-5}	5	10^{-9}
	10^{-6}	6	10^{-8}
Neutral	10^{-7}	7	10^{-7}
Bases	10^{-8}	8	10^{-6}
	10^{-9}	9	10^{-5}
	10^{-10}	10	0.0001
	10^{-11}	11	0.001
	10^{-12}	12	0.01
	10^{-13}	13	0.1
	10^{-14}	14	1.0

atoms to fill its outer orbital with eight electrons. The unique bonding requirements of carbon enable it to join with other carbon atoms to form chains and rings, while still allowing the carbon atoms to bond with hydrogen and other atoms.

Most organic molecules in the body contain hydrocarbon chains and rings as well as other atoms bonded to carbon. Two adjacent carbon atoms in a chain or ring may share one or two pairs of electrons. If the two carbon atoms share one pair of electrons, they are said to have a *single covalent bond;* this leaves each carbon atom free to bond to as many as three other atoms. If the two carbon atoms share two pairs of electrons, they have a *double covalent bond,* and each carbon atom can only bond to a maximum of two additional atoms (fig. 3.8).

The ends of some hydrocarbons are joined together to form rings. In the shorthand structural formulas of these molecules, the carbon atoms are not shown but are understood to be located at the corners of the ring. Some of these cyclic molecules have a double bond between two adjacent carbon atoms. Benzene and related molecules are shown as a six-sided ring with alternating double bonds. Such compounds are called *aromatic.* Since all of the carbons in an aromatic ring are equivalent, double bonds can be shown between any two adjacent carbons in the ring (fig. 3.9).

1. Describe the structure of an atom, and explain how covalent bonds are formed.
2. Write the definition of an acid, and describe the relationship between pH and the H^+ concentration.
3. Describe the structure of water, and explain why ions and molecules with many polar bonds are water-soluble.
4. Describe the structure of a carbon atom, and explain how this structure allows carbon atoms to bond together to form chains and rings.

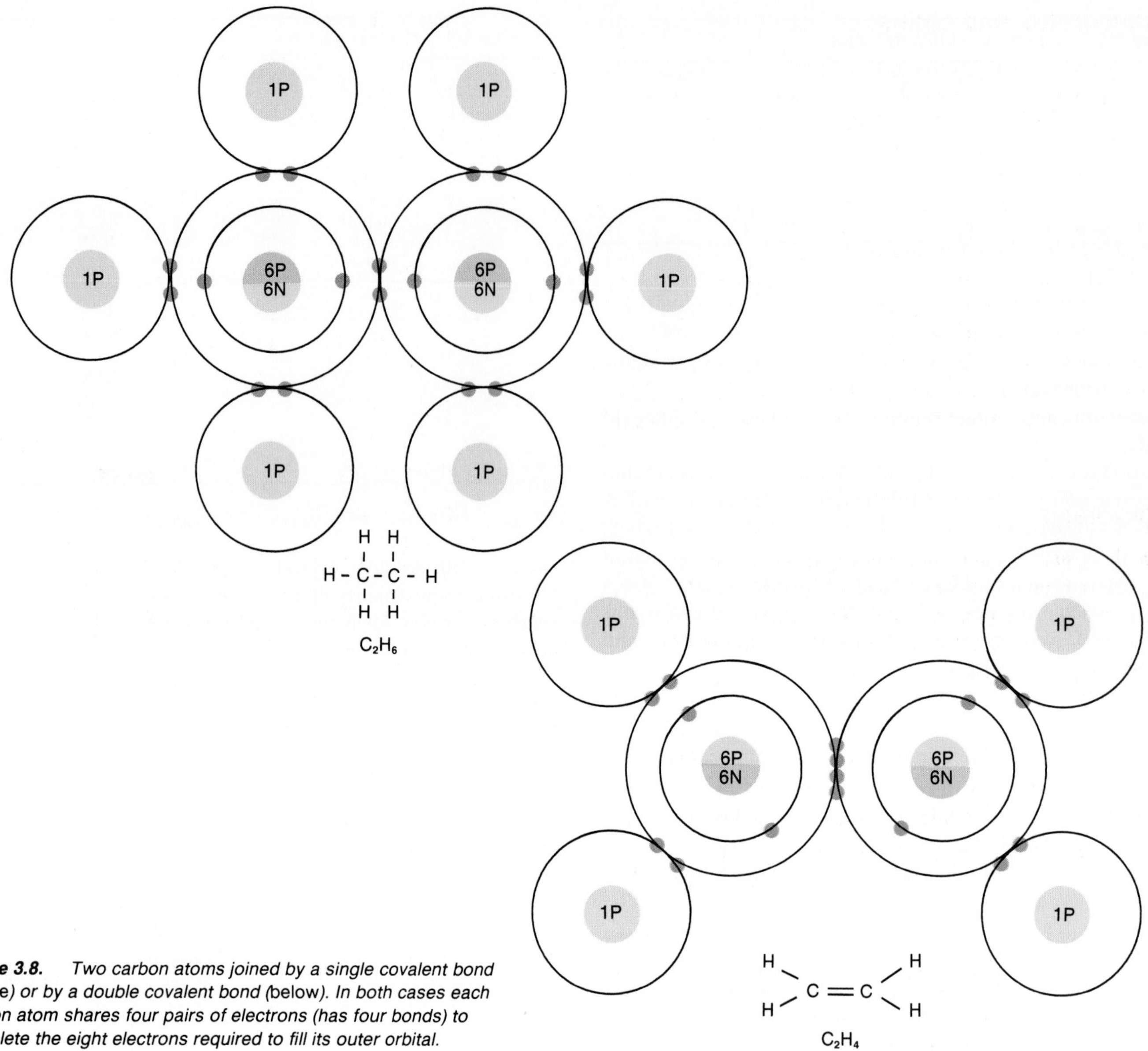

Figure 3.8. *Two carbon atoms joined by a single covalent bond (above) or by a double covalent bond (below). In both cases each carbon atom shares four pairs of electrons (has four bonds) to complete the eight electrons required to fill its outer orbital.*

(a) 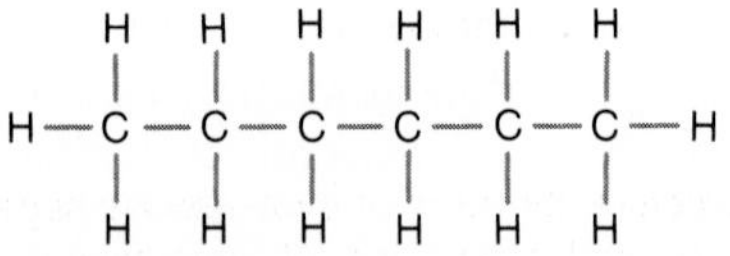 H—C—C—C—C—C—C—H (each C bonded to two H) C_6H_{14} (Hexane)

(b) 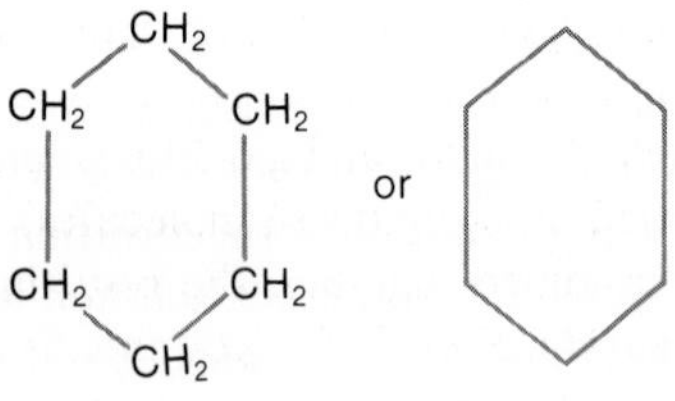 CH_2 ring of six CH_2 groups or [hexagon] C_6H_{12} (Cyclohexane)

(c) 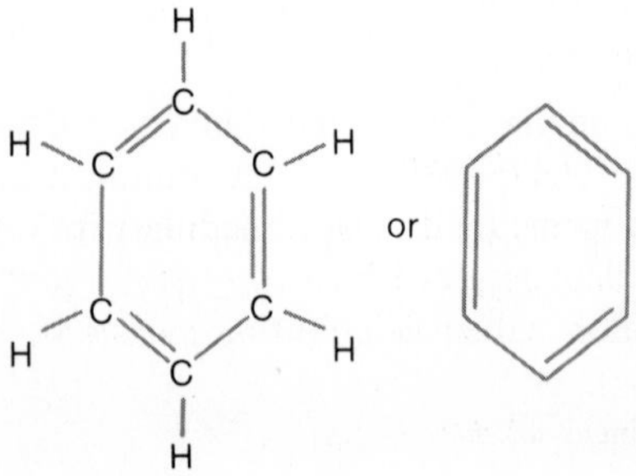 ring of six C—H groups with alternating double bonds or [hexagon with alternating double bonds] C_6H_6 (Benzene)

Figure 3.9. *Hydrocarbons that are (a) linear, (b) cyclic, and (c) aromatic rings.*

Carbohydrates and Lipids

Carbohydrates are a class of organic molecules which consist of monosaccharides, disaccharides, and polysaccharides. All of these molecules are based upon a characteristic ratio of carbon, hydrogen, and oxygen atoms. Lipids are a diverse category of organic molecules which share the physical property of being nonpolar, and thus insoluble in water.

Carbohydrates and lipids are similar in many ways. Both groups of molecules consist primarily of the atoms carbon, hydrogen, and oxygen, and both serve as major sources of energy in the body (comprising most of the calories consumed in food). Carbohydrates and lipids differ, however, in some important aspects of their chemical structures and physical properties. Such differences significantly affect the functions of these molecules in the body.

Carbohydrates

Carbohydrates are organic molecules that contain carbon, hydrogen, and oxygen in the ratio described by their name—*carbo* (carbon) and *hydrate* (water, H_2O). The general formula of a carbohydrate molecule is thus CH_2O; the molecule contains twice the number of hydrogen atoms as it contains carbon or oxygen atoms.

Monosaccharides, Disaccharides, and Polysaccharides

Carbohydrates include simple sugars, or **monosaccharides,**[3] and longer molecules that contain a number of monosaccharides joined together. The suffix *-ose* denotes a sugar molecule; the term *hexose,* for example, refers to a six-carbon monosaccharide with the formula $C_6H_{12}O_6$. This formula is adequate for some purposes, but it does not distinguish between related hexose sugars, which are *structural isomers* of each other. The structural isomers glucose, fructose, and galactose, for example, are monosaccharides that have the same ratio of atoms arranged in slightly different ways (fig. 3.10).

Two monosaccharides can be joined covalently to form a **disaccharide,** or double sugar. Common disaccharides include table sugar, or *sucrose* (composed of glucose and fructose), milk sugar, or *lactose* (composed of glucose and galactose), and malt sugar, or *maltose* (composed of two glucose molecules). When many monosaccharides are joined together, the resulting molecule is called a **polysaccharide.** *Starch,* for example, is a polysaccharide found in many plants, which is formed by the bonding together of thousands of glucose subunits. Animal starch (**glycogen**), found in the liver and muscles, likewise consists of repeating glucose molecules but differs from plant starch in that it is more highly branched (fig. 3.11).

Dehydration Synthesis and Hydrolysis

In the formation of disaccharides and polysaccharides, the separate subunits (monosaccharides) are bonded together covalently by a type of reaction called **dehydration synthesis,** or

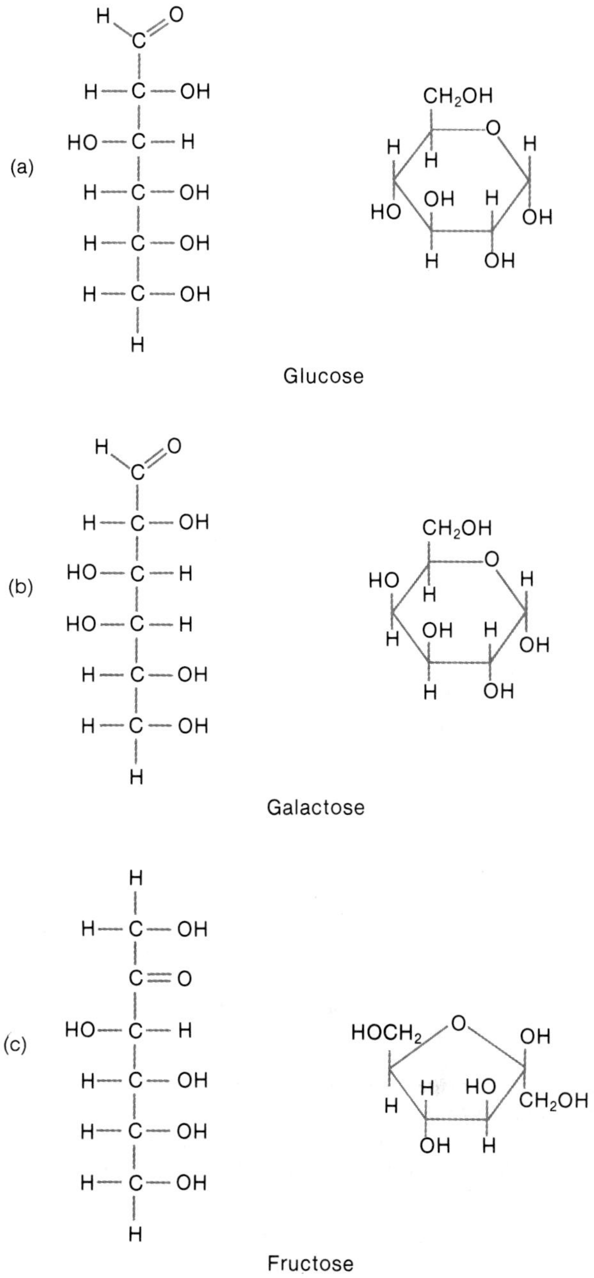

Glucose

Galactose

Fructose

Figure 3.10. *The structural formulas of three hexose sugars—(a) glucose, (b) galactose, and (c) fructose. All three have the same ratio of atoms—*$C_6H_{12}O_6$.

condensation. In this reaction, which requires the participation of specific enzymes (described in a later section), a hydrogen atom is removed from one monosaccharide and a hydroxyl group (OH) is removed from another. As a covalent bond is formed between the two monosaccharides, water (H_2O) is produced. Dehydration synthesis reactions are illustrated in figure 3.12.

When a person eats disaccharides and polysaccharides or when the stored glycogen in the liver and muscles is to be used by tissue cells, the covalent bonds that join monosaccharides

[3]monosaccharide: Gk. *monos,* single; *sakcharon,* sugar

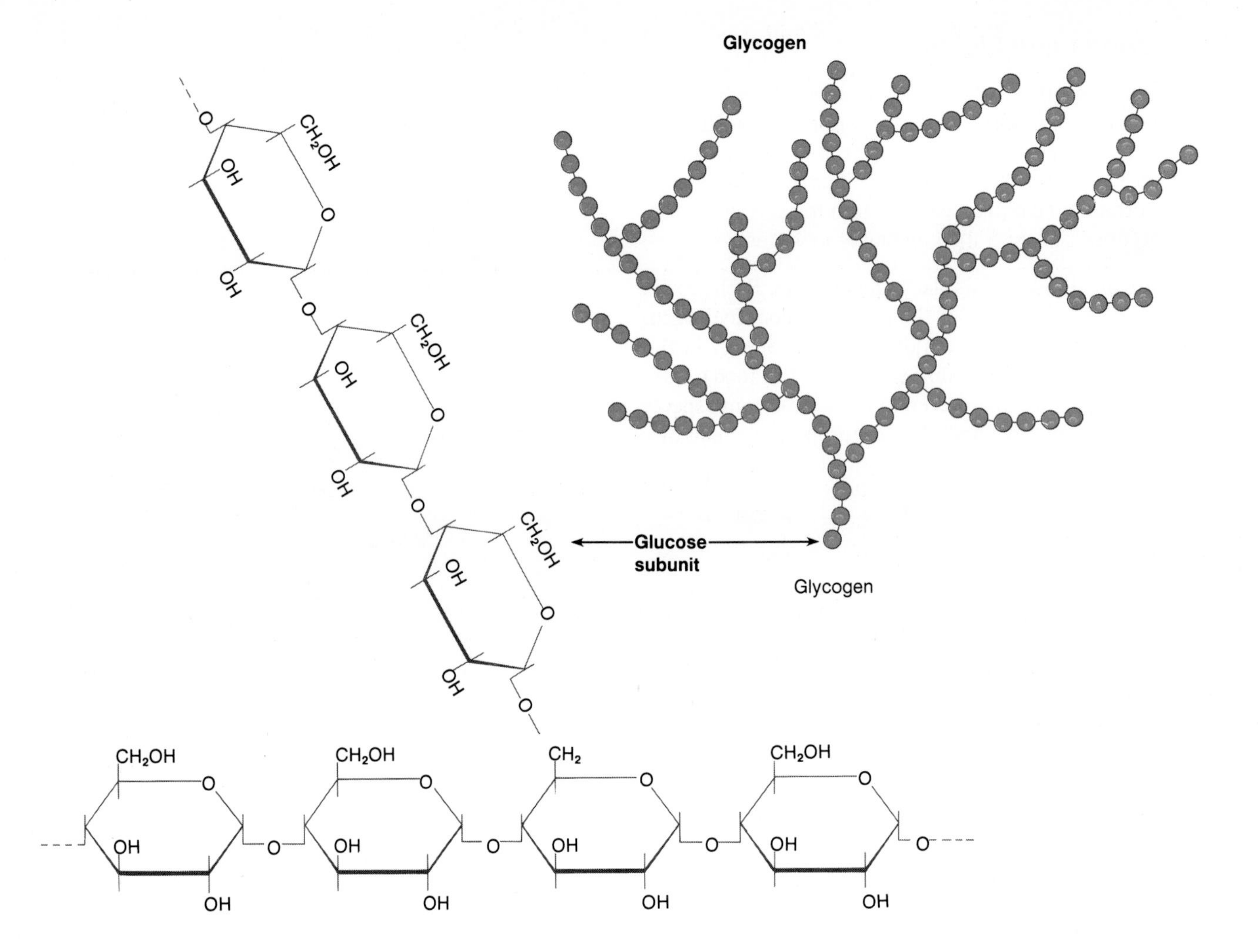

Figure 3.11. Glycogen is a polysaccharide composed of glucose subunits joined together to form a large, highly branched molecule.

into disaccharides and polysaccharides must be broken. These *digestion* reactions occur by means of **hydrolysis.**[4] Hydrolysis is the reverse of dehydration synthesis. A water molecule is split, as implied by the word *hydrolysis,* and the resulting hydrogen atom is added to one of the free glucose molecules as the hydroxyl group is added to the other (fig. 3.13).

When a potato is eaten, the starch within it is hydrolyzed into separate glucose molecules within the intestine. This glucose is absorbed into the blood and carried to the tissues. Some tissue cells may use this glucose for energy. Liver and muscles, however, can store excess glucose by combining these molecules through dehydration synthesis reactions to form glycogen. During fasting or prolonged exercise, the liver can add glucose to the blood through hydrolysis of its stored glycogen.

Dehydration synthesis and hydrolysis reactions do not occur spontaneously; they require the action of specific enzymes. Similar reactions, in the presence of other enzymes, build and break down lipids, proteins, and nucleic acids. In general, therefore, hydrolysis reactions digest molecules into their subunits, and dehydration synthesis reactions build larger molecules by the bonding together of their subunits.

[4]hydrolysis: Gk. *hydor*, water; *lysis*, break

Lipids

The category of molecules known as lipids includes several types of molecules that differ greatly in chemical structure. These diverse molecules are all in the lipid category by virtue of a common physical property—they are all *insoluble in polar solvents* such as water. This is because lipids consist primarily of hydrocarbon chains and rings, which are nonpolar and, thus, hydrophobic. Although lipids are insoluble in water, they can be dissolved in nonpolar solvents such as ether, benzene, and related compounds.

Triglycerides

Triglycerides are a subcategory of lipids that includes fat and oil. These molecules are formed by the condensation of one molecule of *glycerol* (a three-carbon alcohol) with three molecules of *fatty acids.* Each fatty acid molecule consists of a nonpolar

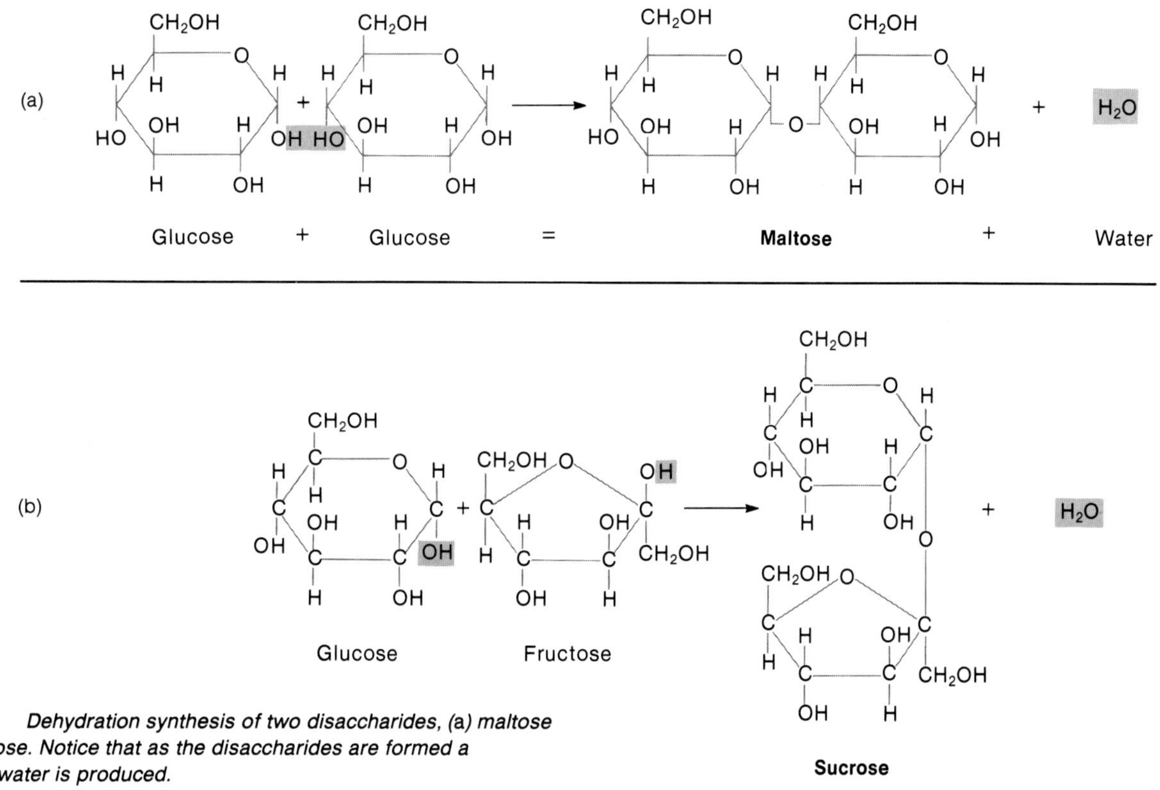

(a)

Glucose + Glucose = **Maltose** + Water

(b)

Glucose Fructose **Sucrose** + H_2O

Figure 3.12. *Dehydration synthesis of two disaccharides, (a) maltose and (b) sucrose. Notice that as the disaccharides are formed a molecule of water is produced.*

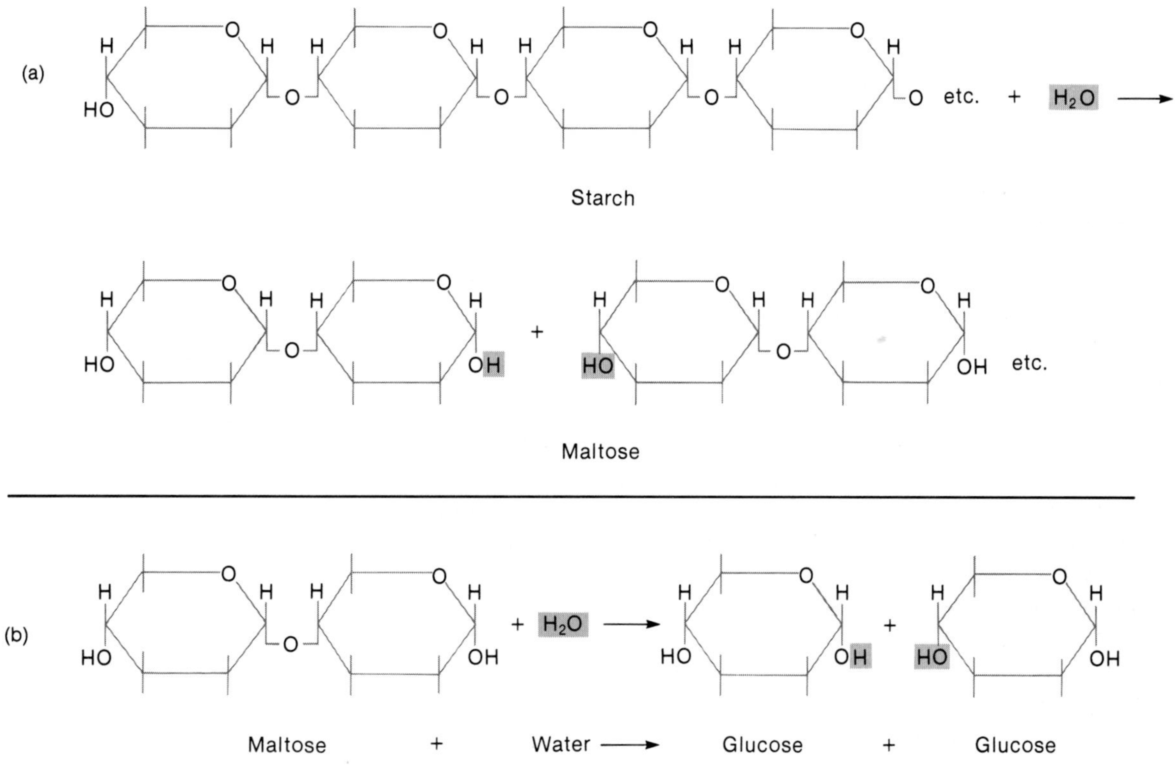

(a)

etc. + H_2O →

Starch

etc.

Maltose

(b)

Maltose + Water → Glucose + Glucose

Figure 3.13. *The hydrolysis of starch (a) into disaccharides (maltose) and (b) into monosaccharides (glucose). Notice that as the covalent bond between the subunits breaks, a molecule of water is split. In this way the hydrogen and hydroxyl from water is added to the ends of the released subunits.*

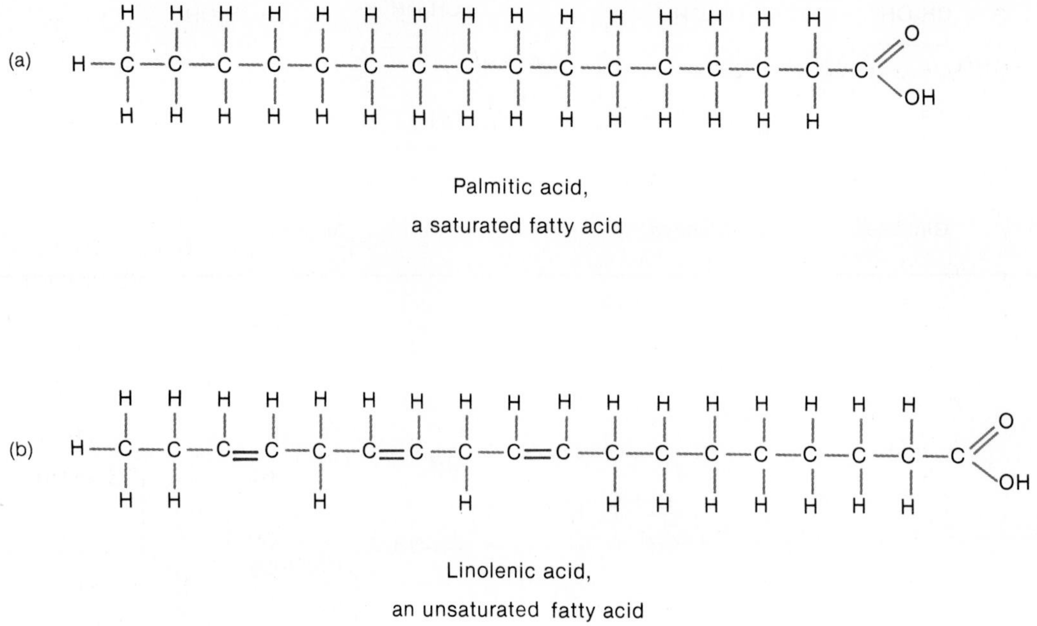

(a)

Palmitic acid,
a saturated fatty acid

(b)

Linolenic acid,
an unsaturated fatty acid

Figure 3.14. *Structural formulas for (a) saturated and (b) unsaturated fatty acids.*

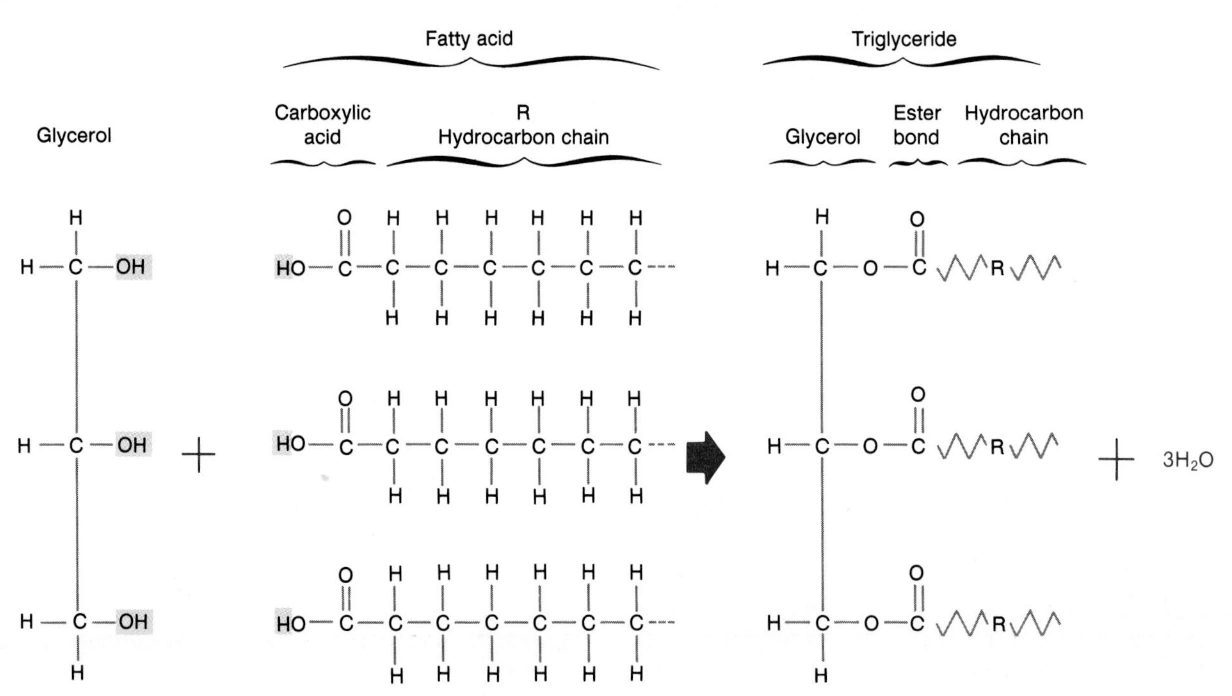

Figure 3.15. *Dehydration synthesis of a triglyceride molecule from a glycerol and three fatty acids. A molecule of water is produced as a bond forms between each fatty acid and the glycerol. Sawtooth lines represent carbon chains, which are symbolized by an* R.

hydrocarbon chain with a *carboxylic acid group* (abbreviated *COOH*) on one end. If the carbon atoms within the hydrocarbon chain are joined by single covalent bonds, so that each carbon atom can also bond to two hydrogen atoms, the fatty acid is said to be *saturated.* If there are a number of double covalent bonds within the hydrocarbon chain, so that each carbon atom can only bond to one hydrogen atom, the fatty acid is said to be *unsaturated.* Triglycerides that contain saturated fatty acids are called **saturated fats;** those that contain unsaturated fatty acids are **unsaturated fats** (fig. 3.14).

Within the adipose cells of the body, triglycerides are formed as the carboxylic acid ends of fatty acid molecules condense with the hydroxyl groups of a glycerol molecule (fig. 3.15). Since the hydrogen atoms from the carboxyl ends of fatty acid molecules

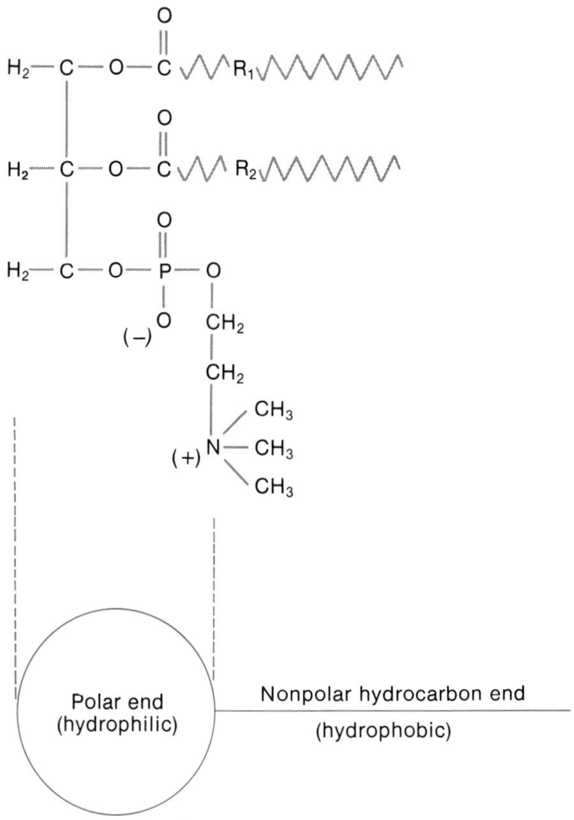

Figure 3.16. *The structure of lecithin, a typical phospholipid (above), and its more simplified representation (below).*

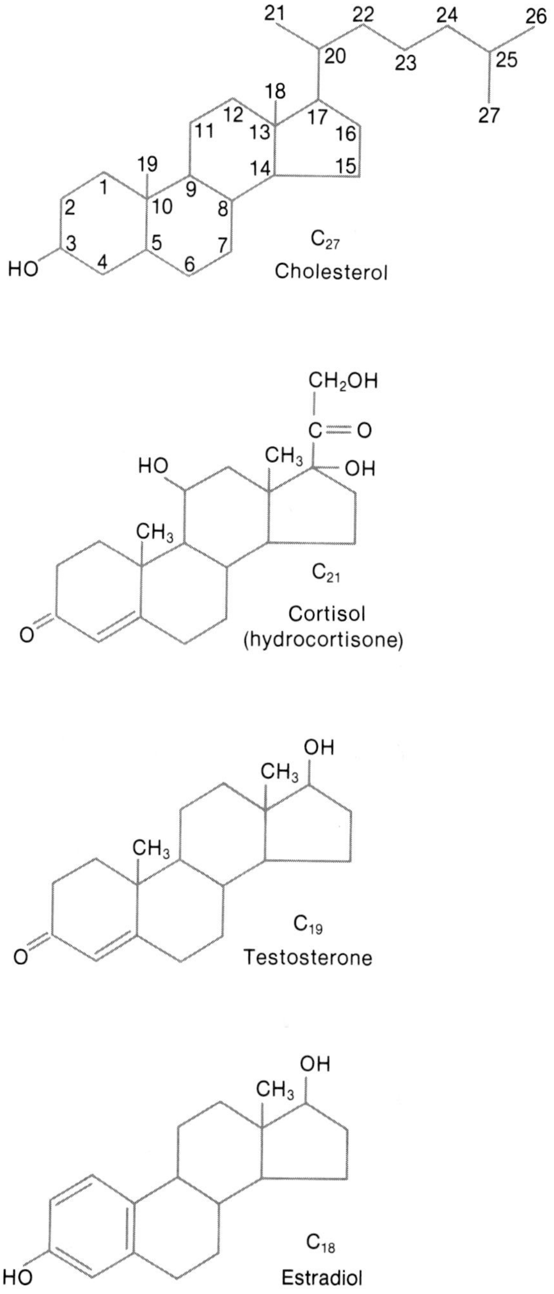

form water molecules during dehydration synthesis, fatty acids that are combined with glycerol can no longer release H^+ and function as acids. For this reason, triglycerides are described as *neutral fats.*

Phospholipids

The class of lipids known as *phospholipids* contains a number of different subclasses, which have in common the fact that they are lipids which contain a phosphate group. The most common type of phospholipid molecule has this structure: the three-carbon alcohol molecule glycerol is attached to two fatty acid molecules; the third carbon atom of the glycerol molecule is attached to a phosphate group, and the phosphate group in turn is bonded to other molecules. If the phosphate group is attached to a nitrogen-containing choline molecule, the phospholipid thus formed is known as *lecithin* (fig. 3.16). Figure 3.16 shows a simple way of illustrating the structure of a phospholipid—the parts of the molecule capable of ionizing and thus becoming charged are shown as a circle, whereas the nonpolar parts of the molecule are represented by lines.

The nonpolar part of a phospholipid is soluble in lipids but not in water; the polar part is soluble in water but not in lipids. These differences in solubility enable phospholipids to form cellular membranes (chapter 4), and to prevent the surface tension of water from collapsing the lungs (chapter 15).

Figure 3.17. *Cholesterol and some steroid hormones derived from cholesterol.*

Steroids

The structure of steroid molecules is quite different from that of triglycerides or phospholipids, and yet steroids are still included in the lipid category of molecules because they are nonpolar and insoluble in water. All steroid molecules have the same basic structure; three six-carbon rings are joined to one five-carbon ring (fig. 3.17). However, different kinds of steroids have different functional groups attached to this basic structure, and they vary in the number and position of the double covalent bonds between the carbon atoms in the rings.

Cholesterol is an important molecule in the body because it serves as the precursor (parent molecule) for the steroid hormones produced by the gonads and adrenal glands. The testes and ovaries (collectively called the *gonads*) secrete **sex steroids,** which include estradiol and progesterone from the ovaries and testosterone from the testes. The adrenal cortex (outer part of the adrenal gland) secretes the **corticosteroids,** including hydrocortisone and aldosterone, among others.

1. Describe the common structure of all carbohydrates, and distinguish between monosaccharides, disaccharides, and polysaccharides.
2. Using dehydration synthesis and hydrolysis reactions, explain how disaccharides and monosaccharides can be interconverted and how triglycerides can be formed and broken down.
3. Explain what a lipid is, and describe the different subcategories of lipids.
4. Explain why a phospholipid molecule is part polar and part nonpolar.

Proteins

Proteins are large molecules composed of amino acid subunits. Since there are twenty different types of amino acids that can be used in constructing a given protein, the variety of protein structures is immense. This variety allows each type of protein to be constructed so that it can perform very specific functions.

The enormous diversity of protein[5] structure results from the fact that there are twenty different building blocks—the amino acids—that can be used to form a protein. These amino acids, as will be described in the next section, are joined together to form a chain that can twist and fold in a specific manner due to chemical interactions between the amino acids. The specific sequence of amino acids in a protein, and thus the specific functional structure of the protein, is determined by genetic information. This genetic information for protein synthesis is contained in another category of organic molecules, the nucleic acids. The structure of nucleic acids, and the mechanisms by which the genetic information they encode directs protein synthesis, is described in chapter 10.

Structure of Proteins

Proteins consist of long chains of subunits called **amino acids.** As their name implies, each amino acid contains an *amino group* (NH_2) on one end of the molecule and a *carboxylic acid group* (COOH) on another end. There are approximately twenty different amino acids, with different structures and chemical properties that are used to build proteins. These differences are due to differences in the *functional groups* of these amino acids, which is abbreviated *R* in the general formula for an amino acid (fig. 3.18). The *R* symbol actually stands for the word *residue,* but it can be thought of as indicating the "rest of the molecule."

When amino acids are joined together by dehydration synthesis, the hydrogen from the amino end of one amino acid combines with the hydroxyl group of the carboxylic acid end of another amino acid. As a covalent bond is formed between the two amino acids, water is produced (fig. 3.19). The bond between adjacent amino acids is called a **peptide bond,** and the compound formed is called a *peptide.* When many amino acids are joined in this way, a chain of amino acids, or **polypeptide,** is produced.

The lengths of polypeptide chains vary greatly. A hormone called *thyrotropin-releasing hormone,* for example, is only three amino acids long, whereas *myosin,* a muscle protein, contains about forty-five hundred amino acids. When the length of a polypeptide chain becomes very long (greater than about a hundred amino acids), the molecule is called a **protein.**

The structure of a protein can be described at four different levels. At the first level, the sequence of amino acids in the protein, called the **primary structure** of the protein, is described. Each type of protein has a different primary structure. All of the billions of *copies* of a given type of protein in the body, however, have the same structure, because the structure of a given protein is coded by the genes. The primary structure of a protein is illustrated in figure 3.20.

Weak interactions (such as hydrogen bonds) between functional (*R*) groups of amino acids in nearby positions in the polypeptide chain cause this chain to twist into a *helix.* The extent and location of the helical structure is different for each protein because of differences in amino acid composition. A description of the helical structure of a protein is termed its **secondary structure** (fig. 3.20).

Most polypeptide chains bend and fold on themselves to produce complex three-dimensional shapes, called the **tertiary structure** of the proteins. Each type of protein has its own characteristic tertiary structure. This is because the folding and bending of the polypeptide chain is produced by chemical interactions between particular amino acids that are located in different regions of the chain.

Most of the tertiary structure of proteins is formed and stabilized by weak chemical interactions (such as hydrogen bonds) between widely spaced amino acids. The tertiary structure of some proteins, however, is made more stable by covalent bonds between sulfur atoms (called *disulfide bonds* and abbreviated S-S) in the functional (*R*) groups of amino acids known as cysteines (fig. 3.21). These strong covalent bonds are the exception. Since most of the tertiary structure is stabilized by weak bonds, this structure can easily be disrupted by high temperature or by changes in pH. Irreversible changes in the tertiary structure of proteins produced by this means are referred to as *denaturation* of the proteins.

Denatured proteins retain their primary structure (the peptide bonds are not broken) but have altered chemical properties. Cooking a pot roast, for example, alters the texture of the

[5]protein: Gk. *proteios,* of the first quality

Functional group

Amino group

Carboxylic acid group

Nonpolar amino acids

Valine

Tyrosine

Polar amino acids

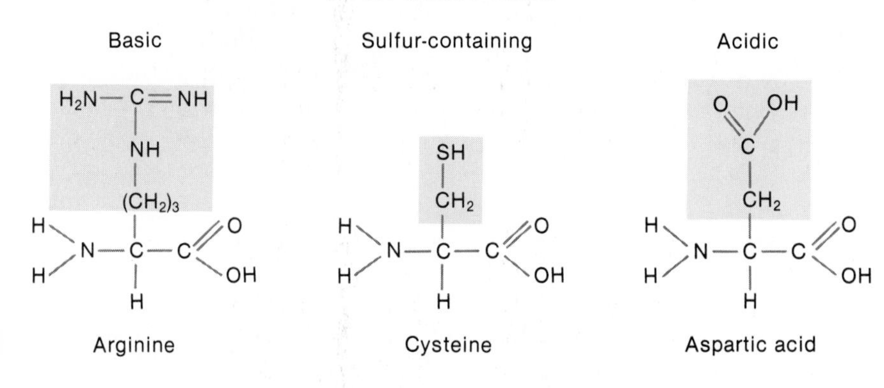

Figure 3.18. *Representative amino acids, showing different types of functional (R) groups.*

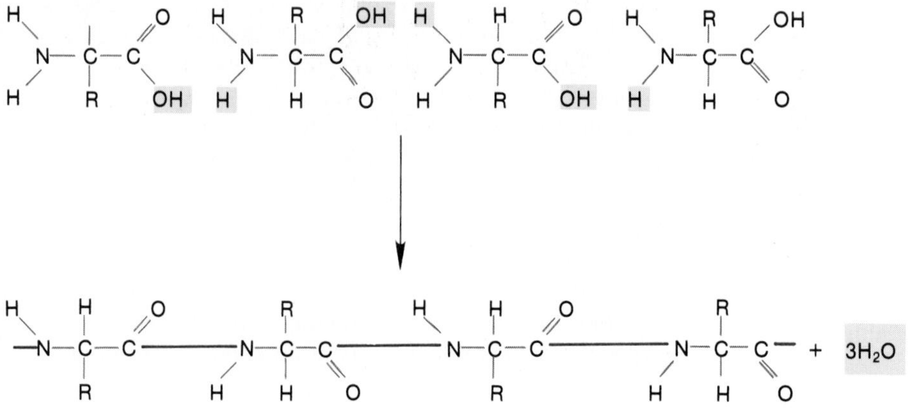

$+ 3H_2O$

Figure 3.19. *The formation of peptide bonds by a dehydration synthesis reaction between amino acids.*

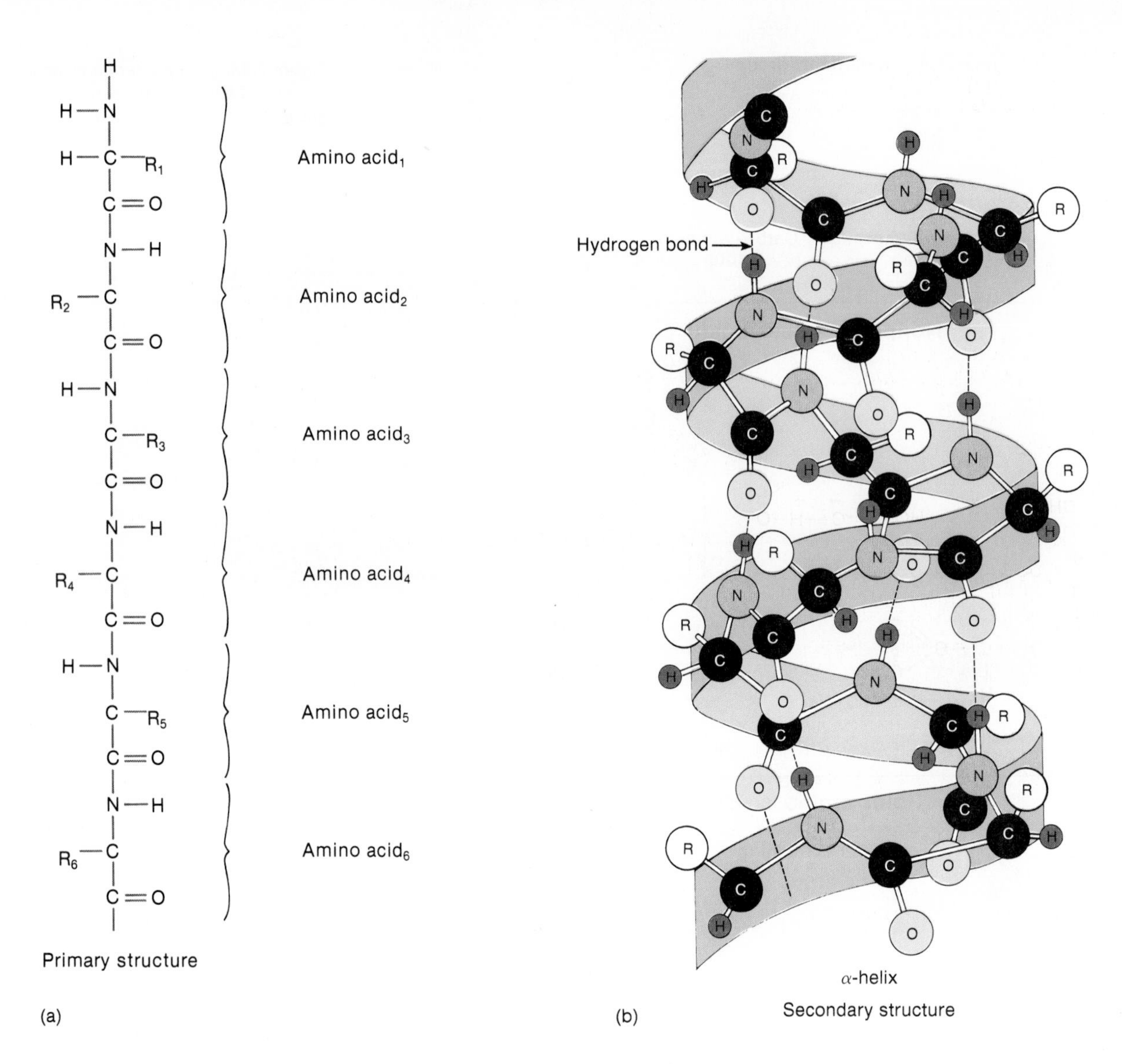

Figure 3.20. *A polypeptide chain, showing (a) its primary structure and (b) secondary structure.*

meat proteins—it doesn't result in an amino acid soup. Denaturation is most dramatically demonstrated by frying an egg. Egg-albumin proteins are soluble in their native state, in which they form the clear, viscous fluid of a raw egg. When denatured by cooking, these proteins change shape, cross-bond with each other, and by this means form an insoluble white precipitate—the egg white.

Some proteins (such as hemoglobin and insulin) are composed of a number of polypeptide chains covalently bonded together. This is the **quaternary structure** of these proteins. Insulin, for example, is composed of two polypeptide chains, one that is twenty-one amino acids long, the other that is thirty amino acids long. Hemoglobin (the protein in red blood cells that carries oxygen) is composed of four separate polypeptide chains. The composition of various body proteins is shown in table 3.4.

Functions of Proteins

Because of their tremendous structural diversity, proteins can serve a wider variety of functions than any other type of molecule in the body. Many proteins, for example, contribute significantly to the structure of different tissues and in this way play a passive role in the functions of these tissues. Examples of such *structural proteins* include collagen (fig. 3.22) and keratin. Collagen is a fibrous protein that provides tensile strength to connective tissues, such as tendons and ligaments. Keratin is found in the outer layer of dead cells in the epidermis, where it serves to prevent water loss through the skin.

Many proteins serve a more active role in the body where specialized structure and function are required. *Enzymes* and *antibodies,* for example, are proteins—no other type of molecule could provide the vast array of different structures needed

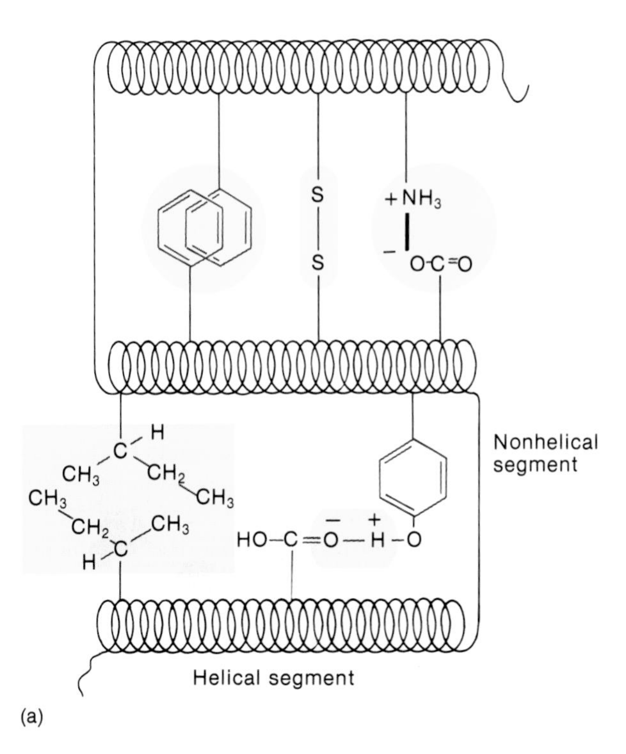

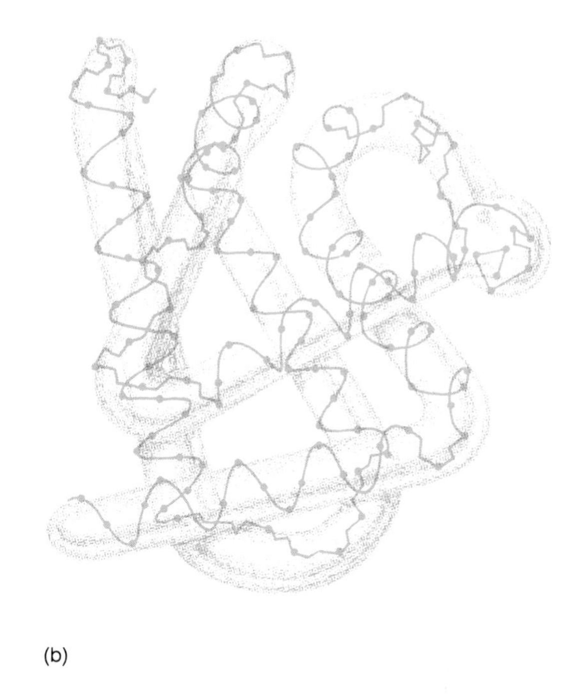

Figure 3.21. The tertiary structure of a protein. (a) Interactions between functional (R) groups of amino acids result in (b) the formation of complex three-dimensional shapes of proteins.

Table 3.4 Composition of selected proteins in the body

Protein	Number of Polypeptide Chains	Nonprotein Component	Function
Hemoglobin	4	Heme pigment	Carries oxygen in the blood
Myoglobin	1	Heme pigment	Stores oxygen in muscle
Insulin	2	None	Hormone-regulating metabolism
Luteinizing hormone	1	Carbohydrate	Hormone that stimulates gonads
Fibrinogen	1	Carbohydrate	Involved in blood clotting
Mucin	1	Carbohydrate	Forms mucus
Blood group proteins	1	Carbohydrate	Produces blood types
Lipoproteins	1	Lipids	Transports lipids in blood

for these functions. Proteins in cell membranes serve as *receptors* for specific regulator molecules (such as hormones) and as *carriers* that transport specific molecules across the membrane. Proteins provide the diversity of shape and chemical properties for the specificity required by these functions.

1. Write the general formula for an amino acid, and describe how amino acids differ from each other.
2. Describe the different levels of protein structure, and explain how these different levels of structure are produced.
3. Describe the different categories of protein function in the body, and explain why proteins can have such diverse functions.

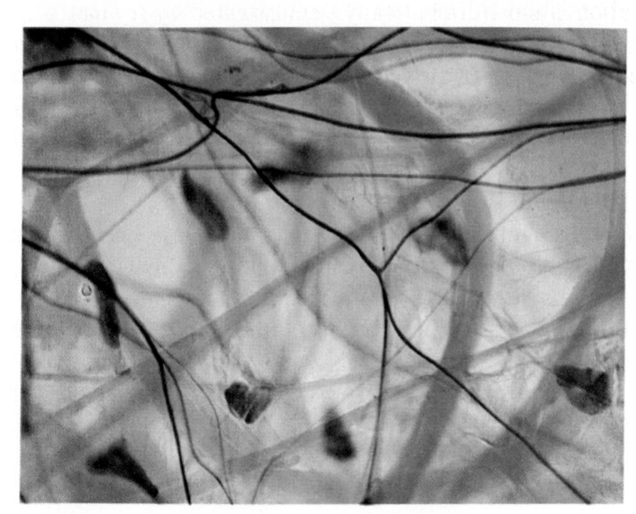

Figure 3.22. A photomicrograph of collagen fibers.

Enzymes as Catalysts

Enzymes are proteins that function as biological catalysts. The catalytic action of enzymes results from the complex structure of these proteins, and the great diversity in the structure of proteins allows different enzymes to be specialized in their actions.

The ability of yeast cells to make alcohol from glucose (a process called *fermentation*[6]) had been known since antiquity, yet no chemist by the mid-nineteenth century could duplicate the trick in the absence of living yeast. Also, yeast and other living cells could perform a vast array of chemical reactions at body temperature that could not be duplicated in the chemical laboratory without adding a substantial amount of heat energy. These observations led many mid-nineteenth-century scientists to believe that chemical reactions in living cells were aided by a "vital force" that operated beyond the laws of the physical world. This "vitalist" concept was squashed along with the yeast cells when a pioneering biochemist, Eduard Buchner, demonstrated that juice obtained from yeast could ferment glucose to alcohol. The yeast juice was not alive—evidently some chemicals in the cells were responsible for fermentation. Buchner didn't know what these chemicals were, so he simply named them **enzymes** (Greek for "in yeast").

In more recent times, biochemists have demonstrated that *all enzymes are proteins* and that enzymes act as *biological catalysts.* (This description must be somewhat modified by recent evidence that RNA may also have limited, but important, catalytic ability.) A catalyst is a chemical that (1) increases the rate of a reaction, (2) is not itself changed at the end of the reaction, and (3) does not change the nature of the reaction or its final result. The same reaction would have occurred in the absence of the catalyst, but it would have progressed at a much slower rate.

In order for a given reaction to occur, the reactants must have sufficient energy. The amount of energy required for a reaction to proceed is called the **energy of activation.** In a large population of molecules, only a small fraction will possess sufficient energy for a reaction. Adding heat will raise the energy level of all the reactant molecules, thus increasing the fraction of the population that has the activation energy. Heat would make reactions go faster, but it would cause protein denaturation and cell death beyond a certain temperature. Catalysts make the reaction go faster at lower temperatures by *lowering the activation energy* required so that a larger fraction of the population of reactant molecules has sufficient energy to participate in the reaction (fig. 3.23).

Since a small fraction of the reactants have the activation energy required for the reaction even in the absence of a catalyst, the reaction could theoretically occur spontaneously at a slow rate. This rate, however, would be much too slow for the needs of a cell. So, from a biological standpoint, the presence or absence of a specific enzyme catalyst acts as a switch—the reaction will occur if the enzyme is present and will not occur if the enzyme is absent.

[6]fermentation: L. *fermentum,* leaven

Mechanism of Enzyme Action

The ability of enzymes to lower the activation energy of a reaction is a result of their structure. Enzymes are proteins—they are thus very large molecules with complex, highly ordered, three-dimensional shapes produced by chemical interactions between their amino acids. Each type of enzyme protein has characteristic ridges, grooves, and pockets that are lined with specific amino acids. The particular pockets that are active in catalyzing a reaction are called the *active sites* of the enzyme.

The reactant molecules, which are the *substrates* of the enzyme, have shapes that allow them to fit into the active sites. The fit may not be perfect at first, but a perfect fit may be induced as the substrate gradually slips into the active site. This induced fit weakens the existing bonds within the substrate molecules and allows them to be more easily broken. New bonds are also more easily formed as substrates are brought close together in the proper orientation. The *enzyme-substrate complex,* formed temporarily in the course of the reaction, then dissociates to yield *products* and the free unaltered enzyme. This model of how enzymes work is known as the **lock-and-key** model of enzyme activity (fig. 3.24).

Naming of Enzymes

Although an international committee has established a uniform naming system for enzymes, the names that are in common use do not follow a completely consistent pattern. With the exception of some older enzyme names (such as pepsin, trypsin, and renin), all enzyme names end with the suffix *-ase.* Classes of enzymes are named according to their job category. *Hydrolases,* for example, promote hydrolysis reactions. Other enzyme categories include *phosphatases,* which catalyze the removal of phosphate groups, *synthetases,* which catalyze dehydration synthesis reactions, and *dehydrogenases,* which remove hydrogen atoms from their substrates.

When tissues become damaged due to diseases, some of the dead cells disintegrate and release their enzymes into the blood. Most of these enzymes are not normally active in the blood because of the absence of their specific substrates, but their enzymatic activity can be measured in a test tube by the addition of the appropriate substrates to samples of plasma. Such measurements are clinically useful, because abnormally high plasma concentrations of particular enzymes are characteristic of certain diseases (table 3.5).

Effects of Temperature and pH

An increase in temperature, as previously described, will increase the rate of non-enzyme-catalyzed reactions because a larger number of reactant molecules will have the activation energy required. A similar relationship between temperature and reaction rate occurs in enzyme-catalyzed reactions. At a temperature of 0° C the reaction rate is unmeasurably slow. As the

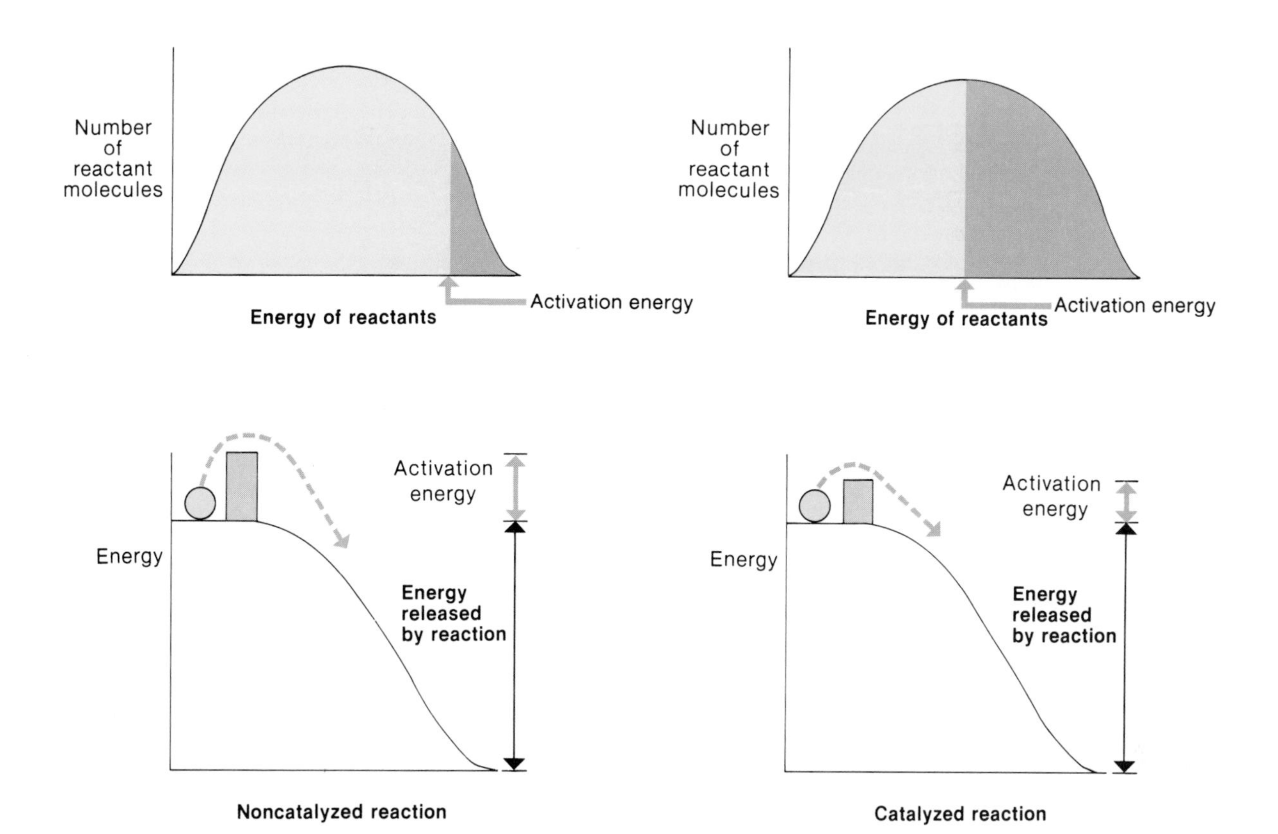

Figure 3.23. *A comparison of a noncatalyzed reaction with a catalyzed reaction. Upper figures compare the proportion of reactant moelcules that have sufficient activation energy to participate in the reaction (color). This proportion is increased in the enzyme-catalyzed reaction, because enzymes lower the activation energy required for the reaction (shown as a barrier on top of an energy "hill" in the bottom figures). Reactants that can overcome this barrier are able to participate in the reaction, as shown by arrows pointing to the bottom of the energy hill.*

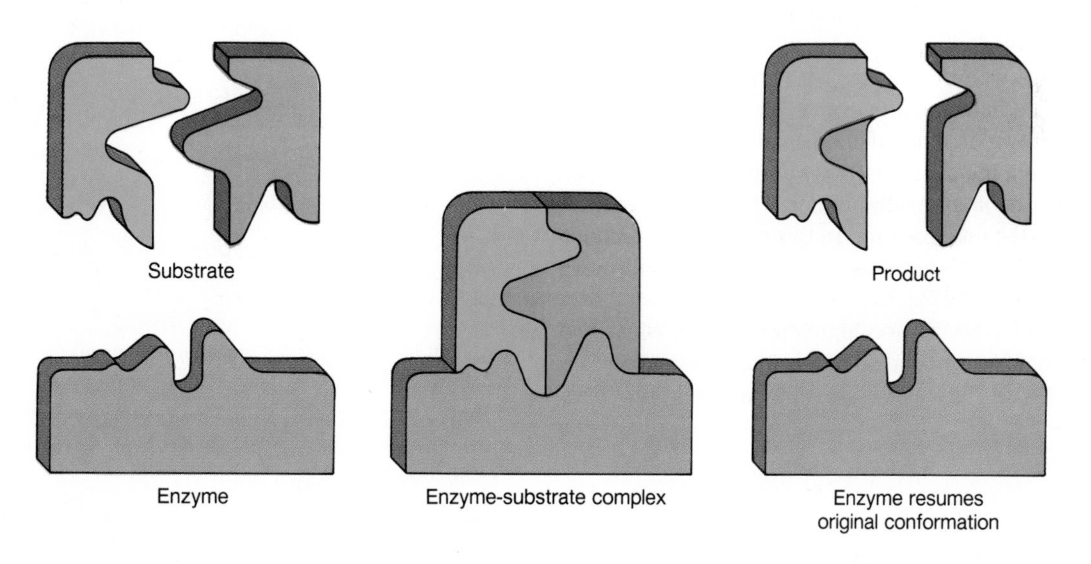

Figure 3.24. *The lock-and-key model of enzyme action.*

temperature is raised above 0° C the reaction rate increases but only up to a point. At a few degrees above body temperature (which is 37° C) the reaction rate reaches a plateau; further increases in temperature actually *decrease* the rate of the reaction (fig. 3.25). This decrease is due to the fact that the tertiary structure of enzymes becomes altered at higher temperatures.

A similar relationship is observed when the rate of an enzymatic reaction is measured at different pH values. Each enzyme characteristically has its peak activity in a very narrow pH range, which is the **pH optimum** for the enzyme. If the pH is changed from this optimum, the reaction rate decreases (fig. 3.26). This decreased enzyme activity is due to changes in the conformation of the enzyme and in the charges of the R groups of the amino acids lining the active sites.

Enzyme	Diseases Associated with Abnormal Plasma Enzyme Concentrations
Alkaline phosphatase	Obstructive jaundice, Paget's disease (osteitis deformans), carcinoma of bone
Acid phosphatase	Benign hypertrophy of prostate, cancer of prostate
Amylase	Pancreatitus, perforated peptic ulcer
Aldolase	Muscular dystrophy
Creatine kinase (or creatine phosphokinase-CPK)	Muscular dystrophy, myocardial infarction
Lactate dehydrogenase (LDH)	Myocardial infarction, liver disease, renal disease, pernicious anemia
Transaminases (GOT and GPT)	Myocardial infarction, hepatitis, muscular dystrophy

Table 3.5 Examples of the diagnostic value of some enzymes found in plasma

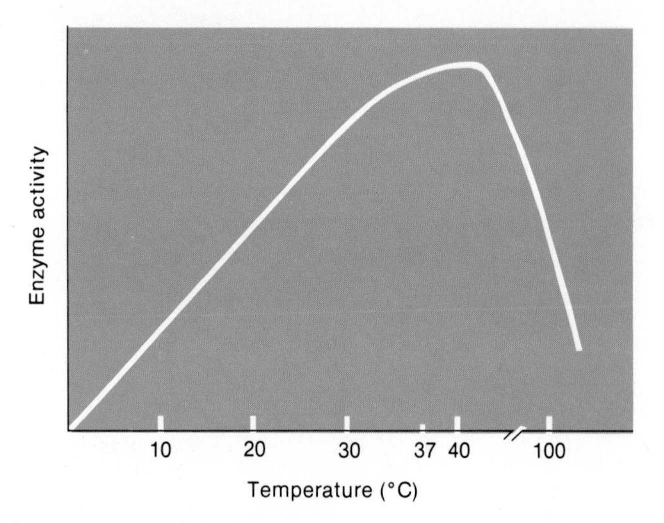

Figure 3.25. *The effect of temperature on enzyme activity, as measured by the rate of the enzyme-catalyzed reaction under standardized conditions.*

The pH optimum of an enzyme usually reflects the pH of the body fluid in which the enzyme is found. The acidic pH optimum of the protein-digesting enzyme *pepsin,* for example, allows it to be active in the strong hydrochloric acid of gastric juice. Similarly, the neutral pH optimum of *salivary amylase* and the alkaline pH optimum of *trypsin* in pancreatic juice allow these enzymes to digest starch and protein, respectively, in other parts of the digestive tract.

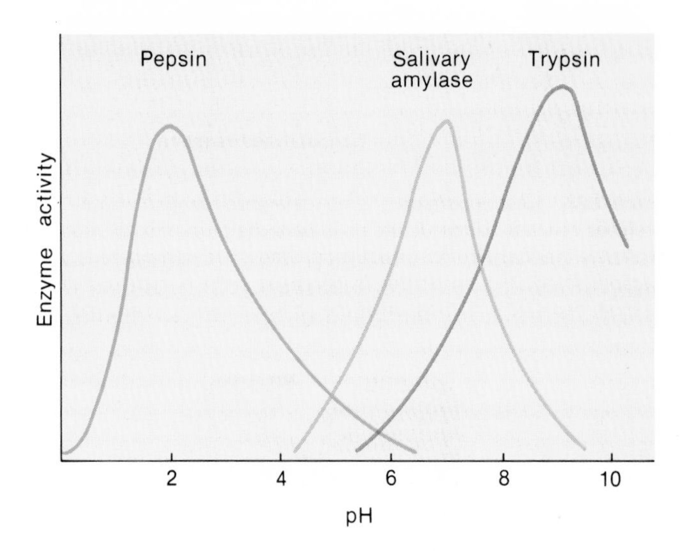

Figure 3.26. *The effect of pH on the activity of three digestive enzymes.*

Metabolic Pathways

The many thousands of different types of enzymatic reactions within a cell do not occur independently of each other. They are, rather, all linked together by intricate webs of interrelationships, the total pattern of which constitutes cellular metabolism.[7] A part of this web that begins with an *initial substrate,* progresses through a number of *intermediates,* and ends with a *final product* is known as a **metabolic pathway.**

The enzymes in a metabolic pathway cooperate in a manner analogous to workers on an assembly line where each contributes a small part to the final product. In this process, the product of one enzyme in the line becomes the substrate of the next enzyme, and so on (fig. 3.27).

Few metabolic pathways are completely linear. Most are branched so that one intermediate at the branch point can serve as a substrate for two different enzymes. Two different products can thus be formed that serve as intermediates of two divergent pathways (fig. 3.28).

1. Explain how catalysts increase the rate of chemical reactions.
2. Draw a labeled picture of the lock-and-key model of enzyme action, and write a formula for this reaction.
3. Explain how enzymes are named.
4. Explain how enzymes cooperate to produce a metabolic pathway, and illustrate this concept with a figure.

Bioenergetics

Living organisms require the constant expenditure of energy to maintain their complex structures and processes. Central to life processes are chemical reactions that are coupled, so that the energy released by one reaction is incorporated into the products of another reaction. The transformation of energy in living systems is largely based on reactions that produce and destroy molecules of ATP.

Bioenergetics[8] refers to the flow of energy in living systems. Organisms maintain their highly ordered structure and life-sustaining activities through the constant expenditure of energy obtained ultimately from the environment. The energy flow in living systems obeys the first and second laws of a branch of physics known as *thermodynamics.*

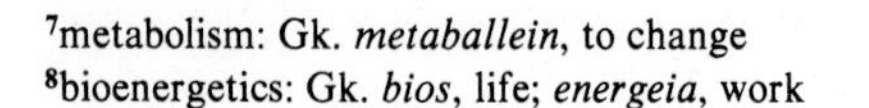
[7]metabolism: Gk. *metaballein,* to change
[8]bioenergetics: Gk. *bios,* life; *energeia,* work

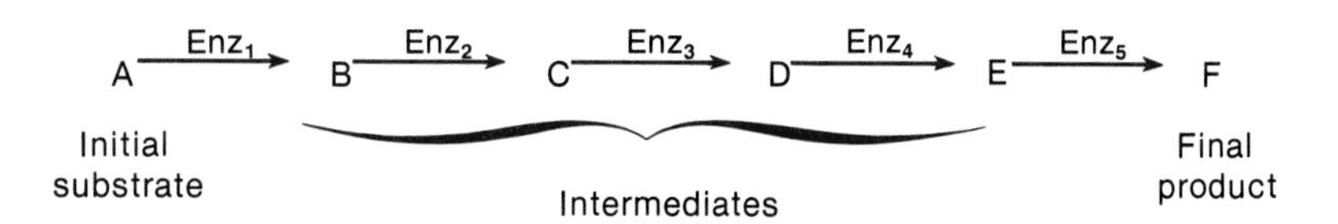

Figure 3.27. A metabolic pathway, where the product of one enzyme becomes the substrate of the next in a multi-enzyme system.

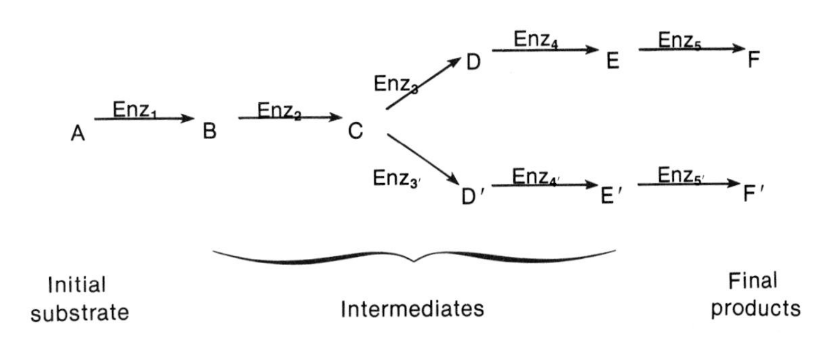

Figure 3.28. A branched metabolic pathway.

According to the **first law of thermodynamics,** energy can be transformed (changed from one form to another), but it can neither be created nor destroyed. This is sometimes called the law of *conservation of energy.* As a result of energy transformations, according to the **second law of thermodynamics,** the universe becomes increasingly disorganized. The term *entropy* is used to describe the degree of disorganization of a system. Energy transformations thus increase the amount of entropy of a system. Only energy that is in an organized state—called *free energy*—can be used to do work. Thus, since entropy increases in every energy transformation, the amount of free energy available to do work decreases. As a result of the increased entropy described by the second law, systems tend to go from states of higher to states of lower free energy.

The chemical bonding of atoms into molecules obeys the laws of thermodynamics. Atoms that are organized into complex organic molecules, such as glucose, have more free energy (less entropy) than six separate molecules each of carbon dioxide and water. Therefore, in order to convert carbon dioxide and water to glucose, energy must be added. Plants perform this feat using energy from the sun in the process of *photosynthesis* (fig. 3.29).

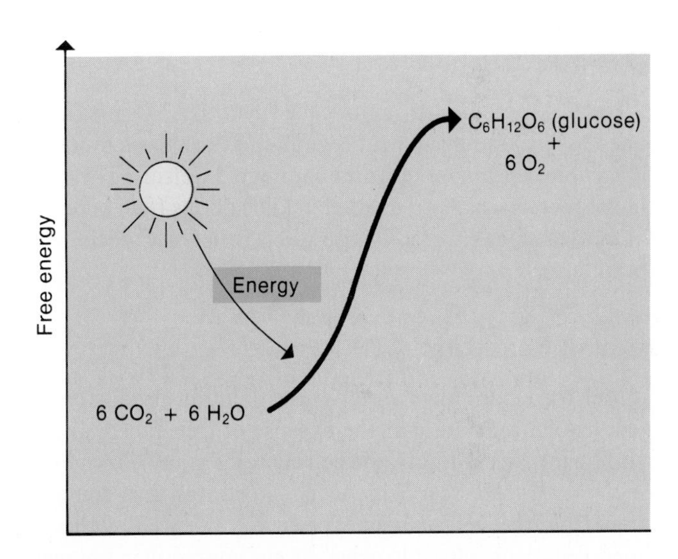

Figure 3.29. A simplified diagram of photosynthesis. Some of the sun's radiant energy is captured by the plants and used to produce glucose from carbon dioxide and water. As the product of this endergonic reaction, glucose has a higher free energy content than the initial reactants.

Endergonic and Exergonic Reactions

Chemical reactions that require an input of energy are known as **endergonic reactions.** Since energy is added to make these reactions "go," the products of endergonic reactions must contain more free energy than the reactants. A portion of the energy added, in other words, is contained within the product molecules. This follows from the fact that energy cannot be created or destroyed (first law of thermodynamics) and from the fact that a more organized state of matter contains more free energy (less entropy) than a less organized state (as described by the second law).

The fact that glucose contains more free energy than carbon dioxide and water can be easily proven by the combustion of glucose to CO_2 and H_2O. This reaction releases energy in the form of heat. Reactions that convert molecules with more free energy to molecules with less—and, therefore, release energy as they proceed—are called **exergonic reactions.**

As illustrated in figure 3.30, the amount of energy released by an exergonic reaction is the same whether the energy is released in a single combustion reaction or in the many small, enzymatically controlled steps that occur in tissue cells. The energy that the body obtains from the consumption of particular foods can, therefore, be measured as the amount of heat energy released when these foods are combusted.

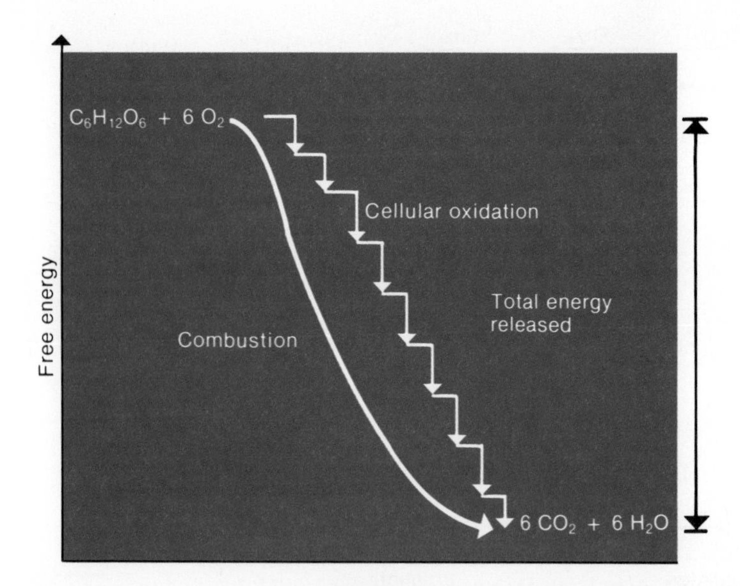

Figure 3.30. *Since glucose contains more energy than carbon dioxide and water, the combustion of glucose is an exergonic reaction. The same amount of energy is released if the glucose is broken dow stepwise within the cell.*

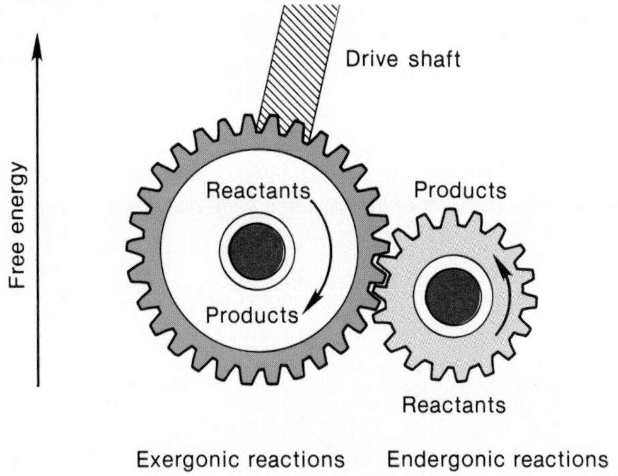

Heat is measured in units called *calories.*[9] One calorie is defined as the amount of heat required to raise the temperature of one cubic centimeter of water one degree Celsius. The caloric value of food is usually indicated in kilocalories (one kilocalorie = 1,000 calories), which is also often called a Calorie with a capital letter *C.*

Coupled Reactions: ATP

In order to remain alive a cell must maintain its highly organized, low entropy state at the expense of free energy from its environment. Accordingly, the cell contains many enzymes that catalyze exergonic reactions, using substrates that come ultimately from the environment. The energy released by these exergonic reactions is used to drive the energy-requiring processes (endergonic reactions) in the cell. Since the cell cannot use heat to drive energy-requiring processes, chemical bond energy that is released by exergonic reactions must be directly transferred to the chemical bonds formed by endergonic reactions. Energy-liberating reactions are thus *coupled* to energy-requiring reactions. This relationship is like two meshed gears; the turning of one (the energy-releasing, exergonic gear) causes turning of the other (the energy-requiring, endergonic gear—fig. 3.31).

The energy released by most exergonic reactions in the cell is used, either directly or indirectly, to drive *one* endergonic reaction (fig. 3.32): the formation of **adenosine triphosphate (ATP)** from adenosine diphosphate (ADP) and inorganic phosphate (abbreviated P_i).

The formation of ATP requires the input of a fairly large amount of energy. Since this energy must be conserved (first law of thermodynamics), the bond that is produced by joining

Free energy
Drive shaft
Reactants
Products
Products
Reactants
Exergonic reactions
Endergonic reactions

Figure 3.31. *A model of the coupling of exergonic and endergonic reactions. The exergonic gear turns the endergonic gear. The reactants of the exergonic reaction (represented by the larger gear) have more free energy than the products of the endergonic reaction because the coupling is not 100% efficient—some energy is lost as heat.*

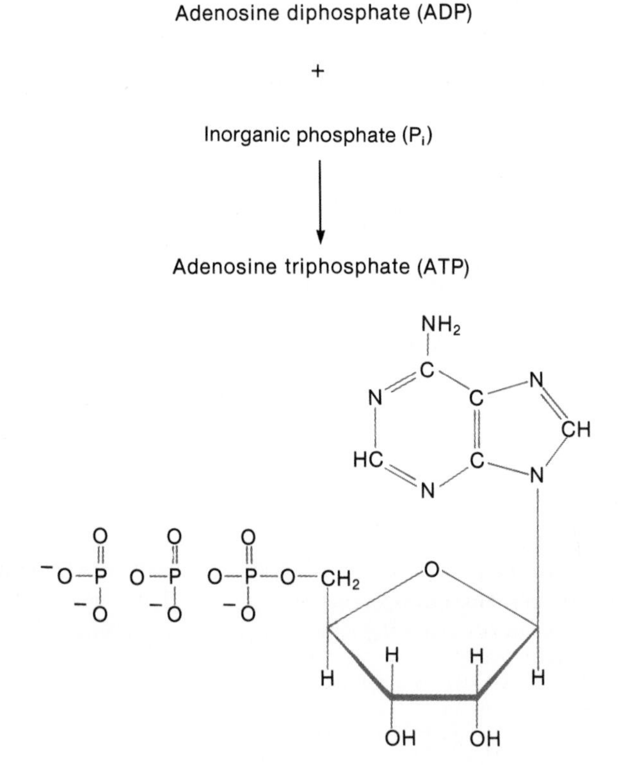

Figure 3.32. *The formation and structure of adenosine triphosphate (ATP).*

P_i to ADP must contain a part of this energy. Thus, when enzymes reverse this reaction and convert ATP to ADP and P_i, a large amount of energy is released. Energy released from the breakdown of ATP is used to power the energy-requiring processes—synthesis reactions, muscle contraction, and so on—in all cells. ATP is thus called the **universal energy carrier** (fig. 3.33).

[9]calorie: L. *calor,* heat

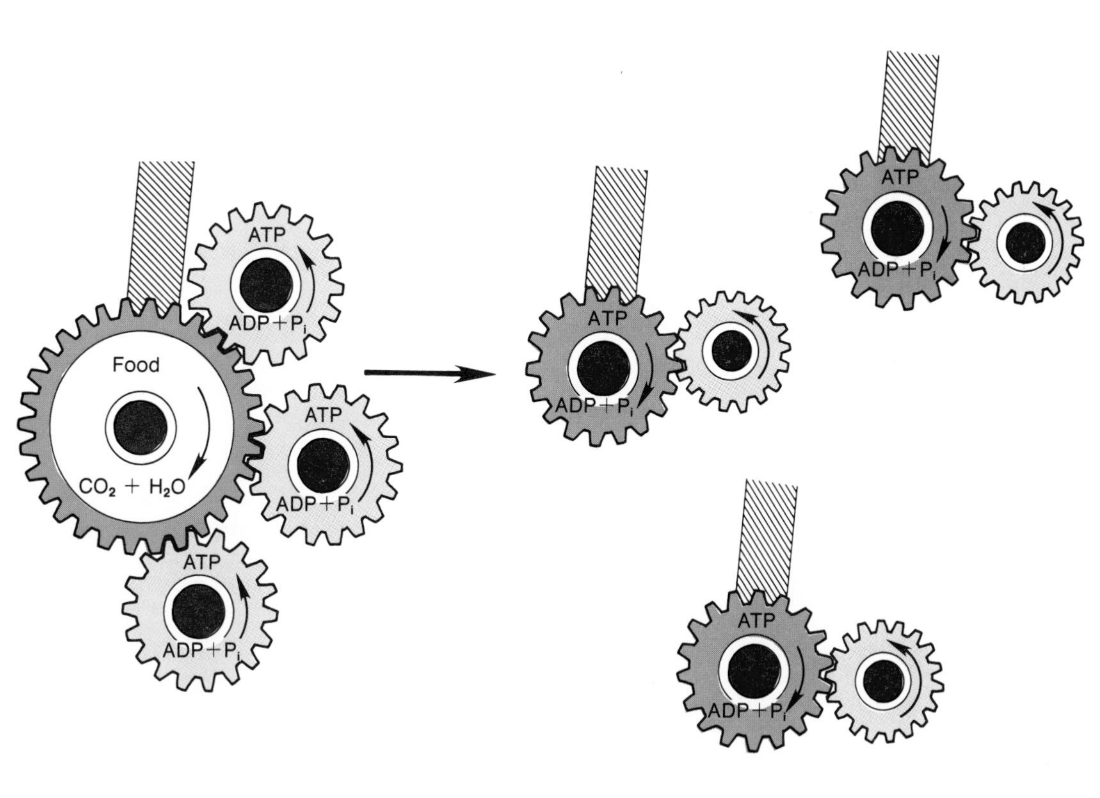

Figure 3.33. *A model of ATP as the universal energy carrier of the cell. Exergonic reactions are shown as gears with arrows going down (reactions produce a decrease in free energy); endergonic reactions are shown as gears with arrows going up (reactions produce an increase in free energy).*

Summary

Atoms, Ions, and Chemical Bonds p. 34

I. Covalent bonds are formed by atoms that share electrons; these are the strongest type of chemical bonds.

II. Ionic bonds are formed by atoms that transfer electrons; these weak bonds join atoms together in an ionic compound.

III. When hydrogen is bonded to an electronegative atom, it gains a slight positive charge and is weakly attracted to another electronegative atom; this weak attraction is a hydrogen bond.

IV. Acids donate hydrogen ions to solution, whereas bases lower the hydrogen ion concentration of a solution.
   A. The pH scale is inversely related to the logarithm of the hydrogen ion concentration.
   B. In a neutral solution the concentration of H^+ is equal to the concentration of OH^-, and the pH is 7.
   C. Acids raise the H^+ concentration and thus lower the pH below 7; bases lower the H^+ concentration and thus raise the pH above 7.

V. Organic molecules contain atoms of carbon joined together by covalent bonds; atoms of nitrogen, oxygen, phosphorus, or sulfur may be present as specific functional groups in the organic molecule.

Carbohydrates and Lipids p. 41

I. Carbohydrates contain carbon, hydrogen, and oxygen, usually in a ratio of 1:2:1.
   A. Carbohydrates consist of simple sugars (monosaccharides), disaccharides, and polysaccharides (such as glycogen).
   B. Covalent bonds between monosaccharides are formed by dehydration synthesis, or condensation; bonds are broken by hydrolysis reactions.

II. Lipids are organic molecules that are insoluble in polar solvents such as water.
   A. Triglycerides (fat and oil) consist of three fatty acid molecules joined to a molecule of glycerol.
   B. Phospholipids (such as lecithin) are phosphate-containing lipids that have a polar group, which is hydrophilic; the rest of the molecule is hydrophobic.
   C. Steroids (including the hormones of the adrenal cortex and gonads) are lipids with a characteristic five-ring structure.

Proteins p. 46

I. Proteins are composed of long chains of amino acids bonded together by covalent peptide bonds.
   A. Each amino acid contains an amino group, a carboxyl group, and a functional group that is different for each of the more than twenty different amino acids.
   B. The polypeptide chain may be twisted into a helix (secondary structure) and bent and folded to form the tertiary structure of the protein.
   C. Because of their great variety of possible structures, proteins serve a wider variety of specific functions than any other type of molecule.

Enzymes as Catalysts p. 50

I. Enzymes are biological catalysts, which increase the rate of chemical reactions.

II. All enzymes are proteins.
   A. The reactants in an enzyme-catalyzed reaction—called the substrates of the enzyme—fit into a specific pocket in the enzyme called the active site.
   B. By forming an enzyme-substrate complex, substrate molecules are brought into proper orientation and existing bonds are weakened; this allows new bonds to be more easily formed.

III. The activity of an enzyme is affected by a variety of factors.
   A. The rate of enzyme-catalyzed reactions increases with increasing temperature, up to a maximum.
   B. Each enzyme has optimal activity at a characteristic pH—called the pH optimum for that enzyme.

IV. Metabolic pathways involve a number of enzyme-catalyzed reactions, in which the product of one enzyme becomes the substrate of the next.

Bioenergetics p. 52

I. The flow of energy in the cell is called bioenergetics.
   A. Chemical reactions that liberate energy are termed exergonic, and those that require energy are endergonic.
   B. Energy is required to produce an organic molecule such as glucose from inorganic molecules.

C. According to the first law of thermodynamics, energy can neither be created nor destroyed.
D. The chemical bond energy in the glucose molecule must thus be released when glucose is broken down into inorganic molecules.

II. In living cells, endergonic reactions are coupled to exergonic reactions.
A. In this way, the energy released by the exergonic reaction can be directly captured as chemical bond energy in the products of the endergonic reaction.
B. The major exergonic reactions in the cell provide energy for the production of ATP from ADP and P_i.
C. ATP is the universal energy carrier; the energy it liberates when it is broken down into ADP and P_i is used to power all the activities of the cell.

Review Activities

Objective Questions

1. Which of the following statements about atoms is *true?*
   (a) They have more protons than electrons.
   (b) They have more electrons than protons.
   (c) They are electrically neutral.
   (d) They have as many neutrons as they have electrons.
2. The bond between oxygen and hydrogen in a water molecule is a(n)
   (a) hydrogen bond
   (b) polar covalent bond
   (c) nonpolar covalent bond
   (d) ionic bond
3. Solution A has a pH of 2, and solution B has a pH of 10. Which of the following statements about these solutions is *true?*
   (a) Solution A has a higher H^+ concentration than solution B.
   (b) Solution B is basic.
   (c) Solution A is acidic.
   (d) All of these are true.
4. Glucose is a
   (a) disaccharide
   (b) polysaccharide
   (c) monosaccharide
   (d) phospholipid
5. Digestion reactions occur by means of
   (a) dehydration synthesis
   (b) hydrolysis
6. Carbohydrates are stored in the liver and muscles in the form of
   (a) glucose
   (b) triglycerides
   (c) glycogen
   (d) cholesterol
7. Lecithin is a
   (a) carbohydrate
   (b) protein
   (c) steroid
   (d) phospholipid
8. The type of bond formed between two molecules of water is a
   (a) hydrolytic bond
   (b) polar covalent bond
   (c) nonpolar covalent bond
   (d) hydrogen bond
9. The carbon-to-nitrogen bond that joins amino acids together is called a
   (a) glycosidic bond
   (b) peptide bond
   (c) hydrogen bond
   (d) double bond
10. Which of the following statements about enzymes is *true?*
    (a) All proteins are enzymes.
    (b) All enzymes are proteins.
    (c) Enzymes are changed by the reactions they catalyze.
    (d) The active sites of enzymes have little specificity for substrates.
11. Which of the following statements about enzyme-catalyzed reactions is *true?*
    (a) The rate of reaction is independent of temperature.
    (b) The rate of all enzyme-catalyzed reactions is decreased when the pH is lowered from 7 to 2.
    (c) The rate of reaction is independent of substrate concentration.
    (d) Under given conditions, the rate of product formation varies directly with enzyme concentration, up to a maximum, at which point the rate cannot be further increased.
12. In a metabolic pathway
    (a) the product of one enzyme becomes the substrate of the next
    (b) the substrate of one enzyme becomes the product of the next
13. Which of the following represents an *endergonic* reaction?
    (a) $ADP + P_i \rightarrow ATP$
    (b) $ATP \rightarrow ADP + P_i$
    (c) glucose + $O_2 \rightarrow CO_2 + H_2O$
    (d) $CO_2 + H_2O \rightarrow$ glucose
    (e) both *a* and *d*
    (f) both *b* and *c*
14. Which of the following statements about ATP is *true?*
    (a) The bond joining ADP and the third phosphate is a high-energy bond.
    (b) The formation of ATP is coupled to energy-liberating reactions.
    (c) The conversion of ATP to ADP and P_i provides energy for biosynthesis, cell movement, and other cellular processes that require energy.
    (d) ATP is the "universal energy carrier" of cells.
    (e) All of these are true.

Essay Questions

1. Compare and contrast nonpolar covalent bonds, polar covalent bonds, and ionic bonds.
2. Give the definition of an acid and base, and explain how these influence the pH of a solution.
3. Using dehydration synthesis and hydrolysis reactions, explain the relationships between starch in an ingested potato, liver glycogen, and blood glucose.
4. "All fats are lipids, but not all lipids are fats." Explain why this statement is true.
5. Explain the relationship between the primary structure of a protein and its secondary and tertiary structures.
6. Explain the relationship between the chemical structure and the function of an enzyme, and describe how various conditions may alter both the structure and the function of an enzyme.
7. Using the first and second laws of thermodynamics, explain how ATP is formed and how it serves as the universal energy carrier.

4

The Cellular Basis of Human Biology

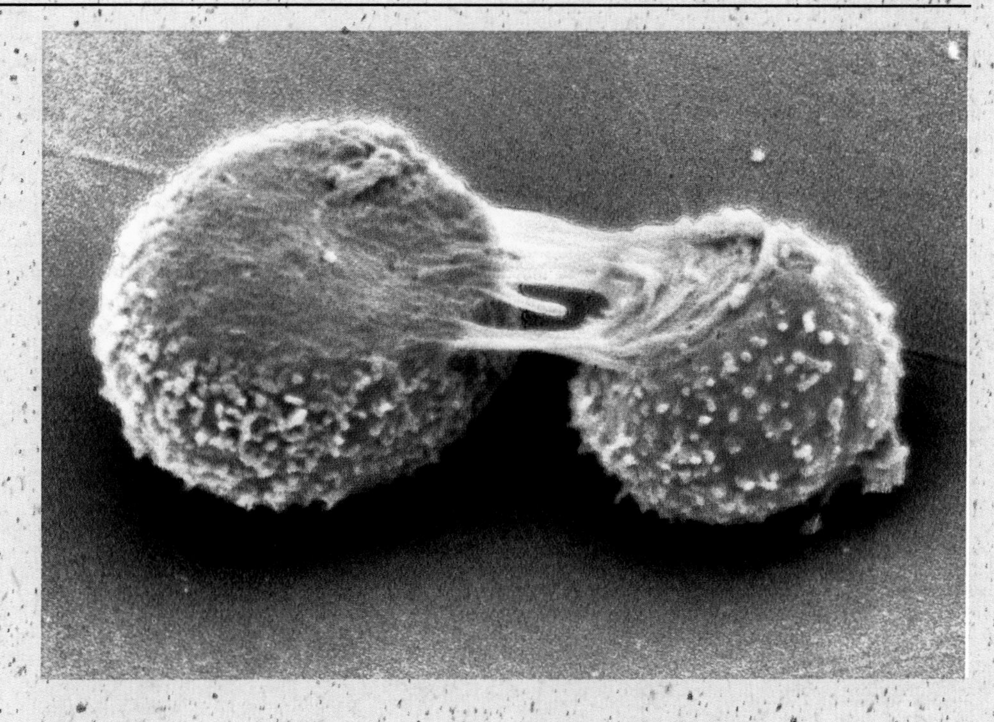

Outline

Objectives

By studying this chapter, you should be able to

1. describe the structure of the cell membrane and explain its functional importance
2. describe how cells move by amoeboid motion, and the structure and significance of cilia and flagella
3. explain the processes of phagocytosis, pinocytosis, receptor-mediated endocytosis, and exocytosis
4. explain the functions of the cytoskeleton, lysosomes, mitochondria, and the endoplasmic reticulum
5. describe the structure and significance of the cell nucleus
6. describe diffusion and explain its physical basis
7. define osmosis and explain the conditions required for osmosis to occur
8. define the terms *isotonic, hypertonic,* and *hypotonic,* and explain their functional significance
9. define active transport and explain its significance

Keys to Pronunciation

cilia: *sil'e-ah*
cytoplasm: *si'to-plazm*
diffusion: *di-fu'zhun*
endoplasmic reticulum: *en''do-plas'mic re-tik'u-lum*
flagella: *fla-jel'ah*
lysosome: *li'so-som*
mitochondria: *mi''to-kon'dre-ah*
organelles: *or''gah-nelz'*
permeable: *per'me-ah-b'l*
phagocytosis: *fag''o-si-to'sis*
pinocytosis: *pin''o-si-to'sis*
pseudopods: *soo'do-pods*

Photo: Hybridoma cells. These are formed by the fusion of cancer cells with normal lymphocytes, the type of white blood cell that produces antibodies. Hybridoma cells are potentially immortal cellular factories that produce specific antibodies.

Cell Membrane and Associated Structures

The cell is the basic unit of structure and function in the body. Many of the functions of cells are performed by particular subcellular structures known as organelles. The cell membrane is an extremely important structure that allows selective communication between the intracellular and extracellular compartments, and that participates in cellular movements.

The cell is so small and so simple in appearance when viewed with the ordinary (light) microscope that it is difficult to conceive that each cell is a living entity unto itself. Equally amazing is the fact that the physiology of our organs and systems is derived from the complex functions of the cells of which they are composed.

As the basic functional unit of the body, each cell is a highly organized molecular factory. Cells come in a great variety of shapes and sizes. This great variation reflects the variation of functions of different cells in the body. All cells, however, share certain characteristics—such as the fact that they are surrounded by a cell membrane—and most cells possess the structures listed in table 4.1. Thus, although no single cell can be considered "typical," the general structure of cells can be shown with a single illustration (fig. 4.1).

[1]aqueous: L. *aqua*, water

For descriptive purposes, a cell can be divided into three principal parts:

1. **Cell (plasma) membrane.** The cell membrane surrounds the cell, gives form to it, and separates the cell's internal structures from the extracellular environment. The cell membrane is selectively permeable, allowing some substances to pass through it while preventing the passage of other substances.
2. **Cytoplasm and organelles.** The cytoplasm is the aqueous[1] content of a cell between the nucleus and the cell membrane. Organelles are subcellular structures within the cytoplasm of a cell that perform specific functions.
3. **Nucleus.** The nucleus is a large, generally spheroid body within a cell that contains the DNA, or genetic material, of a cell and that thus directs the activities of the cell.

Cell Membrane

Because both the intracellular and extracellular environments (or "compartments") are aqueous, a barrier must be present to prevent the loss of cellular molecules that are water-soluble. Since this barrier cannot itself be composed of water-soluble molecules, it makes sense that the cell membrane is composed of lipids.

Table 4.1 Structure and function of cellular components

Component	Structure	Function
Cell (plasma) membrane	Membrane composed of phospholipid and protein molecules	Gives form to cell and controls passage of materials in and out of cell
Cytoplasm	Fluid, jellylike substance in which organelles are suspended	Serves as matrix substance in which chemical reactions occur
Endoplasmic reticulum	System of interconnected membrane-forming canals and tubules	Smooth endoplasmic reticulum metabolizes nonpolar compounds and stores Ca^{++} in striated muscle cells; rough endoplasmic reticulum assists in protein synthesis
Ribosomes	Granular particles composed of protein and RNA	Synthesize proteins
Golgi apparatus	Cluster of flattened, membranous sacs	Synthesizes carbohydrates and packages molecules for secretion; secretes lipids and glycoproteins
Mitochondria	Membranous sacs with folded inner partitions	Release energy from food molecules and transform energy into usable ATP
Lysosomes	Membranous sacs	Digest foreign molecules and worn and damaged cells
Peroxisomes	Spherical membranous vesicles	Contain certain enzymes; form hydrogen peroxide
Centrosome	Nonmembranous mass of two rodlike centrioles	Helps organize spindle fibers and distribute chromosomes during mitosis
Vacuoles	Membranous sacs	Store and excrete various substances within the cytoplasm
Fibrils and microtubules	Thin, hollow tubes	Support cytoplasm and transport materials within the cytoplasm
Cilia and flagella	Minute cytoplasmic extensions from cell	Move particles along surface of cell or move cell
Nuclear membrane	Membrane surrounding nucleus, composed of protein and lipid molecules	Supports nucleus and controls passage of materials between nucleus and cytoplasm
Nucleolus	Dense, nonmembranous mass composed of protein and RNA molecules	Forms ribosomes
Chromatin	Fibrous strands composed of protein and DNA molecules	Controls cellular activity for carrying on life processes

The **cell membrane** (also called the **plasma membrane**) and indeed all of the membranes surrounding organelles within the cell, are composed primarily of phospholipids and proteins. Phospholipids, as described in chapter 3, are polar on the end that contains the phosphate group and nonpolar (and hydrophobic) throughout the rest of the molecule. Since there is an aqueous environment on each side of the membrane, the hydrophobic parts of the molecules "huddle together" in the center of the membrane, leaving the polar ends exposed to water on both surfaces. This results in the formation of a double layer of phospholipids in the cell membrane.

The hydrophobic core of the membrane restricts the passage of water and water-soluble molecules and ions. Certain of these polar compounds, however, do pass through the membrane. The specialized functions and selective transport properties of the membrane are believed to be due to its protein content. Some proteins are found partially submerged on each side of the membrane; other proteins span the membrane completely from one side to the other. Since the membrane is not solid—phospholipids and proteins are free to move—the proteins within the phospholipid "sea" are not uniformly distributed, but rather present a mosaic pattern. This structure is known as the **fluid-mosaic model** of membrane structure (fig. 4.2).

The proteins found in the cell membrane serve a variety of functions, including (1) structural support; (2) transport of molecules across the membrane; (3) enzymatic control of chemical reactions at the cell surface; (4) receptors for hormones and other regulatory molecules that arrive at the outer surface of the membrane; and (5) cellular "markers" (antigens), which identify the blood and tissue type.

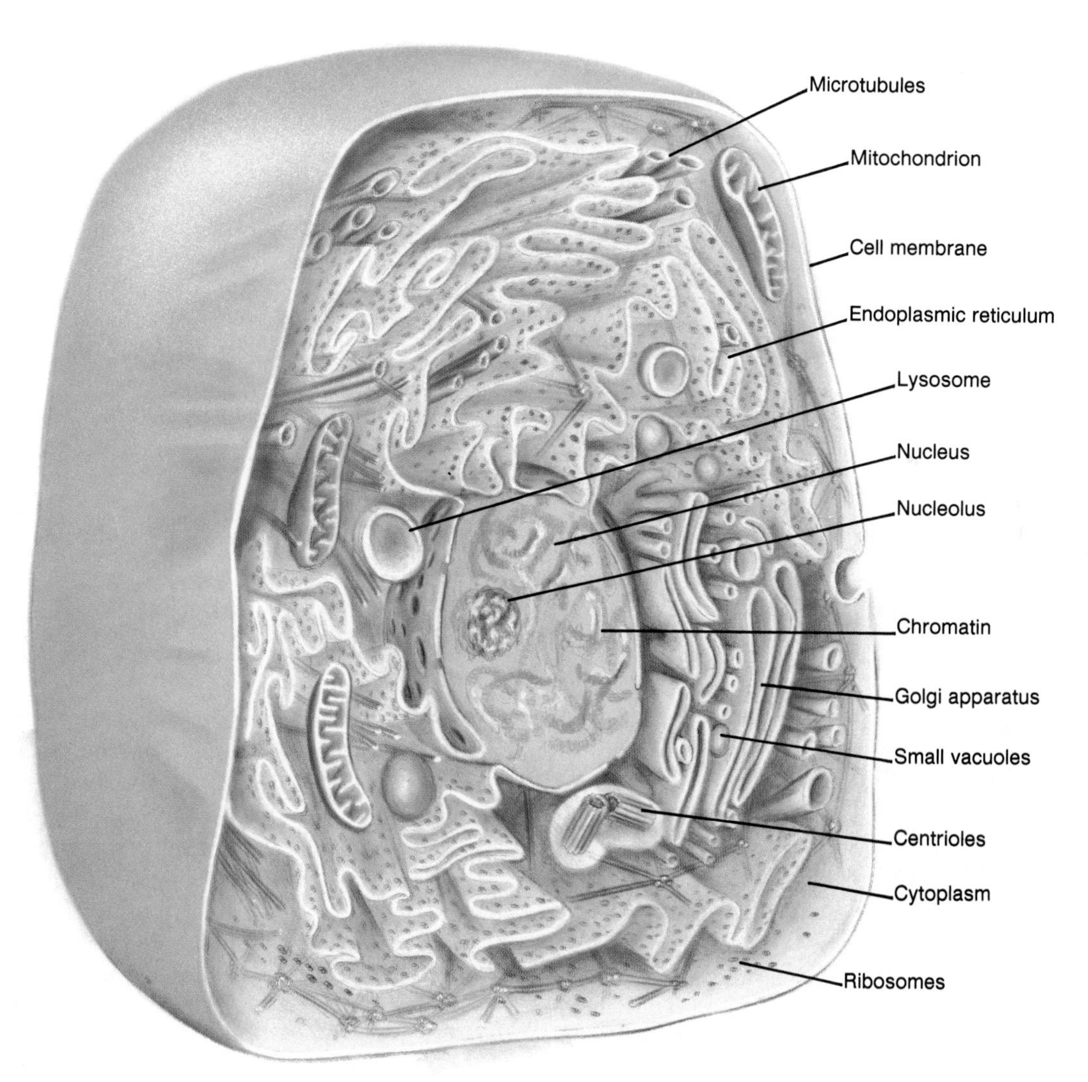

Figure 4.1. *A generalized cell and the principal organelles.*

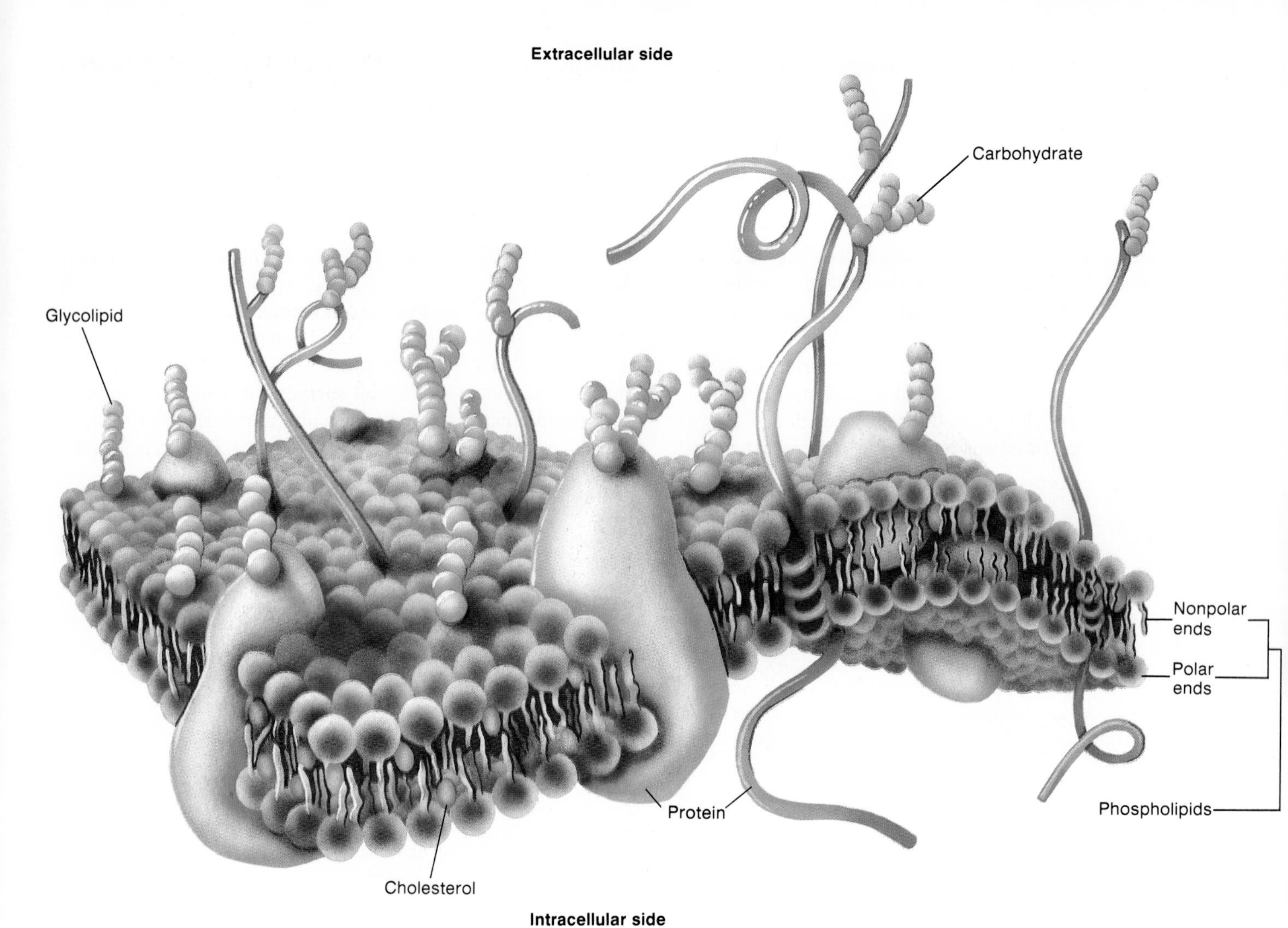

Figure 4.2. *The fluid-mosaic model of the cell membrane. The membrane consists of a double layer of phospholipids, with the phosphates (*open circles*) oriented outward and the hydrophobic hydrocarbons (*wavy lines*) oriented toward the center. Proteins may completely or partially span the membrane. Carbohydrates are attached to the outer surface.*

Cellular Movements

Some body cells—such as certain white blood cells—are able to move like an amoeba (a single-celled animal). This **amoeboid movement** is performed by the extension of parts of the cytoplasm to form *pseudopods*[2] (see fig. 4.4), which attach to a substrate and pull the cell along.

Cilia and Flagella

Cilia[3] (fig. 4.3) are tiny hairlike structures that protrude from the cell and, like the coordinated action of oarsmen in a boat, stroke in unison. Cilia in the human body are found on the apical surface (the surface facing the lumen, or cavity) of stationary epithelial cells in the respiratory and female genital tracts. In the respiratory system, the cilia transport strands of mucus, which are then conveyed by the cilia to a region (the pharynx) where the mucus can either be swallowed or expectorated. In the female genital tract, ciliary movements in the epithelial lining draw the egg (ovum) into the uterine tube and move it toward the uterus.

Sperm are the only cells in the human body that have **flagella.** The flagellum[4] is a single, whiplike structure that propels the sperm through its environment. This movement is required for fertilization, as will be described in chapter 9.

[2]pseudopod: L. *pseudes*, false; *pod*, foot
[3]cilia: L. *cili*, small hair
[4]flagellum: L. *flagrum*, whip

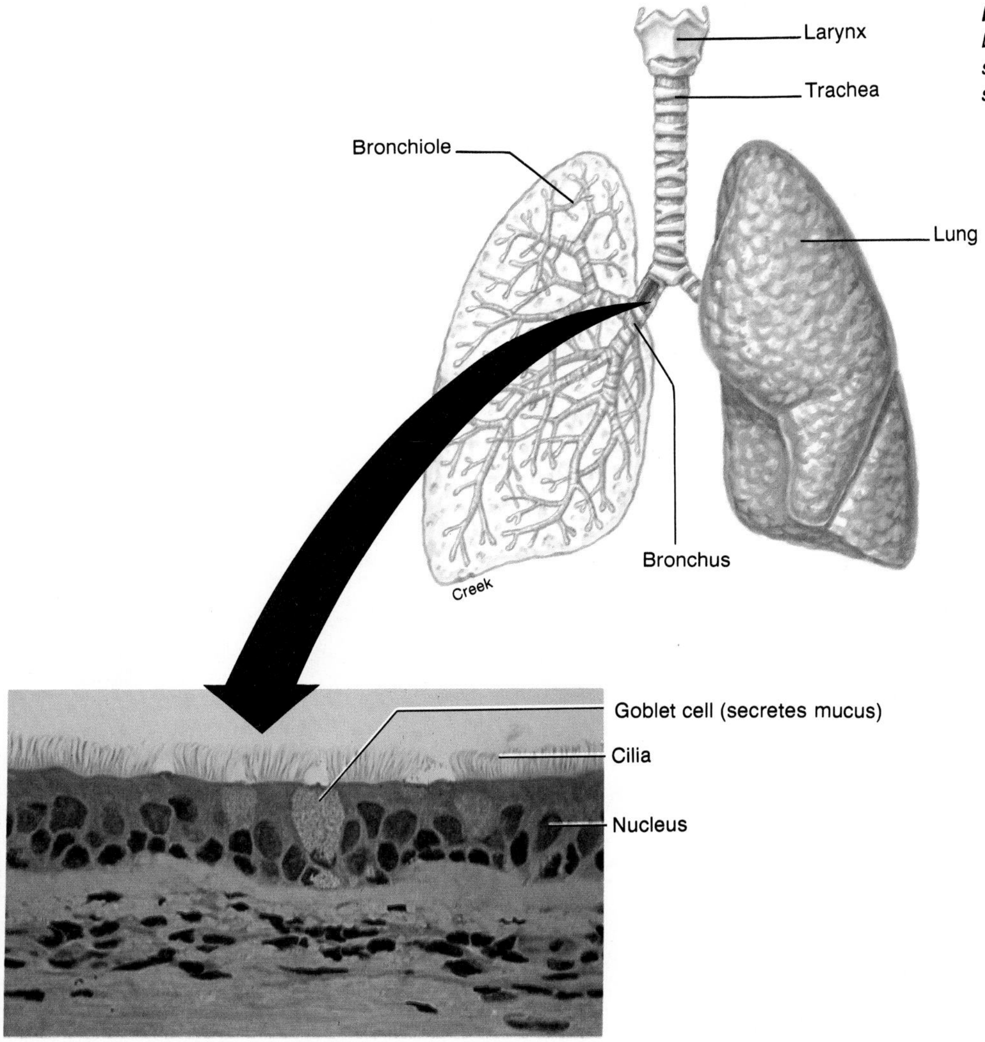

Figure 4.3. The columnar cells that line the bronchus (a passageway of the respiratory system) contain cilia. Goblet cells, which secrete mucus, are also present.

Endocytosis and Exocytosis

Regions of the cell membrane can invaginate (move inward to form a pouch) and pinch off to produce a membrane-enclosed body within the cytoplasm. This process removes regions of cell membrane as it brings part of the extracellular environment into the cell. The process by which part of the extracellular environment is brought into a cell by invagination of the cell membrane is called **endocytosis.**[5] There are three types of endocytosis: phagocytosis, pinocytosis, and receptor-mediated endocytosis.

Phagocytosis and Pinocytosis

Cells that move by amoeboid motion (such as white blood cells)—as well as liver cells, which are not mobile—use pseudopods to surround and engulf particles of organic matter (such as bacteria). This process is a type of cellular "eating" called **phagocytosis,**[6] which serves to protect the body from invading microorganisms and to remove extracellular debris.

Phagocytic cells surround their victim with pseudopods, which join together and fuse (fig. 4.4). After the inner membrane of the pseudopods becomes a continuous membrane around the ingested particle, it pinches off from the cell membrane. The ingested particle is now contained in an organelle called a *food vacuole* within the cell. The particle will subsequently be digested by enzymes contained in a different organelle (the lysosome, described in a later section).

Pinocytosis[7] is a related process performed by many cells. Instead of forming pseudopods, the cell membrane invaginates to produce a deep, narrow furrow. The membrane near the surface of this furrow then fuses, and a small vacuole containing the extracellular fluid is pinched off and enters the cell. In this way a cell can take in large molecules such as proteins which may be present in the extracellular fluid.

Receptor-Mediated Endocytosis

This type of endocytosis involves the smallest area of cell membrane, and it occurs only in response to specific molecules in the extracellular environment. In receptor-mediated endocytosis, the

[5]endocytosis: Gk. *endo*, within; *kytos*, hollow body
[6]phagocytosis: Gk. *phagein*, to eat; *kytos*, hollow body
[7]pinocytosis: Gk. *pinein*, to drink; *kytos*, hollow body

Figure 4.4. *Scanning electron micrographs of phagocytosis, showing the formation of pseudopods (a) and the entrapment of the prey within a food vacuole (b).*

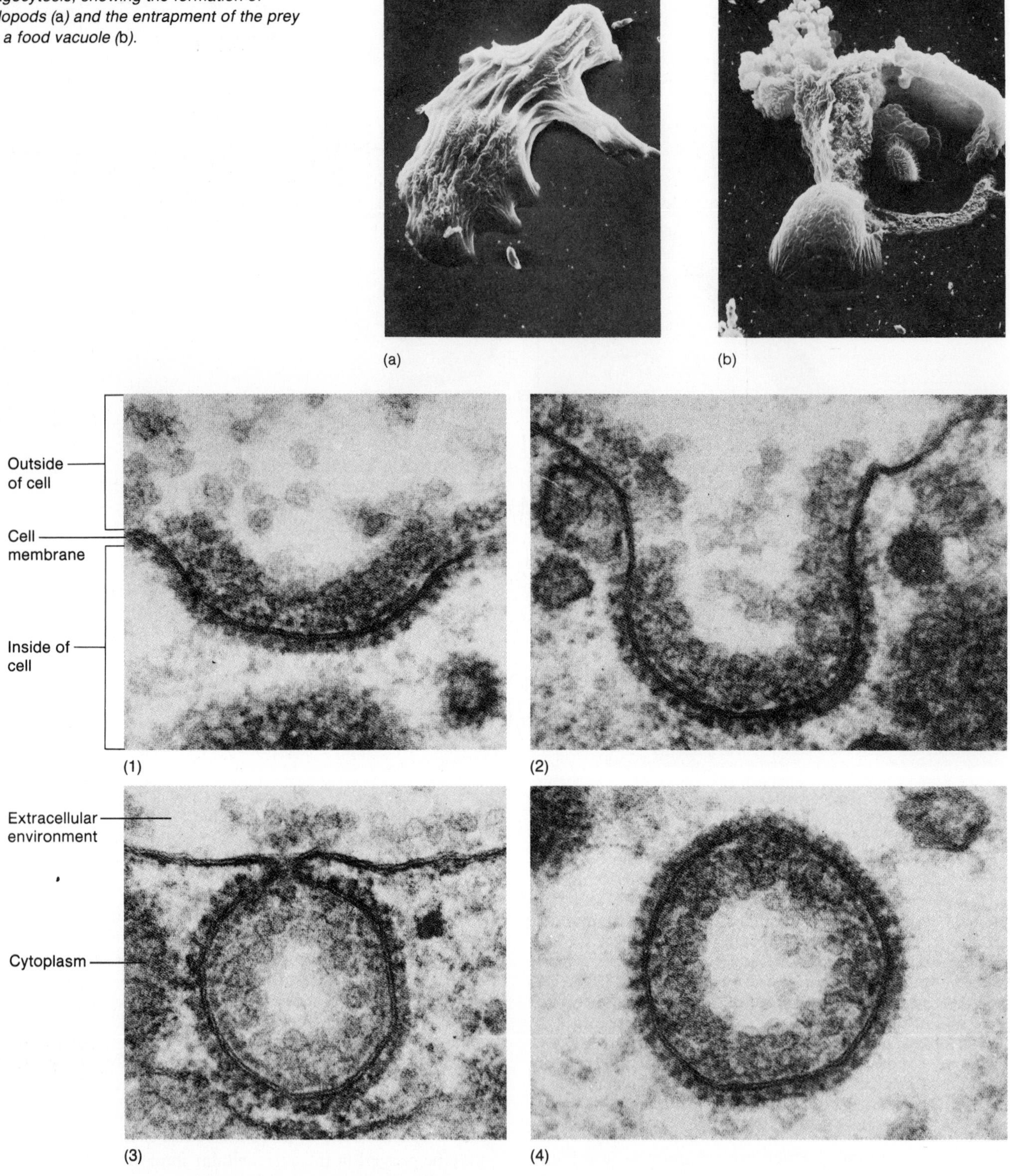

Figure 4.5. *Stages (1–4) of endocytosis, where specific bonding of extracellular particles to membrane receptor proteins is believed to occur.*

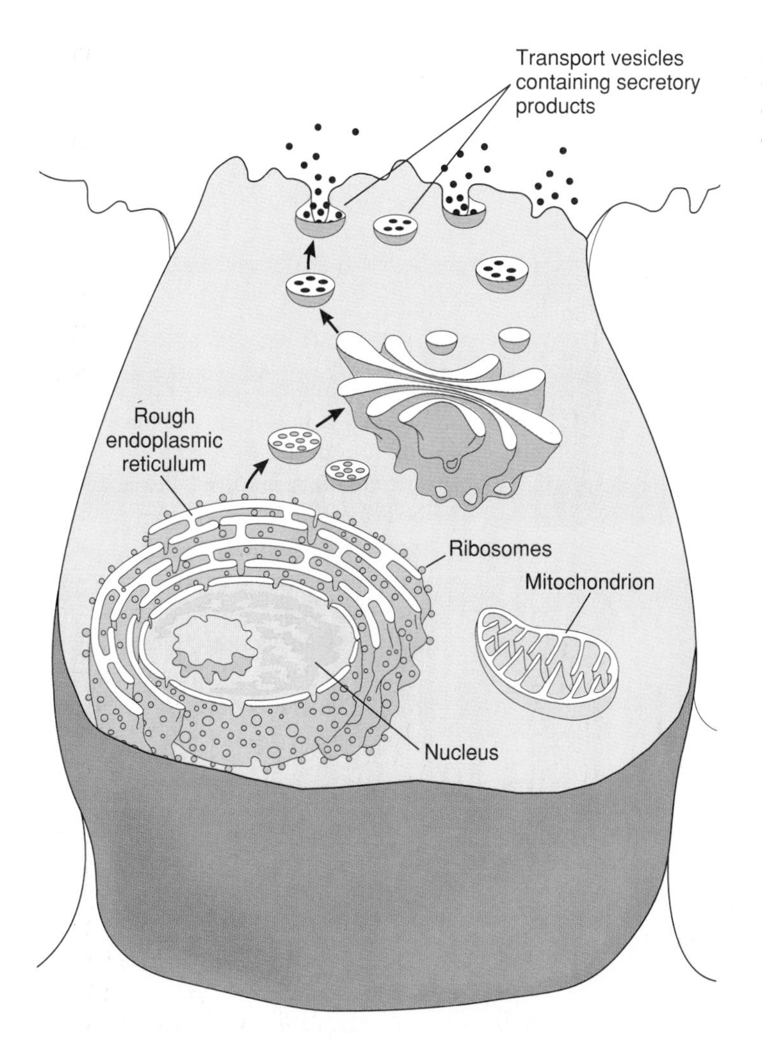

Figure 4.6. Diagrammatic illustration of exocytosis. In this process, vesicles containing a cellular product (represented by dots) fuse with the cell membrane. In this way, the product is secreted.

interaction of very specific molecules in the extracellular environment with specific membrane receptor proteins causes the membrane to invaginate, fuse, and pinch off to form a *vesicle*—a small vacuole (fig. 4.5). Vesicles formed in this way contain extracellular fluid and molecules that could not have passed by other means into the cell. Cholesterol attached to specific proteins in the blood, for example, is taken into cells that line arteries by receptor-mediated endocytosis. (This is in part responsible for atherosclerosis, as will be described in chapter 12.)

Exocytosis

Proteins and other molecules produced within the cell that are destined for export (secretion) are packaged within vesicles. In the process of **exocytosis,**[8] these secretory vesicles fuse with the cell membrane and release their contents into the extracellular environment (fig. 4.6). This process adds new membrane material, which replaces that which was lost from the cell membrane during endocytosis.

Endocytosis and exocytosis account for only part of the two-way traffic between the intracellular and extracellular compartments. Most of this traffic is due to membrane transport processes involving the movement of molecules and ions through the cell membrane, as will be described in a later section.

1. Draw the fluid-mosaic model of the cell membrane, and describe the structure of the membrane.
2. Describe the location and functions of cilia and flagella in the body.
3. Draw a figure showing phagocytosis and pinocytosis, and explain the significance of these processes.
4. Describe the events that occur in receptor-mediated endocytosis, and explain the significance of this process.

[8]exocytosis: Gk. *exo,* outside; *kytos,* hollow body

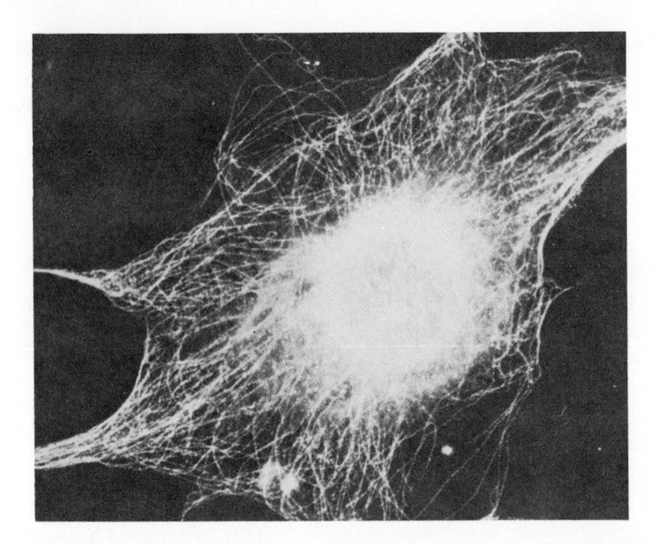

Figure 4.7. A photograph of microtubules forming the cytoskeleton of a cell.

Cytoplasm and Its Organelles

Many of the functions of a cell that are performed in the cytoplasmic compartment result from the activity of specific structures called organelles. Among these are the microtubules and microfilaments of the cytoskeleton; the lysosomes, which contain digestive enzymes; and the mitochondria, where most of the cellular energy is produced.

Cytoplasm and Cytoskeleton

The jellylike matrix within a cell (exclusive of that within the nucleus) is known as **cytoplasm.** When viewed in a microscope without special techniques, the cytoplasm appears to be uniform and unstructured. According to recent evidence, however, the cytoplasm is not a homogenous solution; it is, rather, a highly organized structure in which protein fibers—in the form of *microtubules* and *microfilaments*—are arranged in a complex latticework. These can be seen by fluorescence microscopy with the aid of antibodies against the proteins that compose these structures (fig. 4.7). The interconnected microfilaments and microtubules are believed to provide structural organization for cytoplasmic enzymes and support for various organelles.

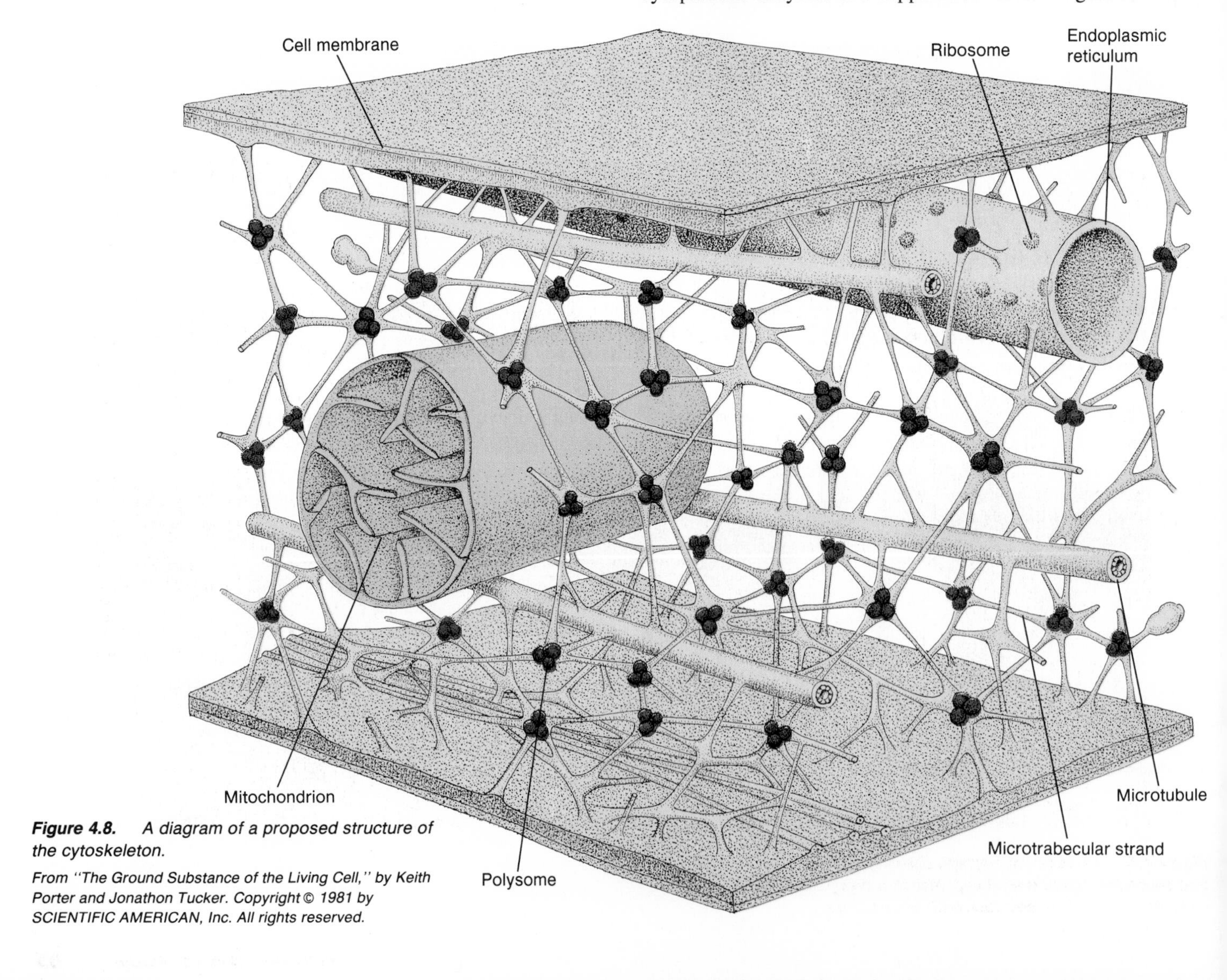

Figure 4.8. A diagram of a proposed structure of the cytoskeleton.

From "The Ground Substance of the Living Cell," by Keith Porter and Jonathon Tucker. Copyright © 1981 by SCIENTIFIC AMERICAN, Inc. All rights reserved.

The latticework of microfilaments and microtubules is thus said to function as a **cytoskeleton** (fig. 4.8). The structure of this "skeleton" is not rigid; it has been shown to be capable of quite rapid reorganization. Contractile proteins—including actin and myosin, which are responsible for muscle contraction—may be able to shorten the length of some microfilaments. The cytoskeleton may thus represent the cellular "musculature." Microtubules, for example, form the *spindle apparatus* that pulls chromosomes away from each other in cell division; they also form the central parts of cilia and flagella. Recent evidence suggests that the rapid movement of organelles within nerve fibers is also dependent upon microtubules, which can move different organelles in opposite directions at the same time.

Lysosomes

After a phagocytic cell has engulfed the proteins, polysaccharides, and lipids present in a particle of "food" (such as a bacterium), these molecules are still kept isolated from the cytoplasm by the membranes surrounding the food vacuole. The large molecules of proteins, polysaccharides, and lipids must first be digested into their smaller subunits (amino acids, monosaccharides, and so on) before they can cross the vacuole membrane and enter the cytoplasm.

The digestive enzymes of a cell are isolated from the cytoplasm and concentrated within membrane-bound organelles called **lysosomes** (fig. 4.9). Partly digested membranes of various organelles and other cellular debris are often observed within lysosomes. This is a result of **autophagy**,[9] a process that destroys worn-out organelles so that they can be continuously replaced. Lysosomes are thus aptly referred to as the "digestive system" of the cell.

Lysosomes have also been called "suicide bags," because a break in their membranes would release their digestive enzymes and thus destroy the cell. This happens normally as part of *programmed cell death,* in which the destruction of tissues is part of embryological development. It also occurs in white blood cells during an inflammation reaction.

Most, if not all, molecules in the cell have a limited life span. They are continuously destroyed and must be continuously replaced. Glycogen and some complex lipids in the brain, for example, are digested normally at a particular rate by lysosomes. If a person, because of some genetic defect, does not have the proper amount of these lysosomal enzymes, the resulting abnormal accumulation of glycogen and lipids could destroy the tissues. Examples of such diseases include Tay-Sachs disease, in which the accumulation of lipids damages the brain, and Gaucher's disease, in which the liver is damaged by accumulation of glycogen.

[9]autophagy: Gk. *autos,* self; *phagein,* to eat

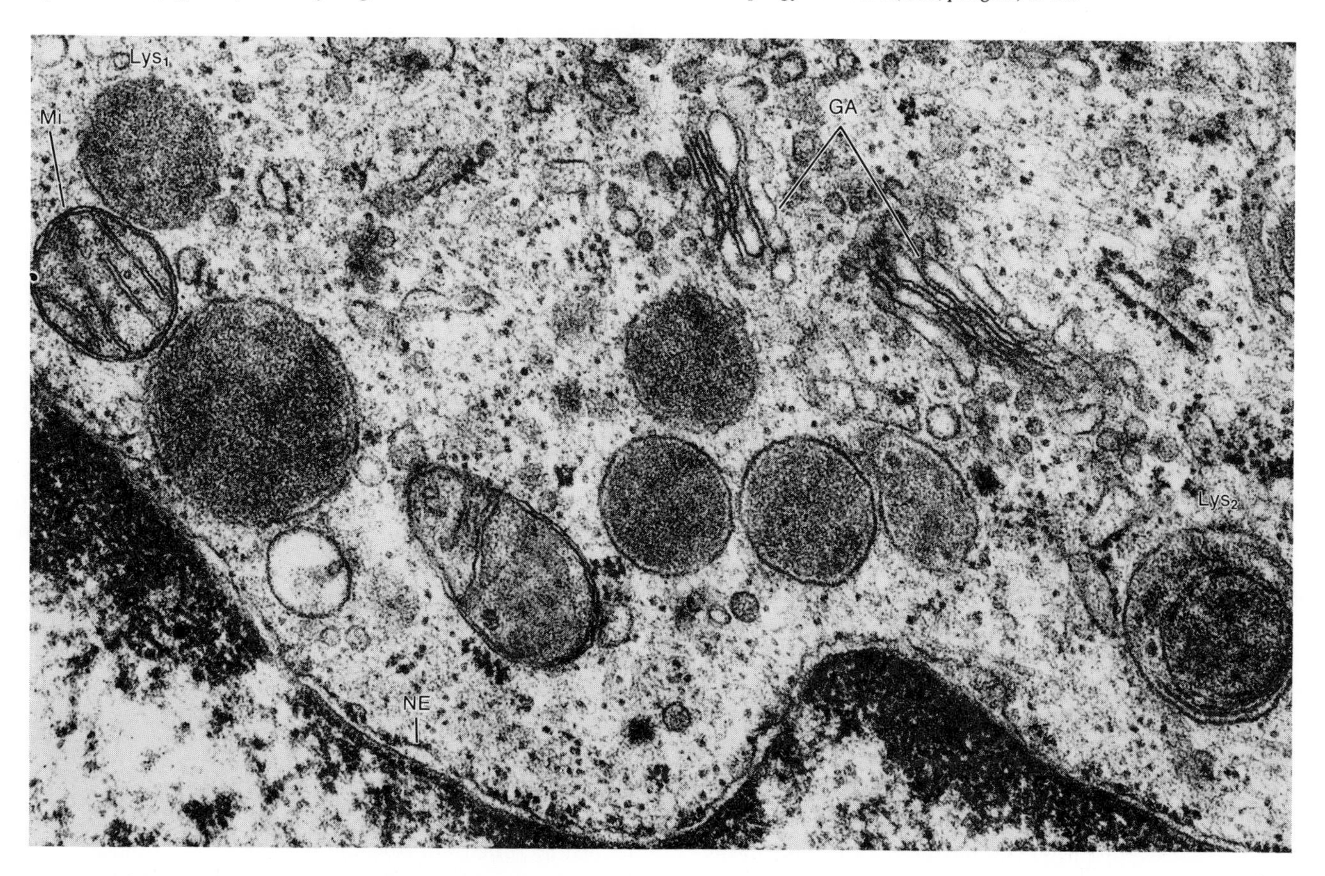

Figure 4.9. *Electron micrograph showing primary lysosomes (Lys_1) and secondary lysosomes (Lys_2). Mitochondria (Mi), Golgi apparatus (GA), and the nuclear envelope (NE) are also seen.*

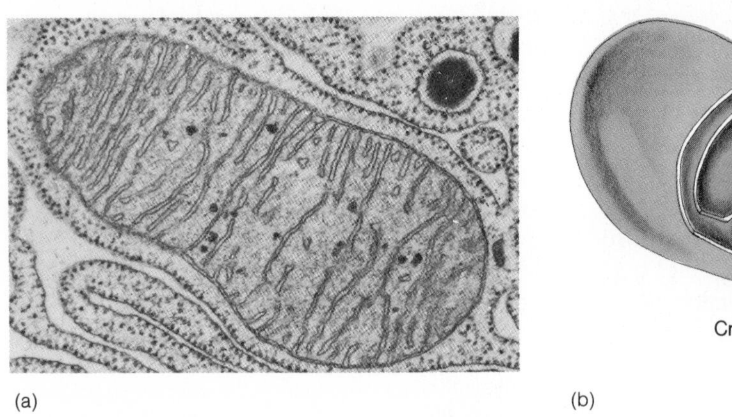

(a)

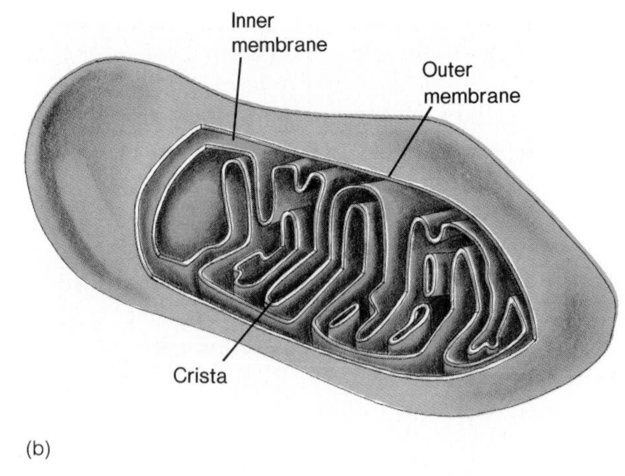

(b)

__Figure 4.10.__ A mitochondrion (a). The outer membrane and the infoldings of the inner membrane—the cristae—are clearly seen. The fluid in the center is the matrix. The structure of a mitochondrion is illustrated in (b).

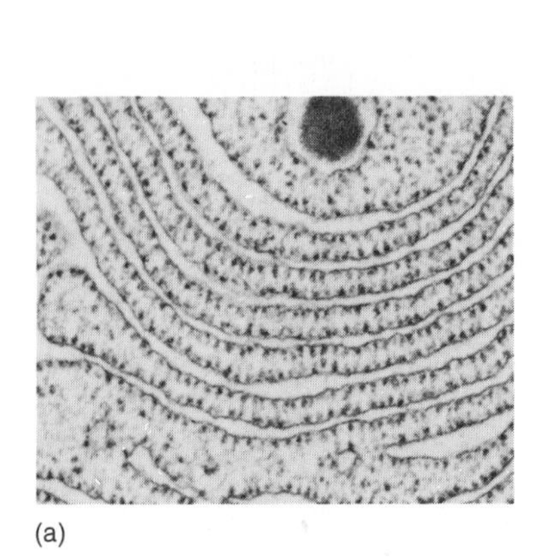

(a)

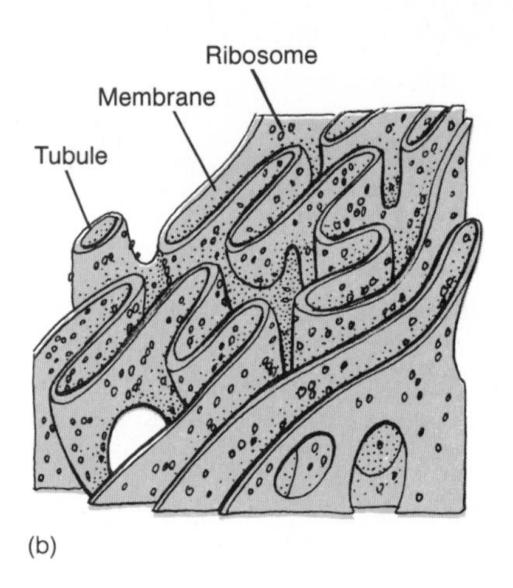

(b)

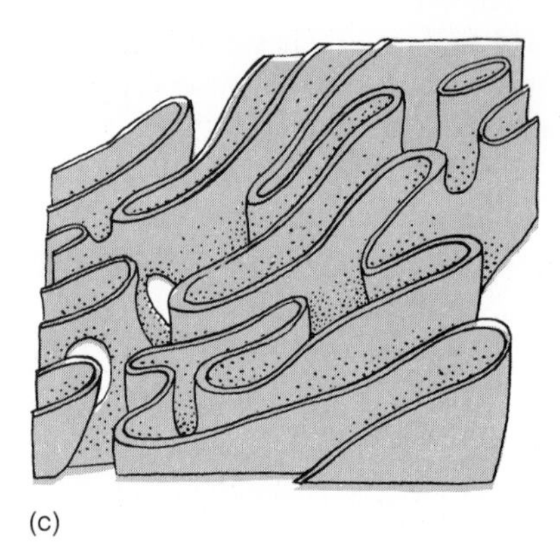

(c)

__Figure 4.11.__ (a) An electron micrograph of endoplamsic reticulum magnified about 100,000 times. (b) Rough endoplasmic reticulum has ribosomes attached to its surface, whereas (c) smooth endoplasmic reticulum lacks ribosomes.

Mitochondria

All cells in the body, with the exception of mature red blood cells, have a hundred to a few thousand organelles called **mitochondria.** Mitochondria serve as sites for the production of most of the cellular energy (chapter 18). For this reason, mitochondria are sometimes called the "powerhouses" of the cell.

Mitochondria vary in size and shape, but all have the same basic structure (fig. 4.10). Each is surrounded by an *outer membrane* that is separated by a narrow space from an *inner membrane.* The inner membrane has many folds, called *cristae,* which extend into the central area (or *matrix*) of the mitochondrion. The cristae and the matrix provide different compartments in the mitochondrion and have different roles in the generation of cellular energy. The detailed structure and function of mitochondria will be described in the context of cellular metabolism in chapter 18.

Endoplasmic Reticulum

Most cells contain a system of membranes known as the endoplasmic reticulum, of which there are two types: (1) a **rough,** or **granular, endoplasmic reticulum;** and (2) a **smooth endoplasmic reticulum** (fig. 4.11). A rough endoplasmic reticulum has ribosomes (small beadlike structures involved in protein synthesis) on its outer surface. A smooth endoplasmic reticulum, in contrast, is not associated with ribosomes. The smooth endoplasmic reticulum is used for a variety of purposes in different cells; it serves as a site for enzyme reactions in steroid hormone production and inactivation, for example, and as a site for the storage of Ca^{++} in skeletal muscle cells. The rough endoplasmic reticulum is found in cells, such as those of exocrine and endocrine glands, that are active secretors of proteins.

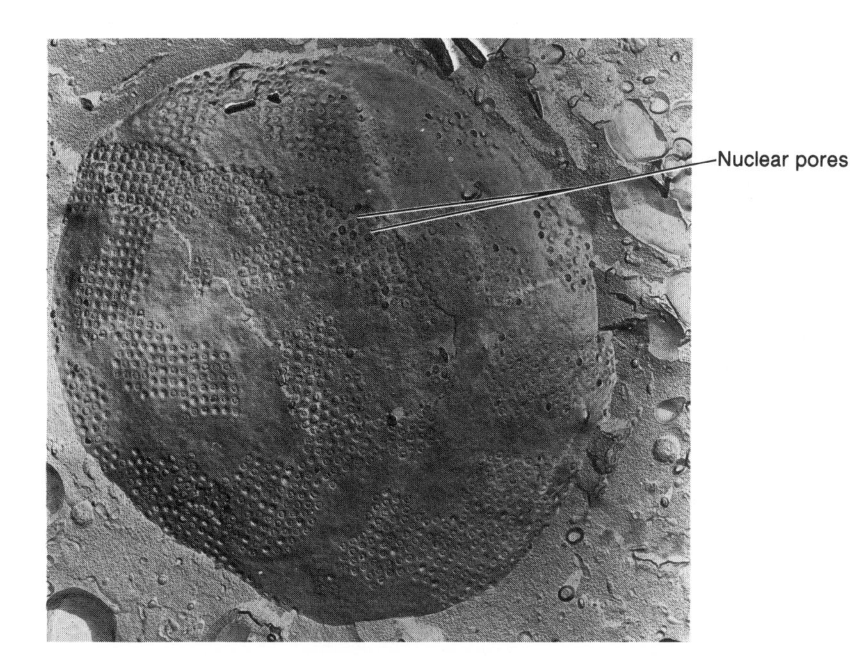

Figure 4.12. An electron micrograph of a freeze-fractured nuclear envelope, showing the nuclear pores.

The smooth endoplasmic reticulum in liver cells contains enzymes used for the inactivation of steroid hormones and many drugs. This inactivation is generally achieved by reactions that convert these compounds to more water-soluble and less active forms, which can be more easily excreted by the kidneys. When people take certain drugs such as alcohol and phenobarbital for a long period of time, a larger dose of these compounds is required to produce a given effect. This phenomenon, called *tolerance,* is accompanied by an increase in the smooth endoplasmic reticulum and thus an increase in the enzymes charged with inactivation of these drugs.

Details of the structure and function of the rough endoplasmic reticulum and its associated ribosomes, and of another organelle called the Golgi apparatus, will be described in conjunction with the topic of protein synthesis in chapter 10. The structure of centrioles and the spindle apparatus, which are involved in DNA replication and cell division, will also be described in chapter 10.

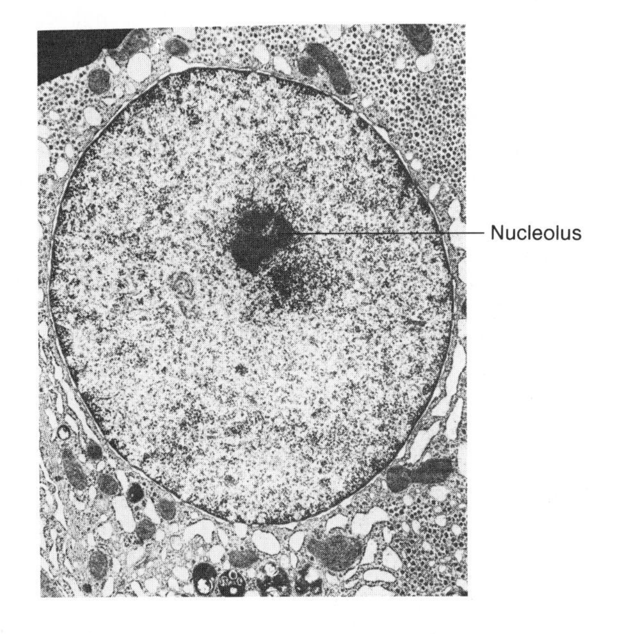

Figure 4.13. *The nucleus of a liver cell showing the nucleolus.*

Cell Nucleus

The nucleus of a cell is the organelle that contains the genes, which are encoded within a molecule known as DNA. DNA contains the code for the production of RNA, which in turn contains the code for the production of proteins. Proteins with enzymatic activity, in turn, regulate the activities of the cell. In this way, the nucleus can be considered the control center of the cell. The chemical nature of DNA and RNA, and the way that these molecules express genetic information, will be described in chapter 10.

Most cells in the body have a single nucleus, although some—such as skeletal muscle cells—are multinucleate. The nucleus is surrounded by a *nuclear envelope* composed of an inner and an outer membrane; these two membranes fuse together to form thin sacs with openings called *nuclear pores* (figs. 4.12 and 4.13). These pores allow RNA to exit the nucleus (where it is formed) and enter the cytoplasm but prevent DNA from leaving the nucleus.

1. Describe the structure of the cell membrane and explain its functional significance.
2. Explain why microtubules and microfilaments can be thought of as the skeleton and musculature of a cell.
3. Describe the contents of lysosomes and explain the significance of autophagy.
4. Describe the structure and function of mitochondria.
5. Distinguish between the structure and function of a rough versus a smooth endoplasmic reticulum.

Diffusion and Osmosis

Net diffusion of a molecule or ion through a cell membrane always occurs in the direction of its lower concentration. Nonpolar molecules can penetrate the phospholipid barrier, and small inorganic ions can pass through channels in the membrane. The net diffusion of water through a membrane is known as osmosis.

The cell (plasma) membrane separates the intracellular environment from the extracellular environment. Proteins and other molecules needed for the structure and function of the cell cannot penetrate, or "permeate," the membrane. The cell membrane is, however, **selectively permeable** to certain molecules and many ions; this allows two-way traffic in nutrients and wastes needed to sustain metabolism and provides electrical currents created by the movements of ions through the membrane.

The mechanisms involved in the transport of molecules and ions through the cell membrane may be divided into two categories: (1) transport that requires the action of specific *carrier proteins* in the membrane (*carrier-mediated transport*); and (2) transport through the membrane that is not carrier-mediated. Carrier-mediated transport includes *facilitated diffusion* and *active transport.* Non-carrier-mediated transport consists of the *simple diffusion* of ions, lipid-soluble molecules, and water through the membrane. The diffusion of water (solvent) through a membrane is called *osmosis.*

Diffusion

Molecules in a gas, as well as molecules and ions dissolved in a solution, are in a constant state of random motion as a result of their thermal (heat) energy. This random motion, called **diffusion,** tends to make the gas or solution evenly mixed, or diffusely spread out, within a given volume. Whenever a *concentration difference,* or *concentration gradient,* exists between two parts of a solution, therefore, random molecular motion tends to abolish the gradient and to make the molecules uniformly distributed (fig. 4.14).

As a result of random molecular motion, molecules in the part of the solution with a higher concentration will enter the area of lower concentration. Molecules will also move in the opposite direction, but not as frequently. As a result, there will be a *net movement* from the region of higher to the region of lower concentration until the concentration difference is abolished. This net movement is called **net diffusion.** Net diffusion is a physical process that occurs whenever there is a concentration difference; when the concentration difference exists across a membrane, diffusion becomes a type of membrane transport.

Diffusion through the Cell Membrane

Since the cell membrane consists primarily of a double layer of phospholipids, molecules that are nonpolar and thus lipid-soluble can easily pass from one side of the membrane to the other. The cell membrane, in other words, does not present a barrier to the diffusion of nonpolar molecules such as oxygen gas (O_2) or steroid hormones. Small, uncharged organic molecules, such as CO_2, ethanol, and urea, are also able to penetrate the double-lipid layers. Net diffusion of these molecules can thus easily occur between the intracellular and extracellular compartments when concentration gradients are present.

The oxygen concentration is relatively high, for example, in the extracellular fluid because oxygen is carried from the lungs to the body tissues by the blood. Since oxygen is converted to water in aerobic cell respiration (chapter 18), the oxygen concentration within the cells is lower than in the extracellular fluid. The concentration gradient for carbon dioxide is in the opposite direction because cells produce CO_2. *Gas exchange* thus occurs by diffusion between the tissue cells and their extracellular environments (fig. 4.15).

Although water is not lipid-soluble, water molecules can diffuse through the cell membrane because of their small size and lack of net charge. In certain membranes the passage of water is restricted to specific channels that can open or close in response to physiological regulation. The net diffusion of water molecules across the membrane (called *osmosis*) occurs when the membrane is permeable to water and when the solution on one side of the membrane is more dilute (has a higher water concentration) than on the other side of the membrane.

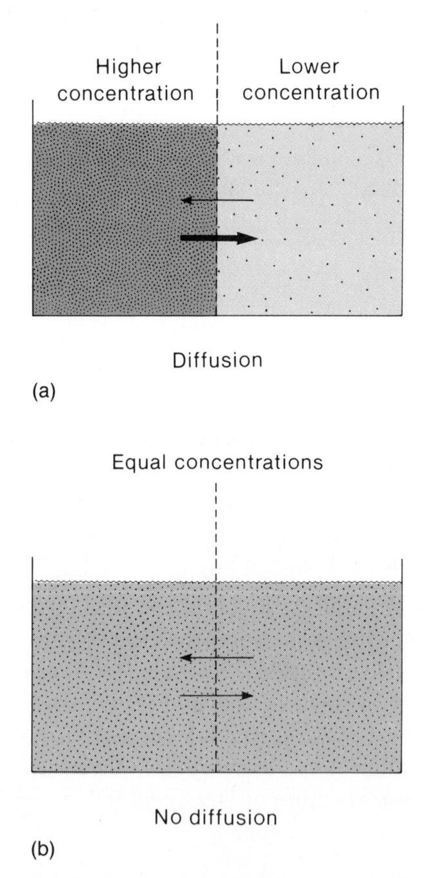

Figure 4.14. *Net diffusion occurs when there is a concentration difference (or concentration gradient) between two regions of a solution (a) provided that the membrane separating these regions is permeable to the diffusing substance. Diffusion tends to equalize the concentration of these solutions (b) and thus to abolish the concentration differences.*

Larger polar molecules, such as glucose, cannot pass through the double phospholipid layers of the membrane and thus require special *carrier proteins* in the membrane for transport (described later). The phospholipid portion of the membrane is similarly impermeable to charged inorganic ions, such as Na^+ and K^+. Passage of these ions through the cell membrane may be permitted by tiny **ion channels** through the membrane that are too small to be seen even with an electron microscope. Many scientists believe that these channels are provided by some of the proteins that span the thickness of the membrane (fig. 4.16).

Osmosis

Suppose that a cylinder is divided into two equal compartments by a membrane partition that can freely move and that one compartment initially contains 180 g/L (grams per liter) of glucose and the other compartment contains 360 g/L of glucose. If the membrane is permeable to glucose, glucose will diffuse from the 360 g/L compartment to the 180 g/L compartment until both compartments contain 270 g/L of glucose.

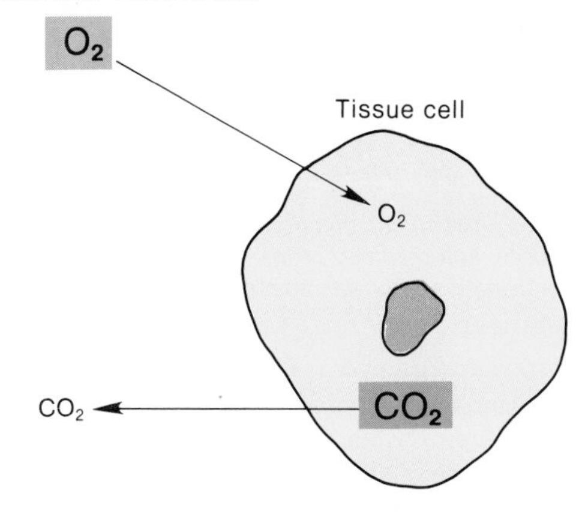

Figure 4.15. *Gas exchange between the intracellular and extracellular compartments occurs by diffusion. The regions of higher concentration are represented by the larger symbols.*

If the membrane is not permeable to glucose but is permeable to water, the same result (270 g/L solutions on both sides of the membrane) will be achieved by the diffusion of water. As water diffuses from the 180 g/L compartment to the 360 g/L compartment, the former solution would become more concentrated while the latter becomes more dilute. This is accompanied by volume changes, as illustrated in figure 4.17.

Osmosis is the net diffusion of water (the solvent) across the membrane. In order for this to occur, the membrane must be *semipermeable*, that is, it must be more permeable to water molecules than to solutes. Like the diffusion of solute molecules, the diffusion of water occurs when the water is more concentrated on one side of the membrane than on the other side; that is, when one solution is more dilute than the other (fig. 4.18). The more dilute solution has a lower concentration of solute but a higher concentration of water molecules. Water thus diffuses from the more dilute solution to the one that has a higher (solute) concentration.

Osmotic Pressure

Osmosis and the movement of the membrane partition could be prevented by an opposing force. If one compartment contained 180 g/L of glucose and the other compartment contained pure water, the osmosis of water into the glucose solution could be prevented by pushing against the membrane with a certain force. This is illustrated in figure 4.19.

The force that would have to be exerted to *prevent* osmosis in this situation is the **osmotic pressure** of the solution. This indirect measurement indicates how strongly the solution "draws" water into it by osmosis. The greater the solute concentration of a solution, the greater its osmotic pressure. Pure water, thus, has an osmotic pressure of zero, and a 360 g/L glucose solution has twice the osmotic pressure of a 180 g/L glucose solution.

Tonicity

Solutions that have the same osmotic pressure are said to be **isotonic** to each other. In physiology and medicine, the blood plasma (the fluid portion of the blood) is used as a reference; isotonic solutions can thus be defined as those that have the same

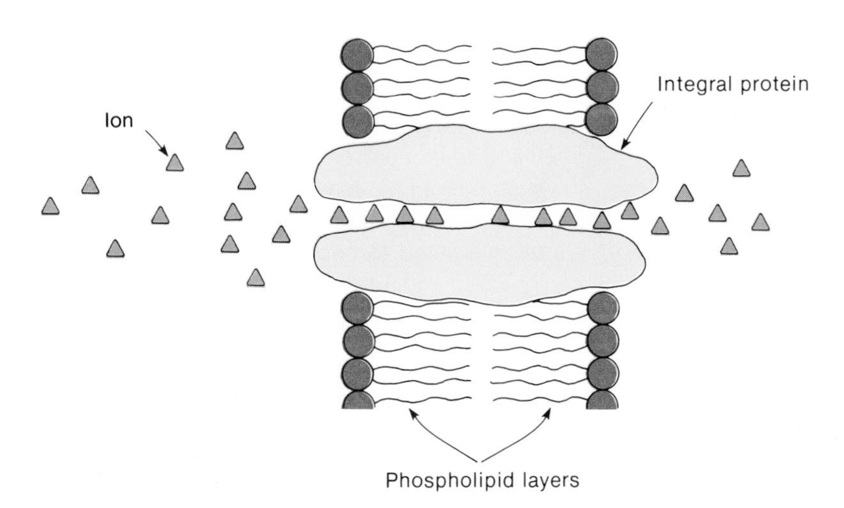

Figure 4.16. *Inorganic ions (such as Na^+ and K^+) may penetrate the membrane through pores within integral proteins that span the thickness of the double phospholipid layers.*

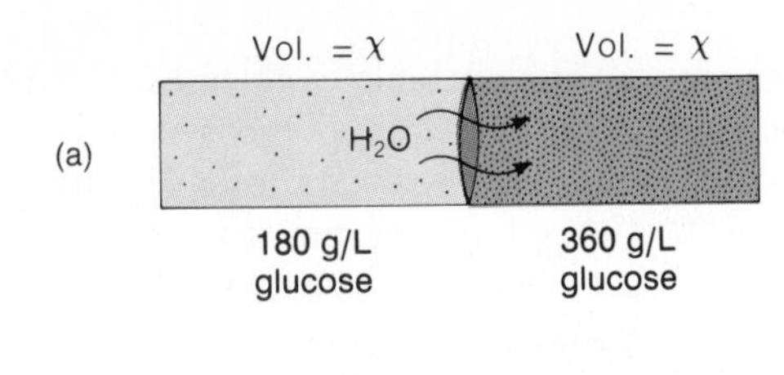

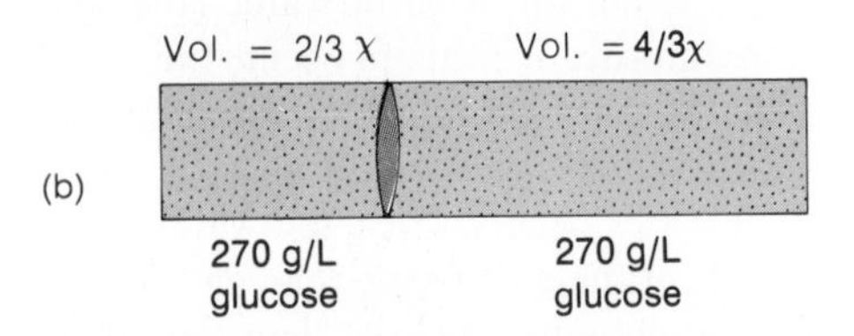

Figure 4.17. *A movable semipermeable membrane (permeable to water but not glucose) separates two solutions of different glucose concentration (a). As a result, water moves by osmosis into the solution of greater concentration until (b) the volume changes equalize the concentrations on both sides of the membrane.*

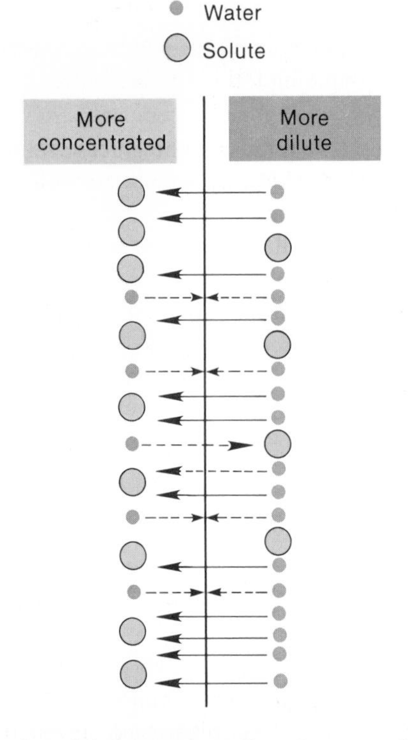

Figure 4.18. *A model of osmosis, or the net movement of water from the solution of lesser solute concentration to the solution of greater solute concentration.*

osmotic pressure as plasma. Cells placed in an isotonic solution will neither gain nor lose water. For this reason, only isotonic solutions are used in hospitals for intravenous infusion or for bathing exposed tissues. Such isotonic solutions include *normal saline* (0.9 g NaCl/100 ml) and *5% dextrose* (5 g glucose/100 ml). Solutions that have a higher osmotic pressure than plasma are **hypertonic**, and those with a lower osmotic pressure than plasma are **hypotonic**. Cells placed in a hypertonic solution will lose water and shrink (fig. 4.20); cells placed in a hypotonic solution will gain water, expand, and possibly will burst.

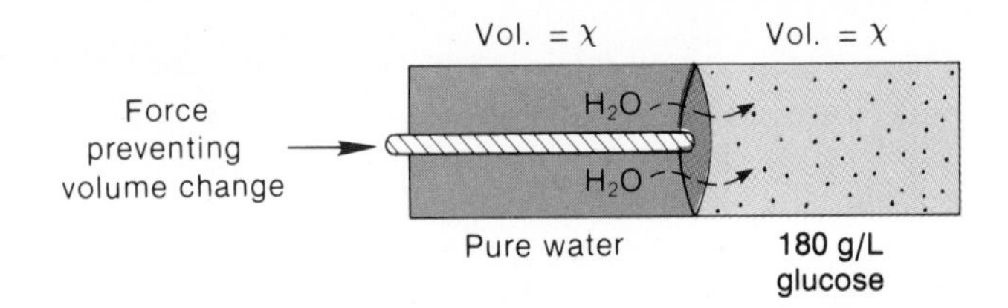

Figure 4.19. *If a semipermeable membrane separates pure water from a 180 g/L glucose solution, water tends to move by osmosis into the glucose solution, thus creating a hydrostatic pressure that pushes the membrane to the left and expands the volume of the glucose solution. The amount of pressure that must be applied to just counteract this volume change is equal to the osmotic pressure of the glucose solution.*

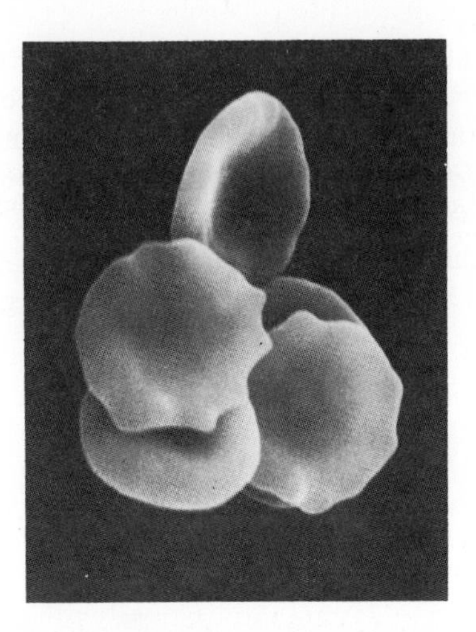

Figure 4.20. *A scanning electron micrograph of normal and crenated red blood cells. The crenated cells have shrunk due to the loss of water, producing a scalloped appearance.*

1. Define simple diffusion, and explain how it occurs.
2. Define the term *osmosis,* and describe the conditions required for it to occur.
3. Define the terms *isotonic, hypotonic,* and *hypertonic,* and explain why hospitals use 5% dextrose and normal saline as intravenous infusions.

Carrier-Mediated Transport

Molecules such as glucose and amino acids are transported across the cell membranes by special protein carriers that are specific and that can be saturated. Carrier-mediated transport in which the net movement is down a concentration gradient, and which is therefore passive, is called facilitated diffusion. Carrier-mediated transport that occurs against a concentration gradient and that requires metabolic energy is called active transport.

In order to sustain metabolism, cells must be able to take up glucose, amino acids, and other organic molecules from the extracellular environment. Molecules such as these, however, are too large and polar to pass through the lipid barrier of the cell membrane by a process of simple diffusion.

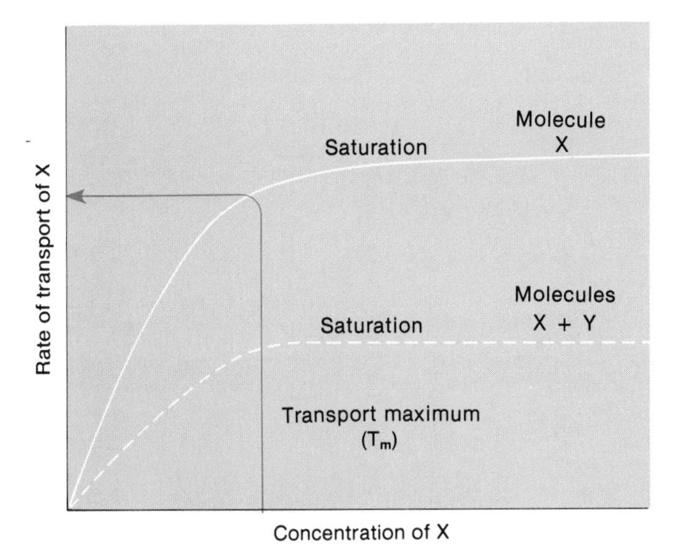

Figure 4.21. *Carrier-mediated transport displays the characteristics of saturation (illustrated by the* transport maximum*) and competition. Molecules* X *and* Y *compete for the same carrier, so that when they are present together the rate of transport of each is less than when either is present separately.*

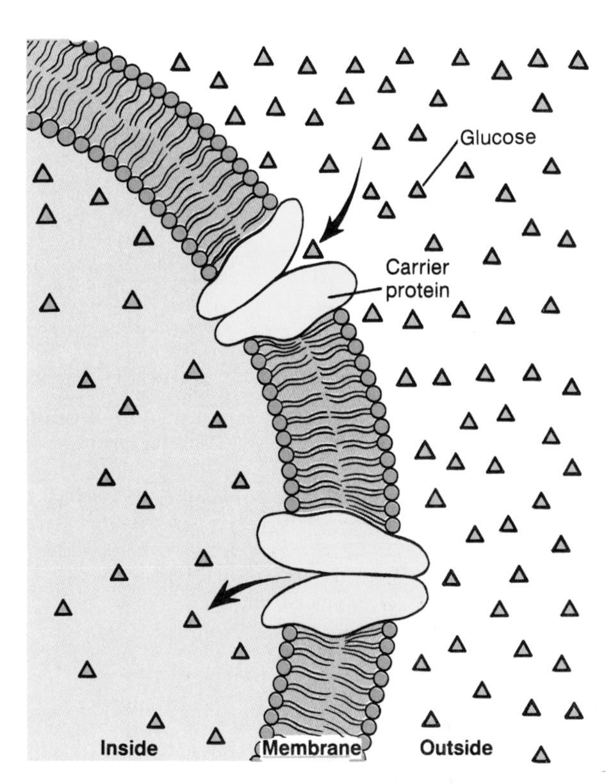

Figure 4.22. *A model of facilitated diffusion, where a molecule is transported across the cell membrane by a carrier protein.*

The transport of glucose, amino acids, and some other molecules is mediated by protein **carriers** within the membrane. Carrier proteins only interact with specific molecules. Glucose carriers, for example, can only interact with glucose and not with closely related monosaccharides. As a further example of specificity, particular carriers for amino acids transport some types of amino acids but not others.

As the concentration of a transported molecule is increased, its rate of transport will also be increased—but only up to a maximum. Beyond this rate, called the *transport maximum* (or T_m), further increases in concentration do not further increase the transport rate. This indicates that the carriers have become saturated (fig. 4.21).

As an example of saturation, imagine a bus stop that is serviced once per hour by a bus that can hold a maximum of forty people (its "transport maximum"). If ten people wait at the bus stop, ten will be transported per hour. If twenty people wait at the bus stop, twenty will be transported per hour. This linear relationship will hold up to a maximum of forty people; if eighty people are at the bus stop, the transport rate will still be forty per hour.

The kidneys transport a number of molecules from the blood filtrate (which will become urine) back into the blood. Glucose, for example, is normally completely reabsorbed so that urine is normally free of glucose. If the glucose concentration of the blood and filtrate is too high (a condition called *hyperglycemia*), however, the transport maximum will be exceeded. In this case, glucose will be found in the urine (a condition called *glycosuria*). This may result from eating too many sweets or from the inadequate action of the hormone *insulin* (in the disease **diabetes mellitus**).

Facilitated Diffusion

The transport of glucose from the blood across the cell membranes of tissue cells occurs by **facilitated diffusion.** Facilitated diffusion, like simple diffusion, is powered by the thermal energy of the diffusing molecules and involves the net transport of substances through a cell membrane from the side of higher to the side of lower concentration. The expenditure of cellular energy in the form of ATP is not required for either facilitated or simple diffusion.

Unlike simple diffusion of nonpolar molecules, water, and inorganic ions through a membrane, the diffusion of glucose through the cell membrane displays the properties of carrier-mediated transport: specificity and saturation. The diffusion of glucose through a cell membrane must therefore be mediated by carrier proteins. One conceptual model of the transport carriers is that they may each be composed of two protein subunits that interact with glucose in a specific way that creates a channel through the membrane (fig. 4.22), so that the glucose can move from the side of higher to the side of lower concentration.

The rate of the facilitated diffusion of glucose into tissue cells depends directly on the plasma glucose concentration. When the plasma glucose concentration is abnormally low—a condition called *hypoglycemia*—the rate of transport of glucose into brain cells may be inadequate for the metabolic needs of the brain. Severe hypoglycemia, as may be produced in a diabetic person by an overdose of insulin, can thus result in the loss of consciousness and even death.

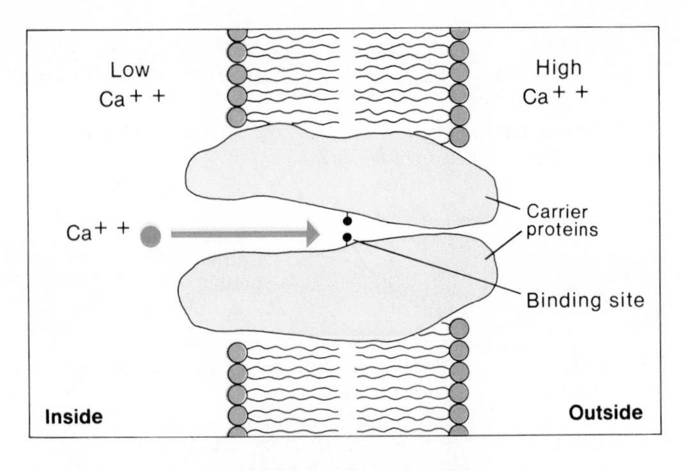

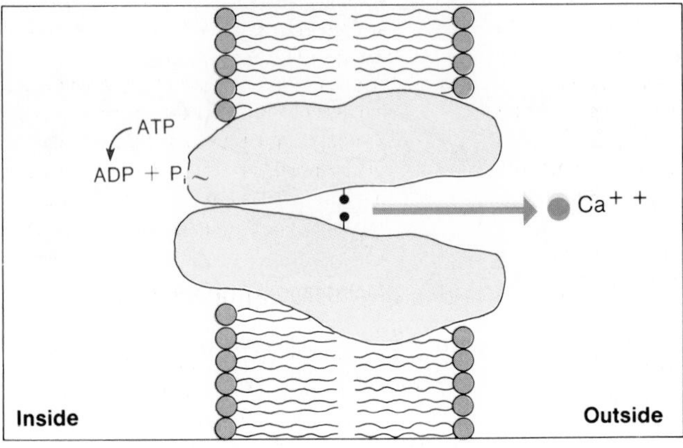

Figure 4.23. A model of active transport, showing the hingelike motion of the integral protein subunits.

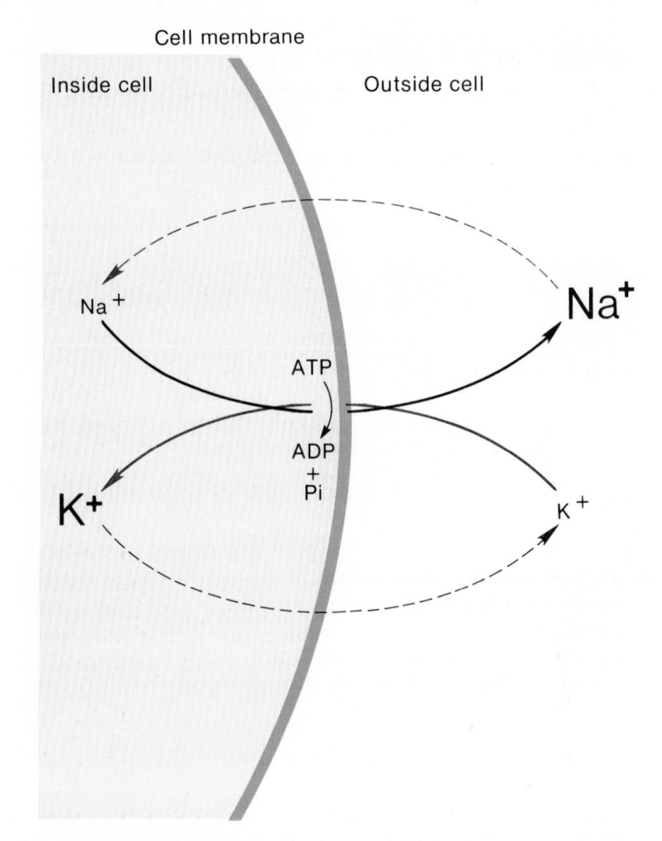

Figure 4.24. The Na^+/K^+ pump actively exchanges intracellular Na^+ for K^+. The carrier itself is an ATPase that breaks down ATP for energy. Dotted lines indicate the direction of passive transport (diffusion); solid arrows indicate the direction of active transport. Relative concentrations are indicated by the size of the symbols.

Active Transport

There are some aspects of cell transport that cannot be explained by simple or facilitated diffusion. The epithelial lining of the intestine, for example, moves glucose from the side of lower concentration in the lumen to the side of higher concentration in the blood. Similarly, all cells extrude Ca^{++} into the extracellular environment and, by this means, maintain an intracellular Ca^{++} concentration that is one thousand to ten thousand times lower than the extracellular Ca^{++} concentration.

The movement of molecules and ions against their concentration gradients, from lower to higher concentrations, requires the expenditure of cellular energy that is obtained from ATP (chapter 3). This type of transport is termed **active transport.** If a cell is poisoned with cyanide, which prevents the production of ATP, active transport is inhibited. This contrasts with passive transport, which can continue even when metabolic poisons kill the cell by preventing the formation of ATP.

Active transport, like facilitated diffusion, is carrier-mediated. These carriers appear to be proteins that span the thickness of the membrane. According to one theory of active transport, the following events may occur: (1) the molecule or ion to be transported bonds to a specific "recognition site" on one side of the carrier protein; (2) this bonding stimulates the breakdown of ATP, which in turn results in the addition of a phosphate to the carrier protein; (3) as a result, the carrier protein undergoes a hingelike motion, which releases the transported molecule or ion on the other side of the membrane. This model of active transport is illustrated in figure 4.23.

The Sodium-Potassium Pump

Active transport carriers are often referred to as "pumps." Although some of these carriers transport only one molecule or ion at a time, other carriers exchange one molecule or ion for another. The most important of the latter type of carriers is the **Na^+/K^+ pump.** This carrier protein, which is also an enzyme that converts ATP to ADP and P_i, actively extrudes three Na^+ ions from the cell as it transports two K^+ into the cell. This transport is energy dependent because Na^+ is more highly concentrated outside the cell and K^+ is more concentrated within the cell. Both ions, in other words, are moved against their concentration gradients (fig. 4.24).

All cells have numerous Na^+/K^+ pumps that are constantly active. This represents an enormous expenditure of energy used to maintain a steep gradient of Na^+ and K^+ across the cell membrane. This steep gradient is needed for a number of functions, including the production of the nerve impulse (chapter 6).

1. List the characteristics of facilitated diffusion that distinguish it from simple diffusion.
2. Describe active transport, and explain how active transport differs from facilitated diffusion.
3. Explain the functional significance of the Na^+/K^+pump.

Summary

Cell Membrane and Associated Structures p. 58

I. The cell membrane is composed predominantly of phospholipids and proteins, arranged as described in the fluid-mosaic model.

II. Cellular movements are achieved by a variety of structures.
   A. White blood cells and some others move by amoeboid motion, in which they extend pseudopods.
   B. Stationary cells that line the respiratory passages and the female reproductive tract have cilia, which beat in a coordinated fashion.
   C. Sperm can move by means of flagella.

III. Endocytosis and exocytosis are processes in which the cell membrane forms pouches.
   A. Endocytosis includes the processes of phagocytosis, pinocytosis, and receptor-mediated endocytosis.
   B. Exocytosis refers to the manner in which a cell secretes a product that was previously packaged in a vesicle; the vesicle fuses with the cell membrane and the product is released into the extracellular environment.

Cytoplasm and Its Organelles p. 64

I. The cytoplasm contains microtubules and microfilaments, which form the cytoskeleton.

II. Lysosomes are organelles that contain digestive enzymes.
   A. Food vacuoles containing structures engulfed by the cell through phagocytosis fuse with lysosomes.
   B. Lysosomes also digest old organelles and are responsible for the turnover of molecules within a cell.

III. Mitochondria are the organelles in which most of the cellular energy (ATP) is produced.

IV. Cells have a rough or a smooth endoplasmic reticulum.
   A. The endoplasmic reticulum is a system of membranous tubes and sacs within the cytoplasm.
   B. The rough endoplasmic reticulum has ribosomes on its surface; it is involved in protein synthesis and secretion.
   C. The smooth endoplasmic reticulum lacks ribosomes; it may be involved in metabolizing particular molecules or (in muscles) in storing calcium ions.

V. The cell nucleus contains the genes (DNA), and is the control center of the cell.

Diffusion and Osmosis p. 68

I. Diffusion occurs as molecules or ions move randomly due to their thermal (heat) energy.
   A. As a result of diffusion, there is a net movement from higher to lower concentration; this net movement stops when the concentration difference is abolished.
   B. In order for a molecule or ion to diffuse across a membrane, there must be a difference in concentration across the membrane and the membrane must be permeable to the diffusing molecule or ion.

II. Osmosis is the net diffusion of water across a membrane.
   A. In order for osmosis to occur, there must be a difference in the concentration of a solute molecule across the membrane, and the membrane must be more permeable to water than to the solute.
   B. The osmotic pressure of a solution is an indirect indication of the tendency of that solution to take in water by osmosis.
   C. Isotonic solutions have the same osmotic pressure as plasma; cells in isotonic solutions will therefore neither gain nor lose water.

Carrier-Mediated Transport p. 70

I. Carrier-mediated transport refers to transport across the cell membrane by proteins; this type of transport displays the properties of specificity and saturation.

II. In facilitated diffusion, carriers move molecules across the membrane from the side of higher to the side of lower concentration.

III. In active transport, carriers move molecules across the membrane from the side of lower to the side of higher concentration.
   A. Active transport requires metabolic energy and the breakdown of ATP.
   B. The Na^+/K^+ pump is an active transport carrier that moves Na^+ out of the cell and K^+ into the cell against their concentration gradients.

Review Activities

Objective Questions

1. According to the fluid-mosaic model of the cell membrane
   (a) protein and phospholipids form a regular, repeating structure
   (b) the membrane is a rigid structure
   (c) phospholipids form a double layer, with the polar parts facing each other
   (d) proteins are free to move within a double layer of phospholipids
2. The organelles that contain digestive enzymes are the
   (a) mitochondria
   (b) lysosomes
   (c) endoplasmic reticulum
   (d) Golgi apparatus
3. The major site of ATP synthesis in a cell is the
   (a) nucleus
   (b) ribosome
   (c) endoplasmic reticulum
   (d) mitochondrion
   (e) lysosome
4. Cells secrete their products through the process of
   (a) phagocytosis
   (b) pinocytosis
   (c) exocytosis
   (d) receptor-mediated endocytosis
5. Structures found on the surface of epithelial cells that line the respiratory passageways are
   (a) cilia
   (b) flagella
   (c) pseudopods
   (d) ribosomes

6. The movement of water across a cell membrane occurs by
   (a) active transport
   (b) facilitated diffusion
   (c) simple diffusion (osmosis)
   (d) all of the above
7. Which of the following statements about the facilitated diffusion of glucose is *true?*
   (a) There is a net movement from the region of low to the region of high concentration.
   (b) Carrier proteins in the cell membrane are required for this transport.
   (c) This transport requires energy obtained from ATP.
   (d) This is an example of osmosis.
8. If a poison such as cyanide stops the production of ATP, which of the following transport processes would cease?
   (a) the movement of Na^+ out of a cell
   (b) osmosis
   (c) the movement of K^+ out of a cell
   (d) all of the above
9. Red blood cells lose water and shrink in
   (a) a hypotonic solution
   (b) an isotonic solution
   (c) a hypertonic solution

Essay Questions

1. The cell membrane is an extremely dynamic structure. Using examples, explain why this statement is true.
2. Describe the conditions required to produce osmosis, and explain why osmosis occurs under these conditions.
3. Explain how simple diffusion can be distinguished from facilitated diffusion and how active transport can be distinguished from passive transport.

5

Tissues, Organs, and Control Systems

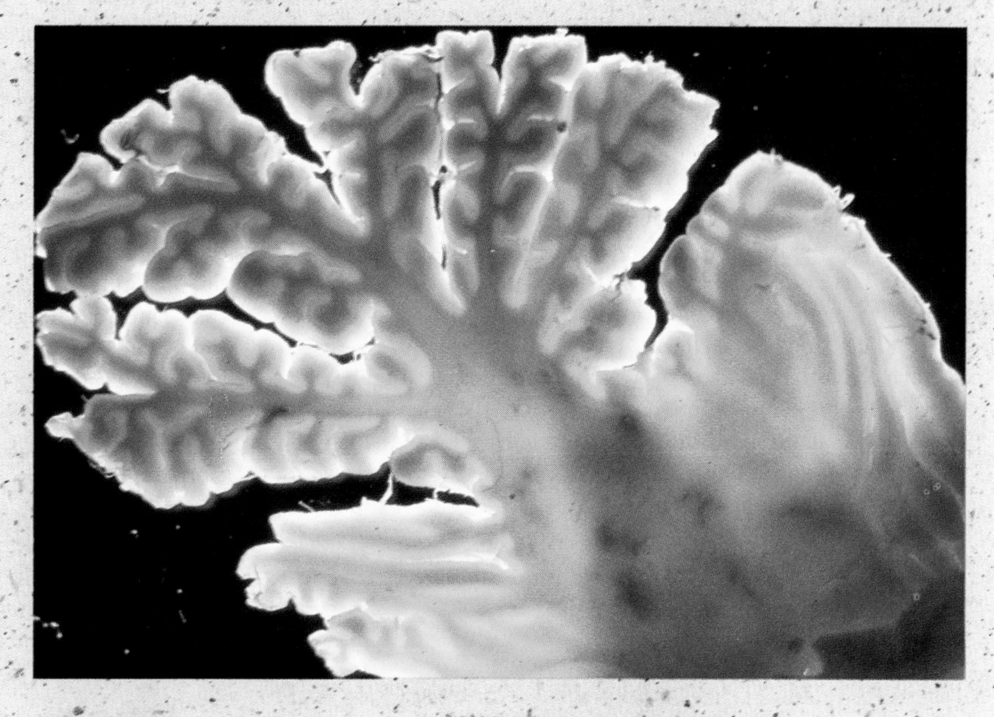

Outline

Objectives

By studying this chapter, you should be able to

1. list the four primary tissues and their subtypes
2. describe the distinguishing features of each primary tissue, and relate the structure of the primary tissue to its functions
3. describe how the primary tissues are organized into organs, using the skin as an example
4. define homeostasis and describe how this concept is used in physiology and medicine
5. explain the nature of negative feedback loops and how these mechanisms act to maintain homeostasis
6. describe the significance of the nervous and endocrine systems
7. explain how negative feedback inhibition helps to regulate the secretion of hormones

Keys to Pronunciation

epithelial: *ep''i-the'le-al*
exocrine: *ek'so-krin*
homeostasis: *ho''me-o-sta'sis*
hormone: *hor'mon*
intercalated: *in-ter'kah-lat-ed*
keratin: *ker'ah-tin*
neuron: *nu'ron*
neuroglia: *nu'rog'le-ah*
squamous: *skwa'mus*

Photo: A slice of the cerebellum, a part of the brain needed for motor coordination.

The Primary Tissues

The organs of the body are composed of four different primary tissues. Each of these tissues has its own characteristic structure and function, and the activities and interactions of these tissues determine the physiology of the organs.

The functions performed by the human body are the result of the performance of the body's cells. Although each cell is itself a living entity, the cells of the body cooperate to form units of structure that share a common function. These units of structure and function composed of many cells are known as the body **tissues**. The entire body is composed of only four types of tissues. These **primary tissues** include (1) muscle, (2) nervous, (3) epithelial, and (4) connective tissues. The tissues of the body, in turn, form larger units of structure and function known as **organs**. A number of different organs, similarly, form larger cooperating structures called the body **systems**. The anatomy and physiology of specific organs and systems will be discussed together with their functions in later chapters. In this section, the common "fabric" of all organs is described.

Muscle

Muscle tissue is specialized for contraction. There are three types of muscles: (1) **skeletal muscle,** (2) **cardiac muscle,** and (3) **smooth muscle.** Skeletal muscle is often called *voluntary muscle,* because we have conscious control of its contraction without special training. Both skeletal and cardiac muscles are **striated;** they have striations, or stripes, that extend across the width of the muscle cell (figs. 5.1 and 5.2), and for this reason they have similar mechanisms of contraction. Smooth muscle (fig. 5.3) lacks these cross-striations and has a different mechanism of contraction.

Skeletal Muscle

Skeletal muscles are generally attached by means of tendons to bones at both ends, so that contraction produces movements of the skeleton. There are, however, exceptions to this pattern; the tongue, superior portion of the esophagus, anal sphincter, and diaphragm are also composed of skeletal muscle.

Since skeletal muscle cells are long and thin they are called **fibers,** or **myofibers** (*myo* = muscle). Within a skeletal muscle the muscle fibers are arranged in bundles, and within these bundles the fibers extend in parallel from one end to the other of the bundle. The parallel arrangement of muscle fibers (seen in fig. 5.1) allows each fiber to be controlled individually: one can thus contract fewer or more muscle fibers and, in this way, vary the strength of the whole muscle's contraction. The ability to vary, or "grade," the strength of skeletal muscle contraction is obviously needed for proper control of skeletal movements.

Cardiac Muscle

Although cardiac muscle is striated, it has a very different appearance from skeletal muscle. Cardiac muscle is found only in the heart, where the **myocardial cells** are short, branched, and intimately interconnected to form a continuous fabric. Special areas of contact between adjacent cells stain darkly to show *intercalated discs* (fig. 5.2), which are characteristic of heart muscle.

The intercalated discs couple myocardial cells together mechanically and electrically. Unlike skeletal muscles, therefore, the heart cannot produce a graded contraction by varying the number of cells stimulated to contract. Because of the way it is constructed, the stimulation of one myocardial cell results in the stimulation of all other cells in the mass and a whole-hearted contraction.

Figure 5.1. *Three skeletal muscle fibers showing the characteristic cross-striations.*

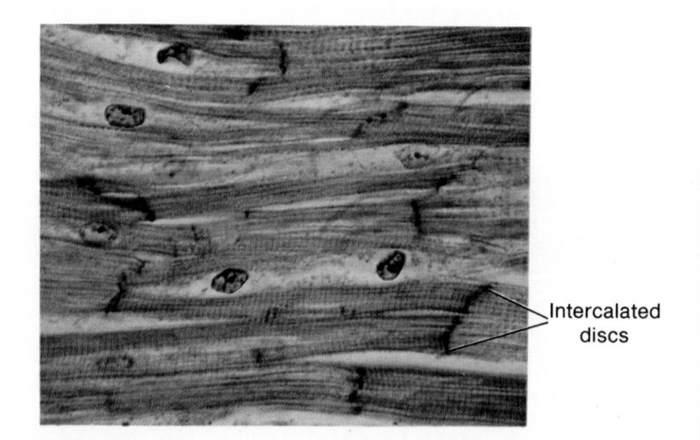

Figure 5.2. *Human cardiac muscle. Notice the striated appearance and dark-staining intercalated discs.*

Smooth Muscle

As implied by the name, smooth muscle cells (fig. 5.3) do not have the cross-striations characteristic of skeletal and cardiac muscle. Smooth muscle is found in the digestive tract, blood vessels, bronchioles (small air passages in the lungs), and in the urinary and reproductive systems. Circular arrangements of smooth muscle in these organs produce constriction of the *lumen* (cavity) when the muscle cells contract. The digestive tract also contains longitudinally arranged layers of smooth muscle. Rhythmic contractions of circular and longitudinal layers of muscle produce *peristalsis,* a process that pushes food from one end of the digestive tract to the other.

Nervous Tissue

Nervous tissue consists of nerve cells, or **neurons,** which are specialized for the generation and conduction of electrical events, and of **neuroglia,** which provide anatomical and functional support to the neurons.

Each neuron consists of three parts (fig. 5.4): (1) a *cell body,* which contains the nucleus and serves as the metabolic center of the cell; (2) *dendrites* (literally, "branches"), which are highly branched cytoplasmic extensions of the cell body that receive input from other neurons or from receptor cells; and (3) an *axon,* which is a single cytoplasmic extension of the cell body, which can be quite long (up to a few feet in length) and is specialized for conducting nerve impulses from the cell body to another neuron or an effector (muscle or gland) cell.

The neuroglia, composed of *neuroglial cells,* do not conduct impulses but instead serve to bind neurons together, modify the extracellular environment of the nervous system, and influence the nourishment and electrical activity of neurons. Neuroglial cells are about five times more abundant than neurons in the nervous system and, unlike neurons, maintain a limited ability to divide by mitosis throughout life.

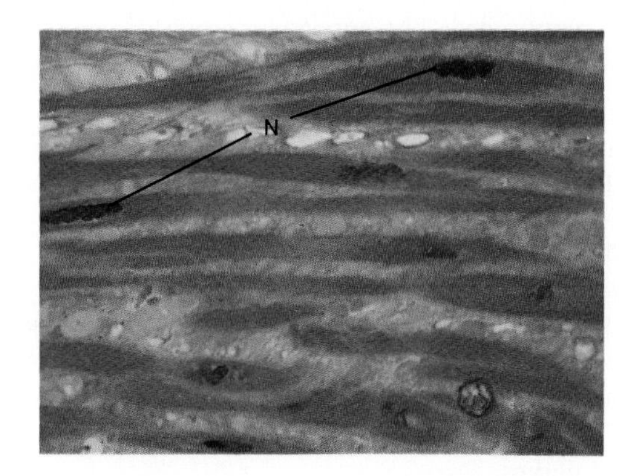

Figure 5.3. *Photomicrograph of smooth muscle cells. Notice that these cells lack striations and contain single, centrally located nuclei.*

Epithelial Tissue

Epithelial tissue consists of cells that form **membranes,** which cover and line the body surfaces, and of **glands** that are derived from these membranes. There are two categories of glands. *Exocrine glands* (*exo* = outside) secrete chemicals through a duct that leads to the outside of the membrane and thus to the outside of the body. *Endocrine glands* (*endo* = within) secrete chemicals called *hormones* into the blood.

Epithelial Membranes

Epithelial membranes are classified according to the number of their layers and the shape of the cells in the upper layer. Epithelial cells that are flattened in shape are *squamous,*[1] those

[1]squamous: L. *squamosus,* scaly

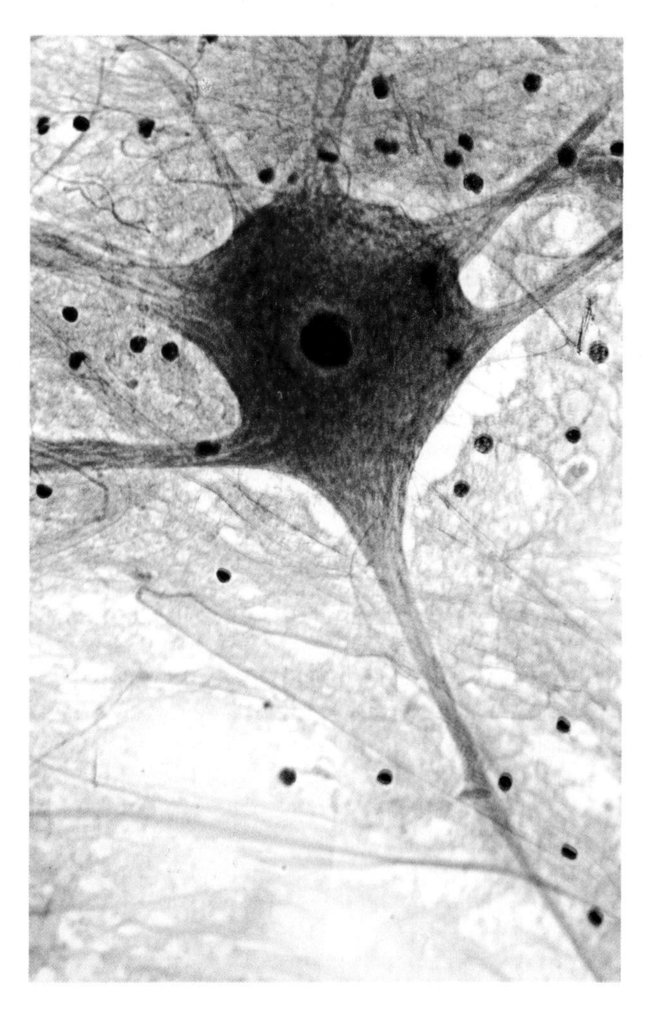

Figure 5.4. *A photomicrograph of a neuron, showing its principal parts.*

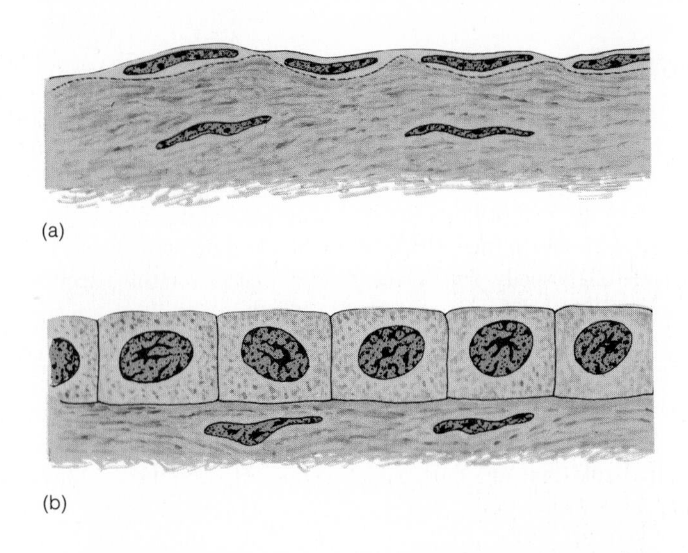

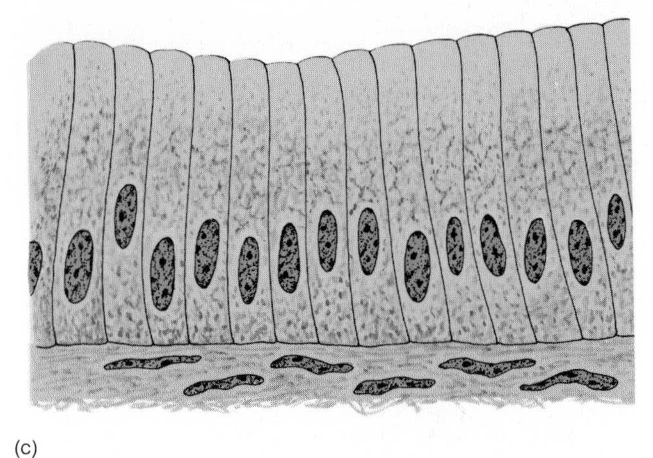

Figure 5.5. *(a) Simple squamous, (b) simple cuboidal, and (c) simple columnar epithelial membranes. The tissue beneath each membrane is connective tissue.*

that are taller than they are wide are *columnar,* and those that are as wide as they are tall are *cuboidal* (fig. 5.5). Those epithelial membranes that are only one cell layer in thickness are known as *simple* membranes; those that are composed of a number of layers are *stratified* membranes.

A simple squamous membrane is adapted for diffusion and filtration; such a membrane lines all blood vessels, where it is known as an *endothelium.*[2] A simple cuboidal epithelium lines the ducts of exocrine glands and part of the tubules of the kidney. A simple columnar epithelium lines the lumen of the stomach and intestine; this epithelium contains specialized unicellular glands, called *goblet cells,* which secrete mucus and are dispersed among the columnar epithelial cells. The columnar epithelial cells in the uterine (fallopian) tubes of females and in the respiratory passages contain numerous cilia, which can move in a coordinated fashion and aid the functions of these organs.

[2]endothelium: Gk. *endon,* within; *thelium,* to cover

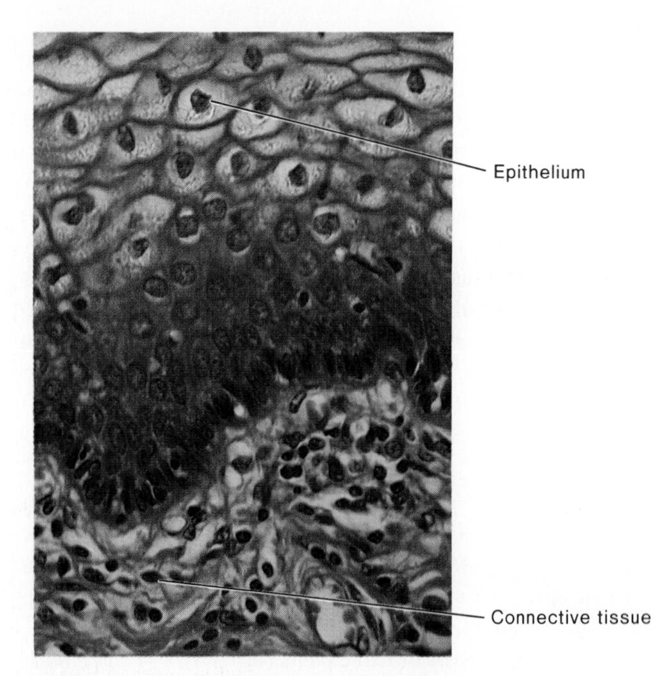

Figure 5.6. *The stratified squamous nonkeratinized epithelial membrane of the vagina.*

The epithelial covering of the esophagus and vagina that provides protection for these organs is a stratified squamous epithelium (fig. 5.6). All layers of this epithelium consist of living cells. The upper layers of the *epidermis* of the skin, in contrast, contain dead cells that are *keratinized,* or *cornified*[3] (fig. 5.7). Since the epidermis is dry and exposed to the potentially desiccating effects of the air, its surface is covered with dead cells that are filled with a water-resistant protein known as *keratin.*[4] This protective layer is constantly flaked off from the surface of the skin and therefore must be constantly replaced by the division of cells in the deeper layers of the epidermis.

The constant loss and renewal of cells is characteristic of epithelial membranes. The entire epidermis is completely replaced every two weeks; the stomach lining is renewed every two to three days. Examination of the cells that are lost, or "exfoliated," from the surface of the female genital tract is a common procedure in gynecology (as in the Pap smear).

In order to form a strong membrane that is effective as a barrier at the body surfaces, epithelial cells are very closely packed and are joined together by structures collectively called **junctional complexes.** There is no room for blood vessels between adjacent epithelial cells. The epithelium must therefore receive nourishment from the connective tissue underneath, which has large intercellular spaces that can accommodate blood vessels and nerves.

[3]cornified: L. *corneus,* horny
[4]keratin: Gk. *keras,* horn

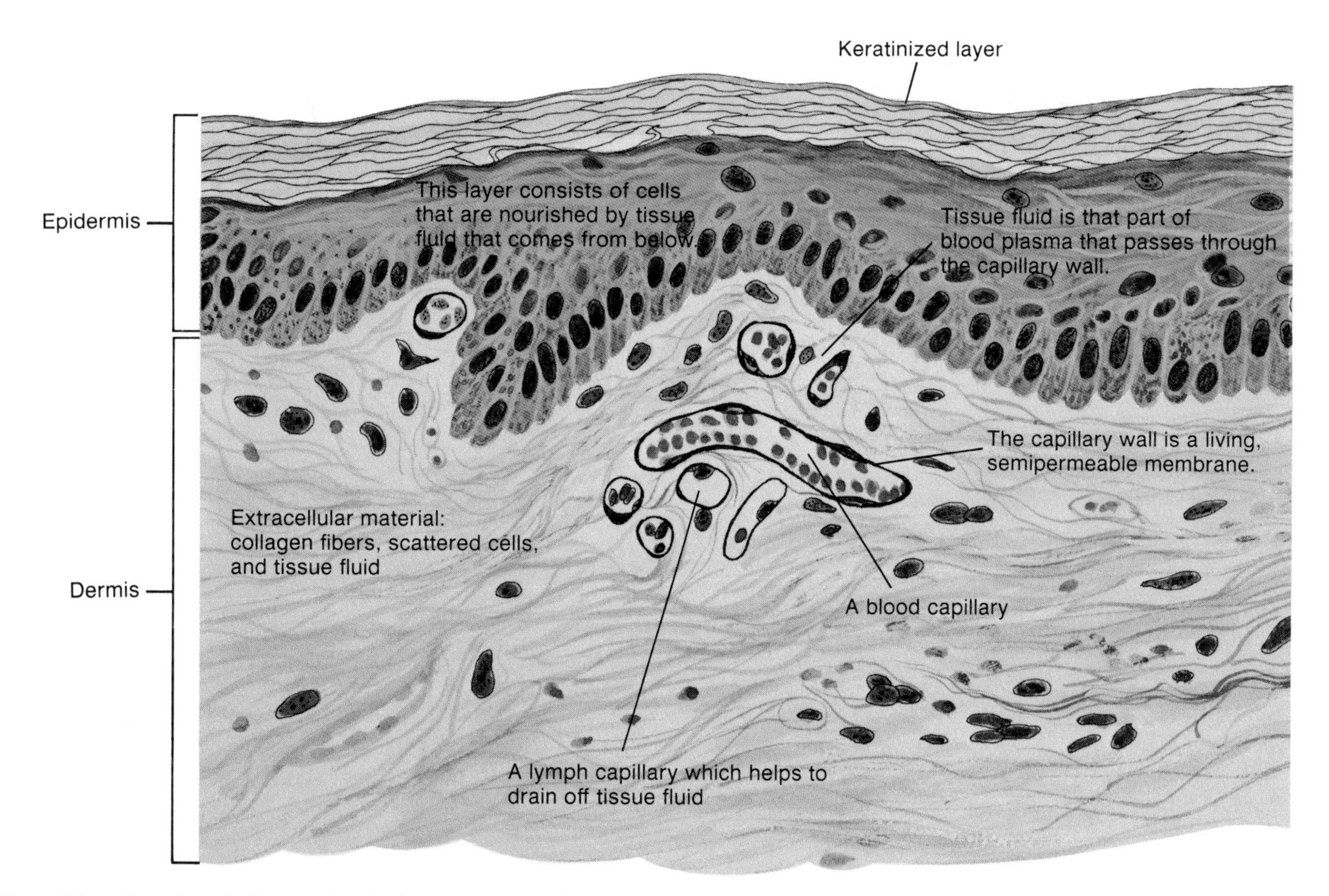

Figure 5.7. *A section of skin showing the loose connective tissue dermis beneath the cornified epidermis. Loose connective tissue contains scattered collagen fibers in a matrix of protein-rich fluid. The intercellular spaces also contain cells and blood vessels.*

Exocrine Glands

Exocrine[5] glands are derived from cells of epithelial membranes that cover and line the body surfaces. The secretions of these cells are expressed to the outside of the epithelial membranes (and hence to the outside of the body) through *ducts.* This is in contrast to endocrine glands, which lack ducts and which therefore secrete into capillaries within the body (fig. 5.8). The structure of endocrine glands will be described in chapter 7.

Examples of exocrine glands in the skin include the lacrimal (tear) glands, sebaceous glands (which secrete oily sebum into hair follicles), and sweat glands. All of the glands that secrete into the digestive tract are also exocrine. This is because the lumen of the digestive tract is a part of the external environment and secretions of these glands go to the outside of the membrane that lines this tract. Mucous glands are located throughout the length of the digestive tract. Other relatively simple glands of the tract include salivary glands, gastric glands, and simple tubular glands in the intestine.

The *liver* and *pancreas* are exocrine (as well as endocrine) glands, derived embryologically from the digestive tract. The exocrine secretion of the pancreas is pancreatic juice, containing digestive enzymes and bicarbonate, which is secreted into the small intestine via the pancreatic duct. The liver produces and secretes bile (an emulsifier of fat) into the small intestine via the gallbladder and bile duct.

Exocrine glands are also prominent in the reproductive system. The female reproductive tract contains numerous mucus-secreting exocrine glands. The male accessory sexual organs—the *prostate* and *seminal vesicles*—are exocrine glands that contribute to the semen. The testes and ovaries (the gonads) are both endocrine and exocrine glands. They are endocrine because they secrete sex steroid hormones into the blood; they are exocrine because they release gametes (ova and sperm) into the reproductive tracts.

[5]exocrine: Gk. *exo,* outside; *krinein,* to separate

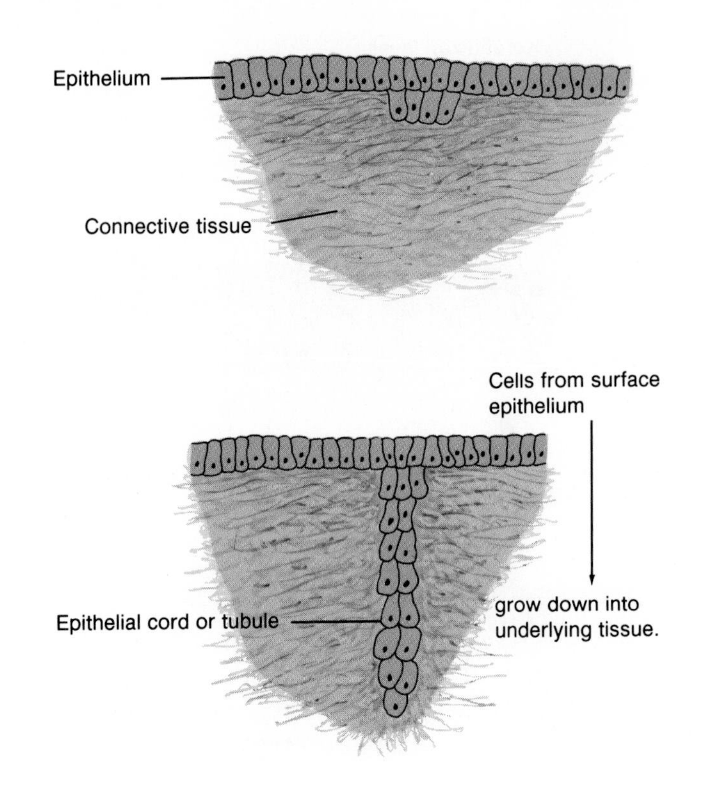

Figure 5.8. *The formation of exocrine and endocrine glands from epithelial membranes.*

Connective Tissue

Connective tissue is characterized by large amounts of extracellular material in the spaces between the connective tissue cells. This extracellular material may be of various types and arrangements and, on this basis, several types of connective tissues are recognized: (1) connective tissue proper; (2) cartilage; (3) bone; and (4) blood. Blood is usually classified as connective tissue because about half its volume is composed of an extracellular fluid known as *plasma.*

Connective tissue proper includes a variety of subtypes. An example of *loose connective tissue* (or *areolar tissue*) is the *dermis* of the skin. This connective tissue consists of scattered fibrous proteins called *collagen*[6] and tissue fluid, which provides abundant space for the entry of blood and lymphatic vessels and nerve fibers. Another type of connective tissue proper is *dense fibrous connective tissue,* which contains densely packed fibers of collagen that may be in an irregular or a regular arrangement. Dense irregular connective tissue contains a meshwork of collagen fibers and forms the tough capsules and sheaths around organs (fig. 5.9). Tendons,[7] which connect muscle to bone, and ligaments,[8] which connect bones together at joints, are examples of dense regular connective tissue. This tissue contains a dense arrangement of collagen fibers that are parallel to each other (fig. 5.10).

[6]collagen: Gk. *kolla,* glue

[7]tendon: L. *tendere,* to stretch

[8]ligament: L. *ligare,* to bind

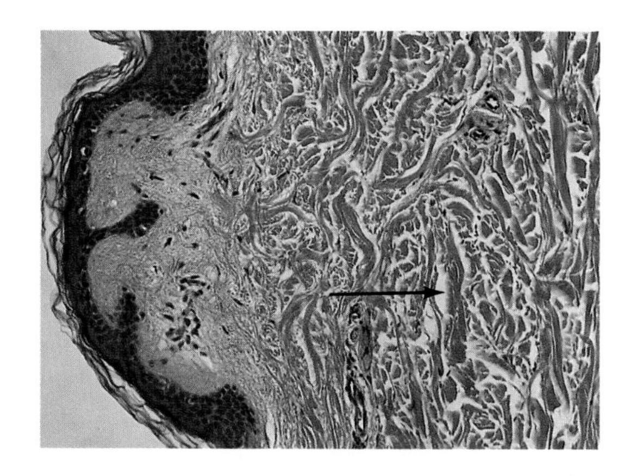

Figure 5.9. A photomicrograph of dense irregular connective tissue. Note the tightly packed, irregularly arranged collagen proteins.

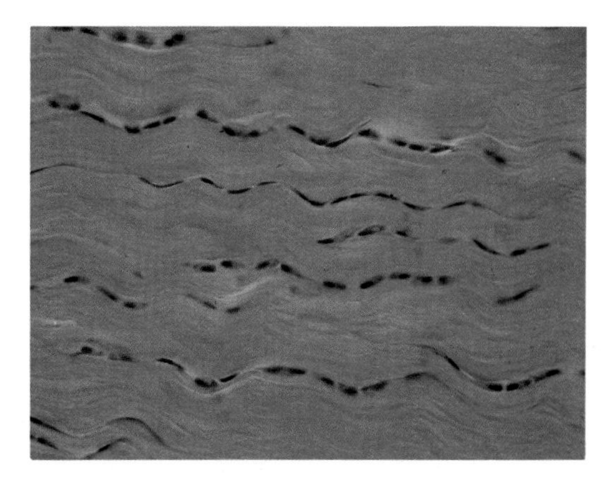

Figure 5.10. A photomicrograph of a tendon showing a dense, regular arrangement of collagen fibers.

Adipose tissue is composed of groupings of adipose cells within loose connective tissue. At the center of each adipose cell is a large droplet of fat, surrounded by the cytoplasm of the adipose cell. In order for this fat to be formed, or for it to be broken down, enzymes in the adipose cells must be activated by specific hormones. This occurs in a manner appropriate to the needs of the body, as described in chapter 18.

Cartilage consists of cells, called *chondrocytes,* surrounded by a semisolid ground substance that imparts elastic properties to the tissue. Cartilage is a type of supportive and protective tissue, and is found at the surface of bones within movable joints, the discs between the bony vertebrae of the backbone, and in other body locations.

Bone is produced as concentric layers of calcified material laid around blood vessels. The cells within the bone remain alive because of small canals within this calcified material, which transport nutrients from the blood vessels. The structure and function of cartilage and bone are described in more detail in chapter 21.

---

1. List the four primary tissues, and describe the distinguishing features of each type.
2. Name the three types of muscle tissue, and distinguish between them.
3. Describe the different types of epithelial membranes, and indicate their locations in the body.
4. Explain why exocrine and endocrine glands are considered to be epithelial tissues, and distinguish between these two types of glands.
5. Describe the different types of connective tissues, and explain how they differ from each other in their content of extracellular material.

Organs and Systems

Organs are generally composed of all four primary tissues which serve the different functions of the organ. The skin is an organ that has numerous functions provided by its constituent tissues. It contains exocrine glands and sensory receptors, for example, and functions as an endocrine gland.

An organ is a structure composed of at least two, and usually all four, types of primary tissues. The largest organ in the body, in terms of its surface area, is the skin. The numerous functions of the skin will serve in this section to illustrate how primary tissues cooperate in the service of organ physiology.

An Example of an Organ: The Skin

The cornified epidermis protects the skin (fig. 5.11) against water loss and against invasion by disease-causing organisms. Invaginations of the epithelium into the underlying connective tissue dermis creates the exocrine glands of the skin. These include sweat glands and sebaceous glands. The secretion of sweat glands cools the body by evaporation and produces odors that, at least in lower animals, serve as sexual attractants. Sebaceous glands secrete oily sebum into hair follicles, where it is transported to the surface of the skin (unless these ducts are blocked, in which case blackheads are produced). Sebum lubricates the cornified surface of the skin, helping to prevent it from drying and cracking (as in chapped lips—there are no sebaceous glands in the lips; one must, therefore, moisten the lips periodically with the tongue).

The skin is nourished by blood vessels within the dermis. In addition to blood vessels, the dermis contains wandering white blood cells and other types of cells that protect against invading

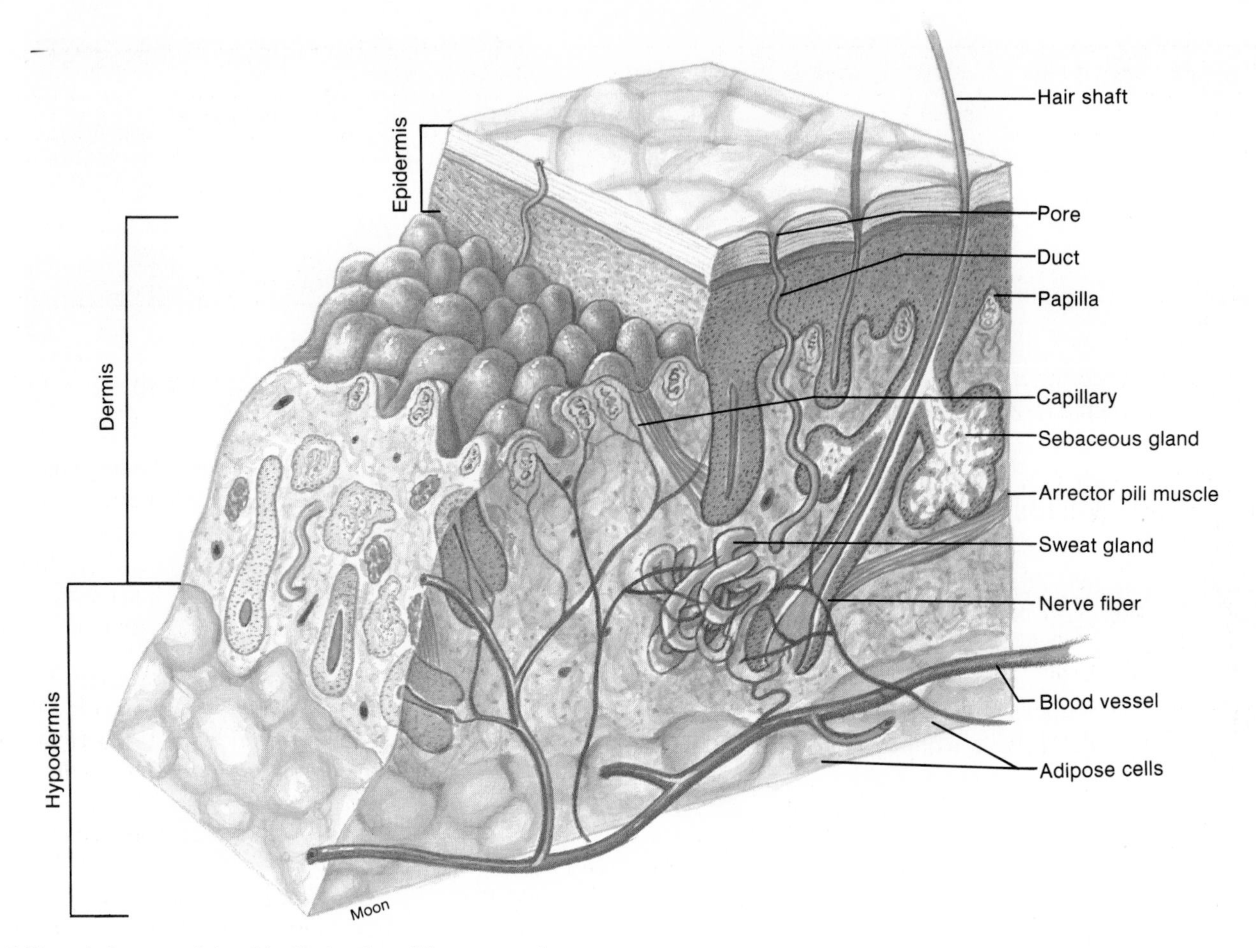

Figure 5.11. *A diagram of the skin. Notice that all four types of primary tissues are present.*

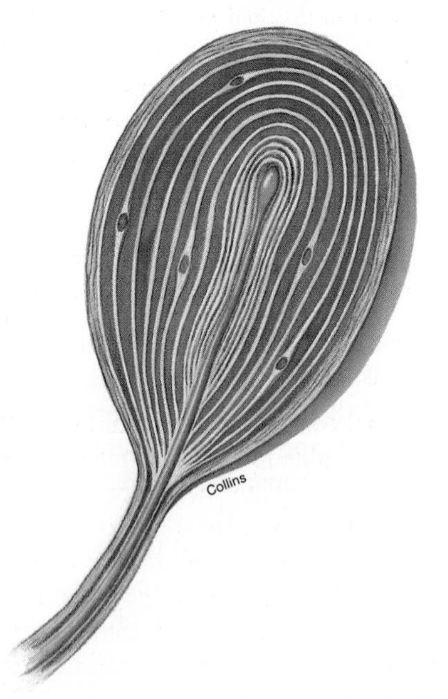

Figure 5.12. *The pacinian corpuscle is a receptor for deep pressure. It consists of epithelial cells and connective tissue proteins that form concentric layers around the ending of a sensory neuron.*

disease-causing organisms, as well as nerve fibers and fat cells. Most of the fat cells, however, are grouped together to form the *hypodermis* (a layer beneath the dermis). Although fat cells are a type of connective tissue, masses of fat deposits throughout the body—such as subcutaneous fat—are referred to as **adipose**[9] **tissue.**

Sensory nerve endings within the dermis mediate the cutaneous sensations of touch, pressure, heat, cold, and pain. Some of these sensory stimuli directly affect the sensory nerve endings. Others act via sensory structures derived from nonneural primary tissues. The pacinian corpuscles in the dermis of the skin (fig. 5.12), for example, monitor sensations of pressure. Motor nerve fibers in the skin stimulate effector organs, resulting in, for example, the secretions of exocrine glands and contractions of the arrector pili muscles, which attach to hair follicles and surrounding connective tissue (producing goose

[9]adipose: L. *adiposus,* fat

Table 5.1 Approximate normal ranges for measurements of some blood values

Measurement	Normal Range	Measurement	Normal Range
Arterial pH	7.35–7.43	Urea	12–35 mg/100 ml
Bicarbonate	21.3–28.5 mEq/L	Amino acids	3.3–5.1 mg/100 ml
Sodium	136–151 mEq/L	Protein	6.5–8.0 g/100 ml
Calcium	4.6–5.2 mEq/L	Total lipids	350–850 mg/100 ml
Oxygen content	17.2–22.0 ml/100 ml	Glucose	75–110 mg/100 ml

bumps). The degree of constriction or dilation of cutaneous blood vessels—and therefore the rate of blood flow—is also regulated by motor nerve fibers.

The epidermis itself is a dynamic structure that can respond to environmental stimuli. The rate of its cell division—and consequently the thickness of the cornified layer—increases under the stimulus of constant abrasion. This produces calluses. The skin also protects itself against the dangers of ultraviolet light by increasing its production of *melanin*[10] pigment, which absorbs ultraviolet light while producing a tan. In addition, the skin is an endocrine gland that produces and secretes vitamin D (derived from cholesterol under the influence of ultraviolet light), which functions as a hormone (chapter 19).

The architecture of most organs is similar to that of the skin. Most are covered by an epithelium immediately over a connective tissue layer. The connective tissue contains blood vessels, nerve endings, scattered cells for fighting infection, and possibly glandular tissue as well. If the organ is hollow—as in the digestive tract or in blood vessels—the lumen is also lined with an epithelium immediately over a connective tissue layer. The presence, type, and distribution of muscular and nervous tissue varies in different organs.

Systems

Organs that are located in different regions of the body and that perform related functions are grouped into **systems.** These include the nervous, endocrine, cardiovascular, lymphatic, respiratory, urinary, muscular, skeletal, integumentary, reproductive, digestive, and immune systems. By means of numerous regulatory mechanisms, these systems work together to maintain the life and health of the entire organism.

1. Describe the location of each type of primary tissue in the skin.
2. Describe the functions of nerve, muscle, and connective tissue in the skin.
3. Describe the functions of the epidermis, and explain why this tissue is called "dynamic."

[10]melanin: Gk. *melas,* black

Homeostasis and Feedback Control

The regulatory mechanisms of the body can be understood in terms of a single, shared function: that of maintaining homeostasis, which is defined as the dynamic constancy of the internal environment. Homeostasis is maintained by effectors which are regulated by sensory information from the internal environment. The activity of effectors is thus controlled by the very changes that these organs help to produce—that is, by feedback control mechanisms.

Over a century ago the French physiologist Claude Bernard observed that the *milieu interieur* ("internal environment") remains remarkably constant despite changing conditions in the external environment. In a book entitled *The Wisdom of the Body* (published in 1932), Walter Cannon coined the term **homeostasis**[11] to describe this internal constancy. Cannon further suggested that mechanisms of physiological regulation exist for one purpose—the maintenance of internal constancy.

The concept of homeostasis has been of inestimable value in the study of anatomy and physiology, because it allows diverse regulatory mechanisms to be understood in terms of their "why" as well as their "how." The concept of homeostasis also provides a major foundation for medical diagnostic procedures. When a particular measurement of the internal environment, such as a blood measurement (table 5.1), deviates significantly from the normal range of values, it can be concluded that homeostasis is not maintained and the person is sick. A number of such measurements, combined with clinical observations, may allow the particular defective mechanism to be identified.

Negative Feedback Loops

In order for internal constancy to be maintained the body must have sensors that are able to detect deviations from a *set point.* The set point is analogous to the temperature set on a house thermostat. In a similar manner, there is a set point for body temperature, blood glucose concentration, the tension on a tendon, and so on. When a sensor detects a deviation from a particular set point, it must relay this information to an **integrating center,** which usually receives information from many

[11]homeostasis: Gk. *homoios,* like; *stasis,* a standing

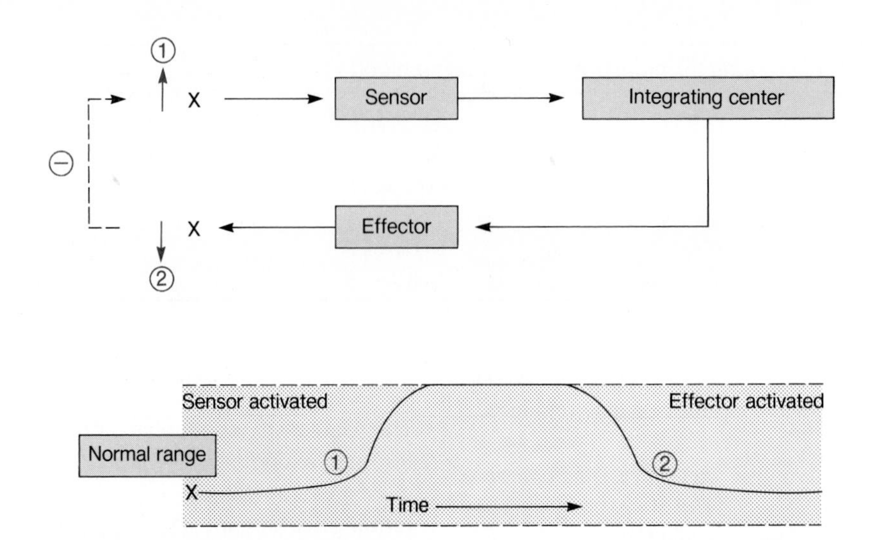

Figure 5.13. A rise in some factor of the internal environment (↑X) is detected by a sensor. Acting through an integrating center, this caused an effector to produce a change in the opposite direction (↓X). The initial deviation is thus reversed, completing a negative feedback loop (shown by dashed arrow and negative sign). The numbers indicate the sequence of changes.

different sensors. The integrating center is often a particular region of the brain or spinal cord, but in some cases it can also be cells of endocrine glands. The relative strengths of different sensory inputs are weighed in the integrating center, and, in response, the integrating center either increases or decreases the activity of particular **effectors,** which are generally muscles or glands.

In response to sensory information about a deviation from a set point, therefore, effectors compensate for this deviation by promoting a reverse change in the internal environment. If the body temperature exceeds the set point of 37°C, for example, effectors act to lower the temperature. If, as another example, the blood glucose concentration falls below normal, the effectors act to increase the blood glucose. Since the activity of the effectors is influenced by the effects they produce, and since this regulation is in a negative, or reverse, direction, this type of control system is known as a **negative feedback loop** (fig. 5.13).

It is important to realize that these negative feedback loops are continuous, ongoing processes. Thus, a particular nerve fiber which is part of an effector mechanism may always display some activity, and a particular hormone, which is part of another effector mechanism, may always be present in the blood. The nerve activity and hormone concentration may decrease in response to deviations of the internal environment in one direction (fig. 5.13), or they may increase in response to deviations in the opposite direction (fig. 5.14). Changes from the normal range in either direction are thus compensated by reverse changes in effector activity.

Homeostasis is best conceived as a state of **dynamic constancy**, rather than a state of absolute constancy. The values of particular measurements of the internal environment fluctuate above and below the set point, which can be taken as the average value within the normal range of measurements (fig. 5.15). This state of dynamic constancy results from greater or lesser activation of effectors in response to sensory feedback, and from the competing actions of antagonistic[12] effectors.

Antagonistic Effectors

Most factors in the internal environment are controlled by several effectors, which often have antagonistic actions. Control by antagonistic effectors is sometimes described as "push-pull," where the increasing activity of one effector is accompanied by decreasing activity of an antagonistic effector. This affords a finer degree of control than could be achieved by simply switching one effector on and off. Normal body temperature, for example, is maintained about a set point of 37° C by the antagonistic effects of sweating, shivering, and other mechanisms (fig. 5.16).

The blood concentrations of glucose, calcium, and other substances are regulated by negative feedback loops that involve hormones which promote opposite effects. While insulin, for example, lowers blood glucose, other hormones raise the blood glucose concentration. The heart rate, similarly, is controlled by nerve fibers that produce opposite effects: stimulation of one group of nerve fibers increases heart rate, and stimulation of another group slows the heart rate.

[12]antagonistic: Gk. *anti,* against; *agonizomai,* to fight

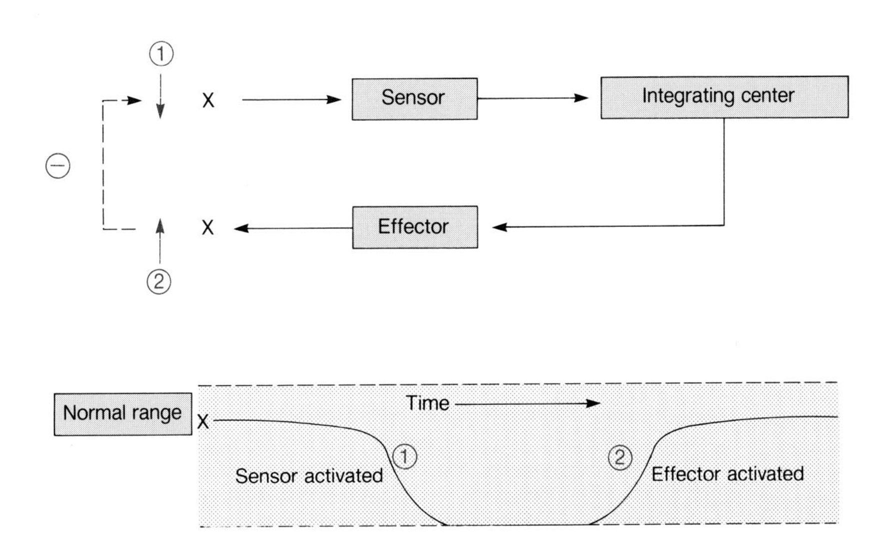

Figure 5.14. *A negative feedback loop that compensates for a fall in some factor of the internal environment (↓X). Compare this figure with figure 5.13.*

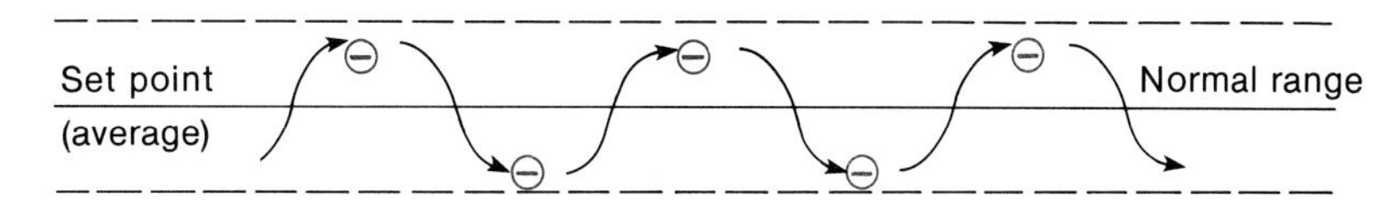

Figure 5.15. *Negative feedback loops (indicated by negative signs) maintain a state of dynamic constancy within the internal environment.*

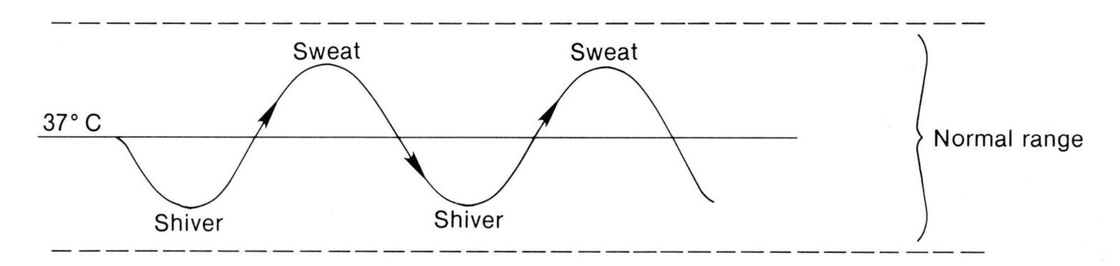

Figure 5.16. *A simplified scheme by which body temperature is maintained within the normal range (with a set point of 37°C) by two antagonistic mechanisms—shivering and sweating. Shivering is induced when the body temperature falls too low and gradually subsides as the temperature rises. Sweating occurs when the body temperature is too high and diminishes as the temperature falls. Most aspects of the internal environment are regulated by the antagonistic actions of different effector mechanisms.*

Neural and Endocrine Regulation

The effectors of most negative feedback loops include the actions of nerves and hormones. In both neural and endocrine[13] regulation, particular chemical regulators released by nerve fibers or endocrine glands stimulate target cells by interacting with specific receptor proteins in these cells. The mechanisms by which this regulation is achieved will be described in later chapters.

The endocrine system functions closely with the nervous system in regulating and integrating body processes and maintaining homeostasis. The nervous system controls the secretion of many endocrine glands, and some hormones in turn affect the function of the nervous system. Together, the nervous and endocrine systems regulate the activities of most of the other systems of the body.

[13]endocrine: Gk. *endon,* within; *krinein,* to separate

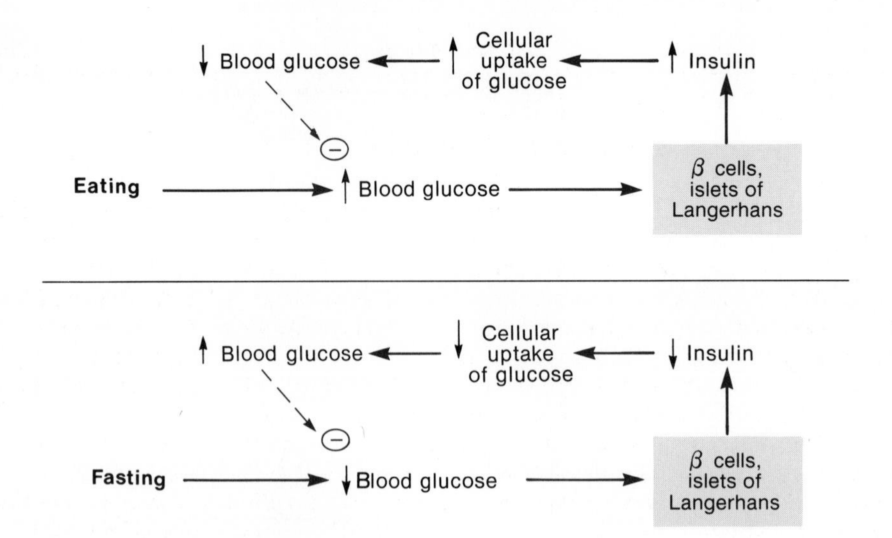

Figure 5.17. The negative feedback control of insulin secretion by changes in the blood glucose concentration. Dashed arrows and negative signs indicate that negative feedback loops compensate the initial changes in blood glucose concentrations produced by eating or fasting.

Regulation by the endocrine system is achieved by the secretion of chemical regulators called **hormones**[14] into the blood. Since hormones are secreted into the blood, they are carried by the blood to all organs in the body. Only specific organs can respond to a particular hormone, however; these are known as the *target organs* of that hormone.

Nerve fibers, or axons (previously described in this chapter), are said to *innervate* the organs that they regulate. When stimulated, these axons produce electrochemical nerve impulses that are conducted from the origin of the axon at the cell body to the end of the axon in the target organ innervated by the axon. These target organs can be muscles or glands, which may function as effectors in the maintenance of homeostasis.

Feedback Control of Hormone Secretion

The details of the nature of the endocrine glands, the interaction of the nervous and endocrine systems, and the actions of hormones, will be explained in later chapters. The purpose of this section is to describe the regulation of hormone secretion in a brief, general manner, because this subject so superbly illustrates the principles of homeostasis and negative feedback regulation.

[14]hormone: Gk. *hormon,* to set in motion

Hormones are secreted in response to specific chemical stimuli. A rise in plasma glucose concentration, for example, stimulates insulin secretion from the islets of Langerhans in the pancreas. Hormones are also secreted in response to transmitter molecules released from nerve endings and to stimulation by other hormones.

The secretion of hormones is controlled by inhibitory as well as by stimulatory influences. The effect of a given hormone's action can inhibit its own secretion. The secretion of insulin—which acts to lower the plasma glucose concentration—is stimulated by a rise in glucose concentration, for example, and is inhibited by a fall in blood glucose. The lowering of blood glucose levels by insulin thus has an inhibitory feedback effect on further insulin secretion. This closed loop control system is called **negative feedback inhibition** (see fig. 5.17).

1. Define homeostasis, and describe how this concept can be used to understand physiological control mechanisms.
2. Describe the meaning of the term negative feedback, and explain how it contributes to homeostasis. Illustrate this concept by drawing a negative feedback loop.
3. Describe positive feedback, and explain how this process functions in the body.
4. Explain how the secretion of a hormone is controlled by negative feedback inhibition. Illustrate this process with an example.

Use of Animals in Research

Political groups who have termed themselves "animal rights activists" accuse scientists of being *speciests*. This term is supposed to be analogous to "racist," and refers to the fact that scientists place human welfare above the welfare of other animals. Most scientists would probably agree that this is true but oversimplified; lower animals also benefit from scientific research which improves animal health and survival of species in changing environments. Further, they would point out that the animal rights activists are also speciests. After all, what do they feed their pet dogs and cats? These animals are carnivores, and their food—even the innocuous-looking dried pellets—must be derived from other animals. Also, the activists fight only for the "rights" of cute animals. Lizards, cockroaches, snails, and worms are as much members of the animal kingdom as dogs and cats, but their "rights" are ignored by these political groups. One never sees activists "liberating" chickens from ranches or lobsters from holding tanks in restaurants. Instead, the activists specifically attack the use of animals in research. While these actions may be well motivated, they are potentially a great threat to human welfare.

Virtually every basic concept in biochemistry, cell biology, and the physiology of the body's organs and systems has been learned through the use of animals in research experiments. These basic concepts provide the foundation for modern medicine—both human and veterinary. Further, laboratory research, with animals used as experimental subjects, has been instrumental in almost all medical advances which have been directly applied to the treatment of human diseases. From early success such as the prevention of polio and the treatment of diabetes mellitus, to current research in AIDS, cancer, heart disease, neural and muscular disorders, and many others, animal experimentation has provided the means by which cures can be sought. Even current knowledge about nutritional requirements, and the body's needs for vitamins and minerals, is based on information gained from animal experimentation. The development of every new prescription item requires the use of laboratory animals. New surgical procedures, the use of prosthetics, and other medical technologies are also the result of animal experimentation.

Although the number of animals used in biomedical research is infinitesimally small compared to the numbers used to provide food, clothing, and other products, the use of animals in research has been the major target of animal rights activists. This is very curious. Whereas alternatives are easily found for meat (many people choose to become vegetarians), the alternatives for research are far more limited. There is no computer bank of information to consult, and no way that the action of drugs or the cure for a disease can be mathematically calculated from first principles, as may be accomplished in some areas of physics. Living systems are far too complex for such easy solutions; in order to learn about them, experiments must be performed. These experiments cannot be done entirely *in vitro* (outside the body), because the test system cannot duplicate the complex interactions that occur in the body. (A drug that is safe when tested on cultured cells may be altered by the body into a carcinogen, for example.) As stated in a 1988 report by the National Research Council, "The chances that alternatives will completely replace animals in the future is nill." The only alternatives to the use of lower animals in basic research, therefore, is to use humans as the experimental subjects or to give up any hope of obtaining further knowledge and finding cures for the diseases that plague humanity.

The Institute of Laboratory Animal Resources estimates that seventeen to twenty-two million laboratory animals are used annually. Of these, 85% are rats and mice. Dogs, cats, and nonhuman primates together account for less than 2%. Researchers use about 200,000 pound animals annually. This may seem like a large number, but it is actually less than 2% of the number destroyed by the pounds themselves. Nevertheless, in response to political pressure from animal rights activists, ten states have currently banned the use of pound animals in laboratory research. This does not eliminate the use of dogs and cats for research, but rather drastically increases the costs of the research because scientists must then buy animals that were raised for this purpose. The manner in which nonrodent laboratory animals are used is federally regulated by the 1966 Animal Welfare Act, and stringent ammendments to this act were passed in 1985.

The National Research Council (NRC), an arm of the National Science Foundation, issued a report in 1988 on the use of laboratory animals in research. In this report, they stated that "humans are morally obliged to each other to improve the human condition." This being so, it is morally reprehensible to refrain from performing experiments that could achieve this goal. The NRC report also stated that "scientists are ethically obliged to ensure the well-being of animals used in research and to minimize their pain and suffering." This seems to be a statement that most people would find reasonable and moral. These positions promote animal welfare while, at the same time, supporting the biomedical research required for the improvement of the human condition.

Summary

The Primary Tissues p. 76

I. There are three types of muscle tissue.
  A. Skeletal muscle is striated, voluntary muscle.
  B. Cardiac muscle is also striated, and its cells are joined together by intercalated discs.
  C. Smooth muscle lacks striations and is found in the internal body organs and the walls of blood vessels.

II. Nervous tissue consists of neurons and neuroglia.
  A. Each neuron is composed of a cell body, dendrites, and a single axon.
  B. Neuroglial cells aid the functions of neurons.

III. Epithelial tissue forms membranes that cover all body surfaces and glands.
  A. Epithelial membranes may be simple or stratified, and their cells may be squamous, cuboidal, or columnar.
  B. Exocrine glands secrete into ducts; endocrine glands lack ducts and secrete hormones into the blood.

IV. Connective tissue is characterized by its abundant extracellular material.
  A. Connective tissue proper includes areolar and dense connective tissue.
  B. Cartilage, bone, dentin, enamel, and blood are specialized types of connective tissue.

Organs and Systems p. 81

I. The skin is an example of an organ, because it is composed of all four primary tissues.
  A. The epidermis and the dermis are examples of epithelial and connective tissue, respectively.
  B. Neurons and blood vessels are found in the dermis, as are the arrector pili muscles.

II. Different organs that cooperate in their functions form the body systems.

Homeostasis and Feedback Control p. 83

I. Homeostasis is the state of dynamic constancy of the internal environment of the body.

II. Homeostasis is maintained by negative feedback loops.
  A. Deviations from a set point are detected by a sensor, which sends information to an integrating center.
  B. The integrating center controls the activity of effectors, which produce changes that compensate for the initial deviation from the set point.

III. The nervous and endocrine systems are the major control systems in the body which help to maintain homeostasis.

IV. The amount of a particular hormone that is secreted is controlled by negative feedback inhibition.
  A. A rise in blood glucose, for example, stimulates insulin secretion, which acts to lower the blood glucose concentration.
  B. As the blood glucose decreases, less insulin is secreted.

Review Activities

Objective Questions

Match the following:

1. Glands are derived from
2. Cells are joined closely together in
3. Cells are separated by large extracellular spaces in
4. Blood vessels and nerves are usually located within

(a) nervous tissue
(b) connective tissue
(c) muscular tissue
(d) epithelial tissue

5. Sweat is secreted by exocrine glands. This means that
   (a) it is produced by epithelial cells
   (b) it is a hormone
   (c) it is secreted into a duct
   (d) it is produced outside the body
6. Which of the following statements about homeostasis is *true?*
   (a) The internal environment is maintained absolutely constant.
   (b) Negative feedback mechanisms act to correct deviations from a normal range within the internal environment.
   (c) Homeostasis is maintained by switching effector actions on and off.
   (d) All of these are true.
7. In a negative feedback loop the effector organ produces changes that are
   (a) similar in direction to that of the initial stimulus
   (b) opposite in direction to that of the initial stimulus
   (c) unrelated to the initial stimulus
8. A hormone called *parathyroid hormone* acts to help raise the blood-calcium concentration. According to the principles of negative feedback, an effective stimulus for parathyroid hormone secretion would be
   (a) a fall in blood calcium
   (b) a rise in blood calcium
9. The act of breathing raises the blood oxygen level, lowers the blood carbon dioxide concentration, and raises the blood pH. According to the principles of negative feedback, sensors that regulate breathing should respond to
   (a) a rise in blood oxygen
   (b) a rise in blood pH
   (c) a rise in blood carbon dioxide concentration
   (d) all of the above

Essay Questions

1. Describe the structure of different epithelial membranes, and explain how their structures relate to their functions.
2. What are the similarities between the dermis of the skin, bone, and blood? What are the major structural differences between these tissues?
3. Explain the role of antagonistic negative feedback processes in the maintenance of homeostasis.
4. Explain, using examples, how the secretion of a hormone is controlled by the effects of that hormone's actions.

6

The Nervous System

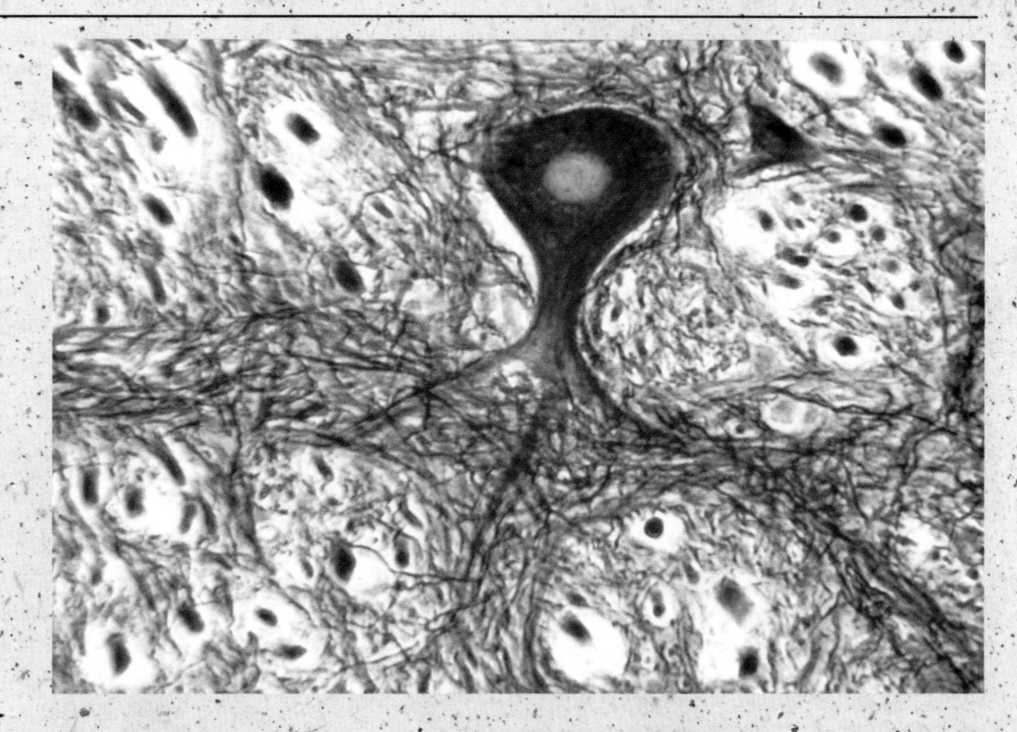

Outline

Objectives

By studying this chapter, you should be able to

1. describe the structure and function of sensory neurons, motor neurons, and association neurons
2. provide the definitions of terms such as *nerve, ganglion, tract,* and *nucleus*
3. list the different types of neuroglial cells and their functions
4. describe how a myelin sheath is formed, and explain its function
5. explain how axons produce an action potential, and describe the characteristics of action potentials
6. describe how synaptic transmission occurs, using acetylcholine as an example of a neurotransmitter
7. list some of the different chemicals that may act as neurotransmitters in the brain
8. distinguish between the CNS and PNS, and describe the structures of the cerebrum
9. list the lobes of the cerebral cortex and their primary functions
10. describe the differences in function between the right and left cerebral hemispheres
11. identify the brain structures involved in language, and the different types of aphasias that may occur when these regions are damaged
12. identify the brain structures involved in emotion and motivation
13. distinguish between short-term and long-term memory, and describe the observations that implicate particular brain regions in these processes

Keys to Pronunciation

acetylcholinesterase: *as''e-til-ko''lin-es'ter-as*
acetylcholine: *as''e-til-ko'len*
adrenergic: *ad''ren-er'jik*
catecholamines: *kat''e-kol'ah-mens*
cholinergic: *ko''lin-er'jik*
depolarization: *de-po''lar-i-za'shun*
enkephalins: *en-kef'ah-lins*
ependyma: *e-pen'di-mah*
exocytosis: *eks'o-si-to'sis*
hypothalamus: *hi''po-thal'ah-mus*
microglia: *mi-krog'le-ah*
myelin: *mi'e-lin*
myoneural: *mi''o-nu'ral*
neuroglial: *nu-rog'le-al*
oligodendrocytes: *ol''i-go-den'dro-sits*
Ranvler: *rahn-ve-a'*
saltatory: *sal'tah-to''re*
sulci: *sul'si*
synapse: *sin'aps*

Photo: Photomicrograph of a neuron in the brain, showing numerous interconnections with other neurons.

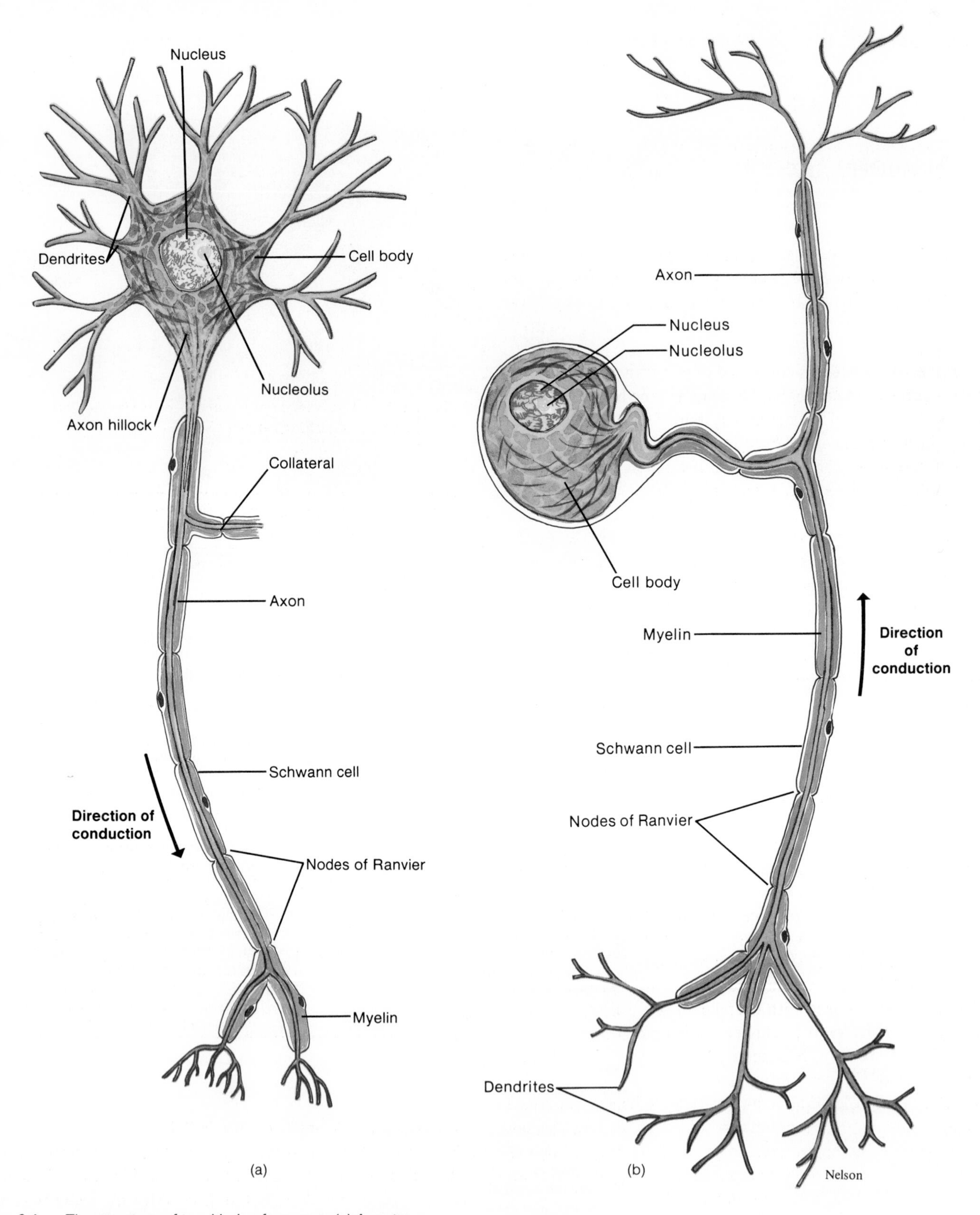

Figure 6.1. *The structure of two kinds of neurons. (a) A motor neuron; (b) a sensory neuron.*

Organization of the Nervous System

The nervous system is composed of neurons, which produce and conduct electrochemical impulses, and of neuroglial cells, which support the functions of neurons. The nervous system is subdivided into the central nervous system and peripheral nervous system, and terms such as "nerve," "ganglion," "tract," and "nucleus" relate to specific structures in the two subdivisions.

The nervous system is divided into the **central nervous system (CNS),** which includes the brain and spinal cord, and the **peripheral nervous system (PNS),** which includes the *cranial nerves* arising from the brain and the *spinal nerves* arising from the spinal cord.

The nervous system is composed of only two principal types of cells, neurons and neuroglia. **Neurons** are the basic structural and functional units of the nervous system. They are specialized to respond to physical and chemical stimuli, conduct electrochemical impulses, and release specific chemical regulators. Through these activities, neurons perform such functions as the perception of sensory stimuli, learning, memory, and the control of muscles and glands. Neurons cannot reproduce themselves by cell division, although some neurons can regenerate a severed portion or sprout small new branches under some conditions.

Neuroglia, or **glial** (*glia* = glue) **cells,** are supportive cells in the nervous system that aid the function of neurons. Glial cells are about five times more abundant than neurons and have limited ability to divide (brain tumors that occur in adults are usually composed of glial cells rather than neurons).

Neurons

Although neurons vary considerably in size and shape, they generally have three principal regions: (1) a cell body; (2) dendrites; and (3) an axon (figs. 6.1 and 6.2). Dendrites and axons can be referred to generically as *processes,* or extensions from the cell body.

The **cell body** is the enlarged portion of the neuron, which contains the nucleus and serves as the metabolic center of the neuron where macromolecules are produced. The cell bodies within the CNS are frequently clustered into groups called *nuclei* (not to be confused with the nucleus of a cell). Cell bodies in the PNS usually occur in clusters called *ganglia* (table 6.1).

Dendrites[1] are thin, branched processes that extend from the cytoplasm of the cell body. Dendrites serve as a receptive area that transmits electrical impulses to the cell body. The **axon**[2] is a longer process that conducts impulses away from the cell body. Axons vary in length from only a millimeter to as long as a meter or more in length. Side branches that may extend from the axon are called *axon collaterals.*

[1]dendrite: Gk. *dendron,* tree branch
[2]axon: Gk. *axon,* axis

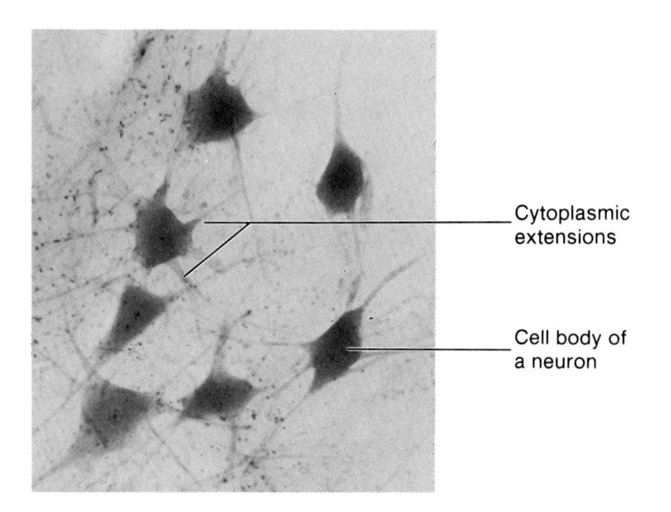

Figure 6.2. *A photomicrograph of neurons from the anterior column of gray matter of the spinal cord (120×).*

Table 6.1 Anatomical terms used in describing the nervous system

Term	Definition
Central nervous system (CNS)	Brain and spinal cord
Peripheral nervous system (PNS)	Nerves and ganglia
Interneuron	Multipolar neuron located entirely within CNS
Sensory neuron	Neuron that transmits impulses from sensory receptor into CNS (afferent fiber)
Motor neuron	Neuron that transmits impulses from CNS to an effector organ (e.g., muscle efferent fiber)
Nerve	Cablelike collection of nerve fibers; may be "mixed" (contain both sensory and motor fibers)
Somatic motor nerve	Nerve that stimulates contraction of skeletal muscles
Autonomic motor nerve	Nerve that stimulates contraction (or inhibits contraction) of smooth muscle and cardiac muscle and secretion of glands
Ganglion	Collection of neuron cell bodies located outside CNS
Nucleus	Groupings of neuron cell bodies within CNS
Tract	Collections of nerve fibers that interconnect regions of CNS

Classification of Neurons and Nerves

Neurons may be classified according to their structure or function. The functional classification is based on the direction that they conduct impulses. **Sensory,** or **afferent, neurons** conduct impulses from sensory receptors into the CNS. **Motor,** or **efferent, neurons** (fig. 6.3) conduct impulses out of the CNS to effector organs (muscles and glands). **Association neurons,** or **interneurons,** are located entirely within the CNS and serve the associative, or integrative, functions of the nervous system.

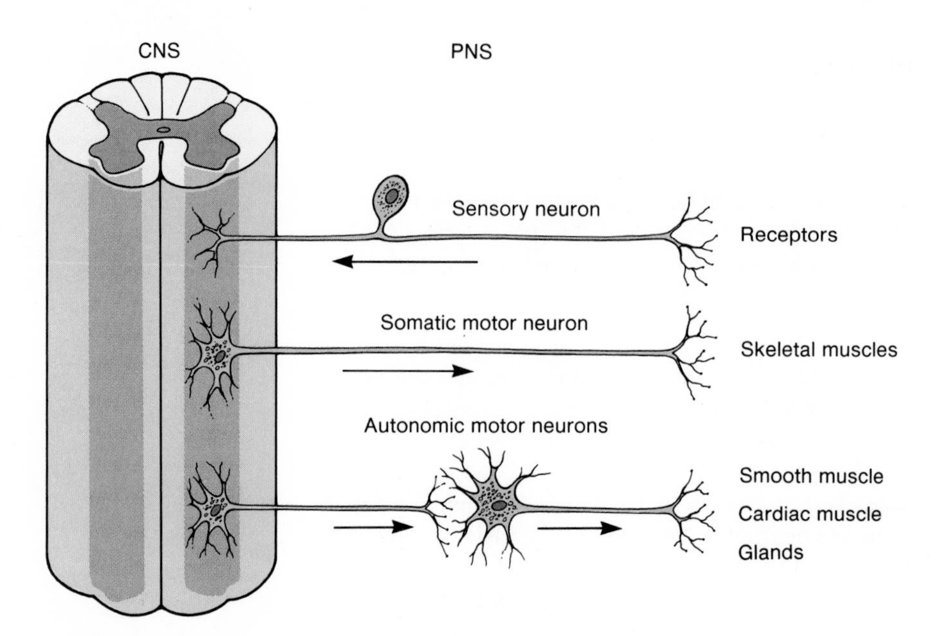

Figure 6.3. *The relationship between sensory and motor fibers of the peripheral nervous system (PNS) and the central nervous system (CNS).*

There are two types of motor neurons: somatic and autonomic. **Somatic motor neurons** provide both reflex and voluntary control of skeletal muscles. **Autonomic motor neurons** innervate the involuntary effectors—smooth muscle, cardiac muscle, and glands. Autonomic motor neurons, together with their central control centers, comprise the *autonomic nervous system,* which will be discussed later in this chapter.

The structural classification of neurons is based on the number of processes that extend from the cell body of the neuron (fig. 6.4). **Bipolar neurons** have two processes, one at either end; this type is found in the retina of the eye. **Multipolar neurons** have several dendrites and one axon extending from the cell body; this is the most common type of neuron (motor neurons are good examples of this type). A **pseudounipolar neuron** has a single short process that divides like a T to form a longer process. Sensory neurons are pseudounipolar—one end of the process formed by the T receives sensory stimuli and produces nerve impulses; the other end of the T delivers these impulses to the brain or spinal cord. The cell bodies of these sensory neurons are located outside the CNS in ganglia.

A **nerve** is a collection of axons outside the CNS. Most nerves are composed of both motor and sensory fibers and are thus called *mixed nerves.* Some of the cranial nerves, however, contain only sensory processes. These are the nerves that serve the special senses of sight, hearing, taste, and smell.

Neuroglia

There are six categories of neuroglial cells:

1. **Schwann cells,**[3] which form sheaths around peripheral axons
2. **Oligodendrocytes,** which form sheaths around axons of the CNS
3. **Microglia,** which are phagocytic cells that migrate through the CNS and remove foreign and degenerated material
4. **Astrocytes,**[4] which may help regulate the passage of molecules from the blood to the brain
5. **Ependyma,** which line the ventricles (cavities) of the brain and the central canal of the spinal cord
6. **Satellite cells,** which support neuron cell bodies within the ganglia of the PNS (table 6.2)

Sheath of Schwann and Myelin Sheath

All axons in the PNS, but not in the CNS, are surrounded by a living sheath of Schwann cells, known as the **sheath of Schwann.** The axons of the CNS, in contrast, lack a sheath of Schwann (because Schwann cells are found only in the PNS). This is significant in terms of nerve regeneration, as will be described in a later section.

Peripheral axons that are less than two micrometers in diameter are surrounded only by a sheath of Schwann. Such axons are said to be *unmyelinated.* Larger axons, in addition to their sheath of Schwann, are also surrounded by a **myelin sheath** and are thus said to be *myelinated.* Myelin is composed of successive wrappings of the cell membrane of Schwann cells. In the process of myelin formation, Schwann cells roll around the axon, much like the way a roll of electrician's tape is used to wrap a wire. Unlike electrician's tape, however, the wrappings are made in the same spot, so that each wrapping overlaps the others. The cytoplasm, meanwhile, becomes squeezed to the outer region of the Schwann cell, much as toothpaste is squeezed to the top of

[3]Schwann: from Theodor Schwann, German histologist, 1810–1882
[4]astrocyte: Gk. *aster,* star; *kytos,* hollow (cell)

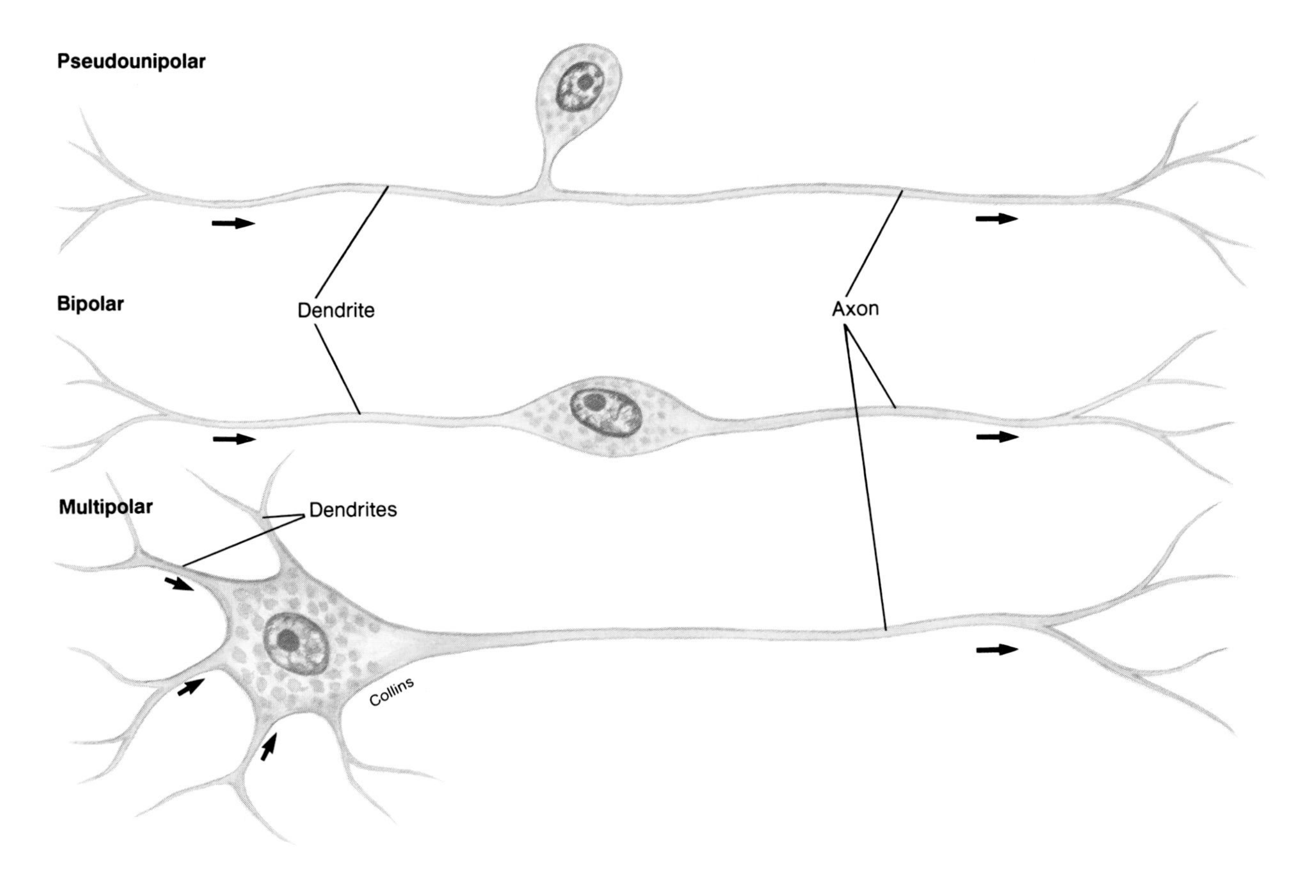

Figure 6.4. Three different types of neurons.

Table 6.2 Some types of neuroglial cells and their functions

Neuroglia	Functions
Schwann cells	Surround axons of all peripheral nerve fibers, forming neurilemmal sheath, or sheath of Schwann; wrap around many peripheral fibers to form myelin sheaths
Oligodendrocytes	Form myelin sheaths around central axons, producing "white matter" of CNS
Astrocytes	Perivascular foot processes, covering capillaries within brain and contributing to the blood-brain barrier
Microglia	Amoeboid cells within CNS that are phagocytic
Ependyma	Form epithelial lining of brain cavities (ventricles) and central canal of spinal cord; cover tufts of capillaries to form choroid plexus—structures that produce cerebrospinal fluid

the tube as the bottom is rolled up (fig. 6.5). The Schwann cells remain alive as their cytoplasm is squeezed to the outside of the myelin sheath. As a result, myelinated axons of the PNS, like their unmyelinated counterparts, are surrounded by a living sheath of Schwann (fig. 6.6).

Each Schwann cell only wraps about 1 mm of axon, leaving gaps of exposed axon between the adjacent Schwann cells. These gaps in the myelin sheath are known as the **nodes of Ranvier.**[5]

[5]nodes of Ranvier: from Louis A. Ranvier, French pathologist, 1835–1922

The successive wrappings of Schwann cell membrane provide insulation around the axon, leaving only the nodes of Ranvier exposed to produce nerve impulses.

The myelin sheaths of the CNS are formed by oligodendrocytes. Unlike a Schwann cell, which forms a myelin sheath around only one axon, each oligodendrocyte has extensions, like the tentacles of an octopus, that form myelin sheaths around several axons (fig. 6.7). The myelin sheaths around axons of the CNS give this tissue a white color; areas of the CNS that contain a high concentration of axons thus form the **white matter.** The **gray matter** of the CNS is composed of high concentrations of cell bodies and dendrites, which lack myelin sheaths.

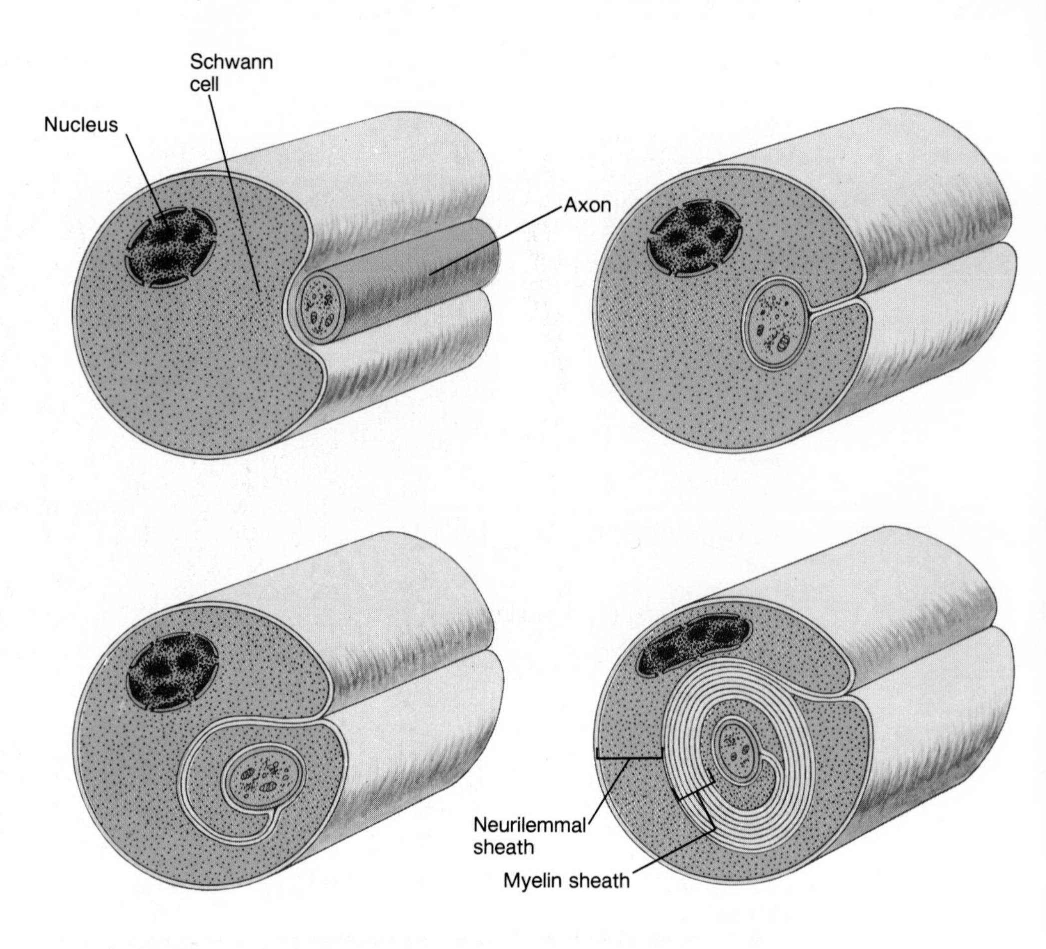

Figure 6.5. The formation of a myelin sheath in a peripheral axon. The myelin sheath is formed by successive wrappings of the Schwann cell membranes, leaving most of the Schwann cell cytoplasm outside the myelin. The neurilemmal sheath of Schwann cells is thus located outside the myelin sheath.

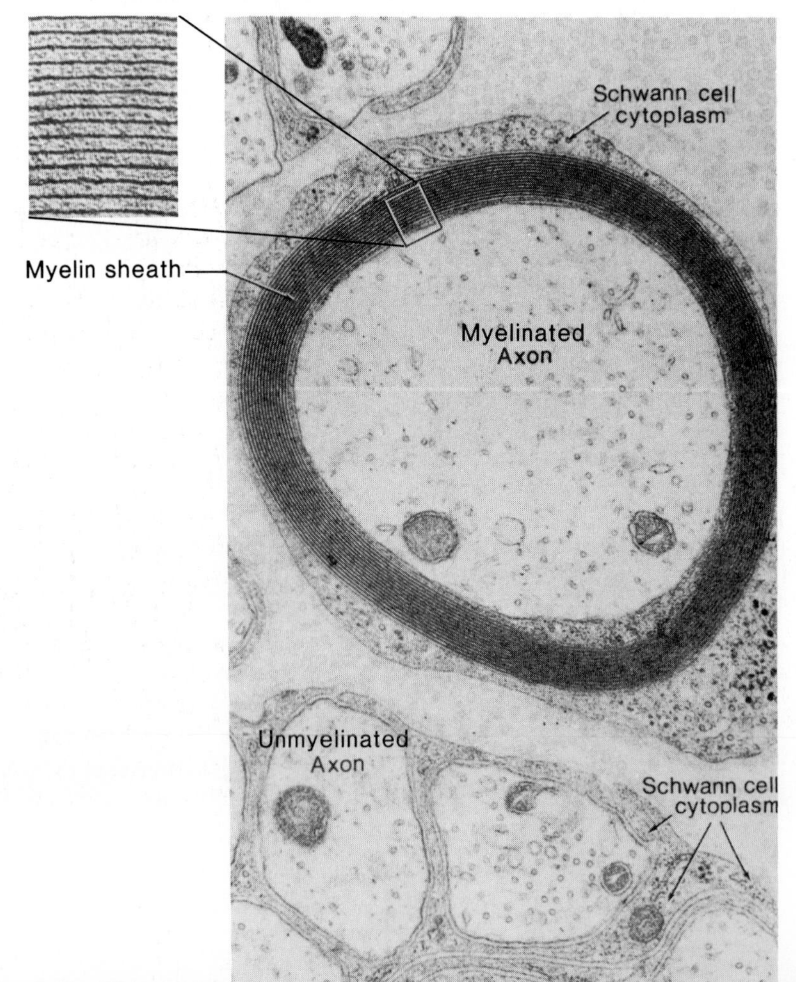

Figure 6.6. An electron micrograph of unmyelinated and myelinated axons.

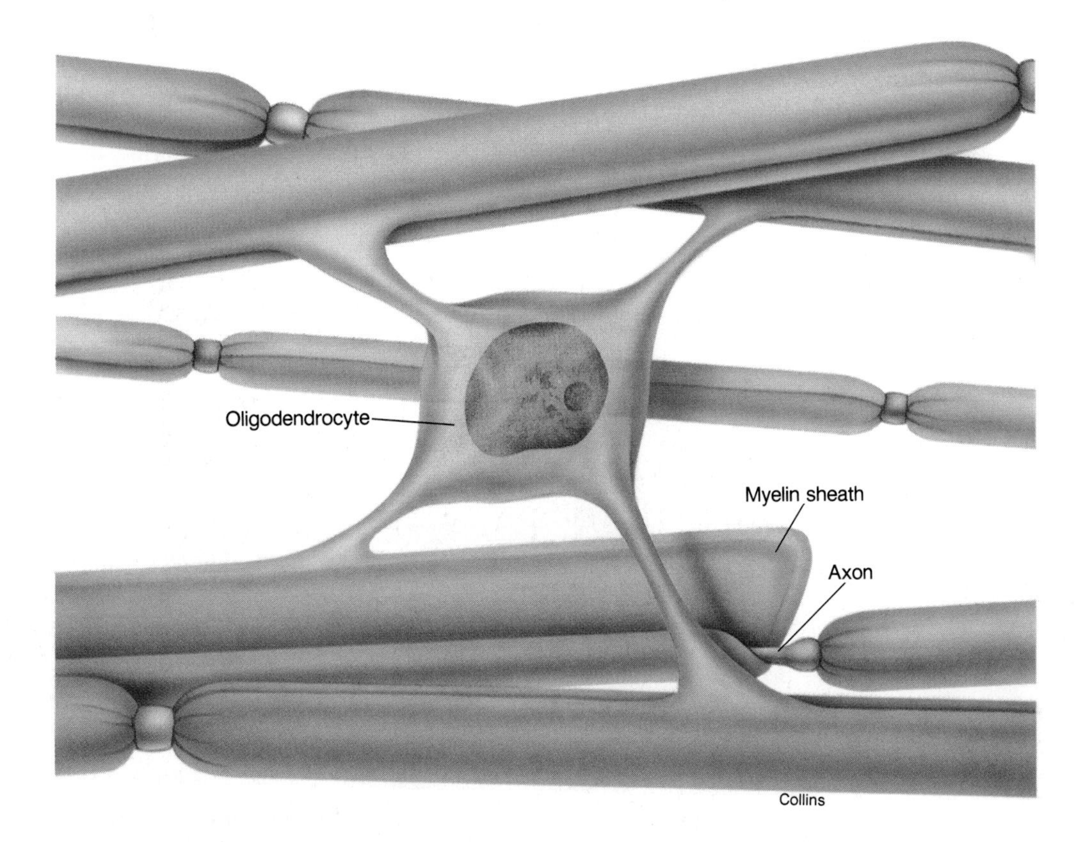

Figure 6.7. Formation of myelin sheaths in the central nervous system by an oligodendrocyte. One oligodendrocyte forms myelin around several axons.

Multiple sclerosis (MS) is a relatively common neurological disease in persons between the ages of twenty and forty. MS is a chronic, degenerating, remitting, and relapsing disease that progressively destroys the myelin sheaths of neurons in multiple areas of the CNS. Initially, lesions form on the myelin sheaths and soon develop into hardened *scleroses* (*skleros* = hardened). Destruction of the myelin sheaths prohibits the normal conduction of impulses, resulting in a progressive loss of functions. Because myelin degeneration is widespread, MS has a greater variety of symptoms than any other neurological disease. This characteristic, coupled with remissions, frequently causes misdiagnosis of this disease.

Regeneration of a Cut Axon

When an axon in a peripheral nerve is cut, the distal portion of the axon that was severed from the cell body degenerates and is phagocytosed by Schwann cells. The Schwann cells then form a *regeneration tube* (fig. 6.8) as the part of the axon that is connected to the cell body begins to grow and exhibit amoeboid movement. The Schwann cells of the regeneration tube are believed to secrete chemicals that attract the growing axon tip, and the regeneration tube helps to guide the regenerating axon to its proper destination. Even a severed major nerve may be surgically reconnected and the function of the nerve largely reestablished if the surgery is performed before tissue death.

Injury in the CNS stimulates the growth of short branches from the axon, but central axons have a much more limited ability to regenerate than peripheral axons. This is believed to be primarily due to the absence of Schwann cells, so that regeneration tubes cannot be formed.

The Blood-Brain Barrier

The blood capillaries in the brain are unique in that their walls lack pores. This is because the cells that comprise the capillary walls are joined by tight junctions, which prevent plasma (the fluid part of blood) from filtering out into the surrounding tissue. The capillaries of the brain, in addition, are surrounded by extensions of a type of neuroglial cell known as astrocytes (fig. 6.9). Astrocytes (*aster* = star) are large cells with numerous cytoplasmic processes that radiate outwards. They are the most abundant of the neuroglial cells in the CNS, constituting up to 90% of the nervous tissue in some areas of the brain.

Unlike other organs, therefore, the brain cannot obtain molecules from the blood plasma by a nonspecific filtering process. Instead, molecules within brain capillaries must be moved

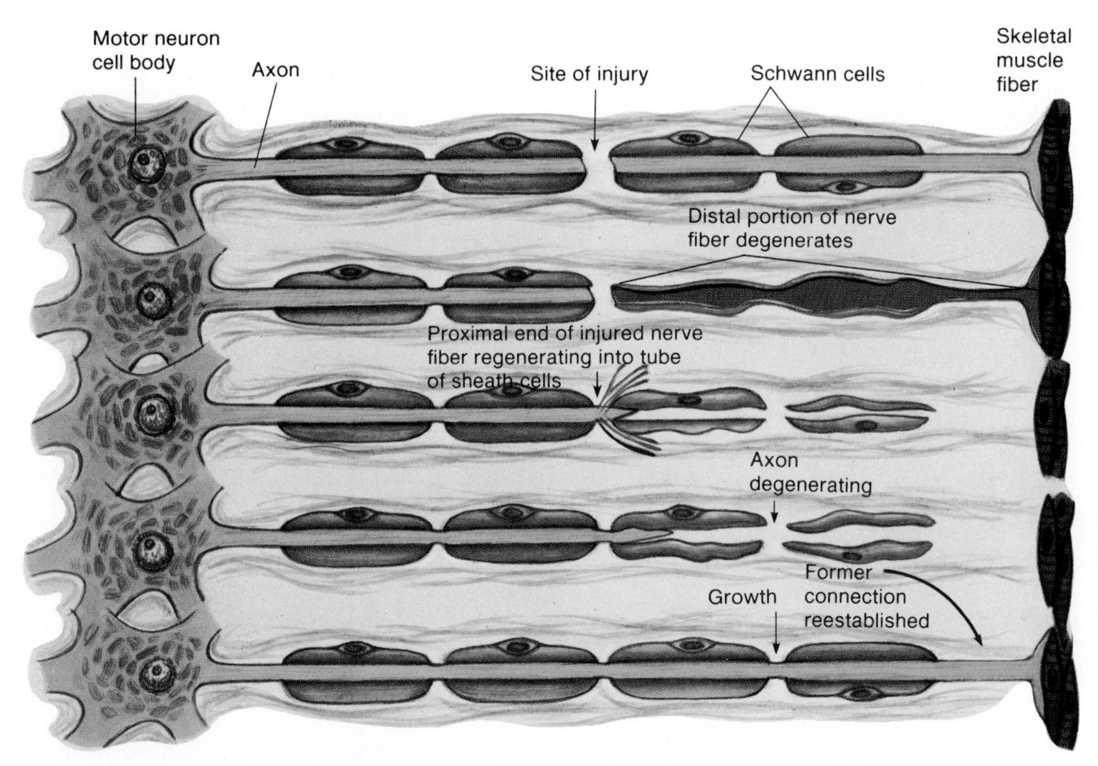

Figure 6.8. *The process of neuron regeneration.* (a) *If a neuron is severed through a myelinated axon, the proximal portion may survive, but* (b) *the distal portion degenerates through phagocytosis. The myelin sheath provides a pathway for* (c *and* d) *the regeneration of an axon, and* (e) *innervation is restored.*

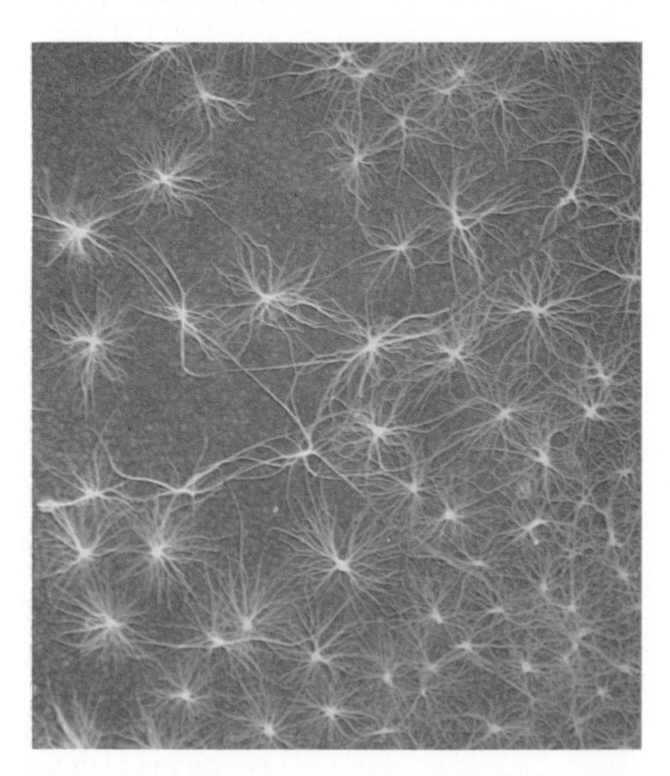

Figure 6.9. *A photomicrograph showing the perivascular feet of astrocytes (a type of neuroglial cell), which cover most of the surface area of brain capillaries.*

through the endothelial cells by diffusion, active transport, endocytosis, and exocytosis (chapter 4). This imposes a very selective **blood-brain barrier.** There is evidence to suggest that the development of tight junctions between adjacent endothelial cells in brain capillaries, and thus the development of the blood-brain barrier, results from the effects of astrocytes on the brain capillaries.

The blood-brain barrier presents difficulties in the chemotherapy of brain diseases because drugs that could enter other organs may not be able to enter the brain. In the treatment of Parkinson's disease, for example, patients who need a chemical called dopamine in the brain must be given a precursor molecule called levodopa (L-dopa). This is because dopamine cannot cross the blood-brain barrier, whereas L-dopa can enter the neurons and be changed to dopamine in the brain.

1. Draw a neuron, label its parts, and describe the functions of these parts.
2. Distinguish between sensory neurons, motor neurons, and interneurons in terms of their structure, location, and function.
3. Describe the structure of the sheath of Schwann and how it functions to promote nerve regeneration. Explain how a myelin sheath in the PNS is formed.
4. Explain the nature of the blood-brain barrier, and describe its significance.

The Nerve Impulse

There is a higher concentration of Na^+ outside cells than inside, and a higher concentration of K^+ inside the cells than outside. In addition to this difference in ion concentration, there is also a difference in charge between the two sides of the cell membrane. All cells are more negatively charged on the inside compared to the outside. This difference in charge is called the membrane potential. The axons of neurons alter their permeabilities to Na^+ and K^+ in response to stimulation. This produces a rapid change in the membrane voltage known as the action potential or nerve impulse.

The cell membrane of all cells acts as a barrier to the movement of ions. Proteins and organic phosphate molecules within the cells possess a negative charge, and are kept in the cell by the cell membrane. This makes the inside of cells more negative than the outside. This negative charge attracts small, positively charged ions from the extracellular fluid, but of these, only K^+ is able to pass through the membrane to an appreciable degree. As a result, there is a higher concentration of K^+ inside cells than in the extracellular fluid. Conversely, the cell membrane is relatively impermeable to Na^+, and the concentration of Na^+ in the extracellular fluid is much higher than it is inside of the cells.

The action of the Na^+/K^+ pumps (chapter 4) contributes to this concentration difference and to the difference in charge across the cell membrane. These active transport carriers pump three Na^+ ions out of the cell for every two K^+ ions that they pump into the cell. As a result, the inside of the cell is made even more negative in comparison to the outside (fig. 6.10).

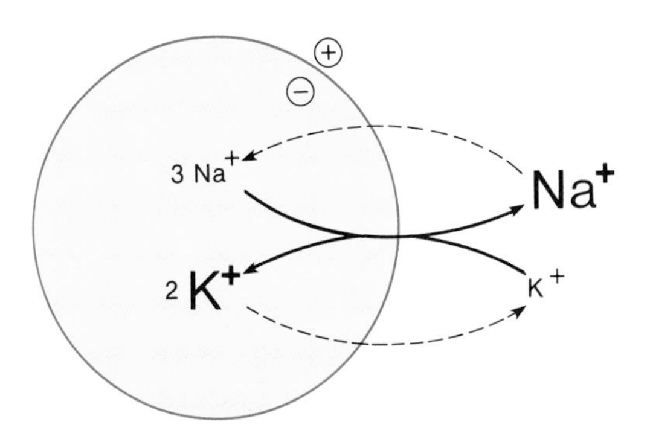

Figure 6.10. *The concentrations of Na⁺ and K⁺ both inside and outside the cell do not change as a result of diffusion* (dotted arrows) *because of active transport* (solid arrows) *by the Na⁺/K⁺ pump. Since the pump transports three Na⁺ for every two K⁺, the pump itself helps to create a charge separation (a potential difference, or voltage) across the membrane.*

The Resting Membrane Potential

In summary, the cell membrane separates two solutions—the cytoplasm on the inside of the cell and the plasma or tissue fluid outside the cell—that have different concentrations of charges. The inside of the cell is more negatively charged than the outside. The difference in charges between two solutions is called a **potential difference.** Since the potential difference occurs across a cell membrane, it can be referred to as a **membrane potential.** Potential differences are measured in units of *voltage.*

Every cell in the body has a membrane potential, but only neurons and muscle cells use this membrane potential to produce electrical impulses. When the neuron or muscle cell is at rest (not producing impulses), the membrane potential is called a **resting membrane potential.** The magnitude of the resting membrane potential can be measured by means of two electrodes connected to an oscilloscope.

In an oscilloscope, electrons from a cathode ray "gun" are sprayed across a fluorescent screen, producing a line of light. Changes in the potential difference between the two recording electrodes produce a deflection of this line. The oscilloscope can be calibrated in such a way that an upward deflection of this line indicates that the inside of the membrane has become less negative (or more positive) compared to the outside of the membrane. A downwards deflection of the line, conversely, indicates that the inside of the cell has become more negative. The oscilloscope can thus function as a fast-responding voltmeter with an ability to display voltage changes as a function of time.

If both recording electrodes were placed outside of the cell, the potential difference between the two would be zero (because there is no charge separation). When one of the two electrodes penetrates the cell membrane, the oscilloscope shows that the intracellular electrode is electrically negative with respect to the extracellular electrode; a membrane potential is recorded. The resting membrane potential of a typical neuron is -70 millivolts (mV). This number indicates the magnitude of the potential difference; the sign in front of the number indicates the polarity of the inside of the cell. (Note that if the inside of the cell were the positive pole, and the potential difference were the same, the membrane potential would be indicated as $+70$ mV.)

If appropriate stimulation causes positive charges to flow into the cell, the line on the oscilloscope would deflect upward; this change is called **depolarization,** because the potential difference between the two recording electrodes is reduced. If the inside of the membrane becomes more negative as a result of stimulation, the line on the oscilloscope would deflect downwards; this is called **hyperpolarization** (fig. 6.11).

The Action Potential

As may be recalled from chapter 4, small inorganic ions must pass through specific channels in the cell membrane. The permeability of the membrane to Na^+, K^+, and other ions is regulated by parts of these channels called **gates.** Gates are believed

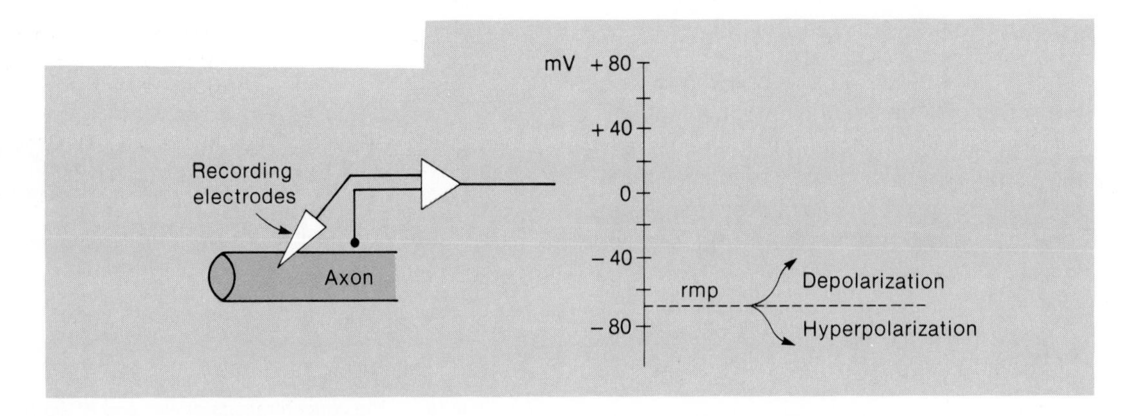

Figure 6.11. *The difference in potential (in millivolts) between an intracellular and extracellular recording electrode is displayed on an oscilloscope screen. The resting membrane potential (rmp) of the axon may be reduced (depolarization) or increased (hyperpolarization).*

to be composed of polypeptide chains that can open or close a membrane channel according to specific conditions. When the gates of specific ion channels are closed, the membrane is not very permeable to that ion, and when the gates are opened, the permeability to that ion is greatly increased.

The gates for Na^+ and K^+ in the axon membrane are open or closed according to the membrane potential. The gates are thus said to be *voltage regulated.* At the resting membrane potential of -70 mV, the gates for the Na^+ channels are closed, although a small amount of Na^+ can leak through. Some of the channels for K^+, however, are open. The resting cell is thus more permeable to K^+ than to Na^+. If the axon membrane is depolarized to a certain level the gated channels can open, and thus allow diffusion of one or the other ion.

Depolarization of a small region of an axon can be experimentally induced by a pair of stimulating electrodes that act as if they inject positive charges into the axon. If a pair of recording electrodes are placed in the same region (one electrode within the axon and one outside), an upward deflection of the oscilloscope line will be observed as a result of this depolarization. If a certain level of depolarization is achieved (from -70 mV to -55 mV, for example) by this artificial stimulation, a sudden and very rapid change in the membrane potential will be observed. This is due to the fact that *depolarization to a threshold level causes the Na^+ gates to open.* Now the permeability properties of the membrane are changed, and Na^+ diffuses down its concentration gradient into the cell.

A fraction of a second after the Na^+ gates open, they close again. At this time, *the depolarization stimulus then causes the K^+ gates to open.* This makes the membrane more permeable to K^+ than it is at rest, and K^+ diffuses out of the cell according to its concentration gradient. The K^+ gates will then close and the permeability properties of the membrane will return to what they were at rest.

Figure 6.12 (bottom) illustrates the movement of Na^+ and K^+ through the axon membrane in response to a depolarization stimulus. Notice that the explosive increase in Na^+ diffusion causes rapid depolarization to 0 mV and then *overshoot* of the membrane potential so that the inside of the membrane actually becomes positively charged (almost $+40$ mV) compared to the outside (fig. 6.12, top). The Na^+ permeability then rapidly decreases as the diffusion of K^+ increases, resulting in *repolarization* to the resting membrane potential. These changes in Na^+ and K^+ diffusion and the resulting changes in the membrane potential that they produce constitute an event called the **action potential,** or **nerve impulse.**

All-or-None Law

The amplitude of action potentials is **all-or-none.** When depolarization is below a threshold value, the voltage-regulated gates are closed and no action potential will be produced. When depolarization reaches a threshold level (-55 mV in the preceding example), a full action potential will be produced, reaching a maximum value of about $+40$ mV. A further increase in the strength of the stimulus will not produce a larger action potential. Since the change from -70 mV to $+40$ mV and back to -70 mV lasts only about three milliseconds (msec), the image of an action potential on an oscilloscope screen looks like a spike. Action potentials are therefore sometimes called *spike potentials.*

Coding for Stimulus Intensity

If one depolarization stimulus is greater than another, the greater stimulus strength is not coded by a greater amplitude of action potentials (because action potentials are all-or-none). The code for stimulus strength in the nervous system is not **AM** (amplitude modulated). When a greater stimulus strength is applied to a neuron, identical action potentials are produced more frequently (more are produced per minute). Therefore, the code for stimulus strength in the nervous system is frequency modulated **(FM).** This is shown in figure 6.13.

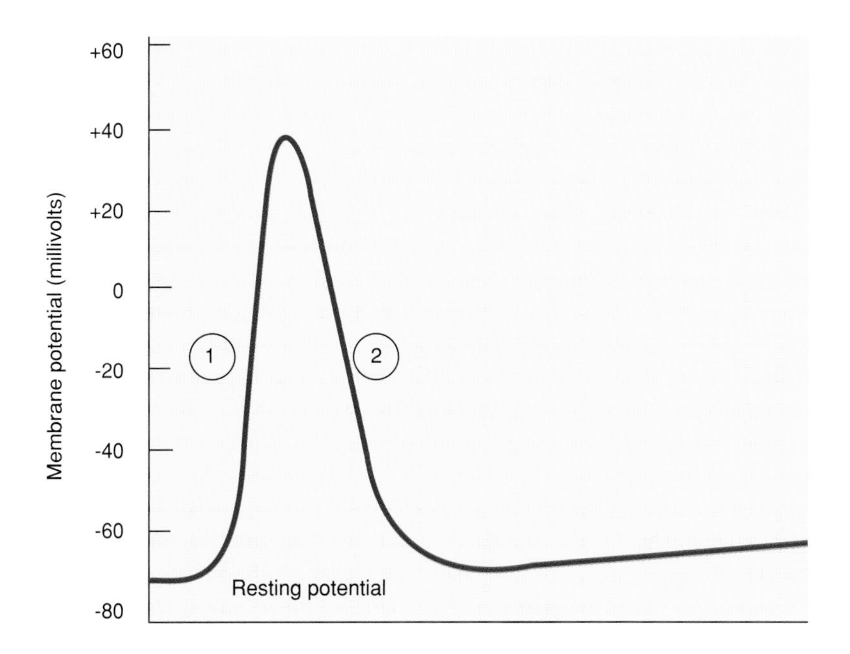

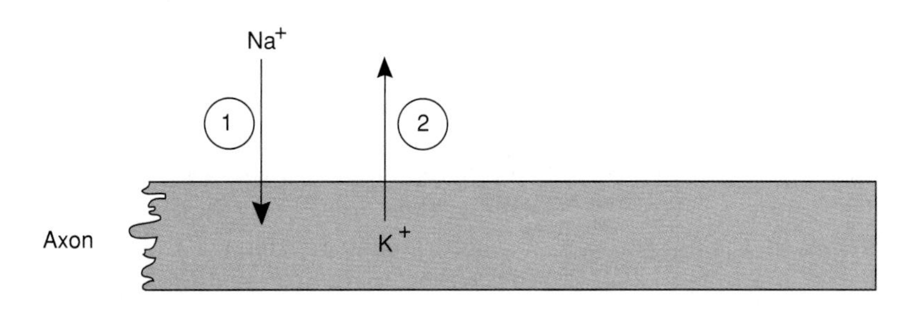

Figure 6.12. *The action potential (top) is produced by diffusion of first Na^+ and then K^+ through the membrane (bottom). Diffusion of Na^+ into the axon produces depolarization, and then diffusion of K^+ out of the axon produces repolarization.*

Keep in mind that action potentials are the means by which information is conveyed between different parts of the nervous system. The brain does *not* receive visual images, sounds, tastes, or smells; it receives action potentials traveling to it along sensory neurons from receptors for these sensory modalities. The brain interprets action potentials in the optic nerve as sight and in the auditory nerve as sound. The intensity of the stimulus—how loud was the thunder?—depends on the frequency of the action potentials in that particular nerve. If these nerves could somehow be switched, we would "see thunder and hear lightning." This short-circuiting of brain pathways can actually occur in a person hallucinating under the influence of a psychoactive drug.

Conduction of Nerve Impulses

When stimulating electrodes artificially depolarize one point of an axon membrane to a threshold level, voltage-regulated gates open and an action potential is produced at that small region of axon membrane containing those gates. For about the first millisecond of the action potential, when the membrane voltage changes from -70 mV to $+40$ mV, a current of Na^+ enters the cell by diffusion due to the opening of the Na^+ gates. Each action potential thus "injects" positive charges (Na^+ ions) into the axon. These positively charged Na^+ ions are conducted to an adjacent region that still has a membrane potential of -70 mV. This

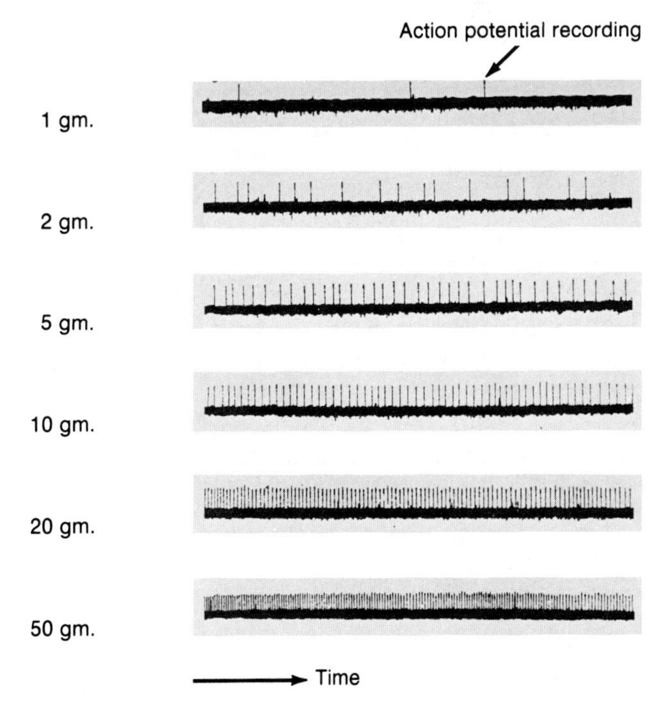

Figure 6.13. *Recordings from a single sensory fiber of a sciatic nerve of a frog stimulated by varying degrees of stretch of its calf muscle. Note that increasing amounts of stretch (indicated by increasing weights attached to the muscle) result in an increased frequency of action potentials.*

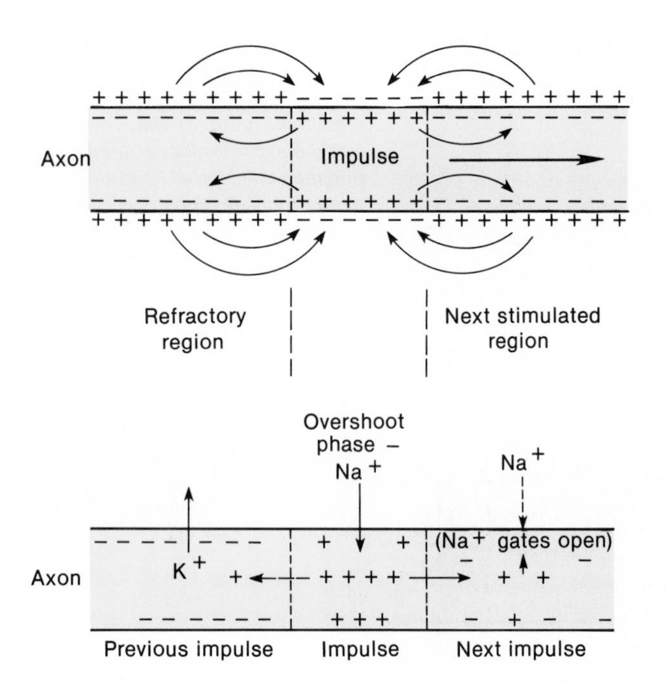

Figure 6.14. The conduction of a nerve impulse (action potential) in an unmyelinated nerve fiber (axon). Each action potential "injects" positive charges that spread to adjacent regions. The region that has previously produced an action potential is refractory. The previously unstimulated region is partially depolarized. As a result, its voltage-regulated Na^+ gates open, repeating the process.

helps to depolarize the adjacent region of axon membrane. When this adjacent region of membrane reaches a threshold level of depolarization, it too produces an action potential as its voltage-regulated gates open (fig. 6.14).

Each action potential thus acts as a stimulus for the production of another action potential at the next region of membrane. In the previous description of action potentials, the stimulus for their production was artificial—depolarization produced by a pair of stimulating electrodes. Now it can be seen that an action potential at one point along an axon is produced by depolarization that results from the production of a preceding action potential. This line of reasoning explains how all action potentials along an axon are produced after the first action potential is generated.

Notice that action potentials are not really conducted, although it is convenient to use that word. Each action potential is a separate, complete event that is repeated, or *regenerated,* along the axon's length. The action potential produced at the end of the axon is thus a completely new event that was produced in response to depolarization from the previous action potential. The last action potential has the same amplitude as the first. Action potentials are thus said to be **conducted without decrement** (without decreasing in amplitude).

The spread of depolarization within an axon is fast compared to the time involved in producing an action potential. Since action potentials are produced at every fraction of a micrometer

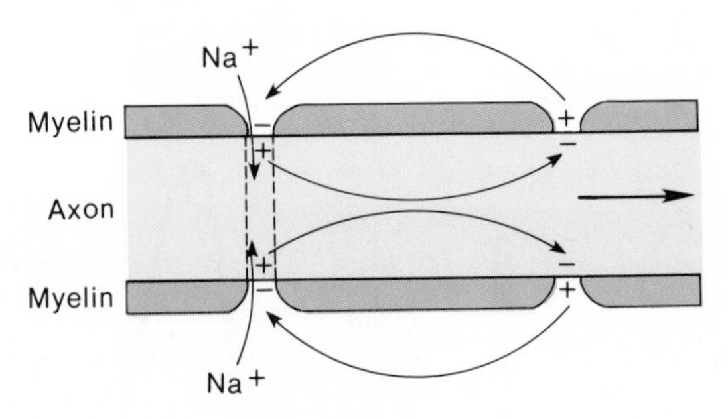

Figure 6.15. The conduction of the nerve impulse in a myelinated nerve fiber. Since the myelin sheath prevents inward Na^+ current, action potentials can only be produced at the interruptions in the myelin sheath, or nodes of Ranvier. This "leaping" of the action potential from node to node is known as saltatory conduction.

in an unmyelinated axon, the conduction rate is relatively slow. The conduction rate is substantially faster if the axon is myelinated.

The myelin sheath provides insulation for the axon, preventing movements of Na^+ and K^+ through the membrane. If the myelin sheath were continuous, therefore, action potentials could not be produced. Fortunately, there are interruptions in the myelin known as the nodes of Ranvier. Studies have shown that Na^+ channels are highly concentrated at the nodes (estimated at 10,000 per square micrometer) and almost absent in the regions of axon membrane between the nodes. Action potentials, therefore, occur only at the nodes of Ranvier (fig. 6.15) and seem to "leap" from node to node; this is called **saltatory conduction.**[6] Saltatory conduction allows a *faster rate of conduction* than is possible in an unmyelinated fiber.

1. Define the meaning of the terms *depolarization* and *repolarization,* and illustrate these processes with a figure.
2. Describe how the permeability of the axon membrane to Na^+ and K^+ is regulated and how changes in permeability to these ions affects the membrane potential.
3. Describe how gating of Na^+ and K^+ in the axon membrane results in the production of an action potential.
4. Define the all-or-none law of action potentials, and describe the effect of increased stimulus strength on action potential production.
5. Describe how action potentials are conducted by unmyelinated nerve fibers, and explain why saltatory conduction in myelinated fibers is more rapid.

Synaptic Transmission

At the synapse of one neuron with another there is usually a small gap separating the cells. Because of this gap, transmission across the synapse must occur chemically. The endings of axons release chemicals called neurotransmitters.

[6]saltatory: L. *saltario,* to leap

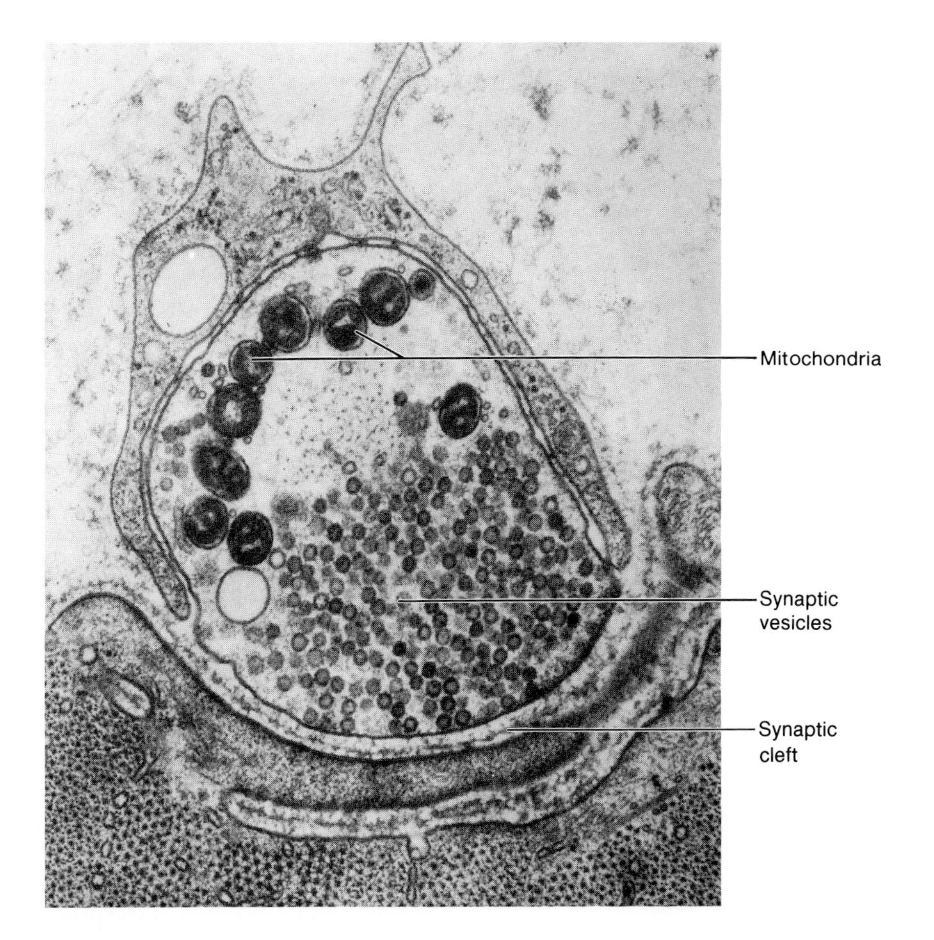

Figure 6.16. *An electron micrograph of a chemical synapse, showing synaptic vesicles at the end of an axon.*

Acetylcholine (ACh) is one such neurotransmitter, released at the neuromuscular junction and at some neuron-neuron synapses. When ACh reaches the postsynaptic cell membrane, it causes a graded depolarization that, in neurons, is known as an excitatory postsynaptic potential (EPSP). The EPSP, in turn, stimulates the production of action potentials in the postsynaptic cell.

A **synapse** is the functional connection between a neuron and a second cell, and serves as the means by which the first (*presynaptic*) neuron can influence the second (*postsynaptic*) cell. In the CNS this postsynaptic cell is also a neuron. In the PNS the postsynaptic cell may be a neuron or an *effector cell* within either a muscle or a gland. Although the physiology of neuron-neuron synapses and neuron-muscle synapses is similar, the latter synapses are often distinguished by the name **neuromuscular junctions.**

Chemical Synapses

Transmission across the majority of synapses in the nervous system is one-way and occurs through the release of chemical neurotransmitters from presynaptic axon endings. These presynaptic endings, which are called **terminal boutons**[7] because of their swollen appearance, are separated from the postsynaptic cell by a **synaptic cleft** so narrow that it can only be seen clearly with an electron microscope (fig. 6.16).

Neurotransmitter molecules within the presynaptic neuron endings are contained within many small, membrane-enclosed **synaptic vesicles.** In order for the neurotransmitter within these vesicles to be released into the synaptic cleft, the vesicle membrane must fuse with the axon membrane in the process of *exocytosis.* The neurotransmitter molecules are then free to diffuse to the postsynaptic cell.

Acetylcholine at the Neuromuscular Junction

The neuromuscular junction—the synapse between a somatic motor neuron and a skeletal muscle fiber—is the most accessible synapse to study and thus the best understood. The presynaptic neuron endings have synaptic vesicles, which each contain about ten thousand molecules of ACh. Once these molecules are released by exocytosis, they quickly diffuse to the membrane of the postsynaptic cell. Here they chemically bond

[7]bouton: Fr. *bouton,* button

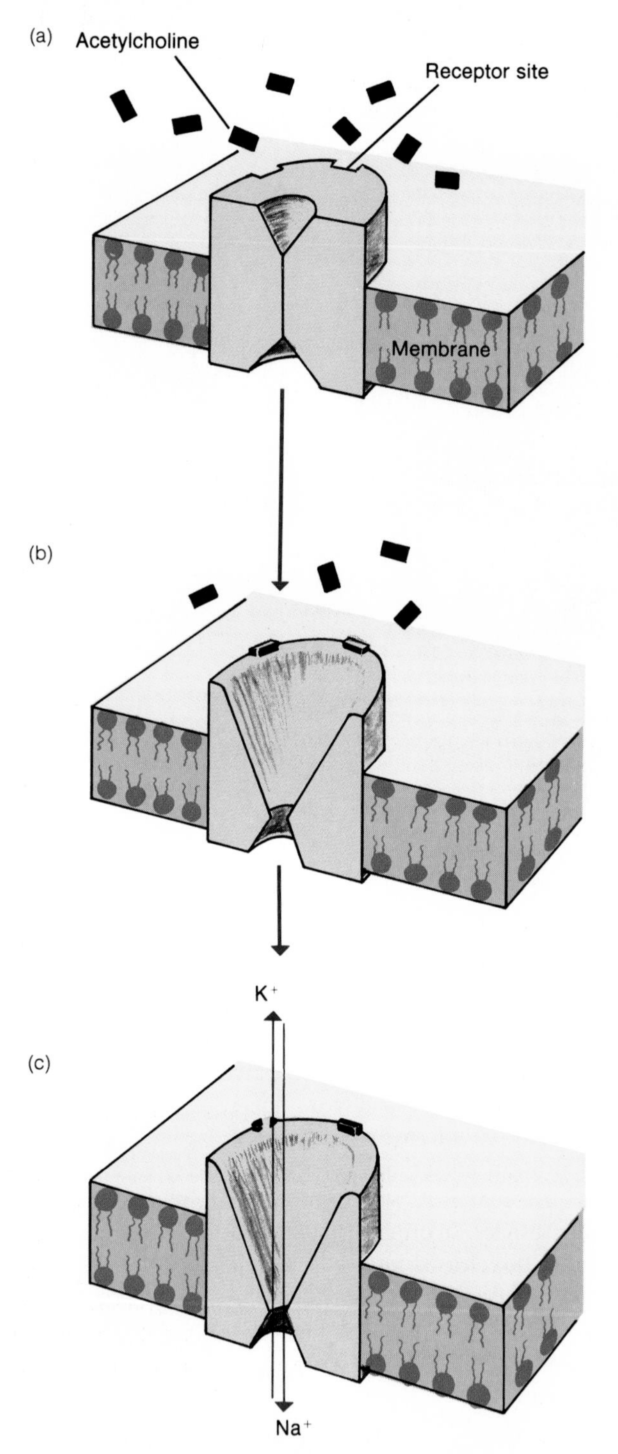

Figure 6.17. (a and b) The binding of acetylcholine to receptor proteins causes the opening of chemically regulated gates in the postsynaptic membrane. (c) This results in the increased diffusion of Na^+ and K^+ through the membrane.

to **receptor proteins** that are built into the postsynaptic membrane. These receptor proteins combine with ACh in a specific manner, analogous to the specific interaction between transport proteins and their substrates (chapter 4).

Acetylcholine is not, however, transported into the postsynaptic cell. Instead, the bonding of ACh to its receptor protein causes changes in membrane structure that result in the opening of ion channel gates for Na^+ and K^+ (fig. 6.17). These gates, located only in the postsynaptic membrane, are called **chemically regulated gates** because they open in response to bonding by a chemical (ACh). Opening of the chemically regulated gates allows Na^+ to enter the muscle cell and depolarize it. This depolarization, in turn, serves as a stimulus for the opening of voltage-regulated gates, so that the muscle cell is stimulated to produce action potentials. As described in chapter 21, these action potentials stimulate the muscle to contract.

Muscle weakness in the disease **myasthenia[8] gravis** is due to the fact that ACh receptors are blocked and destroyed by antibodies secreted by the immune system of the affected person. Paralysis in people who eat shellfish poisoned with saxitoxin, produced by unicellular organisms that cause the red tides, results from the blockage of Na^+ gates.

Acetylcholinesterase

The bond between ACh and its receptor protein exists for only a brief instant. The ACh-receptor complex quickly dissociates but can be quickly reformed as long as free ACh is in the vicinity. In order for activity in the postsynaptic cell to be controlled (in this case, for control of skeletal muscle contraction), free ACh must be inactivated very soon after it is released. The inactivation of ACh is achieved by means of an enzyme called **acetylcholinesterase,** or **AChE,** which is on the postsynaptic membrane or immediately outside the membrane, with its active site facing the synaptic gap (fig. 6.18).

Nerve gas exerts its odious effects by inhibiting AChE in skeletal muscles. Since ACh is not degraded, it can continue to combine with receptor proteins and can continue to stimulate the postsynaptic cell, leading to spastic paralysis. Clinically, drugs such as *neostigmine,* which inhibit AChE, are used to enhance the effects of ACh on muscle contraction when neuromuscular transmission is weak, as in the disease myasthenia gravis.

If any stage in the process of neuromuscular transmission is blocked, muscle weakness—sometimes leading to paralysis and death—may result. The drug *curare,* for example, competes with ACh for attachment to the receptor proteins at the muscle cell membrane. This drug was first used on poison darts by South American Indians because it produced flaccid paralysis of their victims. Clinically, curare is used as a muscle relaxant during anesthesia and to prevent muscle damage during electroconvulsive shock therapy.

[8]myasthenia: Gk. *myos,* muscle; *asthenia,* weakness

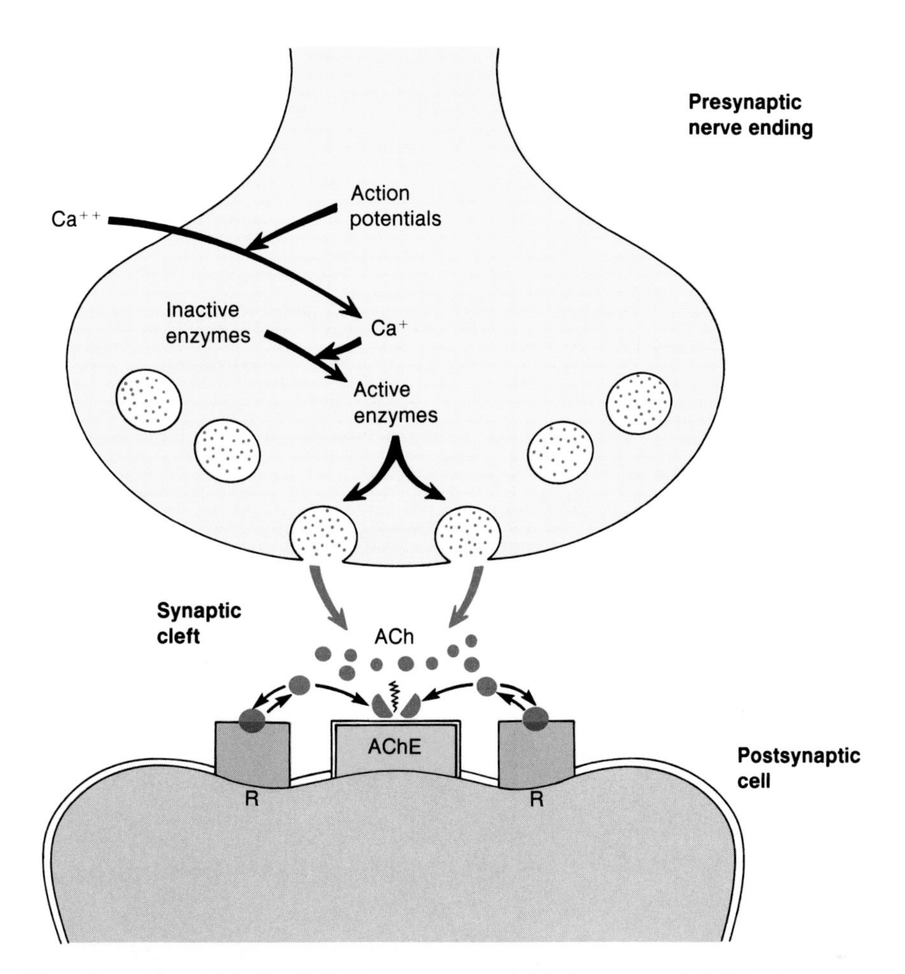

Figure 6.18. *Mechanisms of the release of acetylcholine (ACh) from presynaptic nerve endings and the binding of ACh to receptor proteins (R) in the postsynaptic membrane. Acetylcholine that combines with acetylcholinesterase (AChE) in the postsynaptic membrane is hydrolyzed and thus inactivated.*

Acetylcholine at Synapses between Neurons

Within the central nervous system, the axon terminals of one neuron typically synapse with the dendrites or cell body of another. The dendrites and cell body thus serve as the receptive area of the neuron, and it is in these regions that receptor proteins for neurotransmitters and chemically regulated gates are located. The first voltage-regulated gates are located at the beginning of the axon. It is therefore here that action potentials are first produced (fig. 6.19).

Release of ACh at the synapses results, through the opening of chemically regulated gates, in the production of depolarizations in the dendrites and cell body. These depolarizations in neurons are known as **excitatory postsynaptic potentials (EPSPs).** Unlike action potentials, which are all-or-none, EPSPs can be *graded*—that is, they can vary in their amplitude. Also, unlike action potentials, EPSPs can undergo *summation*: their amplitudes can be added together. If the depolarization is at or above threshold by the time it reaches the beginning of the axon, the EPSP will stimulate the production of action potentials, which can then regenerate themselves along the axon. If, however, the EPSP is below threshold at the axon hillock, no action potentials will be produced in the postsynaptic cell (fig. 6.20).

A neuron may receive synaptic inputs from as many as one thousand other neurons. The excitation of such a neuron requires the summation of many EPSPs before a threshold depolarization is produced at the initial segment of the axon. Summation is thus an extremely important process that can only occur between graded synaptic potentials (action potentials cannot summate, because of their all-or-none properties). Summation of EPSPs allows the postsynaptic cell to "weigh" different combinations of synaptic inputs and respond appropriately.

---

1. Define a synapse, and describe the structures that form a synapse.
2. Describe the events that occur between the release of ACh and the production of action potentials in a skeletal muscle fiber.
3. Describe the function of acetylcholinesterase and its physiological significance.
4. Compare the properties of EPSPs and action potentials, and describe where these events occur in a postsynaptic neuron.
5. Explain how EPSPs produce action potentials in the postsynaptic neuron.

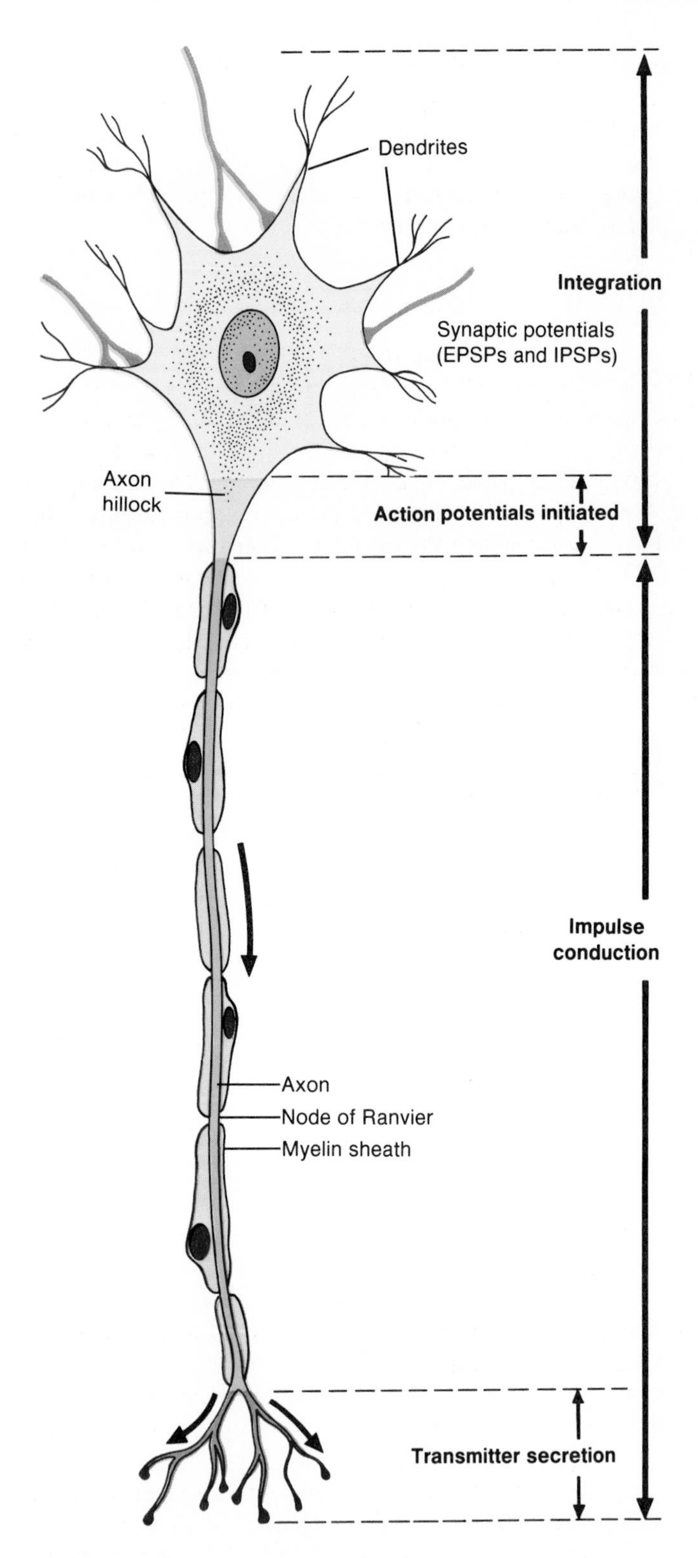

Figure 6.19. *A diagram illustrating the functional specialization of different regions in a "typical" neuron.*

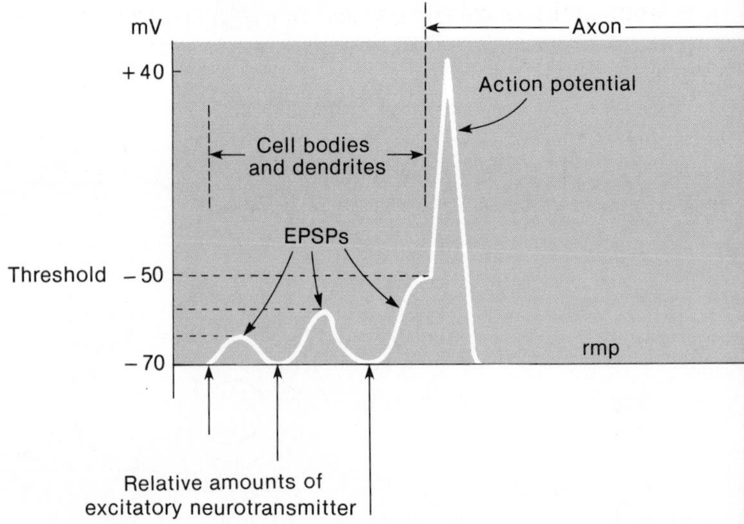

Figure 6.20. *The graded nature of excitatory postsynaptic potentials is shown, in which stimuli of increasing strength produce increasing amounts of depolarization. When a threshold level of depolarization is produced, action potentials are generated in the axon.*

Neurotransmitters in the Brain

In addition to acetylcholine, other chemicals are used as neurotransmitters by neurons in the brain. These include smaller molecules and a variety of polypeptides. Among the polypeptide neurotransmitters is a group of molecules, known as endorphins, which have morphinelike effects. Many drugs, including those used medically and those that are illegally abused, affect the brain by modifying the action of specific neurotransmitters.

Acetylcholine is not only the neurotransmitter of somatic motor neurons and some autonomic neurons of the PNS, it is also an important neurotransmitter within the brain. Since the effects of ACh were discussed previously, this section is devoted primarily to CNS neurotransmitters other than acetylcholine.

Alzheimer's disease is the most common cause of senile dementia, which often begins in middle age and produces progressive mental deterioration. The cause of Alzheimer's disease is not known, but there is evidence that it is associated with a loss of cholinergic neurons (those that use acetylcholine as a neurotransmitter), which terminate in areas of the brain concerned with memory storage.

In addition to acetycholine, there is a great variety of other molecules that serve as neurotransmitters in the CNS. **Catecholamines** are a group of regulatory molecules, derived from the amino acid tyrosine, that include *dopamine, norepinephrine,* and *epinephrine.* Dopamine and norepinephrine function as neurotransmitters; epinephrine and norepinephrine additionally serve as hormones secreted by the adrenal medulla gland (chapter 7).

The catecholamines, together with a related molecule called *serotonin,* are included in a larger category of molecules called *monoamines.* These and other molecules provide the diversity needed for the complex functions of the brain.

Cocaine is a stimulant, related to the amphetamines in its action, which is currently widely abused in the United States. Although early use of this drug produces feelings of euphoria and social adroitness, continued use leads to social withdrawal, depression, dependence upon ever-higher dosages, and serious organic disease that often results in death. It is surprising that the many effects of cocaine on the central nervous system appear to be mediated by one primary mechanism: cocaine blocks the reuptake of the neurotransmitter dopamine from the synaptic gap into the presynaptic axon endings. Cocaine thus results in overstimulation of those neural pathways that use dopamine as a neurotransmitter.

Dopamine and Norepinephrine

Neurons that use **dopamine** as a neurotransmitter are known as *dopaminergic* neurons. These neurons are highly concentrated in the *substantia nigra* (literally, the "dark substance," so-called because it contains melanin pigment) of the brain. Many neurons in the substantia nigra send fibers to the *basal ganglia,* which are large masses of cell bodies deep in the cerebrum that are involved in the coordination of skeletal movements. There is much evidence that **Parkinson's disease**[9] is caused by degeneration of the dopaminergic neurons in the substantia nigra. Parkinson's disease, a major cause of neurological disability in people over sixty years of age, is associated with such symptoms as muscle tremors and rigidity, difficulty in the initiation of movements and in speech, and other severe problems. These patients are frequently given L-dopa, as previously described.

A side effect of L-dopa treatment in some patients with Parkinson's disease is the appearance of symptoms characteristic of **schizophrenia.** This effect is not surprising, in view of the fact that the drugs used to treat schizophrenic patients (chlorpromazine and related compounds) act as specific antagonists of dopamine receptors. As might be predicted from these observations, schizophrenic patients treated with these drugs often develop symptoms of Parkinson's disease. It seems reasonable to suppose, from this evidence, that schizophrenia may be caused, at least in part, by overactivity of the dopaminergic pathways. This hypothesis is strengthened by the additional evidence that the brains of schizophrenic patients appear to have an abnormally high content of dopamine receptors.

Norepinephrine, like ACh, is used as a neurotransmitter in both the PNS and the CNS. Sympathetic neurons of the PNS use norepinephrine as a neurotransmitter at their synapse with smooth muscles, cardiac muscle, and glands. Some neurons in the CNS also appear to use norepinephrine as a neurotransmitter; these neurons seem to be involved in general behavioral arousal. This would help to explain the effects of such drugs as *amphetamines,* which specifically stimulate pathways that use norepinephrine as a neurotransmitter.

Amino Acids as Neurotransmitters

The amino acids **glutamic acid** and **aspartic acid** function as excitatory neurotransmitters in some neurons of the CNS. The amino acid **glycine,** in contrast, is inhibitory; instead of depolarizing the postsynaptic membrane and producing an EPSP, it hyperpolarizes the postsynaptic membrane. This hyperpolarization makes the membrane potential even more negative than it is at rest (changing the membrane potential from -70 mV to, for example, -85 mV). Such hyperpolarization is known as an **inhibitory postsynaptic potential (IPSP).** This inhibitory effect of glycine at synapses within the spinal cord is extremely important in the control of skeletal movements.

The neurotransmitter **GABA (gamma-aminobutyric acid)** is a derivative of the amino acid glutamic acid. GABA is the most prevalent neurotransmitter in the brain; in fact, as many as one-third of all the neurons in the brain use GABA as a neurotransmitter. Like glycine, GABA is an inhibitory neurotransmitter. Also like glycine, some neurons that use GABA are involved in motor control. A deficiency in those neurons that release GABA as a neurotransmitter produces the uncontrolled movements seen in people with *Huntington's chorea.*

In addition to its involvement in motor control, GABA also appears to function as a neurotransmitter involved in mood and emotion. Drugs given to treat extreme anxiety—including the *benzodiazepines* (e.g., Valium)—act by increasing the effectiveness of GABA in the brain. Drugs that antagonize the actions of GABA, conversely, can produce feelings of extreme anxiety.

Endorphins

The ability of opium and its analogues—that is, the *opioids*—to relieve pain (promote analgesia) has been known for centuries. Morphine, for example, has long been used for this purpose. The discovery in 1973 of opioid receptor proteins in the brain suggested that the effects of these drugs might be due to the stimulation of specific neuron pathways. This implied that opioids—like LSD, mescaline, and other mind-altering drugs—might resemble neurotransmitters normally produced by the brain.

---

[9]Parkinson's disease: from James Parkinson, English physician, 1755–1824

These compounds have been identified as a family of chemicals called **endorphins** (for "endogenously produced morphine-like compounds") produced by the brain and pituitary gland. The endorphins include a group of five-amino-acid peptides called **enkephalins,** which may function as neurotransmitters, and a thirty-one-amino-acid polypeptide, produced by the pituitary gland, called *β-endorphin.*

Endorphins have been shown to block the transmission of pain. They also may provide pleasant sensations and thus mediate reward or positive reinforcement pathways. It has been found that blood levels of β-endorphin are increased in exercise. Some people have suggested that the "jogger's high" may thus be due to endorphins. Although evidence for this particular effect is poor, it does appear that endorphins may promote some type of psychic reward system as well as analgesia.

---

1. Explain the relationship between Alzheimer's disease and acetylcholine.
2. Describe the known significance of dopaminergic neurons, and explain how their function relates to Parkinson's disease and schizophrenia.
3. Explain how cocaine and the amphetamines may act in the brain.
4. Explain the effects of glycine and GABA in the brain.
5. Describe the nature and explain the proposed functions of the endorphins.

The Cerebral Hemispheres of the Brain

The brain is composed of an enormous number of association neurons, with accompanying neuroglia, arranged into brain regions and divisions. These neurons receive sensory information, direct the activity of motor neurons, and perform such higher brain functions as learning and memory. Analysis of sensory information, association of that information with past experience and emotional responses, and even self-awareness and consciousness may derive from complex interactions of different brain regions.

The central nervous system, consisting of the brain and spinal cord (fig. 6.21), receives input from *sensory neurons* and directs the activity of *motor neurons,* which innervate muscles and glands. The *association neurons* within the brain and spinal cord are in a position, as their name implies, to associate appropriate motor responses with sensory stimuli and, thus, to maintain homeostasis. Further, the central nervous systems of all vertebrates (and most invertebrates) are capable of at least rudimentary forms of learning and retaining the memories of past experiences. This ability is most highly developed in the human brain, and permits behavior to be modified by experience. Perceptions, learning, memory, emotions, and the self-awareness that forms the basis of consciousness are creations of the brain. Whimsical though it seems, the study of brain physiology is thus the process of the brain studying itself.

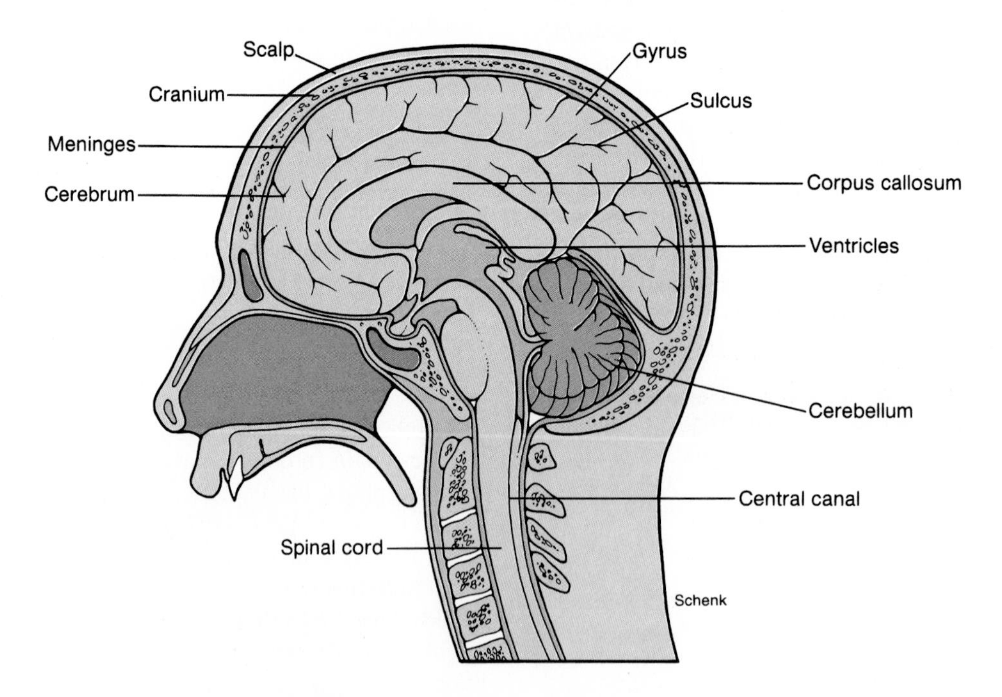

Figure 6.21. *The CNS consists of the brain and the spinal cord, both of which are covered with meninges and bathed in cerebrospinal fluid.*

The **cerebrum** (fig. 6.22) is the largest portion of the brain (accounting for about 80% of its mass) and is believed to be the brain region primarily responsible for higher mental functions. The cerebrum consists of *right* and *left hemispheres,* which are connected internally by a large fiber tract called the *corpus callosum* (see fig. 6.21).

The cerebrum consists of an outer **cerebral cortex,** which is 2–4 mm of gray matter, and underlying white matter. The surface of the cerebrum is folded into convolutions. The elevated folds of the convolutions are called *gyri,*[10] and the depressed grooves are the *sulci.*[11] Each cerebral hemisphere is subdivided by deep sulci, or *fissures,* into five lobes, four of which are visible from the surface (fig. 6.23). These lobes are the *frontal, parietal, temporal,* and *occipital,* which are visible from the surface, and the deep *insula* (table 6.3).

The **frontal lobe** is the anterior portion of each cerebral hemisphere. A deep fissure, called the *central sulcus,* separates the frontal lobe from the **parietal lobe.** The *precentral gyrus,* (see fig. 6.22) involved in motor control, is located in the frontal lobe just in front of the central sulcus. The *postcentral gyrus,* which is located just behind the central sulcus in the parietal lobe, is the primary area of the cortex responsible for the perception of *somatesthetic sensation*—sensation arising from cutaneous, muscle, tendon, and joint receptors.

The precentral (motor) and postcentral (sensory) gyri have been mapped in conscious patients undergoing brain surgery. Electrical stimulation of specific areas of the precentral gyrus was found to cause specific movements, and stimulation of different areas of the postcentral gyrus evoked sensations in specific parts of the body. Typical maps of these regions (fig. 6.24) show an upside-down picture of the body, with the superior regions of cortex devoted to the toes and the inferior regions devoted to the head.

A striking feature of these maps is that the area of cortex responsible for different parts of the body do not correspond to the size of the body parts being served. Instead, the body regions with the highest densities of receptors are represented with the largest areas of the sensory cortex, and the body regions with the greatest number of motor innervations are represented

[10]gyrus: Gk. *gyros,* circle
[11]sulcus: L. *sulcus,* a furrow or ditch

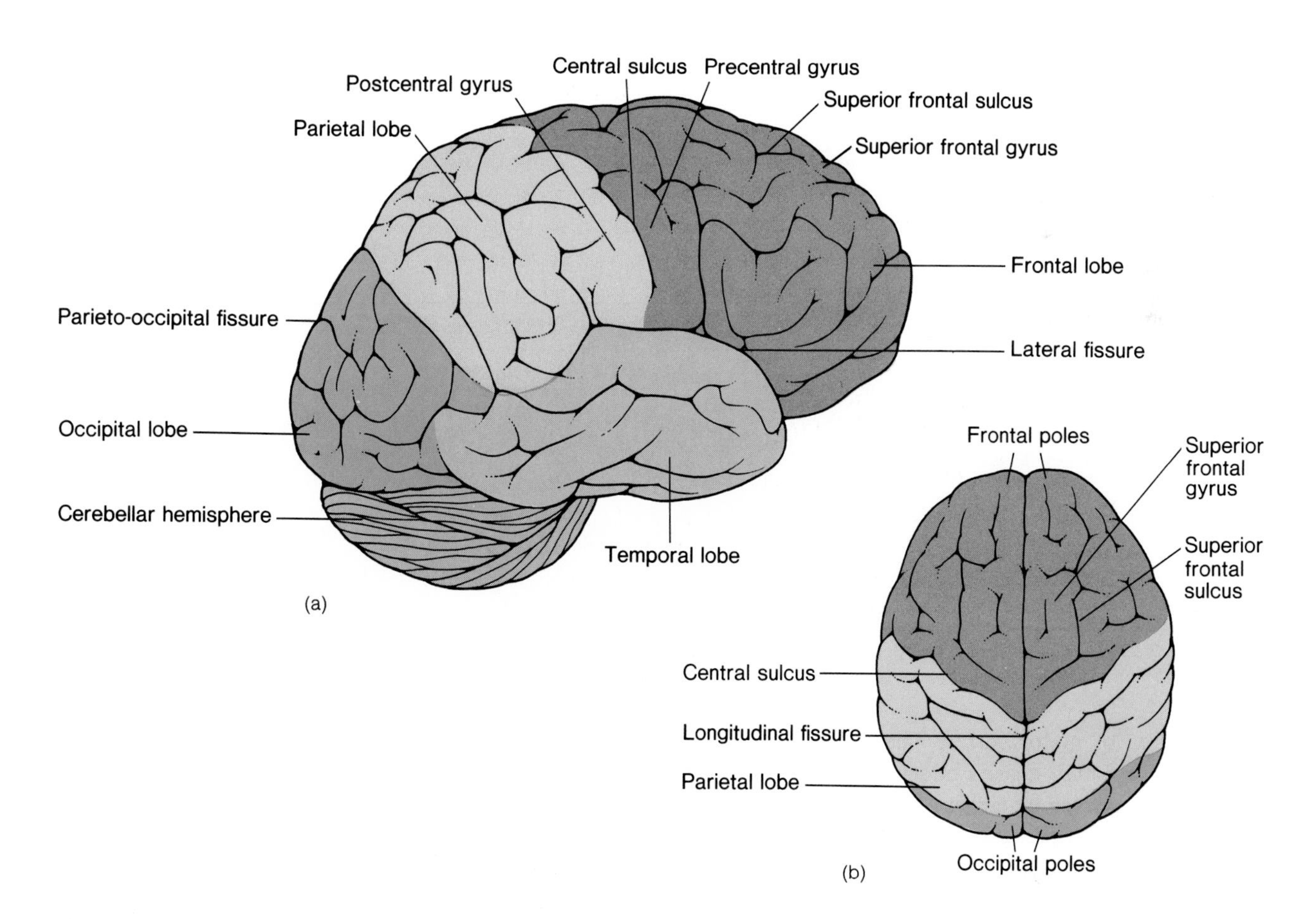

Figure 6.22. *The cerebrum. (a) A lateral view; (b) a superior view.*

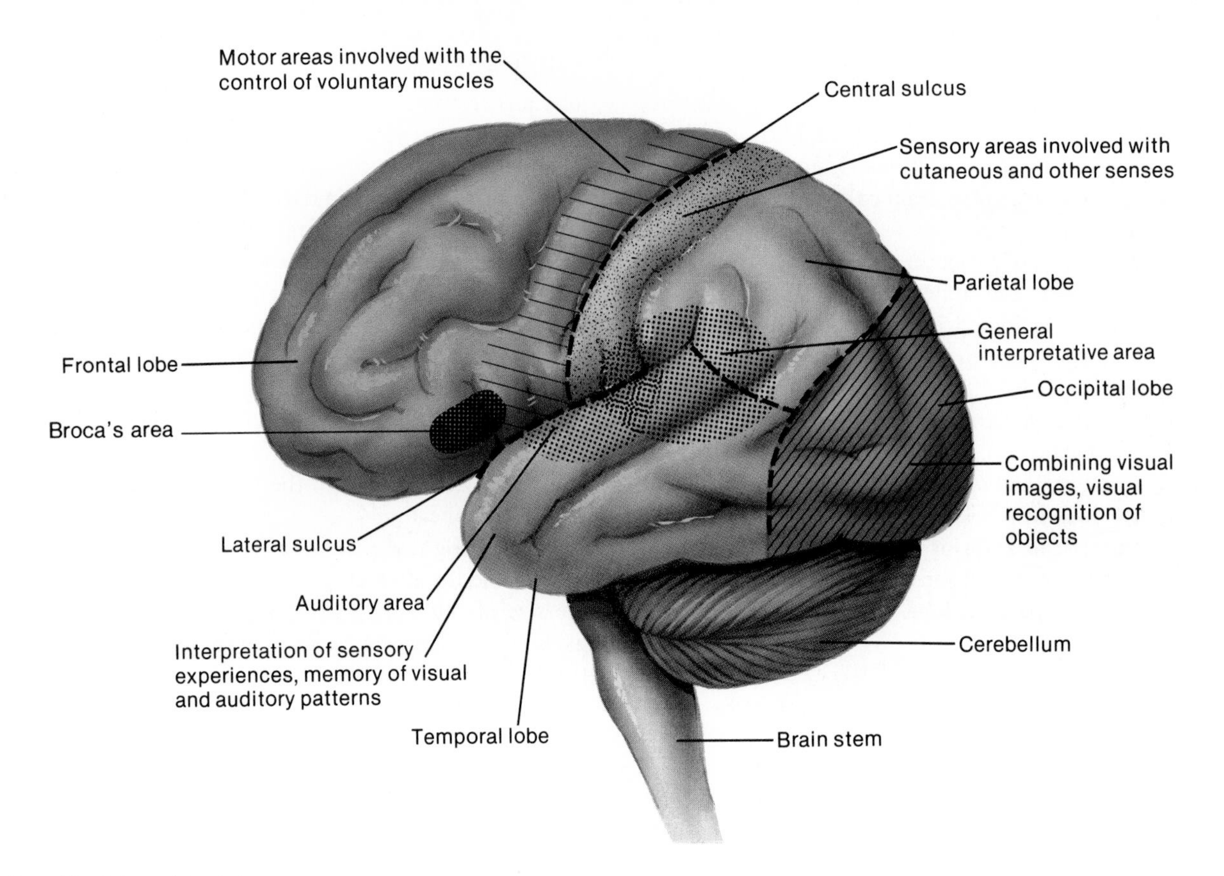

Figure 6.23. *The lobes of the left cerebral hemisphere showing the principal motor and sensory areas of the cerebral cortex. Dotted lines show the division of the lobes.*

Table 6.3 Functions of the cerebral lobes

Lobe	Functions	Lobe	Functions
Frontal	Voluntary motor control of skeletal muscles; personality; higher intellectual processes (e.g., concentration, planning, and decision making); verbal communication	Temporal	Interpretation of auditory sensations; storage (memory) of auditory and visual experiences
Parietal	Somatesthetic interpretation (e.g., cutaneous and muscular sensations); understanding speech and formulating words to express thoughts and emotions; interpretation of textures and shapes	Occipital	Integrates movements in focusing the eye; correlating visual images with previous visual experiences and other sensory stimuli; conscious perception of vision
		Insula	Memory; integrates other cerebral activities

with the largest area of motor cortex. The hands and face, therefore, which have a high density of sensory receptors and motor innervation, are served by larger areas of the precentral and postcentral gyri than is the rest of the body.

The **temporal lobe** contains auditory centers that receive sensory fibers from the inner ear. This lobe is also involved in the interpretation and association of auditory and visual information. The **occipital lobe** is the primary area responsible for vision and for the coordination of eye movements.

Cerebral Lateralization

The two hemispheres of the cerebral cortex control, via motor fibers originating in the precentral gyrus, skilled movements of the contralateral (opposite) side of the body. Also, sensation from the right side of the body projects to the left postcentral gyrus, and vice versa, because of *decussation* (crossing-over) of sensory fibers. Each cerebral hemisphere, however, receives information from both sides of the body because the two hemispheres

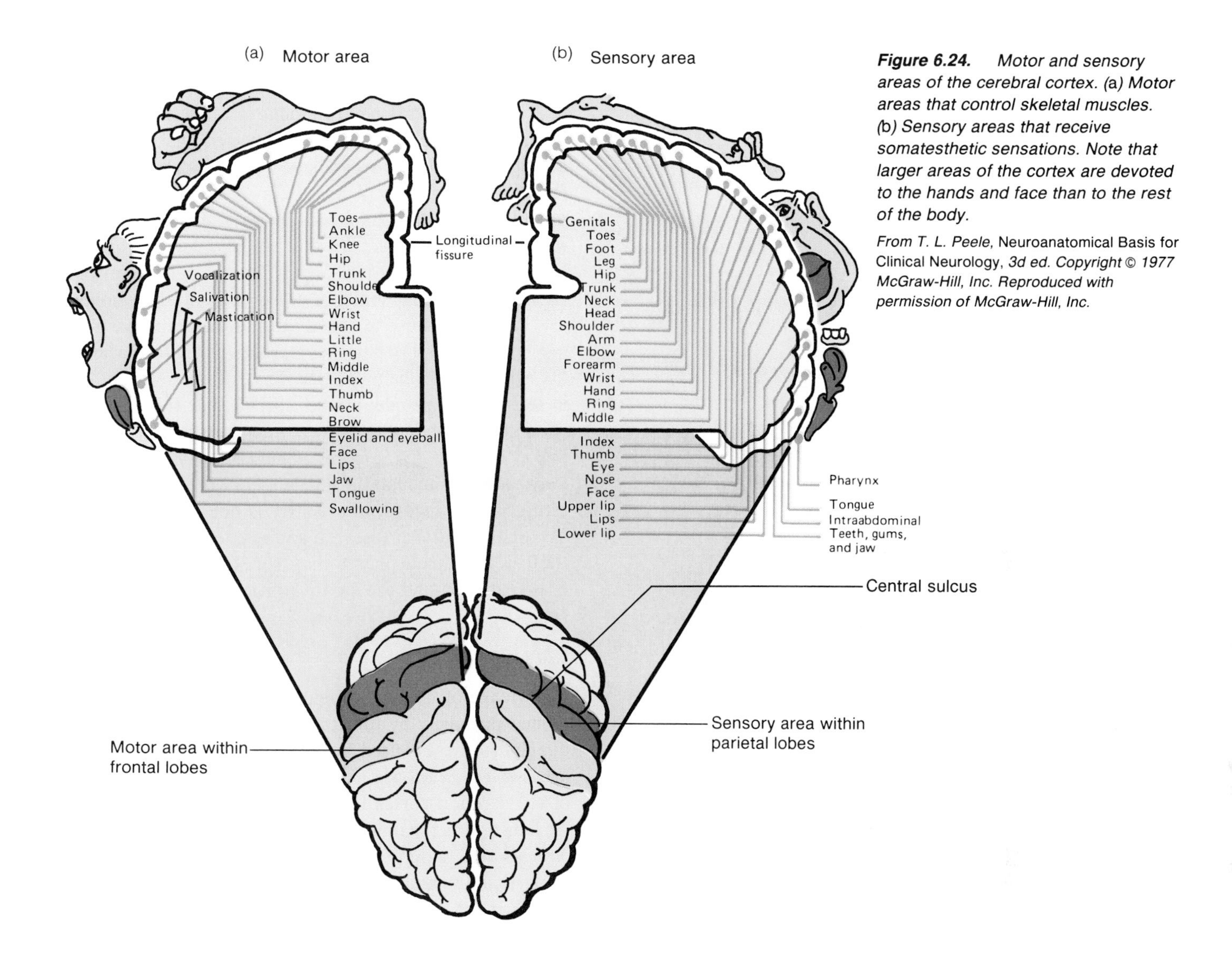

Figure 6.24. *Motor and sensory areas of the cerebral cortex. (a) Motor areas that control skeletal muscles. (b) Sensory areas that receive somatesthetic sensations. Note that larger areas of the cortex are devoted to the hands and face than to the rest of the body.*

From T. L. Peele, Neuroanatomical Basis for Clinical Neurology, *3d ed. Copyright © 1977 McGraw-Hill, Inc. Reproduced with permission of McGraw-Hill, Inc.*

communicate with each other via the *corpus callosum* (see fig. 6.21), a large tract composed of about 200 million fibers.

The corpus callosum has been surgically cut in some victims of severe epilepsy as a way of alleviating their symptoms. These **split-brain** procedures isolate each hemisphere from the other, but to a casual observer, surprisingly, these split-brain patients do not show evidence of any disability as a result of the surgery. However, in specially designed experiments in which each hemisphere is separately presented with sensory images and asked to perform tasks (speech or writing or drawing with the contralateral hand), it has been learned that each hemisphere is good at certain categories of tasks and poor at others (fig. 6.25).

In a typical experiment, the image of an object may be presented to either the right or left hemisphere (by presenting it to either the left or right visual field only) and the person may be asked to name the object. It was found that, in most people, the task could be performed successfully by the left hemisphere but not by the right. Similar experiments have shown that the left hemisphere is generally the one in which most of the language and analytical abilities reside. These results lead to the concept of **cerebral dominance,** which is analogous to the concept of handedness—people generally have better motor competence with one hand than the other. Since most people are right-handed, which is also controlled by the left hemisphere, the left hemisphere was naturally considered to be the dominant hemisphere in most people. Further experiments have shown, however, that the right hemisphere is specialized along different, less obvious lines—rather than one hemisphere being dominant and the other subordinate, the two hemispheres appear to have complementary functions. The term **cerebral lateralization** is used to describe this specialization of function.

Experiments have shown that the right hemisphere does have limited verbal ability; more noteworthy is the observation that the right hemisphere is most adept at *visuospatial tasks.* The right hemisphere, for example, can recognize faces better

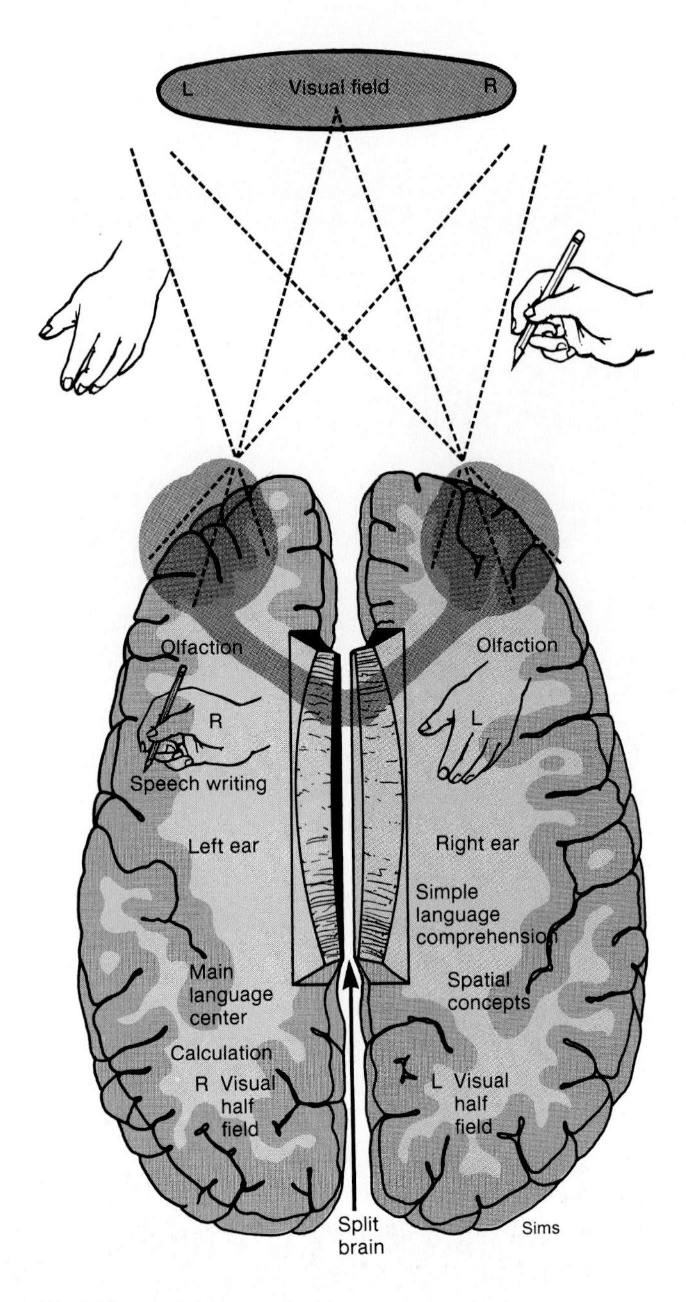

Figure 6.25. Different functions of the right and left cerebral hemispheres, as revealed by experiments with people who have had the tract connecting the two hemispheres (the corpus callosum) surgically split.

than the left, but it cannot describe facial appearances as well as the left. Acting through its control of the left hand, the right hemisphere is better than the left (controlling the right hand) at arranging blocks or drawing cubes. Patients with damage to the right hemisphere, as might be predicted from the results of split-brain research, have difficulty finding their way around a house and reading maps.

Perhaps as a result of the role of the right hemisphere in the comprehension of patterns and part-whole relationships, the ability to compose music, but not to critically understand it, appears to depend on the right hemisphere. Interestingly, damage to the left hemisphere may cause severe speech problems while leaving the ability to sing unaffected.

The lateralization of functions just described—with the left hemisphere specialized for language and analytical ability, and the right hemisphere specialized for visuospatial ability—is true for 97% of all people. It is true for all right-handers (who comprise 90% of all people) and for 70% of all left-handers. The remaining left-handers are split roughly equally into those who have language-analytical ability in the right hemisphere and those in whom this ability is present in both hemispheres.

It is interesting to speculate that the creative ability of a person may be related to the interaction of information between the right and left hemispheres. This interaction may be greater in left-handed people; a study found that the number of left-handers among college art students was disproportionately higher than in the general population. The observation that Leonardo da Vinci and Michelangelo were left-handed is interesting in this regard, but clearly is not scientific proof of any hypothesis. Further research on the lateralization of function of the cerebral hemispheres may reveal much more about both brain function and the creative process.

Language

Knowledge of the brain regions involved in language has been gained primarily by the study of *aphasias*[12]—speech and language disorders caused by damage to the brain. The language areas of the brain are primarily located in the left hemisphere of the cerebral cortex in most people, as previously described. As long ago as the nineteenth century, two areas of the cortex—Broca's area and Wernicke's area (fig. 6.26)—were found to be of particular importance in the production of aphasias.

Damage to **Broca's[13] area,** located in the left inferior frontal gyrus, produces an aphasia in which the person is reluctant to speak, and when speech is attempted, it is slow and poorly articulated. People with *Broca's aphasia,* however, have unimpaired comprehension of speech. It should be noted that this speech difficulty is not simply due to a problem in motor control, since the neural control over the musculature of the tongue, lips, larynx, and so on is unaffected. Damage to **Wernicke's[14] area,** located in the superior temporal gyrus, results in speech that is rapid and fluid but does not convey any information. People with *Wernicke's aphasia* produce speech that has been described as a "word salad." The words used may be real words that are chaotically mixed together, or they may be made-up words. Language comprehension has been destroyed; people with Wernicke's aphasia cannot understand either spoken or written language.

The **angular gyrus,** located at the junction of the parietal, temporal, and occipital lobes, is believed to be a center for the integration of auditory, visual, and somatesthetic information.

[12]aphasia: L. *a,* without; Gk. *phasis,* speech
[13]Broca: from Pierre P. Broca, French neurologist, 1824–1880
[14]Wernicke: from Karl Wernicke, German neurologist, 1848–1905

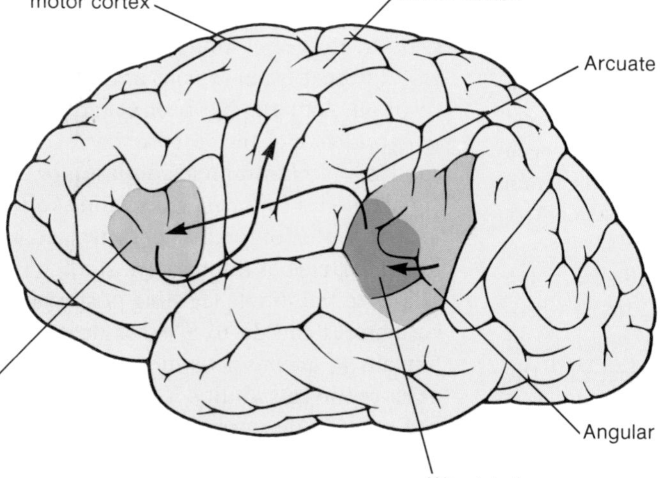

Supplementary motor cortex
Motor cortex
Arcuate fasciculus
Broca's area
Angular gyrus
Wernicke's area

Figure 6.26. Brain areas involved in the control of speech.

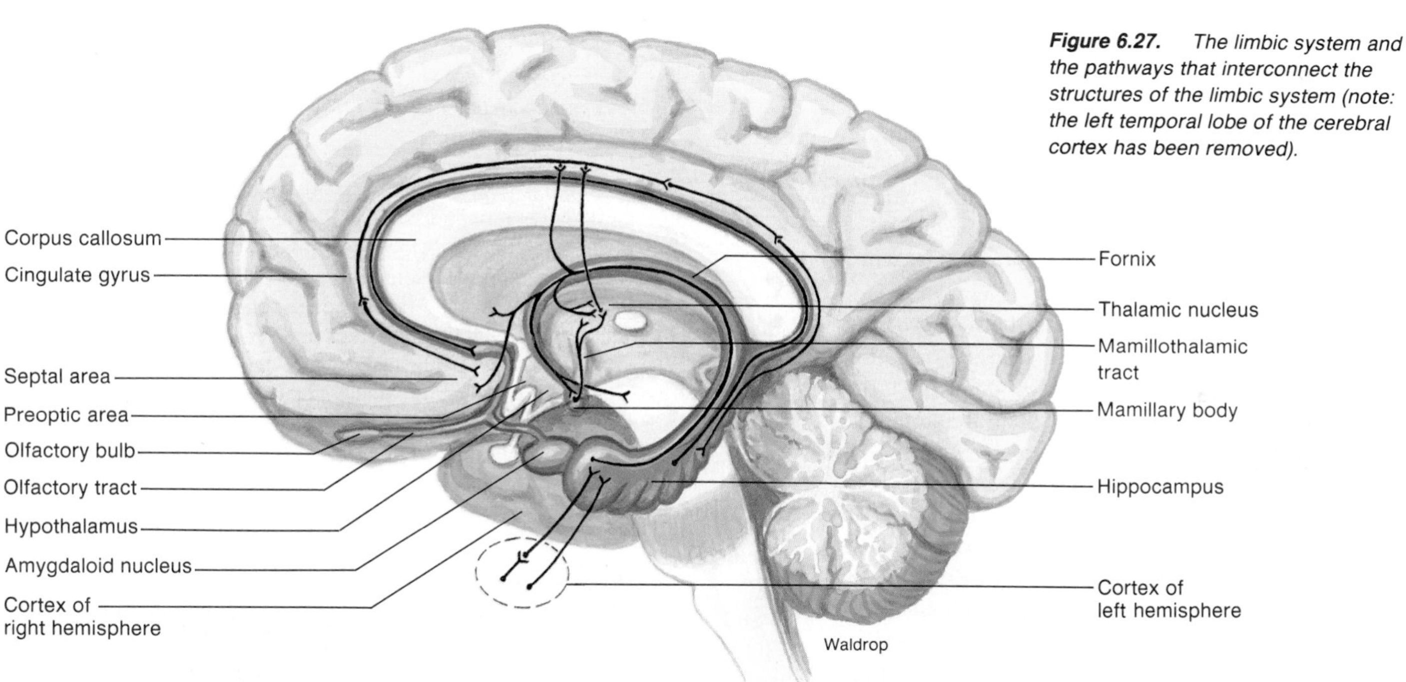

Figure 6.27. The limbic system and the pathways that interconnect the structures of the limbic system (note: the left temporal lobe of the cerebral cortex has been removed).

Damage to the angular gyrus produces aphasias, which suggests that this area projects to Wernicke's area. Some patients with damage to the left angular gyrus can speak and understand spoken language but cannot read or write. Other patients can write a sentence but cannot read it, presumably due to damage to the projections from the occipital lobe (involved in vision) to the angular gyrus.

Recovery of language ability, by transfer to the right hemisphere after damage to the left hemisphere, is very good in children but decreases after adolescence. Recovery is reported to be faster in left-handed people, possibly because language ability is more evenly divided between the two hemispheres in left-handed people. Some recovery usually occurs after damage to Broca's area, but damage to Wernicke's area produces more severe and permanent aphasias.

Emotion and Motivation

The parts of the brain that appear to be of paramount importance in the neural basis of emotional states are the *hypothalamus* (the small brain region directly above the pituitary gland, as described in chapter 7) and the **limbic system.**[15] The limbic system is a group of nuclei and fiber tracts that form a ring around the brain stem. The structures of the limbic system include the *cingulate gyrus* (part of the cerebral cortex), *amygdaloid nucleus* (or *amygdala*), *hippocampus,* and the *septal nuclei* (fig. 6.27).

The limbic system was formerly called the *rhinencephalon,* or "smell brain," because it is involved in the central processing of olfactory[16] information. This function may be the primary

[15]limbic: L. *limbus,* edge or border
[16]olfactory: L. *olfacere,* smell out

one of lower vertebrates, in which the limbic system may constitute the entire forebrain, but it is now known that the limbic system of humans is a center for basic emotional drives. The limbic system was derived early in the course of vertebrate evolution, and its tissue is distinctive from the *neocortex* of the rest of the cerebrum. There are few synaptic connections between the neocortex and the structures of the limbic system, which perhaps helps to explain why we have so little conscious control over our emotions.

Studies of the functions of these regions include electrical stimulation of specific locations, destruction of tissue (producing *lesions*) in particular sites, and surgical removal, or *ablation,* of specific structures. These studies suggest that the hypothalamus and limbic system are involved in the following processes:

Aggression. Stimulation of certain areas of the amygdala produce rage and aggression, and lesions of the amygdala can produce docility in experimental animals. Stimulation of particular areas of the hypothalamus can produce similar effects.

Fear. Fear can be produced by electrical stimulation of the amygdala and hypothalamus, and surgical removal of the limbic system can produce an absence of fear. Monkeys are normally terrified of snakes, for example, but if they have had their limbic system removed they will handle snakes without fear.

Feeding. The hypothalamus contains both a *feeding center* and a *satiety center.* Electrical stimulation of the former produces overeating, and stimulation of the latter will stop feeding behavior in experimental animals.

Sex. The hypothalamus and limbic system are involved in the regulation of the sexual drive and sexual behavior, as shown by stimulation and ablation studies in experimental animals. The cerebral cortex, however, is also critically important for the sex drive in lower animals, and the role of the cerebrum is believed to be even more important for the sex drive in humans.

Reward and punishment system. Electrodes placed in particular sites from the frontal cortex to the hypothalamus can deliver shocks that function as a reward. In rats, this reward is more powerful than food or sex in motivating behavior. Similar studies have been done in some humans, who report a feeling of relaxation and relief from tension, but not ecstasy. Electrodes placed in slightly different positions apparently stimulate a punishment system in experimental animals, who stop their behavior when stimulated in these regions.

Memory

Clinical studies of amnesia (loss of memory) suggest that several different brain regions are involved in memory storage and retrieval. Amnesia has been found to result from damage to the temporal lobe of the cerebral cortex, hippocampus, head of the caudate (in Huntington's disease), or a region of the thalamus (in alcoholics suffering from Korsakoff's syndrome).

A number of researchers now believe that there are two different systems of information storage in the brain. One system relates to the simple learning of stimulus-response that even invertebrates can do to a lesser degree; this may correspond in humans to unconscious memory. Even people with severe amnesia retain this ability. The other system is unique to humans and other higher organisms, and may correspond to conscious memory. Clinical studies also suggest that this latter system of memory can be divided into two major categories: **short-term memory** and **long-term memory.** People with head trauma, for example, and patients with suicidal depression who are treated by *electroconvulsive shock (ECS)* therapy, may lose their memory of recent events but retain their older memories.

Surgical removal of the right and left medial temporal lobes was performed in one patient, designated "H. M.," in an effort to treat his epilepsy. After the surgery he was unable to consolidate any short-term memory. He could repeat a phone number and carry out a normal conversation; he could not remember the phone number if momentarily distracted, however, and if the person to whom he was talking left the room and came back a few minutes later, H. M. would have no recollection of seeing that person or of having had a conversation with that person before. Although his memory of events that occurred before the operation was intact, all subsequent events in his life seemed as if they were happening for the first time.

Surgical removal of the left hippocampus (fig. 6.27) impairs the consolidation of short-term verbal memories into long-term memory, and removal of the right hippocampus impairs the consolidation of nonverbal memories. From these clinical observations, it appears that the hippocampus is an important structure involved in the change from short-term to long-term memory. Experiments in monkeys, however, have failed to duplicate the complete symptoms when only the hippocampus is removed. In these experiments, the amygdaloid nucleus (fig. 6.27) must also be destroyed to produce these effects. The amygdaloid nucleus has been proposed to play an important role in the association of memories with sensory information from other areas of the cortex, and in the association of emotions with both sensory perceptions and memory.

The cerebral cortex is thought to store factual information, with verbal memories lateralized to the left hemisphere and visuospatial information in the right hemisphere. The neurosurgeon Wilder Penfield has electrically stimulated various regions in the brain of awake patients, often evoking visual or auditory memories that were extremely vivid. Electrical stimulation of specific points in the temporal lobe evoked specific memories that were so detailed the patient felt that he was reliving the experience. The medial regions of the temporal lobes, however, cannot be the site where long-term memory is stored, because destruction of these areas in patients being treated for epilepsy did not destroy the memory of events prior to the surgery.

The amount of memory destroyed by ablation (removal) of brain tissue appears to depend more on the amount of brain tissue removed than on the location of the surgery. On the basis of these observations, it was formerly believed that the memory may be diffusely located in the brain; stimulation of the correct location of the cortex then retrieves the memory. According to current thinking, however, particular aspects of the memory—visual, auditory, olfactory, spatial, and so on—are stored in particular areas, and the cooperation of all of these areas is required to produce the complete memory.

1. Identify the location of the sensory and motor cortex, and explain how these areas are organized.
2. Identify the lobes of the cerebral cortex that are involved in analyzing visual and auditory information.
3. Describe the structures of the limbic system and explain its significance.
4. Explain the differences in function of the right and left cerebral hemispheres.
5. Describe the areas of the brain believed to be involved in the production of speech, the relationships between these areas, and the different types of aphasias produced by damage to these areas.
6. Describe the different forms of memory, the brain structures shown to be involved in memory, and some of the experimental evidence on which this information is based.

Cranial and Spinal Nerves

The central nervous system communicates with the body by means of nerves that exit the CNS from the brain (cranial nerves) and spinal cord (spinal nerves). These nerves, together with aggregations of cell bodies located outside the CNS, constitute the peripheral nervous system.

Table 6.4 Summary of cranial nerves

Number and Name	Composition
I Olfactory	Sensory
II Optic	Sensory
III Oculomotor	Mixed
IV Trochlear	Mixed
V Trigeminal	Mixed
VI Abducens	Mixed
VII Facial	Mixed
VIII Vestibulocochlear	Sensory
IX Glossopharyngeal	Mixed
X Vagus	Mixed
XI Accessory	Mixed
XII Hypoglossal	Mixed

As previously mentioned, the peripheral nervous system consists of nerves (collections of axons) and their associated ganglia (collections of cell bodies). The nerves of the PNS may extend from the brain as *cranial nerves* or from the spinal cord as *spinal nerves.*

Cranial Nerves

There are twelve pairs of cranial nerves. Two of these pairs arise from neuron cell bodies located in the forebrain and ten pairs arise from the midbrain and hindbrain. The cranial nerves are designated by Roman numerals and by names. The Roman numerals refer to the order in which the nerves are positioned from the front of the brain to the back. The names indicate structures innervated by these nerves (e.g., facial) or the principal function of the nerves (e.g., occulomotor). A summary of the cranial nerves is presented in table 6.4.

Spinal Nerves

There are thirty-one pairs of spinal nerves. These nerves are grouped into eight cervical, twelve thoracic, five lumbar, five sacral, and one coccygeal according to the region of the vertebral column from which they arise (fig. 6.28).

Each spinal nerve is a mixed nerve composed of sensory and motor fibers. These fibers are packaged together in the nerve, but separate near the attachment of the nerve to the spinal cord. This produces two "roots" to each nerve. The **dorsal root** is

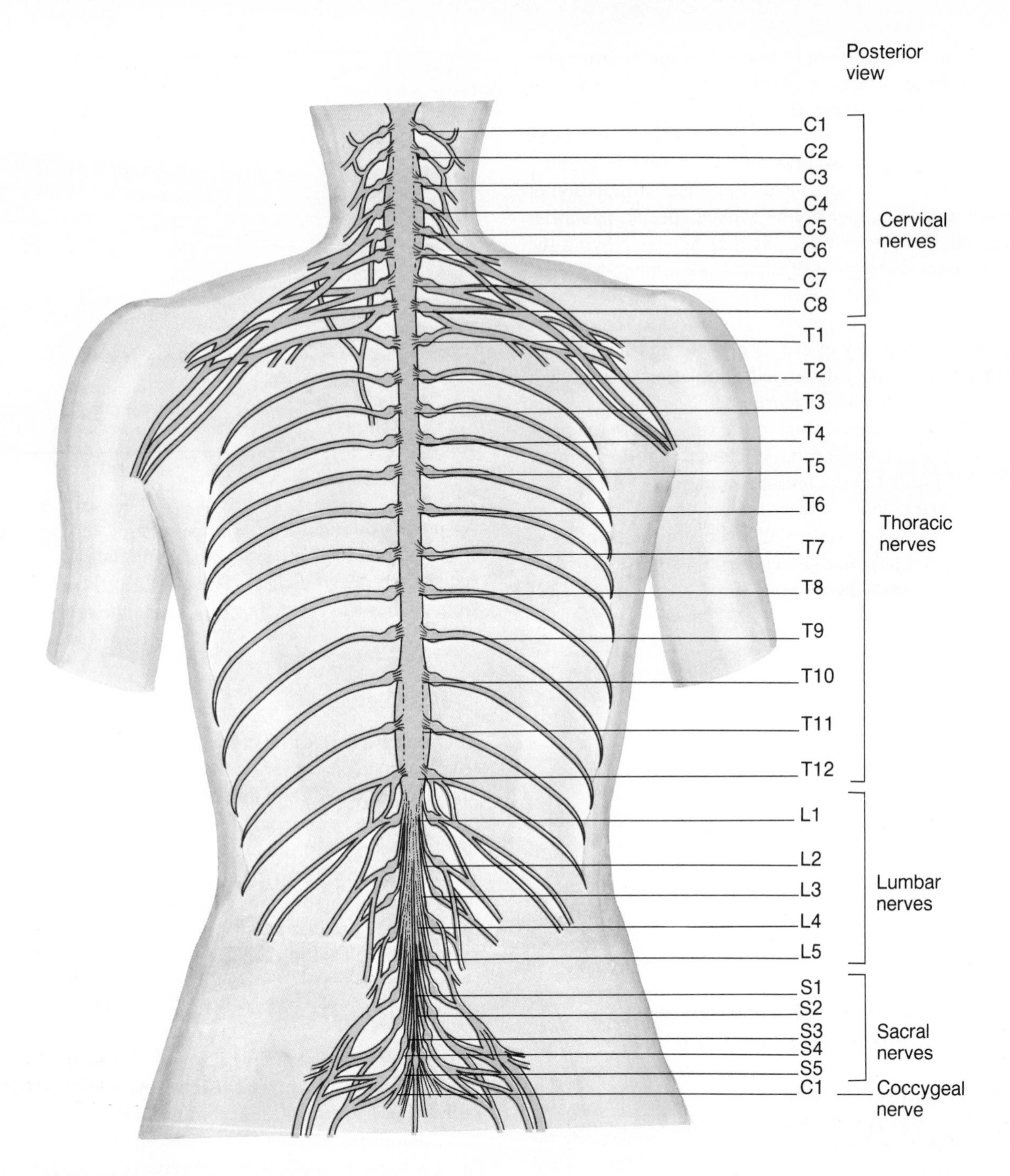

Figure 6.28. *The distribution of the spinal nerves.*

composed of sensory fibers, and the **ventral root** is composed of motor fibers (fig. 6.29). The dorsal root contains an enlargement called the *dorsal root ganglion,*[17] where the cell bodies of the sensory neurons are located. The motor neuron shown in figure 6.29 is a somatic motor neuron, which innervates skeletal muscles; its cell body is not located in a ganglion, but instead is within the gray matter of the spinal cord.

[17]ganglion: Gk. *ganglion,* a swelling

Reflex Arc

The functions of the sensory and motor components of a spinal nerve can most easily be understood by examination of a simple reflex; that is, an unconscious motor response to a sensory stimulus. Figure 6.29 demonstrates the neural pathway involved in a **reflex arc.** Stimulation of sensory receptors evokes action potentials in sensory neurons, which are conducted into the spinal cord. In the example shown, a sensory neuron synapses with an association neuron (or interneuron), which in turn synapses with

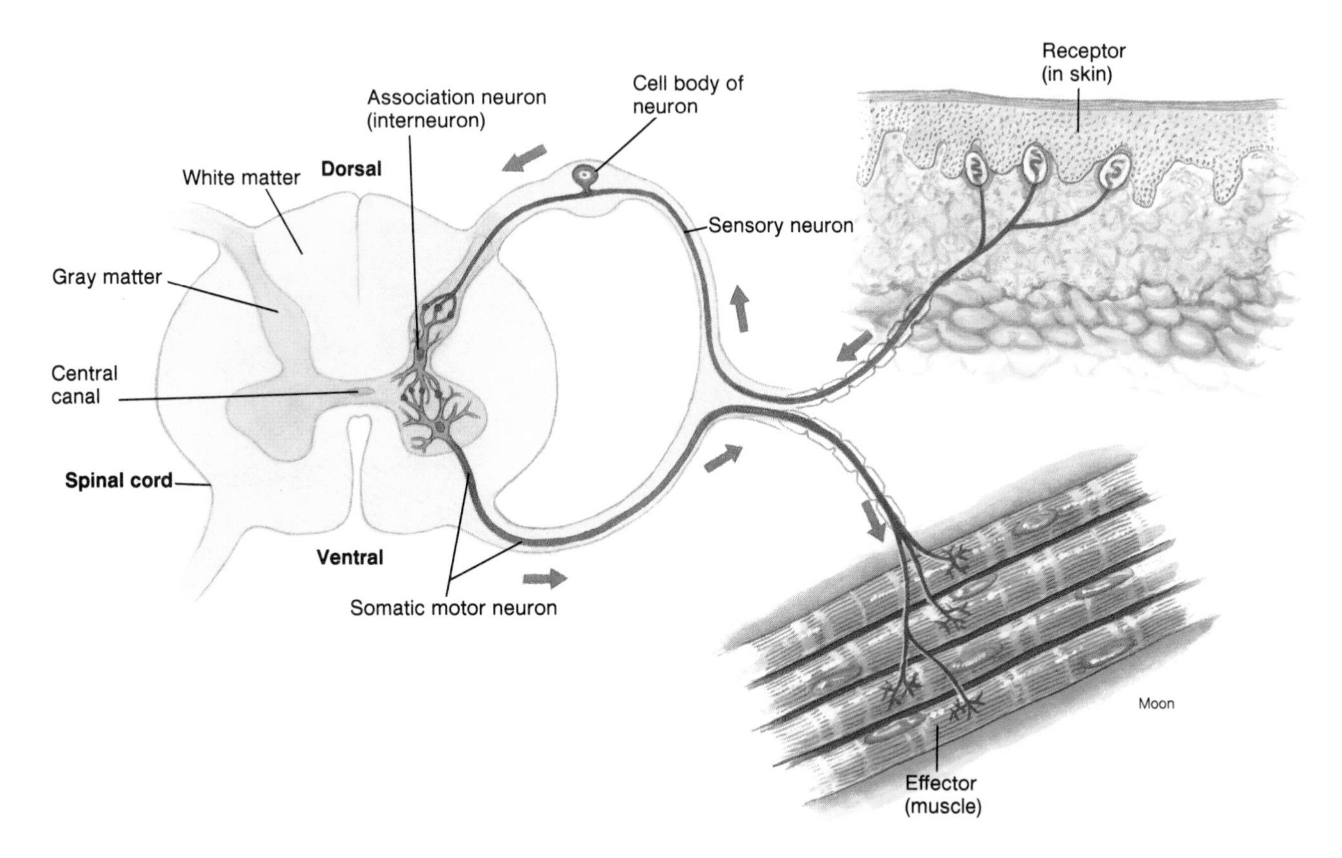

Figure 6.29. *Sensory neuron, association neuron (interneuron), and somatic motor neuron at the spinal cord level.*

a somatic motor neuron. The somatic motor neuron then conducts impulses out of the spinal cord to the muscle and stimulates a reflex contraction. Notice that the brain is not directly involved in this reflex response to sensory stimulation. Some reflex arcs are even simpler than this; in a muscle stretch reflex (the knee-jerk reflex), for example, the sensory neuron makes a synapse directly with a motor neuron. Other reflexes are more complex, involving a number of association neurons and resulting in motor responses on both sides of the spinal cord at different levels.

1. Describe the meaning of the terms *dorsal root, dorsal root ganglion, ventral root,* and *mixed nerve.*
2. Describe the neural pathways and structures involved in a reflex arc.

Autonomic Nervous System

Autonomic nerves are motor nerves that innervate the heart, smooth muscles, and glands. The autonomic nervous system is composed of the sympathetic and parasympathetic divisions. In both of these divisions, an axon exits the CNS and synapses with a second neuron, whose cell body is located in a ganglion. The axon of this postsynaptic neuron, in turn, stimulates the effector organs. Stimulation by sympathetic nerves prepares the body for "fight or flight." Stimulation by parasympathetic nerves generally produces the opposite effects.

Autonomic[18] motor nerves innervate organs whose functions are not usually under voluntary control. The effectors that respond to autonomic regulation include **cardiac muscle** (the heart), **smooth** (visceral) **muscles,** and **glands.** These are part of the organs of the *viscera* (organs within the body cavities) and of blood vessels. The involuntary effects of autonomic innervation contrast with the voluntary control of skeletal muscles by way of somatic motor neurons.

Autonomic Neurons

As previously discussed, neurons of the peripheral nervous system which conduct impulses away from the CNS are known as *motor,* or *efferent,* neurons. There are two major categories of motor neurons: somatic and autonomic. Somatic motor neurons have their cell bodies within the CNS and send axons to skeletal muscles, which are usually under voluntary control.

Unlike somatic motor neurons, which conduct impulses along a single axon from the spinal cord to the neuromuscular junction, autonomic motor control involves two neurons in the

[18]autonomic: Gk. *auto,* self; *nomos,* law

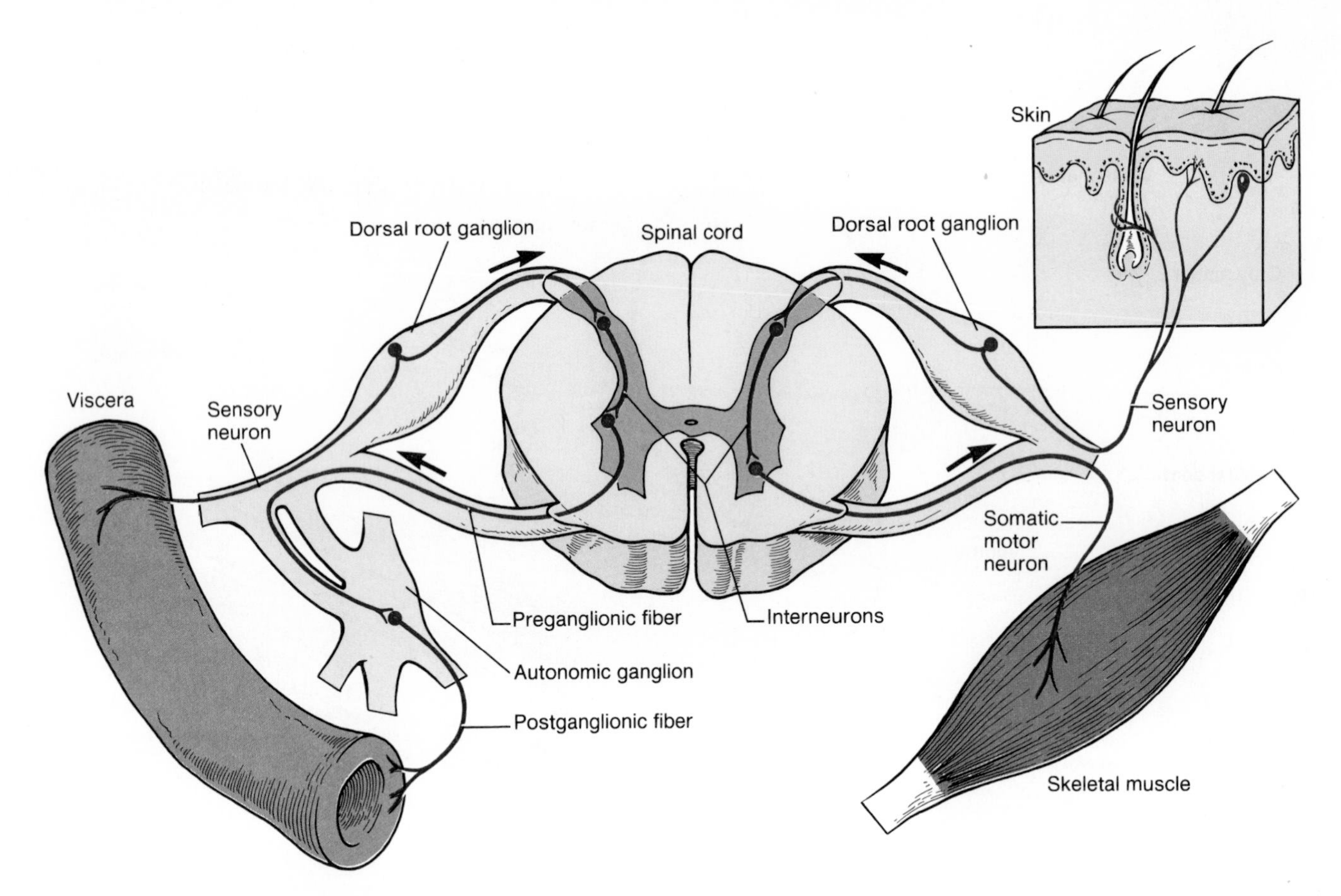

Figure 6.30. A comparison of a somatic motor reflex with an autonomic motor reflex. Although, for the sake of clarity, each is shown on different sides of the spinal cord, both visceral and somatic sensory neurons are found bilaterally.

efferent pathway. The first of these neurons has its cell body in the gray matter of the brain or spinal cord. The axon of this neuron does not directly innervate the effector organ but instead synapses with a second neuron within an *autonomic ganglion.* The first neuron is thus called a **preganglionic neuron.** The second neuron in this pathway, called a **postganglionic neuron,** has an axon that extends from the autonomic ganglion and synapses with the cells of an effector organ (fig. 6.30, right side).

Autonomic ganglia are located in the head, neck, and abdomen; chains of autonomic ganglia also parallel the right and left sides of the spinal cord. The origin of the preganglionic fibers and the location of the autonomic ganglia help to distinguish the **sympathetic** and **parasympathetic** divisions of the autonomic system.

Sympathetic Division

The sympathetic system is also called the thoracolumbar division of the autonomic system because its preganglionic fibers exit the spinal cord from the first thoracic (T1) to the second lumbar (L2) levels. Most sympathetic nerve fibers synapse with postganglionic neurons within a double row of sympathetic ganglia located on either side of the spinal cord (fig. 6.31). Ganglia within each row are interconnected, forming a *sympathetic chain* of ganglia parallel to each side of the spinal cord (fig. 6.32). In addition to the sympathetic chain of ganglia, there are three sympathetic ganglia within the abdomen that are not part of this chain; these are collectively known as *collateral ganglia* (fig. 6.33).

Adrenal Glands

The paired adrenal[19] glands are located above each kidney. Each adrenal is composed of two parts: an outer **cortex**[20] and an inner **medulla.**[21] These two parts are really two functionally different glands with different embryonic origins, different hormones, and different regulatory mechanisms. The adrenal cortex secretes

[19]adrenal: L. *ad,* to; *renes,* kidney
[20]cortex: L. *cortex,* bark
[21]medulla: L. *medulla,* marrow

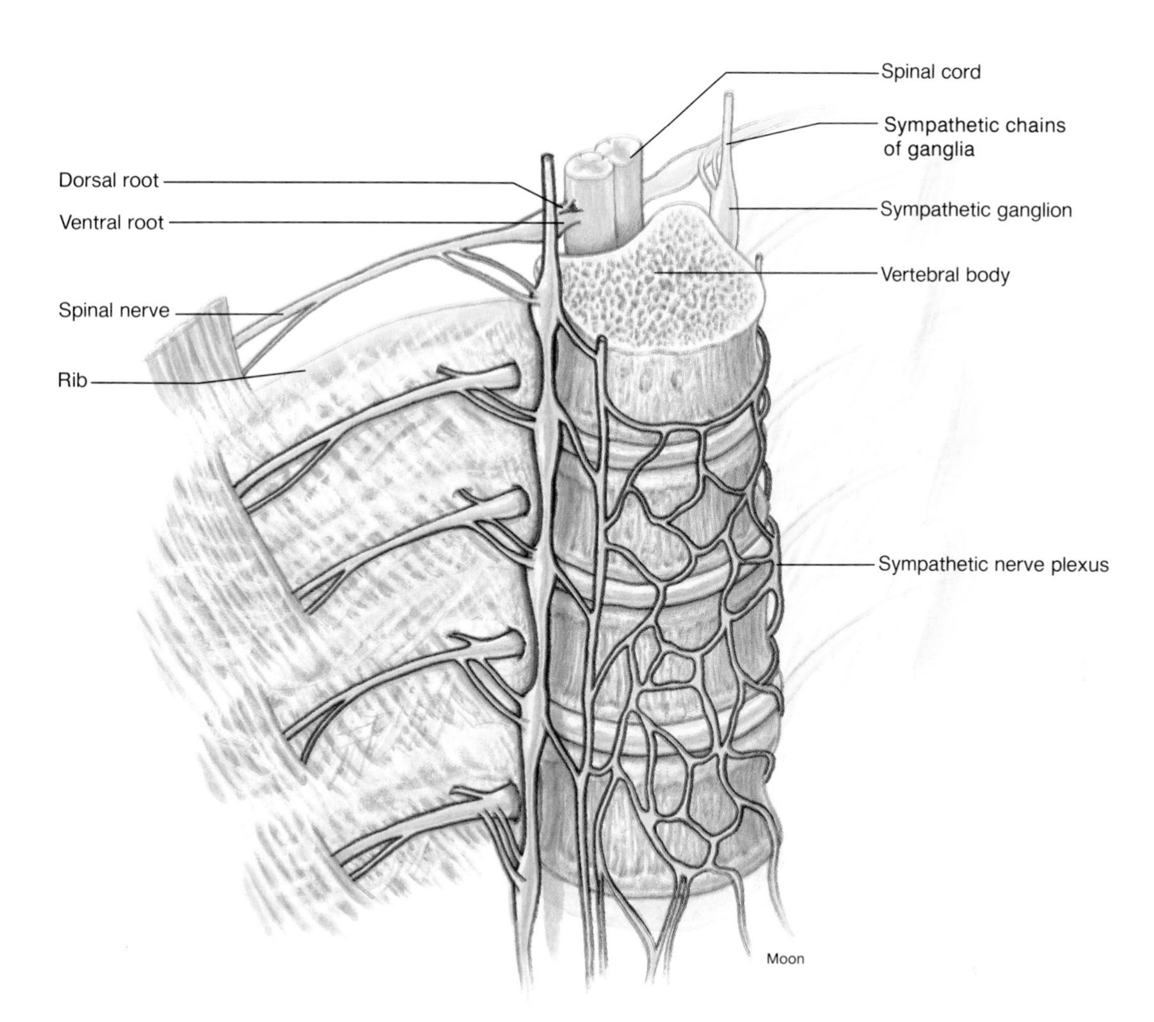

Figure 6.31. *The sympathetic chain of ganglia, showing its relationship to the vertebral column and the spinal cord.*

steroid hormones; the adrenal medulla secretes the hormone **epinephrine** (adrenaline) and, to a lesser degree, **norepinephrine,** when it is stimulated by the sympathetic system.

The adrenal medulla is a modified sympathetic ganglion. The cells of the adrenal medulla are innervated by preganglionic sympathetic fibers (fig. 6.32), and they secrete epinephrine into the blood in response to this neural stimulation. The effects of epinephrine are complementary to those produced by sympathetic nerve stimulation. For this reason, and because the adrenal medulla is stimulated when the sympathetic division is activated, the two are often grouped together as the **sympathoadrenal system.**

Parasympathetic Division

The parasympathetic division is also known as the *craniosacral division* of the autonomic system. This is because its preganglionic fibers originate in the brain and in the sacral levels of the spinal cord. These preganglionic fibers synapse in ganglia that are located next to—or actually within—the organs innervated. For this reason, parasympathetic ganglia—known as *terminal ganglia*—do not form a chain, but are instead scattered at different locations within the body. The parasympathetic ganglia supply the postganglionic fibers that synapse with the effector cells.

Nuclei in the brain stem contribute preganglionic fibers to the very long *tenth cranial,* or *vagus nerves,*[22] which are the major parasympathetic nerves in the body. These preganglionic fibers travel through the neck to the thoracic (chest) cavity, and through the opening in the diaphragm to the abdominal cavity. In each region, some of these fibers leave the main trunks of the vagus nerves and synapse with postganglionic neurons that are located *within* the innervated organs. The preganglionic vagus fibers are thus quite long, and provide parasympathetic innervation to the heart, lungs, esophagus, stomach, pancreas, liver, small intestine, and the upper half of the large intestine. Post-

[22]vagus: L. *vagus,* wandering

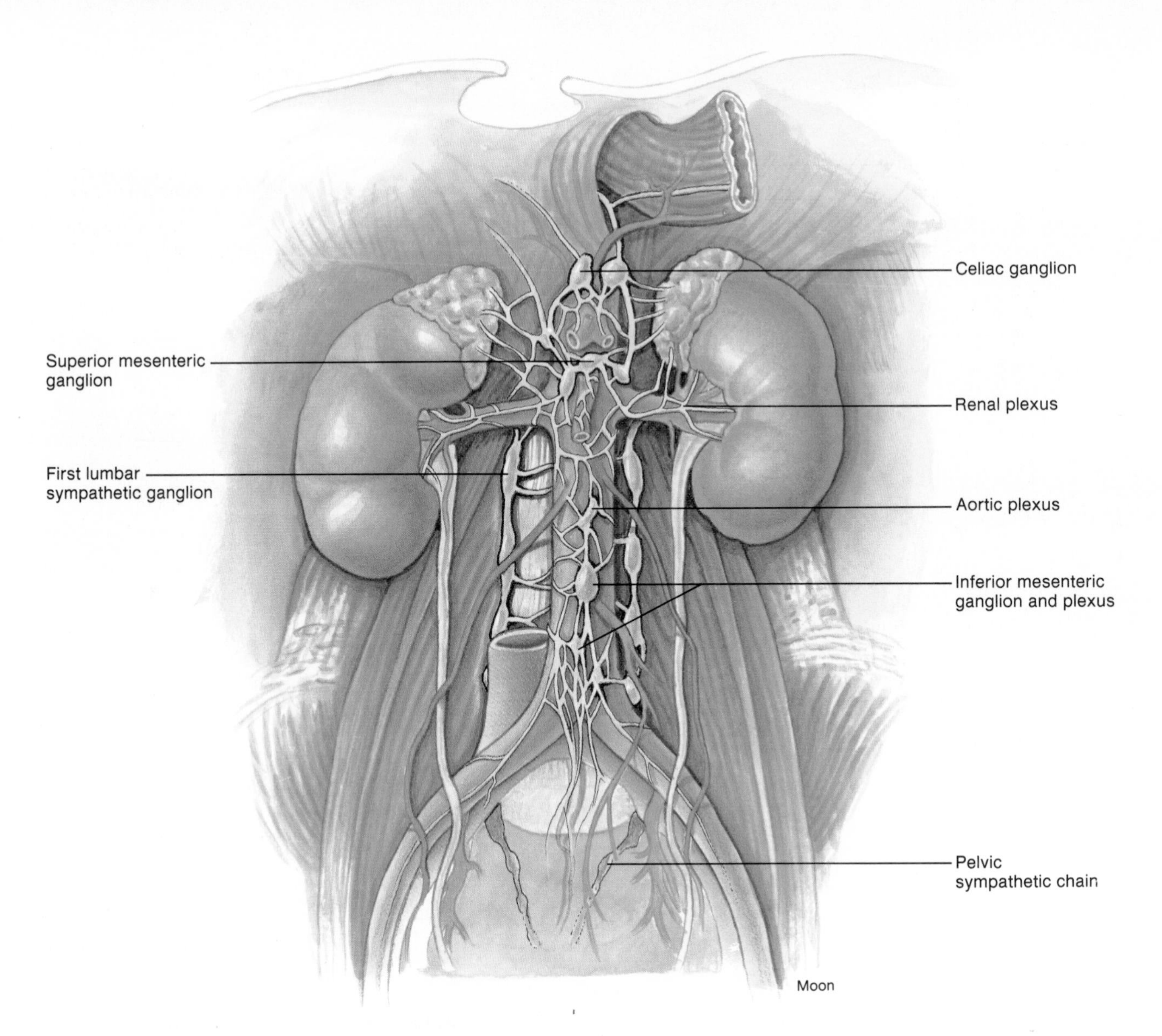

Figure 6.32. *The collateral sympathetic ganglia: the celiac and the superior and inferior mesenteric ganglia.*

ganglionic parasympathetic fibers arise from terminal ganglia within these organs and synapse with effector cells (smooth muscles and glands).

Parasympathetic nerves to the visceral organs thus consist of preganglionic fibers, whereas sympathetic nerves to these organs contain postganglionic fibers. A composite view of the sympathetic and parasympathetic systems is provided in figure 6.33, and these comparisons are summarized in table 6.5.

Functions of the Autonomic Nervous System

The sympathetic and parasympathetic divisions of the autonomic system affect the visceral organs in different ways. Activation of the sympathetic system prepares the body for intense physical activity in emergencies; the heart rate increases, blood glucose rises, and blood is diverted to the skeletal muscles (away from the visceral organs and skin). These and other effects are listed in table 6.6. The theme of the sympathetic system has been aptly summarized in a phrase: **"fight or flight."**

The effects of *parasympathetic nerve* stimulation are in many ways opposite to the effects of sympathetic stimulation. Stimulation of separate parasympathetic nerves can result in slowing of the heart, dilation of visceral blood vessels, and an increased activity of the digestive tract (table 6.6). The different responses of a visceral organ to sympathetic and parasympathetic nerve activity are due to the fact that the postganglionic fibers of these two divisions release different neurotransmitters.

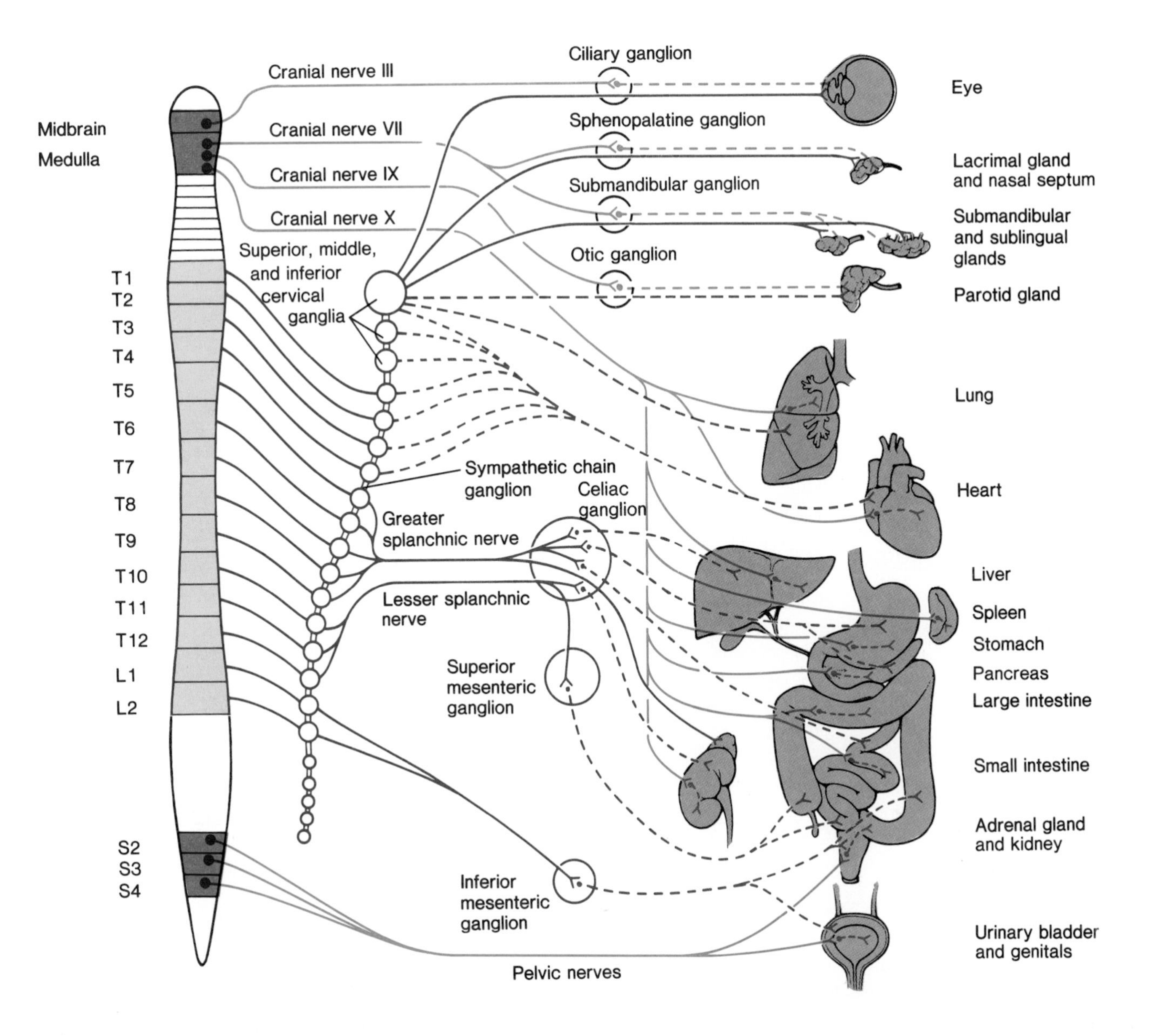

Figure 6.33. *The autonomic nervous system. The sympathetic division is shown in color, the parasympathetic in black. The solid lines indicate preganglionic fibers, and the dotted lines indicate postganglionic fibers.*

Table 6.5 Some comparisons between the structure of the sympathetic and parasympathetic systems

	Sympathetic	**Parasympathetic**
Origin of Preganglionic Outflow	Thoracolumbar levels of spinal cord	Midbrain, hindbrain, and sacral levels of spinal cord
Location of Ganglia	Chain of paravertebral ganglia and prevertebral (collateral) ganglia	Terminal ganglia in or near effector organs
Distribution of Postganglionic Fibers	Throughout the body	Mainly limited to the head and the viscera of the chest, abdomen, and pelvis
Divergence of Impulses from Pre- to Postganglionic Fibers	Great divergence (one preganglionic may activate twenty postganglionic fibers)	Little divergence (one preganglionic only activates a few postganglionic fibers)
Mass Discharge of System as a Whole	Yes	Not normally

Table 6.6 Effects of autonomic nerve stimulation on various visceral effector organs

Effector Organ	Sympathetic Effect	Parasympathetic Effect
Eye		
Iris (radial muscle)	Dilates pupil	———
Iris (sphincter muscle)	———	Constricts pupil
Ciliary muscle	Relaxes (for far vision)	Contracts (for near vision)
Glands		
Lacrimal (tear)	———	Stimulates secretion
Sweat	Stimulates secretion	———
Salivary	Decreases secretion; saliva becomes thick	Increases secretion; saliva becomes thin
Stomach	———	Stimulates secretion
Intestine	———	Stimulates secretion
Adrenal medulla	Stimulates secretion of hormones	———
Heart		
Rate	Increases	Decreases
Conduction	Increases rate	Decreases rate
Strength	Increases	———
Blood Vessels	Mostly constricts; affects all organs	Dilates in a few organs (e.g., penis)
Lungs		
Bronchioles (tubes)	Dilates	Constricts
Mucous glands	Inhibits secretion	Stimulates secretion
Gastrointestinal Tract		
Motility	Inhibits movement	Stimulates movement
Sphincters	Stimulates closing	Inhibits closing
Liver	Stimulates hydrolysis of glycogen	———
Adipose (fat) cells	Stimulates hydrolysis of fat	———
Pancreas	Inhibits exocrine secretions	Stimulates exocrine secretions
Spleen	Stimulates contraction	———
Urinary Bladder	Helps set muscle tone	Stimulates contraction
Piloerector Muscles	Stimulates erection of hair and goose bumps	———
Uterus	If pregnant: contraction If not pregnant: relaxation	———
Penis	Erection; ejaculation	Erection (due to vasodilation)

Functions of the Sympathetic Division

The neurotransmitter released by most postganglionic sympathetic nerve fibers is **norepinephrine** (*noradrenaline*). Transmission at these synapses is thus said to be **adrenergic.**

In view of the fact that the cells of the adrenal medulla are derived from postganglionic sympathetic neurons, it is not surprising that the hormones they secrete (normally about 85% epinephrine and 15% norepinephrine) are similar to the transmitter of postganglionic sympathetic neurons. Epinephrine differs from norepinephrine only slightly in chemical structure and physiological action.

Adrenergic stimulation—by epinephrine in the blood and by norepinephrine released from sympathetic nerve endings—has both excitatory and inhibitory effects. The heart, muscles in the iris of the eye that dilate the pupil, and the smooth muscles of many blood vessels are stimulated to contract. The smooth muscles of the bronchioles (air passages in the lungs) and of some blood vessels, however, are inhibited from contracting; adrenergic chemicals, therefore, cause these structures to dilate.

Drugs that specifically promote the effects of epinephrine and norepinephrine, and drugs that block these effects, are used to treat some common disorders. Certain drugs that promote the adrenergic effects on bronchioles, for example, are used to produce bronchodilation in people suffering from asthma. People with high blood pressure are sometimes treated with the drug *propranolol,* which slows the heartbeat by blocking the adrenergic stimulation of the heart.

Functions of the Parasympathetic Division

All postganglionic parasympathetic neurons are cholinergic—they release acetylcholine as a neurotransmitter. The cholinergic effects of parasympathetic fibers are in some cases excitatory and in others inhibitory. Parasympathetic fibers to the stomach and intestine increase the activity of these organs, whereas the parasympathetic fibers innervating the heart cause slowing of the heart rate. It is useful to remember that the effects of parasympathetic stimulation are, in general, opposite to the effects of sympathetic stimulation.

The parasympathetic effects of ACh are specifically inhibited by the drug **atropine,** or **belladonna,** derived from the deadly nightshade plant (*Atropa belladonna*). Indeed, extracts of this plant were used by women during the Middle Ages to dilate their pupils (atropine inhibits parasympathetic stimulation of the iris). This was done to enhance their beauty (belladonna—beautiful woman). Atropine is used clinically today to dilate pupils during eye examinations, to dry mucous membranes of the respiratory tract prior to general anesthesia, and to inhibit spasmodic contractions of the lower digestive tract.

1. Describe the pathway of impulse conduction from the CNS to a visceral effector organ, such as the heart.
2. Distinguish between the sympathetic and parasympathetic division in terms of the orgin of the preganglionic neurons and the location of the cell bodies of the postganglionic neurons.
3. Describe the general effects of the sympathetic division and indicate the neurotransmitter that produces these effects.
4. Describe the general effects of the parasympathetic division and use this discussion to explain the effects of the drug atropine.

Summary

Organization of the Nervous System p. 91

I. Neurons may have different forms.
   A. Association neurons, located entirely within the CNS, are multipolar.
   B. Sensory neurons are pseudounipolar, conducting impulses from a sensory organ into the CNS.

II. There are several types of neuroglial cells.
   A. Schwann cells are glial cells in the PNS which cover all peripheral axons.
   B. In many cases, the Schwann cells wrap around the axon to form many overlapping layers of membrane around the axon; this is called a myelin sheath.
   C. The gaps between myelin are known as the nodes of Ranvier; this is where impulses are produced in a myelinated axon.
   D. The sheath of Schwann cells aids regeneration of a cut axon in the PNS.
   E. The myelin sheath of axon within the CNS is formed by glial cells called oligodendrocytes.

The Nerve Impulse p. 97

I. All cells have a resting membrane potential.
   A. This is a difference in charge between the inside of the cell and the outside of the cell.
   B. The inside of a cell is negatively charged in comparison to the outside of the cell.
   C. The charge difference is called a potential difference, and is measured in millivolts.

II. The nerve impulse is called an action potential.
   A. In response to depolarization to a threshold value, the voltage-regulated gates that guard the channels for Na^+ and K^+ open.
   B. First, Na^+ diffuses into the axon, and then K^+ diffuses out of the axon.
   C. This causes the membrane potential to switch polarities, to about +40 mV, and then to go back to −70 mV.
   D. Each action potential is an all-or-none event; a stronger stimulus causes more action potentials to be produced per time interval (causes a higher frequency of action potentials).
   E. Action potentials do not decrease in amplitude as they are conducted.
   F. Myelinated axons conduct by saltatory conduction, which is faster than in unmyelinated axons.

Synaptic Transmission p. 100

I. In most synapses, the axon releases a chemical neurotransmitter that stimulates the postsynaptic cell.
   A. Acetylcholine (ACh) is the neurotransmitter of somatic motor axons, which stimulates skeletal muscle cells.
   B. ACh combines with a receptor protein, causing depolarization of the postsynaptic cell.
   C. The ACh released by the axon is inactivated by an enzyme called acetylcholinesterase.

II. Combination of the ACh with its receptor protein opens membrane gates for Na^+ and K^+.
   A. The change in membrane potential is called an excitatory postsynaptic potential (EPSP).
   B. EPSPs are produced in the dendrites and cell bodies of neurons, which are the location of the synapses.
   C. EPSPs, unlike action potentials, are graded events that are capable of summating.
   D. When the EPSP arrives at the axon hillock, it provides the depolarization stimulus needed to produce the action potential.

Neurotransmitters in the Brain p. 104

I. ACh is used by some neurons in the brain; a deficiency in some of these pathways is associated with Alzheimer's disease.

II. Dopamine and norepinephrine are in the chemical family known as catecholamines.
   A. Parkinson's disease is associated with a deficiency in some of the neural pathways that use dopamine as a neurotransmitter; overactivity of dopaminergic pathways is associated with schizophrenia.
   B. Amphetamines act by stimulating neural pathways that use norepinephrine as a neurotransmitter.

III. GABA and glycine are neurotransmitters in the CNS that have inhibitory functions by producing inhibitory postsynaptic potentials (IPSPs).

IV. Endorphins are polypeptides that may act as neurotransmitters in the brain which help to reduce perceptions of pain and produce feelings of euphoria.

The Cerebral Hemispheres of the Brain p. 106

I. The cerebrum consists of an outer gray cortex and underlying white matter.

II. The left cerebral hemisphere in most people is specialized for language and logical thinking, whereas the right hemisphere is devoted more to pattern recognition.

III. Language ability is known to involve specific regions of the left hemisphere, including Broca's and Wernicke's areas.

IV. Emotion and motivation involve the cerebral cortex, a brain region called the hypothalamus, and structures that are known collectively as the limbic system.

V. Memory can be divided into short- and long-term; consolidation of long-term memory requires the function of the hippocampus and amygdaloid nuclei.

Cranial and Spinal Nerves p. 113

I. There are twelve pairs of cranial nerves.

II. There are thirty-one pairs of spinal nerves.
   A. Each spinal nerve is mixed, consisting of the processes of both sensory and motor neurons.

B. The sensory axons enter the spinal cord in the dorsal root, and the motor axons exit the spinal cord in the ventral root of the nerve.

III. A reflex arc involves sensory and motor neurons, and can also involve association neurons.

Autonomic Nervous System p. 115

I. There are two neurons involved in the autonomic motor pathway.
   A. A preganglionic neuron, with its cell body in the CNS, sends its axon to a postganglionic neuron.
   B. The cell body of the postganglionic neuron is located in an autonomic ganglion; its axon may stimulate smooth muscle, cardiac muscle, and glands.

II. The autonomic nervous system is subdivided into the sympathetic and parasympathetic divisions.
   A. In the sympathetic division, the preganglionic neurons are located in the thoracic to the lumbar levels of the spinal cord.
      1. The sympathetic ganglia form a double chain parallel to the spinal cord.
      2. The adrenal medulla also receives preganglionic axons from the sympathetic division, and secretes the hormone epinephrine when the sympathetic division is activated.
   B. In the parasympathetic division, the preganglionic neurons are located in the brain and the sacral level of the spinal cord.
      1. The parasympathetic ganglia are located next to or within the organs that they innervate.
      2. The major parasympathetic nerve is the vagus.

III. The postganglionic sympathetic neurons release norepinephrine as their neurotransmitter, which activates the body for "fight or flight."

IV. Postganglionic parasympathetic neurons release acetylcholine, which usually has actions that are antagonistic to those of norepinephrine from sympathetic nerves.

Review Activities

Objective Questions

1. The neuroglial cells that form myelin sheaths in the peripheral nervous system are
   (a) oligodendrocytes
   (b) satellite cells
   (c) Schwann cells
   (d) astrocytes
   (e) microglia
2. A collection of neuron cell bodies located outside the CNS is called a
   (a) tract
   (b) nerve
   (c) nucleus
   (d) ganglion
3. Which of the following neurons are pseudounipolar?
   (a) sensory neurons
   (b) somatic motor neurons
   (c) neurons in the retina
   (d) autonomic motor neurons
4. Depolarization of an axon is produced by the
   (a) inward diffusion of Na^+
   (b) active extrusion of K^+
   (c) outward diffusion of K^+
   (d) inward active transport of Na^+
5. Repolarization of an axon during an action potential is produced by the
   (a) inward diffusion of Na^+
   (b) active extrusion of K^+
   (c) outward diffusion of K^+
   (d) inward active transport of Na^+
6. As the strength of a depolarizing stimulus to an axon is increased,
   (a) the amplitude of action potentials increases
   (b) the duration of action potentials increases
   (c) the speed with which action potentials are conducted increases
   (d) the frequency with which action potentials are produced increases
7. The conduction of action potentials in a myelinated nerve fiber is
   (a) saltatory
   (b) without decrement
   (c) faster than in an unmyelinated fiber
   (d) all of the above
8. Which of the following is *not* a characteristic of EPSPs?
   (a) They are all-or-none in amplitude.
   (b) They are produced in dendrites and cell bodies.
   (c) They are graded in amplitude.
   (d) They are produced by chemically regulated gates.
9. Which of the following is *not* a characteristic of action potentials?
   (a) They are produced by voltage-regulated gates.
   (b) They are conducted without decrement.
   (c) Na^+ and K^+ diffuse through the membrane at the same time.
   (d) The membrane potential reverses polarity during depolarization.
10. A drug that inactivates acetylcholinesterase
   (a) inhibits the release of ACh from presynaptic endings
   (b) inhibits the attachment of ACh to its receptor protein
   (c) increases the ability of ACh to stimulate muscle contraction
   (d) all of the above
11. The precentral gyrus is
   (a) involved in motor control
   (b) involved in sensory perception
   (c) located in the frontal lobe
   (d) both *a* and *c*
   (e) both *b* and *c*
12. In most people, the right hemisphere controls movement
   (a) of the right side of the body primarily
   (b) of the left side of the body primarily
   (c) of both the right and left sides of the body equally
   (d) of the head and neck only
13. Verbal ability predominates in the
   (a) left hemisphere of right-handed people
   (b) left hemisphere of most left-handed people
   (c) right hemisphere of 97% of all people
   (d) both *a* and *b*
   (e) both *b* and *c*
14. The consolidation of short-term memory into long-term memory appears to be a function of the
   (a) substantia nigra
   (b) hippocampus
   (c) cerebral peduncles
   (d) arcuate fasciculus
   (e) precentral gyrus
15. Parasympathetic ganglia are located
   (a) in a chain parallel to the spinal cord
   (b) in the dorsal roots of spinal nerves
   (c) next to or within the organs innervated
   (d) in the brain
16. Which of the following fibers release norepinephrine?
   (a) preganglionic parasympathetic fibers
   (b) postganglionic parasympathetic fibers
   (c) preganglionic sympathetic fibers
   (d) postganglionic sympathetic fibers
   (e) all of the above

Essay Questions

1. Compare the characteristics of action potentials with those of EPSPs.
2. Explain how action potentials are produced.
3. Explain how action potentials are regenerated along an axon.
4. Describe cerebral lateralization of function, and explain how this knowledge has been gained.
5. Explain why it is believed that there is a difference between short-term and long-term memory, and indicate which brain structures are involved in these two forms of memory.
6. Compare the sympathetic and parasympathetic systems in terms of the location of their ganglia and the distribution of their nerves.
7. Explain the relationship between the sympathetic nervous system and the adrenal glands.
8. Compare the effects of adrenergic and cholinergic stimulation on the cardiovascular and digestive systems.

7

The Endocrine System

Outline

Objectives

By studying this chapter, you should be able to

1. define the terms *hormone* and *endocrine gland*
2. explain how different hormones can exert synergistic, permissive, or antagonistic effects
3. describe the mechanisms of action of different hormones
4. describe the parts of the pituitary gland and the relationship between the pituitary gland and the hypothalamus
5. identify the hormones of the posterior pituitary gland and explain how their secretion is regulated
6. list the hormones of the anterior pituitary and explain how their secretion is regulated by the hypothalamus
7. describe the actions of the thyroid hormones and explain how thyroid secretion is regulated
8. describe the hormones produced by the adrenal cortex and explain how the secretions of the adrenal cortex are regulated
9. describe the actions of epinephrine, secreted by the adrenal medulla, and explain how the secretions of the adrenal medulla are regulated
10. explain why the pancreas is both an exocrine and an endocrine gland and identify the secretions of the islets of Langerhans
11. describe the actions of insulin and glucagon and explain how secretions of these hormones are regulated

Keys to Pronunciation

adenylate cyclase: *ah-den'i-lat si'klas*
antidiuretic: *an''ti-di-u-ret'ik*
circadian: *ser''kah-de'an*
corticosteroids: *kor''ti-ko-ste'roids*
corticotropin: *kor''ti-ko-tro'pin*
endocrine: *en'do-krin*
glucagon: *gloo'kah-gon*
glycogenolysis: *gli''ko-je-nol'i-sis*
gonadotropic: *gon''ah-do-tro'pik*
hypophyseal: *hi''po-fiz'e-al*
hypophysis: *hi-pof'i-sis*
melatonin: *mel''ah-to'nin*
neuroendocrine: *nu''ro-en'do-krin*
phosphodiesterase: *fos''fo-di-es'ter-as*
somatotropin: *so-mah-to-tro'pin*
synergistic: *sin''er-jis'tik*
testosterone: *tes-tos'te-ron*
tetraiodothyronine: *tet''rah-i''o-do-thi'ro-nen*
thymus: *thi'mus*
thyroxine: *thi-rok'sin*
triiodothyronine: *tri''i-o''do-thi'ro-nen*
trophic: *trof'ik*

Photo: Molecular model of oxytocin, a hormone secreted by the posterior pituitary gland.

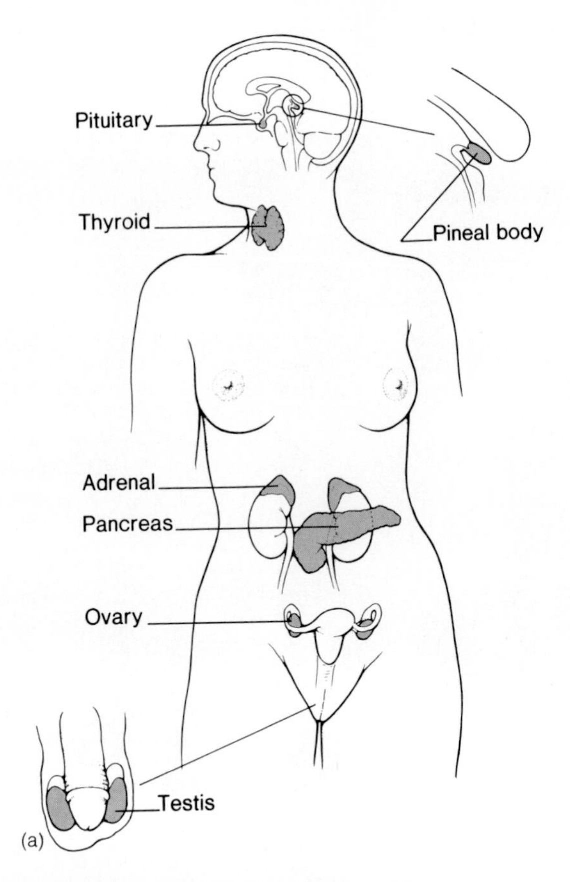

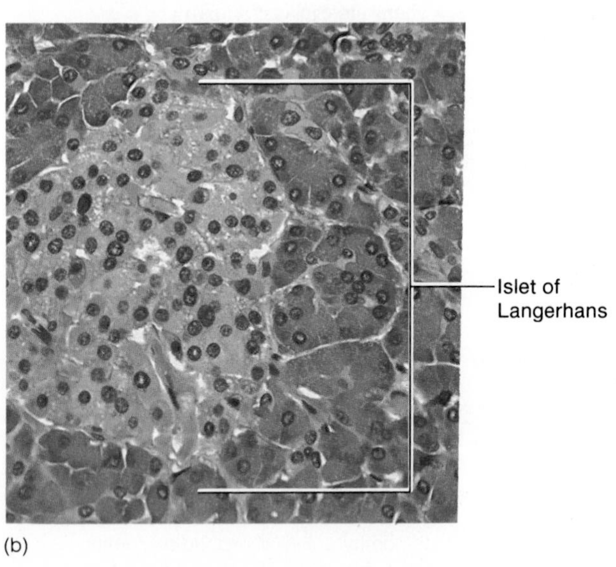

Figure 7.1. *(a) The anatomy of some of th endocrine glands. (b) The islets of Langerhans within the pancreas.*

Hormones: Actions and Interactions

Hormones are regulatory molecules secreted into the blood by endocrine glands. The blood transports hormone molecules to the target organs, where each hormone exerts its characteristic regulatory function through events that occur at a cellular and molecular level. The particular effects of a given hormone are influenced by the concentration of that hormone in the blood and by interactions with effects produced by other hormones.

Endocrine[1] glands, which lack the ducts present in exocrine glands (chapter 5), secrete biologically active chemicals called **hormones[2]** into the blood. Many endocrine glands (fig. 7.1) are organs whose only functions are the production and secretion of hormones. The pancreas, however, functions as both an exocrine and an endocrine gland; the endocrine portion of the pancreas is composed of microscopic structures called the islets of Langerhans (fig. 7.1*b*). In recent years, it has been discovered that many other organs in the body secrete hormones. When these hormones can be demonstrated to have significant physiological functions, the organs that produce these hormones may be categorized as endocrine glands in addition to their other functions. The concept of the **endocrine system** must therefore be extended to include the heart, liver, hypothalamus, and kidneys (table 7.1).

Hormones affect the metabolism of their target organs and, by this means, help regulate (1) total body metabolism, (2) growth, and (3) reproduction. The effects of hormones on body metabolism and growth are discussed in chapter 18; the regulation of reproductive functions by hormones is included in chapter 8.

Chemical Classification of Hormones

Hormones secreted by different endocrine glands are diverse in chemical structure. All hormones, however, can be grouped into three general chemical categories: (1) **catecholamines** (epinephrine and norepinephrine), which are derivatives of an amino acid; (2) **proteins** (table 7.2), which are polypeptides of various lengths; and (3) **steroids,** which are a class of lipids (chapter 3).

Steroid hormones are derived from cholesterol (fig. 7.2). Since they are lipids, they are nonpolar and thus not water-soluble. The gonads—testes and ovaries—secrete *sex steroids;* the adrenal cortex secretes *corticosteroids* such as cortisol, aldosterone, and others.

The thyroid hormones are composed of two derivatives of the amino acid tyrosine bonded together. These hormones are unique because they contain iodine (fig. 7.3). When the hormone contains four iodine atoms, it is called *tetraiodothyronine* (T_4), or *thyroxine.* When it contains three atoms of iodine, it is called *triiodothyronine* (T_3). Although these hormones are not steroids, they are like steroids in that they are relatively small and nonpolar molecules. Steroid and thyroid hormones are active

[1]endocrine: Gk. *endon,* within; *krinein,* to separate
[2]hormone: Gk. *hormon,* to set in motion

Table 7.1 A partial list of the endocrine glands

Endocrine Gland	Major Hormones	Primary Target Organs	Primary Effects
Adrenal cortex	Cortisol Aldosterone	Liver, muscles Kidneys	Glucose metabolism; Na^+ retention, K^+ excretion
Adrenal medulla	Epinephrine	Heart, bronchioles, blood vessels	Adrenergic stimulation
Heart	Atrial natriuretic hormone	Kidneys	Promotes excretion of Na^+ in the urine
Hypothalamus	Releasing and inhibiting hormones	Anterior pituitary	Regulates secretion of anterior pituitary hormones
Intestine	Secretin and cholecystokinin	Stomach, liver, and pancreas	Inhibits gastric motility; stimulates bile and pancreatic juice secretion
Islets of Langerhans (pancreas)	Insulin Glucagon	Many organs Liver and adipose tissue	Insulin promotes cellular uptake of glucose and formation of glycogen and fat; glucagon stimulates hydrolysis of glycogen and fat
Kidneys	Erythropoietin	Bone marrow	Stimulates red blood cell production
Liver	Somatomedins	Cartilage	Stimulates cell division and growth
Ovaries	Estradiol-17β and progesterone	Female genital tract and mammary glands	Maintains structure of genital tract; promotes secondary sexual characteristics
Parathyroids	Parathyroid hormone	Bone, intestine, and kidneys	Increases Ca^{++} concentration in blood
Pineal	Melatonin	Hypothalamus and anterior pituitary	Affects secretion of gonadotropic hormones
Pituitary, anterior	Trophic hormones	Endocrine glands and other organs	Stimulates growth and development of target organs; stimulates secretion of other hormones
Pituitary, posterior	Antidiuretic hormone Oxytocin	Kidneys, blood vessels Uterus, mammary glands	Antidiuretic hormone promotes water retention and vasoconstriction. Oxytocin stimulates contraction of uterus and mammary secretory units
Skin	1,25-Dihydroxyvitamin D_3	Intestine	Stimulates absorption of Ca^{++}
Stomach	Gastrin	Stomach	Stimulates acid secretion
Testes	Testosterone	Prostate, seminal vesicles, other organs	Stimulates secondary sexual development
Thymus	Thymosin	Lymph nodes	Stimulates white blood cell production
Thyroid	Thyroxine (T_4) and triiodothyronine (T_3)	Most organs	Growth and development; stimulates basal rate of cell respiration (basal metabolic rate or BMR)

Table 7.2 Some examples of polypeptide and glycoprotein hormones

Hormone	Structure	Gland	Primary Effects
Antidiuretic hormone	8 amino acids	Posterior pituitary	Water retention and vasoconstriction
Oxytocin	8 amino acids	Posterior pituitary	Uterine and mammary contraction
Insulin	21 and 30 amino acids (double chain)	β cells in islets of Langerhans	Cellular glucose uptake; formation of fat and glycogen
Glucagon	29 amino acids	α cells in islets of Langerhans	Hydrolysis of stored glycogen and fat
ACTH	39 amino acids	Anterior pituitary	Stimulation of adrenal cortex
Parathyroid hormone	84 amino acids	Parathyroid	Increases blood Ca^{++} concentration
FSH, LH, TSH	Glycoproteins	Anterior pituitary	Stimulates growth, development, and secretion of target glands

when taken orally (as a pill); sex steroids are contained in the contraceptive pill, and thyroid hormone pills are taken by people whose thyroid is deficient (who are hypothyroid). Other types of hormones cannot be taken orally because they would be digested into inactive fragments before they were absorbed into the blood.

Common Aspects of Neural and Endocrine Regulation

It may be thought that endocrine regulation is chemical and therefore distinct from the electrical nature of neural control systems. This idea is incorrect. Electrical nerve impulses are in

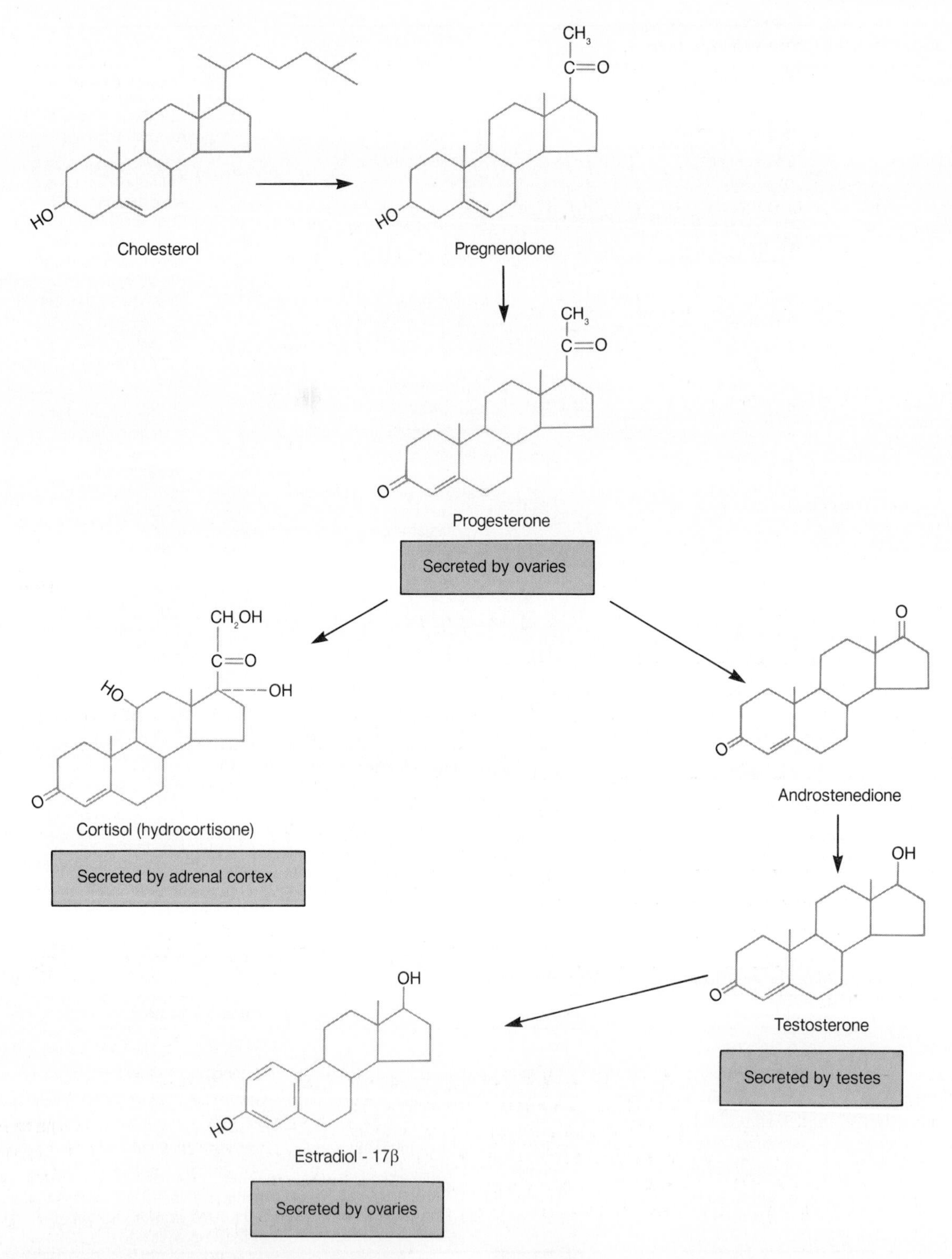

Figure 7.2. *Simplified biosynthetic pathways for steroid hormones. Notice that progesterone (a hormone secreted by the ovaries) is a common precursor in the formation of all other steroid hormones and that testosterone (the major androgen secreted by the testes) is a precursor in the formation of estradiol-17β, the major estrogen secreted by the ovaries.*

fact chemical events produced by the diffusion of ions through the neuron cell membrane (chapter 6). Also, most nerve fibers stimulate the cells they innervate through the release of a chemical neurotransmitter. Neurotransmitters differ from hormones in that they do not travel in the blood but instead diffuse only a very short distance across a synapse. In other respects, however, the actions of neurotransmitters and hormones are very similar.

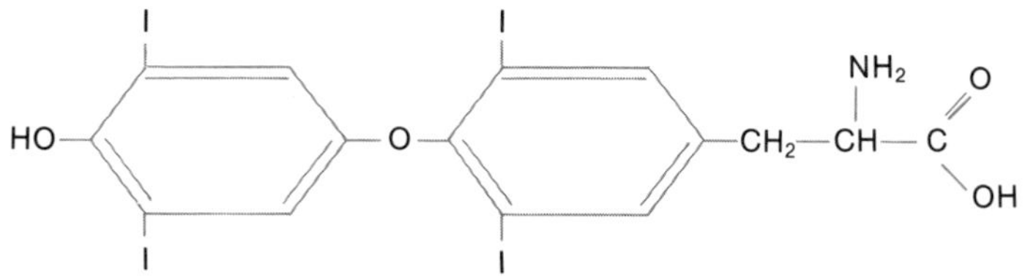

Thyroxine, or tetraiodothyronine (T_4)

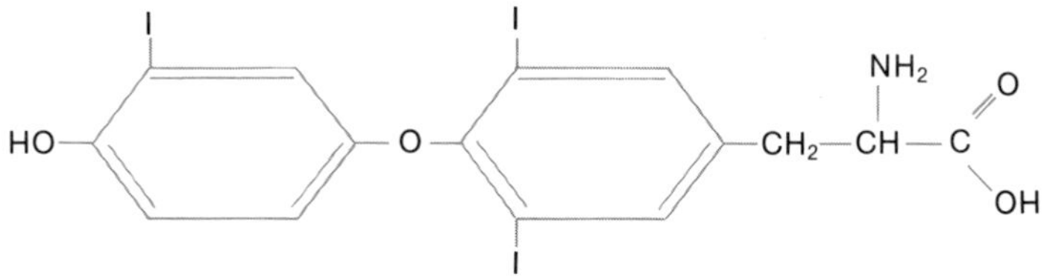

Triiodothyronine (T_3)

Figure 7.3. *The thyroid hormones: thyroxine (T_4) and triiodothyronine (T_3) are secreted in a ratio of 9:1.*

Social Issues

Athletes and Anabolic Steroids

The term **anabolic steroids** refers to male sex hormones (androgens) and their derivatives produced pharmaceutically. These hormones, delivered as pills or injections, may act to enhance the building of muscle mass in people who exercise intensively. This use was endorsed by the Nazis, who envisioned building an army of "supermen." Soviet weight lifters used them in Olympic competition in the 1950s, as did American weight lifters in the 1960s. In 1976, the International Olympic Committee proscribed their use and instituted mandatory urine tests for these compounds. The illicit use of anabolic steroids among Olympic athletes was revealed most dramatically in the 1988 Olympic games at Seoul, in which the gold medal won by Canadian track star Ben Johnson was forfeited when they found stanozolol (an anabolic steroid) in his urine.

Use of anabolic steroids by women causes growth of facial hair, deepening of the voice, and the development of a masculine physique. These effects are certainly predictable from the androgenic nature of these drugs. In a similarly predictable manner, use of anabolic steroids by adolescents can cause premature baldness and put a halt to their growth. Men who take anabolic steroids often develop *gynecomastia*—the development of femalelike mammary tissue. This occurs because the liver can change androgens into estrogens (female sex steroids); the mammary tissue stimulated to grow by these estrogens must be surgically removed. Another common side effect of anabolic steroids is shrinkage of the testes. This effect is produced by androgen-induced inhibition of the pituitary's secretion of gonadotropic hormones (FSH and LH), which are needed to maintain the structure and function of the testes. Production of sperm is thus diminished (causing sterility), and the secretion of testosterone (the major androgen secreted by the testes) is reduced. For example, the testosterone concentration in Ben Johnson's blood at the time his urine was tested was only 15% of the normal value.

In addition to its endocrine effects, the excessive use of anabolic steroids increases the risk of liver damage and liver cancer, kidney disease, and heart disease. These drugs also promote aggressiveness and antisocial behavior, depression, and even psychotic symptoms. Despite all of these negative effects, illegal sales of anabolic steroids total over 100 million dollars annually. Saddest of all, the majority of the users of these drugs are not professional athletes. Rather, they are young men and women who desire an exaggerated growth of muscle mass and who lack the patience to grow into this body image in a slower, more natural, manner. It is very important for people to realize that a highly muscled body is unrelated to health; indeed, it is actually unhealthy if it is achieved through the use of anabolic steroids.

Regardless of whether a particular chemical is acting as a neurotransmitter or as a hormone, in order for it to function in physiological regulation: (1) target cells must have specific **receptor proteins** that combine with the chemical; (2) the combination of the regulator molecule with its receptor proteins must cause a specific sequence of changes in the target cells; and (3) there must be a mechanism to rapidly turn off the action of the regulator; without an "off switch," physiological control is impossible. This last process involves rapid removal and/or chemical inactivation of the regulator molecules.

Effects of Hormone Concentrations on Tissue Response

The effects of hormones are very concentration dependent. Normal tissue responses are only produced when the hormones are present within their normal, or *physiological,* range of concentrations. When some hormones are taken in abnormally high, or *pharmacological,* concentrations (as when they are taken as drugs), they may produce undesirable effects.

Pharmacological doses of hormones, particularly of steroids, can have widespread and often damaging "side effects." People with inflammatory diseases who are treated with high doses of cortisone over long periods of time, for example, may develop characteristic changes in bone and soft tissue structure. Contraceptive pills, which contain sex steroids, have a number of potential side effects that could not have been predicted at the time the pill was first introduced.

Hormone Interactions

A given target tissue is usually responsive to a number of different hormones. These hormones may antagonize each other or work together to produce their effects. The response of a target tissue to a particular hormone is thus affected not only by the concentration of that hormone, but also by the effects of other hormones on that tissue.

When two or more hormones work together to produce a particular result their effects are said to be **synergistic.** In some cases, these effects are additive. Epinephrine and norepinephrine, for example, separately produce an increase in heart rate; if these hormones are given together, the stimulation of heart rate is increased in an additive fashion. The synergistic action of FSH and testosterone is an example of a complementary effect; each hormone separately stimulates a different stage of sperm development during puberty, so that both hormones together are needed at that time to complete sperm development. Likewise, the ability of mammary glands to produce and secrete milk requires the synergistic action of many hormones—estrogen, cortisol, prolactin, oxytocin, and others.

A hormone is said to have a **permissive** effect on the action of a second hormone when it enhances the effect of the second hormone. Prior exposure of the uterus to estrogen, for example, induces the formation of receptor proteins for progesterone, which improves the response of the uterus when it is subsequently exposed to progesterone. Estrogen thus has a permissive effect on the responsiveness of the uterus to progesterone.

In some situations the actions of one hormone antagonize the effects of another. Lactation during pregnancy, for example, is prevented because the high concentration of estrogen in the blood inhibits the secretion and action of prolactin. Another example of antagonism is the action of insulin and glucagon (two hormones from the islets of Langerhans) on adipose tissue; the formation of fat is promoted by insulin, whereas glucagon promotes fat breakdown.

1. Distinguish between the effects of physiological and pharmacological concentrations of a hormone.
2. Explain why men who take anabolic steroids may develop gynecomastia.
3. Explain the ways that the effects of different hormones can interact, and provide examples of these interactions.

Mechanisms of Hormone Action

Each hormone exerts its characteristic effects on a target organ through the actions it has on the cells of these organs. Those hormones that are nonpolar pass through the target cell membrane and act directly within the target cell, whereas those that are polar act on the target cell membrane to cause the production of intracellular second messenger molecules.

Hormones are delivered by the blood to every cell in the body, but only the **target cells** are able to respond to each hormone. In order to respond to a hormone, a target cell must have specific receptor proteins for that hormone. Receptor protein–hormone interaction is highly specific, much like the interaction of an enzyme with its substrate (chapter 3).

The location of a hormone's receptor proteins in its target cells depends on the chemical nature of the hormone. Based on the location of the receptor proteins, hormones can be grouped into three categories: (1) receptor proteins within the nucleus of target cells—*thyroid hormones;* (2) receptor proteins within the cytoplasm of target cells—*steroid hormones;* and (3) receptor proteins in the outer surface of the target cell membrane—*catecholamine* and *protein hormones.* This information is summarized in table 7.3.

Mechanisms of Steroid Hormones and Thyroxine Action

Steroid and thyroid hormones are similar in size and in the fact that they are nonpolar and thus are not very water-soluble. Unlike other hormones, therefore, steroids and thyroid hormones (primarily thyroxine) do not travel dissolved in the aqueous portion of the plasma but instead are transported to their target cells attached to plasma carrier proteins. These hormones then dissociate from the carrier proteins in the blood and easily pass through the lipid component of the target cell's membrane.

Steroid Hormones

Once through the cell membrane, steroid hormones attach to *cytoplasmic receptor proteins* in the target cells. The steroid hormone-receptor protein complex then *translocates* (moves) to the nucleus and attaches by means of the receptor proteins to the chromatin (strands of DNA and protein). The sites of attachment in the chromatin, termed *acceptor sites,* are specific for the target tissue. According to one theory, part of the receptor bonds to a protein while a different part of the receptor bonds to DNA.

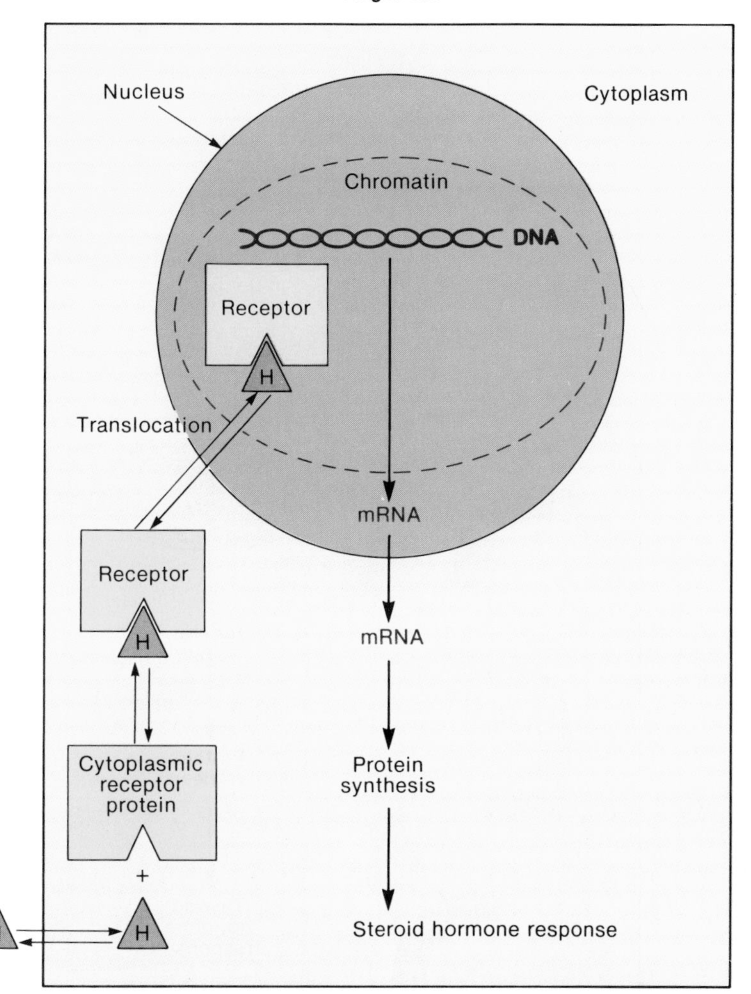

Figure 7.4. The mechanism of the action of a steroid hormone (H) on the target cells.

Table 7.3 Functional categories of hormones, based on the location of their receptor proteins and the mechanisms of their action

Types of Hormones	Secreted by	Location of Receptors	Effects of Hormone-Receptor Interaction
Catecholamines and proteins	All glands except adrenal cortex, gonads, and thyroid	Outer surface of cell membrane	Stimulates production of intracellular "second messenger," which activates previously inactive enzymes
Steroids	Adrenal cortex, testes, ovaries	Cytoplasm of target cells	Stimulates translocation of hormone-receptor complex to nucleus and activation of specific genes
Thyroxine (T_4)	Thyroid	Nucleus of target cells	After conversion to triiodothyronine (T_3), activates specific genes

The attachment of the receptor protein-steroid complex to the acceptor site "turns on" genes. Specific genes become activated by this process and produce messenger RNA (mRNA), which codes for the production of new proteins. (The genetic control of protein synthesis is explained in chapter 10.) Since some of these newly synthesized proteins may be enzymes, the metabolism of the target cell is thus changed in a specific manner (fig. 7.4). Steroid hormones, in short, affect their target cells by stimulating the expression of specific genes.

Thyroxine

The major hormone secreted by the thyroid gland is thyroxine, or **tetraiodothyronine (T_4).** Like steroid hormones, thyroxine travels in the blood attached to carrier proteins. The thyroid also secretes a small amount of **triiodothyronine, or T_3.**

Approximately 99.96% of the thyroxine in the blood is attached to carrier proteins in the plasma; the rest is free. Only the free thyroxine and T_3 can enter target cells; the protein-bound thyroxine serves as a reservoir of this hormone in the blood

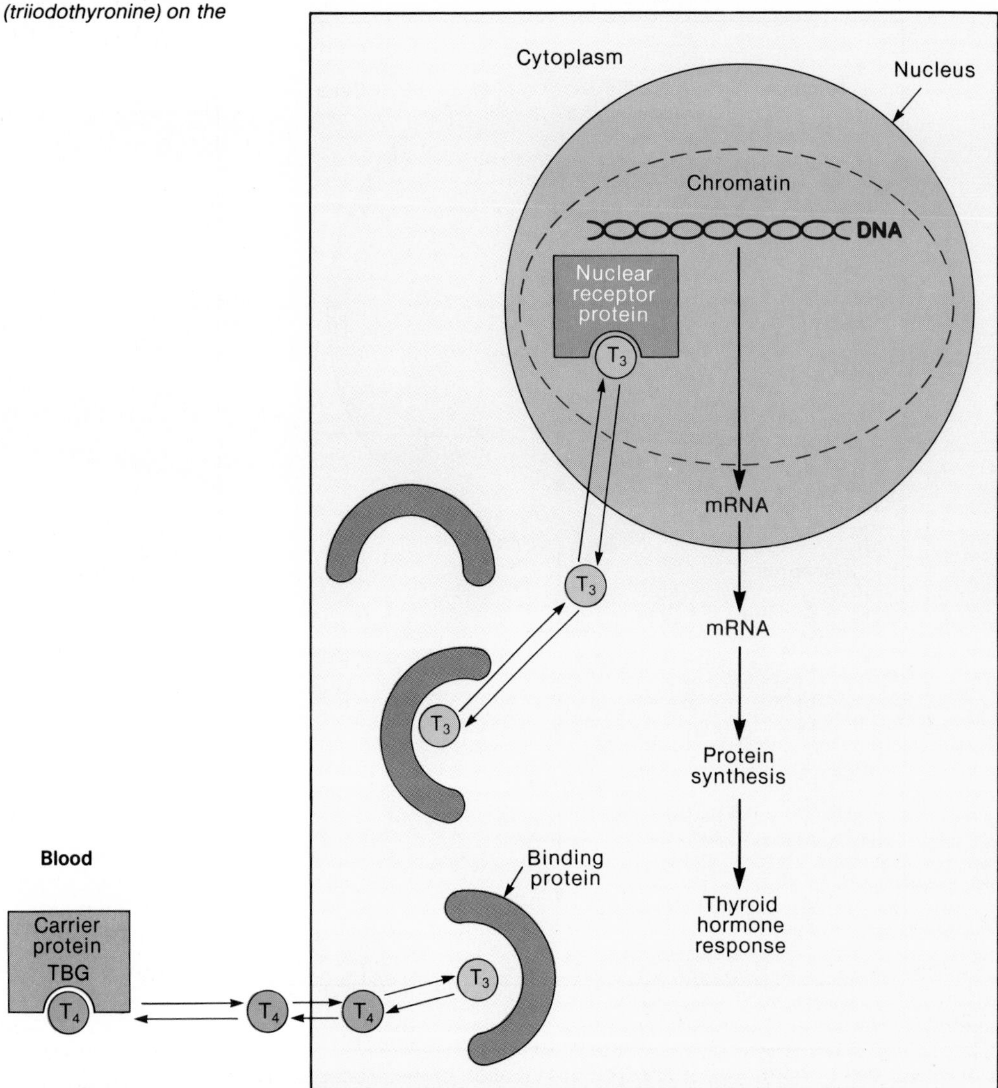

Figure 7.5. *The mechanism of the action of thyroxine (T_4) and T_3 (triiodothyronine) on the target cells.*

(this is why it takes a couple of weeks after surgical removal of the thyroid for the symptoms of hypothyroidism to develop). Once the free thyroxine passes into the target cell cytoplasm, it is enzymatically converted into T_3. It is the T_3 rather than T_4 which is the active form of the hormone within the target cells.

Inactive T_3 receptor proteins are already in the nucleus attached to chromatin. These receptors are inactive until T_3 enters the nucleus from the cytoplasm. The attachment of T_3 to the chromatin-bound receptor proteins activates genes and results in the production of new mRNA and new proteins. This sequence of events is summarized in figure 7.5.

Mechanism of Catecholamine and Protein Hormone Action: Second Messengers

Catecholamine and protein hormones cannot pass through the lipid barrier of the target cell membrane. The effects of these hormones is thus believed to result from interaction of these hormones with receptor proteins in the outer surface of the target cell membrane. Since these hormones do not enter the target cells to exert their effects, other molecules must mediate the actions of these hormones within the target cells. If you think of hormones as "messengers" from the endocrine glands, the intracellular mediators of the hormone's action can be called **second messengers.**

Cyclic AMP

Cyclic adenosine monophosphate (abbreviated **cAMP**) was the first "second messenger" to be discovered and is the best understood. The hormonal effects of epinephrine (adrenaline) and norepinephrine (together known as catecholamines), are due to cAMP production within the target cells. It was later discovered that the effects of many protein hormones are also mediated by cAMP.

The bonding of these hormones to their membrane receptor proteins activates an enzyme called **adenylate cyclase.** This

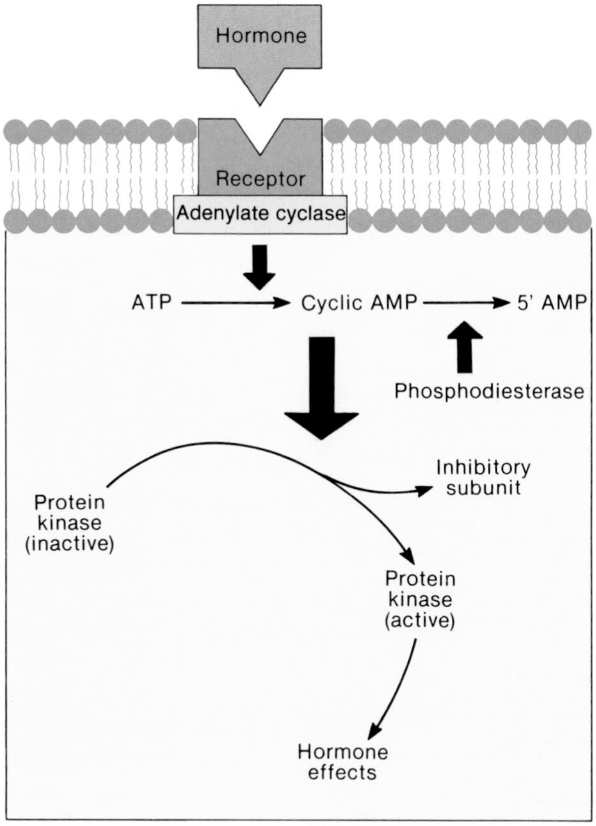

Figure 7.6. Cyclic AMP (cAMP) as a second messenger in the action of catecholamine and protein hormones.

enzyme is built into the cell membrane and, when activated, it catalyzes the following reaction:

$$ATP \rightarrow cAMP + PP_i$$

Adenosine triphosphate (ATP) is thus converted into cAMP plus two inorganic phosphates, abbreviated PP_i. As a result of the interaction of the hormone with its receptor and the activation of adenylate cyclase, therefore, the intracellular concentration of cAMP is increased. Cyclic AMP activates a previously inactive enzyme in the cytoplasm called **protein kinase.** The inactive form of this enzyme consists of two subunits: a catalytic subunit and an inhibitory subunit. The enzyme is produced in an inactive form and becomes active only when cAMP attaches to the inhibitory subunit. Bonding of cAMP to the inhibitory subunit causes it to dissociate from the catalytic subunit, which then becomes active (fig. 7.6). The hormone, in summary—acting through an increase in cAMP production—causes an increase in protein kinase enzyme activity within its target cells.

Active protein kinase catalyzes the attachment of phosphate groups to different proteins in the target cells. This causes some enzymes to become activated, and others to become inactivated. Cyclic AMP, acting through protein kinase, thus modulates the activity of enzymes that are already present in the target cell. This alters the metabolism of the target tissue in a manner characteristic of the actions of that specific hormone (table 7.4).

Table 7.4 Sequence of events that occurs with cyclic AMP as a second messenger

1. The hormones combine with their receptors on the outer surface of target cell membranes.
2. Hormone-receptor interaction stimulates activation of adenylate cyclase on the cytoplasmic side of the membranes.
3. Activated adenylate cyclase catalyzes the conversion of ATP to cyclic AMP (cAMP) within the cytoplasm.
4. Cyclic AMP activates protein kinase enzymes that were already present in the cytoplasm in an inactive state.
5. Activated cAMP-dependent protein kinase transfers phosphate groups to (phosphorylates) other enzymes in the cytoplasm.
6. The activity of specific enzymes is either increased or inhibited by phosphorylation.
7. Altered enzyme activity mediates the target cell's response to the hormone.

---

1. Using diagrams, describe how steroid hormones and thyroxine exert their effects on their target cells.
2. Using a diagram, describe how cyclic AMP is produced within a target cell in response to hormone stimulation. List three hormones that use cAMP as a second messenger.
3. Explain how cAMP functions as a second messenger in the action of some hormones.

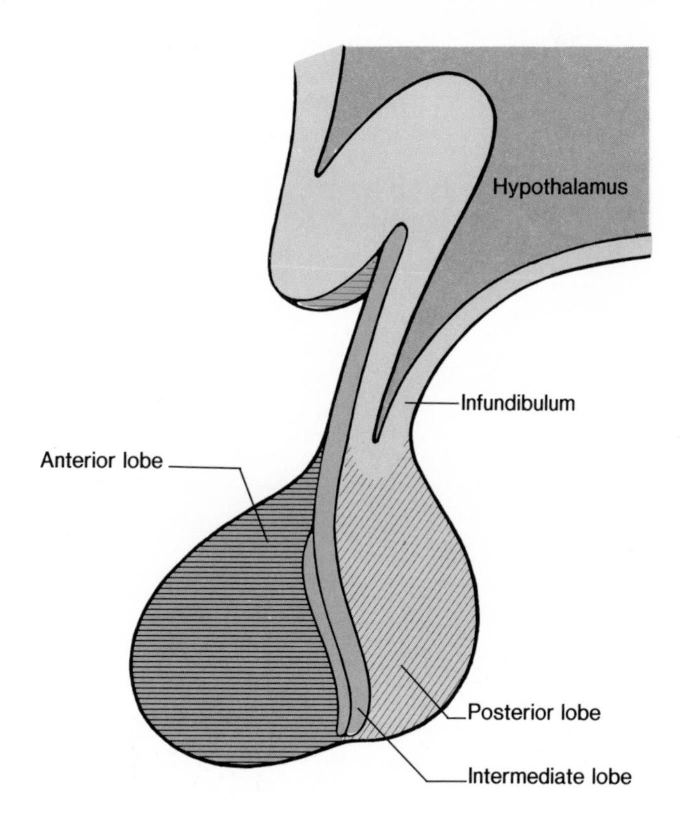

Figure 7.7. *The structure of the pituitary gland as seen in sagittal view.*

Pituitary Gland

The pituitary gland is divided into an anterior portion and a posterior portion. The posterior pituitary secretes hormones that are actually produced by the hypothalamus, whereas the anterior pituitary produces and secretes its own hormones. The anterior pituitary is under the control of the hypothalamus, however, by way of releasing hormones secreted by the hypothalamus into a system of blood vessels. The anterior pituitary responds by secreting specific hormones which act on other endocrine glands.

The pituitary[3] gland is located on the undersurface of the brain below the hypothalamus (chapter 6). The pituitary is a rounded, pea-shaped gland measuring about 1.3 cm (0.5 in.) in diameter, and is attached to the hypothalamus by a stalklike structure called the infundibulum (fig. 7.7).

The pituitary gland is structurally and functionally divided into an anterior lobe and a posterior lobe. These two parts have different embryonic origins. The anterior pituitary is derived from a pouch of epithelial tissue that migrates upwards from the embryonic mouth, whereas the posterior pituitary is formed as a downgrowth of the brain. These parts are illustrated in figure 7.7.

Pituitary Hormones

The hormones secreted by the **anterior pituitary** are called **trophic hormones.** The term *trophic* means "food." Although the anterior pituitary hormones are not food for their target organs, this term is used because high amounts of the anterior pituitary hormones make their target organs hypertrophy (grow), while low amounts cause their target organs to atrophy (shrink). When names are applied to the hormones of the anterior pituitary, therefore, the "trophic" term—conventionally shortened to *tropic* (which has a different meaning—"attracted to"), is incorporated into these names. This is also indicated by the shortened forms of the names for the anterior pituitary hormones, which end in the suffix *-tropin.* The hormones of the anterior pituitary (table 7.5) are:

Growth hormone (GH, or somatotropin). This hormone promotes the movement of amino acids into tissue cells and the incorporation of these amino acids into tissue proteins, thus stimulating growth of organs.

Thyroid-stimulating hormone (TSH, or thyrotropin). This hormone stimulates the thyroid gland to produce and secrete thyroxine (tetraiodothyronine, or T_4).

Adrenocorticotropic hormone (ACTH, or corticotropin). This hormone stimulates the adrenal cortex to secrete steroid hormones such as hydrocortisone (cortisol).

Follicle-stimulating hormone (FSH, or folliculotropin). This hormone stimulates the growth and secretion of ovarian follicles in females and the production of sperm in the testes of males.

Luteinizing hormone (LH, or luteotropin). This hormone and FSH are collectively called **gonadotropic hormones.** In females, LH stimulates ovulation and the conversion of the ovulated ovarian follicle into an endocrine structure called a corpus luteum (chapter 8). In males, LH stimulates the secretion of male sex hormones (mainly testosterone) from the testes.

Prolactin. This hormone is secreted in both males and females. Its best known function is the stimulation of milk production by the mammary glands of women after the birth of their babies. Prolactin plays a supporting role in the regulation of the male reproductive system by the gonadotropins (FSH and LH), and acts on the kidneys to help regulate water and electrolyte balance.

Inadequate growth hormone secretion during childhood causes *pituitary dwarfism.* Inadequate secretion of growth hormone in an adult produces a rare condition called *pituitary cachexia (Simmonds' disease).* One of the symptoms of this disease is premature aging caused by tissue atrophy. Oversecretion of growth hormone during childhood, in contrast, causes *gigantism.* Excessive growth hormone secretion in an adult does not cause further growth in length because the growth discs of cartilage have converted to bone (chapter 21). Instead, hypersecretion of growth hormone in an adult causes *acromegaly.*[4] In acromegaly, the person's appearance gradually changes as a result of thickening of bones and the growth of soft tissues, particularly in the face, hands, and feet (fig. 7.8).

[3]pituitary: L. *pituita,* phlegm (this gland was originally thought to secrete mucus into the nasal cavity)

[4]acromegaly: Gk. *akron,* extremity; *megas,* large

Table 7.5 Hormones secreted by the anterior pituitary

Hormone	Target Tissue	Stimulated by Hormone	Regulation of Secretion
ACTH (adrenocorticotropic hormone)	Adrenal cortex	Secretion of glucocorticoids	Stimulated by CRH (corticotropin-releasing hormone); inhibited by glucocorticoids
TSH (thyroid-stimulating hormone)	Thyroid gland	Secretion of thyroid hormones	Stimulated by TRH (thyrotropin-releasing hormone); inhibited by thyroid hormones
GH (growth hormone)	Most tissue	Protein synthesis and growth; lipolysis and increased blood glucose	Inhibited by somatostatin; stimulated by growth hormone-releasing hormone
FSH (follicle-stimulating hormone) and LH (luteinizing hormone)	Gonads	Gamete production and sex steroid hormone secretion	Stimulated by GnRH (gonadotropin-releasing hormone); inhibited by sex steroids
Prolactin	Mammary glands and other sex accessory organs	Milk production Controversial actions in other organs	Inhibited by PIH (prolactin-inhibiting hormone)
LH (luteinizing hormone)	Gonads	Sex hormone secretion; ovulation and corpus luteum formation	Stimulated by GnRH

The posterior pituitary secretes only two hormones, both of which are produced in the hypothalamus and merely stored in the posterior lobe of the pituitary:

Antidiuretic hormone (ADH, or vasopressin). Antidiuretic hormone stimulates the kidneys to retain water so that less water is excreted in the urine and more water is retained in the blood. This hormone also causes vasoconstriction in experimental animals, although the significance of this effect in humans is controversial.

Oxytocin. This hormone has no known function in males, but in females it is known to stimulate contractions of the uterus during labor and contractions of the mammary gland tissue, which result in the milk-ejection reflex during lactation.

Injections of oxytocin may be given to a woman during labor if she is having difficulties in parturition (childbirth). Increased amounts of oxytocin assist uterine contractions and generally speed up delivery. Oxytocin administration after parturition causes the uterus to regress in size and squeezes the blood vessels, thus minimizing the danger of hemorrhage.

Figure 7.8. The progression of acromegaly in one individual, from age nine (a), sixteen (b), thirty-three (c), and fifty-two (d) years. The coarsening of features and disfigurement are evident by age thirty-three and severe at age fifty-two.

Control of the Posterior Pituitary

The **posterior pituitary** secretes two hormones: antidiuretic hormone (ADH) and oxytocin. These two hormones, however, are actually produced in neuron cell bodies located in specific regions of the hypothalamus. These neurons within the hypothalamus are thus endocrine cells; the hormones they produce are transported along an axon tract (fig. 7.9) to the posterior pituitary, which stores and later secretes these hormones. The posterior pituitary is thus more a storage organ than a true gland.

The secretion of ADH and oxytocin from the posterior pituitary is controlled by **neuroendocrine reflexes.** In nursing mothers, for example, the stimulus of sucking acts via sensory nerve impulses to the hypothalamus to stimulate the reflex secretion of oxytocin. The secretion of ADH is stimulated by neurons in the hypothalamus in response to a rise in blood concentration (chapter 4). These reflexes are discussed in later chapters.

Control of the Anterior Pituitary

At one time the anterior pituitary was called the "master gland" because it secretes hormones that regulate some other endocrine glands (fig. 7.10 and table 7.5). Adrenocorticotropic hormone (ACTH), thyroid-stimulating hormone (TSH), and the gonadotropic hormones (FSH and LH) stimulate the adrenal cortex, thyroid, and gonads, respectively, to secrete their hormones. The anterior pituitary hormones also have a "trophic" effect on their target glands, in that the structure and health of the target glands depend on adequate stimulation by anterior pituitary hormones. The anterior pituitary, however, is not really the master gland, because secretion of its hormones is in turn controlled by hormones secreted by the hypothalamus.

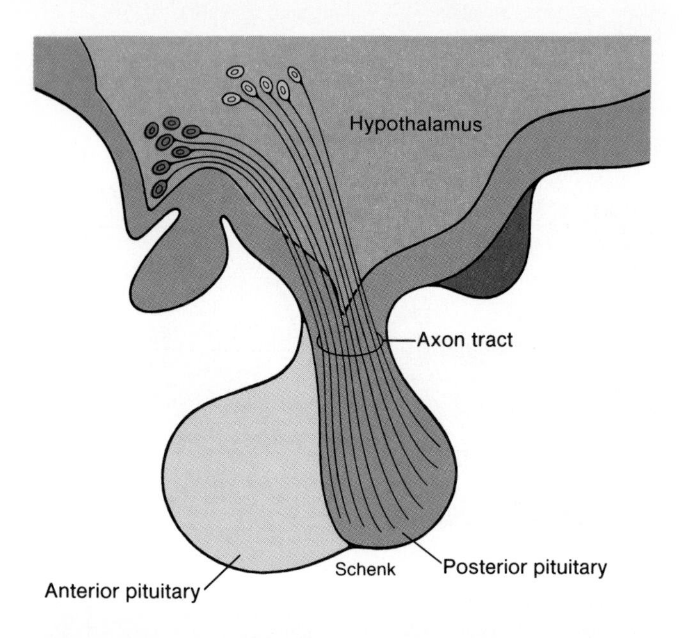

Figure 7.9. The posterior pituitary stores and secretes hormones (vasopressin and oxytocin) produced in neuron cell bodies of the hypothalamus. These hormones are transported to the posterior pituitary by an axon tract.

Releasing and Inhibiting Hormones

Since axons do not enter the anterior pituitary, hypothalamic control of the anterior pituitary is achieved through hormonal rather than neural regulation. Neurons in the hypothalamus produce releasing and inhibiting hormones, which are transported to axon endings in the basal portion of the hypothalamus. This region contains blood capillaries that are drained by venules in the stalk of the pituitary.

The venules that drain the hypothalamus deliver blood to a second capillary bed in the anterior pituitary. Since this second capillary bed receives venous blood from another organ, the vascular link between the hypothalamus and the anterior pituitary comprises a *portal system.* Since an alternate name for the pituitary is *hypophysis,* the vascular link between the hypothalamus and the anterior pituitary is called the **hypothalamo-hypophyseal portal system.**

Neurons of the hypothalamus secrete hormones into this portal system that regulate the secretions of the anterior pituitary (fig. 7.11 and table 7.6). Thyrotropin-releasing hormone (**TRH**) stimulates the secretion of TSH, and corticotropin-releasing hormone (**CRH**) stimulates the secretion of ACTH from the anterior pituitary. A single releasing hormone, gonadotropin-releasing hormone, or **GnRH,** stimulates the secretion of both gonadotropic hormones (FSH and LH) from the anterior pituitary. The secretion of prolactin and of growth hormone from the anterior pituitary is regulated by hypothalamic inhibitory hormones, known as **PIH** (prolactin-inhibiting hormone) and **somatostatin,** respectively.

Table 7.6 Some hypothalamic hormones involved in the control of the anterior pituitary

Hypothalamic Hormone	Structure	Effect on Anterior Pituitary	Action of Anterior Pituitary Hormone
Corticotropin-releasing hormone (CRH)	41 amino acids	Stimulates secretion of adrenocorticotropic hormone (ACTH)	Stimulates secretions of adrenal cortex
Gonadotropin-releasing hormone (GnRH)	10 amino acids	Stimulates secretion of follicle-stimulating hormone (FSH) and luteinizing hormone (LH)	Stimulates gonads to produce gametes (sperm and ova) and secrete sex steroids
Prolactin-inhibiting hormone (PIH)	Controversial (may be dopamine)	Inhibits prolactin secretion	Stimulates production of milk in mammary glands
Somatostatin	14 amino acids	Inhibits secretion of growth hormone	Stimulates anabolism and growth in many organs
Thyrotropin-releasing hormone (TRH)	3 amino acids	Stimulates secretion of thyroid-stimulating hormone (TSH)	Stimulates secretion of thyroid gland
Growth hormone–releasing hormone (GHRH)	44 amino acids	Stimulates growth hormone secretion	Stimulates anabolism and growth in many organs

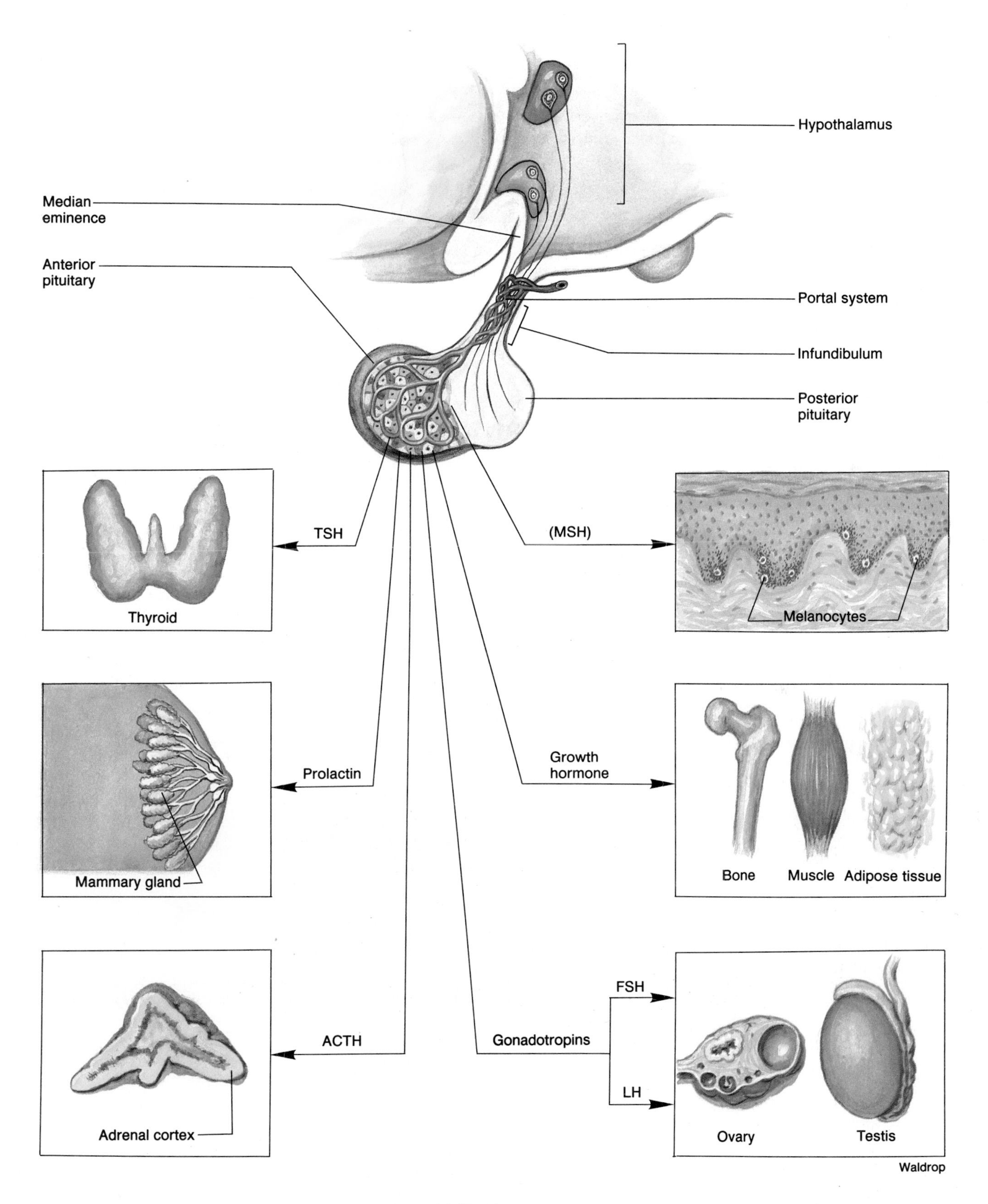

Figure 7.10. *The hormones secreted by the anterior pituitary and the target organs for those hormones.*

Recently, a specific **growth hormone-releasing hormone (GHRH)** that stimulates growth hormone secretion has been identified as a polypeptide consisting of forty-four amino acids. Experiments suggest that a releasing hormone for prolactin may also exist, but no such specific releasing hormone has yet been discovered.

Feedback Control of the Anterior Pituitary

In view of its secretion of releasing and inhibiting hormones, the hypothalamus might be considered the "master gland." The chain of command, however, is not linear; the hypothalamus and anterior pituitary are controlled by the effects of their own actions. In the endocrine system, to use an analogy, the general takes orders from the private. The hypothalamus and anterior pituitary are not master glands because their secretions are controlled by the target glands they regulate.

Anterior pituitary secretion of ACTH, TSH, and the gonadotropins (FSH and LH) is controlled by **negative feedback inhibition** from the target gland hormones. Secretion of ACTH is inhibited by a rise in hormone secretion from the adrenal cortex, for example, and TSH is inhibited by a rise in the secretion of thyroxine from the thyroid. These negative feedback relationships are easily demonstrated by removal of the target glands. Castration (surgical removal of the gonads), for example, produces a rise in the secretion of FSH and LH. In a similar manner, removal of the adrenals or the thyroid would result in an abnormal increase in ACTH or TSH secretion from the anterior pituitary.

These effects demonstrate that under normal conditions the target glands exert an inhibitory effect on the anterior pituitary. This inhibitory effect can occur at two levels: (1) the target gland hormones could act on the hypothalamus and inhibit the secretion of releasing hormones, and (2) the target gland hormones could act on the anterior pituitary and inhibit its response to the releasing hormones. Thyroxine, for example, appears to inhibit the response of the anterior pituitary to TRH and thus acts to reduce TSH secretion (fig. 7.12). Sex steroids, in contrast, reduce the secretion of gonadotropins by inhibiting both GnRH secretion from the hypothalamus and the ability of the anterior pituitary to respond to stimulation by GnRH (see fig. 7.13).

Higher Brain Function and Pituitary Secretion

The feedback effect of estradiol on the secretion of gonadotropic hormones is believed to be exerted at the level of the pituitary gland and hypothalamus. Since the hypothalamus receives neural input from "higher brain centers" (chapter 6), however, it is not surprising that the pituitary-gonad axis can be affected by emotions, so that intense emotions may alter the timing of

Figure 7.11. *Neurons in the hypothalamus secrete releasing hormones (shown as dots) into the blood vessels of the hypothalamo-hypophyseal portal system. These releasing hormones stimulate the anterior pituitary to secrete its hormones into the general circulation.*

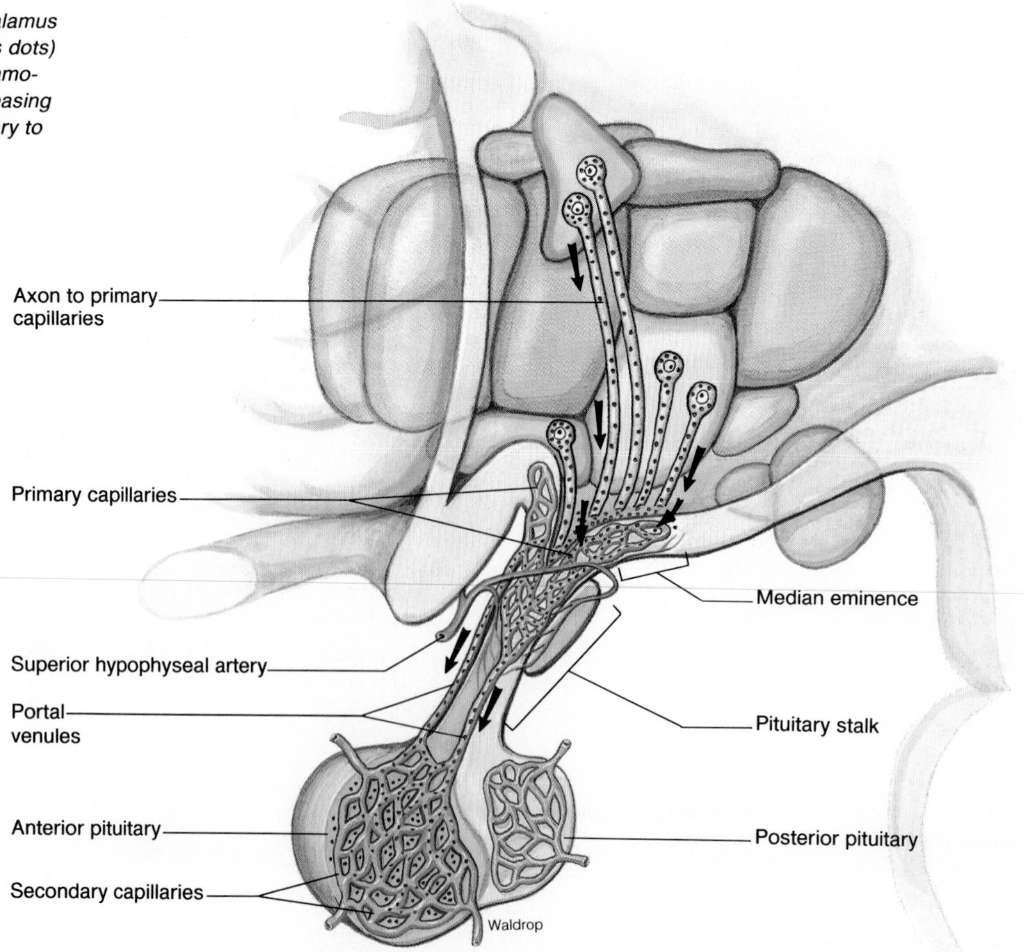

ovulation or menstruation. The influences of higher brain centers on the pituitary-gonad axis also helps to explain the "dormitory effect," in which researchers have noted a tendency for the menstrual cycles to synchronize in women who room together.

This synchronization of menstrual cycles does not occur in a new roommate if her nasal cavity is plugged with cotton, suggesting that the dormitory effect is due to the action of **pheromones.** Pheromones are chemicals excreted by an individual which act through the olfactory sense and modify the physiology or behavior of other members of the same species. Pheromones are important regulatory molecules which help regulate the reproductive cycles and behavior of many lower mammals. The importance of pheromones in human physiology and behavior is much more limited, but—because of our frequent cleansing, use of deodorants, and applications of artificial scents—is difficult to assess.

1. List the hormones secreted by the posterior pituitary. Describe the site of origin of these hormones and the mechanisms by which their secretions are regulated.
2. List the hormones secreted by the anterior pituitary, and describe how the hypothalamus controls the secretion of each hormone.
3. Draw a negative feedback loop showing the control of ACTH secretion. Explain how this system would be affected by (a) an injection of ACTH; (b) surgical removal of the pituitary; (c) an injection of corticosteroids; and (d) surgical removal of the adrenal glands.

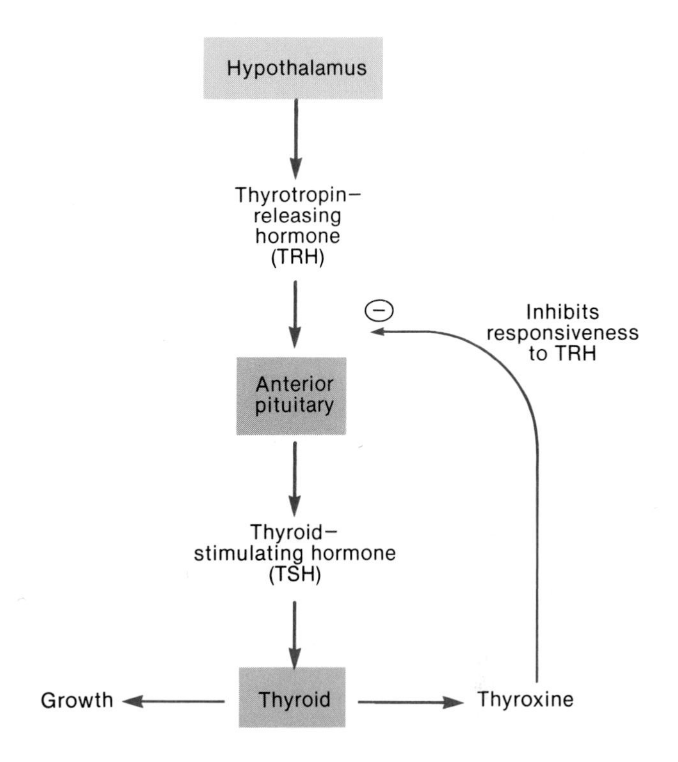

Figure 7.12. *The secretion of thyroxine from the thyroid is stimulated by the thyroid-stimulating hormone (TSH) from the anterior pituitary. The secretion of TSH is stimulated by the thyrotropin-releasing hormone (TRH) secreted from the hypothalamus. This stimulation is balanced by the negative feedback inhibition of thyroxine.*

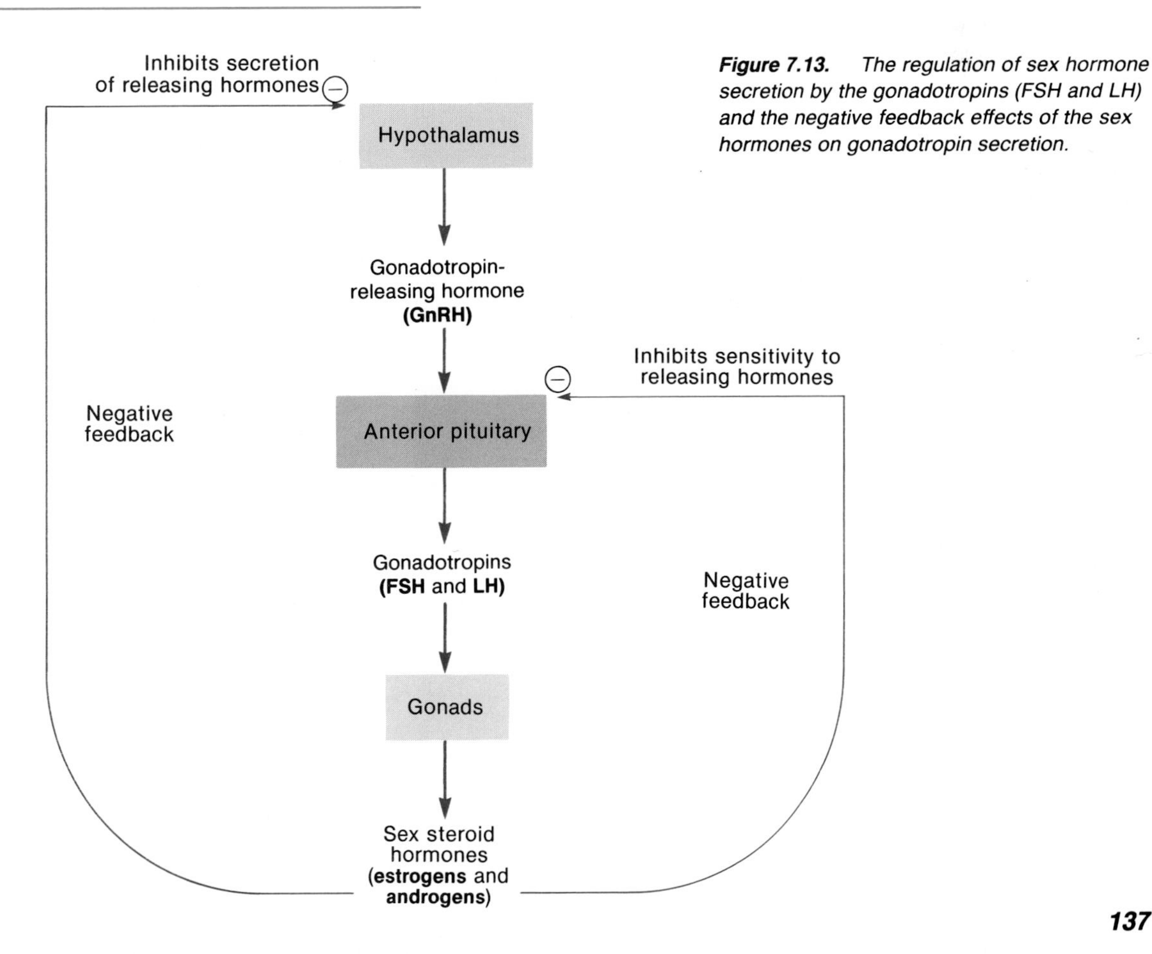

Figure 7.13. *The regulation of sex hormone secretion by the gonadotropins (FSH and LH) and the negative feedback effects of the sex hormones on gonadotropin secretion.*

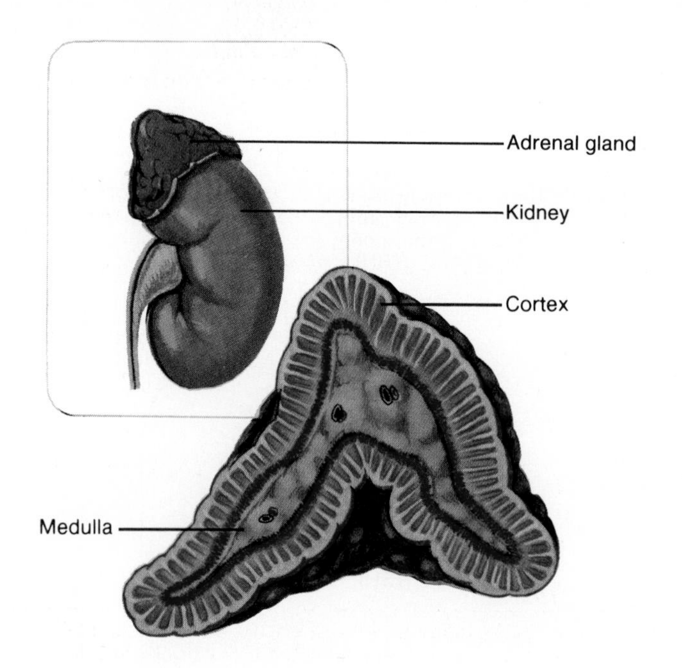

Figure 7.14. *The structure of the adrenal gland.*

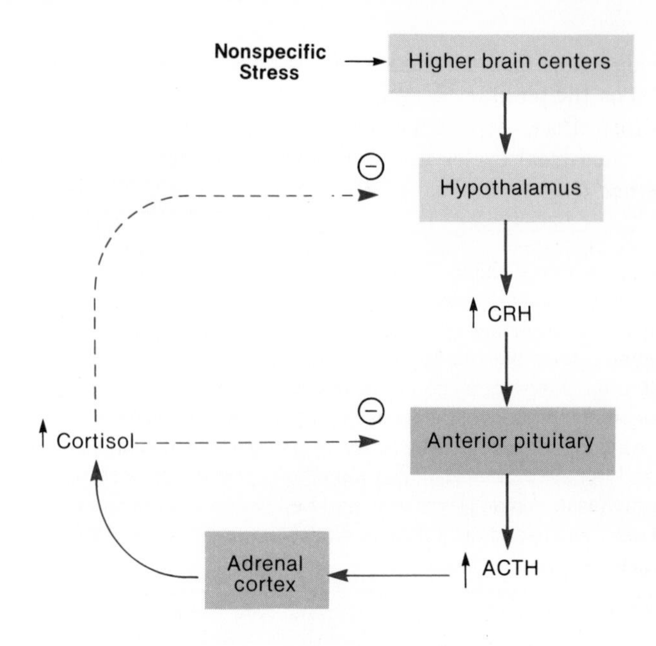

Figure 7.15. *The activation of the pituitary-adrenal axis by nonspecific stress.*

Adrenal Glands

The adrenal cortex and adrenal medulla are structurally and functionally different. The adrenal medulla secretes catecholamine hormones, which complement the sympathetic nervous system in the "fight-or-flight" reaction. The adrenal cortex secretes steroid hormones, which participate in the regulation of mineral balance, energy balance, and reproductive function. The adrenal medulla is stimulated by sympathetic neurons; the adrenal cortex is stimulated by ACTH as a nonspecific response to stress.

The adrenal[5] glands are paired organs that cap the kidneys (fig. 7.14). Each adrenal consists of an outer cortex and inner medulla, which function as separate glands. The differences in structure and function of the adrenal cortex and medulla are related to the differences in their embryonic derivation; the adrenal medulla is derived from embryonic neural tissue (the same tissue that produces the sympathetic ganglia), whereas the adrenal cortex is derived from a different tissue.

As a result of its embryonic derivation, the adrenal medulla secretes epinephrine, and lesser amounts of norepinephrine, into the blood in response to stimulation by sympathetic nerve fibers (chapter 6). The adrenal cortex does not receive neural innervation, and so must be stimulated hormonally (by ACTH secreted from the anterior pituitary).

[5]adrenal: L. *ad,* to; *renes,* kidney

Functions of the Adrenal Cortex

The adrenal cortex secretes steroid hormones called **corticosteroids,** or **corticoids,** for short. There are three functional categories of corticosteroids: (1) **mineralocorticoids,** which regulate Na^+ and K^+ balance; (2) **glucocorticoids,** which regulate the metabolism of glucose and other organic molecules; and (3) **sex steroids,** which are weak androgens (and lesser amounts of estrogens) that supplement the sex steroids secreted by the gonads.

Adrenogenital syndrome is caused by the hypersecretion of adrenal sex hormones, particularly the androgens. Adrenogenital syndrome in young children causes premature puberty and enlarged genitals, especially the penis in a male and the clitoris in a female. An increase in body hair and a deeper voice are other characteristics. This condition in a mature woman can cause the growth of a beard. A related condition, called **congenital adrenal hyperplasia,** can cause masculinization of a female fetus prior to birth.

Aldosterone is the most potent mineralocorticoid, and since it acts to promote retention of salt and water by the body, its secretion is adjusted to maintain a constant blood volume and pressure. The predominant glucocorticoid in humans is cortisol (hydrocortisone). The secretion of cortisol is stimulated by ACTH, which in turn is inhibited by negative feedback from cortisol (fig. 7.15).

Oversecrection of corticosteroids results in **Cushing's syndrome.**[6] This is generally caused by a tumor of the adrenal cortex or by oversecretion of ACTH from the anterior pituitary. Cushing's syndrome is characterized by changes in carbohydrate and protein metabolism, hyperglycemia (high blood glucose), hypertension (high blood pressure), and muscular weakness. Metabolic problems give the body a puffy appearance and can cause structural changes characterized as "buffalo hump" and "moon face." Similar effects are also seen when people with chronic inflammatory diseases receive prolonged treatment with cortisone, which is given to reduce inflammation and inhibit the immune response. **Addison's disease**[7] is caused by inadequate secretion of both glucocorticoids and mineralocorticoids, which results in hypoglycemia, sodium and potassium imbalance, dehydration, hypotension, rapid weight loss, and generalized weakness. A person with this condition who is not treated with corticosteroids will die within a few days.

Stress and the Adrenal Gland

In 1936, Hans Selye discovered that injections of cattle ovaries into rats (1) stimulated growth of the adrenal cortex; (2) caused atrophy (shrinkage) of the spleen, lymph nodes, and thymus; and (3) produced bleeding peptic ulcers. At first he thought that these ovarian extracts contained a specific hormone that caused these effects. He later discovered that injections of a variety of substances, including foreign chemicals such as formaldehyde, could produce the same effects. Indeed, the same pattern of effects occurred when he placed rats in cold environments or when he dropped them into water and made them swim until they were exhausted.

The specific pattern of effects produced by these procedures suggested that these effects were the result of something that the procedures shared in common. Selye reasoned that all of the procedures were *stressful* and that the pattern of changes he observed represented a specific response to any stressful agent. He later discovered that all forms of stress produce these effects because all stressors stimulate the pituitary-adrenal axis. Under stressful conditions, there is increased secretion of ACTH and thus increased secretion of corticosteroids from the adrenal cortex.

Since stress is so very difficult to define, many scientists prefer to define stress operationally as any stimulus that activates the pituitary-adrenal axis. Using this criterion, it has been found that pleasant changes in one's life—such as marriage, a recent promotion, and so on—can be as stressful as unpleasant changes. On this basis, Selye has stated that there is "a nonspecific response of the body to readjust itself following any demand made upon it."

[6]Cushing's syndrome: from Harvey Cushing, U.S. physician, 1869–1939
[7]Addison's disease: from Thomas Addison, English physician, 1793–1860

Table 7.7 Comparison of the hormones from the adrenal medulla

Epinephrine	Norepinephrine
Elevates blood pressure because of increased cardiac output and peripheral vasoconstriction	Elevates blood pressure because of generalized vasoconstriction
Accelerates respiratory rate and dilates respiratory passageways	Similar effect but to a lesser degree
Increases efficiency of muscular contraction	Similar effect but to a lesser degree
Increases rate of glycogen breakdown into glucose, so level of blood glucose rises	Similar effect but to a lesser degree
Increases rate of fatty acid released from fat, so level of blood fatty acids rises	Similar effect but to a lesser degree
Increases release of ACTH and TSH from the adenohypophysis of the pituitary gland	No effect

From Kent M. Van De Graaff, *Human Anatomy,* 2d ed.

Selye termed this nonspecific response the **general adaptation syndrome (GAS).** Stress, in other words, produces GAS. There are three stages in the response to stress: (1) the *alarm reaction,* when the adrenal glands are activated; (2) the *stage of resistance,* in which readjustment occurs; and (3) if the readjustment is not complete, the *stage of exhaustion* may follow, leading to sickness and possibly death.

Excessive stimulation of the adrenal medulla can result in depletion of the body's energy reserves, and high levels of corticosteroid secretion from the adrenal cortex can significantly impair the immune system. It is reasonable to expect, therefore, that prolonged stress can result in increased susceptibility to disease. Indeed, many studies show that prolonged stress results in an increased incidence of cancer and other diseases.

Functions of the Adrenal Medulla

The cells of the adrenal medulla secrete **epinephrine** and **norepinephrine** in an approximate ratio of 4:1, respectively. These hormones are classified as catecholamines and are derived from the amino acid tyrosine.

The effects of these hormones are similar to those caused by stimulation of the sympathetic nervous system, except that the hormonal effect lasts about ten times longer. The hormones from the adrenal medulla increase cardiac output and heart rate, dilate coronary blood vessels, increase mental alertness, increase the respiratory rate, and elevate metabolic rate. A comparison of the effects of epinephrine and norepinephrine are presented in table 7.7.

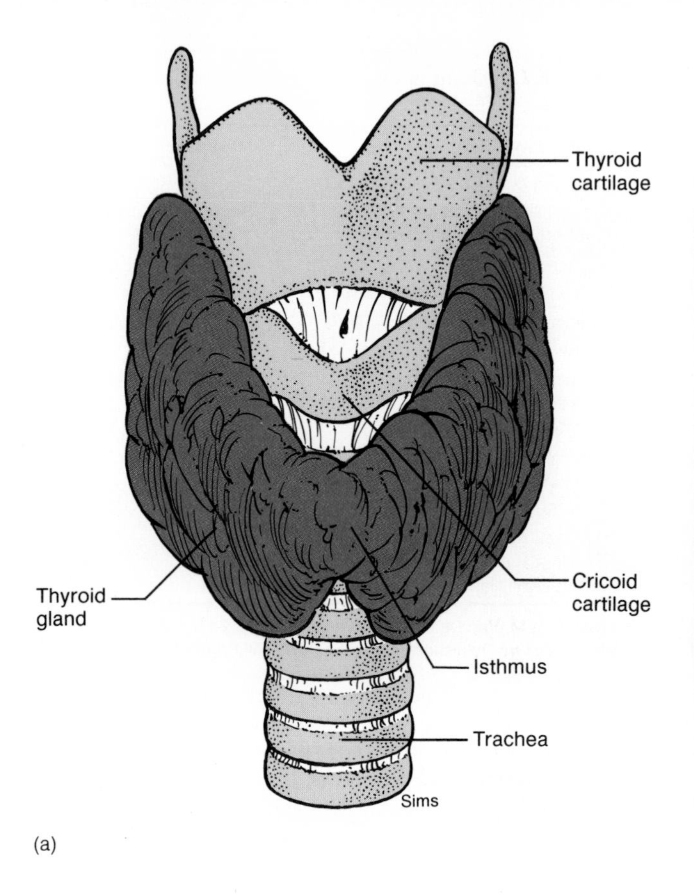

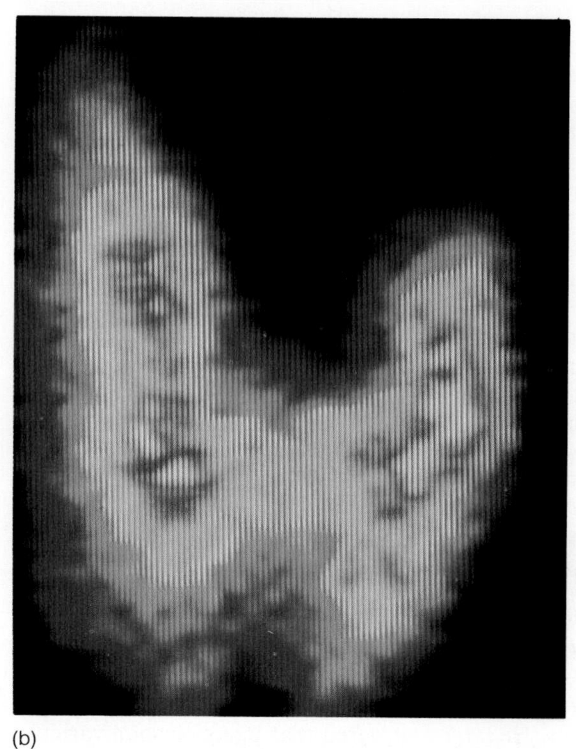

Figure 7.16. *The thyroid gland. (a) Its relationship to the larynx and trachea. (b) A scan of the thyroid gland twenty-four hours after the intake of radioactive iodine.*

The adrenal medulla is innervated by sympathetic nerve fibers. The impulses are initiated from the hypothalamus via the spinal cord when the sympathetic nervous system is stimulated. Many stressors, therefore, activate the adrenal medulla as well as the adrenal cortex. Activation of the adrenal medulla together with the sympathetic nervous system prepares the body for greater physical performance—the *fight-or-flight* response.

---

1. List the categories of corticosteroids. What other endocrine gland(s) produce(s) steroid hormones?
2. List the hormones of the adrenal medulla, and describe their effects.
3. Explain how the secretions of the adrenal cortex and adrenal medulla are regulated.
4. Describe how stress affects the secretions of the adrenal cortex and medulla, and explain how excessive adrenal hormones may produce an increased susceptibility to disease.

Thyroid and Parathyroids

The thyroid secretes thyroxine (T_4) and triiodothyronine (T_3), which are needed for proper growth and development and which are primarily responsible for determining the basal metabolic rate (BMR). Secretions of the thyroid are regulated by TSH from the anterior pituitary. The parathyroid glands secrete parathyroid hormone, which helps to raise the blood Ca^{++} concentration. Low blood Ca^{++} thus stimulates parathyroid hormone secretion in a negative feedback manner.

The thyroid gland is positioned just below the larynx (fig. 7.16). This gland consists of two lobes that lie on either side of the trachea (windpipe). The thyroid is the largest of the endocrine glands, weighing between 20 and 25 g.

On a microscopic level, the thyroid gland consists of many spherical hollow sacs called **thyroid follicles** (fig. 7.17). These follicles are lined with a simple cuboidal epithelium, which synthesizes the principal thyroid hormones. The interior of the follicles contains *colloid,* which is a protein-rich fluid.

Production and Action of Thyroid Hormones

The thyroid follicles actively accumulate iodide (I^-) from the blood and secrete it into the colloid. Once the iodide is in the colloid, it is changed to iodine and attached to specific amino acids (tyrosines) within a protein called **thyroglobulin.** Thyroglobulin contains the thyroid hormones as part of its structure. Upon stimulation by TSH, the cells of the follicle take up a small volume of colloid by pinocytosis, remove the T_3 and T_4 from the thyroglobulin, and secrete the free hormones into the blood.

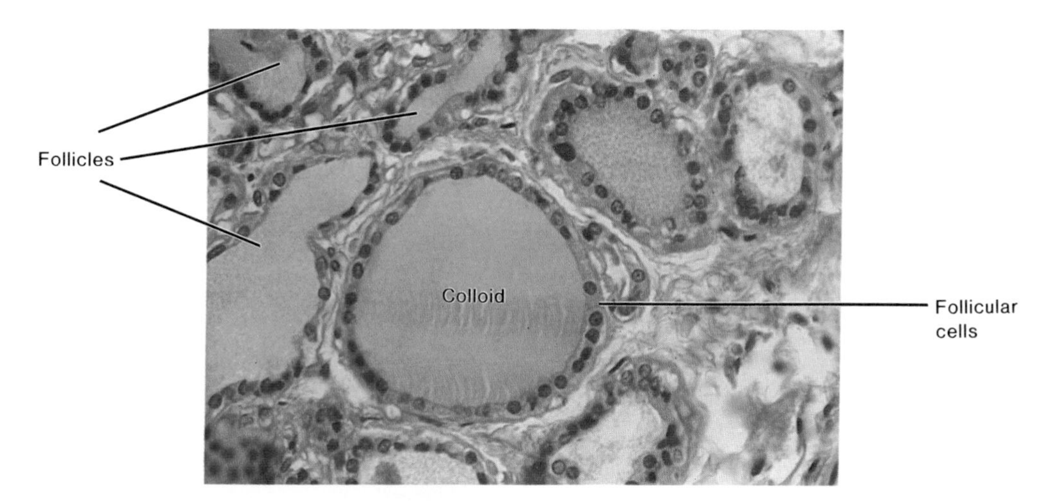

Figure 7.17. *A photomicrograph (magnification ×250) of a thyroid gland, showing numerous thyroid follicles. Each follicle consists of follicular cells surrounding the fluid known as colloid, which contains thyroglobulin.*

Most of the thyroid hormone molecules that enter the blood become attached to plasma carrier proteins. A small percentage of these hormone molecules can detach from the carrier proteins at a certain rate and become free. Only the thyroxine that is free in the plasma can enter the target cells, where it is converted to triiodothyronine and attached to nuclear receptor proteins, as described earlier in this chapter. Through the activation of genes, thyroid hormones stimulate protein synthesis, promote maturation of the nervous system, and increase the rate of energy utilization by the body.

The development of the central nervous system is particularly dependent on thyroid hormones, and a deficiency of these hormones during development can cause serious mental retardation. The basal metabolic rate (BMR)—which is the minimum rate of caloric expenditure when the body is at rest—is determined to a large degree by the level of thyroid hormones in the blood. The physiological functions of thyroid hormones are described in more detail in chapter 18.

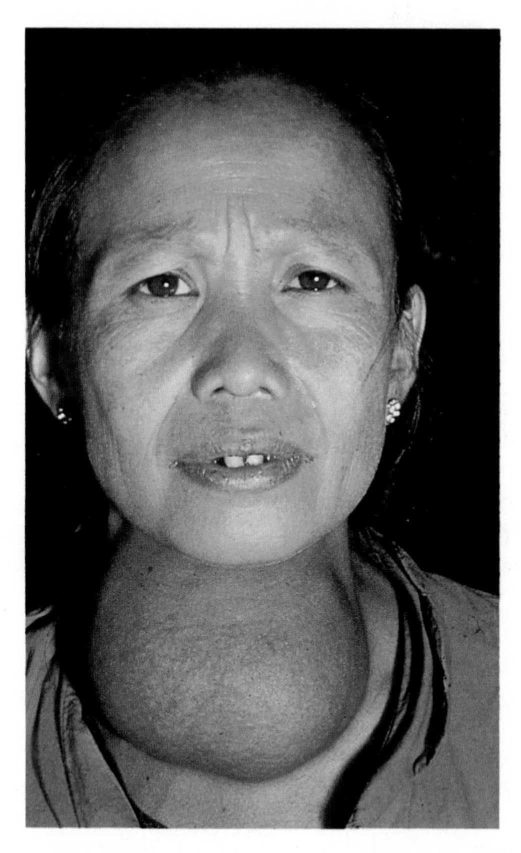

Figure 7.18. *A simple or endemic goiter is caused by insufficient iodine in the diet.*

Diseases of the Thyroid

Thyroid-stimulating hormone (TSH) from the anterior pituitary stimulates the thyroid to secrete thyroxine and exerts a trophic effect on the thyroid gland. This trophic effect is dramatically revealed in people who develop an **iodine-deficiency (endemic) goiter** (fig. 7.18). In the absence of sufficient dietary iodine the thyroid cannot produce adequate amounts of T_4 and T_3. The resulting lack of negative feedback inhibition causes abnormally high levels of TSH secretion, which in turn stimulate the abnormal growth of the thyroid (a goiter). These events are summarized in figure 7.19.

The infantile form of **hypothyroidism** (low thyroid hormone secretion) is known as *cretinism.* An affected child usually appears normal at birth because thyroxine is received from the mother through the placenta. The clinical symptoms of cretinism are stunted growth, thickened facial features, abnormal bone development, mental retardation, low body temperature, and general lethargy. If cretinism is diagnosed early, it can be successfully treated by administering thyroxine.

Hypothyroidism in an adult affects body fluids, causing a characteristic type of edema (accumulation of tissue fluid)

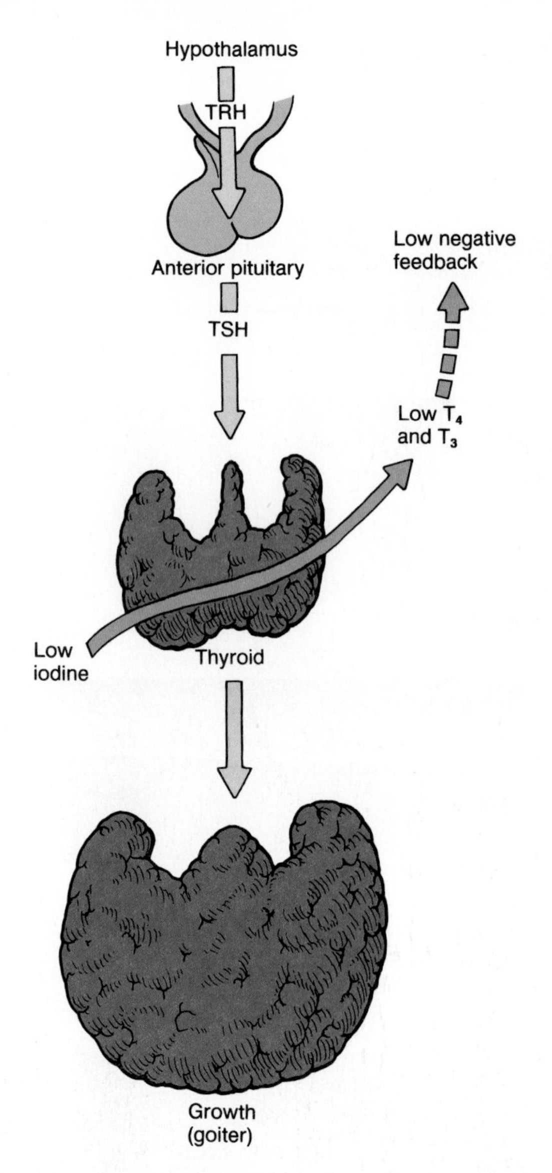

Figure 7.19. The mechanism of goiter formation in iodine deficiency. Low negative feedback inhibition results in excessive TSH secretion, which stimulates abnormal growth of the thyroid.

known as *myxedema.* A person with myxedema has a low metabolic rate, lethargy, and a tendency to gain weight. This condition is treated with thyroxine or with triiodothyronine, which are taken orally (as pills).

Graves' disease,[8] also called **toxic goiter,** involves growth of the thyroid associated with oversecretion of thyroxine. This hyperthyroidism is produced by antibodies that act like TSH and stimulate the thyroid; it is an autoimmune disease (chapter 13). As a consequence of high levels of thyroxine secretion, the metabolic rate and heart rate increase, there is loss of weight, and the autonomic nervous system induces excessive sweating. In about half of the cases, *exophthalmos* (bulging of the eyes) also develops (fig. 7.20) because of edema in the tissues of the eye sockets and swelling of the extrinsic eye muscles.

[8]Graves' disease: from Robert James Graves, Irish physician, 1796–1853

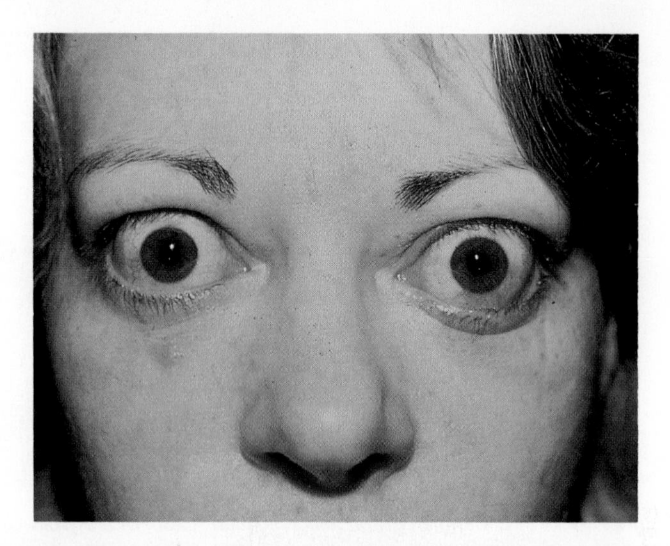

Figure 7.20. Hyperthyroidism is characterized by an increased metabolic rate, weight loss, muscular weakness, and nervousness. The eyes may also protrude.

Thyroid cancer is extremely slow growing, and thus has a more optimistic prognosis than many other forms of cancer. This cancer is first treated by surgical removal of most of the thyroid gland, together with any lymph nodes in the neck that may have been invaded by the cancer. The patient then drinks a solution containing a high concentration of radioactive iodine. Since most of the source of thyroxine has been removed, the pituitary's secretion of TSH is very high. This high level of TSH stimulates any thyroid tissue that is left in the body to take up the radioactive iodine. Only thyroid tissue responds in this manner, and thus only thyroid tissue will concentrate the radioactive iodine and be destroyed by the radiation.

Parathyroid Glands

The small, flattened parathyroid glands are embedded in the posterior surfaces of the thyroid gland (see fig. 7.21). There are usually four parathyroid glands: a *superior* and an *inferior* pair. Each parathyroid gland is a small yellowish brown body measuring 3–8 mm (0.1–0.3 in.) in length, 2–5 mm (0.07–0.2 in.) in width, and about 1.5 mm (0.05 in.) in depth.

The parathyroid glands secrete one hormone called **parathyroid hormone (PTH).** This hormone promotes a rise in blood calcium levels by acting on the bones, kidneys, and intestine (fig. 7.22). Regulation of calcium balance is described in more detail in chapter 19.

1. Describe the structure of the thyroid gland, and list the effect of thyroid hormones.
2. Explain the consequences of an inadequate dietary intake of iodine.
3. Explain why a person who is hypothyroid or hyperthyroid is likely to be overweight or underweight, respectively.

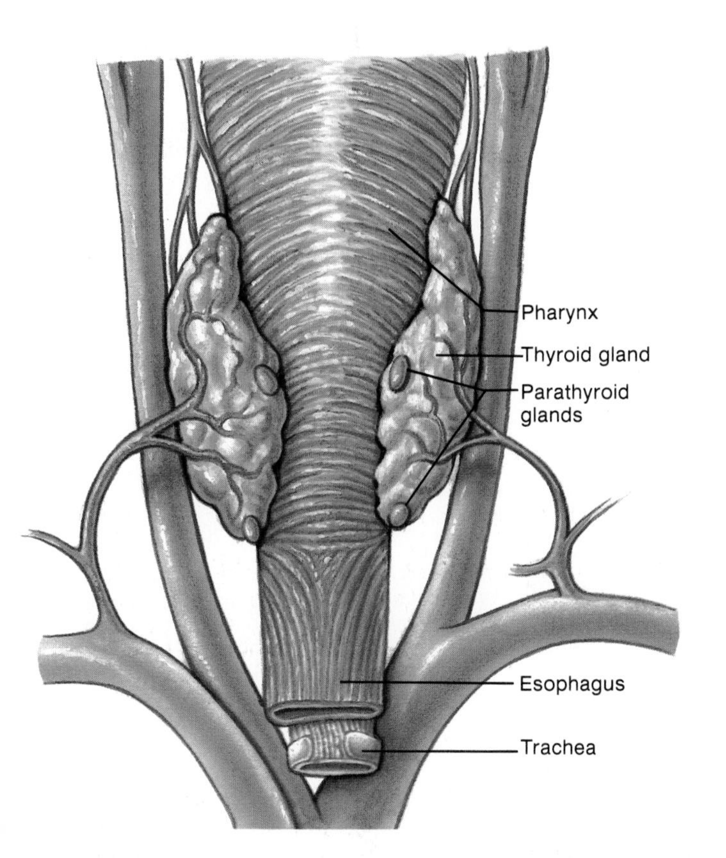

Figure 7.21. *A posterior view of the parathyroid glands.*

Figure 7.22. *Actions of parathyroid hormone. An increased level of parathyroid hormone causes the bones to release calcium, the kidneys to conserve calcium loss through the urine, and the absorption of calcium through the intestinal wall. Negative feedback of increased calcium levels in the blood inhibits the secretion of this hormone.*

Pancreas and Other Endocrine Glands

The islets of Langerhans in the pancreas secrete two hormones, insulin and glucagon, which are critically involved in the regulation of metabolic balance in the body. Insulin promotes the lowering of blood glucose and the storage of energy in the form of glycogen and fat. The hormone glucagon has antagonistic effects that act to raise the blood glucose. Additionally, many other organs secrete hormones that help regulate digestion, metabolism, growth, immune function, and reproduction.

The pancreas is both an endocrine and an exocrine gland. The gross structure of this gland and its exocrine functions in digestion are described in chapter 17. The endocrine portion of the pancreas consists of scattered clusters of cells called the **islets of Langerhans.**[9] (fig. 7.23).

Islets of Langerhans

On a microscopic level, the most conspicuous cells in the islets are the *alpha* and *beta* cells (fig. 7.24). The alpha cells secrete the hormone **glucagon,** and the beta cells secrete **insulin.**[10]

Alpha cells secrete glucagon in response to a fall in the blood glucose concentrations. Glucagon stimulates the liver to convert glycogen to glucose, which causes the blood glucose level to rise. This effect represents the completion of a negative feedback loop. Glucagon also stimulates the breakdown of stored fat and the consequent release of free fatty acids into the blood. This effect helps to provide energy sources for the body during fasting, when blood glucose levels decrease. Glucagon, together with other hormones, also stimulates the conversion of fatty acids to *ketone bodies.* These are four-carbon-long molecules derived from fatty acids, which are produced by the liver and which can serve as immediate sources of energy for a number of organs. Glucagon is thus a hormone that helps to maintain homeostasis during times of fasting, when the body's energy reserves must be utilized (chapter 18).

Beta cells secrete insulin in response to a rise in the blood glucose concentrations after a meal. Insulin promotes the entry of glucose into tissue cells, and the conversion of this glucose into energy storage molecules of glycogen and fat. Insulin also aids the entry of amino acids into cells and the production of cellular protein. The actions of insulin and glucagon are thus antagonistic. After a meal, insulin secretion is increased and glucagon secretion is decreased; fasting, in contrast, causes a rise in glucagon and a fall in insulin secretion.

[9]islets of Langerhans: from Paul Langerhans, German anatomist, 1847–1888
[10]insulin: L. *insula,* island

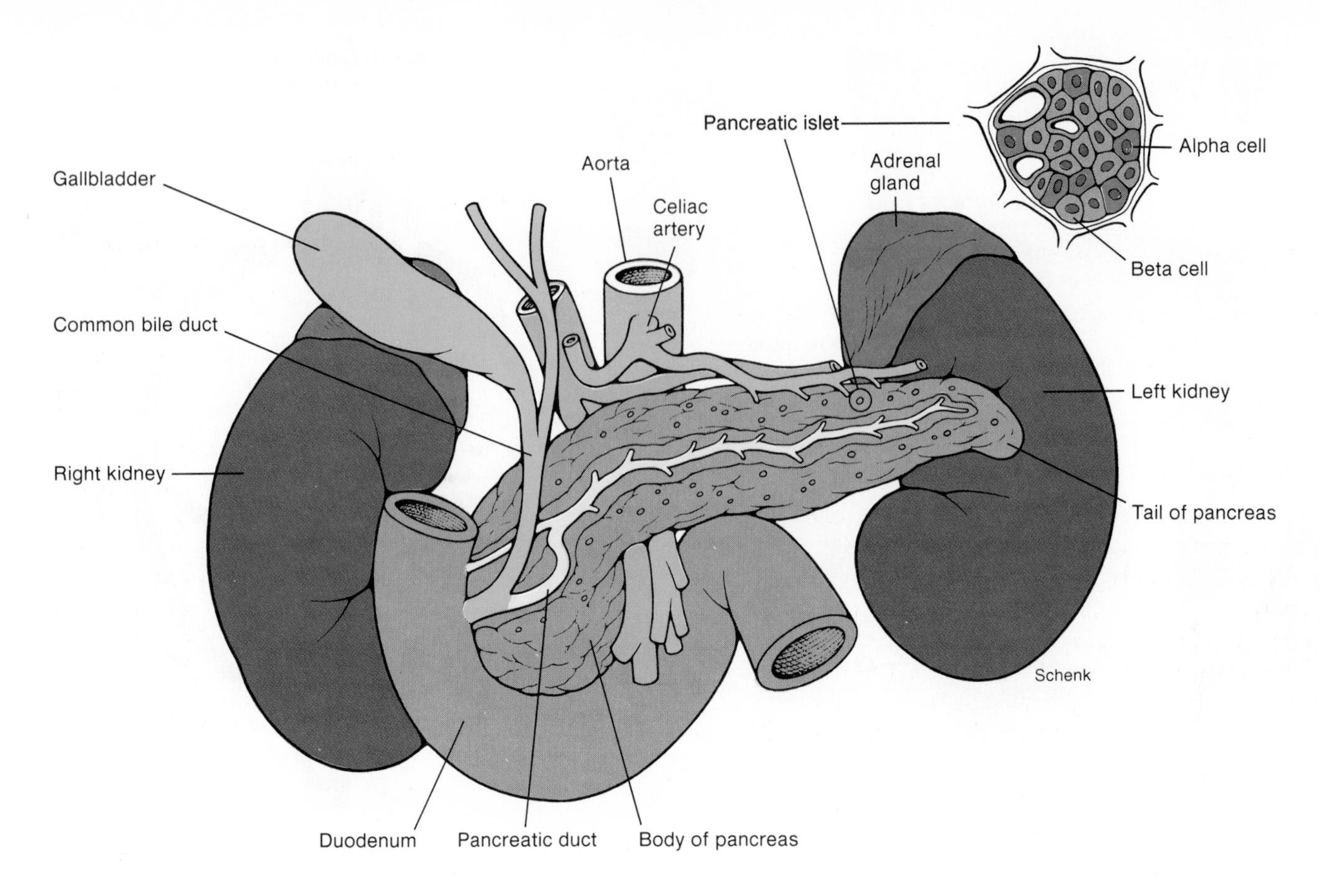

Figure 7.23. *The pancreas and the associated islets of Langerhans.*

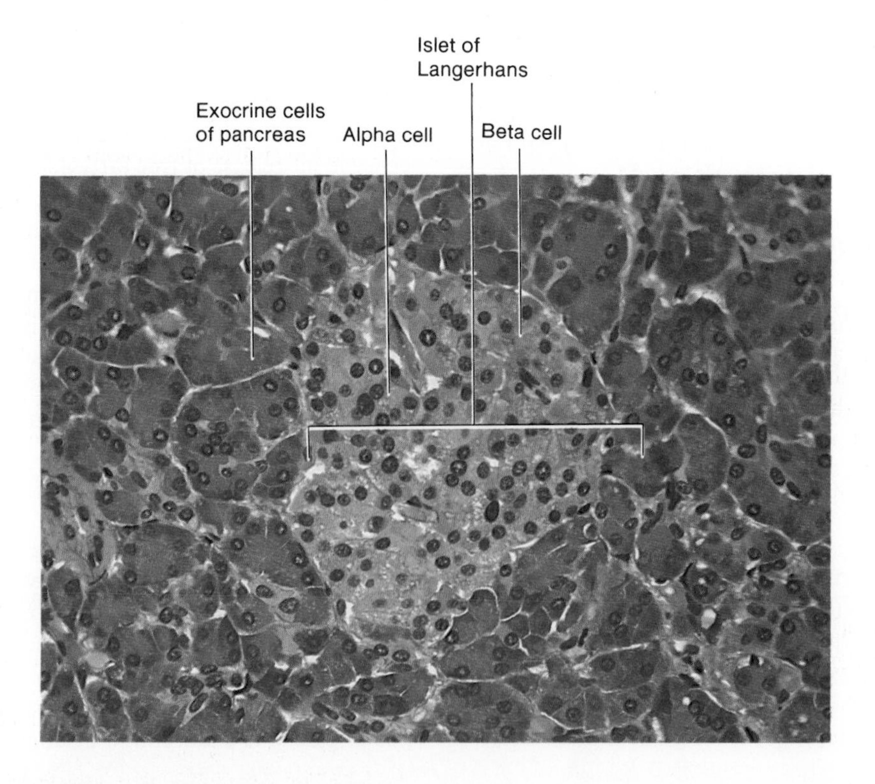

Figure 7.24. *A microscopic view of the pancreas showing an islet of Langerhans and its component cell types.*

Diabetes mellitus[11] is characterized by fasting hyperglycemia and the presence of glucose in the urine. There are two forms of this disease. *Type I*, or *insulin-dependent* diabetes mellitus, is caused by destruction of the beta cells and the resulting lack of insulin secretion. *Type II*, or *non-insulin-dependent* diabetes mellitus (which is the more common form), is caused by decreased tissue sensitivity to the effects of insulin, so that larger amounts of insulin are required to produce a normal effect. Both types of diabetes mellitus are also associated with abnormally high levels of glucagon secretion. The causes and symptoms of diabetes mellitus are described in more detail in chapter 18.

Pineal Gland

The small, cone-shaped pineal[12] gland is located deep within the brain. The pineal gland of a child weighs about 0.2 g and is 5–8 mm (0.2–0.3 in.) long and 9 mm wide. The gland begins to regress in size at about age seven and in the adult appears as a thickened strand of fibrous tissue.

The principal hormone of the pineal is **melatonin.** The secretion of melatonin is inhibited by light and is therefore maximal at night. It has long been suspected that this hormone inhibits the pituitary-gonad axis. Indeed, a decrease in melatonin secretion in many lower vertebrates is responsible for maturation of the gonads during their reproductive season. Melatonin secretion is highest in children of ages one to five and decreases thereafter, reaching its lowest levels at the end of puberty, where concentrations are 75% lower than during early childhood. The secretion of melatonin is, therefore, believed by some researchers to play an important role in the onset of puberty, but this possibility is highly controversial.

Thymus

The thymus is a bilobed organ positioned behind the sternum just above the heart (fig. 7.25). Although the size of the thymus varies considerably from person to person, it is relatively large in newborns and children and sharply regresses in size after puberty. Besides decreasing in size, the thymus of adults becomes infiltrated with strands of fibrous and fatty connective tissue.

The thymus serves as the site of maturation of **T cells** *(thymus-dependent cells)*, which are the white blood cells involved in cell-mediated immunity (chapter 13). In addition to providing T cells, the thymus secretes a number of hormones that are believed to stimulate T cells after they leave the thymus.

Gastrointestinal Tract

The stomach and small intestine secrete a number of hormones that act on the gastrointestinal tract itself and on the pancreas and gallbladder (chapter 17). The effects of these hormones act

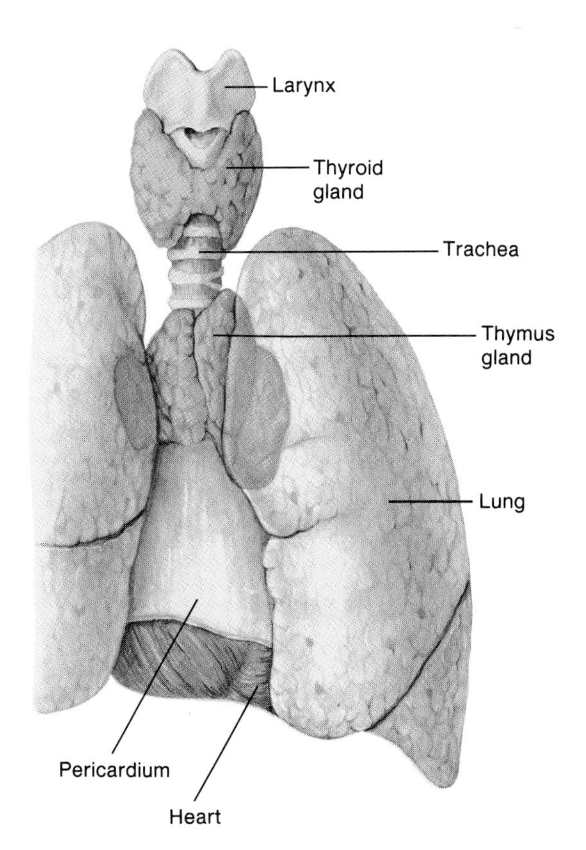

Figure 7.25. The position of the thymus in relation to other anatomical structures.

together with regulation by the autonomic nervous system to coordinate the activities of different regions of the digestive tract and the secretions of pancreatic juice and bile.

Gonads and Placenta

The gonads (**testes** and **ovaries**) secrete **sex steroids.** These include male sex hormones, or **androgens,** and female sex hormones—**estrogens** and **progestogens.** The principal hormones in each of these categories are *testosterone, estradiol-17β,* and *progesterone,* respectively.

The **placenta** is the organ responsible for nutrient and waste exchange between the fetus and mother. The placenta is also an endocrine gland; it secretes large amounts of estrogens and progesterone, as well as a number of polypeptide and protein hormones that are similar to some hormones secreted by the anterior pituitary gland. The physiology of the gonads and placenta are covered in chapters 8 and 9.

1. Describe the structure of the endocrine pancreas, and indicate the sites of origin of insulin and glucagon.
2. Describe how insulin and glucagon secretion is affected by eating and by fasting, and explain the actions of these two hormones.
3. Describe the location of the pineal gland and the possible function of melatonin.
4. Describe the location and function of the thymus gland.
5. Identify the endocrine structures of the reproductive system.

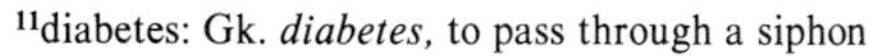

[11] diabetes: Gk. *diabetes,* to pass through a siphon
[12] pineal: L. *pinea,* pine cone

Summary

Hormones: Actions and Interactions p. 124

I. The effects of a hormone in the body depend on its concentration.

II. Hormones can interact in permissive, synergistic, or antagonistic ways.

Mechanisms of Hormone Action p. 128

I. Steroid and thyroid hormones enter their target cells.
   A. Thyroid hormones attach to chromatin-bound receptors located in the nucleus.
   B. Steroid hormones bond to cytoplasmic receptor proteins and translocate to the nucleus.
   C. Attachment of the hormone-receptor protein complex to the chromatin activates genes and thereby stimulates RNA and protein synthesis.

II. Catecholamine and protein hormones bond to receptor proteins on the outer surface of the target cell membrane.
   A. In many cases, this leads to the intracellular production of cyclic AMP, which serves as a second messenger in the action of these hormones.
   B. Cyclic AMP then activates enzymes that were previously inactive in the target cell.

Pituitary Gland p. 132

I. The pituitary secretes eight hormones.
   A. The anterior pituitary secretes growth hormone, thyroid-stimulating hormone, adrenocorticotropic hormone, follicle-stimulating hormone, luteinizing hormone, and prolactin.
   B. The posterior pituitary secretes antidiuretic hormone (also called vasopressin) and oxytocin.

II. The hormones of the posterior pituitary are produced in the hypothalamus and transported to the posterior pituitary by an axon tract.

III. Secretions of the anterior pituitary are controlled by hypothalamic hormones that stimulate or inhibit secretions of the anterior pituitary.

IV. Secretions of the anterior pituitary are also regulated by the negative feedback effects of hormones from the target glands.

V. Higher brain centers, acting through the hypothalamus, can influence pituitary secretion.

Adrenal Glands p. 138

I. The adrenal cortex secretes mineralocorticoids (mainly aldosterone), glucocorticoids (mainly cortisol), and sex steroids (primarily weak androgens).
   A. The glucocorticoids help regulate energy balance; they also can inhibit inflammation and suppress immune function.
   B. The pituitary-adrenal axis is stimulated by stress as part of the general adaptation syndrome.

II. The adrenal medulla secretes epinephrine and lesser amounts of norepinephrine, which complement the action of the sympathetic nervous system.

Thyroid and Parathyroids p. 140

I. The thyroid follicles secrete tetraiodothyronine (T_4, or thyroxine) and lesser amounts of triiodothyronine (T_3).

II. The parathyroids are small structures embedded within the thyroid gland; the parathyroids secrete a hormone that promotes a rise in blood calcium levels.

Pancreas and Other Endocrine Glands p. 143

I. Beta cells in the islets secrete insulin; alpha cells secrete glucagon.
   A. Insulin lowers blood glucose and stimulates the production of glycogen, fat, and protein.
   B. Glucagon raises blood glucose by stimulating the breakdown of liver glycogen; glucagon also promotes the breakdown of fat and the formation of ketone bodies.

II. The pineal gland, located in the brain, secretes melatonin; this hormone may play a role in regulating reproductive function.

III. The thymus is located behind the sternum and above the heart; it is the site of the production of T cell lymphocytes and secretes a number of hormones that may help regulate the immune system.

IV. The gastrointestinal tract secretes a number of hormones that help regulate functions of the digestive system.

V. The gonads secrete sex steroid hormones.

VI. The placenta secretes estrogen, progesterone, and a variety of polypeptide hormones that have actions similar to some anterior pituitary hormones.

Review Activities

Objective Questions

1. Hypothalamic-releasing hormones
   (a) are secreted into capillaries at the base of the hypothalamus
   (b) are transported by portal veins to the anterior pituitary
   (c) stimulate the secretion of specific hormones from the anterior pituitary
   (d) all of the above
2. The hormone primarily responsible for setting the basal metabolic rate and for promoting the maturation of the brain is
   (a) cortisol
   (b) ACTH
   (c) TSH
   (d) thyroxine
3. Which of the following statements about the adrenal cortex is *true?*
   (a) It is not innervated by nerve fibers.
   (b) It secretes some androgens.
   (c) It secretes aldosterone.
   (d) It is stimulated by ACTH.
   (e) All of the above are true.
4. The hormone insulin
   (a) is secreted by alpha cells in the islets of Langerhans
   (b) is secreted in response to a rise in blood glucose
   (c) stimulates the production of glycogen and fat
   (d) both *a* and *b*
   (e) both *b* and *c*

Match the hormone with the primary agent that stimulates its secretion.

5. Epinephrine	(a) TSH
6. Thyroxine	(b) ACTH
7. Corticosteroids	(c) Growth hormone
8. ACTH	(d) Sympathetic nerves
	(e) CRH

9. Steroid hormones are secreted by
   (a) the adrenal cortex
   (b) the gonads
   (c) the thyroid
   (d) both *a* and *b*
   (e) both *b* and *c*
10. The secretion of which of the following hormones would be *increased* in a person with endemic goiter?
    (a) TSH
    (b) thyroxine
    (c) triiodothyronine
    (d) all of the above
11. Which of the following hormones use cAMP as a second messenger?
    (a) testosterone
    (b) cortisol
    (c) insulin
    (d) epinephrine
12. Which of the following terms best describes the type of interaction between the effects of insulin and glucagon?
    (a) synergistic
    (b) permissive
    (c) antagonistic
    (d) cooperative

Essay Questions

1. Compare steroid and polypeptide hormones in terms of their mechanism of action in target organs.
2. Explain the significance of the term *trophic* in regard to the actions of anterior pituitary hormones.
3. Suppose a drug blocks the conversion of T_4 to T_3. Explain what the effects of this drug would be on (*a*) TSH secretion, (*b*) thyroxine secretion, and (*c*) the size of the thyroid gland.
4. Explain why the phrase "master gland" is sometimes used to describe the anterior pituitary, and why this term is misleading.
5. Suppose a person's immune system made antibodies against insulin receptor proteins. Describe the possible effect of this condition on carbohydrate and fat metabolism.

8

Human Reproduction

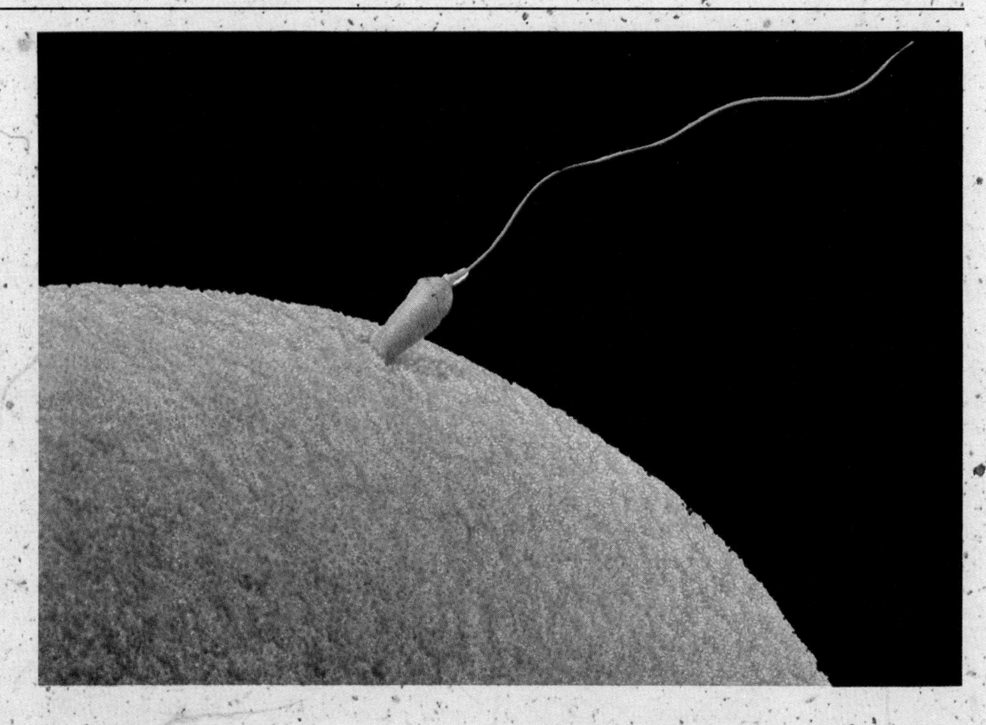

Outline

Objectives

By studying this chapter, you should be able to

1. describe the events that occur during mitosis and meiosis, and the significance of these processes
2. explain how the secretions of FSH and LH are regulated in the male and describe the actions of FSH and LH on the testis
3. describe spermatogenesis and the roles of Sertoli cells in this process
4. explain the physiology of erection and ejaculation and describe the requirements for male fertility
5. describe oogenesis and the stages of ovarian follicle development
6. explain the hormonal interactions involved in the control of ovulation
7. describe the changes that occur in hormone secretion and the function and fate of a corpus luteum during a nonfertile cycle
8. list the phases of the sexual response and describe the events that occur in each phase in males and females
9. explain the nature of different contraceptive procedures, and how they work to prevent pregnancy
10. explain the action of the contraceptive pill and the rhythm method of birth control with reference to the normal female cycle
11. describe the more common sexually transmitted diseases in terms of their causes, symptoms, and treatments

Keys to Pronunciation

androgens: *an'dro-jens*
amenorrhea: *ah-men''o-re'ah*
dihydrotestosterone: *di-hi''dro-tes-tos'ter-on*
endometrium: *en-do-me'tre-um*
epididymis: *ep''i-did'i-mis*
estradiol: *es''trah-di'ol*
fallopian: *fal-lo'pe-an*
gonorrhea: *gon''o-re'ah*
hysterectomy: *his''te-rek'to-me*
meiosis: *mi-o'sis*
menorrhagia: *men''o-ra'je-ah*
menstruation: *men''stroo-a'shun*
mitosis: mi-to'sis
oocyte: *o'o-sit*
oogonium: *o''o-go'ne-um*
progesterone: *pro-jes'te-ron*
seminiferous: *se''mi-nif'er-us*
spermatogonia: *sper''mah-to-go'ne-ah*
testosterone: *tes-tos'te-ron*

Photo: Scanning electron micrograph of a sperm and ovum, in the process of fertilization.

Cellular Reproduction: Mitosis and Meiosis

In sexual reproduction, gametes (sperm and ova) unite to form a new cell which contains twice the number of chromosomes present in each gamete. This new cell is thus said to be diploid, and the gametes are said to be haploid. The new diploid cell formed by the union of gametes must grow by cell division, and this must occur in such a way that the diploid number of chromosomes and their genetic information is preserved from one cell generation to the next. This type of cell division is called mitosis. When a mature individual produces gametes, a different type of cell division—known as meiosis—occurs, in which a diploid parent cell ultimately produces haploid gametes.

"A chicken is an egg's way of making another egg." Phrased in more modern terms, this quote states that organisms exist and function so that the genes they carry can survive beyond the mortal life of individual members of the species. Whether or not one accepts this rather cynical view, it is clear that reproduction is an essential function of life. The incredible complexity of structure and function in living organisms could not be produced in successive generations by chance; mechanisms must exist to transmit the blueprint (genetic code) from one generation to the next. *Sexual reproduction,* in which genes from two individuals are combined in random and novel ways with each new generation, offers the further advantage of introducing great variability into a population. This variability of genetic constitution helps to ensure that some members of a population will survive changes in the environment over evolutionary time.

In sexual reproduction, **germ cells,** or **gametes**[1] (sperm and ova), are formed within the *gonads* (testes and ovaries) by a process of reduction division, or *meiosis*. During this type of cell division, the normal *diploid number* of chromosomes in most human cells—forty-six—is halved, so that each gamete receives the *haploid number* of twenty-three chromosomes. Fusion of a sperm and egg cell (ovum) in the act of **fertilization** results in restoration of the original diploid chromosome number of forty-six in the fertilized egg (the *zygote*). Growth of the zygote into an adult member of the next generation occurs by means of a type of cell division called *mitosis*. When this individual reaches puberty, mature sperm or ova will be formed by meiosis within the gonads so that the life cycle can be continued (fig. 8.1).

Cell Growth and Division

Unlike the life of an organism, which can be pictured as a linear progression from birth to death, the life of a cell follows a cyclical pattern. Each cell is produced as a part of its "parent" cell; when the daughter cell divides, it in turn becomes two new cells. In a sense, then, each cell is potentially immortal as long as its progeny can continue to divide. Some cells in the body divide frequently; the epidermis of the skin, for example, is renewed approximately every two weeks, and the stomach lining is renewed about every two or three days. Other cells, however, such as nerve and striated muscle cells in the adult, do not divide at all. All cells in the body, of course, live only as long as the person lives (some cells live longer than others, but eventually all cells die when vital functions cease).

Prior to cell division, DNA replicates itself by a process that will be described in chapter 10. The replicated DNA forms the short, thick, rodlike structure of chromosomes that are commonly observed in the ordinary (light) microscope. The DNA

[1]gamete: Gk. *gameta,* husband or wife

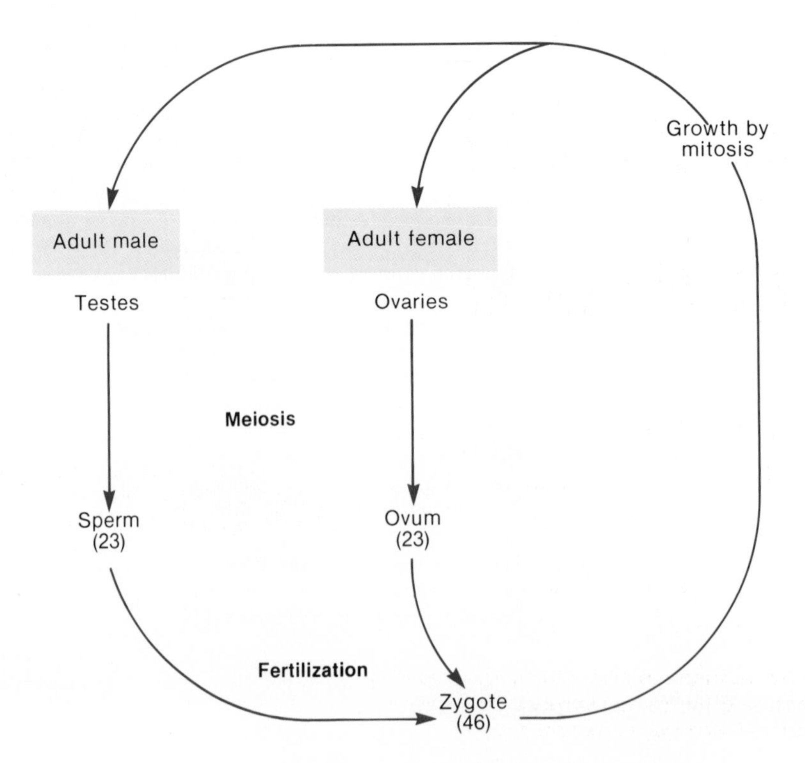

Figure 8.1. The human life cycle. Numbers in parentheses indicate haploid state (23 chromosomes) and diploid state (46 chromosomes).

within these chromosomes is in a "packaged" state, rather than the extended, threadlike state that is active in directing the metabolism of the cell, as will be described in chapter 10.

The chromosomes that were originally present in the ovum (egg cell) of the mother and those originally present in the sperm of the father can be compared and matched into pairs. The matched pairs of chromosomes are called **homologous chromosomes.**[2] Homologous chromosomes do not have identical DNA; one member of the pair may contain a gene that codes for blue eyes, for example, and the other a gene for brown eyes. There are twenty-two homologous pairs of *autosomal chromosomes* and one pair of *sex chromosomes,* described as X and Y. Females have two X chromosomes, whereas males have one X and one Y chromosome (fig. 8.2).

Centrioles

Centrioles are small, rodlike structures located near the nucleus. There are two centrioles, positioned at right angles to each other, at each side ("pole") of the cell. Each centriole is composed of nine evenly spaced bundles of microtubules, with three microtubules per bundle (fig. 8.3).

Centrioles are found only in those cells that are capable of division. During the process of cell division, the centrioles take up positions on opposite sides of the nucleus. The centrioles are involved in the production of, and are attached to, **spindle fibers,** which are also composed of microtubules. The spindle fibers and centrioles help to pull the duplicated chromosomes to opposite poles of the cell during cell division. Cells that lack centrioles, such as mature muscle and nerve cells, cannot divide.

[2]homologous: Gk. *homos,* the same

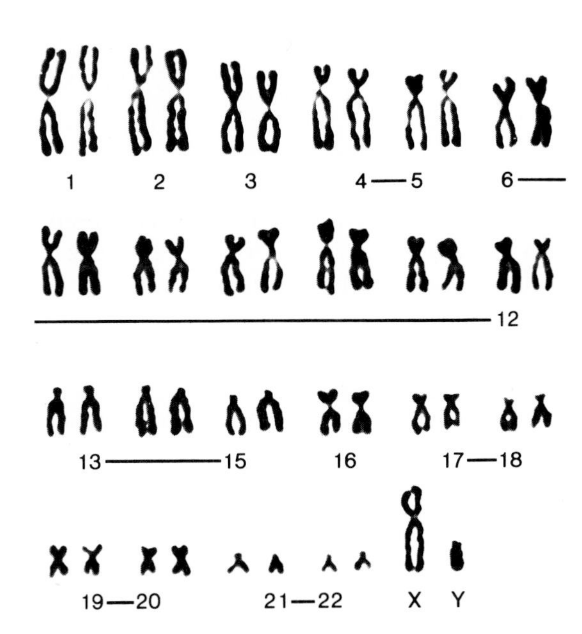

Figure 8.2. A photograph of homologous pairs of chromosomes from a human male cell at metaphase of mitosis (homologous chromosomes have been paired and numbered according to convention).

Mitosis

After the DNA replicates (makes a copy of itself), each chromosome consists of two strands called **chromatids,** which are joined together by a *centromere.* Each chromatid contains a complete DNA molecule that is a copy of the single DNA molecule existing prior to replication. Each chromatid will become a separate chromosome once cell division has been completed.

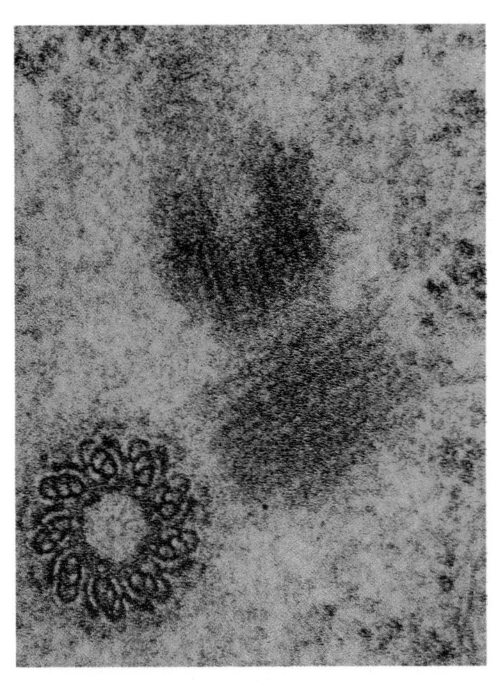

(a)

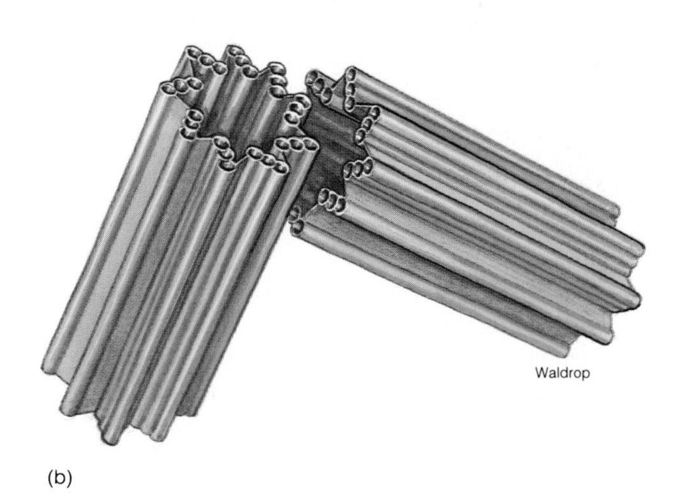

(b)

Figure 8.3. The centrioles. (a) A micrograph of the two centrioles in a centrosome. (b) A diagram showing that the centrioles are positioned at right angles to one another.

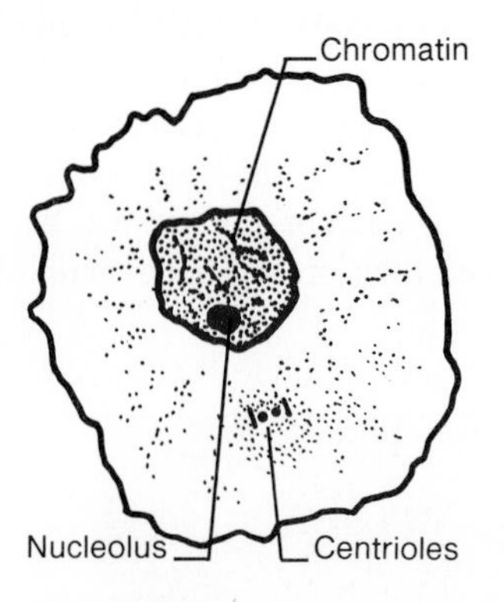

(a) Interphase

The chromosomes are in an extended form and seen as chromatin in the electron microscope.
The nucleus is visible.

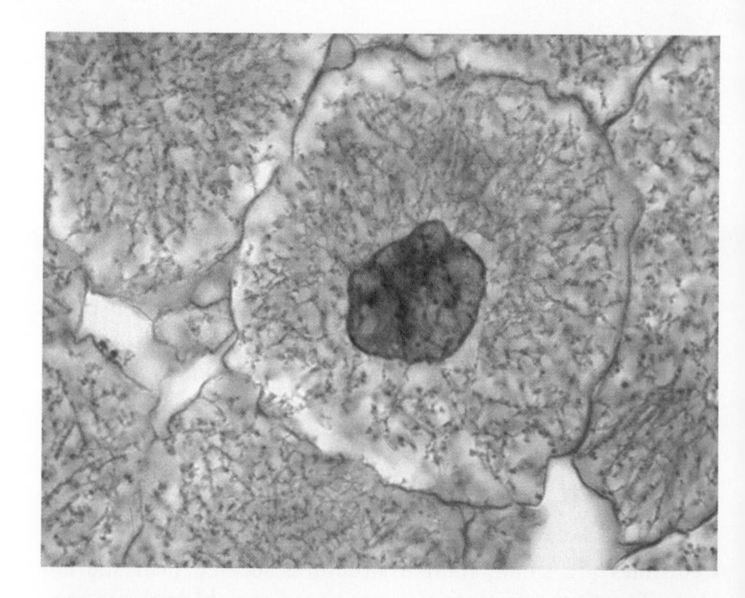

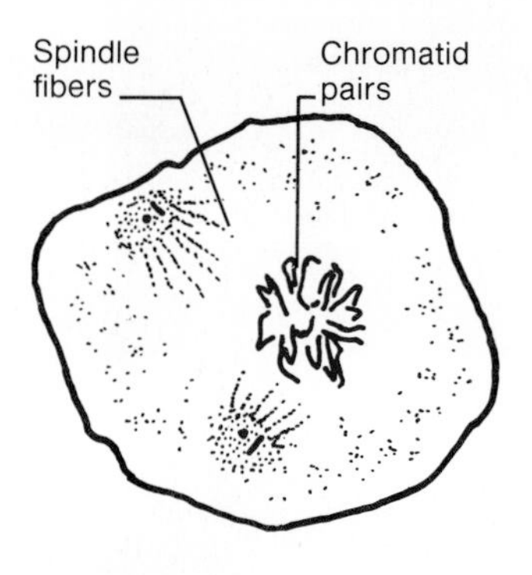

(b) Prophase

The chromosomes are seen and observed to consist of two chromatids joined together by a centromere.
The centrioles move apart towards opposite poles of the cell.
Spindle fibers are produced and extend from each centriole.
The nuclear membrane starts to disappear.
The nucleolus is no longer visible.

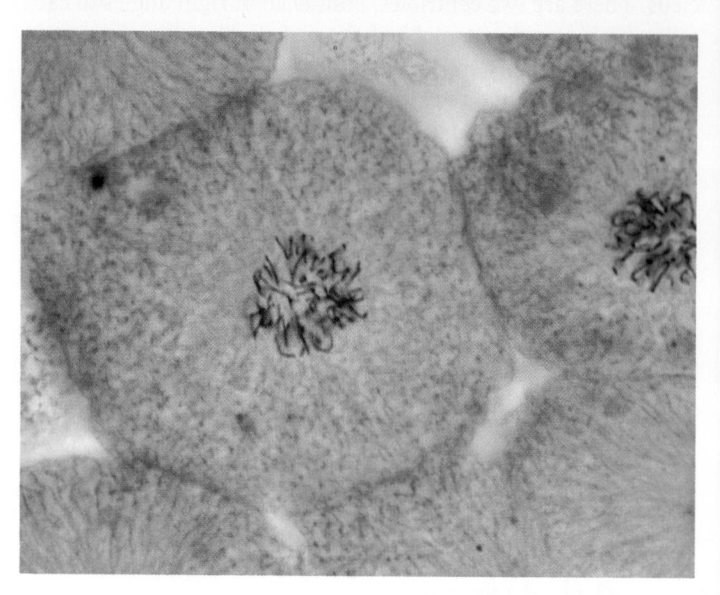

Figure 8.4. *The stages of mitosis (continued on facing page).*

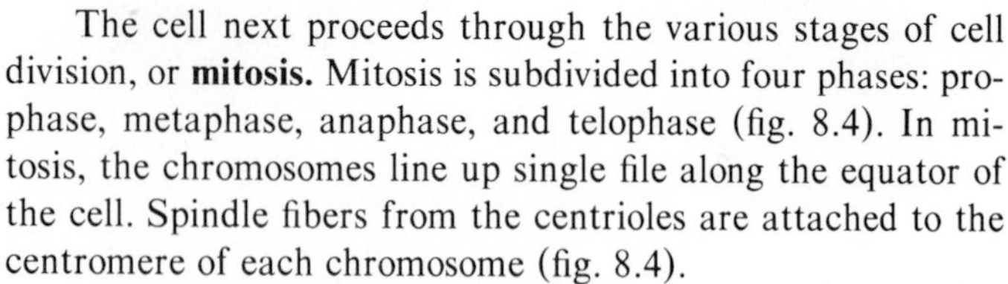

The cell next proceeds through the various stages of cell division, or **mitosis.** Mitosis is subdivided into four phases: prophase, metaphase, anaphase, and telophase (fig. 8.4). In mitosis, the chromosomes line up single file along the equator of the cell. Spindle fibers from the centrioles are attached to the centromere of each chromosome (fig. 8.4).

When the spindle fibers shorten, the centromeres split apart and the two chromatids in each chromosome are pulled to opposite poles. Each pole therefore gets one copy of each of the forty-six chromosomes. Division of the cytoplasm results in the production of two daughter cells, which are genetically identical to each other and to the original parent cell.

Certain types of cells can be removed from the body and grown in nutrient solutions (outside the body, or *in vitro*). Under these artificial conditions the potential longevity of different cell lines can be studied. For unknown reasons, normal connective tissue cells (called fibroblasts) stop dividing *in vitro* after about forty to seventy population doublings. Cells that become transformed into cancer, however, apparently do not age and continue dividing indefinitely in culture. It is ironic that these potentially immortal cells may commit suicide by killing their host.

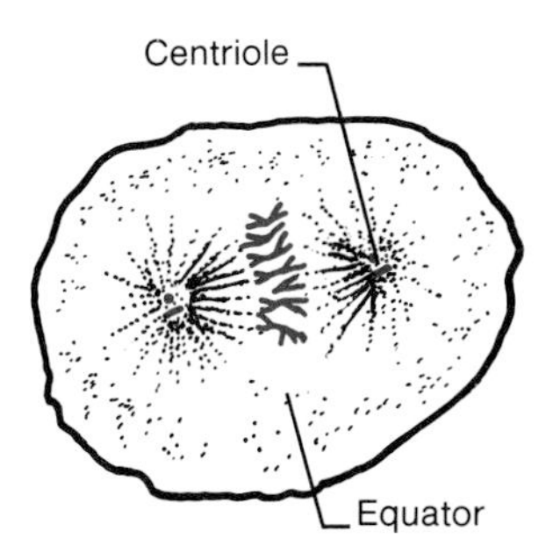

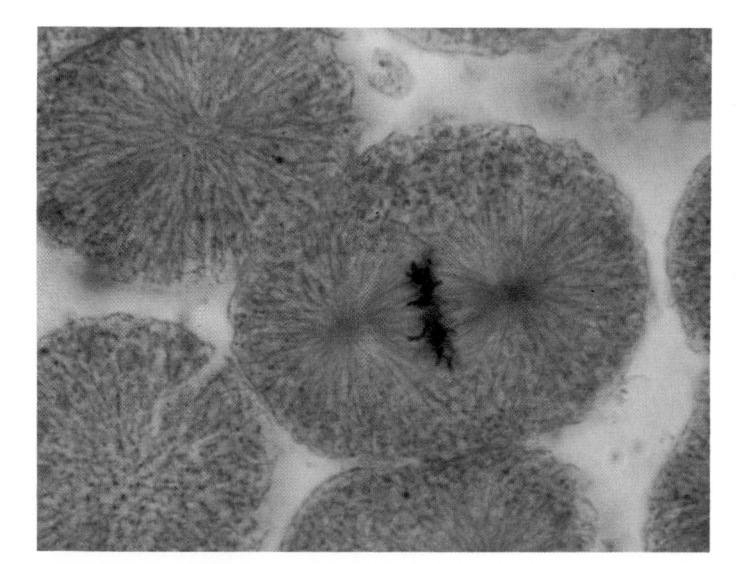

(c) Metaphase

The chromosomes are lined up at the equator of the cell.
The spindle fibers from each centriole are attached
to the centromeres of the chromosomes.
The nuclear membrane has disappeared.

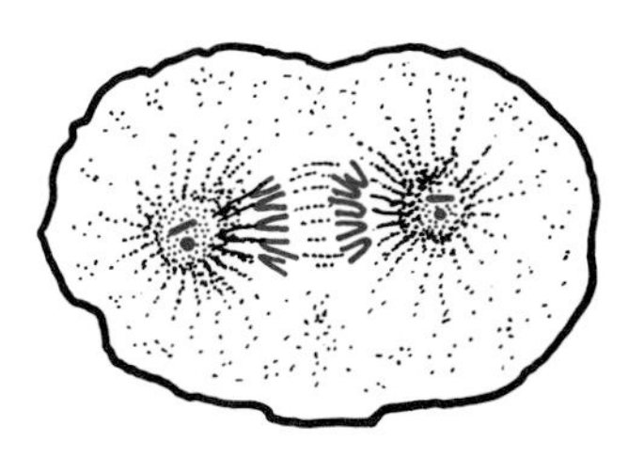

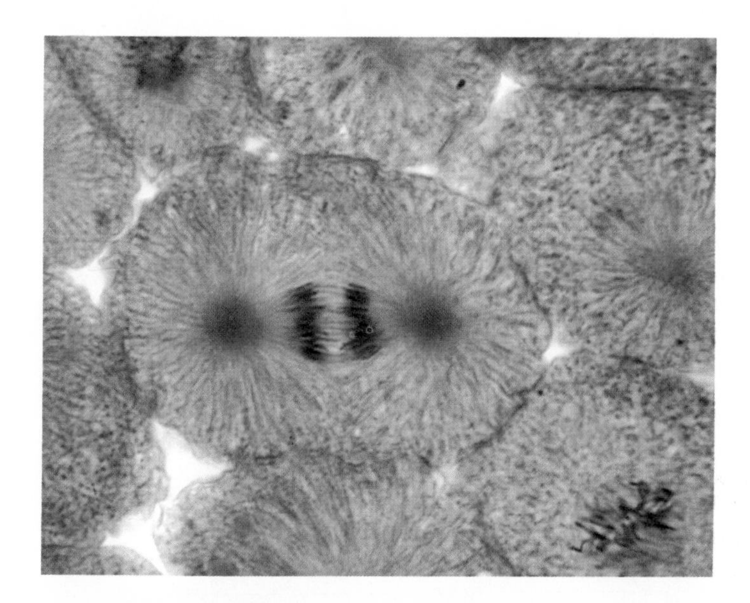

(d) Anaphase

The centromeres split, and the sister chromatids separate
as each is pulled to an opposite pole.

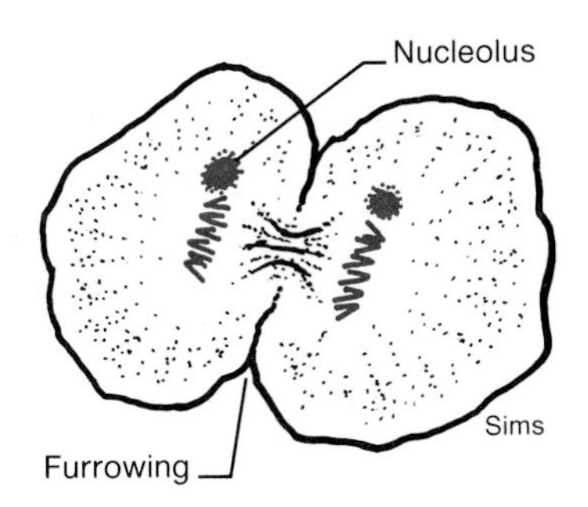

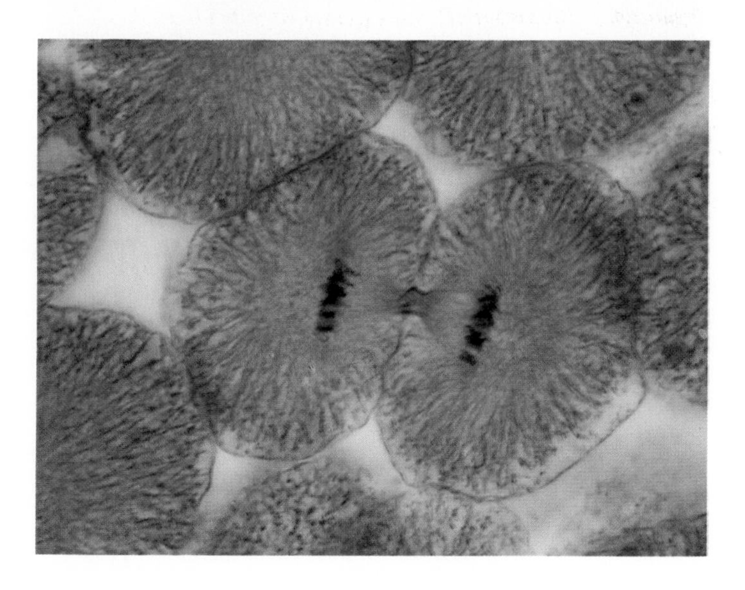

(e) Telophase

The chromosomes become longer, thinner, and less distinct.
New nuclear membranes form.
The nucleolus reappears.
Cell division (cytokinesis) is nearly complete.

Figure 8.4. *continued*

Figure 8.5. *Meiosis, or reduction division. In the first meiotic division the homologous chromosomes of a diploid parent cell are separated into two haploid daughter cells. Each of these chromosomes contain duplicate strands, or chromatids. In the second meiotic division these chromosomes are distributed to two new haploid daughter cells.*

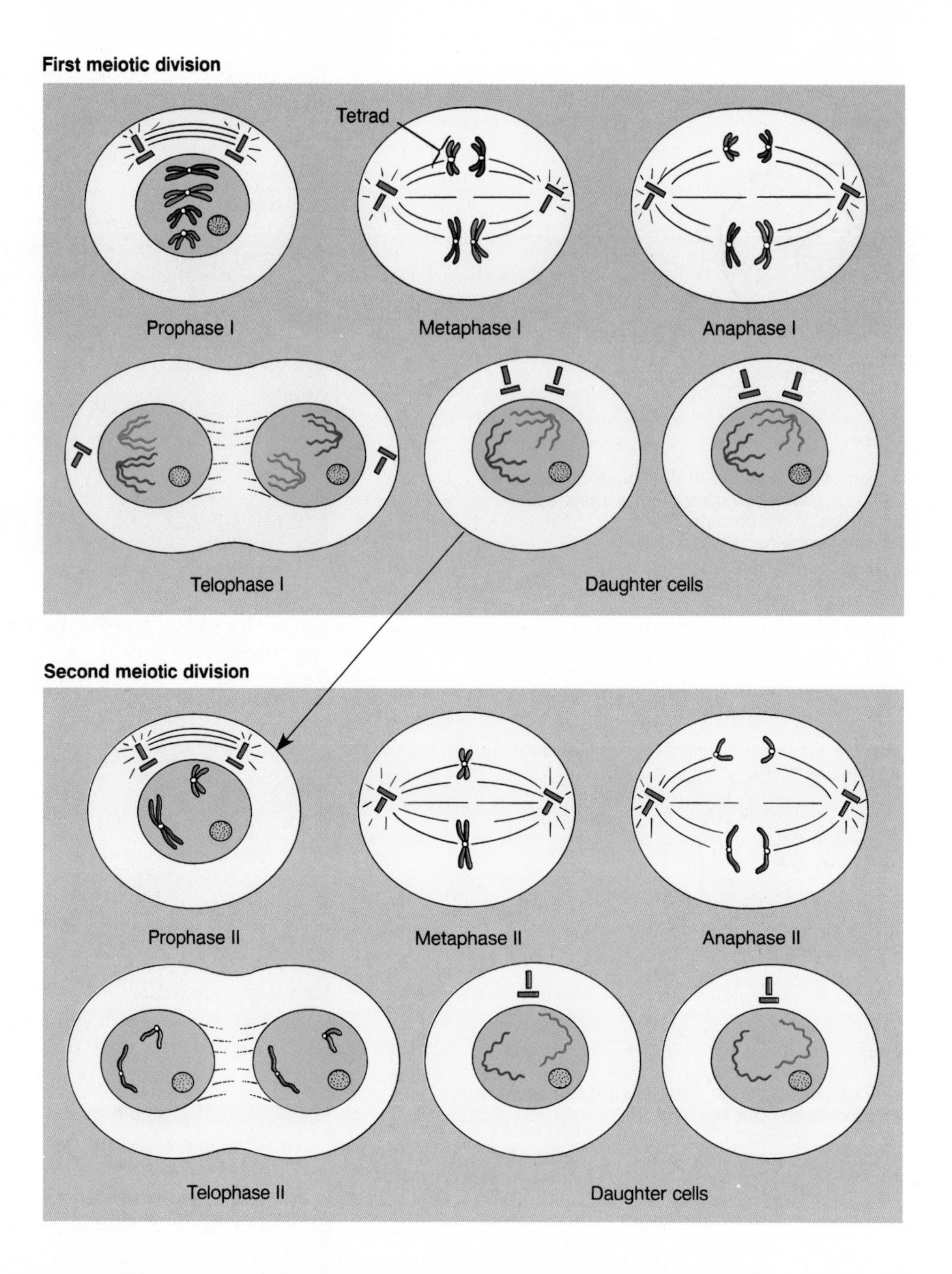

Table 8.1 Stages of meiosis

Stage	Events	Stage	Events
First meiotic division		Second meiotic division	
Prophase I	Chromosomes appear double-stranded. Each strand, called a chromatid, contains duplicate DNA joined together by a structure known as a centromere. Homologous chromosomes pair up side by side.	Prophase II	Chromosomes appear, each containing two chromatids.
Metaphase I	Homologous chromosome pairs line up at equator. Spindle apparatus complete.	Metaphase II	Chromosomes line up single file along equator as spindle formation is completed.
Anaphase I	Homologous chromosomes are separated; each member of a homologous pair moves to opposite poles.	Anaphase II	Centromeres split and chromatids move to opposite poles.
Telophase I	Cytoplasm divides to produce two haploid cells.	Telophase II	Cytoplasm divides to produce two haploid cells from each of the haploid cells formed at telophase I.

Meiosis

When a cell is going to divide, either by mitosis or meiosis, the DNA is replicated (forming chromatids) and the chromosomes become shorter and thicker, as previously described. At this point, the cell has 46 chromosomes (or 23 pairs of homologous chromosomes), which each consist of 2 duplicate chromatids.

Meiosis is a special type of cell division that occurs only in the gonads and is used only in the production of gametes. In meiosis, the homologous chromosomes line up side-by-side along the equator of the cell, rather than single file as in mitosis. The spindle fibers then pull one member of a homologous pair to one pole of the cell, and the other member of the pair to the other pole of the cell. Each of the two daughter cells thus acquires only one chromosome from each of the 23 homologous pairs contained in the parent. The daughter cells, in other words, contain 23 rather than 46 chromosomes. For this reason, meiosis is also known as **reduction division.**

Meiosis, however, consists of two cell divisions. The necessity for this is obvious when you realize that, at the end of the cell division previously described, each daughter cell contains 23 chromosomes—but *each of these consists of two chromatids.* (Since the two chromatids per chromosome are identical, this does not make 46 chromosomes; there are still only 23 *different* chromosomes per cell at this point.) Each of the daughter cells from the first cell division then itself divides, with the duplicate chromatids going to each of two new daughter cells. A grand total of four daughter cells can thus be produced from the meiotic cell division of one parent cell. One parent cell thus produces four sperm (which each contains 23 chromosomes). One parent cell also produces 4 egg cells, but the development of ova is complicated by the fact that three of the four potential egg cells die (this will be described later in this chapter).

The stages of meiosis are subdivided according to whether they occur in the first or the second meiotic cell division. These stages are labeled prophase I, metaphase I, anaphase I, telophase I; and then prophase II, metaphase II, anaphase II, and telophase II (fig. 8.5 and table 8.1).

The reduction of the chromosome number from 46 to 23 is of obvious necessity for sexual reproduction, where the sex cells join and add their content of chromosomes together to produce a new individual. The significance of meiosis, however, goes deeper than simply a reduction in chromosome number. At metaphase I, the pairs of homologous chromosomes can line up with either member facing a given pole of the cell. (Recall that each member of a homologous pair came from a different parent.) Maternal and paternal members of homologous pairs are thus randomly shuffled. When the first meiotic division occurs each daughter cell will thus obtain a complement of 23 chromosomes that are randomly derived from the maternal or paternal contribution to the chromosomes of the parent cell.

In addition to this "shuffling of the deck," or *independent assortment,* of chromosomes, exchanges of parts of homologous chromosomes can occur at metaphase I. That is, pieces of one chromosome of a homologous pair can be exchanged with the other homologous chromosome in a process called *crossing-over* (fig. 8.6). These events together result in **genetic recombination,** and ensure that the gametes produced by meiosis are genetically unique. This provides genetic diversity for organisms that reproduce sexually, and genetic diversity has been shown to promote survival of species over evolutionary time (chapter 2).

1. Describe the cell cycle, indicating which stages are diploid and which are haploid.
2. List the phases of mitosis, and briefly describe the events that occur in each phase.
3. Distinguish between mitosis and meiosis in terms of their final result and their functional significance.
4. Describe in general terms the events that occur during the two meiotic cell divisions, and explain the mechanisms by which genetic recombination occurs during meiosis.

Male Reproductive System

The Leydig cells in the testes are stimulated by LH to secrete testosterone, a potent androgen that acts to maintain the structure and function of the male accessory sexual organs and to promote the development of male secondary sexual characteristics. Spermatogenesis requires the cooperative actions of both FSH and testosterone.

The testes consist of two parts, or "compartments"—the *seminiferous tubules,* where spermatogenesis occurs, and the *interstitial tissue,* which contains androgen-secreting **Leydig cells**[3] (fig. 8.7). About 90% of the weight of an adult testis (which averages 20 g) is comprised of seminiferous tubules. The interstitial tissue is a thin web of connective tissue (containing Leydig cells) between convolutions of the tubules.

There is a strict compartmentation in the testes with regard to FSH and LH action. Cellular receptor proteins for FSH are located exclusively in the seminiferous tubules, where they are confined to the **Sertoli cells** (discussed in a later section). LH receptor proteins are located exclusively in the interstitial Leydig cells. Secretion of testosterone by the Leydig cells is stimulated by LH but not by FSH. Spermatogenesis in the tubules is stimulated by FSH. The apparent simplicity of this compartmentation, however, is an illusion because the two compartments can interact with each other in complex ways.

[3]Leydig cells: from Franz von Leydig, German anatomist, 1821–1908

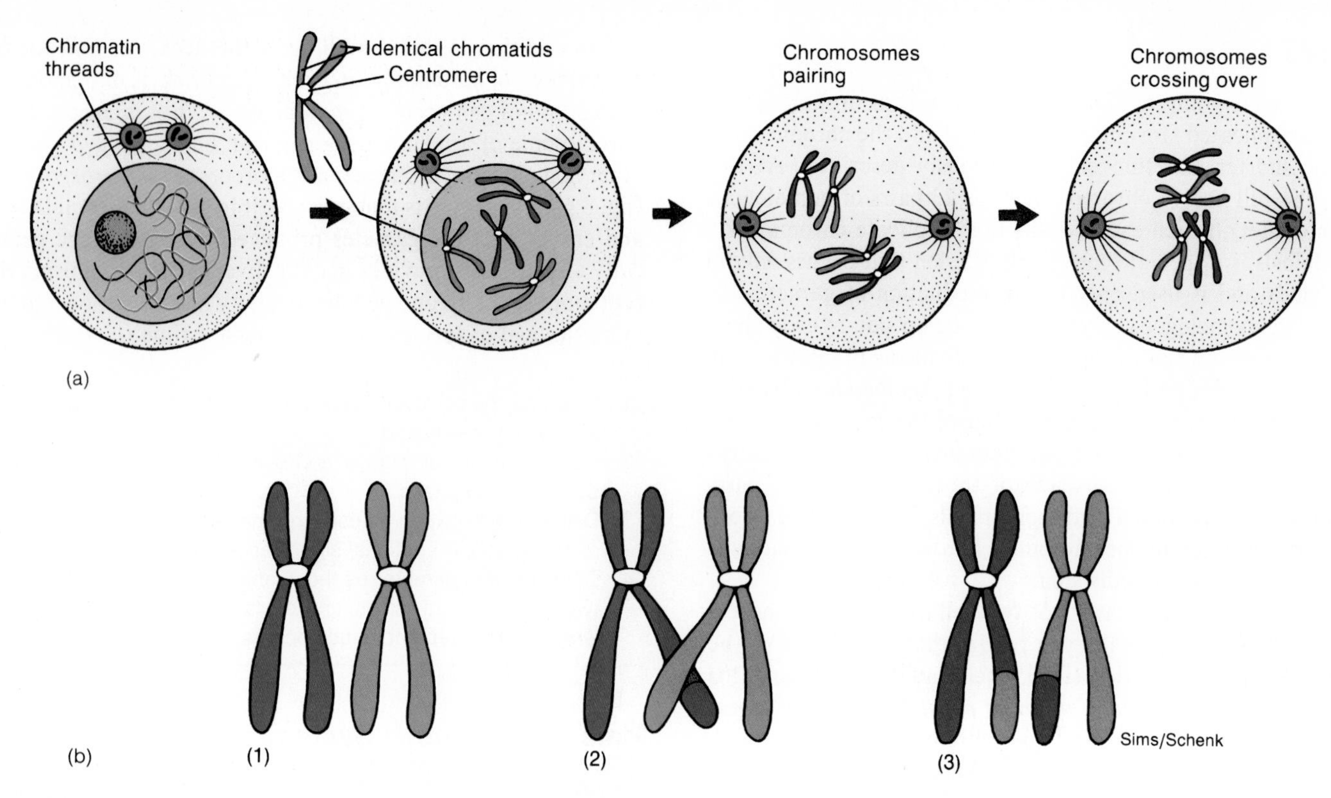

Figure 8.6. *(a) Genetic variation results from the crossing-over, which occurs during the first meiotic prophase. (b) A diagram of the pairing of homologous chromosomes: (1) before the crossing-over; (2) chromatids crossing over; and (3) results of the crossing-over.*

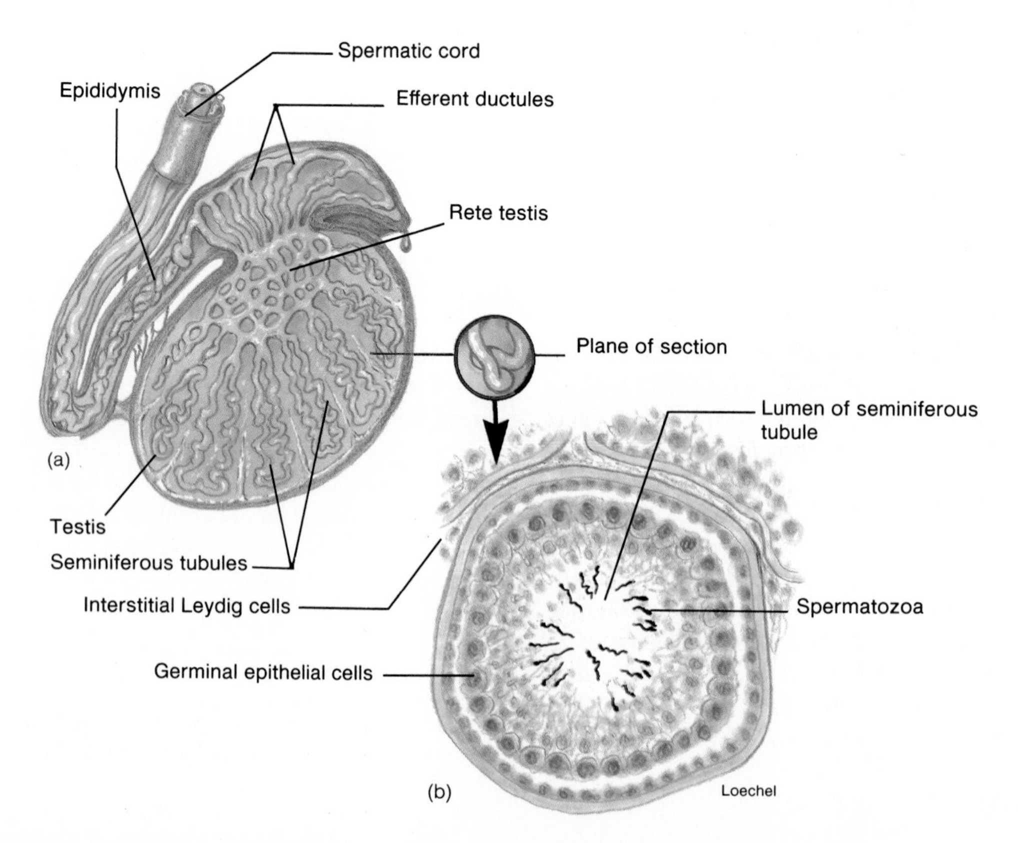

Figure 8.7. *A diagrammatic representation of seminiferous tubules. (a) A sagittal section of a testis; (b) a transverse section of a seminiferous tubule.*

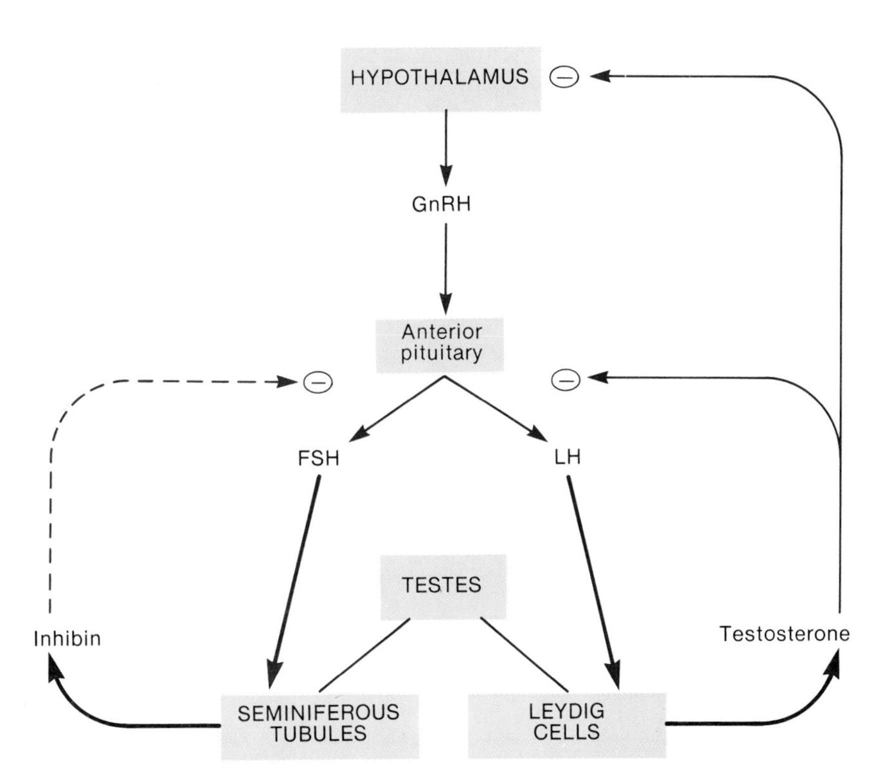

Figure 8.8. Negative feedback relationships between the anterior pituitary and testes. The seminiferous tubules are the targets of FSH action; the interstitial Leydig cells are targets of LH action. Testosterone secreted by the Leydig cells inhibits LH secretion; inhibin secreted by the tubules inhibits FSH secretion.

Control of Gonadotropin Secretion

Castration of a male animal results in an immediate rise in FSH and LH secretion. This demonstrates that hormones secreted by the testes exert negative feedback inhibition of gonadotropin secretion. If testosterone is injected into the castrated animal, the secretion of LH can be returned to the previous (precastration) levels. This provides a classical example of negative feedback—LH stimulates testosterone secretion by the Leydig cells, and testosterone inhibits pituitary secretion of LH (fig. 8.8).

The amount of testosterone that is sufficient to suppress LH, however, is not sufficient to suppress the postcastration rise in FSH secretion. A polypeptide product of the seminiferous tubules, however, has been discovered that specifically suppresses FSH secretion. This hormone, produced by the Sertoli cells, is called **inhibin.** The secretion of inhibin by the tubules is stimulated by FSH, and the secretion of FSH by the anterior pituitary is inhibited, in a negative feedback fashion, by inhibin.

Testosterone Derivatives in Target Tissue

Although testosterone is the major androgen[4] secreted into the blood by the testes, it may not be the active form of the hormone within the cells of particular target organs. In these cases, the testosterone must first be converted to other products within the target cell to exert its effects. In many target organs for testosterone, such as in the prostate, testosterone is converted within the target cells into a derivative called **dihydrotestosterone,** or **DHT** (fig. 8.9). It is then the DHT, rather than the testosterone, which actually stimulates the prostate gland.

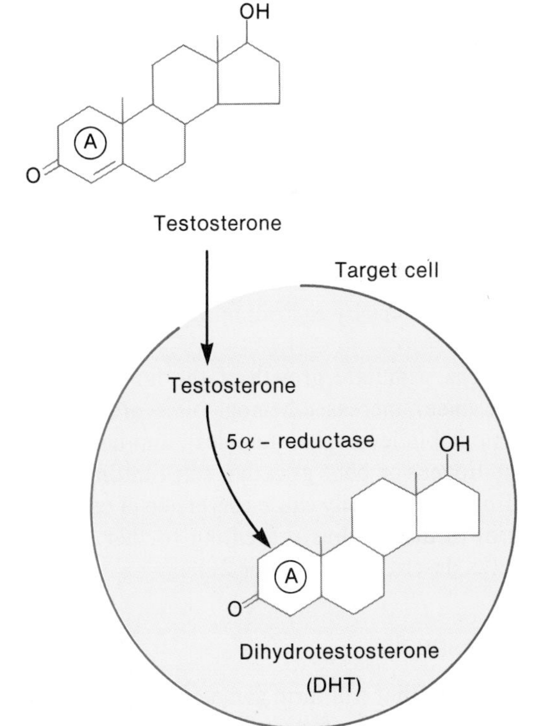

Figure 8.9. The conversion of testosterone, secreted by the interstitial cells of the testes, into dihydrotestosterone (DHT) within the target cells. This reaction involves the addition of a hydrogen (and the removal of the double carbon bond) in the first ring of the steroid.

[4]androgen: Gk. *andros,* male producing

Interestingly, testosterone can also serve as a precursor (parent molecule) for estrogens. In fact, the ovaries of females first produce testosterone (which is not secreted) and then convert it into estrogens. This conversion of testosterone to estrogens also occurs in males, but to a much lesser degree. The testes thus produce small amounts of estrogens whose physiological function, if any, is currently unknown. The brain produces estradiol-17β (the major estrogen) from testosterone, together with DHT and other derivatives of testosterone. These molecules account for the negative feedback effects of testosterone on the hypothalamus.

Testosterone Secretion and Age

The negative feedback effects of testosterone and inhibin help to maintain a constant secretion of gonadotropins in men, resulting in relatively constant levels of androgen secretion from the testes. This contrasts with the cyclic secretion of gonadotropins and ovarian steroids in women. Women experience an abrupt cessation in sex steroid secretion during menopause (as described in a later section). In contrast, the secretion of androgens declines only gradually and to varying degrees in men over fifty years of age. The causes of this age-related change in testicular function are not currently known.

Table 8.2 Summary of some of the actions of androgens in the male

Category	Action
Sex determination	Growth and development of the epididymis, vas deferens, seminal vesicles, and ejaculatory ducts Development of the prostate Development of male external genitalia (penis and scrotum)
Spermatogenesis	At puberty: completion of meiotic division and early maturation of spermatids After puberty: maintenance of spermatogenesis
Secondary sexual characteristics	Growth and maintenance of accessory sexual organs Growth of penis Growth of facial and axillary hair Body growth
Anabolic effects	Protein synthesis and muscle growth Growth of bones Growth of other organs (including larynx) Erythropoiesis (red blood cell formation)

Endocrine Functions of the Testes

Testosterone is by far the major androgen secreted by the adult testis. This hormone, or derivatives of it (such as DHT), is responsible for initiation and maintenance of the body changes associated with puberty in men. Androgens are sometimes called *anabolic steroids* because they stimulate the growth of muscles and other structures (table 8.2). Increased testosterone secretion during puberty is also required for growth of the sex accessory organs—primarily the seminal vesicles and prostate. Removal of androgens by castration results in atrophy of these organs.

Androgens stimulate growth of the larynx (causing lowering of the voice), increased hemoglobin synthesis (males have higher hemoglobin levels than females), and bone growth. The effect of androgens on bone growth is self-limiting, however, because androgens ultimately cause conversion of cartilage to bone in the "growth discs," thus preventing further lengthening of the bones (as described in chapter 21).

Spermatogenesis

Germ cells (those that will form gametes) migrate from the yolk sac of the embryo to the testes during early embryonic development. These become "stem cells" called **spermatogonia** within the outer region of the seminiferous tubules. Spermatogonia are diploid cells that ultimately give rise to mature haploid gametes through the process of meiosis.

Actually, only about 1,000–2,000 stem cells migrate from the yolk sac into the embryonic testes. In order to produce many millions of sperm throughout adult life, these spermatogonia cells duplicate themselves by mitotic division (which produces two diploid cells), and only one of the two cells—now called a **primary spermatocyte**—undergoes meiotic division (fig. 8.10). In this way, spermatogenesis can occur continuously without exhausting the number of spermatogonia.

When a diploid primary spermatocyte completes the first meiotic division, the two haploid cells thus produced are called **secondary spermatocytes.** At the end of the second meiotic division, each of the two secondary spermatocytes produce two haploid **spermatids.** One primary spermatocyte therefore produces four spermatids.

The stages of spermatogenesis are arranged sequentially in the wall of the seminiferous tubule. The epithelial wall of the tubule—called the *germinal epithelium*—is indeed composed of germ cells in different stages of spermatogenesis. The spermatogonia[5] and primary spermatocytes are located toward the outer side of the tubule, whereas spermatids and mature spermatozoa[6] are located on the side of the tubule facing the lumen.

At the end of the second meiotic division, the four spermatids produced by meiosis of one primary spermatocyte are interconnected with each other—their cytoplasm does not completely pinch off at the end of each division. Development of these interconnected spermatids into separate, mature spermatozoa requires the participation of another type of cell in the tubules, the **Sertoli cells**[7] (fig. 8.11).

[5]spermatogonia: Gk. *sperma,* seed; *gone,* generation
[6]spermatozoa: Gk. *sperma,* seed; *zoon,* animal
[7]Sertoli cells: from Enrico Sertoli, Italian histologist, 1842–1910

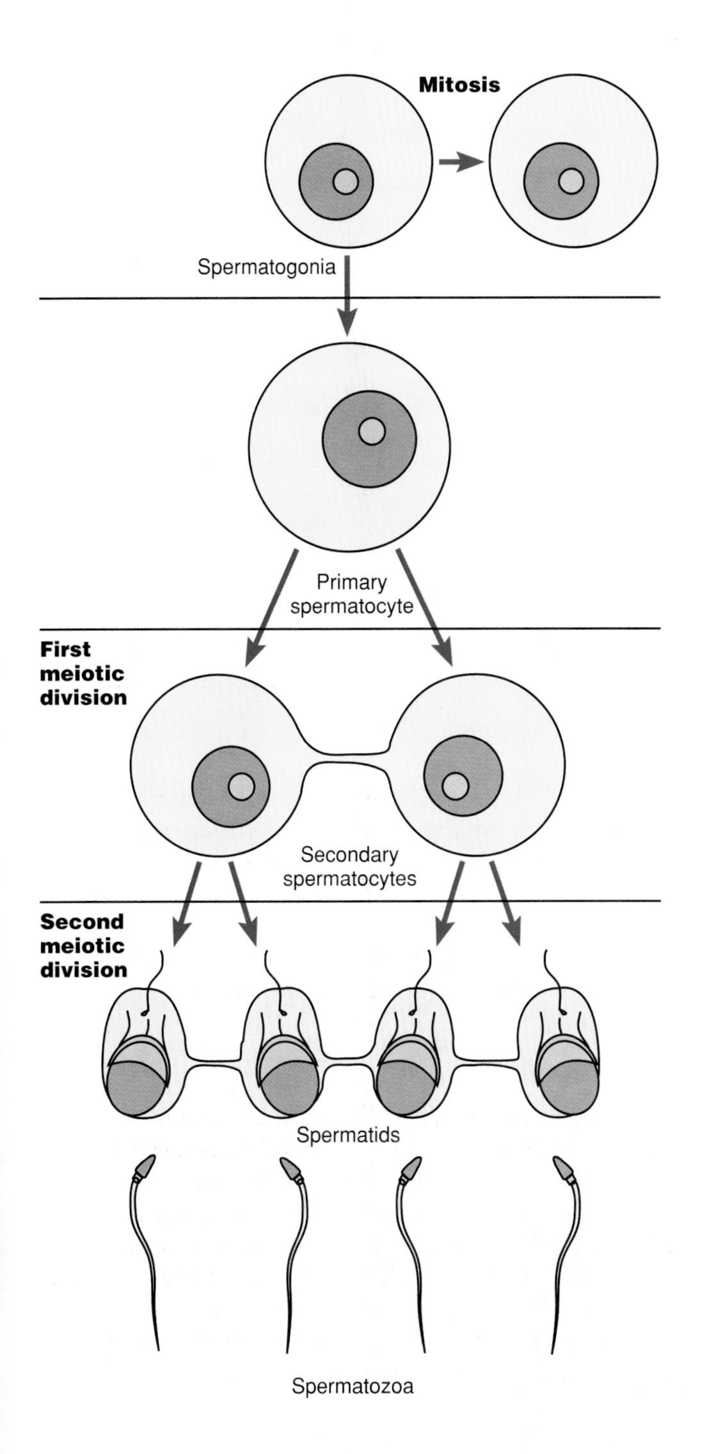

Figure 8.10. Spermatogonia undergo mitotic division to replace themselves and produce a daughter cell that will undergo meiotic division. This cell is called a primary spermatocyte. Upon completion of the first meiotic division, the daughter cells are called secondary spermatocytes. Each of these completes a second meiotic division to form spermatids. Notice that the four spermatids produced by the meiosis of a primary spermatocyte are interconnected. Each spermatid forms a mature spermatozoan.

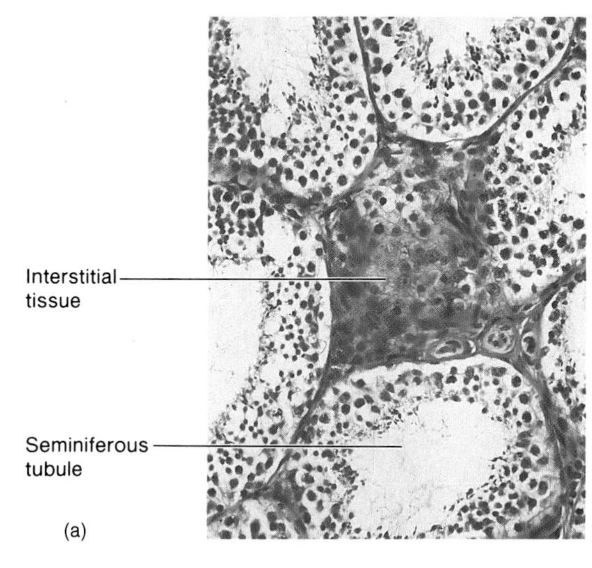

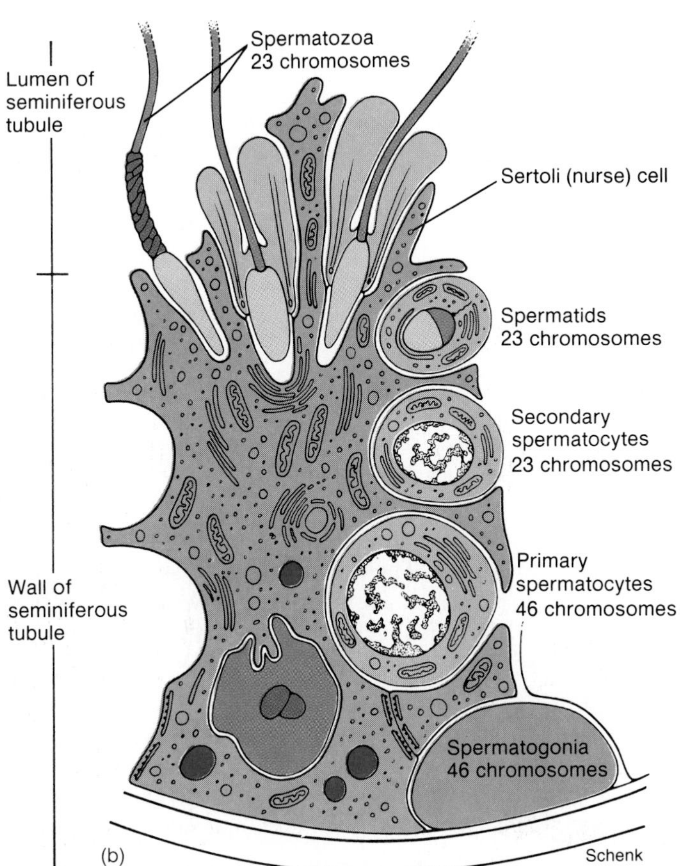

Figure 8.11. Seminiferous tubules. (a) A cross section with surrounding interstitial tissue; (b) the stages of spermatogenesis within the germinal epithelium of a seminiferous tubule, showing the relationship between nurse cells and developing spermatozoa.

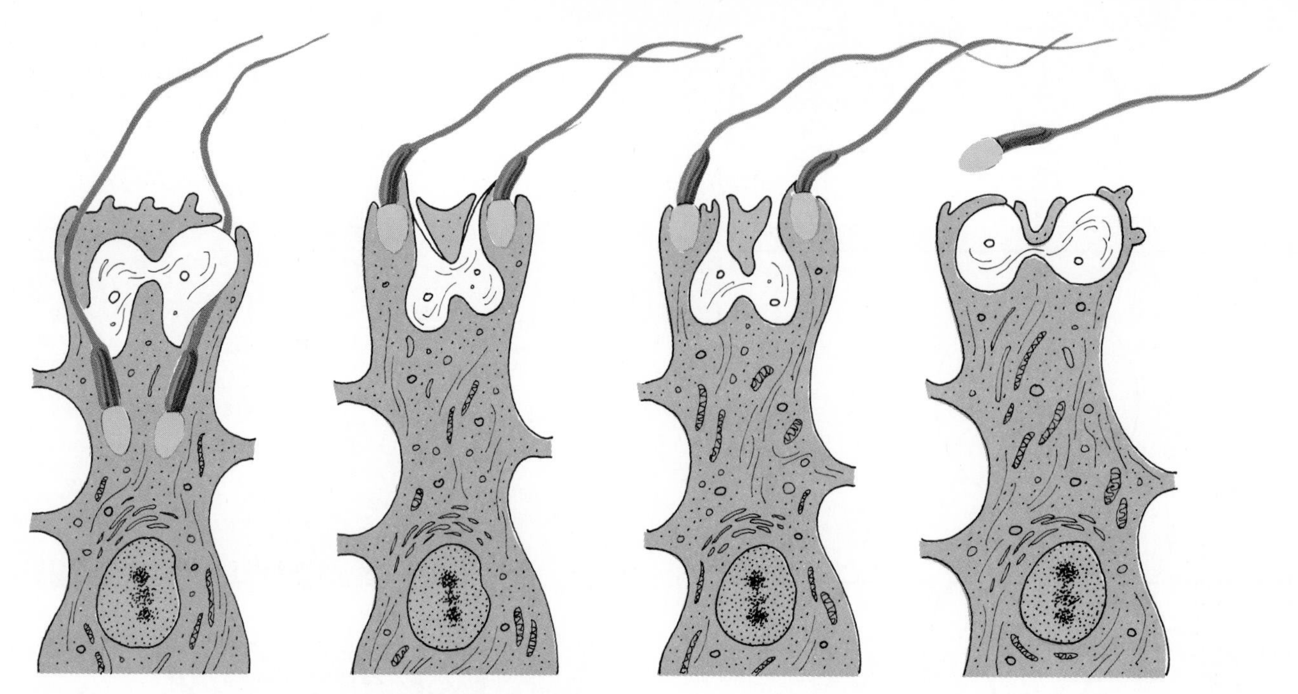

Figure 8.12. Processing of spermatids into spermatozoa. As the spermatids develop into spermatozoa most of their cytoplasm is pinched off as residual bodies and ingested by the surrounding Sertoli cell cytoplasm.

Sertoli Cells

The Sertoli cells are the only nongerminal cell type in the tubules. They form a continuous layer, connected by tight junctions, around the circumference of each tubule. In this way the Sertoli cells comprise a *blood-testis barrier,* because molecules from the blood must pass through the cytoplasm of the Sertoli cells before entering germinal cells. This barrier normally prevents the immune system from becoming sensitized to molecules in the developing sperm, and thus prevents autoimmune destruction of the sperm. The cytoplasm of the Sertoli cells extends through the width of the tubule and envelops the developing germ cells, so that it is often difficult to tell where the cytoplasm of the Sertoli cells and germ cells is separated.

When spermatids are converted to spermatozoa, most of the spermatid cytoplasm is eliminated. This occurs through phagocytosis by Sertoli cells of the "residual bodies" of cytoplasm from the spermatids (fig. 8.12). The mature spermatozoa thus contain very little cytoplasm; all of the cytoplasm of a fertilized ovum is contributed by the egg cell.

Hormonal Control of Spermatogenesis

The very beginning of spermatogenesis is apparently somewhat independent of hormonal control and, in fact, starts during embryonic development. Spermatogenesis is arrested, however, until puberty,[8] when testosterone secretion rises. Testosterone is required for completion of meiotic division and for the early stages of spermatid maturation.

The later stages of spermatid maturation during puberty appear to require stimulation by FSH (fig. 8.13). This FSH effect is mediated by the Sertoli cells which contain FSH receptors and surround and interact with the spermatids. During puberty, therefore, both FSH and testosterone are needed for the initiation of spermatogenesis.

Experiments in rats, and more recently in humans, have revealed that spermatogenesis within the adult testis can be maintained by androgens alone, in the absence of FSH. It appears, in other words, that FSH is needed to initiate spermatogenesis at puberty, but that once spermatogenesis has begun, FSH may no longer be required for this function.

Mature spermatozoa are released into the lumen of the seminiferous tubules. The spermatozoa contain a *head* (with DNA inside), a *body,* and a flagellum *tail* (fig. 8.14). Although the tail will ultimately be capable of flagellar movement, the sperm at this stage are nonmotile (incapable of movement). Motility and other maturational changes occur outside the testes in the epididymis.

Male Sex Accessory Organs

Spermatozoa and tubular secretions are moved within the seminiferous tubules and are drained via the *efferent ductules* into the **epididymis** (fig. 8.7). The epididymis is a single-coiled tube, four to five meters long if stretched out, that receives the tubular products. Spermatozoa enter at the "head" of the epididymis and are drained from its "tail" by a single tube, the **vas deferens.** Sperm can be stored within a widened area of the vas deferens known as the *ampulla.*

[8]puberty: L. *puberty,* grown up

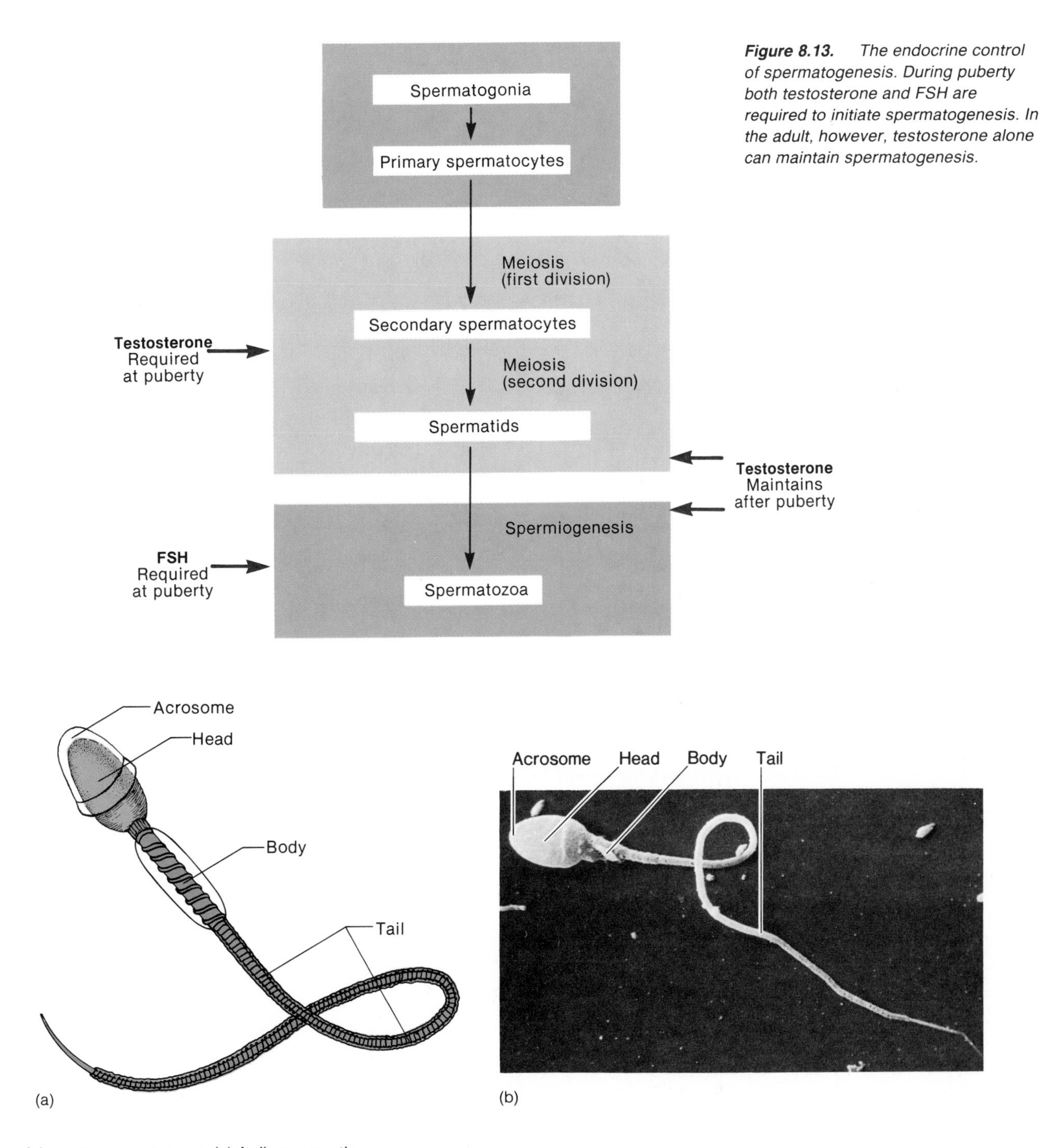

Figure 8.13. *The endocrine control of spermatogenesis. During puberty both testosterone and FSH are required to initiate spermatogenesis. In the adult, however, testosterone alone can maintain spermatogenesis.*

Figure 8.14. *A human spermatozoan. (a) A diagrammatic representation; (b) a scanning electron micrograph.*

(b) From Tissues and Organs: A Text-Atlas of Scanning Electron Microscopy *by Richard G. Kessel and Randy H. Kardon. Copyright © 1979 by W. H. Freeman and Company. Reprinted with permission.*

During their passage through the epididymis, spermatozoa gain motility and undergo other maturational changes so that they are able to fertilize an ovum once they spend some time in the female reproductive tract. Sperm obtained from the seminiferous tubules, in contrast, cannot fertilize an ovum. The epididymis serves as a site for sperm maturation and for the storage of sperm between ejaculations.

The vas deferens carries sperm from the epididymis out of the scrotum and into the body cavity. In its passage, the vas deferens obtains fluid secretions of the **seminal vesicles** and **prostate** gland. This fluid, now called *semen,* is carried by the ejaculatory duct to the *urethra* (fig. 8.15). The urethra is a tube that can carry either semen or urine (chapter 16) through the penis to be expelled from a single opening at the head of the penis (the *glans penis*).

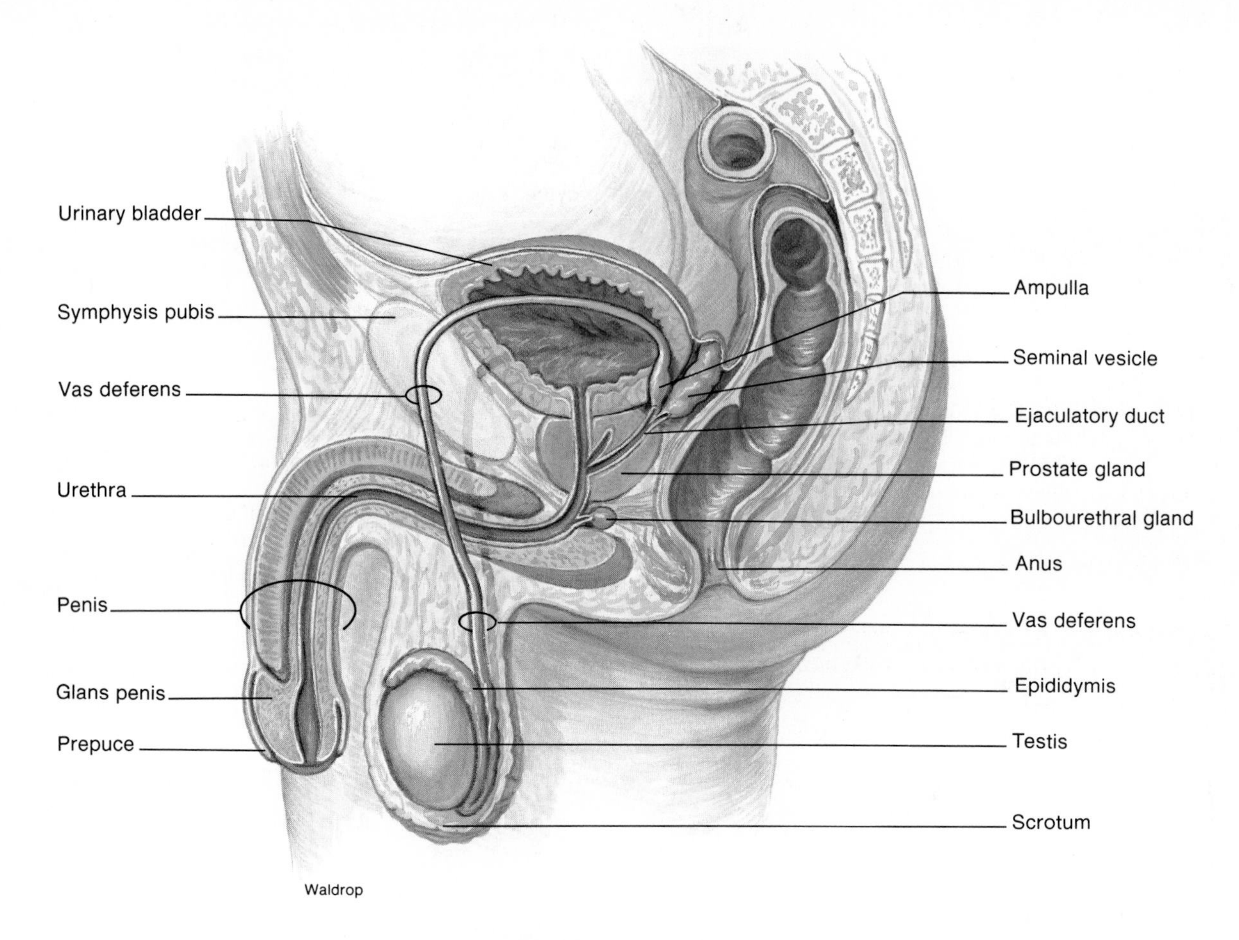

Figure 8.15. *Organs of the male reproductive system in sagittal view.*

The seminal vesicles and prostate are androgen-dependent accessory sexual organs—they atrophy if androgen is withdrawn by castration. The seminal vesicles secrete fluid containing fructose (which serves as an energy source for the spermatozoa), citric acid, coagulation proteins, and prostaglandins (special fatty acids that serve regulatory functions). The prostate secretes a liquefying agent and the enzyme *acid phosphatase,* which is often measured clinically to assess prostate function.

Erection, Emission, and Ejaculation

Erection, accompanied by increases in the length and width of the penis, is achieved as a result of blood flow into the "erectile tissues" of the penis. These erectile tissues include two paired structures—the *corpora cavernosa*—located on the dorsal side of the penis, and one unpaired *corpus spongiosum* on the ventral side (fig. 8.16). The urethra runs through the center of the corpus spongiosum. The erectile tissue forms columns extending the length of the penis, although the corpora cavernosa do not extend all the way to the tip.

Erection is achieved as a result of parasympathetic nerve-induced dilation of small arteries that allows blood to flow into the penis (table 8.3). As the erectile tissues become engorged with blood and the penis becomes turgid,[9] the outflow of blood is partially occluded, thus aiding erection. The term *emission* refers to the movement of semen into the urethra, and *ejaculation* refers to the forcible expulsion of semen from the urethra out of the penis. Emission and ejaculation are stimulated by sympathetic nerves, which cause contractions of the tubular system, seminal vesicles and prostate, and of muscles at the base of the penis.

Erection is controlled by two portions of the central nervous system—the hypothalamus in the brain and the sacral portion of the spinal cord. Conscious sexual thoughts originating in the cerebral cortex act via the hypothalamus to control the sacral region, which in turn increases parasympathetic nerve activity to promote vasodilation and erection in the penis. Conscious thought is not required for erection however, because sensory stimulation of the penis can more directly activate the sacral region of the spinal cord and cause an erection.

[9]turgid: L. *turgeo,* to swell

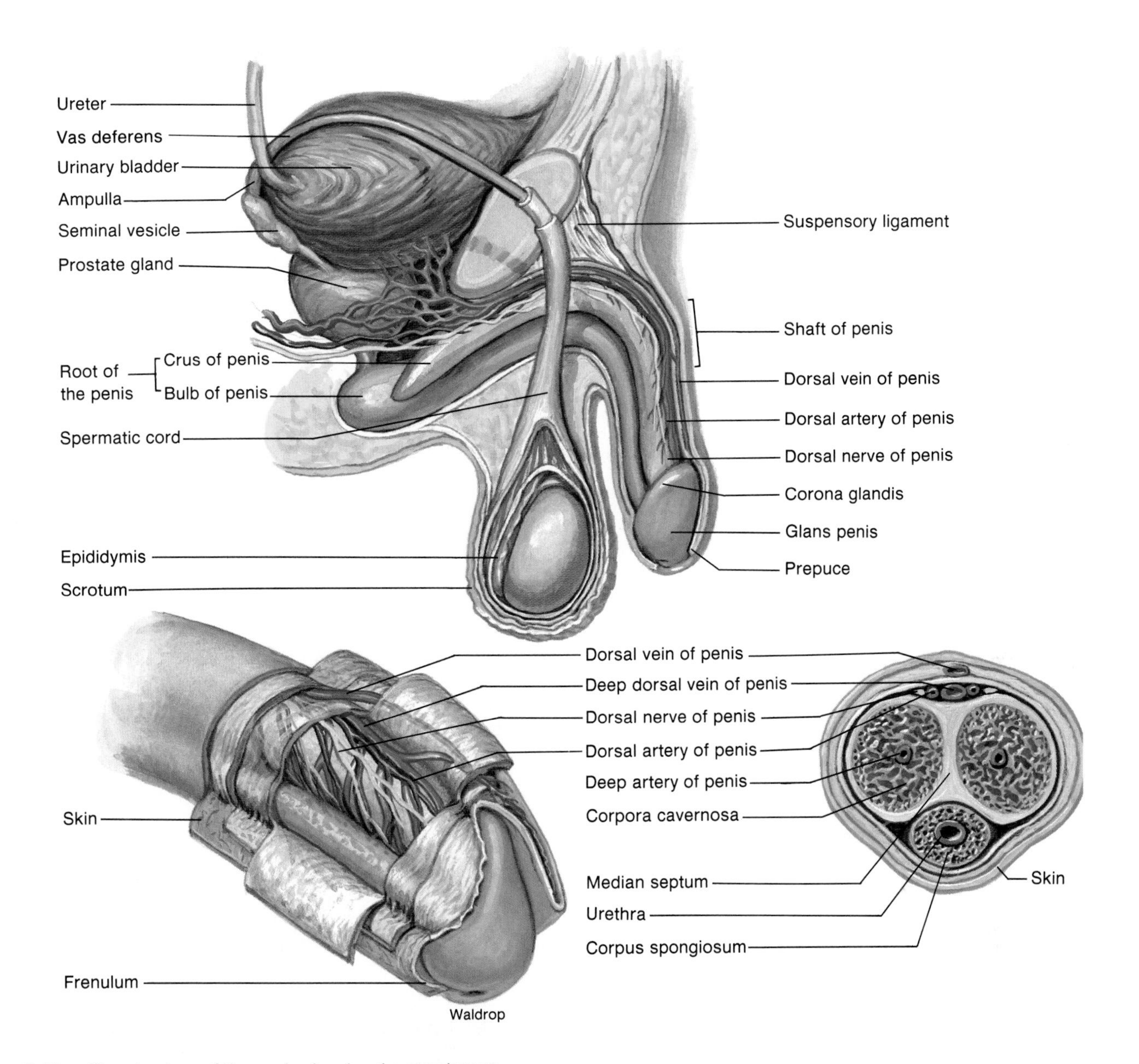

Figure 8.16. *The structure of the penis showing the attachment, blood and nerve supply, and the arrangement of the erectile tissue.*

Table 8.3 Control of erection and ejaculation

Regulation	**Effect**	**Result**
Parasympathetic nerves	Vasodilation produces increased blood flow into erectile tissues—corpora cavernosa and corpus spongiosum. As tissues become turgid, venous outflow is partially occluded, further increasing accumulation of blood.	Erection
Sympathetic nerves	Peristaltic waves of contraction in tubular system—primarily epididymis, vas deferens, and ejaculatory ducts, and contraction of prostate and seminal vesicles.	Ejaculation

Male Fertility

The male ejaculates about 1.5–5.0 ml of semen. The bulk of this fluid (45% to 80%) is produced by the seminal vesicles, and about 15% to 30% is contributed by the prostate. The sperm content in human males averages 60–150 million per milliliter in the ejaculated semen. Some of the values of normal human semen are summarized in table 8.4.

Table 8.4 Some characteristics and reference values used in the clinical examination of semen

Characteristic	Reference Value
Volume of ejaculate	1.5–5.0 ml
Sperm count	40–250 million/ml
Sperm motility	
Percent of motile forms:	
1 hour after ejaculation	70% or more
3 hours after ejaculation	60% or more
Leukocyte count	0–2,000/ml
pH	7.2–7.8
Fructose concentration	150–600 mg/100 ml

From L. Glasser, "Seminal Fluid and Subfertility" in *Diagnostic Medicine*, July/August 1981. Copyright © 1981 Medical Economics Co., Inc.

A sperm concentration below about twenty million per milliliter is termed *oligospermia* (*oligo* = few), and is associated with decreased fertility. A total sperm count below about fifty million per ejaculation is clinically significant in male infertility. In addition to low sperm counts as a cause of infertility, some men and women have antibodies against sperm antigens (this is very common in men with vasectomies). Such antibodies do not appear to affect health, but do reduce fertility.

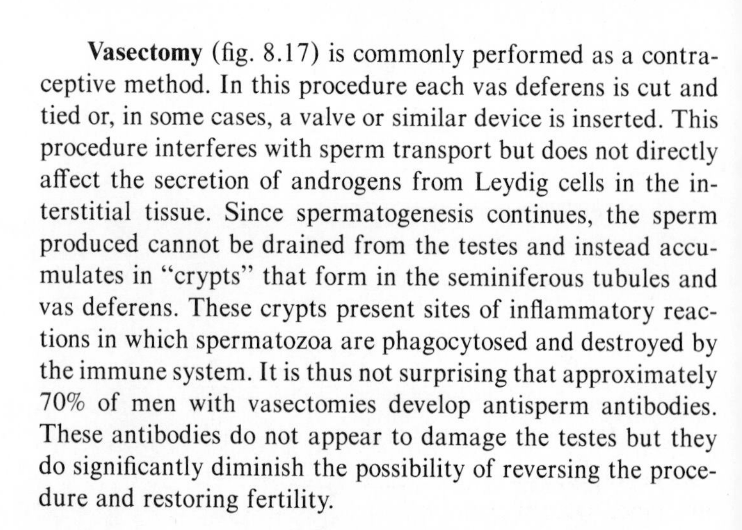

Vasectomy (fig. 8.17) is commonly performed as a contraceptive method. In this procedure each vas deferens is cut and tied or, in some cases, a valve or similar device is inserted. This procedure interferes with sperm transport but does not directly affect the secretion of androgens from Leydig cells in the interstitial tissue. Since spermatogenesis continues, the sperm produced cannot be drained from the testes and instead accumulates in "crypts" that form in the seminiferous tubules and vas deferens. These crypts present sites of inflammatory reactions in which spermatozoa are phagocytosed and destroyed by the immune system. It is thus not surprising that approximately 70% of men with vasectomies develop antisperm antibodies. These antibodies do not appear to damage the testes but they do significantly diminish the possibility of reversing the procedure and restoring fertility.

1. Describe the effects of castration on FSH and LH secretion in the male, and explain the negative feedback control that the testes normally exert on the anterior pituitary.
2. Describe the two compartments of the testes with respect to their (a) structure, (b) function, and (c) response to gonadotropin stimulation.
3. Using a diagram, describe the stages of spermatogenesis. Explain why spermatogenesis can continue throughout life without using up all of the spermatogonia.
4. Describe the hormonal requirements for spermatogenesis.
5. Explain how erection and ejaculation are accomplished.

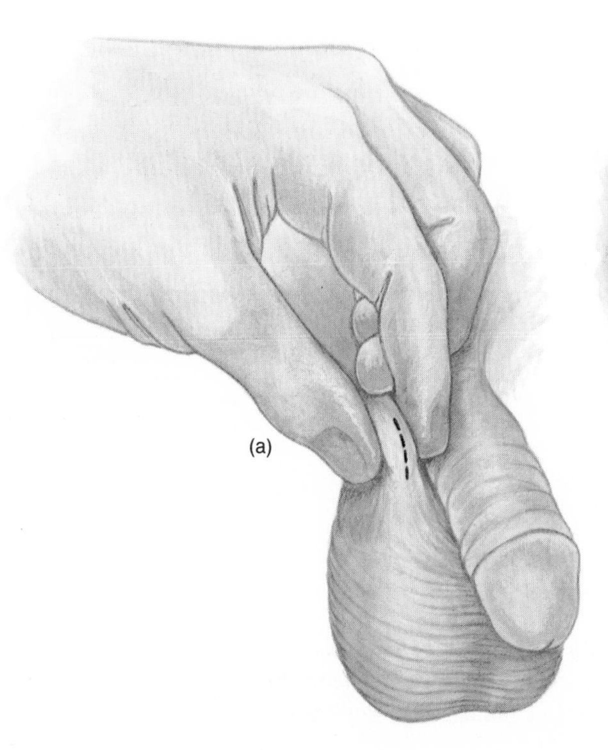

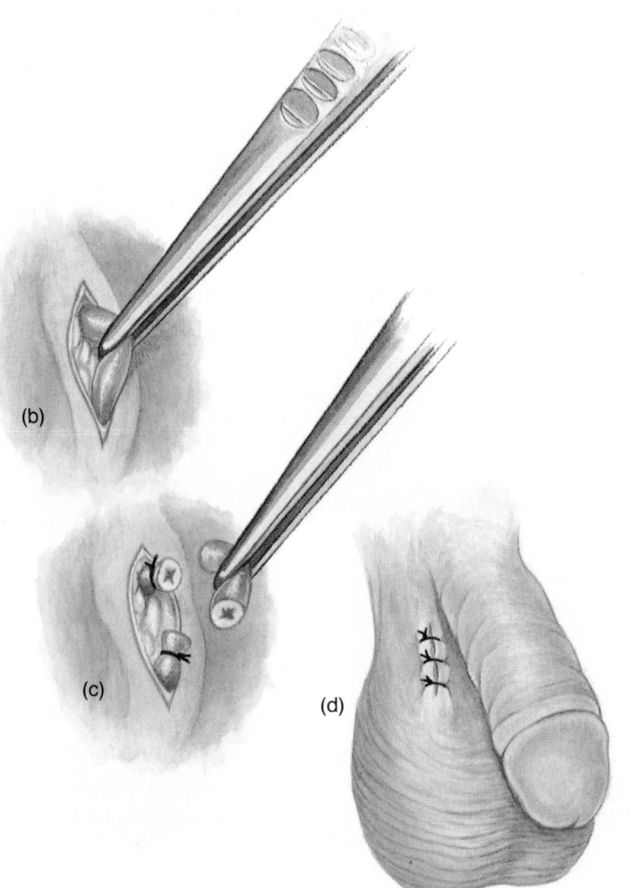

Figure 8.17. *A simplified illustration of a vasectomy, in which a segment of the vas deferens is removed through an incision in the scrotum.*

Female Reproductive System

The ova are contained within the ovary in hollow structures known as follicles. As a result of gonadotropin stimulation, a primary follicle may grow and develop a cavity, thus becoming a secondary follicle. The ovum also undergoes changes during this time, becoming a secondary oocyte. The secondary oocyte contained within a fully mature, or graafian, follicle is expelled from the ovary in the process of ovulation. After ovulation, the empty graafian follicle becomes a corpus luteum.

The two ovaries (fig. 8.18) are located within the body cavity, suspended by means of ligaments from the pelvic girdle. Extensions, or *fimbriae,* of the **uterine,** or **fallopian, tubes**[10] partially cover each ovary. Ova that are released from the ovary—in a process called *ovulation*—are normally drawn into the fallopian tubes by the action of the cilia in the epithelial lining of the tubes. The lumen of each fallopian tube is continuous with the **uterus** (or womb), a pear-shaped muscular organ also suspended within the pelvic girdle by ligaments.

The uterus narrows to form the *cervix* (*cervix* = neck), which opens to the **vagina.** The only physical barrier between the vagina and uterus is a plug of *cervical mucus.* These structures—the vagina, uterus, and fallopian tubes—constitute the sex accessory organs of the female (fig. 8.19). Like the sex accessory organs of the male, the female genital tract is affected by gonadal steroid hormones. Cyclic changes in ovarian secretion, as will be described in the next section, cause cyclic changes in the epithelial lining of the genital tract. The epithelial lining of the uterus, known as the *endometrium,*[11] is most dramatically altered during the ovarian cycle.

The vaginal opening is located immediately posterior to the opening of the urethra, which carries only urine in the female. Both openings are covered by inner **labia minora** and outer **labia majora** (fig. 8.20). The **clitoris,** which is developed from the same embryonic structure as the penis, is located at the anterior margin of the labia minora.

Ovarian Cycle

The germ cells that migrate into the ovaries during early embryonic development multiply, so that by about five months of gestation (prenatal life) the ovaries contain approximately six to seven million parent egg cells, or *oogonia.* The production of new oogonia stops at this point and never resumes again. Toward the end of gestation, the oogonia begin meiosis through prophase I, at which time they are called **primary oocytes.** Meiosis

[10]fallopian tubes: from Gabriele Fallopius, Italian anatomist, 1523–1562

[11]endometrium: Gk. *endon,* within; *metra,* uterus

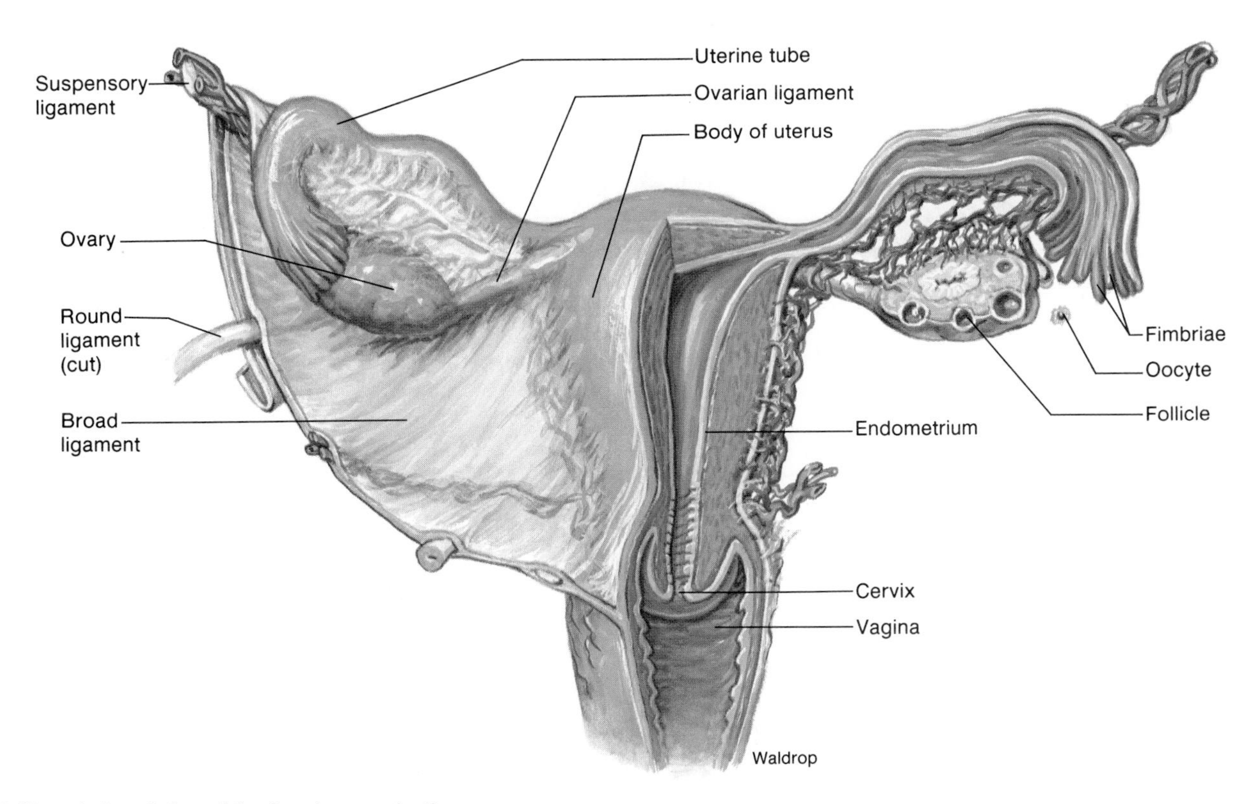

Figure 8.18. *A dorsal view of the female reproductive organs showing the relationship of the ovaries, uterine tubes, uterus, cervix, and vagina.*

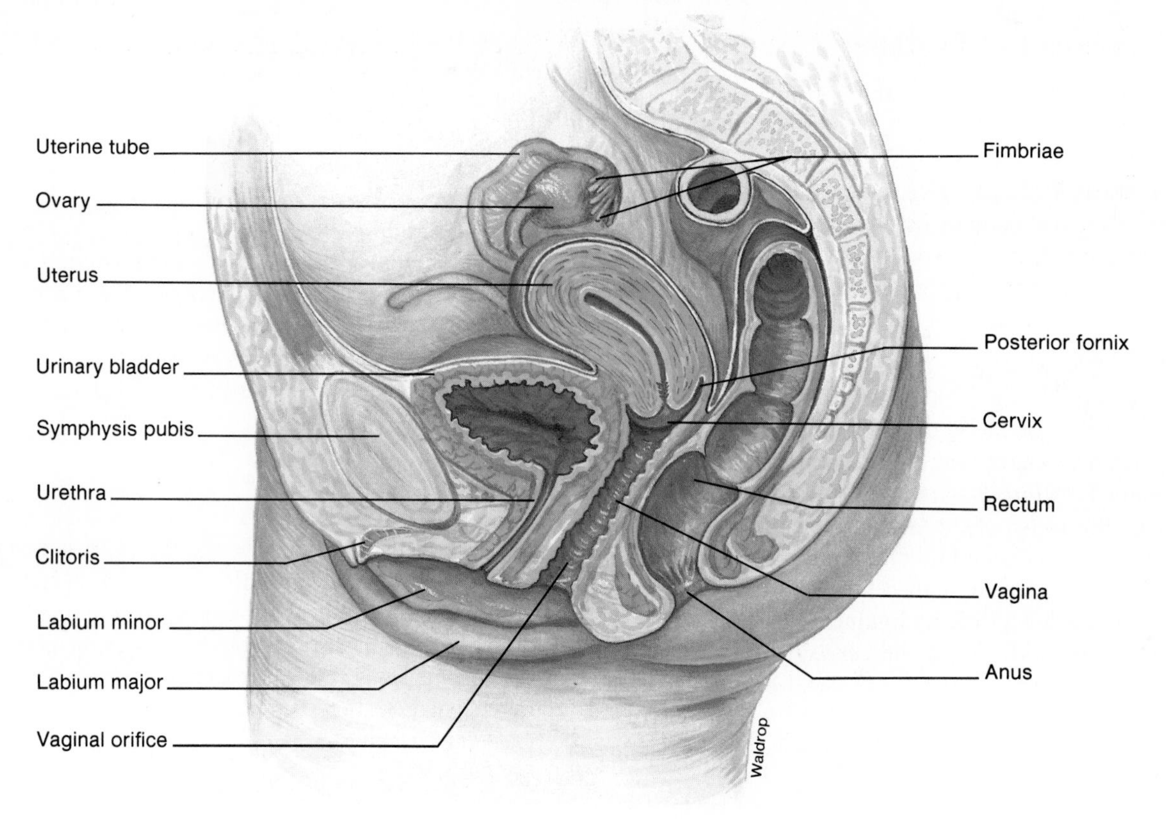

Figure 8.19. *Organs of the female reproductive system.*

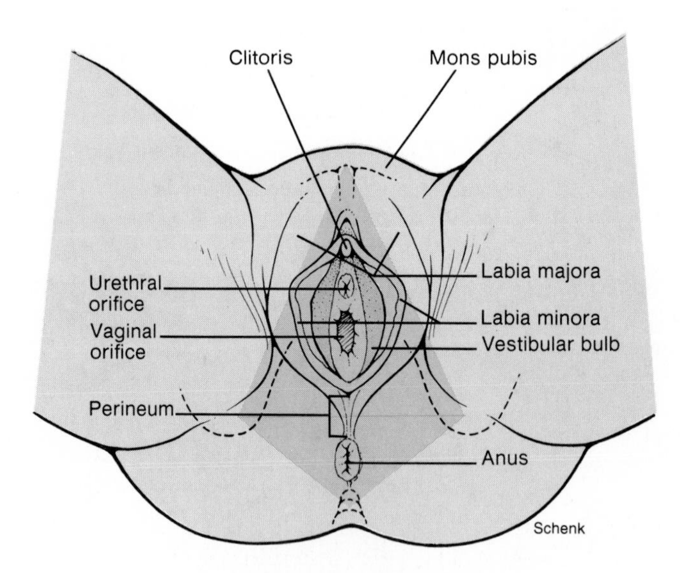

Figure 8.20. *The external female genitalia.*

is arrested at this point until puberty, when particular primary oocytes are stimulated to continue their meiotic division. The number of primary oocytes decreases throughout a woman's lifetime. The ovaries of a newborn girl contain about two million oocytes, but this number is reduced to about 300,000–400,000 by the time the girl enters puberty. Oogenesis ceases entirely at menopause (the time menstruation stops).

Primary oocytes that are not stimulated to complete the first meiotic division are contained within tiny follicles,[12] called **primordial follicles.** In response to gonadotropin stimulation some of these oocytes and follicles get larger, and the follicular cells

[12]follicle: L. diminutive of *follis,* bag

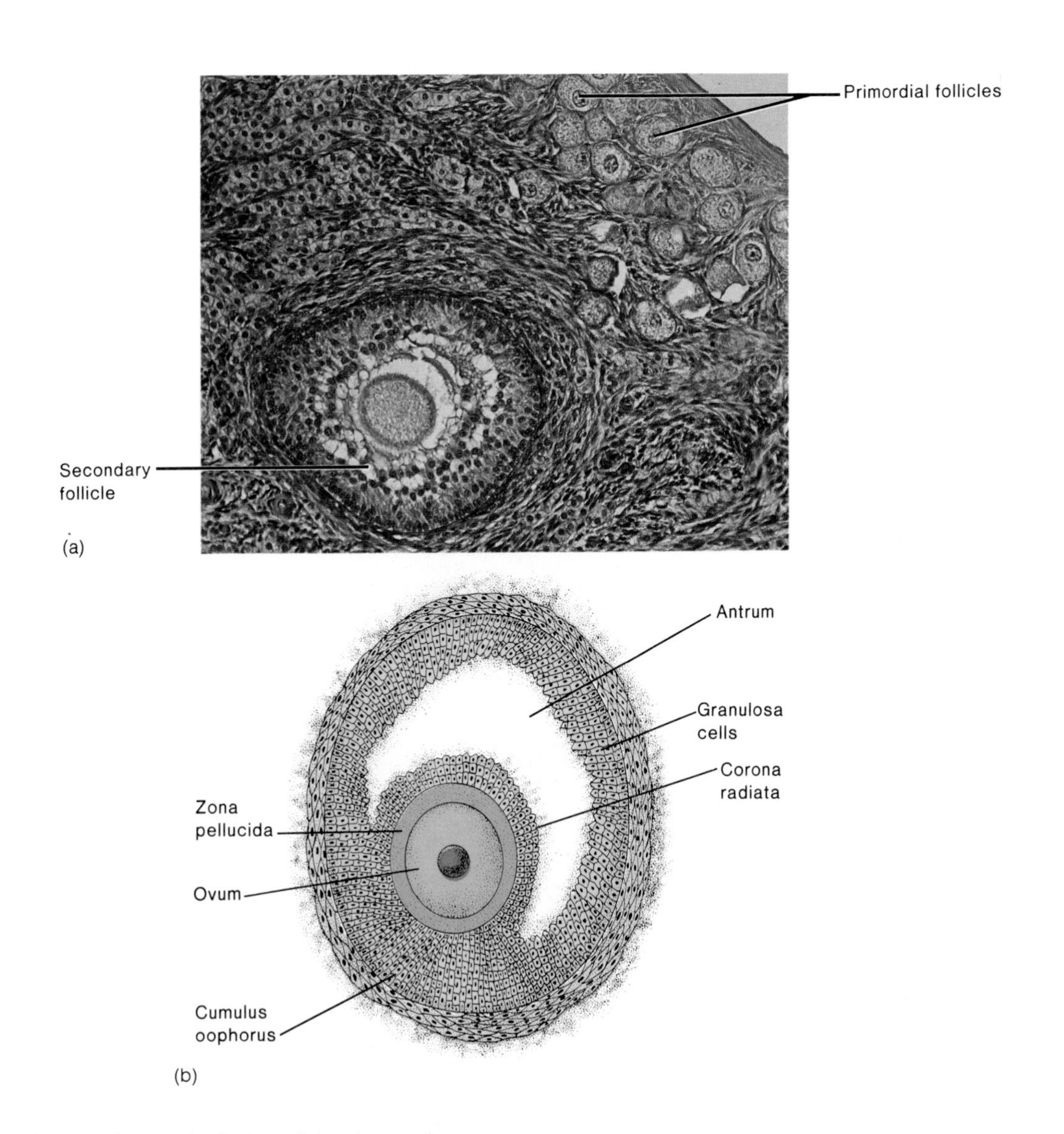

Figure 8.21. (a) A photomicrograph of primordial and secondary follicles. (b) A diagram of the parts of a secondary follicle.

divide to produce numerous small **granulosa cells** that surround the oocyte and fill the follicle. A follicle at this stage in development is called a **primary follicle.**

Some primary follicles will be stimulated to grow still bigger and develop a fluid-filled cavity, called an *antrum,* at which time they are called **secondary follicles** (fig. 8.21). The granulosa cells of secondary follicles form a ring around the circumference of the follicle and form a mound that supports the ovum. This mound is called the *cumulus oophorus.* The ring of granulosa cells surrounding the ovum is the *corona radiata.*[13] Between the oocyte and the corona radiata is a thin gellike layer of proteins and polysaccharides, called the *zona pellucida.*[14] Under the stimulation of FSH from the anterior pituitary, the granulosa cells secrete increasing amounts of estrogen as the follicles grow.

As the follicle develops, the primary oocyte completes its first meiotic division. This does not form two complete cells, however, because only one cell—the **secondary oocyte**—gets all the cytoplasm. The other cell formed at this time becomes a small *polar body* (fig. 8.22), which eventually fragments and disappears. This unequal division of cytoplasm ensures that the ovum will be large enough to become a viable embryo should fertilization later occur. The secondary oocyte then begins the second meiotic division but stops at metaphase II. Meiosis is arrested at metaphase II, before separation of chromosomes and division of cytoplasm occurs, while the ovum is in the ovary. The second meiotic division is only completed by an ovum that has been fertilized.

[13]corona radiata: Gk. *korone,* crown; *radiata,* radiate
[14]zona pellucida: Gk. *zone,* girdle; L. *pellis,* skin

Figure 8.22. (a) A primary oocyte at metaphase I of meiosis. Note the alignment of chromosomes (arrow). (b) A human secondary oocyte formed at the end of the first meiotic division and the first polar body (arrow).

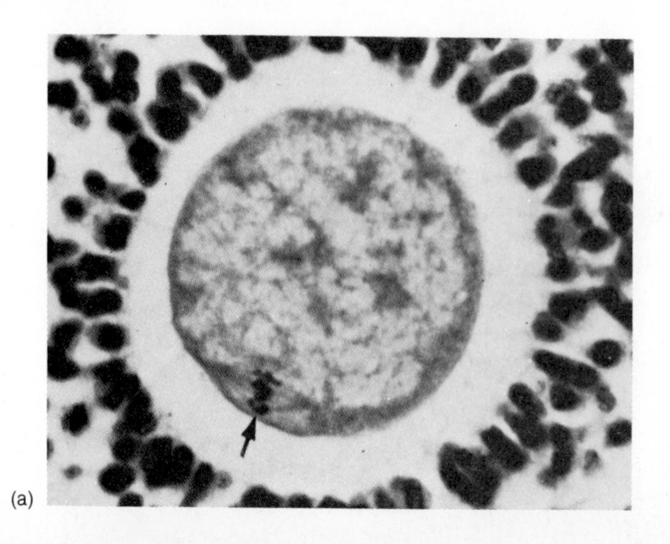

(a)

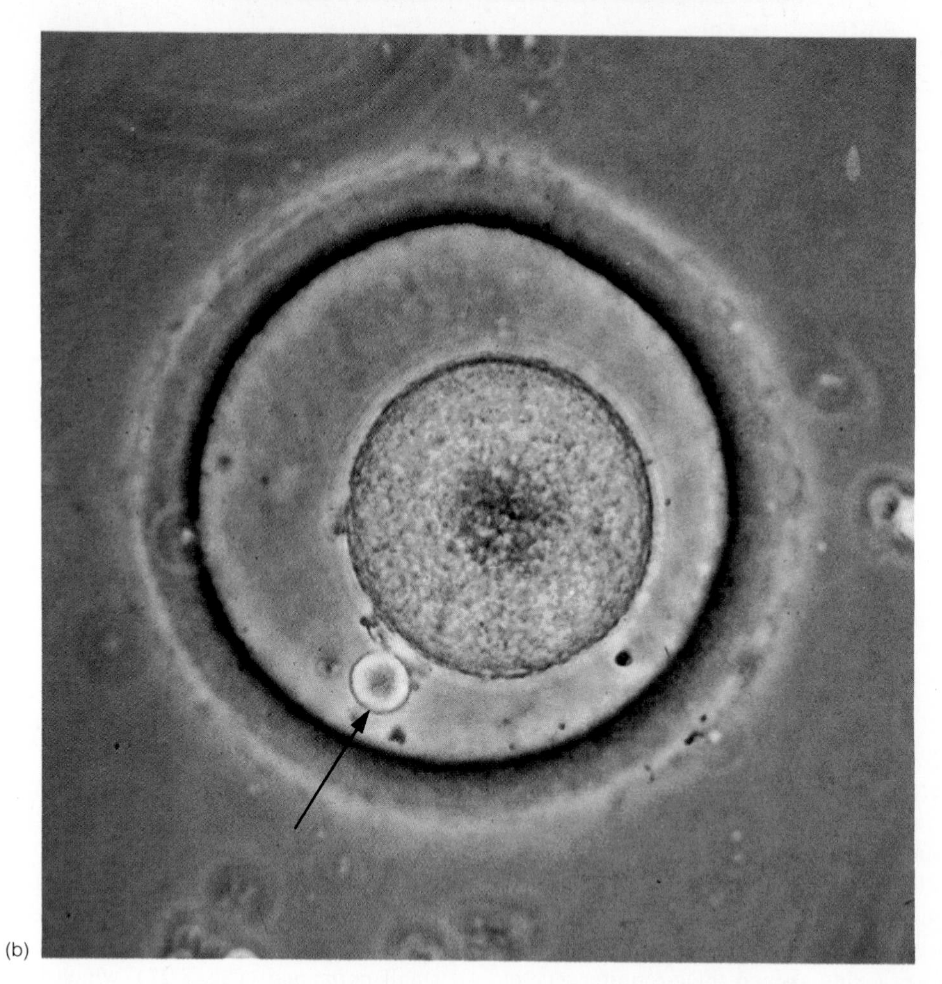

(b)

Ovulation

Usually, by about ten to fourteen days after the first day of menstruation, only one follicle has continued its growth to become a fully mature **graafian follicle**[15] (fig. 8.23); other secondary follicles during that cycle regress in size. The graafian follicle is so large that it forms a bulge on the surface of the ovary. Under proper hormonal stimulation this follicle will rupture—much like the popping of a blister—and extrude its oocyte into the uterine tube in the process of **ovulation** (fig. 8.24).

[15]graafian follicle: from Regnier de Graaf, Dutch anatomist and physician, 1641–1673

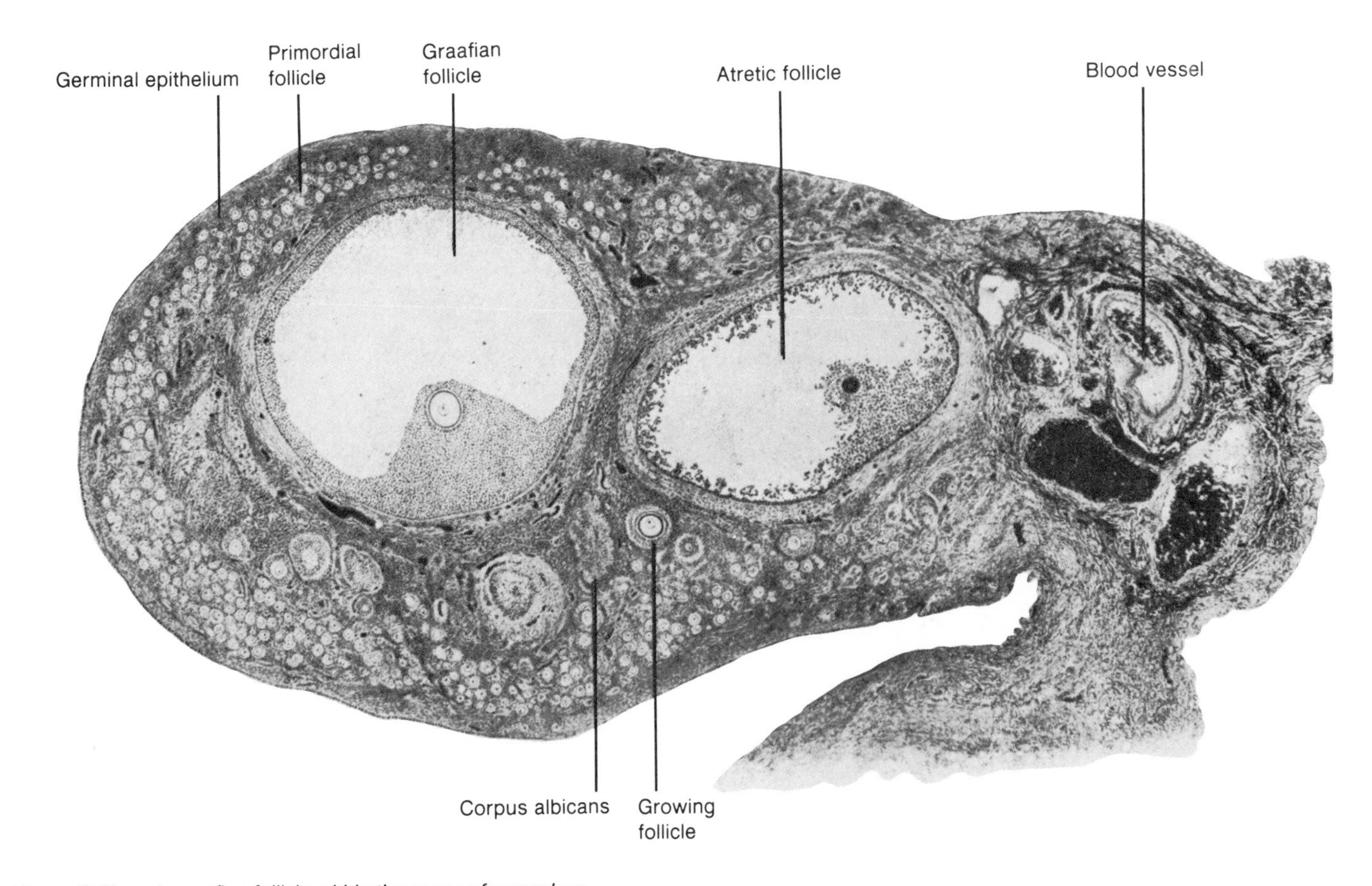

Figure 8.23. *A graafian follicle within the ovary of a monkey.*

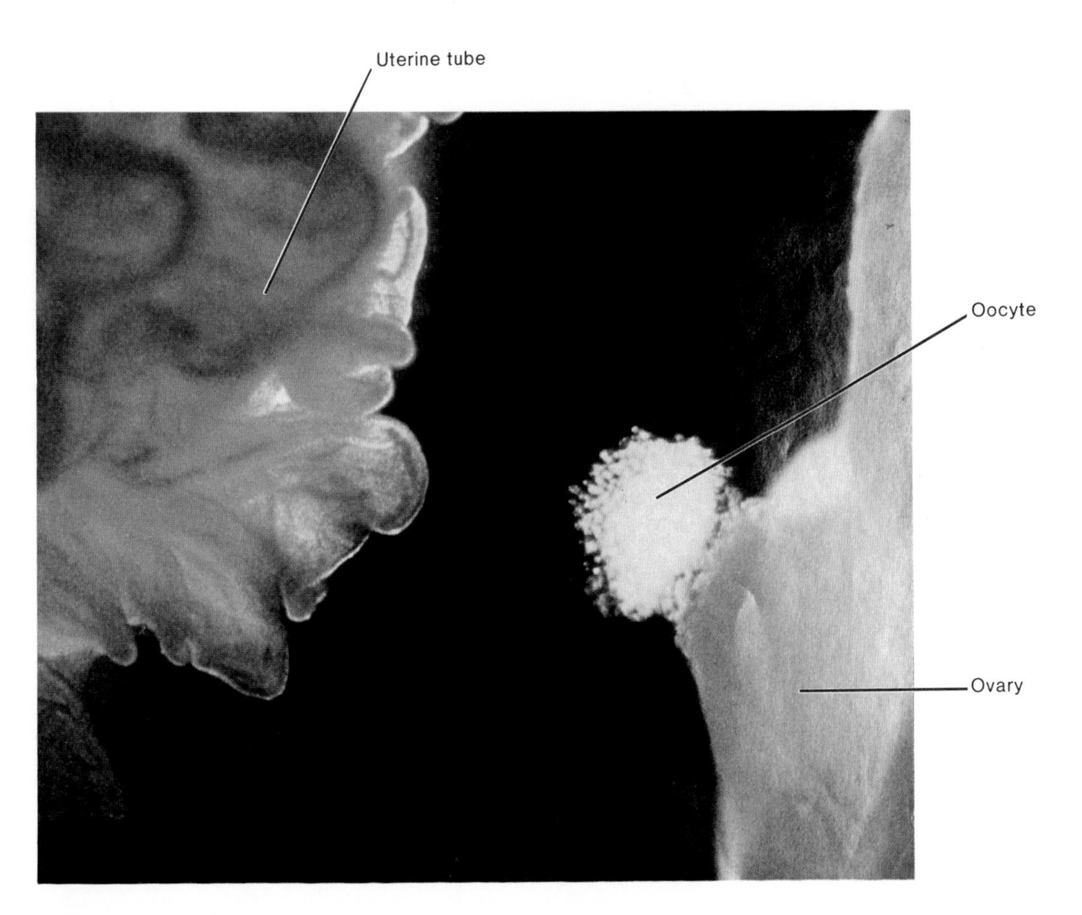

Figure 8.24. *Ovulation from a human ovary.*

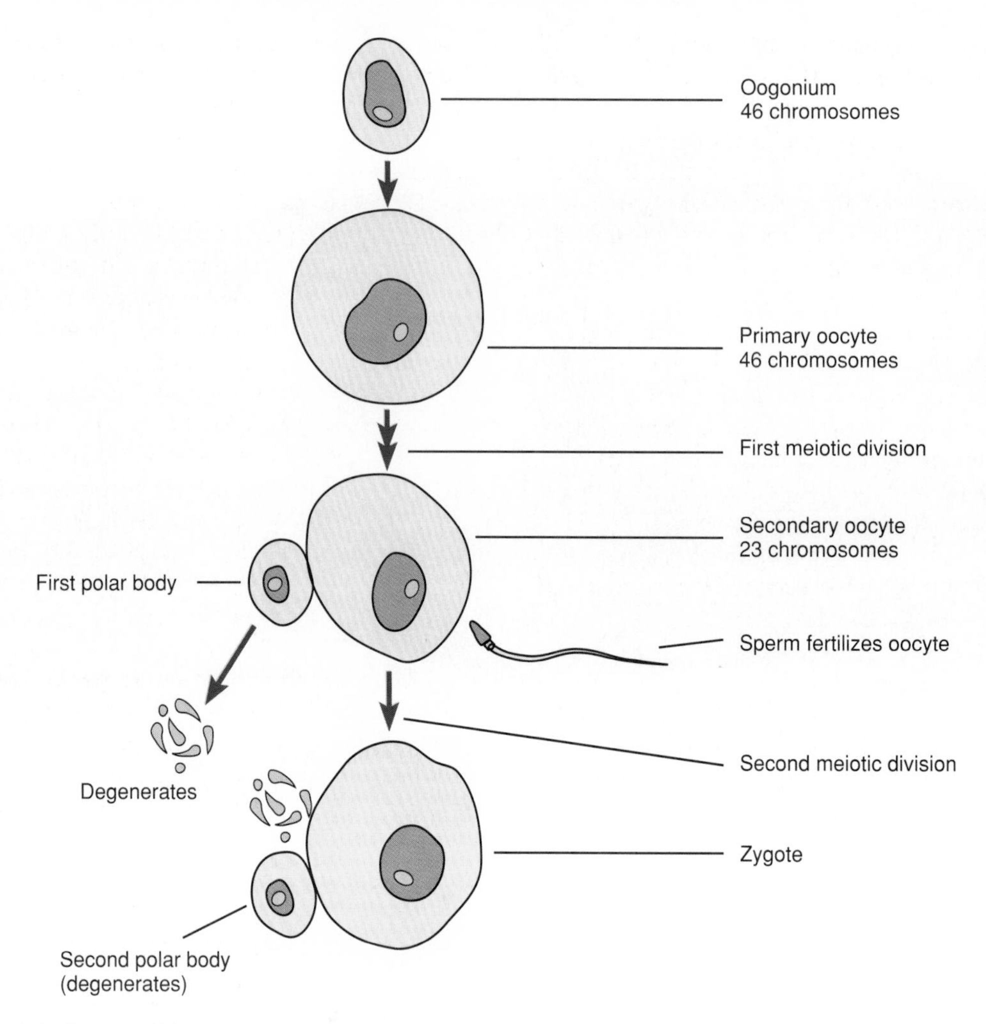

Figure 8.25. *A schematic diagram of the process of oogenesis. During meiosis, each primary oocyte produces a single haploid gamete. If the secondary ooctye is fertilized, it forms a secondary polar body and becomes a zygote.*

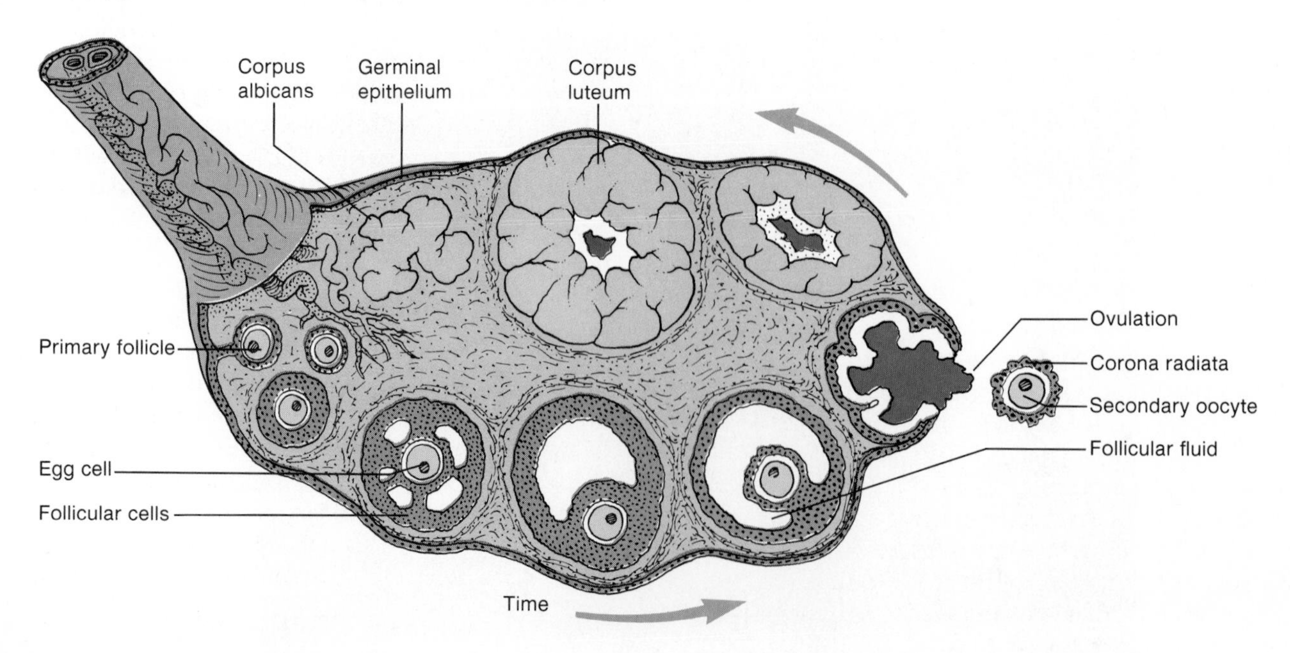

Figure 8.26. *A schematic diagram of an ovary showing the various stages of ovum and follicle development.*

The released cell is a secondary oocyte, surrounded by the zona pellucida and corona radiata. If it is not fertilized, it disintegrates in a couple of days. If a sperm passes through the corona radiata and zona pellucida and enters the cytoplasm of the secondary oocyte, the oocyte will then complete the second meiotic division. In this process the cytoplasm is again not divided equally; most of the cytoplasm remains in the zygote (fertilized egg), leaving another polar body which, like the first, disintegrates (fig. 8.25).

Changes continue in the ovary following ovulation. The empty follicle, under the influence of luteinizing hormone from the anterior pituitary, undergoes structural and biochemical changes to become a **corpus luteum** (*corpus luteum* = yellow body). Unlike the ovarian follicles, which secrete only estrogen, the corpus luteum secretes two sex steroid hormones: estrogen and progesterone. Toward the end of a nonfertile cycle the corpus luteum regresses and is changed into a nonfunctional structure. These cyclic changes in the ovary are summarized in figure 8.26.

Pituitary-Ovarian Axis

The term *pituitary-ovarian axis* refers to the hormonal interactions between the anterior pituitary and the ovaries. The anterior pituitary secretes two gonadotropic hormones—follicle-stimulating hormone (FSH) and luteinizing hormone (LH)—that promote cyclic changes in the structure and function of the ovaries. The secretion of both gonadotropic hormones, as discussed in chapter 7, is controlled by a single releasing hormone from the hypothalamus—called gonadotropin releasing hormone (GnRH)—and by feedback effects from hormones secreted from the ovaries. The nature of these interactions will be described in detail in the next section.

---

1. Compare the structure and contents of a primordial follicle, primary follicle, secondary follicle, and graafian follicle.
2. Define ovulation, and describe the changes that occur in the ovary following ovulation in a nonfertile cycle.
3. Describe oogenesis, and explain why only one mature ovum is produced by this process.
4. Compare the hormonal secretions of the ovarian follicles with those of a corpus luteum.

Menstrual Cycle

Cyclic changes in the secretion of gonadotropic hormones from the anterior pituitary cause the changes observed in the ovaries during a menstrual cycle. FSH stimulates development of the ovarian follicles, and LH promotes ovulation and the development of a corpus luteum from the empty follicle. Growth of the follicles and the development of the corpus luteum, in turn, produces cyclic changes in the secretion of the sex steroids estradiol and progesterone. The rise in secretion first of estradiol and then of progesterone promote growth and development of the endometrium, and the fall in the secretion of these hormones causes menstruation.

Humans, apes, and old-world monkeys have cycles of ovarian activity that repeat at approximately one-month intervals; hence the name **menstrual cycle** (*menstru* = monthly). The term *menstruation* is used to indicate the periodic shedding of the inner two-thirds of the *endometrium* (epithelial lining of the uterus), which becomes thickened prior to menstruation under stimulation by ovarian steroid hormones. In primates (other than new-world monkeys) this shedding of the endometrium is accompanied by bleeding. There is no bleeding, in contrast, when other mammals shed the endometrium; their cycles, therefore, are not called menstrual cycles.

Human females and other primates that have menstrual cycles may permit copulation at any time of the cycle. Nonprimate female mammals, in contrast, are sexually receptive only at a particular time in their cycles (shortly before or shortly after ovulation). These animals are said to have *estrous cycles*. Bleeding occurs in some animals (such as dogs and cats) that have estrous cycles shortly before they permit copulation. This bleeding is a result of high estrogen secretion and is not associated with menstruation.

The bleeding that accompanies menstruation, in contrast, can be experimentally induced by removing the ovaries. This demonstrates that menstrual bleeding is caused by the withdrawal of ovarian hormones. During the normal menstrual cycle, the secretion of ovarian hormones rises and falls in a regular fashion, causing cyclic changes in the endometrium and other sex steroid-dependent tissues.

Phases of the Menstrual Cycle: Pituitary and Ovary

The average menstrual cycle has a duration of about twenty-eight days. Since it is a cycle, there is no beginning or end, and the changes that occur are generally gradual. It is convenient, however, to call the first day of menstruation "day one" of the cycle, because menstrual blood flow is the most apparent of the changes that occur. It is also convenient to divide the cycle into phases based on changes that occur in the ovary and in the endometrium. The ovaries are in the **follicular phase** starting on the first day of menstruation and ending on the day of ovulation. After ovulation, the ovaries are in the **luteal phase** until the first day of menstruation. The phases of the endometrium will be discussed in a later section.

Follicular Phase

Menstruation lasts from day one to day four or five of the average cycle. During this time, the secretions of ovarian steroid hormones are at their lowest ebb, and the ovaries contain only primordial and primary follicles. During the *follicular phase* of the ovaries, which lasts from day one to about day thirteen of the cycle (this is highly variable), some of the primary follicles grow, form an antrum, and become secondary follicles. Toward the end of the follicular phase one follicle in one ovary reaches maturity and becomes a graafian follicle. As follicles grow, the

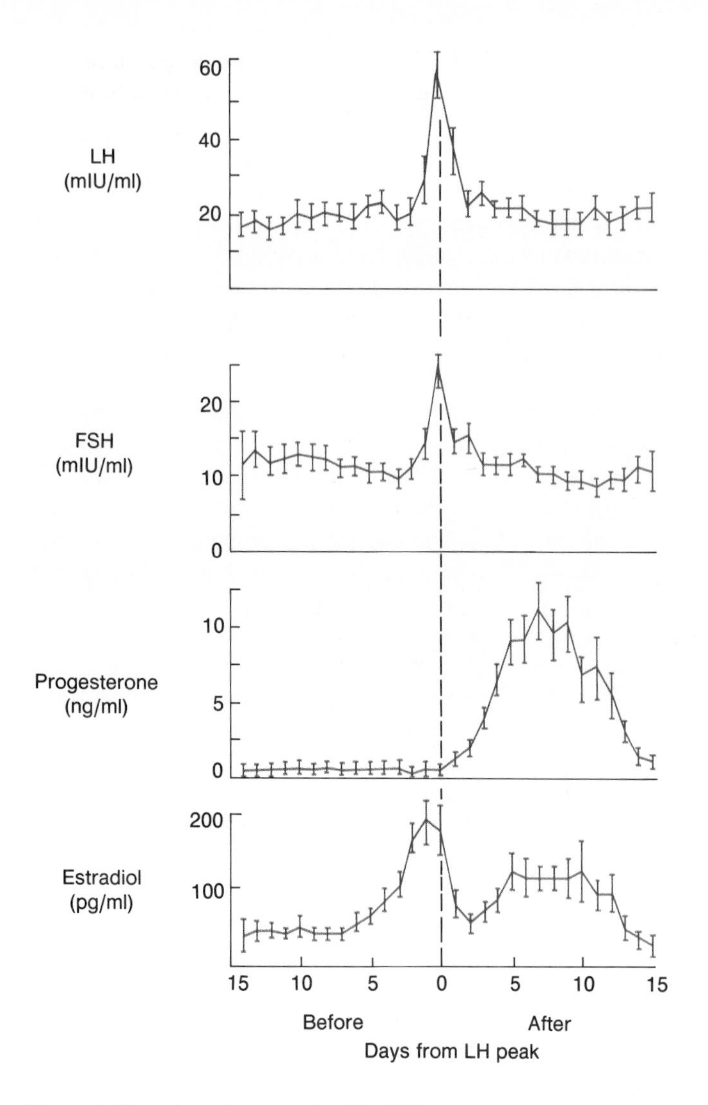

Figure 8.27. *Sample values for LH, FSH, progesterone, and estradiol during the menstrual cycle. The midcycle peak of LH is used as a reference day. (IU = international unit.)*

granulosa cells secrete an increasing amount of **estradiol** (the principal estrogen), which reaches its highest concentration in the blood shortly before ovulation.

The growth of the follicles and the secretion of estradiol are stimulated by, and dependent upon, FSH secreted from the anterior pituitary. The amount of FSH secreted during the early follicular phase is believed to be slightly greater than the amount secreted in the late follicular phase, although this can vary from cycle to cycle (a measure of variance is shown by vertical bars in figure 8.27). FSH stimulates the production of FSH receptors in the granulosa cells, so that the follicles become increasingly sensitive to a given amount of FSH. As a result, the stimulatory effect of FSH on the follicles increases despite the fact that FSH levels in the blood do not increase throughout the follicular phase. Toward the end of the follicular phase, FSH and estradiol also stimulate the production of LH receptors in the graafian follicle. This prepares the graafian follicle for the next major event in the cycle.

The rapid rise in estradiol secretion from the granulosa cells during the follicular phase acts on the hypothalamus to increase the secretion of GnRH. In addition, estradiol augments the ability of the pituitary to respond to GnRH with an increase in LH secretion. As a result of this stimulatory, or **positive feedback,** effect of estradiol on the pituitary, there is an increase in LH secretion in the late follicular phase that culminates in an **LH surge** (fig. 8.27).

The LH surge begins about twenty-four hours before ovulation and reaches its peak about sixteen hours before ovulation. It is this surge that acts to trigger ovulation. Since GnRH stimulates the anterior pituitary to secrete both FSH and LH, there is a simultaneous, though smaller, surge in FSH secretion. Some investigators believe that this midcycle peak in FSH acts as a stimulus for the development of new follicles for the next month's cycle.

Ovulation

Under the influence of FSH stimulation, the graafian follicle grows so large that it becomes a thin-walled "blister" on the surface of the ovary. The growth of the follicle is accompanied by a rapid rate of increase in estradiol secretion. This rapid increase in estradiol, in turn, triggers the LH surge at about day thirteen. Finally, the surge in LH secretion causes the wall of the graafian follicle to rupture at about day fourteen (fig. 8.28 *top*). Ovulation occurs, therefore, as a result of the sequential effects of FSH followed by LH on the ovarian follicles.

By means of the positive feedback effect of estradiol on LH secretion, the follicle in a sense sets the time for its own ovulation. This is because ovulation is triggered by an LH surge, and the LH surge is triggered by increased estradiol secretion, which occurs while the follicle grows. In this way the graafian follicle does not normally ovulate until it has reached the proper size and degree of maturation.

In ovulation, a secondary oocyte, arrested at metaphase II of meiosis, is released into a uterine tube. This oocyte is still surrounded by a zona pellucida and corona radiata as it begins its journey to the uterus. Normally only one ovary ovulates per cycle, with the left and right ovary alternating in successive cycles. Interestingly, if one ovary is removed the remaining ovary does not skip cycles, but ovulates every month. The mechanisms by which this regulation is achieved are not understood.

Luteal Phase

After ovulation, the empty follicle is stimulated by LH to become a new structure, the **corpus luteum** (fig. 8.29). This change in structure is accompanied by a change in function. Whereas the developing follicles secrete only estradiol, the corpus luteum secretes both estradiol and **progesterone.** Progesterone levels in the blood are negligible before ovulation but rise rapidly to reach a peak during the luteal phase at approximately one week after ovulation (see figs. 8.27 and 8.28).

The combined high levels of estradiol and progesterone during the luteal phase exert a **negative feedback inhibition** of FSH and LH secretion. This serves to retard development of

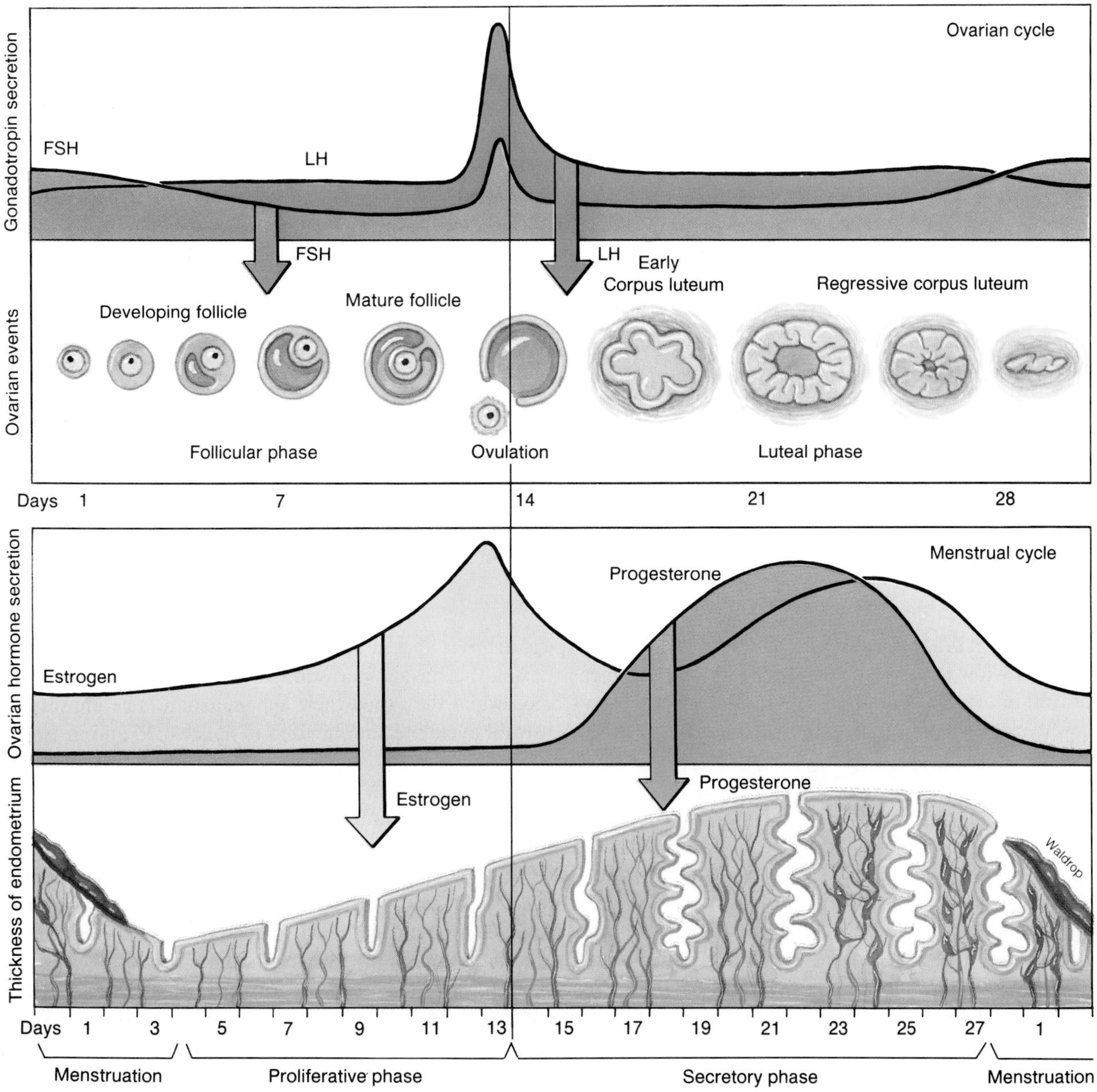

Figure 8.28. *The cycle of ovulation and menstruation.*

new follicles, so that further ovulation does not normally occur during that cycle. In this way multiple ovulations (and possible pregnancies) on succeeding days of the cycle are prevented.

High levels of estrogen and progesterone during the nonfertile cycle do not persist for very long, however. Estrogen and progesterone levels fall during the late luteal phase (starting about day twenty-two), because the corpus luteum regresses and stops functioning. In lower mammals, the decline in corpus luteum function is caused by a hormone secreted by the uterus called *luteolysin.* A similar hormone has not yet been identified in humans, and the cause of corpus luteum regression in humans is not well understood. Breakdown of the corpus luteum can be prevented by high LH secretion, but LH levels remain low during the luteal phase as a result of negative feedback inhibition by ovarian steroids. In a sense, therefore, the corpus luteum causes its own demise.

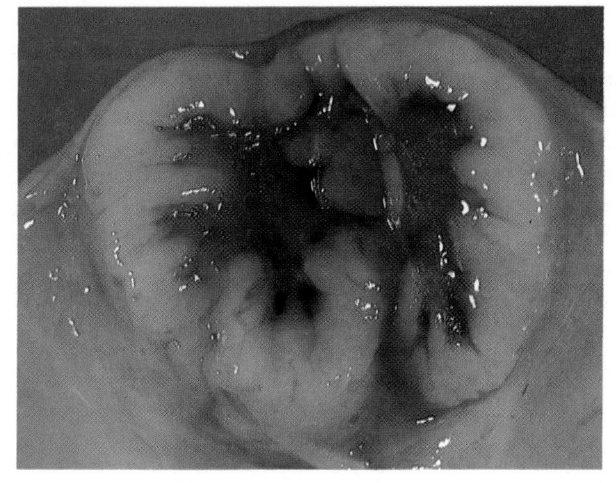

Figure 8.29. *A corpus luteum in a human ovary.*

Table 8.5 Phases of the menstrual cycle

Phase of cycle		Hormonal changes		Tissue changes	
Ovarian	Endometrial	Pituitary	Ovary	Ovarian	Endometrial
Follicular (days 1–4)	Menstrual	FSH and LH secretion low	Estradiol and progesterone remain low	Primary follicles grow	Outer two-thirds of endometrium is shed with accompanying bleeding
Follicular (days 5–13)	Proliferative	FSH slightly higher than LH secretion in early follicular phase	Estradiol secretion rises (due to FSH stimulation of follicles)	Follicles grow; graafian follicle develops (due to FSH stimulation)	Mitotic division increases thickness of endometrium; spiral arteries develop (due to estradiol stimulation)
Ovulatory (day 14)	Proliferative	LH surge (and increased FSH) stimulated by positive feedback from estradiol	Fall in estradiol secretion	Graafian follicle is ruptured and ovum is extruded into fallopian tube	No change
Luteal (days 15–28)	Secretory	LH and FSH decrease (due to negative feedback of steroids)	Progesterone and estrogen secretion increase, then fall	Development of corpus luteum (due to LH stimulation); regression of corpus luteum	Glandular development in endometrium (due to progesterone stimulation)

With the declining function of the corpus luteum, estrogen and progesterone fall to very low levels by day twenty-eight of the cycle. The withdrawal of ovarian steroids causes menstruation and permits a new cycle of ovarian follicle development to progress.

Cyclic Changes in the Endometrium

In addition to a description of the female cycle in terms of the phases of ovarian function, the cycle can also be described in terms of the changes that occur in the endometrium. Three phases can be identified on this basis (fig. 8.28 *bottom*): (1) the proliferative phase; (2) the secretory phase; and (3) the menstrual phase.

The **proliferative phase** of the endometrium occurs while the ovary is in its follicular phase. The increasing amounts of estradiol secreted by the developing follicles stimulates growth (proliferation) of the inner layer of the endometrium. In humans and other primates, spiral arteries develop in the endometrium during this phase. Estradiol may also stimulate the production of receptor proteins for progesterone at this time, in preparation for the next phase of the cycle.

The **secretory phase** of the endometrium occurs when the ovary is in its luteal phase. In this phase, increased progesterone secretion stimulates the development of mucus glands. As a result of the combined actions of estradiol and progesterone, the endometrium becomes thick, vascular, and "spongy" in appearance during the time of the cycle following ovulation. It is therefore well prepared to accept and nourish an embryo if fertilization occurs.

The **menstrual phase** occurs as a result of the fall in ovarian hormone secretion at the end of the late luteal phase. Necrosis (cellular death) and sloughing of the inner layer of the endometrium may be produced by constriction of the spiral arteries. The spiral arteries appear to be responsible for bleeding during menstruation, because lower animals that lack spiral arteries don't bleed when they shed their endometrium. The phases of the menstrual cycle are summarized in figure 8.30 and in table 8.5.

The cyclic changes in ovarian secretion cause other cyclic changes in the female genital tract. High levels of estradiol secretion, for example, cause cornification of the vaginal epithelium (the outer cells die and become filled with keratin). High levels of estradiol also cause the production of a thin, watery cervical mucus, which can be easily penetrated by spermatozoa. During the luteal phase of the cycle, the high levels of progesterone cause the cervical mucus to become thick and sticky after ovulation has occurred. The changes in hormonal secretion after ovulation also affect the basal body temperature, which sharply rises at ovulation and remains elevated throughout the luteal phase. Changes in cervical mucus and basal body temperature are used by many people in the rhythm method of birth control, which will be described in a later section.

The Menopause

The term *menopause* means literally "pause in the menses" and refers to the cessation of ovarian activity that occurs at about the age of fifty. During the postmenopausal years, which account for about a third of a woman's life span, no new ovarian follicles develop and the ovaries cease secreting estradiol. This fall in estradiol is due to changes in the ovaries, not in the pituitary; indeed, FSH and LH secretion by the pituitary is elevated due to the absence of negative feedback inhibition from estradiol.

It is the withdrawal of estradiol secretion from the ovaries that is most responsible for the many debilitating symptoms of

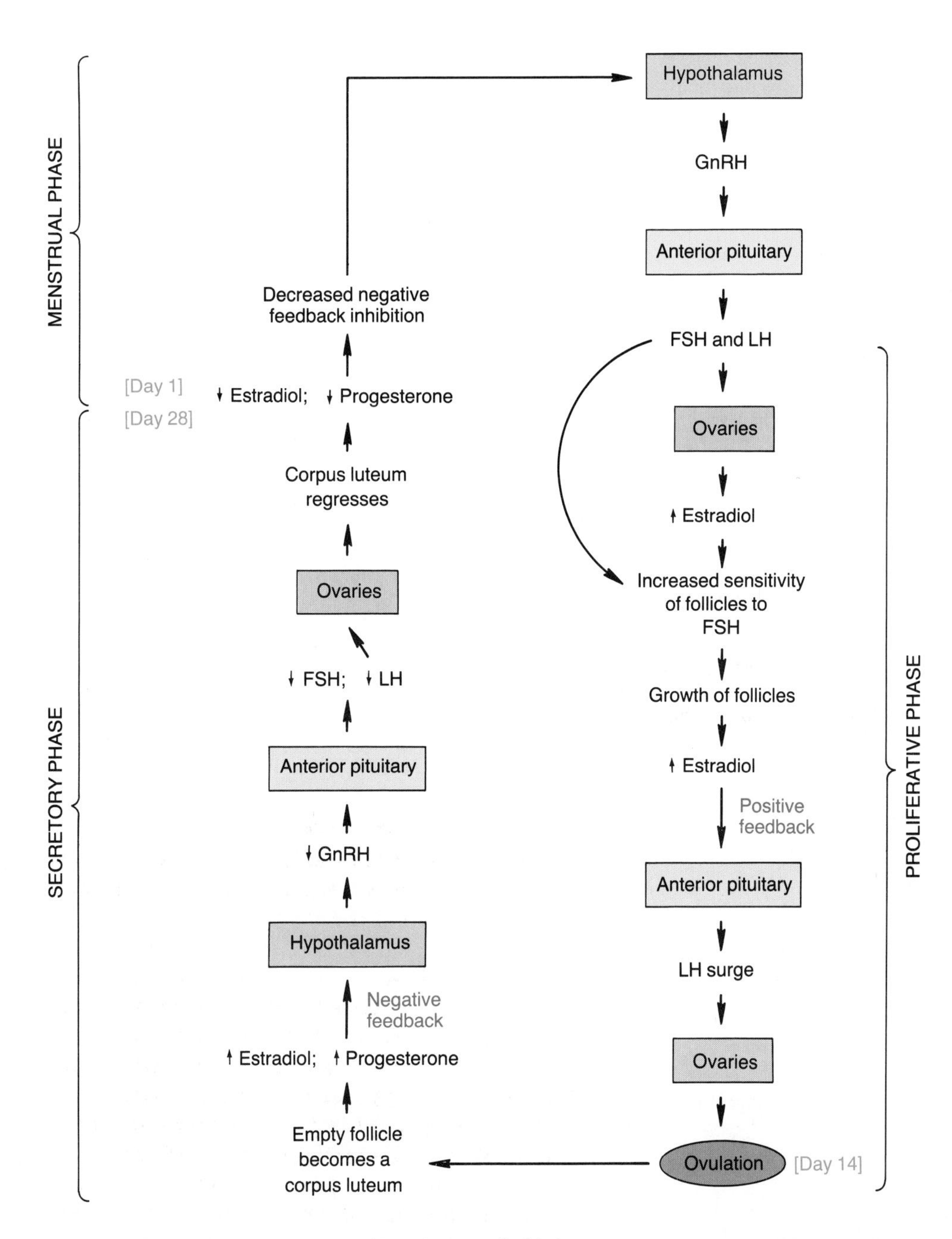

Figure 8.30. *The sequence of events in the endocrine control of the ovarian cycle and their correlation within the phases of the endometrium during the menstrual cycle.*

menopause. These include vasomotor disturbances (which produce "hot flashes"), urogenital atrophy, and the increased development of osteoporosis (see chapter 21). Estrogen replacement therapy helps to alleviate these symptoms, although this treatment is given cautiously because of the increased risk of endometrial cancer. This risk may be reduced by combining estrogen with progesterone, as is done with birth control pills.

Problems Involving the Uterus

Abnormal menstruations are among the most common disorders of the female reproductive system. Abnormal menstruations may be directly related to problems of the reproductive organs and pituitary gland or associated with emotional and psychological stress.

Amenorrhea is the absence of menstruation and can be categorized as normal, primary, or secondary. *Normal amenorrhea* occurs during pregnancy, may occur during lactation, and follows menopause. *Primary amenorrhea* means that a woman has never menstruated although she has passed the age when menstruation normally begins. This is generally accompanied by the failure of the secondary sexual characteristics to develop, and is due to an endocrine disorder.

Secondary amenorrhea is the cessation of menstruation in women who have previously menstruated and who are not pregnant, lactating, or past the menopause. This may be caused by endocrine or psychological factors. It is not uncommon, for example, for young women who are in the process of making major changes in their lives to miss menstrual periods. Secondary amenorrhea is also common in women athletes during periods of intense training. A low percentage of body fat may be a contributing factor. Sickness, fatigue, poor nutrition, or emotional stress may also cause secondary amenorrhea.

Dysmenorrhea is painful or difficult menstruation accompanied by severe menstrual cramps. **Menorrhagia** is excessive bleeding during the menstrual period, and **metrorrhagia** is spotting between menstrual periods. These abnormalities may be caused by a wide variety of factors.

Cancer of the uterus is the most common malignancy of the female reproductive tract. The most common site of uterine cancer is the cervix. Cervical cancer is second only to cancer of the breast in frequency of occurrence and is a disease of young women (age thirty to fifty). If detected early through regular Pap smears, the disease can be cured before it metastasizes (spreads). If the woman is through having children, a *hysterectomy* (surgical removal of the uterus) is usually performed.

Endometriosis is the presence of endometrial tissue at sites other than the inner lining of the uterus. These sites frequently are the ovaries, outer wall of the uterus, abdominal wall, and urinary bladder. The most common symptoms of endometriosis are extreme dysmenorrhea and a feeling of fullness during each menstrual period. Endometriosis can cause infertility. The ectopic (out of place) tissue can be treated chemically or surgically removed. An *oophorectomy,* or removal of the ovaries, may be necessary in extreme cases.

1. Describe the changes that occur in the ovary and endometrium during the follicular phase, and explain the hormonal control of these changes.
2. Describe the hormonal regulation of ovulation.
3. Describe the formation, function, and fate of the corpus luteum, and describe the changes that occur in the endometrium during the luteal phase.
4. Explain the significance of negative feedback inhibition during the luteal phase, and explain the hormonal changes that cause menstruation.

The Sexual Response

Males and females have a similar pattern of physiological changes that occur during sexual activities. These changes fall into a pattern that has four phases: excitement, plateau, orgasm, and resolution. The resolution phase of males, but not females, displays a refractory period.

In their now-classic study published in 1966, Masters and Johnson analyzed 10,000 sexual episodes in 382 women and 312 men. This study laid to rest certain misconceptions about the sexual response, particularly in women, and established sex research as a legitimate scientific field of study.

Masters and Johnson found that, even though males and females differ in their sexual response in important and gender-related ways, the sexual response of the two sexes are more similar than dissimilar. The sexual response of both sexes can be categorized into four phases: *excitement, plateau, orgasm,* and *resolution.* The anatomical and physiological changes that occur during these phases are produced by two processes, which can be scientifically measured: *vasocongestion* and *myotonia.*

Vasocongestion refers to the engorgement of a sexual organ with blood. This process was previously described in relation to erection of the penis, but also occurs in the testes of males and in the vaginal labia, clitoris, and nipples of females. It may even occur in the earlobes of both sexes. Myotonia refers to the development of muscle tension, as a result of the contraction of muscles.

Excitation Phase

During the excitation phase for both sexes there is increased myotonia in the arms and legs, and the heart rate and blood pressure begin to increase. In addition, a *sex flush* may appear as a pinkish rash on the chest and breasts. The nipples erect in both sexes, although this is more intense and evident in females than in males.

During the excitation phase in females the clitoris swells and the labia minora increase to more than twice their previous size. Vaginal lubrication begins and the vaginal walls take on a purplish color. The uterus also increases to about twice its resting size, due to vasocongestion, and it elevates in position. The breasts increase in size by up to 25% in women who have not breast fed a baby. The areolae (pigmented areas around the nipples) also swell in size.

Erection of the penis occurs during the excitation phase in males. The erection may be lost and regained several times. The scrotum contracts and elevates the testes, and the testes begin to increase in size.

Plateau Phase

During the plateau phase, the changes that began in the excitation phase continue to increase in intensity. The heart rate, blood pressure, and breathing rate continue to increase, and the color changes that began during the excitation phase become more pronounced.

In females, during the plateau phase, the clitoris becomes partially hidden behind the labia minora due to a shift in position of the clitoris and continued engorgement of the labia with blood. The clitoris at this point is still sensitive to stimulation, however. Similarly, the erected nipples become partially hidden by continued swelling of the areolae. The labia minora of women who have never been pregnant take on a pink to red color, while those of women who have been pregnant change to a bright red to deep wine color. Masters and Johnson have found that the appearance of this "sex skin" always precedes orgasm if the appropriate stimulation is continued. Engorgement of the outer third of the vagina produces what Masters and Johnson have called the *orgasmic platform,* which is better able to grip the penis.

The changes that occur in males during the plateau phase include the deepening of the red-to-purple color of the glans[16] penis and the continued enlargement and elevation of the testes. Also, a small amount of clear fluid may appear on the glans. This is believed to be a product of the Cowper's glands,[17] and in some cases can contain spermatozoa.

Orgasm

Orgasm,[18] sometimes also called "climax" or "coming," lasts only a few seconds. In both males and females (1) muscle spasms occur involuntarily; (2) the heart rate, blood pressure, and rate of breathing reach their highest values; and (3) the external anal sphincter contracts rhythmically at 0.8 second intervals.

The uterus and the orgasmic platform of the vagina contract several times. The first few contractions occur at 0.8 second intervals and are the most powerful, whereas the others decrease in both frequency and strength. Ejaculation of semen from the penis is accomplished by rhythmic contractions of the urethra within the penis and of other muscles. Interestingly, the first two to three contractions occur at 0.8 second intervals as in the female.

Resolution

The resolution phase refers to the anatomical and physiological return to the preexcitation state. The greatest differences between males and females occur during the resolution phase.

Men enter a *refractory period* immediately after orgasm. In this state, sexual stimulation may cause an erection but orgasm cannot occur. The length of this refractory period is quite variable; it can last from just minutes to days. Women do not enter a refractory period after orgasm, and are thus able to experience multiple orgasms. The research by Masters and Johnson has demonstrated that most women are capable of having multiple orgasms, but numerous surveys have shown that only about 15% regularly experience them. This may be due to lack of appropriate stimulation by their male partners, who are in their refractory period, lack of desire for more than one orgasm, or to other reasons.

1. Define the terms *vasocongestion* and *myotonia,* and give examples of these processes in both the male and female sexual response.
2. Describe the events that occur during orgasm in both males and females, comparing similarities and differences.
3. Explain why women can experience multiple orgasms whereas men, in general, cannot.

[16]glans: L. *glans,* acorn
[17]Cowper's gland: from William Cowper, English anatomist, 1666–1709
[18]orgasm: Gk. *orgasmos,* to swell; to become excited

Table 8.6 Contraceptive use among women exposed to the risk of unwanted pregnancy

Contraceptive Method	No. (Millions)	Percentage of Exposed Women
Exposed users	38.0	92
Sterilization	13.8	33
Female	8.1	20
Male	5.7	14
Oral contraceptive	13.2	32
Condom	6.9	17
Spermicide	2.3	6
Withdrawal	2.3	6
Diaphragm	1.7	4
Periodic abstinence	1.7	4
IUD	1.1	3
Douche	0.6	1
Exposed nonusers	3.3	8

Source: Jacqueline Darroch Forrest and Richard R. Fordyce, "U.S. Women's Contraceptive Attitudes & Practice: How Have They Changed in the 1980's?" *Family Planning Perspectives,* Volume 20: Number 3 (May/June 1988), p. 116. © The Alan Guttmacher Institute.

Contraception and Sexually Transmitted Diseases

Various methods are used to prevent the occurrence of pregnancy in sexually active people. These methods include abstinence of sexual intercourse during the time of a woman's cycle when fertilization is likely to occur; barriers to fertilization, including surgical sterilization (vasecomy or tubal ligation); use of such devices as the condom, diaphragm, and cervical cap; and oral contraceptive pills, which inhibit ovulation. Sexually transmitted diseases may be caused by viruses, bacteria, or protozoa.

According to the latest survey, 92% of sexually active women in the United States ages fifteen to forty-four years old use some method of contraception (fig. 8.31). Sterilization (tubal ligation in women or vasectomy in their male partners) is the most widely used method. Since these surgical procedures are generally not reversible, women who want contraceptive protection now but who may want children in the future use one of the other, reversible, methods. Most popular among these are the oral contraceptives (table 8.6), which are currently used by approximately 13.2 million women.

Sterilization Procedures

The sterilization procedure in a male is called a vasectomy, and has been previously described. The sterilization procedure in a woman is called a **tubal ligation.** In a tubal ligation, the fallopian tubes are either cut and tied or sealed by some other device (fig. 8.32). This procedure can be performed through small incisions into the abdominal cavity under local anesthesia, and generally requires only about twenty minutes.

Tubal ligation and vasectomy are usually performed only for people who do not intend to have children in the future. This is because the chances of being able to reverse these procedures are only about 50%.

Contraceptive Pill

Oral contraceptive pills generally consist of synthetic estrogen and progesterone. These pills are usually taken once each day for three weeks after the last day of a menstrual period. This procedure causes an immediate increase in blood levels of ovarian steroids (from the pill), which is maintained for the normal duration of a monthly cycle. As a result of *negative feedback inhibition* of gonadotropin secretion, *ovulation never occurs*. The entire cycle is like a false luteal phase, with high levels of progesterone and estrogen and low levels of gonadotropins.

Since the contraceptive pills contain ovarian steroid hormones, the endometrium proliferates and becomes secretory just as it does during a normal cycle. In order to prevent an abnormal growth of the endometrium, women stop taking the steroid pills after three weeks (placebo pills are taken during the

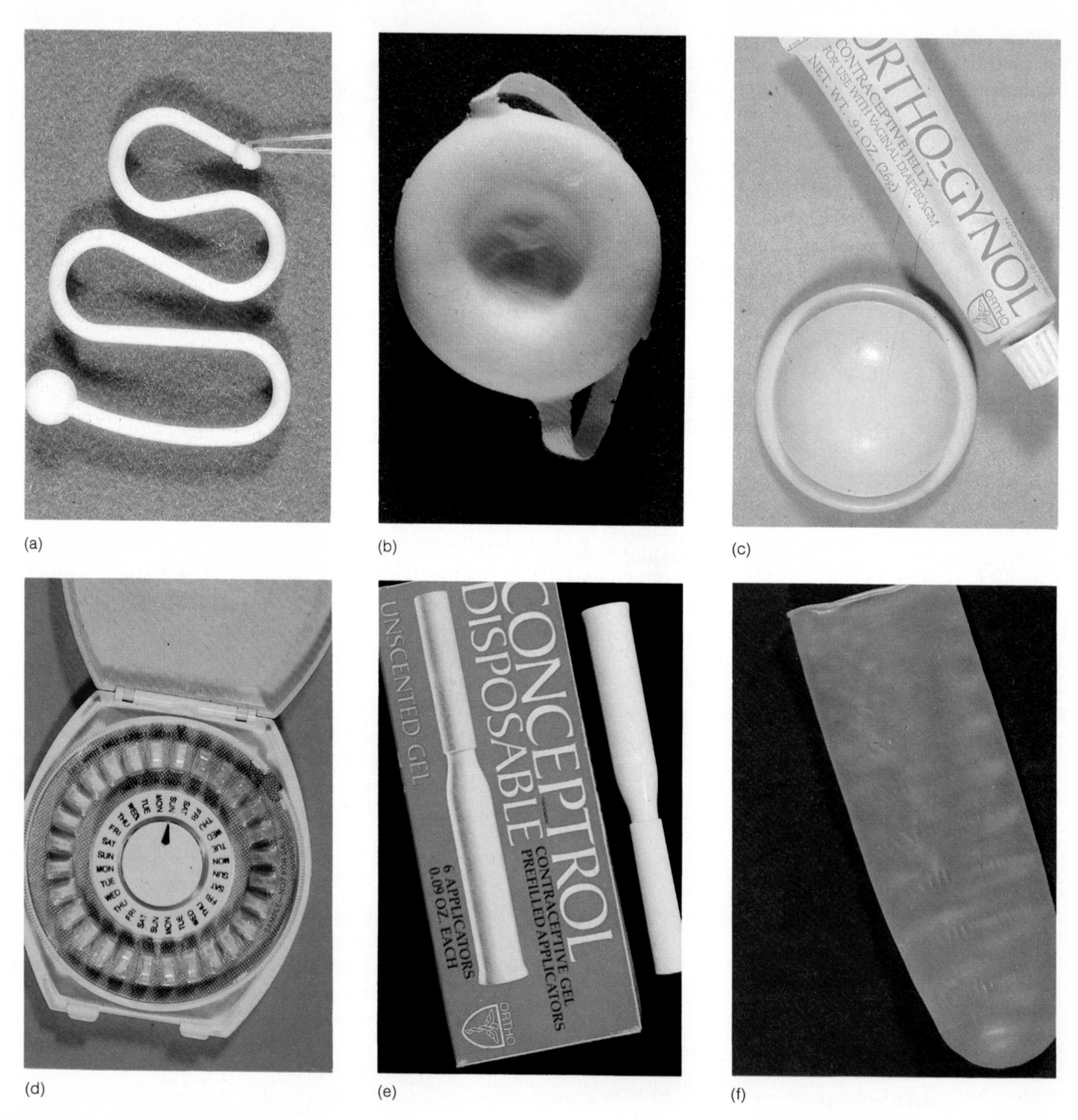

Figure 8.31. *Various types of birth control devices. (a) IUD; (b) contraceptive sponge; (c) diaphragm; (d) birth control pills; (e) vaginal spermicide; (f) condom.*

fourth week). This causes estrogen and progesterone levels to fall, and permits menstruation to occur.

All of the contraceptive pills marketed after 1975 contain less than 50 μg (micrograms) of ethinyl estradiol (the synthetic estrogen used in most pills). This low dose of estrogen provides as good a contraceptive effect as the higher doses previously used, and also has a significantly lower incidence of side effects.

Contraceptive pills have a number of possible negative side effects. Among those proposed or proven are:

1. *Cervical cancer.* Although some studies have found no increased risk of cervical cancer as a result of the use of contraceptive pills, other studies suggest that women who had used oral contraceptives for more than five years had a significantly higher risk. For safety's sake, therefore, women who have used contraceptive pills for more than five years should be tested at least annually for cervical cancer.
2. *Breast cancer.* Most studies have shown that there is no relationship between breast cancer and use of the contraceptive pill. On this basis, the FDA (Food and Drug Administration) has concluded that no change is warranted in how the pills are prescribed, but has recommended further study into this relationship.
3. *Endometrial cancer.* Contrary to what was believed at one time, most studies have demonstrated that contraceptive pills actually exert a protective effect against endometrial cancer, which is the third most common form of cancer among women. The greatest protective effect was found in women who have not had children (these are the women at greatest risk for endometrial cancer).
4. *Cardiovascular disease.* Modern contraceptive pills, containing less than 50 μg of estrogen, do not appear to increase the risk of myocardial infarction (a serious heart injury, described in chapter 12). There is evidence, however, that the pill may increase the risk of myocardial infarction in women who also have other risk factors for coronary disease (including smoking, hypertension, and high blood cholesterol). For these reasons, healthy women thirty-five to forty-four years old can continue to use the contraceptive pill, but women who have particular diseases and those over thirty-five who smoke are advised not to use the birth control pill.
5. *Fertility.* Women who have discontinued the use of oral contraceptives experience a delay in their return to a fertile state. Fertility, however, does eventually return to normal.

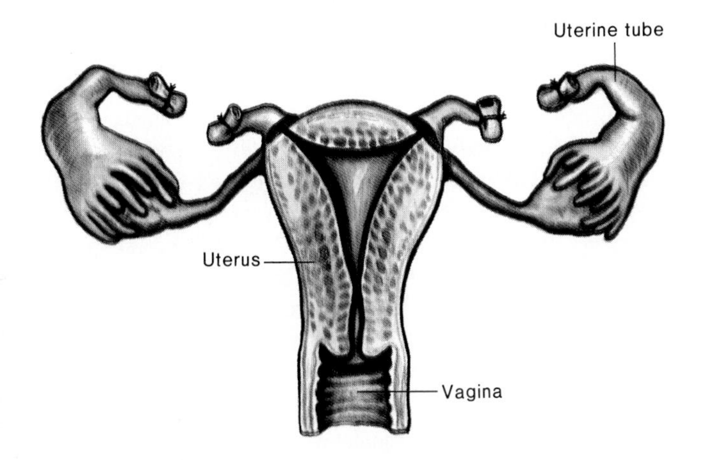

Figure 8.32. *A simplified illustration of a tubal ligation of each uterine tube.*

There are also a number of positive side effects to the use of the birth control pill. These include (1) less menstrual blood loss and less iron-deficiency anemia; (2) lower incidence of menorrhagia, irregular bleeding, dysmenorrhea, and premenstrual syndrome; (3) lowered incidence of benign breast disease; (4) protection against endometrial cancer; (5) lowered incidence of ovarian cysts and ovarian cancer; (6) lowered incidence of salpingitis (inflammation of the fallopian tube); and (7) increased bone density, and thus possible protection against osteoporosis after menopause.

Rhythm Methods

Rhythm methods of birth control, also known as "natural family planning," involve abstinence from sexual relations during the time of the cycle when fertilization can occur. These methods require training as well as high motivation, and for these reasons are not as generally effective at preventing conception as is the birth control pill.

In the temperature method, a woman measures her oral basal body temperature upon waking. Starting one day after the LH peak (at about the time ovulation occurs), the basal body temperature sharply rises as a result of progesterone secretion from the new corpus luteum (fig. 8.33). A woman using this method is required to abstain from sexual intercourse from the

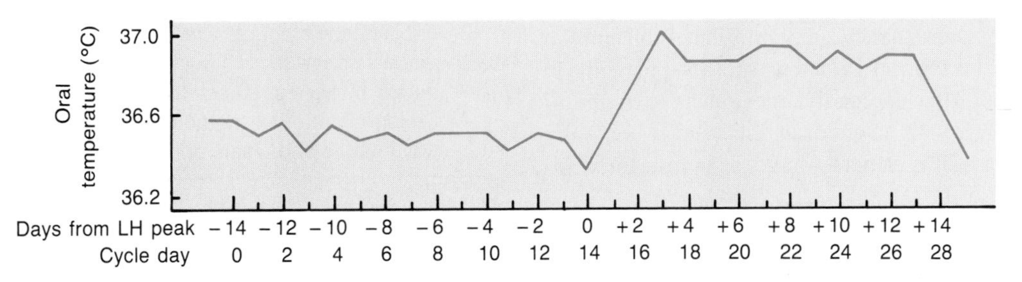

Figure 8.33. *Changes in basal body temperature during the menstrual cycle.*

first day of menstruation through the third consecutive day of elevated basal body temperature (after which the unfertilized ovum has disintegrated). Since this method requires a long period of abstinence, it is not widely used.

In the cervical mucus method, a woman is trained to recognize the changes that occur in the amount and consistency of the cervical mucus which occurs as a result of hormonal changes. This also requires a high degree of motivation and training. The most effective natural family planning techniques involve the use of a combination of methods (basal body temperature and cervical mucus). Studies estimate that overall pregnancy rates among women who use natural family planning techniques are approximately 20%.

Barrier Methods

Although barrier methods—use of devices that physically block the ability of sperm to reach the ovum—are not as successful at preventing pregnancy as the birth control pill, they are quite effective if properly used, and can be made even more effective if they are used together with spermicides. Among these devices are the **condom** (which fits over the man's penis), and the **diaphragm** (a dome-shaped structure made of latex that is inserted into the vagina to block the cervix). These devices must be put on or inserted, respectively, before every act of sexual intercourse. A newer device, the **cervical cap,** can be left on the cervix for up to forty-eight hours.

Condoms, specifically those made of latex, provide benefits other than contraception. They have been shown to help protect against infections by viruses (including the herpesvirus and the AIDS virus), and to protect against transmission of the *Chlamydia trachomatis* bacteria (table 8.7). Use of both condoms and diaphragms is associated with a lower incidence of diseases of the fallopian tubes and cervical cancer. These protective effects may also be related to the ability of barrier contraceptives to reduce the transmission of viruses.

Intrauterine Device

The **intrauterine device (IUD)** is placed in the uterus and is believed to provide contraception by preventing the implantation of the embryo into the endometrium. There are two varieties of IUDs: those that are coated with copper over a plastic frame, and those that release progesterone. Although IUDs are effective contraceptive devices, they are very unpopular. This is because IUDs have a history of serious complications, including perforation of the uterus, inflammation of the fallopian tubes (salpingitis), and ectopic pregnancy (implantation of the embryo at a site other than the uterus). Women who are particularly susceptible to salpingitis—especially those who are under twenty-five years of age, with no children, and who are sexually active with a number of partners—have a significantly increased risk with use of IUDs.

Spermicides

Most spermicides (agents that kill sperm) contain the chemical nonoxynol-9, which interferes with the ability of sperm to swim. These come in a variety of forms, and can be used together with other contraceptive techniques. The most popular method of using a spermicide is the **contraceptive sponge,** in which the spermicide is saturated within a cylindrical piece of polyurethane. The sponge is effective for twenty-four hours and thus does not have to be inserted into the vagina just prior to intercourse.

Sexually Transmitted Diseases

A sexually transmitted disease (STD) is one that is usually passed from one person to another by sexual contact. STDs may be caused by bacteria, viruses, or protozoa (table 8.7). The only contraceptive devices that provide protection against STDs are condoms.

Diseases caused by bacteria and protozoa, which are living cells, are treatable with antibiotics. This is particularly true if the disease is diagnosed early. Sexually transmitted diseases that are viral in origin, however, cannot be cured with any medical treatment presently available. The reasons for this difficulty in treatment relate to the nature of viruses.

Viruses are not living cells. Rather, they are essentially boxes composed of protein that contain genetic information (either DNA or RNA). Most viruses contain DNA, which is injected into a body cell. The viral DNA then commandeers the metabolic machinery of the host cell and directs it to produce new virus particles. An enzyme is then produced that destroys the victim cell and allows the new virus particles to be released, freeing them to infect other cells. The virus that causes AIDS, however, contains RNA instead of DNA. Retroviruses inject their RNA into a body cell (particular white blood cells in the case of the AIDS virus), which is then copied within the host cell to produce viral DNA.

Since a working knowledge of DNA, RNA, and protein synthesis is needed, together with a general understanding of the immune system, to understand the nature of AIDS, details of this topic will be deferred to chapter 13 after these other subjects have been introduced.

---

1. Explain the nature of a vasectomy and a tubal ligation. What are the advantages and disadvantages of these procedures?
2. Describe how the contraceptive pill works. What are its advantages and disadvantages?
3. Explain how a couple who desire to practice family planning would use the rhythm method. What are the advantages and disadvantages of this method?
4. Describe the other methods of contraception, explaining their advantages and disadvantages.

Table 8.7 Common Sexually Transmitted Diseases

	No. of U.S. Cases	Infecting Agent	Female Symptoms	Male Symptoms	Consequences to Women	Treatment Agent	Cure
Viruses							
AIDS	As of July 1987, 38,808 cases reported; 22,328 deaths	HIV (human immunodeficiency virus)	Headache, fever, night sweats, swollen lymph glands, diarrhea, weight loss, fatigue	Same as female	Opportunistic infections, some cancers, death	Azidothymidine, Ribavirin	None
Genital herpes	200,000 to 500,000 new cases per year; 30 to 40 million cases total	Herpes simplex virus II	Often none; blisters in or around vagina; sometimes fever or headache	Often none; sores or clusters of blisters on penis; sometimes fever or headaches	Miscarriage, birth defects, serious infection of newborn; sometimes death of baby	Acyclovir	None
Genital warts	1 million new cases per year	Human papilloma virus	Single warts or clusters of soft growths in and around vagina or anus may be microscopic as well as visible	Warts or clusters on or around penis	Increased risk of cervical cancer	Podophyllin, 5-Fluorouracil, surgical removal	Several treatments may be needed
Bacteria							
Chlamydia	4 million new cases per year	*Chlamydia trachomatis* bacteria	Usually none; sometimes burning at urination, vaginal discharge	Usually none; itching or burning at urination; white discharge	Pelvic inflammatory disease (PID), sterility, ectopic pregnancies	Tetracycline, erythromycin	Yes
Gonorrhea	1.8 million new cases per year	*Neisseria gonorrhoeae* bacteria	Often none; sometimes burning at urination, vaginal discharge, fever, abdominal pain	White discharge from penis, itching or painful urination	PID, sterility, arthritis	Penicillin, ampicillin, amoxicillin	Yes
Syphillis	85,000 new cases per year	*Treponema pallidum* bacteria	Sore (chancre) shortly after infection; fever, sore throat, or rashes	Same as female	Heart disease, brain damage, arthritis, death, damage to babies	Penicillin, erythromycin	Yes
Protozoa							
Trichomonas	Exact number of cases unknown—millions	*Trichomonas vaginalis* protozoa	None or yellow discharge, irritation, unpleasant odor, painful urination	None or prostitis	Urinary tract infections	Metronidazole	Yes

Summary

Cellular Reproduction: Mitosis and Meiosis p. 150

I. In mitosis, the chromosomes duplicate and one of each duplicate pair is distributed to one of the two daughter cells.
   A. In this way, each daughter cell gets a copy of the same diploid set of chromosomes contained in the parent cell.
   B. Mitosis is subdivided into phases: interphase, prophase, metaphase, anaphase, and telophase.
   C. A fertilized egg, or zygote, grows by an increase in cell number due to mitotic cell divisions.

II. Meiosis is a special type of cell division that occurs only in the gonads of mature individuals.
   A. In meiosis, there are two cell divisions that can produce four daughter cells.
   B. The duplicated homologous chromosomes pair up along the equator at metaphase I.
      1. The direction that each homologous chromosome faces at metaphase I is random, so that the maternal and paternal chromosomes can be shuffled; this is called independent assortment.
      2. Parts of each chromosome of a homologous pair can be interchanged; this is called crossing-over.
   C. At telophase I, two cells are produced that have only one of each chromosome; meiosis is thus a reduction division in which haploid cells are produced.
   D. The duplicated strands of each chromosome, called chromatids, are distributed to different daughter cells by the second meiotic division.
   E. In this way, haploid gametes (sperm and ova) are produced.

Male Reproductive System p. 155

I. The testes consist of two parts, or compartments.
   A. The seminiferous tubules are stimulated by FSH to produce sperm.
   B. The interstitial Leydig cells are stimulated by LH to secrete testosterone.

II. Testosterone is the major androgen secreted by the testes.
   A. Testosterone is converted to different products, such as dihydrotestosterone (DHT), which are the active forms of the hormone within particular target organs.
   B. Testosterone is responsible for the development of male secondary sexual characteristics.

III. Spermatogenesis occurs in the seminiferous tubules through meiotic cell division of the diploid spermatogonia.
   A. The Sertoli cells in the tubules participate in the production of spermatozoa in response to FSH stimulation.
   B. Testosterone, as well as FSH, is required for spermatogenesis.

IV. Male sex accessory organs include the epididymis, vas deferens, seminal vesicles, prostate, and ejaculatory duct.

V. Erection of the penis occurs through the engorgement with blood of spongy tissue called the corpora cavernosa and the corpus spongiosum.

VI. The ejaculate contains approximately 300 million spermatozoa.

Female Reproductive System p. 165

I. The ovaries are located in the body cavity, partly covered by the fimbriae at the ends of the fallopian tubes.
   A. The fallopian tubes are continuous with the pear-shaped uterus.
   B. The neck, or cervix, of the uterus is continuous with the vagina.
   C. The external genitalia include the labia majora, labia minora, and clitoris.

II. The ovaries contain hollow structures called follicles, in which the egg cells, or oocytes, are located.
   A. In a secondary follicle, the oocyte is located on a mound of granulosa cells, and there is a fluid-filled cavity or antrum.
   B. Meiosis produces a secondary oocyte that is arrested at metaphase II; the other daughter cell is a tiny polar body that ultimately disintegrates.

III. At ovulation, a mature, or graafian, follicle ruptures to expel the oocyte from the ovary; the empty follicle becomes a corpus luteum.

Menstrual Cycle p. 171

I. The first day of the menstrual cycle is taken as the first day of menstruation; the average cycle lasts twenty-eight days.
   A. During the follicular phase of the cycle, FSH stimulates the ovarian follicles to grow and to secrete increasing amounts of estradiol.
   B. Toward the end of the follicular phase, the rapid rise in blood estradiol stimulates the hypothalamus and anterior pituitary to secrete a very rapidly increasing surge in LH.
      1. This is called a positive feedback effect.
      2. The LH surge stimulates ovulation, so that a secondary oocyte is expelled from the ovary.
      3. The LH surge also causes the empty follicle to change into a corpus luteum.
   C. During the luteal phase of the cycle, the corpus luteum secretes estradiol and progesterone.
      1. Toward the end of the luteal phase, these ovarian hormones exert negative feedback inhibition of FSH and LH secretion.
      2. At the end of the luteal phase, the corpus luteum dies if fertilization does not occur; this causes estradiol and progesterone secretion to fall.

II. The cyclic secretion of ovarian hormones results in cyclic changes in the endometrium of the uterus.
   A. During the proliferative phase, estradiol from the developing follicles causes the endometrium to become thicker.
   B. During the secretory phase, estradiol and progesterone from the corpus luteum cause the endometrium to become glandular and highly vascular.
   C. At the end of the luteal phase of the ovaries, the fall in estradiol and progesterone secretion causes the inner portion of the endometrium to be shed in menstruation.

III. When a woman is past the menopause, the ovaries no longer ovulate and no longer secrete steroid hormones.

The Sexual Response p. 176

I. The sexual response can be divided into four phases, which have characteristics shared by both sexes and those that are unique to each sex.

II. The four phases are excitation, plateau, orgasm, and resolution.
   A. The different responses in each phase display characteristic changes in vasocongestion of different organs and myotonia (muscle tension).
   B. Orgasm in both sexes is characterized by contractions that are spaced at 0.8 second intervals.
   C. Males, but not females, enter a refractory state after orgasm.

Contraception and Sexually Transmitted Diseases p. 177

I. Sterilization (vasectomy or tubal ligation) is the most widely practiced method of contraception, but is generally irreversible.

II. Of the reversible methods of birth control, the contraceptive pill is the most effective and the most widely used.
   A. Contraceptive pills contain derivatives of estrogen and progesterone, which exert negative feedback inhibition of FSH and LH.
      1. This is similar to that which occurs during a normal luteal phase of the ovarian cycle.
      2. As a result, ovulation never occurs.
   B. The rhythm methods utilize the fact that basal body temperature increases, and the cervical mucus becomes more thick and sticky, during the luteal phase of the cycle following ovulation.
   C. The barrier methods of birth control utilize the condom, diaphragm, and the cervical cap.
   D. Spermicides are agents that kill sperm, and can be used with other methods of birth control.

III. Condoms are the only method of birth control that also provide protection against sexually transmitted diseases.
   A. These diseases may be caused by viruses, bacteria, or by protozoa.
   B. At present, the diseases caused by viruses cannot be cured.

Review Activities

Objective Questions

1. The phase of mitosis in which the chromosomes line up at the equator of the cell is called
   (a) interphase
   (b) prophase
   (c) metaphase
   (d) anaphase
   (e) telophase
2. The phase of mitosis in which the chromatids separate is called
   (a) interphase
   (b) prophase
   (c) metaphase
   (d) anaphase
   (e) telophase
3. The phase of meiosis in which independent assortment and crossing-over occur is
   (a) prophase I
   (b) metaphase I
   (c) prophase II
   (d) metaphase II
   (e) interphase I

Match the following:

4. Menstrual phase
5. Follicular phase
6. Luteal phase
7. Ovulation

(a) high estrogen and progesterone; low FSH and LH
(b) low estrogen and progesterone
(c) LH surge
(d) increasing estrogen; low LH and low progesterone

8. In the male
   (a) FSH is not secreted by the pituitary
   (b) FSH receptors are located in the Leydig cells
   (c) FSH receptors are located in the spermatogonia
   (d) FSH receptors are located in the Sertoli cells
9. The secretion of FSH in a male is inhibited by negative feedback effects of
   (a) inhibin secreted from the tubules
   (b) inhibin secreted from the Leydig cells
   (c) testosterone secreted from the tubules
   (d) testosterone secreted from the Leydig cells
10. The approximate number of spermatozoa contained in an ejaculate is
    (a) 100
    (b) 100,000
    (c) 300 million
    (d) 10 million
11. Which of the following is characteristic of the male sexual response but not the female response?
    (a) orgasm
    (b) excitation
    (c) refractory period
    (d) all of these
12. Which of the following may protect against sexually transmitted diseases?
    (a) spermicides
    (b) condoms
    (c) birth control pills
    (d) diaphgragms

Essay Questions

1. Explain why a testis is said to be composed of two separate compartments, and describe the hormonal control of these compartments.
2. Explain the hormonal interactions that control ovulation and make it occur at the proper time.
3. Compare menstrual bleeding and bleeding that occurs during the estrous cycle of a dog, in terms of hormonal control mechanisms and the ovarian cycle.
4. Explain why the effects of the contraceptive pill can be said to duplicate events that occur during a normal luteal phase.
5. Why does menstruation normally occur? Under what conditions does menstruation not occur? Explain.

9

From Conception to Sensescence: Human Development and Aging

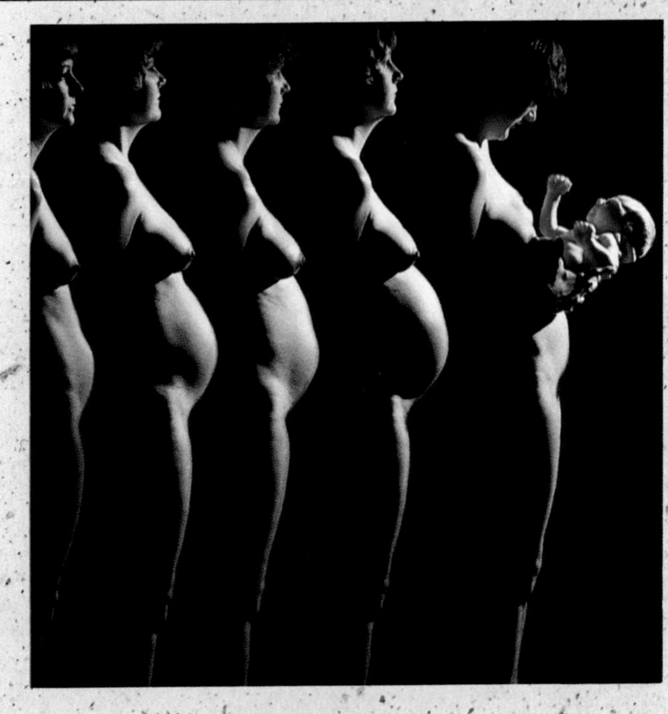

Outline

Objectives

By studying this chapter, you should be able to

1. describe capacitation, the acrosome reaction, and the process of fertilization
2. describe the origin and significance of chorionic gonadotropin
3. give examples of organs and tissues formed from each of the embryonic germ layers
4. correlate the embryonic germ layers with organs that are formed from them
5. explain how the primary sex of the embryo is determined, including the formation of testes or ovaries
6. describe the constituents and functions of the placenta
7. describe, in a general way, the changes that occur during fetal development
8. define the terms *gestation* and *parturition*, and describe the events that occur during labor and delivery
9. distinguish between monozygotic and dizygotic twins
10. explain how the mammary glands and lactation are regulated, and the effects of breast-feeding on the infant and mother
11. distinguish between the stages of childhood, adolescence, and adulthood
12. describe the changes that occur at puberty
13. describe the senescent changes that occur in various body systems

Keys to Pronunciation

adolescence: *ad''o-les'ens*
allantois: *ah-lan'to-is*
amniocentesis: *am''ne-o-sen-te'sis*
amnion: *am'ne-on*
chorion: *ko're-on*
chorionic gonadotropin: *ko''re-on'ik gon''ah-do-tro'pin*
fetus: *fe'tus*
gestation: *jes-ta'shun*
hyaluronidase: *hi''ah-lu-ron'i-das*
lanugo: *lah-nu'go*
menarche: *me-nar'ke*
oxytocin: *ok''si-to'sin*
parturition: *par''tu-rish'un*
placenta: *plah-sen'tah*
prostaglandin: *pros''tah-glan'din*
senescence: *se-nes'ens*
umbilical: *um-bil'i-kal*
zygote: *zi'got*

Photo: Progressive changes occur in a woman's body during the course of pregnancy. These changes are a consequence of, and required for, the development of the fetus and parturition (birth).

Fertilization and Pre-Embryonic Development

When a sperm fertilizes a secondary oocyte, the haploid number of chromosomes from each gamete merge to form a diploid zygote. The zygote divides many times by mitosis to form a blastocyst as the pre-embryo moves along the fallopian tube to the uterus. The blastocyst implants into the uterus, and ensures that its new home is not destroyed by secreting a hormone that prevents menstruation.

During the act of sexual intercourse a man ejaculates an average of 300 million sperm into the vagina. This tremendous number is needed because of the high sperm fatality rate—only about one hundred survive to enter each fallopian tube. During their passage through the female reproductive tract the sperm gain the ability to fertilize an ovum. This process is called **capacitation.**[1] The changes that occur in capacitation are incompletely understood. Experiments have shown, however, that freshly ejaculated sperm are infertile; they must be present in the female tract for at least seven hours before they can fertilize an ovum.

A woman usually ovulates only one ovum a month, amounting to a total number of less than 450 during her reproductive years. Each ovulation releases a secondary oocyte arrested at metaphase of the second meiotic division. The secondary oocyte, as previously described, enters the fallopian tube surrounded by its zona pellucida (a thin transparent layer of protein and polysaccharides) and corona radiata of granulosa cells (fig. 9.1).

[1]capacitation: L. *capacitas,* capable of

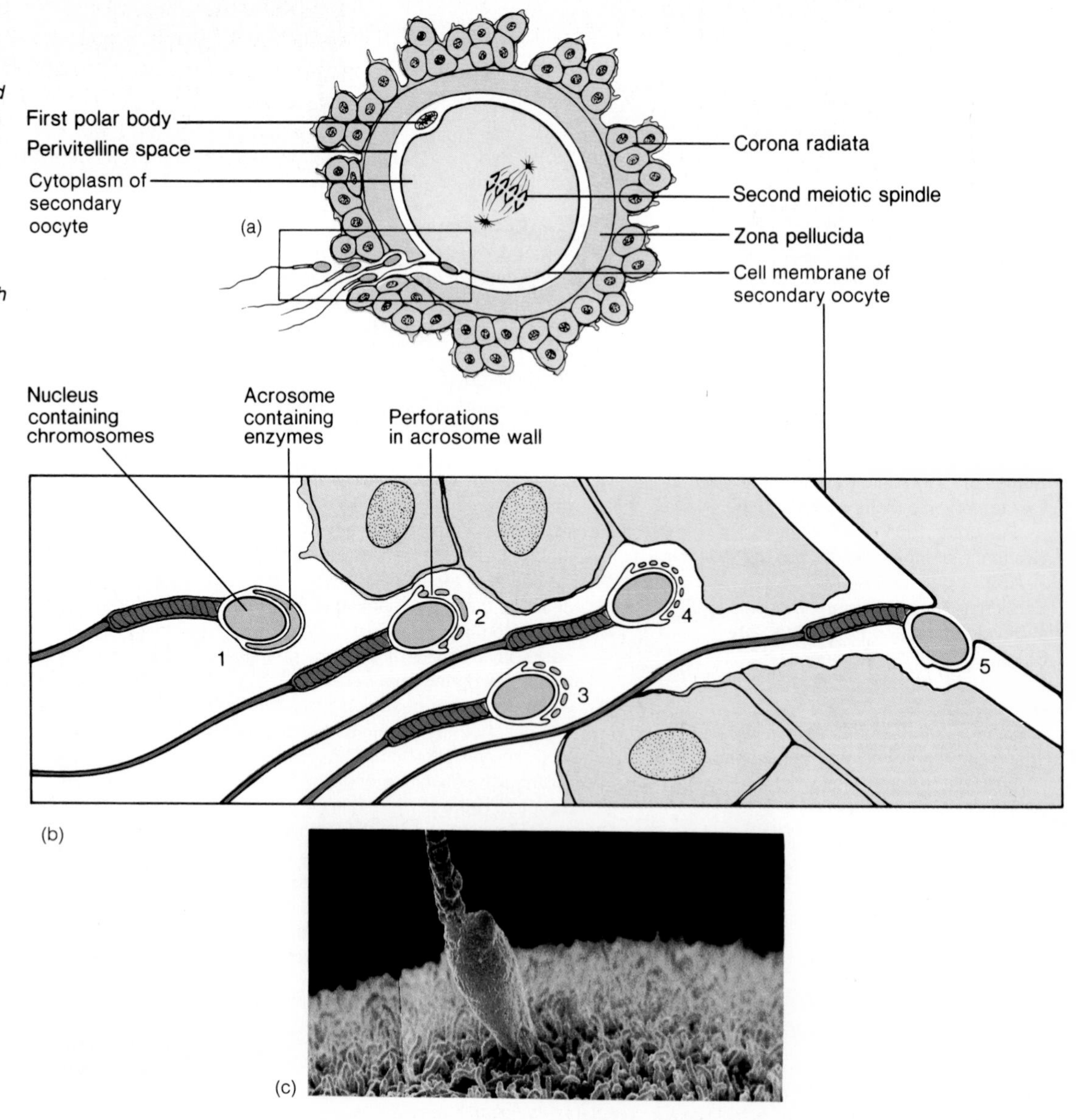

Figure 9.1. *The process of fertilization. (a, b) Diagrammatic representations. As the head of the sperm encounters the gelatinous corona radiata of the egg, which has progressed in meiotic development to the secondary oocyte stage (2), the acrosomal cap ruptures and the sperm digests a path for itself by the action of enzymes released from the acrosome (3, 4). When the cell membrane of the sperm contacts the cell membrane of the egg (5), they become continuous, and the sperm nucleus and other contents move into the egg cytoplasm. (c) A scanning electron micrograph of sperm bound to the egg surface.*

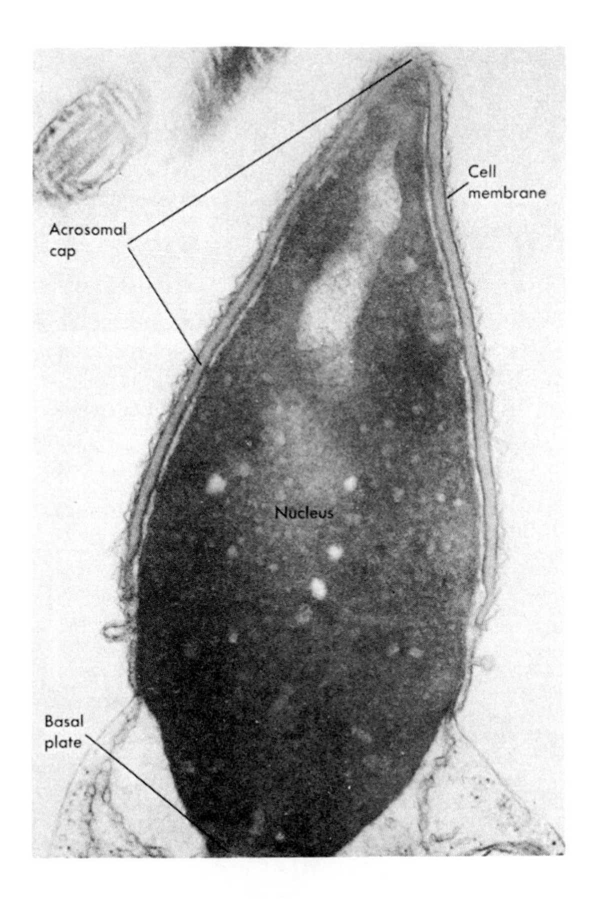

Figure 9.2. *An electron micrograph showing the head of a human sperm with its nucleus and acrosomal cap.*

Fertilization normally occurs in the fallopian tubes. The head of each spermatozoan is capped by an organelle called an *acrosome* (fig. 9.2), which contains digestive enzymes. When sperm meets ovum, an **acrosomal reaction** occurs that exposes the acrosome's digestive enzymes and allows the sperm to penetrate the corona radiata and the zona pellucida. The acrosomal enzymes are not released in this process; rather, the sperm tunnels its way through these barriers by digestion reactions that are localized to its acrosomal cap.

As the first sperm tunnels its way through the zona pellucida, a chemical change in the zona occurs that prevents other sperm from entering. Only one sperm, therefore, is allowed to fertilize one ovum. As fertilization occurs, the secondary oocyte is stimulated to complete its second meiotic division (fig. 9.3). Like the first meiotic division, the second produces one cell that contains all of the cytoplasm—the mature ovum or egg cell—and one polar body. The second polar body, like the first, ultimately fragments and disintegrates.

At fertilization, the sperm enters the cytoplasm of the much larger egg cell. Within twelve hours the nuclear membrane in the ovum disappears, and the haploid number of chromosomes (twenty-three) in the ovum is joined by the haploid number of chromosomes from the sperm. A fertilized egg, or *zygote*,[2] containing the diploid number of chromosomes (forty-six) is thus formed (fig. 9.4).

[2]zygote: Gk. *zygotos*, yolked, joined

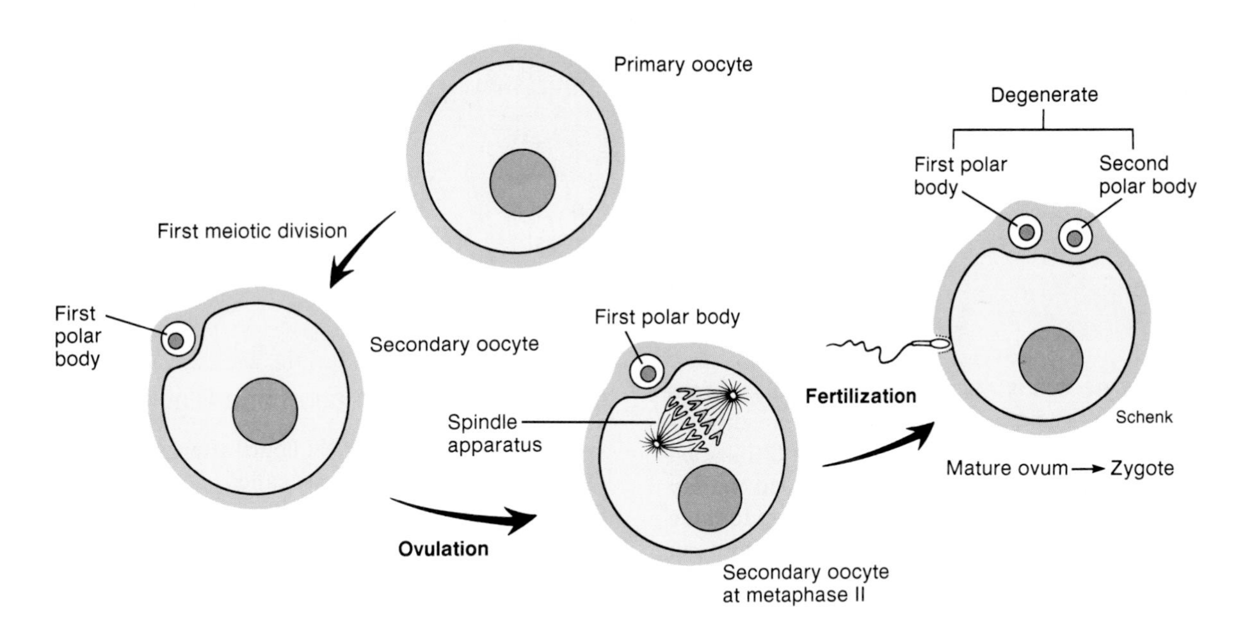

Figure 9.3. *A secondary oocyte, arrested at metaphase II of meiosis, is released at ovulation. If this cell is fertilized, it completes its second meiotic division and produces a second polar body.*

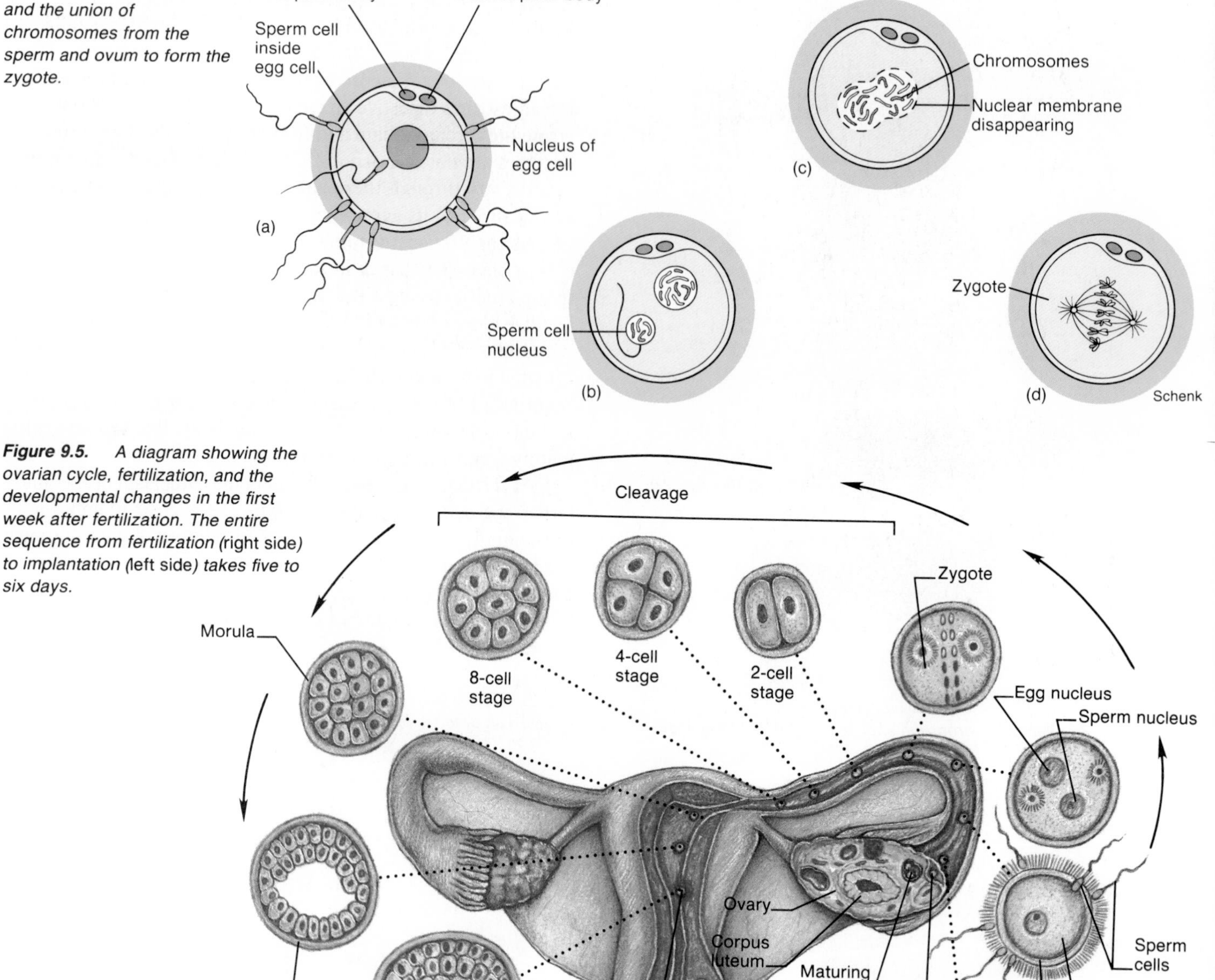

Figure 9.4. *Fertilization and the union of chromosomes from the sperm and ovum to form the zygote.*

Figure 9.5. *A diagram showing the ovarian cycle, fertilization, and the developmental changes in the first week after fertilization. The entire sequence from fertilization (right side) to implantation (left side) takes five to six days.*

A secondary oocyte that is ovulated but not fertilized does not complete its second meiotic division, but instead disintegrates twelve to twenty-four hours after ovulation. Fertilization therefore cannot occur if intercourse takes place beyond one day following ovulation. Sperm, in contrast, can survive up to three days in the female reproductive tract. Fertilization therefore can occur if intercourse is performed within three days prior to the day of ovulation.

Cleavage and Formation of a Blastocyst

At about thirty to thirty-six hours after fertilization the zygote divides by mitosis—a process called *cleavage*—into two smaller cells. The rate of cleavage is thereafter accelerated. A second cleavage, performed about forty hours after fertilization, produces four cells. By about fifty to sixty hours after fertilization a third cleavage occurs, producing a ball of eight cells called a **morula** (= mulberry). This very early embryo enters the uterus three days after ovulation has occurred (fig. 9.5).

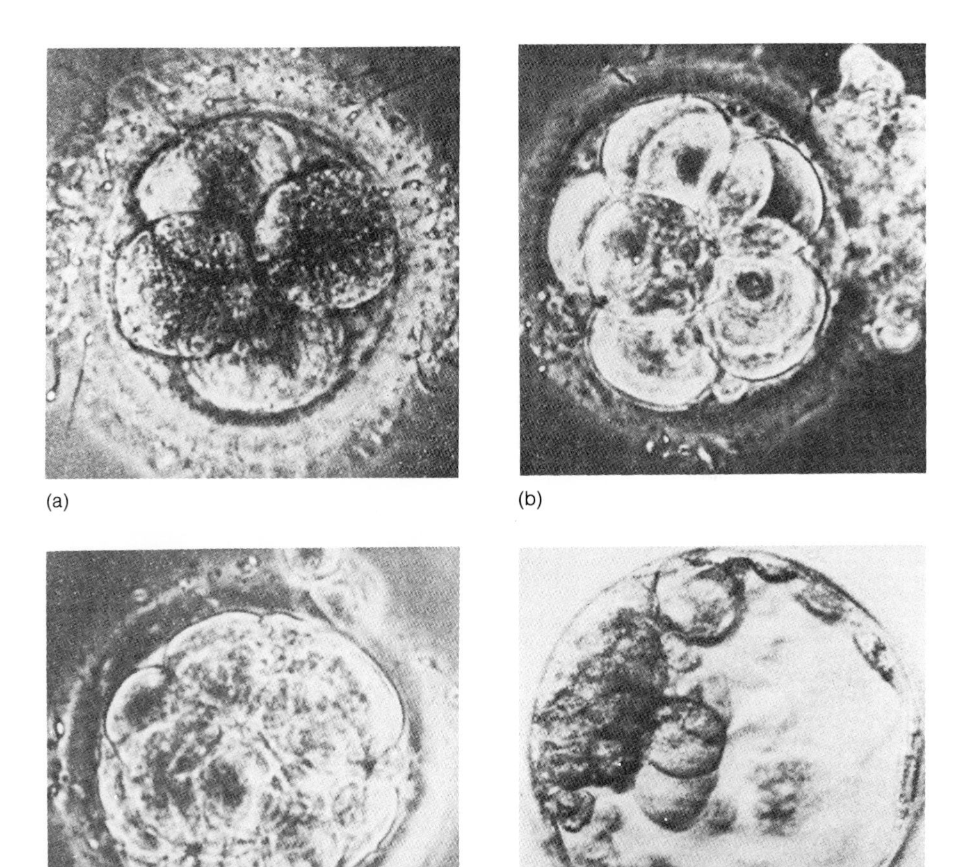

Figure 9.6. Stages of pre-embryonic development of a human ovum fertilized in a laboratory (in vitro) as seen in scanning electron micrographs. (a) 4-cell stages; (b) cleavage at the 16-cell stage; (c) a morula; and (d) a blastocyst.

Cleavage continues so that a morula consisting of thirty-two to sixty-four cells is produced by the fourth day after fertilization. The embryo remains unattached to the uterine wall for the next two days, during which time it undergoes changes that convert it into a hollow structure called a **blastocyst** (fig. 9.6). The blastocyst consists of two parts: (1) an *inner cell mass,* which will become the fetus; and (2) a surrounding *chorion,*[3] which will become part of the placenta.

On the sixth day following fertilization, the blastocyst attaches to the uterine wall, with the side containing the inner cell mass against the endometrium. The trophoblast cells produce enzymes that allow the blastocyst to "eat its way" into the thick endometrium. This begins the process of **implantation,** and by the seventh day the blastocyst is usually completely buried in the endometrium (fig. 9.7).

Implantation

If fertilization does not take place, the corpus luteum begins to decrease its secretion of steroids about ten days after ovulation. This withdrawal of steroids, as previously described, causes

[3]chorion: Gk. *chorion,* external fetal membrane

The process of **in vitro fertilization** is sometimes used to produce pregnancies in women with absent or damaged fallopian tubes, or in women who are infertile for a variety of other reasons. An ovum may be collected following ovulation (as estimated by waiting thirty-six to thirty-eight hours after the LH surge) and placed in a petri dish for two to three days with sperm collected from the husband. The pre-embryos are usually transferred to the woman's uterus at their four-to-eight-cell stage, forty-eight to seventy-two hours after fertilization. The chance of a successful implantation is only 10% to 15% per procedure, but can approach 60% after six such attempted procedures.

Many pre-embryos are often produced *in vitro* and stored in a frozen state. This has raised important moral and legal issues. Do the frozen pre-embryos have legal rights? What becomes of these rights when one is selected for implantation and development, and the rest are kept in storage? In a recent, and severely criticized, decision, a judge in Tennessee awarded custody of seven frozen pre-embryos to the wife in a divorce case. Should the father be forced to pay child support if, in the future, his ex-wife chooses to give birth to one of these frozen pre-embryos? Until recently, these legal issues could have only been imagined in science fiction, or in lawyers' nightmares.

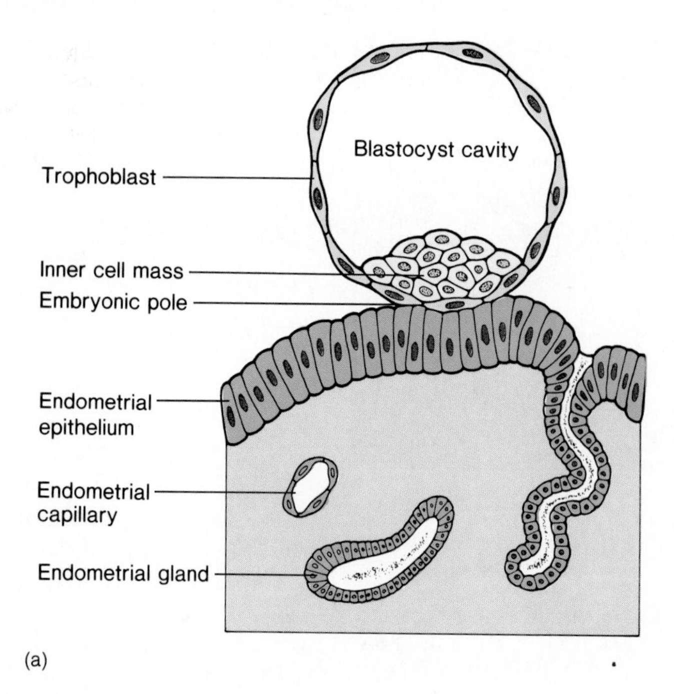

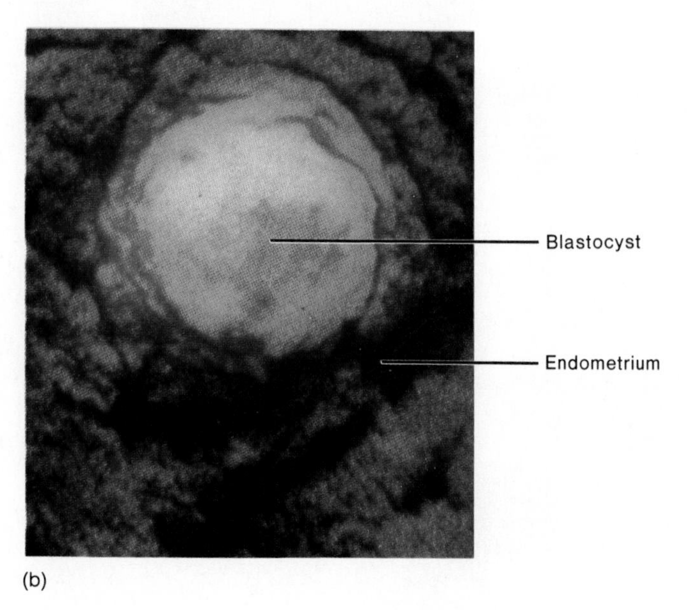

Figure 9.7. *The blastocyst adheres to the endometrium on about the sixth day as shown in diagram (a); (b) is a scanning electron micrograph showing the surface of the endometrium and implantation at twelve days following fertilization.*

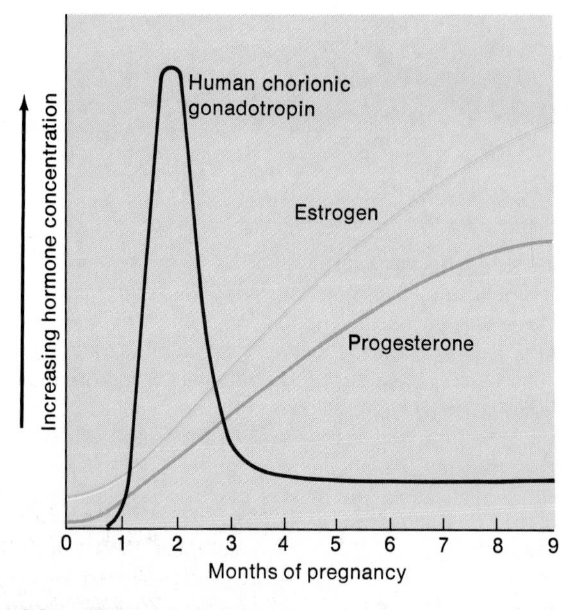

Figure 9.8. *Human chorionic gonadotropin (hCG) is secreted by the chorionic membrane during the first trimester of pregnancy. This hormone maintains the mother's corpus luteum for the first five and a half weeks. After that time the placenta becomes the major sex-hormone-producing gland, secreting increasing amounts of estrogen and progesterone throughout pregnancy.*

menstruation. If fertilization and implantation have occurred, however, menstruation must obviously be prevented to maintain the pregnancy.

Chorionic Gonadotropin

The blastocyst saves itself from being eliminated with the endometrium by secreting a hormone that indirectly prevents menstruation. Even before the sixth day when implantation occurs, the cells of the chorion secrete **chorionic gonadotropin** (**hCG**—the *h* stands for *human*). This hormone is identical to LH in its effects and therefore is able to maintain the corpus luteum past the time when it would otherwise regress. The secretion of estradiol and progesterone from the corpus luteuem is thus maintained and menstruation is normally prevented.

The secretion of hCG declines by the tenth week of pregnancy (fig. 9.8). Actually, this hormone is only required for the first five to six weeks of pregnancy, because the placenta itself becomes an active steroid hormone-secreting gland at this time. Estrogen and progesterone from the placenta are more than sufficient to maintain the endometrium and prevent menstruation for the remainder of the pregnancy.

All **pregnancy tests** assay for the presence of hCG in blood or urine, because this hormone is secreted by the blastocyst but not by the mother's endocrine glands. Modern pregnancy tests detect the presence of hCG by the use of antibodies that specifically bond to hCG. Extremely sensitive techniques utilizing antibodies against a subunit of hCG permit pregnancy to be detected in a clinical laboratory as early as seven to ten days after conception.

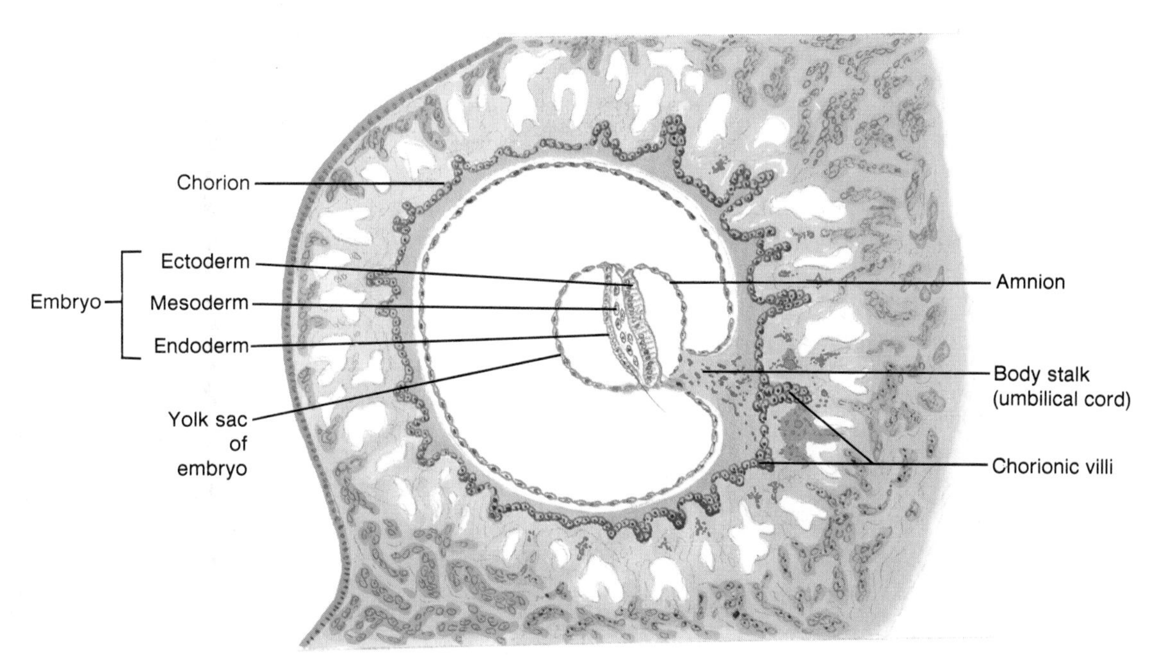

Figure 9.9. *At the completion of implantation, the primary germ layers of the embryo form and the extraembryonic membranes are in the process of developing.*

Table 9.1 Derivatives of germ layers

Ectoderm	Mesoderm	Endoderm
Epidermis of skin and epidermal derivatives: hair, nails, glands of the skin; linings of oral, nasal, anal, and vaginal cavities Nervous tissue; sense organs Lens of eye; enamel of teeth Pituitary gland Adrenal medulla	Muscle: smooth, cardiac, and skeletal Connective tissues: embryonic, connective tissue proper, cartilage, bone, blood Dermis of skin; dentin of teeth Epithelium of blood vessels, lymphatic vessels, body cavities, joint cavities Internal reproductive organs Kidneys and ureters Adrenal cortex	Epithelium of pharynx, auditory canal, tonsils, thyroid, parathyroid, thymus, larynx, trachea, lungs, digestive tract, urinary bladder and urethra, and vagina Liver and pancreas

Formation of Germ Layers

As the blastocyst completes implantation during the second week of development, a number of changes occur. A space, the amniotic cavity, forms between the developing embryo and a surrounding membrane called the amnion (fig. 9.9). The inner cell mass flattens and consists of two layers: an upper **ectoderm,**[4] which borders the amniotic cavity, and a lower **endoderm,**[5] which is next to the blastocyst cavity (fig. 9.9). A short time later, a third layer called the **mesoderm**[6] forms between the endoderm and ectoderm. These three layers are the *primary germ layers.* Once formed, at the end of the second week, the pre-embryonic stage is completed and the embryonic stage begins.

The primary germ layers give rise to the tissue of the later stages and the adult. For example, ectoderm forms the nervous system and the outer layer of the skin (epidermis) and its derivatives (hair and nails). Mesoderm forms the skeleton, muscles, blood, dermis of the skin, and connective tissues. The endoderm produces the lining of the digestive tract and its derivatives (pancreas and liver), the lungs, and the urinary bladder (fig. 9.10 and table 9.1).

[4]ectoderm: Gk. *ecto,* outside; *derm,* skin
[5]endoderm: Gk. *endo,* within; *derm,* skin
[6]mesoderm: Gk. *meso,* middle; *derm,* skin

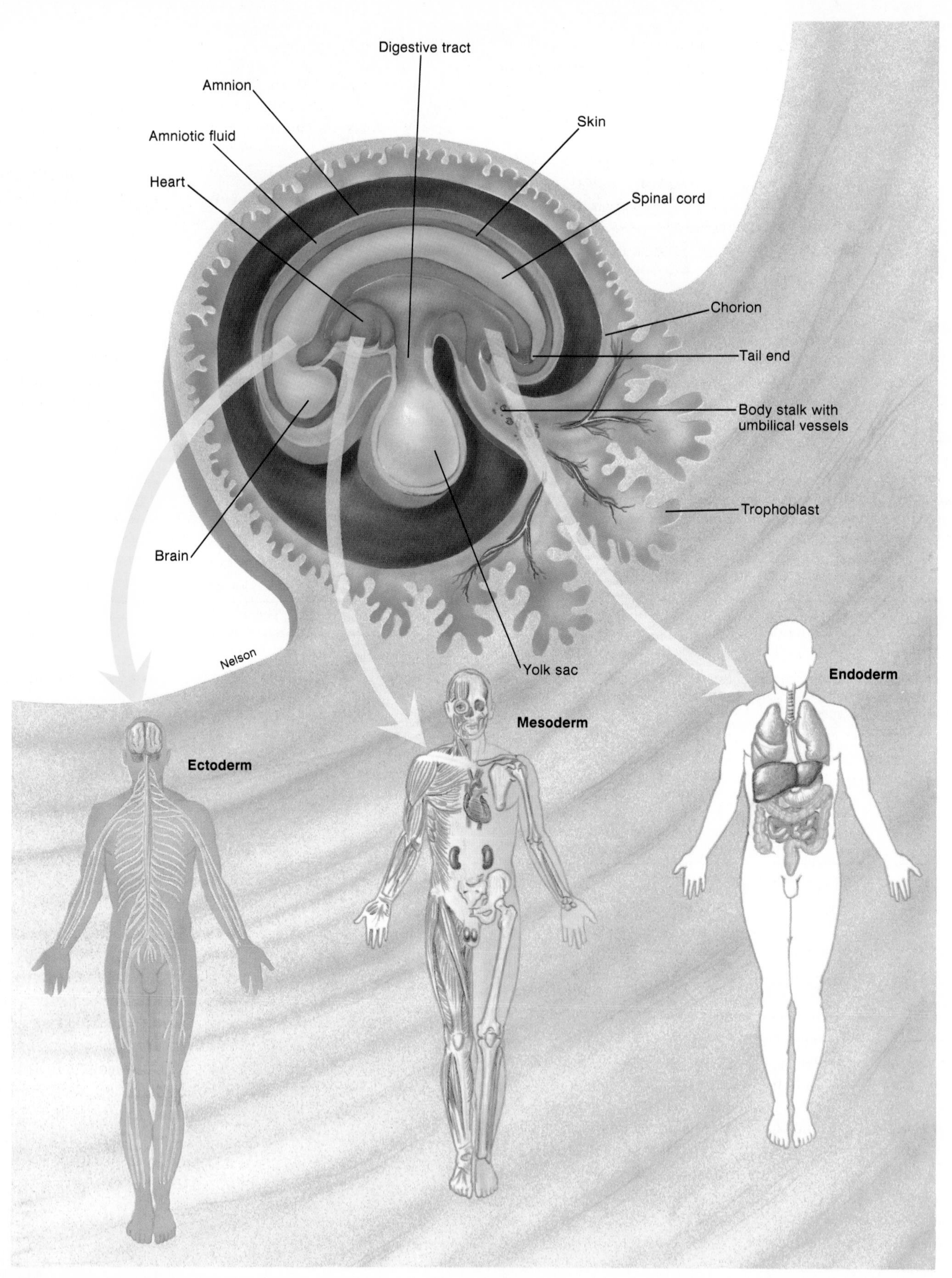

Figure 9.10. *The body systems and the primary germ layers from which they develop.*

In mitotic cell division, the daughter cells have the same genes as the parent cell. Since all the cells of the embryo (or, for that matter, of the adult body) are derived from mitotic cell division of the zygote and its products, all cells should contain the same genes as the zygote. Why, then, do some cells become neurons, others form the epidermis, others skeletal muscles, and so on? The answer is not presently known. We do know that early in embryonic development all cells are *totipotent*—able to become any adult tissue. As development proceeds, different cells become progressively more specialized or *differentiated* to form specific tissues. Experiments with lower organisms suggest that this differentiation is not due to loss of genes, but rather to suppression of the genes for alternate developmental pathways.

The nucleus of a frog intestinal cell, transplanted into the cytoplasm of a frog's ovum and artificially stimulated to divide, has produced an entire frog! This new frog would be the genetic twin of the one that contributed the nucleus. The technique is called *cloning.* Although it makes for great science fiction stories, human beings have not yet been cloned. Even if this were done, the cloned person produced would probably be quite different from the original, because all of the environmental influences on personality development would be different.

1. Describe the significance of the acrosome reaction, and the changes that occur in the secondary oocyte upon fertilization.
2. Describe the structure of the blastocyst, and explain the origin and significance of hCG.
3. List the primary germ layers and some of the adult structures they will eventually form.

Embryonic Development

The type of gonads the embryo will develop—testes or ovaries—depends on the chromosomal sex. Embryos with a Y chromosome develop testes; those without a Y chromosome develop ovaries. If testes develop, their secretions cause the development of male sex accessory organs; if testes are absent, female sex accessory organs develop. The placenta provides the developing embryo with oxygen and nourishment from its mother, and secretes a variety of hormones that aid the pregnancy.

The embryonic stage lasts from the beginning of the third week to the end of the eighth week. During this period of time, the developing organism is called an **embryo.** In addition to developmental changes that occur within the embryo itself, there are also changes that occur in the membranes surrounding the embryo and in the endometrium of the mother.

Sex Determination

Each zygote inherits twenty-three chromosomes from its mother and twenty-three chromosomes from its father. This does not produce forty-six different chromosomes, but rather, twenty-three pairs of homologous chromosomes, as described in chapter 8. Each member of a homologous pair, with the important exception of the sex chromosomes, looks like the other and contains similar genes (such as those coding for eye color, height, and so on). Each cell that contains forty-six chromosomes (that is diploid) has two chromosomes number 1, two chromosomes number 2, and so on through chromosomes number 22. The first twenty-two pairs of chromosomes are called **autosomal chromosomes.**

The twenty-third pair of chromosomes are the **sex chromosomes.** In a female these consist of two X chromosomes, whereas in a male there is one X chromosome and one Y chromosome. The X and Y chromosomes look different and contain different genes. This is the exceptional pair of homologous chromosomes mentioned earlier.

When a diploid cell (with forty-six chromosomes) undergoes meiotic division, its daughter cells receive only one chromosome from each homologous pair of chromosomes. The gametes are therefore said to be haploid. Which of the individual chromosomes of a homologous pair a given gamete obtains is random; it could be either the one originally contributed by the person's mother, or the one originally contributed by the father. This is also true for the sex chromosomes, so that approximately half of the sperm produced will contain an X and approximately half will contain a Y chromosome.

A similar random assortment of maternal and paternal chromosomes into ova will occur in a woman's ovary. Since females have two X chromosomes, however, all of the ova will normally contain one X chromosome. Because all ova contain one X chromosome, whereas some sperm are X-bearing and others are Y-bearing, the *chromosomal sex of the zygote is determined by the sperm.* If a Y-bearing sperm fertilizes the ovum, the zygote will be XY and male; if an X-bearing sperm fertilizes the ovum, the zygote will be XX and female.

Formation of Testes and Ovaries

The gonads of males and females appear similar for the first forty or so days of development following conception. During this time, cells that will give rise to sperm (called *spermatogonia*) and cells that will give rise to ova (called *oogonia*) migrate from the yolk sac to the embryonic gonads. At this stage in development the gonads have the potential to become either testes or ovaries.

Though it has long been recognized that male sex is determined by the presence of a Y chromosome and female sex by the absence of the Y chromosome, the genes involved have only

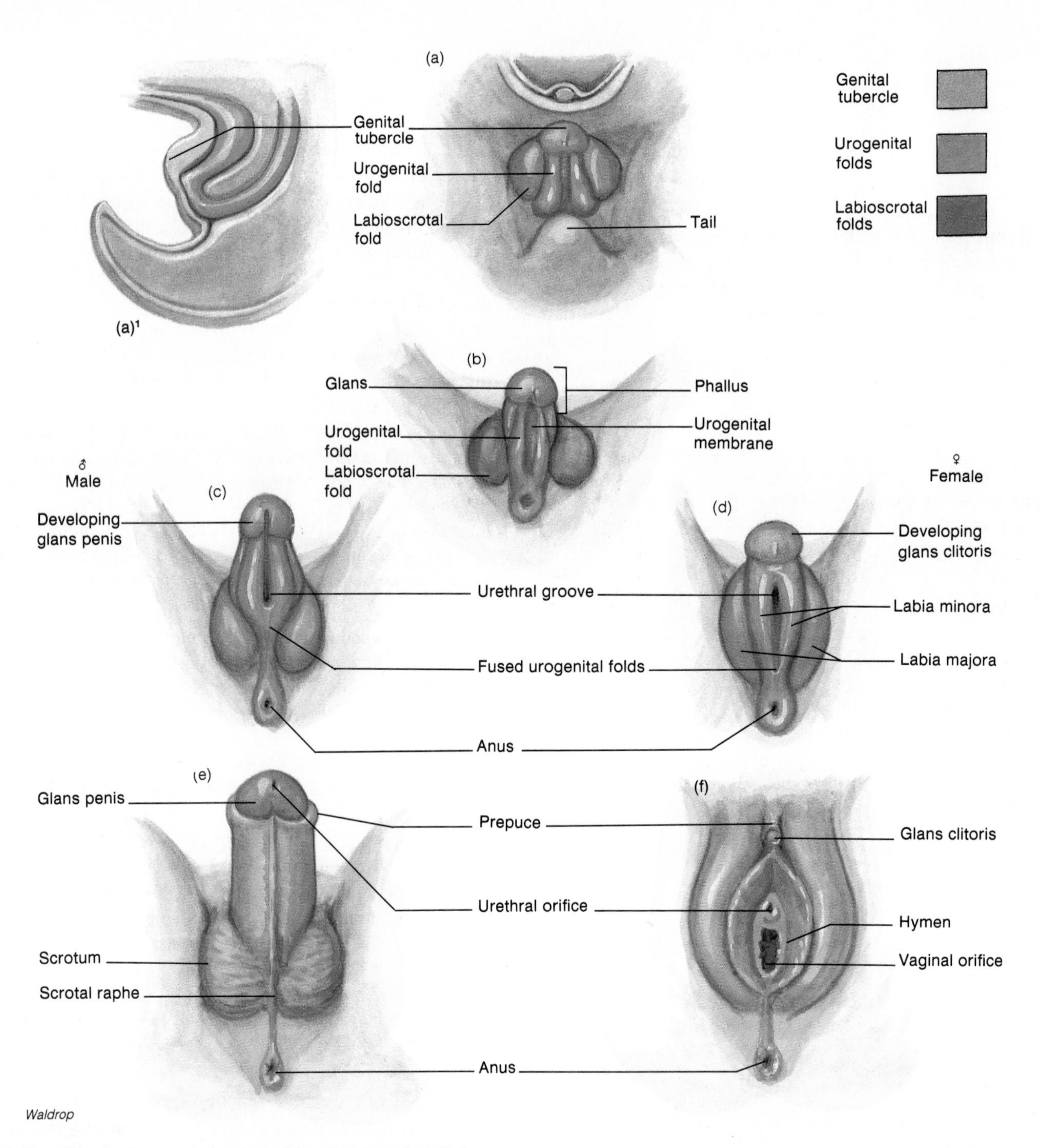

Figure 9.11. *The development of male and female external genitalia. Homologous structures have the same embryonic origins.*

recently been localized. Scientists have discovered that, in rare male babies with XX genotypes, one of the X chromosomes contains a segment of the Y chromosome. This error occurred during the meiotic cell division which formed the sperm. Through this and other observations, it has been shown that the gene for sex determination, called the **testis-determining factor (TDF),** is located on the short arm of the Y chromosome.

Notice that it is normally the presence or absence of the Y chromosome that determines whether the embryo will have testes or ovaries. This point is well illustrated by two genetic abnormalities. In **Klinefelter's syndrome** the affected person has forty-seven instead of forty-six chromosomes because of the presence of an extra X chromosome. These people, with XXY genotypes, develop testes despite the presence of two X chromosomes. Patients with **Turner's syndrome,** who have the genotype XO (and therefore have only forty-five chromosomes) have poorly developed (''streak'') gonads.

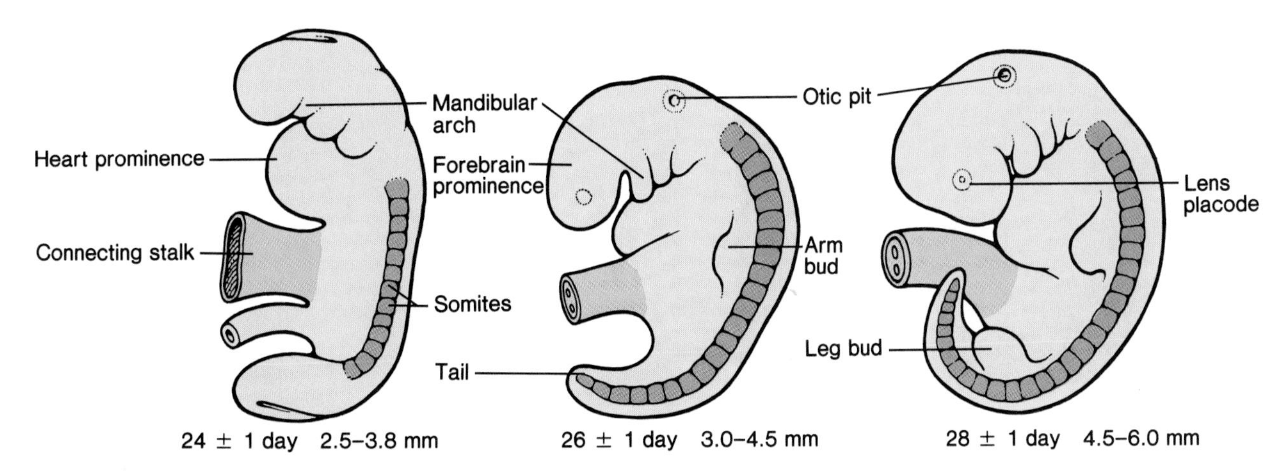

Figure 9.12. *Developmental changes from the third to the fourth week.*

The seminiferous tubules of the testes appear very early in embryonic development (between forty-three and fifty days following conception). Spermatogenesis, however, is arrested until the onset of puberty (this will be described in a later section). At about day sixty-five the Leydig cells appear in the interstitial tissue of the embryonic testes. In contrast to the rapid development of the testes, the functional units of the ovaries—the ovarian follicles—do not appear until about day 105.

The early appearing Leydig cells in the embryonic testes secrete large amounts of testosterone. Testosterone secretion begins as early as nine to ten weeks after conception, reaches a peak at twelve to fourteen weeks, and thereafter declines to very low levels by the end of the second trimester (at about twenty-one weeks). Testosterone secretion during embryonic development in the male serves a very important function (described in the next section); similarly high levels of testosterone will not appear again in the life of the individual until the time of puberty.

As the testes develop they move within the abdominal cavity and gradually descend into the *scrotum.* Descent of the testes is sometimes not complete until shortly after birth. The cooler temperature of the scrotum, which is usually 3–4°C lower than the temperature of the body cavity, is needed for spermatogenesis. This requirement is illustrated by the fact that spermatogenesis does not occur in males with undescended testes—a condition called *cryptorchidism* (*crypt* = hidden; *orchid* = testes).

Sex Accessory Organs and External Genitalia

In the absence of testes, various embryonic structures give rise to the female sex accessory organs (uterus and uterine tubes) and the female external genitalia (vagina, labia, and clitoris). These female structures are not the result of hormonal secretion of the ovaries, but are rather produced by the lack of hormonal secretion by the testes.

Male embryos, in contrast, produce the male sex accessory organs (epididymis, vas deferens, seminal vesicles, prostate, and ejaculatory duct) and the male external genitalia (penis and scrotum) as a result of the masculinizing effect of testosterone secreted by the embryonic testes. The same embryonic structures that produce a penis and scrotum in a male give rise to the vagina, labia, and clitoris in a female. These organs are therefore considered to be *homologous structures* (fig. 9.11).

Stages of Embryonic Development

At three to four weeks, the embryo has grown to about 4 mm in length. Different tissues have begun to specialize, the heart begins to beat and pump blood, and arm and leg buds appear (fig. 9.12). Growth continues from the fifth to the sixth week, in which the embryo is 16–24 mm long. The head becomes larger and the brain begins to develop at this time (fig. 9.13).

During the seventh and eighth weeks, which are the last two weeks of the embryonic stage, the embryo has become 28–40 mm long and has distinctly human characteristics. The body organs are formed, and the nervous system begins coordinating body activity. The eyes are well developed but the lids are stuck together. The nostrils are also developed but plugged with mucus. Since the body systems are developed by the end of the eighth week, the developmental stages that occur after this time constitute *fetal development,* and the embryo is graduated to the category of **fetus.**

Extraembryonic Membranes and the Placenta

The extraembryonic membranes are the yolk sac, amnion, allantois, and chorion (fig. 9.14). The *yolk sac* in humans does not contain nutritive yolk, but does serve various functions that aid the development of the embryo. The *allantois*[7] forms as a small outpouching from the yolk sac and likewise serves some developmental functions. The *amnion*[8] envelops the embryo to form the **amniotic sac,** and the chorion contributes to the placenta. At the time of birth, the extraembryonic membranes separate from the fetus and are expelled from the uterus as the *afterbirth.*

[7]allantois: Gk. *allanto,* sausage; *iodos,* resemblance
[8]amnion: Gk. *amnion,* bowl

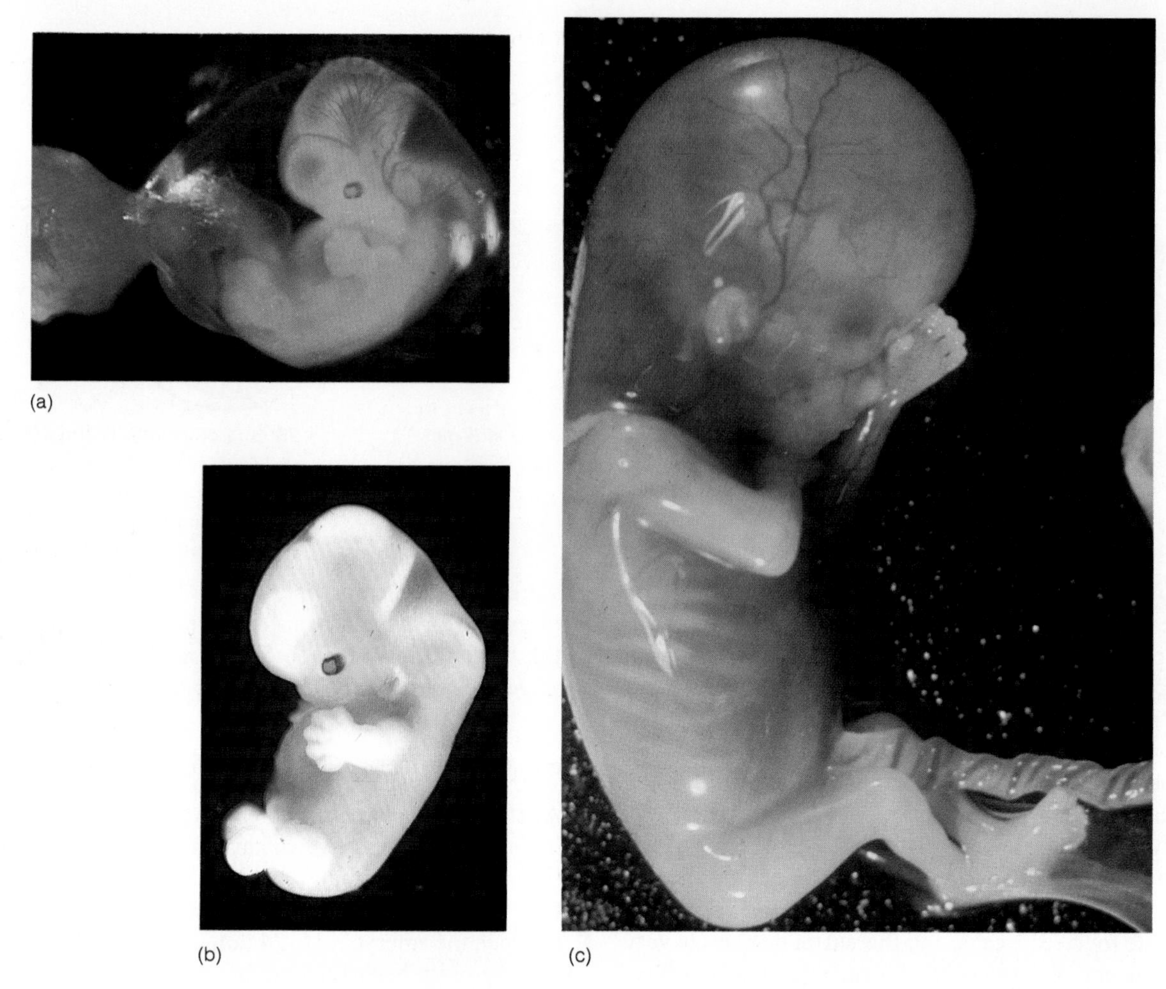

Figure 9.13. *Photographs of a five-week-old (a), six-week-old (b), and eight-week-old (c) embryo. The body systems are developed by the end of the eighth week.*

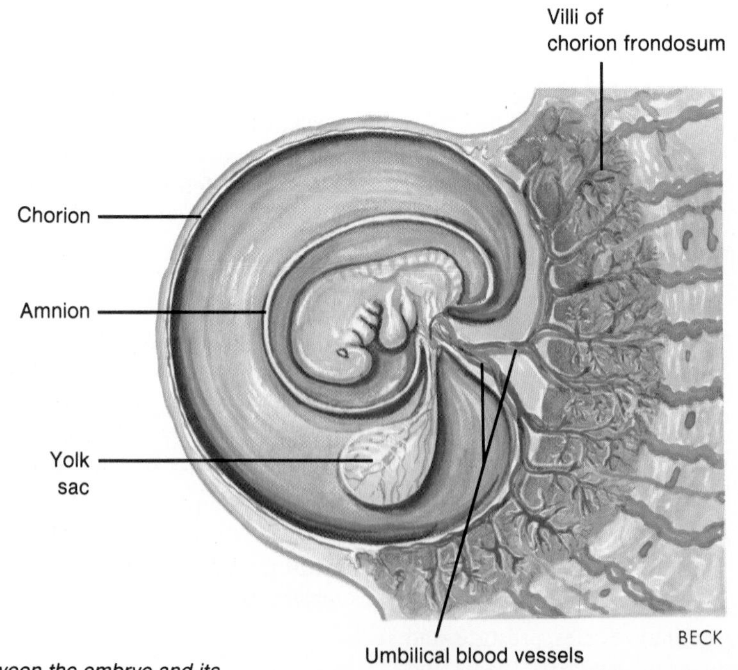

Figure 9.14. *The relationship between the embryo and its extraembryonic membranes.*

Abortion

An estimated 1.5 million *elective abortions* occur in the United States each year. The term "elective" refers to the fact that the abortion is performed due to the wishes of the pregnant woman. This is in contrast to *spontaneous abortions,* which occur with far greater frequency. Spontaneous abortions may be due to biological problems of the embryo or fetus that render it inviable, or to other medical problems associated with the pregnancy.

Elective abortions have engendered more emotional debate than any other biologically based social issue. Biologists can aid this debate by helping to frame proper questions for discussion. For example, the question "Does human life begin at conception?" is often debated, but is not a biologically proper question. A human ovum is genetically human prior to conception, and is as alive before as after fertilization. The genetic constitution of an individual human being is formed at conception and then replicated when each cell of the body is produced. Few people, however, believe that an individual cell (such as a cheek epithelial cell) is equivalent to a human being. This implies that genetic uniqueness cannot be equated with "personhood."

Human life, as opposed to cellular life, involves the presence of consciousness, the ability to experience emotions, and self-awareness. These are higher functions of the human brain. Indeed, a person is considered to be clinically dead when brain functions irreversibly cease. How, then, can one consider a person to be alive if these functions have not yet developed? If the developing organism is not yet a person, an abortion at that stage prevents a person from coming into existence in the future; it cannot be said to end the life of a person who has never existed. According to these arguments, abortions could be considered ethical if performed prior to a particular developmental stage and unethical following that stage.

Unfortunately, arguments on both sides of the abortion issue have tended toward the extremes. Some people who are in favor of legalized abortions ("pro-choice") argue for completely unrestricted access to elective abortions, regardless of the developmental stage of the fetus. Some people who are against legalized abortions ("pro-life") take the position that abortions should not even be performed for victims of rape or incest, calling abortion under any circumstances "murder." This puts a woman who aborts a blastocyst in the same category as one who kills her child. Biological knowledge, in the author's opinion, does not support this contention.

The ethics and legalities of abortion are not simple, and nobody is well served by simplistic and emotionally charged rhetoric. The issues raised by the abortion debate require the best thinking of scientists, philosophers, theologians, and lawmakers, and are not likely to ever be completely settled. In the near future, an "abortion pill" may be available which will enable any woman to obtain an abortion from her family physician. (This pill—known as RU 486—is currently available in France, where it was developed.) The availability of such an abortion pill will surely be opposed by antiabortion groups and be championed by those who are pro-choice. As new scientific advances are made, and new biotechnology emerges, the ethical and legal problems will become even more complex.

Formation of the Placenta and Amniotic Sac

The chorionic membrane, on its side that is in contact with the endometrium, develops leaflike extensions. This part of the chorion is thus known as the *chorion frondosum* (*frond* = leaf). As the blastocyst is implanted in the endometrium and the chorion develops, the cells of the endometrium also undergo changes. These changes, including cellular growth and the accumulation of glycogen, are called the **decidual reaction.** The maternal tissue in contact with the chorion frondosum is called the *decidua basalis.* These two structures—chorion frondosum (fetal tissue) and decidua basalis (maternal tissue)—together form the functional unit known as the **placenta.**[9]

The human placenta is a disc-shaped structure that is continuous at its outer surface with the smooth part of the chorion, which bulges into the uterine cavity. Immediately beneath the chorionic membrane is the amnion, which has grown to envelop the entire fetus (fig. 9.15). The fetus, together with its umbilical cord, is therefore located within the fluid-filled amniotic sac.

The **umbilical cord** attaches near the center of the placenta. It contains two *umbilical arteries,* which carry fetal blood low in oxygen to the placenta, and one *umbilical vein,* which carries blood rich in oxygen from the placenta to the fetus. In about one-fifth of all deliveries, the cord is looped once around the baby's neck. If drawn tightly, the cord may cause death or serious brain damage due to an interruption in the supply of oxygen to the fetus.

Amniocentesis

Amniotic fluid is formed initially as an isotonic secretion, which is later increased in volume and decreased in concentration by urine from the fetus. Amniotic fluid also contains cells that are sloughed off from the fetus, placenta, and amniotic sac. Since all of these cells are derived from the same fertilized ovum, all have the same genetic composition. Many genetic abnormalities can be detected by aspiration of this fluid and examination of the cells thus obtained. This procedure is called **amniocentesis** (fig. 9.16).

[9]placenta: L. *placenta,* a flat cake

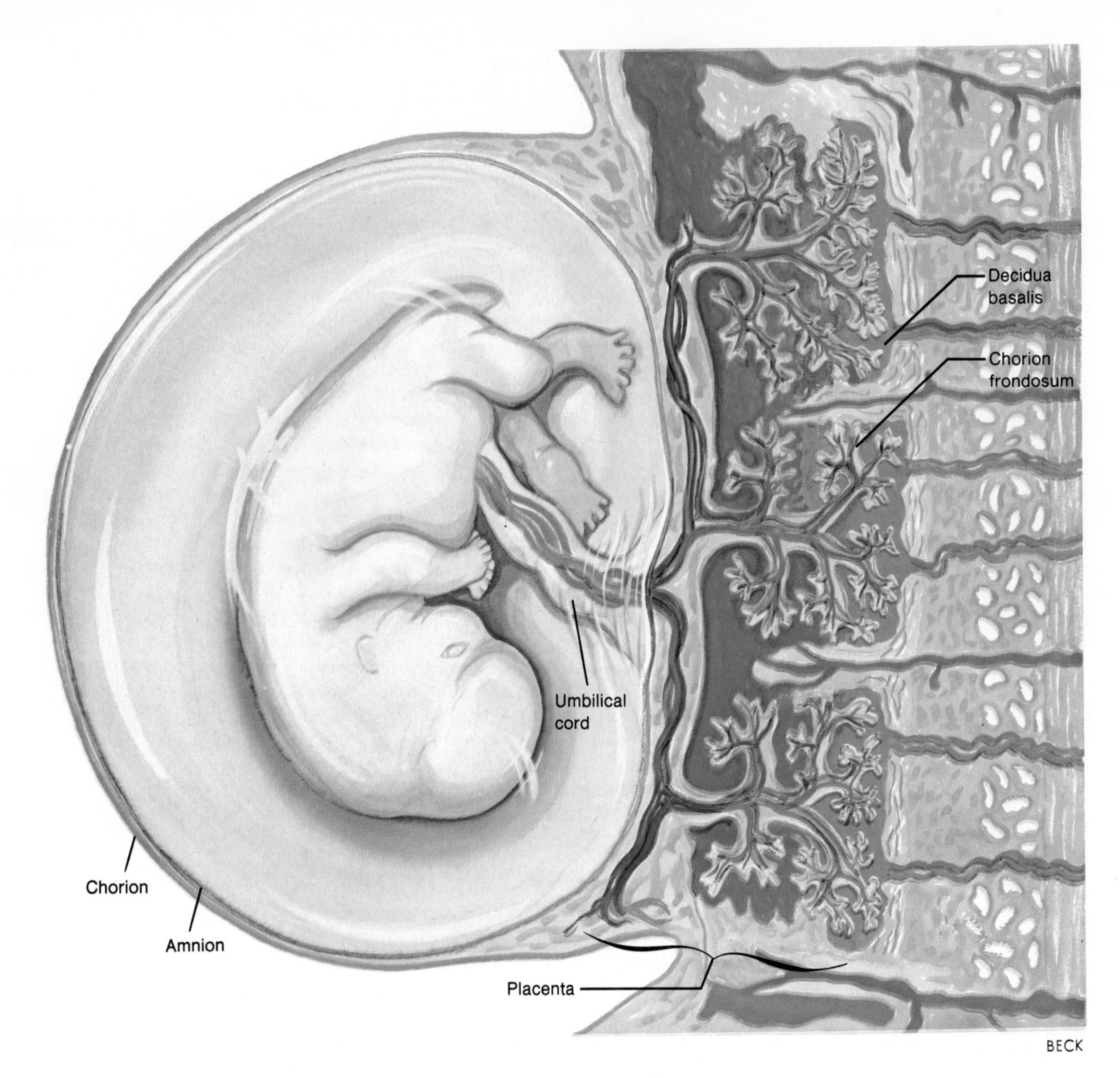

Figure 9.15. *Blood from the fetus is carried to and from the chorion frondosum by umbilical arteries and veins. The maternal tissue between the chorionic villi is known as the decidua basalis, and this tissue, together with the villi, form the functioning placenta.*

Amniocentesis is usually performed at the fourteenth or fifteenth week of pregnancy, when the amniotic sac contains 175–225 ml of fluid. Genetic diseases such as Down's syndrome (where there are three instead of two chromosomes number 21) can be detected by examining chromosomes; diseases such as Tay-Sachs disease, in which there is a defective enzyme involved in the formation of myelin sheaths, can be detected by biochemical techniques.

The amniotic fluid that is withdrawn contains fetal cells at a concentration that is too low to permit direct determination of genetic or chromosomal disorders. These cells must therefore be cultured *in vitro* for at least two weeks before they are present in sufficient numbers for the laboratory tests required. A newer method, called **chorionic villus biopsy,** is now available to detect genetic disorders much earlier than is possible by amniocentesis. In chorionic villus biopsy, a catheter is inserted through the cervix to the chorion, and a sample of a chorionic villus (a fingerlike projection of the chorion) is obtained by suction or cutting. Genetic tests can be performed directly on the villus sample, since this sample contains much larger numbers of fetal cells than does a sample of amniotic fluid. Chorionic villus biopsy can provide genetic information at ten to twelve weeks' gestation; such information obtained by amniocentesis, in comparison, is not generally available before twenty weeks.

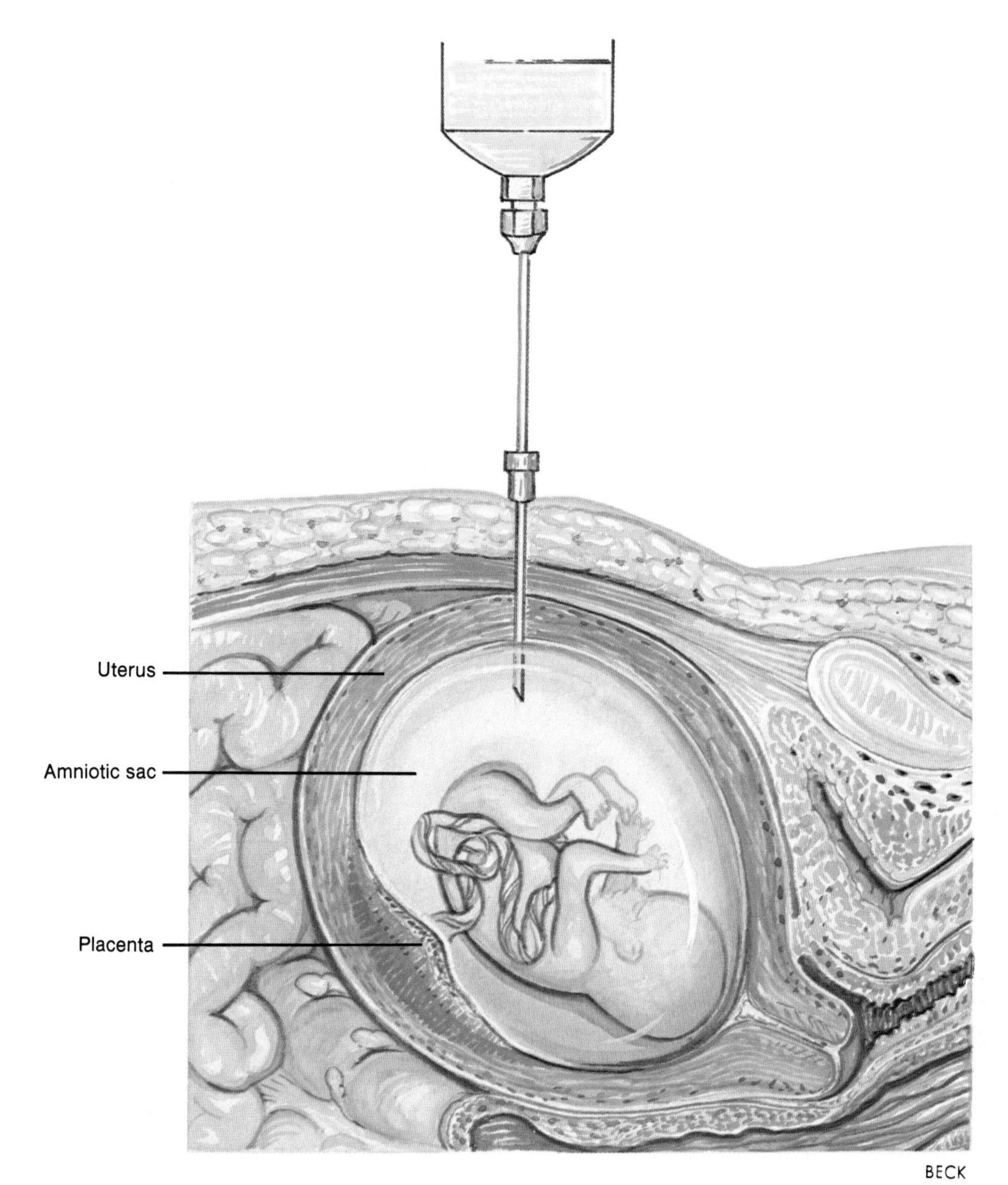

Figure 9.16. *Amniocentesis. In this procedure amniotic fluid, together with suspended cells, is withdrawn for examination. Various genetic diseases can be detected prenatally by this means.*

Exchange of Molecules across the Placenta

The *umbilical artery* delivers fetal blood to vessels within the villi of the chorion frondosum of the placenta. This blood circulates within the villi and returns to the fetus via the *umbilical vein.* Maternal blood is delivered to and drained from the cavities within the decidua basalis that are located between the chorionic villi (fig. 9.17). In this way, maternal and fetal blood are brought close together but never mix within the placenta.

The placenta serves as a site for the exchange of gases and other molecules between the maternal and fetal blood. Oxygen diffuses from mother to fetus, and carbon dioxide diffuses in the opposite direction. Nutrient molecules and waste products likewise pass between maternal and fetal blood; the placenta is, after all, the only link between the fetus and the outside world.

The placenta is not merely a passive conduit for exchange between maternal and fetal blood, however. It has a very high metabolic rate, utilizing about a third of all the oxygen and glucose supplied by the maternal blood. The rate of protein synthesis is, in fact, higher in the placenta than it is in the liver. Like the liver, the placenta produces a great variety of enzymes capable of converting biologically active molecules (such as hormones and drugs) into less active, more water-soluble forms. In this way, potentially dangerous molecules in the maternal blood are often prevented from harming the fetus.

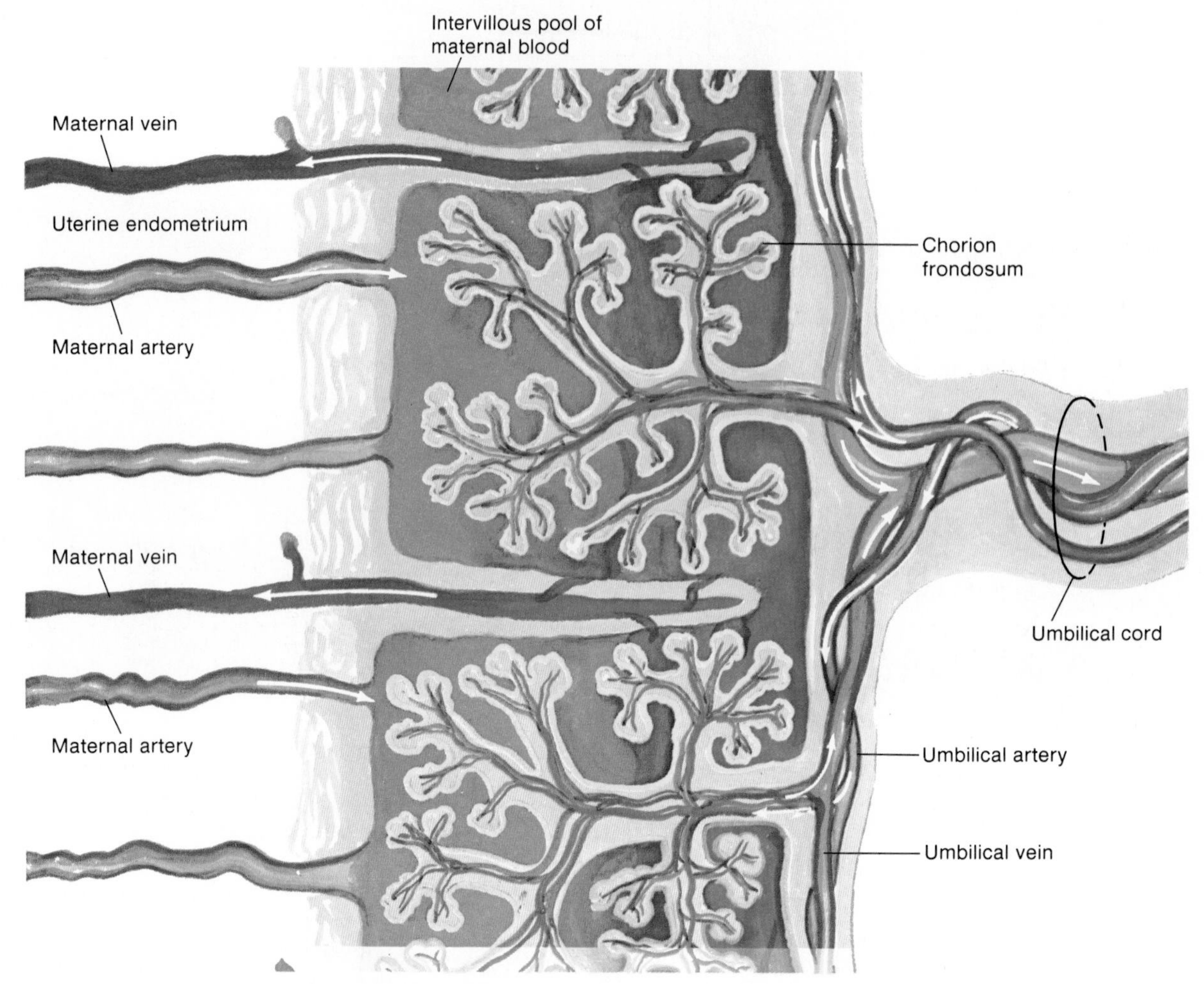

Figure 9.17. *The circulation of blood within the placenta. Maternal blood is delivered to and drained from the spaces between the chorionic villi. Fetal blood is brought to blood vessels within the villi by branches of the umbilical artery and is drained by branches of the umbilical vein.*

Endocrine Functions of the Placenta

The placenta secretes both steroid hormones and protein hormones that have actions similar to those of some anterior pituitary hormones. This latter category of hormones includes **chorionic gonadotropin (hCG)** and **chorionic somatomammotropin (hCS)** (table 9.2). Chorionic gonadotropin has LH-like effects, as previously described; it also has thyroid-stimulating ability, like pituitary TSH. Chorionic somatomammotropin likewise has actions that are similar to two pituitary hormones: growth hormone and prolactin. The placental hormones hCG and hCS thus duplicate the actions of four anterior pituitary hormones.

Pituitary-Like Hormones from the Placenta

The importance of chorionic gonadotropin in maintaining the mother's corpus luteum for the first five and a half weeks of pregnancy has been previously discussed. There is also some evidence that hCG may in some way help to prevent immunological rejection of the implanting embryo. Chorionic somatomammotropin acts together (synergizes with) growth hormone from the mother's pituitary to produce a "diabeticlike" effect in the pregnant woman. The effects of these two hormones stimulate (1) breakdown of fat and increased plasma fatty acid concentration; (2) decreased maternal utilization of glucose and, therefore, increased blood glucose concentrations; and (3) polyuria (excretion of large volumes of urine), thereby producing a degree of dehydration and thirst. This "diabeticlike" effect in the mother helps to spare glucose for the placenta and fetus that, like the brain, use glucose as their primary energy source.

Drug Abuse During Pregnancy

It has been estimated that approximately twelve million Americans have alcohol problems, and of these, about six million have problems sufficiently severe to warrant the label of *alcoholism*. Even though a pregnant woman may not abuse alcohol to the point that she can be considered to be an alcoholic, however, she may ingest enough alcohol to significantly harm the central nervous system of the fetus. This is because the actions of the placenta can raise the level of alcohol in the fetus's blood up to ten times higher than the alcohol concentration in the mother's blood.

A newborn baby may suffer mental retardation and different types of birth defects as a result of the ingestion of alcohol by the mother. This is called **fetal alcohol syndrome,** and may account for a third of all such symptoms among newborns. For these reasons, a woman should refrain from drinking alcoholic beverages completely, or take only very small amounts, during her pregnancy. Mothers who breast feed their baby should also refrain from drinking alcohol; it has been shown that even one drink per day produces a decrease in the motor development of the infant.

A recent study showed that use of marijuana or cocaine during pregnancy can also damage the fetus. In this study, use of marijuana or cocaine resulted in decreases in the birth weight, body length, and head circumference of the newborns. Because the main ingredient of marijuana—tetrahydrocannabinol—is fat-soluble and can cross the placenta, a single episode of use can result in prolonged exposure of the fetus to this drug. (Up to thirty days may be required for complete elimination of the drug after a single use.) Marijuana impairs oxygenation of the blood and has other actions that may be responsible for its effects on the fetus. Cocaine increases maternal heart rate and blood pressure and produces constriction of the uterine arteries, thus decreasing the delivery of oxygen to the fetus. In addition, use of cocaine blunts the appetite of the mother and can lead to undernourishment of the mother and consequently of the fetus.

Table 9.2 Hormones secreted by placenta

Hormones	Effects
Pituitary-like hormones	
Chorionic gonadotropin (hCG)	Similar to LH; maintains mother's corpus luteum for first 5½ weeks of pregnancy; may be involved in suppressing immunological rejection of embryo; also has TSH-like activity
Chorionic somatomammotropin (hCS)	Similar to prolactin and growth hormone; in the mother, hCS acts to promote increased fat breakdown and fatty acid release from adipose tissue and to promote the sparing of glucose use by maternal tissues ("diabeticlike" effects)
Sex steroids	
Progesterone	Helps maintain endometrium during pregnancy; helps suppress gonadotropin secretion; promotes uterine sensitivity to oxytocin; helps stimulate mammary gland development
Estrogens	Help maintain endometrium during pregnancy; help suppress gonadotropin secretion; help stimulate mammary gland development; inhibit prolactin secretion

Steroid Hormones from the Placenta

After the first five and a half weeks of pregnancy, when the corpus luteum regresses, the placenta becomes the major sex steroid–producing gland. The blood concentration of estrogens, as a result of placental secretion, rises to levels more than 100 times greater than those existing at the beginning of pregnancy. The placenta also secretes large amounts of progesterone.

The placenta, however, is an "incomplete endocrine gland" because it cannot produce estrogen and progesterone without the aid of precursors supplied to it by both the mother and the fetus. In order for the placenta to produce estrogens, it needs to cooperate with steroid-producing tissues in the fetus. Fetus and placenta, therefore, form a single functioning system in terms of steroid hormone production.

1. Explain how the sex of the embryo is determined. Why do secretions of the testes rather than of the ovaries play the key role in this process?
2. Describe the formation and components of the placenta, and how these components function to allow the exchange of molecules between the embryo and mother.
3. Describe the structure of the amniotic sac and explain how an amniocentesis is performed.
4. List some of the hormones secreted by the placenta and explain their significance.

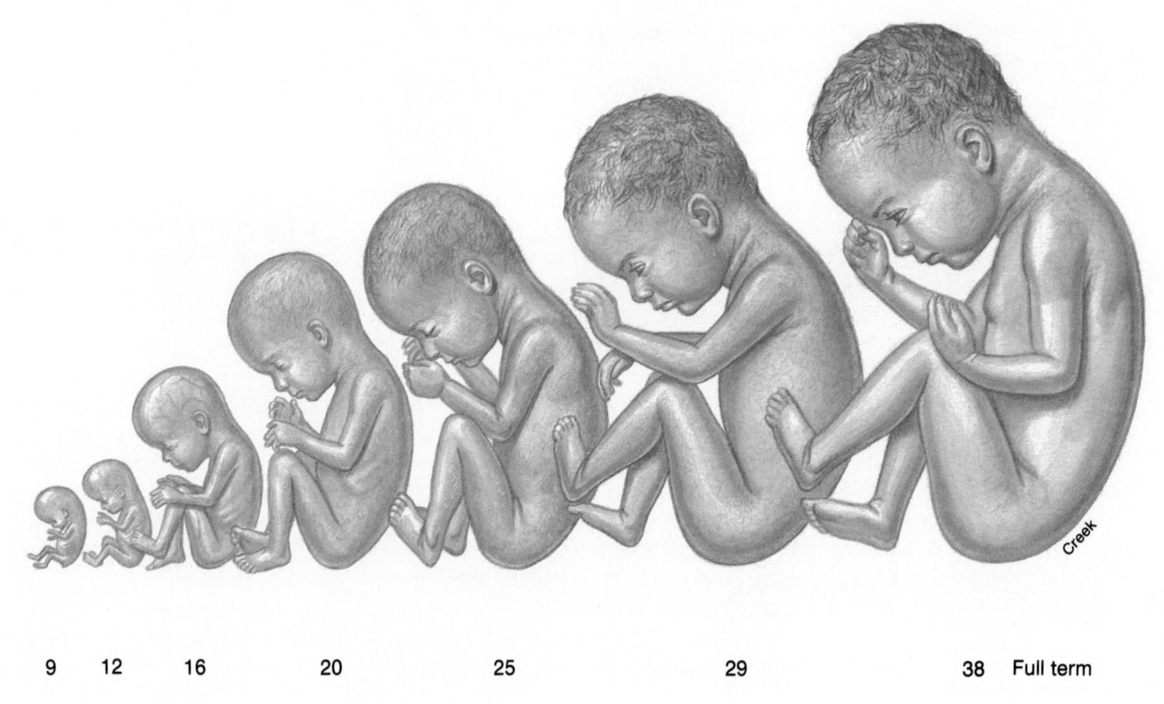

Figure 9.18. *Changes in the external appearance of the fetus from the ninth through the thirty-eighth week.*

Fetal Development and Parturition

The period of fetal development lasts from the beginning of the ninth week until birth, or parturition. Gestation, the period of prenatal development, is usually 266 days. The contractions of the uterus that are required for normal delivery are produced in response to the hormone oxytocin and to chemicals within the uterus called prostaglandins. Production of milk by the mother's mammary glands requires the action of the hormone prolactin, and milk ejection requires the action of oxytocin.

Since most of the tissues and organs of the body appear during the embryonic period, fetal development is primarily limited to body growth. Figure 9.18 shows the changes in external appearance of the fetus from the ninth through the thirty-eighth week.

Stages in Fetal Development

At the beginning of the ninth week, the head is as large as the rest of the body. The eyes are widely spaced, and the ears are set low. Growth of the head slows down during the next three weeks, whereas the growth in body length accelerates. The external genitalia are not developed to the point that the sex can be determined until the twelfth week. By the end of the twelfth week, the fetus is 87 mm long. It can swallow, defecate, and urinate into the amniotic fluid. The nervous system and muscle coordination are developed enough so that the fetus will withdraw its leg if tickled. The fetus begins inhaling through its nose but can take in only amniotic fluid. Figure 9.19 shows the appearance of fetuses at ten and twelve weeks.

Major structural features of a fetus can be detected by *ultrasound* (fig. 9.20). Sound waves are reflected from the interface of tissues with different densities—such as the interface between the fetus and amniotic fluid—and used to produce an image. This technique is so sensitive that it can be used to detect a fetal heartbeat several weeks before it can be detected by a stethoscope.

The facial features of the fetus are formed during the period from thirteen through sixteen weeks. Eyelashes, hair, fingernails, and nipples begin to develop. During the sixteenth week, the fetal heartbeat can be heard by applying a stethoscope to the mother's abdomen. By the end of the sixteenth week, the fetus is 140 mm long and weighs about 200 g (7 oz).

During the period from seventeen to twenty weeks, fetal movements, known as *quickening,* are commonly felt by the mother. The skin of the fetus is covered by a white, cheeselike material known as the **vernix caseosa.**[10] Twenty-week-old fetuses usually have a fine, silklike fetal hair, called **lanugo,**[11] covering the skin. Because of cramped space, the fetus develops the marked spinal flexure and assumes what is commonly called the "fetal position," with the head down in contact with the flexed knees.

During the period from twenty-one to twenty-five weeks, the fetus increases substantially in weight to about 900 g (32 oz). Toward the end of the twenty-ninth week the fetus averages 275 mm (about 11 in.) in length and 1,300 g (46 oz) in

[10]vernix caseosa: L. *vernix,* varnish; *caseus,* cheese
[11]lanugo: L. *lana,* wool

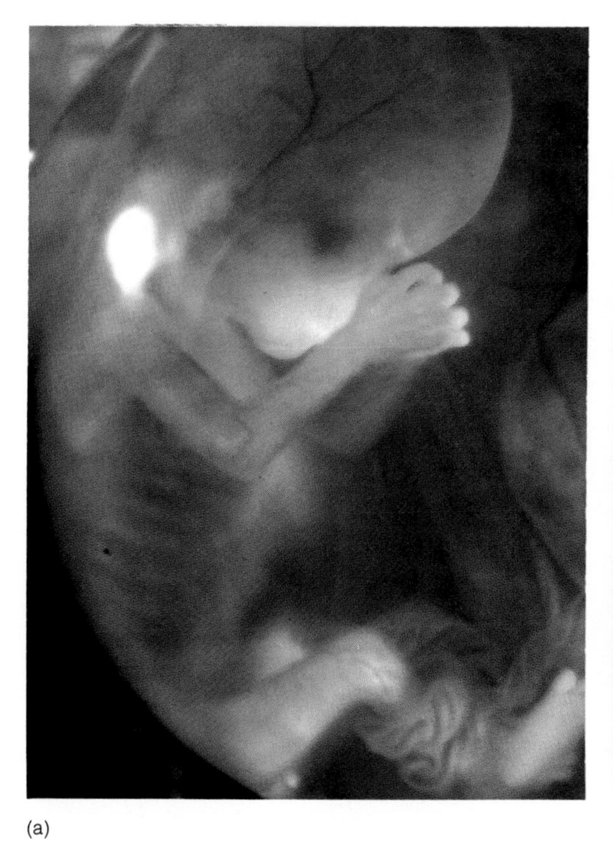

(a)

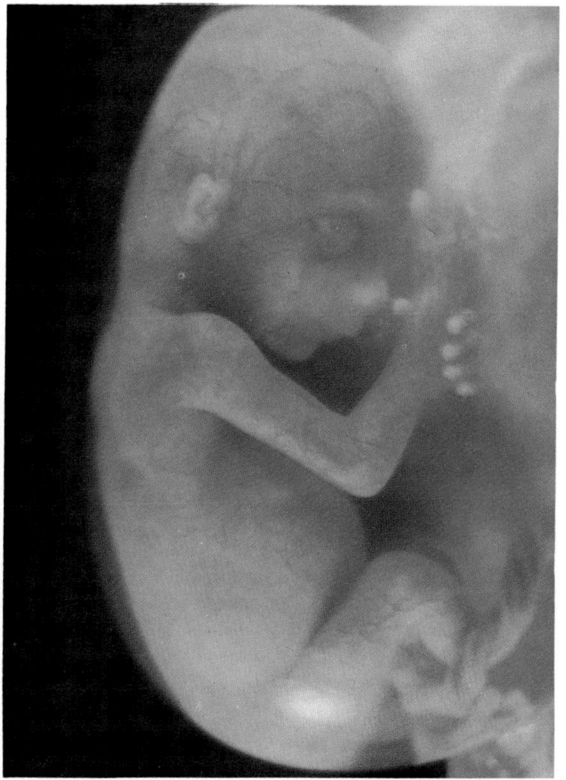

(b)

Figure 9.19. *External appearances of fetuses at (a) ten weeks and (b) twelve weeks.*

weight. The fetus might now survive if born prematurely, but the mortality rate is high. If the fetus is a male, the testes will have begun their descent into the scrotum. As the time of birth approaches, the fetus rotates to a *vertex*[12] position (fig. 9.21).

By the end of thirty-eight weeks, the fetus is considered "full-term." It has reached an average crown-rump length of 36 cm (14 in.), crown-heel length of 50 cm (20 in.), and weight of 3,400 g (7.5 lb). Most fetuses are plump with smooth skin because of the accumulation of subcutaneous fat. The skin is pinkish-blue in color even on fetuses of dark-skinned parents, because skin cells called melanocytes do not produce melanin (the pigment of the skin) until exposed to sunlight.

Labor and Parturition

The time of prenatal development, or the time of pregnancy, is called **gestation.**[13] The human gestational period is usually 266 days or about 280 days from the beginning of the last menstrual period to **parturition,**[14] or birth. Most fetuses are born within 10 to 15 days before or after this time. Parturition is accompanied by a sequence of physiological events called **labor.**

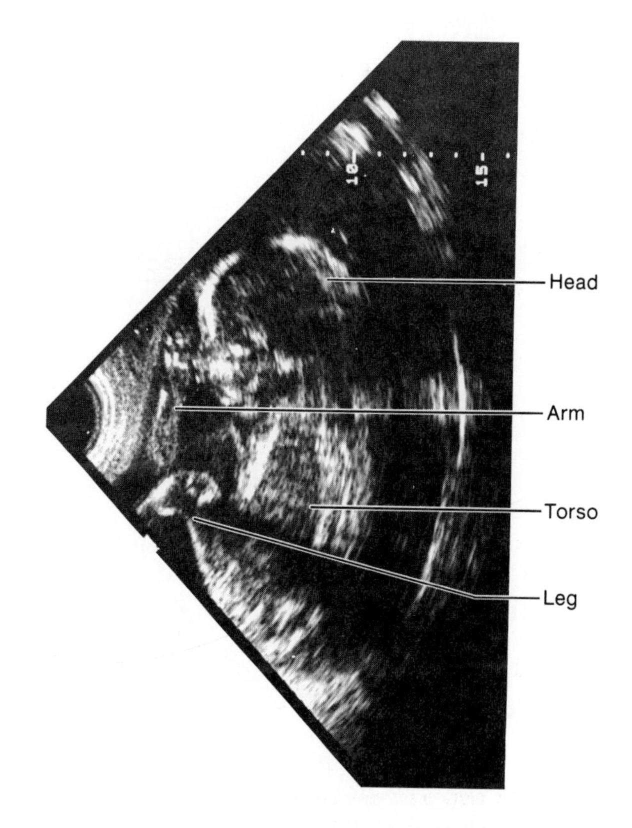

Figure 9.20. *Structures of the human fetus observed through an ultrasound scan.*

[12]vertex: L. *vertex,* summit
[13]gestation: L. *gestatus,* to bear
[14]parturition: L. *parturitio,* giving birth

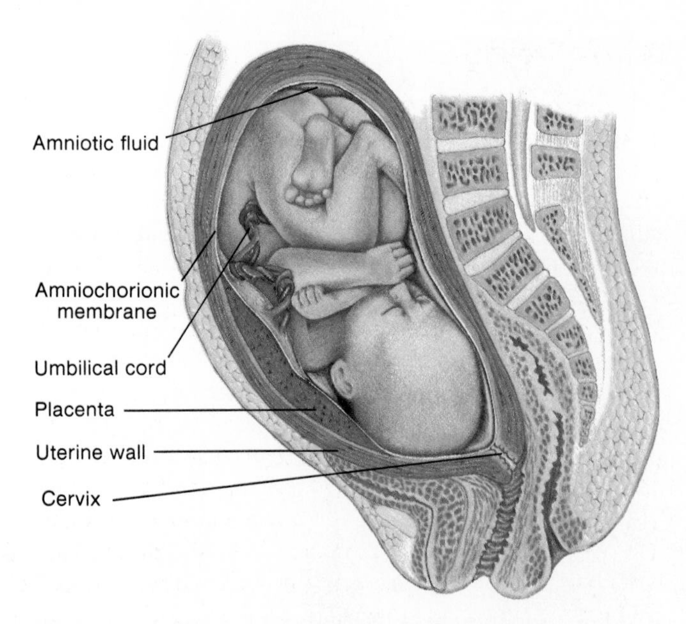

Figure 9.21. *A fetus in vertex position. Toward the end of most pregnancies, the weight of the fetal head causes a rotation of its entire body such that the head is positioned in contact with the cervix of the uterus.*

Powerful contractions of the uterus in labor are needed for childbirth. These uterine contractions are stimulated by two agents: (1) *oxytocin,* a polypeptide hormone produced in the hypothalamus and secreted by the posterior pituitary; and (2) *prostaglandins,* a class of fatty acids with regulatory functions. Prostaglandins are classified as *autocrine* regulators because they act within the same organ in which they are produced. (As distinct from hormones, which travel in the blood to their target organs.) Labor can be induced artificially by injections of oxytocin or by insertion of prostaglandins into the vagina as a suppository.

Stages of Labor

Labor is divided into three stages (fig. 9.22):

1. **Dilation stage.** In this stage the cervix dilates to a diameter of approximately 10 cm. There are regular contractions during this stage and usually a rupturing of the amniotic sac ("bag of waters"). If the amniotic sac does not rupture spontaneously, it is done surgically. The dilation stage may last eight to twenty-four hours.

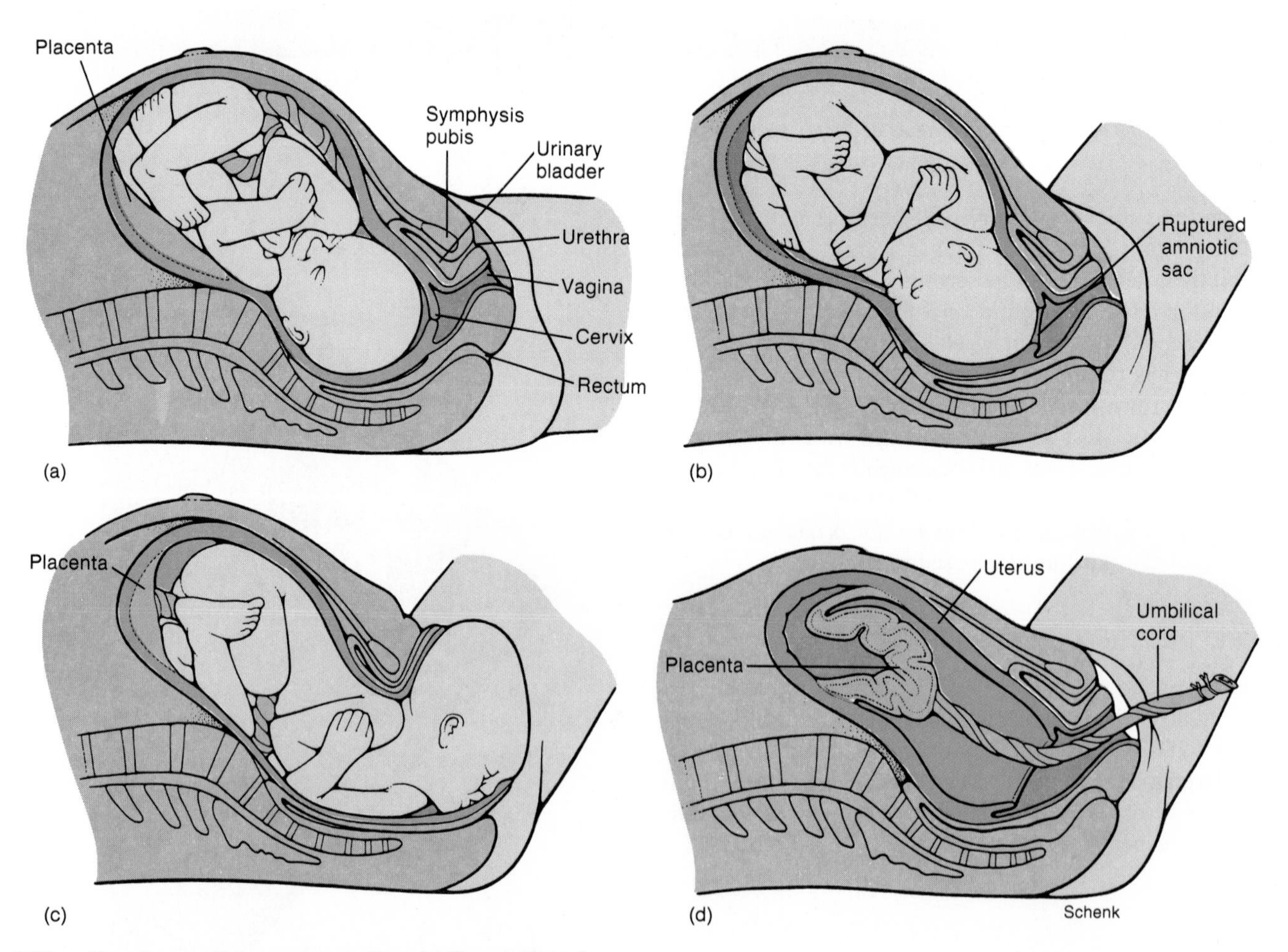

Figure 9.22. *The stages of labor and parturition.* (a) *The position of the fetus prior to labor;* (b) *the ruptured amniotic sac and early dilation of the cervix;* (c) *expulsion state, or the period of parturition;* (d) *the placental stage.*

2. **Expulsion stage.** This is the period of parturition, or actual childbirth. It consists of forceful uterine contractions and abdominal compressions to expel the fetus from the uterus and through the vagina. This stage may require thirty minutes in a first pregnancy, but only a few minutes in subsequent pregnancies.
3. **Placental stage.** Generally within ten to fifteen minutes after parturition, the placenta is separated from the uterine wall and expelled as the *afterbirth.* Forceful uterine contractions occur at this stage, which help to constrict blood vessels and limit hemorrhage. In a normal delivery, blood loss does not exceed 350 ml.

Five percent of newborns are born *breech.* In a breech birth, the fetus has not rotated and the buttocks are the presenting part. The principal concern of a breech birth is the increased time and difficulty of the expulsion stage of parturition. If an infant cannot be delivered breech, a *cesarean section* must be performed. A cesarean section is delivery of the fetus through an incison made into the uterus through the abdominal wall.

Multiple Pregnancy

Twins are born in about one out of eighty-five pregnancies. They can develop in two ways. **Dizygotic twins** (also called *fraternal twins*) are formed when two oocytes are ovulated at the same time and fertilized by two different sperm (fig. 9.23). These are the most common type of twins. **Monozygotic twins** (also called *identical twins*) are produced by the fertilization of a single oocyte by one sperm (fig. 9.24).

Dizygotic twins may be of the same sex or of different sexes, because they bear the same genetic relation to each other as do other siblings. They have two separate amniotic sacs and two chorions, which may become partially fused. Monozygotic twins are genetically identical to each other, because they are derived from the same zygote. Usually toward the end of the first week, the inner cell mass divides to form two embryonic primordia. Monozygotic twins have two amnions but only one chorion and a common placenta. The presence of multiple fetuses, as well as other features of the pregnancy, can be visualized by ultrasound imaging techniques (fig. 9.25).

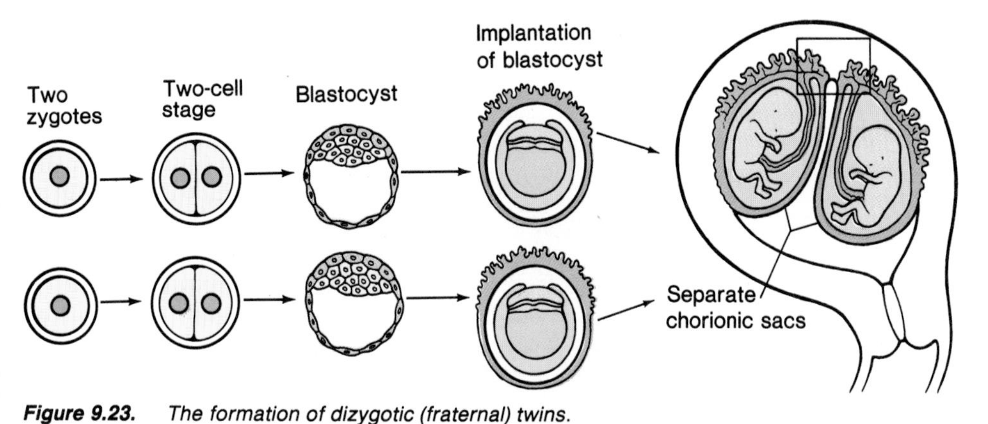

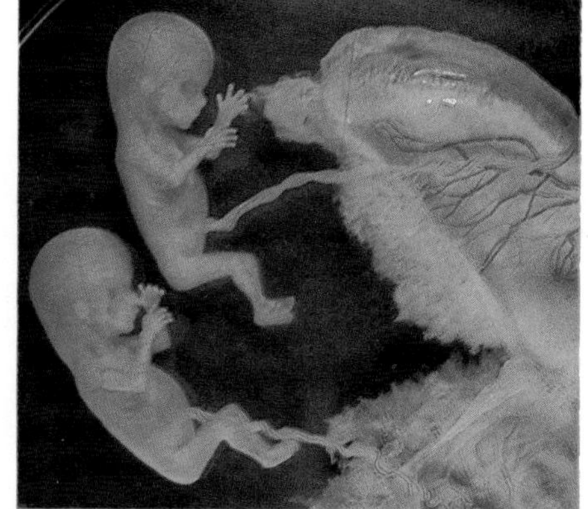

Figure 9.23. *The formation of dizygotic (fraternal) twins.*

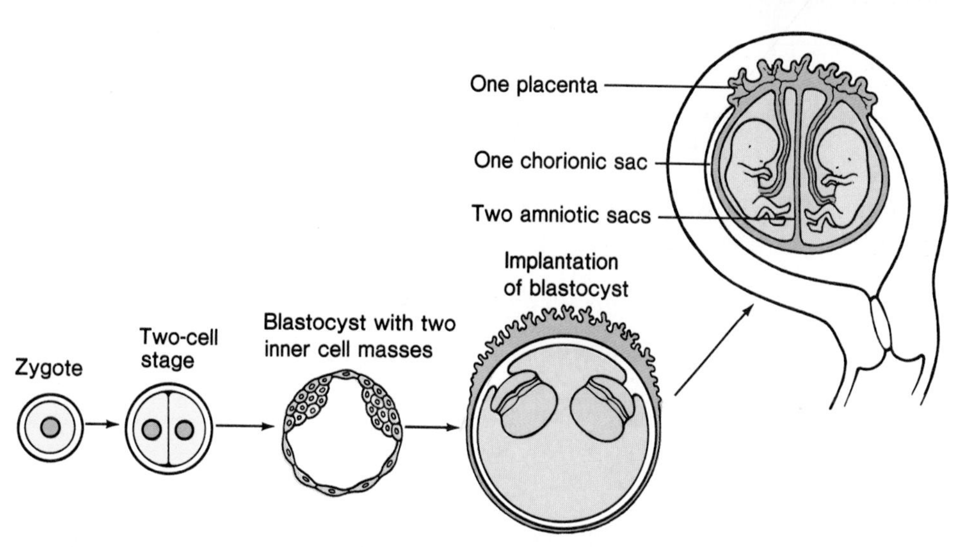

Figure 9.24. *The formation of monozygotic twins. Twins of this type develop from a single zygote and are identical. Such twins have two amnions but one chorion and a common placenta.*

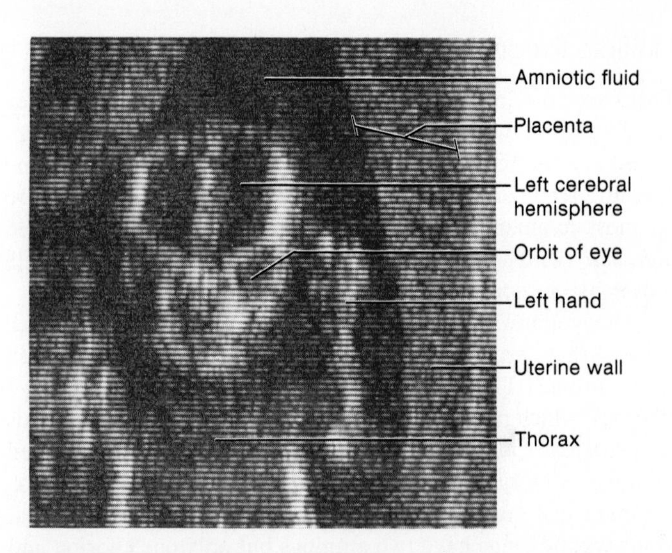

Figure 9.25. *Color-enhanced ultrasonogram of a fetus during the third trimester. The left hand is raised.*

Mammary Glands and Lactation

Each mammary gland is composed of fifteen to twenty lobes, which are divided by adipose tissue. The amount of adipose tissue determines the size and shape of the breast but has nothing to do with the ability of a woman to nurse. Each lobe contains the glandular **alveoli** (fig. 9.26) that secrete the milk of a lactating woman. The clustered alveoli secrete milk into a series of tubules which converge to form a **lactiferous duct** that drains at the tip of the nipple.

The changes that occur in the mammary glands during pregnancy and the regulation of lactation provide excellent examples of hormonal interactions and neuroendocrine regulation (table 9.3). Growth and development of the mammary glands during pregnancy requires the permissive actions of insulin, cortisol, and thyroid hormones. In the presence of adequate amounts of these hormones, high levels of estrogen stimulate the development of the mammary alveoli and progesterone stimulates proliferation of the tubules and ducts (fig. 9.27).

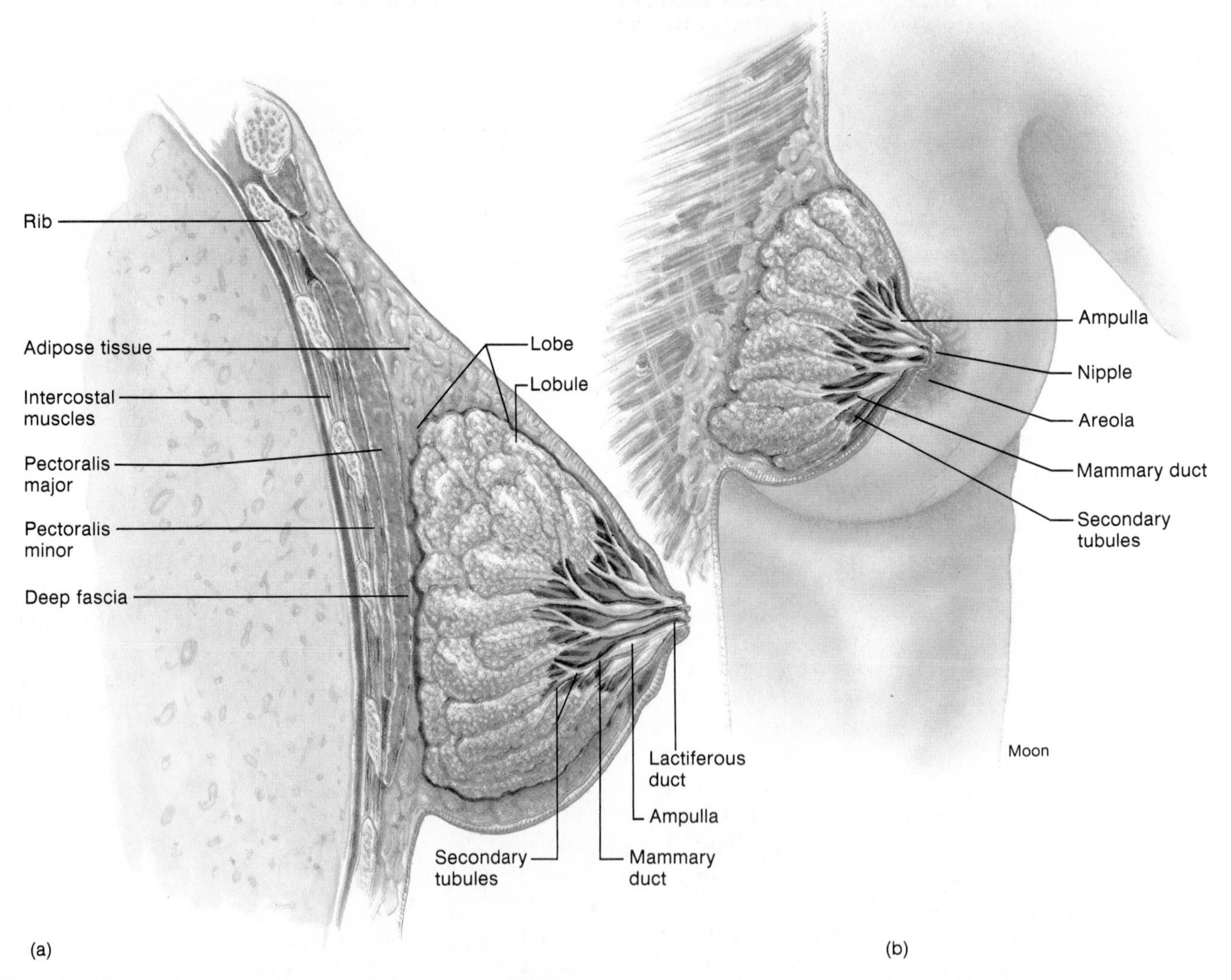

Figure 9.26. *The structure of the breast and mammary glands. (a) A saggital section; (b) an anterior view partially sectioned.*

The production of milk proteins is stimulated after parturition by **prolactin,** a hormone secreted by the anterior pituitary gland. The secretion of prolactin is controlled primarily by *prolactin-inhibiting hormone (PIH)* from the hypothalamus. The secretion of PIH is stimulated by the high levels of estrogen that are present during pregnancy. As a result, prolactin secretion is inhibited during pregnancy. In addition, high levels of estrogen act directly on the mammary glands to block their stimulation by prolactin. During pregnancy, consequently, the high levels of estrogen prepare the breasts for lactation but prevent prolactin secretion and action.

After parturition, when the placenta is eliminated, declining levels of estrogen are accompanied by an increase in the secretion of prolactin. Lactation, therefore, commences. If a woman does not wish to breast-feed her baby, she may be injected with a powerful synthetic estrogen (diethylstilbestrol, or DES), which inhibits further prolactin secretion.

The act of nursing helps to maintain high levels of prolactin secretion via a *neuroendocrine reflex* (fig. 9.28). Sensory endings in the breast, activated by the stimulus of suckling, relay impulses to the hypothalamus and inhibit the secretion of PIH. A proposed prolactin releasing hormone may also be secreted during nursing. Suckling thus results in the reflex secretion of high levels of prolactin, which promotes the secretion of milk from the alveoli into the ducts. In order for the baby to get the milk, however, the action of another hormone is needed.

The stimulus of suckling also results in the reflex secretion of **oxytocin** from the posterior pituitary. This hormone is produced in the hypothalamus and stored in the posterior pituitary; its secretion results in the **milk-ejection reflex,** or milk let-down. Oxytocin stimulates contraction of the lactiferous ducts, as well as of the uterus (this is why women who breast-feed regain uterine muscle tone faster than women who bottle-feed).

Table 9.3 Hormonal factors affecting lactation

Hormones	Major Source	Effects
Insulin, cortisol, thyroid hormones	Pancreas, adrenal cortex, and thyroid	Permissive effects—adequate amounts of these must be present for other hormones to exert their effects on mammary glands
Estrogen and progesterone	Placenta	Growth and development of secretory units (alveoli) and ducts in mammary glands
Prolactin	Anterior pituitary	Production of milk proteins, including casein and lactalbumin
Oxytocin	Posterior pituitary	Stimulates milk-ejection reflex

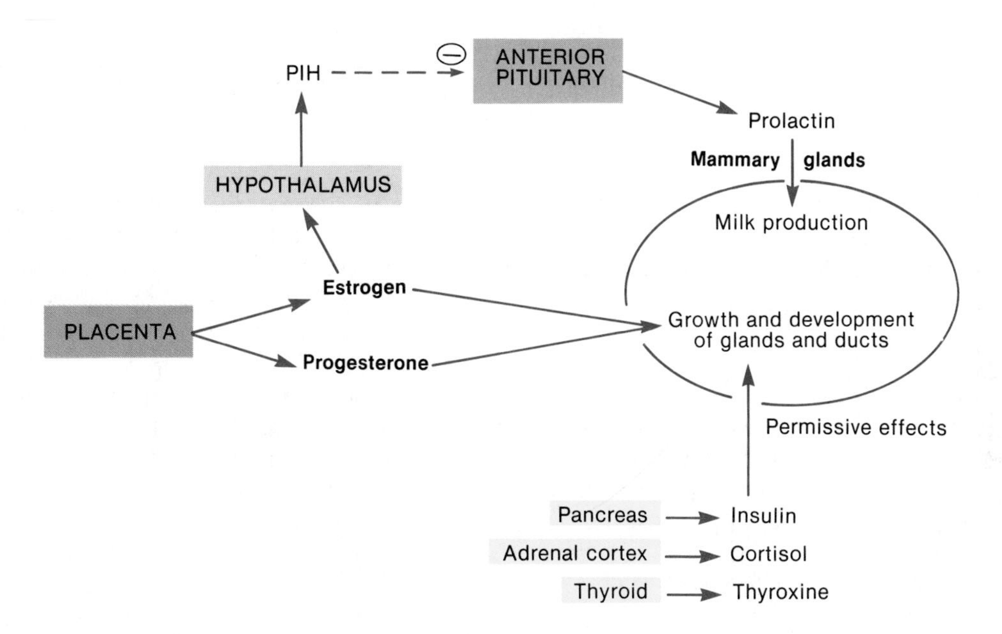

Figure 9.27. *The hormonal control of mammary gland development during pregnancy and lactation. Note that milk production is prevented during pregnancy by estrogen inhibition of prolactin secretion. This inhibition is accomplished by the stimulation of PIH (prolactin-inhibiting hormone) secretion from the hypothalamus.*

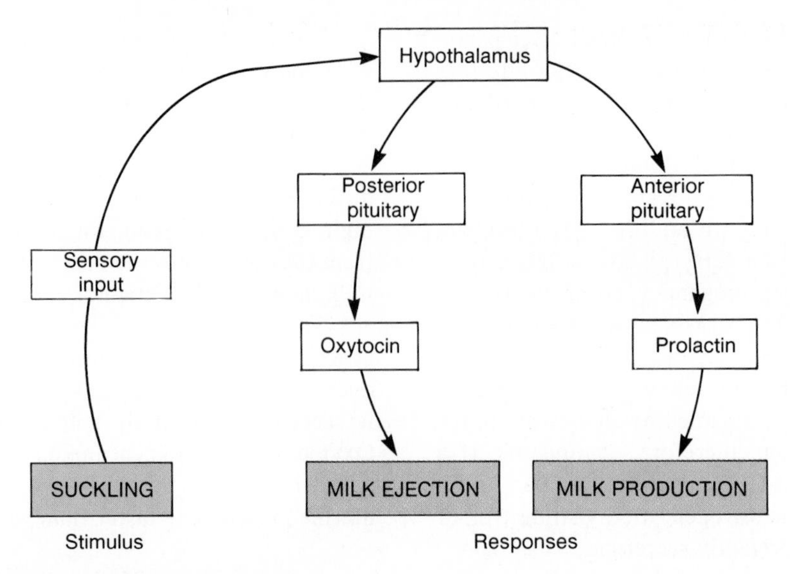

Figure 9.28. Lactation occurs in two stages: milk production (stimulated by prolactin) and milk ejection (stimulated by oxytocin). The stimulus of suckling triggers a neuroendocrine reflex that results in increased secretion of oxytocin and prolactin.

Breast-feeding, acting through reflex inhibition of GnRH secretion, can inhibit the secretion of FSH and LH from the mother's anterior pituitary and thus inhibit ovulation. Breast-feeding is thus a natural contraceptive mechanism that helps to space births. This mechanism appears to be most effective in women with limited caloric intake and those who breast-feed their babies at frequent intervals throughout the day and night. In the traditional societies of the less-industrialized nations, therefore, breast-feeding is an effective contraceptive. Breast-feeding has much less of a contraceptive effect in women who are well nourished, and who breast-feed their babies at more widely spaced intervals.

1. Describe some of the anatomical features of the fetus, and indicate at which week they appear.
2. Explain how contractions of the uterus are produced during labor.
3. List and describe the three stages of labor.
4. Distinguish between monozygotic and dizygotic twins.
5. Explain the hormonal control of lactation.

Postnatal Growth, Aging, and Senescence

During puberty, there is increased secretion of FSH and LH from the pituitary, and increased sex steroid secretion from the gonads. The increase in sex steroids produces the characteristic body changes that occur in adolescents. Age-related changes occur in the body systems during adulthood. The deterioration of function of these systems with advanced age is known as senescence.

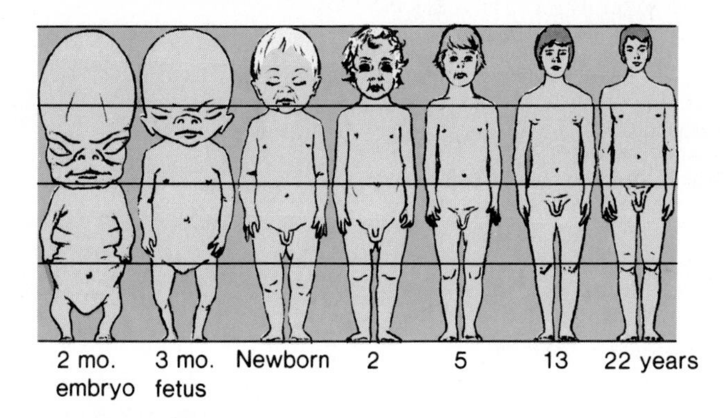

Figure 9.29. The relative proportions of the body from embryo to adult. The head of a newborn accounts for a quarter of the total body length, and the lower appendages make up about one-third. In an adult, the head accounts for about 13% of the total body length, whereas the length of the lower appendages constitutes approximately one-half.

From Human Embryology *by L. Patten (redrawn from Scammon). Copyright © 1933 McGraw-Hill, Inc.*

Human growth is not a steady, linear process. It varies with age, is subject to many individual differences, and is somewhat sex-dependent. The body of a newborn is 6%, and that of a four year old is 25%, of its eventual adult weight. Interestingly, the brain of a newborn is already 25% of its eventual adult weight, and this increases to 90% by the age of four. The relative proportions of the body thus change during growth, as indicated in figure 9.29.

Childhood

Childhood is the period of growth and development between infancy[15] and adolescence, at which time puberty begins. The average weight gain during childhood is 3 to 3.5 kg (7 lb) per

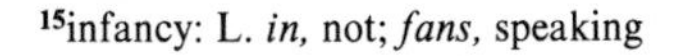

[15]infancy: L. *in,* not; *fans,* speaking

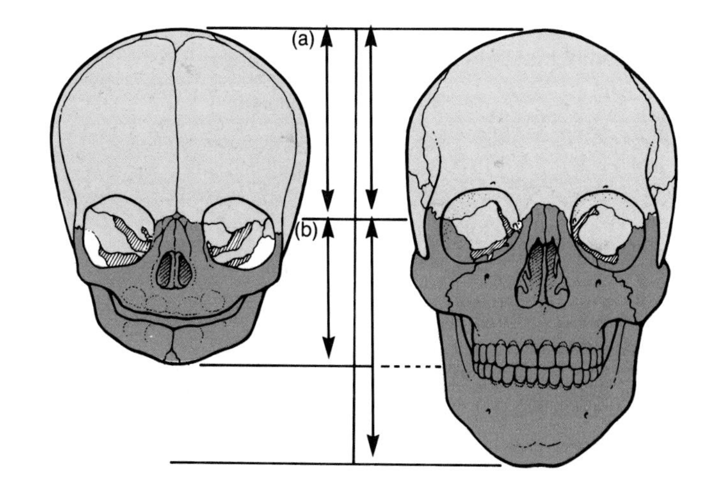

Figure 9.30. Growth of the skull. The height of the cranial vault (distance between planes a and b) is drawn the same in both the infant and adult skulls. Growth of the skull occurs almost exclusively within the bones of the facial region.

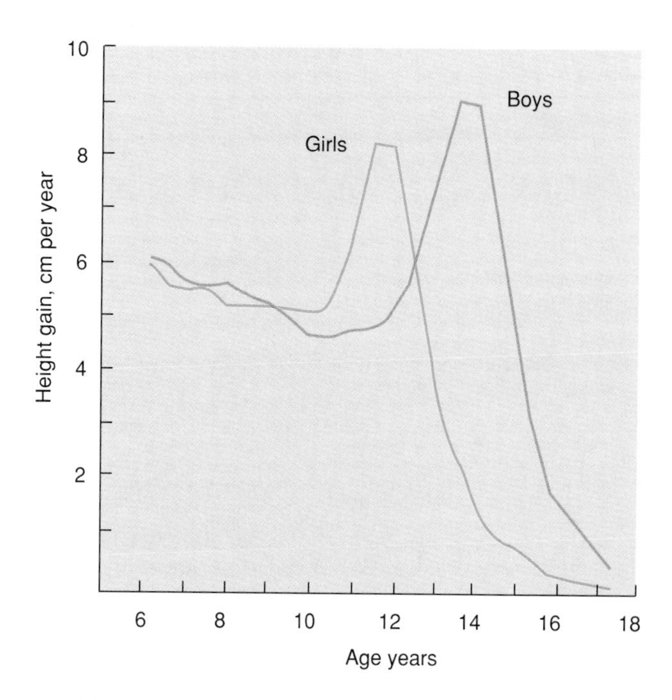

Figure 9.31. The adolescent growth spurt. Notice that the growth spurt for girls occurs approximately two years earlier than for boys.

Table 9.4 Development of secondary sexual characteristics and other changes that occur during puberty in girls

Characteristic	Age of First Appearance	Hormonal Stimulation
Appearance of breast bud	8–13	Estrogen, progesterone, growth hormone, thyroxine, insulin, cortisol
Pubic hair	8–14	Adrenal androgens
Menarche (first menstrual flow)	10–16	Estrogen and progesterone
Axillary (underarm) hair	About two years after the appearance of pubic hair	Adrenal androgens
Eccrine sweat glands and sebaceous glands; acne (from blocked sebaceous glands)	About the same time as axillary hair growth	Adrenal androgens

year. The circumference of the head increases by only about 3 to 4 cm (1.5 in.) during childhood, so that by adolescence the head and brain are virtually adult size.

The facial bones continue to develop during childhood (fig. 9.30). The first permanent teeth generally erupt during the seventh year, and then the deciduous (baby) teeth are shed in approximately the same sequence in which they were acquired. Deciduous teeth are replaced at a rate of about four per year over the next seven years.

Adolescence

Adolescence[16] is the period of rapid growth and development (fig. 9.31) between childhood and adulthood. It begins about the age of ten years in girls and the age of twelve years in boys. The end of adolescence is frequently said to be the age of twenty years, but it is not clearly delineated and varies with the developmental, physical, emotional, mental, or cultural criteria that define an adult.

The Onset of Puberty[17]

Secretion of FSH and LH is high in the newborn, but falls to very low levels a few weeks after birth. Gonadotropin secretion remains low until the beginning of puberty, which is marked by rising levels of FSH followed by LH secretion.

The increased gonadotropin secretion during puberty stimulates a rise in sex hormone secretion from the gonads. Increased secretion of testosterone from the testes and of estradiol from the ovaries during puberty in turn produces changes in body appearance characteristic of the two sexes. Such **secondary sexual characteristics** (tables 9.4 and 9.5) are the physical manifestations of the hormonal changes occurring during puberty.

[16]adolescence: L. *adolescere,* to grow up
[17]puberty: L. *pubertas,* adult form

Table 9.5 Development of secondary sexual characteristics and other changes that occur during puberty in boys

Characteristic	Age of First Appearance	Hormonal Stimulation
Growth of testes	10–14	Testosterone, FSH, growth hormone
Pubic hair	10–15	Testosterone
Body growth	11–16	Testosterone, growth hormone
Growth of penis	11–15	Testosterone
Growth of larynx (voice lowers)	Same time as growth of penis	Testosterone
Facial and axillary (underarm) hair	About two years after the appearance of pubic hair	Testosterone
Eccrine sweat glands and sebaceous glands; acne (from blocked sebaceous glands)	About the same time as facial and axillary hair growth	Testosterone

It has been observed that the age of onset of puberty is related to the amount of body fat and level of physical activity of the child. Ballerinas, and other girls who are very physically active, reach *menarche*—the age when menstrual flow first occurs—at a later average age (15 years old) than in the general population (12.6 years old). This appears to be due to a requirement for a minimum percentage of body fat for menstruation to begin, and may represent a mechanism favored by natural selection for the ability to successfully complete a pregnancy and nurse the baby. Later in life, women who are very lean and physically active may have irregular cycles and *amenorrhea* (cessation of menstruation). This may also be related to the percentage of body fat. In addition, there is evidence that physical exercise may, through activation of neural pathways involving endorphin neurotransmitters (chapter 6), act to inhibit FSH and LH secretion.

Adulthood

Adulthood is the period of life beyond adolescence. An adult has reached maximum physical stature as determined by genetic, nutritional, and environmental factors. Although skeletal maturity is reached in early adulthood, anatomical and physiological changes continue throughout adulthood and are part of the aging process. Changes that occur during aging in the adult are described in the next section.

Aging and Senescence

Aging is a term than has different definitions. *Chronological aging* is simply an indication of the time elapsed from birth. *Biological aging,* which begins with conception and ends with death, is the process of biological changes that occur with time in the life of an individual. These changes permit humans to be at peak performance as they enter early adulthood, but result in a gradual reduction in capabilities from midadulthood onwards.

Senescence[18] refers to the biological aging process, characterized by a gradual deterioration of body structure and function. As senescence progresses through adulthood and old age, the person becomes more vulnerable to diseases. This is a result of characteristic changes in the different body systems with advanced age.

Integumentary System

The skin becomes thin, dry, and inelastic with senescence. Collagen fibers in the dermis become thicker and less elastic, and the amount of adipose tissue in the hypodermis decreases, making it thinner. These changes cause wrinkling of the skin (fig. 9.32). Other changes in the organs of the skin also occur. There is loss of hair on the scalp and on the extremities, a reduction in sweat gland activity, and decreased sebum (oil) production from sebaceous glands. Since elderly people cannot perspire as freely, they are more likely to complain of heat and more likely to suffer heat exhaustion. They also become more sensitive to cold as a result of the loss of adipose tissue.

Skeletal and Muscular Systems

Senescence affects the skeletal system by decreasing skeletal mass and density and increasing porosity and erosion (fig. 9.33). Bones become more brittle and subject to fracture. Articulating surfaces (those at joints) also deteriorate, contributing to *arthritis* (inflammation of the joints). *Osteoporosis* is the most prevalent bone disorder in older people, and is characterized by decreased bone mass and density. Women after menopause are most susceptible to this condition and because of it frequently sustain fractures of the hip, vertebrae, or wrist. The reason for the increased susceptibility is that women start out with a lower bone density than men, and the withdrawal of estrogen at menopause contributes to the development of osteoporosis.

Although the aged experience a general decrease in the strength and endurance of skeletal muscles, the extent of these changes varies considerably among people. Exercise is especially beneficial in retarding these changes as a person approaches old age. Exercise not only strengthens muscles, it also contributes to a healthier circulation. If an elderly person does not maintain muscular strength through exercise, he or she will be more prone to debility and inactivity, thus hastening the muscle atrophy that can occur.

[18]senescence: L. *senis,* old

Figure 9.32. *Senescence of the skin results in a loss of elasticity and the appearance of wrinkles.*

Photograph of child from The 1974 Science Year, *© 1973 Field Enterprises Educational Corporation. By permission of World Book, Inc.*

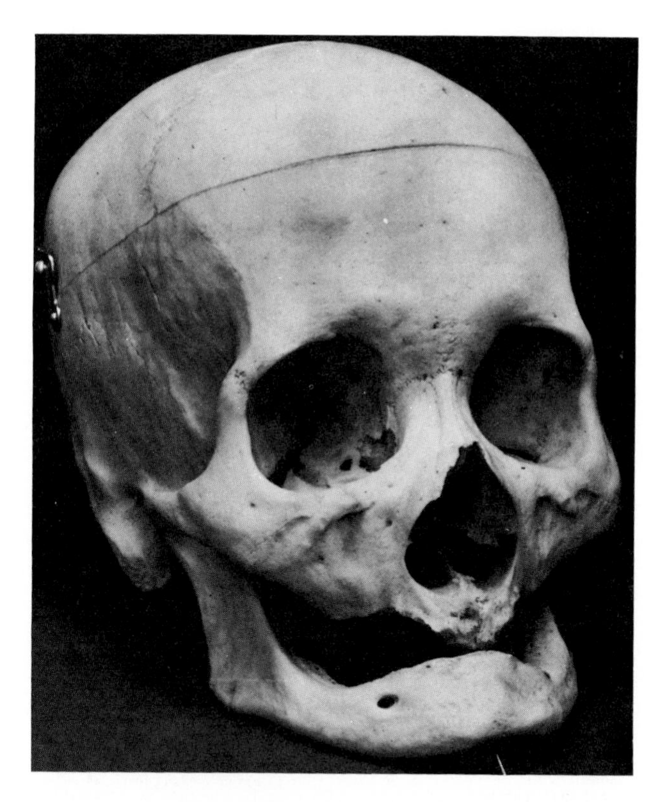

Figure 9.33. *The geriatric skull. Note the loss of teeth and the degeneration of bone, particularly in the facial region.*

Nervous and Sensory Systems

It was once common to hear that people lose 100,000 neurons each day. Recent studies, however, show that such claims are unfounded. The brain does shrink with age, as can be seen in computerized tomography (CT) and magnetic resonance imaging (MRI), but the total number of neurons appears to be the same in elderly people with normal mental ability. Although few neurons are lost as a result of the aging process itself, neurons may be killed by drugs and by interruptions of their vascular supply (as a result of stroke or heart disease). Age-related changes in neurotransmitter function may occur and are partly responsible for such conditions as depression and Alzheimer's disease. For reasons that are not well understood, there is a marked slowing of reaction time in the senescent nervous system.

Most elderly persons develop lipid infiltrates of the cornea of the eyes (chapter 20) in a condition called *arcus senilis.* By the age of sixty-five, 40% of males and 60% of females have vision poorer than 20/70. The pupil of an elderly person's eye cannot dilate fully, so that the amount of light reaching the retina (the receptive part of the eye) may be reduced to 50% of that in youth. The lens loses elasticity (a condition called *presbyopia*), so that a printed page must be held farther from the eye in order for it to be in focus. Bifocals are frequently used to correct this problem. *Cataracts* are the major cause of visual disability among the elderly. In this condition, the normally transparent cornea becomes opaque.

Cardiovascular and Respiratory Systems

Changes in blood vessels include hardening of the arteries (*arteriosclerosis*) and the development of fatty, plaquelike areas in the walls of arteries. The latter condition (described more fully in chapter 12) is called *atherosclerosis* and is responsible for most cases of heart disease and stroke. The maximum blood flow through the coronary arteries (which supply the heart) at the age of sixty is about 35% lower than at the age of thirty. Blood pressure measurements, in contrast, generally increase with age.

After the age of twenty-five, there is a gradual loss of lung elasticity so that by the age of seventy the vital capacity has decreased by about 40% (the vital capacity is the maximum amount of air that can be exhaled after a maximum inhalation). These changes are accompanied by others that decrease the efficiency of the lungs in oxygenating the blood.

With age, however, the oxygen requirements of the body decrease. This is because the *basal metabolic rate (BMR)*—which is the minimum amount of energy that a person consumes in a resting state—declines with age. Young adults with a high BMR may be able to eat almost anything and not get fat, whereas middleaged people with a lower BMR must constantly control food intake to avoid obesity.

Table 9.6 Summary of aging in body systems

Organ System	Principal Senescent Changes
Integumentary system	Degenerative change in collagenous and elastic fibers in dermis; decreased production of pigment in skin and hair follicles; reduced activity of sweat and sebaceous glands Skin tends to become thinner, wrinkled, and dry with pigment spots; hair becomes gray and then white
Skeletal system	Degenerative loss of bone matrix; deteriorating articulations Bones become thinner, more brittle, and more likely to fracture; stature may shorten due to compression of intervertebral discs; susceptibility to joint diseases increases
Muscular system	Loss of skeletal muscle mass; degenerative changes in neuromuscular junctions Loss of muscular strength and motor response
Nervous system	Degenerative changes in neurons; loss of dendrites and synapses; decrease in sensory sensitivity Decreased efficiency in processing and recalling information; decreased ability to communicate; diminished senses of smell, taste, sight, hearing, and touch
Endocrine system	Slightly reduced hormonal secretions Decreased responsiveness to hormones; decreased metabolic rate; reduced ability to maintain homeostasis
Circulatory system	Degenerative changes in cardiac muscle; decreased diameters of lumina of arteries and arterioles; decreased efficiency of immune system Decreased cardiac output; increased resistance to blood flow; increased blood pressure; increased incidence of autoimmune diseases
Respiratory system	Degenerative loss of elastic fibers in lungs; reduced number of functional alveoli Reduced vital capacity; increased dead air space; reduced ability to clear airways by coughing
Digestive system	Decreased motility in GI tract; reduced secretion of digestive enzymes; increased occurrence of periodontal disease Reduced efficiency of digestion
Urinary system	Degenerative changes in kidneys; reduced number of functional nephrons Reduced filtration rate, tubular secretion, and reabsorption
Reproductive system	
Male	Reduced secretion of sex hormones; reduced production of spermatozoa; enlargement of prostate gland Decreased sexual capabilities
Female	Degenerative changes in ovaries; decreased secretion of sex hormones Menopause; regressed secondary sex organs and characteristics

Reproductive System

The senescent changes in the reproductive system in both males and females are generally more important psychologically than physiologically. A decrease in sex hormones in elderly persons may decrease sexual desire and abilities. The testes in a male get smaller and the prostate hypertrophies, which may make urination difficult. Females experience major changes at menopause, when menstruation and ovulation cease.

A summary of the principal changes associated with senescence is provided in table 9.6.

Life Span and Death

Life expectancy is defined as the average number of years lived by people born at a particular date in a particular location. The life expectancy is much longer today than in previous ages. This is primarily due to the fact that medical, public health, and nutritional advances have decreased the death rate among infants and children. The *life span* is the average length of life in a particular culture. The life span in the United States has increased in the past century from forty-seven to seventy-three years, although this varies between the sexes. The life span of males and females is presently seventy years and seventy-six years, respectively. Many scientists believe that humans have a potential life span of between 120 and 150 years if diseases and accidents could be prevented.

Death, in ages past, was determined by cessation of the heartbeat. This definition of death is no longer appropriate, however, because physicians and others can often start a heart that has stopped beating, and surgeons purposely stop the heartbeat during open heart surgery. Human hearts have even been removed and replaced with artificial ones or hearts from people who had recently died. A person with someone else's heart (or kidney, or cornea) is still the same person. Death of that person can only occur when one organ dies—the brain.

Everyone dies for the same reason: the brain has stopped functioning and cannot be revived. All other causes to which death is often attributed (heart disease, cancer, accidents, etc.) must ultimately cause brain death in order for the person to die. This has frequently resulted in very difficult ethical and legal questions concerning the maintenance of life-support technology in people whose bodies are still functional but whose brain is largely destroyed. These questions become even thornier because brain-dead individuals (or babies born *anencephalic*—essentially without a brain) could provide organs for transplantation into people whose lives depend on such transplants.

1. Describe the endocrine events responsible for the changes that occur at puberty, and list some of these changes.
2. At what age does adulthood begin? How is this defined?
3. Describe some of the changes that occur in the organ systems with senescence.

Summary

Fertilization and Pre-Embryonic Development p. 186

I. Fertilization normally occurs in the fallopian tube.
   A. In the acrosome reaction, digestive enzymes on the head of a sperm are exposed that enable it to reach the ovum.
   B. Once the sperm has penetrated a secondary oocyte, the oocyte is stimulated to complete meiotic division.

II. Fertilization produces a diploid zygote.
   A. The zygote divides by mitosis in a process known as cleavage.
   B. A hollow structure called a blastocyst is produced which contains an inner cell mass, which will become the embryo, and a layer of cells that will form the chorion of the placenta.

III. The blastocyst implants into the endometrium of the uterus.
   A. The chorion secretes a hormone called human chorionic gonadotropin (hCG).
   B. hCG stimulates the mother's corpus luteum to continue its secretion of estrogen and progesterone, thus preventing menstruation from occurring.

IV. Three embryonic tissue layers, called primary germ layers, are formed.
   A. These are the ectoderm, mesoderm, and endoderm.
   B. The germ layers will form the adult tissues and organs.

Embryonic Development p. 193

I. The type of gonads an embryo will develop—testes or ovaries—depends on the chromosomal sex.
   A. A female embryo has the genotype XX; since she lacks a Y chromosome, the gonads will become ovaries.
   B. A male embryo has the genotype XY; the Y chromosome causes testes to develop.
   C. Embryonic testes secrete testosterone, which causes the male sex accessory organs and external genitalia to develop.
   D. Female embryos lack testes and testosterone; in their absence, female sex accessory organs and external genitalia develop.

II. The embryonic stage of development lasts until the eighth week, at which time fetal development begins.

III. In addition to forming the body of the embryo, embryonic tissue forms various membranes outside the body of the embryo.
   A. Chief among these extraembryonic membranes are the amnion, which forms the amniotic sac around the embryo, and the chorion.
   B. The part of the chorion in contact with the endometrium becomes part of the placenta.
   C. The placenta includes the chorion and the decidua basalis, which is formed from the endometrium.

IV. The placenta serves to exchange nutrients, gases, and waste products between the fetal and maternal blood.

V. The placenta also secretes a variety of hormones; some of these are similiar to those secreted by a pituitary gland, and others are steroids similar to those secreted by an ovary.

Fetal Development and Parturition p. 202

I. The fetus continues to grow and the body systems continue their development from the eighth to the thirty-eighth week, at which time it is considered full-term.
   A. Labor involves powerful contractions of the uterus, which are stimulated by the posterior pituitary hormone oxytocin and by prostaglandins produced in the uterus.
   B. Labor is divided into three stages: dilation, expulsion, and placental.

II. Multiple pregnancy refers to the development of twins.
   A. Dizygotic twins develop from two different zygotes; they are also known as fraternal twins.
   B. Monozygotic twins develop from one zygtote; they are genetically identical and are thus called identical twins.

III. Mammary glands grow and develop during pregnancy.
   A. After birth of the child, prolactin from the anterior pituitary stimulates the production of milk.
   B. The stimulus of suckling causes reflex secretion of prolactin and oxytocin; oxytocin stimulates milk-ejection as well as contractions of the uterus.

Postnatal Growth, Aging, and Senescence p. 208

I. Childhood is the period of growth and development between infancy and adolescence.

II. Adolescence is the period between childhood and adulthood.
   A. Puberty occurs during adolescence, at which time the gonads secrete increasing amounts of sex steroids in response to increases in gonadotropin secretion.
   B. The rising secretion of sex steroids from the gonads causes the development of secondary sexual characteristics.

III. Adulthood is the period of life beyond adolescence, during which time senescence occurs.
   A. Senescence refers to the deterioration of body structure and function during aging in an adult.
   B. The death of an individual occurs when the brain ceases to function.

Review Activites

Objective Questions

1. A person with the genotype XO has
   (a) ovaries
   (b) testes
   (c) both ovaries and testes
   (d) neither ovaries nor testes
2. An embryo with the genotype XX develops female sex accessory organs because of
   (a) androgens
   (b) estrogens
   (c) absence of androgens
   (d) absence of estrogens
3. The corpus luteum is maintained for the first ten weeks of pregnancy by
   (a) hCG
   (b) LH
   (c) estrogen
   (d) progesterone
4. Fertilization normally occurs in
   (a) the ovaries
   (b) the fallopian tubes
   (c) the uterus
   (d) the vagina
5. The placenta is formed from
   (a) the fetal chorion frondosum
   (b) the maternal decidua basalis
   (c) both of the above
   (d) neither of the above
6. Uterine contractions are stimulated by
   (a) oxytocin
   (b) prostaglandins
   (c) prolactin
   (d) both *a* and *b*
   (e) both *b* and *c*
7. Contraction of the mammary glands and ducts during the milk-ejection reflex is stimulated by
   (a) prolactin
   (b) oxytocin
   (c) estrogen
   (d) progesterone

8. The pre-embryonic stage is completed when the
   (a) blastocyst implants
   (b) placenta forms
   (c) blastocyst reaches the uterus
   (d) primary germ layers form
9. The decidua basalis is
   (a) a component of the umbilical cord
   (b) the embryonic portion of the placenta
   (c) the maternal portion of the placenta
   (d) all of the above
10. During which week following conception does the embryonic heart begin pumping blood?
    (a) fourth
    (b) fifth
    (c) sixth
    (d) eighth
11. Menarche is the
    (a) first menstruation in a girl
    (b) time when menstruation ceases in a woman
    (c) first sperm produced in a boy
    (d) time when pubertal hair appears
12. Senescence refers to the
    (a) improvement of body function that occurs during puberty
    (b) change from an adolescent to an adult
    (c) deterioration of body function with age in an adult
    (d) death of an individual

Essay Questions

1. Explain the meaning of the term *cleavage* in regard to embryonic development, and use this discussion to explain why only one ovum forms from the meiosis of an oogonium.
2. During which days of a female cycle can fertilization occur? Explain why this is true.
3. Describe the structure of the placenta, and explain how this structure allows the exchange of molecules between the mother and the fetus.
4. Explain the physiological significance of hCG, and describe the rationale behind pregnancy testing.
5. Describe the endocrine changes that occur during puberty.
6. Explain why abstinence from alcohol and other drugs is important during pregnancy and during the time a mother is breast-feeding her baby.

10

Gene Action and Genetic Engineering

Outline

Objectives

By studying this chapter, you should be able to

1. describe the structure of DNA, and explain the law of complementary base pairing
2. explain the semiconservative replication of DNA
3. describe the structure of RNA, and identify the different forms of RNA
4. explain how RNA is produced, and the relationship between messenger RNA and a gene
5. identify the nature and significance of codons and anticodons
6. describe the function of the rough endoplasmic reticulum and of the Golgi apparatus
7. identify the components of a virus, and explain how a virus is able to reproduce
8. describe the origin, structure, and significance of plasmids, and explain the function of restriction endonuclease enzymes
9. explain how recombinant DNA is produced in a plasmid, and how it is cloned
10. describe some of the medical uses of genetic engineering
11. describe, in general, how DNA fingerprinting is accomplished, and the uses of this technique
12. describe some of the applications of genetic engineering in agriculture and other industries

Keys to Pronunciation

deoxyribose: *de-ok″se-ri′bos*
endonuclease: *en″do-nu′kle-as*
endoplasmic reticulum: *en″do-plas′mic re-tik′u-lum*
erythropoietin: *e-rith″ro-poi′e-tin*
Escherichia: esh″er-i′ke-a
euchromatin: *u′kro′mah-tin*
eukaryote: *u-kar′e-ot*
Golgi: *gol′je*
hypercholesteremia: *hi″per-ko-les″ter-e′me-ah*
interferon: *in″ter-fer′on*
nucleoli: *nu-kle′o-li*
plasmid: *plaz′mid*
polymerase: *pol″ah-mer-az″*
polymorphism: *pol″e-mor′fizm*
prokaryote: *pro-kar′e-ot*
Pseudomonas: *soo″do-mo′nas*
ribonucleotide: *ri″bo-nu′kle-o-tid*
ribosome: *ri′bo-som*
uracil: *yur′ah-sil*

Photo: Tree clones grown in a test tube illustrate one application of genetic engineering techniques.

The Nucleus and Nucleic Acids

The nucleus contains threads of chromatin, which is a combination of DNA and protein. DNA and RNA are composed of subunits called nucleotides, and together these molecules are known as nucleic acids. DNA is a double-stranded molecule, in which each strand consists of a sugar-phosphate backbone from which units called nitrogenous bases project. The two DNA strands are held together by weak hydrogens bonds that join specific bases in each strand together.

The human life cycle, discussed in chapters 8 and 9, is largely the result of genetic programming. Yet most of the information previously introduced was discovered prior to the time when genetic function was understood. This was also the case with the discovery of the basic laws of evolution by Charles Darwin in 1869 (chapter 2), and the basic laws of inheritance discovered by Gregor Mendel in 1866 (chapter 11). DNA was discovered in 1869 and localized to the chromosomes in 1914, but this information was, at that time, considered unimportant.

As a result of a number of experiments carried out in the 1930s and 1940s, the role of DNA as the "genetic material" was first suspected and then confirmed. Biochemists and physicists then competed to determine the structure of DNA, in the hopes that knowledge about this structure would explain the known behavior of genes. This intense competition among scientists was won by James Watson and Francis Crick, who—in a brief paper published in 1953—successfully described the structure of DNA in such a way that would explain how it could replicate during cell division and provide the basis for the genetic code.

The Watson-Crick model of DNA launched the birth of *molecular genetics,* which is the field of science devoted to the study of gene action at the molecular and cellular levels. In the short time that has elapsed since the birth of this field, there has been a knowledge explosion of awesome magnitude which now permeates all areas of biology and medicine. Although most laymen are more familiar with the revolutionary changes associated with the invention of the microchip in electronics (computers, fax machines, etc.), the revolution in molecular biology is arguably of even greater magnitude. (These two revolutionary areas actually interact; modern molecular biology utilizes computers, and some computer scientists are using biological models of information storage to design computers of the future.)

As a result of the molecular biology revolution, biology is now big business. Advances in this area have already produced significant benefits in medicine, and promise to provide enormous benefits in agriculture and other areas of human endeavor. Molecular genetics, however, is still an extremely young science, and most of its potential is not only unrealized but probably unimagined at the present time. Some of the areas of potential benefits will be discussed later in this chapter, after the basic concepts have been introduced.

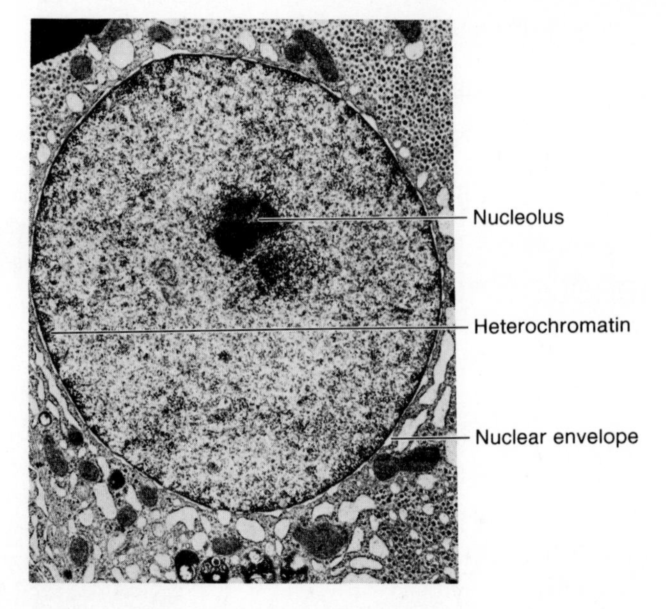

Figure 10.1. *The nucleus of a liver cell, showing the nuclear envelope, heterochromatin, and nucleolus.*

The Cell Nucleus

Most cells in the body have a single nucleus, although some—such as skeletal muscle cells—have many. The nucleus is surrounded by a *nuclear envelope* composed of an inner and an outer membrane; these two membranes fuse together to form thin sacs with openings called *nuclear pores* (fig. 10.1). These pores allow molecules of RNA to exit the nucleus (where they are formed) and enter the cytoplasm but prevent DNA from leaving the nucleus.

Many granulated threads in the nuclear fluid can be seen with an electron microscope. These threads are called *chromatin* and consist of a combination of DNA and protein. There are two forms of chromatin. Thin, extended chromatin—or *euchromatin*—appears to be the active form of DNA in a nondividing cell. Regions of condensed, "blotchy"-appearing chromatin, known as *heterochromatin,*[1] are believed to contain inactive DNA.

One or more dark areas within each nucleus can be seen. These regions, which are not surrounded by membranes, are called *nucleoli.* The DNA within nucleoli contains genes that code for the production of ribosomal RNA (rRNA), an essential component of ribosomes. Ribosomes are organelles involved in protein synthesis, to be discussed in a later section.

[1]chromatin: Gk. *chroma,* color

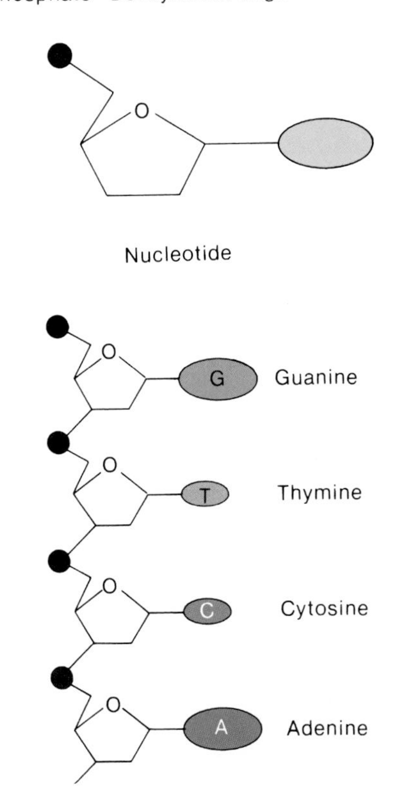

Figure 10.2. *The general structure of a nucleotide and the formation of sugar-phosphate bonds between nucleotides to form a nucleotide chain.*

Deoxyribonucleic Acid (DNA)

Nucleic acids include the macromolecules of **DNA** and **RNA,** which are critically important in genetic regulation, and the subunits from which these molecules are formed. These subunits are known as *nucleotides*. The structure of DNA will be described in this section; the structure of RNA will be described later in relation to its function in genetic expression.

Nucleotides are used as subunits in the formation of long polynucleotide chains. The nucleotides themselves, however, are composed of subunits. Each nucleotide is composed of three parts—a five-carbon sugar, a phosphate group bonded to one end of the sugar, and a *nitrogenous base* bonded to the other end of the sugar (fig. 10.2). The nucleotide bases are cyclic nitrogen-containing molecules with either one or two rings of carbons.

The structure of DNA serves as the basis for the genetic code. One might, therefore, expect DNA to have an extremely complex structure. Actually, although DNA is the largest molecule in the cell, it has a simpler structure than that of most proteins (described in chapter 3). This simplicity of structure deceived some of the early scientists into believing that the protein content of chromosomes, rather than their DNA content, provided the basis for the genetic code.

Sugar molecules in the nucleotides of DNA are a type called **deoxyribose** (hence the name for this nucleic acid). Each deoxyribose sugar can be bonded to one of four possible bases. These bases are **adenine, guanine, cytosine,** and **thymine.** There are thus four different types of nucleotides that can be used to produce the long DNA chains.

When nucleotides combine to form a chain, the phosphate group of one bonds with the deoxyribose sugar of another nucleotide. This forms a sugar-phosphate chain. Since the bases are attached to the sugar molecules, the sugar-phosphate chain looks like a "backbone" from which the bases project. Each of these bases can form a weak type of bond, called a *hydrogen bond,* with another base. These other bases are in turn part of a different chain of nucleotides. Such hydrogen bonding between bases thus produces a *double-stranded* DNA molecule; the two strands are like a staircase, with the paired bases as steps (fig. 10.3).

Actually, the two chains of DNA twist about each other to form a **double helix**—the molecule is like a spiral staircase (fig. 10.4). It has been shown that the number of adenine and guanine bases together in DNA is equal to the number of cytosine and thymine bases. The reason for this is explained by the **law of complementary base pairing**; adenine can only pair with thymine (through two hydrogen bonds), whereas guanine can only pair with cytosine (through three hydrogen bonds). Knowing this rule, we could predict the base sequence of one DNA strand if we knew the sequence of bases in the complementary strand.

Although we can predict which base is opposite a given base in DNA, we cannot predict which bases occur before or after that position within a single polynucleotide chain. Although there are only four bases, the number of possible base sequences along a stretch of several thousand nucleotides (the length of a gene) is almost infinite. Despite the almost infinite possible variety of sequences, almost all of the billions of copies of a particular gene in a person are identical. The mechanisms by which this is achieved will be described in the next section.

---

1. Describe the composition of a nucleotide, and indicate how many different DNA nucleotides are present in a cell.
2. Explain how the nucleotides are arranged into a polynucleotide chain.
3. Explain the law of complementary base pairing, and use this law to explain how the structure of one strand of DNA relates to the structure of the other strand in the molecule.
4. Explain why the structure of DNA is often called a double helix. How are the two strands of DNA held together?

Figure 10.3. The four nitrogenous bases in deoxyribonucleic acid (DNA). Notice that hydrogen bonds can form between guanine and cytosine and between thymine and adenine.

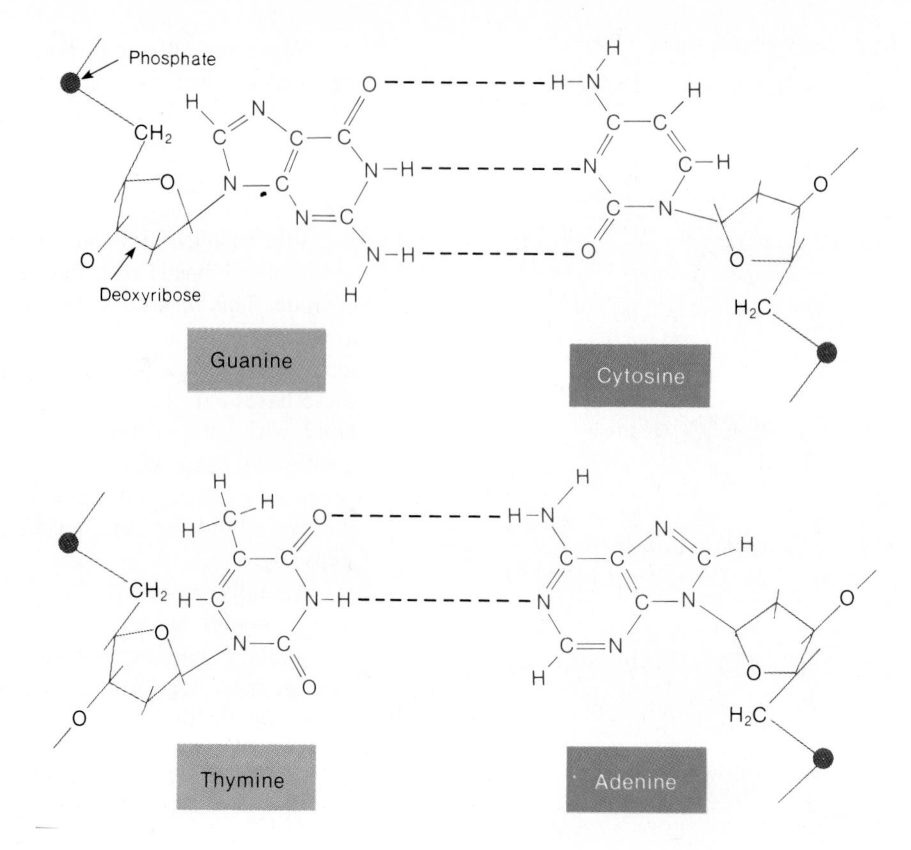

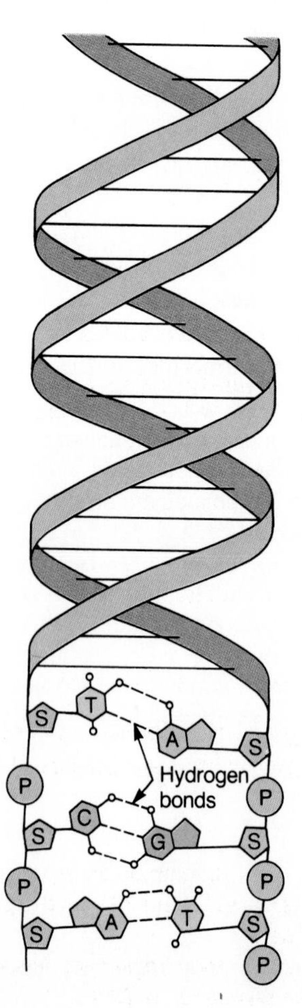

Figure 10.4. *The double helix structure of DNA.*

DNA Synthesis and Cell Division

When a cell is going to divide, each strand of the DNA within its nucleus acts as a template for the formation of a new complementary strand. Because of the law of complementary base pairing, each of the two new double-stranded DNA molecules formed is identical to the original molecule.

Genetic information is required for the life of the cell and for the ability of the cell to perform its functions in the body. Each cell obtains this genetic information from its parent cell through the process of DNA replication and cell division. DNA is the only type of molecule in the body capable of replicating itself, and mechanisms exist within the dividing cell to ensure that the duplicate copies of DNA are properly distributed to the daughter cells.

When a cell is going to divide, each DNA molecule replicates itself, and each of the identical DNA copies thus produced is distributed to the two daughter cells. Replication of DNA requires the action of a specific enzyme known as *DNA polymerase.* This enzyme moves along the DNA molecule, breaking the weak hydrogen bonds between complementary bases as it travels. As a result, the bases of each of the two DNA strands become free to bond to new complementary bases (which are part of nucleotides) that are available within the surrounding environment.

Because of the rules of complementary base pairing, the bases of each original strand will bond to the appropriate free nucleotides: adenine bases pair with thymine-containing nucleotides; guanine bases pair with cytosine-containing nucleotides, and so on. In this way, two new molecules of DNA, each

Figure 10.5. *The replication of DNA. Each new double helix is composed of one old and one new strand. The base sequences of each of the new molecules are identical to that of the parent DNA because of complementary base pairing.*

containing two complementary strands, are formed. The DNA polymerase enzyme links the phosphate groups and deoxyribose sugar groups together to form a second polynucleotide chain in each DNA that is complementary to the first DNA strands. Thus two new double-helix DNA molecules are produced that contain the same base sequence as the parent molecule (fig. 10.5).

When DNA replicates, therefore, each copy is composed of one new strand and one strand from the original DNA molecule. Replication is said to be **semiconservative** (half of the original DNA is "conserved" in each of the new DNA molecules). Through this mechanism, the sequence of bases in DNA—which is the basis of the genetic code—is preserved from one cell generation to the next.

---

1. Explain how the laws of complementary base pairing allow one DNA strand to serve as a template for the production of a new complementary strand.
2. Using colored pencils and simple stick figures, illustrate the semiconservative replication of DNA.

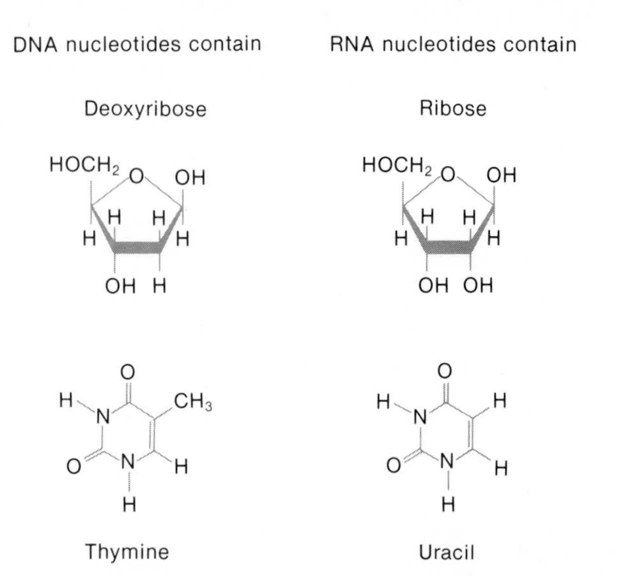

Figure 10.6. Differences between the nucleotides and sugars in DNA and RNA.

Ribonucleic Acid (RNA)

RNA nucleotides contain the bases adenine, guanine, cytosine, or uracil, which can bond to DNA bases by complementary base pairing. In this way, the sequence of bases in one of the strands of DNA serves as a template for the construction of RNA, which can then dissociate from the DNA as a single-stranded molecule. There are different types of RNA that serve different functions in genetic expression.

The genetic information contained in DNA functions to direct the activities of the cell through its production of another type of nucleic acid—*RNA (ribonucleic acid).* Like DNA, RNA consists of long chains of nucleotides joined together by sugar-phosphate bonds. Nucleotides in RNA, however, differ from those in DNA (fig. 10.6) in three ways: (1) a **ribonucleotide** contains the sugar *ribose* (instead of deoxyribose); (2) the base *uracil* is found in place of thymine; and (3) RNA is composed of a single polynucleotide strand (it is not double-stranded like DNA).

There are three kinds of RNA molecules that function in the cytoplasm of cells: *messenger RNA (mRNA), transfer RNA (tRNA),* and *ribosomal RNA (rRNA).* All three types are made within the cell nucleus by using information contained in DNA as a guide.

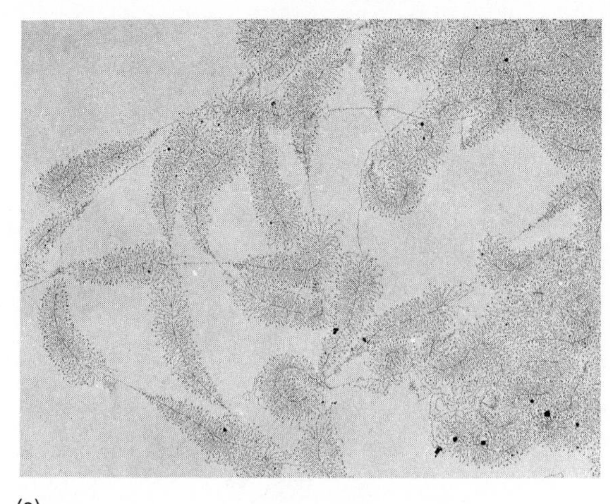

(a)

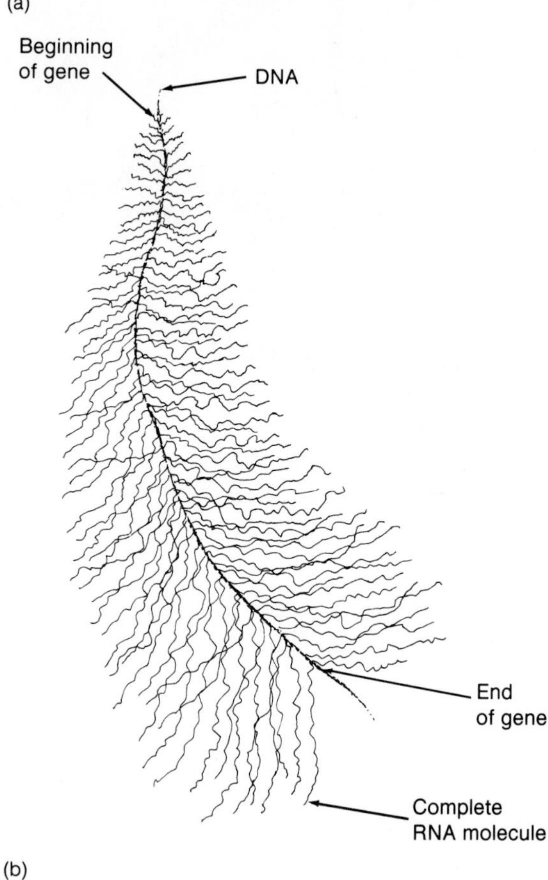

(b)

Figure 10.7. The synthesis of ribosomal RNA on DNA in the nucleolus.

Genetic Transcription: RNA Synthesis

The thin, extended euchromatin is the "working" form of DNA; the more familiar short, stubby form of chromosomes seen during cell division are inactive packages of DNA. The genes do not become active until the chromosomes unravel. Active DNA directs the metabolism of the cell indirectly through its regulation of RNA and protein synthesis.

One gene codes for one polypeptide chain. Each gene is a stretch of DNA that is several thousand nucleotide pairs long. The DNA in a human cell contains three to four billion base pairs—enough to code for at least three million proteins. Since the average human cell contains less than this amount (30,000 to 150,000 different proteins), it follows that only a fraction of the DNA in each cell is used to code for proteins. The remainder of the DNA may be inactive, may be redundant, and may serve to regulate those regions that do code for proteins.

In order for the genetic code to be translated into the synthesis of specific proteins, the DNA code must first be transcribed into an RNA code (fig. 10.7). This is accomplished by DNA-directed RNA synthesis, or **genetic transcription.**

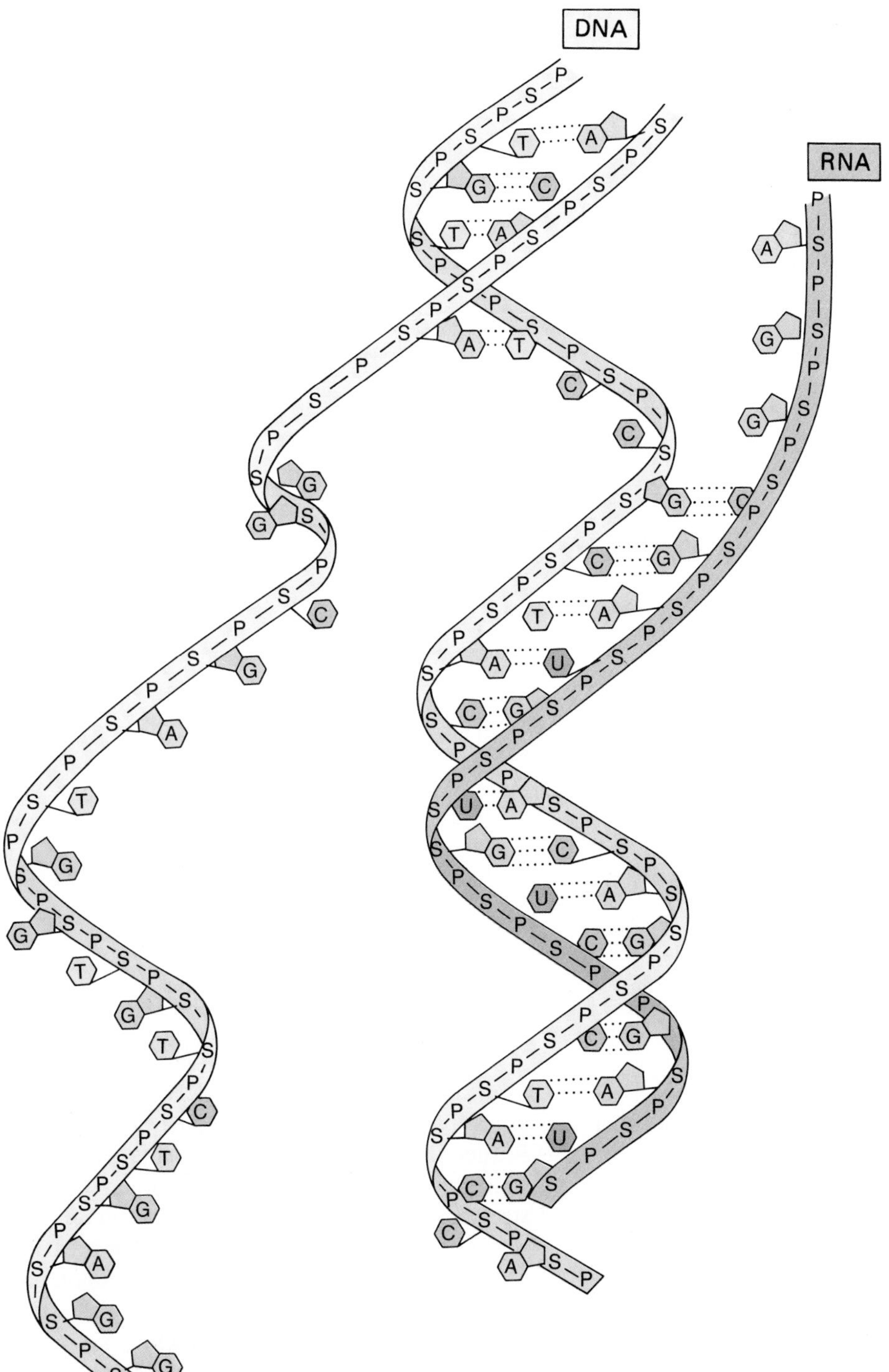

Figure 10.8. RNA synthesis (genetic transcription). Notice that only one of the two DNA strands is used to form a single-stranded molecule of RNA.

In RNA synthesis, the enzyme *RNA polymerase* breaks the weak hydrogen bonds between paired DNA bases. This does not occur throughout the length of DNA, but only in the regions that are to be transcribed (there are base sequences that code for "start" and "stop"). Double-stranded DNA, therefore, separates in these regions so that the freed bases can pair with the complementary RNA nucleotide bases, which are freely available in the nucleus.

This pairing of bases, like that which occurs in DNA replication, follows the law of complementary base pairing: guanine bonds with cytosine (and vice versa), and adenine bonds with uracil (because uracil in RNA is equivalent to thymine in DNA). Unlike DNA replication, however, only *one* of the two freed strands of DNA serves as a guide for RNA synthesis (fig. 10.8). Once an RNA molecule has been produced it detaches from the DNA strand on which it was formed. This process can continue

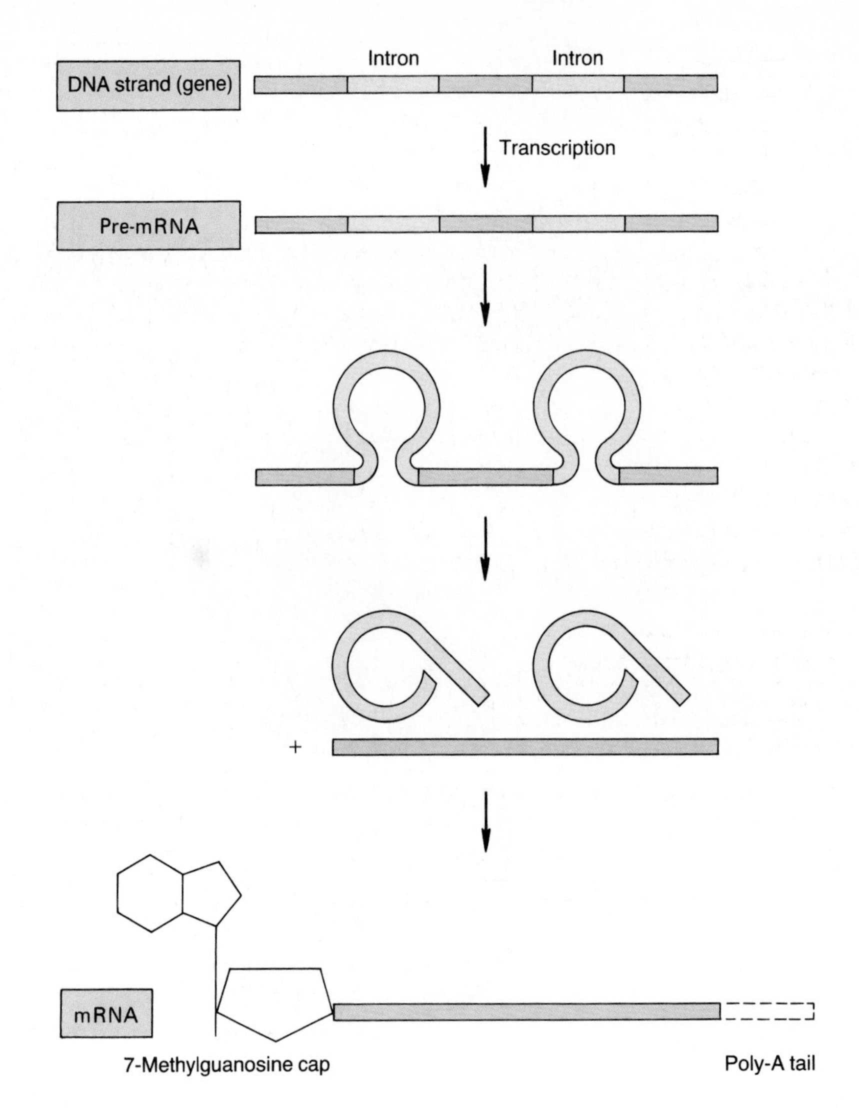

Figure 10.9. *Processing of pre-mRNA into mRNA. Noncoding regions of the gene, called introns, produce excess bases within the pre-mRNA. These excess bases are removed (as lariatlike strands), and the coding regions of mRNA are spliced together.*

indefinitely, producing many thousands of RNA copies of the DNA strand that is being transcribed. When the gene is no longer to be transcribed, the separated DNA strands can then go back together again.

Types of RNA

There are four types of RNA produced within the nucleus by genetic transcription: (1) **precursor messenger RNA (pre-mRNA),** which is altered within the nucleus to form mRNA; (2) **messenger RNA (mRNA),** which contains the code for the synthesis of specific proteins; (3) **transfer RNA (tRNA),** which is needed for decoding the genetic message contained in mRNA; and (4) **ribosomal RNA (rRNA),** which forms part of the structure of ribosomes.

In bacteria, where the molecular biology of the gene is best understood, a gene that codes for one type of protein produces an mRNA molecule that begins to direct protein synthesis as soon as it is transcribed. This is not the case in higher organisms, including humans. In higher cells a pre-mRNA is produced that must be modified within the nucleus before it can enter the cytoplasm as mRNA and direct protein synthesis.

Precursor mRNA is much larger than the mRNA that it forms. This large size of pre-mRNA is, surprisingly, not due to excess bases at the ends of the molecule that must be trimmed. Rather, the excess bases are *within* the pre-mRNA. The genetic code for a particular protein, in other words, is split up by stretches of base pairs that do not contribute to the code. These regions of noncoding DNA within a gene are called *introns*; the coding regions are known as *exons*. As a result, pre-mRNA must be cut and spliced to make mRNA. This cutting and splicing can be quite extensive; a single gene may contain up to fifty introns that must be removed from the pre-mRNA. After cutting and splicing, the mRNA is further modified by the addition of characteristic bases at its ends. These *post-transcriptional modifications* produce the form of mRNA that enters the cytoplasm (fig. 10.9).

1. Describe the structure of RNA, and list the different types of RNA.
2. Explain how RNA is produced within the nucleus according to the information contained in DNA.
3. Explain how precursor mRNA is modified to produce mRNA.

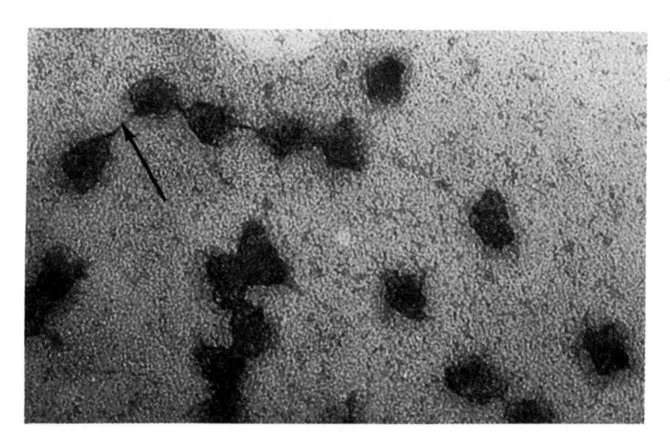

Figure 10.10. *An electron micrograph of polyribosomes. An RNA strand* (arrow) *joins the ribosomes together.*

Table 10.1 Selected DNA base triplets and mRNA codons

DNA Triplet	RNA Codon	Amino Acid
TAC	AUG	"Start"
ATC	UAG	"Stop"
AAA	UUU	Phenylalanine
AGG	UCC	Serine
ACA	UGU	Cysteine
GGG	CCC	Proline
GAA	CUU	Leucine
GCT	CGA	Arginine
TTT	AAA	Lysine
TGC	ACG	Tyrosine
CCG	GGC	Glycine
CTC	GAG	Aspartic acid

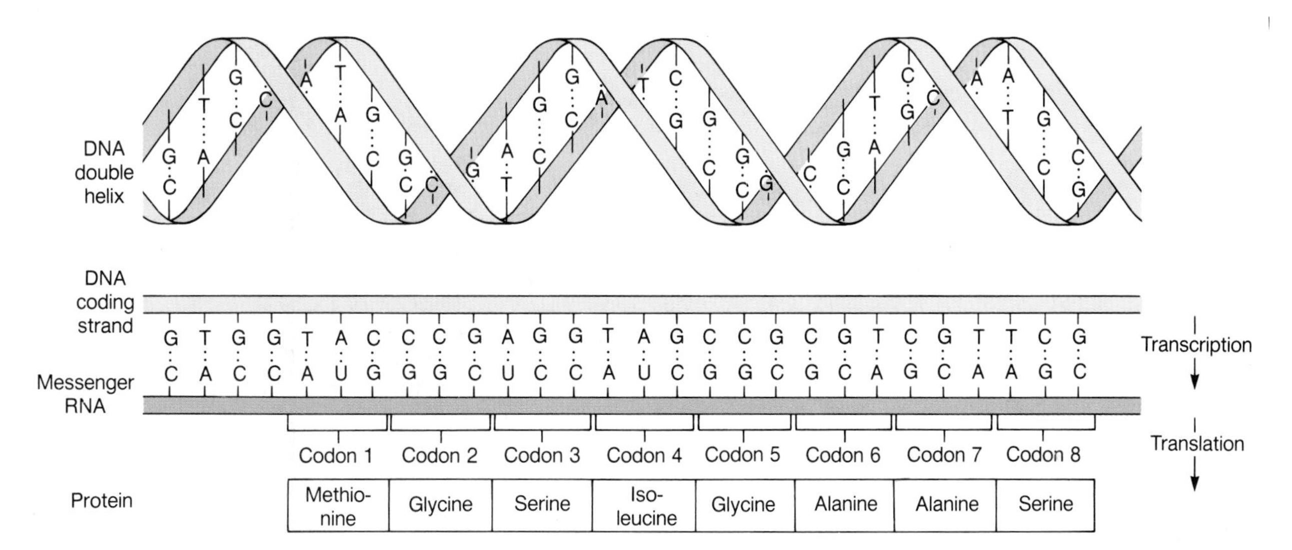

Figure 10.11. *The genetic code is first transcribed into base triplets (codons) in mRNA and then translated into a specific sequence of amino acids in a protein.*

Protein Synthesis and Secretion

In order for a gene to be expressed, it first must be used as a guide, or template, in the production of a complementary strand of messenger RNA. This process is called genetic transcription. The mRNA is then used as a guide to produce a particular type of protein, which has a sequence of amino acids determined by the sequence of base triplets (codons) in the mRNA. This process is called genetic translation. The secretion of proteins by the cell requires the participation of a number of organelles within the cytoplasm of the cell.

When mRNA enters the cytoplasm it attaches to ribosomes, which are seen in the electron microscope as numerous small particles. A **ribosome** is composed of three molecules of ribosomal RNA and fifty-two proteins, arranged to form two subunits of unequal size. The mRNA passes through a number of ribosomes to form a "string-of-pearls" structure called a *polyribosome* (or *polysome,* for short), shown in figure 10.10. The association of mRNA with ribosomes is needed for **genetic translation**—the production of specific proteins according to the code contained in the mRNA base sequence.

Each mRNA molecule contains several hundred or more nucleotides, arranged in the sequence determined by complementary base pairing with DNA during genetic transcription (RNA synthesis). Every three bases, or *base triplet,* is a code word—called a **codon**—for a specific amino acid. Sample codons and their amino acid "translations" are shown in table 10.1 and in figure 10.11. As mRNA moves through the ribosome, the sequence of codons is translated into a sequence of specific amino acids within a growing polypeptide chain.

Transfer RNA

Translation of the codons is accomplished by tRNA and particular enzymes. Each tRNA molecule, like mRNA and rRNA, is single stranded. Although tRNA is single stranded, it bends

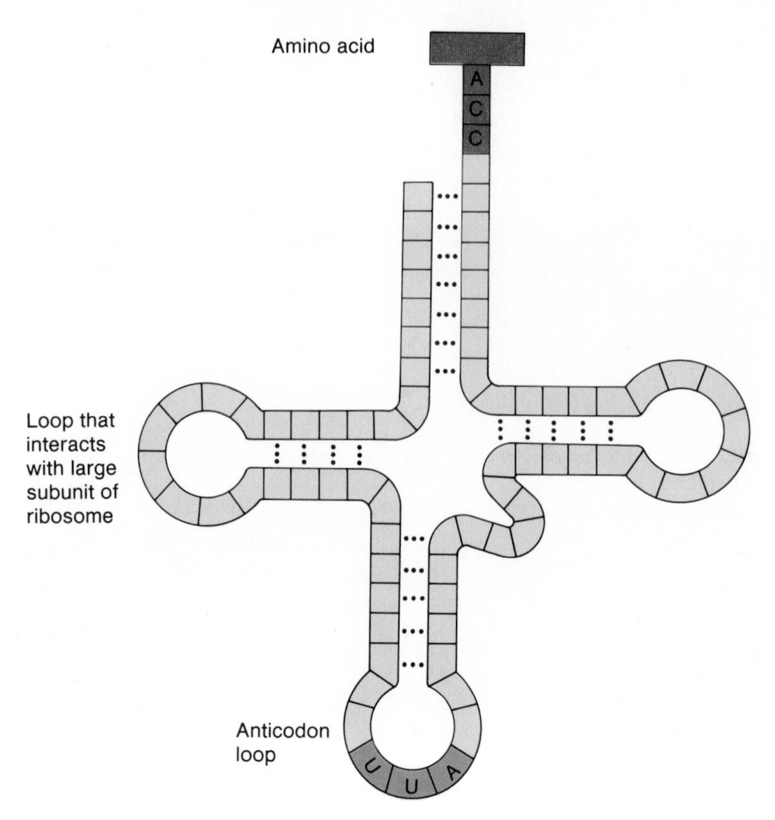

Figure 10.12. *The structure of transfer RNA (tRNA).*

on itself to form three loop regions (fig. 10.12). One of these loops contains the **anticodon**—three nucleotides that are complementary to a specific codon in mRNA.

Enzymes in the cell cytoplasm called *aminoacyl-tRNA synthetase* enzymes join specific amino acids to the ends of tRNA, so that a tRNA with a given anticodon is always bonded to one specific amino acid. There are twenty different aminoacyl-tRNA synthetase enzymes which each join a specific amino acid to a specific tRNA. The cytoplasm of a cell, therefore, contains tRNA molecules that are bonded to specific amino acids, and which are capable of bonding by their anticodon base triplets to specific codons in mRNA.

Formation of a Polypeptide

The anticodons of tRNA bond to the codons of mRNA as the mRNA moves through the ribosome. Since each tRNA molecule carries a specific amino acid, the joining together of these amino acids by peptide bonds creates a polypeptide whose amino acid sequence has been determined by the sequence of codons in mRNA.

When the first and second tRNA bring the first and second amino acids together and a peptide bond forms between them, the first amino acid detaches from its tRNA so that a dipeptide is linked by the second amino acid to the second tRNA. When the third tRNA bonds to the third codon, the third amino acid forms a peptide bond with the second amino acid (which detaches from its tRNA); a tripeptide is then attached by the third amino acid to the third tRNA. The polypeptide chain thus grows as new amino acids are added to its growing tip (fig. 10.13). This growing polypeptide chain is always attached by means of only one tRNA to the strand of mRNA, and this tRNA molecule is always the one that has added the latest amino acid to the growing polypeptide.

As the polypeptide chain grows in length, interactions between its amino acids cause the chain to twist into a helix and to fold and bend on itself. At the end of this process, the new protein detaches from the tRNA as the last amino acid is added. Many proteins are further modified after they are formed; these modifications occur in the rough endoplasmic reticulum and Golgi apparatus.

Function of the Rough Endoplasmic Reticulum

Proteins that are to be used within the cell are produced in polyribosomes that are free in the cytoplasm. If the protein is a secretory product of the cell, however, it is made by mRNA-ribosome complexes located on the rough endoplasmic reticulum. The membranes of this system enclose fluid-filled spaces (cisternae), which the newly formed proteins may enter.

When proteins that are destined for secretion are produced, the first thirty or so amino acids comprise a *leader sequence*. As the polypeptide chain elongates, it is "injected" into the cisterna within the endoplasmic reticulum. The leader sequence is, in a sense, an "address" that directs secretory proteins into the endoplasmic reticulum. Once the proteins are in the cisterna, the leader sequence is removed so the protein cannot reenter the cytoplasm (fig. 10.14).

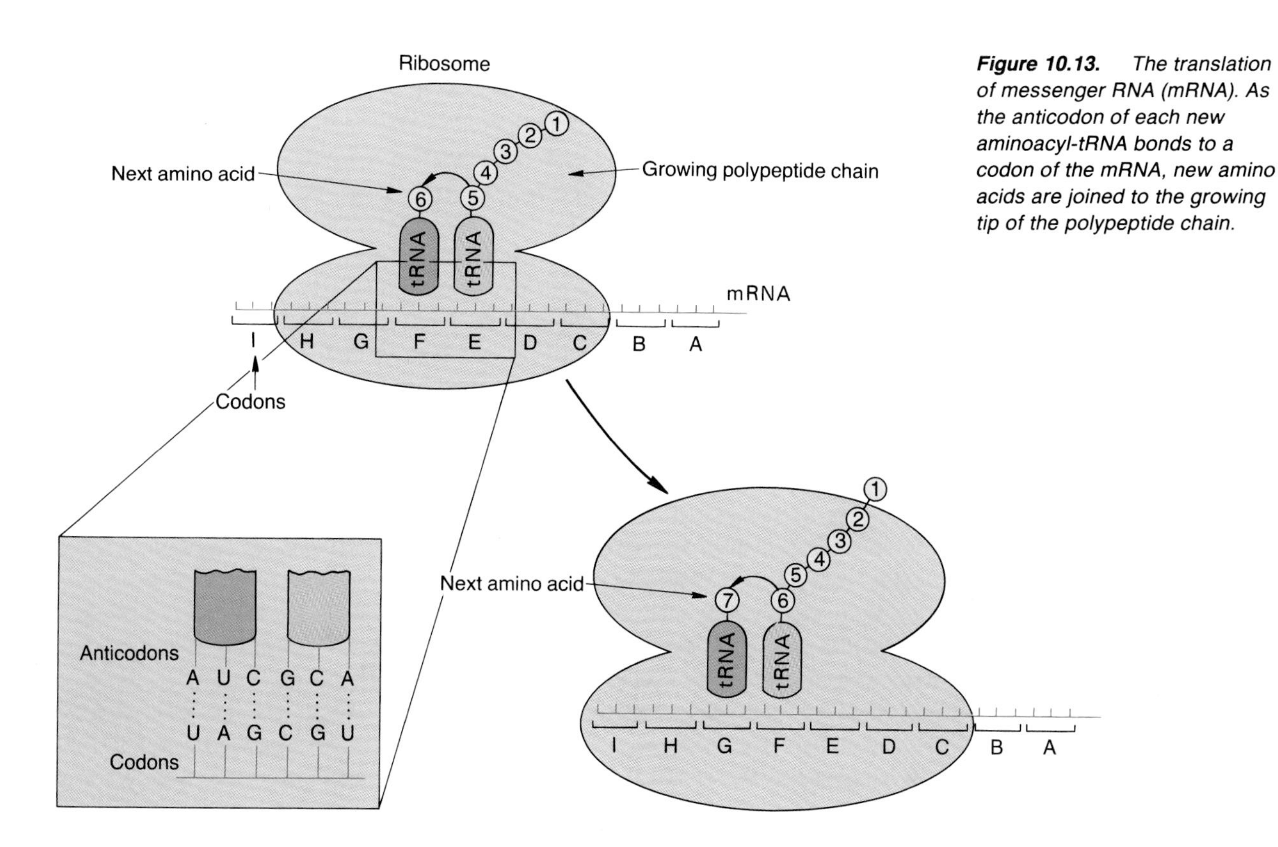

Figure 10.13. *The translation of messenger RNA (mRNA). As the anticodon of each new aminoacyl-tRNA bonds to a codon of the mRNA, new amino acids are joined to the growing tip of the polypeptide chain.*

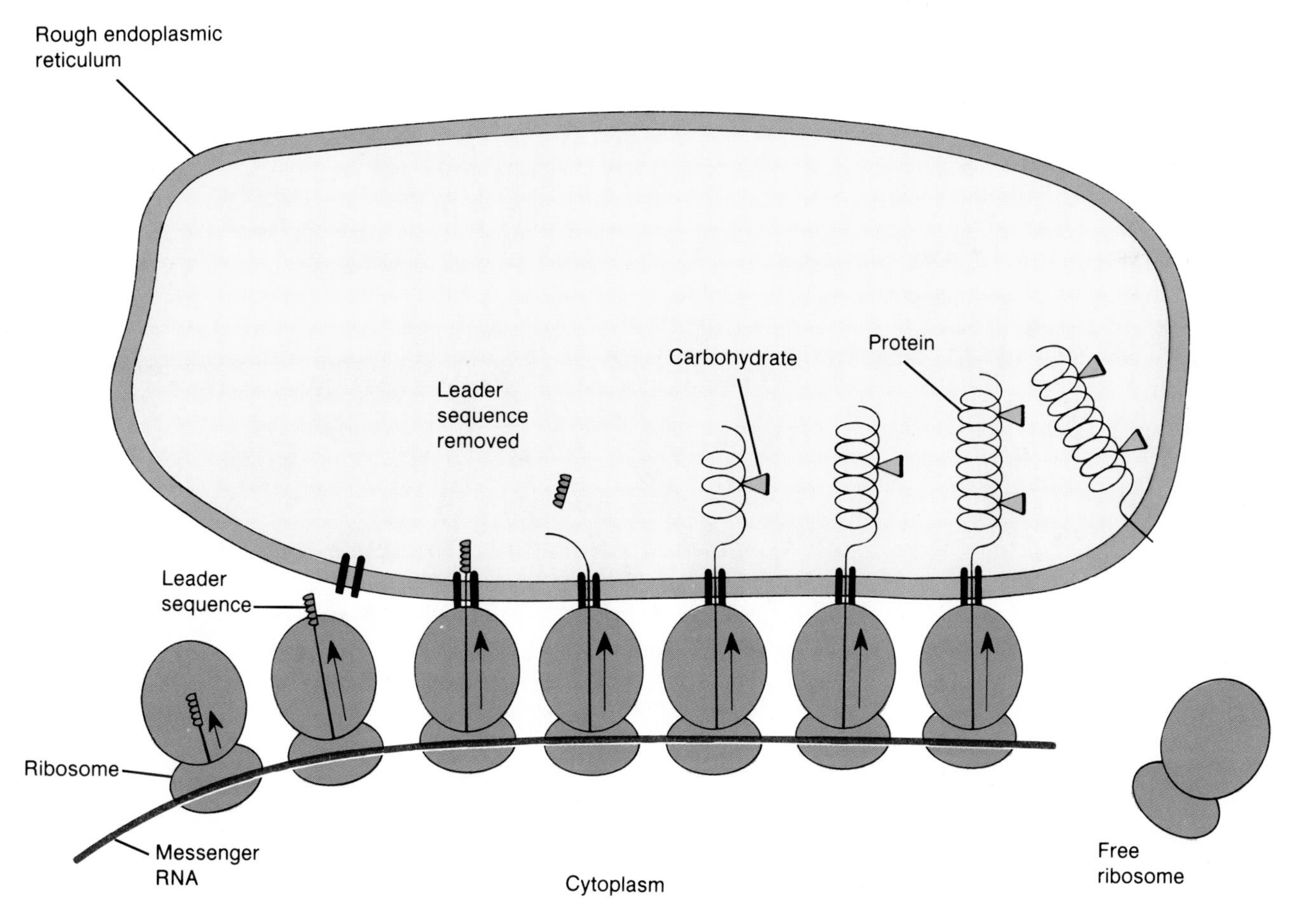

Figure 10.14. *A protein destined for secretion begins with a leader sequence that enables it to be inserted into the endoplasmic reticulum. Once it has been inserted, the leader sequence is removed and carbohydrate is added to the protein.*

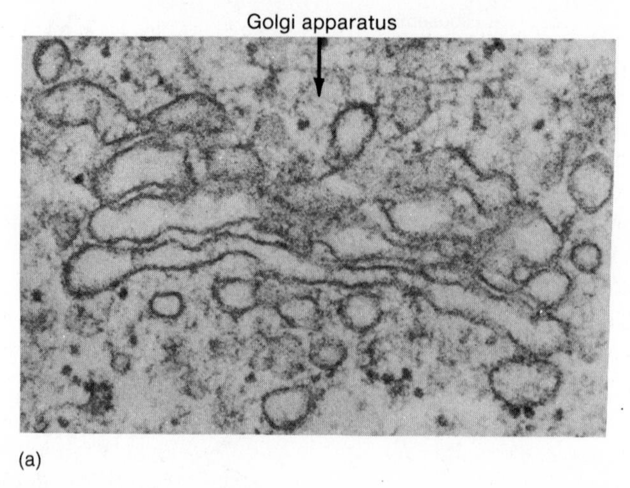

(a)

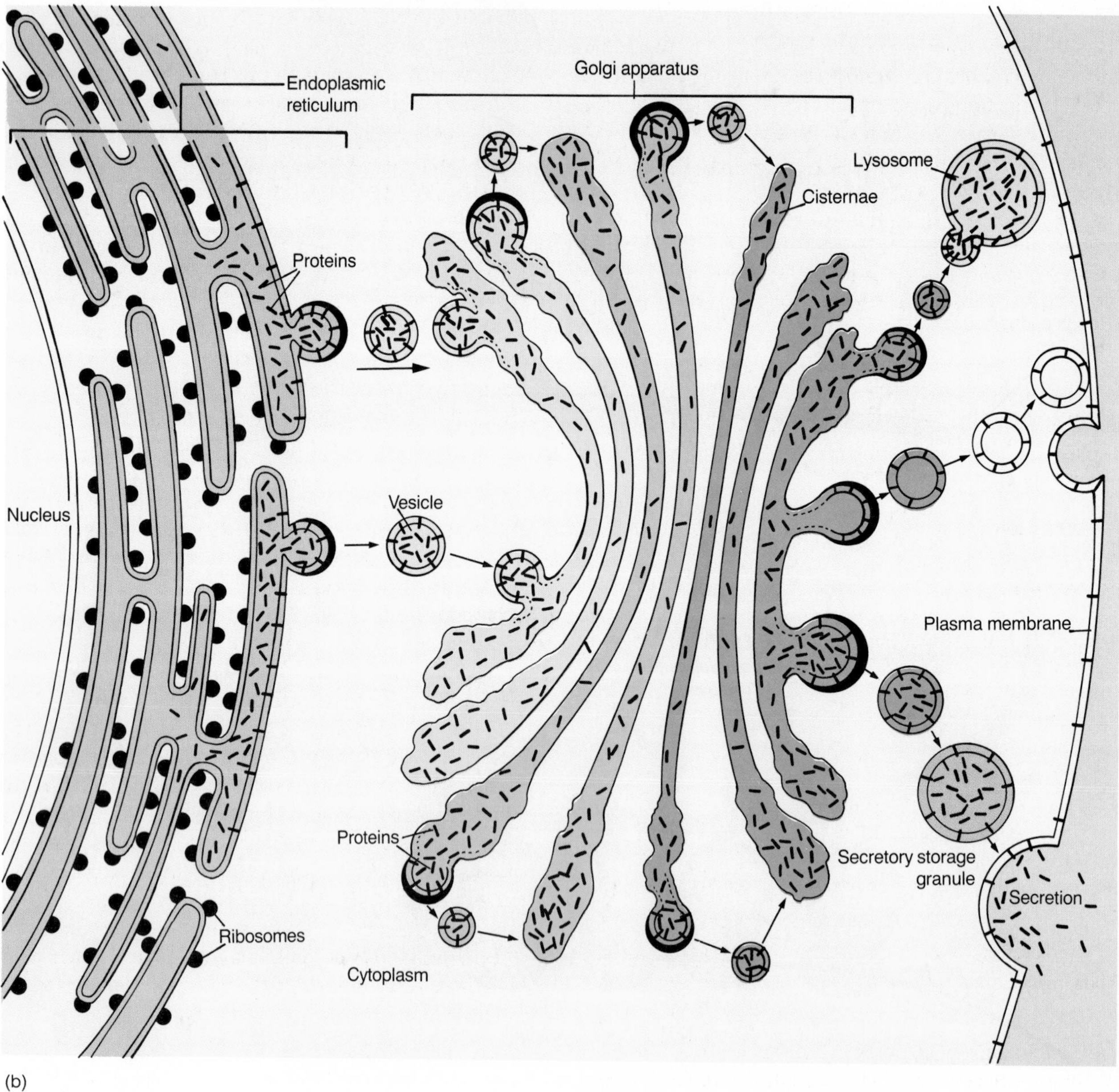

(b)

Figure 10.15. *(a) An electron micrograph of a Golgi apparatus. Notice the formation of vesicles at the ends of some of the flattened sacs. (b) An illustration of the processing of proteins by the rough endoplasmic reticulum and Golgi apparatus.*

The processing of the hormone insulin can serve as an example of the changes that occur within the endoplasmic reticulum. The original molecule enters the cisterna as a single polypeptide composed of 109 amino acids. The first twenty-three amino acids serve as a leader sequence and are quickly removed. The remaining chain folds within the cisterna so that the first and last amino acids in the polypeptide are brought close together. The central region is then enzymatically removed, producing two chains—one is twenty-one amino acids long; the other is thirty amino acids long—which are subsequently joined together by specific bonds between the two polypeptide chains. This is the form of insulin that is normally secreted from the cell.

Function of the Golgi Apparatus

Secretory proteins do not remain trapped within the rough endoplasmic reticulum; they are transported to another organelle within the cell—the **Golgi apparatus.**[2] The Golgi apparatus consists of several flattened sacs. Proteins produced by the rough endoplasmic reticulum are believed to travel in membrane-enclosed vesicles to the sac on one end of the Golgi apparatus. After specialized modifications of the proteins are made within one sac, the modified proteins are passed by means of vesicles to the next sac until the finished products finally leave the Golgi apparatus in vesicles that fuse with the cell membrane (fig. 10.15).

The Golgi apparatus and the rough endoplasmic reticulum contain enzymes that modify the structure of proteins. These changes, combined with the events that occur during genetic transcription and translation, provide numerous possible sites for the regulation of genetic expression.

---

1. Explain how mRNA, rRNA, and tRNA function during the process of protein synthesis.
2. Describe the rough endoplasmic reticulum, and explain how the processing of secretory proteins differs from the processing of proteins that remain within the cell.
3. Describe the structure of the Golgi apparatus, and explain its functions.

Genetic Engineering

Genetic engineering, involving the production of recombinant DNA from different sources, utilizes structures and enzymes present in bacteria and viruses. Various techniques allow DNA from different sources to be cut and spliced together. The enzyme reverse transcriptase, derived from retroviruses, is used to produce DNA from mRNA. Cloned DNA can be used to produce desired genetic products and used as probes for different applications.

The revolution in molecular biology that was noted at the start of this chapter has spawned a large industry in biotechnology. Much of this industry is devoted to applications of **genetic engineering,** also known as **recombinant DNA technology.** The basic techniques for cutting and splicing DNA, so that genes from one organism can be inserted into and used by another organism, originated in laboratories conducting basic scientific research.

Viruses

One of the basic tools of genetic engineering is the use of viruses. In order to understand this new technology, therefore, some information about viruses is required.

A virus is not a cell or composed of cells, and it cannot reproduce by itself. Most scientists, therefore, do not classify viruses among living organisms (it is not a part of the five kingdoms discussed in chapter 1). Despite this, the study of viruses is included in the science of biology. The field of *microbiology,* specifically, deals in part with bacteria (which are small, primitive cells) and with the very much smaller viruses.

A virus can be thought of as a tiny box, made of proteins, which contains genetic information in the form of either DNA or RNA. Viruses are so small that they are bounced around in a fluid because of collisions with randomly moving water molecules. This type of movement is termed *Brownian motion,*[3] and is similar to the way any tiny particle, like a fleck of dust, would move in a solution. The protein component of viruses, however, contains recognition sites that allows them to bind to and interact with the cell membranes of only specific host cells. When such a binding occurs, the virus particle is able to inject its nucleic acids into the cytoplasm of the host cell.

If the virus injects DNA, this viral DNA may use the enzymes in its host cells to replicate itself and then to make the proteins that are part of the viruses. This is a form of parasitism, and results in the production of many new viral particles within the host cell (fig. 10.16). Finally, the viral DNA directs the synthesis of enzymes that break the cell membrane. This kills the host cell, freeing the new viruses to spread out and infect many other new host cells.

If the virus contains RNA instead of DNA it is known as a **retrovirus** (*retro* = backward). The name is appropriate. Whereas the standard sequence of events is the production of RNA from DNA, the retrovirus must reverse this pattern. It accomplishes this by means of an enzyme called **reverse transcriptase,** which the retrovirus injects along with its RNA into the host cell. The name of this enzyme is also apt; it reverses the normal direction of transcription, producing new DNA using the viral RNA as a template! Once this has been accomplished, the virus can replicate itself, destroy its host cell, and infect other cells. Retroviruses cause some types of leukemias and are believed to be responsible for AIDS, as will be described in chapter 13.

---

[2]Golgi: from Camillo Golgi, Italian histologist, 1843–1926

[3]Brownian motion: after Robert Brown, English botanist, 1773–1858

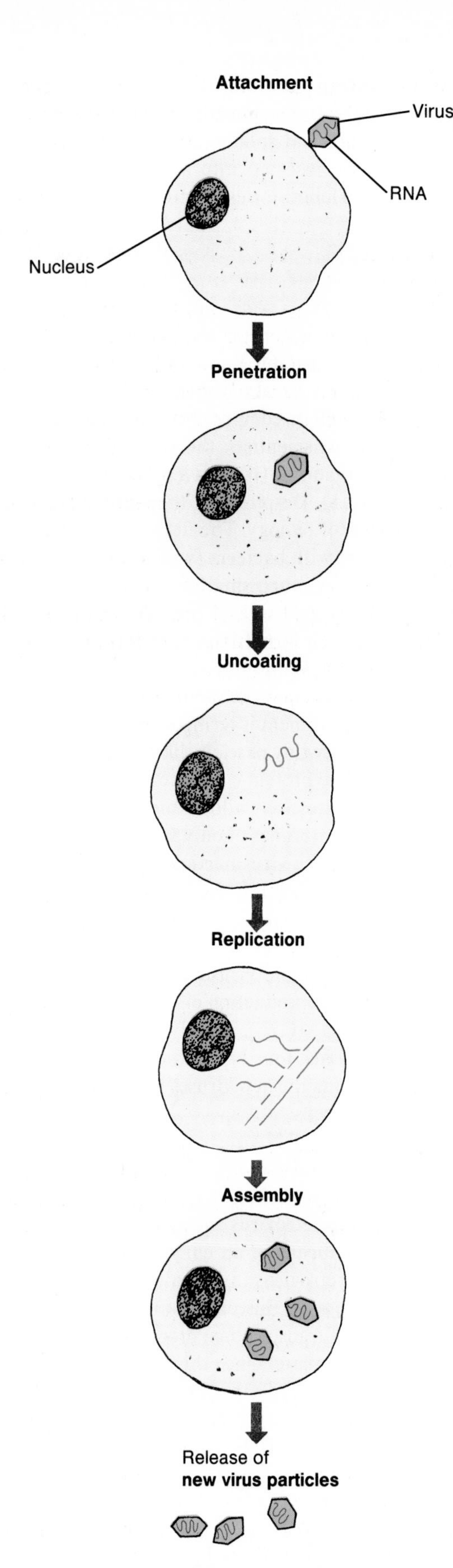

Figure 10.16. *The sequence of events that can occur when human cells are infected with virus particles.*

Bacteria

Bacteria are small cells classified as *prokaryotes,*[4] because they contain only a nuclear area, rather than a distinct nucleus. They therefore differ from *eukaryotes,*[5] which are cells of higher organisms that have a nucleus surrounded by a nuclear membrane. Bacteria have a number of other features that suggest that they are more primitive than eukaryotes. They lack mitochondria (chapter 4), and their DNA consists of only a single, double-stranded molecule in the form of a circle.

Bacteria have some other features that have made them valuable as tools in genetic engineering. Unlike the cells of eukaryotes, some bacteria have **plasmids.** These are small, circular structures of DNA that are located outside the nuclear area. Although cells of higher organisms are impermeable to nucleic acids, bacteria can take up plasmids from a solution. Plasmids have thus served as valuable vehicles, or *vectors,* for the introduction of foreign genes into bacteria.

Bacteria also manufacture more than two hundred different kinds of an enzyme known as **restriction endonuclease.** These enzymes break apart, or cleave, DNA at specific sites which are identified by their base sequence. Such *recognition sequences* may consist of four, six, or eight bases in a row, and different restriction endonuclease enzymes cleave DNA at different and specific sites. These enzymes evolved to protect bacteria from infection by viruses; the enzymes destroy the viral DNA while the bacteria's own DNA is protected. Scientists have used the restriction endonuclease enzymes to cut and splice DNA from different sources.

Production of Recombinant DNA

When a specific restriction endonuclease enzyme interacts with its recognition sequence, it cleaves both strands of a double-stranded DNA molecule. Since a given DNA molecule contains a number of the same recognition sequences, the restriction endonuclease therefore digests DNA into fragments. The lengths of these fragments depend on the distances that the recognition sequences are apart.

When double-stranded DNA is cleaved by a restriction endonuclease, it is cut at an angle. One strand of the fragment produced is four to eight bases longer than the other strand on one end, and the other strand is a corresponding number of bases longer on the opposite end (fig. 10.17). Further, the free bases on one end are always complementary to the free bases on the other end. These free ends could therefore bond with each other, or with the ends of any other DNA cleaved by the same restriction endonuclease. The free ends of the DNA fragments produced by restriction endonuclease digestion are thus known as *sticky ends.*

If, for example, the DNA of a plasmid and the DNA of a human is treated with the same restriction endonuclease enzyme, the fragments of plasmid and human DNA fragments produced

[4]prokaryotes: L. *pro,* before; *karyon,* nucleus
[5]eukaryotes: Gk. *eu,* good; *karyon,* nucleus

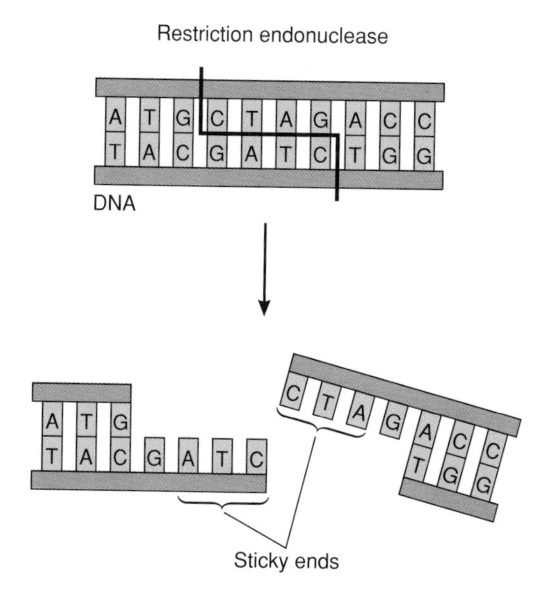

Figure 10.17. Restriction endonuclease cleaves DNA at specific recognition sites. The fragments thus produced have complementary sticky ends.

will have complementary sticky ends. These ends could thus join together by complementary base pairing. Another type of enzyme, known as a *ligase,* bonds the deoxyribose sugars and phosphates together, so that the DNA of the plasmid again forms a circle. This plasmid may, however, contain *recombinant DNA*—part of the DNA is bacterial and part is human (fig. 10.18).

In a less-common variant of this technique, the mRNA that codes for a desired product may be available. In this case, reverse transcriptase can be used to produce the **complementary DNA (cDNA),** analogous to the manner in which retroviruses reproduce. Since it is complementary to the mRNA, the cDNA is a gene. The cDNA is then given artificial sticky ends, as is the plasmid DNA in which it is to be inserted. The vector DNA may be a bacterial plasmid or it may be viral DNA. The DNA of a virus called bacteriophage lambda, for example, has been used for this purpose. The virus then acts as a tiny syringe, injecting the foreign DNA (human, for example) into a bacterial cell.

The first mammalian gene to be introduced into bacteria, converting the bacteria into factories that produced the human product, was the gene for somatostatin (a hormone that regulates the anterior pituitary). This is a polypeptide consisting of only fourteen amino acids. Since that time, numerous other human genes—including the ones that code for insulin and growth hormone—have been successfully introduced into bacteria.

Selection and Cloning

A number of treated plasmids are incubated together with a number of bacterial cells. *Escherichia coli* (*E. coli*), a common bacteria found in the human intestine, are often used as hosts in these procedures. Not all of the plasmids that were treated will have incorporated the foreign gene, however, and not all bacteria incubated with plasmids will take in the plasmids. Those bacteria that do take in plasmids containing the desired gene must be identified and separated from all of the other bacteria in the mixture.

One method of identifying the desired bacteria relies on the presence of two genes in the plasmid that code for antibiotic resistance. Suppose that one gene provides resistance to the antibiotic ampicillin, whereas the other gene codes for tetracycline resistance. The DNA to be grafted (perhaps a fragment of human DNA) is cut and spliced into the middle of the gene for tetracycline resistance, for example (fig. 10.19). This interrupts the code of the gene and significantly decreases the ability of the gene to provide resistance to tetracycline. Now, all bacteria that took in the plasmids can be identified, because they are all resistant to the antibiotic ampicillin. Among these bacteria, the ones that are *not* resistant to tetracycline must contain the desired gene.

The selected bacteria are allowed to reproduce until a large number of genetically identical bacteria, known as a *clone,* is produced. The foreign DNA will be reproduced along with the bacterial DNA, and thus will also be cloned. If the foreign DNA contains a human gene, for example, and if this gene expresses itself (producing mRNA and thus the protein coded by the gene), the bacteria can serve as tiny factories for the production of a product that may be extremely useful for medical and other applications.

Hybridization

There are other uses for cloned DNA. If the plasmids from cloned bacteria are treated with the same restriction endonuclease used to produce the recombinant DNA, the foreign DNA will again be released. These can then be separated into single DNA strands (by breaking the hydrogen bonds with heat) and radioactively labeled, or tagged by some other means. This labeled, single-stranded DNA can then serve as a probe.

For example, if the cloned DNA was derived from some virus of interest (such as the hepatitis B virus), a probe that can detect infection by hepatitis B viruses may be prepared. DNA obtained from a patient's tissue sample can be separated into single strands, and if the hepatitis B virus DNA is present, it will bind by complementary base pairing with the labeled probe DNA. Combination of single-stranded cloned DNA, or of mRNA derived from the cloned DNA, with another single-stranded DNA sample is known as **hybridization.** Hybridization techniques, using mRNA-DNA hybrids, have also been used to determine the location of particular genes on chromosomes.

1. Define the term *recombinant DNA,* and explain how restriction endonuclease enzymes are used to produce recombinant DNA.
2. Define the term *vector* in genetic engineering, and describe how plasmids and viruses can be used as vectors.
3. Explain how bacteria that have incorporated the desired foreign DNA may be identified.
4. Define the term *hybridization,* describe how it is performed, and explain how this technique can be used.

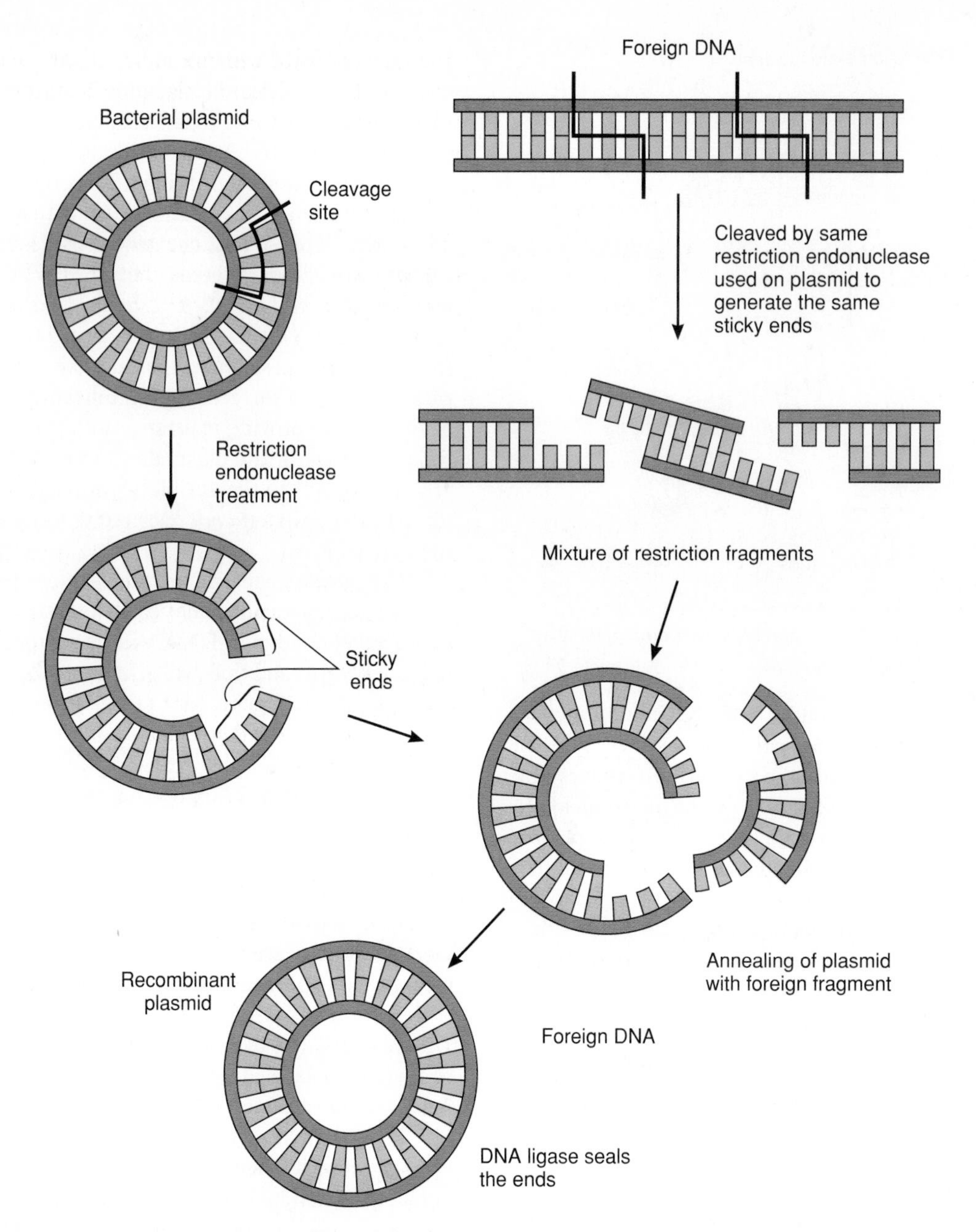

Figure 10.18. *Formation of recombinant DNA in a bacterial plasmid. The recombinant plasmid can then serve as a vector for cloning the foreign DNA.*

Recombinant DNA Applications

Recombinant DNA technology is used for medical, forensic (legal), and agricultural applications. Medical applications include the production of drugs, hormones, and other therapeutic compounds by recombinant bacteria, the development of safer vaccines, and the possible future correction of genetic defects. The technique of DNA fingerprinting can aid law enforcement. Food production may also be greatly increased through various applications of genetic engineering.

As previously indicated, the new recombinant DNA technology has already provided benefits to society, but much of the promise of this technology has yet to be realized. The present and possible future applications of genetic engineering may be grouped into different categories, but these categories actually overlap to a considerable degree.

Medical Applications

Before the development of recombinant DNA technology, people with insulin-dependent diabetes mellitus (chapter 18) had to inject themselves with the hormone insulin derived from the pancreas of pigs or cows. While this insulin works, and has saved the lives of millions of people, some diabetics develop antibodies against the animal-derived insulin which interferes with its effectiveness. Diabetics now can choose to use human insulin,

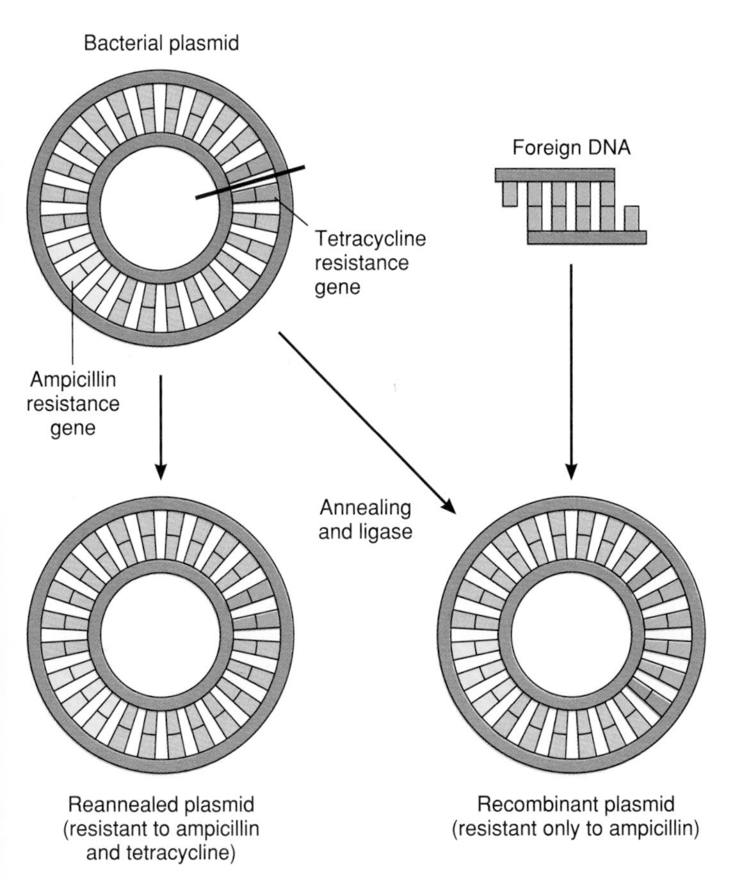

Figure 10.19. *Insertion of foreign DNA into gene that grants resistance to tetracycline in a bacterial plasmid. This breaks up the gene, so that bacteria with the recombinant DNA are not resistant to tetracycline.*

which is readily available in large quantities from genetically engineered bacteria. Indeed, insulin was the first commercially available product (from *Genentech*) of the biotechnology revolution, having been approved for use by the Food and Drug Administration in 1982.

Production of Pharmaceuticals

In addition to insulin, other human hormones, commercially produced using recombinant DNA, are now available. The human growth hormone produced by *Eli Lilly* was approved for use in 1987 in the treatment of children with pituitary dwarfism. Scientists are currently investigating the possibility of using this hormone in the treatment of other conditions, and in the possible treatment of children who are short due to reasons other than lack of growth hormone secretion from their pituitary. Prior to the availability of human growth hormone from recombinant DNA technology, this hormone was in very short supply because it could only be obtained in small amounts from the pituitaries of cadavers.

In 1988, Genentech received permission to market *tissue plasminogen activator (TPA),* a substance that helps to digest blood clots in people in whom the supply of blood in the coronary arteries of the heart is blocked by such clots. In 1989, *Amgen* received permission to produce the hormone *erythropoietin,* which is normally secreted by the kidneys and acts to stimulate red blood cell production in the bone marrow. Patients who have kidney disease often suffer from anemia (low red blood cell count) due to inadequate production of erythropoietin. These patients can now be helped using the erythropoietin produced by cells containing recombinant DNA.

Other recombinant drugs besides hormones have also been approved for use, and a great many products are in development or in the process of being tested in clinical trials. For example, *interleukin-2,* and *interferon,* which are chemicals released by certain white blood cells known as lymphocytes (see chapter 13), are widely used now in the experimental treatment of some types of cancers.

Vaccines

The traditional manner of preparing vaccines involves taking the infectious agent, such as a bacteria or virus, and inactivating it with heat or chemicals so that its ability to cause disease (its *virulence*[6]) is decreased, or attentuated. When these attentuated organisms are injected, they stimulate the immune system (chapter 13) to produce immunity against the active, virulent form of the organism. This works in most cases, but sometimes the infectious agent is not completely inactivated. In these cases, the vaccine itself may cause undesired side effects or even the disease.

Recombinant DNA technology can be used to prepare safer and more effective vaccines for diseases such as malaria, herpes, hepatitis, and pertussis (whooping cough). For example, the gene that codes for one of the coat proteins in a virus may be cloned, so that large amounts of this particular protein can be available for the vaccine. Since the genetic code for the production of the entire virus is lacking in this preparation, the vaccine produced in this way cannot cause the disease. A similar procedure may someday be used to provide immunity against AIDS.

Gene Therapy

There are four thousand known inheritable diseases that are caused by defects in the genetic code. At present, none can be corrected, although the symptoms of a very few can be prevented or reduced. Some preliminary successes have been achieved with diseases that involve blood cells, because these may be curable through transplantation of recombinant bone marrow cells. There are a variety of ways that recombinant DNA can be introduced into such cells, including introduction by altered viruses, treatment with chemicals, and microinjection. It should be noted that such treatment, if successful, would prevent the disease in the treated person but could not be passed on to that person's descendants. This is because the new recombinant genes are located in *somatic cells* (all cells in the body other than sperm or ova), which do not pass genes to the next generation.

Genetic engineering techniques have also been used to identify a previously unknown protein called *dystrophin,* which is not made in patients with *Duchenne's muscular dystrophy.*

[6]virulence: L. *virus,* poison

Social Issues

Sequencing the Human Genome

The human genome (the entire library of human genes) consists of fifty thousand to one hundred thousand genes, each with an average length of about ten thousand base pairs. Since any two humans are far more similar to each other than they are different, their DNA probably differs by only about one base pair in a thousand. These seemingly slight genetic differences, however, may in one case produce a susceptibility and the other case a resistance to a specific disease.

In addition to diseases that result directly from defective genes, there are many other diseases that are partially influenced by genetic endowment. High blood cholesterol and hypertension are inherited to some degree, for example, and these conditions constitute risk factors for heart disease. Future advances in gene therapy may allow "good" genes to be introduced into a person to correct a particular disease. In order for this to occur, however, a more complete knowledge of the human genome is required.

Approximately one thousand human genes have already been mapped to chromosomes, but their positions are not precisely known. The gene library at the Los Alamos National Laboratory now contains about five million base pairs. At current rates of progress, the human genome would not be entirely sequenced until the close of the twenty-first century. The United States and several other countries, however, are mounting an effort to have the entire human genome sequenced in just fifteen years. By the year 2005, if government funding permits, the human genome will be known in its entirety.

This *human genome project,* as it has come to be known, has raised some controversies. Money is the root of the problem; the cost to the United States over the projected fifteen-year period is estimated at $3 billion. Questions concerning where this money will go—to large, national laboratories or to numerous smaller, university-based laboratories, or to some combination of these—has engendered debates over the value of "big" versus "small" science. Once the project is completed, pharmaceutical companies can use the information to develop many new products of great commercial value. If the United States bears most of the financial burden for the human genome project, should it charge other nations who want to use the information obtained? Questions such as these have already resulted in some international friction.

Some people are also concerned with the larger ethical questions that knowledge of the human genome raises. Most people would accept genetic corrections of somatic cells, so that the person does not have the symptoms of the disease; but what of corrections that are inheritable? Should corrections be made in human embryos, so that the genetic disease is actually cured in the treated individuals and all of their progeny? If we know how to cure all genetic diseases, should we? Some diseases may provide a benefit in the carrier state; people who carry the gene for sickle-cell anemia (chapter 11), for example, are protected against malaria. This problem points to another: extensive gene replacement therapy into human embryos could decrease the genetic variation of the human population. As explained in chapter 2, the genetic variability of a population provides the basis for natural selection and evolution. What will then become of the human species over evolutionary time periods?

Perhaps, in the distant future, people will adapt to changing environmental requirements by changing their own genetic constitution. But this raises still other ethical questions, which may become pertinent in the more foreseeable future. Who decides which characteristics are desirable, and on what basis is this decision made? While everyone may agree that the elimination of a particular genetic disease is desirable, the benefits of other changes may be much more debatable. The genetic engineering of humans will not occur in the near future, but when it does, the biological, ethical, and legal questions it will raise will be of supreme importance.

Another genetic disease, *cystic fibrosis* (the most common genetic disease of Caucasians, affecting one in two thousand births), is associated with overproduction of mucus which can block airways and promote bacterial infection and lung disease. Genetic engineering techniques have been used to identify the abnormal gene in this disease. These discoveries may permit improved detection of carriers of these conditions. The inheritance of muscular dystrophy and cystic fibrosis is discussed in chapter 11.

The gene for cystic fibrosis has been localized to the long arm of chromosome number 7. In 1989, scientists achieved a breakthrough by identifying the specific gene involved in cystic fibrosis. They found that this gene spans a length of 250,000 base pairs in DNA, and that the gene in children with cystic fibrosis differs from the normal gene in one base triplet that codes for the amino acid phenylalanine. The scientists have, in addition, succeeded in cloning the gene. This may provide the future basis for experiments that lead to a cure for this disease, and for improved detection of people who are carriers. A prenatal test to determine if a fetus will have cystic fibrosis has already been developed using the knowledge gained through the discovery of the affected gene.

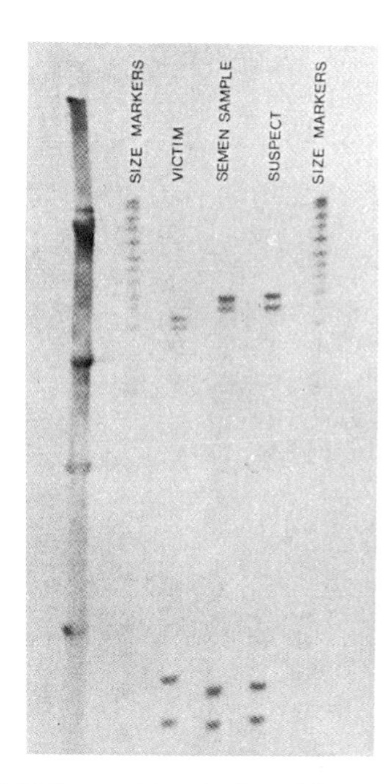

Figure 10.20. *DNA fingerprint analysis matches suspect to a semen sample taken from a rape victim. This figure only shows the results using one probe, whereas three or four probes are required for positive identification.*

Forensic Applications

The new biotechnology has been used in legal, or forensic, applications over the past few years. In some courtroom cases in particular, this new technology provided the critical evidence that has helped convict the suspect. The first case in which this evidence was used was particularly dramatic. A Florida man in 1988 was convicted of rape after a small sample of semen taken from the victim's vagina was matched to a sample from his blood (fig. 10.20). In a recent California case, an accused rapist was aquitted on the basis of the genetic evidence.

The technique used is called **DNA fingerprinting.** In this procedure, a restriction endonuclease is used to cleave the DNA of a sample at its recognition sequence. As previously discussed, the lengths of DNA fragments produced reflect the distance between the recognition sequences. The fragment lengths produced by DNA digestion with a particular restriction endonuclease will often be different in different people because of the differences in their genetic code. This diversity of fragment length produced by digesting DNA from different people with a given restriction endonuclease is called *restriction fragment length polymorphism (RFLP).* The fragments of DNA are separated in an electrical field and visualized by bonding of the single-chain fragments to complementary labeled DNA probes. This produces a characteristic number of bands (generally thirty to forty) by which different samples can be compared.

The particular pattern obtained and the accuracy of the method depends on which restriction endonuclease is used, the nature of the probes, and the amount of DNA in the sample. When properly performed, the chances of two people having the same DNA fingerprints can be as great as one in five billion! (The chances, however, are substantially less than this if the two people are related, or are members of a relatively isolated ethnic population.) The accuracy of these tests has recently been questioned in specific courtroom cases by some scientists and lawyers. Most experts agree that DNA fingerprinting has the potential to greatly aid law enforcement, but caution that the results must be interpreted together with more traditional lines of evidence on a case-by-case basis.

DNA obtained from cells of a hair follicle that cling to a single hair may be sufficient, after it is allowed to produce numerous copies of itself (a process called *gene amplification*), to provide valid DNA fingerprints. The FBI is now involved in developing a central laboratory that can perform these techniques, and hopes to eventually have a DNA fingerprint library similar to their regular fingerprint library. If taken to its logical conclusion, where almost everyone in the country has their DNA fingerprints on file, the ability of law-enforcement personnel to identify criminals may become incredibly effective. One can envision many other applications, both positive and negative, for having a means for positive identification of individuals in the hands of government agencies.

Applications in Food Production

Probably the major areas of benefit from biotechnology in the future will be in agriculture and livestock applications. Like all other technological advances—from the taming of fire to the harnessing of nuclear energy—this application of biotechnology can be a two-edged sword. On the one hand, the ability to better feed the millions of starving people in the world is a noble goal. There are, however, potential negative effects. If the greater availability of food results in a larger world population, and no birth control measures are taken, the new technology would simply postpone the inevitable mass starvations (more people would die of starvation later, rather than less people sooner). Also, the use of bioengineered organisms in agriculture may have undesirable environmental effects, which could be potentially disastrous. These negative possibilities, however, are preventable. In view of the vast potential for good, therefore, recombinant DNA technology will likely continue to revolutionize the production of food.

Genetically Engineered Plants

Why are store-bought tomatoes so tasteless compared to those that are homegrown? It is because the tomatoes grown commercially are not allowed to ripen on the vine; they're picked green, while they are still hard enough to survive the rough handling they must pass through before they get to the store. In the meantime, they're gassed with ethylene (a plant-ripening hormone), which makes them turn red. Genetically engineered tomatoes could be allowed to ripen on the vine and still be tough enough for shipment. Such a tomato has recently been produced by taking the gene that causes the fruit to soften, inserting it into a plasmid and cloning it, and then reinserting it backwards into tomato cells. This results in the production of a backwards mRNA, which bonds to the normal mRNA and thus prevents the expression of the normal gene. Unfortunately, such tomatoes are not yet commercially available.

In broad-leafed plants (scientifically called *dicotyledons*), the vector for gene transfer is a plasmid from the soil bacterium *Agrobacterium tumefaciens.* The plasmid from this bacterium is called the *T_i plasmid,* because it was discovered to be the cause of a tumor (T_i stands for "tumor inducing") in plants. Using this and other vectors, scientists are attempting to create recombinant DNA plants that, for example, produce their own insecticide. This has recently been accomplished by transferring the gene for a protein produced by a particular species of bacteria into tomato plants. The protein is toxic to caterpillars but not to humans. The results of this experiment were very dramatic: the recombinant tomato plants were not damaged whereas a control plot of normal tomatoes were completely defoliated by the caterpillars!

Genetic engineering could also make plants more resistant to viruses. Using a gene that codes for a protein component of tobacco mosaic virus, for example, recombinant tomato and potato plants were made resistant to a wide variety of viruses that normally attack these plants. In another possible use, food plants might be engineered to be resistant to a particular herbicide used to kill weeds, so that the food plants will survive while the weeds will be killed when the herbicide is used.

Perhaps most exciting, plants might be engineered to perform *nitrogen fixation.* Nitrogen fixation refers to the conversion of atmospheric nitrogen gas (N_2) into ammonia (NH_3) and other sources of nitrogen that plants can use to form amino acids. In nature, only plants known as *legumes* can perform nitrogen fixation. The soil in which other plants grow must therefore be supplemented with nitrogen-containing fertilizers.

The reason that legumes (such as alfalfa) can fix nitrogen is because they contain bacteria in their root nodules which have the necessary enzymes. Many companies are working to isolate and clone the genes from the nitrogen-fixing bacteria, *Rhizobium meliloti,* and introduce them into nonlegume plants. If they are successful, the causes of worldwide protein malnutrition (due to lack of sufficient amounts of the essential amino acids) may be eliminated.

Figure 10.21. *Researcher spraying a test field of strawberry plants with recombinant bacteria altered to reduce ice crystal formation. The protective suit that she wears is required by the EPA, even though these bacteria are not harmful to humans.*

San Francisco Chronicle, photographer Mike Maloney.

Genetically Altered Microorganisms in Agriculture

In April of 1987, the first environmental testing of "ice-minus" bacteria was performed (fig. 10.21). This was a species of *Pseudomonas,* a soil bacterium that had been altered by genetic engineering techniques. The normal bacteria produce a protein that stimulates ice crystal formation during a frost; the engineered bacteria had the gene which codes for that protein deleted. Field tests of these altered bacteria indicate that they do, indeed, decrease the frost damage in test plots. Future advances in this area may include the introduction of genes that code for insecticides into the soil bacteria, so that insects which damage plant roots could be destroyed.

Environmental Release of Microorganisms

The field testing of ice-minus bacteria was vigorously opposed by groups of people who were concerned about potential ecological dangers. Some even went so far as to sabotage a testing area. Of greater importance, a judicial order was obtained which delayed the field tests until sufficient evidence could be obtained that this procedure was environmentally safe. Although these concerns appear to have been unnecessary in this particular case, the general questions raised are certainly valid and must be considered before any altered organism is released into the environment.

The ice-minus bacteria were altered in other ways to make them easy to detect once released into the environment. In one method developed by the Monsanto Company, two genes from *E. coli* were transferred to the altered *Pseudomonas* bacteria so that the latter could use lactose, or milk sugar, as an energy source. The altered bacteria could thus be distinguished from the native bacteria, which cannot metabolize lactose. Using these methods to identify the altered bacteria, scientists have found that these bacteria have not spread beyond the test field in which they were sprayed.

Scientists who are in favor of less stringent regulation for environmental release of altered microorganisms have argued that these would be less fit than the native organisms. Since they would be at a selective disadvantage compared to the native organisms, they could not increase in number and invade new territories. In support of this concept, they point out that no problems have occurred in the field tests thus far performed. Considering the huge potential benefit to man through increased agricultural production, the risks, they argue, are negligible.

It should also be noted that microorganisms might be released into the environment for other purposes. For example, bacteria have been developed that can break down particular toxic substances produced by industry. Elimination of environmental toxins—including the massive destructions due to oil spills—by engineered bacteria is a goal that may be realized in the near future. Recombinant bacteria may also play an important role in the extraction and processing of minerals. Indeed, bacterial extraction is used to obtain 10% to 20% of the world's supply of copper, and bacteria are used by Canada in the leaching of uranium.

Scientists who advocate stringent regulation for the environmental release of altered microorganisms point to a number of potential problems. First, the ecology of microorganisms is poorly understood, and it is possible that ecological changes may occur that are not detected. Second, genes can mutate; altered organisms that are less fit when they are released may mutate to a new form that is more fit than the native organisms. If such a mutation grants them the ability to use an environmentally available nutrient that the natives are unable to use, or to live in a particular environment that the natives are unable to tolerate, they would have little competition. Under such circumstances, the engineered organisms could increase in their number and range and could thus drastically alter the ecology in unpredictable ways.

Considering the benefits to be derived (and the money to be made), it is doubtful that any political pressure will halt the applications of biotechnology in agriculture. Such pressure can, however, spur the development of technological solutions to potential hazards and force a slower, more cautious pace in this new agricultural revolution.

Transgenic Animals

In 1987, the U.S. Patent and Trademark Office announced that it would consider granting patents for *transgenic* (recombinant) animals. This announcement was made in response to an application by the University of Washington to patent its sterile oyster, which would be edible throughout the year. The first patent for a transgenic animal, however, was granted to Harvard University in 1988 for its development of a tumor-prone mouse.

In the development of the transgenic mouse, scientists isolated and cloned DNA that contained particular cancer-causing genes (*oncogenes*—chapter 14). They then injected this DNA into fertilized mouse egg cells, which were implanted into surrogate mothers for development. The transgenic mice thus produced have a tendency to develop malignant breast tumors when they become pregnant, and thus serve as excellent experimental models for the investigation of cancer and its cures. Although this is properly considered a medical application, the same type of procedures can be used in producing transgenic farm animals.

Using a similar procedure, scientists have produced transgenic pigs that contain the genes for producing extra amounts of growth hormone. This was done in an effort to produce larger, leaner pigs (growth hormone stimulates growth and protein synthesis, but promotes fat breakdown). The pigs produced in this way were leaner, but only grew larger when fed a high-protein diet. Further, they suffered from a number of ailments that made this experiment less than a complete success.

In another experiment, the genes for producing a human blood clotting factor and a molecule called alpha-1-antitrypsin (a compound that protects the lungs from the destruction associated with emphysema) were inserted into a gene for a milk protein. This DNA was then injected into sheep embryos. When the sheep grew up, the milk they produced contained these human proteins! Although the yields were not high, this is a significant achievement. Many human and animal proteins do not work in the body unless they are combined with particular carbohydrates (in the Golgi apparatus) to produce what are called *glycoproteins*. Bacteria do not attach carbohydrates to

Figure 10.22. *A mouse in a tobacco plant. The tobacco plant was genetically engineered to produce mouse antibodies.*

the proteins they produce, so cloning these genes in bacteria would not produce usable products. In these cases, effective methods require the use of animal cells, as in the sheep experiment.

The development of transgenic animals raises a number of ethical questions. Many people do not want any higher organism to be genetically engineered. But what constitutes a higher organism? Do the oysters previously described fit the category? (They do, if all eukaryotes are to be considered higher organisms.) Those in favor of producing new animals through recombinant DNA technology argue that people have been doing this for thousands of years through breeding techniques (the poodle, for example, is a highly altered, human-engineered derivative of a wolf). The new biotechnology, they argue, simply offers a better, more controlled way of accomplishing what man has previously done through less exact techniques. Most scientists who are working to produce transgenic animals for food production or medical research feel that their efforts are entirely ethical, because they benefit mankind, but feel the line should be drawn sharply at using these same techniques on humans.

Perhaps, at least for some applications, there is an alternative to using transgenic animals for production of needed products. A gene from a higher member of the animal kingdom (a mouse) was recently incorporated into a higher member of the plant kingdom (a tobacco plant). Scientists cloned the gene for producing a specific mouse antibody protein and incorporated this gene into tobacco leaf segments, which were then subsequently used to grow whole tobacco plants. The tobacco plants then manufactured a significant amount of functioning mouse antibodies (fig. 10.22)!

1. List some of the pharmaceuticals that are currently available and those that are in development using recombinant DNA technology.
2. Explain the advantages of using genetic engineering to develop vaccines.
3. Explain the techniques involved in DNA fingerprinting, and the advantages of this procedure.
4. Discuss the human genome project in terms of its goals and possible future uses.
5. Describe how recombinant DNA techniques are being used to improve agriculture, and discuss some of the dangers of this new technology.
6. Explain how recombinant (transgenic) animals have been produced, and discuss the potential uses and ethical considerations of this technology.

Summary

The Nucleus and Nucleic Acids p. 216

I. The DNA is located in the cell nucleus; active DNA is in the form of euchromatin.

II. DNA is composed of a double chain of polynucleotides.

- A. A DNA nucleotide consists of three parts: deoxyribose sugar, phosphate, and a nitrogen-containing base.
- B. There are four types of DNA nucleotides, which each contain one of four possible bases: adenine, guanine, cytosine, and thymine.
- C. The nucleotides are joined together by sugar-phosphate bonds within a polynucleotide chain.
- D. The two chains in DNA are joined together by weak hydrogen bonds between their bases.
- E. According to the law of complementary base pairing, only adenine and thymine can bond together, and only guanine and cytosine can bond together.

DNA Synthesis and Cell Division p. 218

I. Before a cell is to divide, the DNA replicates itself to form two identical copies.

- A. The hydrogen bonds between the two DNA strands break, so that the bases of each strand are free to bond to new complementary bases.
- B. A new complementary DNA strand can be produced using the old strand as a template.
- C. DNA replication is semiconservative: each of the two DNA molecules produced contains one old and one new strand.

Ribonucleic Acid (RNA) p. 220

I. RNA is produced through complementary base pairing with DNA.

- A. Ribonucleotides differ from DNA nucleotides in that they contain ribose instead of deoxyribose, and the base uracil is substituted for thymine.
- B. A region of DNA, which constitutes a gene, codes for the production of a particular RNA molecule.
  1. The weak bonds joining the two DNA strands together in this region break, freeing each strand.
  2. Only one of the two strands is used as a guide.
- C. Ribonucleotides bond by complementary base pairing to the DNA strand, so that the RNA molecule formed is a complementary copy of that DNA strand at that region.
- D. The production of RNA complementary in its base sequence to a gene is called genetic transcription.
- E. Unlike DNA, RNA is composed of only one strand.

II. The complementary RNA copy of a gene, which contains the information for the synthesis of a specific protein, is known as messenger RNA (mRNA).

- A. The gene actually produces a precursor mRNA molecule which is larger than mRNA, because it contains regions within it that do not contribute to the genetic code.
- B. Precursor mRNA is cut and spliced together to make mRNA, which leaves the nucleus and enters the cytoplasm.
- C. Other types of RNA are also made by DNA in the nucleus; these include transfer RNA (tRNA) and ribosomal RNA (rRNA).

Protein Synthesis and Secretion p. 223

I. Messenger RNA codes for the production of proteins in a process called genetic translation.

- A. The mRNA enters ribosomes, which are numerous, tiny organelles composed of protein and rRNA.
- B. Each three bases in mRNA code for a particular amino acid in the protein; these three bases comprise a codon.

II. The cytoplasm contains different transfer RNA molecules.

- A. Each transfer RNA is combined with a specific amino acid.
- B. The tRNA molecules have different anticodons: this is a triplet of bases that is complementary to a codon in mRNA.

III. As the anticodons of tRNA bond to the codons of mRNA, specific amino acids are added to a growing polypeptide chain.

IV. Proteins that are to remain in the cell are produced in free polyribosomes (ribosomes joined together by mRNA); proteins that are to be secreted are produced by ribosomes on a rough endoplasmic reticulum.

V. Proteins to be secreted are inserted into the spaces within the rough endoplasmic reticulum.

- A. These proteins are then transferred to the Golgi apparatus.
- B. The secretory proteins are packaged by the Golgi apparatus into vesicles, which can fuse with the cell membrane and thus release their contents.

Genetic Engineering p. 227

I. Viruses are composed of nucleic acids and protein.

- A. Some viruses contain DNA, which they can inject into a host cell.
- B. The viral DNA takes over the enzymatic machinery of the host cell to produce more virus particles.
- C. Retroviruses inject RNA instead of DNA into the host cell.
- D. The enzyme reverse transcriptase produces viral DNA from this RNA.

II. Bacteria are prokaryotes; they do not have a membrane around their nucleus.

- A. Plasmids are structures composed of circular DNA which are found outside the nucleus of bacterial cells.
- B. Bacteria produce restriction endonuclease enzymes, which cleave DNA at specific recognition sequences.

III. Scientists use reverse transcriptase, plasmids, and restriction endonuclease enzymes for genetic engineering.

- A. Restriction endonuclease cleaves DNA in both strands at its recognition sites, leaving short sticky ends.
- B. Cleavage of a plasmid and a different DNA with the same restriction endonuclease thus allows the foreign DNA to be inserted into the plasmid.
- C. A gene to be inserted into a plasmid may be obtained from a specific mRNA using reverse transcriptase.

IV. Bacteria such as *Escherichia coli* take up plasmids from a solution.

- A. These bacteria and the plasmids then replicate, producing many copies of the foreign gene: this is called cloning.
- B. Various techniques allow the selection and selective cloning of the bacteria that contain the desired gene.

Recombinant DNA Applications p. 230

I. There are numerous medical applications of genetic engineering.

- A. Bacteria containing human genes can be used as factories for the production of many pharmaceuticals.
- B. Safer vaccines can be prepared using cloned DNA.
- C. Gene therapy, involving the correction of inheritable diseases, may someday be possible.

II. The technique of DNA fingerprinting is used as courtroom evidence in legal cases.

III. There are numerous potential applications of genetic engineering in agriculture and livestock production.

- A. Genetically engineered plants that are more nutritious, tastier, or hardier may someday be produced.
- B. Microorganisms that improve plant production, serve as insecticides, or benefit agriculture in other ways may be released into the environment under carefully controlled conditions.

Review Activities

Objective Questions

1. The RNA nucleotide base that pairs with adenine in DNA is
   (a) thymine
   (b) uracil
   (c) guanine
   (d) cytosine
2. Which of the following statements about RNA is *true?*
   (a) It is made in the nucleus.
   (b) It is double stranded.
   (c) It contains the sugar deoxyribose.
   (d) It is a complementary copy of the entire DNA molecule.
3. Which of the following statements about mRNA is *false?*
   (a) It is produced as a larger pre-mRNA.
   (b) It forms associations with ribosomes.
   (c) Its base triplets are called anticodons.
   (d) It codes for the synthesis of specific proteins.
4. The organelle that packages proteins within vesicles for secretion is the
   (a) Golgi apparatus
   (b) rough endoplasmic reticulum
   (c) smooth endoplasmic reticulum
   (d) ribosome
5. If four bases in one DNA strand are A (adenine), G (guanine), C (cytosine), and T (thymine), the complementary bases in the RNA strand made from this region are
   (a) T,C,G,A
   (b) C,G,A,U
   (c) A,G,C,U
   (d) U,C,G,A
6. Which of the following statements about tRNA is *true?*
   (a) It is made in the nucleus.
   (b) It is looped back on itself.
   (c) It contains the anticodon.
   (d) There are over twenty different types of tRNA.
   (e) All of these.
7. The step in protein synthesis during which tRNA, rRNA, and mRNA are all active is known as
   (a) transcription
   (b) translation
   (c) replication
   (d) RNA polymerization
8. The anticodons are located in
   (a) tRNA
   (b) rRNA
   (c) mRNA
   (d) ribosomes
   (e) endoplasmic reticulum
9. When DNA replicates, a DNA molecule formed at the end of this process contains
   (a) one strand from the old DNA, and one strand that is new
   (b) two strands from the old DNA, or two strands that are new
   (c) strands that are each half old and half new
   (d) strands that are a random mixture of old and new DNA
10. The term "double helix" describes the structure of
    (a) mRNA
    (b) tRNA
    (c) rRNA
    (d) DNA
    (e) all of these
11. Reverse transcriptase
    (a) catalyzes DNA replication
    (b) catalyzes RNA synthesis
    (c) catalyzes the production of DNA from mRNA
    (d) breaks up DNA at specific recognition sequences
12. When foreign DNA is joined to a plasmid
    (a) both are digested by the same restriction endonuclease
    (b) complementary sticky ends are formed
    (c) a ligase is used to rejoin the sugar-phosphate bonds
    (d) recombinant DNA is produced
    (e) all of the above

Essay Questions

1. Explain how one DNA molecule serves as a template for the formation of another DNA and why DNA synthesis is said to be semiconservative.
2. What is the genetic code, and how does it affect the structure and function of the body?
3. Why may tRNA be considered the "interpreter" of the genetic code?
4. Compare the processing of cellular proteins with that of proteins that are secreted by a cell.
5. Why is molecular genetics considered to be a revolutionary field? What ethical and legal problems may it pose in the future?

11

Human Genetics

Outline

Objectives

By studying this chapter, you should be able to

1. trace the sequence of events by which a point mutation can result in an inborn error of metabolism
2. explain the specific metabolic disorders that give rise to PKU and albinism
3. define the terms *homozygous, heterozygous, diploid, haploid,* and *allele*
4. define the terms *genotype* and *phenotype,* and explain the relationship between these two
5. predict the probabilities of obtaining offspring with particular genotypes and phenotypes in a simple mendelian cross
6. give examples of human diseases that are inherited as simple mendelian recessive or dominant traits that are carried on autosomal chromosomes
7. give examples of human diseases that are X-linked, and predict the genotypes and phenotypes of offspring in the genetic analysis of these traits
8. explain the causes of sickle-cell anemia and thalassemia, the nature of these diseases, and their patterns of inheritance
9. describe the nature of the blood groups and explain how blood is typed
10. describe the patterns of inheritance of the ABO and Rh blood groups
11. explain how intermediate phenotypes are produced, and the influence of environment on genetic expression
12. identify some disorders caused by abnormal numbers of chromosomes, and explain how these conditions can be produced

Keys to Pronunciation

albinism: *al'bi-nizm*
allele: *ah-lel*
dystrophy: *dis'tro-fe*
erythroblastosis: *e-rith''ro-blas-to'sis*
heterozygous: *het''er-o-zi'gus*
homozygous: *ho''mo-zi'gus*
melanin: *mel'ah-nin*
nondisjunction: *non''dis-junk'shun*
phenotype: *fe'no-tip*
phenylketonuria: *fen''il-ke''to-nu're-ah*
thalassemia: *thal''ah-se'me-ah*

Photo: The faces in a crowd testify to the genetic diversity within the species Homo sapiens.

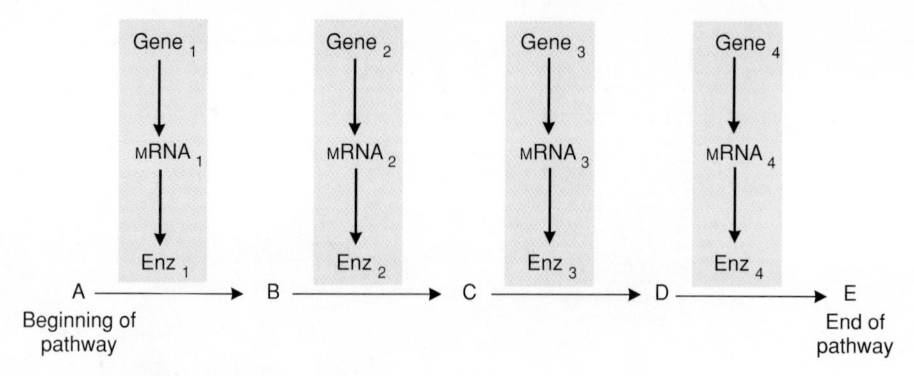

Figure 11.1. A metabolic pathway in which molecule "A" is changed through four separate reactions to molecule "E." Each step is catalyzed by a different enzyme, and each of the enzymes is produced by a different gene.

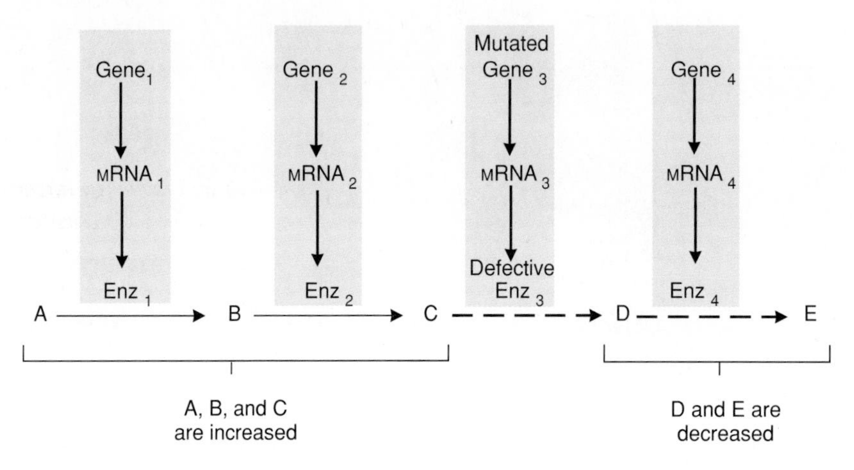

Figure 11.2. A mutated gene can cause the production of a defective enzyme. As a result, this metabolic pathway cannot proceed from "A" to "E."

Gene Action and Inheritance

Genes code for the production of proteins, including enzyme proteins. A defective gene could thus result in the production of a defective enzyme, which interferes in a specific way with metabolism. Phenylketonuria (PKU) and albinism are examples of such genetic diseases which result from a change in just one nucleotide base in a gene.

Francis Crick,[1] the codiscoverer of the structure of DNA, proposed in 1958 what has come to be called the *central dogma* of molecular genetics: DNA makes RNA, and RNA makes protein. The mechanisms by which a gene in DNA codes for the production of RNA (genetic transcription), and by which a specific messenger RNA codes for the production of a specific protein (genetic translation), were described in chapter 10. Since enzymes are proteins that catalyze specific reactions (chapter 3), a defective gene could cause the production of a defective type of enzyme.

Figure 11.1 illustrates a metabolic pathway, as described in chapter 3, in which each step in a series of reactions is catalyzed by a different, specific enzyme. The "central dogma" concept is also shown, with each enzyme coded by a different, specific gene. Since there may be up to 100,000 different human genes, this figure illustrates only a very small part of the total genetic control of cellular metabolism.

Suppose, as illustrated in figure 11.2, that the gene which codes for the third enzyme in the metabolic pathway is defective. This would most likely be due to the inheritance of the defective gene from one or both parents of the affected individual. As a result of the genetic defect, a defective enzyme is produced that is unable to convert its substrate into products. As a consequence, the molecules that would normally be produced after the defective step are not produced, or are produced in abnormally small amounts. The products formed prior to the defective step, in contrast, accumulate to abnormally high amounts. This is an example of one type of inheritable condition, known specifically as an *inborn error of metabolism.*

[1]Crick: Francis Harry Compton Crick (1916–), English biologist

Table 11.1 Selected examples of inborn errors in the metabolism of amino acids, carbohydrates, and lipids

Metabolic Defect	Disease	Abnormality	Clinical Result
Amino acid metabolism	Phenylketonuria (PKU)	Increase in phenylalanine	Mental retardation, epilepsy
	Albinism	Lack of melanin	Susceptibility to skin cancer
	Maple-syrup disease	Increase in leucine, isoleucine, and valine	Degeneration of brain, early death
	Homocystinuria	Accumulation of homocystine	Mental retardation, eye problems
Carbohydrate metabolism	Lactose intolerance	Lactose not utilized	Diarrhea
	Glucose-6-phosphatase deficiency (Gierke's disease)	Accumulation of glycogen in liver	Liver enlargement, hypoglycemia
	Glycogen phosphorylase deficiency (McArdle syndrome)	Accumulation of glycogen in muscle	Muscle fatigue and pain
Lipid metabolism	Gaucher's disease	Lipid accumulation (glucocerebroside)	Liver and spleen enlargement, brain degeneration
	Tay-Sachs disease	Lipid accumulation (ganglioside G_{M_2})	Brain degeneration, death by age 5
	Hypercholestremia	High blood cholesterol	Atherosclerosis of coronary and large arteries

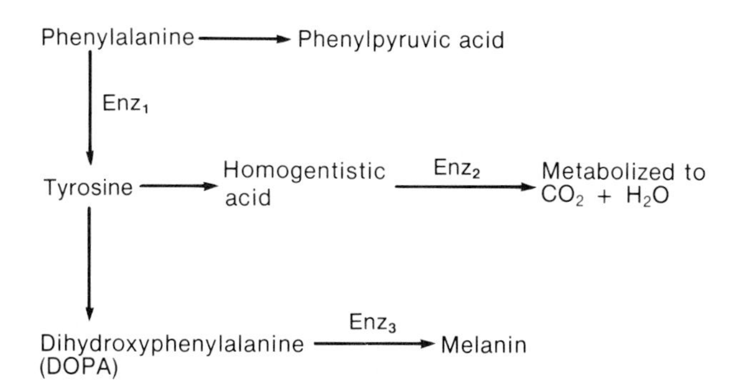

Figure 11.3. *Metabolic pathways for the degradation of the amino acid phenylalanine. Defective enzyme$_1$ produces phenylketonuria (PKU), defective enzyme$_2$ produces alcaptonuria (not a clinically significant condition), and defective enzyme$_3$ produces albinism.*

Inborn Errors of Metabolism

The branched metabolic pathway that begins with the amino acid phenylalanine can serve as an example of an inborn error of metabolism (fig. 11.3). When the enzyme that converts this amino acid to another known as tyrosine is defective, the final products of a divergent pathway accumulate and can be detected in the blood and urine. This disease—*phenylketonuria (PKU)*—can result in severe mental retardation and a shortened life span. Although no inborn error of metabolism is common, PKU occurs so frequently and is so easy to detect that all newborn babies are tested for this defect. If the disease is detected early, brain damage can be prevented by placing the child on an artificial diet low in the amino acid phenylalanine. Besides PKU, there are a large number of other inborn errors of amino acid metabolism, as well as errors in carbohydrate and lipid metabolism (table 11.1).

Point Mutations

Inborn errors of metabolism can vary in severity. Some are invariably fatal, some are potentially serious but can be treated (such as PKU), and some are relatively benign. An example of the latter type of defect is **albinism.**[2] An albino lacks an enzyme needed for the production of the pigment *melanin*[3] (fig. 11.3), which gives color to skin, eyes, and other structures. As a result, an albino has extremely light skin which is more susceptible to the dangerous effects of ultraviolet light from the sun. The eyes of an albino are pink, because the lack of melanin pigment allows the inside of the eyes (which contain blood vessels) to be seen.

Albinism, and indeed most inborn errors of metabolism, are the result of **point mutations.** A point mutation occurs when there is a change in just one nucleotide base in a gene. Suppose that at one point in a gene the base normally present—thymine, for example—is changed to guanine. Since mRNA is produced by complementary base pairing with DNA (chapter 10), the corresponding position in the mRNA will now contain cytosine instead of adenine. The codon of this mRNA which contains the substitute base will thus be different from normal. A codon, it may be recalled, is a sequence of three bases in mRNA which codes for a specific amino acid in a protein. Suppose the normal codon in question is CA*A,* whereas the codon in the person affected with the inborn error of metabolism is now CA*C* (fig. 11.4). This may cause a different amino acid to be placed within a specific location in the protein coded by that mRNA. If that protein is an enzyme, and if the amino acid substitution occurs at a critical location in the enzyme protein, the enzyme will be defective and an inborn error of metabolism will result.

Point mutations are not the only type of mutations that can occur and be inherited. Other, more drastic changes in DNA are also known, and will be described later in this chapter. Point mutations, however, are the most easily understood causes of

[2]albinism: L. *albus,* white
[3]melanin: Gk. *melas,* black

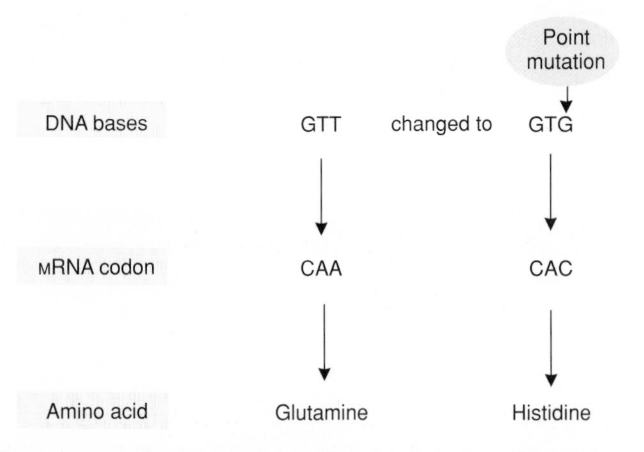

Figure 11.4. *A point mutation in DNA changes one mRNA codon, which may result in a change in one amino acid at one location in the protein produced by that mRNA.*

human genetic diseases. As such, they represent a clear practical application of the recent revolution in knowledge of molecular genetics presented in chapter 10.

The Birth of Genetics

The universe may not only be queerer than we imagine but queerer than we *can* imagine.
J. B. S. Haldane (population geneticist, 1892–1964)

The science of genetics was born before anybody knew of DNA and chromosomes, indeed, before the term "gene" had yet been coined. The basic laws governing inheritance were derived by an Augustinian monk named **Gregor Mendel** (fig. 11.5) based on breeding studies he had performed with the edible pea plant, *Pisum sativum.* Mendel found that each parent contributes an "element" (later to be called a gene) to the offspring, and that these elements could be "antagonistic" (later to be described as a dominance-recessive relationship). These were revolutionary concepts; prior to Mendel, people believed that inheritance was due to a blending of fluids. Such erroneous ideas are still expressed in our language, as in "blood brother" and "blood is thicker than water."

Mendel presented the results of his experiments and his proposed laws of inheritance to the Brünn Society for the Study of Natural Science in 1865. His colleagues responded with boredom to his presentation of peas and math. The proceedings of this meeting were published in 1866; this was Mendel's only publication. Of the 115 copies published and mailed, he received only one response. And that was negative. Mendel died in 1884, a beloved abbot but a scientific unknown.

In 1900, three scientists pursuing independent lines of investigation came to the same conclusions regarding the laws of inheritance as had Gregor Mendel. Mendel's paper was resurrected, and Mendel was given his rightful place as the father of the science of genetics. In 1902, through studies of the behavior of chromosomes during meiosis in grasshopper testes, William Sutton proposed that Mendel's inheritable "elements" are located on the chromosomes.

Figure 11.5. *Gregor Mendel.*

Thomas Hunt Morgan (fig. 11.6) was born the same year that Mendel's paper was published. One of Morgan's uncles was the general who led the Confederate unit known as "Morgan's raiders" during the Civil War; his mother was the granddaughter of Francis Scott Key (the author of our national anthem). Despite a military family tradition, Thomas Hunt Morgan dedicated himself to a career as a biologist and in 1903 became interested in the newly rediscovered mendelian laws of inheritance. He tried and failed to confirm these laws using mice, but succeeded in a spectacular fashion using the fruit fly, or *Drosophila melanogaster.*

In his long and illustrious career, Morgan proved decisively that genes were located on the chromosomes, and showed that males and females differed in their X and Y chromosomes (females have two X chromosomes whereas males have one X and one Y chromosome). He showed why mendelian laws sometimes seem to be violated due to crossing-over of chromosomes (described in chapter 8). Morgan was also able to explain the inheritance of traits such as color-blindness and hemophilia, which are carried by genes located on the X chromosome. For his extensions of Mendel's laws and explanation of the chromosomal basis of inheritance, Morgan was awarded the Nobel prize in 1933.

Thomas Hunt Morgan died in 1945. His death marked the passing of the era in which the science of genetics was devoted to studying the patterns of inheritance. Eight years later, in 1953, the Watson/Crick paper describing the structure of DNA

Figure 11.6. *Thomas Hunt Morgan.*

Table 11.2 Chromosomal locations of some human genetic diseases

Disease	Chromosome	Organ(s) Affected
Huntington's disease	4	Nervous system
Adenomatous polyposis of the colon	5	Large intestine
Cystic fibrosis	7	Lungs
Retinoblastoma	13	Eyes
Polycystic kidney disease	16	Kidneys
Neurofibromatosis	17	Nervous system, muscles, skin, bones
Alzheimer's disease (one form)	21	Nervous system
Muscular dystrophy	X	Skeletal muscles

(chapter 10) was published and the science of molecular genetics was born. First using techniques that had been pioneered by Morgan, and later using the new techniques of genetic engineering, scientists have succeeded in mapping some human genetic diseases to particular chromosomes (table 11.2).

1. Explain the meaning of a "point mutation," and how this can result in the production of an abnormal protein.
2. Explain how PKU and albinism are produced.
3. Why is Gregor Mendel considered to be the father of genetics? What ideas held at the time did he show to be false?

Principles of Mendelian Inheritance

Organisms, including humans, inherit two of each gene. These are carried on each member of a homologous pair of chromosomes. Genes of a pair that specify different conditions (such as tall or short) are known as alleles. A gamete receives only one chromosome from each homologous pair, and thus just one of the alleles. At fertilization, the diploid number of chromosomes is restored and the new individual has a combination of genes (genotype) that is determined by the genetic constitution of the gametes.

The easiest way to understand the basic principles discovered by Mendel is to examine some of Mendel's experiments using pea plants. Although this is not strictly human biology, the basic principles involved in these experiments can be directly applied to the inheritance of some human conditions, as will subsequently be described.

Mendel obtained two separate cultures of the same species of pea plants. One of these was comparatively tall, and the other was short. By placing bags over the flowers and collecting pollen, Mendel could control the reproduction of these plants. He bred the plants in this way to ensure that the tall and short plants always bred true (tall plants always produced tall offspring if interbred, and short plants always produced short offspring). He then mated the tall plants with the short plants.

If inheritance occurred by means of a fluid, the tall and short characteristics should have blended together to produce offspring of intermediate height. This did not occur; all of the offspring were tall! We can borrow from Mendel's later proposals, and from concepts developed still more recently, to explain this outcome.

Each individual pea plant (or human) inherits one set of chromosomes from its mother and one from its father. The cells of the individual are thus *diploid,* containing pairs of homologous chromosomes (as described in chapter 8). A gene determining the height of the pea plant is located on each member of a particular pair of homologous chromosomes. Each individual pea plant, therefore, has two genes for height. Following Mendel's example, we can use the symbol *T* for the tall gene and *t* for the short gene. Tall (T) and short (t) are thus different forms—called different **alleles**[4]—of the gene for height. The term **genotype** describes the genes that are present in each individual. The true-breeding tall plants have the genotype TT, and the true-breeding short plants have the genotype tt. Since both alleles in this case specify the same thing (TT or tt), the parents are said to be **homozygous.**

Gametes (sperm, or pollen, and eggs) are produced by meiosis, which is the type of cell division in which the chromosomal number is halved. A gamete gets only one of each pair of homologous chromosomes. The gametes, therefore, contain only one of the two alleles of each gene present in the diploid cell that underwent meiosis. Gametes are thus *haploid.* If the parent is homozygous (TT, for example), all of the gametes it produces will contain the same allele (T). When the true-breeding tall and short parents are mated, therefore, the offspring must have the genotype Tt.

[4]allele: Gk. *allelon,* of one another

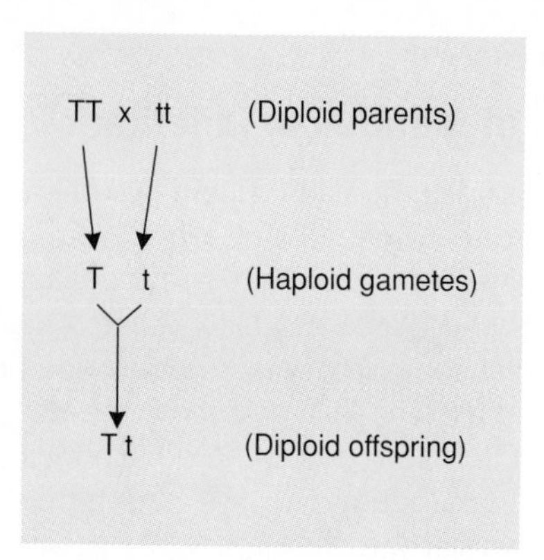

Since the alleles for height specify different conditions, the offspring are termed **heterozygous.** Despite this genotype, these offspring appear indistinguishable from their tall (TT) parents. The appearance, or other result of the action of the genotype, is termed the **phenotype.**[5] The phenotype of the heterozygous offspring, in other words, is the same as their homozygous tall parents. The allele for tall must thus be **dominant,** whereas the allele for short is **recessive.**

Mendel derived the concepts stated above as a result of his next experiment. After obtaining the tall offspring and allowing them to grow to maturity, he interbred them. Using our advantage of viewing his experiments from more than 120 years in the future, we can easily predict the results of this experiment.

A good way to visualize the results of this cross is by use of a device known as a *Punnett square* (fig. 11.7). The different possible genotypes of the gametes of each parent are shown horizontally and vertically, and the possible combinations are represented in boxes where a column and row intersect.

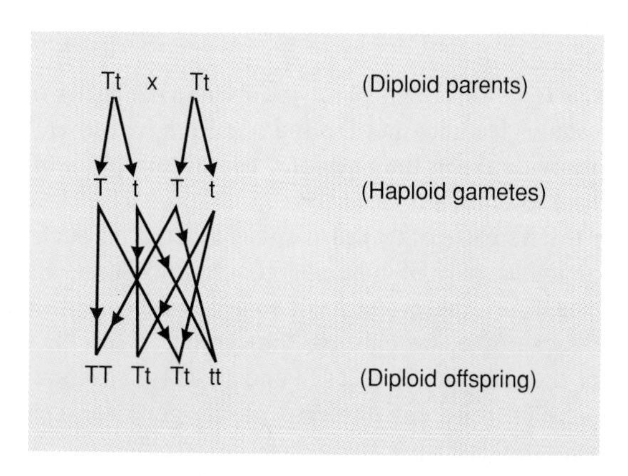

Predicting the Outcome of a Genetic Cross

Since tall is dominant to short (in pea plants), there are three ways that gametes could combine to produce a tall offspring (TT, Tt, and Tt). Only one possible genotype (tt) out of the four total combinations could produce a short phenotype. Phrased another way, the chance that any given offspring from this mating will be tall is 3/4, or 75%, whereas the chance that it will be short is 1/4 (25%).

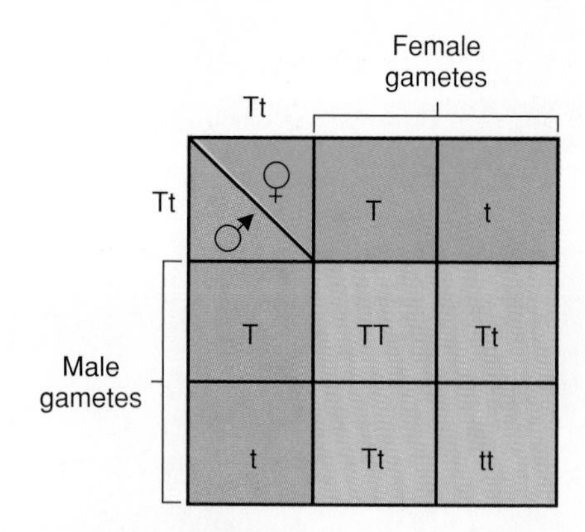

Figure 11.7. *A Punnett square, showing the results of crossing two heterozygous (Tt) pea plants.*

Since nobody can predict which particular sperm (T or t) will fertilize which particular egg (T or t), the above analysis represents only probabilities. Suppose, for example, you have four marbles in a bag; three are black and one is red. If you close your eyes and reach your hand into the bag to pull out one marble, you have a 75% chance that it will be black. Although the chances are against it, however, you could pull out a red marble. If you repeat this with four hundred such bags (drawing just one marble from each bag), the laws of probability are that you would pull out approximately (but probably not exactly) 300 black marbles and 100 red marbles. For example, you might pull out 290 black and 110 red, or 312 black and 88 red, and so on.

The larger the number of times this experiment is repeated, the closer will the combined results approach the predicted 3:1 ratio of black:red (or tall:short). Since the mathematics of probability were unknown to Mendel's audience when he presented his results, it is not surprising that they could not appreciate his findings.

There is one additional and very important aspect of the laws of probability that relates to genetic crosses. If the events involved are random (as in the joining of a sperm with an egg), *the probability of a particular outcome is always the same regardless of the outcomes that preceded it.* Every time a pea plant is born from a Tt × Tt mating, for example, the probability of it being short is 25%. Suppose that you discover one short pea plant offspring as you walk along a row of planted second-generation pea plants. The chances that the next plant in the row is also short is again 25%. This is because the deck is shuffled (the chromosomes assort independently) each time. Lack of understanding of this principle leads to the *gambler's fallacy.* Even if a gambler playing dice has a run of bad (or good) luck, the chances of success or failure at the next throw of the dice should be the same as at the very first throw.

1. Describe the meaning of the terms *diploid* and *haploid.*
2. Describe the meaning of the terms *homozygous* and *heterozygous.*
3. Using Punnett squares, compare the results of these two crosses: (a) Tt × Tt; (b) Tt × TT; and (c) Tt × tt. What is the probability that each of these crosses could produce a tall or a short offspring?
4. Explain the nature of the "gambler's fallacy," and why it is wrong.

[5]phenotype: Gk. *phainein,* to show; *typos,* type

Simple Mendelian Inheritance in Humans

The inheritance of many human conditions follows the same pattern as the inheritance of height in pea plants. Albinism, cystic fibrosis, and Tay-Sachs disease, for example, are inherited as simple mendelian recessive traits. The genes for these conditions are located on autosomal chromosomes. Some conditions, however, are coded by genes located on the X chromosome. Such X-linked genes include those involved in hemophilia, color blindness, and muscular dystrophy. Inheritance of X-linked traits are generally passed from mother to son.

The laws that Mendel discovered in pea plants also apply in human genetics. In this section, some of the human conditions inherited in a simple mendelian manner will be examined.

Some Human Examples of Mendelian Inheritance

The three conditions described in this section share in common the fact that their genetic analysis follows exactly the same pattern as the inheritance of height in pea plants. Of these three conditions, however, only cystic fibrosis and Tay-Sachs disease are of extreme medical significance; the ability to taste PTC has no known consequences to health, and albinism is medically significant only when the albino has too great an exposure to ultraviolet light from the sun.

PTC Tasting

It was discovered by chance that some people could not taste the chemical phenylthiocarbamide (*PTC*), whereas others found it to have a bitter taste. The ability to taste PTC is inherited as a *simple medelian dominant trait.* We can thus use the same analysis as performed in the pea plants. This time, however, we will let T = taster and t = nontaster. Suppose you find that you are a nontaster, but both your parents are tasters. What are the genotypes of you and your parents?

Since you are a nontaster, and since the trait of nontasting is recessive, you must have the homozygous genotype tt. Since you inherited each allele from each parent, each parent must be a carrier for t. In order to be a carrier for t, and yet to be a taster, each parent must thus have the heterozygous genotype Tt. (A **carrier** of a genetic condition is always one who is heterozygous and thus has the normal phenotype.) This being the case, what are the chances that your brother is a taster or a nontaster? (The answers are 75% and 25%, respectively.)

Albinism

Albinism is inherited as a simple mendelian recessive trait. If we let A = normal and a = albino, we can predict the results of matings just as with Mendel's pea plants. Suppose an albino marries a person with normal pigment and they produce four children. Three of these four children have normal pigmentation and one is an albino. What are the genotypes of the parents?

We immediately know the genotype of one of the parents: the albino parent must be aa. But what of the other parent? A person with normal pigmentation must have the *A* allele, but could have either the AA or Aa genotypes. If this parent had the AA genotype, the cross between the parents would look like this:

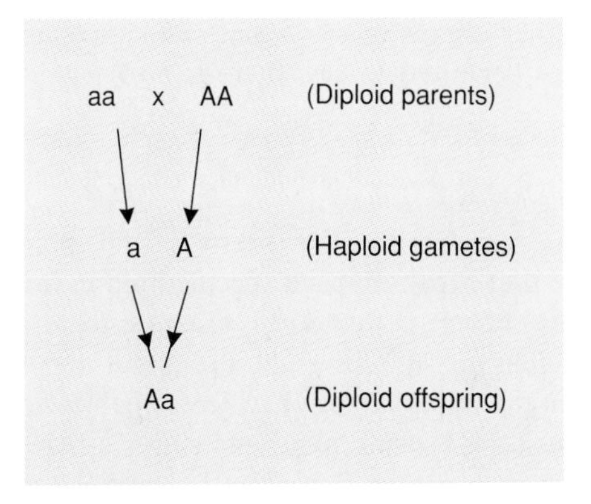

All of the children possible from this mating would have normal pigmentation (because A is dominant to a) but would be carriers of albinism. Since we know that these parents produced an albino child, we know that the parent with normal pigmentation must have the genotype Aa. This cross would be as follows:

From this genetic analysis, we would predict that the chances of this couple producing normal or albino children would be 50% each. But, since this couple had four children, one might think that two of the four would be normal and two albino. This is not true. The genetic analysis tells us only that the probability of a given child being normal or albino is 50%. It is quite possible, as in this example, that three children could have been born with normal pigmentation and one with the albino condition.

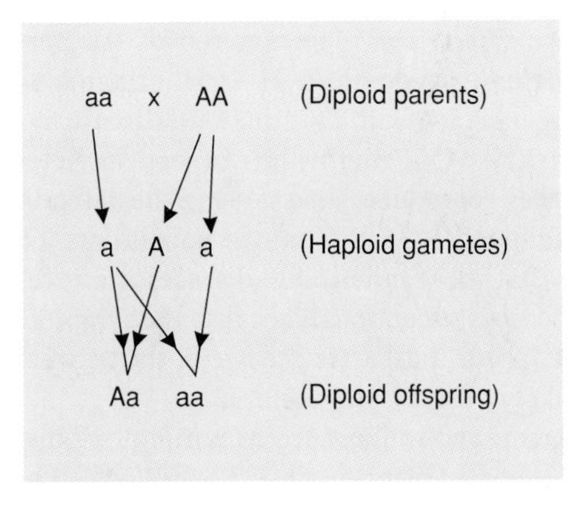

Cystic Fibrosis

Cystic fibrosis is the most common fatal inheritable disorder among Caucasians. Cystic fibrosis is inherited, like albinism, as a simple mendelian recessive trait. If C = normal and c = cystic fibrosis, a person must have the genotype cc to inherit the disease. Approximately one out of twenty whites are heterozygous carriers (Cc) for this condition, and about one in two thousand are children with this disease who are homozygous (cc). Most children who inherit cystic fibrosis die before the age of thirteen.

It has recently been demonstrated that people with cystic fibrosis have a defect in the ability of chloride ions (Cl^-) to pass through cell membranes. This results in a very thick, sticky mucus that blocks the airways of the lungs. It also affects the liver and other organs in ways that can cause death. As with most serious genetic diseases, there is no known cure for this disease.

Tay-Sachs Disease

Tay-Sachs disease[6] is an inborn error of metabolism in which the enzyme that breaks down a specific lipid in the brain is defective. This enzyme is found in lysosomes, the organelles that are responsible for digesting old structures and turning over molecules in the cell (chapter 4). Affected babies appear normal at birth, but soon become blind and suffer other symptoms of nervous system degeneration. Death invariably occurs after about the age of four or five. This disease is rare in the general population, but is carried in a heterozygous state by about one in twenty-eight Ashkenazi Jews (those with Central or Eastern Europe ancestry).

The heterozygous carrier state for Tay-Sachs disease can now be diagnosed by routine blood tests. If both husband and wife find that they are heterozygous for this condition, they stand a 25% chance of giving birth to a baby with Tay-Sachs disease. They can thus elect to not have children. Such knowledge and its application has led to a dramatic reduction in the incidence of this disease in recent years.

Huntington's Disease

Unlike cystic fibrosis, albinism, and Tay-Sachs disease, which are inherited as recessive traits, the gene for Huntington's disease[7] is *dominant* to the normal gene. People with Huntington's disease suffer from degeneration of the nervous system leading eventually to death. If H = Huntington's disease, and h = normal, a person with Huntington's disease most likely has the genotype Hh. One might thus expect the heterozygotes to die before they reproduce, thus causing the defective gene to be rapidly eliminated from the human population. This is indeed the reason that most inheritable diseases are recessive. Huntington's disease is exceptional, because the symptoms of the disease do not appear until after the age of thirty, when the person has most likely already had children.

One person who suffered from Huntington's disease was the folksinger Woody Guthrie (fig. 11.8). Woody Guthrie wrote many famous songs (including "This Land is Your Land") before developing the symptoms of the disease. Since the disease was not well known at the time, physicians had difficulty diagnosing it (some thought the symptoms were due to alcoholism). Woody Guthrie served as a model for many future folksingers and songwriters—including Bob Dylan, who composed a song in his honor and sang it to Guthrie before he died. People with this disease can thus live normal, very productive lives until the symptoms progress too far. It is certainly understandable that many of those who might inherit the disease, including Woody Guthrie's son, singer Arlo Guthrie, refuse to take tests that could determine if they will later develop the disease.

Figure 11.8. *Woody Guthrie, a folksinger who inherited Huntington's disease.*

X-Linked Inheritance

You may have noticed that, in the previous genetic examples, the sex of the parents was not specified. This is because the traits previously considered are determined by genes located on the *autosomal* chromosomes (all chromosomes other than X or Y). Since both males and females have a homologous pair of all autosomal chromosomes, the genetic analysis need not specify whether Aa (for example) is the mother or father. The results are the same either way.

If a trait is determined by a gene located on the X chromosome, however, the situation is different. This is because females have two X chromosomes whereas males have only one X chromosome. (The other chromosome in males is the Y chromosome, which appears to function only in determining the sex of the embryo, as described in chapter 9.) Inheritance that is carried on the X chromosome is referred to as **X-linked** or **sex-linked** inheritance.

[6]Tay-Sachs disease: after Warren Tay, English physician (1843–1927), and Bernard Sachs, American neurologist (1858–1944)

[7]Huntington's disease: after George Huntington, American physician (1850–1916)

Figure 11.9. *Queen Victoria (seated in front center) together with some of her descendants. Alexandra (standing behind Victoria to her left) was a carrier for hemophilia who married Czar Nicholas II.*

Hemophilia

Hemophilia is the inability to form blood clots properly. There are a variety of inheritable forms of this disease, but the most famous form is a genetic defect found in the royal families of Europe. This disease has been traced to Queen Victoria (fig. 11.9) of England (1819–1901), who was normal but a carrier of this condition. Thus far, ten of her male descendants inherited this disease (fig. 11.10).

Why were all of the hemophiliac descendants of Queen Victoria male? A genetic analysis of one such mating will provide the answer. Alexandra, a granddaughter of Queen Victoria, was a carrier for hemophilia. Let H = normal and h = hemophilia. Alexandra thus had the genotype Hh, but since these genes are carried on the X chromosome, we will show this genotype as X^HX^h. Alexandra married Czar Nicholas II of Russia, who was normal. We can show his genotype as X^HY. Please notice that the Y chromosome does *not* contain a gene for blood clotting. A male thus has only one allele for this condition. He is either normal or hemophilic; males *cannot* be heterozygous carriers of X-linked traits.

Nicholas and Alexandra had four daughters and one son. All of the daughters had normal blood clotting; the son—Alexis—was hemophilic. The genetic analysis is presented in a Punnett square in figure 11.11. From this analysis, we can see that all of the daughters would have normal blood clotting, although the probability of a daughter being a carrier of this condition is 50%. The chances that a son would inherit hemophilia is 50% (but the chances that a hemophilic child from this mating is male is 100%).

Color Blindness

Color vision is produced by specific receptors called *cones* in the retina of the eyes. There are three types of cones that respond best to the colors red, green, and blue, respectively. This selectivity is due to a slightly different pigment in each type of cone. The gene that codes for the pigment in the blue cones is located on an autosomal chromosome. The genes for the red and green cones are located on the X chromosome.

In the most common form of **red-green color blindness,** the gene for green cones is defective, so that green cannot be distinguished from red (fig. 11.12). In a less-common form of this disorder, the gene for the red cones is defective, so that red cannot be distinguished from green. Both types of red-green color blindness together are found in approximately 8% of males and 0.04% of females. The lack of both red and green cones in a given individual would abolish color vision entirely; this is an extremely rare condition.

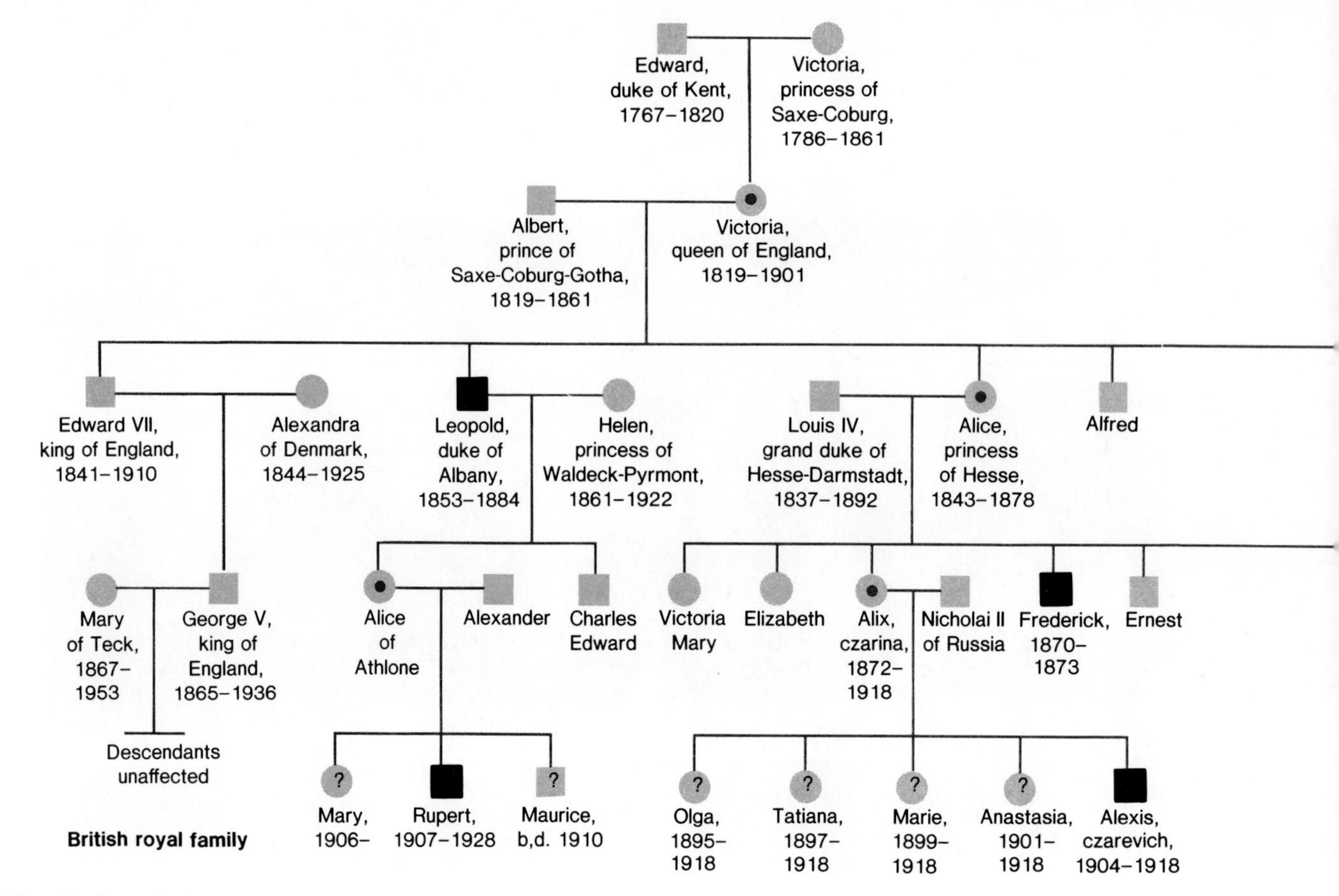

Figure 11.10. *Pedigree for hemophilia among the royal families of Europe. Note the key above the figure.*

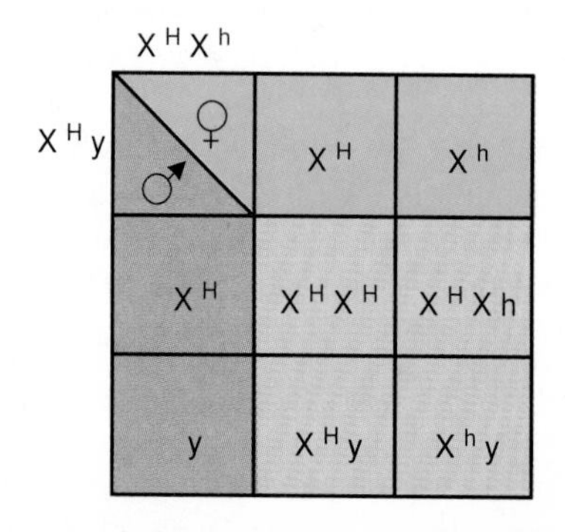

Figure 11.11. *A Punnett square showing the inheritance of hemophilia. Notice that, among the male children, there is a 50% chance of contracting the disease. None of the daughters would have hemophilia, but there is a 50% chance that a daughter could be a carrier.*

We can perform the same type of genetic analysis with color blindness as with hemophilia. One difference between these conditions is that there are far more color-blind females than there are hemophilic females. This is probably due to the fact that hemophilia is fatal for mature females undergoing menstruation (at least in the past, before the clotting factors were

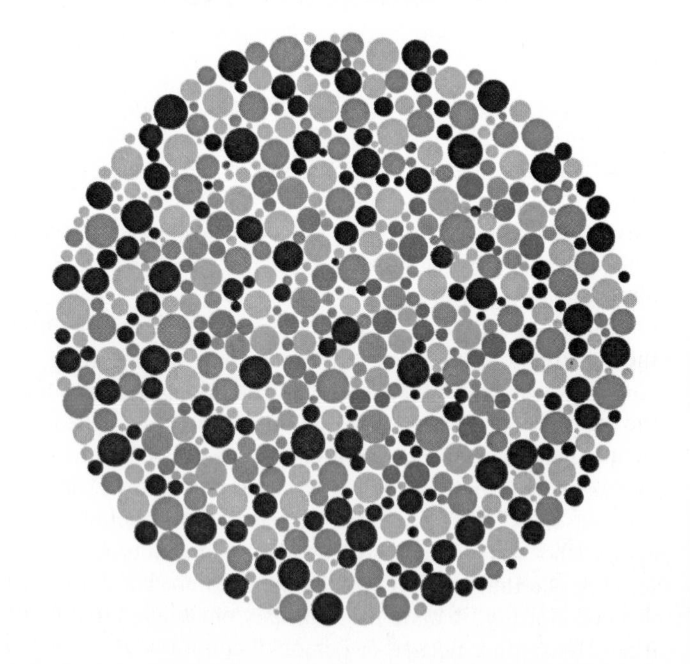

Figure 11.12. *A test for red-green color blindness. People with this X-linked trait cannot read the number embedded in the colored dots. The above has been reproduced from Ishihara's Tests for Colour Blindness published by Kanehara & Co., Ltd., Tokyo, Japan, but tests for color blindness cannot be conducted with this material. For accurate testing, the original plates should be used.*

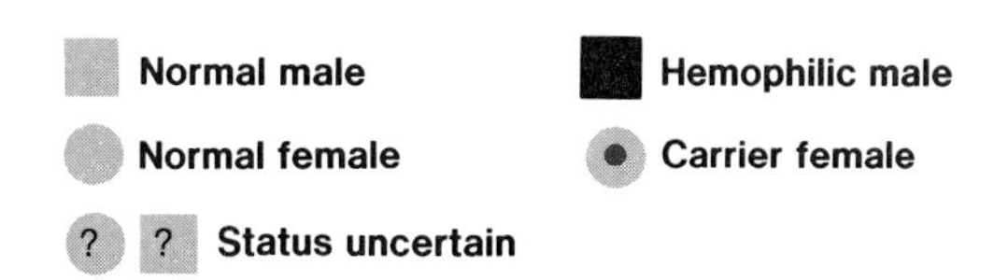

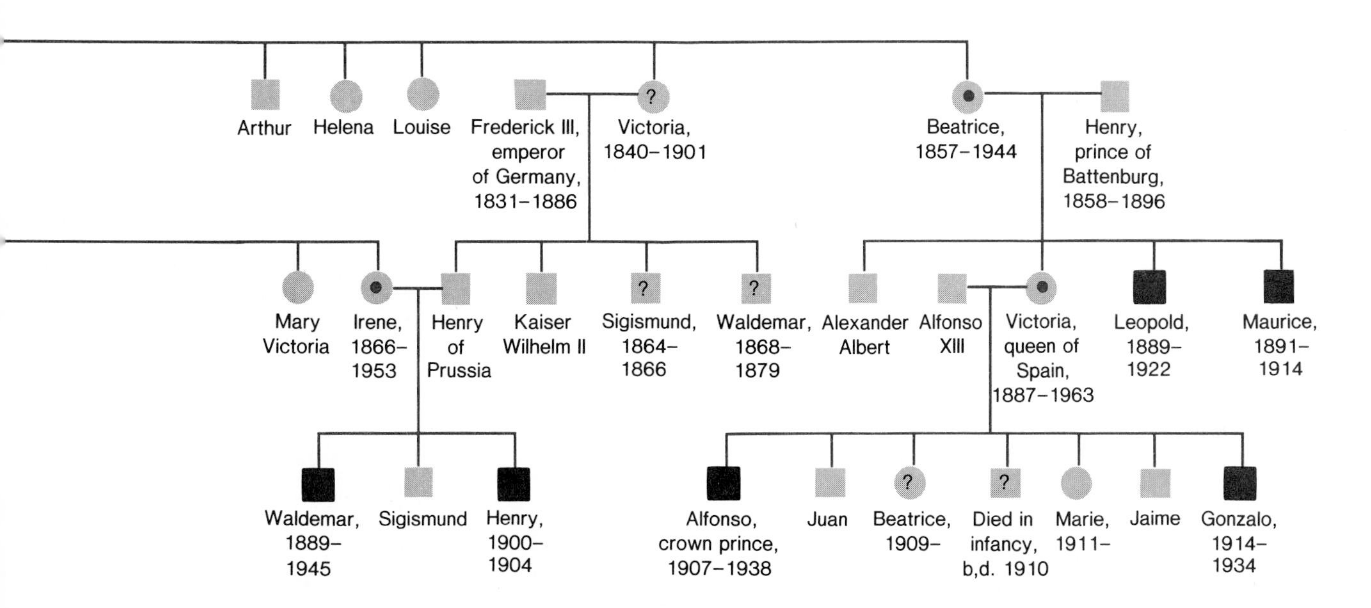

medically available). Color blindness, on the other hand, is relatively benign. (People with red-green color blindness learn to use other cues to distinguish traffic lights, for example.)

Muscular Dystrophy

Muscular dystrophy[8] **(MD)** is a family of diseases that involves muscle wasting. The most common and serious form of this disease is called *Duchenne*[9] muscular dystrophy, which is inherited on the X chromosome. Since it is an X-linked trait, it is found almost entirely in males with a frequence of one in three thousand five hundred births. Most who inherit the disease die in their young adulthood.

Duchenne muscular dystrophy can involve the heart and also the brain, causing mental retardation. The skeletal muscles lose their normal structure, developing connective tissue fibers in place of lost muscle cells, and thus weaken. It has recently been discovered that patients with MD lack a particular protein called *dystrophin.* Using the techniques of DNA analysis described in chapter 10, scientists have found that the gene that codes for this protein is huge, being composed of two to three million base pairs and containing about sixty exons. Finding a cure for this disease through genetic engineering techniques is thus a formidable task for scientists dedicated to this endeavor.

[8]dystrophy: Gk. *dys,* abnormal; *trephein,* to nourish

[9]Duchenne: after Guillaume Benjamin Amand Duchenne, French neurologist (1806–1875)

1. If one prospective parent finds that he or she is not a carrier for an autosomal recessive trait, need the other prospective parent be tested for the carrier state? Explain.
2. If two parents who are carriers for cystic fibrosis have a child with this disease, what are the chances that their next child will be born with the disease? Explain.
3. Can a man be a carrier for an X-linked trait? Explain. Explain why he cannot pass any of his X-linked genes to his sons.
4. Explain why X-linked traits are generally passed from the mother to the son. How could a girl be born hemophilic?

Inherited Diseases of Hemoglobin

Hemoglobin contains two pairs of polypeptide chains, called alpha and beta, which are coded by two genes. A point mutation in one of these genes produces a single amino acid substitution which results in sickle-cell anemia. This disease is inherited as an autosomal recessive trait. Thalassemia is a related disorder which has a greater variety of causes. The heterozygous states for both disorders confer a resistance to malaria.

The inherited diseases involving the protein hemoglobin follow the same patterns previously discussed, with the diseases inherited as simple mendelian recessive traits. These diseases are probably the most widely distributed and best characterized of the genetic disorders.

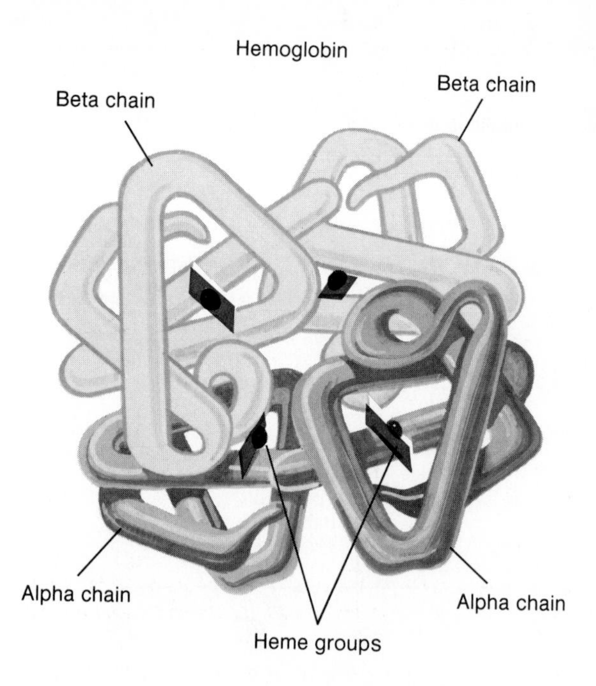

Figure 11.13. *The structure of hemoglobin, showing the two alpha and two beta polypeptide chains. The heme groups are the parts that bind to oxygen.*

In order to understand these genetic diseases, some knowledge of the structure of hemoglobin is required. Hemoglobin is a protein found in the red blood cells. Indeed, the red color of blood is produced by the hemoglobin molecules. There are approximately 280 million molecules of hemoglobin within each red blood cell. Each hemoglobin molecule consists of four polypeptide chains and four organic structures known as the heme groups (fig. 11.13). The heme groups are the parts of the hemoglobin that bind to oxygen. This enables the red blood cells to pick up oxygen in the lungs, transport it in the blood, and release it to the tissues (chapter 15).

It is the protein component of hemoglobin that is altered in genetic disorders. The four polypeptide chains constitute two identical pairs. In normal hemoglobin, called *hemoglobin A* (the *A* stands for "adult") there are two polypeptide chains called the *alpha chains,* and two called the *beta chains.* A specific genetic disease involving hemoglobin will affect either the alpha or the beta polypeptide chains.

Sickle-Cell Anemia

The production of an abnormal form of hemoglobin known as *hemoglobin S* is caused by an autosomal recessive gene, found predominately among the black population. Suppose S = normal and s = sickle cell. Approximately 9% of the black population of the United States is heterozygous (Ss) for this condition and 0.2% has the disease and is thus homozygous (ss).

When a person is homozygous for this condition, the abnormal hemoglobin S forms a crystallinelike structure when the oxygen levels of the blood are low. This causes the red blood cells to assume their characteristic sickle shape (fig. 11.14). These sickled cells are less flexible than normal, and thus cannot easily pass through narrow blood channels known as capillaries. As a result, the flow of blood and oxygen to different organs is reduced, leading to organ damage.

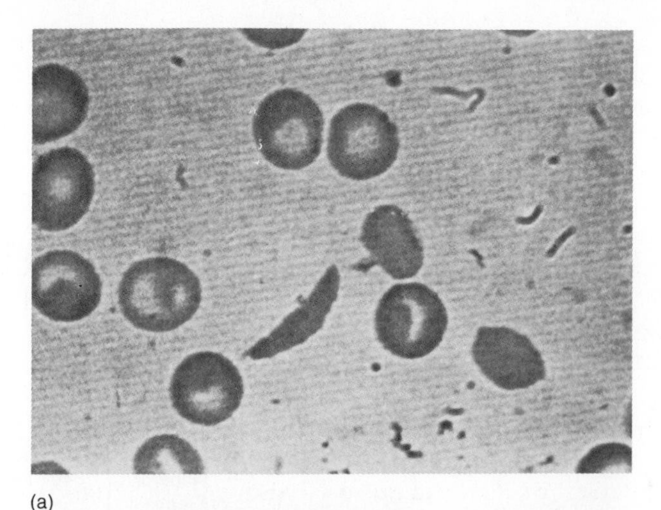

(a)

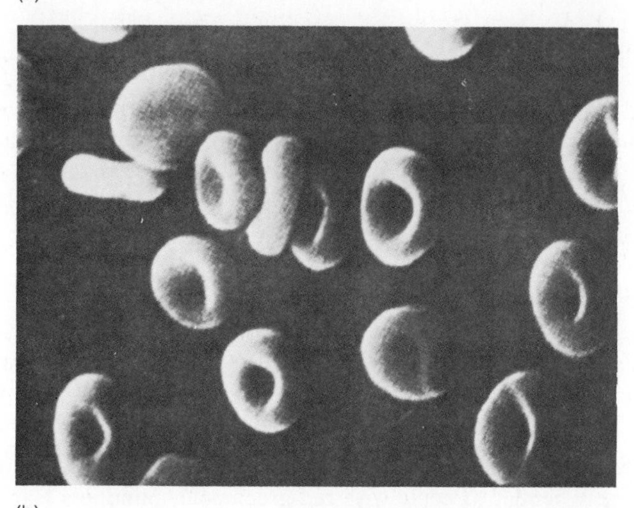

(b)

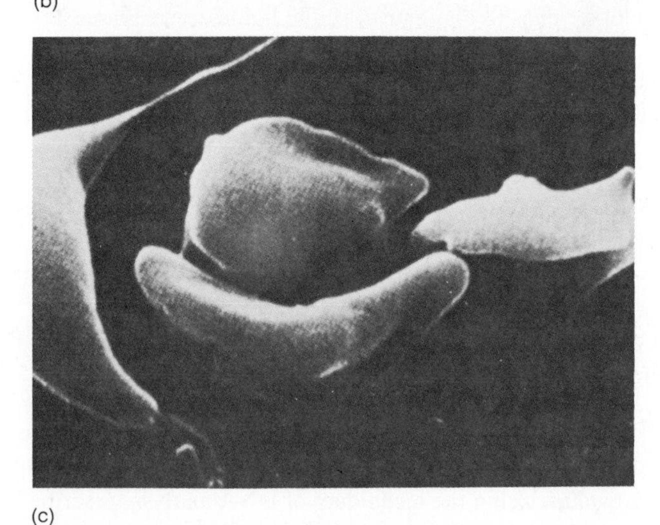

(c)

Figure 11.14. *(a) A sickled red blood cell as seen in the light microscope. (b) Normal cells. (c) Sickled red blood cells as seen in the scanning electron microscope.*

A person who is heterozygous for sickle-cell anemia does not have the disease but is said to have *sickle-cell trait.* This is because the abnormal allele does produce abnormal hemoglobin, and the normal allele produces normal hemoglobin. The red blood cells thus contain a combination of these two forms of hemoglobin. When the oxygen concentration of the blood is very low, as in high altitude, people with sickle-cell trait may experience some symptoms of the disease. Moreover, their hemoglobin will be insoluble under these conditions, which provides a basis for the diagnosis of the carrier condition.

Sickle-cell anemia was the first genetic disease that was understood at the molecular level. It was discovered that a point mutation is present in the gene coding for the beta chains of hemoglobin: the sequence GAG is changed to GTG. This point mutation causes a change in one codon in the messenger RNA produced by this gene, which results in the substitution of the amino acid valine for glutamic acid at the sixth position of the beta chains.

Mutations occur as a result of random changes in DNA. Those changes that are completely harmful are removed by natural selection (chapter 2). One therefore might wonder why sickle-cell anemia is so prevalent in the black population. The answer was first suggested by the geographical distribution of the sickle-cell gene: it is most frequent in regions that have a high incidence of malaria (fig. 11.15). This relationship was subsequently confirmed by experimental evidence showing that the parasite that causes malaria, which resides in the host's red blood cells, cannot live in red blood cells containing hemoglobin S. People who are heterozygous for hemoglobin S (who have the genotype Ss), are thus at a selective advantage in an area where death from malaria is more common than death from sickle-cell anemia.

Since the carrier state for sickle-cell anemia can easily be detected from a simple blood test, the birth of children with the disease can be prevented. If both parents are normal, but one parent is a carrier (with the genotype Ss) and one is not a carrier (has the genotype SS), there is absolutely no chance that they can produce a child with sickle-cell anemia. If both parents are carriers, however, the chances of producing a child with this disease is 25%. In such a situation, the couple may elect to not have children. This analysis, it should be noted, is true for every disease that is inherited as a simple recessive trait, but is useful only for those diseases that can be detected in a carrier state. At present, only sickle-cell disease and a few others can be so detected.

Thalassemia

Thalassemia[10] is a family of hemoglobin disorders that occurs principally among people of Mediterranean ancestry. It is similar to sickle-cell anemia in that people who are heterozygous for this condition display a resistance to infection with malaria. Also, as in sickle-cell anemia, it is the beta chains of hemoglobin that are usually affected.

[10]thalassemia: Gk. *thalassa,* sea (referring to the Mediterranean); *haima,* blood

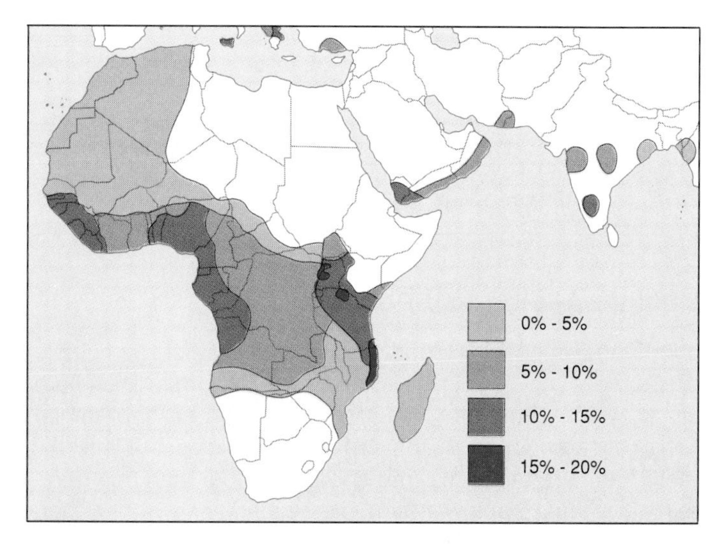

Figure 11.15. *The frequency distribution of the gene for sickle-cell anemia. It is most prevalent in the equatorial region, where malaria is common.*

The gene that codes for the alpha chains of hemoglobin are present in two copies on chromosome number 16. The presence of duplicate copies of this gene help to explain why disorders of the alpha chain rarely occur. The gene that codes for the beta chain of hemoglobin is found as only one copy on chromosome number 11. Farther along on the same chromosome are two copies of a gene that codes for a different polypeptide chain. This different polypeptide chain, called the *gamma chain,* is found in the hemoglobin of all fetuses (forming *hemoglobin F,* or fetal hemoglobin). Normally, the gamma gene is turned off shortly before birth, causing a decrease in hemoglobin F production as the production of normal hemoglobin A is increased.

Surprisingly, if a mutation causes the deletion of the entire gene for the beta chains of hemoglobin, the type of thalassemia thus produced is not dangerous. This is because the gamma genes are activated, and fetal hemoglobin is produced in the adult as a compensation for this condition. More serious effects are produced when the beta gene is present but defective, because in these cases the gamma genes are not as fully activated. There are over two hundred known point mutations in the gene for the beta chains (resulting in the substitution of one amino acid for another) which can cause thalassemias of different degrees of severity.

A different type of mutation, called a **frame-shift mutation,** produces severe forms of thalassemia. In a frame-shift mutation, one nucleotide base pair in the gene is deleted. Since the DNA code is read off as base triplets, such a mutation causes all triplets downstream of the deletion to read differently. For example, suppose the normal base sequence is AGTCCGATG. The base triplets would be AGT, CCG, and ATG. If the first T (thymine) is deleted, the base triplets are changed to AGC, CGA, and TG __ . These altered codons result in numerous changes in the amino acid sequence of the beta chains of hemoglobin.

1. Explain the cause of sickle-cell disease, starting from the DNA code and ending with organ damage.
2. Explain why sickle-cell disease and thalassemia have not been eliminated by natural selection from the human population.
3. Describe the meaning of the term *frame-shift mutation,* and compare it with a point mutation.

Multiple Alleles and Polygenic Traits

Inheritance of the ABO blood groups is an example of multiple alleles, because one gene can exist in three forms. These are designated by the letters A, B, and O, which describe molecules called antigens on the surface of red blood cells. Polygenic traits are those that are coded by more than one gene pair, and thus display intermediate conditions rather than either-or phenotypes. Skin color is an example of a polygenic trait.

Analysis of the patterns of heredity in humans and other organisms can be more involved than those that have previously been described. This does not mean that the simple mendelian laws are disobeyed. Rather, the more complex patterns of inheritance are extensions of these laws which take additional facts into account. These include the fact that there can be more than two alleles for a particular gene, and that more than one gene pair may influence a particular trait.

Multiple Alleles: Inheritance of Blood Types

The immune system is able to recognize particular molecules and respond by producing *antibody* proteins which can combine with these molecules. The molecules that can stimulate the immune system in this way are known as *antigens.* An antibody made in response to a specific antigen can only bind to that particular antigen. Although the immune system is discussed more completely in chapter 13, the terminology of antibodies and antigens is required in order to understand blood typing.

The red blood cells contain, on the surface of their cell membranes, molecules that can act as antigens. These are the **blood group antigens.** The major blood group antigens are known as the *ABO system.* Another important red blood cell antigen is known as the *Rh factor,* and will be discussed separately.

The ABO System

The antigens of the ABO system are coded by a gene identified as *I.* Unlike all of the other genes that we have considered up to this point, the I gene can exist in three, rather than two, alleles. These **multiple alleles** are: I^A, I^B, and I^O. For simplicity, we will show these alleles as A, B, and O. The *A* allele produces the A antigen, the *B* allele produces the B antigen, and the *O* allele produces neither the A nor the B antigen ("O" can be thought of as indicating zero).

The blood type of a person indicates the antigens present on the red blood cell surface. A person who is *blood type A* thus has the A antigens on his or her red blood cells. This is the person's phenotype. The genotypes that can produce this phenotype are AA or AO. This person inherited the *A* allele from one parent and either the *A* or the *O* allele from the other parent.

Table 11.3 The ABO Blood Group

Possible Genotypes	Antigens on RBCs	Antibodies in Plasma
AA; AO	A	Anti-B
BB; BO	B	Anti-A
AB	AB	Neither anti-A nor anti-B
OO	O	Both anti-A and anti-B

Similarly, a person who is *blood type B* has the genotye BB or BO. A person who is *blood type AB* has both the *A* and *B* antigens on the surface of the red blood cells, and thus has the genotype AB. Notice that there is no dominance-recessive relationship between A and B—they are said to be **codominant.** A person with *blood type O* has neither the *A* nor the *B* antigens on his or her red blood cells and must have the genotype OO (table 11.3).

A person's blood plasma (the fluid part of the blood) contains antibodies against the ABO antigens that are absent from the person's blood. A person who is blood type A thus has antibodies against type B and type AB blood. A person who is blood type B has antibodies against type A and type AB blood. A person who is blood type AB has no antibodies against these antigens, and one who is blood type O has both anti-A and anti-B antibodies. It is not clear what causes these antibodies to be produced; one possibility is that exposure to closely related bacterial antigens causes the production of antibodies that cross-react with the blood group antigens. A person does not develop antibodies against his or her own blood group antigens due to tolerance mechanisms.

What would happen if a person with type A blood did have antibodies against type A red blood cells? This can easily be seen by mixing the blood from a type A person with the serum from a person who is type B. The type B blood does have antibodies against the type A red blood cells. These antibodies combine with the antigens on the red blood cell surface and in this way cause the red blood cells to stick together (fig. 11.16). This clumping of red blood cells due to antigen-antibody reactions is known as **agglutination.** Such a reaction could happen in the blood vessels of a type A person who was erroneously given type B blood in a transfusion. The agglutinated particles could then block small blood vessels, leading to organ damage. Further, the antigen-antibody reactions at the red blood cell surface could lead to widespread hemolysis (destruction of red blood cells).

Before blood transfusions are given, therefore, the blood must be typed. This is done by mixing the person's blood on a glass slide with anti-A and anti-B antibodies contained in two fluids known as antisera (fig. 11.16). In determining the blood type, you are what you clump. An agglutination with anti-A antiserum but not anti-B means that you are type A. Agglutination with anti-B but not anti-A means that you are type B. If agglutination occurs with both antisera you are type AB, and if it occurs with neither you are type O.

Since the type O red blood cells will not agglutinate when mixed with the plasma of a type A, type B, or type AB person, blood type O is sometimes called the *universal donor.* This is only valid in emergency situations when small amounts of blood

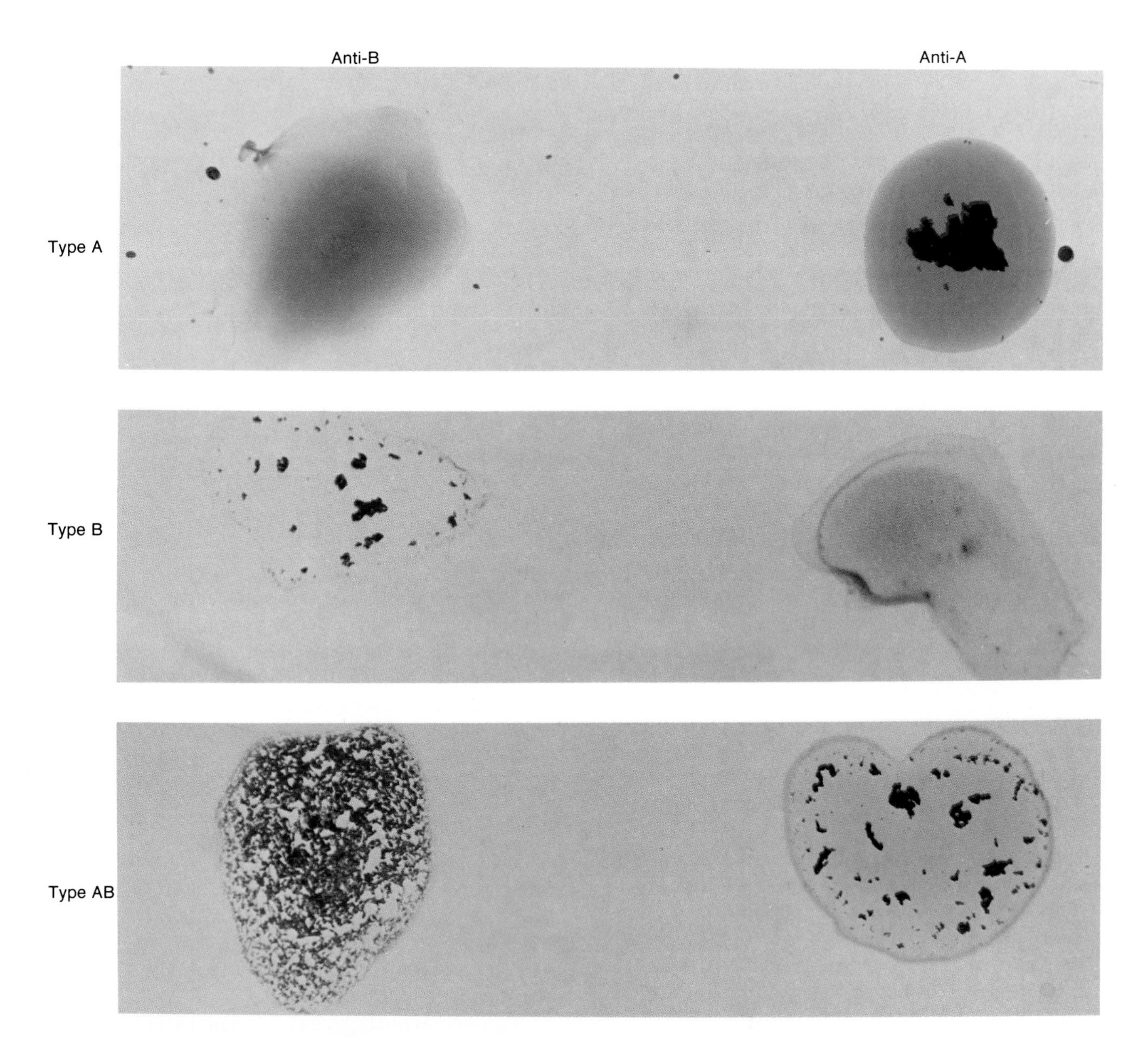

Figure 11.16. *The agglutination (clumping) of red blood cells occurs when cells with A-type antigens are mixed with anti-A antibodies and when cells with B-type antigens are mixed with anti-B antibodies. No agglutination would occur with type O blood (not shown).*

are transfused, however, because (1) there are other blood group antigens on red blood cells; and (2) if any plasma from a type O person is given along with the red blood cells, the anti-A and anti-B antibodies in the type O plasma can agglutinate the recipient's red blood cells. Type AB is sometimes called the *universal recipient,* because the plasma of people with type AB blood lacks antibodies to agglutinate red blood cells with the *A* or *B* antigens. The universal recipient concept is also subject to the limitations mentioned for the universal donor.

Rh Factor

Another antigen present on the red blood cells of most people is known as the **Rh factor.** Although first discovered in rhesus monkeys (from which the name is derived), humans have similar antigens. The Rh antigen is inherited as a simple mendelian dominant trait. If + indicates presence of the Rh antigen and − indicates absence of the Rh antigen, a person who is Rh positive may have the genotypes ++ or +−, whereas a person who is Rh negative must have the genotype −−.

Suppose an Rh negative woman marries an Rh positive man and becomes pregnant with an Rh positive fetus. The fetus thus has an antigen on its red blood cells that is foreign to the mother. Since the maternal and fetal bloods do not usually mix in the placenta, the mother will not become exposed to this foreign antigen while carrying the fetus. At the time of birth, however, the immune system of the mother can become exposed and sensitized to this foreign antigen. The next time the woman is pregnant with an Rh positive fetus, the antibodies that she has made against the Rh antigen may cross the placenta and attack the fetal red blood cells. This can result in immune destruction of the fetal red blood cells, a condition called **hemolytic disease of the newborn (HDN).** A baby born with this condition must receive massive blood transfusions.

Fortunately, this disease can be prevented if the Rh negative mother receives an injection of antibodies against the Rh factor within seventy-two hours after the birth of her first and subsequent Rh positive babies. These antibodies (called *Rh immune globulin*) destroy the Rh antigens from the fetus before they can sensitize the mother's immune system.

Polygenic Traits

The term **polygenic traits** refers to traits that are determined by more than one pair of genes. When only one pair of genes determine a trait, and one allele is dominant to another, the phenotype can be described as "either-or": either tall or short (as in Mendel's peas), normal or albino, and so on. When more than one pair of genes are involved in determining a trait, in contrast, intermediate conditions can be produced. The more pairs of genes that are involved in determining the trait, the more intermediate conditions can be produced. It is the production of these intermediate conditions that gives the illusion of a blending of characteristics, and thus confused people prior to Mendel's discoveries.

The Dihybrid Cross

Suppose we consider two pairs of genes, each of which is located on a different pair of homologous chromosomes. One pair of homologous chromosomes contains the alleles Aa and another pair contains the alleles Bb. The genotype of this person could thus be described as AaBb (the diploid condition). Since these chromosomes assort independently at metaphase I of meiosis (chapter 8), the gametes produced by this individual could have the following combination of alleles: AB, Ab, aB, and ab (fig. 11.17). If two individuals with this genotype mated, there are sixteen possible combinations of gametes, as shown by the Punnett square in figure 11.17.

Now, if A is dominant to a, and B is dominant to b, these sixteen combinations will produce four different phenotypes. Individuals with the genotypes AABB, AaBb, AaBB, and AABb will all have phenotypes determined by the dominant genes A and B. There are nine combinations out of the sixteen that produce this phenotype (fig. 11.17). Individuals with the genotypes AAbb and Aabb (three out of sixteen) will have phenotypes determined by the dominant gene A and the recessive gene b. Those with the genotypes aaBB and aaBb (again, three out of sixteen) will have the phenotypes determined by the recessive gene a and the dominant gene B. Finally, there is only one out of the sixteen combinations that will have the genotype aabb, and thus have the phenotype produced by the recessive genes a and b. The four different phenotypes produced by this cross are thus said to be present in a *9:3:3:1 ratio.*

AaBb

AaBb

♂ \ ♀	AB	aB	Ab	ab
AB	AABB	AaBB	AABb	AaBb
aB	AaBB	aaBB	AaBb	aaBb
Ab	AABb	AaBb	AAbb	Aabb
ab	AaBb	aaBb	Aabb	aabb

Figure 11.17. *A Punnett square showing a dihybrid cross of two individuals who each have the genotype AaBb.*

Human Skin Color and Height

If there was absolutely no dominance-recessive relationship in the preceding example, then AA would have a different phenotype than Aa, and BB would appear different from Bb. In that case there would be nine rather than four different possible phenotypes from the interbreeding of AaBb individuals: AABB, AABb, AaBB, AaBb, Aabb, aaBB, aaBb, AAbb, and aabb. If the two gene pairs independently affect the same characteristic (such as skin color, for example), there would be seven intermediate shades between the extremes of AABB and aabb.

One can thus see that there are more intermediate conditions produced when more gene pairs contribute to a trait and when there is less of a dominance relationship between gene pairs. In the genetic determination of skin color, three or four gene pairs with incomplete dominance are believed to be involved. A person who is as genetically dark as possible could thus be shown to have the genotype AABBCCDD, and one who is genetically as light as possible would have the genotype aabbccdd. Different combinations of these genes would yield a large number of intermediate colors.

Skin color is produced by a combination of factors. Genetic endowment is believed to contribute approximately 70% to the skin color, whereas environmental influences (tanning from the sun) contributes the other 30%. Darker skin, whether produced by genetic influences or from tanning, is due to an increased content of the pigment melanin in the skin. Melanin can absorb ultraviolet light, and thus prevent this high-energy light from damaging the skin. Light-skinned people are thus more susceptible to the damaging effects of ultraviolet light—including the development of skin cancer—than are darker-skinned people.

Unlike the situation with Mendel's pea plants, the height of humans shows many intermediate variations. To some degree, these variations are obviously based on genetic differences—tall parents are more likely to have tall children, and so on. Like the inheritance of skin color, the genetic determination of height in humans must be due to the presence of more than one gene pair determining this trait. The height of a person, however, is strongly influenced by nongenetic factors, such as nutrition and emotional health during the growing years. For these reasons, the children of immigrants are very often taller than their parents.

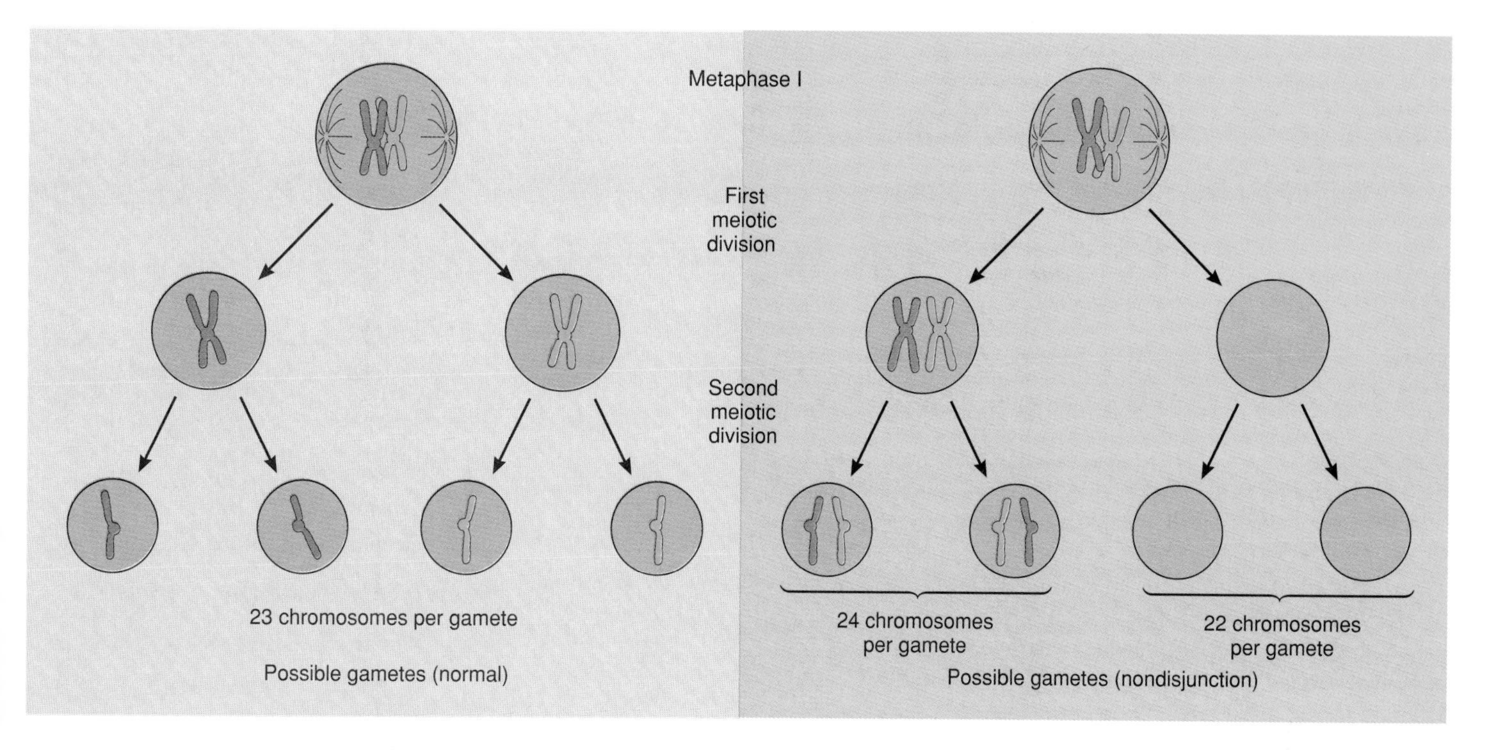

Figure 11.18. *Nondisjunction of one pair of homologous chromosomes can result in the production of gametes that contain both members of the homologous pair, and gametes that have neither member.*

Skin color and height are thus the result of the interaction of genetic endowment and environmental influences. Other qualities of a human—including intelligence and behavior patterns—are likewise the product of both genetic and nongenetic influences. The degree to which genetic ("nature") and nongenetic ("nurture") factors influence human behavior and potential is the subject of continuous controversy. Studies with identical twins separated at birth and reared with different families have revealed that genetic influences are stronger than was once suspected. Nevertheless, all agree that nurture plays a major role in determining how a person's genetic potential will be expressed.

1. Suppose a woman who is type A positive has a type O negative child, and accuses a man who is type B positive of being the father. Is this possible? Explain.
2. Explain what happens if type A blood is given to a type O person. What might happen if type O red blood cells are given to a type A person? Explain.
3. If a tall father and a short mother have a child, will the child be tall, short, or intermediate in height? What things could influence the results of this cross?

Diseases Caused by Abnormal Numbers of Chromosomes

During meiosis, when gametes are produced, two homologous chromosomes may not separate. This is known as nondisjunction of chromosomes. Nondisjunction of autosomal chromosome number 21 produces trisomy 21, or Down's syndrome. Nondisjunction of the sex chromosomes produces Klinefelter's syndrome (XXY), Turner's syndrome (XO), and the XYY genotype.

During meiosis, when gametes are produced, homologous chromosomes may not separate. Instead of one member of a pair of homologous chromosomes going to one of the two daughter cells, and the other member to the other daughter cell, both may go into the same daughter cell. This event, known as **nondisjunction** of chromosomes, results in the production of abnormal gametes. One gamete will thus get both chromosomes of a homologous pair, while the other gamete produced in this meiotic division gets none (fig. 11.18).

If an ovum that has both homologous chromosomes is fertilized by a normal sperm that contains one of these chromosomes, the zygote produced will have three members instead of two of this chromosome. The total number of chromosomes in this zygote will thus be forty-seven instead of forty-six (fig. 11.18). If the gamete that received neither member of the homologous pair of chromosomes at meiosis contributes to the zygote, the zygote will receive a total of forty-five instead of forty-six chromosomes. Such gross abnormalities can be detected by microscopic examination of the fetal chromosomes (fig. 11.19). Samples for this examination are usually obtained by the technique of amniocentesis (chapter 9).

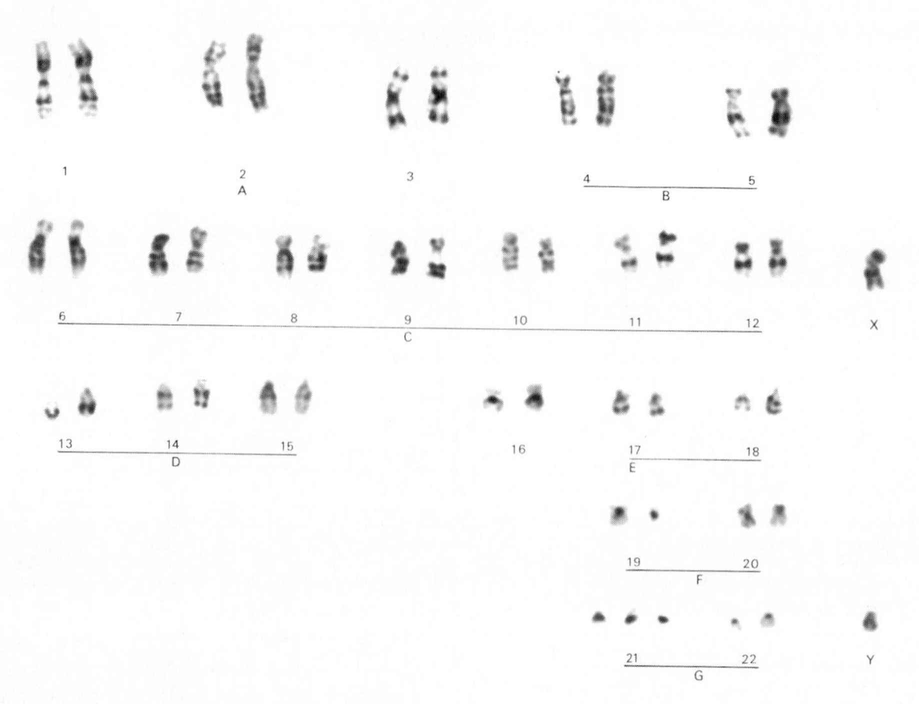

Figure 11.19. The chromosomes in a diploid cell of a person with Down's syndrome. Notice that there are three chromosomes number 21.

Figure 11.20. Children with Down's syndrome.

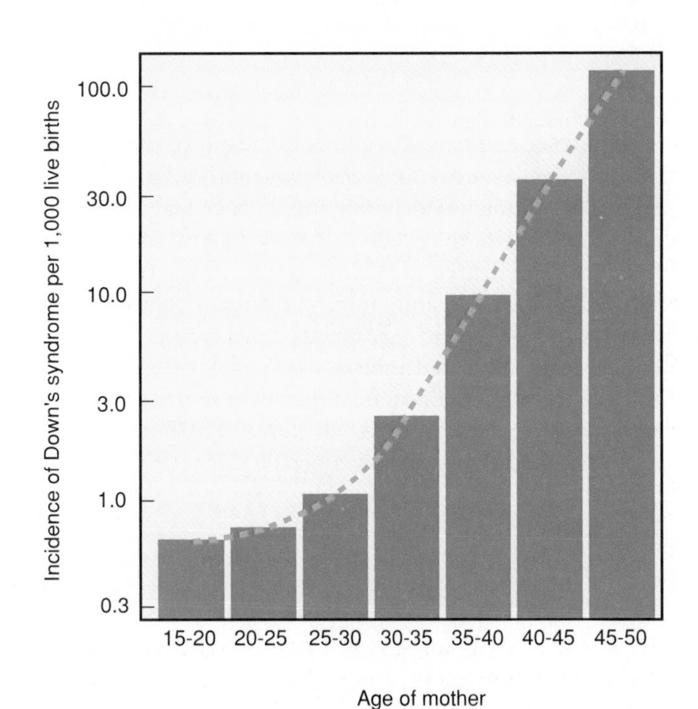

Figure 11.21. The relationship between the age of the mother and the frequency of Down's syndrome births.

The most famous disease caused by the nondisjunction of chromosomes is **Down's syndrome**[11] (**Mongolism,** fig. 11.20). Since an individual with this condition has three, instead of two, of autosomal chromosome number 21, this condition is more properly known as **trisomy 21.** People with this disease suffer from mental retardation and a shortened life span. The average incidence of trisomy 21 is one out of 750 births. The incidence of trisomy 21, and indeed of other chromosomal abnormalities, however, increases with the age of the mother (fig. 11.21).

The sex chromosomes can also be subject to nondisjunction. The X chromosomes may not separate during meiosis, so that the egg cell receives both X chromosomes. Upon fertilization by a Y-bearing sperm, the zygote will have the genotype XXY.

[11]Down's syndrome: after John Langdon Haydon Down, English physician (1828–1896)

Social Issues

Identification of Genetic Abnormalities

The ability to detect genetic abnormalities of a fetus raises a number of ethical questions. If an abnormality is present, should the fetus be aborted? What type of abnormality might justify an abortion? Clearly, there are no absolute answers. If a couple who are each carriers of Tay-Sachs disease, for example, decide to take a chance and have a child, they know before conception that the odds are 25% of having a child with the disease. Since there is currently an *in utero* test for Tay-Sachs, they can tell if they won or lost their gamble. Do they then have the right to abort the fetus? If they don't, the baby will suffer from blindness and neural disorders and certainly die by the age of five.

What about disorders that are not immediately fatal, but that decrease the quality of life and require extensive medical attention, such as sickle-cell anemia? Or those, such as Down's syndrome, that cause mental retardation? Most parents who have children with these conditions state that they are glad to have these children. Still, if prospective parents do not feel that they can emotionally or financially shoulder the burden of a child with these genetic disorders, should they have the right to abort the fetus after diagnosis? These questions are tied to the more general controversy over elective abortions discussed in chapter 9.

When it was first observed that the frequency of the XYY genotype is higher in the prison than in the general population a controversy was raised. Some people thought that children with this genotype should be closely watched, and thus treated differently from other children. Most scientists, however, argued against such special treatment. There are many people with the XYY genotype in the general population who do not become criminals, they pointed out, and special treatment of children with this genotype may actually increase their sense of isolation and promote criminal behavior. It is presently illegal—and most believe, immoral—to stigmatize a person because of a genetic abnormality.

Controversies such as these can only become more common and convoluted as more is learned of the human genome. Suppose, for example, that a specific gene is identified that increases a person's susceptibility to developing cancer when exposed to a particular chemical. Should a person with this cancer-prone gene be barred from working at a job in which this chemical is used? What if this chemical is prevalent in all fields of work that this person is able to perform? In these circumstances, if you were that person, would you even want to know that you had this gene?

Such hard choices must already be made by some people; remember that many who had a parent with Huntington's disease have refused diagnostic tests to see if they have inherited this condition. As more is learned of human genetics and more diagnostic tests become available, more decisions such as these will have to be made.

This is **Klinefelter's syndrome.**[12] As described in chapter 9, the presence of the Y chromosome causes the embryo to become male. The testes formed, however, are very abnormal, and people suffering from this condition are infertile. Mental retardation appears to be a common symptom of people suffering from syndromes involving the nondisjunction of chromosomes. Klinefelter's syndrome occurs with a frequency of about one out of five hundred births.

The nondisjunction of X chromosomes could have a different effect: both X chromosomes could go into the polar body rather than into the egg cell (chapter 9). The egg cell would thus be missing an X chromosome and would contain twenty-two rather than twenty-three chromosomes. Upon fertilization with an X-bearing sperm, the zygote would have the genotype XO, with a total of forty-five instead of forty-six chromosomes. This is called **Turner's syndrome,**[13] and occurs with a frequency of one out of five thousand female births (as described in chapter 9, the absence of the Y chromosome results in the production of a female embryo).

Nondisjunction of the Y chromosomes during the formation of sperm can produce a sperm with two Y chromosomes. When this sperm fertilizes a normal ovum, which contains one X chromosome, a zygote with the genotype XYY is produced. This condition occurs with a frequency of about one in one thousand male births. Men with the XYY genotype grow to be large individuals and suffer from mild mental retardation. Interestingly, the incidence of the XYY genotye among jail prisoners is about twenty times higher than in the general population.

Mapping the Human Genome

It has been estimated that there are 50,000 to 100,000 genes in the human genome. Only about 4,600 of these, however, are known and listed in *Mendelian Inheritance in Man,* an encyclopedia of human genes. By 1968, there were about 68 genes known to be located on the X chromosome, and one gene was localized to a specific autosomal chromosome. Now, using the techniques described in chapter 10, the chromosomal locations of over 1,500 human genes are known. In most instances, moreover, these genes can be localized to specific regions of particular chromosomes. Over 600 of these genes have now been cloned and sequenced.

The end result of this process will eventually be a complete encyclopedia of the human genome. Since there are three billion base pairs in the haploid genome of one human, this is certainly a formidable task. If the base pairs are indicated with one letter (such as A for adenine), it would require at least thirteen sets of the *Encyclopedia Britannica* to print this haploid genome. And this would not express the complete diploid genotype of that person, because a person is heterozygous for many of these genes. Further, such a printout would not take the genetic variation between individuals into account.

[12]Klinefelter's syndrome: after Harry Fitch Klinefelter, Jr., American physician (1912–   )

[13]Turner's syndrome: after Henry Hubert Turner, American endocrinologist (1892–   )

A group of international scientists formed the *Human Genome Organization (HUGO)* in 1988. As expressed by one scientist, this organization will be a "U.N. for the human genome." Once the project is complete, it should be possible to develop tests that detect the carrier state of most, if not all, genetic diseases, and to detect such diseases in fetuses *in utero.* Knowledge of the complete DNA sequences of the human genome will greatly facilitate the development of gene therapies for these conditions. And such diagnostic and therapeutic benefits are not limited to diseases that are purely genetic in nature. Most cancers, for example, are now believed to involve interactions between environmental agents that function as carcinogens and the human genome (chapter 14). Complete knowledge of the human genome will thus greatly improve diagnosis, and quite possibly treatment, of many forms of cancer.

1. Explain nondisjunction, and how Down's syndrome is produced.
2. Explain how the XXY, XYY, and XO genotypes are produced, and indicate the sex of these individuals.

Summary

Gene Action and Inheritance p. 240

I. Each enzyme protein is coded by a different gene.
  A. When a gene is defective, it produces a defective enzyme.
  B. PKU and albinism are produced in this way.
II. A single base change in DNA constitutes a point mutation.
  A. This causes a different sequence of three bases to be present in the DNA, causing a change in a codon in mRNA.
  B. A changed codon in mRNA results in the substitution of one amino acid for another in the protein.
III. Gregor Mendel is the father of genetics; the classical patterns of heredity were further clarified by Thomas Hunt Morgan.

Principles of Mendelian Inheritance p. 243

I. A pea plant (or a person) inherits two sets of genes called alleles on the two sets of homologous chromosomes.
  A. If both alleles specify the same condition, the person is homozygous for that condition; if the alleles are different, the person is heterozygous.
  B. The combination of alleles in an organism is termed its genotype; the physical result of that genotype is called the phenotype.
  C. If the organism is heterozygous, and one allele is dominant to the other, the organism will have the phenotype of the dominant allele and will be a carrier for the recessive allele.
II. The probabilities of obtaining a particular combination of alleles in the offspring from a particular cross can be determined using a Punnett square.
  A. If two individuals who are each heterozygous for a recessive trait mate, the offspring stand a 75% chance of inheriting the dominant phenotype and a 25% chance of inheriting the recessive phenotype.
  B. The chances of obtaining a particular combination of alleles from a cross remain the same each time a prediction is made.

Simple Mendelian Inheritance in Humans p. 245

I. Many human conditions are inherited as simple mendelian traits that are determined by genes located on the autosomal chromosomes.
  A. The ability to taste PTC is inherited as a dominant trait.
  B. Albinism, cystic fibrosis, and Tay-Sachs disease are inherited as autosomal recessive traits.
  C. Huntington's disease is inherited as a dominant trait.
II. Some traits are inherited on the X chromosome.
  A. Since a boy gets his X chromosome from his mother, X-linked traits are usually passed from mother to son.
  B. Hemophilia, color blindness, and Duchenne muscular dystrophy are X-linked traits.

Inherited Diseases of Hemoglobin p. 249

I. A hemoglobin molecule contains four separate polypeptide chains that are grouped into two identical pairs.
  A. In sickle-cell anemia, a point mutation in DNA causes a specific amino acid substitution in the beta chains.
  B. The abnormal hemoglobin, called hemoglobin S, produces sickling of the red blood cells and the symptoms of sickle-cell anemia.
  C. The heterozygous condition for sickle-cell anemia confers resistance to malaria.
II. Thalassemia is a family of hemoglobin disorders that usually affect the beta chains of hemoglobin.
  A. If the gene for the beta chains is completely deleted, a different gene is activated that produces polypeptide chains normally found only in fetal hemoglobin.
  B. A frame-shift mutation is the deletion of one base, causing all base triplets downstream of this deletion to have a different translation.

Multiple Alleles and Polygenic Traits p. 252

I. The ABO blood group represents multiple alleles of one gene.
  A. The blood type refers to the type of antigen molecules present on the surface of the red blood cells.
  B. A type A person may have the genotype AA or AO; a type B person may have the genotype BB or BO.
  C. There is no dominance-recessive relationship between A and B; they are said to be codominant.
  D. Antibodies are present in a person's blood plasma that can agglutinate red blood cells with all other antigens than those of that person.
II. The Rh factor is an antigen present on the red blood cells of most people.
  A. The Rh negative condition is inherited as a simple mendelian recessive trait.
  B. If an Rh negative mother gives birth to an Rh positive baby, her subsequent babies could develop hemolytic disease of the newborn.
III. Phenotypes showing intermediate conditions between two extremes are produced by more than one pair of alleles.
  A. Two gene pairs that are located on different chromosomes assort independently, producing a phenotypic ratio of 9:3:3:1 if these gene pairs display dominance-recessive relationships.
  B. Skin color is an example of a polygenic trait, with three or four gene pairs contributing to a person's skin color.
  C. Environmental influences (such as tanning of the skin) also affect the phenotype.

Diseases Caused by Abnormal Numbers of Chromosomes p. 255

I. Nondisjunction of chromosomes refers to the abnormal condition in which homologous chromosomes do not separate during meiosis.
  A. A gamete produced in this way will thus have either one extra chromosome or it will be missing a chromosome.
  B. If this gamete contains two of the autosomal chromosome number 21, at fertilization it will form an individual with trisomy 21 (Down's syndrome).
II. Scientists are attempting to compile a complete encyclopedia of the human genome.

Review Activities

Objective Questions

1. Suppose the enzyme that catalyzes the reaction A ⟶ B is defective. Which of the following statements is *true*?
   (a) Molecule A will decrease in amount.
   (b) Molecule B will increase in amount.
   (c) This represents an inborn error of metabolism.
   (d) All of these.
2. Suppose the normal DNA code AGCTCCGAT is changed to AGCGCCGAT.
   (a) A point mutation has occurred
   (b) The mRNA will have one codon that is changed
   (c) One amino acid may be substituted for another in the protein
   (d) All of these.
3. The scientist who pioneered the use of *Drosophila* in learning about genetics was
   (a) Gregor Mendel
   (b) Francis Crick
   (c) Thomas Hunt Morgan
   (d) J. B. S. Haldane
4. One parent is a carrier for this disease and one parent is homozygous normal. If one of their children gets the disease, the disease may be
   (a) albinism
   (b) PKU
   (c) cystic fibrosis
   (d) Huntington's disease
5. Suppose that both parents are heterozygous for sickle-cell anemia. The probability that one of their children will have this disease is
   (a) zero
   (b) 25%
   (c) 50%
   (d) 75%
   (e) 100%
6. Suppose the couple in question number 5 have a child with sickle-cell anemia. The probability that their second child will have this disease is
   (a) zero
   (b) 25%
   (c) 50%
   (d) 75%
   (e) 100%
7. Suppose that a man finds that he is not a carrier for Tay-Sachs disease, but that his wife is a carrier. What is the probability that they can produce a child with this disease?
   (a) zero
   (b) 25%
   (c) 50%
   (d) 75%
   (e) 100%
8. A child with hemophilia has normal parents. Therefore
   (a) the father was a carrier for this disease
   (b) his mother was a carrier for this disease
   (c) his sisters stand a 50% chance of having this disease
   (d) his brothers stand a 25% chance of having this disease
   (e) all of these
9. A person who has type A blood
   (a) has anti-A antibodies in his or her plasma
   (b) can donate blood to a type O person
   (c) could not be the parent of a type O child
   (d) has antibodies in his or her plasma against the B antigen
10. Suppose two parents who are each Rh positive have a child who is Rh negative. The probability of having another Rh negative child is
    (a) zero
    (b) 25%
    (c) 50%
    (d) 75%
    (e) 100%
11. The universal blood donor is
    (a) type A
    (b) type B
    (c) type AB
    (d) type O
12. A child with Down's syndrome
    (a) has forty-seven chromosomes
    (b) probably has an older mother
    (c) has three autosomal chromosomes number 21
    (d) all of these

Essay Questions

1. Why weren't Gregor Mendel's theories accepted during his time? What conclusions can you draw from this in regard to the progress of human knowledge?
2. Explain, in a step-by-step fashion, how a point mutation can cause an inborn error of metabolism.
3. Could a couple who are both carriers of cystic fibrosis have two children in a row with this disease? Explain your answer.
4. King Henry VIII beheaded wives who didn't produce male children. Was it their fault? Explain.
5. Can a girl be color blind or hemophilic? If yes, how?
6. Evaluate the "nature vs. nurture" controversy, using skin color and height as examples.

12

The Circulatory System

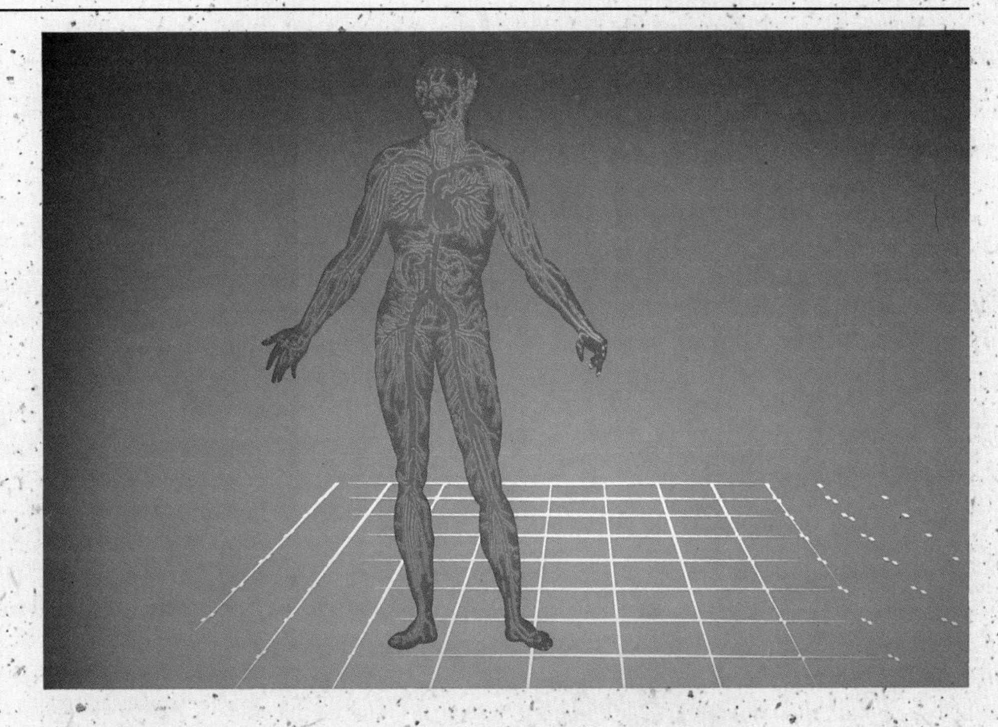

Outline

Objectives

By studying this chapter, you should be able to

1. describe the appearance and functions of erythrocytes and the different categories of leukocytes
2. describe the structure of the heart and the pattern of blood flow through the heart
3. identify the path of blood flow in the pulmonary and systemic circulations
4. identify the heart valves, and describe their actions
5. explain how the heart sounds are produced and the nature of heart murmurs
6. explain the electrical activity that results in the heartbeat, and describe the pathway of impulse conduction
7. identify the waves of the electrocardiogram and the causes of these waves
8. define the terms *bradycardia, tachycardia, flutter,* and *fibrillation*
9. explain the significance of the cardiac output, and identify the major factors that affect this measurement
10. describe how blood flow through the heart and skeletal muscles is regulated, and how these change during exercise
11. explain the nature and significance of atherosclerosis, and describe its dangers
12. explain how the arterial blood pressure is regulated and how the blood pressure is measured
13. identify the structures and functions of the lymphatic system

Keys to Pronunciation

aorta: *a-or'tah*
atherosclerosis: *ath''er-o''skle-ro'sis*
atrioventricular: *a''tre-o-ven-trik'u-lar*
atrium: *a'tre-um*
capillary: *kap'i-lar''e*
diastole: *di-as'to-le*
erythrocyte: *e-rith'ro-sit*
interstitial: *in''ter-stish'al*
ischemia: *is-ke'me-ah*
leukocyte: *loo'ko-sit*
lymph: *limf*
mitral: *mi'tral*
Purkinje: *pur-kin'je*
sinoatrial: *si''no-a'tre-al*
systemic: *sis-tem'ik*
systole: *sis'to-le*
tachycardia: *tak''e-kar'de-ah*
venule: *ven'ul*

Photo: Image of the human circulatory system.

Functions and Components of the Circulatory System

Blood serves numerous functions, including the transport of respiratory gases, nutritive molecules, metabolic wastes, and hormones. Blood is transported through the body in a system of vessels from and to the heart. The heart consists of four chambers; two of these receive blood from veins, and two pump blood into arteries. The arterial and venous systems are continuous by way of tiny vessels known as blood capillaries.

A unicellular organism can provide for its own maintenance and continuity by performing the wide variety of functions needed for life. By contrast, the complex human body is composed of trillions of specialized cells that demonstrate a division of labor. Cells of a multicellular organism depend on one another for the very basics of their existence. The majority of the cells of the body are stationary and thus incapable of procuring food and oxygen or even moving away from their own wastes. Therefore, a highly specialized and effective means of transporting materials within the body is needed.

The blood serves this transportation function. An estimated 60,000 miles of vessels throughout the body of an adult ensure that nutrients reach each of the trillions of living cells. The blood can also, however, serve to transport disease-causing viruses, bacteria, and their toxins. To safeguard against this, the circulatory system has protective mechanisms: the white blood cells and lymphatic system. In order to perform its various functions, the circulatory system works together with the respiratory, urinary, digestive, endocrine, and integumentary systems in maintaining homeostasis.

Functions of the Circulatory System

The functions of the circulatory system can be divided into three areas: transportation, regulation, and protection.

Transportation. All of the substances essential for cellular metabolism are transported by the circulatory system. These substances can be categorized as follows:

a. Respiratory. Red blood cells, called **erythrocytes,** transport oxygen to the tissue cells. In the lungs, oxygen from the inhaled air attaches to hemoglobin molecules within the erythrocytes and is transported to the cells. Carbon dioxide produced by cell metabolism is carried by the blood to the lungs for elimination in the exhaled air.

b. Nutritive. The digestive system is responsible for the breakdown of food so that it can be absorbed through the wall of the intestine and into the blood vessels. The blood then carries these absorbed products of digestion through the liver and to the cells of the body.

c. Excretory. Metabolic wastes, excessive water and ions, as well as other molecules in plasma (the fluid portion of blood) are filtered through the capillaries of the kidneys and excreted in urine.

Regulation. The blood carries hormones and other regulatory molecules from their site of origin to distant target tissues.

Protection. The circulatory system protects against injury and foreign microbes or toxins introduced into the body. The clotting mechanism protects against blood loss when vessels are damaged, and **leukocytes** (white blood cells) provide immunity from many disease-causing agents.

Major Components of the Circulatory System

The circulatory system is frequently divided into the **cardiovascular system,** which consists of the heart and blood vessels, and the **lymphatic system,** which consists of lymph vessels and lymph nodes.

The **heart** is a four-chambered, double pump. Its pumping action creates the pressure head needed to push blood in the vessels to the lungs and body cells. At rest, the heart of an adult pumps about five liters of blood per minute. It takes about one minute for blood to be circulated through the body and back to the heart.

Blood vessels form a tubular network that permits blood to flow from the heart to all the living cells of the body and then back to the heart. *Arteries* carry blood away from the heart, whereas *veins* return blood to the heart. Arteries and veins are continuous with each other through smaller blood vessels.

Arteries branch extensively to form a "tree" of ever smaller vessels. Those that are microscopic in diameter are called *arterioles.* Conversely, microscopic-sized veins, called *venules,* deliver blood into ever larger vessels that empty into the large veins. Blood passes from the arterial to the venous system in *capillaries,* which are the thinnest and most numerous blood vessels. All exchanges of fluid, nutrients, and wastes between the blood and tissues occur across the walls of capillaries.

As blood plasma passes through capillaries, the pressure of the blood forces some of this fluid through the capillary walls. Fluid derived from plasma that passes out of capillary walls into the surrounding tissues is called *tissue fluid,* or *interstitial fluid.* Some of this fluid returns directly to capillaries, and some enters into **lymph vessels** located in the connective tissues around the blood vessels. Fluid in lymph vessels is called *lymph.* This fluid is returned to the venous blood at particular sites. **Lymph nodes,** positioned along the way, cleanse the lymph prior to its return to the venous blood. The lymphatic system is thus considered a part of the circulatory system and is discussed at the end of this chapter.

---

1. Name the components of the circulatory system that function in oxygen transport, in the transport of nutrients, and in protection.
2. Define and describe the function of arteries, veins, and capillaries.
3. Define the terms *interstitial fluid* and *lymph.* Describe their relationships to blood plasma.

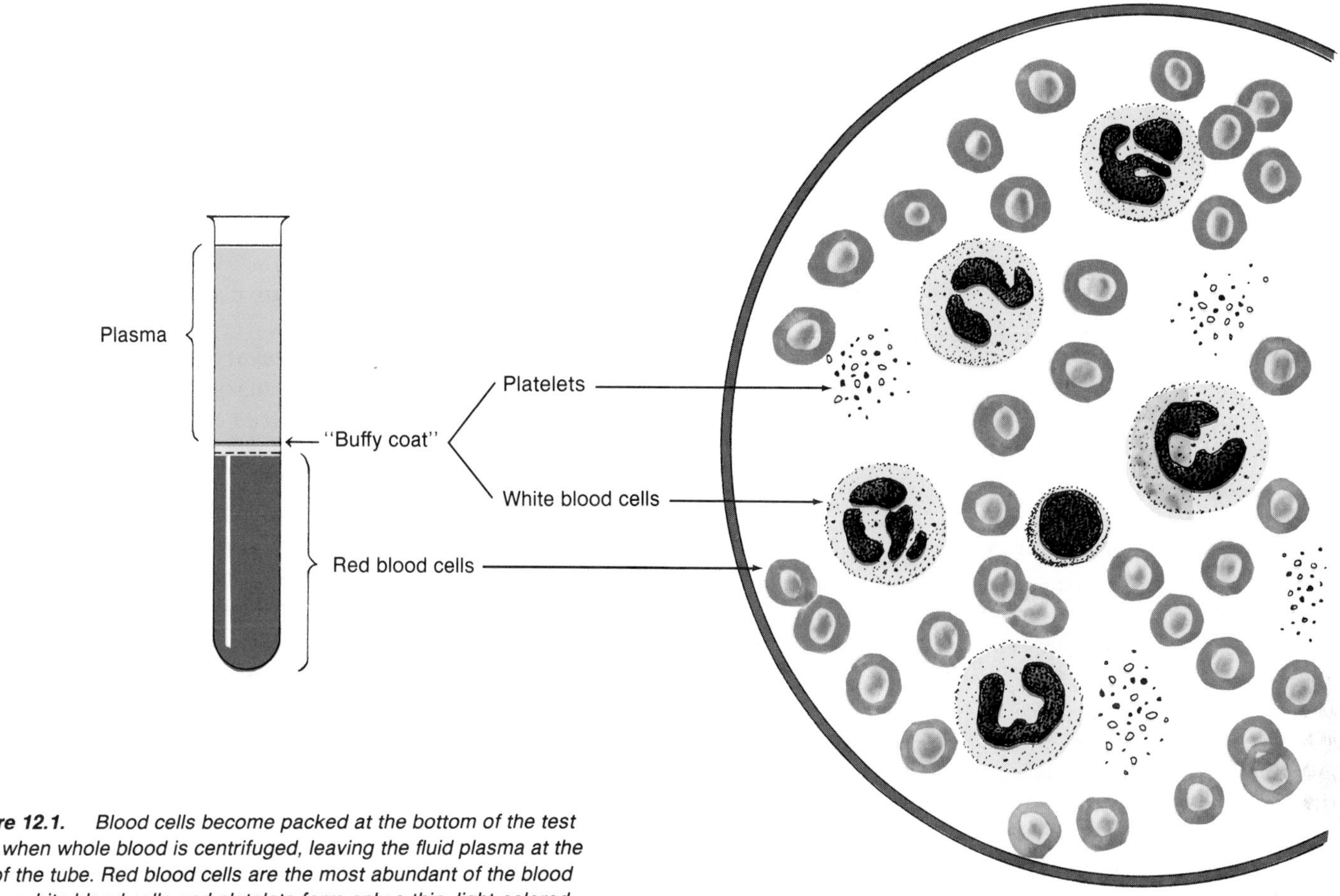

Figure 12.1. *Blood cells become packed at the bottom of the test tube when whole blood is centrifuged, leaving the fluid plasma at the top of the tube. Red blood cells are the most abundant of the blood cells—white blood cells and platelets form only a thin, light-colored "buffy coat" at the interface between the packed red blood cells and the plasma.*

Composition of the Blood

Blood consists of formed elements that are suspended and carried in the plasma. These formed elements and their major functions include erythrocytes (oxygen transport), leukocytes (immune defense), and platelets (blood clotting). Plasma contains different types of proteins and many water-soluble molecules.

Blood leaving the heart is referred to as *arterial blood.* Arterial blood, with the exception of that going to the lungs, is bright red in color due to a high concentration of oxyhemoglobin (the combination of oxygen with hemoglobin—chapter 15) in the erythrocytes. *Venous blood* is blood returning to the heart and, except for the venous blood from the lungs, has a darker color due to hemoglobin that is no longer combined with oxygen.

Blood is composed of a cellular portion, called **formed elements,** and a fluid portion, called **plasma.** When a blood sample is centrifuged, the heavier formed elements are packed into the bottom of the tube, leaving plasma at the top (fig. 12.1). The formed elements constitute approximately 45% of the total blood volume (the *hematocrit*), and the plasma accounts for the remaining 55%.

Plasma

Plasma is a straw-colored liquid consisting of water and dissolved solutes. Sodium ion is the major solute of the plasma in terms of its concentration. In addition to Na^+, plasma contains many other salts and ions, as well as organic molecules such as metabolites, hormones, enzymes, antibodies, and other proteins.

There are three types of plasma proteins: albumins, globulins, and fibrinogen. **Albumins** account for most (60%–80%) of the plasma proteins and are the smallest in size. They are produced by the liver and serve to draw water from the surrounding tissue fluid into the capillaries by osmosis. This action is needed to maintain blood volume and pressure. **Globulins** are divided into three subtypes: **alpha globulins, beta globulins,** and **gamma globulins.** The alpha and beta globulins are produced by the liver and function to transport lipids and fat-soluble vitamins in the blood. Gamma globulins are antibodies produced by lymphocytes (one of the white blood cells) and function in immunity. **Fibrinogen** protein is an important clotting factor produced by the liver. During the process of clot formation, fibrinogen is converted into insoluble threads of *fibrin.* The fluid from clotted blood, which is called **serum,** thus does not contain fibrinogen but is otherwise identical to plasma.

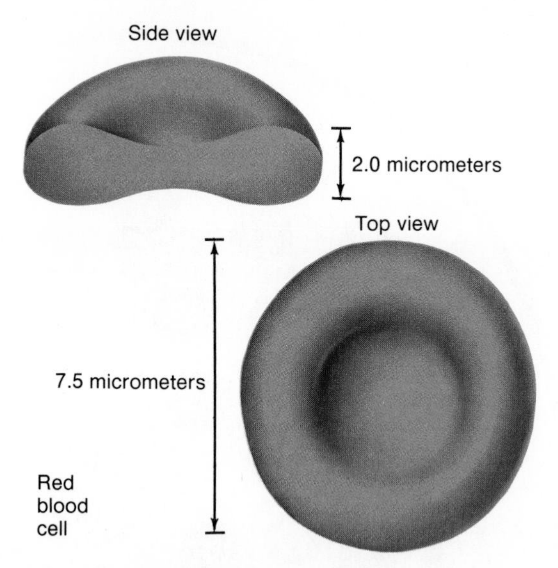

Figure 12.2. The structure of an erythrocyte.

The Formed Elements of Blood

The formed elements of blood include erythrocytes, or red blood cells, and leukocytes, or white blood cells. Erythrocytes are by far the most numerous of these two types: a cubic millimeter of blood contains 5.1 million to 5.8 million erythrocytes in males and 4.3 million to 5.2 million erythrocytes in females. The same volume of blood, in contrast, contains only 5,000 to 9,000 leukocytes.

Erythrocytes

Erythrocytes are about 7 μm in diameter and 2.2 μm thick, and resemble rounded tablets with their middle pinched in. Their unique shape relates to their function of transporting oxygen and provides an increased surface area through which gas can diffuse (fig. 12.2). Erythrocytes lack a nucleus and mitochondria. Because of these deficiencies, erythrocytes have a circulating life span of only about 120 days before they are destroyed by phagocytic cells in the liver, spleen, and bone marrow.

Each erythrocyte contains approximately 280 million *hemoglobin* molecules, which give blood its red color. Each hemoglobin molecule consists of four polypeptide chains (described in chapter 11), and an iron-containing pigment, called heme. The iron group of heme is able to combine with oxygen in the lungs and release oxygen in the tissues.

> Anemia is present when there is an abnormally low hemoglobin concentration and/or red blood cell count. The most common cause of this condition is a deficiency in iron (**iron-deficiency anemia**), which is an essential component of the hemoglobin molecule. In **pernicious anemia,** there is inadequate availability of vitamin B_{12}, which is needed for red blood cell production. This results from atrophy of the lining of the stomach, which normally secretes a substance, called *intrinsic factor,* that is needed for absorption of vitamin B_{12} obtained in the diet. **Aplastic anemia** is anemia due to destruction of the bone marrow, which may be caused by chemicals (including benzene and arsenic) and by X rays.

Leukocytes

Leukocytes differ from erythrocytes in several ways. Leukocytes contain nuclei and mitochondria and can move in an amoeboid fashion (erythrocytes are not able to move independently). Because of their amoeboid ability, leukocytes can squeeze through pores in capillary walls and get to a site of infection, whereas erythrocytes usually remain confined within blood vessels.

Leukocytes, which are almost invisible under the microscope unless they are stained, are classified according to their stained appearance. Those leukocytes that have granules in their cytoplasm are called **granular leukocytes,** and those that do not are called **agranular** (or nongranular) **leukocytes.** The granular leukocytes are also identified by their oddly shaped nuclei, which in some cases are contorted into lobes separated by thin strands. The granular leukocytes are therefore known as *polymorphonuclear leukocytes* (abbreviated PMN).

The stain used to identify leukocytes is usually a mixture of a pink-to-red stain called eosin and a blue-to-purple stain called a "basic stain." Granular leukocytes with pink-staining granules are therefore called *eosinophils,* and those with blue-staining granules are called *basophils.* Those with granules that have little affinity for either stain are *neutrophils.* Neutrophils are the most abundant type of leukocyte, comprising 50%–70% of the leukocytes in the blood.

There are two types of agranular leukocytes: lymphocytes and monocytes. *Lymphocytes* are usually the second most numerous type of leukocyte; they are small cells with round nuclei and little cytoplasm. *Monocytes,* in contrast, are the largest of the leukocytes and generally have kidney- or horseshoe-shaped nuclei. In addition to these two cell types, there are smaller numbers of large lymphocytes, or *plasma cells.* Plasma cells produce and secrete large amounts of antibodies.

Platelets

Platelets, or **thrombocytes,** are the smallest of the formed elements and are actually fragments of large cells found in bone marrow. (This is why the term *formed elements* is used rather than *blood cells* to describe erythrocytes, leukocytes, and platelets.) The fragments that enter the circulation as platelets lack nuclei but, like leukocytes, are capable of amoeboid movement. The platelet count per cubic millimeter of blood is 130,000 to 360,000. Platelets survive about five to nine days and then are destroyed by the spleen and liver.

Platelets play an important role in blood clotting. They constitute the major portion of the mass of the clot, and molecules in their cell membranes serve to activate the clotting factors in plasma that result in threads of fibrin, which reinforce the platelet plug.

The appearance of the formed elements of the blood is shown in figure 12.3, and a summary of the characteristics of these formed elements is presented in table 12.1.

1. Define the terms *plasma* and *serum,* and identify their components.
2. Identify each of the formed elements of the blood and their major functions.
3. Define the term *anemia* and describe some of its causes.
4. Describe the classification of the white blood cells, and list the different types of these cells.

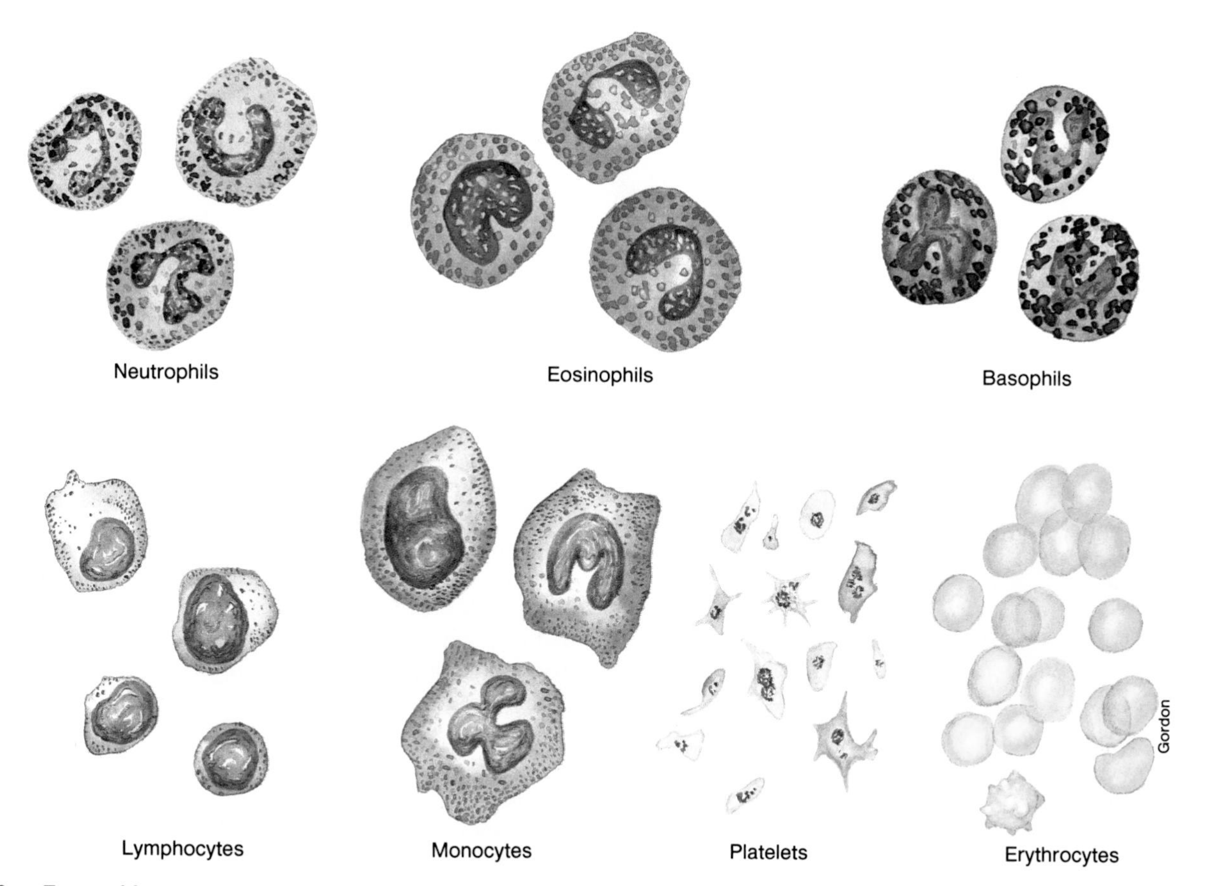

Figure 12.3. *Types of formed elements in blood.*

Table 12.1 Formed elements of the blood

Component	Description	Number Present	Function
Erythrocyte (red blood cell)	Biconcave disc without nucleus; contains hemoglobin; survives 100–120 days	4,000,000 to 6,000,000/mm³	Transports oxygen and carbon dioxide
Leukocytes (white blood cells)		5,000 to 10,000/mm³	Aid in defense against infections by microorganisms
Granulocytes	About twice the size of red blood cells; cytoplasmic granules present; survive 12 hours to 3 days		
1. Neutrophil	Nucleus with 2–5 lobes; cytoplasmic granules stain slightly pink	54%–62% of white cells present	Phagocytic
2. Eosinophil	Nucleus bilobed; cytoplasmic granules stain red in eosin stain	1%–3% of white cells present	Helps to detoxify foreign substances; secretes enzymes that break down clots
3. Basophil	Nucleus lobed; cytoplasmic granules stain blue in hematoxylin stain	Less than 1% of white cells present	Releases anticoagulant heparin
Agranulocytes	Cytoplasmic granules absent; survive 100–300 days		
1. Monocyte	2–3 times larger than red blood cell; nuclear shape varies from round to lobed	3%–9% of white cells present	Phagocytic
2. Lymphocyte	Only slightly larger than red blood cell; nucleus nearly fills cell	25%–33% of white cells present	Provides specific immune response (including antibodies)
Platelet (thrombocyte)	Cytoplasmic fragment; survives 5–9 days	130,000 to 360,000/mm³	Clotting

Table 12.2 Summary of the pulmonary and systemic circulations

	Source	Arteries	O_2 Content of Arteries	Veins	Termination
Pulmonary Circulation	Right ventricle	Pulmonary arteries	Low	Pulmonary veins	Left atrium
Systemic Circulation	Left ventricle	Aorta and its branches	High	Superior and inferior venae cavae and their branches	Right atrium

Structure of the Heart

The heart contains four chambers: two atria, which receive venous blood, and two ventricles, which eject blood into arteries. The right ventricle pumps blood to the lungs, where the blood becomes oxygenated, and the left ventricle pumps oxygenated blood to the entire body. The flow of blood from atria to ventricles, and then from ventricles to arteries, is assured by the action of two pairs of one-way valves within the heart.

The heart is divided into four chambers. The right and left **atria** (singular, *atrium*[1]) receive blood from the venous system; the right and left **ventricles**[2] pump blood into the arterial system. The right atrium and ventricle (sometimes called the *right pump*) are separated from the left atrium and ventricle (the *left pump*) by a muscular wall, or *septum.* This septum normally prevents mixture of the blood from the two sides of the heart.

The muscle cells in the heart—called *myocardial cells*—are striated (striped) in appearance, as are the muscle cells of skeletal muscle (chapter 5). Unlike the cells of skeletal muscle, however, the myocardial cells are joined together both mechanically and electrically. This occurs because of special junctions, known as *intercalated discs,* between adjacent myocardial cells. Intercalated discs are actually a type of electrical synapse that permits action potentials (the electrical signal of nerves and muscles—chapter 6) to move from one cell to the next. The right and left atria form one such electrically connected unit known as a *myocardium.* The right and left ventricles form a separate myocardium.

Pulmonary and Systemic Circulations

Blood that has become partially depleted of its oxygen and increased in carbon dioxide, as a result of gas exchange across tissue capillaries, returns by way of veins to the right atrium. This blood then enters the right ventricle, which pumps it into the *pulmonary arteries.* The pulmonary arteries branch to transport blood to the lungs, where gas exchange occurs between the lung capillaries and the air sacs (alveoli) of the lungs. Oxygen diffuses from the air to the capillary blood, while carbon dioxide diffuses in the opposite direction.

The blood that returns to the left atrium by way of the *pulmonary veins* is therefore enriched in oxygen and partially depleted in carbon dioxide. The path of blood from the heart (right ventricle), through the lungs, and back to the heart (left atrium) completes one circuit: the **pulmonary circulation.**

Oxygen-rich blood in the left atrium enters the left ventricle and is pumped into a very large artery—the *aorta.* The aorta ascends for a short distance, makes a U-turn (the aortic arch), and then descends through the thoracic (chest) and abdominal cavities. Arterial branches from the aorta supply oxygen-rich blood to all of the organ systems and are thus part of the **systemic circulation.**

As a result of cellular metabolism, the oxygen concentration is lower and the carbon dioxide concentration is higher in the tissues than in the capillary blood. Blood that drains into the systemic veins is thus partially depleted in oxygen and increased in carbon dioxide content. These veins ultimately empty into two large veins—the *superior* and *inferior venae cavae*[3]—that return the oxygen-poor blood to the right atrium. This completes the systemic circulation: from the heart (left ventricle), through the organ systems, and back to the heart (right atrium). The characteristics of the systemic and pulmonary circulations are summarized in table 12.2 and illustrated in figure 12.4.

The Heart Valves

Although adjacent myocardial cells are joined together mechanically and electrically by intercalated discs, the atria and ventricles are separated into two functional units by a sheet of connective tissue located between them. Embedded within this sheet of tissue are one-way **atrioventricular (AV) valves** (fig. 12.5). The AV valve located between the right atrium and right ventricle has three flaps, and is therefore called the *tricuspid valve.* The AV valve between the left atrium and left ventricle has two flaps and is thus called the *bicuspid valve;*[4] this is also known as the *mitral valve.*[5]

The AV valves allow blood to flow from the atria to the ventricles, but normally prevent the backflow of blood into the atria. Opening and closing of these valves occur as a result of pressure

[1]atrium: L. *atrium,* chamber
[2]ventricle: L. *ventriculus,* diminutive of *venter,* belly
[3]vena cava: L. *vena,* vein; *cava,* empty
[4]bicuspid: L. *bi,* two; *cuspis,* tooth point
[5]mitral: L. *mitra,* like a bishop's mitre

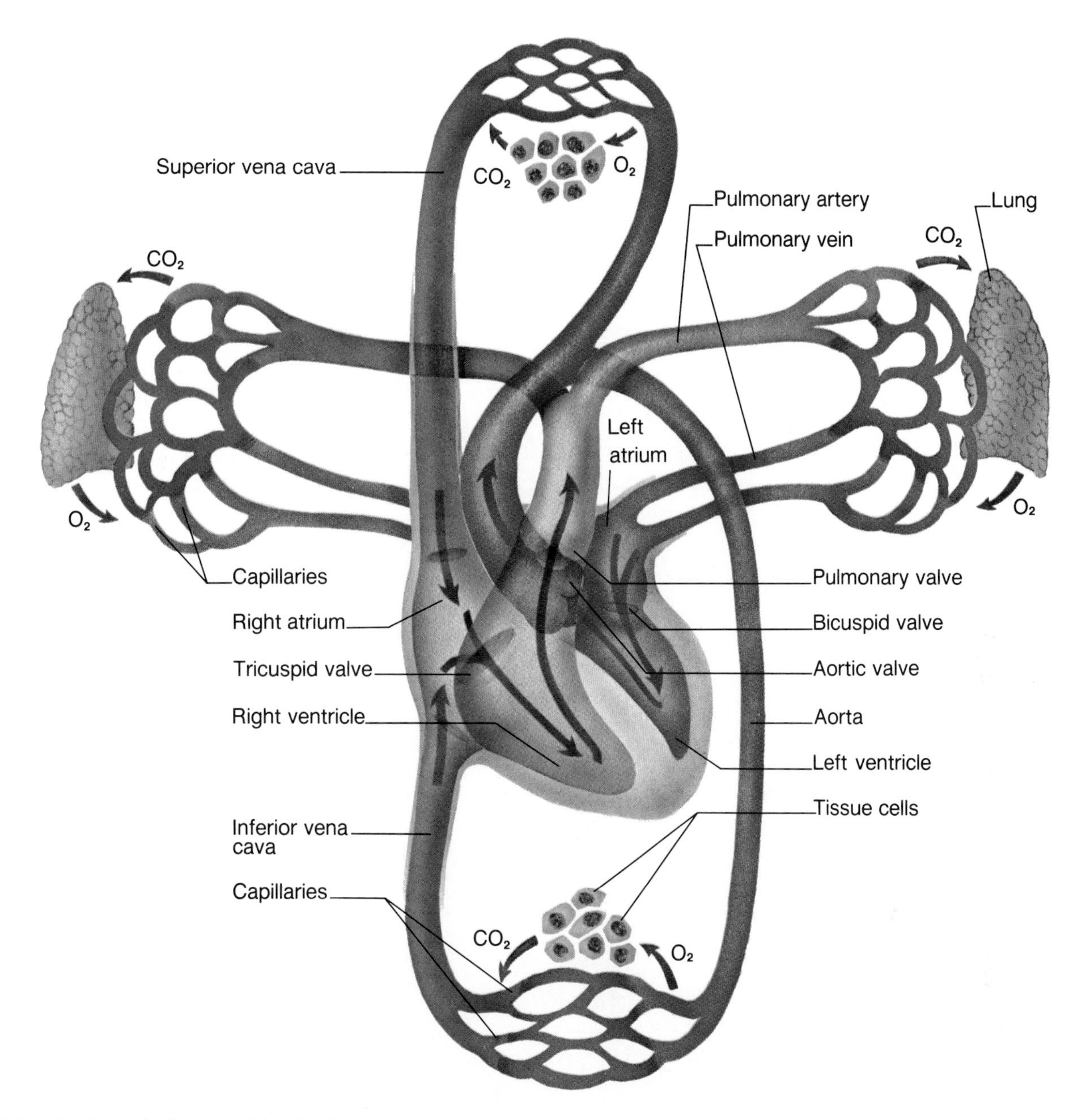

Figure 12.4. *A schematic diagram of the circulatory system.*

differences between the atria and ventricles. When the ventricles are relaxed, the return of blood through the veins to the atria causes the pressure in the atria to exceed that in the ventricles. The AV valves therefore open, allowing blood to enter the ventricles. As the ventricles contract, the intraventricular pressure rises above the pressure in the atria and pushes the AV valves closed.

One-way **semilunar valves**[6] are located at the origins of the pulmonary artery and aorta. These valves open during ventricular contraction, allowing blood to enter the pulmonary and systemic circulations. During ventricular relaxation, when the pressure in the arteries is greater than the pressure in the ventricles, the semilunar valves snap shut and thus prevent the backflow of blood into the ventricles.

[6]semilunar: L. *semi,* half; *luna,* moon

---

1. Using a flow diagram (arrows), describe the pathway of the pulmonary circulation. Indicate the relative amounts of oxygen and carbon dioxide in the vessels involved.
2. Use a flow diagram to describe the systemic circulation, and indicate the relative amounts of oxygen and carbon dioxide in the blood vessels.
3. Name the AV valves and the valves of the pulmonary artery and aorta. Describe how these valves ensure a one-way flow of blood.

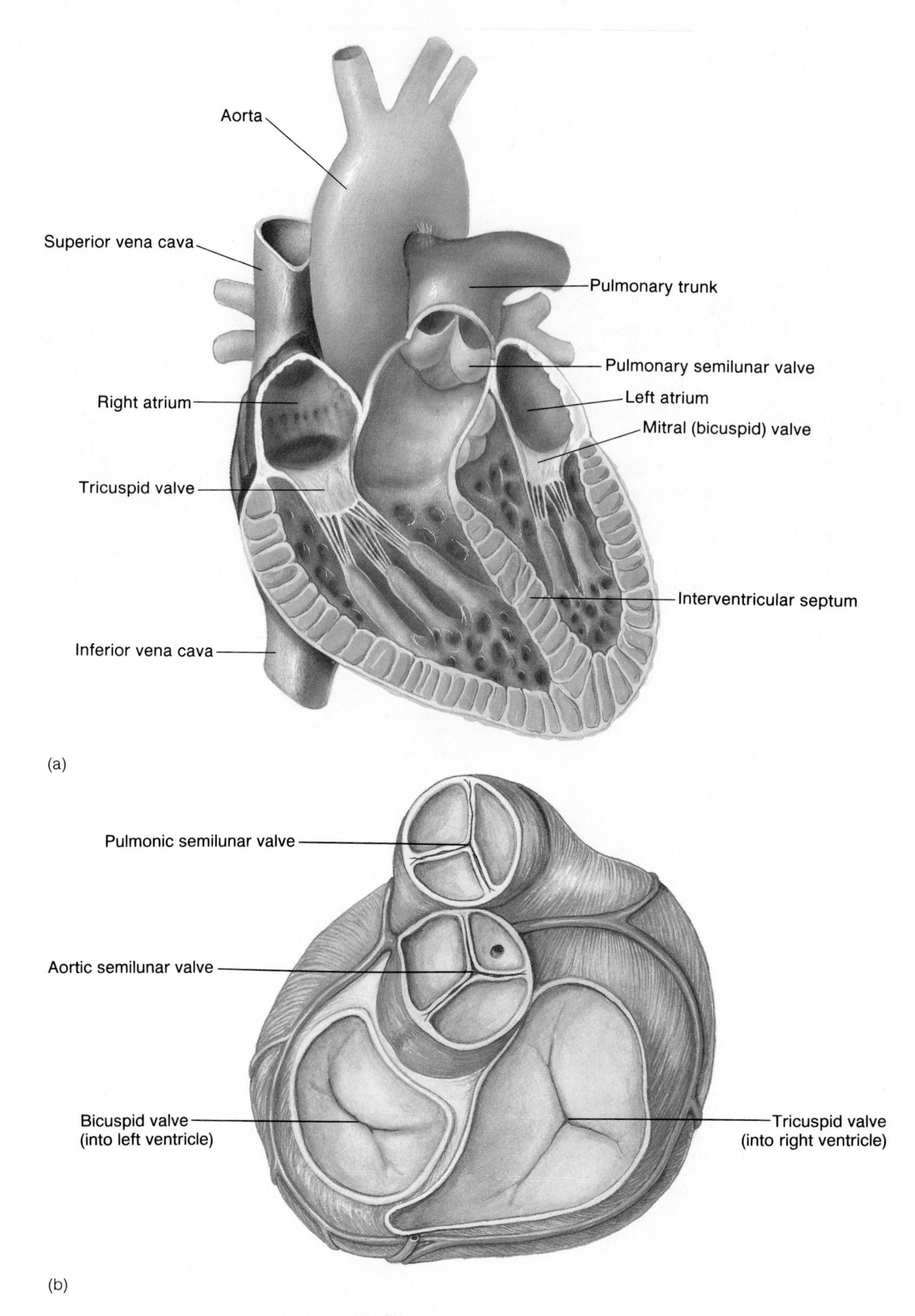

Figure 12.5. *A diagram of the structure of the heart (a) and of the aortic and pulmonary semilunar valves (b).*

The Cardiac Cycle and the Heart Sounds

Simultaneous contraction of both ventricles sends blood through the pulmonary and systemic circulations. Contraction of the ventricles causes a rise in pressure that closes the AV valves; relaxation of the ventricles produces a fall in pressure that closes the semilunar valves. The closing of first the AV valves and then the semilunar valves produces the "lub-dub" sounds heard with a stethoscope.

The *cardiac cycle* refers to the repeating pattern of contraction and relaxation of the heart. The phase of contraction is called **systole,**[7] and the phase of relaxation is called **diastole.**[8] When these terms are used alone they refer to contraction and relaxation of the ventricles. It should be noted, however, that the atria also contract and relax. Atrial contraction occurs toward the end of diastole, when the ventricles are relaxed; when the ventricles contract during systole, the atria are relaxed.

The heart thus has a two-step pumping action. The right and left atria contract almost simultaneously, followed about 0.1–0.2 seconds later by contraction of the right and left ventricles. During the time when both the atria and ventricles are relaxed the venous return of blood fills the atria. The buildup of pressure that results causes the AV valves to open and blood to flow from atria to ventricles. It has been estimated that the ventricles are about 80% filled with blood even before the atria contract. Contraction of the atria adds the remaining 20% of the total amount of blood in the ventricles just prior to ventricular contraction.

It is interesting that the blood contributed by contraction of the atria does not appear to be essential for life. Elderly people who have atrial fibrillation (a condition in which the atria fail to contract) do not appear to have a higher mortality than those who have normally functioning atria. People with atrial fibrillation, however, become fatigued more easily during exercise because the reduced filling of the ventricles compromises the ability of the heart to increase its output during exercise.

Contraction of the ventricles in systole ejects about two-thirds of the blood that they contain. The amount ejected by a contraction of each ventricle is called the *stroke volume.* The ventricles then fill with blood during the next cycle. At an average **cardiac rate** of seventy-five beats per minute, each cycle lasts 0.8 second; 0.5 second is spent in diastole, and systole takes 0.3 second.

Heart Sounds

Closing of the AV and semilunar valves produces sounds that can be heard at the surface of the chest with a stethoscope. These sounds are often described phonetically as *lub-dub.* The "lub," or **first sound,** is produced by closing of the AV valves when contraction of the ventricles begins. The "dub," or **second sound,** is produced by closing of the semilunar valves when the ven-

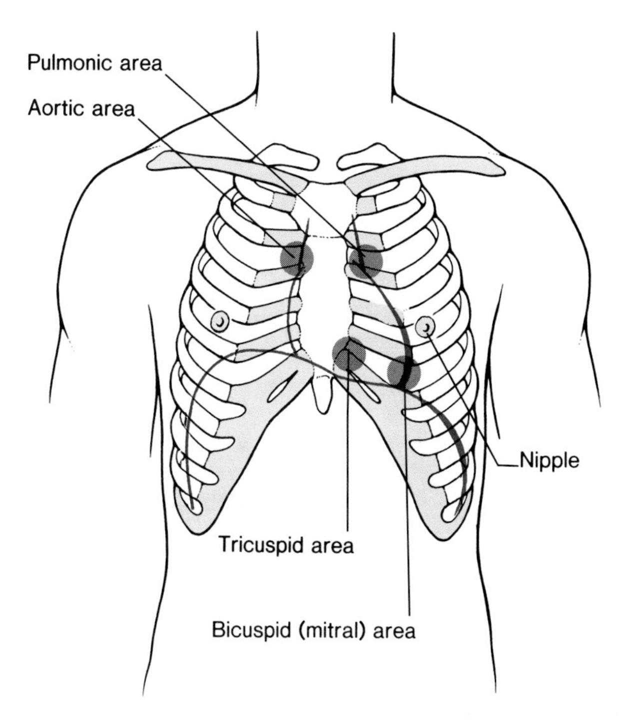

Figure 12.6. *Routine stethoscope positions for listening to the heart sounds.*

tricles begin to relax. The first sound is thus heard at systole, and the second sound is heard at the beginning of diastole. The standard stethoscope positions on the chest for hearing the heart sounds are shown in figure 12.6

Heart Murmurs

Murmurs are abnormal heart sounds produced by abnormal patterns of blood flow in the heart. Many murmurs are caused by defective heart valves. Defective heart valves may be congenital (inherited), or more commonly, they can occur as a result of ***rheumatic endocarditis,*** associated with rheumatic fever. In this disease the valves become damaged by antibodies made in response to an infection caused by streptococcus bacteria (the same bacteria that produce strep throat). Many people have small defects that produce detectable murmurs but do not seriously compromise the pumping ability of the heart. Larger defects, however, may have dangerous consequences and thus may require surgical correction.

The lungs of a fetus are collapsed, and blood is routed away (shunted) from the pulmonary circulation by an opening in the interatrial septum called the **foramen ovale** and by a connection between the pulmonary trunk and aorta called the **ductus arteriosus** (fig. 12.7). These shunts normally close after birth, but when they remain open (are *patent*), murmurs can result. When other defects are also present, a significant amount of oxygen-depleted blood from the right side of the heart may enter the left side and thus the systemic circulation through these defects. This lowers the oxygen concentration of the blood ejected into the systemic circulation. Since blood low in oxygen imparts a bluish tinge to the skin, the baby may be born *cyanotic* (blue).

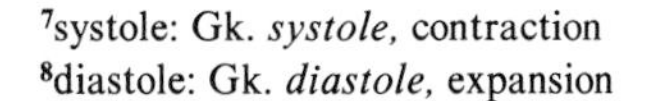

[7]systole: Gk. *systole,* contraction
[8]diastole: Gk. *diastole,* expansion

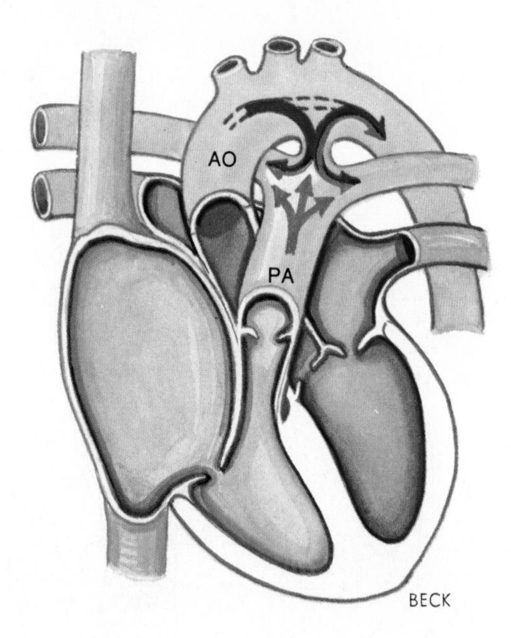

Figure 12.7. *The flow of blood through a patent (open) ductus arteriosus.*

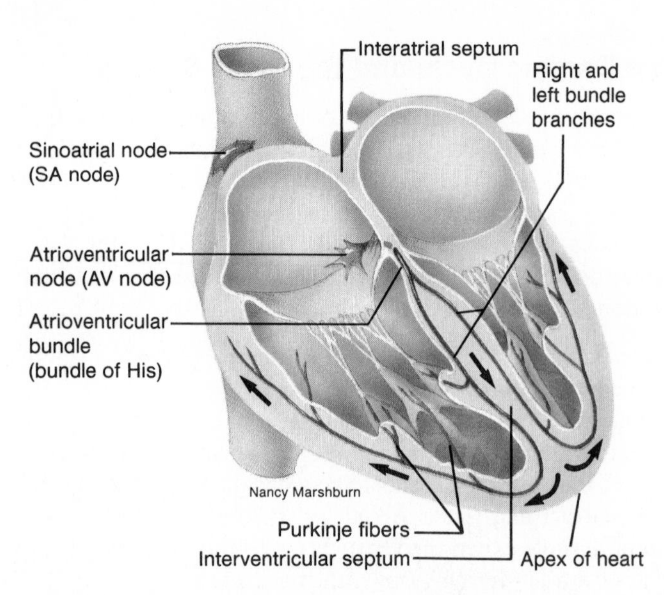

Figure 12.8. *The conduction system of the heart.*

1. Using a figure or an outline, describe the sequence of events that occurs during the cardiac cycle. Indicate when atrial and ventricular filling occur and when atrial and ventricular contraction occur.
2. Describe how the heart sounds are produced.
3. Explain the nature of heart murmurs.

Electrical Activity of the Heart and the Electrocardiogram

The pacemaker region of the heart exhibits a spontaneous excitation which causes electrical impulses and results in the automatic beating of the heart. Electrical impulses are conducted by myocardial cells in the atria and are transmitted to the ventricles by special conducting tissue. Electrocardiogram waves correspond to the electrical events in the heart.

If the heart of a frog is removed from the body, and all neural innervations are cut, it will still continue to beat as long as the myocardial cells remain alive. The automatic nature of the heartbeat is referred to as *automaticity.* Experiments with isolated myocardial cells, and clinical experience with patients who have specific heart disorders, have revealed that there are many regions within the heart that are capable of originating action potentials (electrical impulses) and functioning as a pacemaker. In a normal heart, however, only one region demonstrates spontaneous electrical activity and by this means functions as a pacemaker. This pacemaker region is called the **sinoatrial node,** or **SA node.** The SA node is located in the right atrium near the opening of the superior vena cava, and is about 5 mm wide, 15 mm long, and 1.5 mm deep.

Some other regions of the heart can, under abnormal conditions, fire automatically and assume a pacemaker function. The rate of spontaneous excitation of these cells, however, is slower than that of the SA node. The potential pacemaker cells are, therefore, stimulated by impulses from the SA node before they can stimulate themselves. If impulses from the SA node are prevented from reaching these areas (through blockage of conduction), they will fire at their own rate and function as pacemakers. A pacemaker other than the SA node is called an *ectopic pacemaker.*

Conducting Tissues of the Heart

Action potentials that originate in the SA node spread to adjacent myocardial cells of the right and left atria through the junctions between these cells. Since the myocardium of the atria is separate from the myocardium of the ventricles, however, the impulse cannot be conducted directly from the atria to the ventricles. Specialized conducting tissue, composed of modified myocardial cells, is thus required.

Once the impulse spreads through the atria, it passes to the **atrioventricular node (AV node),** which is located on the bottom portion of the interatrial septum (fig. 12.8). From here, the impulse continues through the **atrioventricular bundle,** or **bundle of His,**[9] beginning at the top of the interventricular septum. This conducting tissue descends along the interventricular septum until dividing into right and left bundle branches, which are continuous with the **Purkinje fibers**[10] within the ventricular walls. Stimulation of these fibers causes both ventricles to contract simultaneously and eject blood into the pulmonary and systemic circulations.

[9]bundle of His: from Wilhelm His, Jr., Swiss physician, 1863–1934
[10]Purkinje fibers: from Johannes E. von Purkinje, Bohemian anatomist, 1787–1869

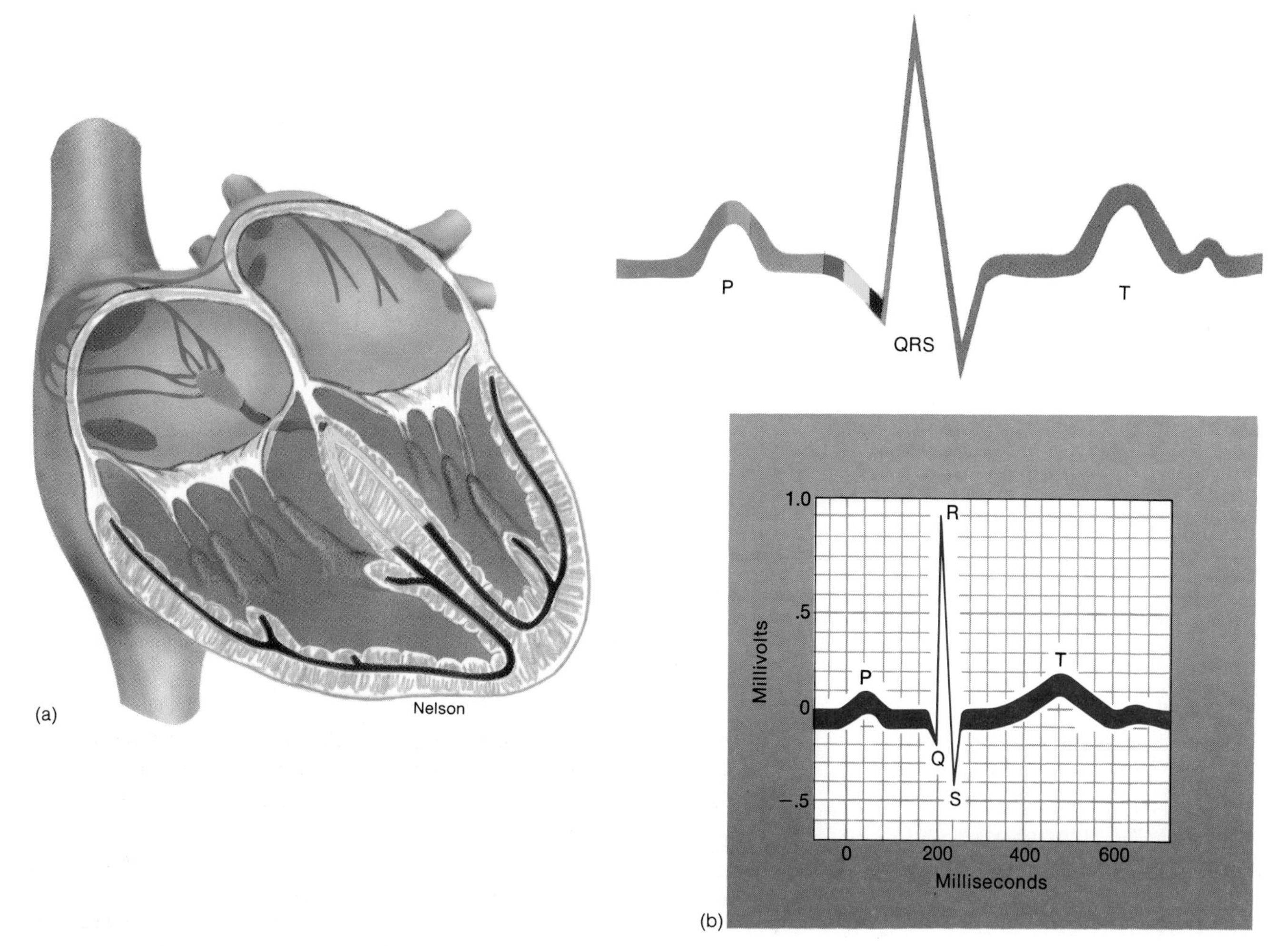

Figure 12.9. *The electrocardiogram indicates the conduction of electrical impulses through the heart (a) and measures and records both the intensity of this electrical activity (in millivolts) and the time intervals involved (b).*

The Electrocardiogram

A pair of surface electrodes placed directly on the heart will record a repeating pattern of potential (voltage) changes. As impulses spread from the atria to the ventricles, the voltage measured between these two electrodes will vary in a way that provides a picture of the electrical activity of the heart. The nature of this picture can be varied by changing the position of the recording electrodes on the heart; different positions provide different perspectives, enabling an observer to gain a more complete picture of the electrical events.

The body is a good conductor of electricity because tissue fluids contain a high concentration of ions that move (creating a current) in response to potential differences. Potential differences generated by the heart are thus conducted to the body surface, where they can be recorded by surface electrodes placed on the skin. The recording thus obtained is called an **electrocardiogram (ECG** or **EKG)** (fig. 12.9).

Electrodes are placed on the arms and legs and on the chest (fig. 12.10). These provide *ECG leads* which measure the potential difference between two designated electrodes (lead one, for example, measures the voltage between the left and right arms). There are a total of twelve standard ECG leads that "view" the changing pattern of the heart's electrical activity from different perspectives. This is important because certain abnormalities are best seen with particular leads and may not be visible at all with other leads.

Each cardiac cycle produces three distinct ECG waves, designated P, QRS, and T. These waves represent changes in potential between two regions on the surface of the heart which are produced by the composite effects of action potentials in many myocardial cells. The spread of electrical waves through the right and left atria produces the **P wave**; the spread of impulses through the ventricles produces the **QRS wave**; and repolarization (electrical recovery) of the ventricles produces the **T wave** (fig. 12.11).

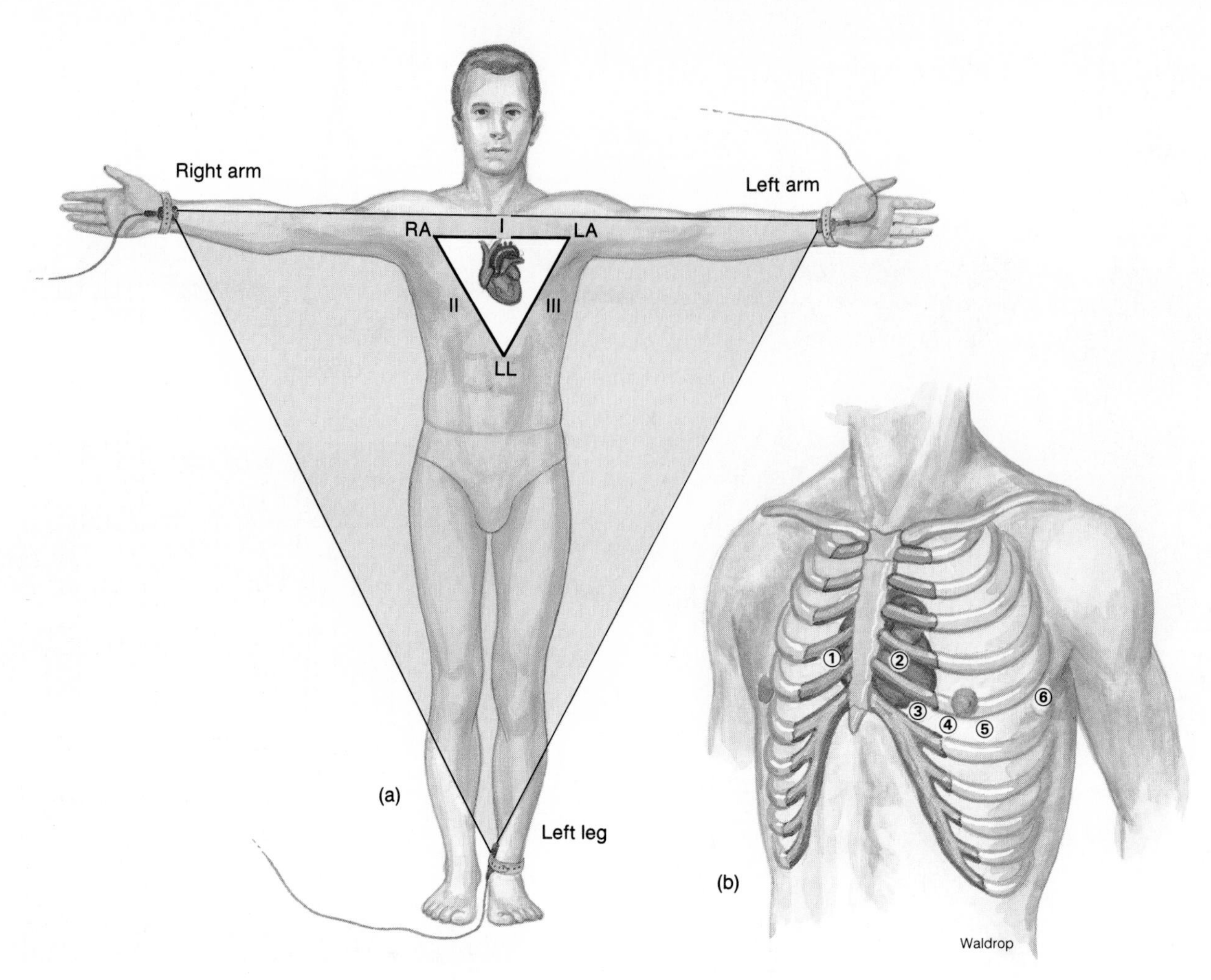

Figure 12.10. *The placement of the bipolar limb leads and the exploratory electrode for the unipolar chest leads in an electrocardiogram (ECG). (RA = right arm, LA = left arm, LL = left leg.)*

Arrhythmias Detected by the Electrocardiograph

Arrhythmias, or abnormal heart rhythms, can be detected and described by the abnormal ECG tracings they produce. Although proper clinical interpretation of electrocardiograms requires more information than is presented in this chapter, some knowledge of abnormal rhythms is interesting in itself and is useful in gaining an understanding of normal physiology.

Since a heartbeat occurs whenever a normal QRS complex is seen, the cardiac rate (beats per minute) can be easily obtained from examination of the ECG recording. A cardiac rate slower than 60 beats per minute indicates **bradycardia**; a rate faster than 100 beats per minute is described as **tachycardia** (fig. 12.12).

Both bradycardia and tachycardia can occur normally. Endurance-trained athletes, for example, commonly have a slower heart rate than the general population. This *athlete's bradycardia* occurs as a result of greater inhibitory activity of the vagus nerve (a parasympathetic nerve) fibers to the SA node and is a beneficial adaptation. Activation of the sympathetic nervous system, during exercise or emergencies ("fight or flight"), causes a normal tachycardia to occur.

Abnormal tachycardia occurs when a person is at rest. This may result from abnormally fast pacing by the atria, due to drugs or to the development of abnormally fast ectopic pacemakers—cells located outside the SA node that assume a pacemaker function. This abnormal atrial tachycardia thus differs from normal "sinus" (SA node) tachycardia. *Ventricular tachycardia* results when abnormally fast ectopic pacemakers in the ventricles cause them to beat rapidly and independently of the atria. This is very dangerous because it can quickly degenerate into a lethal condition known as ventricular fibrillation.

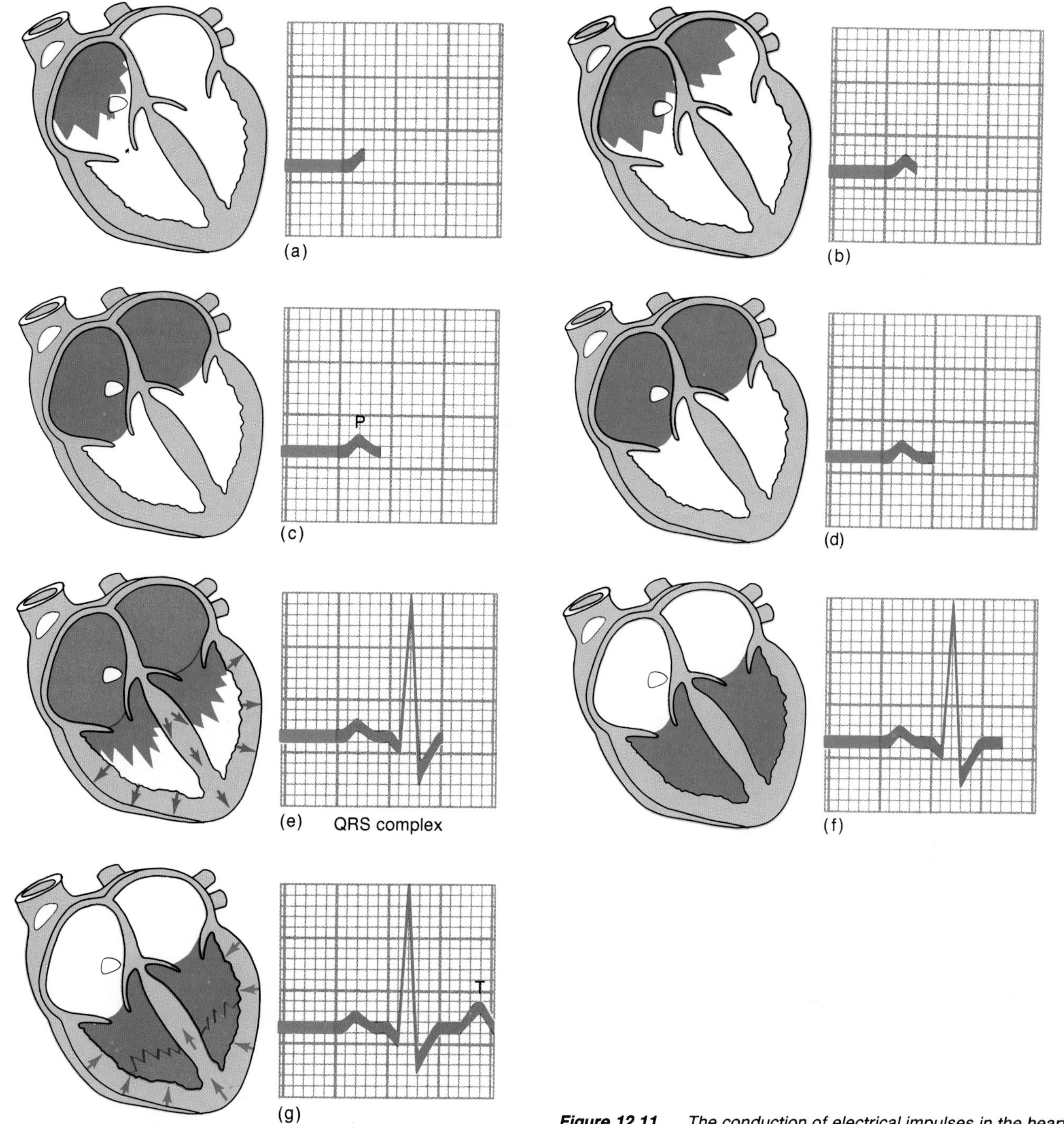

Figure 12.11. The conduction of electrical impulses in the heart, as indicated by the electrocardiogram (ECG).

Flutter and Fibrillation

Extremely rapid rates of electrical excitation and contraction of either the atria or the ventricles may occur. In *flutter,* the contractions are very rapid (200–300 per minute) but are coordinated. In *fibrillation,* contractions of different groups of myocardial fibers occur at different times so that a coordinated pumping action of the chambers is impossible.

Fibrillation is caused by a continuous recycling of electrical waves, known as **circus rhythms,** through the myocardium. Recycling of electrical waves along continuously changing pathways produces an uncoordinated contraction and stops the pumping action of the heart. Fibrillation can sometimes be stopped by a strong electric shock delivered to the chest. This procedure is called **electrical defibrillation.** The electric shock

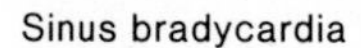

Sinus bradycardia

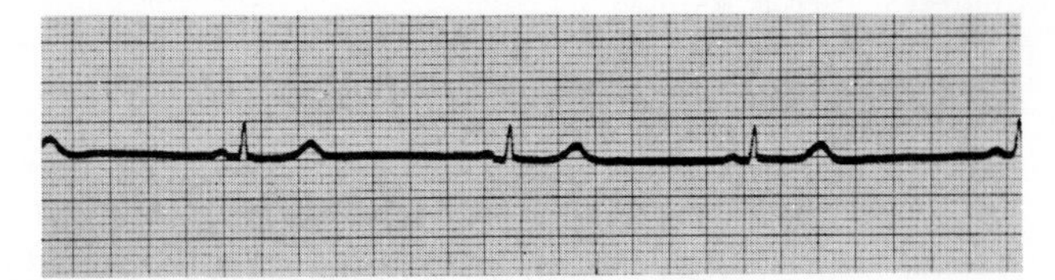

Sinus tachycardia

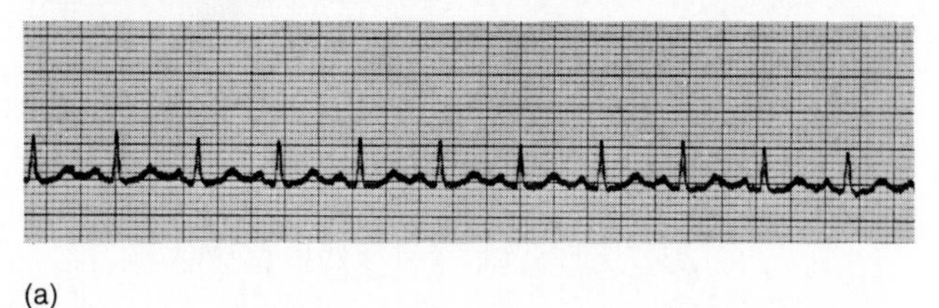

(a)

Ventricular tachycardia

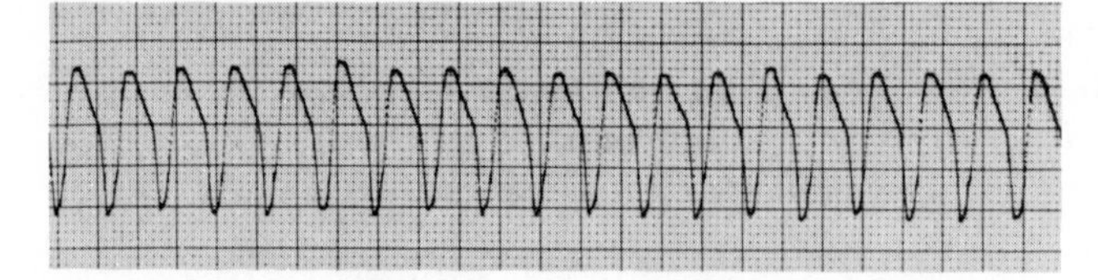

Ventricular fibrillation

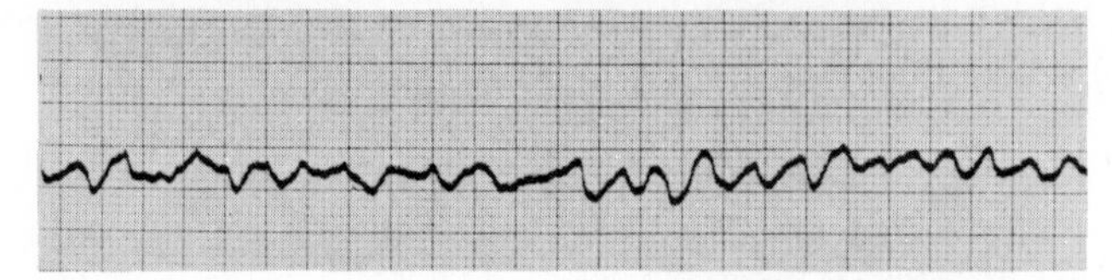

(b)

Figure 12.12. *In (a) the heartbeat is paced by the normal pacemaker—the SA node. This can be abnormally slow (bradycardia—46 beats per minute in this example) or fast (tachycardia—136 beats per minute in this example). Compare the pattern of tachycardia in (a) with the tachycardia in (b), which is produced by an ectopic pacemaker in the ventricles. This dangerous condition can quickly lead to ventricular fibrillation, also shown in (b).*

depolarizes the heart and prevents it from being able to produce action potentials for a short time. Conduction of circus rhythms thus stops, and—within a couple of minutes—the SA node can begin to stimulate contraction in a normal fashion. This does not correct the initial problem that caused circus rhythms and fibrillation, of course, but it does keep the person alive long enough to take other corrective measures.

A number of abnormal conditions, including a blockage in conduction of the impulse along the bundle of His, require the insertion of an **artificial pacemaker.** These are battery-powered devices, about the size of a locket, which may be placed in permanent position under the skin. The electrodes from the pacemaker are guided by means of a fluoroscope through a vein to the right atrium, through the tricuspid valve, and into the right ventricle. The electrodes are fixed in contact with the wall of the ventricle. When these electrodes deliver shocks—either at a continuous pace or on demand (when the heart's own impulse doesn't arrive on time)—both ventricles are depolarized and contract and then repolarize and relax just as they do in response to normal stimulation.

1. Identify and describe the action of the heart's normal pacemaker. Explain the meaning of the term "ectopic pacemaker."
2. Describe the pathway of impulse conduction in the heart.
3. Draw an ECG and label the waves. Indicate the electrical events in the heart that produce these waves.
4. Describe normal and pathological examples of bradycardia and tachycardia. Also describe how fibrillation is produced.

Cardiac Output and Blood Flow

The cardiac output is the volume of blood pumped per minute by each ventricle. This volume can be adjusted by variations in the cardiac rate and the stroke volume. During exercise, the cardiac output is greatly increased to provide an increased blood flow through the skeletal muscles. Blood flow through the heart and skeletal muscles is further increased during exercise as a result of vasodilation in these organs.

The heart functions as a pump, with each ventricle ejecting about 80 ml of blood per beat (the *stroke volume*) and beating about 70 times per minute (the *cardiac rate*). If the stroke volume and cardiac rate are multiplied together, a **cardiac output** of about 5,600 ml/min is obtained for each ventricle. Since the circulatory system is closed—so that the amount of blood returning to the heart per minute must equal the amount ejected—this value for the cardiac output also represents the total rate of blood flow through the circulation.

Although the heartbeat is automatic, the rate of beat is adjusted by the nervous and endocrine systems. Sympathetic nerves (chapter 6) to the SA node stimulate an increased rate of beat. This is produced directly by the release of norepinephrine from the sympathetic nerve endings. The adrenal medulla secretes epinephrine and lesser amounts of norepinephrine (chapter 7) into the blood as hormones. These hormones support the action of the sympathetic nerves in stimulating an increased cardiac rate during emergency situations (the "fight-or-flight" response). In contrast, parasympathetic nerve fibers that innervate the SA node cause a slowing of the rate of beat.

The stroke volume can be adjusted by variations in (1) the strength with which the ventricles contract; and (2) the volume of blood available within the ventricles to be ejected. Epinephrine and norepinephrine stimulate the ventricles to contract more strongly, thus ejecting more blood per beat. Variations in the total amount of blood in the body affect the stroke volume; a decrease in blood volume (as may occur during dehydration or in hemorrhage), produces a decrease in the stroke volume. Factors that affect the ability of veins to return blood to the heart, in addition, influence the volume of blood in the ventricles and thus the stroke volume and cardiac output.

Venous Return

The **venous return** is a shorthand phrase that refers to the flow of blood in the veins to the heart. The average pressure in the veins is only 2 mm Hg, compared to a much higher average arterial pressure of about 100 mm Hg. These pressures represent the hydrostatic pressure that the blood exerts on the walls of the vessels, and the numbers indicate the differences from atmospheric pressure.

The low venous pressure is insufficient to return blood to the heart, particularly from the lower limbs. Veins, however, pass between skeletal muscle groups that produce a massaging action as they contract (fig. 12.13). As the veins are squeezed by contracting skeletal muscles, a one-way flow of blood to the heart is ensured by the presence of **venous valves.** The ability of these valves to prevent the flow of blood away from the heart was demonstrated in the seventeenth century by William Harvey (fig. 12.14). After applying a tourniquet to a subject's arm, Harvey found that he could push the blood in a bulging vein toward the heart but not in the reverse direction.

The effect of the massaging action of skeletal muscles on venous blood flow is often described as the **skeletal muscle pump.** The rate of venous return to the heart is dependent, in large part, on the action of skeletal muscle pumps. When these pumps are less active, as when a person stands still or is bedridden, blood accumulates in the veins and causes them to bulge. When a person is more active, blood returns to the heart at a faster rate and less is left in the venous system.

The accumulation of blood in the veins of the legs over a long period of time, as may occur in people with occupations that require standing still all day, can cause the veins to stretch to the point where the venous valves are no longer efficient. This can produce **varicose**[11] **veins.** During walking the movements of the foot activate muscle pumps. This effect can be produced in bedridden people by upward and downward manipulations of the feet.

[11]varicose: L. *varicosus,* dilated vein

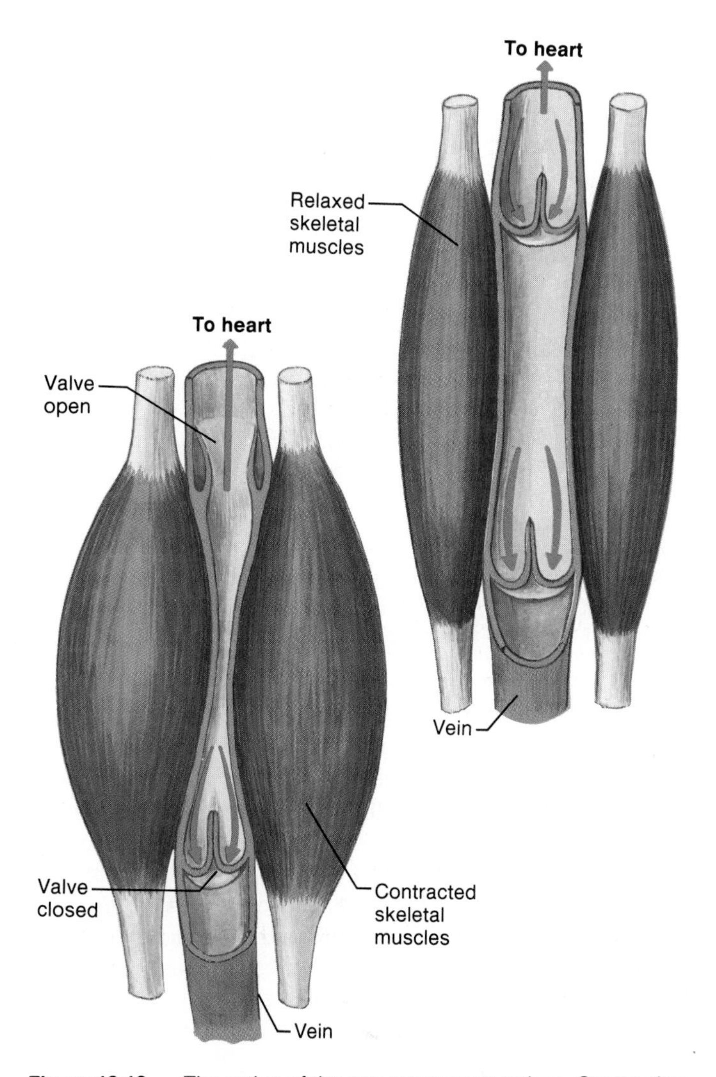

Figure 12.13. *The action of the one-way venous valves. Contraction of skeletal muscles helps to pump blood toward the heart, but is prevented from pushing blood away from the heart by closure of the venous valves.*

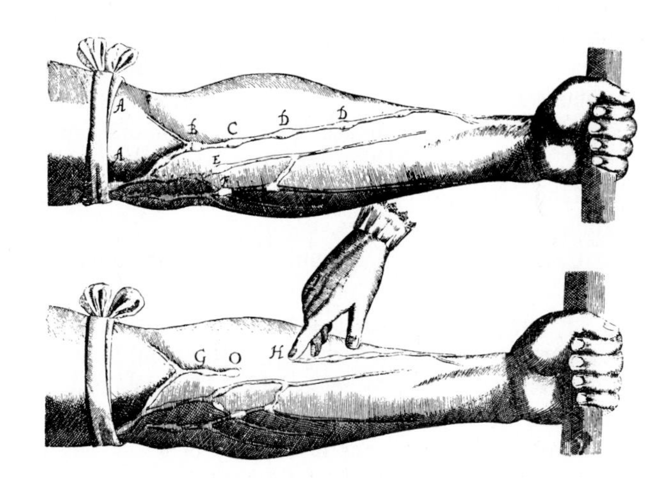

Figure 12.14. *A classical demonstration by William Harvey of the existence of venous valves that prevent the flow of blood away from the heart.*

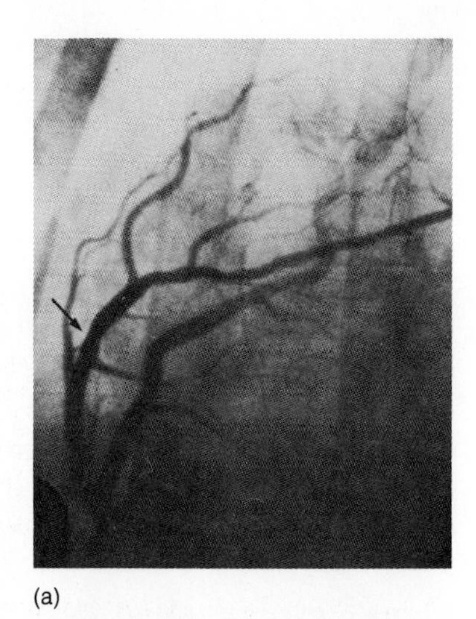

(a)

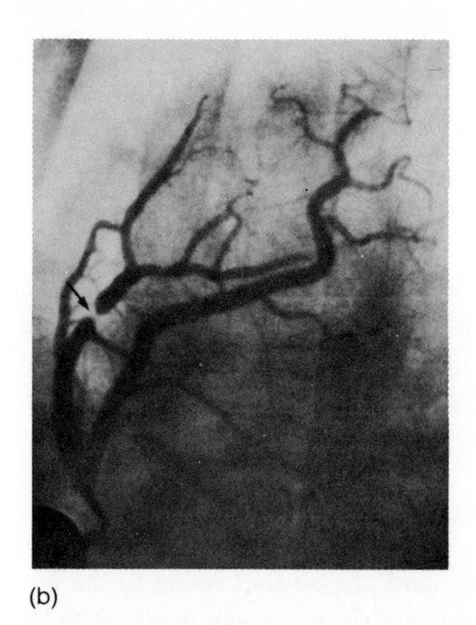

(b)

Figure 12.15. *An angiogram of the left coronary artery in a patient (a) when the ECG was normal and (b) when the ECG showed evidence of myocardial ischemia. Notice that a coronary artery spasm—see arrow in (b)—appears to accompany the ischemia.*

of myocardial ischemia. Notice that a coronary artery spasm—see arrow in (b)—appears to accompany the ischemia.

Action of the skeletal muscle pumps aid the return of venous blood from the lower limbs to the large abdominal veins. Movement of venous blood from abdominal to thoracic veins, however, is aided by an additional mechanism: breathing. When a person inhales, the diaphragm—a muscular sheet separating the thoracic and abdominal cavities—contracts. As the diaphragm contracts it changes from dome-shaped to a more flattened form and thus protrudes more into the abdomen. This has two effects; it increases the pressure in the abdomen, thus squeezing the abdominal veins, and it decreases the pressure in the thoracic cavity. The pressure difference in the veins created by this inspiratory movement forces blood into the thoracic veins that return the venous blood to the heart.

Regulation of Coronary Blood Flow

The coronary arteries supply an enormous number of capillaries in the heart, which are packed within the myocardium at a density of about 2,500–4,000 per cubic millimeter of tissue. Each myocardial cell, as a consequence, is within 10 μm of a capillary (compared to an average distance in other organs of 60–80 μm). The exchange of gases by diffusion between myocardial cells and capillary blood thus occurs very quickly.

Contraction of the myocardium squeezes the coronary arteries. Unlike blood flow in all other organs, flow in the coronary vessels thus decreases in systole and increases during diastole. The myocardium, however, contains large amounts of *myoglobin,* a pigment related to hemoglobin. Myoglobin in heart muscle stores oxygen during diastole and releases its oxygen during systole. In this way, the myocardial cells can receive a continuous supply of oxygen even though coronary blood flow is temporarily reduced during systole.

Under abnormal conditions, the blood flow to the myocardium may be inadequate, resulting in *ischemia* (inadequate blood flow). This can result from blockage by atheromas (plaques, as described in a later section) and/or blood clots or from muscular spasm of a coronary artery (fig. 12.15). Blockage of a coronary artery can be visualized by a technique called *selective coronary arteriography.* In this procedure, a catheter (plastic tube) is inserted into a brachial or femoral artery all the way to the opening of the coronary arteries in the aorta, and radiographic contrast material is injected. The picture thus obtained is called an **angiogram.** If the occlusion is sufficiently great, a coronary bypass operation may be performed. In this procedure a length of blood vessel, usually taken from the saphenous vein in the leg, is sutured to the aorta and to the coronary artery at a location beyond the site of the occlusion (fig. 12.16).

Regulation of Blood Flow through Skeletal Muscles

Blood flow in a skeletal muscle decreases when the muscle contracts and squeezes its arterioles; indeed, blood flow stops entirely when the muscle contracts beyond about 70% of its maximum. Pain and fatigue result when the muscle cells do not receive adequate amounts of oxygen from the blood (chapter 21), and thus occur much more quickly when a strong contraction is sustained than when rhythmic contractions and relaxations are performed.

Skeletal muscles have such a large mass that they receive 20% to 25% of the total blood flow in the body at rest. As exercise progresses, the small arteries and arterioles within the muscles dilate. This vasodilation results primarily from the effects of increased cellular metabolism within the muscle. As the metabolic rate of a muscle increases, chemical changes occur within the tissue—CO_2 increases, pH decreases, and the amount of oxygen within the tissue is reduced. These chemical changes

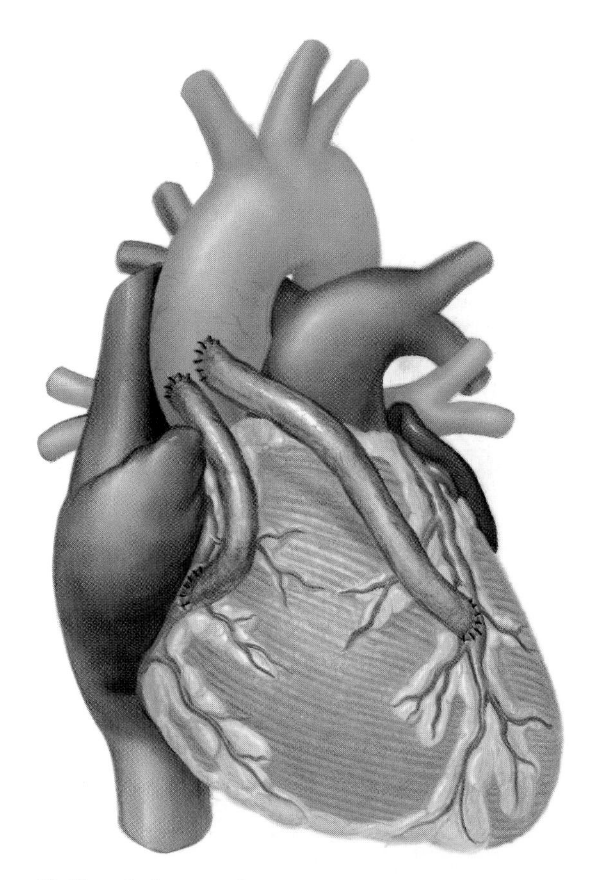

Figure 12.16. *A diagram of coronary artery bypass surgery.*

Age	Maximum Cardiac Rate
Table 12.3 Relationship between age and average maximum cardiac rate	
20–29	190 beats/min
30–39	160 beats/min
40–49	150 beats/min
50–59	140 beats/min
60 and above	130 beats/min

act directly on the arterioles to promote vasodilation. As a result, the percentage of the total blood flow in the body that goes through skeletal muscles can increase from 20% at rest to as high as 85% during exercise.

Circulatory Changes during Exercise

During exercise the cardiac output of an average person can increase five-fold, from about 5 L per minute to about 25 L per minute. This is primarily due to an increase in cardiac rate. The cardiac rate, however, can only increase up to a maximum value (table 12.3), which is determined mainly by the person's age. In very well-trained athletes the stroke volume can also increase significantly, allowing these individuals to achieve a cardiac output during strenuous exercise as much as six or seven times greater than their resting values.

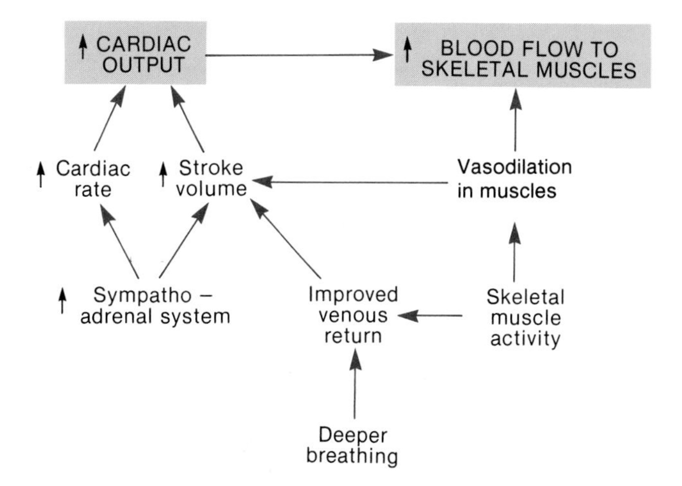

Figure 12.17. *Cardiovascular adpatations to exercise.*

In most people the stroke volume can only increase 10% to 35% during exercise. The fact that the stroke volume can increase at all during exercise may at first be surprising, in view of the fact that the heart has less time to fill with blood between beats when it is pumping faster. The shorter filling time, however, is compensated by the more rapid rate of venous return. The venous return is aided by the improved action of the skeletal muscle pumps and by increased respiratory movements during exercise (fig. 12.17).

Endurance-trained athletes commonly have a slower resting cardiac rate and an increased resting stroke volume compared to nonathletes. This increased resting stroke volume is believed to be due to an increase in blood volume. Indeed, studies have shown that the blood volume can increase by about 500 ml after only eight days of training. These adaptations enable the trained athlete to produce a larger increase in cardiac output and achieve a higher cardiac output during exercise. This large cardiac output is the major factor in the improved oxygen delivery to skeletal muscles that occurs as a result of endurance training.

1. Describe blood flow and oxygen delivery to the heart during systole and diastole.
2. Define *myocardial ischemia,* and describe the factors that prevent ischemia from occurring in the normal heart.
3. Describe how vasodilation of arterioles in skeletal muscles is produced during exercise, and explain the significance of this effect.
4. Explain how the stroke volume can increase during exercise, and the effect of endurance training on the resting stroke volume.

Atherosclerosis

Atherosclerosis is the most dangerous type of hardening of the arteries, accounting for most incidents of heart disease and stroke. High blood cholesterol, and a low ratio of HDL to LDL (two types of protein carriers that transport cholesterol in the blood), contribute to the risk of atherosclerosis.

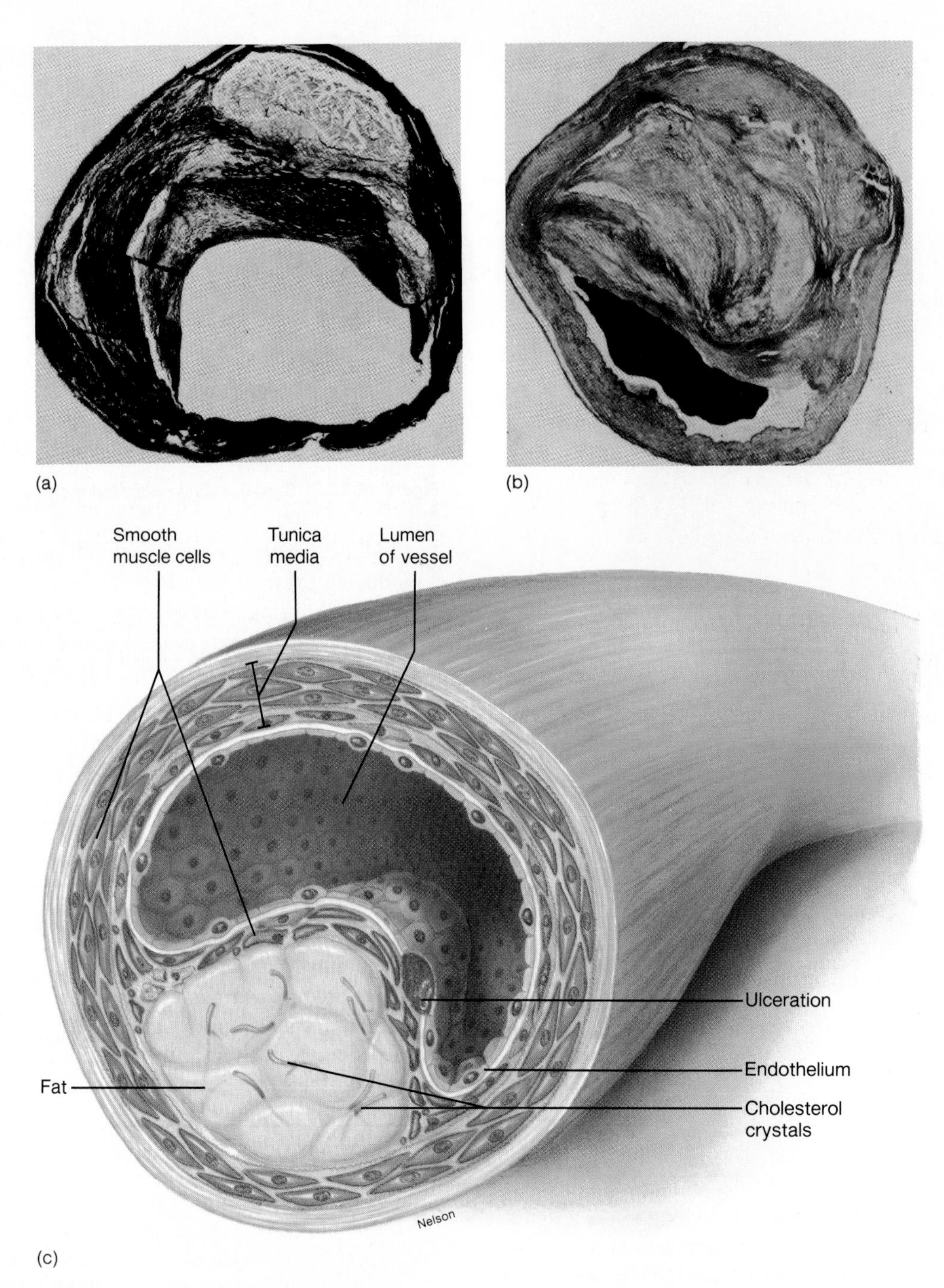

Figure 12.18. *(a) The lumen (cavity) of a human cornary artery is partially occluded by an atheroma and (b) almost completely occluded by a thrombus. (c) The structure of an atheroma is diagrammed.*

Atherosclerosis[12] is the most common form of arteriosclerosis (hardening of the arteries) and, through its contribution to heart disease and stroke, is responsible for about 50% of the deaths in the United States. In atherosclerosis, localized *plaques,* or **atheromas,** protrude into the lumen of the artery and thus reduce blood flow. The atheromas additionally serve as sites for *thrombus* (blood clot) formation, which can further occlude the blood supply to an organ (fig. 12.18).

[12]atherosclerosis: Gk. *athere,* gruel; *skleros,* hard

It is currently believed that atheromas begin as "fatty streaks," which are gray-white areas that protrude into the lumen of arteries, particularly at arterial branch points. These are universally present in the aorta and coronary arteries by the age of ten, but progress to more advanced stages at different rates in different people. As this progression occurs, smooth muscle cells and others fill with lipids to give a "foamy cell" appearance. Later, damage to the endothelium—the simple squamous epithelium (chapter 5) that lines the artery—causes blood platelets to stick to underlying connective tissue. Platelets are believed to release chemicals that cause smooth muscle cells to move into this area and proliferate, resulting in a tumorlike growth. The intercellular space of the atheroma later becomes filled with lipids and cholesterol crystals, and then is hardened by deposits of calcium. The damage to the endothelial lining also promotes the formation of blood clots at the atheromas, which further occlude the artery.

Risk factors in the development of atherosclerosis include advanced age, smoking, hypertension, and high blood cholesterol concentrations. It is currently believed that an intact and properly functioning endothelium protects against atherosclerosis. If the endothelium in a particular region of an artery is removed, or if it is damaged in some way, growth factors, or *mitogens* (chemicals that stimulate mitosis), may stimulate proliferation of the smooth muscle cells and the growth of an atheroma. These growth factors may be derived from blood platelets, endothelial cells, and/or monocytes.

LDL and HDL Cholesterol

There is good evidence that high blood cholesterol is associated with an increased risk of atherosclerosis. This high blood cholesterol can be produced by a diet rich in cholesterol and saturated fat, or it may be the result of an inherited condition known as *familial hypercholesteremia.* This condition is inherited as a single dominant gene; individuals who inherit two of these genes have extremely high cholesterol concentrations (regardless of diet) and usually suffer heart attacks during childhood.

Lipids, including cholesterol, are carried in the blood attached to protein carriers. Cholesterol is carried to the arteries by plasma proteins called **low-density lipoproteins (LDL).** These particles, produced by the liver, consist of a core of cholesterol surrounded by a layer of phospholipids (to make the particle water-soluble) and a protein. Cells in various organs contain receptors for the protein in LDL. When LDL attaches to its receptors, the cell engulfs the LDL and utilizes the cholesterol for different purposes. Most of the LDL in the blood is removed in this way by the liver.

Once LDL has passed through the endothelium of an artery, it may stimulate monocytes to enter the area and engulf the cholesterol (thereby becoming "foam cells"). The monocytes may then be stimulated by LDL to secrete a growth factor that either begins or contributes to the development of an atheroma. A high blood concentration of LDL favors these events. Recent evidence shows that people who eat a diet high in cholesterol and saturated fat, and people with familial hypercholesteremia, have a high blood LDL concentration because their tissues (principally the liver) have a low number of LDL receptors. With fewer LDL receptors the liver is less able to remove the LDL from the blood, the blood LDL concentration is raised, and the risk of atherosclerosis is greatly increased.

Excessive cholesterol may be released from cells and travel in the blood as **high-density lipoproteins (HDL),** which are removed by the liver. The cholesterol in HDL is not taken into the artery wall, and therefore this cholesterol does not contribute to atherosclerosis. Indeed, a high proportion of cholesterol in HDL as compared to LDL is beneficial, since it indicates that cholesterol may be traveling away from the blood vessels to the liver. The concentration of HDL-cholesterol appears to be higher and the risk of atherosclerosis lower in people who exercise regularly. The HDL-cholesterol concentration, for example, is higher in marathon runners than in joggers and is higher in joggers than in inactive men. Women in general have higher HDL-cholesterol concentrations and a lower risk of atherosclerosis than men.

Ischemic Heart Disease

A tissue is said to be **ischemic** when it receives an inadequate supply of oxygen because of an inadequate blood flow. The most common cause of myocardial ischemia is atherosclerosis of the coronary arteries. The adequacy of blood flow is relative—it depends on the metabolic requirements of the tissue for oxygen. An obstruction in a coronary artery, for example, may allow sufficient blood flow at rest but may produce ischemia when the heart is stressed by exercise or emotional conditions.

Myocardial ischemia is associated with increased concentrations of blood lactic acid (chapter 18) produced by the ischemic tissue. This condition causes pain, which may be referred to the left shoulder and arm, as well as other areas. This referred pain is called **angina pectoris.** People with angina frequently take nitroglycerin or related drugs that help relieve the ischemia and pain. These drugs are effective because they stimulate vasodilation, which improves circulation to the heart and decreases the work that the heart must perform to eject blood into the arteries.

Myocardial cells cannot be sustained in the absence of oxygen for more than a few minutes. If ischemia continues for more than a few minutes, *necrosis* (cellular death) may occur in the areas most deprived of oxygen. A sudden, irreversible injury of this kind is called a **myocardial infarction,** or **MI.** The lay term "heart attack," though imprecise, usually refers to a myocardial infarction.

---

1. Describe the development of atherosclerosis and the dangers of this condition.
2. Explain how cholesterol is carried in the plasma and how the concentrations of cholesterol carriers is related to the risk of atherosclerosis.
3. Explain how angina pectoris is produced and the relationship of this symptom to conditions in the heart.

Blood Pressure

The pressure of the arterial blood is regulated by the cardiac output and the degree of constriction or dilation of the arterial vessels. Changes in these factors normally function to maintain a correct arterial pressure. As the heart goes through systole and diastole, the arterial pressure rises and falls. These systolic and diastolic blood pressures are measured indirectly through the use of a blood pressure cuff.

Resistance to flow in the arterial system is greatest in the arterioles because these vessels have the smallest diameters. Blood flow rate and pressure is thus reduced in the capillaries, which are located downstream of the high resistance imposed by the arterioles. The blood pressure and flow rate within the capillaries are further reduced by the fact that their total cross-sectional area is much greater than the cross-sectional areas of arteries and arterioles (fig. 12.19). (Although each capillary is much narrower than each arteriole, there are many more capillaries than there are arterioles.) Variations in the diameter of arterioles due to vasoconstriction and vasodilation simultaneously affect both blood flow through capillaries downstream and the *arterial blood pressure* upstream from the arterioles.

Vasoconstriction of arterioles increases the resistance to blood flow and raises arterial blood pressure. Blood pressure can also be raised by an increase in the cardiac output. This may be due to elevations in cardiac rate or stroke volume, which in turn are affected by other factors. The three most important variables affecting blood pressure are the cardiac rate, stroke volume (determined primarily by the blood volume), and *total peripheral resistance.*

The total peripheral resistance refers to the total resistances to blood flow through all of the arteries in the body. The resistance to blood flow through each vessel is increased by vasoconstriction and decreased by vasodilation. If there is a net vasoconstriction in the arterial supply to different organs, therefore, the total peripheral resistance and blood pressure will be

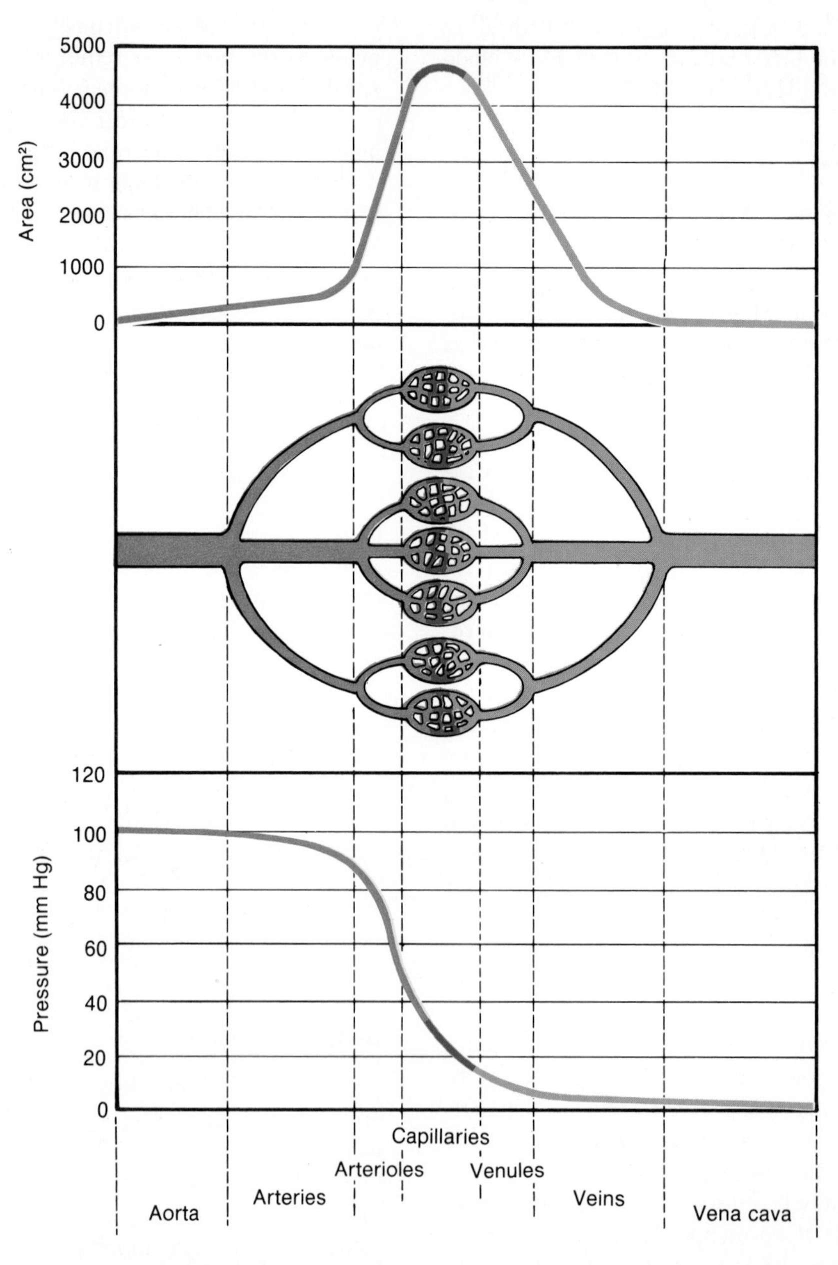

Figure 12.19. *As blood passes from the aorta to the smaller arteries, arterioles, and capillaries, the cross-sectional area increases as the pressure decreases.*

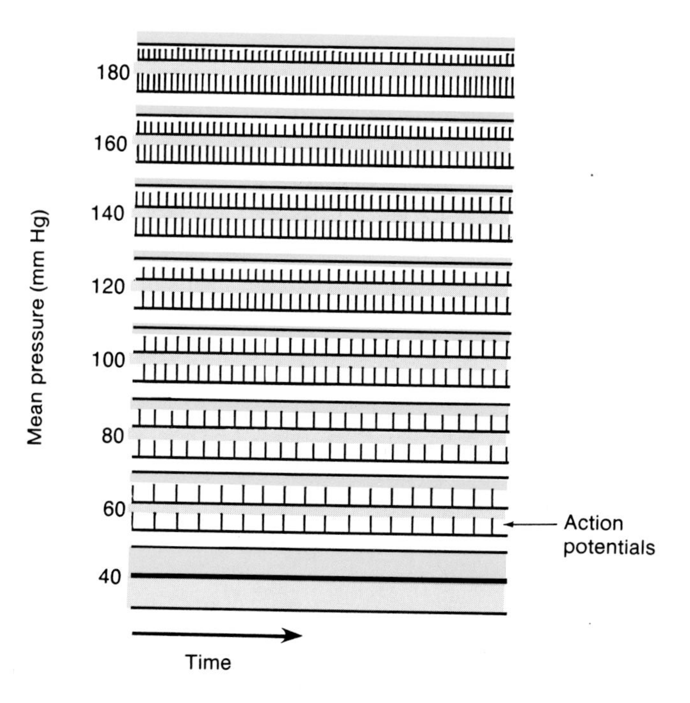

__Figure 12.20.__ Action potential frequency in sensory nerve fibers from baroreceptors in the carotid sinus and aortic arch. As the blood pressure increases, the baroreceptors become increasingly stretched. This results in an increase in the frequency of action potentials that are transmitted to the cardiac and vasomotor control centers in the medulla oblongata.

increased. If there is a net vasodilation, the total peripheral resistance and blood pressure will be decreased. All of these effects, it should be understood, are the results that would occur in the absence of compensations. A decrease in total peripheral resistance, for example, might be compensated by an increase in cardiac output to maintain a constant blood pressure.

Baroreceptor Reflex

In order for blood pressure to be maintained within limits, specialized receptors for pressure are needed. These baroreceptors[13] are stretch receptors located in the *aortic arch* and in the *carotid sinuses.*[14] An increase in pressure causes the walls of these arterial regions to stretch and stimulate the activity of sensory nerve endings (fig. 12.20). A fall in pressure below the normal range, in contrast, causes a decrease in the frequency of action potentials produced by these sensory nerve fibers.

Sensory nerve activity from the baroreceptors ascends to the medulla oblongata of the brain (chapter 6), which directs the autonomic nervous system to respond appropriately. **Vasomotor control centers** in the medulla control vasoconstriction/vasodilation and hence help regulate total peripheral resistance. **Cardiac control centers** in the medulla regulate the cardiac rate (fig. 12.21).

[13]baroreceptors: Gk. *baros,* pressure; L. *receiver,* to receive
[14]carotid: Gk. *karotikos,* stupefying (because the ancient Greeks knew that finger pressure to the carotid sinuses could cause loss of consciousness)

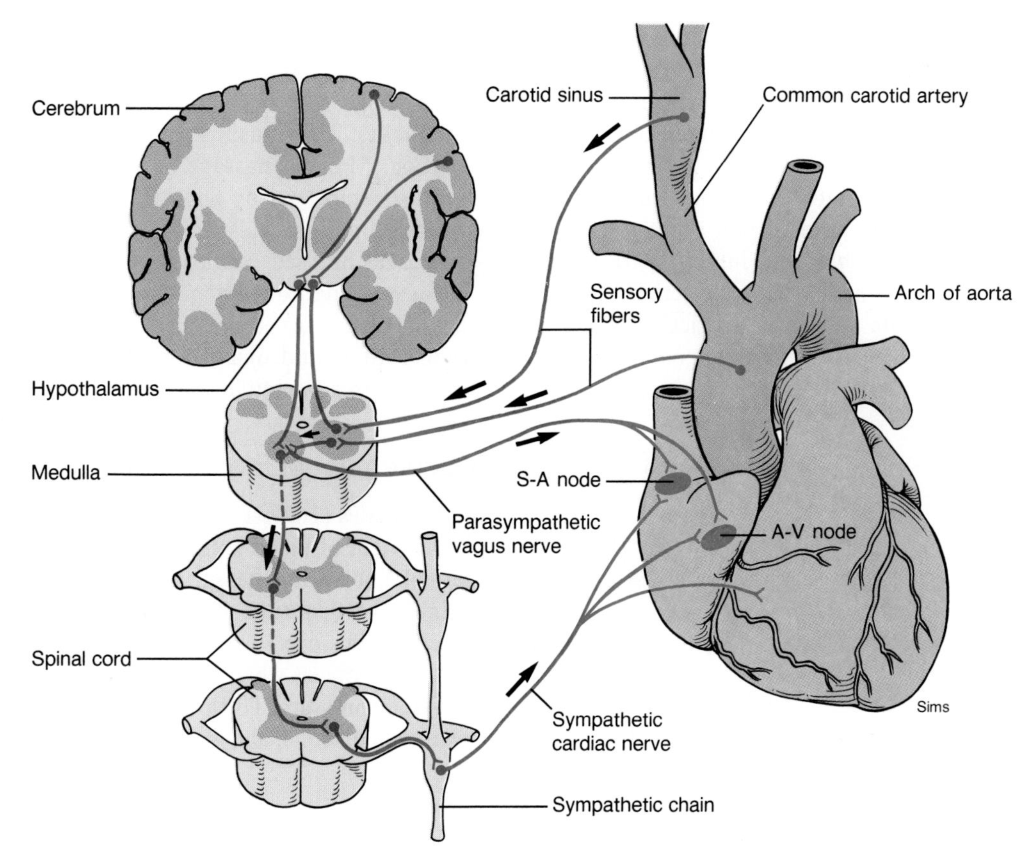

__Figure 12.21.__ The baroreceptor reflex. Sensory stimuli from baroreceptors in the carotid sinus and the aortic arch, acting via control centers in the medulla oblongata, affect the activity of sympathetic and parasympathetic nerve fibers in the heart.

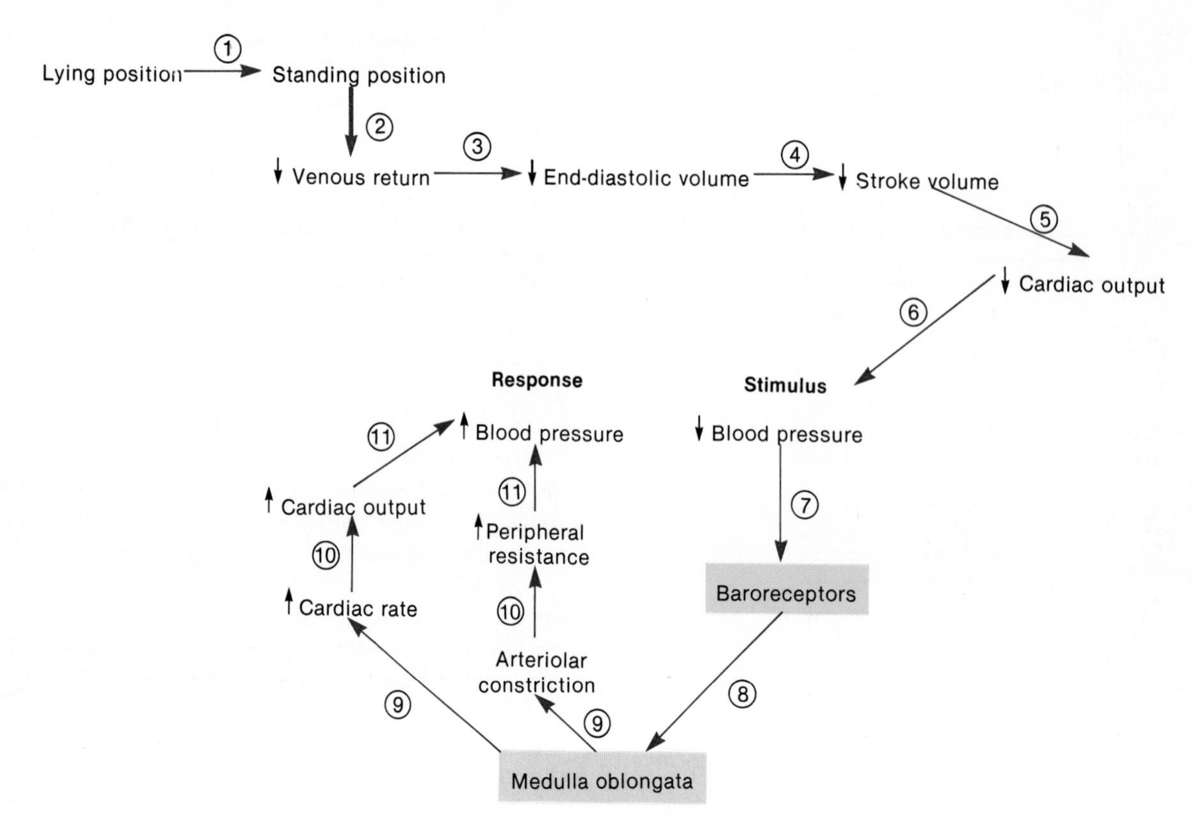

Figure 12.22. Compensations for the upright posture induced by the baroreceptor reflex. Numbers indicate the sequence of cause-and-effect steps.

When a person goes from a lying to a standing position, there is a shift of 500–700 ml of blood from the veins of the thoracic cavity to veins in the lower extremities, which expand to contain the extra volume of blood. This pooling of blood reduces the venous return and cardiac output. The resulting fall in blood pressure is almost immediately compensated by the baroreceptor reflex. A decrease in baroreceptor sensory information traveling to the medulla oblongata inhibits parasympathetic activity and promotes sympathetic nerve activity. This results in increased cardiac rate and vasoconstriction, which help to maintain an adequate blood pressure upon standing (fig. 12.22).

Since the baroreceptor reflex may require a few seconds to be fully effective, many people feel dizzy and disoriented if they stand up too rapidly. If the baroreceptor sensitivity is abnormally reduced, perhaps by atherosclerosis, an uncompensated fall in pressure may occur upon standing. This condition—called **postural,** or **orthostatic, hypotensior**)otension = low blood pressure)—can make a person feel e. emely dizzy or even faint because of inadequate blood flow to the brain.

The baroreceptor reflex can also mediate the opposite response. When the blood pressure rises above an individual's normal range, the baroreceptor reflex activates parasympathetic nerves. This causes a slowing of the cardiac rate and vasodilation. Manual massage of the carotid sinus, a procedure sometimes employed by physicians to reduce tachycardia and lower blood pressure, also evokes this reflex. Such carotid massage should be used cautiously, however, because the intense vagus-nerve-induced slowing of the cardiac rate could cause loss of consciousness (as occurs in emotional fainting). Manual massage of both carotid sinuses simultaneously can even cause cardiac arrest in susceptible people.

Measurement of Blood Pressure

Stephen Hales (1677–1761) accomplished the first documented measurement of blood pressure by inserting a cannula into the artery of a horse and measuring the height to which blood would rise in the vertical tube. This height varied between two levels, bouncing from one to the other. The highest pressure in the artery is produced when the heart contracts at systole. This is the **systolic pressure.** The lowest pressure occurs when the heart is resting in diastole, and is thus known as the **diastolic pressure.**

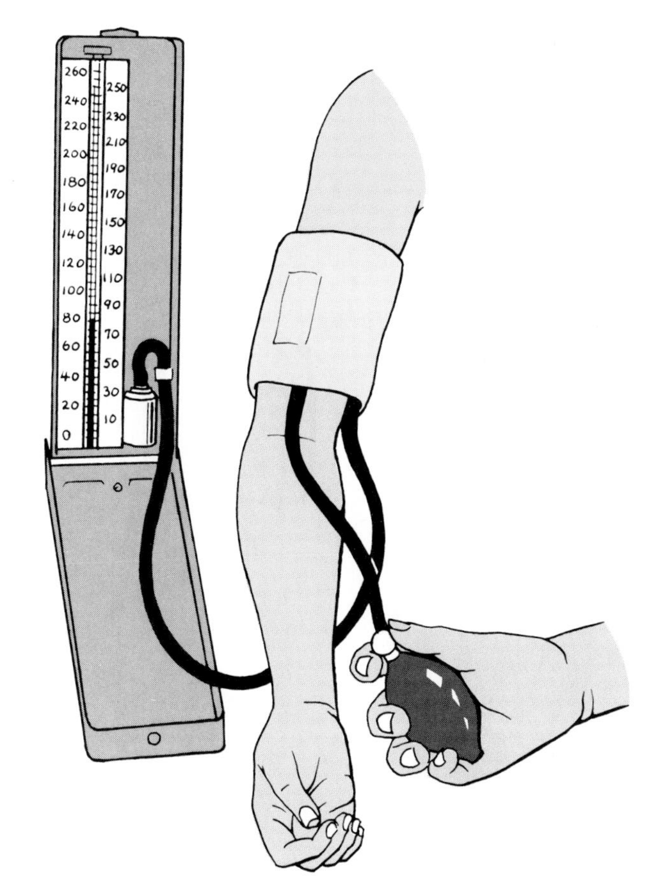

Figure 12.23. *The use of a pressure cuff and sphygmomanometer to measure blood pressure.*

Modern clinical blood pressure measurements, fortunately, are less direct than Hale's method. The indirect, or **auscultatory,[15] method** of blood pressure measurement is based on the correlation of blood pressure with arterial sounds. In the auscultatory method, an inflatable rubber bladder within a cloth cuff is wrapped around the upper arm and a stethoscope is applied over the brachial artery (fig. 12.23). The artery is normally silent before inflation of the cuff because blood travels in a smooth *laminar flow* through the arteries. The term laminar means layered—blood in the central axial stream moves the fastest, and blood flowing closer to the artery wall moves more slowly. There is little transverse movement between these layers that would produce mixing.

The laminar flow that normally occurs in arteries produces little vibration and is thus silent. When the artery is pinched, however, blood flow through the constriction becomes turbulent, which causes the artery to vibrate and produce sounds (much like the sounds produced by water through a kink in a garden hose). The tendency of the cuff pressure to constrict the artery is opposed by the blood pressure. Thus, in order to constrict the artery, the cuff pressure must be greater than the diastolic blood pressure. If the cuff pressure is also greater than the systolic blood pressure the artery would be pinched off and silent. *Turbulent flow* and its associated sounds only occur, therefore, when the cuff pressure is greater than the diastolic pressure but less than the systolic blood [illegible]sure.

Suppose that a pe[illegible] has a systolic pressure of 120 mm Hg and a diastolic pressure of 80 mm Hg (the average normal values). This is commonly recorded as 120/80 (systolic/diastolic). When the cuff pressure is between 80 and 120 mm Hg, the artery will be closed during diastole and open during systole. As the artery begins to open with every systole, turbulent flow of blood through the constriction will create vibrations that are heard as the **sounds of Korotkoff,[16]** as shown in figure 12.24. These are usually "tapping" sounds because the artery becomes constricted, blood flow stops, and silence resumes with every diastole. It should be understood that the sounds of Korotkoff are *not* "lub-dub" sounds produced by closing of the heart valves (those sounds can only be heard on the chest, not on the brachial artery).

Initially, the cuff is usually inflated to produce a pressure greater than the systolic pressure so that the artery is pinched off and silent. The pressure in the cuff is read from an attached meter called a *sphygmomanometer.*[17] A valve is then turned to allow the release of air from the cuff, causing a gradual decrease in cuff pressure. When the cuff pressure is equal to the systolic pressure, the **first sound** of Korotkoff is heard as blood passes in a turbulent flow through the constricted opening of the artery.

Korotkoff sounds will continue to be heard at every systole as long as the cuff pressure remains greater than the diastolic pressure. When the cuff pressure becomes equal to or less than the diastolic pressure, the sounds disappear because the artery remains open, laminar flow occurs, and the vibrations of the artery stop (fig. 12.25). The **last sound** of Korotkoff thus occurs when the cuff pressure is equal to the diastolic pressure. Normal blood pressure values are shown in table 12.4.

Hypertension

Approximately 20% of all adults in the United States have *hypertension*—blood pressure in excess of the normal range for the person's age and sex. Hypertension that is a result of (secondary to) known disease processes is logically called **secondary hypertension.** Secondary hypertension comprises only about 10%

[15]auscultatory: L. *auscultare,* to listen to

[16]sounds of Korotkoff: from Nicolai S. Korotkoff, Russian physician, 1874–1920

[17]sphygmomanometer: Gk. *sphygmos,* pulse; *manos,* thin; *metro,* measure

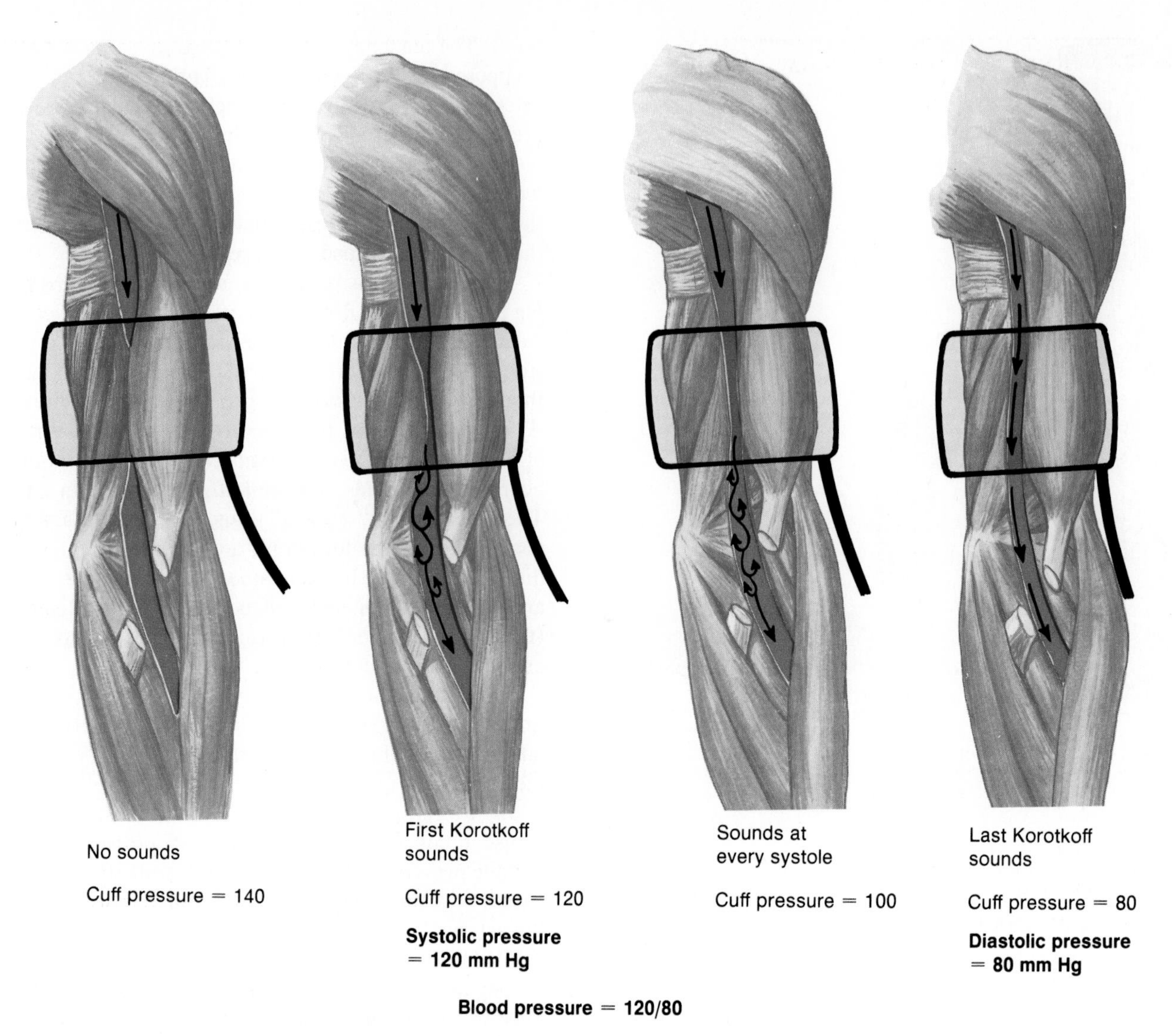

Figure 12.24. *Korotkoff sounds are produced by the turbulent flow of blood through the partially constricted brachial artery. This occurs when the cuff pressure is greater than the diastolic pressure but less than the systolic pressure.*

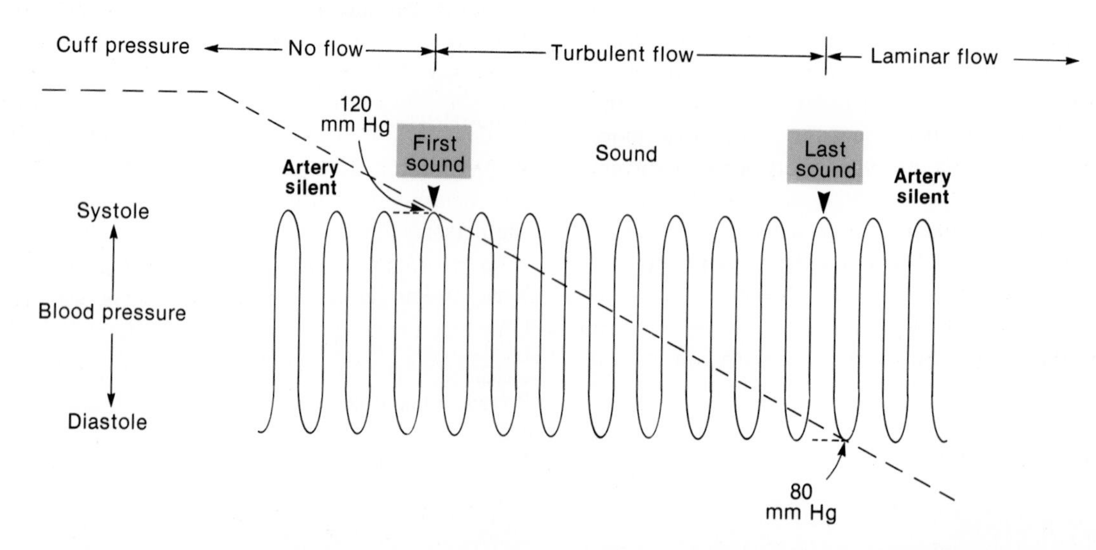

Figure 12.25. *The indirect, or auscultatory, method of blood-pressure measurement. The first Korotkoff sound is heard when the cuff pressure is equal to the systolic blood pressure, and the last sound is heard when the cuff pressure and diastolic blood pressures are equal.*

Table 12.4 Normal arterial blood pressure at different ages

Age	Systolic		Diastolic		Age	Systolic		Diastolic	
	Men	Women	Men	Women		Men	Women	Men	Women
1 day	70				16 years	118	116	73	72
3 days	72				17 years	121	116	74	72
9 days	73				18 years	120	116	74	72
3 weeks	77				19 years	122	115	75	71
3 months	86				20–24 years	123	116	76	72
6–12 months	89	93	60	62	25–29 years	125	117	78	74
1 year	96	95	66	65	30–34 years	126	120	79	75
2 years	99	92	64	60	35–39 years	127	124	80	78
3 years	100	100	67	64	40–44 years	129	127	81	80
4 years	99	99	65	66	45–49 years	130	131	82	82
5 years	92	92	62	62	50–54 years	135	137	83	84
6 years	94	94	64	64	55–59 years	138	139	84	84
7 years	97	97	65	66	60–64 years	142	144	85	85
8 years	100	100	67	68	65–69 years	143	154	83	85
9 years	101	101	68	69	70–74 years	145	159	82	85
10 years	103	103	69	70	75–79 years	146	158	81	84
11 years	104	104	70	71	80–84 years	145	157	82	83
12 years	106	106	71	72	85–89 years	145	154	79	82
13 years	108	108	72	73	90–94 years	145	150	78	79
14 years	110	110	73	74	95–106 years	145	149	78	81
15 years	112	112	75	76					

From Diem, K., and Lentner, C., eds., *Documenta Geigy Scientific Tables,* 7th ed., J. R. Geigy S. A., Basle, Switzerland, 1970. With permission.

of the hypertensive population. Hypertension that is the result of complex and poorly understood processes is not-so-logically called **primary,** or **essential, hypertension.**

Essential Hypertension

The vast majority of people with hypertension have essential hypertension. An increased total peripheral resistance is a universal characteristic of this condition. Cardiac rate and the cardiac output are elevated in many, but not all, of these cases. Sustained high stress (acting via the sympathetic nervous system) and high salt intake appear to act together in the development of hypertension. There is some evidence that Na^+ enhances the vascular response to sympathetic stimulation. Further, sympathetic nerves can cause constriction of the renal blood vessels and thus decrease the excretion of salt and water. There appears to be a genetic basis for these responses, so that essential hypertension tends to run in families.

As an adaptive response to prolonged high blood pressure, the arterial wall becomes thickened. This can lead to arteriosclerosis and results in an even greater increase in total peripheral resistance, thus raising blood pressure still further.

Dangers of Hypertension

If other factors remain constant, blood flow increases as arterial blood pressure increases. People with hypertension thus have adequate flow of blood to their organs until the hypertension causes vascular damage. Hypertension, as a result, is usually without symptoms until a dangerous amount of vascular damage is produced.

Hypertension is dangerous because (1) high arterial pressure makes it more difficult for the ventricles to eject blood; this increases the amount of work that the heart must perform and may result in pathological changes in the heart; (2) high pressure may damage cerebral blood vessels, leading to cerebrovascular accident (stroke); and (3) it contributes to the development of atherosclerosis, which can itself lead to heart disease and stroke as previously described.

Treatment of Hypertension

Hypertension is usually treated by restricting salt intake and by taking drugs that act in a variety of ways. Most commonly, these drugs are *diuretics*[18] that increase urine volume. Since urine is derived from blood (chapter 16), this decreases blood volume

[18]diuretic: Gk. *dia,* completely; *ouresis,* urination

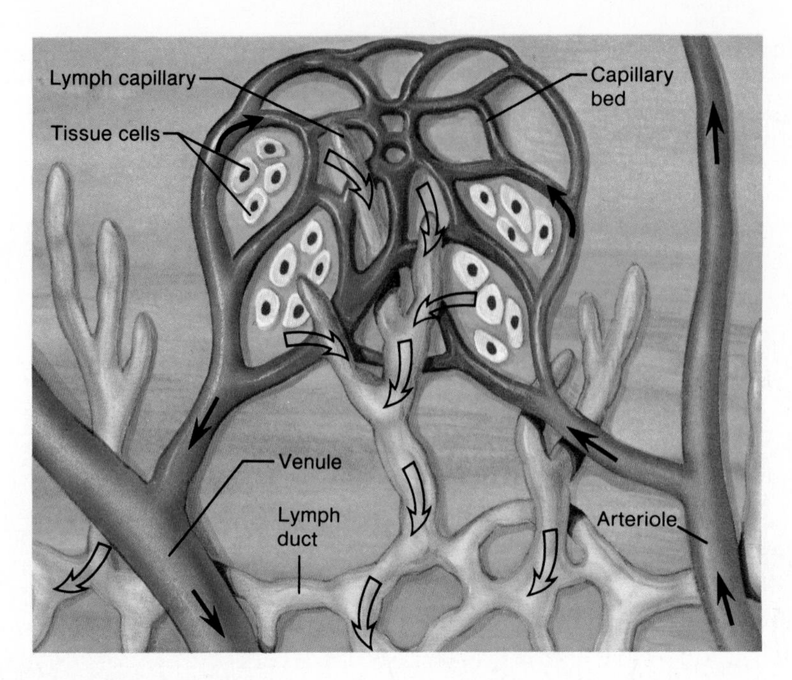

Figure 12.26. A schematic diagram showing the structural relationship of a capillary bed and a lymph capillary.

and pressure. Sympathetic-blocking drugs are also often used; drugs that block receptors for norepinephrine (such as propranolol) decrease cardiac rate. Various vasodilators may also be used to decrease total peripheral resistance.

1. Describe the significance of arterioles in influencing blood flow through capillaries and arterial blood pressure.
2. Describe how the baroreceptor reflex helps to compensate for a fall in blood pressure.
3. Describe how the sounds of Korotkoff are produced and how these sounds are used to measure blood pressure.

Lymphatic System

Lymphatic vessels absorb excess tissue fluid and transport this fluid—now called lymph—to ducts that drain into veins. Lymph nodes, and lymphoid tissue in the thymus, spleen, and tonsils produce lymphocytes, which are white blood cells involved in immunity.

The lymphatic system has three basic functions: (1) it transports interstitial (tissue) fluid, which was initially formed as a blood filtrate, back to the blood; (2) it transports absorbed fat from the small intestine to the blood; and (3) its cells—called *lymphocytes*—help provide immunological defenses against disease-causing agents.

The smallest vessels of the lymphatic system are the *lymph capillaries* (fig. 12.26). Lymph capillaries are microscopic, closed-ended tubes that form vast networks in the intercellular spaces within most organs. Because the walls of lymph capillaries are porous, various substances—including interstitial fluid, proteins, microorganisms, and absorbed fat (in the intestine—chapter 17)—can easily enter. The fluid within the lymphatic system is referred to as **lymph.**

From merging lymph capillaries, the lymph is carried into larger *lymphatic vessels* (fig. 12.27). The lymphatic vessels eventually empty into one of two principal ducts: the *thoracic duct* and the *right lymphatic duct.* These ducts drain the lymph into the left and right subclavian veins (under the clavicles, or collarbones), respectively. Thus tissue fluid, which is formed by filtration of plasma out of blood capillaries, is ultimately returned back to the cardiovascular system (fig. 12.28).

Before the lymph is returned to the cardiovascular system, it is filtered through **lymph nodes.** Lymph nodes contain phagocytic cells, which help to remove pathogens, and *germinal centers,* which are the sites of lymphocyte production. The tonsils, thymus, and spleen—together called *lymphoid organs*—likewise contain germinal centers and are sites of lymphocyte production. Lymphocytes are the cells of the immune system that respond in a specific fashion to antigens, and their functions are described as part of the immune system in chapter 13.

1. Describe the structure and function of lymphatic capillaries, and explain how lymph originates.
2. Trace the flow of lymph from its origin to its final destination.
3. Describe the functions of lymph nodes and lymphoid organs.

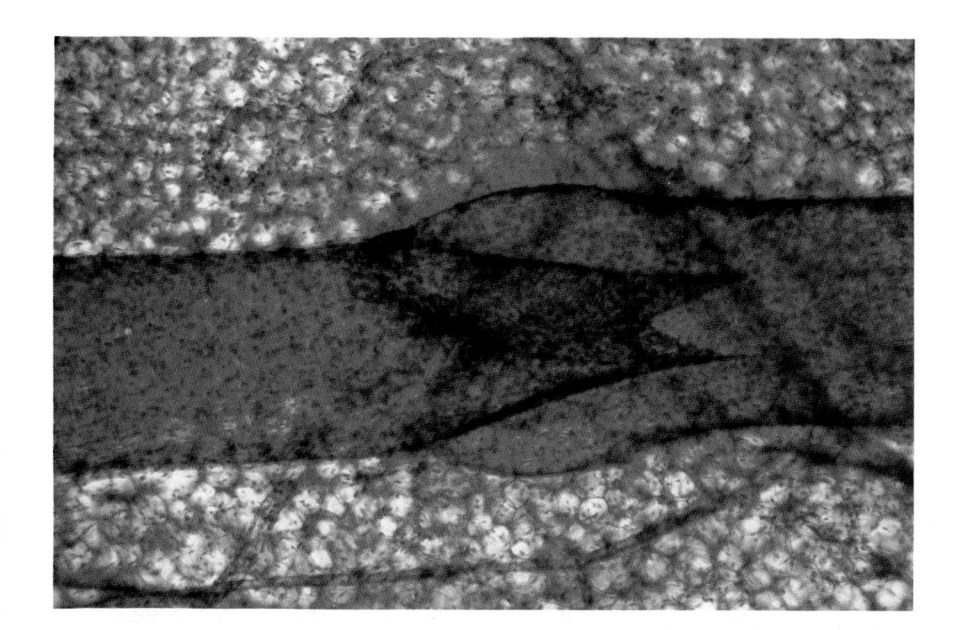

Figure 12.27. Photomicrograph of a lymphatic vessel.

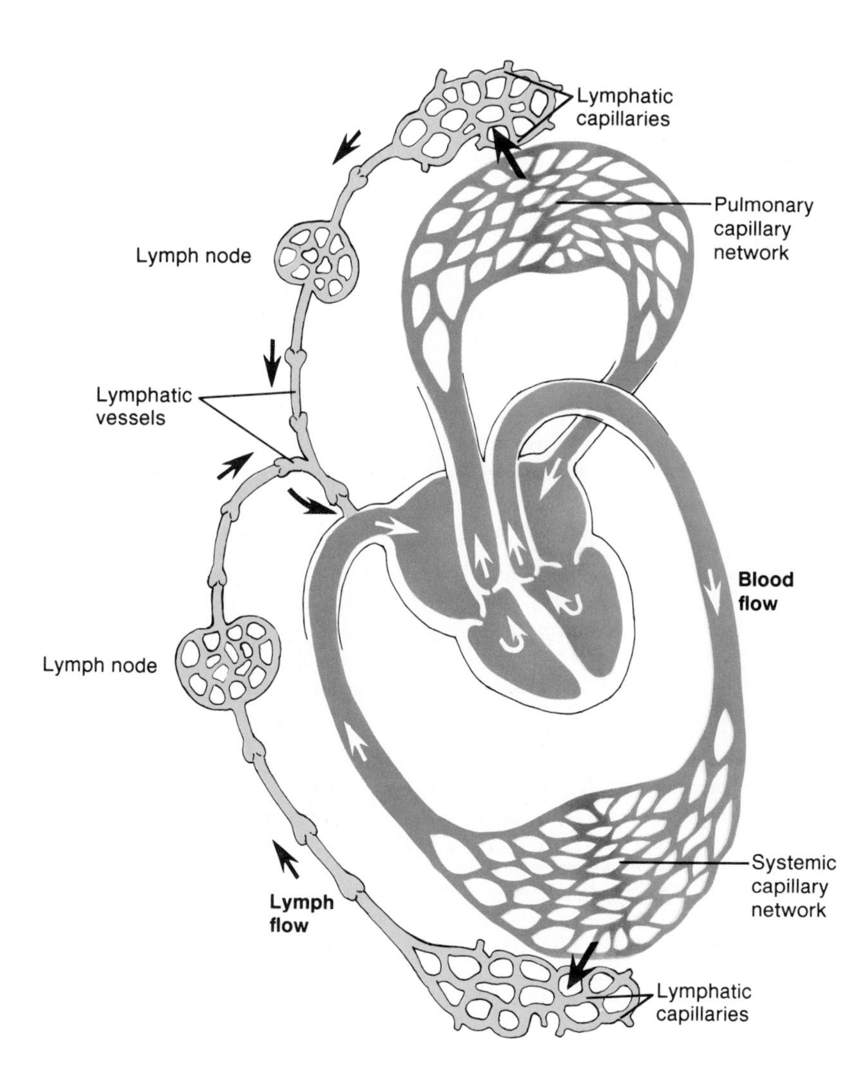

Figure 12.28. *The schematic relationship of the circulatory and lymphatic systems. Lymphatic vessels transport lymph fluid from interstitial spaces to the venous bloodstream.*

Summary

Functions and Components of the Circulatory System p. 262

I. The circulatory system serves to deliver molecules needed by the tissue cells for nutritional and regulatory purposes, and to remove waste products.

II. The heart functions as a pump to circulate the blood.
   - A. Arteries carry blood away from the heart.
   - B. Veins return blood to the heart.
   - C. Capillaries, located between arteries and veins, are the smallest and most numerous of the blood vessels.
   - D. Capillaries have very thin walls, so that molecules can move across these walls and be exchanged with the tissue cells.

Composition of the Blood p. 263

I. Plasma is the fluid portion of the blood; serum is the fluid from clotted blood.

II. The formed elements of blood include erythrocytes, leukocytes, and platelets.
   - A. Erythrocytes (red blood cells) contain hemoglobin and function to transport oxygen.
   - B. Leukocytes (white blood cells) are divided into different categories.
     1. The agranular leukocytes include lymphocytes and monocytes.
     2. Granular leukocytes are the neutrophils, eosinophils, and basophils.
   - C. Platelets function in blood clotting.

Structure of the Heart p. 266

I. The heart is composed of two upper chambers, the atria, and two lower chambers, the ventricles.

II. There are two separate circulations: pulmonary and systemic.
   - A. The right ventricle pumps blood into the pulmonary arteries that deliver blood to the lungs.
   - B. The left ventricle pumps blood into the large aorta, which branches to supply oxygenated blood to all of the organs in the body.

III. There are two pairs of valves in the heart.
   - A. The atrioventricular (AV) valves allow blood to go from the atria to the ventricles.
   - B. The semilunar valves are located at the origin of the pulmonary artery and aorta.

The Cardiac Cycle and the Heart Sounds p. 269

I. The cardiac cycle consists of two parts: systole and diastole.
   - A. The phase of ventricular contraction is systole; the phase of ventricular relaxation is diastole.
   - B. The atria contract just before the contraction of the ventricles.

II. When the ventricles contract and relax, two sounds are heard.
   - A. The first sound—lub—is produced when the AV valves close at systole.
   - B. The second sound—dub—is produced when the semilunar valves close as the ventricles relax in diastole.

Electrical Activity of the Heart and the Electrocardiogram p. 270

I. The heartbeat is automatic, because electrical impulses (action potentials) are produced spontaneously by the pacemaker SA node.
   - A. Impulses spread from the SA node through both right and left atria.
   - B. Impulses excite the AV node, which conducts these impulses in the bundle of His.
   - C. The Purkinje fibers carry the impulses into the ventricular muscle, stimulating the ventricles to contract.

II. The pattern of electrical conduction in the heart is revealed by the electrocardiogram (ECG).
   - A. The P wave is produced by electrical excitation of the atria.
   - B. The QRS wave is produced by electrical excitation of the ventricles.
   - C. The T wave is produced by repolarization of the ventricles.

III. Abnormal patterns of electrical conduction are detected by the ECG.

Cardiac Output and Blood Flow p. 274

I. The cardiac output equals the cardiac rate times the stroke volume.
   - A. The cardiac rate is increased by stimulation from a sympathetic nerve, and decreased by inhibition from the vagus nerve.
   - B. The stroke volume is also increased by sympathetic nerve stimulation, which increases the strength of contraction of the ventricles.

II. The venous return is the return of blood in the veins to the atria.
   - A. The massaging action of contracting skeletal muscles on the veins helps to push blood back to the heart; this is the skeletal muscle pump.
   - B. Breathing movements of the diaphragm also aid the flow of blood from veins in the abdomen to the thorax.

III. The coronary arteries are constricted during systole, so that blood flow is greater during diastole.

IV. Blood flow through skeletal muscles is increased during exercise because the arterioles dilate in response to local chemical changes (such as increased carbon dioxide) in the muscle tissue.

V. During exercise, the cardiac output may increase from 5 L/min (at rest) to as high as 25 L/min or more.
   - A. This is primarily due to an increase in the cardiac rate.
   - B. The stroke volume may also increase somewhat, particularly in highly trained athletes.

Atherosclerosis p. 277

I. Atherosclerosis is a type of hardening of the arteries in which localized plaques, or atheromas, are present.
   - A. These plaques are produced by proliferation of smooth muscle cells, deposits of lipids, and hardening due to calcification.
   - B. Atherosclerosis is responsible for most heart disease and stroke (cerebrovascular accident).

II. High blood cholesterol is a major risk factor in the development of atherosclerosis.

Blood Pressure p. 279

I. The arterial blood pressure varies directly with the cardiac output and the total peripheral resistance.
   - A. An increase in cardiac rate or blood volume, therefore, can raise the blood pressure.
   - B. The total peripheral resistance is increased by vasoconstriction and decreased by vasodilation of the arterioles.

II. The blood pressure is maintained within limits by the action of the baroreceptor reflex.
   - A. If the pressure falls, baroreceptors are stimulated.
   - B. This activation sends impulses to the medulla oblongata of the brain.
   - C. The medulla oblongata, in turn, stimulates sympathetic nerve activity, which promotes an increase in blood pressure.

III. Arterial pressure is measured using a pressure cuff.
   - A. Blood normally flows through arteries in a smooth, laminar flow; the artery is thus quiet.
   - B. When the pressure in the cuff is greater than the diastolic pressure of the blood, a constriction is produced in the artery.
   - C. Sounds that are produced by the turbulent flow of blood through the constricted artery are called the sounds of Korotkoff.

IV. Hypertension is a risk factor in the development of atherosclerosis.

Lymphatic System p. 286

I. Tissue fluid that enters lymphatic vessels is known as lymph.
   - A. Lymphatic capillaries drain away excessive tissue fluid.
   - B. This fluid is carried by lymphatic vessels to two large ducts, which empty the lymph into veins.

II. Lymph nodes, and lymphatic tissue in the tonsils, spleen, and thymus, produce lymphocytes which function in immunity.

Review Activities

Objective Questions

1. All arteries in the body contain oxygen-rich blood with the exception of
   (a) the aorta
   (b) the pulmonary arteries
   (c) the renal artery
   (d) the coronary arteries
2. The "lub," or first heart sound, is produced by closing of
   (a) the aortic semilunar valve
   (b) the pulmonary semilunar valve
   (c) the tricuspid valve
   (d) the bicuspid valve
   (e) both AV valves
3. The first heart sound is produced at
   (a) the beginning of systole
   (b) the end of systole
   (c) the beginning of diastole
   (d) the end of diastole
4. The QRS wave of an ECG is produced by
   (a) excitation of the atria
   (b) repolarization of the atria
   (c) excitation of the ventricles
   (d) repolarization of the ventricles
5. The normal pacemaker of the heart is
   (a) the SA node
   (b) the AV node
   (c) the bundle of His
   (d) the Purkinje fibers
6. An ischemic injury to the heart that results in death of some myocardial cells is
   (a) angina pectoris
   (b) a myocardial infarction
   (c) fibrillation
   (d) heart block
7. Platelets
   (a) form a plug by sticking to each other
   (b) release chemicals that stimulate vasoconstriction
   (c) provide molecules needed for the formation of a clot
   (d) all of the above
8. Which of the following cells is *not* a granular leukocyte?
   (a) neutrophils
   (b) basophils
   (c) eosinophils
   (d) lymphocytes
9. Which of the following statements about tissue fluid is *true*?
   (a) It is formed from blood plasma.
   (b) Most tissue fluid returns directly to the blood capillaries.
   (c) Some tissue fluid enters lymphatic capillaries.
   (d) Excessive accumulation of tissue fluid is called edema.
   (e) All of these.
10. Which of the following would cause a decrease in the cardiac rate?
    (a) increased activity of a sympathetic nerve
    (b) increased activity of a parasympathetic nerve
    (c) increased secretion of epinephrine from the adrenals
    (d) all of these
11. The greatest resistance to blood flow occurs in the
    (a) large arteries
    (b) medium-sized arteries
    (c) arterioles
    (d) capillaries
12. The sounds of Korotkoff are produced by
    (a) closing of the semilunar valves
    (b) closing of the AV valves
    (c) the turbulent flow of blood through an artery
    (d) elastic recoil of the aorta
13. If a person's blood pressure is recorded as 118/76
    (a) the systolic pressure is 76 mm Hg
    (b) the first sound was heard at a pressure of 76 mm Hg
    (c) the diastolic pressure is 76 mm Hg
    (d) this person has hypertension
14. Blood flow in the coronary circulation is
    (a) increased during systole
    (b) increased during diastole
    (c) constant throughout the cardiac cycle

Essay Questions

1. Trace the path of a drop of blood from the lungs to the right atrium of the heart.
2. Explain how the lub-dub sounds of the heart are produced and when they occur in the cardiac cycle.
3. Explain how the rate of beat of the heart is regulated.
4. Could atherosclerotic plaque be cleaned out of an artery by reaming it, as with a pipe cleaner? Explain.
5. Describe the changes in the cardiovascular system that occur during exercise.
6. Explain the mechanism that produces a rapid pulse when a person suffers from an abnormally low blood pressure.

13

The Immune System

Outline

Objectives

By studying this chapter, you should be able to
1. describe some of the mechanisms of nonspecific immunity
2. describe the structure of antibodies and the nature of antigens
3. explain how antigen-antibody reactions lead to the destruction of an invading pathogen
4. describe the events that occur during a local inflammation
5. describe the process of active immunity and explain how the clonal selection theory may account for this process
6. describe the mechanisms of passive immunity and give natural and clinical examples of this form of immunization
7. explain the meaning of the term *monoclonal antibodies* and describe some of their clinical uses
8. explain how T lymphocytes are classified and describe the function of the thymus gland
9. define the term *lymphokines* and list some of these molecules and their functions
10. describe the nature of the HIV virus, how it affects the immune system, and the dangers of AIDS
11. describe how helper T lymphocytes are activated and their function in the immune system
12. explain the possible mechanisms responsible for tolerance of self-antigens
13. define the term *autoimmune disease,* and provide examples of different autoimmune diseases
14. distinguish between immediate and delayed hypersensitivity

Keys to Pronunciation

antigenic: *an-ti-jen'ik*
antitoxin: *an''ti-tok'sin*
chemotaxis: *ke''mo-tak'sis*
chimera: *ki-me'rah*
diapedesis: *di''ah-pe'de'sis*
endogenous pyrogen: *en-doj'e-nus pi'ro-jen*
glomerulonephritis: *glo-mer''u-lo-ne-fri'tis*
histamine: *his'tah-men*
hybridoma: *hi''bri-do'mah*
immunoglobulin: *im''u-no-glob'u-lin*
interferon: *in''ter-fer'on*
lymphokines: *lim'fo-kins*
macrophages: *mak''ro-faj'es*
monoclonal: *mon''o-klon'al*
periarteritis: *per''e-ar''te-ri'tis*
phagocytosis: *fag''o-si-to'sis*
Pneumocystis: *nu''mo-sis'tis*
thymopoietin: *thi''mo-poi'e-tin*
vaccination: *vak''si-na'shun*

Photo: Not everyone smiles when getting an injection, but all are grateful for the benefits provided by immunizations.

Defense Mechanisms

Nonspecific immune protection is provided by such mechanisms as phagocytosis, fever, and the release of molecules called interferons. Specific immunity involves the functions of lymphocytes and is directed at specific molecules, or parts of molecules, known as antigens.

The immune system includes all of the structures and processes that provide a defense against potential pathogens[1] (disease-causing agents). These defenses can be grouped into *nonspecific* and *specific* categories.

Nonspecific defense mechanisms are inherited as part of the structure of each organism. Epithelial membranes that cover the body surfaces, for example, restrict infection by most pathogens. The strong acidity of gastric juice (pH 1–2) also helps to kill many microorganisms before they can invade the body. These external defenses are backed by internal defenses, such as phagocytosis, which function in both a specific and nonspecific manner (table 13.1).

Nonspecific Immunity

Invading pathogens, such as bacteria, that have crossed epithelial barriers enter connective tissues. These invaders—or chemicals, called *toxins,*[2] secreted from them—may enter blood or lymphatic capillaries and be carried to other areas of the body. The invasion and spread of infection is fought in two stages: (1) nonspecific immunological defenses are employed; if these are sufficiently effective, the pathogens may be destroyed without progression to the next step; (2) lymphocytes may be recruited, and their specific actions used to reinforce the nonspecific immune defenses.

[1]pathogen: Gk. *pathema,* suffering; *gen,* to produce
[2]toxin: Gk. *toxikon,* poison

Phagocytosis[3]

There are three major groups of phagocytic cells: (1) **neutrophils** (described in chapter 12); (2) the cells of the **mononuclear phagocyte system;** this includes *monocytes* in the blood and *macrophages* (derived from monocytes) in the connective tissues; and (3) **organ-specific phagocytes** in the liver, spleen, lymph nodes, lungs, and brain (table 13.2).

The *Kupffer cells* in the liver, together with phagocytic cells in the spleen and lymph nodes, are **fixed phagocytes.** This term refers to the fact that these cells are immobile ("fixed") in the channels within these organs. As blood flows through the liver and spleen and as lymph percolates through the lymph nodes, foreign chemicals and debris are removed by phagocytosis and chemically inactivated within the phagocytic cells. Invading pathogens are very effectively removed in this manner, so that blood is usually sterile after a few passes through the liver and spleen.

Connective tissues contain a resident population of all leukocyte types. Neutrophils and monocytes in particular can be highly mobile within connective tissues as they scavenge for invaders and cellular debris. These leukocytes are recruited to the site of an infection by a process known as **chemotaxis**[4]—movement toward chemical attractants. Neutrophils are the first to arrive at the site of an infection; monocytes arrive later and can be transformed into macrophages as the battle progresses.

If the infection is sufficiently large, new phagocytic cells from the blood may join those already in the connective tissue. These new neutrophils and monocytes are able to squeeze through the tiny gaps between adjacent endothelial cells in the capillary wall and enter the connective tissues. This process, called **diapedesis,**[5] is illustrated in figure 13.1.

[3]phagocytosis: Gk. *phagein,* to eat; *kytos,* hollow (cell)
[4]chemotaxis: Gk. *chemeia,* alchemy; *taxis,* orderly arrangement
[5]diapedesis: Gk. *dia,* through; *pedesis,* a leaping

Table 13.1 Structures and defense mechanisms of nonspecific immunity

	Structure	Mechanisms
External	Skin	Anatomic barrier to penetration by pathogens; secretions have lysozyme (enzyme that destroys bacteria)
	Digestive tract	High acidity of stomach
		Protection by normal bacterial population of colon
	Respiratory tract	Secretion of mucus; movement of mucus by cilia; alveolar macrophages
	Genitourinary tract	Acidity of urine
		Vaginal lactic acid
Internal	Phagocytic cells	Ingest and destroy bacteria, cellular debris, denatured proteins, and toxins
	Interferons	Inhibit replication of viruses
	Complement proteins	Promote destruction of bacteria and other effects of inflammation
	Endogenous pyrogen	Secreted by leukocytes and other cells; produces fever

Phagocytic cells engulf particles in a manner similar to the way an amoeba eats. The particle becomes surrounded by cytoplasmic extensions called pseudopods,[6] which ultimately fuse together. The particle thus becomes surrounded by a membrane derived from the plasma membrane (fig. 13.2) and contained within an organelle analogous to a food vacuole in an amoeba. This vacuole then fuses with lysosomes (organelles that contain digestive enzymes), so that the ingested particle and the digestive enzymes remain separated from the cytoplasm by a continuous membrane. Often, however, lysosomal enzymes are released before the food vacuole has completely formed. When this occurs, free lysosomal enzymes may be released into the infected area and contribute to inflammation.

Table 13.2 Phagocytic cells and their locations

Phagocyte	Location
Neutrophils	Blood and all tissues
Monocytes	Blood and all tissues
Tissue macrophages (histiocytes)	All tissues (including spleen, lymph nodes, bone marrow)
Kupffer cells	Liver
Alveolar macrophages	Lungs
Microglia	Central nervous system

Interferons

In 1957, researchers demonstrated that cells infected with a virus produced polypeptides that interfered with the ability of a second, unrelated strain of virus to infect other cells in the same

[6]pseudopod: Gk. *pseudes,* false; *pous,* foot

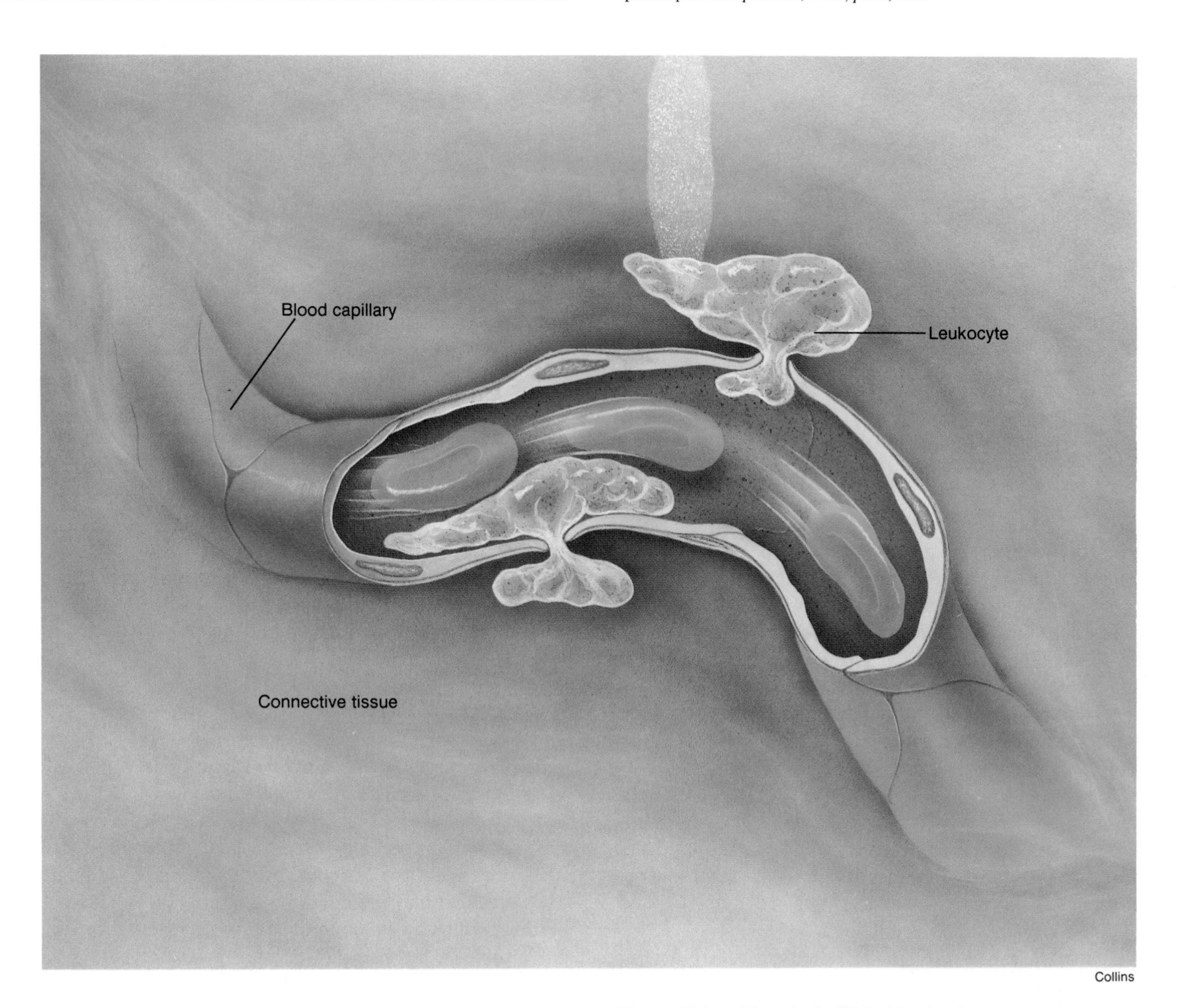

Figure 13.1. *Diapedesis. White blood cells squeeze through openings between capillary endothelial cells to enter surrounding connective tissues.*

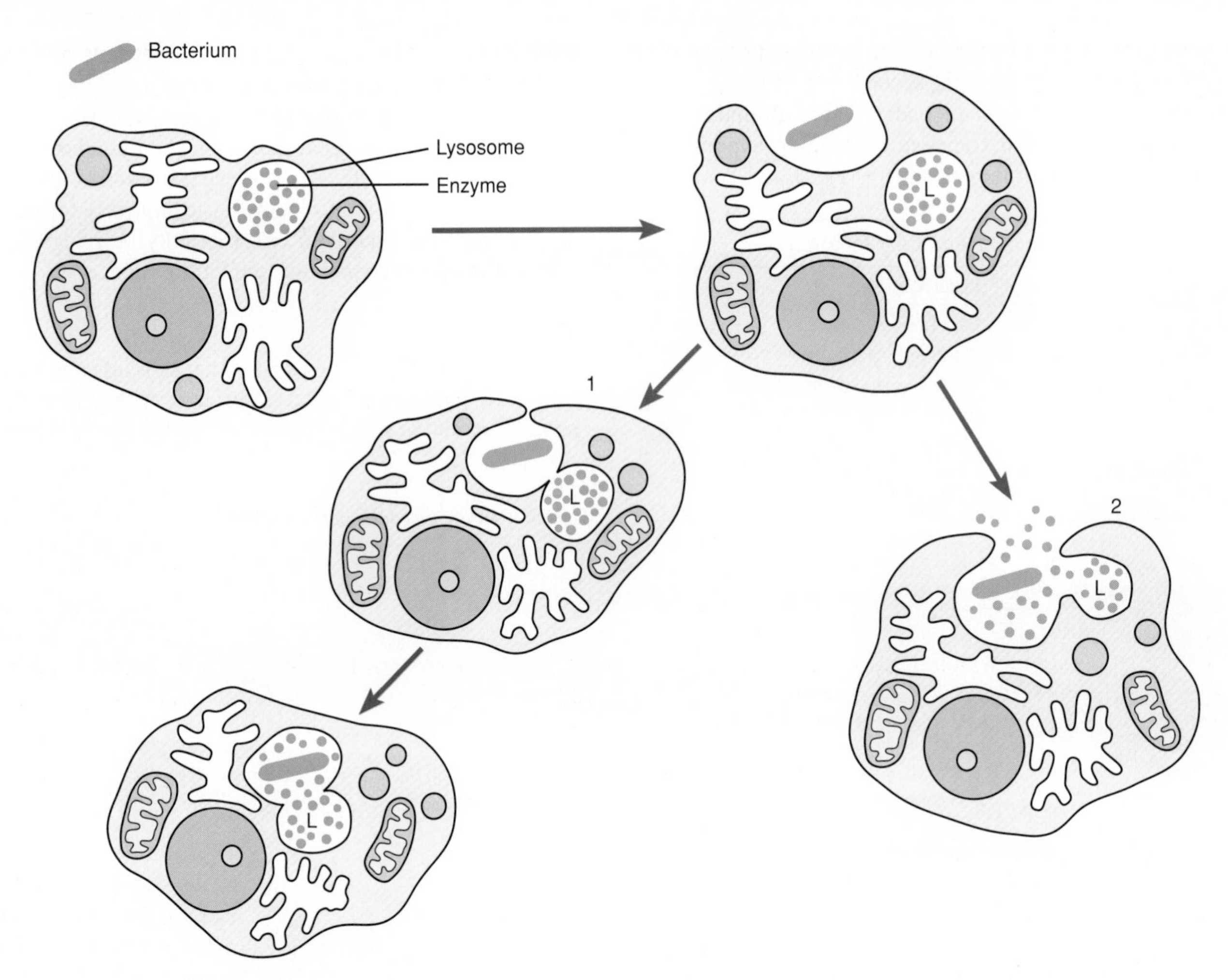

Figure 13.2. *Phagocytosis by a neutrophil or macrophage. A phagocytic cell extends its pseudopods around the object to be engulfed (such as a bacterium). Dots represent lysosomal enzymes (L = lysosomes). If the pseudopods fuse to form a complete food vacuole, lysosomal enzymes are restricted to the organelle formed by the lysosome and food vacuole. If the lysosome fuses with the vacuole before fusion of the pseudopods is complete, lysosomal enzymes are released into the infected area of tissue.*

culture. These **interferons,**[7] as they were called, thus produced a nonspecific, short-acting resistance to viral infection. This discovery produced a great deal of excitement, but further research in this area was hindered by the fact that human interferons could only be obtained in very small quantities and that animal interferons had little effect in humans. In 1980, however, technological breakthroughs allowed researchers to introduce human interferon genes into bacteria (chapter 10) so that bacteria could act as interferon factories.

Leukocytes and many other cells make their own characteristic types of interferons. These polypeptides act as messengers that protect other cells in the vicinity from viral infection. The viruses are still able to penetrate these other cells, but the ability of the viruses to replicate (described in chapter 10) and assemble new virus particles is inhibited.

Lymphocytes called *T lymphocytes* (described in a later section) release interferons in response to viral infections and perhaps as part of their immunological surveillance against cancer. Interferons may destroy cancer cells directly and indirectly by the activation of T lymphocytes and *natural killer cells* (cells related to T lymphocytes). The actions of these cells in combating cancer is described in chapter 14.

Specific Immunity

In 1890, a scientist named von Behring[8] demonstrated that a guinea pig which had been previously injected with a safe dose of diphtheria toxin could survive subsequent injections of otherwise lethal doses of that toxin. Further, von Behring showed that this immunity could be transferred to a second, nonexposed animal by injections of serum from the immunized guinea pig. He concluded that the immunized animal had chemicals in its serum—which he called **antibodies**—that were responsible for the immunity. He also showed that these antibodies conferred

[7]interferon: L. *inter,* between; *ferio,* to strike

[8]von Behring: Emil von Behring, German bacteriologist (1854–1917)

Table 13.3 Comparison of B and T lymphocytes

Characteristic	B Lymphocytes	T Lymphocytes
Site where processed	Bone marrow	Thymus
Type of immunity	Humoral (secretes antibodies)	Cell-mediated
Subpopulations	Memory cells and plasma cells	Cytotoxic (killer) T cells, helper cells, suppressor cells
Presence of surface antibodies	Yes	Not detectable
Receptors for antigens	Present—are surface antibodies	Present—are related to immunoglobulins
Life span	Short	Long
Tissue distribution	High in spleen, low in blood	High in blood and lymph
Percent of blood lymphocytes	10%–15%	75%–80%
Transformed by antigens to	Plasma cells	Small lymphocytes
Secretory product	Antibodies	Lymphokines
Immunity to viral infections	Enteroviruses, poliomyelitis	Most others
Immunity to bacterial infections	*Streptococcus, Staphylococcus,* many others	Tuberculosis, leprosy
Immunity to fungal infections	None known	Many
Immunity to parasitic infections	Trypanosomiasis, maybe to malaria	Most others

immunity only to subsequent diphtheria infections; the antibodies were *specific* in their actions. It was later learned that antibodies are proteins produced by a particular type of lymphocyte.

Antigens

Antigens are molecules that stimulate antibody production and combine with these specific antibodies. Most antigens are large molecules (such as proteins), and they are foreign to the blood and other body fluids (although there are exceptions to both descriptions). The ability of a molecule to function as an antigen depends not only on its size but also on the complexity of its structure. Proteins are therefore more antigenic than polysaccharides, which have a simpler structure. Plastics used in artificial implants are composed of large molecules but are not very antigenic because of their simple, repeating structures.

A large, complex, foreign molecule can have a number of different **antigenic determinant sites,** which are areas of the molecule that stimulate production of and combine with different antibodies. Most naturally occurring antigens have many antigenic determinant sites and stimulate the production of different antibodies with specificities for these sites.

Haptens

Many small organic molecules are not antigenic by themselves but can become antigens if they bond to proteins (and thus become antigenic determinant sites on the proteins). This was discovered by Karl Landsteiner,[9] the same man who discovered the ABO blood groups (chapter 11). By bonding these small molecules—which Landsteiner called **haptens**—to proteins in the laboratory, new antigens could be created for research or diagnostic purposes. The bonding of foreign haptens to a person's own proteins can also occur in the body; by this means, derivatives of penicillin, for example, that would otherwise be harmless can produce fatal allergic reactions in susceptible people.

[9]Karl Landsteiner: Austrian-born pathologist and immunologist, who immigrated to America (1868–1943)

Lymphocytes

Leukocytes, erythrocytes, and blood platelets (chapter 12) are all ultimately derived from ("stem from") unspecialized cells in the bone marrow. These *stem cells* produce the specialized blood cells, and they replace themselves by mitosis so that the stem cell population is not exhausted. Lymphocytes produced in this manner seed the thymus, spleen, and lymph nodes, producing self-replacing lymphocyte colonies in these organs.

The lymphocytes that are seeded in the thymus become **T lymphocytes.** These cells have an immunological function that is different from other lymphocytes. The thymus, in turn, seeds other organs; about 65% to 85% of the lymphocytes in blood and most of the lymphocytes in lymph nodes are T lymphocytes. T lymphocytes, therefore, come from or had an ancestor that came from the thymus gland.

Most of the lymphocytes that are not T lymphocytes are called **B lymphocytes.** The letter *B* stands for "bone marrow–derived" because it is believed that the B lymphocytes in mammals are processed in the bone marrow.

Both B and T lymphocytes function in specific immunity. The B lymphocytes combat bacterial and some viral infections by secreting antibodies into the blood and lymph. They are therefore said to provide **humoral[10] immunity.** T lymphocytes attack host cells that have become infected with viruses or fungi, transplanted human cells, and cancerous cells. The T lymphocytes do not secrete antibodies; they must come into close proximity or have actual physical contact with the target cell in order to destroy it. T lymphocytes are therefore said to provide **cell-mediated immunity** (table 13.3).

[10]humoral: L. *humor,* a fluid in the body

Figure 13.3. *B lymphocytes have antibodies on their surface that function as receptors for specific antigens. The interaction of antigens and antibodies on the surface stimulates cell division and the maturation of the B cell progeny into memory cells and plasma cells. Plasma cells produce and secrete large amounts of the antibody (note the extensive rough endoplasmic reticulum in these cells.)*

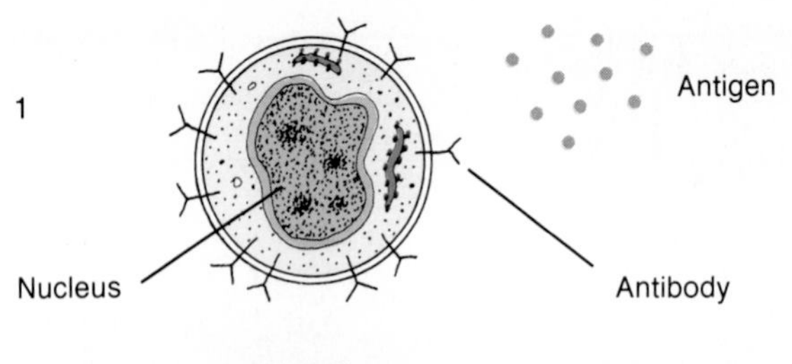

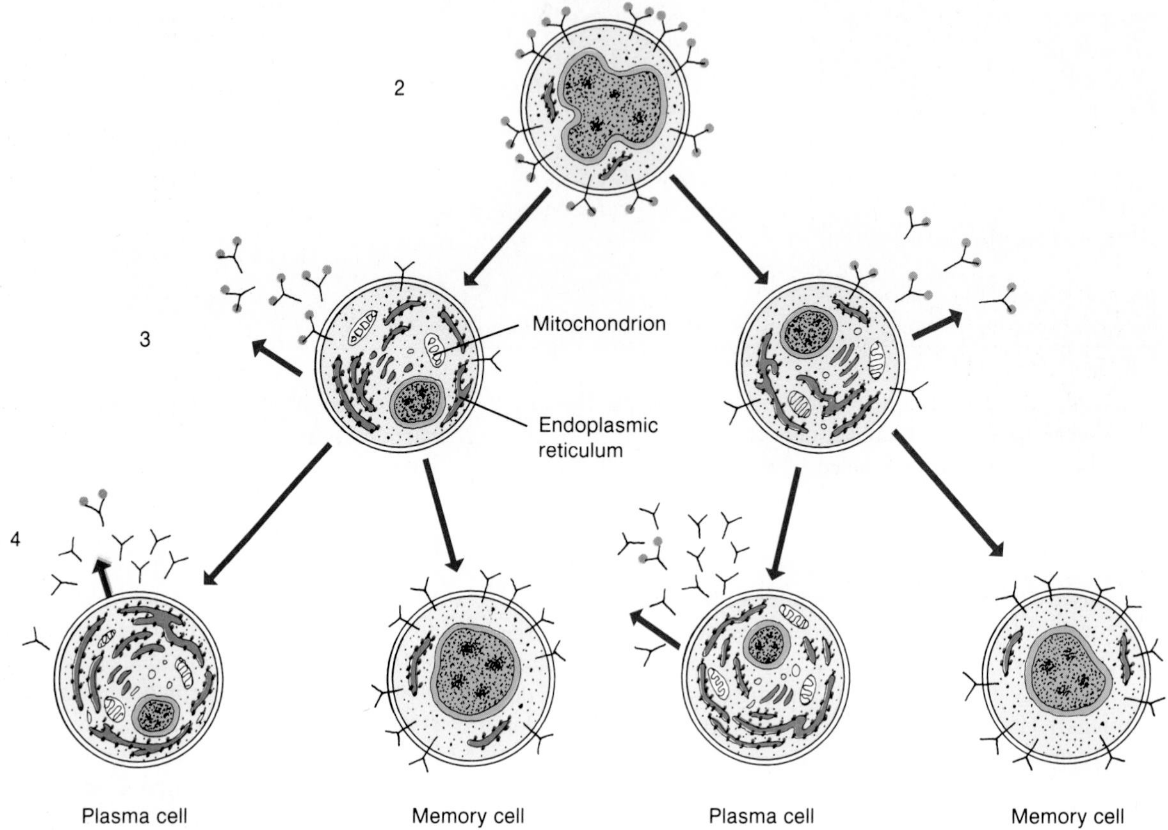

1. List the phagocytic cells in blood and lymph, and indicate which organs contain fixed phagocytes.
2. Describe the actions of interferons.
3. Describe the characteristics of antigens.
4. Describe the meaning of the term *hapten,* and give an example.
5. Distinguish between B and T lymphocytes in terms of their origins and immune functions.

Functions of B Lymphocytes

The specific antibody that a particular B lymphocyte produces will bind only to a specific antigen. When that antigen interacts with its appropriate antibody at the surface of that B cell it stimulates the B cell to divide many times, producing inactive memory cells and active, antibody-secreting cells called plasma cells. Binding of these secreted antibodies to antigens stimulates events that help the immune system to destroy the invading pathogens.

Exposure of a B lymphocyte to the appropriate antigen results in cell growth followed by many cell divisions. Some of the progeny become **memory cells,** which are indistinguishable from the original cell; others are transformed into **plasma cells** (fig. 13.3). Plasma cells are protein factories that produce about two thousand antibody proteins per second in their brief (five to seven day) life span.

Antibodies

Antibody proteins are also known as **immunoglobulins.** These constitute the gamma globulin class of plasma proteins, as described in chapter 12. Since antibodies are specific in their actions, it could be predicted that different types of antibodies have different structures. An antibody against measles, for example, does not confer immunity to poliomyelitis and, therefore, must have a slightly different structure than an antibody against polio.

All antibody molecules consist of four interconnected polypeptide chains. Two longer, higher molecular weight chains (the *H chains*) are joined to two shorter, lighter *L chains.* Research

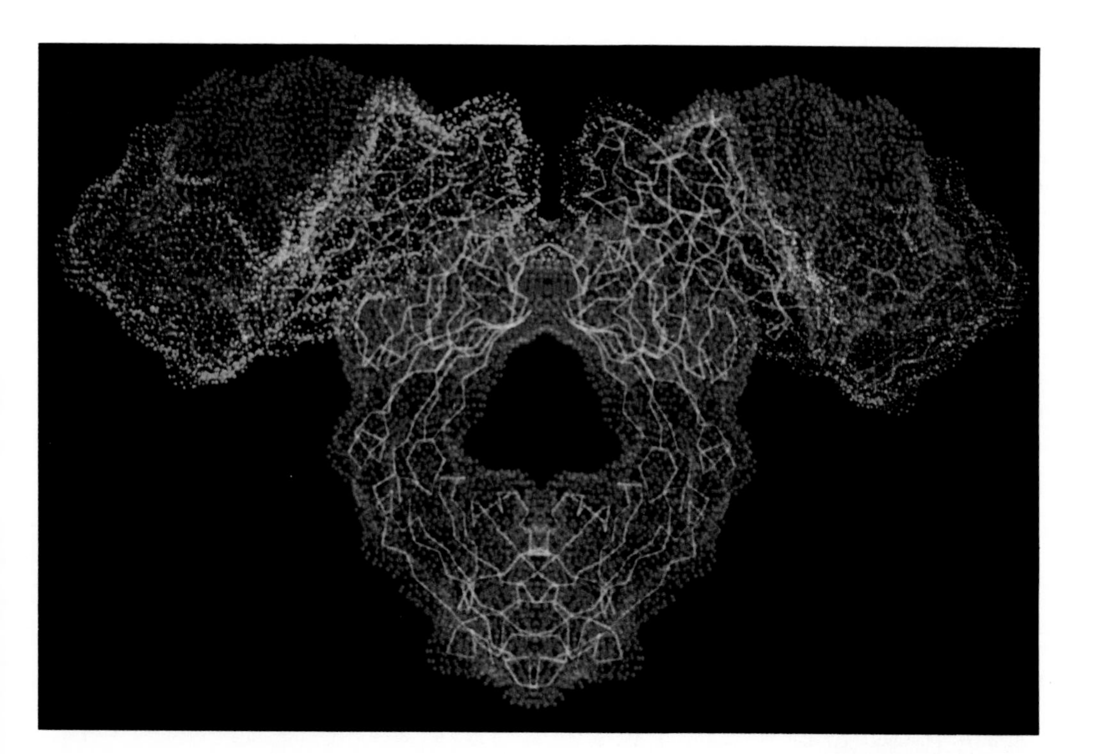

Figure 13.4. A computer-generated model of an antibody molecule. An antigen can combine with each of the two side branches located in the upper part of the antibody.

has shown that these four chains are arranged in the form of a Y. The top of the Y is the antigen-binding portion. This is shown in figure 13.4. One antibody molecule can thus bind to two antigen molecules.

B lymphocytes have antibodies on their cell membrane that serve as **receptors** for antigens. Combination of antigens with these antibody receptors stimulates the B cell to divide and produce more of these antibodies, which are secreted. Exposure to a given antigen thus results in increased amounts of the specific type of antibody that can attack that antigen. This provides active immunity, as described in the next major section.

The combination of antibodies with antigens does not itself produce destruction of the antigens or of the pathogenic organisms that contain these antigens. Antibodies, rather, serve to identify the targets for immunological attack and to activate nonspecific immune processes that destroy the invader. Bacteria that are buttered with antibodies, for example, are better targets for phagocytosis by neutrophils and macrophages.

When antibodies combine with antigens in the blood plasma, a system of plasma proteins known as **complement** become activated. Some of these activated complement proteins form pores in the cell membrane of the invading bacteria (fig. 13.5). These pores allow the osmotic influx of water, so that the target cell swells and bursts. Notice that the complement proteins, not the antibodies directly, kill the cell; antibodies only serve as activators of this process.

When complement proteins are activated, they have a number of effects. These include (1) *chemotaxis*—the activated complement proteins attract phagocytic cells to the site of infection; (2) increased activity of phagocytic cells due to complement activation; and (3) the release of *histamine* from

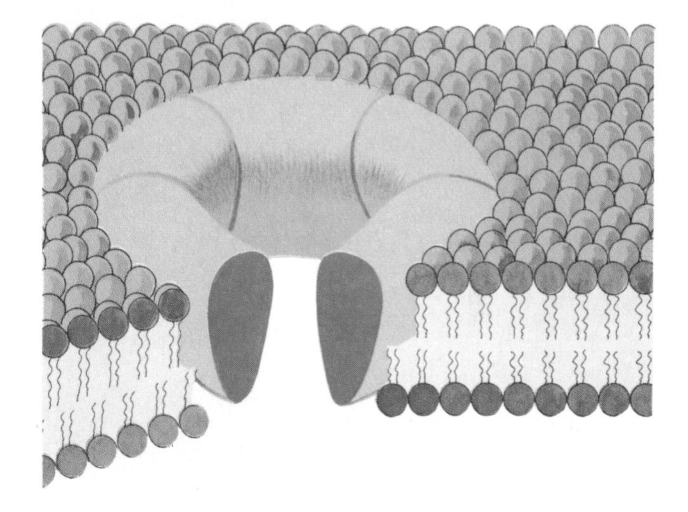

Figure 13.5. Complement proteins (illustrated as a doughnut-shaped ring) puncture the membrane of the cell to which they are attached (fixed). This aids destruction of the cell.

mast cells (a connective tissue cell type) and basophils. Histamine stimulates vasodilation, which increases blood flow to the infected area, and stimulates an increase in capillary permeability. The latter effect can result in the leakage of plasma proteins into the surrounding tissue fluid, producing a local edema.

Local Inflammation

Aspects of the nonspecific and specific immune responses and their interactions are well illustrated by the events that occur when bacteria enter a break in the skin and produce a local inflammation (table 13.4). The inflammatory reaction is initiated

Table 13.4 Summary of events that occur in a local inflammation when a break in the skin permits entry of bacteria

Category	Events
Nonspecific immunity	Bacteria enter through break in anatomic barrier of skin. Resident phagocytic cells—neutrophils and macrophages—engulf bacteria. Nonspecific activation of complement proteins occurs.
Specific immunity	B cells are stimulated to produce specific antibodies. Phagocytosis is enhanced by antibodies attached to bacterial surface antigens. Specific activation of complement proteins occurs, which stimulates phagocytosis, chemotaxis of new phagocytes to the infected area, and secretion of histamine from tissue mast cells. Diapedesis allows new phagocytic leukocytes (neutrophils and monocytes) to invade the infected area. Vasodilation and increased capillary permeability (as a result of histamine secretion) produce redness and edema.

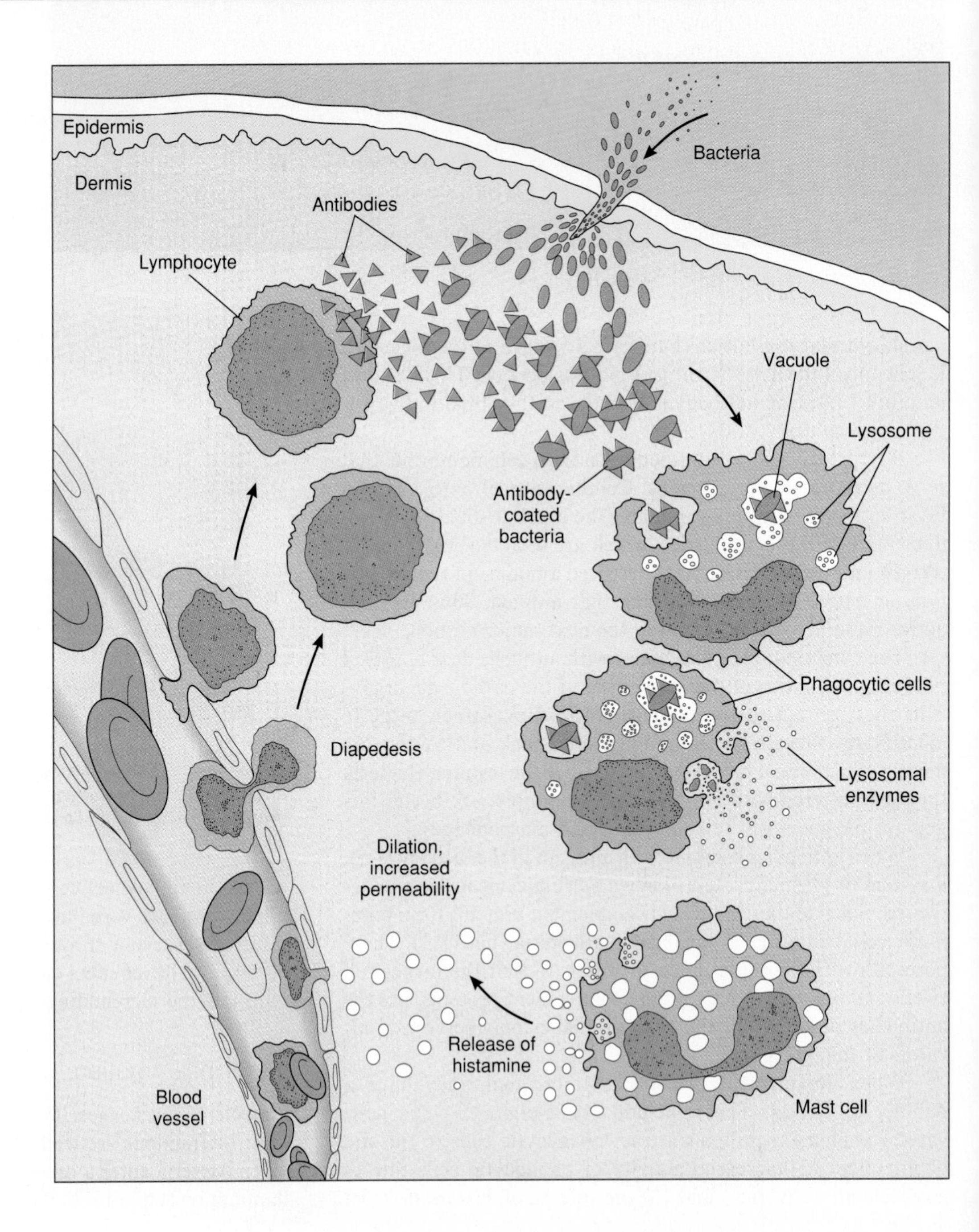

Figure 13.6. *The entry of bacteria through a cut in the skin produces a local inflammatory reaction. In this reaction, antigens on the bacterial surface are coated with antibodies and ingested by phagocytic cells. Symptoms of inflammation are produced by the release of lysosomal enzymes and by the secretion of histamine and other chemicals from tissue mast cells.*

by the nonspecific mechanisms of phagocytosis and complement activation. Activated complement further increases this nonspecific response by attracting new phagocytes to the area and by increasing the activity of these phagocytic cells.

After some time, B lymphocytes are stimulated to produce antibodies against specific antigens that are part of the invading bacteria. Attachment of these antibodies to antigens in the bacteria greatly increases the previously nonspecific response. This occurs because of greater activation of complement, which directly destroys the bacteria and which—together with the antibodies themselves—promotes the phagocytic activity of neutrophils, macrophages, and monocytes (fig. 13.6).

As inflammation progresses, the release of lysosomal enzymes from macrophages causes the destruction of leukocytes and other tissue cells. These effects, together with those produced by histamine and other chemicals released from mast cells, produce the characteristic symptoms of a local inflammation: *redness and warmth* (due to vasodilation); *swelling* (edema); and *pus* (the accumulation of dead leukocytes). If the infection continues, leukocytes and macrophages may release a chemical called *endogenous pyrogen.* This chemical stimulates the hypothalamus of the brain to raise its internal "thermostat" (the center that controls body temperature), producing a fever.

---

1. Describe the chemical structure of antibodies, and indicate which parts of these molecules bind to antigens.
2. Explain the significance of complement proteins when the immune system attacks a cell containing foreign antigens.
3. Describe the events that occur during a local inflammation.

Active and Passive Immunity

When a person is first exposed to the antigens in a pathogen, too few antibodies may be produced to combat the disease. In the process, however, the lymphocytes which have specificity for those antigens are stimulated to divide many times and produce a clone. This is active immunity, and can protect the person from getting the disease upon subsequent exposures. A person may be passively immunized to a disease by obtaining antibodies produced by another organism.

It was first known in the mid-eighteenth century that the fatal effects of smallpox could be prevented by inducing mild cases of the disease. This was accomplished at that time by rubbing needles into the pustules of people who had mild forms of smallpox and injecting these needles into healthy people. This method of immunization sometimes produced fatal cases of smallpox, and was outlawed as a result.

Acting on the observation that milkmaids who contracted cowpox—a disease similar to smallpox but less *virulent* (pathogenic)—were immune to smallpox, an English physician, named Edward Jenner,[11] inoculated a healthy boy with cowpox. When the boy recovered, Jenner inoculated him with an otherwise deadly amount of smallpox, from which he also proved to be immune. (This was fortunate for both the boy—who was an orphan—and Jenner; Jenner's fame spread, and as the boy grew into manhood he proudly gave testimonials on Jenner's behalf.) This experiment, performed in 1796, marked the first widespread immunization program.

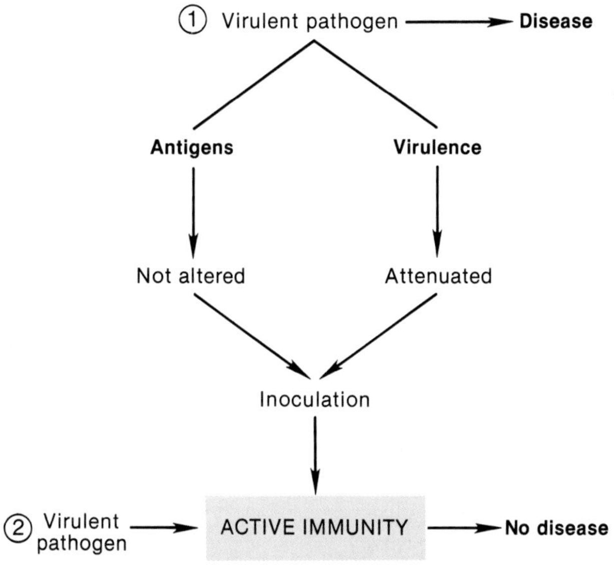

A similar, but more sophisticated, demonstration of the effectiveness of immunizations was performed by Louis Pasteur[12] almost a century later. Pasteur isolated the bacteria that cause anthrax and heated them until their ability to cause disease was greatly reduced (their virulence was *attenuated*), but the nature of their antigens was not significantly changed (fig. 13.7). He then injected these attenuated bacteria into twenty-five cattle, leaving twenty-five unimmunized. Several weeks later, before a gathering of scientists, he injected all fifty cattle with the completely active anthrax bacteria. All twenty-five of the unimmunized cattle died—all twenty-five of the immunized cattle survived.

Figure 13.7. *Active immunity to a pathogen can be gained by exposure to the fully virulent form or by inoculation with a pathogen whose virulence (ability to cause disease) has been attenuated (reduced) but whose antigens are the same as in the virulent form.*

Active Immunity

When a person is exposed to a particular pathogen for the first time, there is a variable latent period before measurable amounts of specific antibodies appear in the blood. This sluggish **primary response** may not be sufficient to protect the individual against the disease caused by the pathogen. Antibody concentrations in the blood during this primary response reach a plateau in a relatively short time and decline after a few weeks.

[11]Edward Jenner: English physician (1749–1823)

[12]Louis Pasteur: French chemist and bacteriologist (1822–1895)

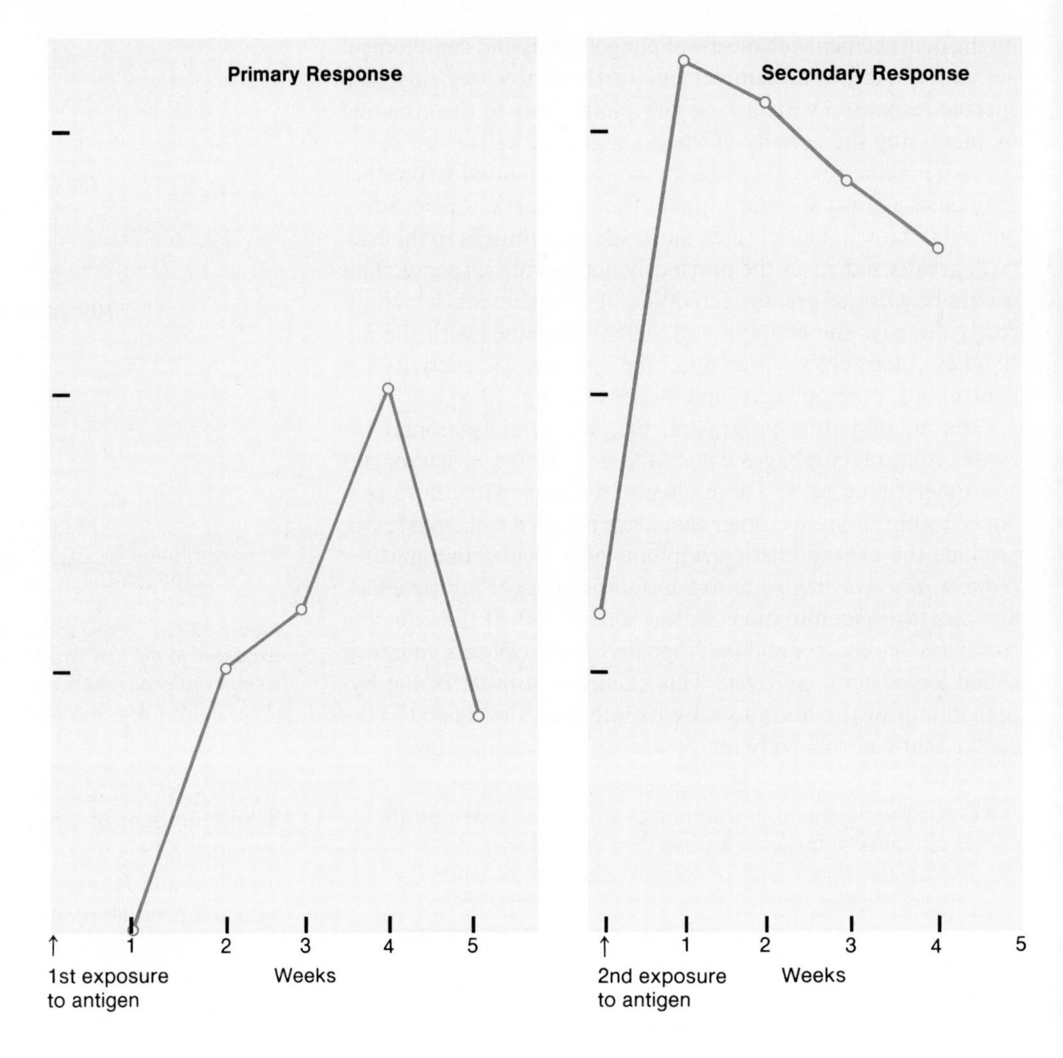

Figure 13.8. A comparison of antibody production in the primary response (upon first exposure to an antigen) to antibody production in the secondary response (upon subsequent exposure to the antigen). The greater secondary response is believed to be due to the development of lymphocyte clones produced during the primary response.

A subsequent exposure of the same individual to the same antigen results in a **secondary response** (fig. 13.8). Compared to the primary response, antibody production during the secondary response is much more rapid. High antibody concentrations in the blood are reached more quickly and are maintained for a longer time than in the primary response. This rapid rise in antibody production is usually sufficient to prevent the disease.

Clonal Selection Theory

The immunization procedures of Jenner and Pasteur were effective because the people who were inoculated produced a secondary rather than a primary response when exposed to the virulent pathogens. This protection is not simply due to accumulations of antibodies in the blood, because secondary responses occur even after antibodies produced by the primary response have disappeared. Immunizations, therefore, seem to produce a type of "learning" in which the ability of the immune system to combat a particular pathogen is improved by prior exposure.

The mechanisms by which secondary responses are produced are not completely understood; the **clonal selection theory,** however, appears to account for most of the evidence. According to this theory, B lymphocytes *inherit* the ability to produce particular antibodies (and T lymphocytes inherit the ability to respond to particular antigens). One B lymphocyte can only produce one type of antibody, with specificity for one antigen. Since this ability is genetic rather than acquired, some lymphocytes can, for example, respond to measles and produce antibodies against it even if the person has never been previously exposed to this disease.

The inherited specificity of each lymphocyte is reflected in the antigen receptor proteins on the surface of the lymphocyte's plasma membrane. Exposure to measles antigens thus stimulates these specific lymphocytes to divide many times until a large population of genetically identical cells—a clone—is produced. Some of these cells become plasma cells that secrete antibodies for the primary response; others become memory cells that can be stimulated to secrete antibodies during the secondary response (fig. 13.9).

Notice that according to the clonal selection theory (table 13.5), antigens do not induce lymphocytes to make the appropriate antibodies. Rather, antigens *select* lymphocytes (through interaction with surface receptors) that are already able to make antibodies against that antigen. This is analogous to evolution

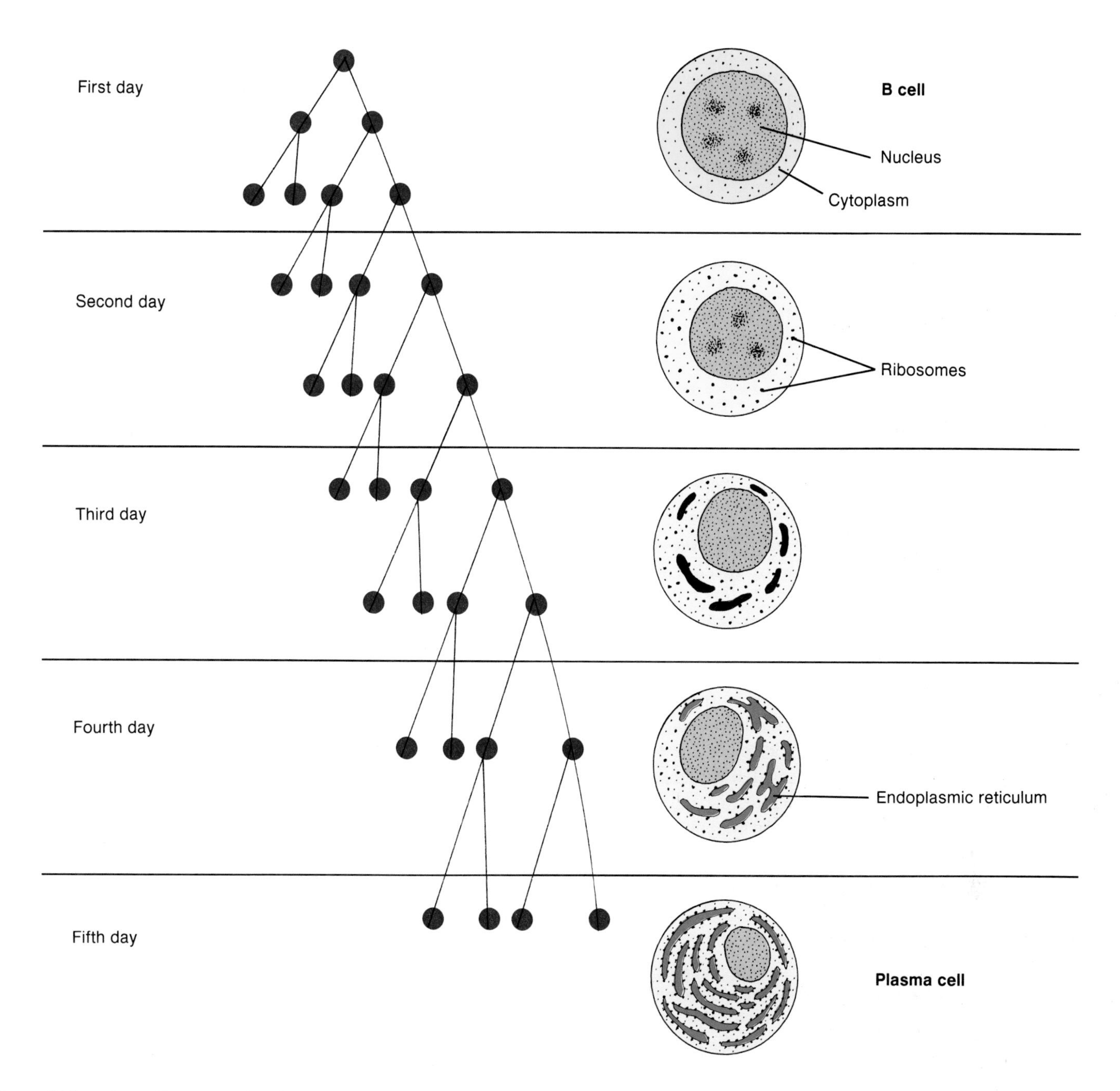

Figure 13.9. *According to the clonal selection theory, exposure to an antigen stimulates the production of lymphocyte clones and the maturation of some members of B cell clones into antibody-secreting plasma cells.*

Table 13.5 Summary of the clonal selection theory (with regard to B cells)

Process	Results
Lymphocytes inherit the ability to produce specific antibodies.	Prior to antigen exposure, lymphocytes are already present in the body that can make the appropriate antibodies.
Antigens interact with antibody receptors on the lymphocyte surface.	Antigen-antibody interaction stimulates cell division and the development of lymphocyte clones containing memory cells and plasma cells that secrete antibodies.
Subsequent exposure to the specific antigens produce a more efficient response.	Exposure of lymphocyte clones to specific antigens results in greater and more rapid production of specific antibodies.

by natural selection (chapter 2). An environmental agent (in this case, antigens) acts on the genetic diversity already present in a population of organisms (lymphocytes) to cause increasing numbers of the individuals that are selected.

Vaccinations

The development of a secondary response provides **active immunity** against the specific pathogens. The development of active immunity requires prior exposure to the specific antigens, during which time a primary response is produced and the person may get sick. Some parents, for example, deliberately expose their children to others who have measles, chicken pox, and mumps so that their children will be immune to these diseases in later life.

Clinical immunization programs induce primary responses by inoculating people with pathogens whose virulence has been attenuated or destroyed (such as Pasteur's heat-inactivated anthrax bacteria) or by using closely related strains of microorganisms that are antigenically similar but less pathogenic (such as Jenner's cowpox inoculations). The name for these procedures—**vaccinations** (after the Latin word *vacca* = cow)—reflects the history of this technique. All of these procedures cause the development of lymphocyte clones that can combat the virulent pathogens by producing secondary responses.

The first successful polio vaccine (the Salk[13] vaccine) was composed of viruses that had been inactivated by treatment with formaldehyde. These viruses were injected into the body, in contrast to the currently used oral (Sabin[14]) vaccine. The oral vaccine contains active viruses that have attenuated virulence. These viruses invade the epithelial lining of the intestine and multiply but do not invade nerve tissue. The immune system can, therefore, become sensitized to polio antigens and produce a secondary response if polio viruses that attack the nervous system are later encountered.

There is always a danger when vaccines are prepared using attenuated viruses or toxins that some virulence may remain and cause disease in vaccinated people. One example is the commonly used vaccine for *pertussis* (whooping cough). This disease is responsible for about one million infant deaths annually worldwide, and is caused by a toxin released from a species of bacteria. Though the vaccine against pertussis is relatively effective, it occasionally produces severe side effects. Scientists have recently produced a protein subunit of pertussis toxin using genetic engineering techniques which appears to confer immunity without having the virulent actions of the toxin. In the future, production of specific proteins from cloned DNA (chapter 10) may provide other vaccines that are safer and more effective than those prepared by traditional methods.

Passive Immunity

The term **passive immunity** refers to the immune protection that can be produced by transferring antibodies produced by another person or by an animal. The person or animal that produced the antibodies to particular antigens was actively immunized as explained by the clonal selection theory. The person who receives these ready-made antibodies is thus passively immunized to the same antigens. Passive immunity occurs naturally in the transfer of immunity from a mother to her infant, and can be produced artificially by injections of antibodies to protect against certain diseases.

The ability to mount an immune response—called **immunological competence**—does not fully develop until shortly after birth. The fetus, therefore, cannot immunologically reject its mother. The immune system of the mother is fully competent but does not usually respond to fetal antigens for reasons that are not completely understood. Some antibodies from the mother do cross the placenta and enter the fetal circulation, however, and these serve to confer passive immunity to the fetus.

The fetus and the newborn baby are, therefore, immune to the same antigens as the mother. Since the baby did not itself produce the lymphocyte clones needed to form these antibodies, such passive immunity disappears when the infant is five to six months old. If the baby is breast-fed it can receive additional antibodies in its mother's first milk (the *colostrum*).

Passive immunizations are used clinically to protect people who have been exposed to extremely virulent infections or toxins, such as snake venom, tetanus, and others. In these cases the affected person is injected with *antiserum* (serum-derived antibodies), also called *antitoxin,* from an animal or human that has been previously exposed to the pathogen. The animal develops the lymphocyte clones and active immunity and thus has a high concentration of antibodies in its blood. Since the person who is injected with these antibodies does not develop active immunity, he or she must be actively immunized later or again be injected with antitoxin upon subsequent exposures. A comparison of active and passive immunity is shown in table 13.6.

Monoclonal Antibodies

In addition to their use in passive immunity, antibodies are also commercially prepared for use in research and clinical laboratory tests. In the past, antibodies were obtained by chemically purifying a specific antigen and then injecting this antigen into animals. Since an antigen typically has many different antigenic determinant sites, however, the antibodies obtained by this method had different specificities. This could result in some degree of cross-reaction with closely related antigen molecules.

In the preparation of monoclonal antibodies, an animal (mice are commonly used) is injected with an antigen and then subsequently killed. B lymphocytes are then obtained from the spleen and placed in thousands of different *in vitro* incubation vessels. These cells soon die, however, unless they are hybridized with cancerous multiple myeloma cells. The fusion of a B

[13]Salk vaccine: from Jonas Salk, American immunologist (1914–)
[14]Sabin vaccine: from Albert B. Sabin, American virologist (1906–)

Table 13.6 Comparison of active and passive immunity

Characteristic	Active Immunity	Passive Immunity
Injection of person with	Antigens	Antibodies
Source of antibodies	The person inoculated	Natural—the mother; artificial—injection with antibodies
Method	Injection with killed or attenuated pathogens or their toxins	Natural—transfer of antibodies across the placenta; artificial—injection with antibodies
Time to develop resistance	5 to 14 days	Immediately after injection
Duration of resistance	Long (perhaps years)	Short (days to weeks)
When used	Before exposure to pathogen	Before or after exposure to pathogen

lymphocyte with a cancerous cell produces a hybrid that undergoes cell division and produces a clone, called a *hybridoma.* Each hybridoma secretes large amounts of identical, **monoclonal antibodies.** From among the thousands of hybridomas produced in this way, the one that produces the desired antibody is cultured while the rest are disgarded.

The availability of large quantities of pure monoclonal antibodies has resulted in the development of much more sensitive clinical laboratory tests (of pregnancy, for example). In the future, monoclonal antibodies against specific tumor antigens may aid the diagnosis of cancer. Even more exciting, drugs that can kill normal as well as cancerous cells might be aimed directly at a tumor by combining these drugs with monoclonal antibodies against specific tumor antigens. The production of monoclonal antibodies and their potential use in cancer treatments are discussed more completely in chapter 14.

The commercial production of monoclonal antibodies for medical diagnosis and treatment is part of the new biotechnology revolution. The commercial applications of monoclonal antibody production is now a multimillion dollar industry that, together with the applications of genetic engineering (chapter 10), hold much promise for medical, social, and economic progress.

1. Describe three methods used to induce active immunity.
2. Explain the characteristics of the primary and secondary immune responses, and draw graphs to illustrate your discussion.
3. Explain the clonal selection theory and how this theory accounts for the secondary response.
4. Describe passive immunity, and give natural and clinical examples of this type of immunization.

Functions of T Lymphocytes

T lymphocytes are activated by specific antigens and respond by secreting chemicals called lymphokines. Through their secretion of lymphokines, T cells provide numerous functions that assist all aspects of the immune system. These functions include cell-mediated destruction by killer T cells and supporting roles by helper and suppressor T cells. T cells are activated by foreign antigens only when those antigens are presented to the T cells by macrophages.

The thymus processes lymphocytes in such a way that their functions become quite distinct from those of B cells. Unlike B cells, the T lymphocytes provide specific immune protection without secreting antibodies. This is accomplished in different ways by the three subpopulations of T lymphocytes, which will be described shortly.

Thymus Gland

The thymus gland extends from below the thyroid in the neck into the thoracic cavity. This organ grows during childhood but gradually regresses after puberty. Lymphocytes from the bone marrow seed the thymus and become transformed into T cells. These lymphocytes in turn enter the blood and seed lymph nodes and other lymphoid organs, where they divide to produce new T cells when stimulated by antigens.

Small T lymphocytes that have not yet been stimulated by antigens have very long life spans—months or perhaps years. Still, new T cells must be continuously produced to provide efficient cell-mediated immunity. Since the thymus atrophies after puberty, this organ may not be able to provide new T cells in later life. Colonies of T cells in the lymph nodes and other organs are apparently able to produce new T cells under the stimulation of various **thymus hormones.**

The hormones that are secreted by the thymus may promote the transformation of lymphocytes into T cells, and the maturation of T lymphocytes. There is some experimental evidence suggesting that the administration of thymus hormones may be able to restore cell-mediated immunity in some cases where T cell function has declined. This decline occurs in some congenital and acquired diseases as well as naturally in the course of aging in conjunction with an increased susceptibility to viral infections and cancer.

Killer, Helper, and Suppressor T Lymphocytes

The **killer T lymphocytes** destroy specific target cells that are identified by specific antigens on their surface. In order to effect this *cell-mediated* destruction, the T lymphocytes must be in actual contact with their target cells (in contrast to B cells, which kill at a distance). Although the mechanisms by which these lymphocytes kill their targets are not completely understood, there is evidence that this is accomplished by the secretion of certain molecules at the region of contact. Among these molecules, specific polypeptides called *perforins* have been identified which enter the cell membrane of the target cell and form cylindrical channels through the membrane. This is analogous to the channels formed by complement proteins previously discussed, and can result in osmotic destruction of the target cell.

The killer lymphocytes defend against viral and fungal infections and are also responsible for transplant rejection reactions and for immunological surveillance against cancer. Although most bacterial infections are fought by B lymphocytes, some are the targets of cell-mediated attack by killer T lymphocytes. This is the case with the *tubercle bacilli* that cause tuberculosis. Injections of an extract of these bacteria under the skin produce inflammation after a latent period of forty-eight to seventy-two hours. This reaction is cell mediated rather than humoral, as shown by the fact that it can be induced in an unexposed guinea pig by an infusion of lymphocytes, but not of serum, from an exposed animal.

The **helper T lymphocytes** and **suppressor T lymphocytes** indirectly participate in the specific immune response by regulating the responses of the B cells and the killer T cells. The activity of B cells and killer T cells is increased by helper T lymphocytes and decreased by suppressor T lymphocytes. The amount of antibodies secreted in response to antigens is thus affected by the relative numbers of helper to suppressor T cells that develop in response to a given antigen.

As a result of advances in recombinant DNA technology (genetic engineering) that allow the production of monoclonal antibodies, it is now possible for clinical laboratories to distinguish between the different subcategories of lymphocytes by means of antigen "markers" on their surfaces. Counting the lymphocytes in each of these subcategories provides far more information about diseases and their causes than was previously available. Tests of this sort have provided valuable information about the effects of the AIDS virus.

Lymphokines

The helper T lymphocytes secrete a number of polypeptides that serve to regulate many aspects of the immune system. These products are called **lymphokines.** When a lymphokine is first discovered, it is named according to its biological activity (e.g., *B cell stimulating factor*). Since each lymphokine has many different actions, however, such names can be misleading. Scientists have thus agreed to use the name *interleukin,* followed by a number, to indicate a lymphokine once its amino acid sequence has been determined.

Interleukin-1, for example, is secreted by macrophages and other cells and can activate the T cell system. *Interleukin-2* is released by helper T lymphocytes and stimulates proliferation and activation of killer T lymphocytes. *Interleukin-4* is secreted by T lymphocytes and is required for the proliferation and clone development of B cells. A number of additional lymphokines are currently known, and more will probably be discovered with continued research.

Activation of T Lymphocytes

Unlike B cells, T cells do not make antibodies and thus do not have antibodies on their surface to serve as receptors for antigens. The T cells do, however, have specific receptors for antigens on their membrane surface, and these T cell receptors have recently been identified as molecules closely related to antibodies. The T cell receptors differ from the antibody receptors on B cells in another, and very important, characteristic: they *cannot bond to free antigens.* In order for a T lymphocyte to respond to a foreign antigen, the antigen must be presented to the T lymphocyte on the membrane of an *antigen-presenting cell.* The chief antigen-presenting cells are macrophages, which engulf the antigens by phagocytosis and then move these antigens to the cell membrane surface. In this position, the foreign antigens can stimulate helper T lymphocytes.

1. List the different types of T lymphocytes and describe their functions.
2. Describe the origin and significance of lymphokines, and provide examples of some of the lymphokines.
3. Describe the nature of the HIV and how it causes some of the symptoms of AIDS.
4. Explain how the drug AZT works in the treatment of AIDS, and describe how a vaccine for AIDS might be developed.
5. Explain the role of macrophages in the activation of T lymphocytes.

Tolerance, Autoimmunity, and Allergy

There are cases where the immune system itself, rather than foreign invaders, can be the cause of disease. Autoimmune diseases are produced when the immune system attacks self-antigens. Allergy is produced by antigen-antibody reactions that cause the release of histamine and similarly acting molecules.

The immune system of a person exists to combat foreign invasion by organisms and their products which can cause disease. In order to accomplish this, however, the immune system activates processes that are, by their nature, destructive. A person's immune system must therefore not attack the molecules produced by that person's genes. This does, however, occasionally occur. When it does, *autoimmune diseases* are produced. Also, the immune system may respond to benign[15] foreign antigens in such a way as to cause health problems. This produces the symptoms of *allergy.*

Tolerance

The ability to produce an immune response against foreign, **non-self** antigens, while tolerating (not producing an immune response against) **self**-antigens occurs during the first month or

[15]benign: L. *benignus,* kind

Acquired Immune Deficiency Syndrome (AIDS)

Acquired immune deficiency syndrome (AIDS) has, to date, caused more than 65,000 deaths in the United States alone. This figure is misleading, however. An estimated ten million people worldwide are believed to be infected, and since AIDS has been shown to have a latency period of approximately eight years, most will display symptoms of the disease in the near future. The U.S. General Accounting Office (GAO) estimates that there will be 300,000 to 485,000 cases of AIDS in the United States by the end of 1991. People at high risk include homosexual and bisexual men (through anal intercourse) and intravenous drug users (through sharing of needles with infected individuals). Intravenous drug users account for one-third of AIDS cases in the United States and Europe; half of the estimated 200,000 intravenous drug users in New York City are believed to be infected. People at lesser risk include those who received blood transfusions prior to 1985 (before blood was tested for AIDS) and the spouses of those at high risk. In Haiti and the countries of central Africa, heterosexual contact is believed to be the primary route of infection.

Most scientists now believe that AIDS is caused by a virus, known as the *human immunodeficiency virus (HIV),* shown in figure 13.10. This is a retrovirus, which injects RNA into its host cell. As described in chapter 10, the virus contains the enzyme reverse transcriptase, which produces viral DNA using the viral RNA as a template. Once the viral DNA is produced, it is incorporated into the DNA of the host cell, and later may direct the synthesis of proteins required to produce new virus particles inside the host cell (fig. 13.11). These bud from the infected cell and are free to infect other cells.

The HIV specifically attacks helper T lymphocytes (fig. 13.12). Laboratory tests have been developed to identify these cells and count the relative proportions of helper T cells to suppressor T cells. Helper T lymphocytes have a particular antigen marker on their surface known as the CD4 antigen, which can be distinguished from a different antigen (known as CD8) on suppressor and killer T lymphocytes. A healthy person has a *CD4/CD8 ratio* greater than 2.0; a person with AIDS has a ratio reduced to 0.5 or less, indicating a great reduction in helper T lymphocytes. A different laboratory test detects the presence of antibodies against HIV in the plasma of an infected person. This test serves to identify people with this infection, although several months may elapse between the time of infection and the time when the person tests antibody-positive for HIV.

AIDS is a progressive disease. Early, less severe symptoms are sometimes referred to as *ARC: AIDS-related complex.* The decreased immunological function that results from the loss of helper T lymphocytes, however, makes the person susceptible to serious opportunistic infections, including *Pneumocystis carinii* pneumonia. Many people with AIDS develop a previously rare form of cancer known as *Kaposi's sarcoma.* The AIDS virus also attacks the nervous system, producing symptoms of neurological damage in approximately one-third of AIDS patients.

Since HIV is a retrovirus, it requires the activity of reverse transcriptase in order to reproduce. The drug *AZT,* an analog of the nucleotide base thymine, inhibits the ability of reverse transcriptase to produce viral DNA from RNA. This drug, however, can

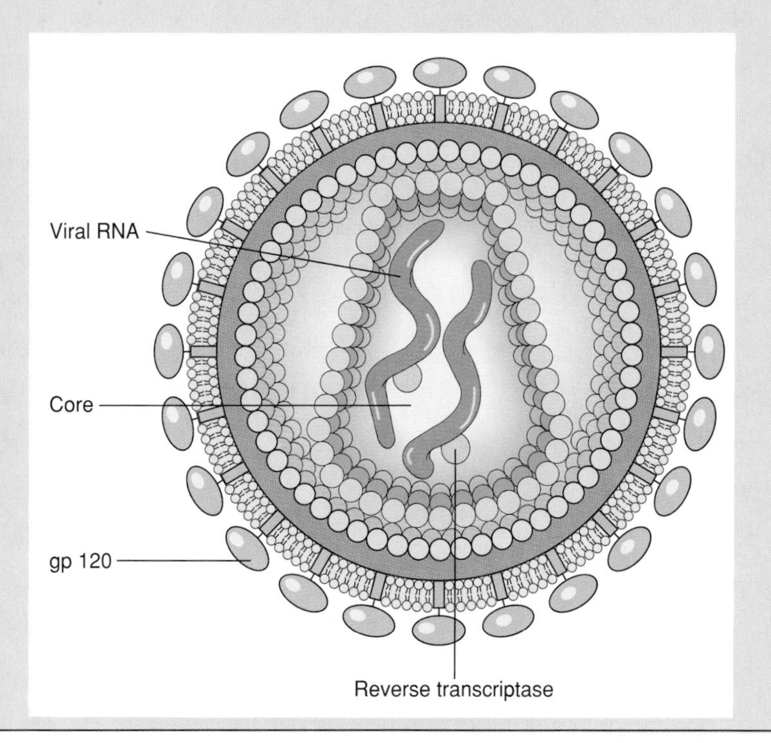

Figure 13.10. *A diagram of HIV, the virus that causes AIDS.*

(continued)

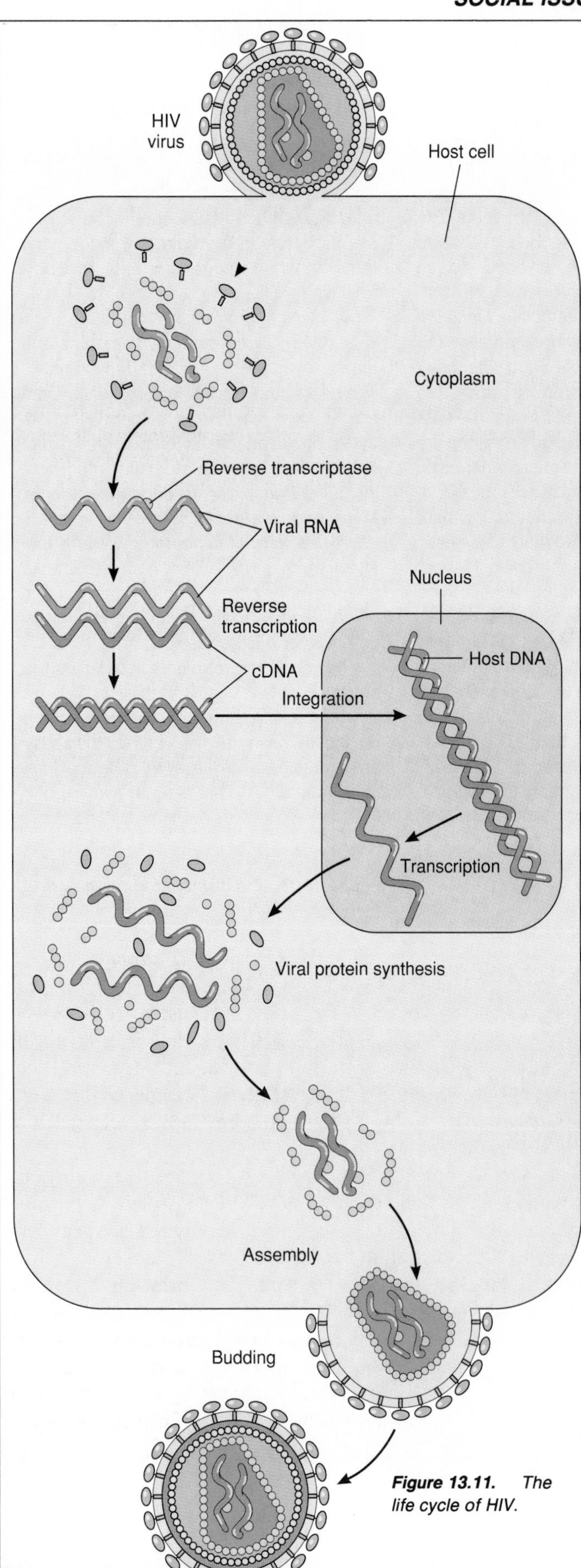

Figure 13.11. *The life cycle of HIV.*

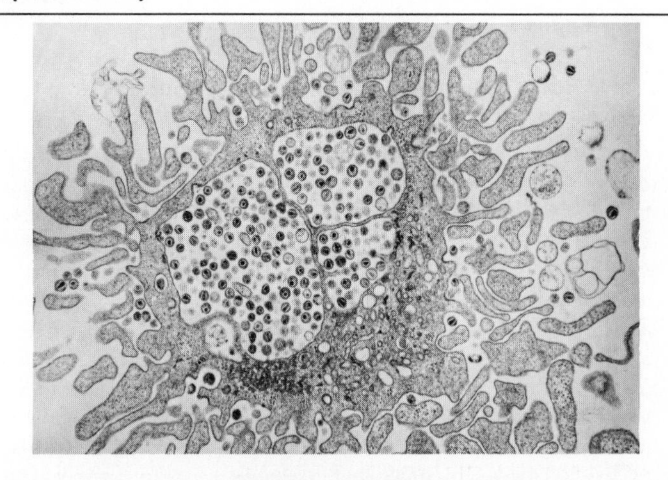

Figure 13.12. *A helper T lymphocyte infected with HIV.*

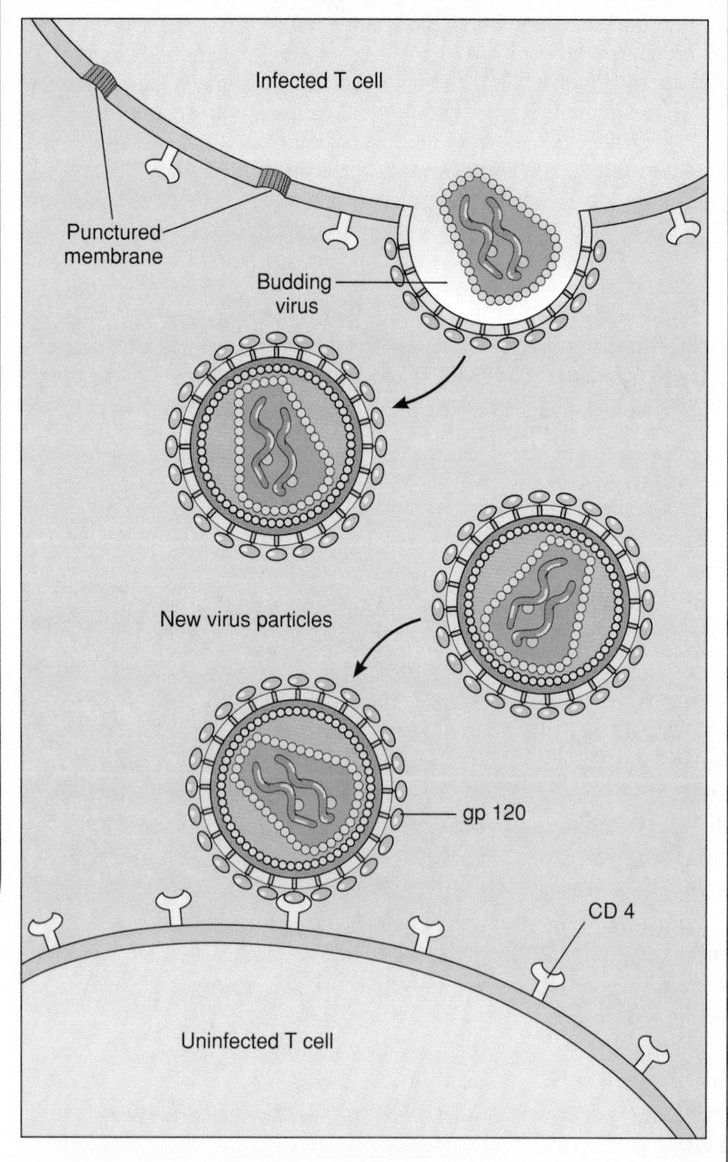

Figure 13.13. *Interaction of the gp120 particle of HIV and the CD4 receptor of a T lymphocyte leads to infection.*

(continued)

produce anemia and severe gastrointestinal disorders. Other drugs that similarly inhibit reverse transcriptase activity, but which have less toxicity, are therefore under investigation. One such drug that shows promise is *DDI* (dideoxyinosine).

Scientists have recently shown that AZT can delay the onset of symptoms in people infected with HIV who do not yet display more advanced symptoms of AIDS. Moreover, people who are in the early stages of AIDS can tolerate higher, more effective doses of AZT. This exciting discovery was soon followed by another; children treated with AZT can be prevented from experiencing the mental deterioration caused by HIV infection of the brain. In fact, AZT was shown to reverse the loss of I.Q. and symptoms of autism in infected children. The FDA has moved quickly in response to these findings, so that people without overt AIDS may benefit from AZT. Unfortunately, the cost of AZT treatment is still quite high, even though it was reduced 20% by Burroughs Wellcome, the manufacturer of this drug.

AZT and similar drugs cannot cure AIDS; they can only decrease the ability of the virus to reproduce. Scientists are attempting to develop a vaccine that could actively immunize people to HIV. If successful, vaccinated people would develop T cell and B cell clones that recognize and attack the AIDS-causing virus. This possibility is strengthened by a recent report of successful vaccination of rhesus monkeys with a vaccine against the *simian immunodeficiency virus (SIV)*. This virus is closely related to HIV, and causes symptoms very similar to AIDS in monkeys. The vaccine was prepared by inactivating whole SIV viruses with the chemical formalin. Nine monkeys were vaccinated with these attenuated viruses and then given active SIV in doses that proved deadly to unvaccinated animals. All nine vaccinated monkeys were disease-free at the end of one year, and eight out of the nine appeared to be free of SIV infection as a result of the vaccination.

A vaccine containing whole HIV viruses (analogous to the attenuated SIV viruses used in the monkey experiments) is undesirable, because these retroviruses would likely mutate to disease-causing forms within the body. A more promising approach—one that has been used successfully in the development of a hepatitis B vaccine—is to inoculate a person with one of the proteins found on the surface of the viral particle. A particular target in this research is the protein known as *gp120*, which stands for a glycoprotein with a molecular weight of 120,000. This protein sticks out from the surface of the virus particle (fig. 13.13) and thus could be recognized by the immune system.

This approach to the development of an AIDS vaccine is made difficult by the fact that the genes for the gp120 proteins of the HIV virus mutate frequently, so that a single infected individual may harbor many variants of the HIV virus. Scientists are attempting to locate regions of this protein that are less variable and that can serve as antigens for the development of a vaccine. Other approaches are also being investigated. Since the gp120 viral protein must combine with a receptor protein on the surface of helper T cells for the virus to be taken into these cells, attempts are being made to block this interaction. Some genetic engineering companies have produced this helper T cell receptor, called *CD4* (fig. 13.13), in the hopes that injecting the free receptor could "tie up" the virus particles before they can be taken into the T lymphocytes.

Although a cure for AIDS is not yet in sight, current knowledge of the immune system and the development of recombinant DNA technology at least offer the hope that a vaccine may be found in the future. When found, it would owe its existence to the modern biotechnology revolution.

so of postnatal life when immunological competence is established. If a fetal mouse of one strain receives transplanted antigens from a different strain, therefore, it will not recognize tissue transplanted later in life from the other strain as foreign and, as a result, will not immunologically reject the transplant.

The ability of an individual's immune system to recognize and tolerate self-antigens requires continuous exposure of the immune system to those antigens. If this exposure begins when the immune system is weak—such as in fetal and early postnatal life—tolerance is more complete and long lasting than when exposure occurs later in life. Some self-antigens, however, are normally hidden from the blood. An exposure to these self-antigens results in antibody production just as if these proteins were foreign. Antibodies made against self-antigens are called **autoantibodies.**

The reasons why autoantibodies are not normally produced—that is, why tolerance to self-antigens occurs—are not well understood. There are two general types of theories that have been proposed to account for immunological tolerance: (1) *clone deletion;* and (2) *immunological suppression.* According to the clone deletion theory, tolerance is achieved by destruction of the lymphocytes that recognize self-antigens. Presumably this occurs primarily during fetal life, when those lymphocytes that have receptors on their surface for self-antigens are recognized and destroyed. There is evidence that clonal deletion occurs in the thymus and is the major mechanism responsible for T cell tolerance.

According to the immunological suppression theory, the lymphocytes that make autoantibodies are present throughout life but are normally inhibited from attacking self-antigens. This can be due to the effects of suppressor T lymphocytes and/or antibodies that block the actions of autoantibodies. An alteration in the ratio of suppressor to helper T lymphocytes in later life, or a shift in the interactions among autoantibodies and their antibody blockers, therefore, might result in the production of autoantibodies. Immunological suppression, also called *clonal anergy,* is the major mechanism responsible for B cell tolerance.

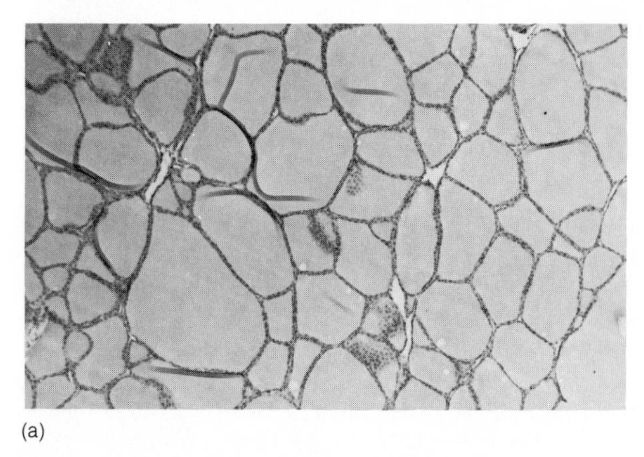

(a)

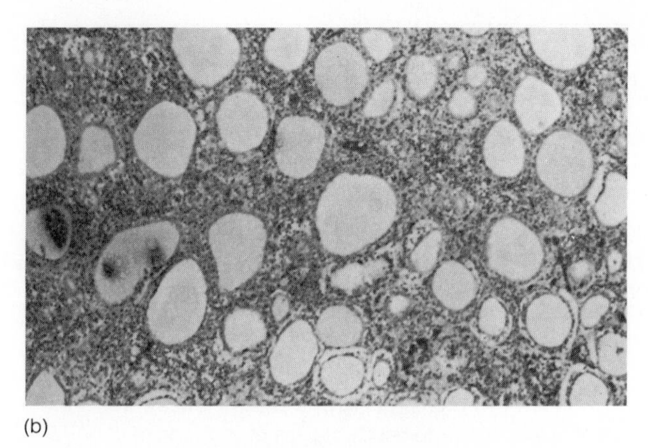

(b)

Figure 13.14. Autoimmune thyroiditis in a rabbit, induced experimentally by injection with thyroglobulin. Compare the picture of a normal thyroid (a) with that of the diseased thyroid (b). The grainy appearance of the diseased thyroid is due to the infiltration of large numbers of lymphocytes and macrophages.

A spontaneous mutation in mice leads to the development of a strain that suffers from severe combined immunodeficiency (SCID). This condition is similar to a rare congenital condition in humans, and results in the absence of B and T lymphocytes. Grafts in SCID mice are therefore not rejected. This inability to reject transplants has recently been exploited by reconstituting a human immune system in the SCID mice, using lymphocytes from peripheral human blood or human fetal liver, thymus, and lymph node grafts. This technique may provide a means for studying the function of the immune system, and for studying diseases of the human immune system. Since the HIV virus, for example, only infects human (and chimpanzee) lymphocytes, the mouse-human chimera[16] may provide an animal model for experimental investigation of AIDS.

The ability of the normal immune system to tolerate self-antigens while it identifies and attacks foreign antigens provides a specific defense against invading pathogens. In every individual, however, this system of defense against invaders at times commits domestic offenses. This can result in diseases that range in severity from the sniffles to sudden death.

Autoimmunity

Autoimmune diseases are those produced by failure of the immune system to tolerate self-antigens. This failure results in the production of autoantibodies that can cause inflammation and organ damage. Such autoimmune destruction may occur as a result of the following mechanisms.

An antigen that does not normally circulate in the blood may become exposed to the immune system. Thyroglobulin protein that is normally trapped within the thyroid follicles (chapter 7), for example, can stimulate the production of autoantibodies that cause the destruction of the thyroid (fig. 13.14); this occurs in *Hashimoto's thyroiditis.* Similarly, autoantibodies developed against lens protein in a damaged eye may cause the destruction of a healthy eye (in *sympathetic ophthalmia*).

A self-antigen, which is otherwise tolerated, may be altered by combining with a foreign hapten. The disease *thrombocytopenia* (low platelet count), for example, can be caused by the autoimmune destruction of thrombocytes (platelets). This occurs when drugs such as aspirin, sulfonamide, antihistamines, digoxin, and others combine with platelet proteins to produce new antigens. The symptoms of this disease usually stop when the person stops taking these drugs.

Antibodies may be produced that are directed against other antibodies. Such interactions may be necessary for the prevention of autoimmunity, as previously suggested, but imbalances may actually cause autoimmune diseases. *Rheumatoid arthritis,* for example, is an autoimmune disease associated with the abnormal production of one group of antibodies that attack other antibodies. This results in an inflammation reaction of the joints characteristic of the disease.

Antibodies produced against foreign antigens may cross-react with self-antigens. Autoimmune diseases of this sort can occur, for example, as a result of *streptococcus* bacterial infections. Antibodies produced in response to antigens in this bacterium may cross-react with self-antigens in the heart, and inflammation induced by such autoantibodies can produce heart damage (including valve defects).

There are a number of diseases that are caused by inflammation when antibodies combine with free antigens in the plasma. These diseases can produce widespread systemic effects. In viral hepatitis B, for example, the combination of viral antigens and antibodies can cause widespread inflammation of arteries (*periarteritis*). Note that the arterial damage is not caused by the hepatitis virus itself but by the inflammatory process. Another example of this sort of an autoimmune disease is *systemic lupus erythematosus (SLE).* People with SLE produce antibodies against their own DNA and nuclear proteins. This can result in inflammation throughout the body, leading to damage to the kidneys and other organs.

[16]chimera: Gk. *chimaira,* a mythological fire-spouting monster with a lion's head, goat's body, and serpent's tail

Table 13.7 Allergy: comparison between immediate and delayed hypersensitivity reactions

Characteristic	Immediate Reaction	Delayed Reaction
Time for onset of symptoms	Within several minutes	Within one to three days
Lymphocytes involved	B cells	T cells
Immune effector	IgE antibodies	Cell-mediated immunity
Allergies most commonly produced	Hay fever, asthma, and most other allergic conditions	Contact dermatitis (such as to poison ivy and poison oak)
Therapy	Antihistamines and adrenergic drugs	Corticosteroids (such as cortisone)

Allergy

The term *allergy,* usually used synonymously with *hypersensitivity,* refers to particular types of abnormal immune responses to antigens, which are called *allergens* in these cases. There are two major forms of allergy: (1) **immediate hypersensitivity,** which is due to an abnormal B lymphocyte response to an allergen that produces symptoms within seconds or minutes; and (2) **delayed hypersensitivity,** which is an abnormal T cell response that produces symptoms within about forty-eight hours after exposure to an allergen. A comparison between these two types of hypersensitivity is provided in table 13.7.

Immediate Hypersensitivity

Immediate hypersensitivity can produce such symptoms as allergic rhinitis (chronic runny or stuffy nose), conjunctivitis (red eyes), allergic asthma, atopic dermatitis (hives), and others. These symptoms result from the production of antibodies of the **immunoglobulin E (IgE)** subclass, instead of the normal antibodies (which are of the subclass designated IgG).

Unlike the normal antibodies produced in response to an antigen, IgE antibodies do not circulate in the blood but instead attach to tissue mast cells. When the person is again exposed to the same allergen, the allergen bonds to the antibodies attached to the mast cells. This stimulates the mast cells to secrete various chemicals, including **histamine** and others (fig. 13.15). These chemicals produce the symptoms of the allergic reactions.

The itching, sneezing, tearing, and runny nose of persons suffering from hay fever are produced largely by histamine and can be treated effectively by antihistamine drugs. Food allergies, causing diarrhea and colic, are mediated primarily by prostaglandins (a type of regulatory fatty acid—chapter 3) and can be treated with aspirin, which inhibits prostaglandin synthesis (these are the only allergies that respond to aspirin). Asthma, produced by smooth muscle constriction in the bronchioles (airways) in the lungs, is due to the release of molecules related to prostaglandins called leukotrienes. Since there are no antileukotriene drugs presently available, asthma is treated with epinephrinelike compounds (which cause bronchodilation) and corticosteroids. Corticosteroids are the hormones of the adrenal cortex and their derivatives, such as hydrocortisone and cortisone.

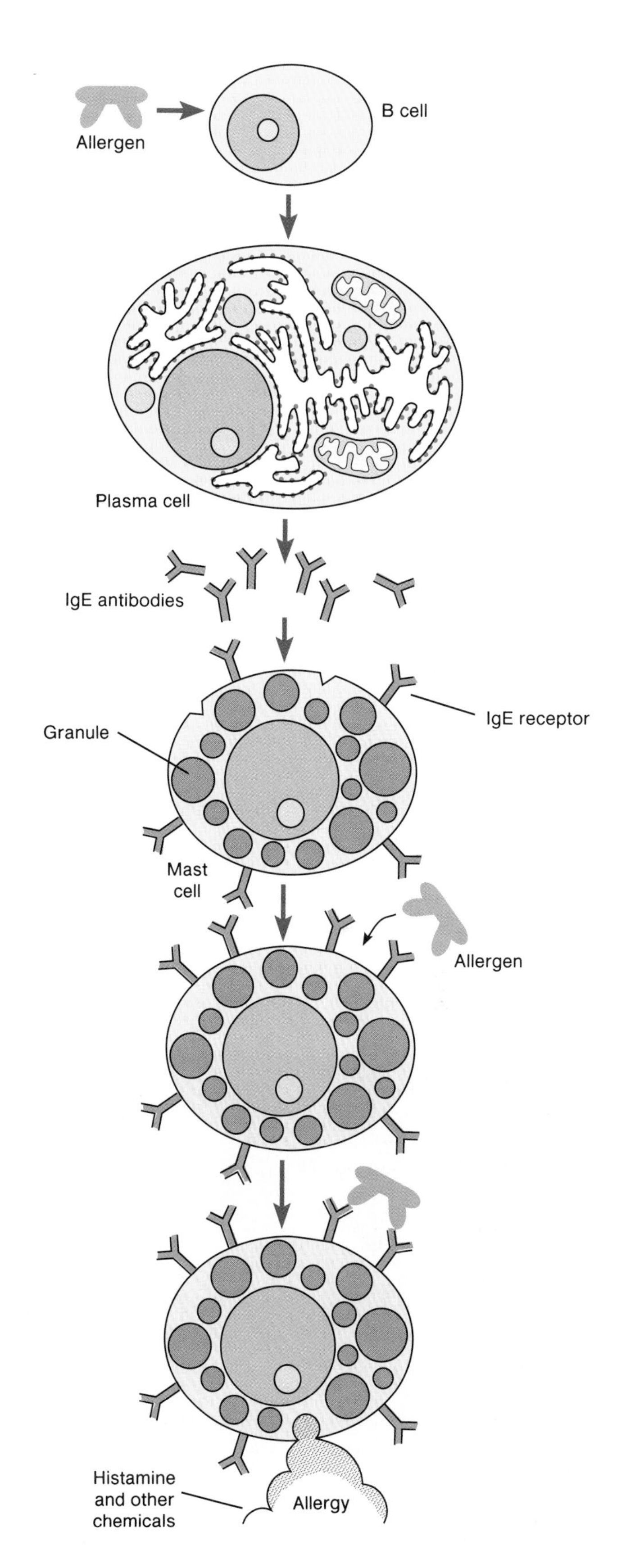

Figure 13.15. *Allergy is produced when antibodies of the IgE subclass attach to tissue mast cells. The combination of these antibodies with allergens cause the mast cell to secrete histamine and other chemicals that produce the symptoms of allergy.*

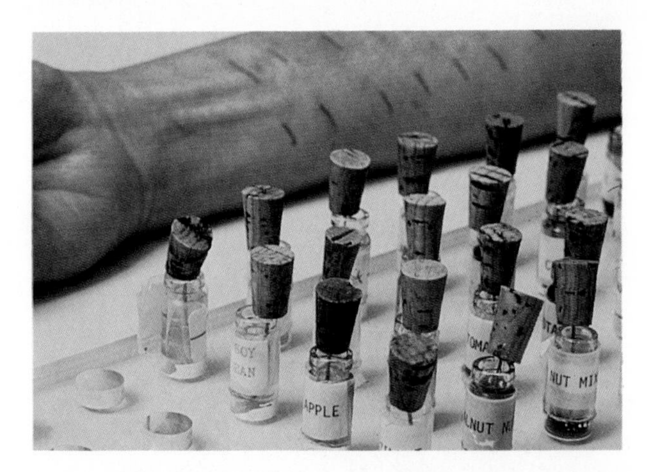

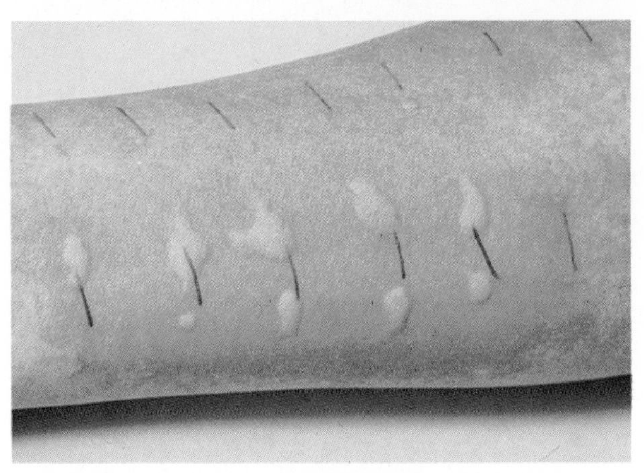

Figure 13.16. *Skin test for allergy. If an allergen is injected into the skin of a sensitive individual, a typical flare-and-wheal response occurs within several minutes.*

(From Scanning Electron Microscopy in Biology *by Kessel and Shih Springer-Verlag, Inc.)*

Immediate hypersensitivity to a particular antigen is commonly tested by injecting various antigens under the skin (fig. 13.16). Within a short time a *flare-and-wheal reaction* is produced if the person is allergic to that antigen. This reaction is due to the release of histamine and other chemical mediators: the flare is due to vasodilation, and the wheal results from local edema.

Allergens that provoke immediate hypersensitivity include various foods, bee stings, and pollen grains. The most common allergy of this type is seasonal hay fever, which may be provoked by ragweed (*Ambrosia*) pollen grains. People with chronic allergic rhinitis and asthma due to an allergy to dust or feathers are usually allergic to a tiny mite (fig. 13.17) that lives in dust and eats the scales of skin that are constantly shed from the body. Actually, most of the antigens from the dust mite are not in its body but rather in its feces, which are tiny particles that can enter the nasal mucosa much like pollen grains.

Delayed Hypersensitivity

As the name implies, a longer time is required for the development of symptoms in delayed hypersensitivity (hours to days) than in immediate hypersensitivity. This may be due to the fact that immediate hypersensitivity is mediated by antibodies, whereas delayed hypersensitivity is a cell-mediated T lymphocyte response. Since the symptoms are caused by the secretion of lymphokines, rather than by the secretion of histamine, treatment with antihistamines provides little benefit. At present, corticosteroids are the only drugs that can effectively treat delayed hypersensitivity.

One of the best-known examples of delayed hypersensitivity is **contact dermatitis,** caused by poison ivy, poison oak, and poison sumac. The skin tests for tuberculosis—the tine test and

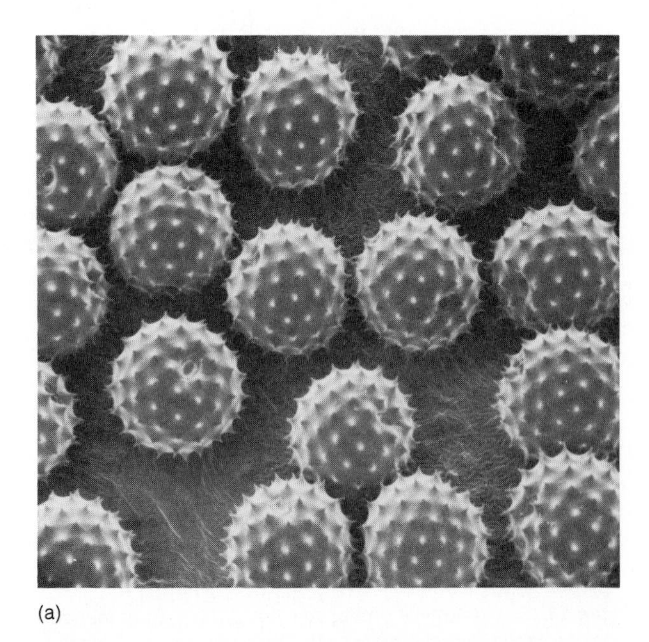

(a)

(b)

Figure 13.17. *(a) A scanning electron micrograph of ragweed (*Ambrosia*), which is responsible for hay fever. (b) A scanning electron micrograph of the house dust mite (*Dermatophagoides farinae*), which lives in dust and is often responsible for yearlong allergic rhinitis and asthma.*

([a] From Scanning Electron Microscopy in Biology *by R. G. Kessel and C. Y. Shih. © Springer-Verlag, 1976.)*

the Mantoux test—also rely on delayed hypersensitivity reactions. If a person has been exposed to the tubercle bacillus and has, as a result, developed T cell clones, skin reactions appear within a few days after the tubercle antigens are rubbed into the skin with small needles (tine test) or are injected under the skin (Mantoux test).

1. Explain the possible mechanisms responsible for autoimmune diseases.
2. Distinguish between immediate and delayed hypersensitivity.
3. Describe the sequence of events by which allergens can produce symptoms of runny nose, skin rash, and asthma.

Summary

Defense Mechanisms p. 292

I. Nonspecific defense mechanisms include barriers to penetration of the body and internal defenses.
   A. Phagocytic cells engulf invading pathogens.
   B. Interferons are polypeptides secreted by cells infected with viruses that help protect other cells from viral infections.
II. Specific immune responses are directed against antigens.
   A. Antigens are molecules or parts of molecules that are usually large, complex, and foreign.
   B. A given molecule can have a number of antigenic determinant sites that stimulate the production of different antibodies.
III. Specific immunity is a function of lymphocytes.
   A. B lymphocytes secrete antibodies and provide humoral immunity.
   B. T lymphocytes provide cell-mediated immunity.

Functions of B Lymphocytes p. 296

I. Antibodies are produced by B lymphocytes.
   A. Each antibody has two regions that combine with specific antigens.
   B. The combination of antigens with antibodies promotes phagocytosis.
II. Antigen-antibody complexes activate a system of proteins called the complement system.
   A. This results in complement fixation, where complement proteins attach to a cell membrane and promote the destruction of the cell.
   B. Free complement proteins promote phagocytosis and chemotaxis and stimulate the release of histamine from tissue mast cells.
III. Specific and nonspecific immune mechanisms cooperate in the development of a local inflammation.

Active and Passive Immunity p. 299

I. A primary response is produced when a person is first exposed to a pathogen; a subsequent exposure results in a secondary response.
   A. During the primary response antibodies are produced slowly, and the person is likely to get sick.
   B. During the secondary response antibodies are produced quickly, and the person has resistance to the pathogen.
   C. The secondary response is believed to be due to the development of lymphocyte clones in response to the first exposure.
II. Passive immunity is provided by antibodies made by a different organism.
   A. This occurs naturally from mother to fetus.
   B. Injections of antiserum provide passive immunity to some pathogenic organisms and toxins.
III. Monoclonal antibodies are made by hybridomas, which are formed artificially by the fusion of B lymphocytes with cancer cells.

Functions of T Lymphocytes p. 303

I. The thymus processes T lymphocytes and secretes hormones that are believed to be required for the function of T lymphocytes.
II. There are three subcategories of T lymphocytes.
   A. Killer T lymphocytes kill target cells by a mechanism that does not involve antibodies but that does require close contact between the killer T cell and the target cell.
   B. Killer T lymphocytes are responsible for transplant rejection and for the immunological defense against fungal and viral infections.
   C. Helper T lymphocytes stimulate, and suppressor T lymphocytes suppress, the function of B lymphocytes and killer T lymphocytes.
   D. The T lymphocytes secrete a family of compounds called lymphokines, which promote the action of lymphocytes and macrophages.
III. T lymphocytes cannot be activated by free antigens; the antigens must be presented to the T cells on the surface of macrophages.
IV. AIDS is caused by the human immunodeficiency virus (HIV).
   A. This is a retrovirus, which injects RNA into its host cell.
   B. Viral RNA directs the production of viral DNA; this requires the action of reverse transcriptase.
   C. HIV kills helper T lymphocytes, resulting in susceptibility to infections and certain forms of cancer.

Tolerance, Autoimmunity, and Allergy p. 304

I. Tolerance to self-antigens may be due to the destruction of lymphocytes that can recognize the self-antigens, or it may be due to suppression of the immune response to self-antigens.
II. Autoimmune diseases are caused by the production of autoantibodies against self-antigens.
III. There are two types of allergic responses, which are characterized as immediate hypersensitivity and delayed hypersensitivity.
   A. Immediate hypersensitivity results when an allergen provokes the production of antibodies in the IgE class, which attach to tissue mast cells and stimulate the release of histamine from the mast cells.
   B. Delayed hypersensitivity is a cell-mediated response of T lymphocytes, which secrete the lymphokines that produce the allergic symptoms.

Review Activities

Objective Questions

1. Which of the following offers a nonspecific defense against viral infection?
   (a) antibodies
   (b) leukotrienes
   (c) interferon
   (d) histamine

Match the cell type with its secretion.

2. killer T cells	(a) antibodies
3. mast cells	(b) lymphokines
4. plasma cells	(c) lysosomal enzymes
5. macrophages	(d) histamine

6. Mast cell secretion during an immediate hypersensitivity reaction is stimulated when antigens combine with
   (a) IgG antibodies
   (b) IgE antibodies
   (c) IgM antibodies
   (d) IgA antibodies

7. During a secondary immune response
   (a) antibodies are made quickly and in great amounts
   (b) antibody production lasts longer than in a primary response
   (c) lymphocyte clones are believed to develop
   (d) all of the above
8. Which of the following cells aids the activation of T lymphocytes by antigens?
   (a) macrophages
   (b) neutrophils
   (c) mast cells
   (d) natural killer cells
9. Which of the following statements about T lymphocytes is *false?*
   (a) Some T cells promote the activity of B cells.
   (b) Some T cells suppress the activity of B cells.
   (c) Some T cells secrete interferon.
   (d) Some T cells produce antibodies.
10. Delayed hypersensitivity is mediated by
   (a) T cells
   (b) B cells
   (c) plasma cells
   (d) natural killer cells
11. Active immunity may be produced by
   (a) having a disease
   (b) receiving a vaccine
   (c) receiving gamma globulin injections
   (d) both *a* and *b*
   (e) both *b* and c

Essay Questions

1. Explain how antibodies help destroy invading bacterial cells.
2. Explain how T lymphocytes interact with macrophages and the infected cells in fighting viral infections.
3. Explain the possible roles of helper and suppressor T lymphocytes in (a) defense against infections and (b) tolerance to self-antigens.
4. Describe the clonal selection theory, and use this theory to explain how active immunity is produced.
5. Distinguish between the causes of autoimmune diseases and allergy, and give examples of each disease category.

14

Cancer

Outline

Objectives

By studying this chapter, you should be able to
1. describe how cancerous cells differ from normal cells
2. distinguish between benign and malignant tumors
3. explain the nature and significance of metastasis
4. explain how chemical carcinogens and ionizing radiation may cause cancer
5. describe the role of diet in cancer
6. explain the cancer risks associated with cigarette smoking
7. explain the nature and significance of oncogenes
8. explain how oncogenes may transform normal cells to cancerous cells
9. explain the nature of proto-oncogenes
10. explain how the existence of anti-oncogenes was determined, and the possible actions of these genes
11. explain how the immune system might recognize tumor cells, and which parts of the immune system are most active in immunological surveillance against cancer
12. describe the relationship between aging and stress and the risk of cancer
13. describe the mechanism by which most antitumor drugs work
14. describe how lymphocytes and lymphokines are being used in the experimental treatment of cancer
15. explain how monoclonal antibodies are produced, and how they are used in the experimental treatment of cancer

Keys to Pronunciation

angiogenesis: *an''je-o-jen'e-sis*
carcinogen: *kar-sin'o-jen*
carcinogenic: *kar''si-no-jen'ik*
carcinoma: *kar''si-no'mah*
differentiation: *dif''er-en''she-a'shun*
interferon: *in''ter-fer'on*
ionizing: *i'on-izing*
leukemia: *loo-ke'me-ah*
lymphokine: *lim'fo-kin*
lymphoma: *lim-fo'mah*
metastasis: *me-tas'tah-sis*
mitogen: *mi'to-jen*
monoclonal: *mon''o-klon'al*
neoplasm: *ne'o-plazm*
oncogene: *ong'ko-jen*
oncogenic: *ong'ko-jen'ik*
oncology: *ong-kol'o-je*
retinoblastoma: *ret''i-no-blas-to'mah*
tumor: *too'mor*

Photo: Cancer cells grown in vitro *from a human breast tumor. These cells are proliferating in an uncontrolled fashion, just as they do* in vivo.

The Nature of Cancer

A cancer is an uncontrolled growth of cells that produces a tumor, or neoplasm. Cancerous cells lose their tissue specializations and do not respond to control mechanisms that normally limit cell division. Further, cancers can spread away from the primary tumor and establish secondary tumors in a process called metastasis.

Oncology[1] is the field of biomedicine devoted to the study and treatment of tumors. A *tumor,*[2] or **neoplasm** ("new growth"), is produced by an abnormal proliferation of cells through mitotic cell division. Tumors are described as *benign* when they are relatively slow growing and limited to a specific location; examples include warts, moles, and polyps. Tumors are *malignant*[3] when they grow rapidly and undergo **metastasis.**[4] Metastasis refers to the dispersion of tumor cells and the resulting seeding of new tumors in different locations. These tumors interfere with the normal function of the organs in which they are located, causing illness and often death. The term **cancer**[5]—coined by Hippocrates[6] because the tumors spread through the body like the movements of a crab—usually designates malignant tumors.

Normal cells within tissues are specialized in their structure and function. These specialized, or *differentiated,* cells developed from the undifferentiated cells of the early embryo, as described in chapter 9. When cells become cancerous, they lose much of their specializations (fig. 14.1), so that they cannot perform the functions for which they are responsible in the organ. Like the less differentiated embryonic cells, cancer cells undergo rapid rates of mitosis. When normal cells are grown in a laboratory dish, for example, they will divide until they spread out in the bottom of the dish, but will stop dividing at that point. The inhibition of cell division in this case is known as *contact inhibition,* because it is induced by the contact of normal cells with each other. Also, the division of normal cells is inhibited by chemical factors released within the tissue. Such inhibition is absent in a cancer. When cancerous cells are grown in a dish they continue to grow indefinitely, forming multiple layers of cells.

[1]oncology: Gk. *onkos,* mass, tumor; *log,* study of
[2]tumor: L. *tumere,* to swell
[3]malignant: L. *malignans,* acting maliciously
[4]metastasis: Gk. *meta,* beyond; *stastis,* stand
[5]cancer: L. *cancer,* a crab
[6]Hippocrates: after Hippocrates of Cos, late fifth-century B.C. Greek physician, who is regarded as the father of medicine

The loss of specialization when cells become cancerous sometimes results in the production of antigens (chapter 13) that are identical to those produced by embryonic cells. The genes that produced these antigens in the embryo or fetus were inactivated when the cells became differentiated, and are abnormally activated again when the cells become cancerous. This provides the basis for some laboratory tests for cancer. **Carcinoembryonic antigen (CEA),** for example, is a type of molecule normally produced by the fetal gut, pancreas, and liver, but not by adult organs. A positive CEA test may indicate cancer of the colon (large intestine). Similarly, the liver of a fetus, but not of an adult, normally produces **alpha fetoprotein (AFP).** Cancer of the liver may be indicated by a positive AFP test in an adult.

Most tumors are believed to arise from the uncontrolled proliferation of a single cell. They are thus clones, analogous to the clones of lymphocytes normally produced during active immunity (chapter 13). Unlike the normal lymphocyte clones, which are subject to extensive controls, the proliferation of tumor cells is uncontrolled. As the number of cells in the tumor increases, the likelihood of a mutation occurring during mitosis becomes greater. Such a mutation is called a *somatic mutation* (fig. 14.2), because it occurs in a body cell rather than a gamete (and thus it cannot be passed to the next generation). Through

(a)

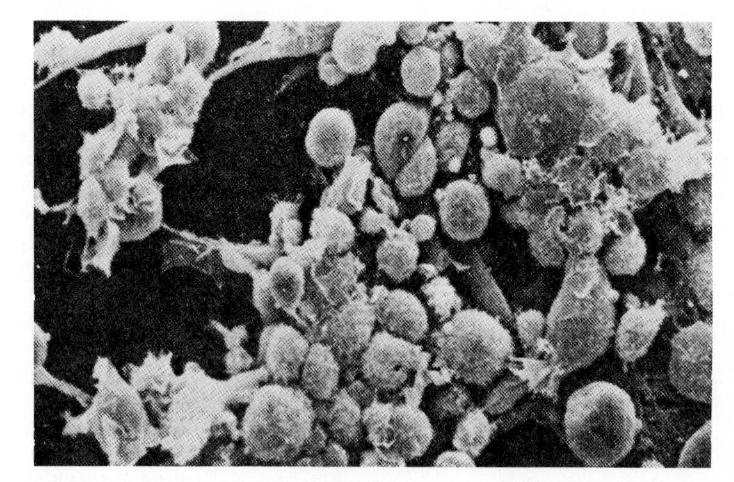

(b)

Figure 14.1. Appearance of normal fibroblasts (a) on a glass plate compared to fibroblasts that have become transformed (by means of a virus) into cancerous cells (b).

such mutations, the cells of a tumor usually display some genetic differences even though they are ultimately derived from one cell. These genetic differences within a tumor add difficulty to the development of cancer therapies.

Types of Cancers

There are four major types of cancers, which are classified as follows:

1. **Carcinoma.**[7] These are tumors derived from epithelial tissues, and include cancers of the skin, glands, breasts, and most internal organs. Carcinomas comprise about 80%–90% of all cancers.
2. **Sarcoma.**[8] These are tumors of the connective tissues, such as bones and adipose tissue, and of muscles.
3. **Leukemia.** Leukemias involve uncontrolled proliferation of white blood cells. These types of cancers are described in more detail later.
4. **Lymphoma.** Lymphomas are cancers of the lymphoid organs (chapter 12), particularly of the spleen and lymph nodes. The most common type of lymphoma is called *Hodgkin's lymphoma.*

[7]carcinoma: Gk. *karkinos,* crab, cancer
[8]sarcoma: Gk. *sarko,* flesh; *oma,* tumor

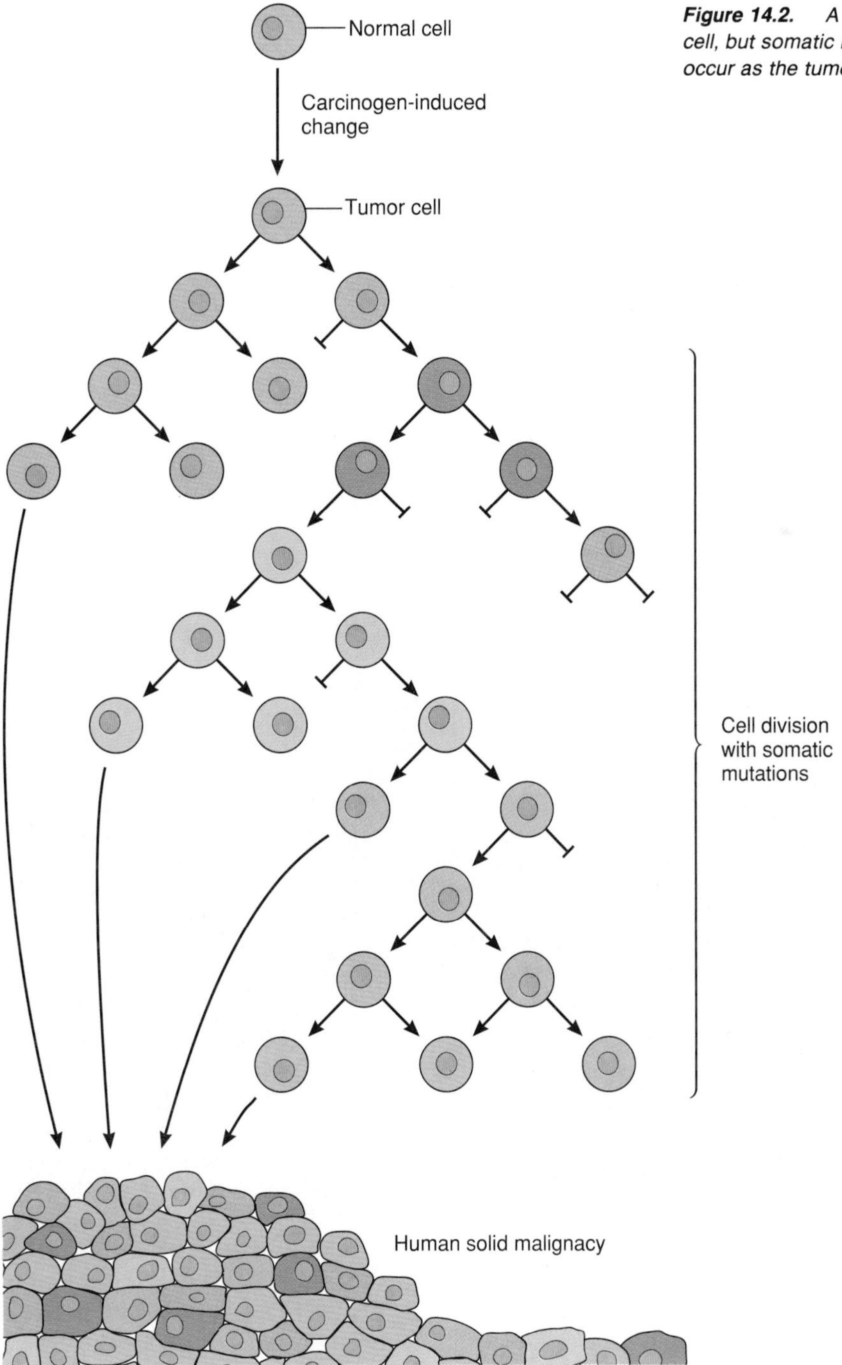

Figure 14.2. *A tumor develops from one cell, but somatic mutations during cell division occur as the tumor grows.*

Table 14.1 The most common forms of cancer

	Conventional Treatment	Chances of Surviving	Experimental Treatments*
Lung			
Small Cell 38,000 new cases 28,000 deaths	Chemotherapy.	If diagnosed at early stage, patient has a 33% chance of surviving 10–16 months. Overall, a patient has a 5%–10% chance of surviving 2 years.	Drugs or antibodies to suppress tumor's growth. Chemotherapy plus radiation followed by surgery.
Other Lung 114,000 new cases 111,000 deaths	Surgery and radiation.	If tumor can be removed by surgery, patient has a 40%–80% chance of surviving 5 years. If not, a 2%–30% chance.	Chemotherapy.
Colorectal			
Rectum 42,000 new cases 8,000 deaths	Surgery, sometimes followed by radiation.	If diagnosed early, patient has a 75%–100% chance of surviving 5 years. At later stages, a 30%–75% chance.	Chemotherapy. Biological therapy using interferon, interleukin-2, and/or monoclonal antibodies.
Colon 105,000 new cases 53,500 deaths	Surgery.	If diagnosed early, patient has a 75%–100% chance of surviving 5 years. At later stages, a 10%–75% chance.	Chemotherapy. Biological therapy using interferon, interleukin-2 and/or monoclonal antibodies.
Breast 136,000 new cases 42,300 deaths	In early stages, surgery sometimes with radiation. In later stages, surgery with radiation plus chemotherapy or hormonal therapy.	If diagnosed early, patient has at least an 85% chance of surviving 5 years. At later stages a 10%–66% chance.	Chemotherapy or hormonal therapy used at early stages. High-dose chemotherapy.
Prostate 99,000 new cases 28,000 deaths	In early stages, surgery or radiation. In later stages, surgery and/or hormonal therapy to block activity of male hormones that stimulate cancer's growth.	If diagnosed early, patient has a 65%–77% chance of surviving 5 years. At later stages, a 21%–48% chance. Since half the men diagnosed with prostate cancer are 70 or older, many of those who get prostate cancer will die of other diseases.	New drugs and synthetic hormones to block the activity of male hormones. Radioactive pellets implanted in prostate. Chemotherapy using experimental drugs.
Uterus			
Cervix† 12,900 new cases 7,000 deaths	In early stages, surgery. In later stages, radiation.	If diagnosed early, patient has an 80%–90% chance of surviving 5 years. At later stages, a 15%–80% chance.	Chemotherapy using experimental drugs. Radiation.
Endometrium 34,000 new cases 3,000 deaths	In early stages, hysterectomy. In later stages, hysterectomy, radiation, and sometimes hormonal therapy.	If diagnosed early, patient has an 80%—100% chance of surviving 5 years. At later stages a 5%–60% chance.	Chemotherapy and hormonal therapy using experimental drugs.
Bladder 46,400 new cases 10,400 deaths	In early stages, surgery or radiation. In later stages, chemotherapy and/or radiation.	If diagnosed early, patient has a 75%–90% chance of surviving 5 years. At later stages, a 2%–55% chance.	High-dose chemotherapy before surgery. Standard-dose chemotherapy injected directly into the bladder. Bacteria injected in bladder to stimulate immune response.
Non-Hodgkin's Lymphoma 31,700 new cases 16,500 deaths	Chemotherapy	Overall patient has a 50%–60% chance of surviving 5 years. Most patients have advanced disease when they are first seen.	Chemotherapy and local radiation. Total body irradiation. High-dose chemotherapy. Monoclonal antibodies.
Pancreas 27,400 new cases 24,500 deaths	Radiation and/or chemotherapy.	Most patients die within 1 year. Fewer than 2% survive for 5 years.	High-dose chemotherapy. Radiation. Surgery.
Melanoma (Skin)‡ 27,300 new cases 5,800 deaths	Surgery.	If diagnosed early, a patient has a 70%–100% chance of surviving 5 years. At later stages, a 10%–65% chance.	Surgery plus interferon, interleukin-2, and/or killed melanoma cells to stimulate immune response.

*Treatments listed are given both individually and in various combinations—and usually after conventional therapy.
†Not included in this figure are the more than 50,000 yearly cases of cervical dysplasia, a precancerous condition usually treated by local surgery or cryotherapy.
‡Non-melanoma skin cancers are much more common—more than 500,000 new cases each year—but much less serious: 90%–95% of patients are cured by surgery.
Research by Mary James. Reprinted from In Health.

All together, there are over 100 different cancers. The most common of these are listed in table 14.1. Approximately one in four Americans will have cancer, and about 80% of these will die from it. Cancer is the second leading cause of death in the United States (after heart disease), resulting in 400,000 deaths annually. And the incidence of cancer has been steadily in-

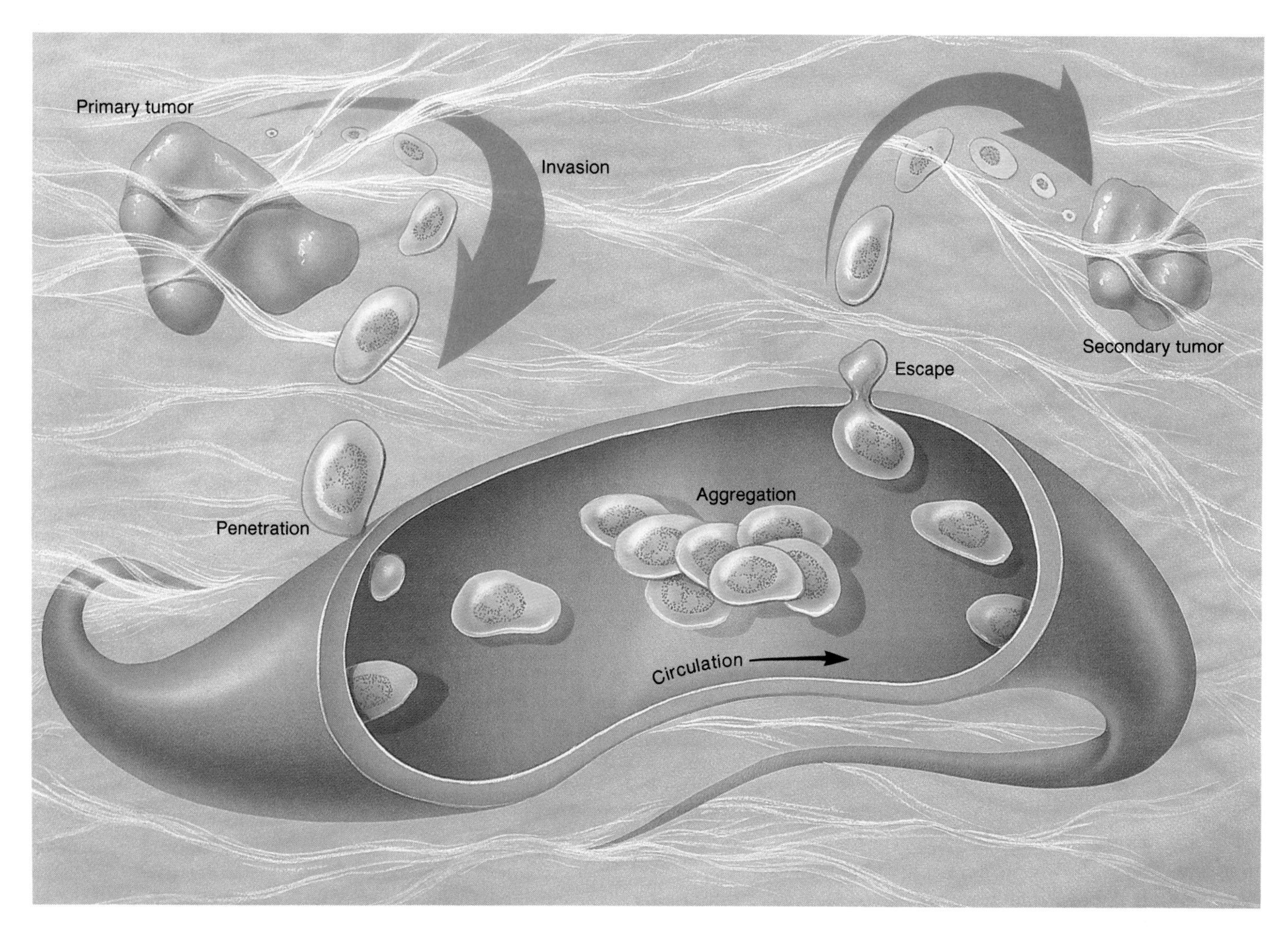

Figure 14.3. The steps involved in the metastasis of a primary tumor and the establishment of a secondary tumor.

creasing: 157 people per 100,000 died of cancer in 1950, but this number increased to 162 in 1975 and 171 in 1985. Over 120,000 women are diagnosed with breast cancer each year; statistically, the chances of a woman in the United States developing breast cancer are about 1 in 14. The prognosis for some forms of cancer, however, has improved substantially with newer forms of treatment. These will be discussed in a later section.

Physicians often describe the course of a particular form of cancer as occurring in stages. The staging of cancer allows a more accurate prognosis to be made and determines the form of treatment. In general, the stages are as follows:

Stage 1. Cancer can be diagnosed in a pathology microscope slide. The tumor is localized to one region and can be surgically removed.

Stage 2. Metastasis has occurred to lymph nodes located near the primary tumor.

Stage 3. Metastasis has occurred and secondary tumors have formed in organs located near the site of the primary tumor.

Stage 4. Secondary tumors have spread throughout the body.

Metastasis

The hallmark of a cancer, as opposed to a benign tumor, is its ability to metastasize to other sites. This is apparently a very difficult thing to do. A cancerous cell must be able to break away from the mass of cells of which it is a part, enter a blood or lymphatic vessel, travel through this vessel, and establish itself at a new site. There is evidence that only some cells in a tumor have the ability to metastasize. Because of the genetic diversity within a tumor, some cells will, however, have this ability. In this way, secondary tumors are produced.

In order for a primary tumor to metastasize, it must be able to digest the collagen protein fibers that form the basement membrane under epithelia (chapter 5). In this way, the tumor can invade the underlying connective tissue where lymphatic and blood vessels are located (fig. 14.3). Cells from this tumor may then spread to another site. In order for a tumor to become established, blood vessels must supply the growing mass of cells. Growth of new blood vessels is termed *angiogenesis,*[9] and may

[9]angiogenesis: Gk. *angeion,* vessel; *genesis,* production

occur as a result of stimulation by particular molecules released from the cancerous cells. Finally, the cancerous cells may respond to various growth factors, which stimulate cell division. One of these growth factors is known to be secreted by blood platelets.

Leukemia

Leukemia is not a "typical" cancer, in that the primary neoplasm is not a solid tumor. Rather, this form of cancer is characterized by an uncontrolled proliferation of white blood cells (leukocytes) by either the bone marrow or lymphoid organs. Although leukemia is statistically uncommon (fifty out of one million Americans), it is a very important disease that can serve as an example of cancer.

White blood cells are produced in the bone marrow and in the lymphoid organs (spleen, lymph nodes, and thymus). In the process of leukocyte production, a parent cell that is less specialized undergoes cell division in these organs. The daughter cells then undergo division and differentiation to produce the mature white blood cells. Normally, only the mature leukocytes are released into the blood. In a person with leukemia, however, immature forms of leukocytes are characteristically found in the circulating blood.

There are two forms of leukemia: chronic and acute. *Chronic leukemia* usually affects middle-aged and older people and has a more optimistic prognosis. *Acute leukemia* is usually a disease of children and young adults; survival from the time of diagnosis is only a matter of weeks to months for some leukemias. Sadly, leukemia is the most common malignancy of childhood.

In acute leukemia, immature leukocytes continue to multiply wherever they are found—in the bone marrow, blood vessels, and in organs that they invade. The continued cell division of leukocytes that have invaded secondary sites destroys the normal tissue of these organs. In chronic leukemia, the cells of the primary tumors continue to divide, but the cells that they produce do mature and cease cell division. These cells, however, may also invade various organs and cause damage.

Investigations with cats showed that leukemia in these animals can be produced by a virus, specifically a retrovirus (chapter 10). More recently, scientists have provided evidence of a viral cause of human leukemia as well. The mechanisms by which viruses might cause cancer will be described in a later section.

---

1. Define the term *cancer,* and distinguish between benign and malignant tumors.
2. Explain how cancerous cells differ from normal cells.
3. Define the term *metastasis,* and describe the events that occur during this process.
4. Describe the nature of leukemia, and distinguish between acute and chronic forms of this disease.

Causes of Cancer

Most cancers are caused by environmental factors, including chemical carcinogens, ionizing radiation, and viruses. The development of cancer proceeds in two major stages, initiation and promotion, which have different characteristics. The avoidance of smoking, and the choice of a proper diet, are the major preventative measures that can be taken to reduce the risk of cancer.

Based on statistical analysis of the incidence of cancer in different groups of people, authorities have estimated that approximately 70%–90% of all cancers are caused by environmental factors. Such environmental factors include *chemical carcinogens, ionizing radiation,* and *viruses.* There is also evidence that some people have a genetically caused predisposition to developing particular types of cancers. The development of cancer thus depends on both environmental and genetic influences.

Environmental Causes of Cancer

The first observation that chemicals can act as **carcinogens,** or cancer-causing agents, was made by an English physician named Percivall Pott in 1775, who noted that chimney sweeps exposed to soot had a tendency to develop cancer of the scrotum. In 1915, Japanese pathologists observed that skin cancer can be produced by putting coal tar on the ears of rabbits. Subsequent experiments showed that other molecules could produce skin cancer when rubbed on animal skins. In 1935, two Japanese pathologists noted that ingestion of particular dyes could cause liver cancer in experimental rabbits. These observations provided the framework for subsequent studies of *carcinogenesis,* or cancer development, in response to particular chemicals. The number of chemicals that are known to be carcinogens is now quite long, and growing longer as new chemicals are tested for carcinogenesis.

Chemical Carcinogens and Ionizing Radiation

Although rabbit ears were the first method of testing the carcinogenesis of particular chemicals, we now have a more humane testing method. This is the *Ames test,* developed by Bruce Ames in 1979. It has been established that chemical carcinogens produce cancer by inducing mutations during cell division. Since bacteria undergo cell division at a more rapid rate than mammalian cells, the Ames test assesses the mutagenic potential (ability to cause mutations) of potential carcinogens in bacterial systems.

Ionizing radiation can cause changes in the structure of DNA, and thus, like chemicals, can cause mutations. *Ionizing radiation* is energy produced by decaying radioactive materials and by X-rays. This energy is known as ionizing radiation because it can knock electrons out of their orbitals in an atom, thus

Cancer Prevention versus Cure

Cancer is probably the most feared of all diseases. Because of this, many people react to it irrationally. People with cancer are sometimes shunned, as those with leprosy were in past ages. This is a manifestation of denial, of wishing to distance oneself from the disease. Such an attitude is not only destructive to the cancer patient, but—considering the probability of developing cancer—it is potentially self-defeating. People with some forms of cancer can recover from the disease, especially with early diagnosis. This recovery is aided by a positive mental attitude, for reasons that relate to the immune system (as will be discussed later). People with cancer can thus benefit from continued socialization with their friends and relatives.

The fact that most cancers are environmentally caused means that they are potentially preventable. By refraining from smoking, eliminating alcohol or drinking in only moderate amounts, and eating correctly, one can minimize the risk of cancer. One cannot, however, reduce this risk to zero. Exposure to carcinogens is an unavoidable fact of life in developed countries, and some people might be particularly sensitive to carcinogens because of a genetic predisposition. The risk of contracting cancer is analogous to the risk of dying in an automobile accident. You can minimize this risk by driving defensively and wearing seat belts, but you may still get into an accident due to factors beyond your control. So it is with cancer.

There are two opposing attitudes regarding cancer that are both irrational. One is summarized by the statement "everything causes cancer, so I'm not going to worry about it." This statement is untrue: only particular things cause cancer. Most of these can be avoided, and by doing this the chances of getting cancer can be reduced. The other irrational attitude may be described by this statement: "If person *X* got cancer, he must have done something wrong." This statement is based on the understandable wish that one can avoid cancer by positive action—eating right, for example. This, however, is a most perfidious attitude, because it places the blame for the disease on its sufferer. It should always be remembered that one can do everything right and still get cancer.

So, what should one do? The most rational course of action is to avoid those things that are known to cause cancer and to support biomedical research into the causes of, and possible cures for, cancer. This research involves the use of animals in experiments; a cure for cancer can not be found without animal experimentation. As will be made clear in subsequent sections of this chapter, animal research has already yielded many new insights into the basic biology of cancer and cell control mechanisms involved in cancer. It is from such research that cures will eventually be found.

producing an ion and free electrons. The fact that ionizing radiation can cause cancer was dramatically shown by the increased incidence of cancer in survivors of Hiroshima and Nagasaki in Japan after World War II, and by the increased incidence of cancer in people living in southern Utah during the 1950s as a result of aboveground atom bomb testing.

Various studies have shown that there are two stages in cancer development. The first stage is known as **initiation.** This stage is believed to be due to one or more mutations, and was found to be irreversible. This is followed by a *latent period,* during which no changes are apparent. The second stage is known as **promotion**; little is known about the events that occur during this stage, but unlike the initiation stage, promotion is reversible. *Complete carcinogens* are both initiators and promoters; they can, by themselves, cause cancer. This includes most chemical carcinogens, ionizing radiation, and tumor-producing viruses. There are, however, some *incomplete carcinogens* which function only as initiators. Additional agents that act as promoters are thus required for cancer to develop in response to incomplete carcinogens.

Viruses and Cancer

The fact that viruses might cause cancer was first noted in mice, where a type of leukemia was shown to be virally induced. In 1964, investigators demonstrated that leukemia in cats was caused by a virus. A number of human cancers are known to be associated with infection by specific viruses. Leukemias and lymphomas, for example, are now believed to be caused, at least in part, by RNA-containing viruses (retroviruses). One of the more interesting associations is that of *Epstein-Barr virus (EBV)* with *Burkitt's lymphoma.* The EBV is a DNA-containing virus found worldwide, and is the cause of a common benign condition called infectious mononucleosis. In equatorial Africa, however, EBV has been found to cause Burkitt's lymphoma in children. Nobody knows why EBV causes this form of cancer in Africa but not in the rest of the world. This illustrates that particular environmental agents may produce cancer only when other predisposing conditions are present.

Diet and Cancer

Studies of the rates of cancer in different countries suggest that different aspects of the diet may have either preventative or causative roles in cancer. For example, the rate of breast and colon cancer in the underdeveloped countries is less than one-fifth that in the United States. Similarly, a high degree of correlation has been found between per capita fat intake and the national mortality from breast and colon cancer. Studies of immigrant populations have consistently shown that the immigrants take on the cancer risks of their adopted country in the first or second generation. This indicates that enviromental factors, particularly diet, determine which forms of cancer will be most prevalent in a particular society.

These statistics, however, are only suggestive, since many differences exist besides diet in the environment of different countries. Nevertheless, other types of scientific studies suggest that as much as 35% of the risk of developing cancer is due to diet. Improper diet is thus second only to cigarette smoking as the most significant and easily avoidable risk factor in cancer.

Ingestion of alcohol and coffee has been suggested by some studies to increase the risk of particular forms of cancer, although this is controversial. Also, the method of food preparation may be related to cancer risk. Certain preservatives, for example, may be carcinogenic. The process of charbroiling meats is known to produce carcinogens from the products of fat combustion. It is thus prudent to minimize consumption of alcohol, coffee, preserved meats, and barbecued steaks. (For a variety of reasons, the fat should be trimmed from all meats before cooking.)

Some of the specific dietary factors that may have either preventative or causative roles in cancer are as follows:

1. **Vitamin A and Carotene.** A study of Norwegian men showed that those whose dietary intake of vitamin A was above average had less than half the rate of lung cancer as those whose intake of vitamin A was below average. Similar studies in other countries have confirmed that dietary intake of vitamin A is inversely related to cancer risk, suggesting a preventative role for this vitamin. This effect may be due not only to vitamin A, but also to its parent molecule, called beta-carotene, which is produced by plants.
2. **Selenium.** Selenium is an essential trace element in the diet, functioning as a cofactor for a particular enzyme that helps to protect cells against oxidative damage. Further, selenium has been shown to decrease the ability of carcinogens to cause mutations in the Ames test. Statistical studies have shown an inverse relationship between the concentration of selenium in the soil and the incidence of cancer in different regions. This association suggests a preventative effect, but such an effect has not yet been proven.
3. **Dietary Fiber.** The term *fiber* refers to undigestible material in food, particularly the cellulose present in plants. Good sources for dietary fiber are thus fruits, vegetables, and whole grains. The association of fiber with cancer has been shown by statistical studies. The incidence of colon cancer, for example, is up to eight times higher in the United States than in the underdeveloped countries where fiber intake is higher. Further studies have confirmed that ingestion of fiber can reduce the risk of colon cancer.
4. **Dietary Fat and Cholesterol.** There is much evidence linking a high-fat diet with an increased risk of colon and breast cancer. In a Greek study, the intake of meat (the most important source of fat) was associated with an increased risk of colorectal cancer. In Japan and other Far Eastern countries, where the fat intake is lower than in the United States, for example, the incidence of breast cancer is one-fifth that of the U.S. This is not the result of genetic differences, because the offspring of Japanese immigrants to the United States have the same rates of breast cancer as the general population.
5. **Excess Calories.** When more energy, measured in Calories, is ingested than is used by the body, the excess energy is stored primarily as fat. This is true regardless of the food source (carbohydrates, fat, or protein). If this positive energy balance is maintained over a period of time, obesity will result. Obesity is statistically associated with increased risk of endometrial cancer and of cancer of the gallbladder.
6. **Vitamin E.** Vitamin E functions as an antioxidant in the body, and has been shown to decrease the rate of mutations. It therefore makes sense that this vitamin may help to prevent cancer, but the statistical evidence for such an effect is not conclusive.
7. **Vitamin C.** Although some studies have suggested that vitamin C may have a preventative effect on cancer, most studies have failed to confirm this effect.

From this information, certain recommendations can be made in regards to diet. One should eat more fruits, vegetables, and whole grains and less meat. Considering the association between a high-fat diet, blood cholesterol, and atherosclerosis (chapter 12), such a diet would help to reduce the risk of both the number one and number two killers: heart disease and cancer.

Smoking and Cancer

The 1964 Surgeon General's Report identified cigarette smoking as the single most important cause of preventable mortality. Since that time, the evidence has been accumulating to support this statement, and to confirm the relationship between smoking and cancer. It has been estimated that approximately 30% of all cancer deaths can be attributed to smoking. Smoking has been found to be the major cause of cancer of the lungs, larynx, oral cavity, and esophagus. Smoking also contributes to the risk of cancers of the bladder, pancreas, and kidneys. There are some studies that have also suggested an increased risk of stomach and uterine cervical cancer in smokers. The statistics are very clear: about 96% of people who die of lung cancers were smokers. About 30% of female bladder cancers and 40% of male bladder cancers can be attributed to cigarette smoking alone. The risk

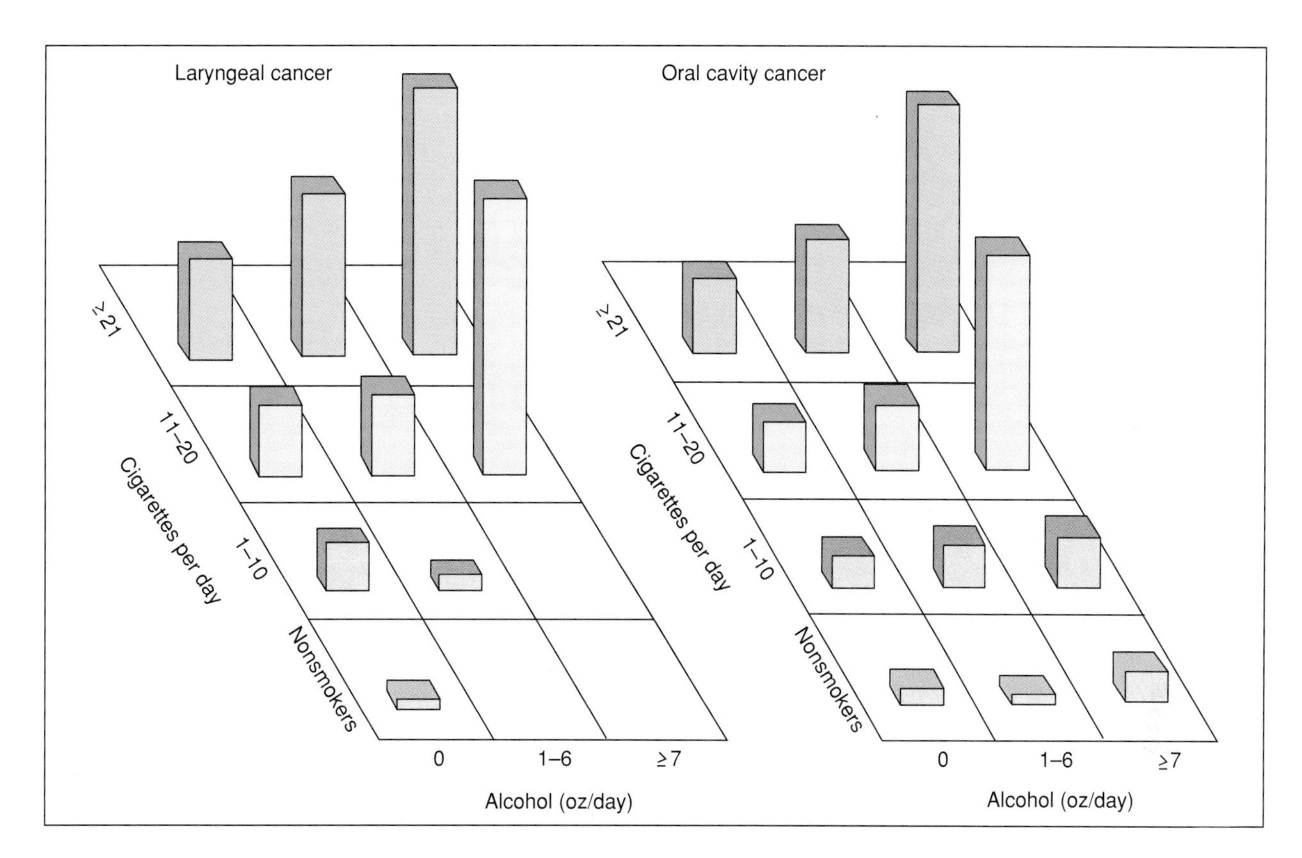

Figure 14.4. *Relative risk of developing laryngeal and oral cavity cancer as a function of cigarettes smoked and ounces of alcohol ingested per day. Notice that cigarettes and alcohol have additive effects on the cancer risk.*

of lung cancer mortality in a person who smokes more than a pack a day is fifteen to twenty-five times greater than in nonsmokers!

Lung cancer has been the leading cause of death in men since the early 1950s, following the increased use of cigarettes by men after World War I. Death from lung cancer did not start to rise in women until the early 1960s, because the prevalence of smoking among women was substantially less than among men until after World War II. The rate of increase in lung cancer mortality among women, however, has grown so high that it may soon become greater than the mortality from breast cancer. The implications from the different statistical studies are supported by analysis of cigarette smoke. Smoke from cigarettes has been found to contain over two dozen carcinogens which act as initiators of cancer.

Drinking alcohol appears to act synergistically with cigarettes to increase the risk of laryngeal and oral cancers (fig. 14.4). In one study, for example, the relative risk of oral cancer in a nonsmoker who drank at least 7 oz of alcohol per day was 2.5. The risk in someone who drank the same amount but also smoked one to ten cigarettes per day, in comparison, was 5.1. This risk increased to 20.5 in those who smoked eleven to twenty cigarettes per day, and further increased to 24 in those who smoked more than twenty cigarettes daily.

In 1986, the Surgeon General's Report stated that the involuntary inhalation of cigarette smoke in nonsmokers (known as *passive smoking*) can cause lung cancer. Although the risks from passive smoking are less than for active smoking, greater numbers of people are affected. It has recently been estimated that about 4,700 nonsmokers die annually of lung cancer due to passive smoking. Such studies are supported by analysis of "sidestream" smoke, which escapes into a room (versus "mainstream" smoke, which is inhaled). The sidestream smoke contains a lower concentration of carcinogens than does mainstream smoke, but the effects can accumulate over time.

---

1. Explain the nature and significance of the Ames test.
2. Distinguish between the initiation and promotion stages of cancer development.
3. Identify the probable cause of Burkitt's lymphoma. What might the geographic distribution of this form of cancer indicate?
4. Identify the dietary factors that are believed to help prevent cancer. Also, identify the dietary factors that may contribute to the risk of cancer.
5. In which forms of cancer does smoking comprise the major risk factor? In which forms of cancer does smoking comprise a less, but still significant, risk factor? What other health problems, besides cancer, are associated with smoking?

Smoking and Public Policy

Considering the great health hazards of smoking, to both the smoker and others in the same room, a nonsmoker may well wonder why anyone smokes. Many people began to smoke before the health hazards were fully appreciated and later found it difficult to quit. Tobacco is an addictive drug, as should be evident by the commercial success of various enterprises devoted to helping smokers quit their habit. But why should young people today start smoking? Some people believe that they are encouraged to do so by advertising, and thus advocate the elimination of cigarette advertisements. Tobacco, they argue, is a dangerous, addictive drug much like marijuana or cocaine, and should be legally treated as such. Others, however, feel that a ban on tobacco advertisements would be an infringement of basic rights. The tobacco companies, of course, have a great financial interest in this argument. Since many smokers are giving up their habit every year, these companies must recruit new smokers to maintain their present level of income.

In view of the documented health hazards of passive smoking, there have been increasing efforts to ban smoking in offices and public places since the middle 1980s. The federal government has banned smoking on all domestic airline flights . Some cities have banned smoking in restaurants. Forty-two states have passed legislation banning smoking in public transportation, hospitals, elevators, indoor recreational facilities, schools, and libraries. Many smokers have argued that such restrictions are violations of their personal freedom. The attitude of those in favor of smoking restrictions, on the other hand, may best be summarized by the old saying: "Your right to swing your arm stops at my nose."

It should be noted that cigarette smoking is also associated with health hazards other than cancer. These include (1) chronic bronchitis and emphysema (deaths from these conditions are twenty times more frequent in people who smoke heavily); (2) cardiovascular disease (smoking is a risk factor in atherosclerosis); (3) pregnancy (average birth weight of infants from mothers who smoke is 6 oz less than from nonsmoking mothers); and (4) peptic ulcers. In view of all of these health hazards, it is certain that legislators will attempt to impose increasing restrictions on tobacco use and advertisements, while competing interests will mount aggressive campaigns to oppose these restrictions.

Oncogenes and Anti-Oncogenes

Oncogenes are cancer-causing genes carried by viruses. Normal cells have similar genes, known as proto-oncogenes, which are believed to have regulatory functions in normal cells. These proto-oncogenes, however, may mutate to cancer-causing forms. Normal cells are also believed to have tumor-suppressing genes, or anti-oncogenes.

The association of a virus with a specific tumor was first observed by a scientist named Peyton Rous in 1911. This virus causes a sarcoma in chickens, and so is known as the Rous sarcoma virus. In later years, a mouse leukemia virus and a feline leukemia virus were discovered. While there are a number of human tumors associated with specific DNA-containing viruses, it is not known if they are the direct causes of the tumors. Rather, they may contribute in part to the development of the cancers. Retroviruses (which contain RNA), however, may directly cause particular cancers. In humans, this has been shown to be true for adult T cell leukemia, caused by the *human T cell lymphotropic virus (HTLV-1)*. This virus is a close relative of the human immunodeficiency virus (HIV) which causes AIDS, as discussed in chapter 13.

The ability of specific viruses to induce cancers has been shown to be due to specific genes, called **oncogenes** (from the Greek word *onkos,* for cancer). These oncogenes are designated with three letters, often preceded by the letter *v* (for virus). The oncogene in the Rous sarcoma virus, for example, is *v-src.* In order for retroviruses to transform their host cells, their RNA must be used to produce complementary DNA (cDNA) by the enzyme reverse transcriptase (chapter 10). This cDNA, which contains the oncogene, is next incorporated into the host's DNA (fig. 14.5). When the cDNA is transcribed, messenger RNA is produced which codes for the protein product of the oncogene. It is this oncogene protein that causes the transformation of a normal cell into a cancerous one.

Proto-Oncogenes

Over thirty different viral oncogenes are now known. In addition to *v-src,* there is *v-myc* (after avian myelocytomatosis), *v-ras* (for rat sarcoma), and many others. The description of different oncogenes was soon followed by an amazing discovery. Oncogenes are not viral genes after all; they don't code for the structure of the virus. Rather, oncogenes are copies of genes from host cells that were picked up by the ancestors of the virus. Further experiments have shown that normal human DNA contains regions that appear to be the same as the viral oncogenes! In tribute to this discovery, the Nobel prize was awarded to Drs. J. Michael Bishop and Harold E. Varmus in 1989.

The normal human cellular genes that are the same as the viral oncogenes isolated from the tumors of lower animals are called **proto-oncogenes.** Proto-oncogenes are distinguished from their viral counterparts by the letter *c* (for cellular). For example, there is a *c-src,* a *c-myc,* and a *c-ras.* The functions of these genes in the normal cell are not well understood. They are, however, believed to be involved in the regulation of cell division. Three proto-oncogenes, for example, appear to code for known *mitogens* (chemicals that stimulate mitosis).

According to the oncogene theory, cancer may be produced in two different ways: (1) a cell may be infected by a virus that

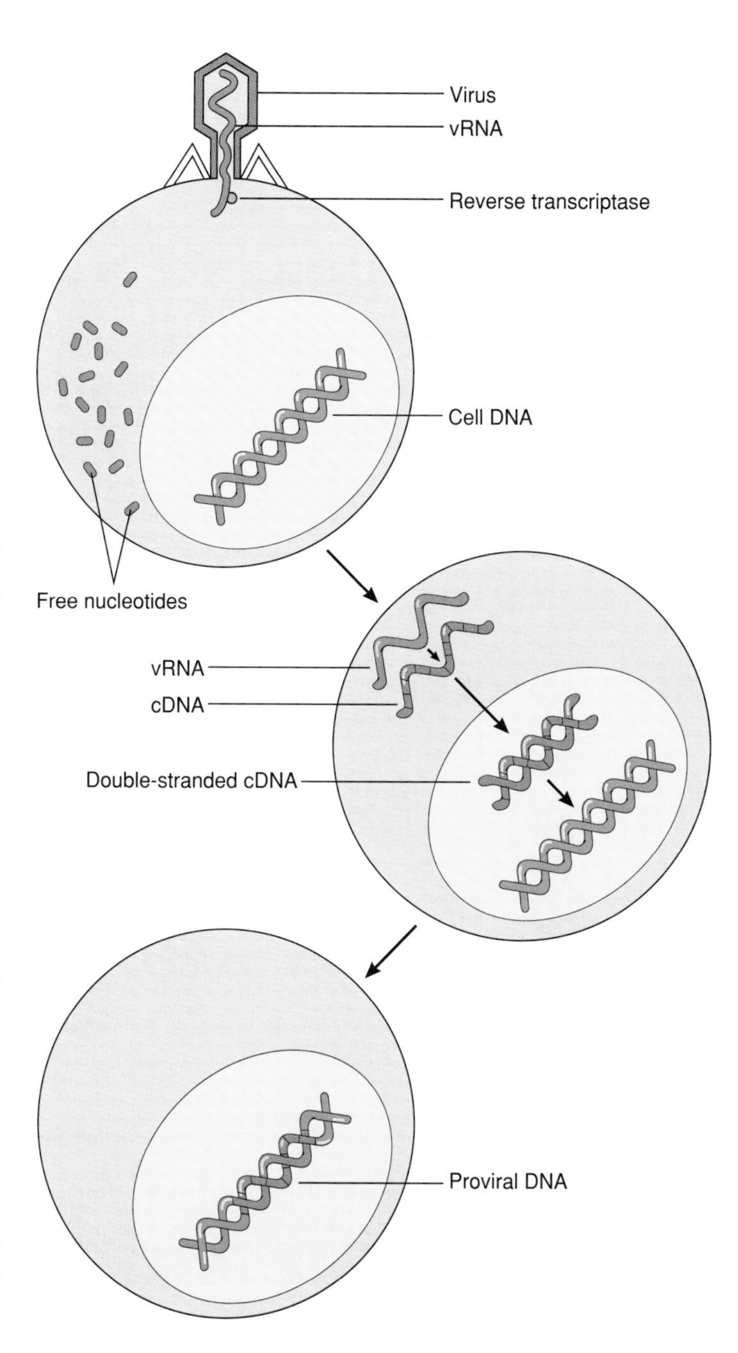

Figure 14.5. A retrovirus injects its viral RNA (vRNA) into the host cell. Through the action of reverse transcriptase, complementary DNA (cDNA) is produced. These viral genes can become inserted into the DNA of the host cell. The inserted viral genes form what is known as a provirus.

contains an oncogene; or (2) a proto-oncogene within a cell may mutate to become an oncogene. The latter process may be stimulated by chemical carcinogens or ionizing radiation.

Genetic Changes in Cancer

The mutation of a body (somatic) cell, in which a proto-oncogene changes into an oncogene, can apparently be produced in different ways. In some cases, a simple point mutation

Table 14.2 The loss of particular regions of specific chromosomes is found in many types of cancer

Tumor	Site of chromosomal deletion*
Retinoblastoma	13q
Osteosarcoma	13q
Wilms	11p
Rhabdomyosarcoma	11p
Hepatoblastoma	11p
Hepatocellular carcinoma	11p
Bladder	11p
Renal carcinoma	3p
Lung small cell	3p, 13q, 17p
Other types	3p (13q, 17p less frequent)
Breast	11p, 13q
Stomach	13q
Colon	5q, 17p, 18q, 22, others
Insulinoma	11
Phaeochromocytoma	1p, 22
Medullary thyroid carcinoma	1p
Meningioma	22
Acoustic neuroma	22
Melanoma	Various

*p = short arm of chromosome
q = long arm of chromosome

Reprinted by permission from *Nature*, 335:400–401. Copyright © 1988 Macmillan Magazines Ltd.

(change in one DNA base—chapter 10) may begin the multistep process of transformation to cancer. In many cases, cancers are also associated with gross changes in parts of chromosomes and even entire chromosomes. In an attempt to explain the development of colorectal cancer, for example, the following process has been proposed: (1) first, a mutation occurs in the *c-ras* proto-oncogene; (2) then, a segment of chromosome number 5 is lost; and (3) deletions of parts of chromosomes number 17 and 18 occur in advanced stages of the tumor. The loss of particular parts of chromosomes associated with specific cancers is shown in table 14.2.

In addition to point mutation, a proto-oncogene may become oncogenic (cancer-causing) when the region of the chromosome on which it is located is moved to a different location. **Chromosomal translocation**—the movement of pieces of chromosomes from one chromosome to another—is a process that occurs in a number of cancers. In Burkitt's lymphoma, for example, part of chromosome 8, which contains the *c-myc* gene, is transferred to chromosome number 14 (fig. 14.6). In chronic myelogenous leukemia, the proto-oncogene known as *c-abl* is moved when translocation occurs between chromosomes number 9 and 22. A list of some of the chromosomal translocations known to occur in cancer is provided in table 14.3.

Another means by which a proto-oncogene may become oncogenic is by a process known as **gene amplification.** This occurs when there are multiple copies of the gene within a single cell, so that more gene product is made than normally occurs. Twenty or more copies of the *c-myc* gene per tumor cell were found in

Table 14.3 The chromosomal translocations that occur in different cancers	
Chromosome Translocation*	**Cancer**
(8;21)	Acute myelogenous leukemia, acute myeloblastic leukemia
(9;22)	Chronic myelogenous leukemia
(15;17)	Acute promyelocytic leukemia
(11;19)	Acute monocytic leukemia, acute myelomonocytic leukemia
(1;19)	Pre–B cell leukemia
(8;14), (8;22), (2;8)	Burkitt lymphoma, acute lymphocytic leukemia of the B cell type
(11;14) (13;32)	Chronic lymphocytic leukemia, diffuse large- and small-cell lymphoma, multiple myeloma
(14;18)	Follicular lymphoma
(4;11)	Acute lymphocytic leukemia
(11;14) (13;13)	Acute lymphocytic leukemia of the T cell type

*Translocations occur between the chromosome numbers separated by a semicolon, and enclosed with parentheses.

a patient with promyelocytic leukemia, for example. Other genes that have been shown to be amplified include the *c-myb* gene in colon carcinoma, the *c-erb-B* gene in epidermal cancer cells, and the *c-myc* gene in lung cancer. Generally, gene amplification is only seen in the later stages of a cancer when the prognosis is poor. Indeed, some physicians have attempted to determine the stages of breast cancer by the degree of amplification of the oncogene associated with this condition.

The Actions of Oncogenes

The protein coded by the *src* oncogene was isolated in 1977 and found to be an enzyme that is a *protein kinase.* This enzyme moves to the cell membrane, where it adds phosphate groups to proteins within the membrane (fig. 14.7). There is evidence to suggest that this action may make the cell more sensitive than normal to mitogens (growth-stimulating chemicals) which are always present. The product of the *ras* oncogene, for example, has been shown to make the cell more sensitive to the effects of a mitogen known as epidermal growth factor.

Although most oncogene products thus far identified are protein kinase enzymes that act at the cell membrane, an interesting exception has recently been identified. The oncogene known as *v-jun,* from the avian sarcoma virus, appears to be identical to the active part of a previously known cellular protein that stimulates genetic transcription (RNA synthesis). This protein binds to specific regions of DNA and activates genes. This is the first evidence that an oncogene can directly alter the expression of other genes.

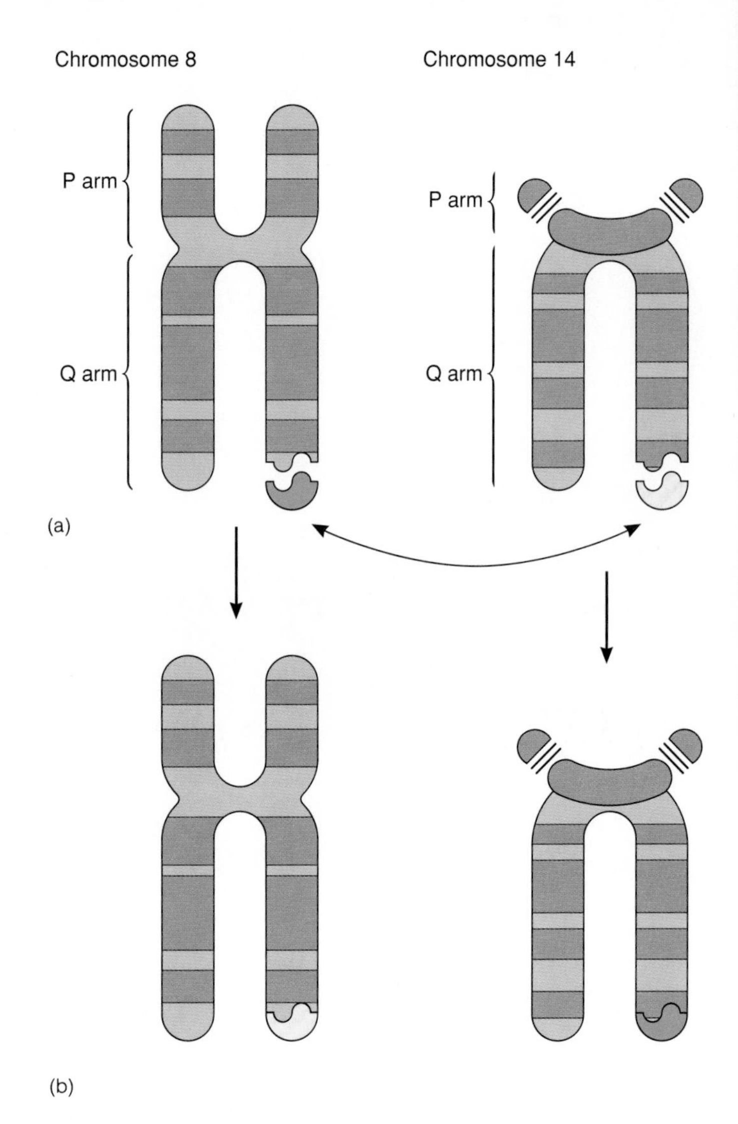

Figure 14.6. *Appearances of chromosomes 8 and 14 before (a) and after (b) chromosomal translocation occurs in Burkitt's lymphoma. Notice that pieces from the long arm (the Q arm) of each chromosome are reciprocally traded.*

Anti-Oncogenes

The fact that some people have a genetic predisposition for developing particular cancers has been known for some time. Various lines of evidence suggest that this predisposition is not due to the presence of proto-oncogenes, which all people have, but rather to the presence of other genes. Those genes that predispose a person to a particular cancer could act in a variety of possible ways, but until recently, no particular mechanism was demonstrated.

Retinoblastoma is a rare cancer of the retinal layer of the eyes. Before the invention of the ophthalmoscope (the instrument for examining the interior of the eyes) by Hermann von Helmholtz in 1850, this cancer was invariably fatal. Today,

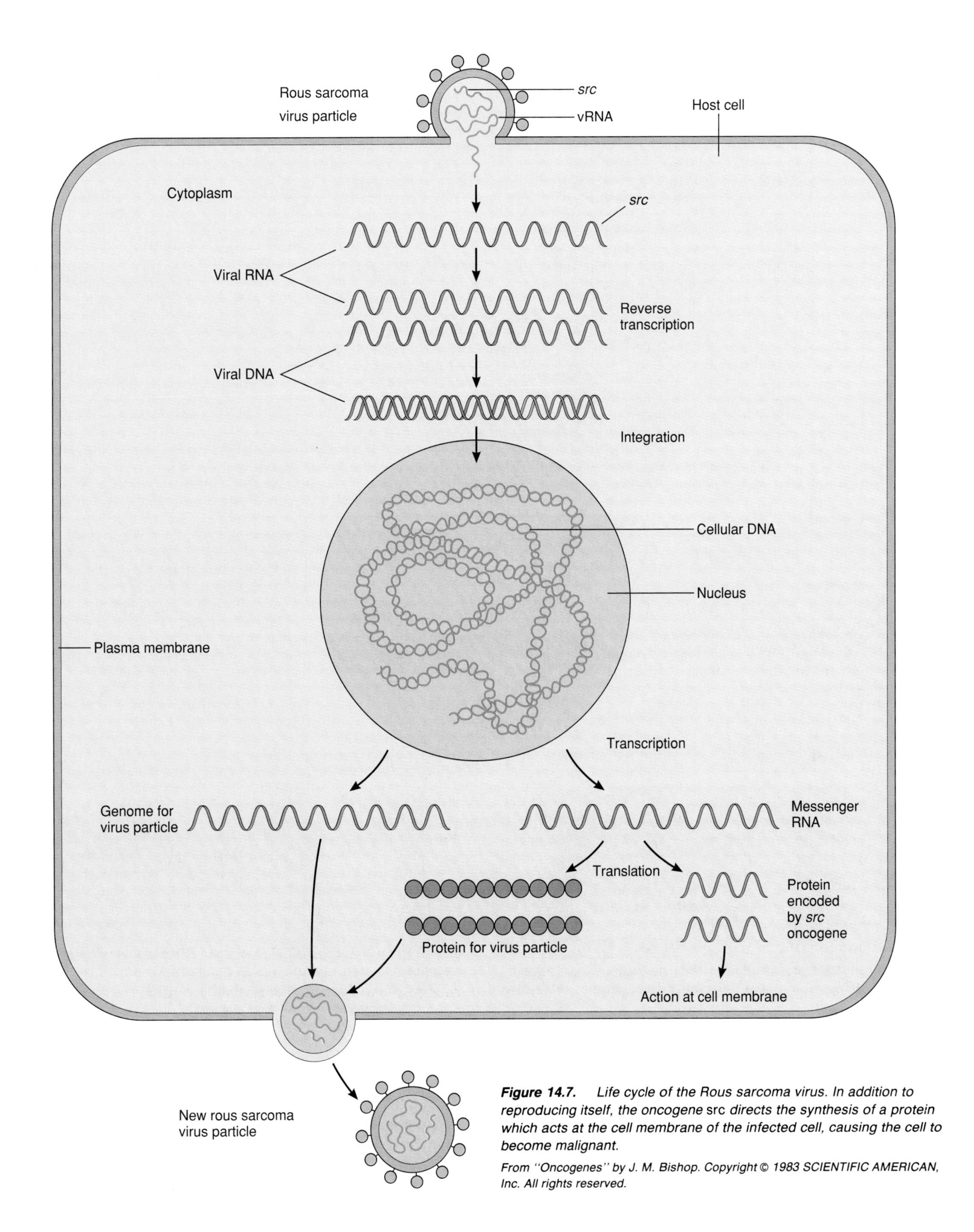

Figure 14.7. *Life cycle of the Rous sarcoma virus. In addition to reproducing itself, the oncogene* src *directs the synthesis of a protein which acts at the cell membrane of the infected cell, causing the cell to become malignant.*

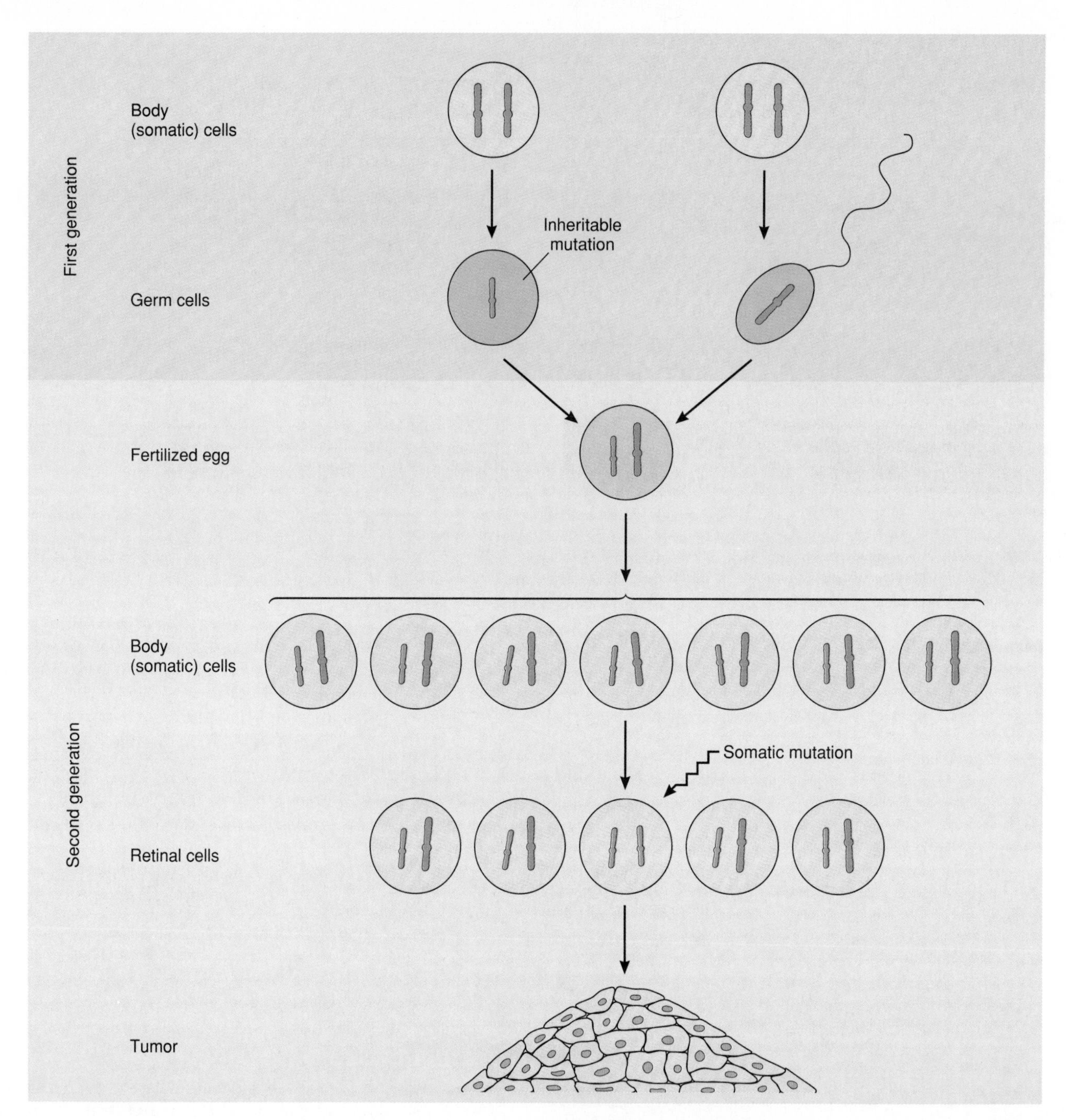

Figure 14.8. *The inheritance of a predisposition to develop retinoblastoma. This disease is prevented by having at least one good anti-oncogene, which is located on chromosome 13. If a baby inherits one mutant anti-oncogene, and the other also becomes deleted as a result of a somatic mutation, the child will develop the cancer.*

From "Finding the Anti-Oncogene" by R. Weinberg.

however, the child can be saved by early diagnosis and the removal of the affected eye. In 1986, scientists discovered a gene, located on chromosome number 13, which *prevents* the development of this cancer. It has thus been called an **anti-oncogene,** or a **tumor-suppressing gene.** Experiments demonstrated that the addition of normal retinoblastoma genes to tumor cells in tissue culture can convert them back to normal cells.

An inherited form of this cancer, it was discovered, is due to a mutation in which the retinoblastoma gene is completely absent. The fetus inherits the defective gene from one parent and the normal gene from the other parent (fig. 14.8). In the many cell divisions that occur during the course of eye development, however, the normal gene may mutate. The baby thus has a genetic predisposition to develop retinoblastoma; when the baby has two defective retinoblastoma genes, the cancer develops.

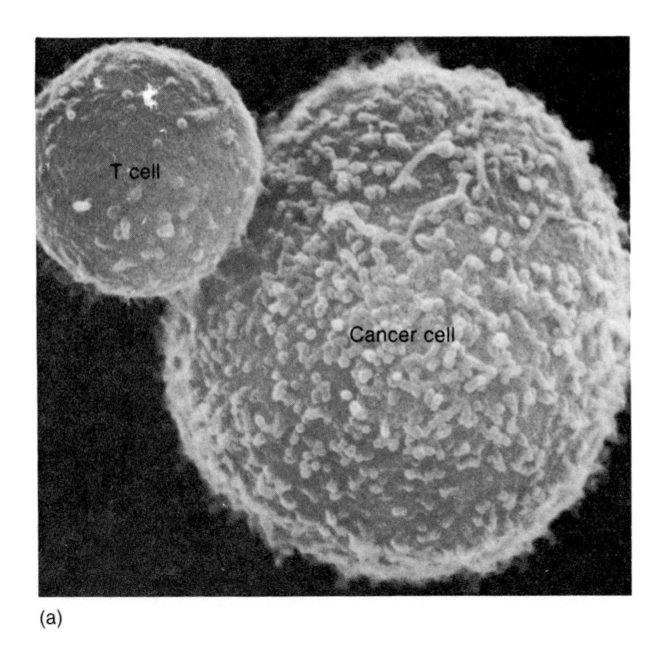

(a)

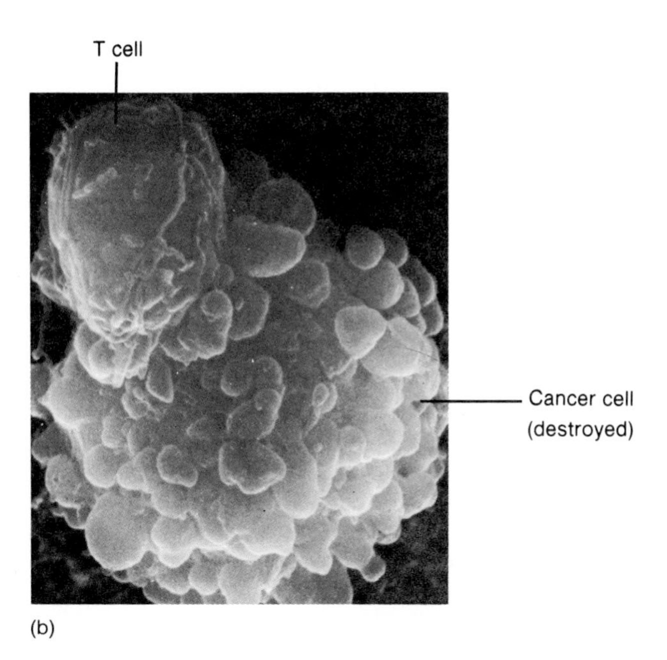

(b)

Figure 14.9. *A killer T cell (a) contacts a cancer cell (the larger cell), in a manner that requires specific interaction with antigens on the cancer cell. The killer T cell releases lymphokines, including toxins that cause the death of the cancer cell (b).*

There is evidence that the retinoblastoma anti-oncogene produces a protein that prevents the development of tumors in normal cells. When the cell is infected by particular cancer-causing viruses, the oncogene of such viruses produces a protein that combines with and inactivates the protein produced by the retinoblastoma anti-oncogene. This may represent an important mechanism by which these particular oncogenes cause cancer. If a cancer is indeed the result of infection by a specific virus, and if the oncogene of this virus can be blocked by larger amounts of the anti-oncogene protein, a new means of cancer therapy will have been found. Much more research is needed to explore this exciting possibility.

1. Explain the significance of oncogenes and proto-oncogenes, and explain the relationship between these two.
2. Use the oncogene theory to explain how a chemical carcinogen may cause cancer.
3. Define the terms *chromosomal translocation* and *genetic amplification.* How are these processes related to cancer?
4. Describe how proteins coded by oncogenes might act to transform a normal cell to a cancerous one.
5. Define the term *anti-oncogene,* and explain how the disease retinoblastoma is believed to be produced.

Cancer and the Immune System

The immune system may be able to recognize cancerous cells and destroy them before they can produce a tumor. This is known as immunological surveillance, and is primarily the function of killer T lymphocytes and natural killer (NK) cells. Stress is known to suppress the activity of the immune system, and thus may contribute to the development of cancer.

As tumors become cancerous, they become less specialized. In this process, as previously described, new molecules are produced by the cells due to activation of previously inactive cellular genes, or to the addition of oncogenes from viruses. These abnormally active genes can produce proteins that travel to the cell membrane. As a result, the cells may contain surface antigens that the immune system is able to recognize as foreign. These antigens can thus stimulate the immune destruction of the tumor cells.

Tumor antigens activate the immune system, initiating an attack primarily by killer T lymphocytes (fig. 14.9) and natural killer cells (described below). The concept of **immunological surveillance** against cancer was introduced in the early 1970s to describe the proposed role of the immune system in fighting cancer. According to this concept, tumor cells originate frequently in the body but are normally recognized and destroyed by the immune system before they can cause cancer.

There is now evidence that immunological surveillance does prevent some types of cancer. This explains why, for example, AIDS victims (with a depressed immune system) have a higher incidence of Kaposi's sarcoma. It is not clear, however, why all types of cancers do not appear with high frequency in AIDS patients and others whose immune system is suppressed. For these reasons, the generality of the immunological surveillance system concept is currently controversial.

Natural Killer Cells

Researchers observed that a strain of hairless mice, which genetically lacks a thymus and T lymphocytes, does not suffer from a high incidence of tumor production. This surprising observation led to the discovery of **natural killer (NK) cells,** which are lymphocytes that are related to, but distinct from, T lymphocytes. Unlike killer T cells, NK cells destroy tumors in a nonspecific fashion and do not require prior exposure for sensitization to the tumor antigens. The NK cells thus provide a first line of cell-mediated defense, which is subsequently backed up by a specific response through the action of killer T cells. Killer T cells and NK cells have been shown to interact; the activity of NK cells is stimulated by interferon, released from T lymphocytes.

Effects of Aging and Stress

It is known that cancer risk increases with age. According to one theory, this is due to the fact that aging lymphocytes accumulate genetic errors over the years that decrease their effectiveness. The secretion of thymus hormones also decreases with age, in parallel with a decrease in cell-mediated immune competence. Both of these changes and perhaps others not yet discovered could increase susceptibility to cancer.

Numerous experiments have demonstrated that tumors grow faster in experimental animals subject to stress than in unstressed control animals. This is generally believed to result from the fact that stressed animals, including humans, have increased secretion of corticosteroid hormones (chapter 7), which act to suppress the immune system (this is why cortisone is given to people who receive organ transplants and to people with chronic inflammatory diseases). Some recent experiments, however, suggest that the stress-induced suppression of the immune system may also be due to other factors that do not involve the adrenal cortex.

There are a number of new, experimental approaches to the treatment of cancer that involve use of lymphocytes and lymphokines (the molecules secreted by T lymphocytes). These will be discussed in the next section. Future advances in cancer therapy may also incorporate methods of strengthening the immune system together with methods that directly destroy tumors. In the meantime, it is clear that the reduction of stress—as accompanies a positive mental attitude (through "laugh therapy," for example)—strengthens the immune system and thus improves the prognosis of cancer.

1. Describe the concept of immunological surveillance against cancer. Does the disease AIDS support this concept?
2. Explain the possible relationship between stress and the development of cancer.

Cancer Therapies

Conventional cancer therapy includes surgery, chemotherapy, and radiation therapy. Most antitumor drugs destroy cells that are in the process of DNA replication and cell division. Experimental treatments for cancer include the use of interferon and interleukin-2, which are lymphokines that can strengthen the immune system. Monoclonal antibodies that are targeted against tumor antigens offer additional hope for new cancer therapies.

Therapies for cancer are generally classified as *conventional*—including surgery, chemotherapy, and radiation therapy—and *experimental.* Conventional treatments have increased the overall odds of surviving cancer from 39% in 1950 to 50% in 1985. For some forms of cancer, the chances of survival (as defined by being disease-free after five years) are quite good—testis cancer (90%), Hodgkin's lymphoma (74%), and colon cancer (74%), for example. Survival is improved when the diagnosis is made early. Many other forms of cancer, unfortunately, have a significantly less optimistic prognosis. This is why further research is so critically important, and why so much hope is invested in experimental treatments.

Chemotherapy

If diagnosed early, there are fifteen forms of cancer (out of over one hundred) that can potentially be cured by standard chemotherapy (table 14.4). Most antitumor drugs act in a similar manner—they kill dividing cells. Younger tumors, in which 30%–100% of the cells are undergoing mitosis, are more susceptible to these drugs than are older tumors, which have a lower proportion of their cells undergoing DNA synthesis and division. Older tumors are also more likely to have mutant cells that are resistant to the drugs. In these cases, surgery and radiation may be used together with chemotherapy.

Some normal tissues also have rapid rates of cell division. This is the case with the cells at the hair roots, the epithelium that lines the gastrointestinal tract, and the tissue that forms blood cells. Antitumor drugs thus also attack these cells, resulting in negative side effects. For this reason, periods in which chemotherapy is given are usually short and followed by rest periods, which enables the normal tissue to recover. It should be understood that the normal tissue can recover, once the cancer is destroyed.

There are a number of different types of antitumor drugs that inhibit DNA synthesis and cell division. Some—including those developed from mustard gas (HN_2), a poison gas used in World War I—directly act on DNA to prevent its strands from separating during replication. Other drugs interfere with the production of nucleotides, which are the subunits required for DNA synthesis (chapter 10). Still others interfere with the formation of the spindle fibers, which are needed to separate the chromosomes during cell division. In addition to these types of drugs, many others—including antibiotics, corticosteroids, and antiestrogens—may be used in particular cases.

Unfortunately, conventional chemotherapy may be relatively ineffective for many tumors. Sometimes, mutant cells have developed that are resistant to the drug. Larger amounts of the drug might theoretically work, but cannot be used because of its toxicity to healthy tissues. Also, most of these drugs are not able to penetrate the blood-brain barrier (chapter 6), and thus are ineffective in treating tumors that have invaded the brain or spinal cord. Many patients with cancer that is incurable by conventional methods, therefore, participate in clinical trials of experimental methods of cancer therapy.

Table 14.4 Beneficial effect of chemotherapy on survival with particular types of cancer

Type of Cancer	Percentage of Patients Reaching Normal Life Expectancy	
	Estimated % prior to chemotherapy	Estimated % with chemotherapy
Wilms' tumor	30	90
Choriocarcinoma	20	80
Rhabdomyosarcoma	10	60
Advanced Hodgkin's disease	5	58*
Ewing's sarcoma	10	55
Burkitt's lymphoma	5	55
Retinoblastoma	20	55
Acute lymphoblastic leukemia	0	50
Mycosis fungoides	5	50*
Osteosarcoma	20	50*
Diffuse histiocytic lymphoma	0	42
Metastatic embryonal testicular carcinoma	0	32–70*
Acute myelocytic leukemia	0	20*
Ovarian cancer	0	20*
Nodular mixed lymphoma	0	20*

*Longer observation required to be certain of this percentage.

From *The Merck Manual of Diagnosis and Therapy,* Edition 14, p. 2375, edited by Robert Berkow. Copyright 1982 by Merck & Co., Inc. Used with permission.

Lymphocytes and Lymphokines

As discussed in chapter 13, T lymphocytes secrete a group of chemicals called *lymphokines,* which have regulatory effects on the immune system. Some of these lymphokines also have additional actions in the body. **Interferons** were the first lymphokines to be identified and to be genetically engineered. These are a family of glycoproteins, produced by a variety of cells in addition to lymphocytes, whose best-known effect is to protect against viral infections. In addition to their antiviral action, and their ability to stimulate the immune system, interferons act to inhibit cell division. They thus have a direct antitumor effect.

The production of human interferons by genetically engineered bacteria (chapter 10) has made large amounts of these substances available for the experimental treatment of cancer. Thus far, interferons have proven to be a useful addition to the treatment of particular forms of cancer, including some types of lymphomas, renal carcinoma, melanoma, Kaposi's sarcoma, and breast cancer. They have not, however, proved to be the "magic bullet" against cancer (a termed coined by Paul Ehrlich[10]) as had previously been hoped.

A team of scientists headed by Steven Rosenberg at the National Cancer Institute has pioneered the use of another lymphokine that is now available through genetic engineering techniques. This important lymphokine is called **interleukin-2 (IL-2),** and is known to activate both killer T lymphocytes and B lymphocytes (chapter 13). These investigators removed some of the blood from cancer patients who could not be successfully treated by conventional means and isolated a population of their lymphocytes. They treated these lymphocytes with IL-2, to produce *lymphokine-activated killer (LAK) cells,* and then reinfused these cells, together with IL-2 and interferons, into the patients. Depending on the combinations and dosages, they obtained remarkable success (but not a complete cure of all cancers) in many of these patients.

Rosenberg's group next identified a subpopulation of lymphocytes that had invaded solid tumors in mice. The number of these *tumor-infiltrating lymphocyte (TIL) cells* was increased by tissue culture, and then reintroduced into the mice with excellent effects. As a result of these findings, the investigators have recently used these techniques to treat an experimental group of people who have metastatic melanoma, a cancer that claims the lives of six thousand Americans annually. The patients were first given conventional chemotherapy and radiation therapy. They were then treated with their own TIL cells and interleukin-2. Some of the preliminary results of this treatment seem very promising (see fig. 14.10). It should be noted, however, that the long-term effects of this treatment are not known at the present time, and more research is needed to clarify the optimum procedures for this therapy.

The National Institutes of Health (NIH) granted permission in 1990 to attempt the first gene therapy of cancer. In one approach, the human gene for **tumor necrosis factor,** a lymphokine shown to be effective against cancer in mice, will be inserted into cultured TIL cells taken from a patient's melanoma. The genetically-altered TIL cells will then be stimulated to proliferate with IL-2 for thirty to forty days and infused into the same patient. This marriage of genetic engineering and immunology has exciting promise, but it is too early to tell if it will beget new cancer treatments.

[10]Paul Ehrlich: German bacteriologist, 1854–1915

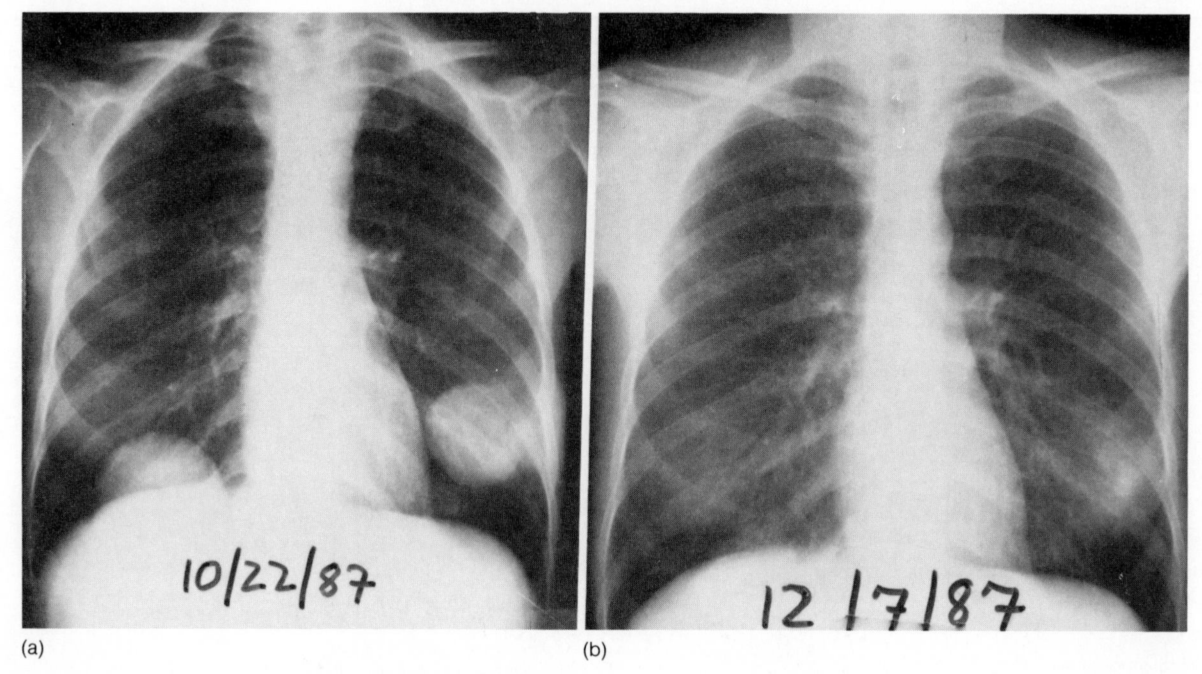

Figure 14.10. *Secondary tumors in the lungs of a patient (a) are significantly regressed (b) as a result of treatment with tumor-infiltrating lymphocytes and interleukin-2.*

Monoclonal Antibodies

As described in chapter 13, unlimited amounts of specific, homogenous antibodies—called **monoclonal antibodies**—can now be produced that will combine with almost any antigen. This revolutionary breakthrough was the result of the discovery that an antibody-secreting B lymphoycte could be fused with a cancerous myeloma cell (fig. 14.11). Differentiated cells, such as B lymphocytes, can divide in tissue culture only a limited number of times. Such a culture of cells, therefore, would quickly die, and the antibodies produced by that particular B cell would be available in only very small amounts. Tumor cells, in contrast, can continue to divide indefinitely. By fusing these two together, the B cell becomes immortal. Since one B cell only makes one kind of antibody, all of its progeny produced by cell division will only make that type of antibody. This discovery has produced a large biotechnology industry that has aided many fields of medicine.

One hope of research scientists has been to discover a particular antigen that is present on the surface of all cancerous cells and is absent on the surface of all normal cells. If such an antigen is ever found, monoclonal antibodies could be made against it. In that case, these antibodies would indeed be magic bullets against cancer. Unfortunately, there are no antigens yet known that are found on all cancer cells and are completely unique to these cells. Still, some progress has been made, and there is evidence that monoclonal antibodies may be useful in the fight against cancer.

Taken together, the experimental use of lymphokines, lymphocytes, and monoclonal antibodies comprise a very hopeful means of exploiting the body's own defense system in the fight against cancer. In the future, perhaps knowledge of antioncogene action will also provide tools for combating cancer. We are fortunate that knowledge has progressed so rapidly in the areas of immunology, cancer biology, and molecular biology, and that an active biotechnology industry is available. The ability to cure cancer, however, is still beyond us. This will require a much more complete knowledge of basic cell function than we now have, and thus is dependent upon continued high levels of biomedical research.

---

1. Explain how most antitumor drugs act, and why a person treated with these drugs may experience the loss of hair and other negative side effects.
2. Describe the significance of interferon, interleukin-2, and lymphocytes in the experimental treatment of cancer.
3. Explain how monoclonal antibodies are produced, and how they might be used in the treatment of cancer.

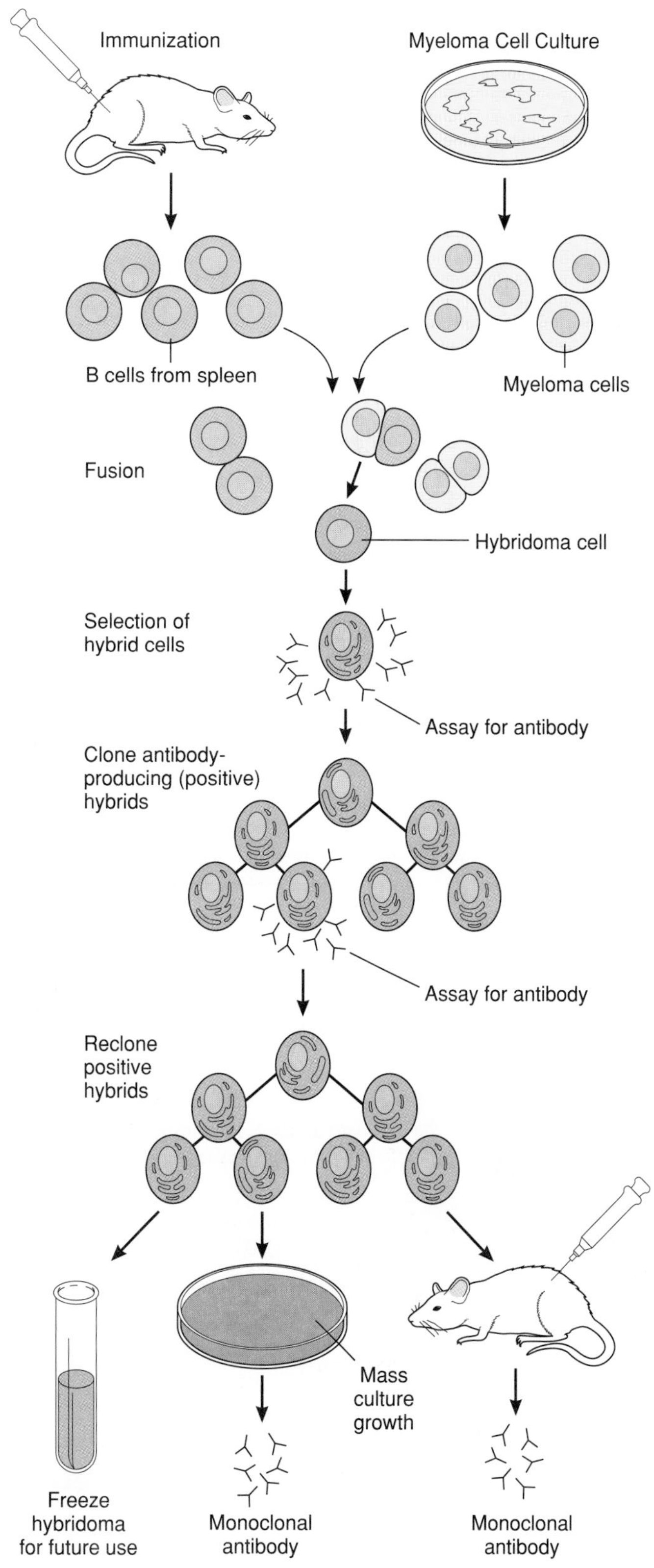

Immunization

Myeloma Cell Culture

B cells from spleen

Myeloma cells

Fusion

Hybridoma cell

Selection of hybrid cells

Assay for antibody

Clone antibody-producing (positive) hybrids

Assay for antibody

Reclone positive hybrids

Mass culture growth

Freeze hybridoma for future use

Monoclonal antibody

Monoclonal antibody

Figure 14.11. *Formation of a hybridoma which produces unlimited amounts of one specific type of antibody.*

Summary

The Nature of Cancer p. 314

I. Tumors result from uncontrolled proliferation of a single cell that has lost its specialization (differentiation) and that divides without the normal inhibitions.

II. There are several categories of cancers, depending on the tissues involved; the extent of cancer development is often marked by stages.

III. Leukemia involves uncontrolled proliferation of leukocyte-forming cells; it is divided into an acute and a chronic form, which have different characteristics.

Causes of Cancer p. 318

I. Chemical carcinogens and ionizing radiation are important environmental causes of cancer.
   A. The development of cancer is divided into two stages: initiation and promotion.
   B. The initiation of cancer may involve one or more mutations; the promotion stage appears to be reversible.
   C. Complete carcinogens are those that produce both initiation and promotion of tumor growth.

II. Viruses have been shown to cause certain cancers in lower animals, and there is evidence that they may also contribute to cancer in humans.

III. Vitamin A, vitamin E, fiber, and selenium may help to prevent cancer; dietary fat and excess calories may contribute to cancer.

IV. Cigarette smoking is the single most important preventable cause of cancer.

Oncogenes and Anti-Oncogenes p. 322

I. Some retroviruses are known to cause specific cancers in lower animals.
   A. The genes that cause cancer in these viruses are known as oncogenes.
   B. The oncogenes of viruses are believed to be derived from mammalian genes, which are known as proto-oncogenes.
   C. The human proto-oncogenes may become activated to cancer-causing forms by point mutations, by chromosomal translocations, or by genetic amplification.

II. The proteins produced by oncogenes may cause cancer by acting at the plasma membrane to make the cell more sensitive to mitogens (chemicals that stimulate mitosis).

III. Anti-oncogenes are genes that suppress the formation of tumors.

Cancer and the Immune System p. 327

I. As cells become cancerous and lose their specialization, new antigens appear on the surface of the cancer cells.
   A. The immune system may be able to recognize these tumor antigens and destroy the cancerous cells; this is known as immunological surveillance against cancer.
   B. Killer T cells and natural killer (NK) cells are the primary agents of immunological surveillance.

II. The risk of cancer increases with age and with stress.

Cancer Therapies p. 328

I. Conventional therapy involves surgery, chemotherapy, and radiation therapy, which kills cells undergoing DNA synthesis and cell division.

II. Lymphokines and lymphocytes are used in the experimental treatment of cancer.
   A. Interferon and interleukin-2 are lymphokines (chemicals secreted by T lymphocytes) which have been used in the fight against cancer.
   B. In addition, lymphokine-activated killer (LAK) cells and tumor-infiltrating lymphocyte (TIL) cells have also been used.

III. Monoclonal antibodies may be used to target specific tumor antigens in the future treatment of cancer.

Review Activities

Objective Questions

1. Which of the following statements is *false?*
   (a) A tumor is a neoplasm.
   (b) A tumor is derived from the division of a single cell.
   (c) All of the cells in a tumor are genetically identical.
   (d) Cancerous cells are less differentiated than normal cells.
2. The most common kinds of cancers are
   (a) carcinomas
   (b) sarcomas
   (c) leukemias
   (d) lymphomas
3. Which of the following statements about metastasis of a tumor is *true?*
   (a) Some cells break away from the primary tumor.
   (b) Metastatic tumors produce enzymes that digest collagen.
   (c) Metastatic tumors secrete chemicals that promote the growth of new blood vessels that supply the tumor.
   (d) All of these.
4. Which of the following statements is *true* about an incomplete carcinogen?
   (a) It causes initiation of the tumor only.
   (b) It causes promotion of the tumor only.
   (c) It first causes initiation, then it causes promotion.
   (d) It first causes promotion, then it causes initiation.
5. Burkitt's lymphoma
   (a) is believed to be caused by a virus
   (b) results whenever a person is exposed to the virus
   (c) occurs worldwide
   (d) none of these
6. Which of the following dietary substances might be effective in preventing cancer?
   (a) vitamin A
   (b) vitamin E
   (c) selenium
   (d) all of these
7. Dietary fiber is believed to help prevent
   (a) cancer of the esophagus
   (b) cancer of the stomach
   (c) cancer of the pancreas
   (d) cancer of the colon
8. Smoking is believed to be the major contributor to
   (a) cancer of the lungs
   (b) cancer of the larynx
   (c) cancer of the oral cavity
   (d) cancer of the esophagus
   (e) all of these
9. Leukemias and lymphomas are believed to be often caused by
   (a) DNA-containing viruses
   (b) retroviruses
   (c) chemical carcinogens
   (d) ionizing radiation
10. Oncogenes
    (a) are genes that cause cancer
    (b) are present in some viruses
    (c) are derived from normal mammalian genes
    (d) all of these
11. A proto-oncogene may become activated when
    (a) a point mutation occurs
    (b) an anti-oncogene is inactivated
    (c) it is moved to a different chromosomal location
    (d) all of these
12. The gene *c-src*
    (a) is found in a virus
    (b) is found only in cancer patients
    (c) is found in normal human cells
    (d) requires reverse transcriptase in order to be expressed

Essay Questions

1. Explain the theoretical basis for the CEA test for cancer.
2. Explain the physiological reason why a positive mental attitude may contribute to a cancer cure.
3. Can cancer be completely prevented, so that a cure is not necessary? Explain your answer.
4. Describe the ways that immune mechanisms are currently used in the experimental treatment of cancer.

15

The Respiratory System

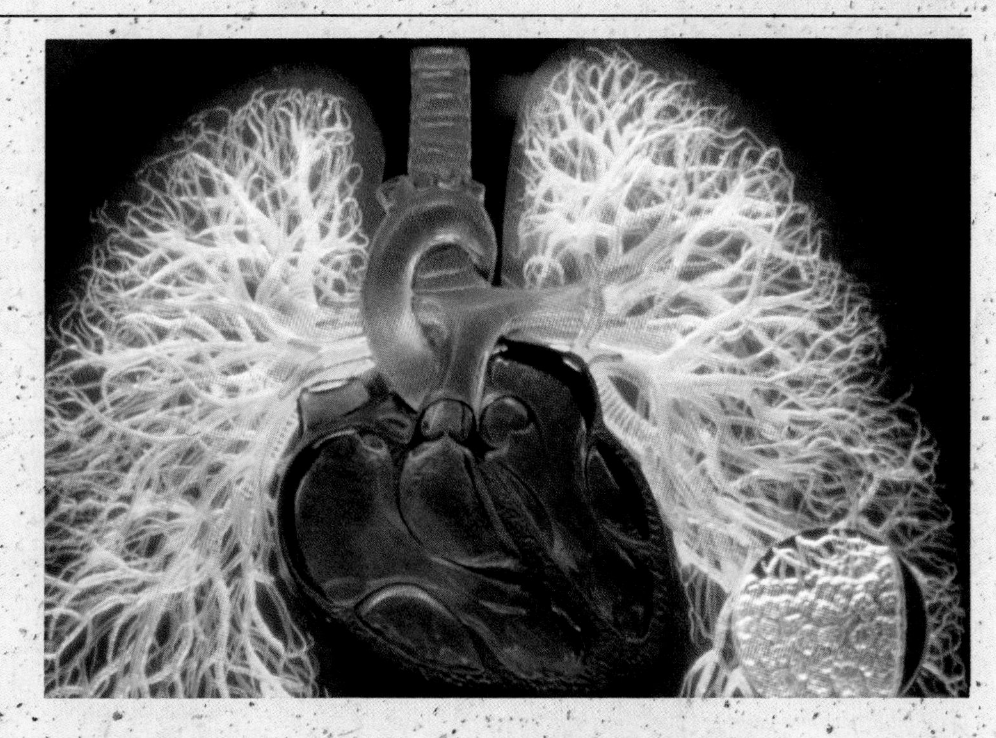

Outline

Objectives

By studying this chapter, you should be able to

1. describe the structures that comprise the lungs and explain their functions
2. explain how the intrapulmonary and intrapleural pressures vary during ventilation
3. define the term *elasticity* and explain how this lung property affects ventilation
4. explain the significance of surface tension and the role of pulmonary surfactant in normal lung function
5. explain how inspiration and expiration are accomplished in unforced breathing
6. describe the nature of some pulmonary disorders, including asthma, bronchitis, emphysema, and fibrosis
7. identify the parts of the brain involved in the regulation of breathing
8. identify the location of the chemoreceptors that regulate breathing, and explain the nature of this regulation
9. describe the nature of deoxyhemoglobin and oxyhemoglobin, and explain how carbon monoxide acts as a poison
10. describe the loading and unloading reactions, and explain how the extent of the unloading reaction is increased during exercise
11. explain how carbon dioxide is transported in the blood, and define the terms *hyperventilation* and *hypoventilation*
12. define the maximal oxygen uptake and anaerobic threshold, and explain how these are affected by endurance training
13. explain the compensations of the respiratory system to life at a high altitude

Keys to Pronunciation

alveolus: *al-ve'o-lus*
anaerobic: *ah"a-er-o'bik*
apnea: *ap-ne'ah*
asthma: *az'mah*
bronchiole: *brong'ke-ol*
bronchus: *brong'kus*
carboxyhemoglobin: *kar-bok"se-he"mo-glo'bin*
chemoreceptors: *ke"mo-re-cep'tors*
dyspnea: *disp'ne-ah*
emphysema: *em"fi-se'mah*
erythropoietin: *e-rith"ro-poi'e-tin*
hyperpnea: *hi"perp-ne'ah*
hypoxemia: *hi"pok-se'me-ah*
oxyhemoglobin: *ok"se-he"mo-glo'bin*
pleura: *ploo'rah*
pneumothorax: *nu"mo-tho'raks*
polycythemia: *pol"e-si-the'me-ah*
surfactant: *sur-fak'tant*
trachea: *tra'ke-ah*

Photo: The anatomical relationship between the lungs and heart are clearly shown in this model.

The Structure of the Respiratory System

The respiratory system is divided into a respiratory zone, where gas exchange between air and blood occur, and a conducting zone, which conducts the air to the respiratory zone. The exchange of gases between air and blood occurs across the walls of tiny air sacs called alveoli.

The term *respiration*[1] includes three separate but related functions: (1) **ventilation** (breathing); (2) **gas exchange,** which occurs between the air and blood in the lungs and between the blood and other tissues of the body; and (3) **oxygen utilization** by the tissues in energy-liberating reactions of cell respiration (discussed in chapter 18).

Ventilation is the mechanical process that moves air into and out of the lungs. Since air in the lungs has a higher oxygen concentration than in the blood, oxygen diffuses from air to blood. Carbon dioxide, conversely, moves from the blood to the air within the lungs by diffusing down its concentration gradient. As a result of this gas exchange, blood leaving the lungs (in the pulmonary veins) contains a higher oxygen and a lower carbon dioxide concentration than the blood delivered to the lungs in the pulmonary arteries.

Gas exchange between the air and blood occurs entirely by diffusion through lung tissue. This diffusion occurs very rapidly because there is a high surface area within the lungs and a very short distance between blood and air. The fact that blood in the pulmonary veins is almost completely saturated with oxygen is testimony to the high efficiency of normal lung function.

[1]respiration: L. *re,* back; *spirare,* to breathe

Respiratory and Conducting Zones

Gas exchange in the lungs occurs across about 300 million tiny (0.25–0.50 mm in diameter) air sacs, known as **alveoli.** The enormous number of these structures provides a high surface area—60–80 square meters, or about 760 square feet—for diffusion of gases. The diffusion rate is further increased by the fact that each alveolus[2] is only one cell-layer thick, so that the total "air-blood barrier" is only two cells across (an alveolar cell and a capillary endothelial cell).

Alveoli are polyhedral in shape and are usually clustered together, like the units of a honeycomb (fig. 15.1). These clusters of alveoli occur at the ends of *respiratory bronchioles,* which are the very thin air tubes that end blindly in alveolar sacs. The alveoli and respiratory bronchioles comprise most of the mass of the lungs.

The air passages of the respiratory system are divided into two functional zones. The **respiratory zone** is the region where gas exchange occurs, and it therefore includes the respiratory bronchioles (because they contain separate outpouchings of alveoli) and the terminal clusters of alveolar sacs (fig. 15.1). The **conducting zone** includes all of the anatomical structures through which air passes before reaching the respiratory zone (fig. 15.2). These air-conducting passages begin with the *trachea,*[3] or windpipe. The trachea divides into the primary *bronchi* (singular: *bronchus*[4]), which enter the lungs and subdivide into ever-smaller bronchioles (fig. 15.3).

[2]alveolus: L. diminutive of *alveus,* cavity
[3]trachea: L. *trachia,* rough air vessel
[4]bronchus: Gk. *bronchos,* windpipe

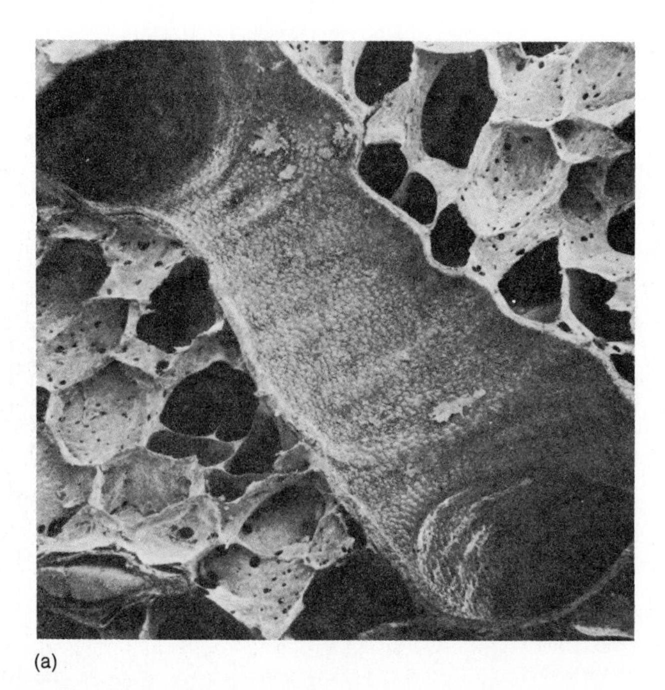
(a)

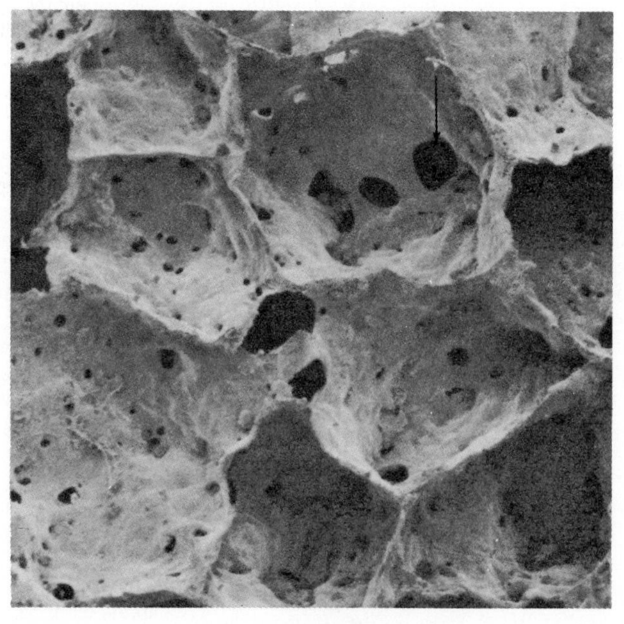
(b)

Figure 15.1. *(a) Scanning electron micrograph showing lung alveoli and a small bronchiole. (b) The alveoli under higher power (the arrow points to an alveolar pore through which air can pass from one alveolus to another).*

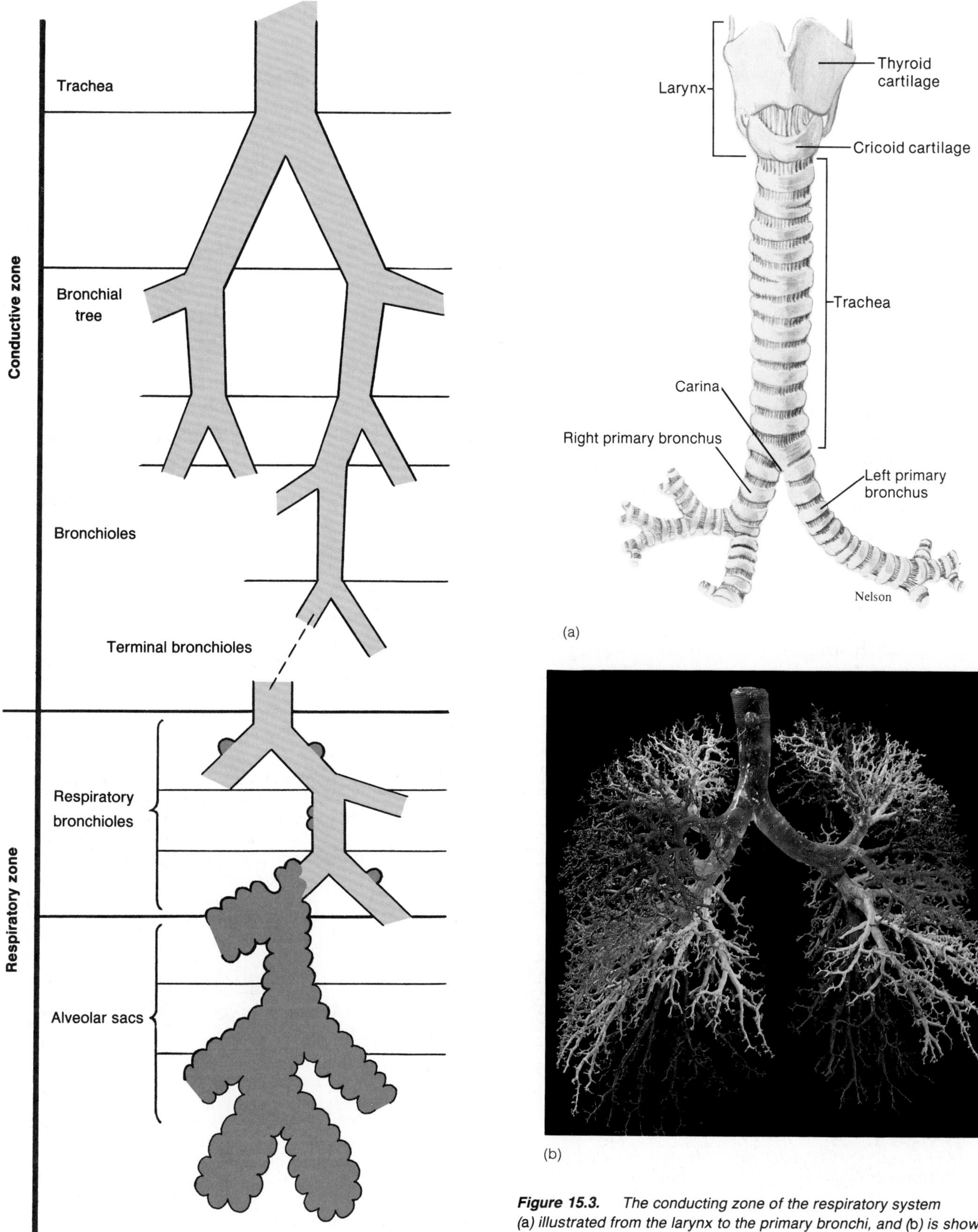

Figure 15.2. *The conducting and respiratory zones of the respiratory system.*

Figure 15.3. *The conducting zone of the respiratory system (a) illustrated from the larynx to the primary bronchi, and (b) is shown by a plastic cast of the airway from the trachea to the terminal bronchioles.*

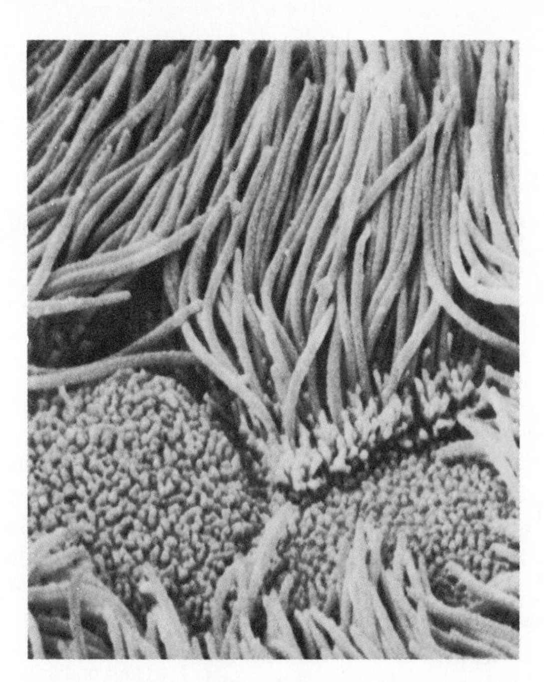

Figure 15.4. *A scanning electron micrograph of a bronchial wall showing cilia, which help to cleanse the lung by moving trapped particles.*

If the trachea becomes occluded, it may be necessary to create an emergency opening into this tube so that ventilation can still occur. A **tracheotomy** is the process of surgically opening the trachea, and a **tracheostomy** is the procedure of inserting a tube into the trachea to permit breathing and to keep the passageway open. A tracheotomy should be performed only by a competent physician, however, as there is great risk of cutting a nerve or jugular vein, which are also in this area.

In addition to conducting air into the respiratory zone, the air passages serve additional functions: *warming* and *humidification* of the inspired air and *filtration* and *cleaning.* Mucus secreted by cells of the conducting zone serves to trap small particles in the inspired air and thereby performs a filtration function. This mucus is moved along at a rate of 1–2 centimeters per minute by cilia projecting from the tops of epithelial cells that line the conducting zone (fig. 15.4). There are about three hundred cilia per cell that beat in a coordinated fashion to move mucus toward the mouth, where it can either be swallowed or expectorated.

The importance of this cleansing function is evidenced by the disease called *black lung,* which occurs in miners who inhale too much carbon dust and therefore develop pulmonary fibrosis (as described in a later section). The alveoli themselves are normally kept clean by the phagocytic action of macrophages (chapter 13) that reside within them. The cleansing action of cilia and macrophages in the lungs has been shown to be diminished by cigarette smoke.

Thoracic Cavity

The *diaphragm,* a dome-shaped sheet of striated muscle, divides the body cavity into two parts. The area below the diaphragm, or the *abdominal cavity,* contains the liver, pancreas, gastrointestinal tract, spleen, genitourinary tract, and other organs. Above the diaphragm, the chest, or *thoracic cavity,* contains the heart, large blood vessels, trachea, esophagus, and thymus gland in the central region and is filled elsewhere by the right and left lungs.

The *parietal pleural*[5] *membrane* is a wet epithelial membrane that lines the inside of the thoracic wall. The *visceral pleural membrane* covers the surface of the lungs. The lungs normally fill the thoracic cavity so that the visceral pleural membrane covering the lungs is pushed against the parietal pleural membrane lining the thoracic wall. There is thus, under normal conditions, little or no air between the visceral and parietal pleural membranes. There is, however, a "potential space"—called the *intrapleural space*—that can become a real space if the visceral and parietal pleural membranes separate when a lung collapses. The normal position of the lungs in the thoracic cavity is shown in the radiograph in figure 15.5.

1. Describe the structures involved in gas exchange in the lungs and how this process occurs.
2. Describe the structures and functions of the conducting zone of the respiratory system.
3. Describe the position of the lungs in the thoracic cavity and the relationship between the visceral and parietal pleural membranes.

Ventilation

During inspiration, air enters the lungs because the pressure within the lungs (intrapulmonary pressure) is lower than the atmospheric pressure. The lower intrapulmonary pressure results from an increase in lung volume, caused by expansion of the thoracic cavity. During expiration, air is forced out of the lungs because the intrapulmonary pressure is greater than the atmospheric pressure. This results from a decrease in the volume of the lungs and thoracic cavity.

Movement of air from the conducting zone to the terminal bronchioles occurs as a result of the pressure difference between the two ends of the airways. The pressure differences in the pulmonary[6] system are induced by changes in lung volumes, which result from changes in the volume of the thoracic cavity due to muscle contraction and relaxation.

Intrapulmonary and Intrapleural Pressures

The wet visceral and parietal pleural membranes are normally against each other, so that the lungs are stuck to the chest wall in the same manner that two wet pieces of glass stick to each

[5]pleura: Gk. *pleura,* side or rib
[6]pulmonary: Gk. *pleumon,* lung

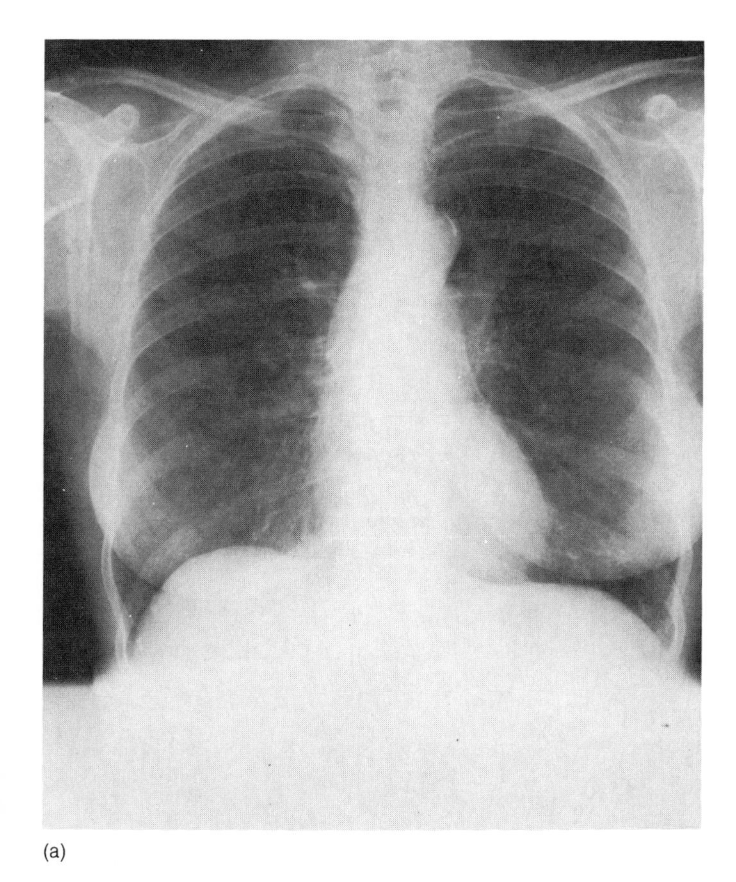

(a)

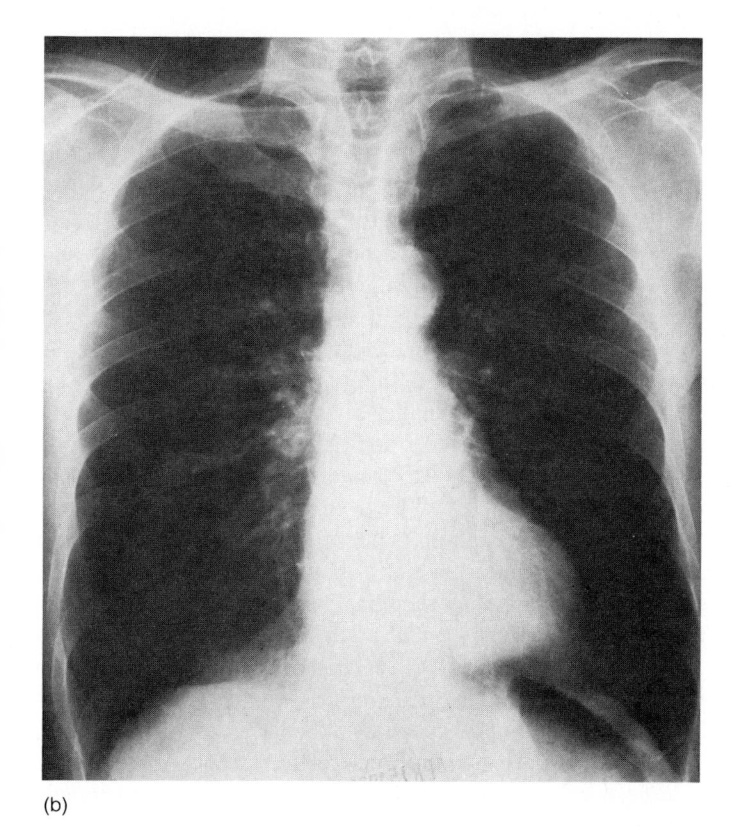

(b)

Figure 15.5. *Radiographic (X ray) views of the chest of a normal female (a) and a normal male (b).*

other. The *intrapleural space* between the two wet membranes contains only a thin layer of fluid secreted by the pleural membranes. The pleural cavity in a healthy, living person is thus potential rather than real; it can become real only in abnormal situations when air enters the intrapleural space. Since the lungs normally remain against the chest wall, they get larger and smaller together with the thoracic cavity during ventilation.

Air enters the lungs during inspiration because the pressure of the atmosphere is greater than the **intrapulmonary,** or **alveolar, pressure.** In order for inspiration to occur, therefore, the intrapulmonary pressure must fall below the atmospheric pressure. A pressure below that of the atmosphere is called a *subatmospheric pressure.* During quiet inspiration, for example, the intrapulmonary pressure may become 3 mm Hg less than the pressure of the atmosphere. This subatmospheric pressure is commonly shown as -3 mm Hg. Expiration, conversely, occurs when the intrapulmonary pressure is greater than the atmospheric pressure. During quiet expiration, for example, the intrapulmonary pressure may rise to $+3$ mm Hg over the atmospheric pressure.

Changes in intrapulmonary pressure occur as a result of changes in lung volume. This follows from **Boyle's law,**[7] which states that *the pressure of a given quantity of gas is inversely proportional to its volume.* An increase in lung volume during inspiration decreases intrapulmonary pressure to subatmospheric levels. Air therefore goes in. A decrease in lung volume raises the intrapulmonary pressure above that of the atmosphere, thus pushing air out. These changes in lung volume occur as a consequence of changes in thoracic volume, as will be described in a later section.

Lung Elasticity and Surface Tension

The term **elasticity** refers to the tendency of a structure to return to its initial size after being distended (stretched). The lungs are very elastic, and since they are normally stuck to the chest wall, they are always in a state of elastic tension. This tension increases during inspiration when the lungs are stretched and is reduced by elastic recoil during expiration. The elasticity of the lungs and of other thoracic structures thus aids in pushing the air out during expiration.

> The elastic nature of lung tissue is revealed when air enters the intrapleural space (as a result of an open chest wound, for example). This condition is called a **pneumothorax,**[8] which is shown in figure 15.6. As air enters the intrapleural space, the intrapleural pressure rises until it is equal to the atmospheric pressure. When the intrapleural pressure is the same as the intrapulmonary pressure, the lung collapses away from the chest wall as a result of elastic recoil. Since each lung is packaged separately by pleural membranes, a pneumothorax usually causes collapse of only one of the two lungs, leaving the other fully functional.

[7]Boyle's law: after Robert Boyle, British physicist, 1627–1691
[8]pneumothorax: Gk. *pneumon,* breath; L. *thorax,* chest

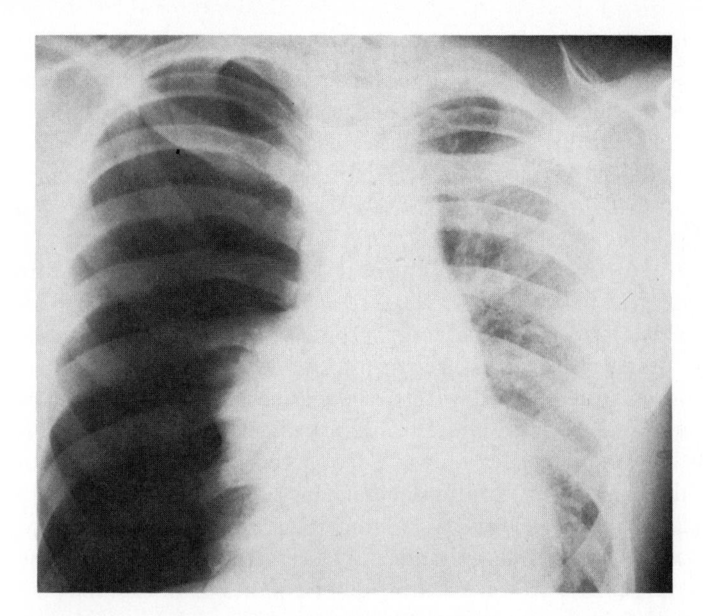

Figure 15.6. A pneumothorax of the right lung. The right side of the thorax appears uniformly dark because it is filled with air; the space between the ribs is also greater than on the left due to release from the elastic tension of the lungs.

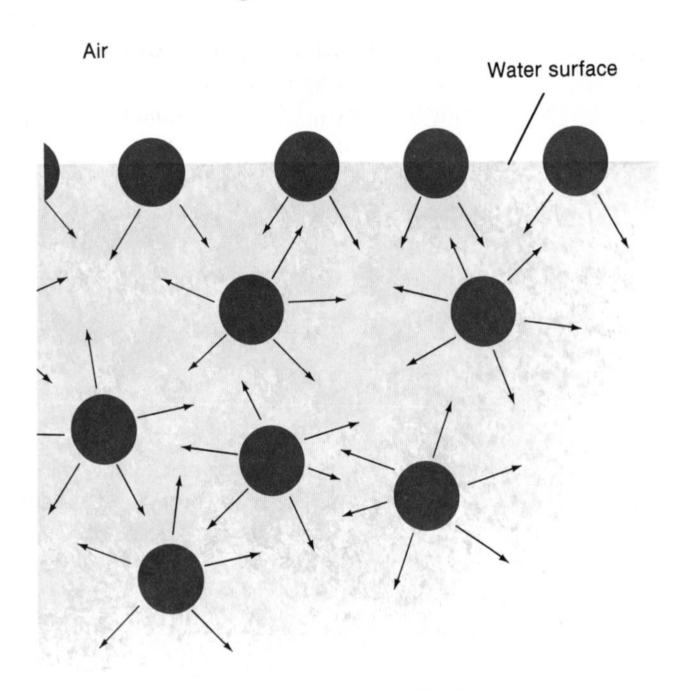

Figure 15.7. Water molecules at the surface have a greater attraction for other water molecules than for air. The surface molecules are thus attracted to each other and pulled tightly together by the attractive forces of water underneath. This produces surface tension.

The forces that act to resist the expansion of the lungs include elastic resistance and the **surface tension** that is exerted by fluid in the alveoli. Although the alveoli are relatively dry, they do contain a very thin film of fluid, much like soap bubbles.

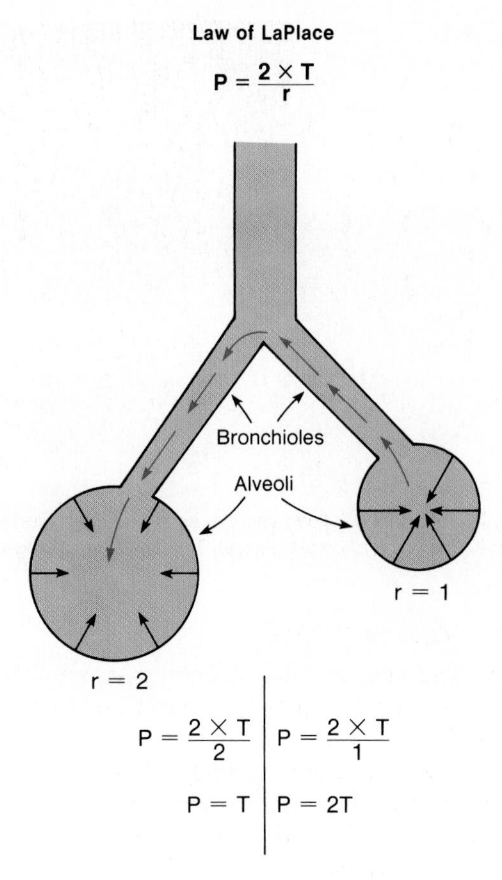

Figure 15.8. In the absence of pulmonary surfactant, smaller alveoli would collapse and empty their air into larger alveoli as a result of their surface tension. This is normally prevented by surfactant.

Surface tension is created by the fact that water molecules at the surface are attracted more to other water molecules than to air. As a result, the surface water molecules are pulled tightly together by attractive forces from underneath (fig. 15.7).

The surface tension of an alveolus produces a force that is directed inward and, as a result, creates pressure within the alveolus. The pressure in a smaller alveolus would be greater than in a larger alveolus as a result of their surface tension. The greater pressure of the smaller alveolus would then cause it to empty its air into the larger one (fig. 15.8). This does not normally occur because of the action of a type of molecule present within the alveoli.

Alveolar fluid contains a phospholipid (chapter 3), known as dipalmitoyl lecithin, which functions to lower surface tension. This compound is called **lung surfactant** (a contraction of the term *surface active agent.*) The ability of this lung surfactant to lower surface tension is greater in smaller than in larger alveoli. Surfactant thus prevents the smaller alveoli from collapsing and emptying their air into the large alveoli. Even after a forceful expiration, the alveoli remain open and a *residual volume* of air remains in the lungs. Since the alveoli do not collapse, less surface tension has to be overcome to inflate them at the next inspiration.

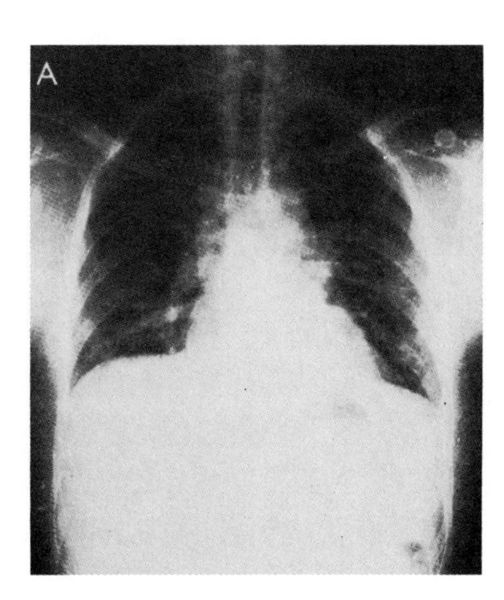

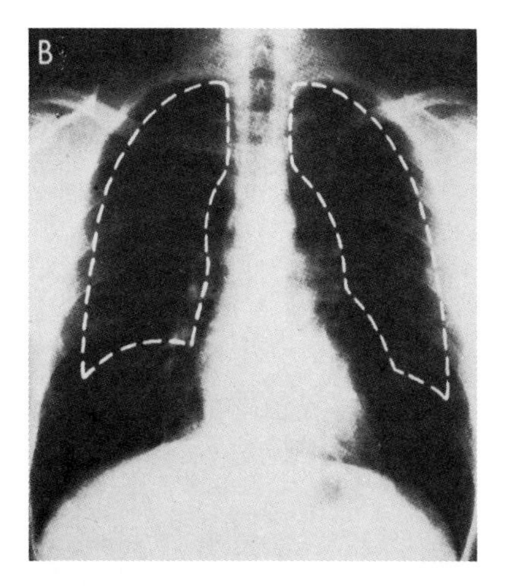

Figure 15.9. *A change in lung volume, as shown by radiographs, during expiration (a) and inspiration (b). The increase in lung volume during full inspiration is shown by comparison with the lung volume in full expiration (dashed lines).*

Respiratory Distress Syndrome

Since surfactant does not start to be produced until about the eighth month, premature babies are sometimes born with lungs that lack sufficient surfactant, and their alveoli are collapsed as a result. This condition is called **respiratory distress syndrome.** It is also called **hyaline membrane disease,** because the high surface tension causes plasma fluid to leak into the alveoli, producing a glistening "membrane" appearance (and pulmonary edema). This condition does not occur in all premature babies; the rate of lung development depends on hormonal conditions (thyroxine and hydrocortisone primarily) and on genetic factors.

Even under normal conditions, the first breath of life is a difficult one because the newborn must overcome great surface tension forces in order to inflate its partially collapsed alveoli. The effort required for the first breath is fifteen to twenty times that required for subsequent breaths, and an infant with respiratory distress syndrome must duplicate this effort with every breath. Fortunately, many babies with this condition can be saved by mechanical ventilators that keep them alive long enough for their lungs to mature and manufacture sufficient surfactant.

Mechanics of Breathing

Breathing, or **pulmonary ventilation,** refers to the movement of air into and out of the respiratory system. This movement of air occurs as a result of differences between the atmospheric and the intrapulmonary pressures; air goes in when the intrapulmonary pressure is subatmospheric, and air goes out when the intrapulmonary pressure rises above that of the atmosphere. As previously discussed, these pressure changes within the lungs occur as a consequence of changes in thoracic volume.

The thorax must be sufficiently rigid so that it can protect vital organs, yet must be flexible to function as a bellows during the ventilation cycle. The rigidity is provided by the bony composition of the rib cage. The rib cage is pliable, however, because the ribs are separate from one another and because most ribs (the upper ten of the twelve pairs) are attached to the sternum by cartilages. The structure of the rib cage and associated cartilages provide continuous elastic tension, so that when stretched by muscle contraction during inspiration, the rib cage can return passively to its resting dimensions when the muscles relax. This elastic recoil is greatly aided by the elasticity of the lungs.

Pulmonary ventilation consists of two phases, inspiration and expiration. Inspiration (inhalation) and expiration (exhalation) are accomplished by alternately increasing and decreasing the volumes of the thorax and lungs (fig. 15.9).

Inspiration and Expiration

The thoracic cavity increases in size during inspiration. This is accomplished by contractions of the **diaphragm**[9] and the **external intercostal** muscles between the ribs (fig. 15.10). A contraction of the dome-shaped diaphragm downward increases the thoracic volume vertically. A simultaneous contraction of the external intercostals moves the ribs upwards and outwards, to further increase the volume of the thorax. Other thoracic muscles which can expand the thoracic cavity still further become involved in forced (deep) inspiration.

Quiet expiration is a passive process. After becoming stretched by contractions of the diaphragm and thoracic muscles, the thorax and lungs recoil as a result of their elastic tension when the respiratory muscles relax. The decrease in lung volume raises the pressure within the alveoli above the atmospheric pressure and pushes the air out.

[9]diaphragm: Gk. *dia,* across; *phragma,* fence

Figure 15.10. The position of the principal muscles of inspiration. Contraction of these muscles enlarges the size of the thoracic cavity.

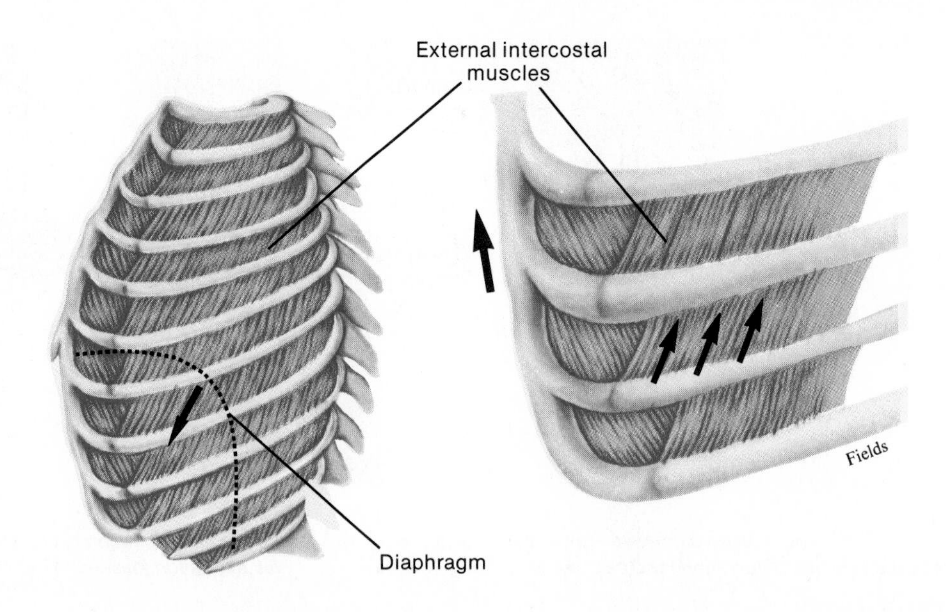

Table 15.1 Summary of the mechanisms involved in normal, quiet ventilation and forced ventilation

	Inspiration	Expiration
Normal, quiet breathing	Contraction of the diaphragm and external intercostal muscles increases the thoracic and lung volume, decreasing intrapulmonary pressure to about −3 mm Hg.	Relaxation of the diaphragm and external intercostals, plus elastic recoil of lungs, decreases lung volume and increases intrapulmonary pressure to about +3 mm Hg.
Forced ventilation	Inspiration, aided by contraction of accessory muscles such as the scalenes and sternocleidomastoid, decreases intrapulmonary pressure to −20 mm Hg or less.	Expiration, aided by contraction of abdominal muscles and internal intercostal muscles, increases intrapulmonary pressure to +30 mm Hg or more.

During forced expiration the *internal intercostal* muscles contract and depress the rib cage. The abdominal muscles may also aid expiration, because when they contract they force abdominal organs up against the diaphragm and further decrease the volume of the thorax. The events that occur during inspiration and expiration are summarized in table 15.1 and shown in figure 15.11.

Pulmonary Disorders

People with pulmonary disorders frequently complain of **dyspnea,**[10] which is a subjective feeling of "shortness of breath." Some of the terms used to define ventilation are defined in table 15.2.

Asthma

The obstruction of air flow through the bronchioles may occur as a result of excessive mucus secretion, inflammation, and/or contraction of the smooth muscles in the bronchioles. **Asthma**[11] results from bronchiolar constriction, which increases airway resistance and makes breathing difficult. Constriction of the bronchiolar smooth muscles is stimulated by molecules called *leukotrienes* (a type of fatty acid related to prostaglandins) and to a lesser degree by histamine, released by leukocytes and mast cells. This can be provoked by an allergic reaction (chapter 13) or by stimulation from parasympathetic nerve endings.

Epinephrine and related compounds promote bronchodilation. Asthmatics can, therefore, take epinephrine as an inhaled spray to relieve the symptoms of an asthma attack. Bronchodilation occurs as a result of the combination of epinephrine with its receptor proteins on bronchiole smooth muscle cells. These receptors are known as adrenergic receptors, of a subtype designated β (beta). It has been learned that there are two subtypes of beta receptors for epinephrine and that the subtype in the heart (called β_1) is different from the one in the bronchioles (β_2). Drugs have been developed that selectively stimulate the β_2 adrenergic receptors, causing bronchodilation, without stimulating the heart to the extent that epinephrine does.

Emphysema

Alveolar tissue is destroyed in **emphysema,**[12] resulting in fewer but larger alveoli (fig. 15.12). This produces a decreased surface area for gas exchange and a decreased ability of the bronchioles to remain open during expiration. Collapse of the

[10]dyspnea: Gk. *dys,* bad; *pnoe,* breathing
[11]asthma: Gk. *asthma,* panting

[12]emphysema: Gk. *emphysan,* blow up, inflate

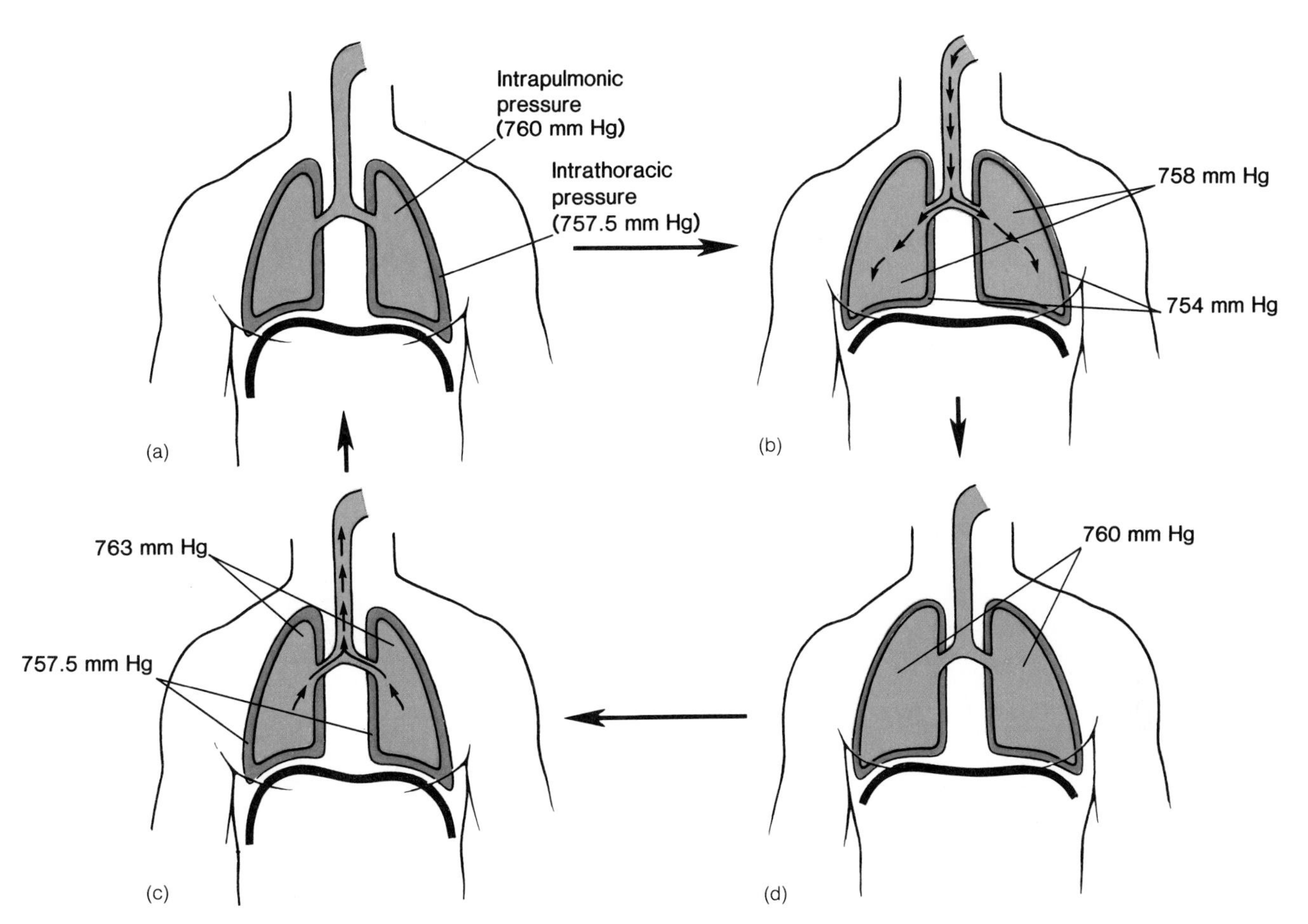

Figure 15.11. *The mechanics of pulmonary ventilation. (a) The lungs and pleural cavities prior to inspiration. (b) During inspiration, the diaphragm is contracted and the rib cage elevated; intrapleural pressure is reduced and air inflates the lungs. (c) Inspiration is completed as intrapulmonary pressure is equal to atmospheric pressure. (d) As the muscles of inspiration are relaxed, the rib cage recoils and the intrapleural and intrapulmonary pressures are raised until the intrapulmonary pressure equals the atmospheric pressure (a).*

Table 15.2 Definitions of some terms used to describe ventilation

Term	Definition
Air spaces	Alveolar ducts, alveolar sacs, and alveoli
Airways	Structures that conduct air from the mouth and nose to the respiratory bronchioles
Anatomical dead space	Volume of the conducting airways to the zone where gas exchange occurs
Apnea	Cessation of breathing
Dyspnea	Unpleasant, subjective feeling of difficult or labored breathing
Eupnea	Normal, comfortable breathing at rest
Hyperventilation	Alveolar ventilation that is excessive in relation to metabolic rate; results in abnormally low alveolar CO_2
Hypoventilation	An alveolar ventilation that is low in relation to metabolic rate; results in abnormally high alveolar CO_2
Pneumothorax	Presence of gas in the pleural space (the space between the visceral and parietal pleural membranes) causing lung collapse
Torr	Synonymous with millimeters of mercury (760 mm Hg = 760 torr)

bronchioles as a result of the compression of the lungs during expiration produces *air trapping,* which further decreases the efficiency of gas exchange in the alveoli.

There are different types of emphysema. The most common type occurs almost exclusively in people who have smoked cigarettes heavily over a period of years. A component of cigarette smoke apparently stimulates the macrophages and leukocytes to secrete protein-digesting enzymes that destroy lung tissues. This is why the chances of getting emphysema are over twenty times greater for smokers than for nonsmokers.

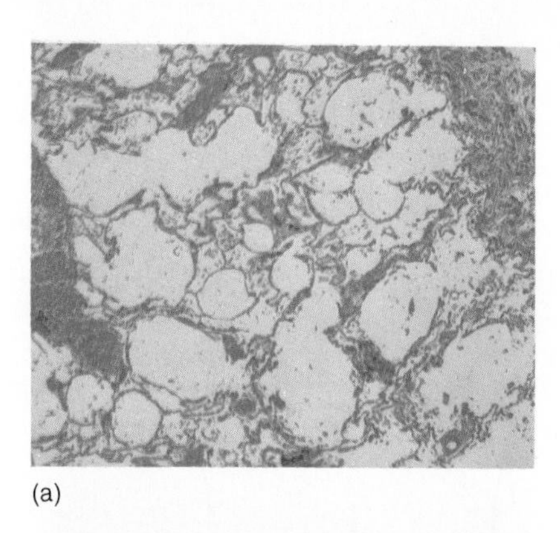

(a)

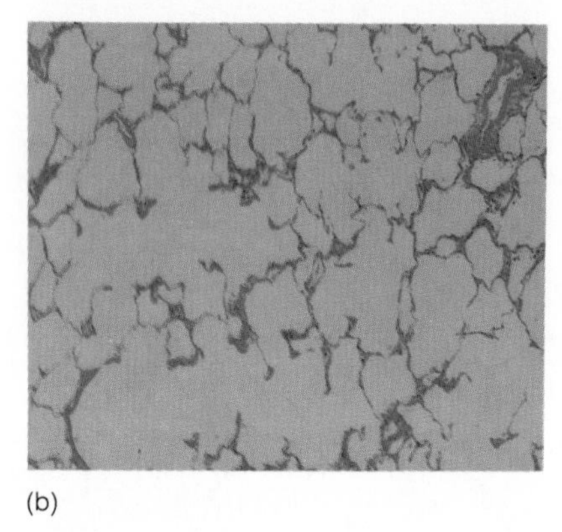

(b)

Figure 15.12. *Photomicrographs of tissue from a normal lung (a) and from the lung of a person with emphysema (b). In emphysema, lung tissue is destroyed, resulting in the presence of fewer and larger alveoli.*

Four out of five people who die from lung disease (*not* counting cancer) are smokers. The number of these deaths (over 71,000 Americans per year at last count), furthermore, increased by 33% from 1979 to 1986. The 33% figure is an overall rate for both men and women; the death rate from lung disease among women actually increased by 80% during this time period. Women are thus catching up to men in their incidence of emphysema, as they are for their incidence of lung cancer (chapter 14). Like the statistics for lung cancer, the delayed increase in emphysema among women compared to men can also be attributed to the fact that women began smoking in large number more recently than did men.

Chronic bronchitis and emphysema, the most common causes of respiratory failure, together are called **chronic obstructive pulmonary disease (COPD).** In addition to the more direct aspects of these conditions, other pathological changes may occur. These include edema, inflammation, hyperplasia (increased cell number), zones of pulmonary fibrosis, pneumonia, pulmonary emboli (traveling blood clots), and heart failure. Patients with severe emphysema may eventually develop *cor pulmonale*—pulmonary hypertension with hypertrophy and the eventual failure of the right ventricle.

Pulmonary Fibrosis

Under certain conditions, for reasons that are poorly understood, lung damage leads to pulmonary fibrosis instead of emphysema. In this condition the normal structure of the lungs is disrupted by the accumulation of fibrous connective tissue proteins. Fibrosis can result, for example, from the inhalation of particles less than 6 μm in size, which are able to accumulate in the respiratory zone of the lungs. Included in this category is *anthracosis,* or black lung, which is produced by the inhalation of carbon particles from coal dust.

1. Describe how the intrapulmonary and intrapleural pressures change during inspiration and expiration, and explain the reasons for these changes in terms of Boyle's law.
2. Define the term *elasticity,* and explain how this lung property affects inspiration and expiration.
3. Explain the significance of lung surfactant in normal physiology and in the respiratory distress syndrome.
4. Describe the actions of the diaphragm and external intercostal muscles during inspiration, and explain how quiet expiration is produced.
5. Describe the nature of asthma and emphysema.

Gas Exchange in the Lungs

Gas exchange between the alveolar air and the blood in pulmonary capillaries results in an increased concentration of oxygen, and a decreased concentration of carbon dioxide, in the blood leaving the lungs. This blood enters the systemic arteries, where blood gas measurements are taken to assess the effectiveness of lung function.

The atmosphere is an ocean of gas that exerts pressure on all objects within it. The amount of this pressure can be measured with a glass U-tube filled with fluid. One end of the U-tube is exposed to the atmosphere, while the other side is continuous with a sealed vacuum tube. Since the atmosphere presses on the exposed side, but not on the side connected to the vacuum tube, atmospheric pressure pushes fluid in the U-tube up on the vacuum side to a height determined by the atmospheric pressure and the density of the fluid. Water, for example, would be pushed up to a height of 33.9 feet (10,332 mm) at sea level, whereas mercury (Hg)—which is more dense—is raised to a height of 760 mm. As a matter of convenience, therefore, devices used to measure atmospheric pressure (barometers) use mercury rather than water. The atmospheric pressure at sea level is thus said to be equal to *760 mm Hg* (or *760 Torr*), which is also described as a pressure of *one atmosphere* (fig. 15.13).

According to **Dalton's law,**[13] the total pressure of a gas mixture (such as air) is equal to the sum of the pressures that each gas in the mixture would exert independently. The pressure exerted by a particular gas in a mixture is known as its **partial pressure.** Applying Dalton's law, the partial pressure of a particular gas can be calculated by multiplying the percent of the gas in the mixture by the total pressure of the mixture. Since

[13]Dalton's law: after John Dalton, English chemist, 1766–1844

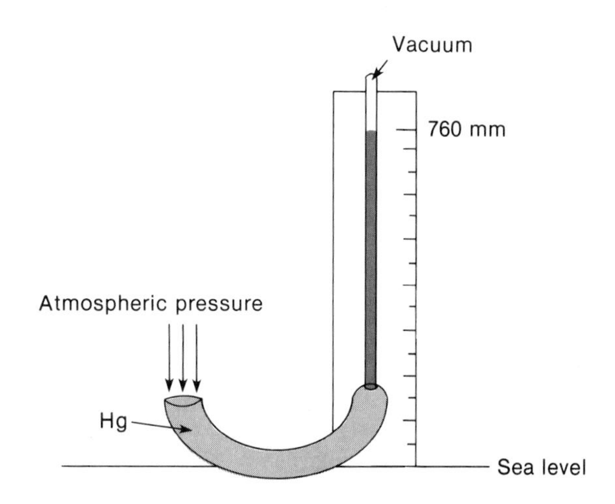

Figure 15.13. Atmospheric pressure at sea level can push a column of mercury to a height of 760 millimeters. This is also described as 760 torr, or one atmospheric pressure.

Table 15.3 The effect of altitude on P_{O_2}

Changes in P_{O_2} at Various Altitudes

Altitude (feet above sea level)	Atmospheric pressure (mm Hg)	P_{O_2} in air (mm Hg)	P_{O_2}in alveoli (mm Hg)	P_{O_2} in arterial blood (mm Hg)
0	760	159	105	100
2,000	707	148	97	92
4,000	656	137	90	85
6,000	609	127	84	79
8,000	564	118	79	74
10,000	523	109	74	69
20,000	349	73	40	35
30,000	226	47	21	19

oxygen comprises about 21% of the atmosphere, for example, its partial pressure (abbreviated P_{O_2}) is 21% of 760, or about 159 mm Hg. Since nitrogen comprises about 78% of the atmosphere, its partial pressure is equal to $0.78 \times 760 = 593$ mm Hg. These two gases thus contribute about 99% of the total pressure of 760 mm Hg.

Effect of Altitude

As one goes to a higher altitude, the total atmospheric pressure and the partial pressure of the constituent gases decrease (table 15.3). At Denver, for example (5,000 feet above sea level), the atmospheric pressure is decreased to 619 mm Hg, and the P_{O_2} is therefore reduced to $619 \times 0.21 = 130$ mm Hg. At the peak of Mount Everest (at 29,000 feet), the P_{O_2} is only 42 mm Hg. As one descends below sea level, as in ocean diving, the total pressure increases by one atmosphere for every 33 feet. At 33 feet therefore, the pressure equals $2 \times 760 = 1{,}520$ mm Hg. At 66 feet, the pressure equals three atmospheres.

The P_{O_2} of the air within the lung alveoli is always a little less than the calculated P_{O_2} of the air that is breathed. This is partly due to the fact that the lungs are not 100% efficient, and partly to the fact that water vapor in the air has a dilution effect that was not previously considered.

Partial Pressures of Gases in Blood

The enormous surface area of alveoli and the short distance between alveolar air and capillary blood allow quick diffusion of gases between air and blood. This function is further aided by the fact that each alveolus is surrounded by so many capillaries that they form an almost continuous sheet of blood around the alveoli (fig. 15.14).

The amount of a gas that can be dissolved in a fluid depends on (1) the solubility of the gas in the fluid, which is a physical constant; (2) the temperature of the fluid—more gas can be dissolved in cold water than warm water; and (3) the partial pressure of the gas. Since the temperature of the blood does not vary significantly, *the concentration of a gas dissolved in a fluid (such as plasma) depends directly on its partial pressure in the gas mixture.* When water—or plasma—is brought into equilibrium with air at a P_{O_2} of 100 mm Hg, for example, the fluid will

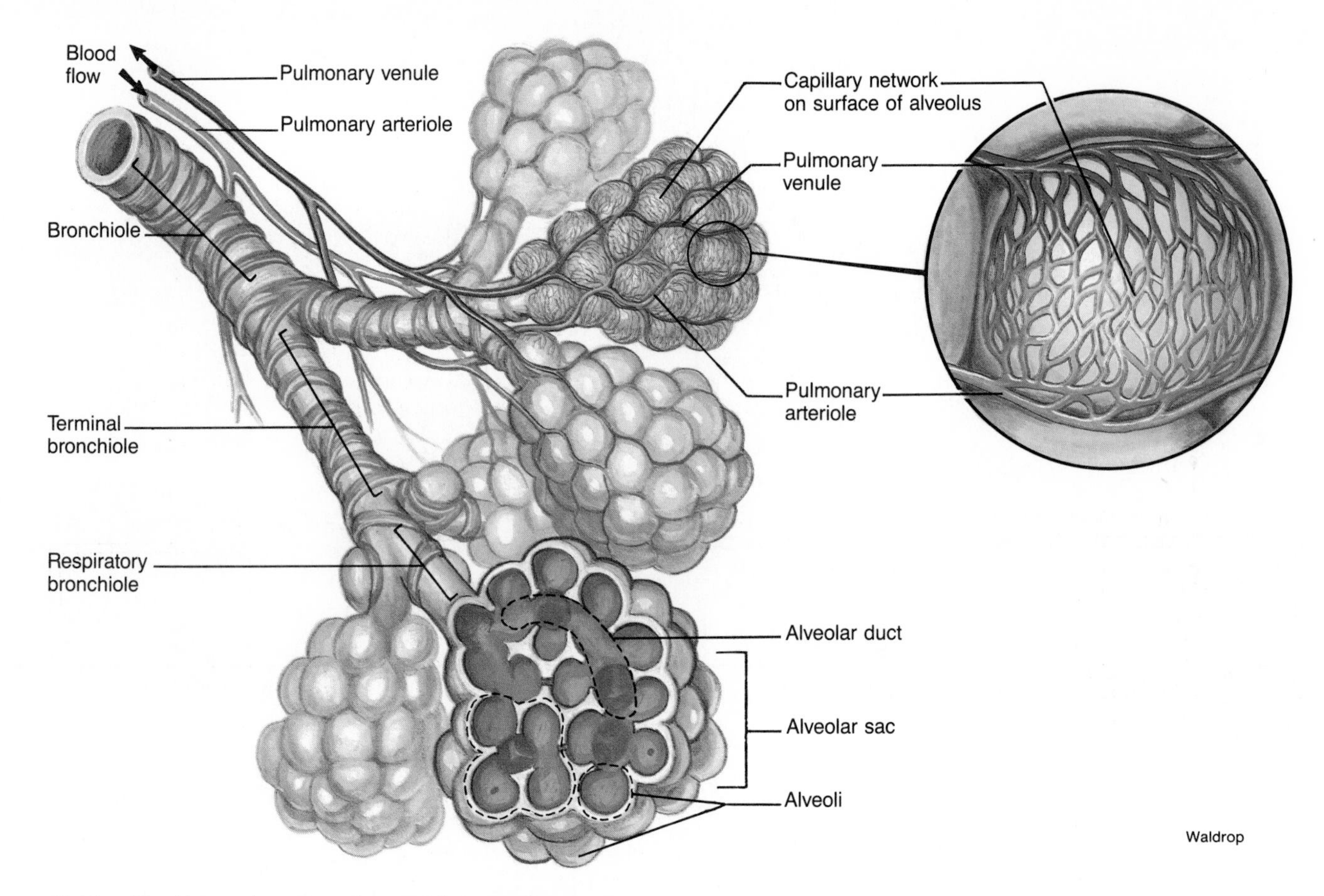

Figure 15.14. The high surface area of contact between the pulmonary capillaries and the alveoli allows rapid exchange of gases between the air and blood.

contain 0.3 ml O_2 per 100 ml of fluid at body temperature. If the P_{O_2} of the gas were reduced by half, the amount of dissolved oxygen would also be reduced by half.

The amount of oxygen dissolved in the blood plasma is thus directly proportional to the partial pressure of oxygen in the alveolar air. Since blood leaving the lungs returns to the left atrium and is pumped by the left ventricle into the systemic circulation (chapter 12), the blood from a systemic artery is commonly used for clinical blood-gas measurements. (This differs from all other blood tests, in which the blood from a systemic vein is used.) Since the amount of dissolved oxygen is proportional to the P_{O_2}, the dissolved oxygen in arterial blood plasma is clinically indicated using P_{O_2} units (fig. 15.15).

Disorders Caused by High Blood-Gas Concentrations

The total atmospheric pressure increases by one atmosphere (760 mm Hg) for every 10 m (33 feet) below sea level. If one dives 10 m below sea level, therefore, the amounts of dissolved gases in the plasma are twice that at sea level. At 66 feet they are three times, and at 100 feet they are four times the values at sea level. The increased amounts of nitrogen and oxygen dissolved in the blood plasma under these conditions can have serious effects on the body.

Oxygen Toxicity

Although breathing 100% oxygen at one or two atmospheres pressure can be safely tolerated for a few hours, higher partial oxygen pressures can be very dangerous. *Oxygen toxicity* develops rapidly when the P_{O_2} rises above about 2.5 atmospheres. This is apparently caused by the oxidation of enzymes and other destructive changes that can damage the nervous system and lead to coma and death. For these reasons, deep-sea divers commonly use gas mixtures in which oxygen is diluted with inert gases such as nitrogen (as in ordinary air) or helium.

Hyperbaric oxygen—oxygen at greater than one atmosphere pressure—is often used to treat conditions such as carbon monoxide poisoning, circulatory shock, and gas gangrene. Before the dangers of oxygen toxicity were realized these hyperbaric oxygen treatments sometimes resulted in tragedy. Particularly tragic were the cases of **retrolental fibroplasia,** in which damage to the retina of the eyes (chapter 20) and blindness resulted from hyperbaric oxygen treatment of premature babies who had hyaline membrane disease.

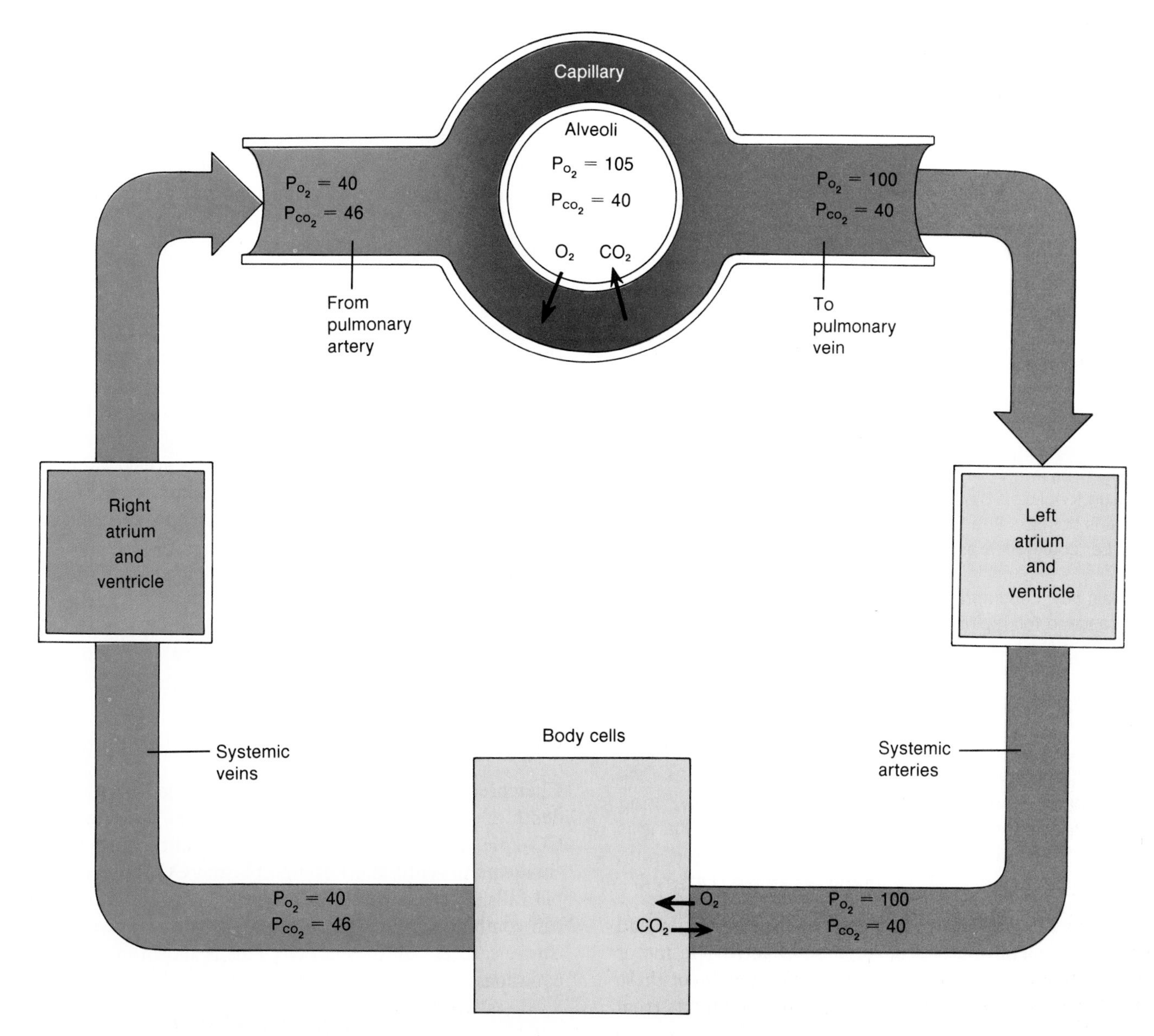

Figure 15.15. *The P_{O_2} and P_{CO_2} of blood as a result of gas exchange in the lung alveoli and gas exchange between systemic capillaries and body cells.*

Nitrogen Narcosis

Although at sea level nitrogen is physiologically inert, larger amounts of dissolved nitrogen can have dangerous effects. Since it takes time for the nitrogen to dissolve, these effects usually don't appear until the person has remained submerged over an hour. *Nitrogen narcosis* resembles alcohol intoxication; depending on the depth of the dive, the diver may experience "rapture of the deep" or may become so drowsy that he or she is totally incapacitated.

Decompression Sickness

The fact that more of a gas is dissolved at higher than at lower gas pressures can easily be demonstrated with a champagne bottle. When the bottle is corked, there is a high pressure in the gas above the champagne within the bottle. At this high pressure, there is a large amount of carbon dioxide dissolved in the fluid (and no bubbles can be seen within the champagne). When the cork is loosened, the higher pressure within the bottle is relieved as gas rushes out and the cork pops. Now that the air in the champagne is at a lower pressure, less carbon dioxide can remain dissolved. Much of the CO_2 thus comes out of solution, producing the bubbles seen in a glass of champagne.

A similar effect, involving nitrogen gas instead of CO_2, may occur in a deep-sea diver who ascends too rapidly. The amount of nitrogen dissolved in the plasma decreases as the diver ascends to sea level, due to a progressive decrease in the gas pressure. If the diver surfaces slowly, a large amount of nitrogen can diffuse through the alveoli and be eliminated in the expired breath. If decompression occurs too rapidly, however, bubbles of nitrogen gas (N_2) can form in the blood and block small blood channels, producing muscle and joint pain as well as more serious damage. These effects are known as *decompression sickness,* or the bends.

The cabins of airplanes that fly long distances at high altitudes (30,000 to 40,000 feet) are pressurized so that the passengers and crew are not exposed to the very low atmospheric pressures of these altitudes. If a cabin were to become rapidly depressurized at high altitude, much less nitrogen could remain dissolved at the greatly lowered pressure. People in this situation, like the divers who ascend too rapidly, would thus experience decompression sickness.

1. Describe how the P_{O_2} of air is calculated and how this value is affected by altitude and deep-sea diving.
2. Explain why systemic arterial blood, rather than venous blood, is used for clinical blood-gas measurements.
3. What is normal value for arterial P_{O_2} and P_{CO_2}? How would this number change if a person climbed a mountain to 10,000 feet?
4. Explain how decompression sickness is produced in divers who ascend too rapidly.

Regulation of Breathing

Breathing is accomplished by contraction and relaxation of particular muscles under the direction of motor neurons. The neural control of involuntary breathing is regulated by a respiratory center in the medulla oblongata of the brain. Changes in the blood gases and pH influence the respiratory center, causing compensatory changes in the breathing pattern.

Inspiration and expiration are produced by the contraction and relaxation of skeletal muscles in response to activity in motor neurons (chapter 6) from the spinal cord. The activity of these motor neurons, in turn, is controlled by descending tracts from the brain. The automatic control of breathing is regulated by nerve fibers that descend to the spinal cord from the medulla oblongata. The voluntary control of breathing is a function of the cerebral cortex and involves nerve fibers that descend in a different pathway.

The separation of the voluntary and involuntary pathways is dramatically illustrated in the condition called **Ondine's curse,** in which neurological damage abolishes the automatic but not the voluntary control of breathing. People with this condition must consciously force themselves to breathe and be put on artificial respirators when they sleep.

Brain Stem Respiratory Centers

A loose aggregation of neurons in the *medulla oblongata* forms the **rhythmicity center** that controls automatic breathing. The rhythmicity center consists of interacting groups of neurons that fire either during inspiration or expiration. The cycle of inspiration and expiration is thus built into the neural activity of the medulla. The rhythmicity center in the medulla directly controls nerves that innervate the diaphragm and other respiratory muscles.

The automatic control of breathing is also influenced by input from receptors sensitive to the chemical composition of the blood. There are two groups of *chemoreceptors* that respond to changes in the dissolved plasma CO_2, pH (H^+), and dissolved plasma O_2 of the arterial blood. These are the **central chemoreceptors** in the medulla oblongata and the **peripheral chemoreceptors.** The peripheral chemoreceptors are contained within small nodules associated with the aorta and the carotid arteries. The peripheral chemoreceptors include the **aortic bodies,** located around the aortic arch, and the **carotid bodies,** located just beyond the branching of each common carotid into the internal and external carotid arteries (fig. 15.16). The aortic and carotid bodies should not be confused with the aortic and carotid sinuses (chapter 12), which are located within these arteries and contain receptors that monitor the blood pressure.

Effects of Blood Gases and pH on Ventilation

Chemoreceptor input to the brain stem modifies the rate and depth of breathing so that, under normal conditions, plasma CO_2, pH, and O_2 remain relatively constant. If hypoventilation (inadequate ventilation) occurs, plasma CO_2 quickly rises and pH falls. The fall in pH is due to the fact that carbon dioxide can combine with water to form carbonic acid, which in turn can release H^+ to the solution. This is shown in the following equations:

$$CO_2 + H_2O \rightarrow H_2CO_3$$

$$H_2CO_3 \rightarrow HCO_3^- + H^+$$

The rate and depth of breathing are normally adjusted to maintain a relatively constant plasma CO_2 and pH; proper oxygenation of the blood is a by-product of this regulation. If a person does not breathe adequately, for example, the plasma CO_2 will rise and the pH will fall. The fall in plasma pH stimulates chemoreceptors in the aortic and carotid bodies. The rise in plasma CO_2 also causes a fall in the pH of cerebrospinal fluid (CSF), which stimulates chemoreceptors in the medulla oblongata (fig. 15.17). As a result of chemoreceptor stimulation, breathing will be increased so that plasma CO_2 and pH will be brought back to the normal levels. In the process, more oxygen will also enter the blood.

If the plasma CO_2 rises above a normal level, the person is not breathing adequately to "blow off" the proper amount of carbon dioxide. By definition, this condition is called **hypoventilation.** If the arterial CO_2 is much below normal, conversely, the person must be breathing excessively. This condition is called **hyperventilation.**

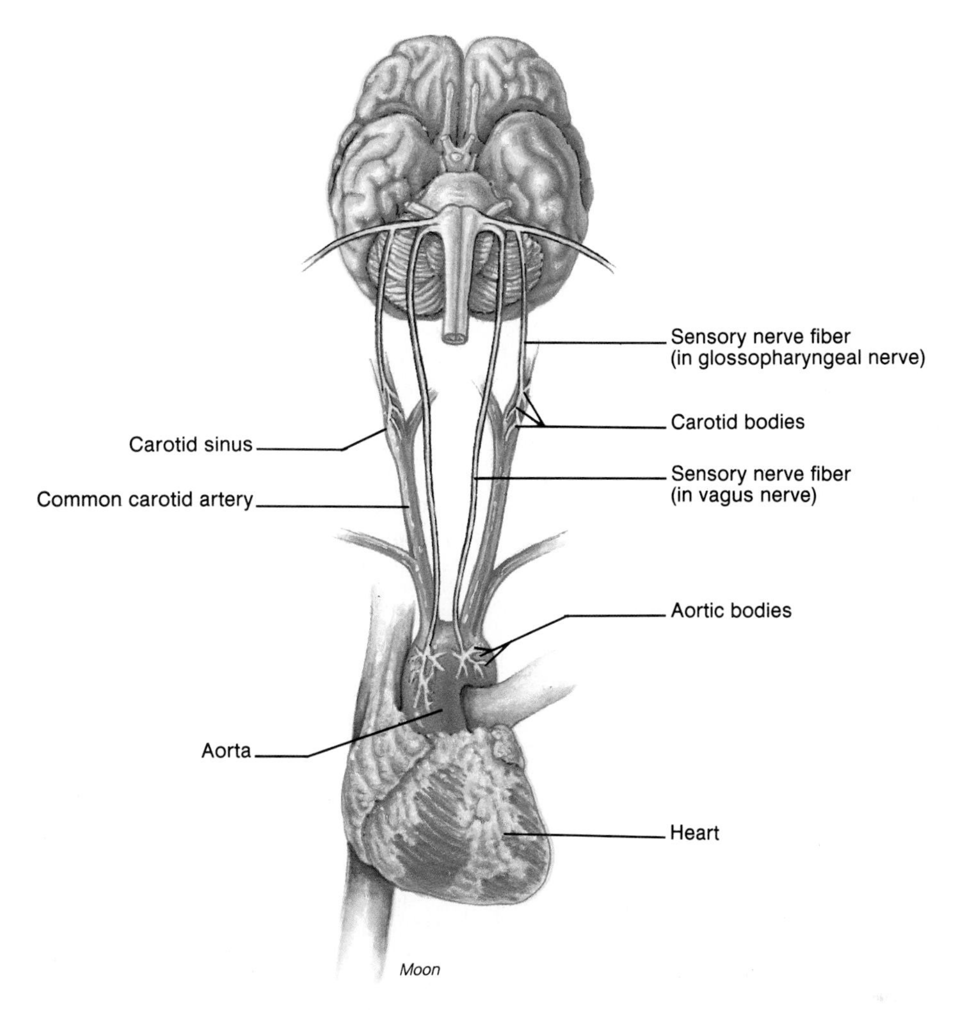

Figure 15.16. *The peripheral chemoreceptors (aortic and carotid bodies) regulate the brain stem respiratory centers by means of sensory nerve stimulation.*

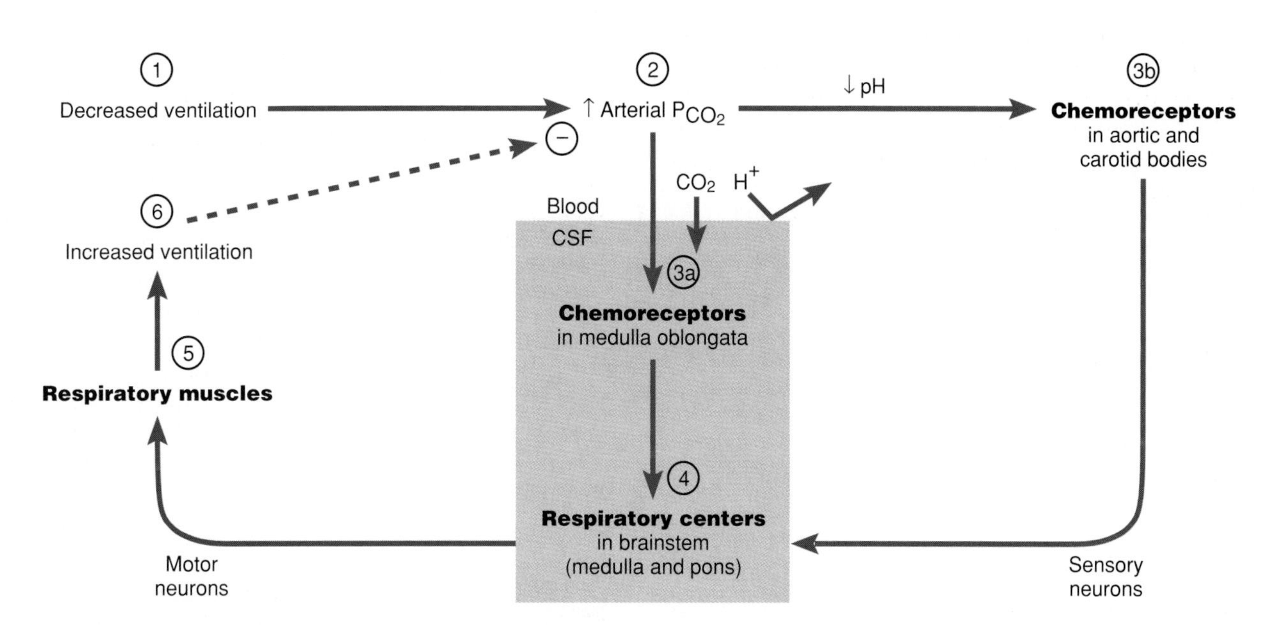

Figure 15.17. *The control of ventilation by blood CO_2 and pH. The box represents the blood-brain barrier, which allows CO_2 to pass into the cerebrospinal fluid but prevents the passage of H^+.*

People who hyperventilate during psychological stress are sometimes told to breathe into a paper bag so that they rebreathe their expired air that is enriched in CO_2. This procedure helps to raise their plasma CO_2 back up to the normal range. This is needed because low blood CO_2 causes cerebral vasoconstriction. In addition to producing dizziness, the cerebral ischemia that results can cause further hyperventilation. Breathing into a paper bag can thus raise the blood CO_2 and stop the hyperventilation.

The plasma O_2 concentration has a secondary role in the regulation of breathing. If a person goes hiking or skiing at a high elevation, the low P_{O_2} of the air produces a low plasma O_2 concentration. This is sensed by chemoreceptors in the carotid bodies, which, as a result, become more sensitive to the plasma CO_2 and pH. Increased breathing is thus stimulated indirectly by the low plasma O_2. (This is why people who first arrive at a high elevation commonly hyperventilate.) At very high elevations, plasma O_2 may drop to half-normal concentrations; at this point, the low plasma O_2 can directly stimulate breathing. Breathing in people with severe emphysema may also be stimulated by a low plasma O_2. However, it should be remembered that normal ventilation at sea level is controlled by the plasma CO_2 and pH rather than by the plasma O_2.

There are a variety of disease processes that can produce cessation of breathing during sleep, or *sleep apnea.* **Sudden infant death syndrome (SIDS)** is an especially tragic form of sleep apnea that claims the lives of about ten thousand babies annually in the United States. Victims of this condition are apparently healthy two-to-five-month-old babies who die in their sleep without apparent reason—hence, the layman's term of *crib death.* These deaths seem to be caused by failure of the respiratory control mechanisms in the brain stem and/or by failure of the carotid bodies to be stimulated by reduced arterial oxygen.

1. Identify the parts of the brain that serve as the control centers for breathing.
2. Distinguish between the central and peripheral chemoreceptors, and give their locations.
3. What chemical changes in the blood are most important in stimulating breathing? What role does oxygen play in the regulation of breathing?
4. Define the terms *hyperventilation* and *hypoventilation.*

Oxygen and Carbon Dioxide Transport

Hemoglobin without oxygen, or deoxyhemoglobin, can bond to oxygen to form oxyhemoglobin. This "loading" reaction occurs in the capillaries of the lungs. The dissociation of oxyhemoglobin, or "unloading" reaction, occurs in the tissue capillaries. Carbon dioxide from the tissues is carried largely as bicarbonate in the blood, and then reconverted to carbon dioxide gas in the lungs.

If the lungs are functioning properly, blood leaving in the pulmonary veins and traveling in the systemic arteries has a plasma oxygen concentration of 0.3 ml of O_2 per 100 ml of blood. The total oxygen concentration of blood, however, is much higher, due to its hemoglobin[14] content. If the oxygen and hemoglobin concentration are normal, arterial blood contains about 20 ml of O_2 per 100 ml of blood (fig. 15.18).

Hemoglobin

Most of the oxygen in the blood is contained within the red blood cells, where it is chemically bonded to **hemoglobin.** Each hemoglobin molecule consists of (1) a protein globin part, composed of four polypeptide chains (discussed in chapter 11); and (2) four nitrogen-containing, disc-shaped organic pigment molecules, called *hemes.*

Each of the four polypeptide chains is combined with one heme group. In the center of each heme group is one atom of iron, which can combine with one molecule of oxygen (O_2). One hemoglobin molecule can thus combine with four molecules of oxygen. Since there are about 280 million hemoglobin molecules per red blood cell, each red blood cell can carry over a billion molecules of oxygen.

Normal hemoglobin can share electrons and bond to oxygen to form **oxyhemoglobin.** When oxyhemoglobin dissociates to release oxygen to the tissues, the hemoglobin is called **deoxyhemoglobin.**

Carboxyhemoglobin is an abnormal form of hemoglobin, in which the heme is combined with *carbon monoxide* instead of oxygen. Since the bond with carbon monoxide is 210 times stronger than the bond with oxygen, carbon monoxide tends to displace oxygen in hemoglobin and remains attached to hemoglobin. The transport of oxygen to the tissues is thus reduced in carbon monoxide poisoning.

According to federal standards, the percent of carboxyhemoglobin in the blood of active nonsmokers should not be higher than 1.5%. Studies have shown, however, that these values in many cities can range as high as 3% or more in nonsmokers and as high as 10% or more in smokers. Although these high levels may not cause immediate problems in healthy people, long-term effects on health might occur. People with respiratory or cardiovascular diseases would be particularly vulnerable to the negative effects of carboxyhemoglobin on oxygen transport. Fortunately, the carbon monoxide will be gradually eliminated from the blood once a person moves to an environment with cleaner air.

The *oxygen-carrying capacity* of whole blood is determined by its concentration of normal hemoglobin. If the hemoglobin concentration is below normal—a condition called **anemia**[15]—the oxygen concentration of the blood is reduced below normal.

[14]hemoglobin: Gk. *haima,* blood; L. *globus,* globe
[15]anemia: Gk. *a,* negative; *haima,* blood

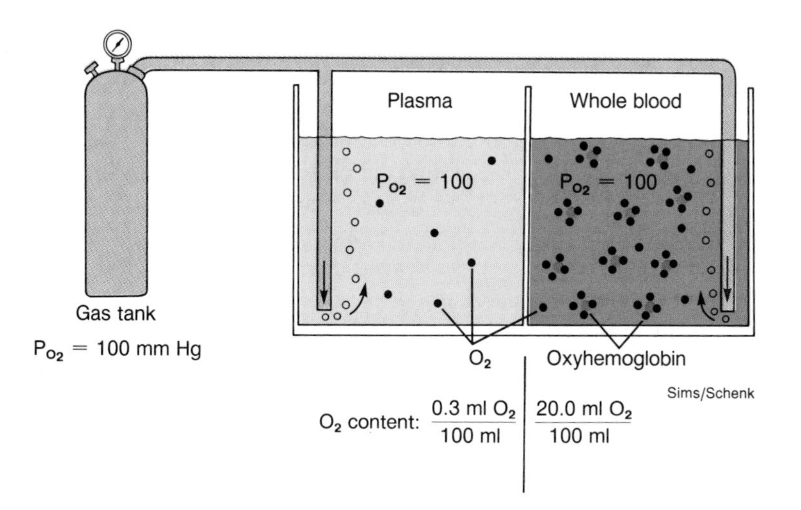

Figure 15.18. Plasma and whole blood that are brought into equilibrium with the same gas mixture have the same P_{O_2} and thus the same amount of dissolved oxygen molecules (shown as black dots). The oxygen content of whole blood, however, is much higher than that of plasma because of the binding of oxygen to hemoglobin.

Social Issues

Air Pollution and Lung Function

Pollutants in the air include particulate matter (as in soot, or the exhaust from a diesel engine), incompletely combusted organic molecules (giving rise to the eye irritation characteristic of smog), and gases such as carbon monoxide (CO) and ozone (O_3). The nature of air pollution is discussed in chapter 22, but the health hazards of air pollution relate most specifically to lung function.

The constituents of air pollution may contribute to pulmonary fibrosis and emphysema, although the extent of this contribution is controversial. The particulate matter in smoke and the organic molecules within smog also irritate the air passages of the lungs. This irritation promotes a neural reflex that causes constriction of the bronchioles. Bronchoconstriction is also evoked by ozone, a prominent constituent of smog. When the bronchioles constrict, the resistance to air flow becomes greater, and more effort is required to move air through the airways. The ability to perform exercise is impaired, because more time is required to move a given volume of air into and out of the lungs. A number of studies have suggested that it is actually healthier to refrain from exercising on a smoggy day.

The carbon monoxide present in smog causes the production of abnormal amounts of carboxyhemoglobin. As previously described, this form of hemoglobin cannot carry oxygen; a person with 15% carboxyhemoglobin, for example, has 15% less oxygen in the blood than if no carboxyhemoglobin were present. (Fifteen percent is not an unreasonable value for a cigarette smoker who drives professionally in a smoggy city.)

Studies have shown that the carboxyhemoglobin concentrations are higher in many cities (Los Angeles, New York, and others) than in more rural areas. According to federal standards, everyone in these cities suffers from a degree of carbon monoxide poisoning. An interesting footnote to this is that laboratory technicians in such a city who take blood-gas measurements, including the percent carboxyhemoglobin, can tell if the blood sample came from an out-of-towner!

Within a smoggy city, the carboxyhemoglobin concentrations are higher in people who live closer to major traffic arteries. This illustrates the fact that automobile emissions are the most significant contributor to air pollution. Recognition of this fact has prompted the California state government to institute progressive restrictions on allowable automobile emissions, and similar restrictions are currently being considered by the federal government. These regulations also require the automobile industry to produce a higher proportion of smaller, more fuel-efficient cars, and to produce a certain number of cars that run on a gasoline-methyl alcohol ("gasohol") mixture. A surtax on gasoline is considered by some legislators as another possible means of reducing gasoline consumption and thus automobile emissions.

It should be understood that cleaning our air will be expensive in the short run. But then, the ill-effects of hypoxemia (low blood oxygen), and the increased risk of pulmonary diseases are not without costs. Considering the health risks of air pollution, clean air is a bargain at any price.

Conversely, when the hemoglobin concentration is increased above the normal range—as occurs in **polycythemia**[16] (high red blood cell count)—the oxygen-carrying capacity of blood is increased accordingly. This can occur as an adaptation to life at a high altitude.

The production of hemoglobin and red blood cells in bone marrow is controlled by a hormone called **erythropoietin,**[17] produced primarily by the kidneys. The secretion of erythropoietin—and thus the production of red blood cells—is stimulated when the delivery of oxygen to the kidneys and other organs is

[16]polycythemia: Gk. *polys,* much; *kytos,* cell; *haima,* blood

[17]erythropoietin: Gk. *erythros,* red; *poiesis,* a making

lower than normal. Red blood cell production is also promoted by androgens, which explains why the hemoglobin concentration in men averages higher than in women.

The Loading and Unloading Reactions

Deoxyhemoglobin and oxygen combine to form oxyhemoglobin; this is called the **loading reaction.** Oxyhemoglobin, in turn, dissociates to yield deoxyhemoglobin and free oxygen molecules; this is the **unloading reaction.** The loading reaction occurs in the lungs and the unloading reaction occurs in the systemic capillaries.

Loading and unloading can thus be shown as a reversible reaction:

$$\text{Deoxyhemoglobin} + O_2 \underset{\text{tissues}}{\overset{\text{lungs}}{\rightleftharpoons}} \text{Oxyhemoglobin}$$

The extent that the reaction will go in each direction depends on two factors: (1) the oxygen concentration of the plasma; and (2) the *affinity,* or bond strength, between hemoglobin and oxygen. High plasma O_2 drives the equation to the right (favors the loading reaction); at the high plasma O_2 of the pulmonary capillaries almost all the deoxyhemoglobin molecules combine with oxygen. Low plasma O_2 in the systemic capillaries drives the reaction in the opposite direction to promote unloading.

The affinity (bond strength) between hemoglobin and oxygen also influences the loading and unloading reactions. A very strong bond would favor loading but inhibit unloading; a weak bond would hinder loading but improve unloading. The bond strength between hemoglobin and oxygen is normally strong enough so that 97% of the hemoglobin leaving the lungs is in the form of oxyhemoglobin; this is known as the *percent saturation* of the blood. The bond is sufficiently weak, however, so that adequate amounts of oxygen are unloaded in the tissues. Under normal resting conditions about 20% of the oxygen carried by the arterial blood is unloaded to the tissues.

The bond strength between hemoglobin and oxygen is weakened as the temperature increases. The blood going through muscle capillaries when the muscles are warmed during exercise thus can release a higher percentage of the oxygen it carries. Also, the affinity of hemoglobin for oxygen is decreased when there is a fall in pH (increased acidity). This also occurs during exercise, when the muscles produce increased amounts of carbon dioxide (forming carbonic acid) and lactic acid. During exercise, therefore, a higher percentage of the oxygen carried by arterial blood is delivered to the tissues than is the case at rest.

Transport of Carbon Dioxide

Carbon dioxide is carried by the blood in three forms: (1) as *dissolved CO_2*; (2) attached to an amino acid in hemoglobin; and (3) as *bicarbonate,* which accounts for most of the CO_2 carried by the blood.

Carbon dioxide is able to combine with water to form carbonic acid, as described in a previous section. This reaction occurs spontaneously in the plasma at a slow rate, but occurs much more rapidly within the red blood cells due to the action of the enzyme **carbonic anhydrase.** Since this enzyme is confined to the red blood cells, most of the carbonic acid is produced there rather than in the plasma. The buildup of carbonic acid concentrations within the red blood cells favors the dissociation of these molecules into H^+ (protons, which contribute to the acidity of a solution) and HCO^- (bicarbonate). The equation describing this reaction has been previously introduced.

Bicarbonate produced within the red blood cells can then diffuse into the plasma and be carried by the blood to the lungs. When blood reaches the pulmonary capillaries, HCO_3^- enters the red blood cells from the plasma, and combines with H^+ to form carbonic acid:

$$HCO_3^- + H^+ \rightarrow H_2CO_3$$

In the pulmonary capillaries, carbonic anhydrase catalyzes the conversion of carbonic acid to carbon dioxide and water:

$$H_2CO_3 \xrightarrow[\text{low } P_{CO_2}]{\text{carbonic anhydrase}} H_2O + CO_2$$

In summary (fig. 15.19), the carbon dioxide produced by the tissue cells is converted in the red blood cells to carbonic acid. With the buildup of carbonic acid concentrations, the carbonic acid dissociates into bicarbonate and H^+. A reverse of these events occurs in the pulmonary capillaries to convert carbonic acid to CO_2 gas, which is eliminated in the expired breath.

1. Describe the effect of plasma O_2 on the loading and unloading reactions.
2. Explain the nature of carbon monoxide poisoning.
3. Describe the changes that allow an increase in oxygen unloading to the muscle during exercise.
4. Explain how carbon dioxide is transported in the blood and eliminated in the exhaled air.

Effects of Exercise and High Altitude on Respiratory Function

The arterial blood gases and pH do not significantly change during moderate exercise because ventilation increases to keep pace with increased metabolism. Endurance training increases the ability of muscles to extract oxygen from the blood. Adaptations occur at high altitude in both the control of ventilation and in the oxygen transport abilities of blood to permit an adequate delivery of oxygen to the tissues.

Changes in ventilation and oxygen delivery occur during exercise and during acclimatization to a high altitude. These changes help to compensate for the increased metabolic rate during exercise and for the decreased arterial O_2 at a high altitude.

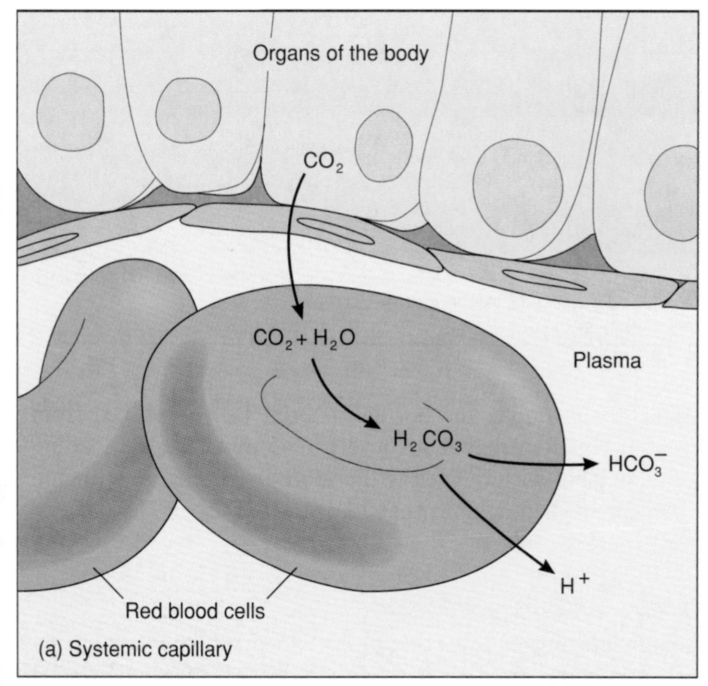

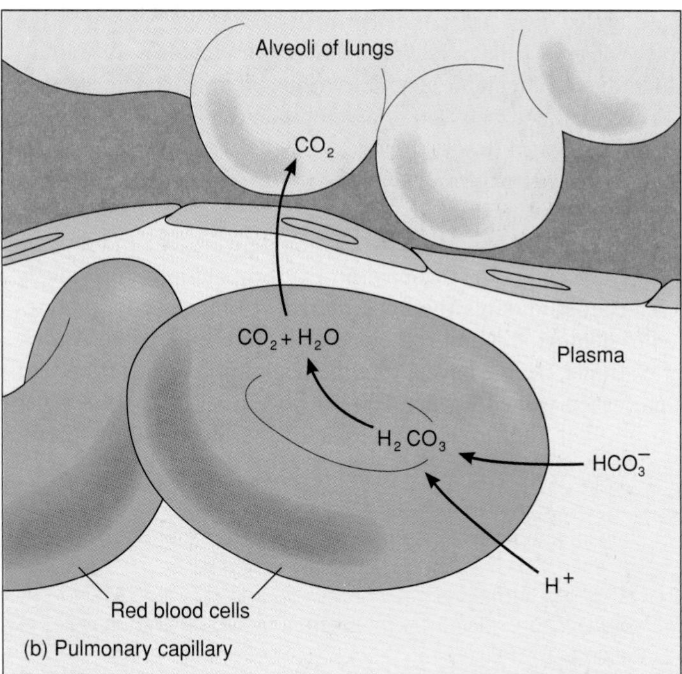

Figure 15.19. *The transport of carbon dioxide. In systemic capillaries (a), CO_2 is converted into carbonic acid and then bicarbonate. In the pulmonary capillaries (b), bicarbonate is converted into carbonic acid and then back to CO_2.*

Ventilation during Exercise

Immediately upon exercise, the rate and depth of breathing increase many times over their resting values. This increased ventilation, particularly in well-trained athletes, is exquisitely matched to the simultaneous increase in oxygen consumption and carbon dioxide production by the exercising muscles. The arterial O_2, CO_2, and pH thus remain surprisingly constant during exercise.

It is tempting to suppose that ventilation increases during exercise as a result of the increased CO_2 production by the exercising muscles, which stimulate the chemoreceptors. Ventilation increases together with increased CO_2 production, however, so that measurements of plasma CO_2 during exercise are not significantly higher than at rest. The mechanisms responsible for the increased ventilation during exercise must therefore be more complex.

Various theories have been proposed to explain the increased ventilation that occurs during exercise. It has been suggested, for example, that sensory nerve activity from the exercising limbs may stimulate the respiratory muscles, either through spinal reflexes or via the respiratory control center in the medulla oblongata. Alternatively, input from the cerebral cortex may stimulate the respiratory control center to modify ventilation. These theories are useful in explaining the immediate increase in ventilation that occurs at the beginning of exercise.

Rapid and deep ventilation continues after exercise has stopped, however, suggesting that chemical factors—specifically, the plasma CO_2 and pH—in the blood may also stimulate ventilation during exercise. The evidence suggests that all of these mechanisms may be involved in the *hyperpnea,* or increased ventilation, of exercise. (Note that hyperpnea differs from hyperventilation in that the plasma CO_2 remains in the normal range during hyperpnea but is decreased in hyperventilation.)

Anaerobic Threshold and Endurance Training

At the beginning of exercise, the cardiovascular system may not be able to deliver adequate amounts of blood and oxygen to the muscles. As a result of the time lag required to make proper cardiovascular adjustments, a "stitch in the side"—probably due to insufficient oxygen delivery to the diaphragm—may develop. After the cardiovascular adjustments have been made, a person may experience a "second wind" when the muscles receive sufficient oxygen for their needs.

Continued heavy exercise can cause a person to reach the **anaerobic threshold,** which is the maximum rate of oxygen consumption that can be attained before blood lactic acid levels rise. (Lactic acid is the product of anaerobic respiration—metabolism in the absence of oxygen, as described in chapter 18.) This occurs when 50%–60% of the *maximal oxygen uptake* (maximum rate of oxygen consumption) has been reached. The anaerobic threshold is higher in endurance-trained athletes than

Table 15.4 Changes in respiratory function during exercise

Variable	Change	Comments
Ventilation	Increased	This is not hyperventilation because ventilation is matched to increased metabolic rate. Mechanisms responsible for increased ventilation are not well understood.
Blood gases	No change	Blood-gas measurements during light, moderate, and heavy exercise show little change because ventilation is increased to match increased muscle oxygen consumption and carbon dioxide production.
Oxygen delivery to muscles	Increased	Although the total oxygen content of the blood does not increase during exercise, there is an increased rate of blood flow to the exercising muscles.
Oxygen extraction by muscles	Increased	Increased oxygen consumption lowers the plasma O_2 and lowers the affinity of hemoglobin for oxygen (due to the effect of increased temperature). More oxygen, as a result, is unloaded to the tissues. This effect is enhanced by endurance training.

it is in other people, because endurance-trained athletes have a higher maximal oxygen uptake. They therefore produce less lactic acid at a given level of exercise, and are less subject to fatigue (lactic acid contributes to fatigue, as described in chapter 21). The effects of exercise and endurance training on respiratory function are summarized in table 15.4.

Acclimatization to High Altitude

When a person who is from a region near sea level moves to a significantly higher elevation, several adjustments in the respiratory system must be made to compensate for the decreased plasma O_2 concentration at the higher altitude. These adjustments include changes in ventilation, in the hemoglobin affinity for oxygen, and in the total hemoglobin concentration.

Reference to table 15.3 indicates that at an altitude of 8,000 feet, for example, the P_{O_2} of arterial blood is 74 mm Hg (compared to 100 mm Hg at sea level). The amount of oxygen attached to hemoglobin, and thus the total oxygen content of blood, is significantly decreased. People may thus experience rapid fatigue at surprising low levels of physical activity. Compensations made by the respiratory system gradually reduce the amount of fatigue caused by a given amount of exertion at high altitudes.

Changes in Ventilation

A decrease in plasma O_2 at higher elevations stimulates the carotid bodies to increase ventilation. These changes cause *hyperventilation,* which decreases the arterial CO_2 below its normal value. Hyperventilation may cause dizziness or a general feeling of "light-headedness," but it does not cause a significant increase in the blood oxygen levels. These remain low as a result of the low P_{O_2} of the atmosphere. In the Peruvian Andes, for example, the normal arterial P_{O_2} is reduced from 100 mm Hg (its value at sea level) to 45 mm Hg. The loading of hemoglobin with oxygen is therefore incomplete, producing a percent oxyhemoglobin that is decreased from 97% (at sea level) to 81%.

Changes in Oxygen Transport

Though less oxygen is carried by the blood, a higher proportion of the amount carried is released to the tissues. Normal arterial blood at sea level unloads only about 20% of its oxygen to the tissues at rest. This percentage unloading increases at high altitude due to chemical changes within the red blood cells that weaken the bond between hemoglobin and oxygen. As a result of the increase in oxygen unloading, people are able to perform better after two or three days at high altitude than they could when they first arrived.

In response to inadequate oxygen delivery, the kidneys secrete the hormone erythropoietin. Erythropoietin, as previously described, stimulates the bone marrow to increase its production of hemoglobin and red blood cells. In the Peruvian Andes, for example, people have a total hemoglobin concentration that is increased from 15 g per 100 ml (at sea level) to 19.8 g per 100 ml. The total oxygen content of the blood is thus correspondingly increased.

---

1. Does moderate exercise have a significant effect on the blood values of O_2, CO_2, and pH? Explain your answer.
2. Define the terms *anaerobic threshold* and *maximal oxygen uptake,* and explain how these are affected by endurance training.
3. Describe the changes that occur in the respiratory system during acclimatization to a high altitude.

Summary

The Structure of the Respiratory System p. 334

I. Alveoli are small, numerous, thin-walled air sacs that provide an enormous surface area for gas diffusion.
   A. The region of the lungs where gas exchange with the blood occurs is known as the respiratory zone.
   B. The trachea, bronchi, and bronchioles that deliver air to the respiratory zone comprise the conducting zone.

II. The thoracic cavity and lungs are covered by wet pleural membranes, which are against each other.

Ventilation p. 336

I. The intrapleural and intrapulmonary pressures vary during ventilation.
   A. The intrapulmonary pressure is subatmospheric during inspiration and greater than the atmospheric pressure during expiration.
   B. Pressure changes in the lungs are produced by variations in lung volume.

II. The mechanics of ventilation are influenced by the physical properties of the lungs.
   A. The elasticity of the lungs refers to their tendency to recoil after distension; this elastic recoil aids expiration.
   B. The surface tension within alveoli would cause collapse of these alveoli without the action of pulmonary surfactant; this indeed occurs in respiratory distress syndrome.

III. Inspiration and expiration are accomplished by contraction and relaxation of striated muscles.
   A. During quiet inspiration, the diaphragm and external intercostal muscles contract and thus increase the volume of the thorax.
   B. During quiet expiration, these muscles relax, and the elastic recoil of the lungs and thorax causes a decrease in thoracic volume.

IV. Asthma results from bronchoconstriction; emphysema and chronic bronchitis are frequently referred to as chronic obstructive pulmonary disease.

Gas Exchange in the Lungs p. 342

I. The partial pressure of a gas in a gas mixture is equal to the total pressure times the percent composition of that gas in the mixture.
   A. Since the total pressure of a gas mixture decreases with altitude above sea level, the partial pressures of the constituent gases likewise decrease with altitude.
   B. The amount of gas that can be dissolved in a fluid is directly proportional to the partial pressure of that gas in contact with the fluid.

II. Abnormally high plasma concentrations of gases can cause a variety of disorders, including oxygen toxicity, nitrogen narcosis, and decompression sickness.

Regulation of Breathing p. 346

I. Neural centers in the medulla oblongata directly control the muscles of respiration.

II. Breathing is affected by chemoreceptors sensitive to the plasma CO_2, pH, and O_2 of the blood.
   A. An increase in plasma CO_2 and a decrease in pH are usually of greater importance than changes in the plasma O_2 in the regulation of breathing.
   B. The chemoreceptors in the aortic and carotid bodies are sensitive to changes in plasma CO_2 indirectly, because of consequent changes in blood pH.
   C. A decrease in the plasma O_2 concentration can stimulate breathing at high altitudes.

Oxygen and Carbon Dioxide Transport p. 348

I. Hemoglobin is composed of four polypeptide chains and four heme groups that contain a central atom of iron.
   A. When the iron is not attached to oxygen, the hemoglobin is called deoxyhemoglobin; when it is attached to oxygen, it is called oxyhemoglobin.
   B. If the iron is attached to carbon monoxide, the hemoglobin is called carboxyhemoglobin; this molecule cannot bond to oxygen.
   C. Deoxyhemoglobin combines with oxygen in the lungs (the loading reaction) and breaks its bonds with oxygen in the tissue capillaries (the unloading reaction).

II. The extent of oxygen unloading is increased by a rise in temperature or a fall in pH; both of these effects occur in skeletal muscles during exercise.

III. Red blood cells contain an enzyme called carbonic anhydrase, which catalyzes the reversible reaction whereby carbon dioxide and water are used to form carbonic acid.

IV. An abnormally high plasma CO_2 indicates hypoventilation; an abnormally low plasma CO_2 indicates hyperventilation.

Effects of Exercise and High Altitude on Respiratory Function p. 350

I. During exercise there is increased ventilation, or hyperpnea, which is matched to the increased metabolic rate.
   A. During heavy exercise the anaerobic threshold may be reached at about 50%–60% of the maximal oxygen uptake; at this point, lactic acid is released into the blood by the muscles.
   B. Endurance training enables the muscles to utilize oxygen more effectively, so that greater levels of exercise can be performed before the anaerobic threshold is reached.

II. Acclimatization to a high altitude involves changes that help to deliver oxygen more effectively to the tissues despite reduced plasma O_2 concentrations.
   A. Hyperventilation occurs in response to the low plasma O_2.
   B. The bond strength between hemoglobin and oxygen becomes weaker, so that the blood can unload a higher percentage of the oxygen it carries.
   C. The kidneys produce the hormone erythropoietin, which stimulates the bone marrow to increase its production of red blood cells, so that more oxygen can be carried by the blood.

Review Activities

Objective Questions

1. Gas exchange within the lungs occurs across the walls of the
   (a) trachea
   (b) bronchi
   (c) bronchioles
   (d) alveoli
   (e) all of these
2. During inhalation, the chest expands because
   (a) the lungs inflate with air, pushing the chest out
   (b) contraction of skeletal muscles causes the chest to expand
   (c) the atmospheric pressure falls below the pressure in the lungs
   (d) none of these
3. Which of the following statements about intrapulmonary and intrapleural pressure is *true*?
   (a) The intrapulmonary pressure is always subatmospheric.
   (b) The intrapleural pressure is always greater than the intrapulmonary pressure.
   (c) The intrapulmonary pressure is greater than the intrapleural pressure.
   (d) The intrapleural pressure equals the atmospheric pressure.

Match the pulmonary disorder to its cause:

4. collapse of the lung in an open chest wound
5. collapse of the lungs due to lack of surfactant
6. loss of alveoli, with larger remaining alveoli
7. increased resistance to air flow through bronchioles

(a) asthma
(b) fibrosis
(c) respiratory distress syndrome
(d) pneumothorax
(e) emphysema

8. If a gas tank contained 100% oxygen at one atmosphere pressure, the P_{O_2} of the air in the tank would be
   (a) 100 mm Hg
   (b) 160 mm Hg
   (c) 760 mm Hg
   (d) none of these
9. If a person has 50% carboxyhemoglobin
   (a) he has breathed too much carbon dioxide
   (b) his lungs are not functioning correctly
   (c) he is anemic
   (d) his tissues are suffering from hypoxia

Match the following terms with their definition:

10. difficulty breathing
11. cessation of breathing
12. normal breathing during exercise
13. excessive breathing, with low plasma CO_2

(a) hyperventilation
(b) dyspnea
(c) apnea
(d) hyperpnea

Essay Questions

1. Using a flow diagram (with arrows) to show cause and effect, explain how contraction of the diaphragm produces inspiration.
2. Explain the course of events that occurs when a child, who is trying to get his way by holding his breath, is forced to take a breath.
3. Explain why a person gets dizzy as a result of hyperventilation, and why breathing into a paper bag can be beneficial.
4. Does a person normally hyperventilate while engaging in moderate exercise? Explain.
5. Why don't deep-sea divers breathe 100% oxygen? Why do they have to ascend slowly rather than quickly back to the surface?
6. Explain the physiological mechanisms that enable a person who has lived at a high altitude for a period of time to adjust to this altitude.

16

The Urinary System

Outline

Objectives

By studying this chapter, you should be able to

1. identify the organs of the urinary system, and describe the path of urine flow from its origin to its point of excretion
2. describe the gross structure of the kidney
3. identify the parts of the renal nephron, the different regions of the nephron tubule, and the location of the nephron in the kidney
4. explain the process of glomerular filtration
5. define the terms *reabsorption* and *secretion* in regards to the nephron tubule
6. define the term *renal plasma clearance,* and explain the significance of this concept
7. explain why glucose does not normally appear in the urine, and how glycosuria occurs in a person with diabetes mellitus
8. explain how salt and water are reabsorbed in the proximal tubule
9. describe the action and significance of the countercurrent multiplier system
10. explain how antidiuretic hormone (ADH) and aldosterone act on the kidneys
11. identify different ways that drugs can act as diuretics, and the uses for these drugs
12. describe some of the diseases of the kidneys, and the tests that are performed in a urinalysis

Keys to Pronunciation

aldosterone: *al''do-ster'on*
antidiuretic: *an''ti-di''u-ret'ik*
calyces: *kal'i-sez*
calyx: *ka'liks*
creatinine: *kre-at'i-nin*
diabetes: *di''ah-be'tez*
dialysis: *di-al'i-sis*
diuretic: *di''u-ret'ik*
glomerulonephritis: *glo-mer'u-lo-ne-fri'tis*
glomerulus: *glo-mer'u-lus*
glycosuria: *gli''ko-su're-ah*
Henle: *hen'le*
hyperkalemia: *hi''per-kah-le'me-ah*
micturition: *mik''tu-rish'un*
nephron: *nef'ron*
thiazide: *thi'ah-zid*
ureter: *u're-ter*
urethra: *u-re'thrah*
urinalysis: *u''ri-nal'i-sis*

Photo: Arterial blood within a kidney. The tufts of capillaries shaped like balls are the glomeruli. These produce a blood filtrate which becomes urine.

The Kidneys and Urinary System

Each kidney contains many tiny tubules. Each of these tubules receives a blood filtrate, similar to tissue fluid, which is modified as it passes through different regions of the tubule and is thereby changed into urine. The tubules and associated blood vessels thus form the functioning units of the kidneys known as nephrons.

The primary function of the kidneys is regulation of the extracellular fluid (plasma and tissue fluid) environment in the body. This function is accomplished through the formation of urine, which is a modified filtrate of plasma. In the process of urine formation, the kidneys regulate (1) the volume of blood plasma and thus contribute significantly to the regulation of blood pressure; (2) the concentration of waste products in the blood; (3) the concentration of electrolytes (Na^+, K^+, HCO_3^-, and other ions) in the plasma; and (4) the pH of plasma. In order to understand how these functions are performed by the kidneys, a knowledge of kidney structure is required.

Structure of the Urinary System

The right and left kidneys are located in the abdominal cavity below the diaphragm and the liver. Each kidney in an adult weighs approximately 160–175 g, and is about 10–12 cm long and 5–6 cm wide—about the size of a fist. Urine produced in the kidneys is drained into a cavity known as the *renal pelvis*[1] (the term "renal" always refers to the kidneys). From this cavity the urine is channeled via two long ducts—the *ureters*—to the single *urinary bladder* (fig. 16.1).

A slice through the kidney shows two distinct regions (fig. 16.2): the outer **cortex,** and the deeper region, or **medulla.** The medulla is composed of eight to fifteen conical *renal pyramids.*

As the ureter enters the kidney, it expands to create the cavity of the renal pelvis. The apex of each renal pyramid in the medulla extends into a branch of the pelvis known as a *minor calyx.*[2] Several minor calyces (the plural of calyx) unite to form a *major calyx.* The minor and major calyces funnel the urine into the cavity of the pelvis (fig. 16.3), from which the urine is drained into the ureter and channeled to the urinary bladder.

The **urinary bladder** is a storage sac for urine, and its shape is determined by the amount of urine it contains. As the urinary bladder fills, it changes from a triangular to an oval shape as it bulges upward into the abdominal cavity. The urinary bladder is drained from below by a tube called the **urethra.** In females, the urethra is 4 cm (1.5 in.) long and opens into the space between the labia minora (chapter 8). In males, the urethra is about 20 cm (8 in.) long and opens at the tip of the penis, from which it can discharge either urine or semen.

[1]pelvis: Gk. *pyelos,* a trough or basin
[2]calyx: Gk. *kalyx,* cup

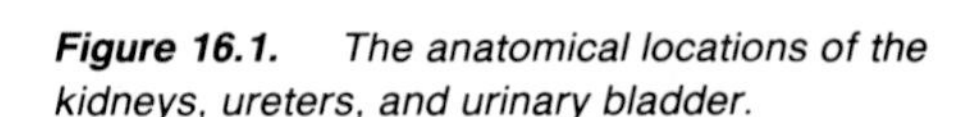
Figure 16.1. *The anatomical locations of the kidneys, ureters, and urinary bladder.*

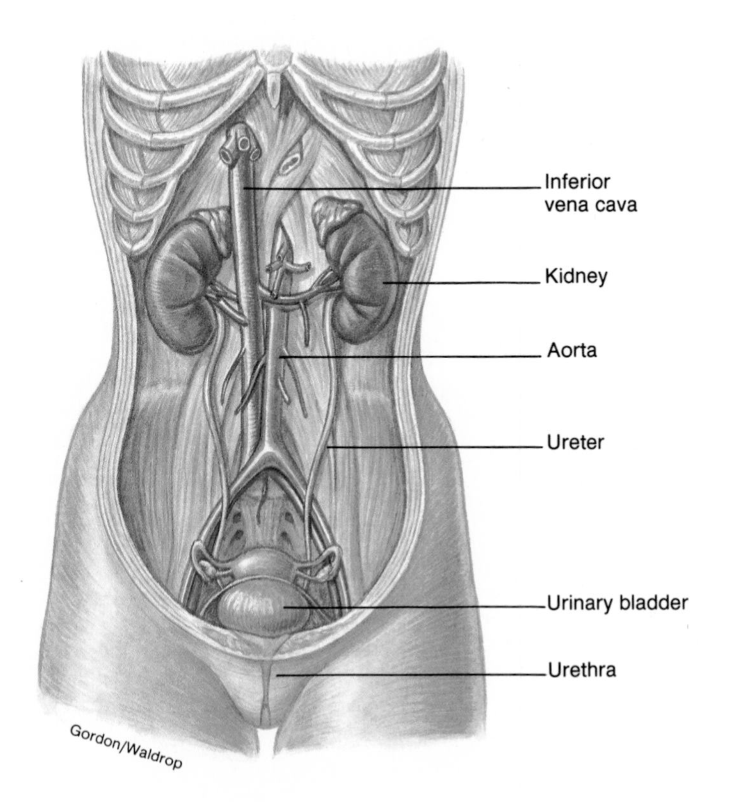

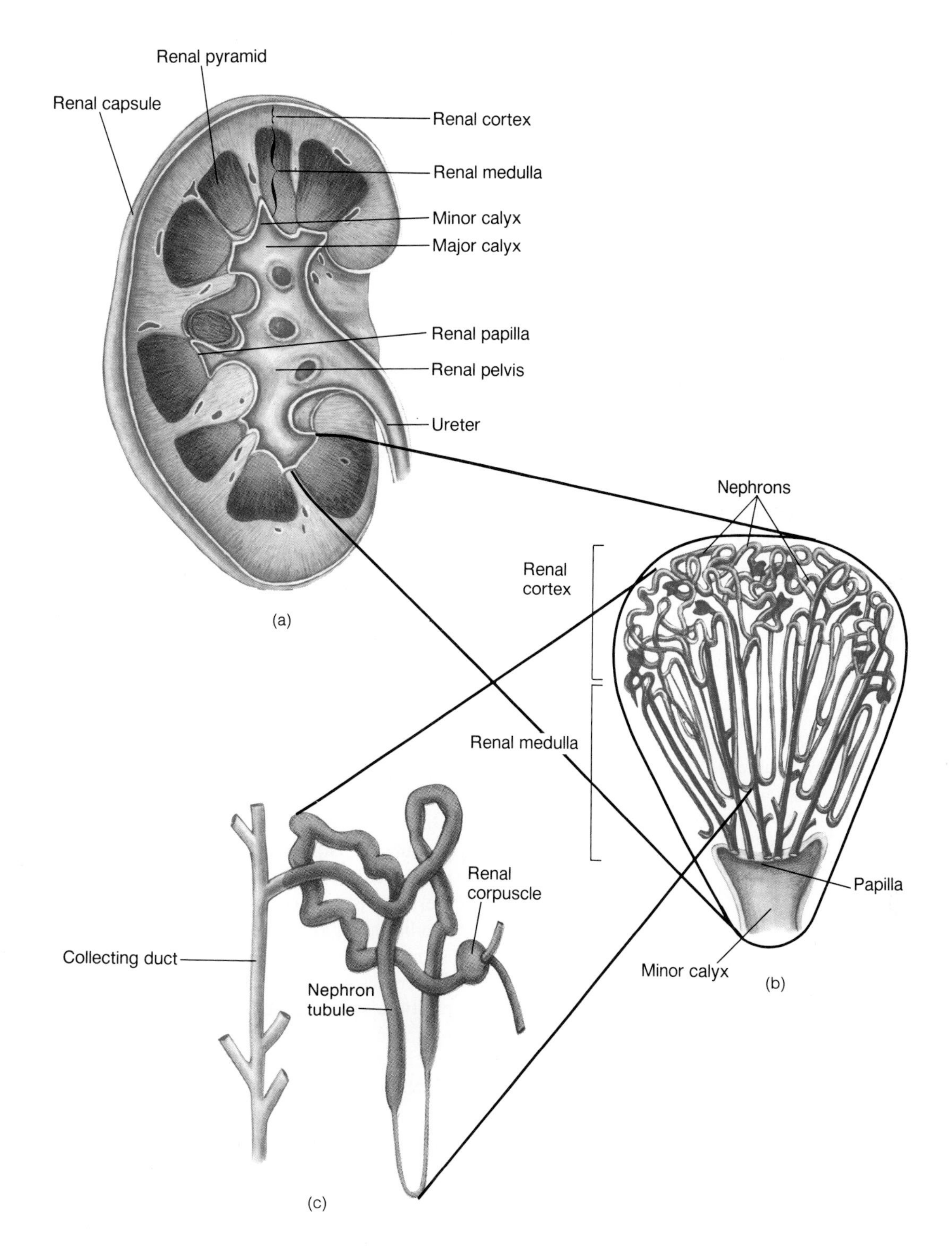

Figure 16.2. *The internal structures of a kidney.* (a) *A coronal section showing the structure of the cortex, medulla, and renal calyces.* (b) *A diagrammatic magnification of a renal pyramid and cortex to depict the tubules.* (c) *A diagrammatic view of a single nephron.*

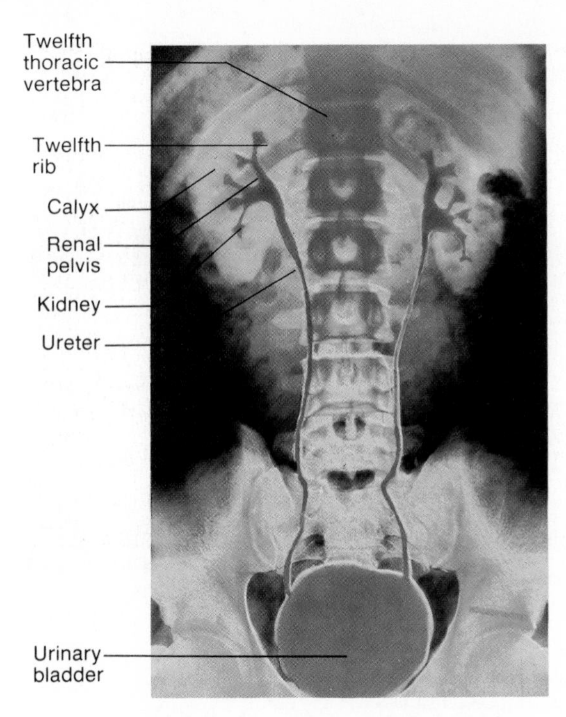

Figure 16.3. *A false color X ray of the calyces and renal pelvises of the kidneys, the ureters, and the urinary bladder.*

> **Kidney stones** are composed of crystals and proteins that grow in the renal medulla until they break loose and pass into the urine collection system. Small stones that are anchored in place are not usually noticed, but large stones in the calyces or pelvis may obstruct the flow of urine. When a stone breaks loose and passes into a ureter it produces a steadily increasing sensation of pain, which often becomes so intense that the patient requires narcotic drugs. Most kidney stones contain crystals of calcium oxalate, but stones may also be composed of crystals of calcium phosphate, uric acid, or cystine. These substances are dissolved in normal urine but form crystals when people develop kidney stones.

Micturition Reflex

The urethra has two muscular sphincters (muscles that are circularly arranged around an opening). The upper sphincter is composed of smooth muscle and is called the *internal urethral sphincter*; the lower sphincter is composed of voluntary skeletal muscle and is called the *external urethral sphincter.* The actions of these sphincters are regulated in the process of urination, which is also known as **micturition.**[3]

Micturition is controlled by a reflex center located in the second, third, and fourth sacral levels of the spinal cord (chapter 6). Filling of the urinary bladder activates stretch receptors which send impulses to this micturition center. As a result, parasympathetic neurons are activated which stimulate rhythmic contractions of the urinary bladder and cause relaxation of the internal urethral sphincter. At this point, a sense of urgency is perceived by the brain, but there is still voluntary control over the external urethral sphincter.

When urination is consciously allowed to occur, descending motor tracts from the brain to the micturition center cause inhibition of somatic motor fibers (chapter 6) to the external urethral sphincter. This muscle, as a result, relaxes, and urine is expelled.

[3]micturition: L. *micturire,* to urinate

Microscopic Structure of the Kidney

The **nephron**[4] is the functional unit of the kidney that is responsible for the formation of urine. Each kidney contains more than a million nephrons. A nephron consists of **tubules** and associated small blood vessels. Fluid formed by capillary filtration enters the tubules and is subsequently modified by transport processes; the resulting fluid that leaves the tubules is urine.

Renal Blood Vessels

Arterial blood enters the kidney through the *renal artery,* which divides into smaller arteries (fig. 16.4) that pass into the renal medulla. These, in turn, give rise to still smaller arteries that enter the cortex of the kidney. Finally, the small arteries in the cortex subdivide into numerous **afferent arterioles** (fig. 16.5), which are microscopic in size. The afferent arterioles deliver blood into capillary networks, called **glomeruli,** which produce a blood filtrate that enters the urinary tubules. The blood remaining in the glomerulus[5] leaves through an **efferent arteriole,** which delivers this blood into another capillary network, the **peritubular capillaries,** that surrounds the tubules. This arrangement of blood vessels, in which a capillary bed (the glomerulus) is drained by an arteriole and is delivered to a second capillary bed located downstream (the peritubular capillaries), is unique. Everywhere else in the body, capillary beds are drained by venules which return the blood to the heart. Blood from the peritubular capillaries, however, is finally drained into veins and returned to the general circulation.

Nephron Tubules

The tubular portion of a nephron consists of a glomerular capsule, a proximal convoluted tubule, a descending limb of the loop of Henle, an ascending limb of the loop of Henle, and a distal convoluted tubule (fig. 16.5).

The **glomerular (Bowman's)**[6] **capsule,** which is the beginning of the nephron tubule, surrounds the glomerulus in the cortex of the kidney. One way to mentally picture this arrangement is to imagine the capsule as a large, soft balloon. Now imagine that you push your fist partway into the balloon, so that the balloon surrounds your fist up to your wrist. In this analogy, the balloon is the capsule and your fist is the glomerulus. Fluid,

[4]nephron: Gk. *nephros,* kidney
[5]glomerulus: L. diminutive of *glomus,* ball
[6]Bowman's capsule: from Sir William Bowman, English anatomist, 1816–1892

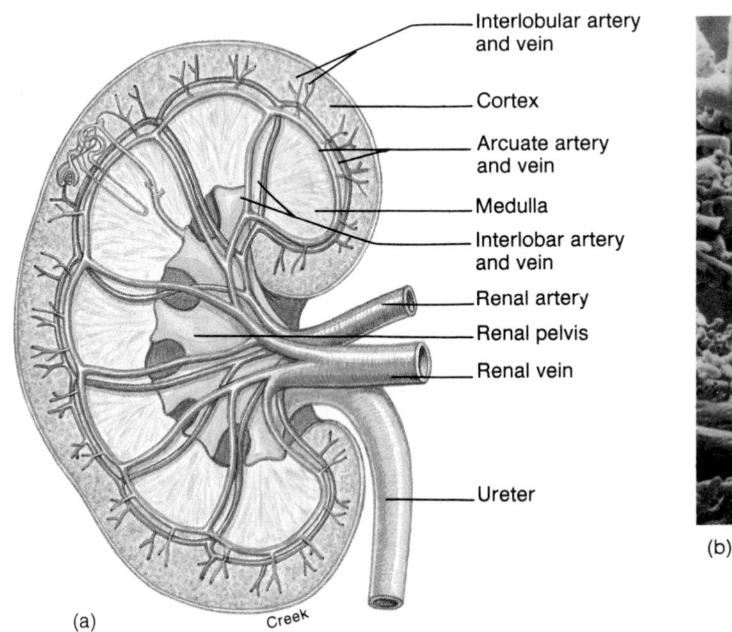

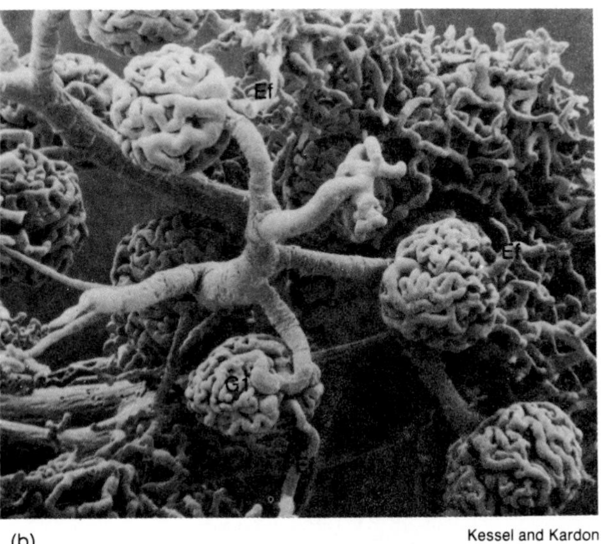

Figure 16.4. *The vascular structure of the kidneys. (a) An illustration of the major arterial supply and (b) a scanning electron micrograph of the glomeruli.*

(b) From Tissues and Organs: A Text-Atlas of Scanning Electron Microscopy *by Richard G. Kessel and Randy H. Kardon. Copyright © 1979 by W. H. Freeman and Company. Reprinted by permission.*

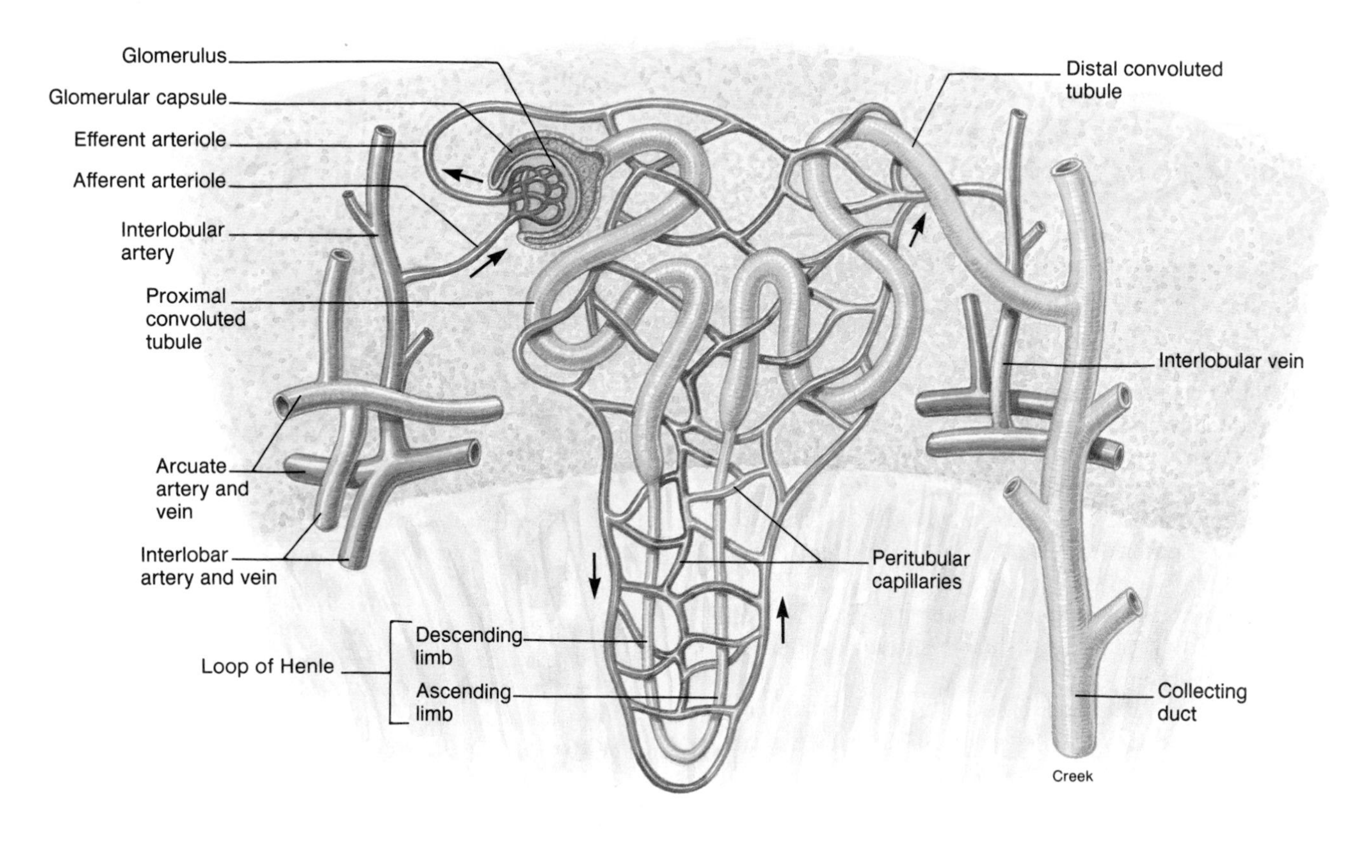

Figure 16.5. *A simplified illustration of blood flow from a glomerulus to an efferent arteriole, to the peritubular capillaries, and to the venous drainage of the kidneys.*

Social Issues

Development and Costs of New Drugs

The kidneys produce and secrete a hormone known as erythropoietin (EPO). As explained in chapter 15, this hormone stimulates the bone marrow to produce more red blood cells and hemoglobin, so that the blood can carry more oxygen when required. People with kidney disease undergoing dialysis treatment usually do not have adequate amounts of EPO, and therefore commonly suffer from anemia. Until very recently, there was little that could be done to help such patients.

Recently, a bioengineering company (Amgen) developed a method of producing human EPO using genetic engineering techniques (chapter 10). The medical community is excited at the prospect of being able to treat its kidney patients with this new therapy, but some legislators have expressed dismay at its high cost. Since many of the patients who will be treated with EPO are on medicare, it is ultimately the taxpayers who will bear much of the financial burden.

Critics of the cost of EPO have pointed out that EPO costs more, gram for gram, than gold. Unlike gold, however, it has been claimed that one gram of EPO can treat 8,000 patients for one week and eliminate the need for about 1,000 blood transfusions. Amgen defends the high cost of EPO as a necessary result of the high cost of research and development of this drug. Further, the development of similar EPO drugs by other companies will reduce the sales of EPO by the company that was the initial pioneer and will bring the prices down. This scenario holds for the development of all new pharmaceutical agents and explains their high price when first introduced. A recent study found that the average pharmaceutical takes twelve years and costs $231 million to develop.

Pharmaceuticals that are designed to treat rare diseases are less profitable than those which treat common ailments. In order to stimulate the research and development of drugs for rare diseases, the federal government passed the *Orphan Drug Act.* This allows the government to subsidize the development of such drugs. Development of recombinant human growth hormone, for example, was subsidized under the Orphan Drug Act. This seemed to be reasonable, as long as the growth hormone was used only to treat children with the rare condition of pituitary dwarfism. More recently, however, there have been suggestions that growth hormone might be used for other conditions; if so, the number of patients treated and thus the profitability of the drug would be greatly increased. If the orphan drug stops being an orphan, should the government be reimbursed for its subsidy?

The development of pharmaceuticals using the techniques of genetic engineering has thus opened new areas of legal controversy. In addition to direct government subsidies for development of some drugs, the government pays for recovery of the costs of drug development through such governmental programs as medicare. Taxpayers should therefore be knowledgeable about these controversies; our health (financial and otherwise) depends on it.

derived as a filtrate of blood plasma from the glomerulus, enters the capsule through tiny slits. The process of filtration will be described in the next section.

Filtrate in the glomerular capsule next passes into the lumen (cavity) of the **proximal convoluted tubule.** The wall of the proximal convoluted tubule contains millions of microvilli; these microvilli serve to increase the surface area for reabsorption. In the process of reabsorption, salt, water, and other molecules needed by the body are transported from the lumen, through the tubular cells, and into the peritubular capillaries which surround the nephron tubules.

The glomerulus, glomerular capsule, and proximal convoluted tubule are located in the renal cortex. Fluid then passes from the proximal convoluted tubule to the **loop of Henle.**[7] This fluid is carried into the medulla in the **descending limb** of the loop and returns to the cortex in the **ascending limb** of the loop. Back in the cortex, the tubule becomes coiled again and is called the **distal convoluted tubule.**

The distal convoluted tubules of several nephrons drain into a **collecting duct** (fig. 16.6). Fluid is then drained by the collecting duct from the cortex to the medulla as the collecting duct passes through a renal pyramid. This fluid, now called urine, is emptied into a minor calyx. Urine is then funneled through the renal pelvis and out of the kidney in the ureter.

Polycystic kidney disease is a condition inherited as an autosomal dominant trait (chapter 11) which affects one in 600–1,000 people. This disease is thus more common than such genetic diseases as sickle-cell anemia, cystic fibrosis, or muscular dystrophy. In 50% of the people who inherit the defective gene (located on the short arm of chromosome 16), progressive renal failure develops during middle age to the point that dialysis or kidney transplants are required. The cysts that develop are expanded portions of the renal tubule.

1. Describe the "theme" of kidney function in a single sentence, and list the components of this functional theme.
2. Trace the course of blood flow through the kidney from the renal artery to the renal vein.
3. Trace the course of tubular fluid from the glomerular capsules to the ureter.
4. Draw a diagram of the tubular component of a nephron. Label the segments, and indicate which parts are in the cortex and which are in the medulla.

[7]loop of Henle: from Friedrich G. J. Henle, German anatomist, 1809–1885

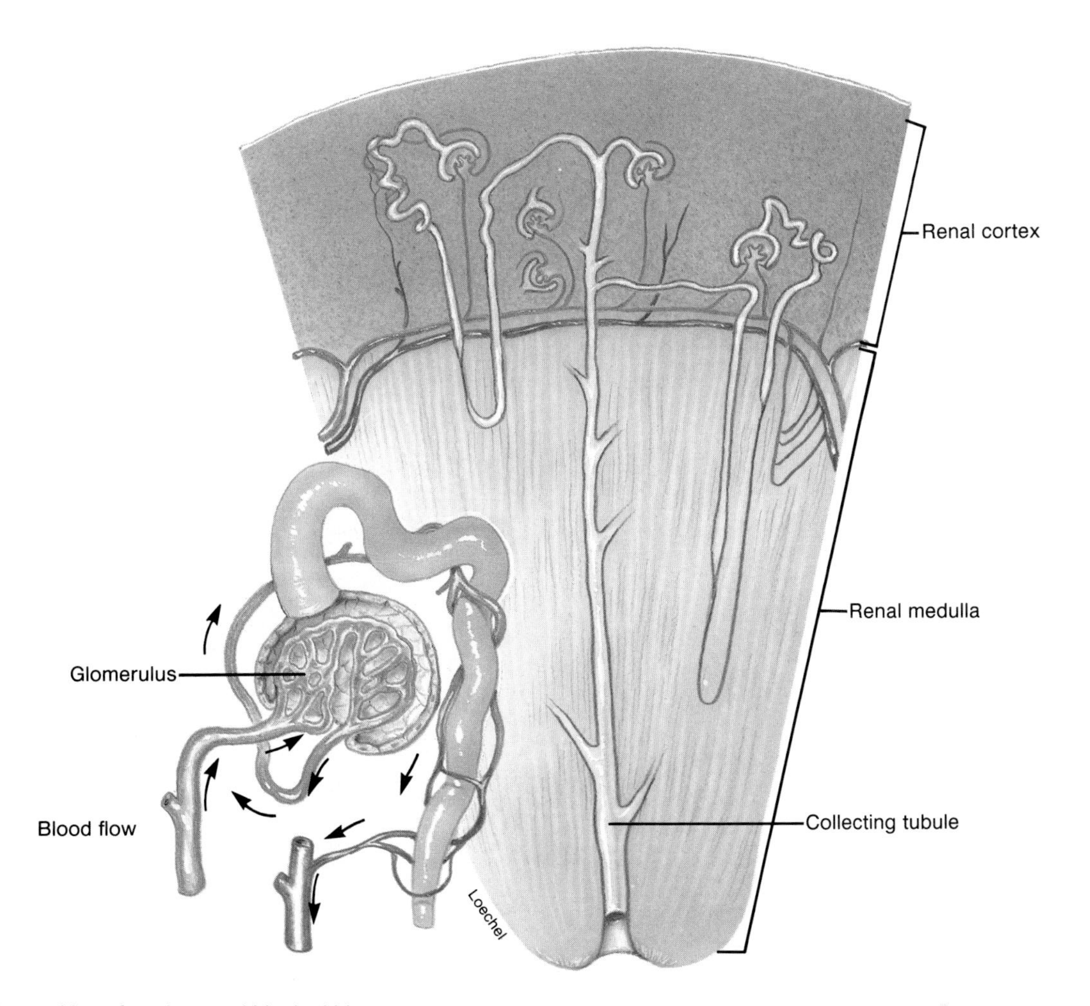

Figure 16.6. *The position of nephrons within the kidney.*

Glomerular Filtration, Reabsorption, and Secretion

Water and dissolved solutes—but not proteins—can pass from the blood plasma into the glomerular capsule. The volume of this filtrate produced per minute by both kidneys is called the glomerular filtration rate (GFR). Molecules needed by the body can be reabsorbed from the filtrate back into the blood. Some waste products can be eliminated by transport through the tubule wall.

The glomerular capillaries have extremely large pores. As a result of these large pores, glomerular capillaries are up to four hundred times more permeable to plasma water and dissolved solutes than are the capillaries of skeletal muscles. Although the pores of glomerular capillaries are large, they are still small enough to prevent the passage of red blood cells, white blood cells, and platelets into the filtrate.

Before the filtrate can enter the interior of the glomerular capsule it must pass through the capillary pores, the basement membrane (a thin layer of glycoproteins immediately outside the endothelial cells), and the inner layer of the glomerular capsule (fig. 16.7). The inner layer of the glomerular capsule contains narrow slits through which filtered molecules pass to enter

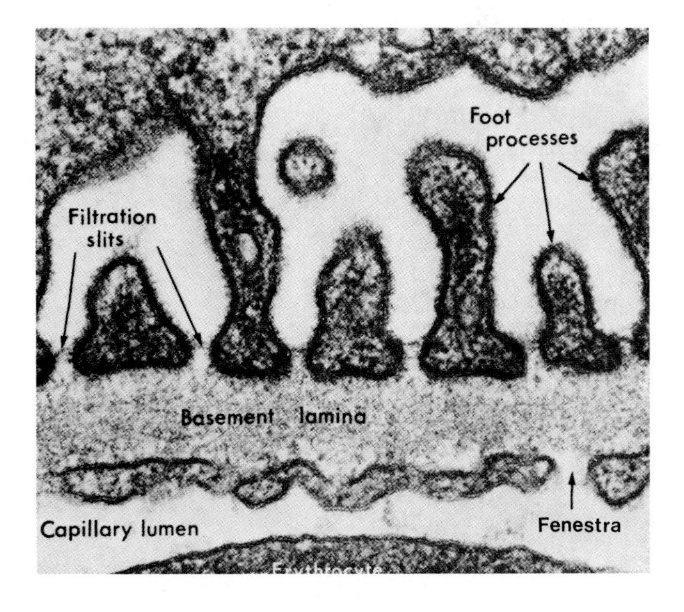

Figure 16.7. *An electron micrograph of the "filtration barrier" between the capillary lumen and the cavity of the glomerular (Bowman's) capsule.*

the interior of the glomerular capsule (fig. 16.8). All molecules dissolved in the blood plasma can enter the capsule except for plasma proteins, which are too large to pass through the filtration barriers.

Glomerular Filtration

Glomerular filtration is the process by which fluid enters into the capsules. The fluid that enters the glomerular capsule is called *glomerular filtrate* (fig. 16.9), and is formed under pressure (the hydrostatic pressure of the blood). This is similar to the formation of tissue fluid by other capillary beds in the body (chapter 12). Because glomerular capillaries are extremely permeable and have a high surface area, an extraordinarily large volume of filtrate is formed. The **glomerular filtration rate (GFR)** is the volume of filtrate produced per minute by both kidneys. The GFR averages 115 ml per minute in women and 125 ml per minute in men. This is equivalent to 7.5 L per hour or 180 L per day (about 45 gallons)! Since the total volume of blood is only about 5.5 L, this means that the total blood volume is filtered into the urinary tubules every forty minutes. Most of the filtered water must obviously be returned immediately to the vascular system, or a person would literally urinate to death within minutes!

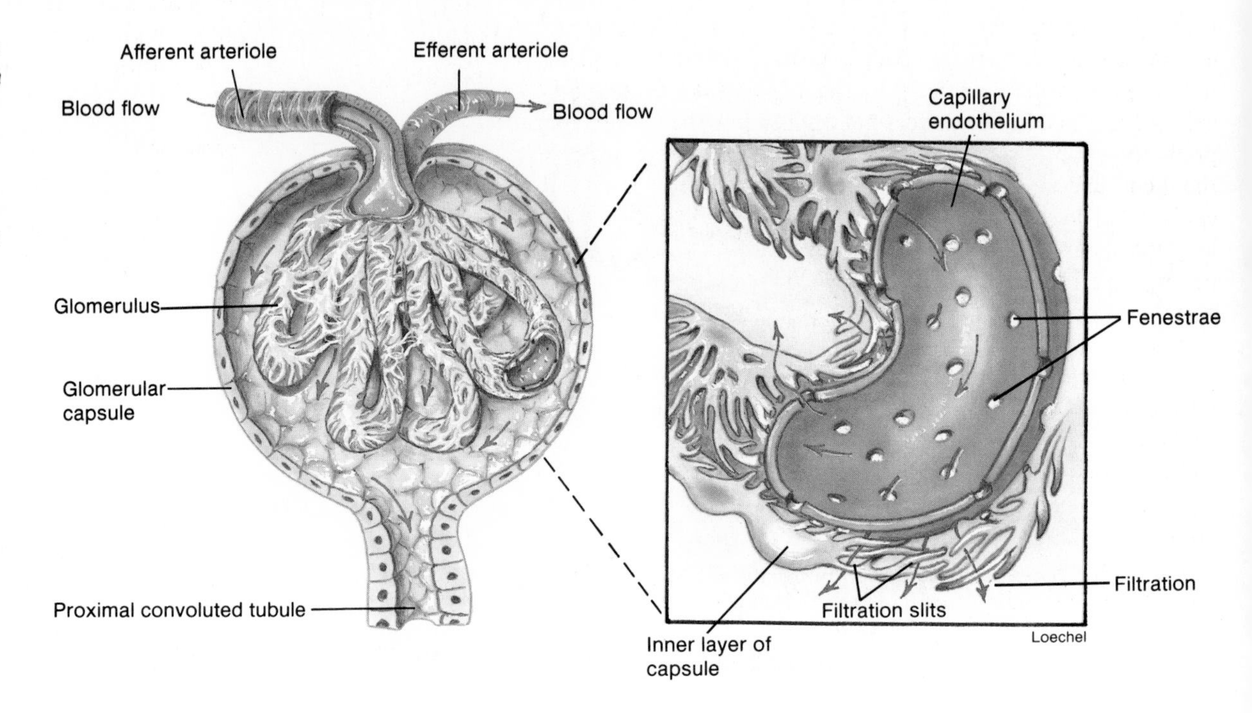

Figure 16.8. *Fluid derived from plasma is filtered through the large pores (called fenestrae) in the glomerular capillaries and then through the filtration slits of the capsule.*

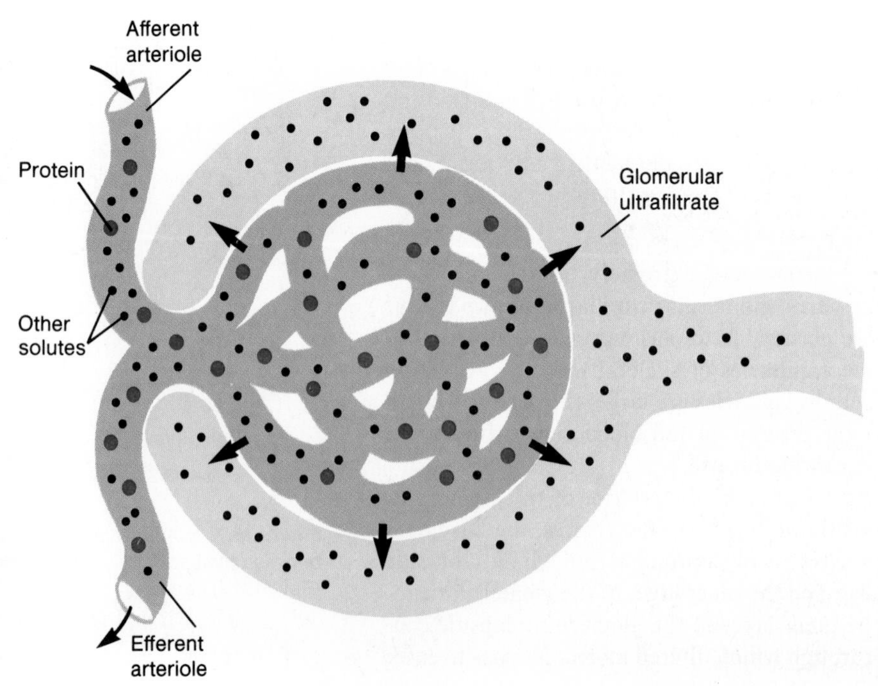

Figure 16.9. *The formation of glomerular filtrate. Proteins (large circles) are not filtered, but smaller plasma solutes (dots) easily enter the glomerular filtrate.*

Measurements of the plasma concentration of **creatinine** are often used clinically as an index of kidney function. Creatinine is a waste product of a molecule called creatine, normally found in muscles. Since creatinine is released into the blood at a constant rate and since its excretion is closely matched to the GFR, an abnormal decrease in GFR causes the plasma creatinine concentration to rise. A simple measurement of blood creatinine concentration can thus indicate if the GFR is normal and thereby provide information about the health of the kidneys.

Reabsorption

It should be understood that the molecules within the glomerular filtrate are on their way out of the body. If nothing happens to them, they will be excreted by the urinary system. The process of filtration, however, does not discriminate between waste products and molecules needed by the body. Any molecule in the blood plasma that is small enough to enter the glomerular capsule will do so. Most of the filtered water, and molecules such as glucose, amino acids, vitamins, and many ions which the body needs, must be reclaimed. This is where the selection is made between waste products to be excreted and molecules to be retained by the body.

In the process of **reabsorption,** a particular molecule in the filtrate within the nephron tubules is moved through the tubule wall (fig. 16.10), released into the tissue space outside the tubule, and transported into the surrounding peritubular capillaries (fig. 16.11). The transport of these molecules out of the filtrate is often the function of specific membrane carriers, so that this transport can be selective. Once these molecules are in the peritubular capillaries, they are transported eventually into the renal vein and back to the general circulation.

Reabsorption of Glucose

The reabsorption of glucose and amino acids is an energy-requiring process, which occurs primarily in the proximal convoluted tubules. This is an example of active transport, in which the expenditure of energy in the form of ATP (chapter 4), is required. This form of cellular transport, in addition, requires the use of specific carrier proteins in the cell membrane of the proximal tubule epithelial cells. As described in chapter 4, membrane transport that involves the use of carrier proteins is known as carrier-mediated transport.

Carrier-mediated transport displays the property of *saturation.* This means that when the transported molecule (such as glucose) is present in sufficiently high concentrations, all of the carriers become "busy," and the transport rate reaches a maximal value. The concentration of transported molecules

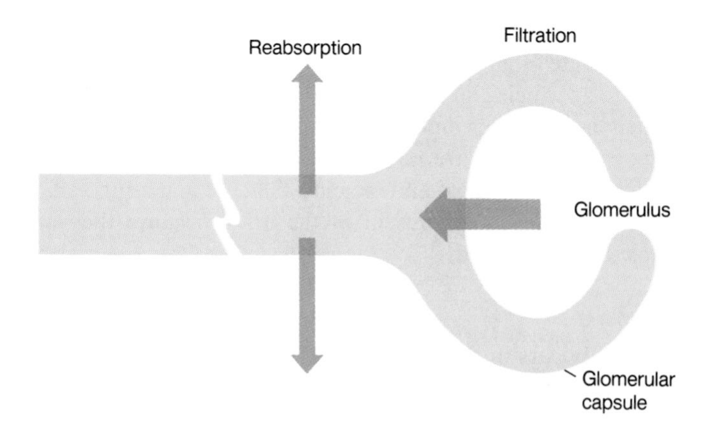

Figure 16.10. *Plasma water and its dissolved solutes (except proteins) enter the glomerular filtrate, but most of these filtered molecules are reabsorbed. The term reabsorption refers to the transport of molecules out of the tubular filtrate back into the blood.*

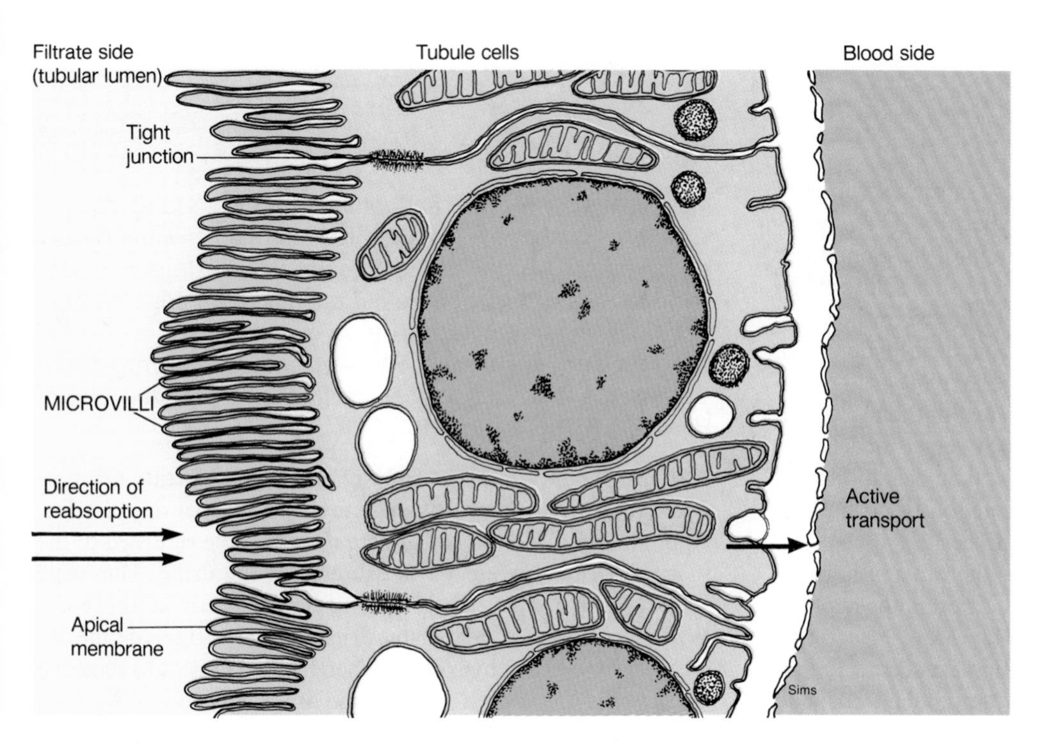

Figure 16.11. *An illustration of the appearance of proximal tubule cells in the electron microscope. Molecules that are reabsorbed pass through the tubule cells from the apical membrane (facing the filtrate) to the membrane on the opposite side (facing the blood).*

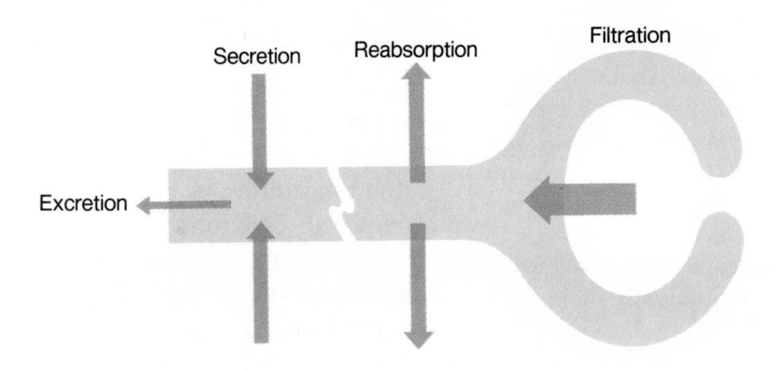

Figure 16.12. Secretion refers to the active transport of substances from the peritubular capillaries into the tubular fluid. This transport is in a direction opposite to that of reabsorption.

needed to just saturate the carriers and achieve the maximal transport rate is called the **transport maximum** (abbreviated T_m).

For example, the transport maximum for glucose is achieved when the rate of glucose delivery to the tubules is 375 mg/min. The rate at which glucose is normally delivered to the tubules is much less than this amount, and thus the carriers for glucose in the renal tubules are not normally saturated. The same is true for the transport of amino acids. Glucose and amino acids are thus not normally present in the urine because they are completely reabsorbed.

Glycosuria

Glucose appears in the urine—a condition called **glycosuria**—when more glucose passes through the tubules than can be reabsorbed. This occurs when the plasma glucose concentration reaches 180–200 mg per 100 ml. Glucose is normally absent from urine because plasma glucose concentrations remain below this threshold value. When the plasma concentration of glucose is abnormally high (*hyperglycemia*), however, the threshold can be exceeded and glucose can "spill over" into the urine. Fasting hyperglycemia and glycosuria indicate the disease **diabetes mellitus** (chapter 18).

Secretion

One of the major functions of the kidneys is the excretion of waste products such as urea and other molecules. These molecules are filtered through the glomerulus into the glomerular capsule along with water, salt, and other plasma solutes. The filtered waste products, therefore, can be excreted in the urine as long as they are not reabsorbed. It should be understood, however, that only a fraction of the blood arriving in the glomeruli gets filtered into the capsules. The rest of the blood sent to the kidneys simply remains in the blood vessels and leaves the kidney in the renal vein. Waste products in this unfiltered blood would thus escape from being excreted in the urine if it were not for the operation of a different process.

In addition to the process of filtration, there is another way in which a substance can be put into the fluid within the nephron tubules. This process is called **secretion** (fig. 16.12). Secretion is the opposite of reabsorption. Molecules that are secreted move out of the peritubular capillaries, across the tubular cells, and into the fluid within the tubule. This latter movement is an active, energy-requiring transport process. In this way, molecules that were not filtered out of the blood in the glomerulus, but instead passed through the efferent arterioles to the peritubular capillaries, can still be excreted in the urine.

Many antibiotics, like penicillin, are secreted by the renal tubules and thus are rapidly eliminated from the body. Because antibiotics are excreted so rapidly by the kidneys, large amounts must be administered to be effective. Drugs and hormones which are not water-soluble, however, can bind to plasma proteins and thus escape elimination in the urine. Many of these drugs and hormones are inactivated in the liver by chemical transformations that make them water-soluble, and are then rapidly cleared from the blood by active secretion in the nephrons.

1. Describe the formation and composition of glomerular filtrate.
2. Explain the process and significance of tubular reabsorption and secretion.
3. Explain how a person with diabetes mellitus can have glucose in the urine.

Reabsorption of Salt and Water

Most of the filtered salt and water is reabsorbed across the wall of the proximal tubule. The reabsorption of water occurs by osmosis, in which water follows the active extrusion of NaCl from the tubule and into the surrounding capillaries. Most of the remaining water in the filtrate is reabsorbed across the wall of the collecting duct. This also occurs by osmosis, and is a result of transport processes that operate in the loop of Henle.

Although about 180 L per day of glomerular filtrate are produced, the kidneys normally excrete only 1–2 L per day of urine. Approximately 99% of the filtrate must thus be returned to the vascular system, while 1% is excreted in the urine. The urine volume, however, varies according to the needs of the body. When a well-hydrated person drinks water, urine volume increases. In severe dehydration, when the body needs to conserve water, only 400 ml per day of urine are produced.

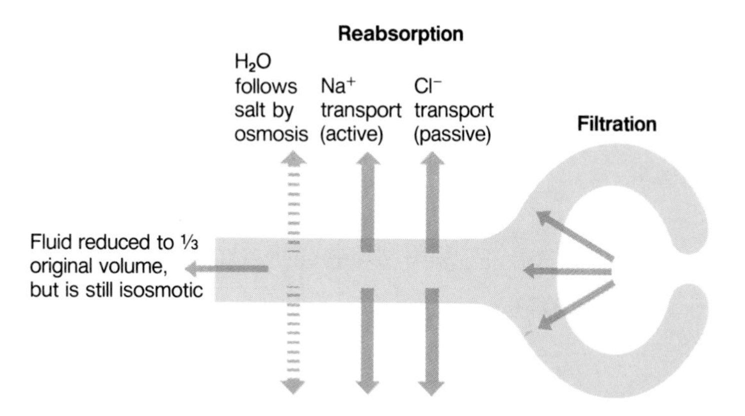

Figure 16.13. Mechanisms of salt and water reabsorption in the proximal tubule. Sodium is actively transported out of the filtrate, and chloride follows passively by electrical attraction. Water follows the salt out of the tubular filtrate into the peritubular capillaries by osmosis.

Regardless of the body's state of hydration, it is clear that most of the filtered water must be returned to the vascular system to maintain blood volume and pressure. It is important to realize that the transport of water always occurs passively by *osmosis;* there is no such thing as active transport of water. A concentration gradient must thus be created between tubular fluid and blood that favors the osmosis of water into the blood.

Reabsorption in the Proximal Tubule

Since all plasma solutes, with the exception of proteins, are able to freely enter the glomerular filtrate, the concentration and osmotic pressure of the filtrate is essentially the same as that of plasma. The filtrate is thus isotonic to the plasma (chapter 4). Osmosis cannot occur unless the concentrations of plasma and filtrate are altered by active transport processes. This is achieved by the active transport of Na^+ from the filtrate to the peritubular blood.

The transport of Na^+ from the tubular fluid to the tissue fluid surrounding the epithelial cells of the proximal tubule creates a difference in charge (potential difference) across the wall of the tubule. This electrical gradient favors the passive transport of Cl^- toward the higher Na^+ concentration in the tissue fluid. Chloride ions, therefore, passively follow sodium ions out of the filtrate to the tissue fluid. As a result of the accumulation of NaCl, the concentration and osmotic pressure of the tissue fluid surrounding the proximal tubule is increased above that of the tubular fluid.

An osmotic gradient is thus created beween the tubular fluid and the tissue fluid surrounding the proximal tubule. Since the cells of the proximal tubule are permeable to water, water moves by osmosis from the tubular fluid into the epithelial cells and then out of the epithelial cells into the tissue fluid. The salt and water which were reabsorbed from the tubular fluid can then move passively into the surrounding peritubular capillaries and in this way be returned to the blood (fig. 16.13).

Approximately 65% of the salt and water in the original glomerular filtrate is reabsorbed across the proximal tubule and returned to the vascular system. The reabsorption of salt and water in the proximal tubule occurs constantly regardless of the person's state of hydration. This reabsorption is very costly in terms of energy expenditures, accounting for as much as 6% of the calories consumed by the body at rest.

The Loop of Henle

Water cannot be actively transported across the tubule wall, and osmosis of water cannot occur if the tubular fluid and surrounding tissue fluid are isotonic to each other. In order for water to be reabsorbed by osmosis, the surrounding tissue fluid must be hypertonic (chapter 4). The osmotic pressure of the tissue fluid in the renal medulla is, in fact, raised to over four times that of plasma. This results partly from the fact that the tubule bends; the geometry of the loop of Henle allows interaction to occur between the descending and ascending limbs. Since the ascending limb is the active partner in this interaction, its properties will be described before those of the descending limb.

Ascending Limb of the Loop of Henle

Salt (NaCl) is actively moved out of the ascending limb into the surrounding tissue fluid (fig. 16.14*a*). Unlike the epithelial walls of the proximal tubule, however, the walls of the ascending limb of the loop of Henle are *not permeable to water.* The tubular fluid thus becomes increasingly dilute as it ascends toward the cortex, whereas the tissue fluid around the loops of Henle in the medulla becomes increasingly more concentrated. By means of these processes, the tubular fluid that enters the distal tubule in the cortex is made hypotonic, whereas the tissue fluid in the medulla is made hypertonic.

Descending Limb of the Loop of Henle

The deeper regions of the medulla, around the tips of the loops of Henle, have a very high salt concentration. In order to reach this high a concentration, the salt pumped out of the ascending limb must accumulate in the tissue fluid of the medulla. This occurs as a result of the properties of the descending limb.

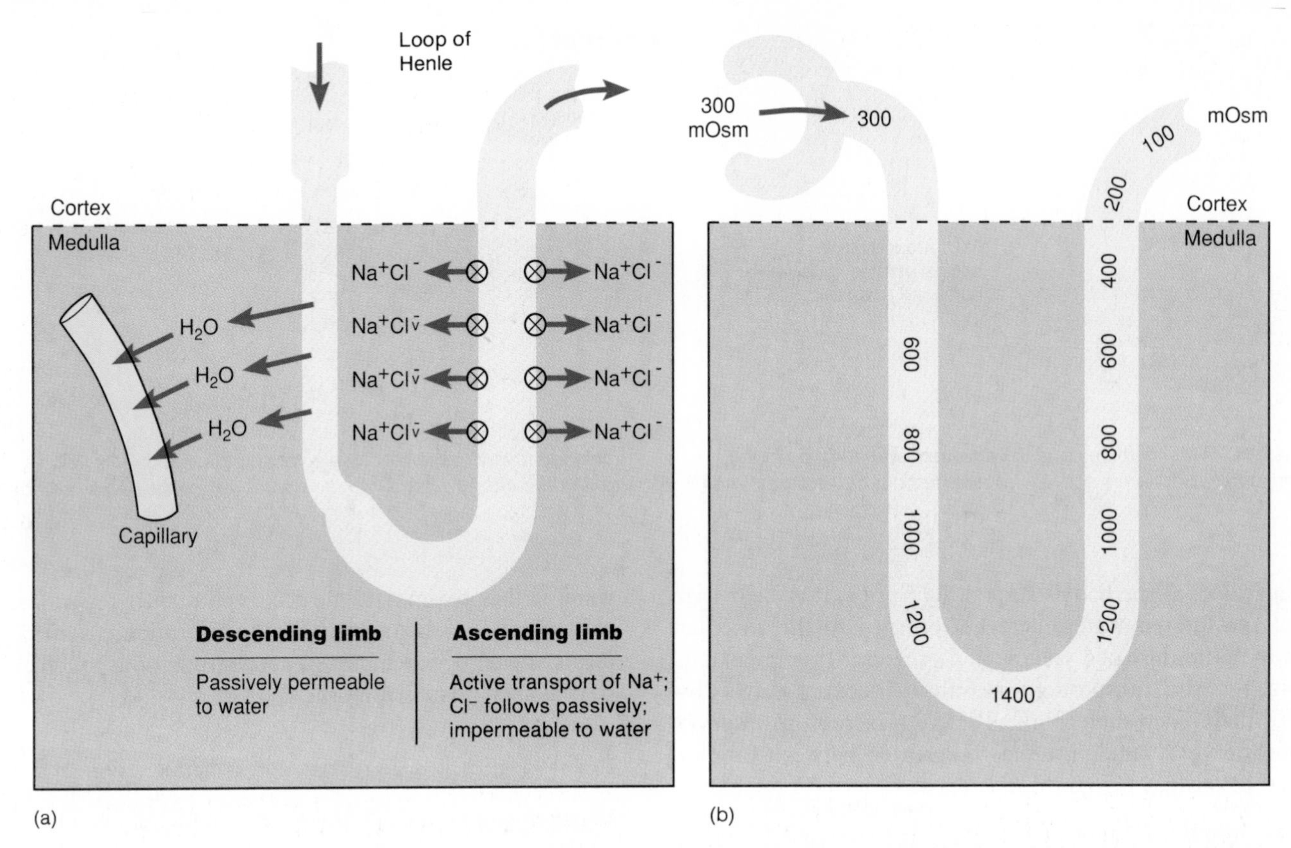

Figure 16.14. The countercurrent multiplier system. The extrusion of sodium chloride from the ascending limb makes the surrounding tissue fluid more concentrated. This concentration is multiplied by the fact that the descending limb is passively permeable so that its fluid increases in concentration as the surrounding tissue fluid becomes more concentrated. The transport properties of the loop and their effect on tubular fluid concentration is shown in (a). The values of these changes in concentration (in units of total moles per liter) are shown in (b).

The descending limb does not actively transport salt, and indeed is believed to be impermeable to the passive diffusion of salt. It is, however, permeable to water. Since the surrounding tissue fluid is hypertonic to the filtrate in the descending limb, water is drawn out of the descending limb by osmosis and enters blood capillaries. The concentration of tubular fluid is thus increased, and its volume is decreased, as it descends toward the tips of the loops.

As a result of these passive transport processes in the descending limb, the fluid that "rounds the bend" at the tip of the loop is just as concentrated as the surrounding tissue fluid. There is, therefore, a higher salt concentration arriving in the ascending limb than there would be if the descending limb simply delivered isotonic fluid. Salt transport by the ascending limb is increased accordingly, so that the "saltiness" of the tissue fluid is multiplied (fig. 16.14*b*).

Countercurrent Multiplication

Countercurrent flow (flow in opposite directions) in the ascending and descending limbs and the close proximity of the two limbs allow interaction to occur. A *positive feedback* mechanism is created; the more salt the ascending limb extrudes, the more concentrated will be the fluid that returns to it from the descending limb. This positive feedback mechanism multiplies the concentration of tissue fluid and descending limb fluid and is thus called the **countercurrent multiplier system.**

The countercurrent multiplier system traps some of the salt that enters the loop of Henle in the tissue fluid of the renal medulla. This system thus results in a gradually increasing concentration of renal tissue fluid from the cortex to the inner medulla. The "payoff" for all this effort is described in the next section.

Collecting Duct: Effect of Antidiuretic Hormone (ADH)

As a result of the countercurrent multiplier system, the tissue fluid of the renal medulla is made very hypertonic. The collecting ducts must transport their fluid through this hypertonic environment in order to empty their contents of urine into the renal pelvis. Whereas the fluid surrounding the collecting ducts in the medulla is hypertonic, the fluid that passes into the collecting ducts in the cortex is hypotonic.

The walls of the collecting ducts are *permeable to water but not to salt.* Since the surrounding tissue fluid in the renal medulla is very hypertonic, as a result of the countercurrent

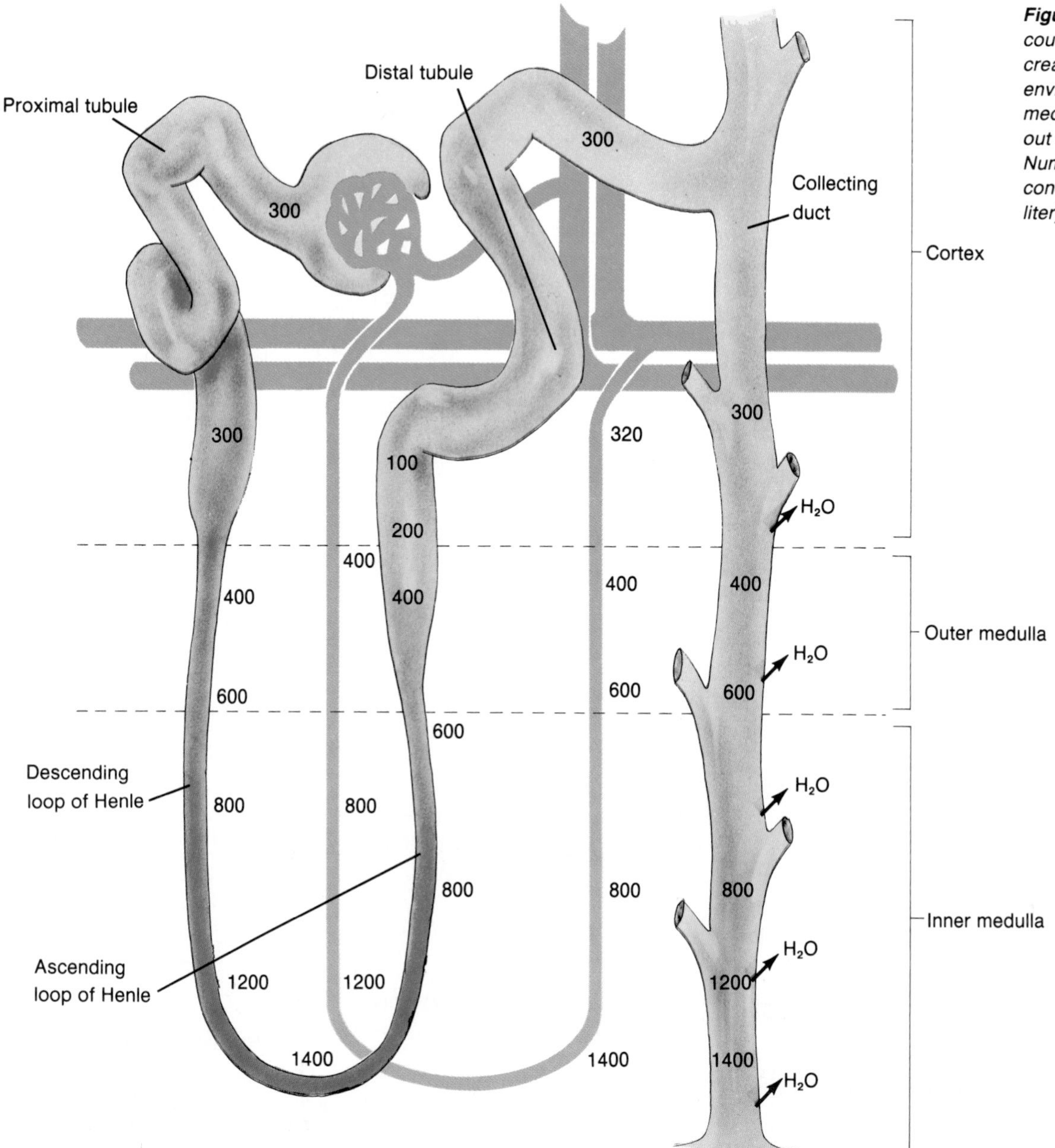

Figure 16.15. *The countercurrent multiplier system creates a concentrated environment in the renal medulla. This acts to draw water out of the collecting duct. Numbers indicate solute concentration (in total moles per liter).*

multiplier system, water is drawn out of the collecting ducts by osmosis. This water does not dilute the surrounding tissue fluid because it is transported by capillaries to the general circulation. In this way, most of the water remaining in the filtrate is returned to the vascular system (fig. 16.15).

The osmotic gradient created by the countercurrent multiplier system provides the force for water reabsorption through the collecting ducts. The rate of this reabsorption, however, is determined by the permeability of the collecting duct cell membranes to water. The permeability of the collecting duct to water, in turn, is determined by the concentration of **antidiuretic hormone (ADH)** in the blood. ADH opens pores in the membrane, allowing water to pass.

ADH is produced by neurons in the hypothalamus and is secreted from the posterior pituitary gland (chapter 7). The secretion of ADH is stimulated when osmoreceptors in the hypothalamus respond to an increase in blood osmotic pressure (concentration). During dehydration, therefore, when the plasma becomes more concentrated, increased secretion of ADH promotes increased permeability of the collecting ducts to water. In severe dehydration only the minimal amount of water needed to eliminate the body's wastes is excreted. Under these conditions about 99.8% of the initial glomerular filtrate is reabsorbed. Conversely, when a person drinks an excessive amount of water, the secretion of ADH is inhibited. Less water is therefore reabsorbed, so that a larger volume of more dilute urine is excreted.

Diabetes insipidus (not to be confused with diabetes mellitus) is a disease produced by the inadequate secretion or action of ADH. The collecting ducts are thus not very permeable to water and, therefore, a large volume (5–10 L per day) of dilute urine is produced. The dehydration that results causes intense thirst, but a person with this condition has difficulty drinking enough to compensate for the large volumes of water lost in the urine. A similar, but less intense, effect is produced in healthy people by ingestion of caffeine and alcohol, which act directly on the hypothalamus to inhibit ADH secretion.

Effect of Aldosterone on the Kidneys

In addition to ADH, there is another hormone that has important regulatory functions in the kidneys. This hormone is a steroid, known as **aldosterone,** which is secreted by the adrenal cortex. Aldosterone acts primarily on the distal convoluted tubules of the nephrons, and has three important effects. Aldosterone stimulates (1) reabsorption of Na^+ from the distal tubule into the surrounding capillaries; (2) reabsorption of water, which follows the Na^+; and (3) secretion of K^+ from the blood into the fluid within the distal tubule.

The ability of aldosterone to promote sodium and water retention is significant but of lesser importance than its effect on potassium. Indeed, aldosterone is absolutely essential for the elimination of potassium in the urine. This is because all of the K^+ that arrives in the proximal tubule is automatically reabsorbed; the only way K^+ can be eliminated in the urine is through secretion. Secretion of K^+ into the filtrate occurs at the distal tubule (fig. 16.16), and is stimulated by aldosterone.

Since food contains potassium, the kidneys must eliminate a corresponding amount of potassium to prevent the blood potassium concentration from getting too high (a condition called *hyperkalemia*).[8] This is necessary because hyperkalemia can cause heart failure. The same dangerous condition can occur if there is inadequate secretion of aldosterone from the adrenal cortex in *Addison's disease*.[9]

1. Describe how salt and water are reabsorbed in the proximal tubule, and the significance of this reabsorption.
2. Describe the transport of salt and water in the ascending and descending limbs of the loop of Henle, and explain how these two regions of the nephron interact.
3. What is the purpose of the countercurrent multiplier system?
4. Explain how ADH helps the body to conserve water, and describe how variations in ADH secretion affect the volume and concentration of urine.
5. Explain the actions of aldosterone. What effect would a drug that blocks aldosterone action have on the blood K^+ concentration?

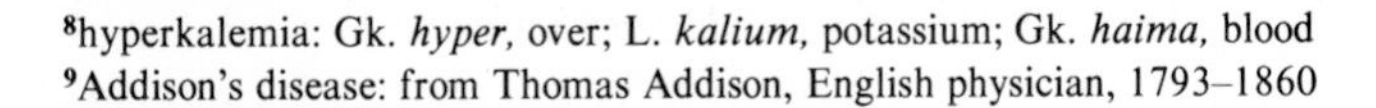

[8]hyperkalemia: Gk. *hyper,* over; L. *kalium,* potassium; Gk. *haima,* blood
[9]Addison's disease: from Thomas Addison, English physician, 1793–1860

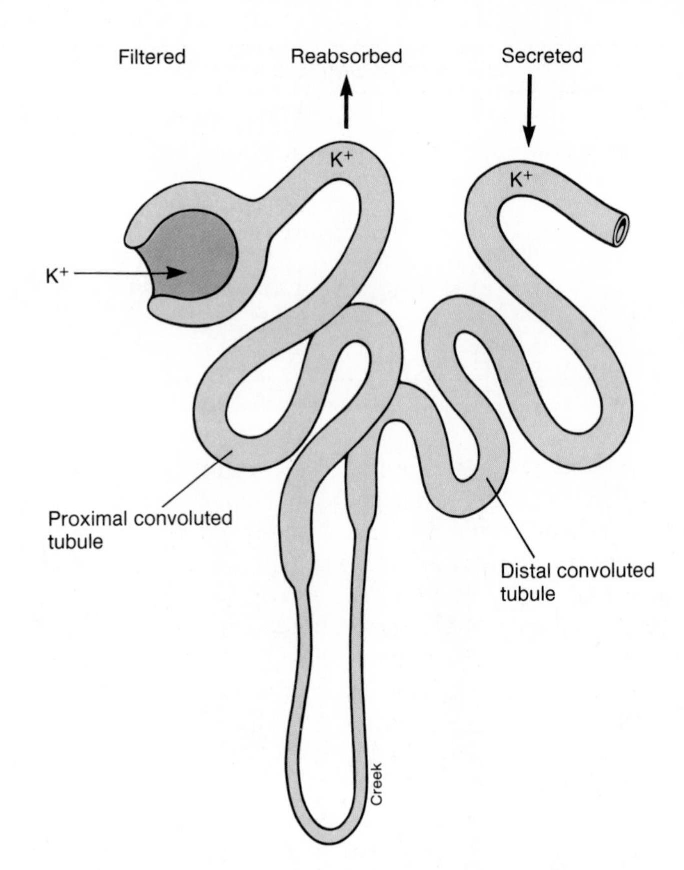

Figure 16.16. *Potassium is almost completely reabsorbed in the proximal tubule, but under aldosterone stimulation, it is secreted into the distal tubule. All of the K^+ in urine is derived from secretion rather than from filtration.*

Clinical Applications

Different types of diuretic drugs act on specific segments of the nephron tubule to inhibit the reabsorption of water and thus to lower blood volume. A knowledge of the mechanism of action of diuretics promotes better understanding of the physiology of the nephron. Clinical analysis of the urine, similarly, is meaningful only when the mechanisms that produce normal urine composition are understood.

The importance of kidney function in maintaining homeostasis (chapter 5), and the ease with which urine can be collected and tested, make the clinical study of renal function and urine composition particularly significant. Further, the ability of the kidneys to regulate blood volume is exploited clinically in the management of high blood pressure.

Use of Diuretics

People who need to lower their blood volume because of hypertension, congestive heart failure, or edema take medications that increase the volume of urine excreted. Such medications are

Table 16.1 Actions of different classes of diuretics

Category of Diuretic	Example	Mechanism of Action	Major Site of Action
Carbonic anhydrase inhibitors	Acetazolamide	Inhibits reabsorption of bicarbonate	Proximal tubule
Loop diuretics	Furosemide	Inhibits sodium transport	Thick segments of ascending limbs
Thiazides	Hydrochlorothiazide	Inhibits sodium transport	Last part of ascending limb and first part of distal tubule
Potassium-sparing diuretics	Spironolactone	Inhibits action of aldosterone	Last part of distal tubule and cortical collecting duct
	Triamterene	Inhibits Na^+/K^+ exchange	Last part of distal tubule and cortical collecting duct

called **diuretics.** There are a number of diuretic drugs in clinical use that act on the renal nephron in different ways (table 16.1).

The most powerful diuretics, inhibiting salt and water reabsorption by as much as 25%, are the drugs that act to inhibit active salt transport out of the ascending limb of the loop of Henle. Examples of these loop diuretics include *furosemide* and *ethacrynic acid.* The thiazide diuretics, like *hydrochlorothiazide,* are more commonly used for hypertension and inhibit salt and water reabsorption by as much as 8%. The thiazides act by inhibiting salt transport in the first segment of the distal convoluted tubule.

When extra solutes are present in the filtrate, they increase the osmotic pressure of the filtrate and in this way decrease the osmotic reabsorption of water throughout the nephron. *Mannitol* is sometimes used clinically for this purpose. A similar osmotic effect can occur in diabetes mellitus due to the presence of glucose in the filtrate and urine. This extra solute causes the excretion of excessive amounts of water in the urine and can result in severe dehydration of the person with uncontrolled diabetes.

The previously mentioned diuretics can result in the excessive secretion of K^+ into the filtrate and its excessive elimination in the urine. For this reason, potassium-sparing diuretics are sometimes used. *Spironolactones* are aldosterone antagonists, which compete with aldosterone for receptor proteins in the cells of the distal tubule. These drugs, therefore, block the aldosterone stimulation of Na^+ reabsorption and K^+ secretion.

Renal Function Tests and Kidney Disease

Renal function can be tested in a variety of ways, including measurement of the GFR. These tests aid the diagnosis of kidney diseases such as glomerulonephritis and renal insufficiency.

Glomerulonephritis

Inflammation of the glomeruli, or glomerulonephritis, is currently believed to be an *autoimmune disease* which involves the person's own antibodies (as described in chapter 13). These antibodies appear to be produced in response to streptococcus infections (such as strep throat). A variable number of glomeruli are destroyed in this condition, and the remaining glomeruli become more permeable to plasma proteins. Leakage of proteins into the urine decreases the plasma protein concentration, resulting in edema (chapter 12).

Renal Insufficiency

When nephrons are destroyed—as in chronic glomerulonephritis, infection of the renal pelvis (*pyelonephritis*), and loss of a kidney—or when kidney function is reduced by either arteriosclerosis or blockage by kidney stones, a condition of *renal insufficiency* may develop. This can cause hypertension, due primarily to the retention of salt and water, and *uremia* (high plasma urea concentrations). The inability to excrete urea is accompanied by elevated plasma H^+ (acidosis) and elevated K^+, which are more immediately dangerous than the high urea. Uremic coma results from these associated changes.

Patients with uremia or the potential for developing uremia are often placed on **dialysis** machines (fig. 16.17). The term *dialysis* refers to the separation of molecules on the basis of size by their ability to diffuse through an artificial semipermeable membrane. Urea and other wastes can easily pass through the membrane pores, whereas plasma proteins are left behind (just as occurs across glomerular capillaries). The plasma is thus cleansed of these wastes as they pass from the blood into the dialyzing fluid. Unlike the tubules, however, the dialysis membrane cannot reabsorb Na^+, K^+, glucose, and other needed molecules. These substances are prevented from diffusing through the membrane by including them in the dialysis fluid. More recent techniques include the use of the patient's own peritoneal membranes (which line the body cavity) for dialysis.

Urinalysis

The urine sample obtained as part of a physical examination is used for a variety of tests. There are two major types of tests: (1) chemical analysis of the urine; and (2) microscopic examination of urine sediment obtained after centrifuging the urine sample.

A urine sample is commonly analyzed chemically for such solutes as glucose (to detect diabetes mellitus), protein (to detect glomerulonephritis), bilirubin (which is increased in jaundice),

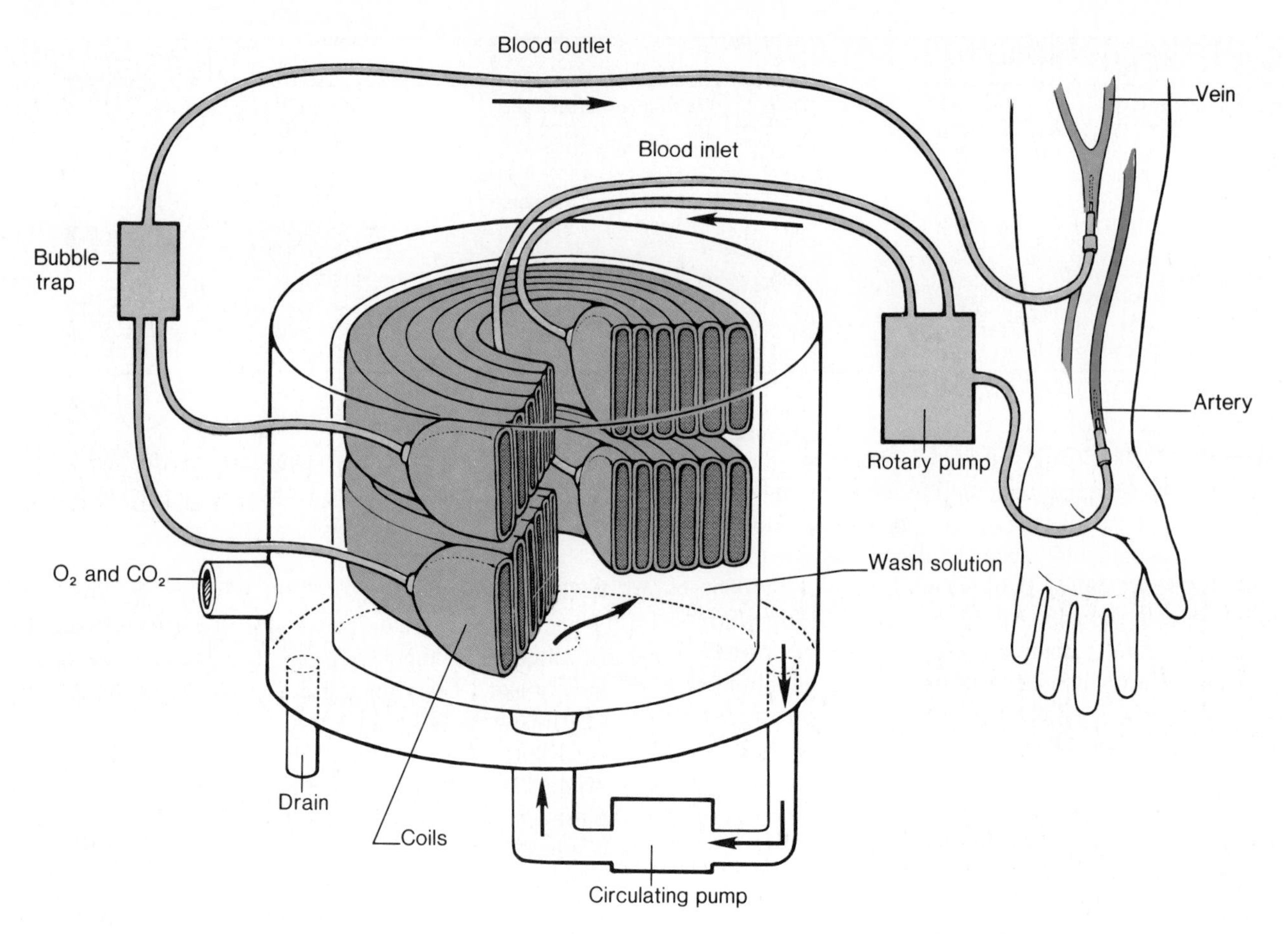

Figure 16.17. *A diagram of a hemodialysis machine.*

blood (perhaps due to a kidney stone), and others. Urine sediment usually contains a variety of epithelial cells that are normally flaked off from the urinary tract (fig. 16.18), as well as various crystals. The appearance of bacteria and white blood cells in large numbers indicates a urinary tract infection. If a person has inflamed glomeruli, the proteins that leak through may stick together and be pushed out, much like toothpaste through a tube. This forms cylindrical structures in the urine sediment known as *casts*. A large number of casts is a clinically significant finding.

The dangers of kidney disease highlight the central importance of the kidneys in maintaining homeostasis within the body. The function of one damaged kidney can be taken over by the other one, if the other is not affected by disease. Often, however, renal dialysis or a kidney transplant may be required.

1. List the different categories of clinical diuretics, and explain how each exerts its diuretic effect.
2. Explain why most diuretics can cause excessive loss of K^+, and explain how this is prevented by the potassium-sparing diuretics.
3. Define uremia and describe its dangers. Under what conditions might uremia occur?
4. Explain the significance of protein and casts in the analysis of a urine sample.

Summary

The Kidneys and Urinary System p. 356

I. Urine is drained from the cavity of the kidney, called the renal pelvis, into a ureter.
   A. Each ureter from each kidney carries urine to the urinary bladder.
   B. The urinary bladder is drained by a single urethra.
   C. Urination, or micturition, is controlled by a center located in the sacral region of the spinal cord.

II. Each kidney contains more than a million functional units known as nephrons, which are comprised of tubules and blood vessels.
   A. Afferent arterioles in the renal cortex end in tufts of capillaries called glomeruli.
   B. Each glomerulus is surrounded by a glomerular capsule, which is the beginning segment of a nephron tubule.
   C. The fluid in the glomerular capsule passes, in turn, through the proximal tubule, loop of Henle, distal tubule, and collecting duct.

Glomerular Filtration, Reabsorption, and Secretion p. 361

I. Glomerular filtrate is formed from plasma and enters through slits in the glomerular capsule.
   A. This filtrate is identical in composition to plasma except that it has a much lower protein concentration.
   B. The glomerular filtration rate averages about 120 ml/min, which is equivalent to 180 L/day.

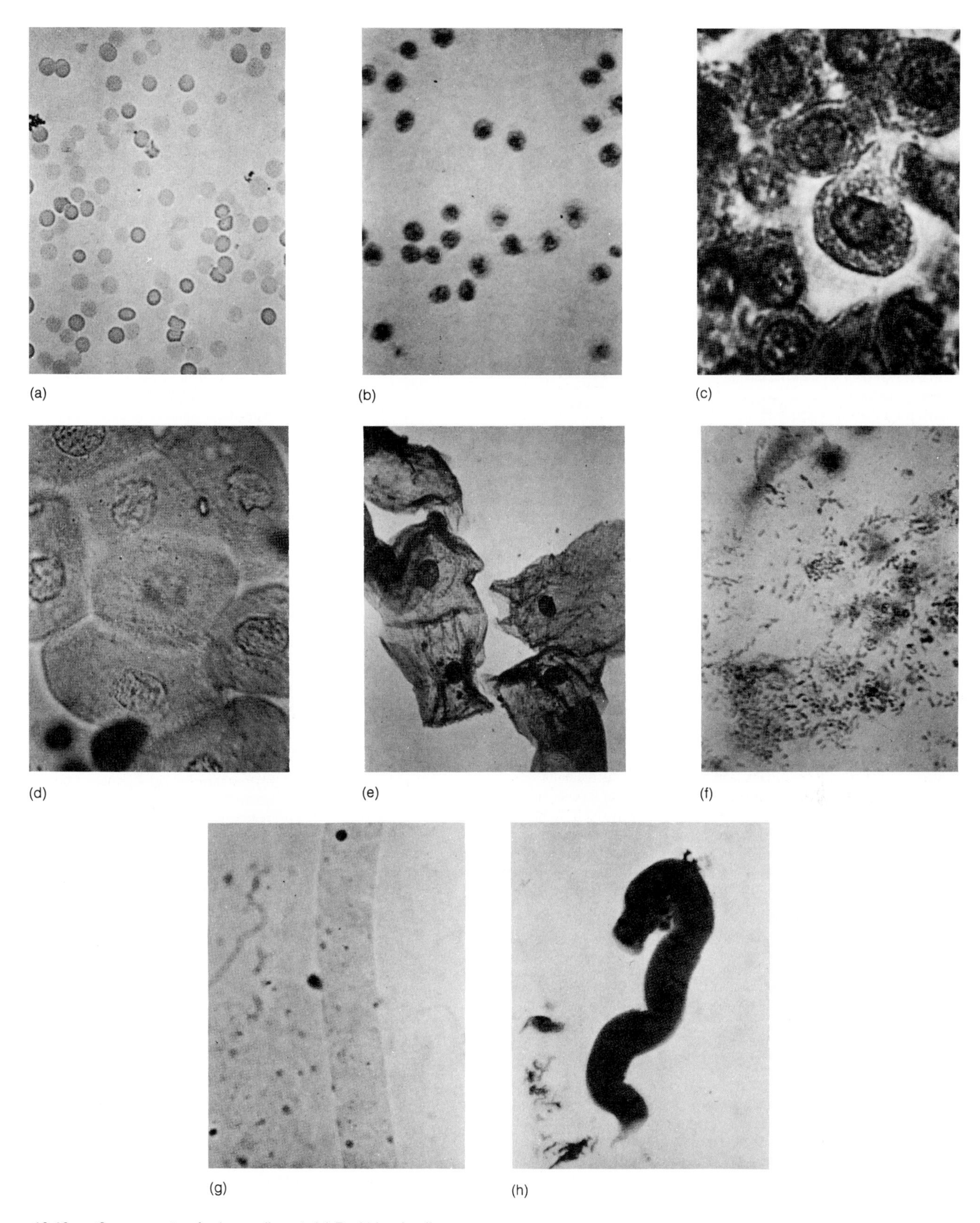

Figure 16.18. *Components of urine sediment. (a) Red blood cells; (b) white blood cells; (c) renal tubule epithelial cells; (d) bladder epithelial cells; (e) urethral epithelial cells; (f) bacteria; (g) a hyaline cast; (h) a waxy cast.*

II. Substances in the filtrate that are needed by the body are transported through the tubule wall and into surrounding capillaries.
   A. This process is called reabsorption, and it usually requires the action of specific membrane carriers.
   B. Carriers can become saturated; when this occurs, the molecules that they normally reabsorb will appear in the urine.
III. Molecules in the unfiltered blood can be transported across the tubule wall into the filtrate; this is called secretion.

Reabsorption of Salt and Water p. 364

I. The proximal tubule actively transports Na^+ across the tubule wall and into the blood.
   A. Cl^- passively follows the Na^+, and water follows the NaCl.
   B. In this way, about 65% of the initial volume of the glomerular filtrate is immediately reabsorbed.
II. The ascending limb of the loop of Henle extrudes salt into the surrounding tissue fluid of the renal medulla.
   A. Water cannot follow the salt, so that the fluid that ascends to the distal tubule is made dilute (hypotonic).
   B. The surrounding tissue fluid is made concentrated (hypertonic) by the action of the ascending limb.
   C. The descending limb of the loop is permeable to water; water therefore leaves by osmosis.
   D. The interaction of the ascending and descending limbs produces a greatly multiplied saltiness of the renal medulla, which becomes very hypertonic; this is the countercurrent multiplier system.
III. The walls of the collecting duct are permeable to water, so that water can be drawn out by osmosis due to the hypertonic renal medulla.
   A. The permeability of the collecting duct to water is stimulated by antidiuretic hormone (ADH).
   B. ADH secretion is stimulated by dehydration; under these conditions, more water is reabsorbed and less is excreted in the urine.
IV. Aldosterone stimulates the reabsorption of Na^+ and water and the secretion of K^+ into the distal tubule.

Clinical Applications p. 368

I. Diuretic drugs act on different segments of the nephron tubules to partially inhibit the reabsorption of salt and water; in this way, they can lower blood volume and blood pressure.
II. Glomerulonephritis is inflammation of the glomeruli; people with renal insufficiency for different reasons may undergo dialysis treatment.
III. Urinalysis includes the chemical analysis of urine for such solutes as glucose, and the microscopic analysis of sediment from centrifuged urine samples.

Review Activities

Objective Questions

1. Which of the following statements about the renal pyramids is *false?*
   (a) They are located in the medulla.
   (b) They contain glomeruli.
   (c) They contain collecting ducts.
   (d) They empty urine into the calyces.

Match the following:

2. Active transport of sodium; water follows passively
3. Active transport of sodium; impermeable to water
4. Passively permeable to water
5. Passively permeable to water if ADH is present

   (a) proximal tubule
   (b) descending limb
   (c) ascending limb
   (d) distal tubule
   (e) collecting duct

6. Antidiuretic hormone promotes the retention of water by stimulating the
   (a) active transport of water
   (b) active transport of chloride
   (c) active transport of sodium
   (d) permeability of the collecting duct to water
7. Aldosterone stimulates sodium reabsorption and potassium secretion in the
   (a) proximal convoluted tubule
   (b) descending limb of the loop
   (c) ascending limb of the loop
   (d) distal convoluted tubule
   (e) collecting duct
8. About 65% of the glomerular ultrafiltrate is reabsorbed in the
   (a) proximal tubule
   (b) distal tubule
   (c) loop of Henle
   (d) collecting duct
9. Diuretic drugs that act in the loop of Henle
   (a) inhibit active sodium transport
   (b) result in the increased flow of filtrate to the distal convoluted tubule
   (c) cause the increased secretion of potassium into the tubule
   (d) promote the excretion of salt and water
   (e) all of the above
10. The appearance of glucose in the urine
   (a) occurs normally
   (b) indicates the presence of kidney disease
   (c) occurs only when the transport carriers for glucose become saturated
   (d) is a result of hypoglycemia

Essay Questions

1. Explain how glomerular filtrate is produced and why it has a low protein concentration.
2. Explain how the countercurrent multiplier system works, and describe its functional significance.
3. Explain the mechanisms whereby diuretic drugs may cause an excessive loss of potassium; also explain how the potassium-sparing diuretics work.
4. Explain how the kidneys select between waste molecules to be excreted and molecules that the body needs to retain.

17

The Digestive System

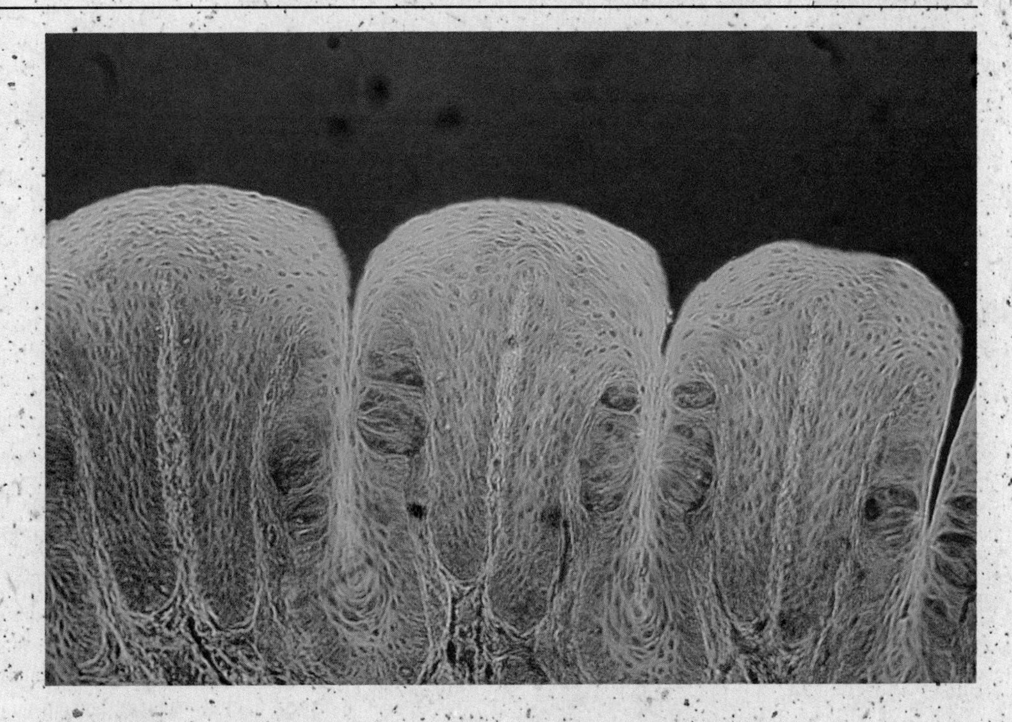

Outline

Objectives

By studying this chapter, you should be able to

1. describe the two major functions of the digestive system
2. describe peristalsis and explain the significance of the lower esophageal sphincter
3. describe the structure and secretions of the gastric mucosa
4. explain the roles of HCl and pepsin in digestion, and explain why the stomach does not normally digest itself
5. describe the structure and function of the villi, microvilli, and crypts of Lieberkühn in the small intestine
6. describe the location and functions of the brush border enzymes of the intestine
7. identify the major functions of the liver
8. describe the pattern of blood flow to the liver, and explain how this enables the liver to modify the composition of the blood
9. describe the composition and functions of bile, and explain how bile is kept separate from blood in the liver
10. describe the composition and functions of pancreatic juice
11. explain how gastric secretion is regulated during the different phases of digestion
12. explain the steps involved in the digestion and absorption of carbohydrates and proteins
13. describe the roles of bile and pancreatic lipase in fat digestion, and trace the steps involved in the absorption of lipids

Keys to Pronunciation

acini: *as'i-ni*
amylase: *am'i-las*
bilirubin: *bil'i-roo'bin*
canaliculi: *kan''ah-lik'u-li*
cecum: *se'kum*
cholecystokinen: *ko''le-sis''to-kin'in*
chylomicrons: *ki''lo-mi'krons*
cirrhosis: *sir-ro'sis*
diverticulum: *di''ver-tik'u-lum*
duodenum: *du''o-de'num* or *du-od'e-num*
esophageal hiatus: *e-sof'ah-je'al hi-a'tus*
esophagus: *e-sof'ah-gus*
haustra: *hows'tra*
ileum: *il'e-um*
jaundice: *jawn'dis*
jejunum: *je-joo'num*
secretin: *se-kre'tin*
trypsin: *trip'sin*
viscera: *vis'er-ah*

Photo: Photomicrograph of the surface of the tongue. The raised areas are called papillae, and the barrel-shaped structures on the sides of the papillae are the taste buds.

Introduction to the Digestive System

Within the lumen of the digestive tract, large food molecules are broken down by hydrolysis reactions into their subunits. These subunits pass through the mucosa (inner layer) of the intestine to enter the blood or lymph in a process called absorption.

Unlike plants, which can form organic molecules using the inorganic molecules of carbon dioxide, water, and ammonia, humans and other animals must obtain their basic organic molecules from food. Some of the ingested food molecules are needed for their energy (caloric) value, which is obtained by chemical reactions in the cells (chapter 18). The rest of the ingested molecules are used to make additional tissue.

Most of the organic molecules that are ingested are similar to the molecules that form the structure of human tissues. These are generally large molecules (*polymers*), which are composed of subunits (*monomers*). Within the gastrointestinal tract, the process of **digestion** occurs in which these large molecules are broken down into their monomers by *hydrolysis* reactions (discussed in chapter 3 and reviewed in fig. 17.1). The monomers thus formed are transported across the wall of the intestine into the blood and lymph; this process is called **absorption.** Digestion and absorption are the primary functions of the digestive system.

The lumen of the gastrointestinal tract is continuous with the environment because it is open at both ends (mouth and anus). Indigestible materials, such as cellulose from plant walls, pass from one end to the other without crossing the epithelial lining of the digestive tract (that is, without being absorbed). In this sense, these indigestible materials never enter the body, and the harsh conditions required for digestion thus occur *outside* the body.

In *planaria* (a type of flatworm) the gastrointestinal tract has only one opening—the mouth is also the anus. Each cell that lines the gastrointestinal tract is thus exposed to food, absorbable digestion products, and waste products. The two-ended digestive tract of higher organisms, in contrast, allows one-way transport, which is ensured by wavelike muscle contractions and

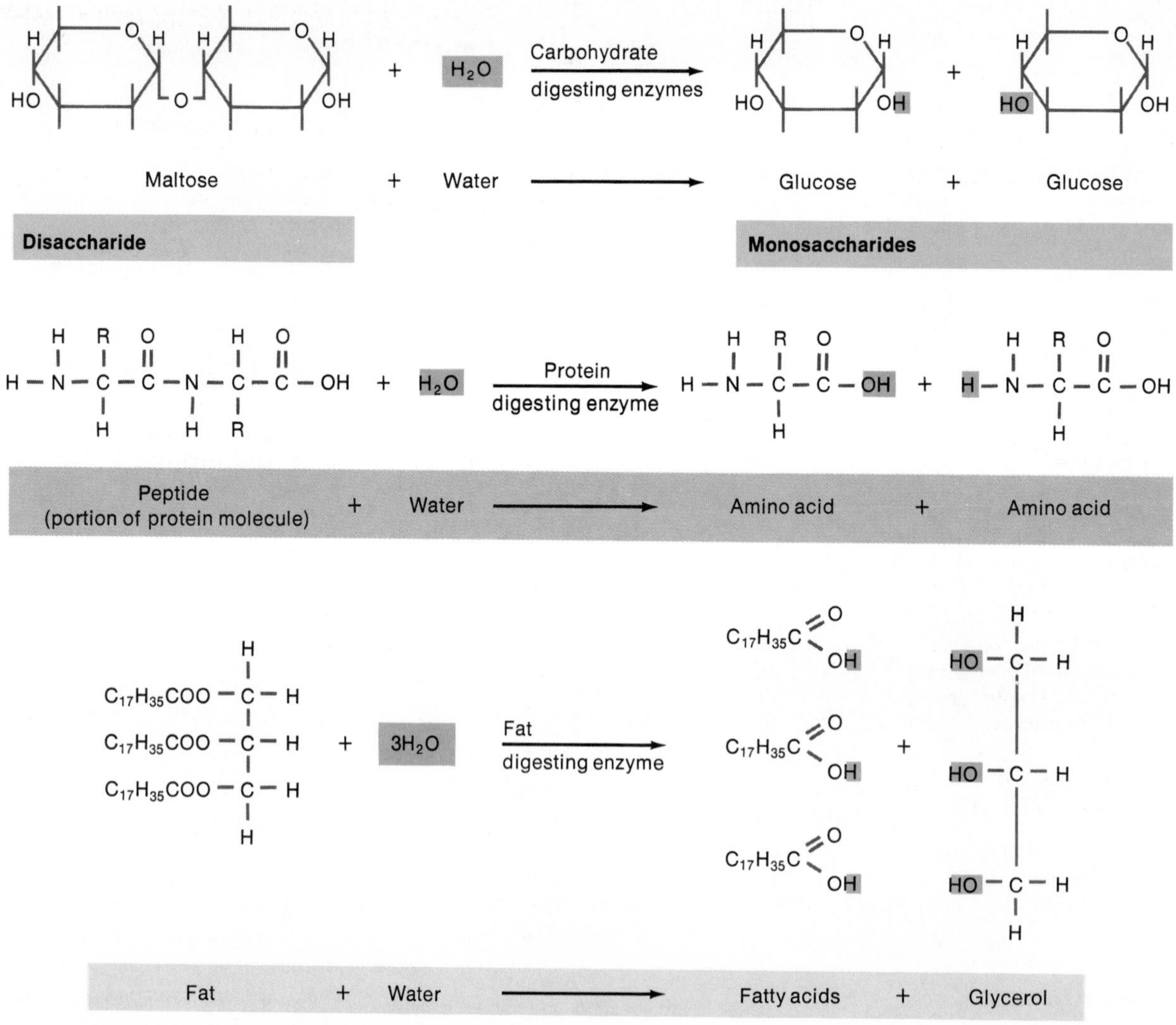

Maltose + Water → Glucose + Glucose (Carbohydrate digesting enzymes)
Disaccharide → Monosaccharides

Peptide (portion of protein molecule) + Water → Amino acid + Amino acid (Protein digesting enzyme)

Fat + $3H_2O$ → Fatty acids ($C_{17}H_{35}COOH$) + Glycerol (Fat digesting enzyme)

Figure 17.1. *The digestion of food molecules occurs by means of hydrolysis reactions.*

by the action of sphincter muscles. This one-way transport allows different regions of the gastrointestinal tract to be specialized for different functions, as a "dis-assembly line."

The digestive system can be divided into a tubular *gastrointestinal (GI) tract* and *accessory organs.* The GI tract is approximately 9 m (30 ft) long and extends from the mouth to the anus. It traverses the thoracic cavity and enters the abdominal cavity through the diaphragm. The organs of the GI tract include the *oral (buccal) cavity, esophagus, stomach, small intestine,* and *large intestine* (fig. 17.2). The accessory digestive organs include the *teeth, tongue, salivary glands, liver, gallbladder,* and *pancreas.* The term *viscera* is frequently used to refer to the abdominal organs of digestion, but this term can actually be used to indicate any organ in the thoracic and abdominal cavities.

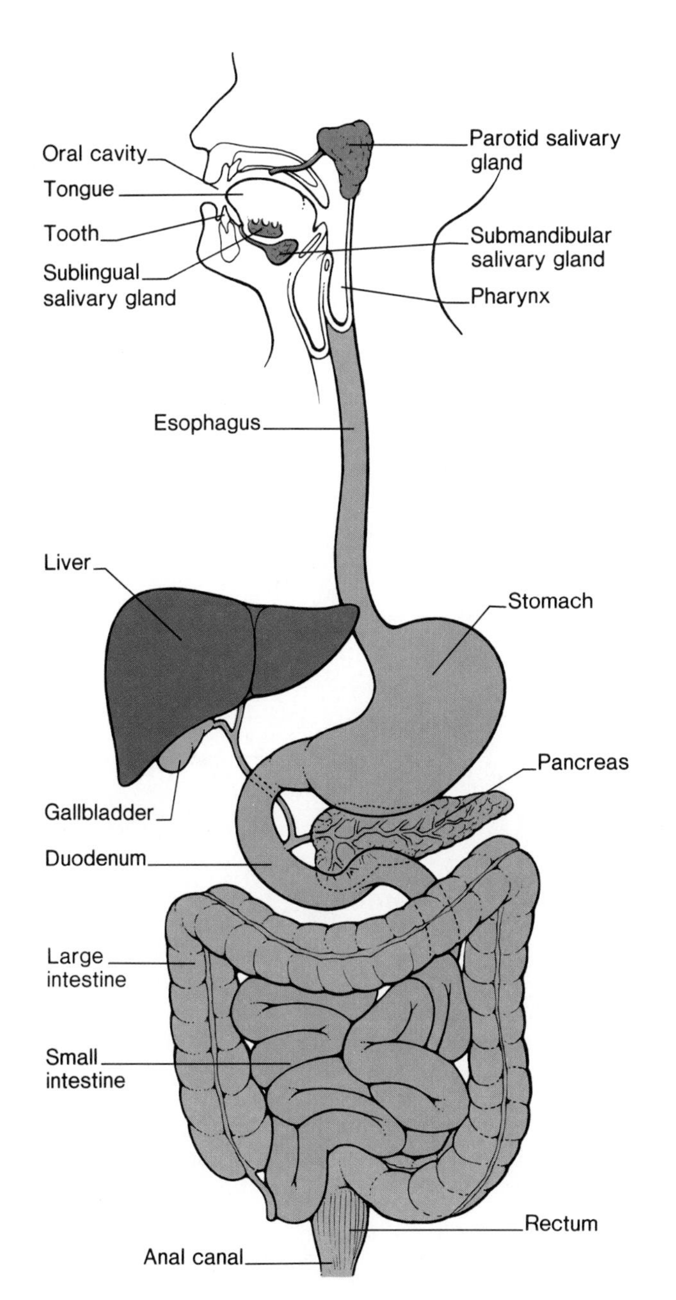

Figure 17.2. *The digestive system, including the digestive tract and the accessory digestive organs.*

Layers of the Gastrointestinal Tract

The GI tract is composed of four layers. Each layer contains a dominant tissue type that performs specific functions in the digestive process. The four layers of the GI tract, from the inside out, are the **mucosa, submucosa, muscularis,** and **serosa** (fig. 17.3*a*).

Mucosa

The mucosa surrounds the lumen of the GI tract. It consists of a simple columnar *epithelium* supported by the *lamina propria,* which is a thin layer of connective tissue. The lamina propria contains numerous lymph nodules that are important in protecting against disease (fig. 17.3*b*). External to the lamina propria is a thin layer of smooth muscle called the *muscularis mucosa.* This is the muscle layer that causes portions of the GI tract to have numerous small folds, thus greatly increasing the surface area. Specialized goblet cells in the mucosa secrete mucus throughout most of the GI tract.

Submucosa

The relatively thick submucosa is a highly vascular layer of connective tissue serving the mucosa. Absorbed molecules that pass through the columnar epithelial cells of the mucosa enter into blood vessels of the submucosa. In addition to blood vessels, the submucosa contains glands and nerve fibers (fig. 17.3*b*).

Muscularis

The muscularis (also called the muscularis externa) is responsible for contractions and peristaltic movement through the GI tract. **Peristalsis** refers to the rhythmic, wavelike contractions that move food through the digestive tract. The muscularis has an inner circular and an outer longitudinal layer of smooth muscle. Contractions of these layers move the food through the tract and physically pulverize and churn the food with digestive enzymes.

Serosa

The outer serosa layer completes the wall of the GI tract. It is a binding and protective layer consisting of loose connective tissue covered with a layer of simple squamous epithelium.

Nerve Supply to the Digestive System

The GI tract is innervated by the sympathetic and parasympathetic divisions of the autonomic nervous system (chapter 6). The vagus nerve is the source of parasympathetic activity. Stimulation of these parasympathetic fibers increases peristalsis and the secretions of the GI tract, thus promoting digestive functions. Sympathetic nerve stimulation acts antagonistically to the effects of parasympathetic nerves by reducing peristalsis and

Figure 17.3 *(a) An illustration of the major layers of the intestine. The insert shows how folds of mucosa form projections called* villi. *(b) An illustration of a cross section of the intestine showing layers and glands.*

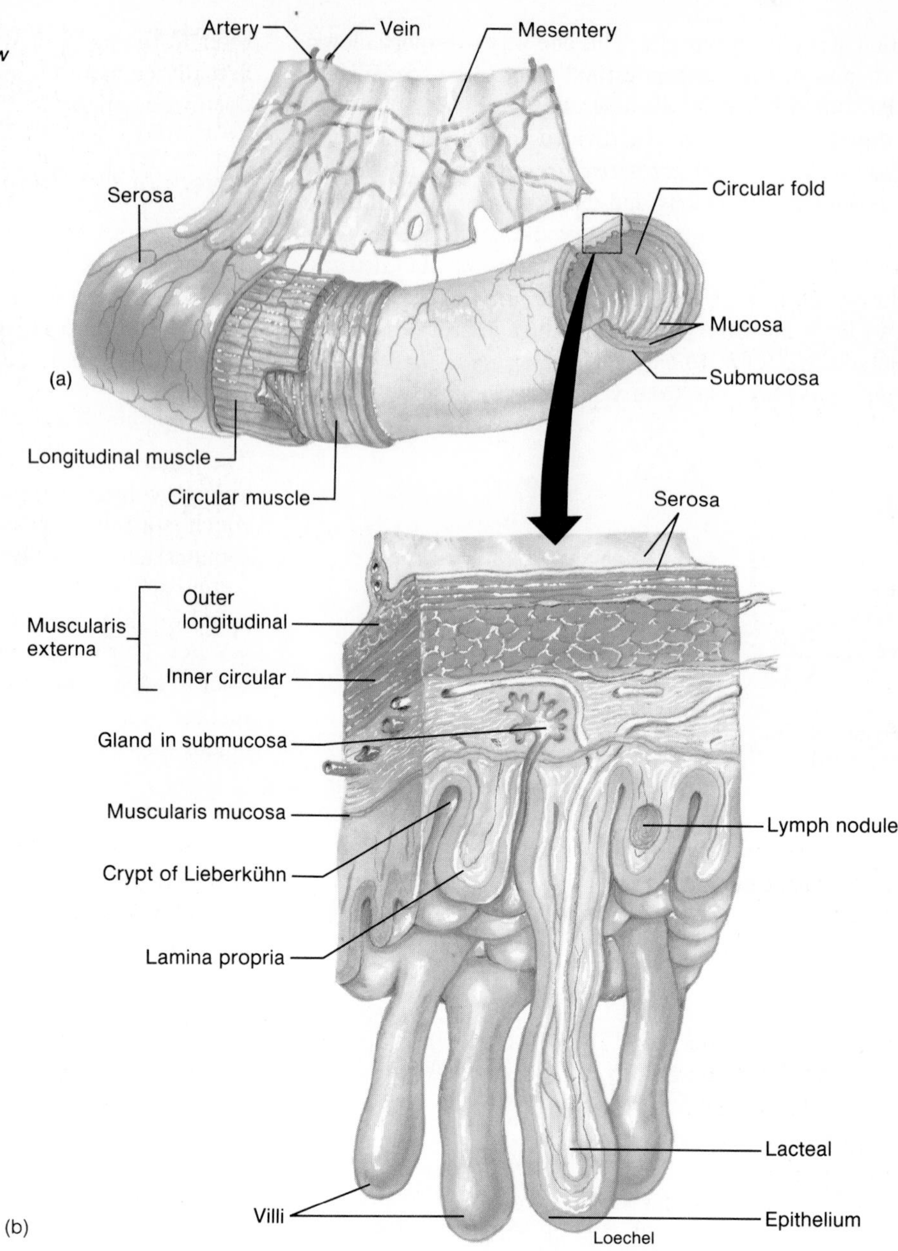

secretions and stimulating the contraction of sphincter muscles along the GI tract. This inhibits the activity of the digestive system.

---

1. Define the terms *digestion* and *absorption,* and indicate which molecules are absorbed.
2. Describe the structure and function of the mucosa, submucosa, and muscularis layers of the gastrointestinal tract.
3. Explain the actions of autonomic nerves on the gastrointestinal tract.

Esophagus and Stomach

The mucosa of the stomach contains gastric glands with parietal cells, which secrete hydrochloric acid, and chief cells, which secrete a protein-digesting enzyme named pepsin. The stomach partially digests proteins and functions to store its contents, called chyme, for later processing by the intestine.

Swallowed food is passed from the esophagus[1] to the stomach, where it is churned and mixed with hydrochloric acid and pepsin, a protein-digesting enzyme. The mixture thus produced is

[1]esophagus: Gk. *oisein,* to carry; *phagema,* food

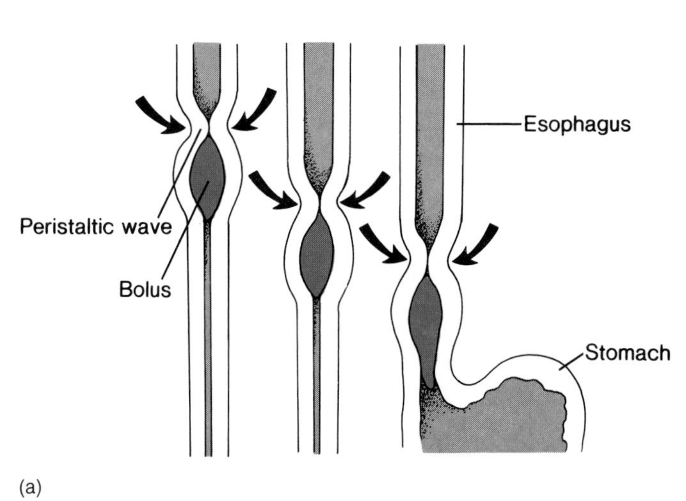

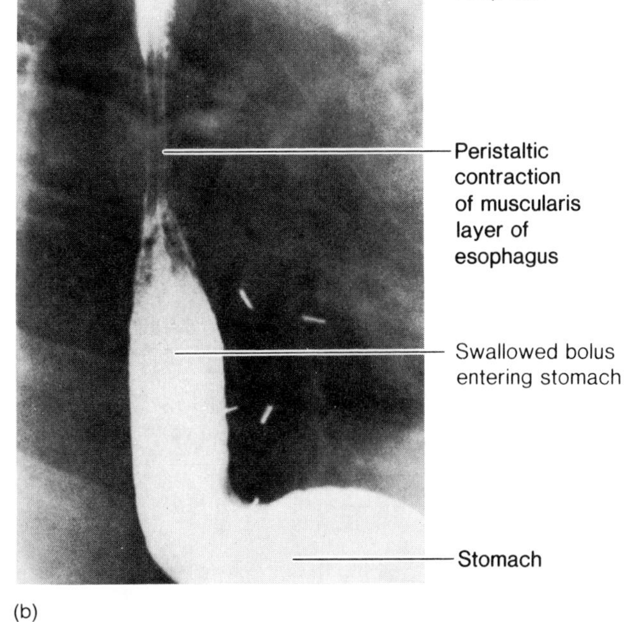

Figure 17.4. *(a) A diagram and (b) an X ray of peristalsis in the esophagus.*

pushed by muscular contractions of the stomach past the pyloric sphincter, which guards the junction of the stomach with the small intestine.

Esophagus

The **esophagus** is the portion of the GI tract that transports swallowed food to the stomach. It is a muscular tube approximately 25 cm (10 in.) long, located posterior to the trachea.

The esophagus descends through the neck and thoracic cavity and passes through an opening in the diaphragm called the *esophageal hiatus,* located just above the stomach. The walls of the esophagus contain either striated or smooth muscle, depending on the location. The upper third of the esophagus contains only voluntary, striated muscle, whereas the terminal portion contains only involuntary smooth muscle.

A **hiatus hernia** is present when part of the stomach protrudes above the diaphragm through the esophageal hiatus. This condition is quite common, and often does not produce noticeable symptoms or require special treatment. Sometimes, however, chest pain may be produced. If the hiatus hernia causes gastroesophageal reflux (movement of gastric contents into the esophagus), however, medical treatment may be needed. When the lumen of the esophagus is strangulated (constricted) by the hernia, surgical correction is required.

Swallowed food is pushed from one end of the esophagus to the other by the wavelike muscular contraction of peristalsis (fig. 17.4). These contractions progress to the *gastroesophageal junction* as they empty the contents of the esophagus into the stomach.

The terminal portion of the esophagus contains a thickening of the circular muscle fibers in its wall. This portion is referred to as the **lower esophageal sphincter.** The muscle fibers of this region constrict after food passes into the stomach to help prevent the stomach contents from regurgitating into the esophagus. Regurgitation would occur because the pressure in the abdominal cavity is greater than the pressure in the thoracic cavity as a result of movements of the diaphragm during respiration. The lower esophageal sphincter must remain closed, therefore, until food is pushed through it by peristalsis into the stomach.

The lower esophageal sphincter is not a true sphincter muscle, and it does at times permit the acidic contents of the stomach to enter the esophagus. This can create a burning sensation commonly called **heartburn,** although the heart is not involved. Infants under one year old have difficulty controlling their lower esophageal sphincter and thus often "spit up" following meals. Certain mammals, such as rodents, have a true gastroesophageal sphincter and cannot regurgitate, which is why poison grains can kill mice and rats effectively.

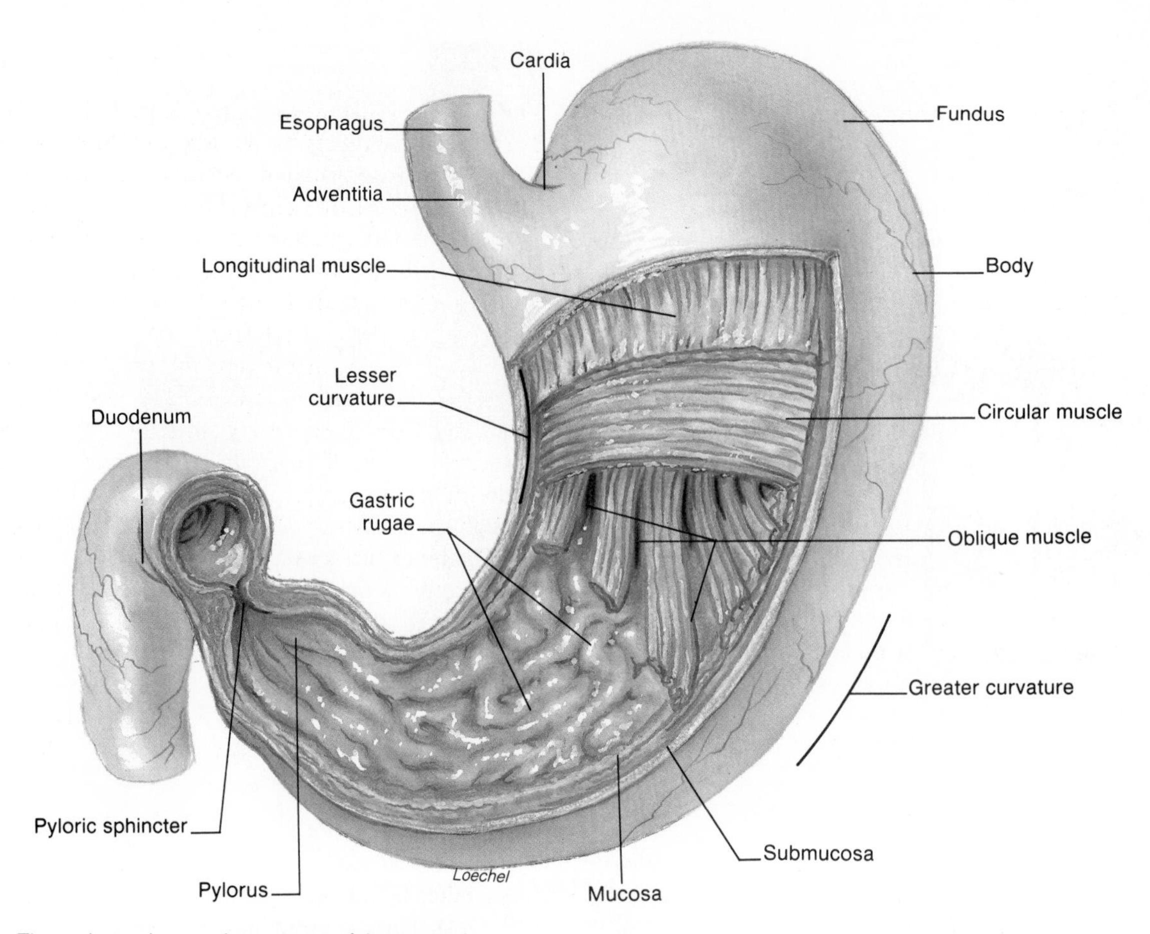

Figure 17.5. *The major regions and structures of the stomach.*

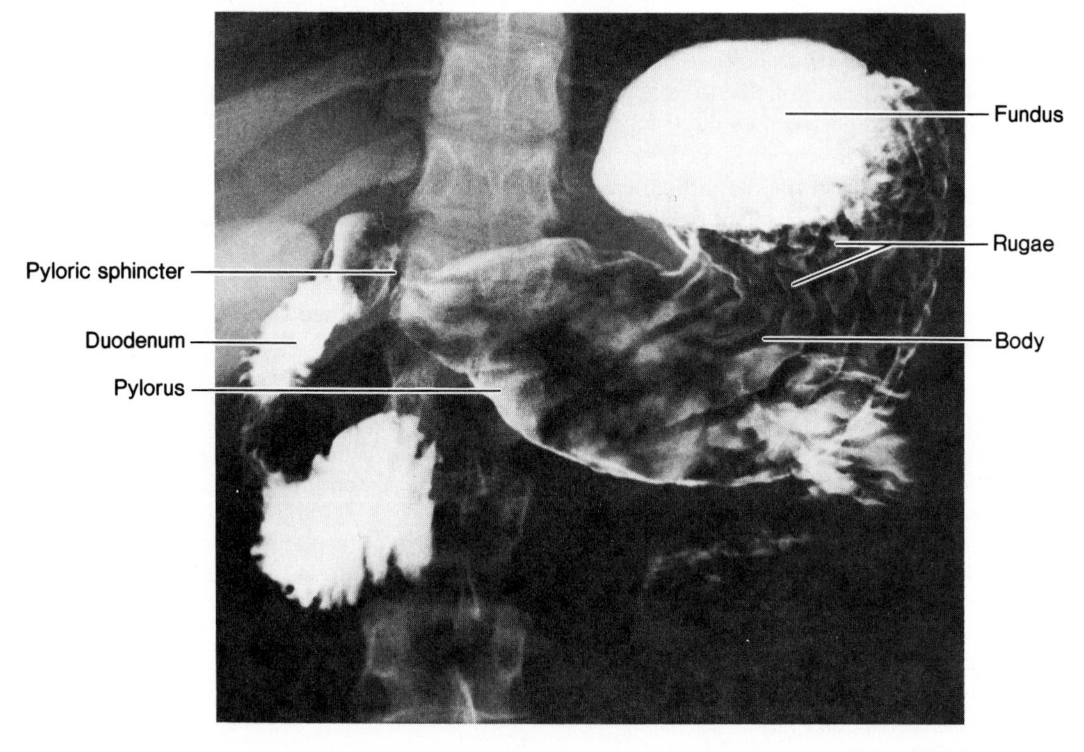

Figure 17.6. *An X ray of a stomach.*

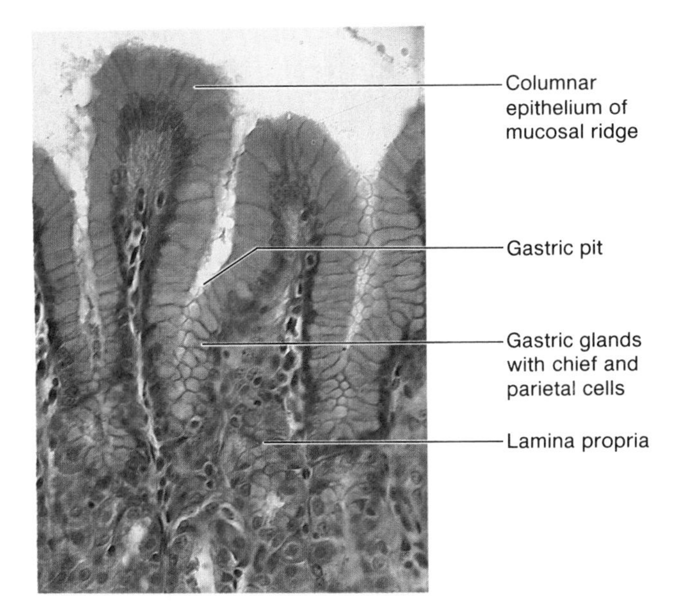

Figure 17.7. *Microscopic structures of the mucosa of the stomach.*

Stomach

The stomach is the most distensible part of the GI tract. It is a J-shaped pouch that is continuous with the esophagus and empties into the first portion of the small intestine (the duodenum). The functions of the stomach are to store food; to initiate the digestion of proteins; and to move the food into the small intestine as a pasty material called **chyme.**

Swallowed food is delivered from the esophagus to the *cardiac region* of the stomach (figs. 17.5 and 17.6). An imaginary horizontal line drawn through the cardiac region divides the stomach into an upper *fundus* and a lower *body,* which comprises about two-thirds of the stomach. The last portion of the stomach is called the *pyloric region,* or *antrum.* Contractions of the stomach push partially digested food from the antrum, through the *pyloric sphincter,* and thus into the duodenum.

The inner surface of the stomach is thrown into long folds called *rugae,* which can be seen with the unaided eye. Microscopic examination of the gastric mucosa shows that it is likewise folded. The openings of these folds into the stomach lumen are called **gastric pits.** The cells that line the folds deeper in the mucosa secrete various products into the stomach; these form the exocrine **gastric glands** (figs. 17.7 and 17.8).

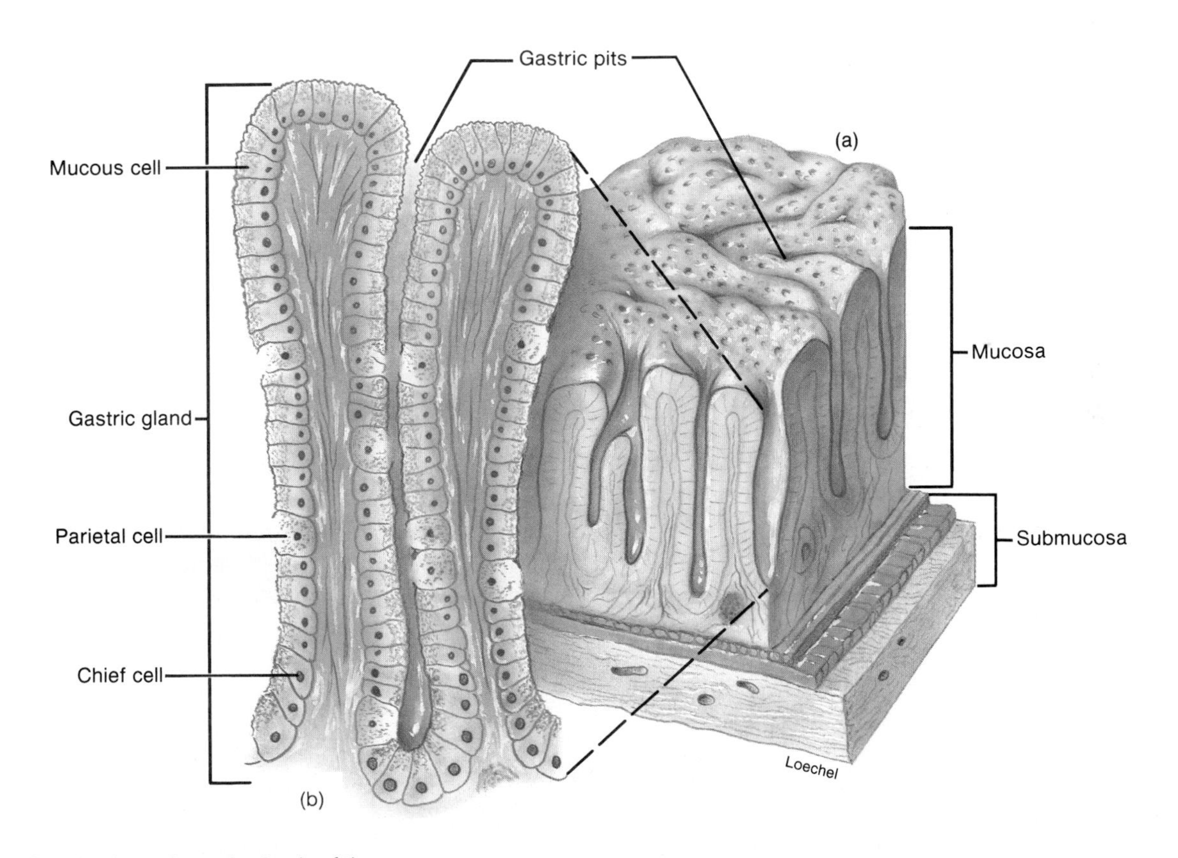

Figure 17.8. *Gastric pits and gastric glands of the mucosa. (a) Gastric pits are the openings of the gastric glands. (b) Gastric glands consist of mucous cells, chief cells, and parietal cells, each of which produces a specific secretion.*

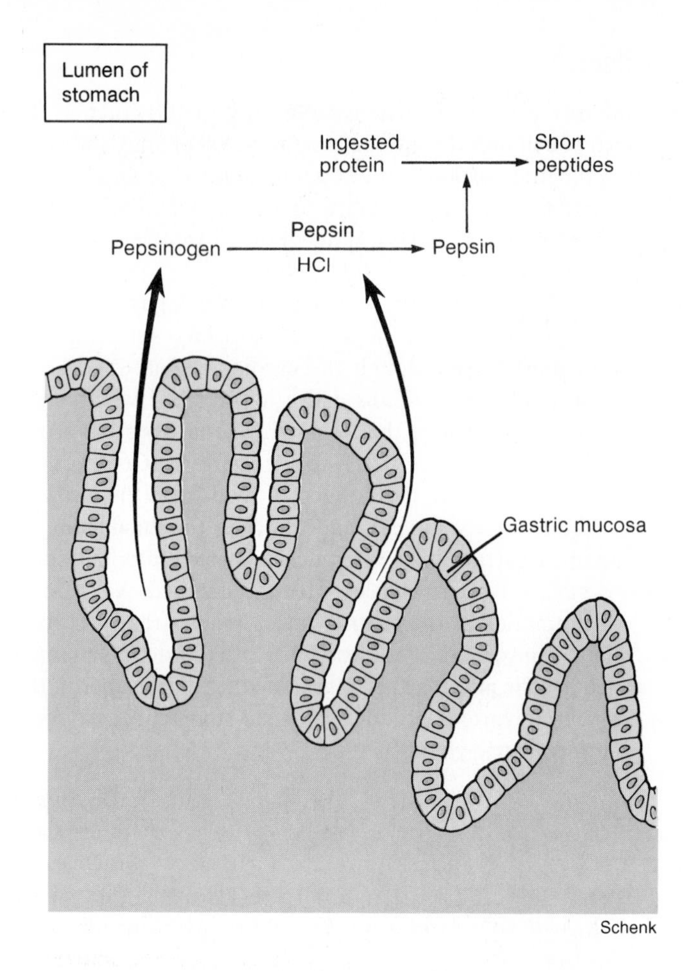

Figure 17.9. The gastric mucosa secretes the inactive enzyme pepsinogen and hydrochloric acid (HCl). Through a process of self-digestion, the active enzyme pepsin is produced. Pepsin digests proteins into shorter polypeptides.

There are several different types of cells in the gastric glands that secrete different products. The major cells and their products include (1) **goblet cells,** which secrete mucus; (2) **parietal cells,** which secrete hydrochloric acid (HCl); and (3) **chief cells,** which secrete pepsinogen (an inactive form of the protein-digesting enzyme pepsin).

Pepsin and Hydrochloric Acid Secretion

The secretion of hydrochloric acid by parietal cells makes gastric juice very acidic, with a pH less than 2. This strong acidity serves three functions: (1) ingested proteins are denatured at low pH—that is, their three-dimensional structure (chapter 3) is altered so that they become more digestible; (2) under acidic conditions, weak pepsinogen enzymes partially digest each other—this frees the active pepsin enzyme as small peptide fragments are removed (fig. 17.9); and (3) pepsin is more active under acidic conditions, becoming better able to break the peptide bonds within the food proteins.

Digestion and Absorption in the Stomach

Proteins are only partially digested in the stomach by the action of pepsin. Carbohydrates and fats are not digested at all in the stomach. The complete digestion of food molecules occurs later, when chyme enters the small intestine. Even patients with complete gastrectomies (removal of the stomach) can still adequately digest and absorb their food.

The only commonly ingested substances that can be absorbed across the stomach wall are alcohol and aspirin. Absorption occurs as a result of the lipid solubility of these molecules. The passage of aspirin through the gastric mucosa has been shown to cause bleeding, which may be significant if large amounts of aspirin are taken.

The only function of the stomach that appears to be essential for life is the secretion of a polypeptide called *intrinsic factor.* This polypeptide is needed for the absorption of vitamin B_{12} in the terminal portion of the small intestine, and vitamin B_{12} is required for maturation of red blood cells in the bone marrow. A patient with a gastrectomy must thus take intrinsic factor together with vitamin B_{12} to prevent the development of **pernicious anemia.**

Gastritis and Peptic Ulcers

The gastric mucosa is normally resistant to the harsh conditions of low pH and pepsin activity in gastric juice. The reasons for its resistance to pepsin digestion are not well understood. Its resistance to hydrochloric acid appears to be due to three interrelated mechanisms: (1) the stomach lining is covered with a thin layer of alkaline mucus, which may offer limited protection; (2) the epithelial cells of the mucosa are joined together by tight junctions, which prevent acid from leaking into the submucosa; and (3) the epithelial cells that are damaged are exfoliated (shed) and replaced by new cells. This latter process results in the loss and replacement of about one-half million cells a minute, so that the entire epithelial lining is replaced every three days.

These three mechanisms provide a barrier that prevents self-digestion of the stomach. Breakdown of the gastric mucosa barrier may occur at times, however, perhaps as a result of the detergent action of bile salts that are regurgitated from the small intestine through the pyloric sphincter. When this occurs, acid can leak through the mucosal barrier to the submucosa, which causes direct damage and stimulates inflammation. The histamine released from mast cells (chapter 13) during inflammation may stimulate further acid secretion and result in further damage to the mucosa. The inflammation that occurs during these events is called **acute gastritis.**

Once the mucosal barrier is broken, the acid can produce craterlike holes and even complete perforations of the stomach wall. These are called gastric ulcers. More commonly encountered, however, are ulcers of the first 3–4 cm of the duodenum. Ulcers of the stomach and duodenum are known as **peptic ulcers.** It has been estimated that about 10% of American men and 4% of American women will develop peptic ulcers.

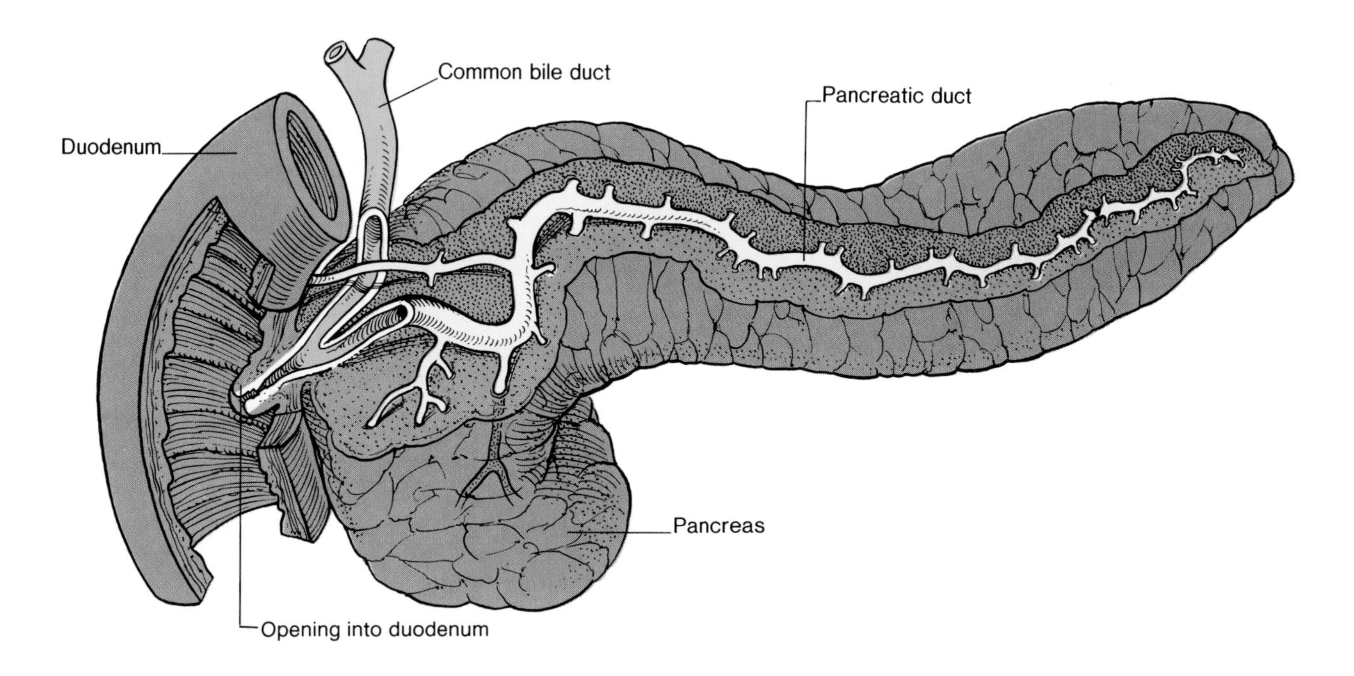

Figure 17.10. *The duodenum and associated structures.*

The duodenum is normally protected against gastric acid by the buffering action of bicarbonate in alkaline pancreatic juice. People who develop duodenal ulcers, however, produce excessive amounts of gastric acid that cannot be neutralized by pancreatic juice. This excessive gastric acid secretion is apparently due to excessive parasympathetic nerve stimulation of the stomach. People with gastritis and peptic ulcers must avoid substances that stimulate acid secretion, including caffeine and alcohol, and often must take antacids.

---

1. Describe the structure and function of the lower esophageal sphincter. Explain how heartburn is produced.
2. List the secretory cells of the gastric mucosa and the products they secrete.
3. Explain the roles of hydrochloric acid in the stomach and how pepsinogen is activated.
4. Explain the cause of peptic ulcers and why they are more likely to be duodenal than gastric in location.

Small Intestine

The mucosa of the small intestine has folds, called villi, which project into the lumen. The cells that line these villi, further, have foldings of their plasma membrane called microvilli. This arrangement greatly increases the surface area for absorption. It also improves digestion, because the digestive enzymes of the small intestine are embedded within the cell membrane of the microvilli.

The *small intestine* is that portion of the GI tract between the pyloric sphincter of the stomach and the opening into the large intestine. It is called "small" because of its smaller diameter compared to that of the large intestine. The small intestine is approximately 3.7 m (12 ft) long in a living person, but it will measure nearly twice this length in a cadaver when the muscle wall is relaxed. The first 20–30 cm (10 in.) from the pyloric sphincter is the **duodenum**[2] (figs. 17.5 and 17.10). The next two-fifths of the small intestine is the **jejunum,** and the last three-fifths is the **ileum.** The ileum empties into the cecum of the large intestine.

The products of digestion are absorbed across the epithelial lining of the intestinal mucosa. Absorption of carbohydrates, lipids, protein, calcium, and iron occurs primarily in the duodenum and jejunum. Bile salts, vitamin B_{12}, water, and electrolytes are absorbed primarily in the ileum. Absorption occurs at a rapid rate as a result of the large mucosal surface area in the small intestine, which is provided by folds. The mucosa and submucosa form large folds called the *plicae*[3] *circularis,* which can be observed with the unaided eye. The surface area is further increased by the microscopic folds of mucosa, called villi, and by the foldings of the cell membrane of epithelial cells (which can only be seen with an electron microscope), called microvilli.

Villi and Microvilli

Each **villus**[4] is a fingerlike fold of mucosa that projects into the intestinal lumen (fig. 17.11). The villi are covered with columnar epithelial cells and interspersed among these are the

[2]duodenum: L. *duodeni,* twelve each (length of twelve fingers across)
[3]plica: L. *plicatus,* folded
[4]villus: L. *villosus,* shaggy

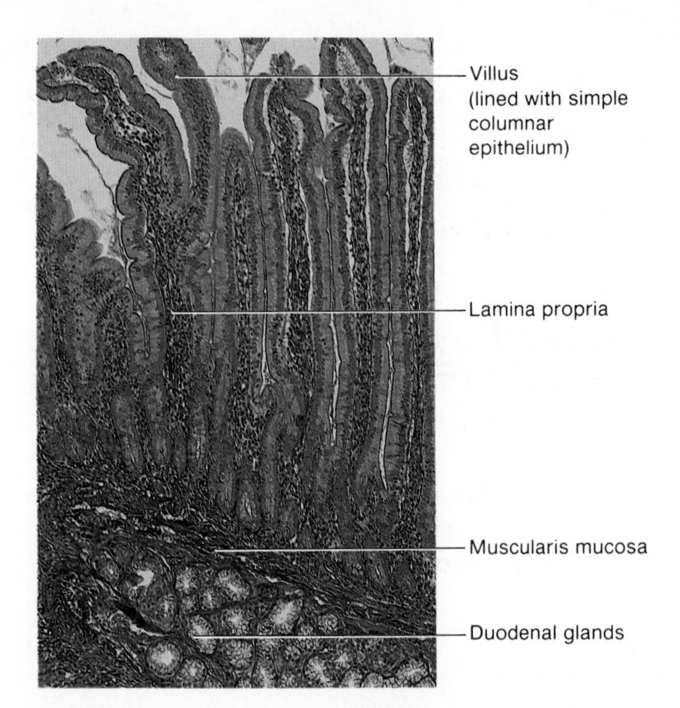

Figure 17.11. The histology of the duodenum.

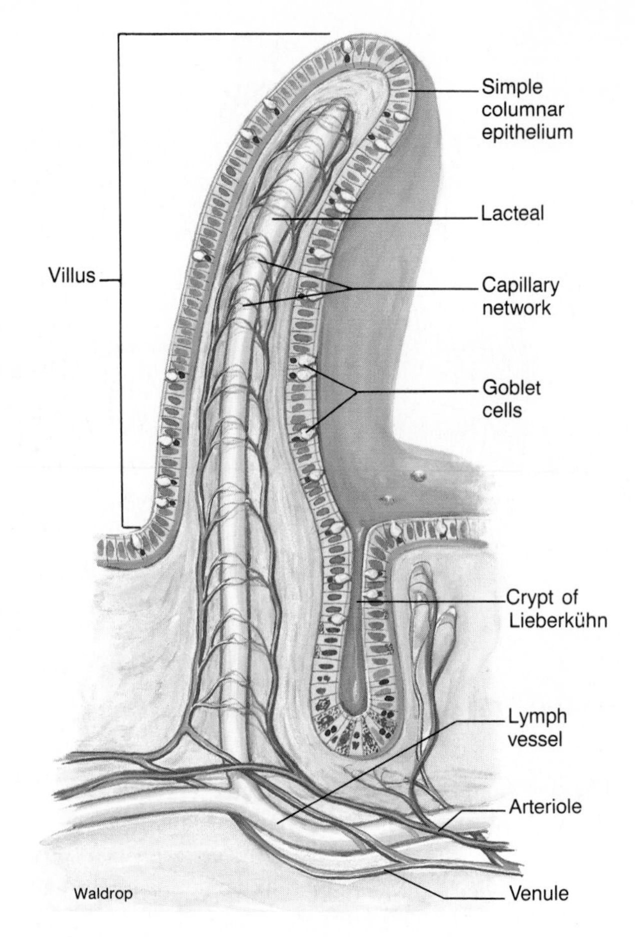

Figure 17.12. A diagram of the structure of an intestinal villus.

mucous-secreting goblet cells. The lamina propria forms a connective tissue core of each villus and contains numerous lymphocytes, blood capillaries, and a lymphatic vessel called the *central lacteal*[5] (fig. 17.12). Absorbed monosaccharides and amino acids are secreted into the blood capillaries; absorbed fat enters the central lacteals.

Epithelial cells at the tips of the villi are continuously shed and are replaced by cells that are pushed up from the bases of the villi. The epithelium at the base of the villi invaginates downwards at various points to form narrow pouches that open through pores to the intestinal lumen. These structures are called the **crypts of Lieberkühn**[6] (fig. 17.13).

Microvilli are fingerlike projections formed by foldings of the cell membrane, which can only be clearly seen in an electron microscope. In a light microscope, the microvilli produce a somewhat vague **brush border** on the edges of the columnar epithelial cells. The term *brush border* is thus often used synonymously with microvilli in descriptions of the intestine (fig. 17.14).

Intestinal Enzymes

In addition to providing a large surface area for absorption, the cell membranes of the microvilli contain digestive enzymes. These enzymes are not secreted into the lumen, but instead remain attached to the cell membrane with their active sites exposed to the chyme. These **brush border enzymes** hydrolyze (digest) disaccharides, polypeptides, and other substrates into smaller molecules (table 17.1).

[5]lacteal: L. *lacteus,* milk

[6]crypts of Lieberkühn: from Johann N. Lieberkühn, German anatomist, 1711–1756

The ability to digest milk sugar, or lactose, depends on the presence of a brush border enzyme called lactase. This enzyme is present in most children under the age of four but becomes inactive in most adults. This can result in **lactose intolerance.** The presence of large amounts of undigested lactose in the intestine causes diarrhea, gas, cramps, and other unpleasant symptoms. Yogurt is better tolerated than milk because it contains lactase produced by the yogurt bacteria, which becomes activated in the duodenum and digests lactose.

Intestinal Contractions and Motility

Like cardiac muscle, intestinal smooth muscle is capable of spontaneous electrical activity and automatic, rhythmic contractions. The rate at which this automatic activity occurs is influenced by autonomic nerves. Contraction is stimulated by parasympathetic (vagus nerve) innervation and is reduced by sympathetic nerve activity.

The small intestine has two major types of contractions: peristalsis and segmentation. Peristalsis is much weaker in the small intestine than in the esophagus and stomach. **Intestinal motility**—the movement of chyme through the intestine—is relatively slow and is due primarily to the fact that the pressure at the pyloric end of the small intestine is greater than at the other end.

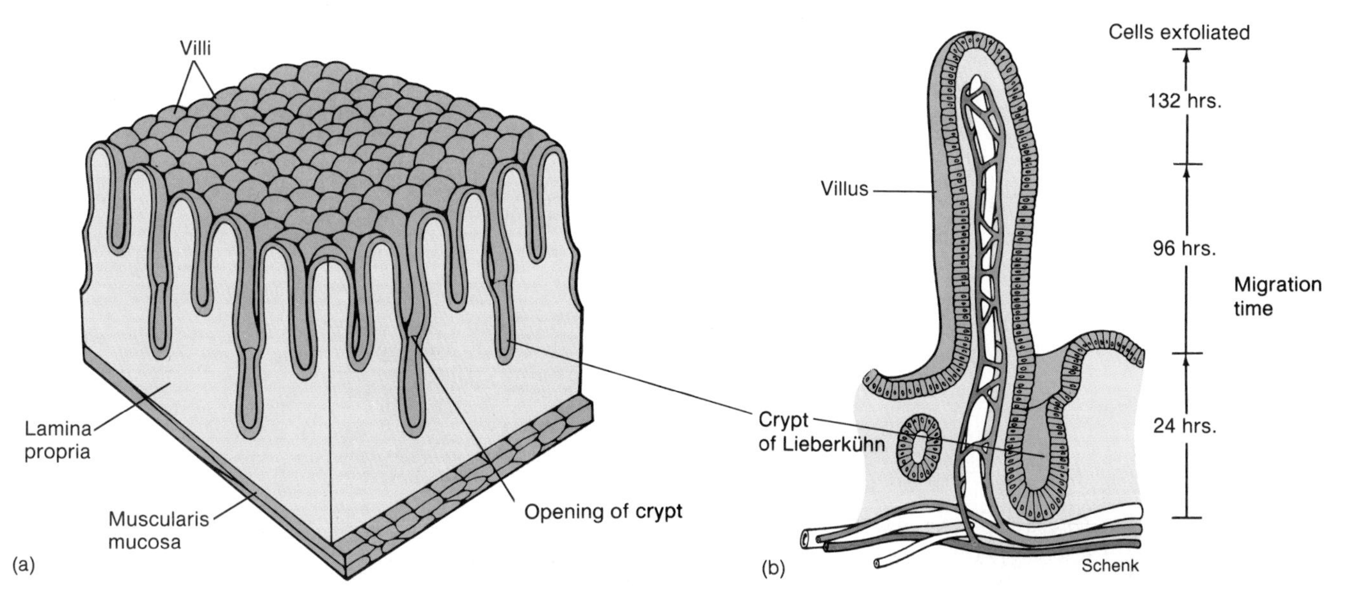

Figure 17.13. *(a) Intestinal villi and crypts of Lieberkühn. The crypts serve as sites for production of new epithelial cells. The time required for migration of these new cells to the tip of the villi is shown in (b). Epithelial cells are exfoliated from the tips of the villi.*

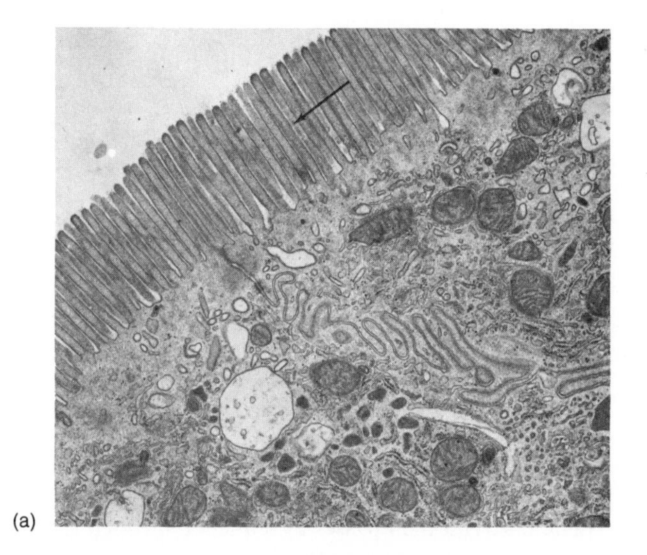

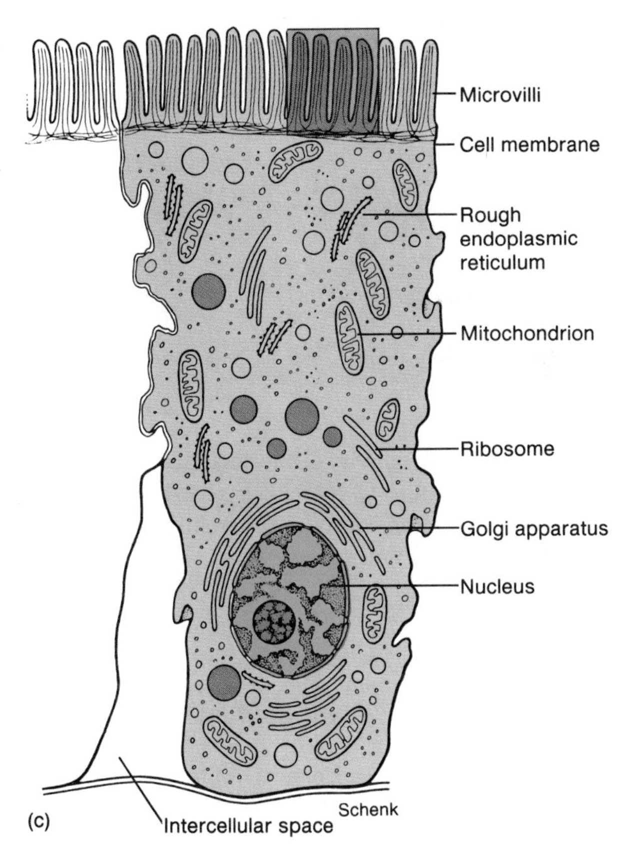

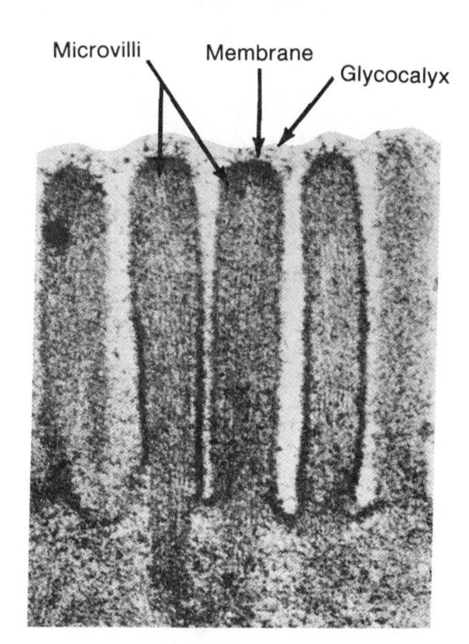

Figure 17.14. *(a) Electron micrograph of microvilli, forming the brush border of the small intestine. (b) Individual microvilli seen at higher magnification. (c) Illustration of an intestinal epithelial cell showing microvilli.*

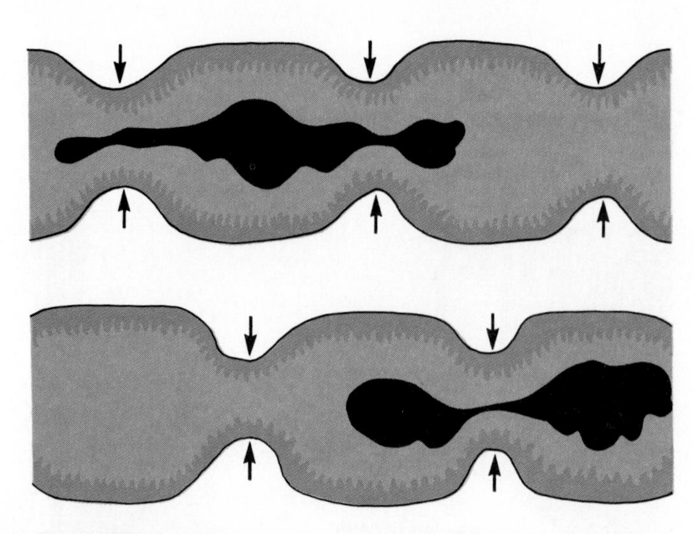

Figure 17.15. Segmentation of the small intestine. Simultaneous contractions of many segments of the intestine help to mix the chyme with digestive enzymes and mucus.

Table 17.1 Brush border enzymes attached to the cell membrane of microvilli in the small intestine

Category	Enzyme	Comments
Disaccharidase	Sucrase	Digests sucrose to glucose and fructose; deficiency produces gastrointestinal disturbances
	Maltase	Digests maltose to glucose
	Lactase	Digests lactose to glucose and galactose; deficiency produces gastrointestinal disturbances (lactose intolerance)
Peptidase	Aminopeptidase	Produces free amino acids, dipeptides, and tripeptides
	Enterokinase	Activates trypsin (and indirectly other pancreatic juice enzymes); deficiency results in protein malnutrition
Phosphatase	Ca^{++}, Mg^{++}—ATPase	Needed for absorption of dietary calcium; enzyme activity regulated by vitamin D
	Alkaline phosphatase	Removes phosphate groups from organic molecules; enzyme activity may be regulated by vitamin D.

The major contractile activity of the small intestine is **segmentation.** This term refers to muscular constrictions of the lumen, which occur simultaneously at different intestinal segments (fig. 17.15). This action serves to mix the chyme more thoroughly.

1. Describe the intestinal structures that increase surface area, and explain the function of the crypts of Lieberkühn.
2. Define the term *brush border enzymes,* and list examples. Explain why many adults cannot tolerate milk.
3. Describe how smooth muscle contraction in the small intestine is regulated, and explain the function of segmentation.

Large Intestine

The large intestine absorbs water and electrolytes from the chyme it receives from the small intestine and, in a process regulated by the action of sphincter muscles, passes waste products out of the body through the rectum and anal canal.

Chyme from the ileum passes into the *cecum,*[7] which is a blind-ending pouch at the beginning of the large intestine, or *colon.* Waste material then passes in sequence through the ascending colon, transverse colon, descending colon, sigmoid colon, rectum, and anal canal (fig. 17.16). Waste material (feces) is excreted through the anus.

As in the small intestine, the mucosa of the large intestine contains many scattered lymphocytes and lymphatic nodules and is covered by columnar epithelial cells and mucous-secreting goblet cells. There are no villi, however, in the large intestine. The outer surface of the colon bulges outwards to form pouches, or **haustra** (fig. 17.17). Occasionally, the muscularis of the haustra may become so weakened that the wall forms a more elongated outpouching, or diverticulum (*divert* = turned aside). Inflammation of these structures is called *diverticulitis.*

[7]cecum: L. *caecum,* blind pouch

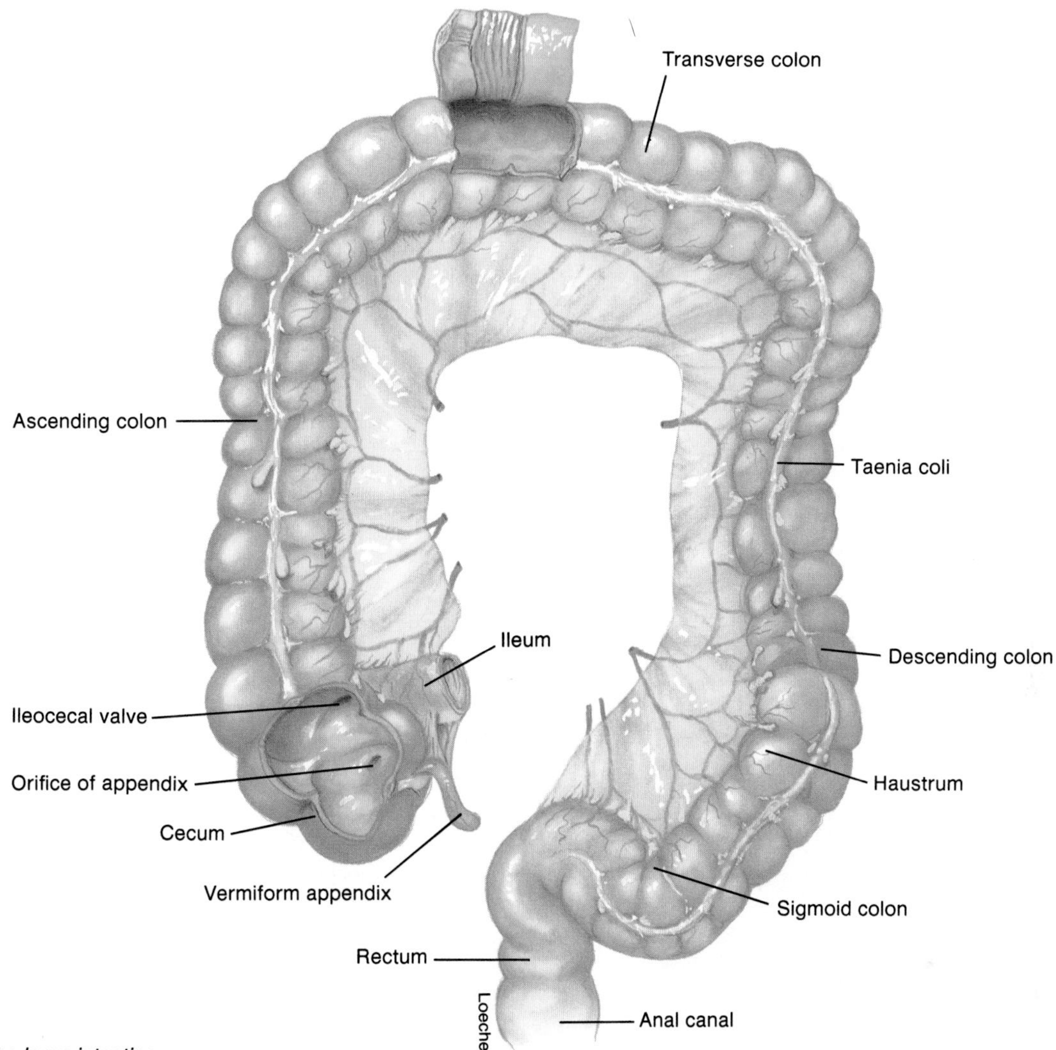

Figure 17.16. *The large intestine.*

The *vermiform appendix*[8] is a short, thin outpouching of the cecum. It does not function in digestion, but like the tonsils, it contains numerous lymphatic nodules (fig. 17.18) and is subject to inflammation—a condition called **appendicitis.** This is commonly detected in its later stages by pain in the lower right quadrant of the abdomen. Rupture of the appendix can cause inflammation of the surrounding body cavity—*peritonitis.* This dangerous event may be prevented by surgical removal of the inflamed appendix (appendectomy).

Fluid and Electrolyte Absorption in the Intestine

Most of the fluid and electrolytes (ions) in the lumen of the digestive tract are absorbed by the small intestine. Although a person may drink only about 1.5 L/day of water, the small intestine receives 7–9 L/day as a result of the fluid secreted into the digestive tract by the salivary glands, stomach, pancreas, liver, and gallbladder. The small intestine absorbs most of this fluid and passes about 1.5–2.0 L/day of fluid to the large intestine. The large intestine absorbs about 90% of this remaining volume, leaving less than 200 ml/day of fluid to be excreted in the feces.

Diarrhea is characterized by excessive fluid excretion in the feces. There are three different mechanisms, illustrated by three different diseases, which can cause diarrhea. In *cholera,* severe diarrhea results from a chemical called *enterotoxin,* released from the infecting bacteria. Enterotoxin stimulates active Na^+ and water secretion. In *celiac sprue,* a disease produced in susceptible people by eating foods that contain gluten (proteins from grains such as wheat), diarrhea results from inadequate absorption of fluid due to damage to the intestinal mucosa. In *lactose intolerance,* diarrhea is produced by the increased osmotic pressure of the contents of the intestinal lumen as a result of the presence of undigested lactose.

[8]vermiform appendix: L. *vermiformis,* wormlike; *appendix,* attachment

Defecation

After electrolytes and water have been absorbed, the waste material that is left passes to the rectum,[9] leading to an increase in rectal pressure and the urge to defecate. If the urge to defecate is denied, feces are prevented from entering the anal canal by the anal sphincter. In this case the feces remain in the rectum and may even back up into the sigmoid colon. The **defecation reflex** normally occurs when the rectal pressure rises to a particular level that is determined, to a large degree, by habit. At this point the anal sphincter relaxes to admit feces into the anal canal.

During the act of defecation the longitudinal rectal muscles contract to increase rectal pressure and the anal sphincter muscle relaxes. Excretion is aided by contractions of abdominal and pelvic skeletal muscles, which raise the intra-abdominal pressure and help push the feces from the rectum through the anal canal and out the anus.[10]

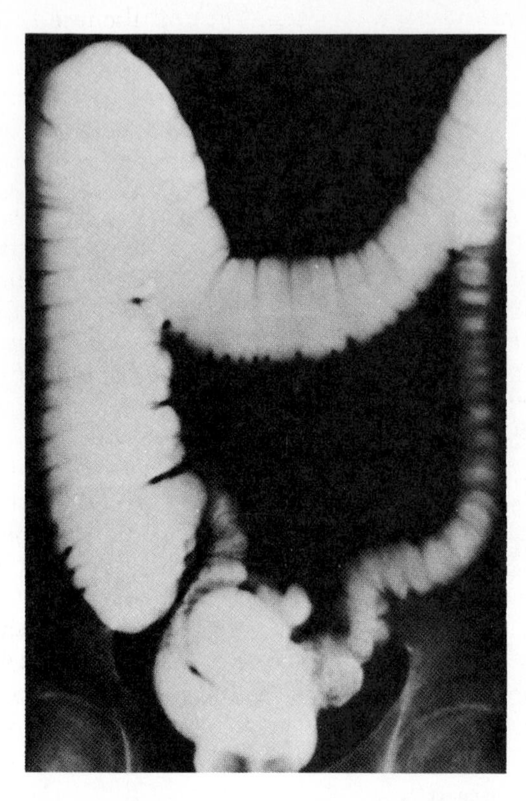

Figure 17.17. A radiograph after a barium enema showing the haustra of the large intestine.

---

1. Describe the functions of the large intestine, and explain how diarrhea may be produced.
2. Describe the structures and mechanisms involved in defecation.

Liver, Gallbladder, and Pancreas

The primary digestive function of the liver is the production of bile, which is secreted into ducts that drain to the gallbladder and duodenum. The gallbladder simply stores and concentrates bile, which contains bile salts that aid fat digestion and absorption. Blood from the intestine, which contains the absorbed products of digestion, flows to the liver where it is chemically modified. The exocrine portion of the pancreas secretes bicarbonate and digestive enzymes into the small intestine.

[9]rectum: L. *rectum,* straight tube
[10]anus: L. *anus,* ring

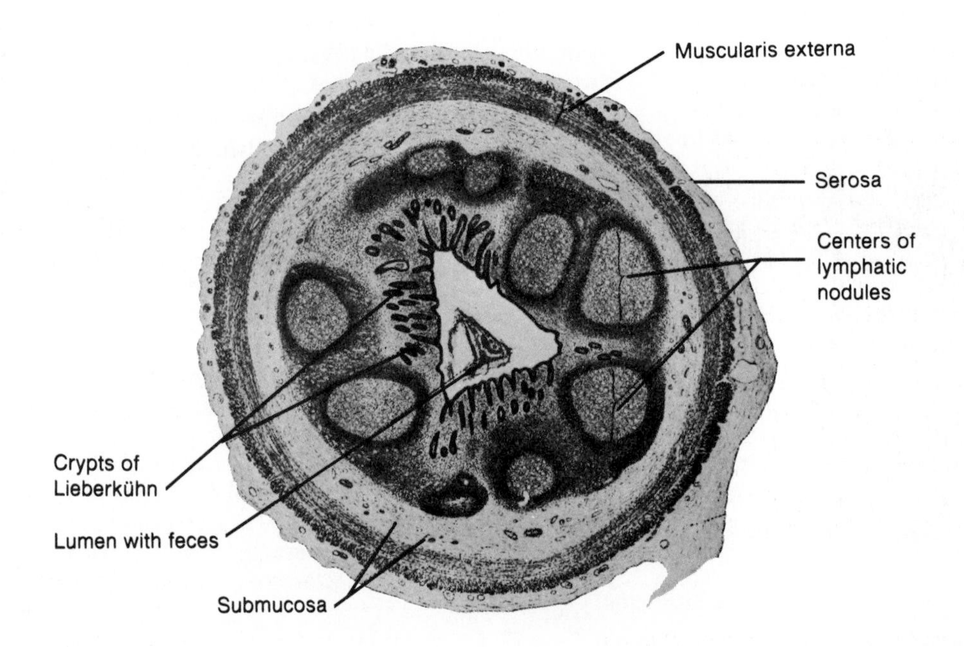

Figure 17.18. The microscopic appearance of a cross section of the human appendix.

The liver, which is the largest internal organ, lies immediately beneath the diaphragm in the abdominal cavity. Attached to the inferior surface of the liver is the pear-shaped gallbladder. This organ is approximately 10 cm long by 3.5 cm wide. The pancreas is located behind the stomach along the posterior abdominal wall.

Structure of the Liver

Although the liver is the largest internal organ, it is, in a sense, only one to two cells thick. This is because the liver cells, or **hepatocytes,** form **plates** that are one to two cells thick and separated from each other by large capillary spaces, called **sinusoids** (fig. 17.19). The sinusoids are lined with phagocytic cells, which are spread apart from each other to produce wide pores. The plate structure of the liver and the high permeability of the sinusoids allow each hepatocyte to have direct contact with the blood.

In **cirrhosis,** large portions of liver tissue are destroyed and replaced with permanent connective tissue and "regenerative nodules" of hepatocytes. These regenerative nodules don't have the platelike structure of normal liver tissue and are therefore less functional. One indication of this decreased function is the entry of ammonia (normally converted by the liver to urea) into the general circulation. Cirrhosis may be caused by chronic alcohol abuse, viral hepatitis, and other agents that attack liver cells.

Portal System

The products of digestion that are absorbed into blood capillaries in the intestine do not directly enter the general circulation. Instead, this blood is delivered first to the liver. Capillaries in the digestive tract drain into the *hepatic portal vein,*[11] which carries this blood to capillaries in the liver. (The term *hepatic* refers to the liver.) It is not until the blood has passed through this second capillary bed that it enters the general circulation through the *hepatic vein* that drains the liver. The term **portal system** is used to describe this unique pattern of circulation: capillaries → vein → capillaries → vein. In addition to receiving venous blood from the intestine, the liver also receives arterial blood via the *hepatic artery.*

Liver Lobules

The hepatic plates are arranged into functional units called **liver lobules** (fig. 17.20). In the middle of each lobule is a *central vein,* and at the periphery of each lobule are branches of the hepatic portal vein and of the hepatic artery, which open into the sinusoids *between* hepatic plates. Arterial blood and portal venous blood, containing molecules absorbed in the GI tract, thus mix as the blood flows within the sinusoids from the periphery of the lobule to the central vein. The central veins of different liver lobules converge to form the hepatic vein, which carries blood away from the liver.

Bile is produced by the hepatocytes and secreted into thin channels called **bile canaliculi,** located *within* each hepatic plate (fig. 17.20). These bile canaliculi are drained at the periphery of each lobule by *bile ducts,* which in turn empty into *hepatic ducts* that carry bile away from the liver. Since blood travels in the sinusoids and bile travels in the opposite direction within the hepatic plates, blood and bile do not mix in the liver lobules.

In addition to the normal constituents of bile, a wide variety of exogenous compounds (drugs) are secreted by the liver into the bile ducts. The liver can thus "clear" the blood of particular

[11]portal: L. *porta,* gate

Portal vein
Hepatic artery
Bile duct
(a)

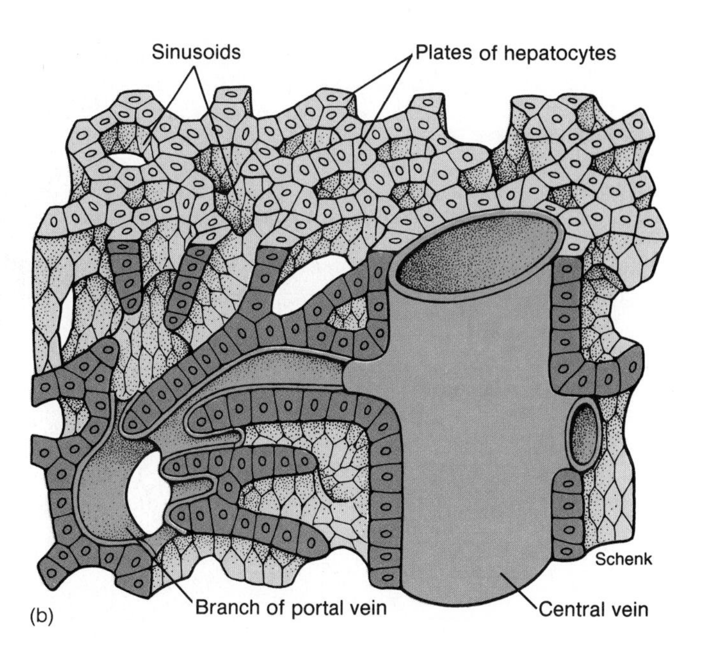

Figure 17.19. *The structure of the liver. (a) A scanning electron micrograph of the liver. Hepatocytes are arranged in plates so that blood that passes through sinusoids (b) will be in contact with each liver cell.*

(a) From: Tissues and Organs: A Text-Atlas of Scanning Electron Microscopy *by Richard G. Kessel and Randy Kardon. Copyright © 1979 by W. H. Freeman and Company. Reprinted with permission.*

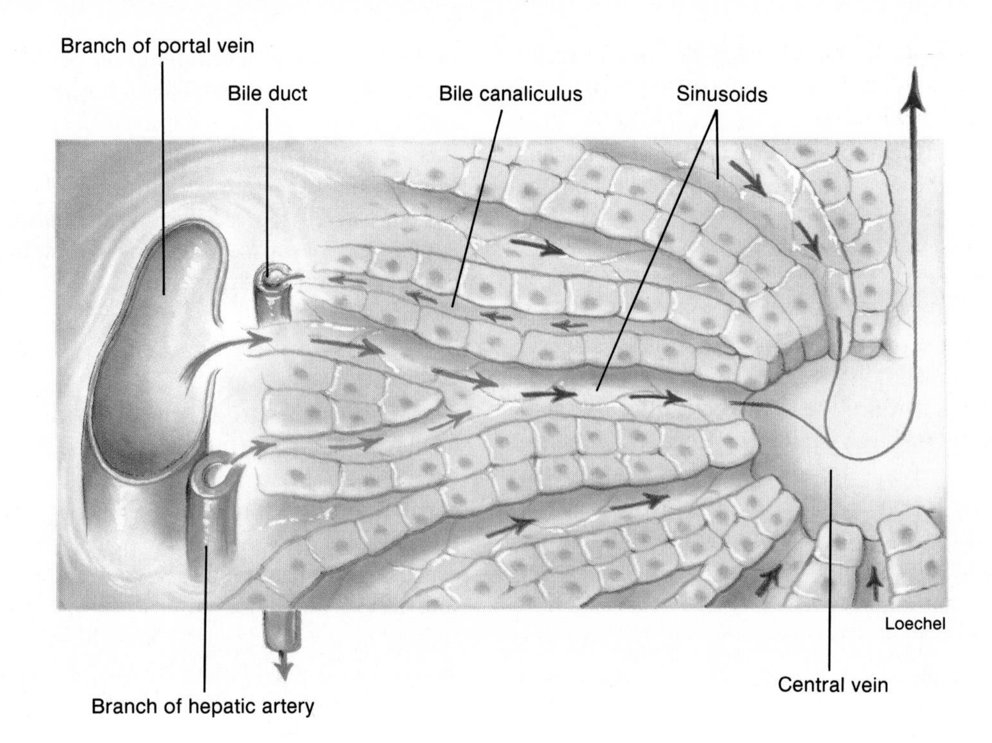

Figure 17.20. *The flow of blood and bile in a liver lobule. Blood flows within sinusoids from a portal vein to the central vein (from the periphery to the center of a lobule). Bile flows within hepatic plates from the center to bile ducts at the periphery of a lobule.*

Table 17.2 Summary of the major categories of liver functions

Functional Category	Actions
Detoxification of blood	Phagocytosis
	Chemical alteration of biologically active molecules (hormones and drugs)
	Production of urea, uric acid, and other molecules that are less toxic than parent compounds
	Excretion of molecules in bile
Carbohydrate metabolism	Conversion of blood glucose to glycogen and fat
	Production of glucose from liver glycogen and from other molecules (amino acids, lactic acid) by gluconeogenesis
	Secretion of glucose into the blood
Lipid metabolism	Synthesis of triglyceride and cholesterol
	Excretion of cholesterol in bile
	Production of ketone bodies from fatty acids
Protein synthesis	Production of albumin
	Production of plasma transport proteins
	Production of clotting factors (fibrinogen, prothrombin, and others)
Secretion of bile	Synthesis of bile salts
	Conjugation and excretion of bile pigment (bilirubin)

compounds by removing them from the blood and excreting them into the intestine with the bile. Molecules that are cleared from the blood by secretion into the bile are excreted in the feces; this is analogous to renal clearance of blood through excretion in the urine (chapter 16).

Functions of the Liver

As a result of its very large and diverse enzymatic content, its unique structure, and the fact that it receives venous blood from the intestine, the liver has a wider variety of functions than any other organ in the body. A summary of the major categories of liver function is presented in table 17.2.

Bile Production and Secretion

The liver produces and secretes 250–1,500 ml of bile per day. The major constituents of bile include bile salts, bile pigment (bilirubin), phospholipids (mainly lecithin), cholesterol, and inorganic ions.

Bile pigment, or **bilirubin,** is produced in the spleen, liver, and bone marrow as a derivative of the heme groups (minus the iron) from hemoglobin. *Free bilirubin* is secreted by these organs into the blood, which transports the bilirubin to the liver. The liver can then remove some of this bilirubin from the blood by secreting it into the bile. In order to do this, enzymes in the liver are required to combine the bilirubin with another molecule (glucuronic acid). Bilirubin combined in this way is referred to as *conjugated bilirubin.*

Jaundice[12] is a yellow staining of the tissues produced by high blood concentrations of either free or conjugated bilirubin. Jaundice due to high blood levels of conjugated bilirubin in adults may result when bile excretion is blocked by gallstones. Since free bilirubin is derived from heme, jaundice due to high blood levels of free bilirubin is usually caused by an excessively high rate of red blood cell destruction. *Physiological jaundice of the newborn* is due to high levels of free bilirubin in otherwise healthy neonates. This type of jaundice may be caused by the rapid fall in blood hemoglobin concentrations that normally occurs at birth, or in premature infants it may be caused by inadequate amounts of hepatic enzymes that are needed to conjugate bilirubin and thus excrete it in the bile.

Newborn infants with jaundice are usually treated by *phototherapy,* in which they are placed under blue light in the 400–500 nm wavelength range. This light is absorbed by bilirubin in cutaneous vessels and results in the conversion of bilirubin to a more polar form, which is soluble in plasma without having to be conjugated. The more water-soluble form of bilirubin can then be excreted in the bile and urine.

The **bile salts** are derivatives of cholesterol that have two to four polar (charged) groups on each molecule. In aqueous solutions these molecules "huddle" together to form aggregates known as **micelles.** The nonpolar parts are located in the central region of the micelle (away from water), whereas the polar groups face water around the periphery of the micelle. Lecithin, cholesterol, and other lipids in the small intestine enter these micelles in a process that aids the digestion and absorption of fats (described in a later section).

Detoxification of the Blood

The liver can remove biologically active molecules such as hormones and drugs from the blood by (1) excretion of these compounds in the bile; (2) phagocytosis by cells, which line the sinusoids; and (3) chemical alteration of these molecules within the hepatocytes.

Ammonia, for example, is a very toxic molecule produced by the action of bacteria in the intestine. The liver has the enzymes needed to convert ammonia into less toxic *urea* molecules, which are secreted by the liver into the blood and excreted by the kidneys in the urine. Similarly, the liver converts toxic purines (from nucleic acids) into *uric acid.*

Secretion of Glucose, Triglycerides, and Ketone Bodies

The liver helps to regulate the blood glucose concentration by either removing glucose from or adding glucose to the blood, according to the needs of the body. After a carbohydrate-rich meal, the liver can remove some glucose from the hepatic portal blood and convert it into glycogen and triglycerides. During fasting, the liver secretes glucose into the blood. The liver also contains the enzymes required to convert fatty acids into ketone bodies, which are secreted into the blood in large amounts during fasting. These processes are controlled by hormones and are explained in more detail in chapter 18.

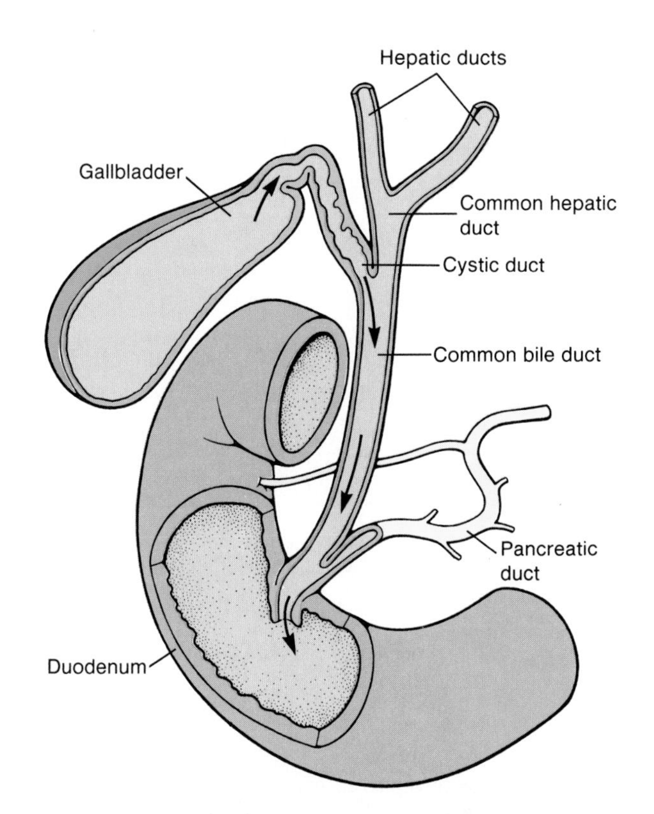

Figure 17.21. *The pancreatic duct joins the common bile duct to empty its secretions into the duodenum.*

Production of Plasma Proteins

Plasma albumin and most of the plasma globulins (with the exception of antibodies) are produced by the liver. Albumin comprises about 70% of the total plasma protein. The globulins produced by the liver have a wide variety of functions, including the transport of cholesterol and triglycerides, transport of steroid and thyroid hormones, inhibition of trypsin activity, and blood clotting.

Gallbladder

The gallbladder is a saclike organ attached to the inferior surface of the liver. This organ stores and concentrates bile, which drains to it from the liver by way of the *cystic*[13] *duct.* A sphincter valve at the neck of the gallbladder allows a storage capacity of about 35 to 50 ml. When the gallbladder fills with bile, it expands to the size and shape of a small pear. Contraction of the muscle layer of the gallbladder ejects bile through the cystic duct into the *common bile duct,* which conveys bile into the duodenum (fig. 17.21).

Bile is continuously produced by the liver and drains through the hepatic and common bile ducts to the duodenum. When the small intestine is empty of food, a sphincter at the end of the common bile duct closes, and bile is forced up to the cystic duct and then to the gallbladder for storage.

[12]jaundice: L. *galbus,* yellow
[13]cystic: Gk. *kystis,* pouch

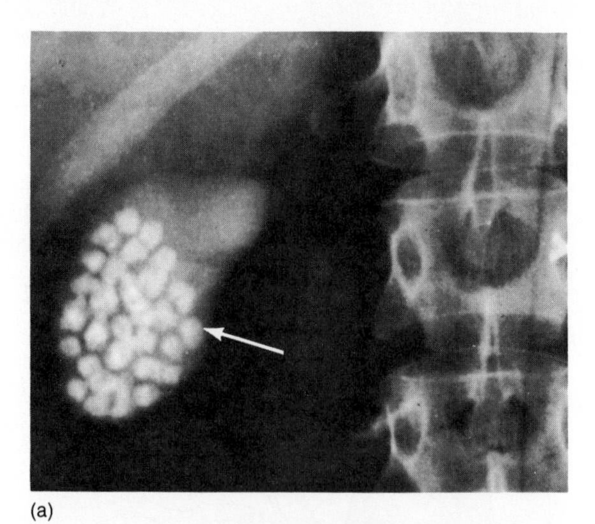

(a)

(b)

Figure 17.22. *(a) An X ray of a gallbladder that contains gallstones. (b) A posterior view of a gallbladder that has been removed and cut open to reveal its gallstones. A dime is placed in the photo to show relative size.*

Approximately twenty million Americans have **gallstones,** which can produce painful symptoms by obstructing the cystic or common bile ducts. Gallstones commonly contain cholesterol as their major component, and become hardened by the precipitation of inorganic salts (fig. 17.22). In order for gallstones to be produced, the liver must secrete enough cholesterol to create a supersaturated solution, and some substance must serve as a nucleus for the formation of cholesterol crystals. Gallstones may sometimes be dissolved by treatment with a bile salt, or they may have to be removed surgically.

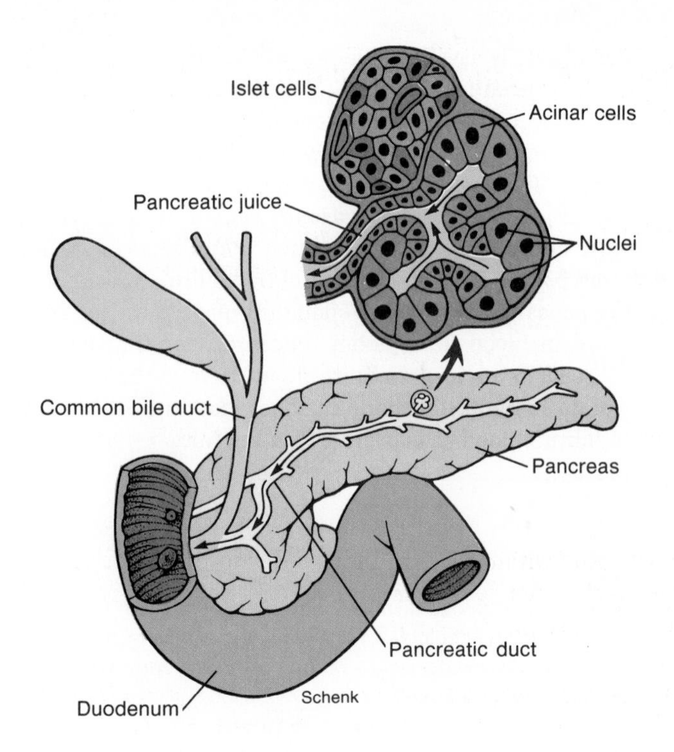

Figure 17.23. *The pancreas is both an exocrine and an endocrine gland. Pancreatic juice—the exocrine product—is secreted by acinar cells into the pancreatic duct. Scattered "islands" of cells, called the islets of Langerhans, secrete the hormones insulin and glucagon into the blood.*

Pancreas

The pancreas is a soft, glandular organ that has both exocrine and endocrine functions. The endocrine function is performed by clusters of cells, called the **islets of Langerhans,**[14] that secrete the hormones insulin and glucagon into the blood (chapter 18). As an exocrine gland, the pancreas secretes a fluid called **pancreatic juice** through the pancreatic duct (fig. 17.23) into the duodenum. Within the tissue of the pancreas are the exocrine secretory units, called *acini.*[15] Each acinus consists of a cluster of glandular cells surrounding a lumen, into which the pancreatic juice is secreted.

Pancreatic Juice

Pancreatic juice contains water, bicarbonate, and a wide variety of digestive enzymes that are secreted into the duodenum. These enzymes include (1) **amylase,** which digests starch; (2) **trypsin,** which digests protein; and (3) **lipase,** which digests triglycerides (fat). It should be noted that the complete digestion of food molecules in the small intestine requires the action of both pancreatic enzymes and brush border enzymes.

Most pancreatic enzymes are produced as inactive molecules, which helps to minimize the risk of self-digestion within the pancreas. The inactive form of trypsin, called trypsinogen,

[14]islets of Langerhans: from Paul Langerhans, German anatomist, 1847–1888
[15]acinus: L. *acinus,* grape

Table 17.3 Summary of the physiological effects of gastrointestinal hormones

Secreted by	Hormone	Effects
Stomach	Gastrin	Stimulates parietal cells to secrete HCl Stimulates chief cells to secrete pepsinogen Maintains structure of gastric mucosa
Small intestine	Secretin	Stimulates water and bicarbonate secretion in pancreatic juice Potentiates actions of cholecystokinin on pancreas
Small intestine	Cholecystokinin (CCK)	Stimulates contraction of the gallbladder Stimulates secretion of pancreatic juice enzymes Potentiates action of secretin on pancreas Maintains structure of exocrine pancreas (acini)
Small intestine	Gastric inhibitory peptide (GIP)	Inhibits gastric emptying Inhibits gastric acid secretion Stimulates secretion of insulin from endocrine pancreas (islets of Langerhans)

is activated within the small intestine into the active form of the enzyme. Active trypsin then converts the other pancreatic enzymes into their active forms.

1. Describe the structure of liver lobules, and trace the pathways for the flow of blood and bile in the lobules.
2. Describe the composition and function of bile, and trace the flow of bile from the liver and gallbladder to the duodenum.
3. List the different functions of the liver.
4. Explain how the liver helps maintain a constant blood glucose concentration.
5. Describe the endocrine and exocrine structures and functions of the pancreas, and explain why the pancreas does not normally digest itself.

Regulation of the Digestive System

The activities of different regions of the GI tract are coordinated by the actions of the vagus nerve and various hormones secreted by the stomach and intestine. The stomach begins to increase its secretion in anticipation of a meal, and further increases its activities in response to the presence of chyme. The entry of chyme into the duodenum stimulates secretion of hormones that promote contractions of the gallbladder, secretion of pancreatic juice, and inhibition of gastric activity.

The activities of the GI tract are, to a large degree, automatic. Neural and endocrine control mechanisms, however, can stimulate or inhibit these automatic functions to help coordinate the different stages of digestion. The sight, smell, or taste of food, for example, can stimulate salivary and gastric secretions via activation of the vagus nerve (the tenth cranial nerve—chapter 6). This helps to "prime" the digestive system in preparation for a meal. Stimulation of the vagus, in this case, originates in the brain and is a conditioned reflex (as Pavlov demonstrated by training dogs to salivate in response to a bell).

The GI tract is both an endocrine gland and a target for the action of various hormones. Indeed, the first hormones to be discovered were gastrointestinal hormones. In 1902, Bayliss and Starling[16] discovered that the duodenum produced a chemical regulator, which they named **secretin**; in 1905, these scientists proposed that secretin was but one of many yet undiscovered chemical regulators produced by the body. They coined the term *hormones* for this new class of regulators. Other investigators in 1905 discovered that an extract from the stomach antrum (pyloric region) stimulated gastric secretion. The hormone **gastrin** was thus the second hormone to be discovered.

The chemical structures of gastrin, secretin, and the duodenal hormone **cholecystokinin** (*CCK*) were determined in the 1960s. More recently, a fourth hormone produced by the small intestine, **gastric inhibitory peptide** (*GIP*), has been added to the list of proven GI tract hormones. The effects of these hormones are summarized in table 17.3.

Regulation of Gastric Function

Gastric motility and secretion are, to some extent, automatic. Waves of contraction that serve to push chyme through the pyloric sphincter, for example, are initiated spontaneously in the stomach. The secretion of hydrochloric acid (HCl) and pepsinogen, likewise, can be stimulated in the absence of neural and hormonal influences by the presence of cooked or partially digested protein in the stomach. The effects of autonomic nerves and hormones are superimposed on this automatic activity. The control of gastric function is conveniently divided into three phases: (1) the cephalic phase; (2) the gastric phase; and (3) the intestinal phase.

Cephalic Phase

The cephalic phase of gastric regulation refers to control by the brain via the vagus nerves. As previously discussed, various conditioned stimuli can evoke gastric secretion. This conditioning in humans is, of course, more subtle than that shown in response to a bell by Pavlov's[17] dogs. A conversation about appetizing

[16]Starling: Ernest Henry Starling, English physiologist, 1866–1927
[17]Pavlov: Ivan Petrovich Pavlov, Russian physiologist, 1849–1936

food was actually shown, in one study, to be a more potent stimulus for gastric acid secretion than was the sight and smell of food!

Gastric Phase

The arrival of chyme into the stomach stimulates the gastric phase of regulation. In this phase, the secretions of the stomach are stimulated by two factors: (1) distension (stretch) of the stomach, which is determined by the amount of chyme; and (2) the chemical nature of the chyme. While intact proteins have little stimulatory effect, the presence of short polypeptides and amino acids in the stomach stimulates the gastric secretion of pepsinogen and HCl. These stimuli also cause the stomach to secrete the hormone gastrin, which itself promotes the further secretion of pepsinogen and HCl.

Intestinal Phase

The intestinal phase of gastric regulation refers to the *inhibition* of gastric activity when chyme enters the small intestine. Investigators in 1886 demonstrated that the addition of olive oil to a meal inhibits gastric emptying, and in 1929 it was shown that the presence of fat inhibits gastric juice secretion.

The arrival of chyme into the duodenum produces a neural reflex that inhibits the stomach's motility and secretion. The presence of fat in the chyme also stimulates the duodenum to secrete a hormone that inhibits gastric function. This inhibitory hormone is believed to be gastric inhibitory peptide (GIP), although the ability of GIP to inhibit gastric acid secretion is currently controversial.

The inhibitory neural and endocrine mechanisms during the intestinal phase prevent the further passage of chyme from the stomach to the duodenum. This gives the duodenum time to process the load of chyme that it has previously received. Since this inhibitory reflex is stimulated by fat in the chyme, a breakfast of bacon and eggs takes longer to pass through the stomach—and makes one feel "fuller" for a longer time—than does a breakfast of pancakes and syrup.

Regulation of Pancreatic Juice and Bile Secretion

The arrival of chyme into the duodenum also stimulates reflex secretion of pancreatic juice and bile. The secretion of pancreatic juice and bile is stimulated by both neural reflexes initiated in the duodenum and by secretion of the duodenal hormones cholecystokinin (CCK) and secretin.

Secretion of Pancreatic Juice

The secretion of pancreatic juice is stimulated by both secretin and CCK, which are hormones produced by the duodenum. Secretion of secretin and CCK, however, occurs in response to different stimuli and these two hormones have different effects on the composition of pancreatic juice. The release of secretin occurs in response to a fall in duodenal pH below 4.5. Secretin stimulates the production of bicarbonate by the pancreas. Since bicarbonate neutralizes the acidic chyme and since secretin is released in response to the low pH of chyme, this completes a negative feedback loop. The secretion of CCK occurs in response to the fat content of chyme in the duodenum. CCK stimulates the production of pancreatic enzymes such as trypsin, lipase, and amylase.

Secretion of Bile

The liver secretes bile continuously, but this secretion is greatly augmented following a meal. This increased secretion is due to the release of secretin and CCK from the duodenum. Secretin is the major stimulator of bile secretion by the liver, and CCK enhances this effect. The arrival of chyme in the duodenum also causes the gallbladder to contract and eject bile. Contraction of the gallbladder occurs in response to neural reflexes from the duodenum and in response to stimulation by CCK.

1. Describe the events that occur during the cephalic and gastric phases of regulation of the activities of the stomach.
2. Describe the mechanisms involved in the intestinal phase of gastric regulation, and explain why a fatty meal takes longer to leave the stomach than a meal low in fat.
3. Explain the regulation of pancreatic juice and bile secretion.

Digestion and Absorption of Food

Carbohydrates and proteins are digested into their monomers of monosaccharides and amino acids, respectively. These are absorbed through the intestinal mucosa into blood capillaries. Fat is emulsified by the action of bile salts, digested into fatty acids and monoglycerides, and absorbed into the intestinal epithelial cells. Once in the cells, triglycerides are resynthesized and secreted into lymphatic capillaries.

The caloric (energy) value of food is found predominantly in its content of carbohydrates, lipids, and proteins. In the average American diet, carbohydrates account for approximately 50% of the total calories, protein accounts for 11% to 14%, and lipids make up the balance. These food molecules consist primarily of long combinations of subunits (monomers), which must be digested by hydrolysis reactions into the free monomers before absorption can occur. The characteristics of the major digestive enzymes are summarized in table 17.4.

Digestion and Absorption of Carbohydrates

Carbohydrates include polysaccharides (such as starch), and disaccharide and monosaccharide sugars (chapter 3). Most of the ingested carbohydrates are in the form of starch, which is a long polysaccharide composed of glucose subunits. The most commonly ingested sugars are the disaccharides sucrose (table sugar, consisting of glucose and fructose) and lactose (milk sugar, consisting of glucose and galactose). The digestion of starch begins in the mouth with the action of **salivary amylase.** This enzyme cleaves some of the bonds between adjacent glucose molecules, but most people don't chew their food long enough

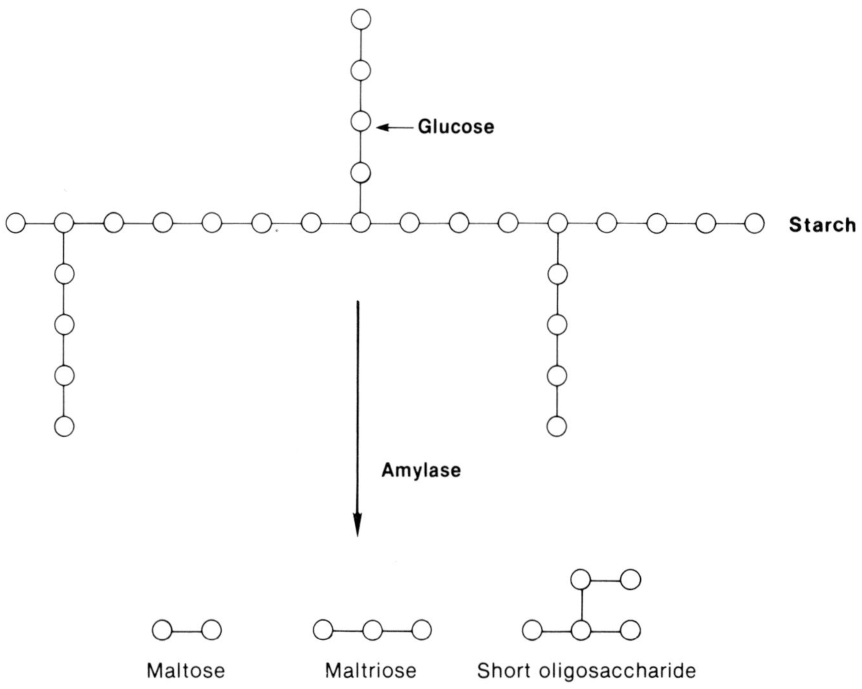

Glucose

Starch

Amylase

Maltose Maltriose Short oligosaccharide

Figure 17.24. Pancreatic amylase digests starch into maltose, maltriose, and short oligosaccharides containing branch points in the chain of glucose molecules.

Table 17.4 Summary of the sources and activities of the major digestive enzymes

Region or Source					
Organ	Source	Substrate	Enzymes	Optimum pH	Products
Mouth	Saliva	Starch	Salivary amylase	6.7	Maltose
Stomach	Gastric glands	Protein	Pepsin	1.6–2.4	Shorter polypeptides
Duodenum	Pancreatic juice	Starch	Pancreatic amylase	6.7–7.0	Maltose, maltriose, and oligosaccharides
		Polypeptides	Trypsin, chymotrypsin, carboxypeptidase	8.0	Amino acids, dipeptides, and tripeptides
		Triglycerides	Pancreatic lipase	8.0	Fatty acids and monoglycerides
	Epithelial membranes	Maltose	Maltase	5.0–7.0	Glucose
		Sucrose	Sucrase	5.0–7.0	Glucose + fructose
		Lactose	Lactase	5.8–6.2	Glucose + galactose
		Polypeptides	Aminopeptidase	8.0	Amino acids, dipeptides, tripeptides

for sufficient digestion to occur in the mouth. The digestive action of salivary amylase stops when the food enters the stomach because this enzyme is inactivated at the low pH of gastric juice.

The digestion of starch, therefore, occurs mainly in the duodenum as a result of the action of **pancreatic amylase.** This enzyme cleaves the straight chains of starch to produce short chains that are two or three glucose units long. In addition, short, branched chains of glucose molecules, called *oligosaccharides,*[18] are also produced (fig. 17.24).

These shorter products of carbohydrate digestion are further hydrolyzed to their monosaccharides (mainly glucose) by brush border enzymes, located on the microvilli of the epithelial cells in the small intestine. Glucose is then secreted from the epithelial cells into blood capillaries within the villi.

Digestion and Absorption of Proteins

Protein digestion begins in the stomach with the action of pepsin. Pepsin digestion results in the liberation of some amino acids, but the major products are short-chain polypeptides. This activity helps to produce a more homogenous chyme, but it is not essential for the complete digestion of protein that occurs in the small intestine.

Most protein digestion occurs in the duodenum and jejunum. Particular pancreatic juice enzymes, including trypsin, cleave peptide bonds within the protein. Other enzymes present in pancreatic juice, and enzymes on the brush border of the small intestine, remove amino acids from the ends of polypeptide chains. As a result of the action of these enzymes, polypeptide chains are digested into free amino acids. It is usually only the free amino acids that can be absorbed into the blood (fig. 17.25).

[18]oligosaccharide: Gk. *oligos,* little; *sakcharon,* sugar

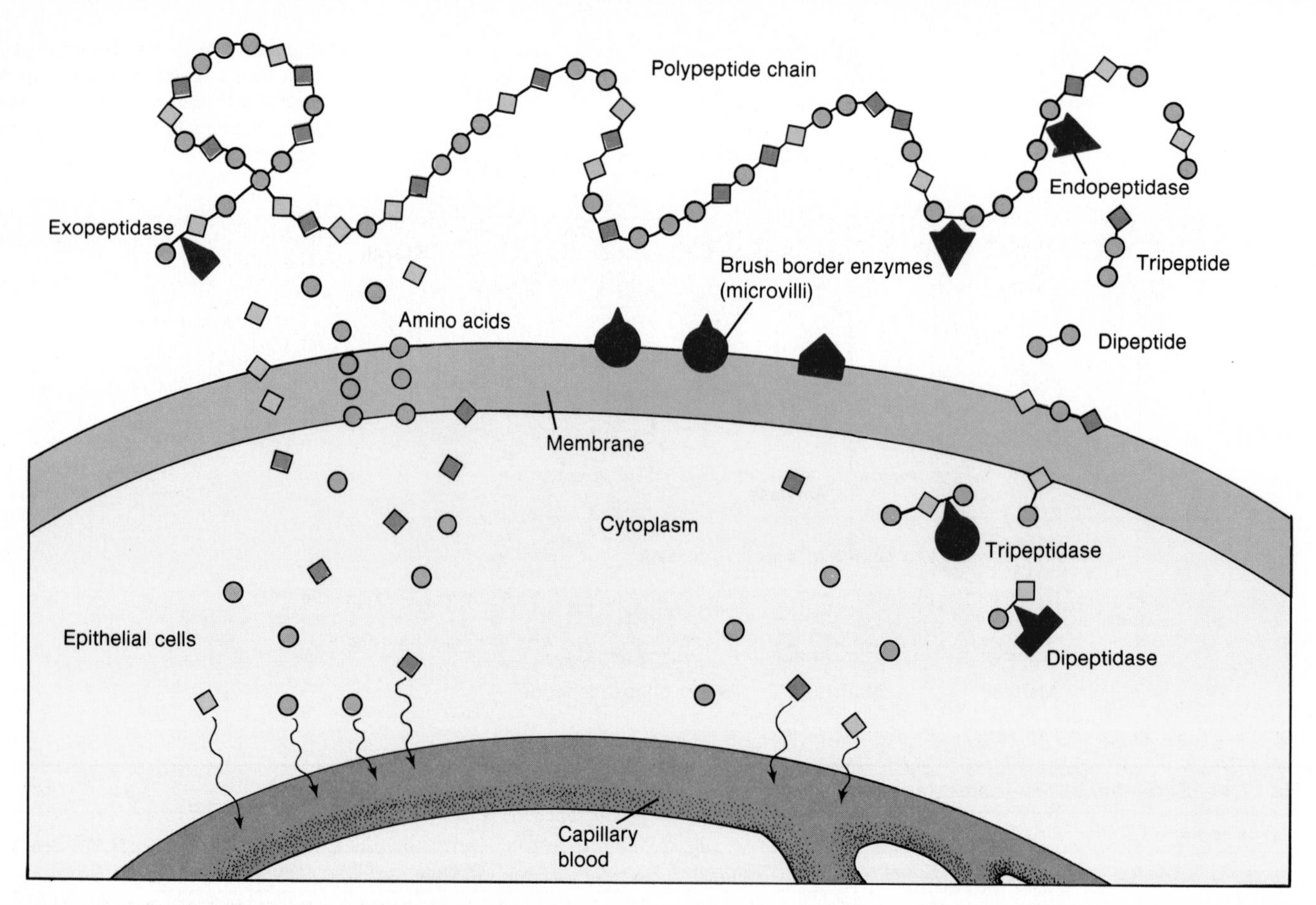

Figure 17.25. *Polypeptide chains are digested into free amino acids, dipeptides, and tripeptides by the action of pancreatic juice enzymes and brush border enzymes. The amino acids, dipeptides, and tripeptides enter duodenal epithelial cells. Dipeptides and tripeptides are hydrolyzed into free amino acids within the epithelial cells, and these products are secreted into capillaries that carry them to the hepatic portal vein.*

Triglyceride

Glycerol **Fatty acids**

Lipase + 2 **HOH**

Monoglyceride + **Free fatty acids**

Figure 17.26. *Pancreatic lipase digests fat (triglycerides) by cleaving off the first and third fatty acids. This produces free fatty acids and monoglycerides. Sawtooth structures indicate hydrocarbon chains in the fatty acids.*

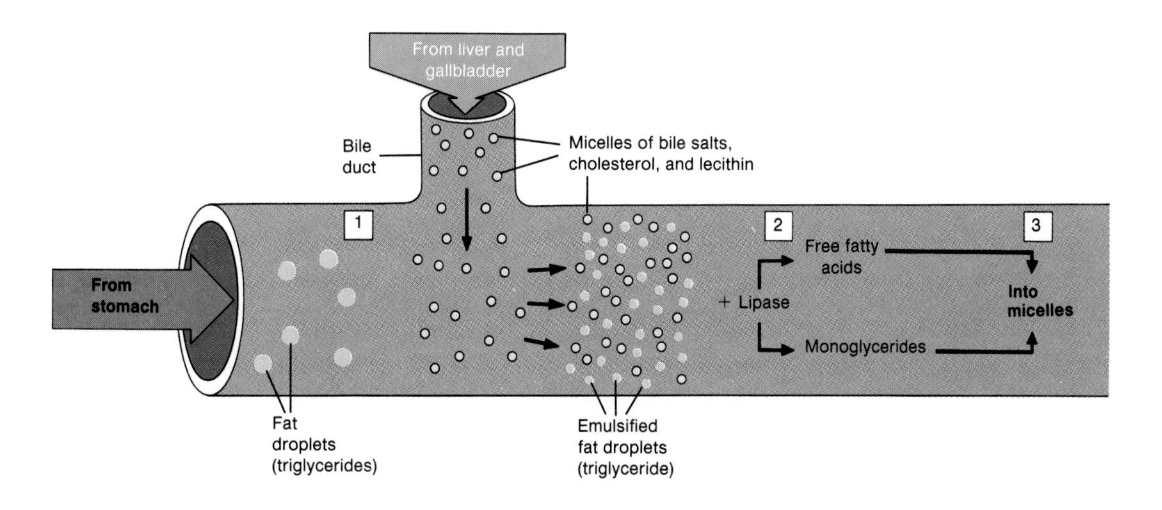

Figure 17.27. *Steps in the digestion of fat (triglycerides) and the entry of fat digestion products (fatty acids and monoglycerides) into micelles of bile salts secreted by the liver into the duodenum.*

Newborn babies appear to be capable of absorbing a substantial amount of undigested proteins (hence they can absorb antibodies from their mother's first milk); in adults, however, only the free amino acids enter the hepatic portal vein. Foreign food protein, which would be attacked by the immune system, does not normally enter the blood. (Chicken proteins, for example, do not circulate in the blood following a chicken dinner.) An interesting exception to this is the protein toxin of *Clostridium botulinum* bacteria (that causes botulism), which is resistant to digestion and which can thus be absorbed from contaminated food.

Digestion and Absorption of Lipids

Although the salivary glands and stomach produce lipid-digesting enzymes (lipases), there is very little fat digestion until the fat in chyme arrives in the duodenum in the form of fat globules. Through mechanisms previously described, the arrival of fat in the duodenum serves as a stimulus for the secretion of bile. Bile salts act to break up large fat droplets into much finer droplets. This process, called **emulsification,** results in the formation of tiny *emulsification droplets* of triglycerides. Note that emulsification is not chemical digestion—the bonds joining glycerol and fatty acids (the subunits of triglycerides) are not broken when fat is emulsified.

Digestion of Lipids

The emulsification of fat aids digestion because the smaller and more numerous emulsification droplets present a greater surface. Fat digestion occurs at the surface of the droplets through the enzymatic action of **pancreatic lipase.** Through hydrolysis reactions, lipase removes two of the three fatty acids from each triglyceride molecule and thus liberates *free fatty acids* and *monoglycerides* (fig. 17.26).

Free fatty acids and monoglycerides are more polar than the undigested lipids and are able to move more easily into the mixtures (known as *micelles*) of bile salts, lecithin, and cholesterol (fig. 17.27). These micelles then move to the brush border of the intestinal epithelium where absorption occurs.

Absorption of Lipids

Free fatty acids and monoglycerides can leave the micelles and pass through the membrane of the microvilli to enter the intestinal epithelial cells. These products are used to *resynthesize* triglycerides within the epithelial cells. This is different from the absorption of amino acids and monosaccharides, which pass through the epithelial cells without being altered.

Triglycerides, and other lipids such as cholesterol, are then combined with protein inside the epithelial cells to form small particles called **chylomicrons.** These tiny combinations of lipid and protein are secreted into the lymphatic capillaries of the intestinal villi (fig. 17.28). Absorbed lipids thus pass through the lymphatic system, which eventually carries these lipids into large veins (chapter 12). The absorption of lipids is thus significantly different from that of amino acids and monosaccharides, which enter the blood directly.

Lipid Transport in Blood

Once the chylomicrons are in the blood, their triglyceride content is removed by the enzyme **lipoprotein lipase,** which is attached to the lining of blood vessels. This enzyme digests triglycerides and thus provides free fatty acids and glycerol for use by the tissue cells. The remaining *remnant particles,* containing cholesterol, are taken up by the liver.

Figure 17.28. *Fatty acids and monoglycerides from the micelles within the small intestine are absorbed by epithelial cells and converted intracellularly into triglycerides. These are then combined with protein to form chylomicrons, which enter the lymphatic vessels (lacteals) of the villi. These lymphatic vessels transport the chylomicrons to the thoracic duct, which empties them into the venous blood.*

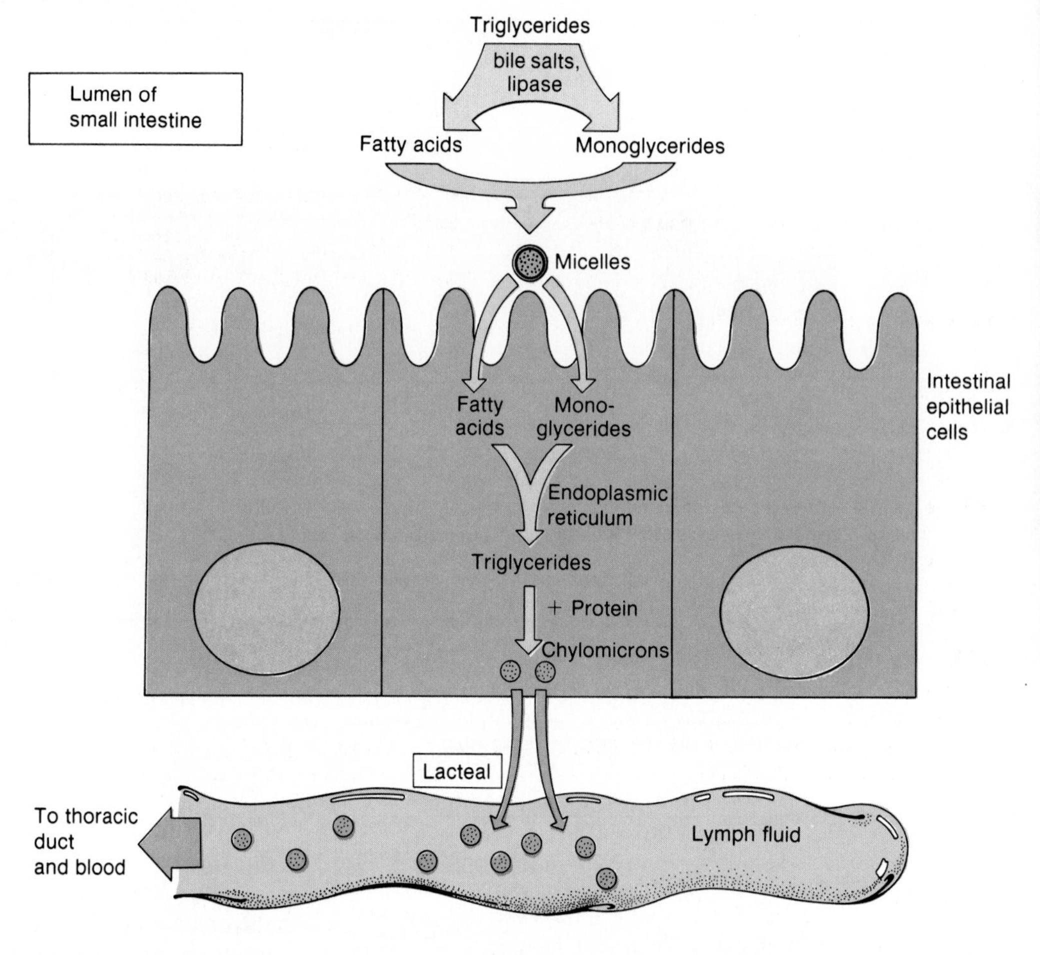

Table 17.5 Summary of the characteristics of the lipid carrier proteins (lipoproteins) found in plasma				
Lipoprotein class	**Origin**	**Destination**	**Major lipids**	**Functions**
Chylomicrons	Intestine	Many organs	Triglycerides, other lipids	Delivers lipids of dietary origin to body cells
Very-low density lipoproteins (VLDL)	Liver	Many organs	Triglycerides, cholesterol	Delivers endogenously produced triglycerides to body cells
Low-density lipoproteins (LDL)	Intravascular removal of triglycerides from VLDL	Blood vessels, liver	Cholesterol	Delivers endogenously produced cholesterol to various organs
High-density lipoproteins (HDL)	Liver and intestine	Liver and steroid hormone producing glands	Cholesterol	Removes and degrades cholesterol

Cholesterol and triglycerides produced by the liver are combined with other proteins and secreted into the blood as *very-low-density lipoproteins (VLDL).* Once the triglycerides are removed, the VLDL is converted to *low-density lipoproteins (LDL),* which transport cholesterol to various organs, including blood vessels. Excess cholesterol is returned from these organs to the liver by *high-density lipoproteins (HDL).* The characteristics of these lipoproteins are summarized in table 17.5. Knowledge of these lipoproteins is required for proper understanding of the roles of LDL and HDL in heart disease (as discussed in chapter 12).

1. List the enzymes involved in carbohydrate digestion, indicating their locations in the GI tract and their actions.
2. List the enzymes involved in protein digestion, indicating their locations in the GI tract and their actions.
3. Describe how bile aids both the digestion and absorption of fats. Explain how the absorption of fat differs from the absorption of amino acids and monosaccharides.
4. Trace the fate of triglyceride and cholesterol in a chylomicron within an intestinal epithelial cell.

Summary

Introduction to the Digestive System p. 374

I. The digestion of food molecules involves the hydrolysis of these molecules into their subunits.

II. The layers of the GI tract are, from the inside outward, mucosa, submucosa, muscularis, and serosa.

Esophagus and Stomach p. 376

I. Peristaltic waves of contraction push food through the lower esophageal sphincter into the stomach.

II. The lining of the stomach is thrown into folds, or rugae, and the mucosa is formed into gastric pits and gastric glands.
  A. The parietal cells of the gastric glands secrete HCl, and the chief cells secrete pepsinogen.
  B. In the acidic environment of gastric juice, pepsinogen is converted into the active protein-digesting enzyme called pepsin.
  C. Some digestion of protein occurs in the stomach; the most important function of the stomach is the secretion of intrinsic factor, which is needed for the absorption of vitamin B_{12} in the intestine.

Small Intestine p. 381

I. Regions of the small intestine include the duodenum, jejunum, and ileum; the common bile duct and pancreatic duct empty into the duodenum.

II. Villi project into the lumen, and at the bases of the villi the mucosa forms narrow pouches called the crypts of Lieberkühn.
  A. New epithelial cells are formed in the crypts.
  B. The membrane of intestinal epithelial cells is folded to form microvilli; this is called the brush border of the mucosa and serves to increase surface area.

III. Digestive enzymes, called brush border enzymes, are located in the membranes of the microvilli.

Large Intestine p. 384

I. The large intestine is divided into the cecum, colon, rectum, and anal canal.
  A. The vermiform appendix is attached to the cecum.
  B. The colon consists of ascending, transverse, descending, and sigmoid portions.

II. The large intestine absorbs water and electrolytes.

III. Defecation occurs when the anal sphincter relaxes and contraction of other muscles raises the rectal pressure.

Liver, Gallbladder, and Pancreas p. 386

I. The liver is composed of functional units called lobules.
  A. Liver lobules consist of plates of hepatic cells separated by capillary sinusoids.
  B. Blood flows from the periphery of each lobule, through the sinusoids, and out the central vein.
  C. Bile flows within the hepatocyte plates, in canaliculi, to the bile ducts.
  D. Bile consists of a pigment called bilirubin, bile salts, cholesterol, and other molecules.
  E. The liver detoxifies the blood by excreting substances in the bile, by phagocytosis, and by chemical inactivation.
  F. The liver modifies the plasma concentrations of proteins, glucose, triglycerides, and ketone bodies.

II. The gallbladder serves to store and concentrate the bile, and it releases bile through the cystic duct and common bile duct to the duodenum.

III. The pancreas is both an exocrine and an endocrine gland.
  A. The endocrine portion is known as the islets of Langerhans and secretes the hormones insulin and glucagon.
  B. The exocrine acini of the pancreas produce pancreatic juice, which contains various digestive enzymes and bicarbonate.

Regulation of the Digestive System p. 391

I. The regulation of gastric function occurs in three phases.
  A. In the cephalic phase, the activity of higher brain centers, acting via the vagus nerve, stimulates gastric juice secretion.
  B. In the gastric phase, the secretion of HCl and pepsin is controlled by the gastric contents.
  C. In the intestinal phase, the activity of the stomach is inhibited by neural reflexes from the duodenum and by gastric inhibitory peptide (GIP), secreted by the duodenum.

II. The secretion of the hormones secretin and cholecystokinin (CCK) from the duodenum regulate pancreatic juice and bile secretion.

Digestion and Absorption of Food p. 392

I. The digestion of starch begins in the mouth through the action of salivary amylase.
  A. Pancreatic amylase digests starch into shorter molecules within the small intestine.
  B. Complete digestion into monosaccharides is accomplished by brush border enzymes.

II. Protein digestion begins in the stomach by the action of pepsin.
  A. Pancreatic juice contains protein-digesting enzymes, including trypsin.
  B. The brush border contains digestive enzymes that help to complete the digestion of proteins into amino acids.

III. Lipids are digested in the small intestine after being emulsified by bile salts.
  A. Fatty acids and monoglycerides enter particles called micelles, composed of bile salts.
  B. Once the products of fat digestion are inside the mucosal epithelial cells, these subunits are used to resynthesize triglycerides.
  C. Triglycerides together with proteins are secreted into the central lacteals and transported by the lymphatic system to the blood.

Review Activities

Objective Questions

1. Intrinsic factor
  (a) is secreted by the stomach
  (b) is a polypeptide
  (c) promotes absorption of vitamin B_{12} in the intestine
  (d) helps prevent pernicious anemia
  (e) all of the above

2. Intestinal enzymes such as lactase are
  (a) secreted by the intestine into the chyme
  (b) produced by the crypts of Lieberkühn
  (c) produced by the pancreas
  (d) attached to the cell membrane of microvilli in the epithelial cells of the mucosa

3. Which of the following statements about gastric secretion of HCl is *false?*
  (a) HCl is secreted by parietal cells.
  (b) HCl hydrolyzes peptide bonds.
  (c) HCl is needed for the conversion of pepsinogen to pepsin.
  (d) HCl is needed for maximum activity of pepsin.

4. Most digestion occurs in the
   (a) mouth
   (b) stomach
   (c) small intestine
   (d) large intestine
5. Which of the following statements about trypsin is *true?*
   (a) Trypsin is derived from trypsinogen by the digestive action of pepsin.
   (b) Active trypsin is secreted into the pancreatic acini.
   (c) Trypsin is produced in the crypts of Lieberkühn.
   (d) Trypsin is active within the small intestine.
6. During the gastric phase, the secretion of HCl and pepsinogen is stimulated by
   (a) vagus nerve stimulation that originates in the brain
   (b) polypeptides in the gastric lumen and by gastrin secretion
   (c) secretin and cholecystokinin from the duodenum
   (d) all of the above
7. The first organ to receive the blood-borne products of digestion is the
   (a) liver
   (b) pancreas
   (c) heart
   (d) brain
8. Which of the following statements about hepatic portal blood is *true?*
   (a) It contains absorbed fat.
   (b) It contains ingested proteins.
   (c) It is mixed with bile in the liver.
   (d) It is mixed with blood from the hepatic artery in the liver.
9. Absorption of salt and water is the principal function of which region of the GI tract?
   (a) esophagus
   (b) stomach
   (c) duodenum
   (d) jejunum
   (e) large intestine
10. Cholecystokinin (CCK) is a hormone that stimulates the
   (a) production of bile
   (b) release of pancreatic enzymes
   (c) contraction of the gallbladder
   (d) both *a* and *b*
   (e) both *b* and *c*

Essay Questions

1. Explain how the gastric secretion of HCl and pepsin is regulated during the cephalic, gastric, and intestinal phases.
2. Describe how pancreatic enzymes become activated in the lumen of the intestine, and explain the need for these mechanisms.
3. What is the function of bicarbonate in pancreatic juice? Explain why ulcers are more likely to be located in the duodenum than in the stomach.
4. Explain why the pancreas is considered to be both an exocrine and an endocrine gland. Given this information, predict what effects tying of the pancreatic duct would have on pancreatic function.
5. Explain how jaundice is produced when (*a*) the person has gallstones, (*b*) the person has a high rate of red blood cell destruction, and (*c*) the person has liver disease.

18

The Metabolism of the Body

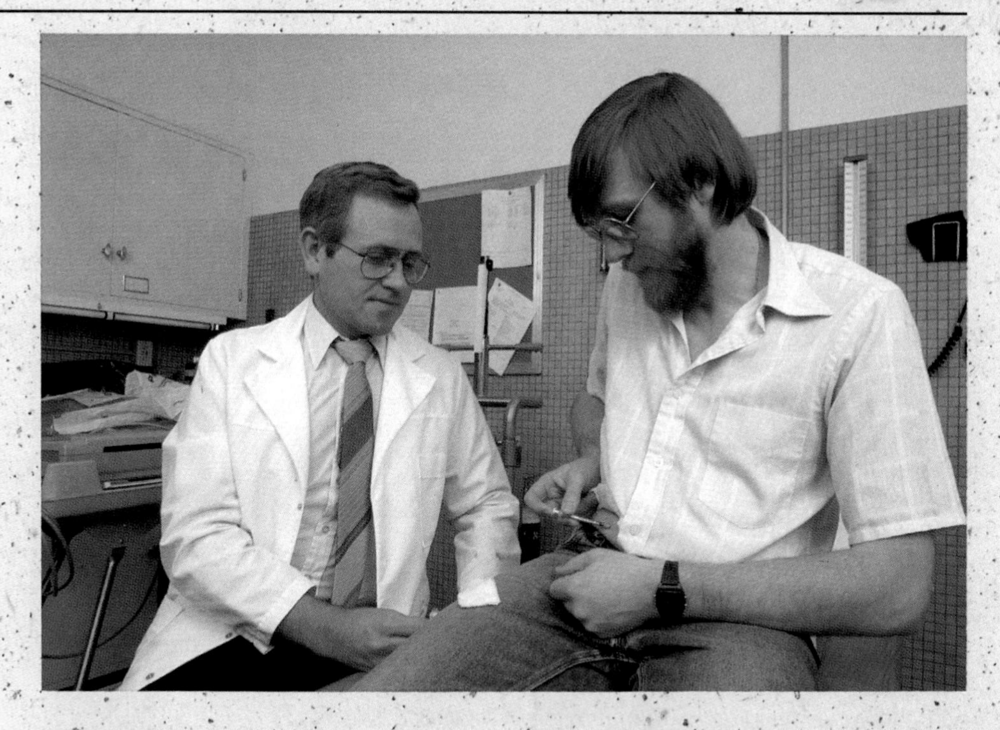

Outline

Objectives

By studying this chapter, you should be able to

1. define oxidation-reduction reactions, and explain how NAD and FAD function in these reactions
2. describe the pathway of glycolysis and the products of this pathway
3. explain the significance of anaerobic respiration in the body
4. describe the Krebs cycle in general terms, and explain its significance
5. explain the function of the electron transport system, and the significance of oxidative phosphorylation
6. explain the role of oxygen in aerobic cell respiration
7. explain, in general terms, how fat can be used to obtain energy within the body cells
8. describe the pathway by which carbohydrates can be converted into fat
9. explain how amino acids can be used by the body cells to obtain energy, and how amino acids can be interconverted
10. define the terms *anabolism* and *catabolism*
11. describe how the secretions of insulin and glucagon are regulated
12. explain the effects of insulin and glucagon on the body
13. distinguish between type I and type II diabetes mellitus
14. explain how reactive hypoglycemia is produced
15. describe the effects of thyroxine and growth hormone on body metabolism

Keys to Pronunciation

aerobic: *a-er-o'bik*
anabolism: *ah-nab'o-lizm*
anaerobic: *an''a-er-o'bik*
catabolism: *ca-tab'o-lizm*
cytochromes: *si'to-kromz*
diabetes mellitus: *di''ah-be'tez mel'i-tus*
glucagon: *gloo'kah-gon*
gluconeogenesis: *gloo''ko-ne''o-jen'e-sis*
glycolysis: *gli-kol'i-sis*
lipogenesis: *lip''o-jen'i-sis*
lipolysis: *li-pol'i-sis*
myxedema: *mik''se-de'mah*
nicotinamide: *nik''o-tin'ah-mid*
phosphorylation: *fos''for-i-la'shun*
thyroxine: *thi-rok'sin*

Photo: A man with diabetes mellitus is injecting himself with the hormone insulin. In the past, this insulin was derived from cattle and pigs. Today, however, human insulin is available as a result of genetic engineering techniques.

Energy Flow Within the Body

Many reactions involved in cell respiration are coupled in such a way that the energy released from one reaction is used to power another reaction. Electrons or hydrogen atoms can also be transferred from one molecule to another. The molecule that loses electrons or hydrogens is oxidized, while the one that accepts them is reduced.

As discussed in chapter 3, some chemical reactions liberate energy (are exergonic) while other reactions require the input of energy in order to proceed (and are thus termed endergonic)[1]. When larger molecules are broken down into smaller molecules, energy is liberated. If these larger molecules are combusted, for example, the products carbon dioxide (CO_2) and water (H_2O) are produced, and heat energy is released. Similar reactions also occur within living cells. Within a cell, these exergonic reactions are used to "drive" endergonic reactions. The energy-releasing and energy-requiring reactions are *coupled.*

As explained in chapter 3, the reaction

$$ADP + P_i + \text{energy} \rightarrow ATP$$

is an endergonic reaction which is driven by energy released from exergonic reactions. According to the law of conservation of energy, the chemical bond joining the last phosphate group in ATP (adenosine triphosphate) must contain some of the energy that was required for the formation of this bond. (The rest of the energy released by the exergonic reactions escapes as heat.) ATP is thus a "high energy" compound, which functions as the **universal energy carrier** of the cell. When the cell requires energy for any of its activities, it obtains this energy by the breakdown of ATP:

$$ATP \rightarrow ADP + P_i + \text{energy}$$

Coupled Reactions: NAD and FAD

When an atom or a molecule gains electrons, it is said to become **reduced;** when it loses electrons, it is said to become **oxidized.** Reduction and oxidation are always coupled reactions: an atom or a molecule cannot become oxidized unless it donates electrons to another, which therefore becomes reduced.

[1]endergonic: Gk. *endon,* within; *ergon,* work

Notice that the term *oxidation* does not imply that oxygen participates in the reaction. This term is derived from the fact that oxygen has a great tendency to accept electrons. This property of oxygen is exploited by cells; oxygen acts as the final electron acceptor in a chain of oxidation-reduction reactions that provides energy for ATP production.

Oxidation-reduction reactions in cells often involve the transfer of hydrogen atoms rather than of free electrons. Since a hydrogen atom contains one electron (and one proton in the nucleus), a molecule that loses hydrogen becomes oxidized and one that gains hydrogen becomes reduced. In many oxidation-reduction reactions, pairs of electrons—either as free electrons or as a pair of hydrogen atoms—are transferred from the molecule being oxidized to the molecule being reduced.

Two molecules that serve important roles in the transfer of hydrogens are **nicotinamide adenine dinucleotide (NAD),** which is derived from the vitamin niacin (vitamin B_3), and **flavin adenine dinucleotide (FAD),** which is derived from the vitamin riboflavin (vitamin B_2). These molecules are coenzymes that function as *hydrogen carriers,* because they accept hydrogens (becoming reduced) in one enzyme reaction and donate hydrogens (becoming oxidized) in a different enzyme reaction (fig. 18.1). The oxidized form of these molecules is written simply as NAD and FAD.

Each FAD can accept the two electrons and can bind the two protons from two hydrogen atoms. The reduced form of FAD is therefore written as $FADH_2$. Each NAD can also accept two electrons but can only bind one proton. The reduced form of NAD, therefore, is shown as $NADH + H^+$ (the H^+ represents a free proton). For the sake of simplicity, however, the reduced form of NAD will be shown as $NADH_2$ in this chapter. Both $NADH_2$ and $FADH_2$ can then donate their hydrogens to other molecules (fig. 18.1).

Cell Respiration

All of the reactions in the body that involve energy transformation are collectively termed **metabolism.**[2] Metabolism may be divided into two categories: *anabolism*[3] and *catabolism.*[4] Catabolic reactions release energy, usually by the breakdown of

[2]metabolism: Gk. *metabole,* change
[3]anabolism: Gk. *anabole,* a raising up
[4]catabolism: Gk. *katabole,* a casting down

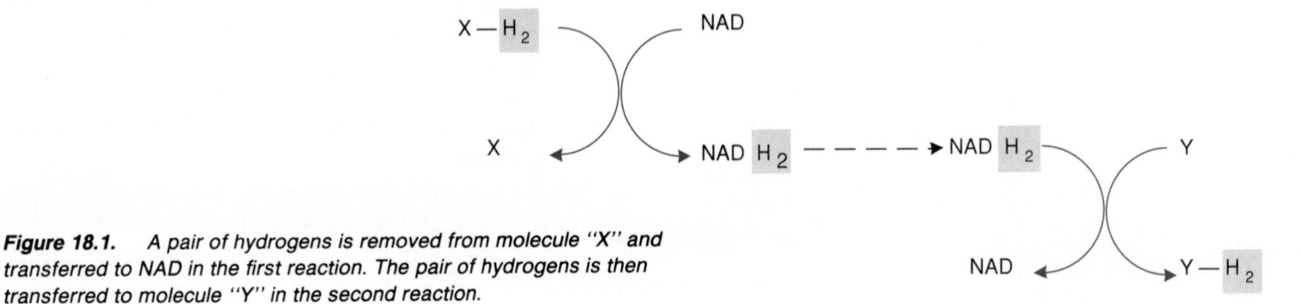

Figure 18.1. *A pair of hydrogens is removed from molecule "X" and transferred to NAD in the first reaction. The pair of hydrogens is then transferred to molecule "Y" in the second reaction.*

larger organic molecules into smaller molecules. Anabolic reactions require the input of energy and include the synthesis of large, energy-storage molecules such as glycogen, fat, and protein.

The catabolic reactions that break down glucose, fatty acids, and amino acids serve as the primary sources of energy for the cellular synthesis of ATP. These metabolic pathways are known collectively as *cellular respiration.* When oxygen is utilized, these processes are called **aerobic cell respiration.** The final products of aerobic respiration are carbon dioxide, water, and energy (a part of which is trapped in the chemical bonds of ATP). The overall equation for aerobic respiration, therefore, is identical to the equation that describes combustion (fuel + $O_2 \rightarrow CO_2 + H_2O$ + energy).

Notice that the term *respiration* refers to chemical reactions that liberate energy for the production of ATP. The oxygen used in aerobic respiration by tissue cells is obtained from the blood; the blood, in turn, becomes oxygenated in the lungs by the process of breathing (described in chapter 15). Breathing is thus needed for, but is different from, aerobic respiration.

1. Define the terms *oxidation* and *reduction.*
2. Describe how NAD and FAD can function to transfer hydrogens from one molecule to another.
3. Compare aerobic respiration with a combustion reaction.

Glycolysis and Anaerobic Respiration

In cellular respiration, chemical reactions that transform food molecules into simpler molecules liberate energy, some of which is used to produce ATP. The complete combustion of a molecule requires the presence of oxygen; some energy is liberated, however, during incomplete combustion in the absence of oxygen. Anaerobic respiration provides a net gain of 2 ATP per glucose through the process of glycolysis and the conversion of pyruvic acid into lactic acid.

Unlike combustion, the conversion of glucose to carbon dioxide and water within the cells occurs in small, enzymatically catalyzed steps. Oxygen is used only at the last step (this will be described in a later section). Since a small amount of the chemical bond energy of glucose is released at early steps in the metabolic pathway, some cells in the body can obtain energy for ATP production in the temporary absence of oxygen. This process is called **anaerobic[5] respiration.**

Glycolysis

Both the anaerobic and the aerobic respiration of glucose begin with a metabolic pathway known as **glycolysis.[6]** Glycolysis is the metabolic pathway by which glucose—a six-carbon sugar—is converted into two molecules of *pyruvic acid.* Even though each pyruvic acid molecule is roughly half the size of a glucose, one must not think of glycolysis as simply the breaking in half of glucose. Glycolysis is a metabolic pathway involving many enzymatically controlled steps.

Each pyruvic acid molecule contains three carbons, three oxygens, and four hydrogens. The number of carbon and oxygen atoms in one molecule of glucose—$C_6H_{12}O_6$—can thus be accounted for in the two pyruvic acid molecules. Since the two pyruvic acids together account for only eight hydrogens, however, it is clear that four hydrogen atoms are removed during the intermediate steps in glycolysis. Each pair of these hydrogen atoms are combined with NAD to form $NADH_2$. Starting from one glucose molecule, therefore, glycolysis results in the conversion of two molecules of NAD into two molecules of $NADH_2$.

Glycolysis is exergonic, and a portion of the energy that is released is used to drive the reaction ADP + $P_i \rightarrow$ ATP. At the end of the glycolysis pathway there is a net gain of two ATP per glucose molecule, as indicated in the overall equation for glycolysis:

$$\text{Glucose} + 2\text{NAD} + 2\text{ADP} + 2\text{P}_i \rightarrow 2 \text{ pyruvic acid} + 2\text{NADH}_2 + 2\text{ATP}$$

Although the overall equation for glycolysis is exergonic, glucose must be "activated" at the beginning of the pathway before energy can be obtained. This activation requires the addition of two phosphate groups derived from two molecules of ATP. Energy from the reaction ATP $\rightarrow$ ADP + P_i is therefore consumed at the beginning of glycolysis. This is shown as an "up-staircase" in figure 18.2. Notice that the P_i is not shown in these reactions in figure 18.2; this is because the phosphate is not released but instead is added to the intermediate molecules of glycolysis. At later steps in glycolysis, four molecules of ATP are produced as energy is liberated (the "down-staircase" in figure 18.2). The two molecules of ATP used in the beginning, therefore, represent an energy investment; the net gain of two ATP (and two $NADH_2$) by the end of the pathway represent an energy profit.

Anaerobic Respiration

In order for glycolysis to continue, there must be adequate amounts of NAD available to accept hydrogen atoms. The $NADH_2$ that is produced in glycolysis, therefore, must become oxidized by donating its hydrogens to another molecule.

When oxygen is *not* available in sufficient amounts, the $NADH_2$ produced in glycolysis donates its hydrogens to pyruvic acid. This addition of two hydrogen atoms to pyruvic acid produces *lactic acid* (fig. 18.3). This metabolic pathway, by which glucose is converted through pyruvic acid to lactic acid, is called **anaerobic respiration.**

Anaerobic respiration yields a net gain of two ATP (produced by glycolysis) per glucose molecule. A cell can survive anaerobically as long as it can produce sufficient energy for its needs in this way and as long as lactic acid concentrations do

[5]anaerobic: Gk. *an,* without; *aer,* air; *bios,* life
[6]glycolysis: Gk. *glyco,* sugar; *lysis,* breaking

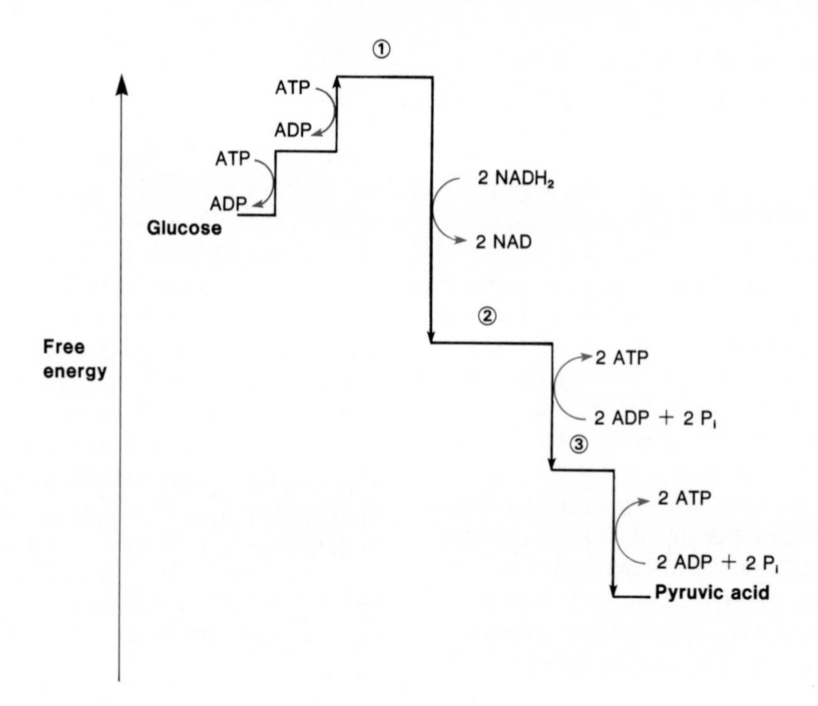

Figure 18.2. *The energy expenditure and gain in glycolysis. Notice that there is a "net profit" of 2 ATP and 2 NADH$_2$ per glucose molecule in glycolysis. Molecules listed by number are* (1) *fructose-1, 6-diphosphate,* (2), *1, 3-diphosphoglyceric acid, and* (3) *3-phosphoglyceric acid.*

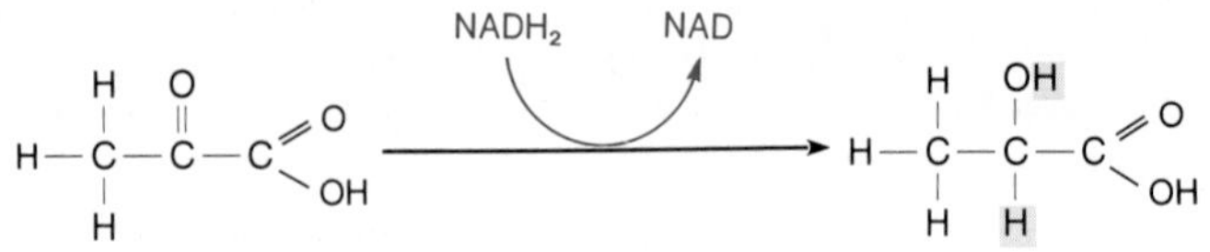

NADH$_2$ NAD

Pyruvic acid → Lactic acid

Figure 18.3. *When pyruvic acid gains two hydrogens from NADH$_2$ it is converted into lactic acid.*

not become excessive. Some tissues are better adapted for anaerobic respiration than others—skeletal muscles survive longer than cardiac muscle, which in turn can survive under anaerobic conditions longer than can the brain.

Except for red blood cells, which can only respire anaerobically (thus sparing the oxygen they carry), anaerobic respiration provides only a temporary sustenance for tissues that have energy requirements in excess of their aerobic ability. Anaerobic respiration can only occur for a limited period of time (longer for skeletal muscles, shorter for the heart, and shortest for the brain) when the *ratio of oxygen supply to oxygen need* falls below a critical level. Anaerobic respiration is, in a sense, an emergency procedure that provides some ATP until the emergency (oxygen deficiency) has passed.

The term "emergency" is actually overstated in the case of skeletal muscles, in which anaerobic respiration is a normal, daily occurrence that does not harm the tissue or the individual. Anaerobic respiration does not normally occur in the heart, however, and when it does occur a potentially dangerous condition may be present.

Ischemia refers to inadequate blood flow to an organ, such that the rate of oxygen delivery is insufficient to maintain aerobic respiration. Inadequate blood flow to the heart, or *myocardial ischemia,* may occur if the coronary blood flow is occluded by atherosclerosis, a blood clot, or by an artery spasm (chapter 12). People with myocardial ischemia often experience *angina pectoris,* severe pain in the chest and left arm area. This pain is associated with increased blood levels of lactic acid, which are produced by anaerobic respiration by the heart muscle. The degree of ischemia and angina can be decreased by vasodilator drugs such as nitroglycerin and amyl nitrite, which improve blood flow to the heart and also decrease the work of the heart by dilating peripheral blood vessels.

1. Define the term *glycolysis.* Explain why there is a net gain of two ATP in this process.
2. Define the term *anaerobic respiration* in terms of its initial substrates and final products. Explain the significance of lactic acid formation at the end of this process.
3. Explain why anaerobic respiration occurs. In which tissue(s) is anaerobic respiration normal? In which tissue is it abnormal?

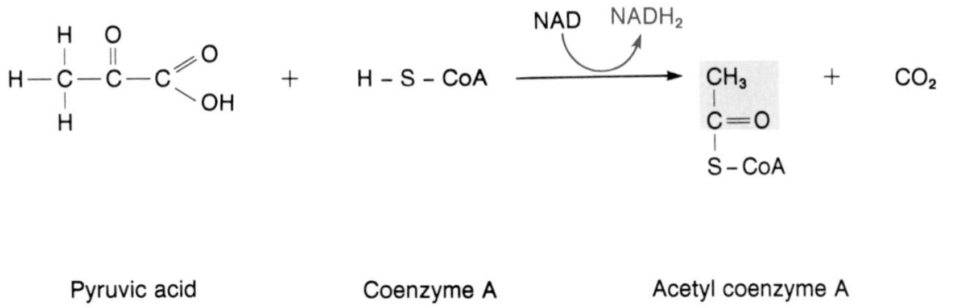

$$\text{Pyruvic acid } (H_3C{-}CO{-}COOH) + \text{H-S-CoA (Coenzyme A)} \xrightarrow{NAD \rightarrow NADH_2} \text{Acetyl coenzyme A } (CH_3{-}C{=}O{-}S{-}CoA) + CO_2$$

Figure 18.4. *The formation of acetyl coenzyme A in aerobic respiration.*

Aerobic Respiration

In the aerobic respiration of glucose, pyruvic acid is formed by glycolysis and then converted into acetyl coenzyme A. This begins a cyclic metabolic pathway called the Krebs cycle. As a result of these pathways, a large number of $NADH_2$ and $FADH_2$ are generated. These molecules provide electrons for an energy-generating process that drives the formation of many ATP. A total of 38 ATP can be produced by the aerobic respiration of a single molecule of glucose.

Aerobic respiration is equivalent to combustion in terms of its final products (CO_2 and H_2O) and in terms of the total amount of energy liberated. In aerobic respiration, however, the energy is released in small, enzymatically controlled oxidation reactions, and a portion (38%–40%) of the energy released in this process is captured in the high-energy bonds of ATP.

The aerobic respiration of glucose begins with glycolysis. Glycolysis in both anaerobic and aerobic respiration results in the production of two molecules of pyruvic acid, two molecules of ATP, and two molecules of $NADH_2$ per glucose. In aerobic respiration, however, the hydrogens in $NADH_2$ are *not* donated to pyruvic acid and lactic acid is not formed, as happens in anaerobic respiration. Instead, the pyruvic acids will move to a different cellular location and undergo a different reaction. The $NADH_2$ produced by glycolysis will eventually be oxidized, but that occurs later in the story.

The enzymes that catalyze glycolysis are located in the cell cytoplasm. In aerobic respiration, pyruvic acid leaves the cytoplasm and enters the interior (the matrix) of mitochondria (chapter 4). Once pyruvic acid is inside a mitochondrion, carbon dioxide is enzymatically removed from each three-carbon-long pyruvic acid to form a two-carbon-long organic acid—acetic acid. The enzyme that catalyzes this reaction combines the acetic acid with a coenzyme (derived from the vitamin pantothenic acid) called *coenzyme A*. The combination thus produced is called **acetyl coenzyme A,** abbreviated **acetyl CoA** (fig. 18.4).

Glycolysis converts one glucose molecule into two molecules of pyruvic acid. Since each pyruvic acid molecule is converted into one molecule of acetyl CoA and one CO_2, two molecules of acetyl CoA and two molecules of CO_2 are derived from each glucose. It should be noted that the oxygen in CO_2 is derived from pyruvic acid, not from oxygen gas.

The Krebs Cycle

Once acetyl CoA is formed, the acetic acid subunit (two carbons long) is combined with oxaloacetic acid (four carbons long) to form a molecule of citric acid (six carbons long). Coenzyme A acts only as a transporter of acetic acid from one enzyme to another (similar to the transport of hydrogen by NAD). The formation of citric acid begins a cyclic metabolic pathway known as the **citric acid cycle.** Most commonly, however, this cyclic pathway is named after its principal discoverer, and is called the **Krebs cycle.**[7] A simplified illustration of this pathway is shown in figure 18.5.

Through a series of reactions involving the elimination of two carbons and four oxygens (as two CO_2 molecules) and the removal of hydrogens, citric acid is eventually converted to oxaloacetic acid, which completes the cyclic metabolic pathway. In this process the following occur: (1) one ATP is produced; (2) three molecules of NAD are reduced to $NADH_2$; and (3) one molecule of FAD is reduced to $FADH_2$.

The production of $NADH_2$ and $FADH_2$ by each "turn" of the Krebs cycle is far more significant, in terms of energy production, than the single ATP produced directly by the cycle. This is because $NADH_2$ and $FADH_2$ eventually donate their electrons to an energy-generating process that results in the formation of many molecules of ATP.

Electron Transport and Oxidative Phosphorylation

Built into the foldings, or cristae, of the inner mitochondrial membrane are a series of molecules that serve in **electron transport** during aerobic respiration. This electron transport chain of molecules consists of a flavoprotein (derived from the vitamin

[7]Krebs: from Sir Hans A. Krebs, British biochemist, 1900–1981

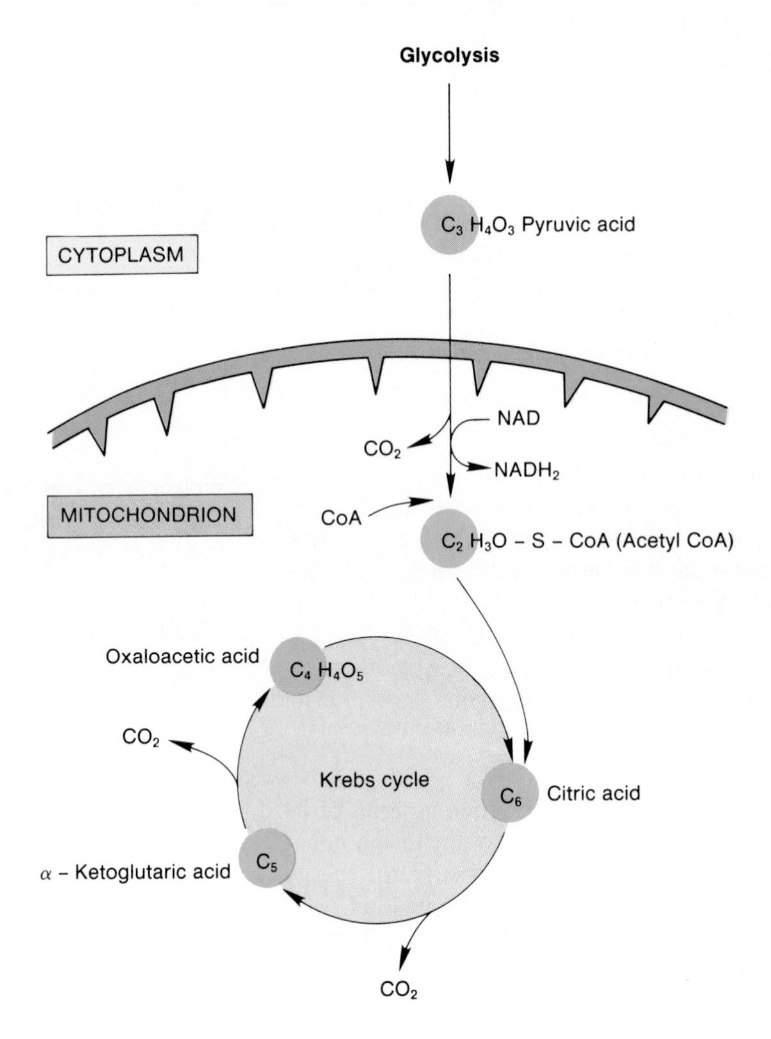

Figure 18.5. *A simplified diagram of the Krebs cycle, showing how the original four-carbon-long oxaloacetic acid is regenerated at the end of the cyclic pathway. Only the numbers of carbon atoms in the Krebs cycle intermediates are shown, the numbers of hydrogens and oxygens are not accounted for in this simplified scheme.*

riboflavin), coenzyme Q (derived from vitamin E), and a group of iron-containing pigments called *cytochromes.* The last of these cytochromes is cytochrome a_3, which donates electrons to oxygen in the final oxidation-reduction reaction (fig. 18.6). These molecules of the electron transport system are fixed in position within the inner mitochondrial membrane in such a way that they can pick up electrons from $NADH_2$ and $FADH_2$ and transport them in a definite sequence and direction.

In aerobic respiration, $NADH_2$ and $FADH_2$ become oxidized by transferring their pairs of electrons (from the two hydrogen atoms) to the electron transport system of the cristae. This reduces the first molecule of the electron transport chain, and leaves two protons (two H^+) free in the mitochondrion (these will be used later in the story). The oxidized forms of NAD and FAD are thus regenerated and able to continue to "shuttle" electrons from hydrogen atoms in the Krebs cycle to the electron transport chain.

One reduced cytochrome transfers its electron pair to the next cytochrome in the chain. The molecules of the electron transport chain thus move pairs of electrons from one to the other in a bucket brigade fashion. This is an exergonic process, and the energy derived is used to phosphorylate ADP to ATP. The production of ATP in this manner is thus appropriately termed **oxidative phophorylation.**

When one molecule of glucose is respired aerobically to carbon dioxide and water, a *grand total of 38 ATP* are produced. This number includes the 2 ATP produced during glycolysis, 2 ATP produced by two turns of the Krebs cycle, and 34 ATP produced through oxidative phosphorylation by the electron transport chain. Since anaerobic respiration of glucose yields only 2 ATP per glucose, it is easy to see why the cells of the body always "choose" to respire aerobically when oxygen is available.

Function of Oxygen

If the last cytochrome remained in a reduced state, it would be unable to accept more electrons. Electron transport would then progress only to the next-to-last cytochrome. This process would continue until all of the components of the electron transport chain remained in the reduced state. At this point, the electron transport system would stop functioning and no ATP could be produced within the mitochondria. With the electron transport system incapacitated, the Krebs cycle would stop. Respiration would then become anaerobic.

Oxygen, from the air we breathe, allows electron transport to continue by functioning as the **final electron acceptor** of the electron transport chain. Oxygen accepts electrons from the last cytochrome in the chain so that electron transport and oxidative

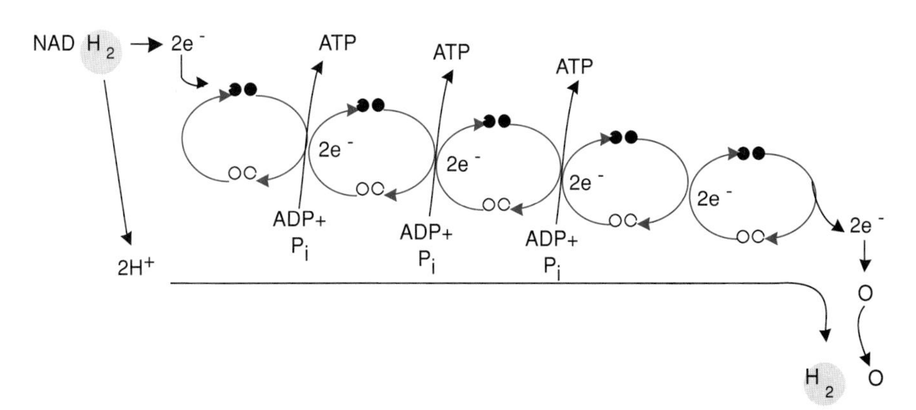

Figure 18.6. *The electron transport chain and the formation of ATP by oxidative phosphorylation. Molecules of the electron transport chain that lack a pair of electrons (are oxidized) are represented by open circles. Those that have received a pair of electrons (are reduced) are represented by filled circles.*

phosphorylation can continue. At the very last step of aerobic respiration, therefore, oxygen becomes reduced by the two electrons that were passed to the chain from $NADH_2$ or $FADH_2$. This reduced oxygen binds two protons (H^+), and a molecule of water is formed. Since the oxygen atom is part of a molecule of oxygen gas (O_2), this last reaction can be shown as follows:

$$O_2 + 4\,e^- + 4\,H^+ \rightarrow 2\,H_2O$$

> *Cyanide* is a very rapid lethal poison (1–15 minutes), which produces symptoms of rapid heart rate, hypotension, coma, and ultimately death if not quickly treated. Cyanide produces these effects because it has one very specific action: it blocks the transfer of electrons from cytochrome a_3 to oxygen. The effects of this deadly poison are thus the same as would occur if oxygen were completely removed; aerobic cell respiration and the production of ATP by oxidative phosphorylation comes to a halt.

1. Describe the fate of pyruvic acid in aerobic respiration, and explain how this differs from its fate in anaerobic respiration.
2. Draw a simplified Krebs cycle, using C_2 for acetic acid, C_4 for oxaloacetic acid, C_5 for alpha-ketoglutaric acid, and C_6 for citric acid. List the high-energy products that are produced at each turn of the Krebs cycle.
3. Explain the function of oxygen in the body and how a lack of sufficient oxygen in an organ can make respiration become anaerobic.

Metabolism of Fat and Protein

Triglycerides can be hydrolyzed into glycerol and fatty acids. Fatty acids can be converted into numerous molecules of acetyl CoA, which enter Krebs cycles and generate a large amount of ATP. Amino acids derived from proteins may also be used for energy. This involves removal of the amine group and the conversion of the remaining molecule into either pyruvic acid or one of the Krebs cycle molecules.

When food intake by the body occurs at a faster rate than energy consumption, the cellular concentration of ATP rises. Cells, however, do not store extra energy in the form of extra ATP. When cellular ATP concentrations rise, because more energy (from food) is available than can be immediately used, high ATP concentrations inhibit glycolysis. Under conditions of high cellular ATP, when glycolysis is inhibited, glucose is instead converted into glycogen and fat. The interconversion of glucose, glycogen, and fat is depicted in figure 18.7, with the heavy arrows indicating the pathways that are favored when a person's intake of carbohydrates exceeds the energy requirements.

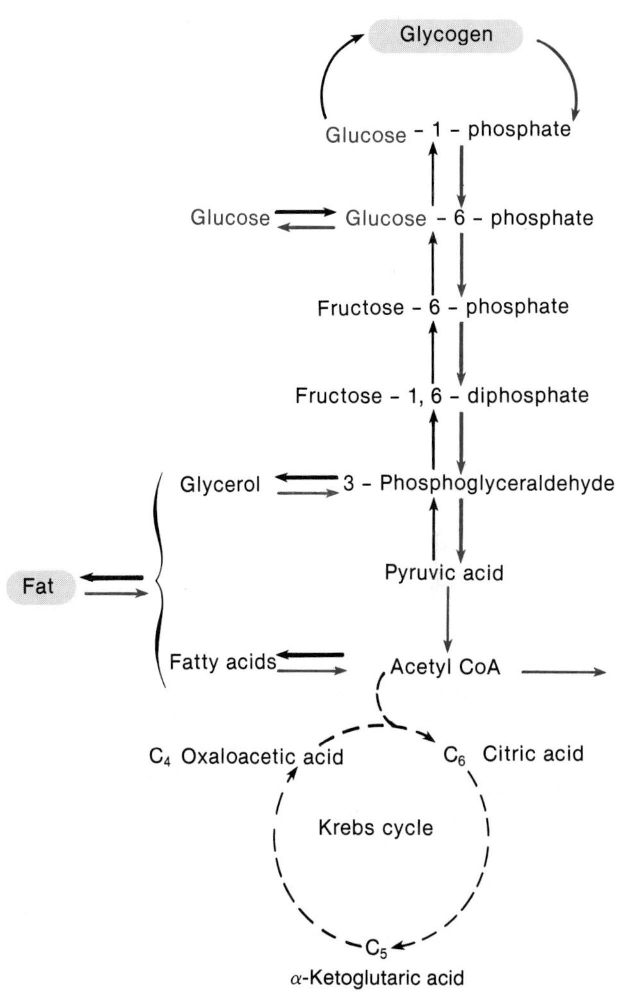

Figure 18.7. *The metabolic pathways by which glucose can be converted into glycogen and fat. Reverse reactions are indicated by lighter arrows.*

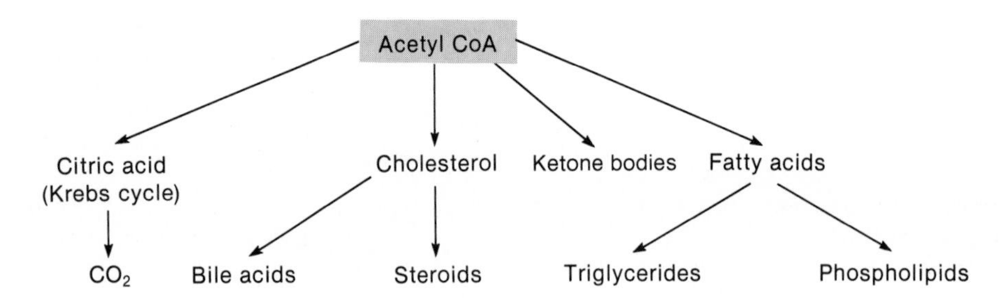

Figure 18.8. *Divergent metabolic pathways for acetyl coenzyme A.*

Lipid Metabolism

When glucose is going to be converted into fat, glycolysis occurs and pyruvic acid is converted into acetyl CoA. Some of the molecules produced during glycolysis, however, do not complete their conversion to pyruvic acid, and acetyl CoA does not enter a Krebs cycle. The two-carbon acetic acid subunits of acetyl CoA molecules can instead be used to produce a variety of lipids, including steroids such as cholesterol as well as ketone bodies and fatty acids (fig. 18.8). Acetyl CoA may thus be considered a branch point at which a number of different possible metabolic pathways may progress.

In the formation of fatty acids, a number of acetic acid (two-carbon) subunits are joined together to form the fatty acid chain. Six acetyl CoA molecules, for example, will produce a fatty acid that is twelve carbons long. When three of these fatty acids condense with one glycerol, a triglyceride (fat) molecule is formed. The formation of fat, or **lipogenesis,** occurs primarily in adipose tissue and in the liver when the concentration of blood glucose is elevated following a meal.

Breakdown of Fat (Lipolysis)

When fat stored in adipose tissue is going to be used as an energy source, enzymes hydrolyze (break down) triglycerides into *glycerol* and *free fatty acids,* a process called **lipolysis.** These molecules (primarily the free fatty acids) enter the blood and can be used by the liver, skeletal muscles, and other organs for aerobic respiration.

Free fatty acids serve as the major energy source derived from triglycerides. Most fatty acids consist of a long hydrocarbon chain with a carboxylic acid group (COOH) at one end. In a process known as **β-oxidation** (β is the Greek letter *beta*), enzymes remove a two-carbon acetic acid molecule from the acid end of a fatty acid. This results in the formation of acetyl CoA and a new fatty acid molecule that is now two carbons shorter. The process of β-oxidation continues until the entire fatty acid molecule is converted to acetyl CoA.

Since each acetyl CoA molecule can start a Krebs cycle, and since many acetyl CoA molecules are liberated by the breakdown of a fatty acid, more energy is obtained from the respiration of fat than of glucose. The aerobic respiration of one fatty acid, for example, might yield over 130 ATP, and the complete respiration of a triglyceride (with three fatty acids and a glycerol) can result in the production of over 400 ATP molecules!

Ketone Bodies

There is a continuous turnover of triglycerides in adipose tissue, even when a person is not losing weight. New triglycerides are produced, while others are hydrolyzed into glycerol and fatty acids. This turnover ensures that the blood normally contains a sufficient amount of fatty acids for aerobic respiration by skeletal muscles, the liver, and other organs.

Some of the fatty acids enter an alternate pathway. This pathway begins with the production of two-carbon acetyl CoA molecules from fatty acids, followed by the conversion of two molecules of acetyl CoA into four-carbon-long acidic derivatives. These derivatives of fatty acids are known as **ketone bodies,** and include acidic molecules and *acetone,* which is a three-carbon-long molecule identical to the solvent in nail polish remover.

Ketone bodies, which can be used for energy by many organs, are found in the blood under normal conditions. Under conditions of fasting or of diabetes mellitus, however, the increased liberation of free fatty acids from adipose tissue results in an elevated production of ketone bodies by the liver. The secretion of abnormally high amounts of ketone bodies into the blood produces **ketosis,** which is one of the signs of fasting or an uncontrolled diabetic state. A person in this condition may also have a sweet-smelling breath due to the presence of acetone, which is volatile and leaves the blood in the exhaled air.

Amino Acid Metabolism

Nitrogen is ingested primarily as proteins, enters the body as amino acids, and is excreted mainly as urea in the urine. In childhood, the amount of nitrogen excreted may be less than the amount ingested because amino acids are incorporated into proteins during growth. Growing children are thus said to be in a state of *positive nitrogen balance.* People who are starving or suffering from prolonged wasting diseases, in contrast, are in a state of *negative nitrogen balance*; they excrete more nitrogen than they ingest because they are breaking down their tissue proteins.

Healthy adults maintain a state of nitrogen balance, in which the amount of nitrogen excreted is equal to the amount ingested. This does not imply that the amino acids ingested are unnecessary; on the contrary, they are needed to replace the approximately 400 grams of protein that is "turned over" each

Social Issues

Alcohol Consumption and Health

The type of alcohol present in beer, wine, and distilled liquor is chemically known as *ethanol*. It is a small, two-carbon long molecule that, unlike most substances ingested in the diet, can be absorbed from the stomach as well as from the intestine. Once absorbed, it is transported by the blood within the hepatic portal vein to the liver. Some of the alcohol in this blood is metabolized by the liver, and the rest is passed into the general circulation. Since the liver is the major site of alcohol metabolism, blood containing alcohol must make many passes through the liver before the blood is cleared of this substance.

Within the liver cells, ethanol is eliminated by conversion into acetyl CoA, which is also two carbons long. The hepatic enzyme that catalyzes this reaction is known as *alcohol dehydrogenase*. Skeletal muscles lack this enzyme; forcing someone who is drunk to walk, therefore, cannot cause alcohol to be metabolized faster. Neither can drinking coffee help a person to recover from alcohol intoxication, because alcohol dehydrogenase is not affected by caffeine. The amount of time required to clear the blood of a given amount of alcohol depends only on the amount of alcohol dehydrogenase enzymes present in the person's liver. The activity of this enzyme is determined in part by heredity, but can be increased through frequent drinking of alcoholic beverages.

Since acetyl CoA can enter a Krebs cycle and provide energy, alcohol provides calories that are empty of nutrients. Once the cells have produced the amount of ATP they need through aerobic respiration, the remainder of the acetyl CoA derived from alcohol is channeled into fatty acids within the liver. Through this and other mechanisms, fat deposits accumulate in the liver of heavy drinkers. This can cause liver damage, leading to fibrosis. Fibrosis of the liver is reversible if the heavy drinker abstains from alcohol. Further liver damage, however, can lead to the irreversible condition of cirrhosis (described in chapter 17). The damaged liver decreases its production of proteins and other molecules that are needed by the body for a wide variety of functions.

Scientists have recently discovered that the human stomach contains alcohol dehydrogenase, and is thus capable of metabolizing some of the alcohol that a person drinks before this alcohol can be absorbed. Further, they discovered that the amount of this gastric enzyme is lower in women than in men. This may, in part, explain why the blood alcohol level is higher in women than in men after drinking a given amount of alcohol. The higher blood concentration of alcohol helps to explain why women are more susceptible than men to the effects of alcohol. In particular, the greater absorption of alcohol from the stomach in women, and the consequent higher concentration of alcohol in the hepatic portal vein, helps to explain why women are more susceptible to liver damage from alcohol.

Heavy drinking may cause increased breakdown of proteins in skeletal and cardiac muscles. In extreme cases, this can lead to serious heart damage. Alcohol also interferes with utilization of glucose by the brain. Since the brain requires blood glucose as its energy source, depression of brain activity results. This may in part explain the depressant effects of alcohol on higher brain functions and loss of inhibitions in the inebriated person. In advanced cases of alcoholism, irreversible brain damage may result. There is an additional danger if a pregnant woman engages in heavy alcohol consumption. The alcohol can easily pass through the placenta and interfere with glucose utilization by the fetus, causing damage to its developing central nervous system.

A heavy drinker commonly experiences a number of nutritional problems. In addition to loss of the body's proteins, there is a depletion of B vitamins (chapter 19). This is due in part to the fact that thiamine and niacin are required for metabolism of alcohol, so that less is available for other metabolic activities. Also, these water-soluble vitamins may be lost in the urine due to the diuretic action (chapter 16) of alcohol. A number of needed minerals, including potassium, magnesium, and zinc, are also excreted from the body in excessive amounts. The loss of B vitamins and minerals may be compensated by eating well, but depletion of protein, and deficiencies in iron and folic acid, are found even in heavy drinkers whose intake of food appears adequate. The effects of these nutritional disorders are superimposed upon the damages produced more directly by alcohol.

There are other health-related consequences of heavy alcohol consumption. For example, dangerous interactions can occur between alcohol and other drugs. Doses of drugs that are otherwise safe can become lethal when taken with alcohol; doses of drugs that should be effective may not work correctly in a heavy drinker. A heavy drinker undergoing surgery, for example, requires much higher amounts of the general anesthetic than someone who is a nondrinker. This is why anesthesiologists routinely ask the drinking habits of a patient prior to surgery, and why honest answers to these questions are very important.

In addition to its harmful effects on the body, the consumption of alcohol is responsible for a high percentage of automobile, boating, hunting, and job-related accidents. Alcohol abuse is thus a major health hazard in most of the world.

day. When more amino acids are ingested than are needed to replace proteins, the excess amino acids are not stored as additional protein (one cannot build muscles simply by eating large amounts of protein). Rather, the amine groups can be removed, and the "carbon skeletons" of the organic acids that are left can be used for energy or converted to carbohydrate and fat.

Oxidative Deamination

If there are more amino acids than are needed for protein synthesis, the amine group from an amino acid may be removed and excreted as *urea* in the urine (fig. 18.9). The metabolic process that removes amine groups from amino acids—leaving an organic acid and ammonia (which is converted to urea)—is known as **oxidative deamination.** The organic acids left behind after the amine group has been removed may then be used in the Krebs cycle as a source of energy (fig. 18.10).

Depending on which amino acid is deaminated, the organic acid left over may be either pyruvic acid or one of the Krebs cycle acids. These can be respired for energy, converted to fat, or converted to glucose. In the latter case, the amino acids are eventually changed to pyruvic acid, which is used to form glucose in a process that is essentially the reverse of glycolysis. The production of glucose from amino acids is known as **gluconeogenesis.** Gluconeogenesis from amino acids occurs primarily in the liver, and is extremely important in maintaining blood glucose levels during prolonged fasting and exercise. The possible interrelationships between amino acids, carbohydrates, and fat are illustrated in figure 18.11.

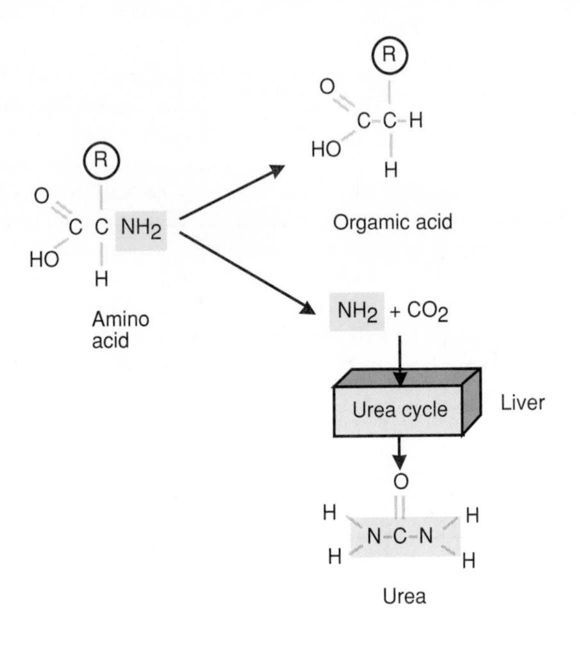

Figure 18.9. *Oxidative deamination of an amino acid produces an organic acid (pyruvic acid or Krebs cycle acid) and urea.*

Transamination

An adequate amount of all twenty amino acids is required to build proteins for growth and for replacement of the proteins that are turned over. Fortunately, only eight (in adults) or nine (in children) amino acids cannot be produced by the body and so must be obtained in the diet; these are the *essential amino acids* (table 18.1). Arginine is a "semiessential" amino acid because it can be produced by the body, but not in sufficient quantities to classify it as a nonessential amino acid. The remaining amino acids are "nonessential" only in the sense that the body can produce them if it is given a sufficient amount of the essential amino acids and of carbohydrates.

The body can produce the nonessential amino acids by a type of reaction known as **transamination.** In this process, pyruvic acid or a Krebs cycle acid is converted into an amino acid. In order to do this, the amino group (NH_2) of a different amino acid must be removed and transferred to the new molecule (fig. 18.12). When an amino group is added to pyruvic acid, for example, the amino acid known as alanine is produced. Of course, the old amino acid, which donated its amine group, is no longer

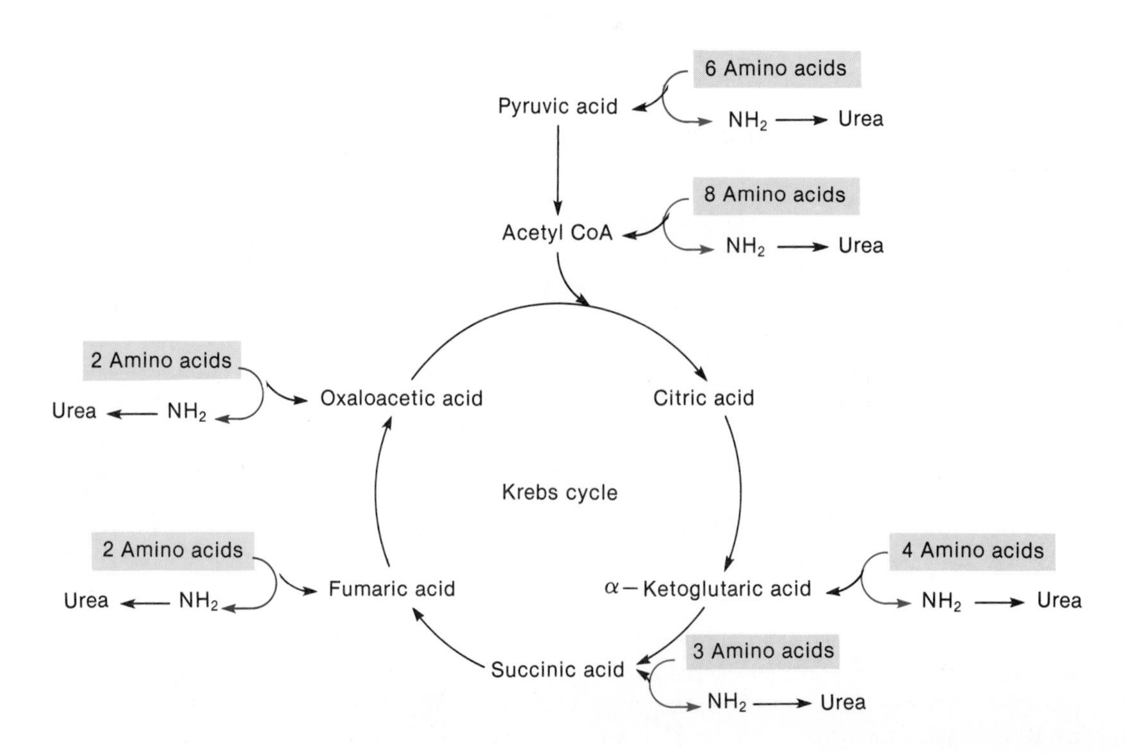

Figure 18.10. *Pathways by which amino acids can be catabolized for energy. These pathways are indirect for some amino acids, which must be transaminated into other amino acids before being converted into keto acids by deamination.*

an amino acid. Transamination, therefore, does not result in a net gain in the number of amino acids within the cell. Rather, it provides the cell with the different types of amino acids required for protein synthesis.

1. Construct a flow chart to show the metabolic pathway by which glucose can be converted to fat. Indicate only the major steps involved.
2. Define the terms *lipolysis* and *β-oxidation,* and explain in general terms how fat can be used for energy.
3. Describe transamination and deamination, and explain their functional significance.

Table 18.1 The essential, semiessential, and nonessential amino acids

Essential Amino Acids	Semiessential Amino Acid	Nonessential Amino Acids
Lysine	Arginine	Aspartic acid
Tryptophan		Glutamic acid
Phenylalanine		Proline
Threonine		Glycine
Valine		Serine
Methionine		Alanine
Leucine		Cysteine
Isoleucine		
Histidine (children)		

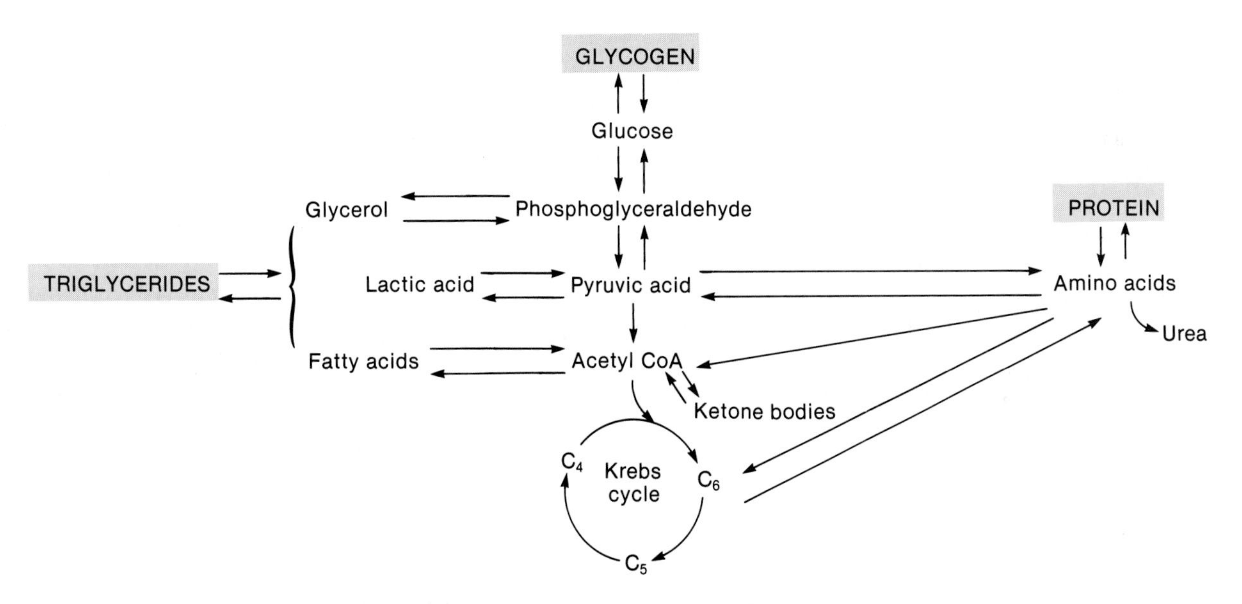

Figure 18.11. *Simplified metabolic pathways showing how glycogen, fat, and protein can be interconverted.*

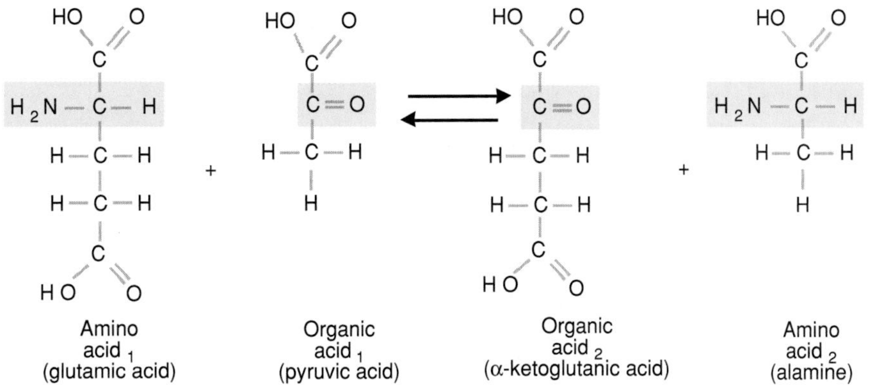

Amino acid $_1$ (glutamic acid) + Organic acid $_1$ (pyruvic acid) ⇌ Organic acid $_2$ (α-ketoglutaric acid) + Amino acid $_2$ (alamine)

Figure 18.12. *Transamination of amino acids.*

Regulation of Energy Metabolism

The blood plasma contains circulating glucose, fatty acids, amino acids, and other molecules that can be used by the body tissues for cell respiration. These circulating molecules may be derived from the food or from breakdown of the body's glycogen and fat reserves, or from body protein. The building of the body's energy reserves following a meal, and the utilization of these reserves between meals, is regulated by the action of a number of hormones.

The molecules that can be oxidized for energy by the processes of cell respiration may be derived from the **energy reserves** of glycogen, fat, or protein. Glycogen and fat function primarily as energy reserves, whereas this represents a secondary, emergency function for proteins. Although body protein can provide amino acids for energy, this requires the breakdown of proteins needed for muscle contraction, structural strength, enzymatic activity, and other functions. Alternatively, the molecules used for cell respiration can be derived from the products of digestion that are absorbed through the small intestine. Since these molecules—glucose, fatty acids, amino acids, and others—are carried by the blood to the tissue cells for use in cell respiration, they can be called **circulating energy carriers** (fig. 18.13).

Because of differences in cellular enzyme content, different organs have different *preferred energy sources.* The brain has an almost absolute requirement for blood glucose as its energy source, for example. A fall in the plasma concentration of glucose below about 50 mg per 100 ml can thus "starve" the brain and have disastrous consequences. Resting skeletal muscles, in contrast, use fatty acids as their preferred energy source. Similarly, ketone bodies (derived from fatty acids), lactic acid, and amino acids can be used to different degrees as energy sources by various organs. The plasma normally contains adequate concentrations of all of these circulating energy carriers to meet the energy needs of the body.

Hormonal Regulation of Metabolism

The absorption of energy carriers from the intestine is not continuous; it rises to high levels following meals (the *absorptive state*) and tapers toward zero between meals (the *postabsorptive* or *fasting* state). Despite this, the plasma concentration of glucose and other energy carriers does not remain high during periods of absorption and does not normally fall below a certain level during periods of fasting. During the absorption of digestion products from the intestine, energy carriers are removed from the blood and deposited as energy reserves from which withdrawals can be made during times of fasting (fig. 18.14). This assures that there will be an adequate plasma concentration of energy carriers to sustain tissue metabolism at all times.

The rate of deposit and withdrawal of energy carriers into and from the energy reserves and the conversion of one type of molecule into another are regulated by the actions of hormones. The balance between anabolism and catabolism is determined by the antagonistic effects of hormones such as insulin, glucagon, growth hormone, thyroxine, and others. The specific metabolic effects of these hormones are illustrated in figure 18.15.

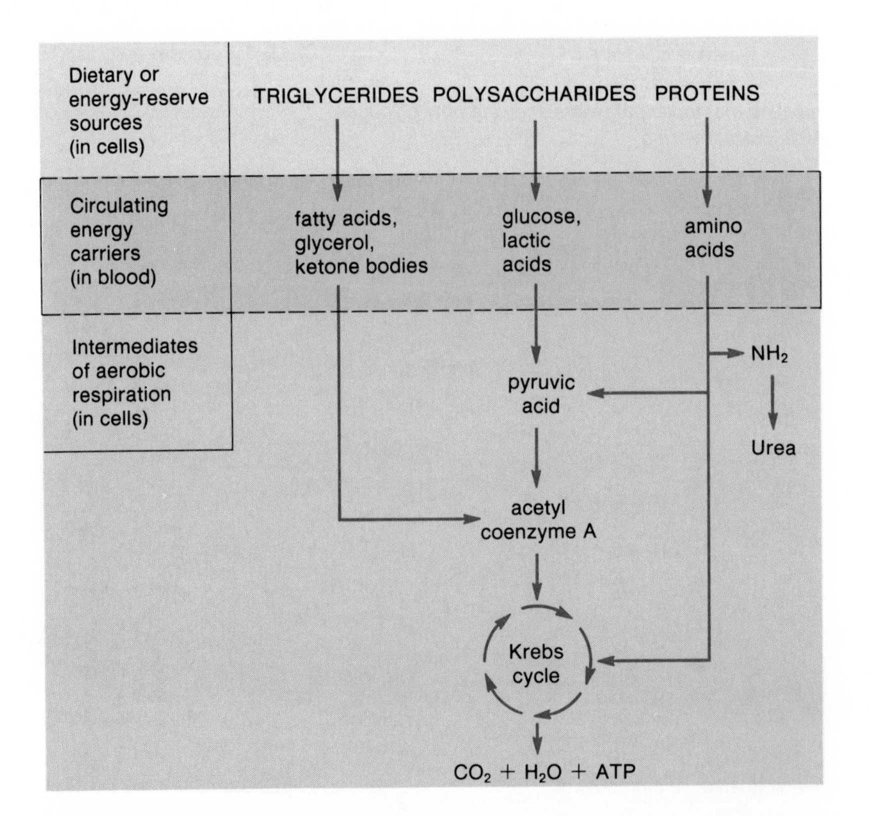

Figure 18.13. *A schematic flowchart of energy pathways in the body.*

1. Define the terms *energy reserves* and *circulating energy carriers,* and give examples.
2. Which hormones promote an increase in blood glucose? Which promote a decrease? List the hormones that stimulate fat synthesis (lipogenesis) and fat breakdown (lipolysis).

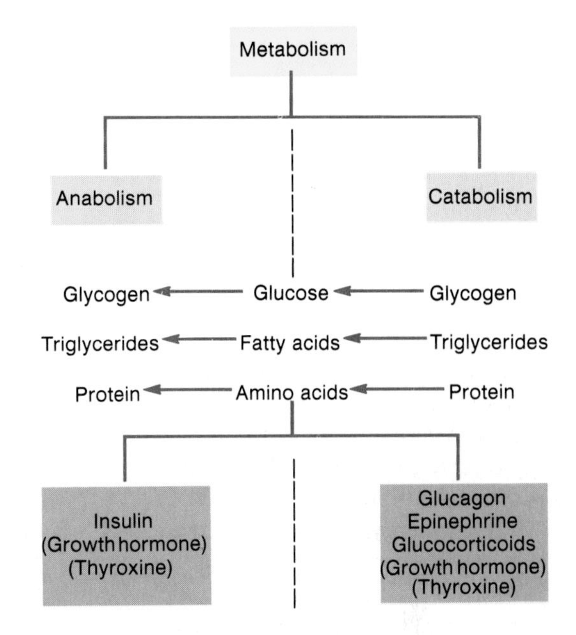

Figure 18.14. *The balance of metabolism can be tilted toward anabolism (synthesis of energy reserves) or catabolism (utilization of energy reserves) by the combined action of various hormones. Growth hormone and thyroxine have both anabolic and catabolic effects.*

Regulation by the Islets of Langerhans

Insulin and glucagon are two hormones secreted by the islets of Langerhans in the pancreas. Insulin stimulates the entry of blood glucose into tissue cells; it thus increases the storage of glycogen and fat while causing a fall in the blood glucose concentration. Glucagon causes a rise in blood glucose by stimulating the breakdown of liver glycogen and the secretion of glucose into the blood.

Scattered within a "sea" of pancreatic exocrine tissue are islands of hormone-secreting cells. These **islets of Langerhans** contain two major cell types that secrete different hormones (fig. 18.16). The most numerous are the *beta cells;* these cells comprise 60% of each islet and secrete the hormone **insulin.** The *alpha cells* comprise about 25% of each islet and secrete the hormone **glucagon.**

Insulin and glucagon secretion is largely regulated by the plasma concentrations of glucose and, to a lesser degree, of amino acids. Since the plasma concentration of glucose and amino acids

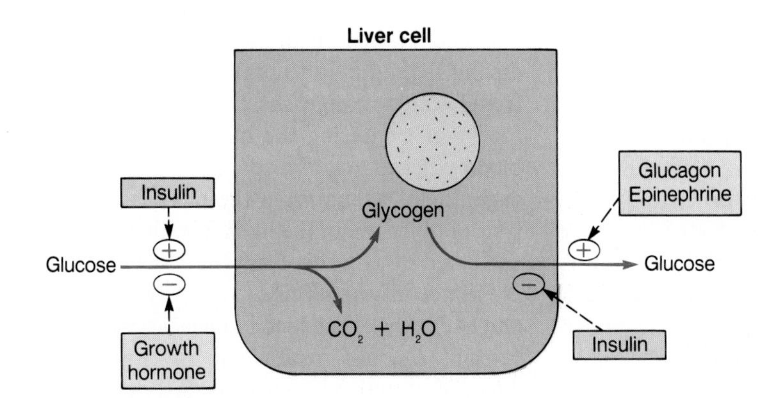

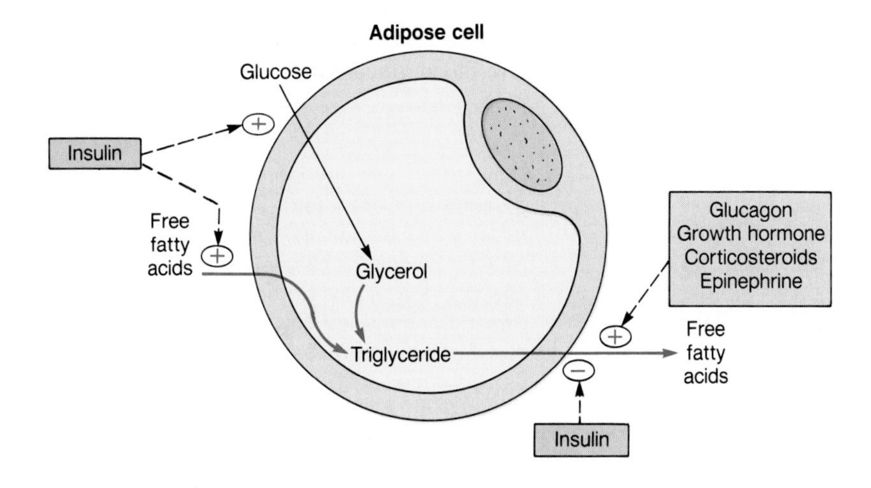

Figure 18.15. *Different hormones participate both synergistically and antagonistically in the regulation of metabolism.*

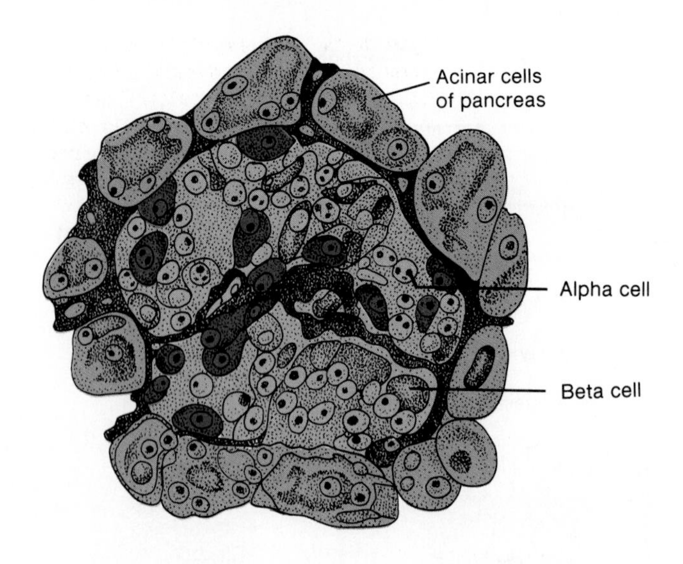

Figure 18.16. *The cellular composition of a normal pancreatic islet.*

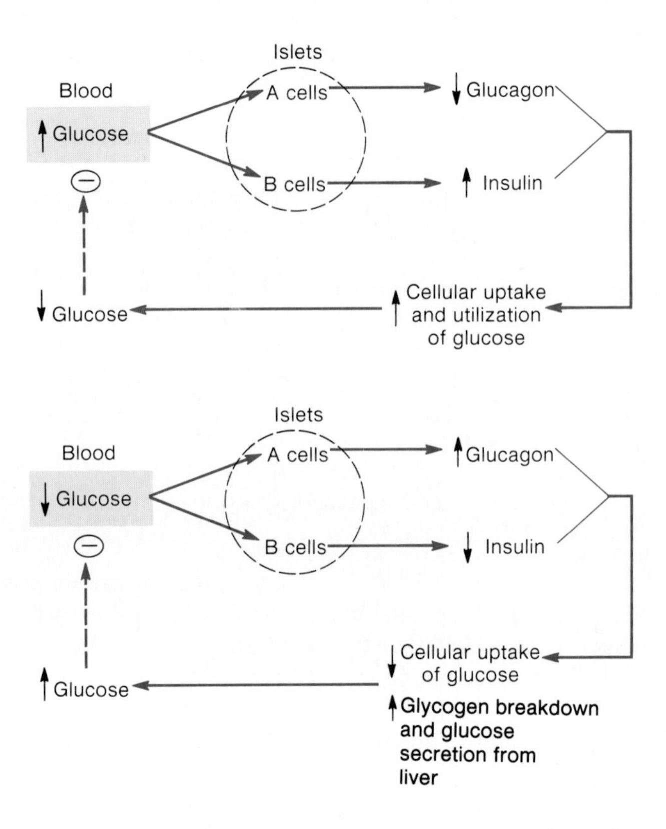

Figure 18.17. *The control of insulin secretion, and the effects of insulin on the metabolism of glucose and glycogen.*

rises during the absorption of a meal and falls during fasting, the secretion of insulin and glucagon likewise changes between the absorptive and postabsorptive states. These changes in insulin and glucagon secretion, in turn, cause changes in plasma glucose and amino acid concentrations and thus help to maintain homeostasis via negative feedback loops (fig. 18.17).

Absorptive State

During the absorption of a carbohydrate meal, the plasma glucose concentration rises. This rise in plasma glucose (1) stimulates the beta cells to secrete insulin and (2) inhibits the secretion of glucagon from the alpha cells. Insulin acts to *stimulate the cellular uptake of plasma glucose.* A rise in insulin secretion therefore lowers the plasma glucose concentration.

The lowering of the plasma glucose concentration by insulin is a side effect of its anabolic actions. By promoting the cellular uptake of glucose into liver and adipose cells, insulin stimulates the production of the energy reserves of glycogen and fat. In addition, insulin secretion is increased in response to particular amino acids, and insulin aids the cellular uptake of amino acids into skeletal muscles and other organs. Through these effects, insulin serves as the major anabolic hormone of the body.

Postabsorptive State

Between meals (the postabsorptive state), the plasma glucose concentration falls. At this time, therefore, (1) insulin secretion decreases and (2) glucagon secretion increases. These changes in hormone secretion prevent the cellular uptake of blood glucose into organs such as the muscles, liver, and adipose tissue and promote the release of glucose from the liver (through the actions of glucagon). A negative feedback loop is therefore completed (fig. 18.17), which helps to retard the fall in plasma glucose concentration that occurs during fasting.

The low insulin and high glucagon secretion from the islets during fasting help to prevent the blood glucose concentration from falling to dangerously low levels. Since glucose derived from food is not entering the blood from the intestine at this time, blood glucose must be spared for the brain. The hormone glucagon helps to maintain the blood glucose by inducing hydrolysis of glycogen in the liver. The liver thus secretes glucose into the blood during this time (fig. 18.18).

Since insulin levels are low, organs such as skeletal muscles and liver cannot use blood glucose very well as an energy source. Fortunately, low insulin and high glucagon promotes breakdown of stored fat, releasing fatty acids into the blood as an energy source. As previously discussed, ketone bodies (derived from fatty acids) will also be released at this time, providing alternate energy sources for many organs and thus helping to spare blood glucose for the brain.

The ability of the beta cells to secrete insulin, as well as the ability of insulin to lower blood glucose, is measured clinically by the **oral glucose tolerance test** (fig. 18.19). In this procedure, a person drinks a glucose solution and blood samples are taken periodically for plasma glucose measurements. In a normal person the rise in blood glucose produced by drinking this solution is reversed to normal levels within two hours following glucose ingestion. People with *diabetes mellitus,* in contrast, maintain a high blood glucose concentration (hyperglycemia) during the oral glucose tolerance test.

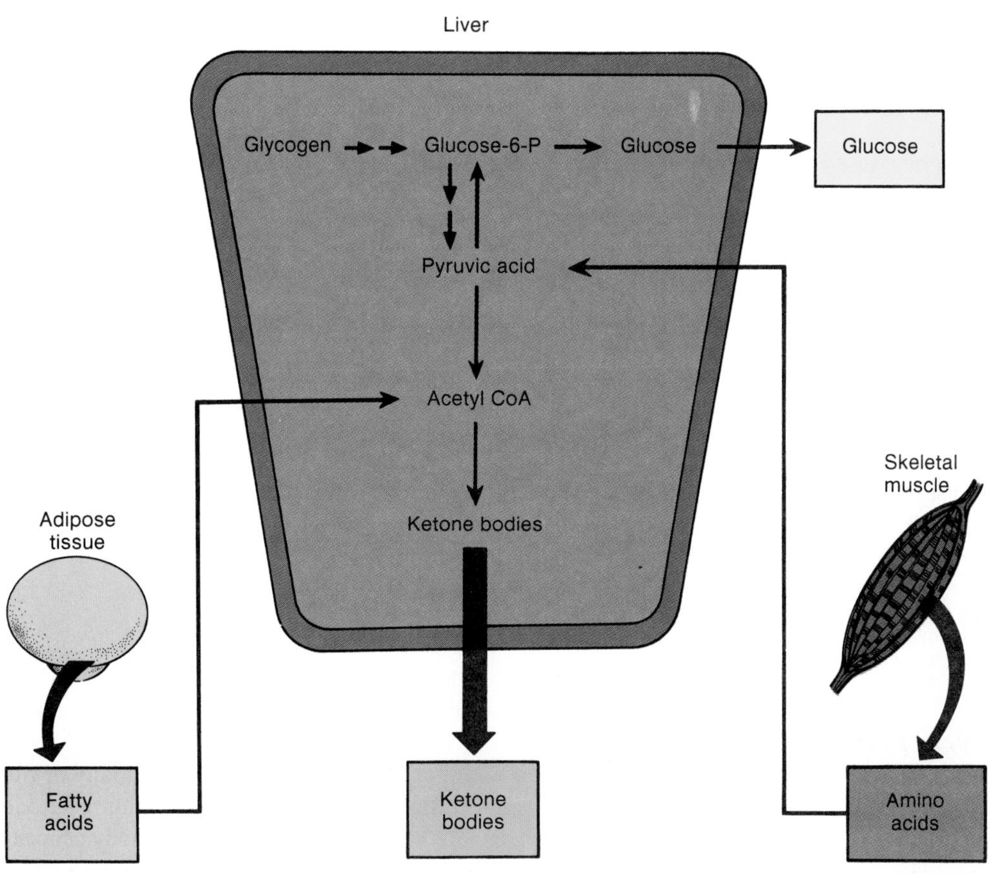

Figure 18.18. Increased glucagon secretion and decreased insulin secretion during fasting favors catabolism. These hormonal changes result in elevated release of glucose, fatty acids, ketone bodies, and amino acids into the blood. Notice that the liver secretes glucose that is derived both from the breakdown of liver glycogen and from the conversion of amino acids in gluconeogenesis.

A summary of the metabolic effects of insulin and glucagon during the absorptive and postabsorptive states is provided in figure 18.20.

1. Describe how the secretions of insulin and glucagon change during the absorptive and postabsorptive states.
2. Explain how changes in insulin and glucagon secretion help to maintain the blood glucose concentration.
3. Explain how insulin and glucagon affect fat metabolism during the absorptive and postabsorptive states.

Diabetes Mellitus

Defects in the actions of insulin produce the metabolic disturbances characteristic of diabetes mellitus. A person with type I diabetes requires injections of insulin; a person with type II diabetes can control the condition by other methods. In both types, hyperglycemia results from the inability of glucose to move out of the blood and into the tissue cells.

Chronic high blood glucose, or hyperglycemia, is the hallmark of the disease **diabetes mellitus.**[8] The name of this disease is derived from the fact that glucose "spills over" into the urine when

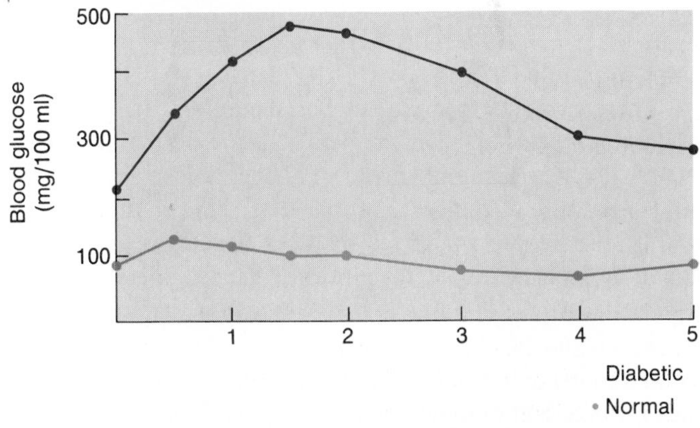

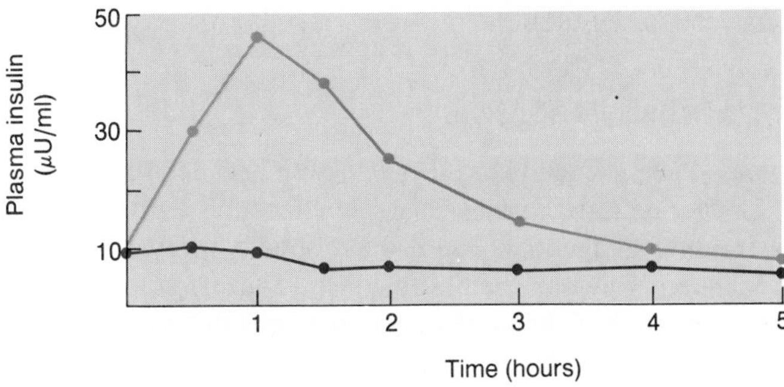

Figure 18.19. Changes in blood glucose and plasma insulin concentrations after the ingestion of 100 grams of glucose in an oral glucose tolerance test.

[8]diabetes mellitus: Gk. *dia,* through; *bainein,* to go; *meli,* honey

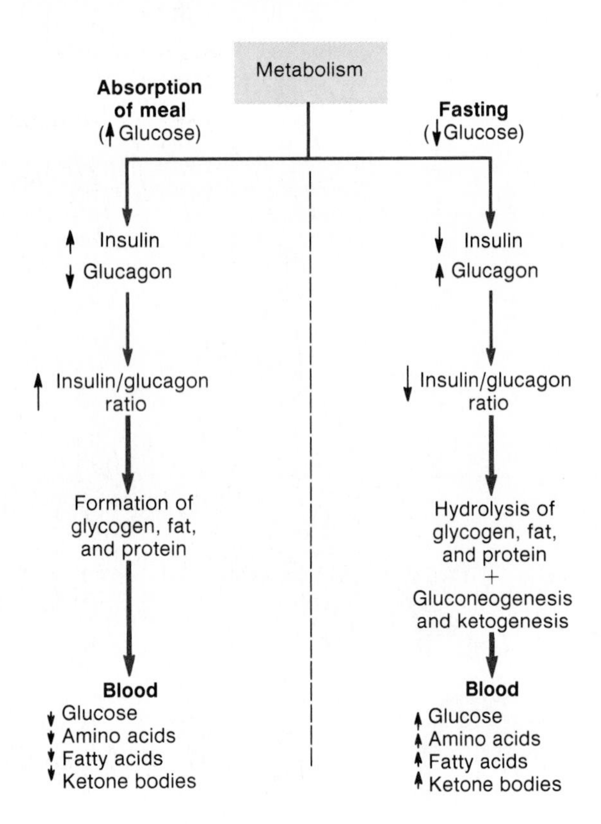

Figure 18.20. *The inverse relationship between insulin and glucagon secretion during the absorption of a meal and during fasting. Changes in the insulin: glucagon ratio tilts metabolism toward anabolism during the absorption of food and toward catabolism during fasting.*

the blood glucose concentration is too high. The hyperglycemia of diabetes mellitus results from the inadequate secretion or action of insulin.

There are two forms of diabetes mellitus. In **type I,** or **insulin-dependent,** diabetes the beta cells are destroyed and secrete little or no insulin. This form of the disease accounts for only about 10% of the cases of diabetes in the country. About 90% of the people who have diabetes have **type II,** or **non-insulin-dependent,** diabetes mellitus. Type I diabetes has also been called *juvenile-onset diabetes,* because this condition is usually diagnosed in people under the age of thirty. Type II, or *maturity-onset, diabetes* is usually diagnosed in people over the age of thirty. Some comparisons of these two forms of diabetes mellitus are shown in table 18.2.

Type I Diabetes Mellitus

Type I diabetes mellitus results when the beta cells of the islets of Langerhans are destroyed by the effects of a virus or other environmental agent, which may act directly or by causing the production of autoantibodies (chapter 13) that attack the beta cells. Removal of the insulin-secreting cells in this way causes hyperglycemia and the appearance of glucose in the urine. Without insulin, glucose cannot enter the adipose cells; the rate of fat synthesis thus lags behind the rate of fat breakdown and large amounts of free fatty acids are released from the adipose cells.

In a person with uncontrolled type I diabetes, many of the fatty acids released from adipose cells are converted into ketone bodies in the liver. This causes an elevated ketone body concentration in the blood (ketosis), and, if the blood pH is lowered by the ketone bodies, it may also result in *ketoacidosis.* During this time, the glucose and excess ketone bodies that are eliminated in the urine also cause the excessive excretion of water. This can produce severe dehydration, which, together with ketoacidosis and associated disturbances, may lead to coma and death (fig. 18.21).

Table 18.2 Comparison of juvenile-onset and maturity-onset diabetes mellitus

Characteristics	Juvenile-Onset (Type I)	Maturity-Onset (Type II)
Usual age at onset	Under 20 years	Over 40 years
Development of symptoms	Rapid	Slow
Percent of diabetic population	About 10%	About 90%
Development of ketoacidosis	Common	Rare
Association with obesity	Rare	Common
Beta cells of islets	Destroyed	Usually not destroyed
Insulin secretion	Decreased	Normal or increased
Autoantibodies to islet cells	Present	Absent
Usual treatment	Insulin injections	Diet; oral stimulators of insulin secretion

Type II Diabetes Mellitus

The effects produced by insulin, or any hormone, depend on the concentration of that hormone in the blood and on the sensitivity of the target tissue to given amounts of the hormone. Tissue responsiveness to insulin, for example, varies under normal conditions. For reasons that are incompletely understood, exercise increases insulin sensitivity and obesity decreases insulin sensitivity of the target tissues. The islets of an obese person, therefore, must secrete high amounts of insulin to maintain the blood glucose concentration in the normal range. Conversely, people who are thin and exercise regularly require lower insulin secretion to maintain the proper blood glucose concentration.

Type II diabetes is usually slow to develop, is hereditary, and occurs most often in people who are overweight. Unlike type I diabetes mellitus, most people who have type II diabetes have normal or even elevated levels of insulin in their blood. Despite

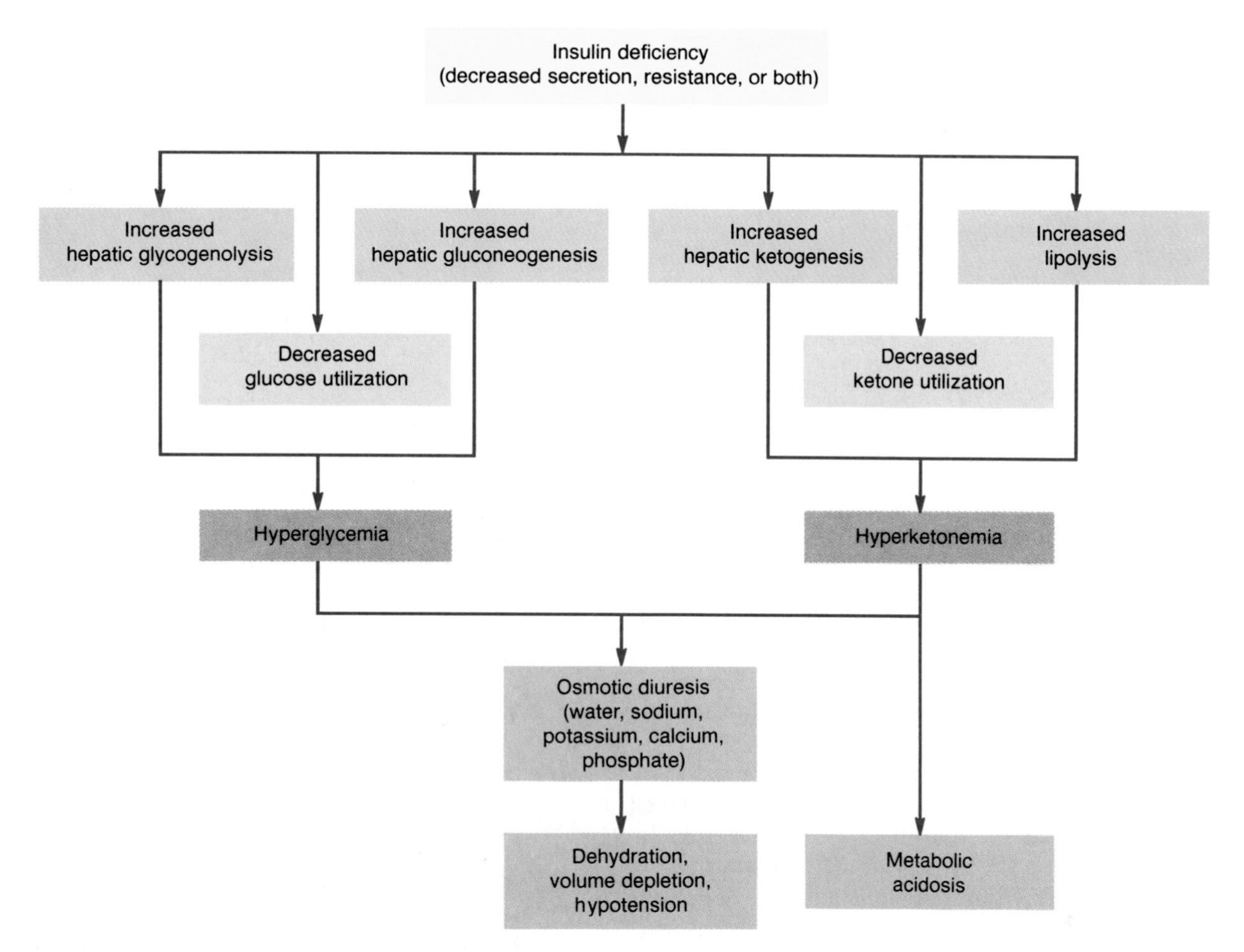

Figure 18.21. *The sequence of events by which an insulin deficiency may lead to coma and death.*

this, people with type II diabetes have hyperglycemia if untreated. This must mean that, even though the insulin levels may be in the normal range, the amount of insulin secreted is inadequate.

Much evidence has been obtained to show that people with type II diabetes have an abnormally low tissue sensitivity to insulin. This is true even if the person is not obese, but the problem is compounded by the decreased tissue sensitivity that accompanies obesity. There is also evidence, however, that the beta cells are not functioning correctly: whatever amount of insulin they secrete is inadequate to the task.

Since obesity decreases insulin sensitivity, people who are genetically predisposed to insulin resistance may develop symptoms of diabetes when they gain weight. Conversely, this type of diabetes mellitus can usually be controlled by increasing tissue sensitivity to insulin through diet and exercise. If this is not sufficient, oral drugs are available that increase insulin secretion and also stimulate tissue responsiveness to insulin.

It is uncommon for people with type II diabetes to develop ketoacidosis. The hyperglycemia itself, however, can be dangerous on a long-term basis. Diabetes is the second leading cause of blindness in the United States, and people with diabetes frequently have circulatory problems that increase the tendency to develop gangrene and increase the risk of atherosclerosis. The causes of damage to the eyes and to blood vessels are not well understood.

Hypoglycemia

People with type I diabetes mellitus are dependent upon insulin injections to prevent hyperglycemia and ketoacidosis. If inadequate insulin is injected, the person may enter a coma as a result of the ketoacidosis, electrolyte imbalance, and dehydration that develop. An overdose of insulin, however, can also produce a coma as a result of the hypoglycemia (abnormally low blood glucose levels) produced. The physical signs and symptoms of diabetic and hypoglycemic coma are sufficiently different (table 18.3) to allow hospital personnel to distinguish between these two types.

Less severe symptoms of hypoglycemia may be produced by an oversecretion of insulin from the islets of Langerhans after a carbohydrate meal. This **reactive hypoglycemia** is caused by

Table 18.3 Comparison of coma due to diabetic ketoacidosis and to hypoglycemia

	Diabetic Ketoacidosis	Hypoglycemia
Onset	Hours to days	Minutes
Causes	Insufficient insulin; other diseases	Excess insulin; insufficient food; excessive exercise
Symptoms	Excessive urination and thirst; headache, nausea, and vomiting	Hunger, headache, confusion, stupor
Physical Findings	Deep, labored breathing; breath has acetone odor; blood pressure decreased, pulse weak; skin is dry	Pulse, blood pressure, and respiration are normal; skin is pale and moist
Laboratory Findings	Urine: glucose present, ketone bodies increased Plasma: glucose and ketone bodies increased, bicarbonate decreased	Urine: normal Plasma: glucose concentration low, bicarbonate normal

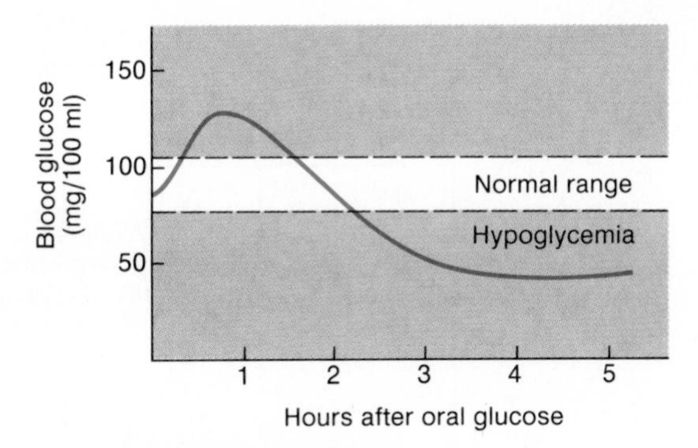

Figure 18.22. An idealized oral glucose tolerance test in a person with reactive hypoglycemia. The blood glucose concentration falls below the normal range within five hours of glucose ingestion as a result of excessive insulin secretion.

an exaggerated response of the beta cells to a rise in blood glucose and is most commonly seen in adults who are genetically predisposed to type II diabetes. People with reactive hypoglycemia, therefore, must limit their intake of carbohydrates and eat smaller meals at frequent intervals, rather than two or three larger meals per day.

The symptoms of reactive hypoglycemia include tremor, hunger, weakness, blurred vision, and impaired mental ability. The appearance of some of these symptoms, however, does not necessarily indicate reactive hypoglycemia and a given level of blood glucose does not always produce these symptoms. For these reasons a number of tests must be performed, including the oral glucose tolerance test, to confirm the diagnosis of reactive hypoglycemia. In the glucose tolerance test, reactive hypoglycemia is shown when the initial rise in blood glucose produced by the ingestion of a glucose solution triggers excessive insulin secretion, so that the blood glucose levels fall below normal within five hours (fig. 18.22).

1. Distinguish between type I and type II diabetes mellitus.
2. Explain how weight loss and exercise may help to control type II diabetes.
3. Describe the nature of reactive hypoglycemia, and how the symptoms of this condition are produced.

Regulation by Thyroxine and Growth Hormone

Both thyroxine and growth hormone are required for proper protein synthesis, and thus both are needed for growth and development of the body. Thyroxine, in addition, promotes cell respiration and thus influences the basal metabolic rate.

The anabolic effects of insulin are antagonized by glucagon, as previously described, and by the actions of a variety of other hormones. Growth hormone and thyroxine antagonize the action of insulin on carbohydrate and lipid metabolism. The actions of insulin, thyroxine, and growth hormone, however, cooperate in the stimulation of protein synthesis.

Thyroxine

The thyroid secretes thyroxine, also called tetraiodothyronine (T_4), in response to stimulation by thyroid-stimulating hormone (TSH) from the anterior pituitary (chapter 7). Thyroxine (1) regulates the rate of cell respiration and (2) contributes to proper growth and development, particularly during early childhood.

Thyroxine and Cell Respiration

Thyroxine stimulates the rate of cell respiration in almost all cells in the body, and thus influences the metabolic rate of the body. The metabolic rate under resting and carefully defined conditions is known as the **basal metabolic rate (BMR).** The BMR indicates the "idling speed" of the body, and is directly determined by the blood concentration of thyroxine. Indeed, measurements of BMR were used clinically to evaluate thyroid function prior to the development of direct chemical determinations of thyroid hormones in the blood.

Thyroxine in Growth and Development

By increasing the rate of cell respiration, thyroxine stimulates the increased consumption of circulating energy carriers such as glucose, fatty acids, and other molecules. These effects, however, are produced by the synthesis of specific enzyme proteins in response to thyroxine. As a result of its stimulation of protein synthesis throughout the body, thyroxine is considered to be an anabolic hormone like insulin and growth hormone.

Because of its stimulation of protein synthesis, thyroxine is needed for growth of the skeleton and, most importantly, for the proper development of the central nervous system. Recent evidence has demonstrated the presence of receptor proteins for

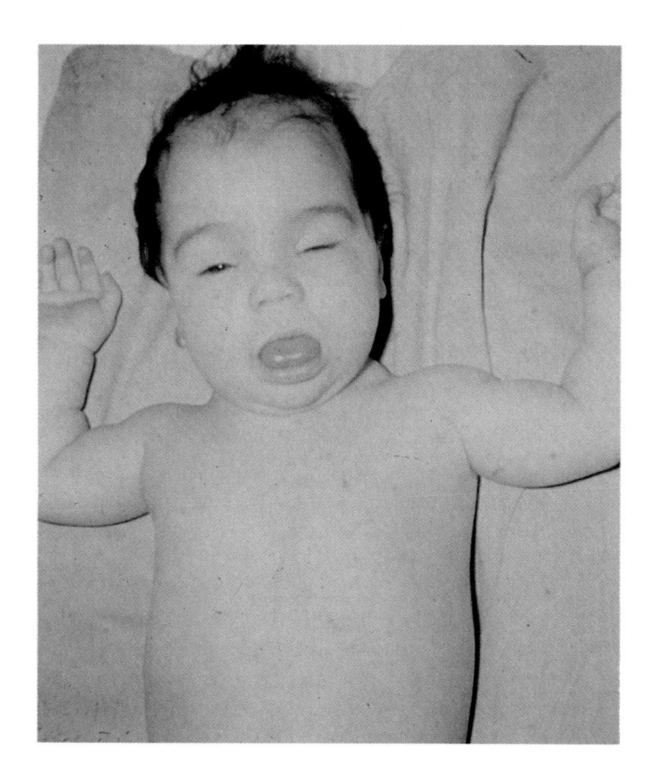

Figure 18.23. *Cretinism is a disease of infancy caused by an underactive thyroid gland.*

Characteristics	Hypothyroid	Hyperthyroid
Growth and Development	Impaired growth	Accelerated growth
Activity and Sleep	Decreased activity; increased sleep	Increased activity; decreased sleep
Temperature Tolerance	Intolerance to cold	Intolerance to heat
Skin Characteristics	Coarse, dry skin	Smooth skin
Perspiration	Absent	Excessive
Pulse	Slow	Rapid
Gastrointestinal Symptoms	Constipation; decreased appetite; increased weight	Frequent bowel movements; increased appetite; decreased weight
Reflexes	Slow	Rapid
Psychological Aspects	Depression and apathy	Nervous, "emotional"
Plasma T_4 Levels	Decreased	Increased

Table 18.4 Comparison of hypothyroidism and hyperthyroidism

thyroxine in the neurons and astrocytes of the brain. This need for thyroxine is particularly great during the time the brain is undergoing its greatest rate of development, from the end of the first trimester of prenatal life to six months after birth. Hypothyroidism during this time may result in **cretinism** (fig. 18.23). Unlike dwarfs, who have normal thyroxine secretion but a low secretion of growth hormone, cretins suffer from severe mental retardation. Treatment with thyroxine soon after birth, particularly before one month of age, has been found to restore development of intelligence as measured by I.Q. tests administered five years later.

Hypothyroidism and Hyperthyroidism

As might be predicted from the effects of thyroxine, people who are hypothyroid have an abnormally low basal metabolic rate (BMR) and experience weight gain and lethargy. There is also a decreased ability to adapt to cold stress when there is a thyroxine deficiency (table 18.4). Another symptom of hypothyroidism is **myxedema**—accumulation of mucoproteins and fluid in subcutaneous connective tissues. Hypothyroidism due to lack of iodine is accompanied by excessive TSH secretion (chapter 7), which stimulates abnormal growth of the thyroid (goiter).

A goiter can also be produced by another mechanism. In **Graves' disease,**[9] antibodies are produced that have TSH-like effects on the thyroid. Unlike TSH secretion, the production of these antibodies is not controlled by negative feedback. These antibodies thus cause excessive stimulation of the thyroid. A goiter is thus produced that is associated with a hyperthyroid state. Hyperthyroidism produces a high BMR accompanied by weight loss, nervousness, irritability, and an intolerance to heat.

[9]Graves' disease: after Robert James Graves, Irish physician, 1796–1853

Growth Hormone

The anterior pituitary secretes growth hormone in larger amounts than any other of its hormones. As its name implies, growth hormone stimulates growth in children and adolescents. The continued high secretion of growth hormone in adults, particularly under the conditions of fasting and other forms of stress, implies that this hormone can have important metabolic effects even after the growing years have ended.

Effects of Growth Hormone on Metabolism

The fact that growth hormone secretion is increased during fasting and also during absorption of a protein meal reflects the complex nature of this hormone's action. Growth hormone has both anabolic and catabolic effects; it promotes protein synthesis (anabolism) and it also stimulates the catabolism of fat and the release of fatty acids from adipose tissue. Growth hormone is similar to insulin in its former effect and similar to glucagon in its latter effect.

Growth hormone stimulates the cellular uptake of amino acids and protein synthesis in many organs of the body. These actions are useful during a protein-rich meal; amino acids are removed from the blood and used to form proteins, and the plasma concentration of glucose and fatty acids is increased to provide alternate energy sources. The anabolic effect of growth hormone on protein synthesis is particularly important during the growing years, when it contributes to increases in bone length and in the mass of many soft tissues.

Effects of Growth Hormone on Body Growth

The stimulatory effects of growth hormone on skeletal growth results from stimulation of mitosis in the discs of cartilage present in the long bones of growing children and adolescents (chapter 21). Part of this growing cartilage is then converted to bone, and the cycle of cartilage growth and partial conversion to bone is repeated, enabling the bone to grow in length. This skeletal growth stops when all of the growth discs of cartilage are converted to bone at the end of the growing years. The pituitary of an adult continues to secrete growth hormone, but increases in the length of bones cannot occur without the cartilage growth discs.

An excessive secretion of growth hormone in children can produce **gigantism** in which people may grow up to eight feet tall. An excessive growth hormone secretion that occurs after the growth discs in bones have disappeared, however, cannot produce increases in height. The oversecretion of growth hormone in adults results in an elongation of the jaw and deformities in the bones of the face, hands, and feet. This condition, called **acromegaly,**[10] is accompanied by the growth of soft tissues and coarsening of the skin (see fig. 7.8). It is interesting that athletes who (illegally) take growth hormone supplements to increase their muscle mass may also experience body changes similar to acromegaly.

An inadequate secretion of growth hormone during the growing years results in **dwarfism.** An interesting variant of this is *Laron dwarfism,* in which there is a genetic insensitivity of the target tissues to the effects of growth hormone.

[10]acromegaly: Gk. *akron,* extremity; *melos,* limb

An adequate diet, particularly of proteins, is required for the action of growth hormone. This helps to explain the common observation that many children are significantly taller than their parents, who may not have had an adequate diet in their youth. Children with protein malnutrition (**kwashiorkor**) have low growth rates, despite the fact that their growth hormone levels are abnormally elevated. If these children later eat an adequate diet, growth rates and the secretion of growth hormone can become normal.

1. Explain the actions of thyroxine on the basal metabolic rate, and explain why people who are hypothyroid have a tendency to gain weight and are less resistant to cold stress.
2. Distinguish between dwarfism and cretinism, and between gigantism and acromegaly.
3. Explain how growth hormone stimulates skeletal growth.

Summary

Energy Flow Within the Body p. 400

I. Metabolism is divided into anabolism and catabolism.

II. NAD and FAD are molecules that transport hydrogen from one reaction to another.

III. Aerobic cell respiration refers to the complete combustion to molecules such as glucose to carbon dioxide and water; this also yields energy, which is used to produce ATP.

Glycolysis and Anaerobic Respiration p. 401

I. In glycolysis, a molecule of glucose is converted through a series of reactions into two molecules of pyruvic acid.
   A. This also reduces two molecules of NAD to $NADH_2$.
   B. During glycolysis, there is a net gain of 2 ATP per glucose molecule.

II. In anaerobic respiration, the pyruvic acids formed by glycolysis are converted into lactic acid molecules.
   A. Anaerobic respiration occurs only when there is an insufficient supply of oxygen for aerobic respiration.
   B. A total of 2 ATP per glucose is obtained through glycolysis.
   C. Anaerobic respiration in skeletal muscles is normal; the heart only respires anaerobically when it does not receive a normal supply of blood.

Aerobic Respiration p. 403

I. When adequate amounts of oxygen are present, pyruvic acid enters the mitochondria and is converted into a two-carbon acetyl CoA molecule.

II. The combination of acetyl CoA with oxaloacetic acid (four carbons long) produces a six-carbon-long molecule called citric acid.
   A. This begins a cyclic metabolic pathway known as the citric acid or Krebs cycle.
   B. In the course of the Krebs cycle, 2 ATP, 3 $NADH_2$ and one $FADH_2$ are produced.

III. The electrons from the hydrogen atoms in $NADH_2$ and $FADH_2$ are transferred to the molecules of the electron transport chain.
   A. Two pairs of electrons from the pairs of hydrogen atoms are passed from one member of the electron transport chain to the next.
   B. As electrons move along the electron transport chain of molecules, energy is released and is used to power the formation of ATP; this process is called oxidative phosphorylation.
   C. At the last step of electron transport, the electrons are transferred to an oxygen atom; when 2 H^+ are also combined, water is formed.

Metabolism of Fat and Protein p. 405

I. Triglycerides (fat) are formed from glycerol and three fatty acids.

II. The breakdown of fat liberates free fatty acids into the blood for use by tissue cells; ketone bodies are formed in the liver as derivatives of fatty acids.

III. Amino acids from protein can have their amine groups removed in the process of oxidative deamination.
   A. The amine groups are incorporated into molecules of urea for excretion.
   B. The organic acids formed from deamination of amino acids can be used in cell respiration for energy.
   C. Amino acids that are not among the essential group can be formed by transamination reactions.

Regulation of Energy Metabolism p. 410

I. Molecules present in the blood which can be used for cell respiration are called circulating energy carriers; glycogen, fat, and protein within cells can function as energy reserves.

II. The absorptive state refers to the condition when molecules obtained from digestion of food enter the blood from the intestine.

III. The postabsorptive state refers to the periods between meals when circulating energy carriers can only be obtained from the energy reserves within the tissue cells.

Regulation by the Islets of Langerhans p. 411

I. Insulin secretion is stimulated when blood glucose rises following a meal; this also inhibits the secretion of glucagon.
   A. A rise in insulin stimulates the cellular uptake of blood glucose, causing a fall in the blood glucose concentration back to normal.
   B. Insulin stimulates the conversion of glucose to glycogen and fat, thus promoting anabolism.

II. During the postabsorptive state, when blood glucose falls, insulin secretion is inhibited and glucagon secretion is increased.
   A. Glucagon stimulates the hydrolysis of liver glycogen, allowing the liver to secrete glucose into the blood.
   B. Glucagon, in the presence of low insulin levels, also stimulates lipolysis, and thus results in the release of free fatty acids from adipose tissues.

Diabetes Mellitus p. 413

I. In type I diabetes, the beta cells of the islets are destroyed and little or no insulin is secreted.

II. In type II diabetes, the islets secrete inadequate amounts of insulin and the target organs have a lowered sensitivity to the effects of insulin.

III. Reactive hypoglycemia occurs when excessive insulin is secreted following a carbohydrate meal.

Regulation by Thyroxine and Growth Hormone p. 416

I. Thyroxine stimulates cell respiration, and thus acts to set the basal metabolic rate (BMR).
   A. A person who is hypothyroid thus has a slow metabolism and tends to gain weight and be intolerant to cold.
   B. Thyroxine also stimulates protein synthesis, and is needed for proper development of the brain; inadequate thyroxine can result in cretinism.

II. Growth hormone promotes protein synthesis, inhibits the cellular utilization of glucose, and promotes the breakdown of fat.
   A. Inadequate growth hormone during the growing years can result in dwarfism.
   B. Excessive growth hormone during the growing years causes gigantism; after this time, it can cause acromegaly.

Review Activities

Objective Questions

1. The net gain of ATP per glucose molecule in anaerobic respiration is ______ ; the net gain in aerobic respiration is ______ .
   (a) 2;4
   (b) 2;38
   (c) 38;2
   (d) 24;30
2. In anaerobic respiration, the hydrogens from $NADH_2$ become part of
   (a) pyruvic acid
   (b) lactic acid
   (c) citric acid
   (d) oxygen
3. When organs respire anaerobically, there is an increased blood concentration of
   (a) oxygen
   (b) glucose
   (c) lactic acid
   (d) ATP
4. The oxygen in the air we breathe
   (a) functions as the final electron acceptor of the electron transport chain
   (b) combines with hydrogen to form water
   (c) combines with carbon to form CO_2
   (d) both *a* and *b*
   (e) both *a* and *c*
5. In terms of the number of ATP molecules directly produced, the major energy-yielding process in the cell is
   (a) glycolysis
   (b) the Krebs cycle
   (c) oxidative phosphorylation
   (d) gluconeogenesis
6. Ketone bodies are derived from
   (a) fatty acids
   (b) glycerol
   (c) glucose
   (d) amino acids
7. Which of the following organs can secrete glucose into the blood?
   (a) the liver
   (b) the skeletal muscles
   (c) both of these organs
8. The formation of glucose from amino acids is called
   (a) glycogenesis
   (b) glycogenolysis
   (c) glycolysis
   (d) gluconeogenesis
9. Which of the following organs has an almost absolute requirement for blood glucose as its energy source? The
   (a) liver
   (b) brain
   (c) skeletal muscles
   (d) heart
10. When amino acids are used as an energy source
    (a) oxidative deamination occurs
    (b) pyruvic acid or one of the Krebs cycle acids (keto acids) are formed
    (c) urea is produced
    (d) all of the above occur
11. During the absorption of a carbohydrate meal, when the blood glucose concentration is increasing
    (a) insulin secretion decreases and glucagon secretion increases
    (b) insulin secretion increases and glucagon secretion decreases
    (c) the secretion of insulin and glucagon increase
    (d) the secretion of insulin and glucagon decrease
12. Formation of fat from glucose is promoted by which of the following hormones?
    (a) insulin
    (b) glucagon
    (c) thyroxine
    (d) growth hormone
    (e) all of these
13. Protein synthesis is promoted by all of the following hormones *except*
    (a) insulin
    (b) glucagon
    (c) thyroxine
    (d) growth hormone
    (e) all of these
14. The basal metabolic rate is determined primarily by
    (a) hydrocortisone
    (b) insulin
    (c) growth hormone
    (d) thyroxine
15. Ketoacidosis in untreated diabetes mellitus is due to
    (a) excessive fluid loss
    (b) hypoventilation
    (c) excessive eating and obesity
    (d) excessive fat catabolism

Essay Questions

1. Explain the advantages and disadvantages of anaerobic respiration.
2. What purpose is served by the formation of lactic acid during anaerobic respiration? How is this purpose achieved during aerobic respiration?
3. Describe the metabolic pathway by which fat can be used as a source of energy, and explain why the metabolism of fatty acids can yield more ATP than the metabolism of glucose.
4. Explain how energy is obtained from the metabolism of amino acids. Why does a starving person have high concentrations of urea in the blood?
5. Compare the metabolic effects of fasting to the state of uncontrolled type I diabetes. Explain the hormonal similarities of these conditions.
6. Describe how thyroxine affects cell respiration, and explain why a hypothyroid person has a tendency to gain weight and has a reduced tolerance to cold.

19

Nutrition

Outline

Objectives

By studying this chapter, you should be able to

1. define the term *basal metabolic rate*
2. estimate your energy expenditure per day, and determine how many kilocalories you would have to lose or gain to achieve your ideal body weight
3. Describe the significance of the essential amino acids and how they can be obtained in the diet and in a vegetarian diet
4. distinguish between the different nutritional categories of carbohydrates
5. explain the significance of the recommended daily allowance (RDA)
6. define the terms *enriched, fortified*, and *supplemented*
7. list some common misconceptions in regards to nutrition, and explain the errors involved in these
8. list the different sources for the B vitamins, and describe some of the diseases associated with deficiencies in these vitamins
9. list the dietary sources of vitamin C, and describe the association between vitamin C deficiency and scurvy
10. describe the functions of each of the fat-soluble vitamins, and indicate the deficiency diseases associated with these vitamins
11. explain how the body produces vitamin D, and the functions of vitamin D

Keys to Pronunciation

anabolic: *an''ah-bol'ik*
anorexia: *an''o-rek'se-ah*
calorie: *kal'o-re*
cholecalciferol: *ko''le-kal-sif'er-ol*
niacin: *ni'ah-sin*
nutrient: *nu'tre-ent*
nutrition: *nu-trish'un*
pellagra: *pe-la'grah*
phenylalanine: *fen''il-al'ah-nin*
pyridoxine: *per''i-dok'sen*
riboflavin: *ri''bo-fla'vin*
thiamine: *thi'ah-min*
tocopherol: *to-kof'er-ol*
tryptophan: *trip'to-fan*
vitamin: *vi'tah-min*

Photo: Good nutrition provides adequate calories to maintain a proper body weight, essential amino acids, and the vitamins and minerals needed for specific functions in the body.

Energy Value of Food

Since the body consumes energy in its metabolism, energy must be taken into the body in the form of food. The rate of the body's energy consumption is the metabolic rate; this is composed of a basal component and a component produced by muscular activity. Weight is gained when the caloric intake is greater than the caloric expenditure, and is lost when the opposite occurs.

Living tissue is maintained by the constant expenditure of energy. This energy is obtained directly from ATP and indirectly from the cell respiration of glucose, fatty acids, ketone bodies, amino acids, and other organic molecules (chapter 18). These molecules are ultimately obtained from food, but they can also be obtained from the glycogen, fat, and protein stored in the body.

The energy value of food is commonly measured in *kilocalories,* which are also called "big calories" and spelled with a capital letter (*C*alories). One kilocalorie is equal to 1,000 calories.[1] One calorie is defined as the amount of heat required to raise one cubic centimeter of water from 15° to 16°C. This can be directly measured by combusting a sample of food in a device known as a calorimeter (fig. 19.1). As described in chapter 18, the same amount of energy is released within cells by the aerobic respiration of that food, but only some of the energy escapes as heat; the rest is transferred to the high-energy bonds of ATP.

Metabolic Rate and Caloric Requirements

The total rate of body metabolism, or the **metabolic rate,** can be measured by either the amount of heat generated by the body or by the amount of oxygen consumed by the body per minute. This rate is influenced by a variety of factors. For example, the metabolic rate is increased by physical activity and by the digestion and absorption of food.

Temperature is also an important factor determining metabolic rate. Body temperature is sensed by cells in the hypothalamus of the brain, which also contains a *temperature control center.* In response to deviations from a "set point" for body temperature (chapter 5), the control areas of the hypothalamus can direct physiological responses that help to correct the deviations and maintain a constant body temperature. Ambient temperatures that are either too hot or too cold are thus accompanied by physiological responses (such as shivering or sweating) that increase the metabolic rate.

Basal Metabolic Rate

The metabolic rate of an awake, relaxed person 12–14 hours after eating in a room at comfortable temperature is known as the **basal metabolic rate (BMR).** The BMR is determined primarily by the age, sex, and body surface area of the person, but is also strongly influenced by the level of thyroid secretion. A person with hyperthyroidism has an abnormally high BMR, and a person with hypothyroidism has a low BMR (chapter 18). An interesting recent finding is that the BMR may be influenced by genetic inheritance, and that at least some families that are prone to obesity may have a genetically determined low BMR. They are thus more "fuel efficient." This ability may have evolved over many generations in response to chronic food shortage, and, in the words of one scientist, may represent a "thrifty gene rendered detrimental by progress."

Effect of Physical Activity on Metabolic Rate

The differences in energy requirements among most people, however, are due primarily to differences in physical activity. Average daily energy expenditures may range from about 1,300 to 5,000 kilocalories per day. The average values for active young adults is about 2,900 kilocalories per day for a man and 2,100 kilocalories per day for a woman.

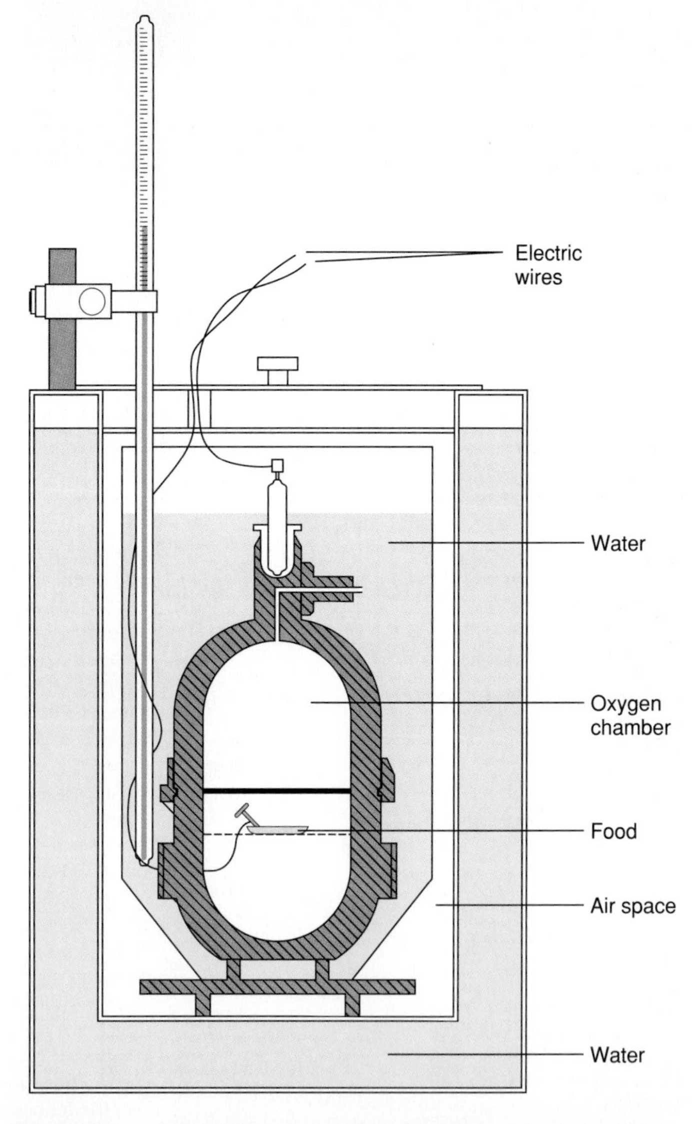

Figure 19.1. *A calorimeter to measure the amount of calories in a sample of food. When the food is combusted, the temperature of the water in the inner chamber is raised and measured by the thermometer.*

[1]calorie: L. *calor,* heat

Table 19.1 gives the *resting metabolic rate* (which is approximately 10% higher than the BMR) for men and women. In order to estimate your resting metabolic rate, first divide your weight in pounds by a factor of 2.2 to obtain your weight in kilograms. Then obtain your metabolic rate per minute from the table. Finally, multiply this last figure by a factor of 1,440 (the number of minutes in twenty-four hours) to obtain your resting metabolic rate per day.

In order to estimate your actual energy expenditure per day you must increase the value of your resting metabolic rate by an activity factor. Add 30% to your resting value if you are inactive, 50% if your day is spent in light activity, 75% if you do extensive amounts of walking, housework, and so on, or 100% if you perform hours of strenuous activity (such as athletics, dancing, and so on). Table 19.2 provides estimates of the caloric expenditures of different types of physical activity.

Table 19.1 Resting metabolic rate*

			Weight in kg†						
Men	**Women**	**Body Fat, %**	**50**	**55**	**60**	**65**	**70**	**75**	**80**
			kcal per minute						
Thin		5	0.99	1.06	1.12	1.19	1.26	1.32	1.39
Average		10	0.94	1.01	1.08	1.14	1.21	1.28	1.34
Plump	Thin	15	0.89	0.96	1.03	1.09	1.16	1.23	1.30
Fat	Average	20	0.84	0.91	0.98	1.05	1.11	1.18	1.25
	Plump	25	0.80	0.86	0.93	1.00	1.07	1.13	1.20
	Fat	30	—	0.81	0.88	0.95	1.02	1.08	1.15

*These values are slightly higher than basal because the subject is not in the strictly postabsorptive condition (12 hours or more from the last meal). (Adapted from Durnin and Passmore.)

†To convert to pounds multiply by 2.2.

From L. J. Bogert, et al., *Nutrition and Physical Fitness,* 9th ed. Copyright © 1973 W. B. Saunders, Orlando, FL.

Table 19.2 Energy consumed (in kilocalories per minute) by different types of activities

	Weight in Pounds			
Activity	**105–115**	**127–137**	**160–170**	**182–192**
Bicycling				
10 mph	5.41	6.16	7.33	7.91
Stationary, 10 mph	5.50	6.25	7.41	8.16
Calisthenics	3.91	4.50	7.33	7.91
Dancing				
Aerobic	5.83	6.58	7.83	8.58
Square	5.50	6.25	7.41	8.00
Gardening, weeding, and digging	5.08	5.75	6.83	7.50
Jogging				
5.5 mph	8.58	9.75	11.50	12.66
6.5 mph	8.90	10.20	12.00	13.20
8.0 mph	10.40	11.90	14.10	15.50
9.0 mph	12.00	13.80	16.20	17.80
Rowing Machine				
Easily	3.91	4.50	5.25	5.83
Vigorously	8.58	9.75	11.50	12.66
Skiing				
Downhill	7.75	8.83	10.41	11.50
Cross-country, 5 mph	9.16	10.41	12.25	13.33
Cross-country, 9 mph	13.08	14.83	17.58	19.33
Swimming, crawl				
20 yards per minute	3.91	4.50	5.25	5.83
40 yards per minute	7.83	8.91	10.50	11.58
55 yards per minute	11.00	12.50	14.75	16.25
Walking				
2 mph	2.40	2.80	3.30	3.60
3 mph	3.90	4.50	5.30	5.80
4 mph	4.50	5.20	6.10	6.80

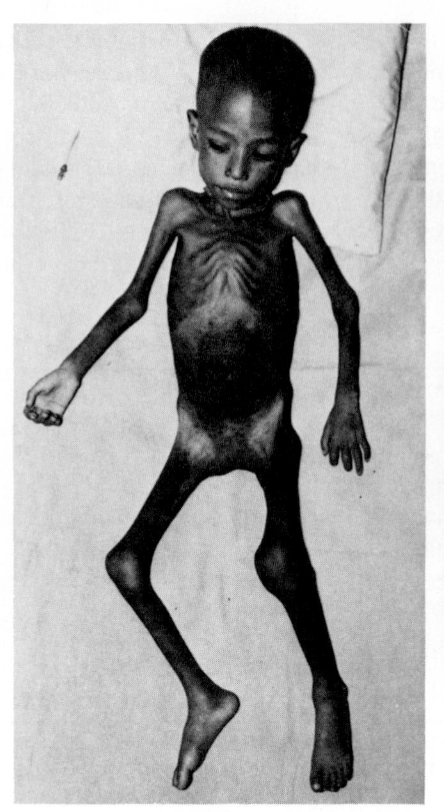

(a)

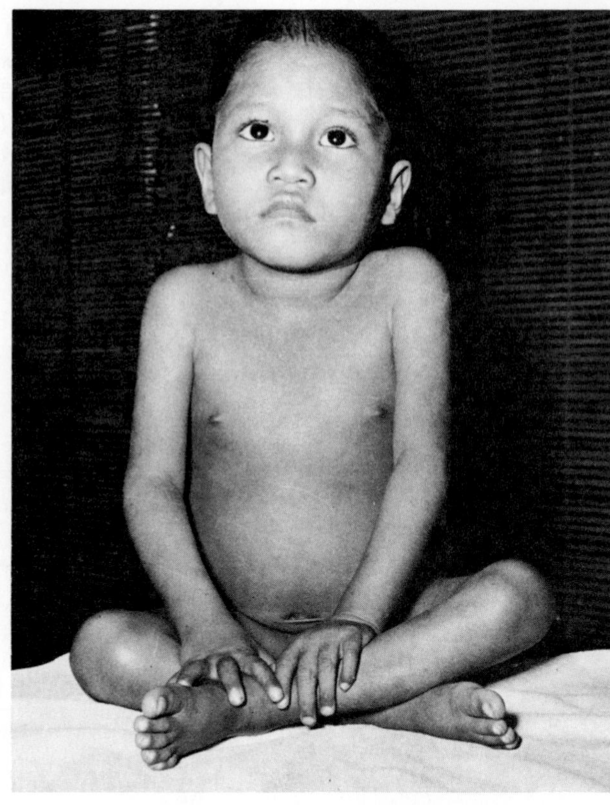

(b)

Figure 19.2. A starving three-year-old girl (a) found abandoned in Burma was hospitalized and provided with an adequate diet; (b) is the same girl three months later.

Body Weight

The weight for a person of a given sex and height is determined in part by the *lean body mass* of the person, which can vary depending on bone structure and degree of muscle development. Body weight is also, of course, influenced by the amount of fat contained in adipose tissue. While variations in lean body mass do not appear to affect health, abnormally high or low amounts of body fat are clinically significant. When the caloric intake is greater than the energy expenditures, excess calories are stored primarily as fat. This is true regardless of the source of the calories—carbohydrates, protein, or fat—because these molecules can be converted to fat in adipose tissue by the metabolic pathways described in chapter 18.

Ideal body weights, as compiled by life insurance companies, are indicated in table 19.3. It should be noted that these values assume that the person is wearing shoes with a one-inch heel. Therefore, add one inch to your barefoot height when using these tables. Also, it is assumed that the person is dressed. Thus, if you know your weight when naked, add two to four pounds to account for clothing.

People are considered to be underweight or overweight when they weigh 10% less or 10% more than their ideal body weight. In severe starvation, the body weight can become far lower than the ideal (fig. 19.2). The condition of *obesity* is present when a person's body weight is 15% to 25% greater than their ideal body weight.

Obesity is a risk factor for cardiovascular diseases, renal disease, diabetes mellitus, gallbladder disease, the development of kidney stones, and some malignancies (particularly endometrial and breast cancer). Obesity in childhood is due to an increase in both the size and number of adipose cells; weight gain in adulthood is due mainly to an increase in adipose cell size, although the number of these cells may also increase in extreme weight gains. When weight is lost, the size of the adipose cells decreases, but the number of adipose cells does not decrease. It is thus important to prevent further increases in weight in all overweight people but particularly so in children.

Dieting and Weight Loss

Food is so plentiful in our society that obesity is, unfortunately, common. For the obese person, a weight reduction program is essential for health. Many people who are not obese, but who are only overweight, may also benefit from weight reduction. The health benefits of weight loss for all who are above their ideal weight, however, has not been proven. Indeed, some people may be healthier when they are somewhat above the ideal weight provided in the tables. To a degree, the slim body idealized by our society is a cultural quirk. This can actually be damaging to health, as in the teenage girl with *anorexia*[2] *nervosa* who continues to consider herself "fat" even when she is dangerously underweight.

[2]anorexia: Gk. *anorexia,* lack of appetite

Table 19.3 Desirable body weights, according to sex, age, height, and body frame

	Height Feet	Height Inches	Small Frame	Medium Frame	Large Frame
Men	5	2	128–134	131–141	138–150
	5	3	130–136	133–143	140–153
	5	4	132–138	135–145	142–156
	5	5	134–140	137–148	144–160
	5	6	136–142	139–151	146–164
	5	7	138–145	142–154	149–168
	5	8	140–148	145–157	152–172
	5	9	142–151	148–160	155–176
	5	10	144–154	151–163	158–180
	5	11	146–157	154–166	161–184
	6	0	149–160	157–170	164–188
	6	1	152–164	160–174	168–192
	6	2	155–168	164–178	172–197
	6	3	157–172	167–182	176–202
	6	4	162–176	171–187	181–207

Weights at ages 25–59 based on lowest mortality. Weight in pounds according to frame (in indoor clothing weighing 5 lbs., shoes with 1″ heels).

	Height Feet	Height Inches	Small Frame	Medium Frame	Large Frame
Women	4	10	102–111	109–121	118–131
	4	11	103–113	111–123	120–134
	5	0	104–115	113–126	122–137
	5	1	106–118	115–129	125–140
	5	2	108–121	118–132	128–143
	5	3	111–124	121–135	131–147
	5	4	114–127	124–138	134–151
	5	5	117–130	127–141	137–155
	5	6	120–133	130–144	140–159
	5	7	123–136	133–147	143–163
	5	8	126–139	136–150	146–167
	5	9	129–142	139–153	149–170
	5	10	132–145	142–156	152–173
	5	11	135–148	145–159	155–176
	6	0	138–151	148–162	158–179

Weights at ages 25–59 based on lowest mortality. Weight in pounds according to frame (in indoor clothing weighing 3 lbs., shoes with 1″ heels).

It should be understood that the goal of weight reduction should be loss of fat, not loss of muscle tissue and water. Fat is produced when more calories are taken in as food than are consumed by the body per day in cell respiration. Fat is lost when the opposite occurs—when the caloric intake per day is less than the caloric expenditure. This can be achieved by reducing the caloric intake and/or by increasing the caloric expenditure through exercise. The stores of body fat are analogous to a bank account; if you take out more than you put in, your savings (energy reserves in the form of fat) must decrease. It may not be easy to do, but it really is that simple. Fat may be deposited more in some part of the body than in others, and thus some parts of the body may seem to be responding to the diet earlier than other parts. This has given rise to the myth of "cellulite." Science has shown that there is no such thing; fat is fat, and a program in which weight is lost will eventually cause a visually apparent reduction in fat from all parts of the body.

There are 3,500 kilocalories in a pound of fat. In order to lose a pound of fat, therefore, you must have a negative cash (calorie) flow of 3,500 kilocalories. If you eat 500 kilocalories per day less than you use, therefore, you will lose one pound of fat in a week.

The source of the calories in the diet is irrelevant to the above calculations. More fat will *not* be lost if the calories are ingested in the form of protein compared to carbohydrates or fat, as many people erroneously believe. High-protein diets, in fact, are dangerous for a variety of reasons. The best diets are those that are balanced with all four food groups (described in a later section), but which are simply reduced in total calories. This can be achieved by substituting low calorie foods for those with higher calories in each food group. A 100 g (3 1/2 oz) portion of skinned chicken, for example, has 170 kilocalories compared to almost 400 kilocalories in a portion of the same weight of a fatty steak. One cup of skim milk has 85 kilocalories compared to 150 kilocalories in a cup of whole milk. As a further example of substitution, a 100 g portion of a boiled potato has 65 kilocalories compared to 270 kilocalories in a comparable weight of french fries.

An examination of the above comparisons reveals that the caloric difference between the substituted foods is due to their fat (lipid) content. This is because lipids have far more calories per unit weight than do carbohydrates or proteins. The body obtains 9 kilocalories of energy per gram of fat, but only 4 kilocalories per gram of either carbohydrates or protein.

A weight-reduction diet should thus be low in fat, but some fat is required to supply the fat-soluble vitamins. People who eat less than 1,200–1,500 kilocalories per day may have to supplement their diet with a multivitamin/mineral tablet that provides the recommended daily allowance (RDA) of vitamins and minerals. Nutritionists, however, advise against taking amounts of vitamins and minerals above the recommended dosages, as will be described in a later section.

1. Define the basal metabolic rate and the conditions required for its measurement.
2. How many calories must you eat to maintain your weight? How could you gain or lose weight? Explain.
3. Calculate how many calories you would have to eat per day to acheive a weight loss of ten pounds over a period of two months.

Nutritional Values of Foods

In addition to calories, food provides needed essential amino acids and fatty acids, vitamins, minerals, and fiber. A balanced diet is required to obtain the recommended daily allowances of all the nutrients. Food packages contain information regarding the U.S. RDA for each nutrient, the composition of the food by weight, and the additives that may be present.

In addition to providing the energy required to maintain a person's ideal body weight, food must also provide a variety of substances called *nutrients.* These include vitamins, a number of elements (minerals), the essential amino acids, and one essential fatty acid. Food that has caloric value but which contains little or no nutrients is said to provide *empty calories.*

Anabolic Requirements

In addition to providing the body with energy, food also supplies the raw materials for synthesis reactions—collectively termed **anabolism**—that occur constantly within the cells of the body. Anabolic reactions include those that synthesize DNA and RNA, protein, glycogen, triglycerides, and other polymers. These anabolic reactions must occur constantly to replace those molecules that are catabolyzed into their subunit monomers.

Exercise and fasting, acting through changes in hormonal secretion, cause an increase in the catabolism of stored glycogen, fat, and body protein. These molecules are also broken down at a certain rate in a person who is neither exercising nor fasting. For this reason, new monomers must be obtained from food to prevent a continual decline in the amount of protein, glycogen, and fat stored in the body.

The *turnover rate* of a particular molecule is the rate at which it is broken down and resynthesized. For example, the average daily turnover for protein is 150 g/day, but since many of the amino acids derived from body proteins can be reused in protein synthesis, a person needs only about 35 g/day of protein in the diet. It should be noted that this is an average figure and will differ in people as a result of differences in size, sex, age, genetics, and physical activity. The average daily turnover of fat is about 100 g/day, but very little is required in the diet (other than that which supplies sufficient fat-soluble vitamins and essential fatty acids), because fat can easily be produced from excess carbohydrates.

The minimal amounts of dietary protein and fat required to meet the turnover rate are only adequate if they supply sufficient amounts of the essential amino acids and fatty acids. These molecules are termed *essential* because the body cannot make them and thus is forced to obtain them in the diet for proper protein and fat synthesis. The eight **essential amino acids** are lysine, methionine, valine, leucine, isoleucine, tryptophan, phenylalanine, and threonine. Linoleic acid is the only **essential fatty acid;** the body can make all of the others given this one and sufficient carbohydrates and proteins in the diet to provide the carbon "skeletons" for the molecules.

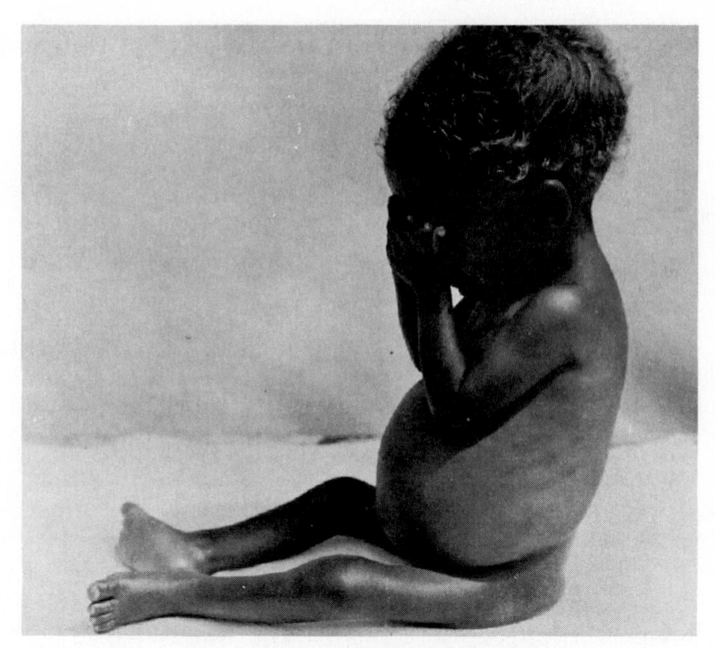

Figure 19.3. *Kwashiorkor in a two-year-old African child.*

Prolonged protein malnutrition that is associated with a swollen abdomen is called *kwashiorkor* (fig. 19.3). The swelling of the abdomen is caused by edema (abnormal accumulation of tissue fluid) due to a low blood protein concentration.

Types of Carbohydrates and Proteins

Carbohydrates are nutritionally divided into **simple carbohydrates,** which include the sugars, and **complex carbohydrates,** which include starch and fiber. The most common food sugars are the disaccharides sucrose (table sugar, composed of the monosaccharides glucose and fructose), and lactose (milk sugar, composed of glucose and galactose). Honey is a concentrated solution of glucose and fructose, and thus is exactly the same as table sugar after the sucrose has been digested. It should be recalled that starch (a polysaccharide composed of repeating glucose subunits) and disaccharides are completely digested into their monosaccharides before they are absorbed.

Fiber is a term that refers to indigestible plant material, primarily the molecule *cellulose,* which provides structural support for the plant cells. Since the human digestive tract cannot digest cellulose, it passes through into the feces. One form of fiber, which is provided by wheat bran, softens the stools. This helps to prevent constipation and helps to prevent infection and possibly cancer of the intestine. Another form of fiber, known as *soluble fiber,* is provided by oat and rice bran, as well as by fruits and many vegetables. Soluble fiber may help to lower blood cholesterol concentrations, but this is currently controversial.

The nutritional quality of protein sources in food depends on its content of the essential amino acids. *High quality proteins* are rich in all of the essential amino acids and thus are able to sustain the body, whereas *low quality proteins* are those

Table 19.4 The recommended daily allowances (RDA) for nutrients

Age (years)	Weight (kg)	Weight (lbs)	Height (cm)	Height (in)	Protein (g)	(RE) Vitamin A	(μg) Vitamin D	(mg) Vitamin E	(mg) Vitamin C	(mg) Thiamin	(mg) Riboflavin	(mg equiv.) Niacin	(mg) Vitamin B_6	(μg) Folacin	(μg) Vitamin B_{12}	(mg) Calcium	(mg) Phosphorus	(mg) Magnesium	(mg) Iron	(mg) Zinc	(μg) Iodine
Infants																					
0.0–0.5	6	13	30	24	kg × 2.2	420	10	3	35	0.3	0.4	6	0.3	30	0.5	360	240	50	10	3	40
0.5–1.0	9	20	71	28	kg × 2.0	400	10	4	35	0.5	0.6	8	0.6	45	1.5	540	360	70	15	5	50
Children																					
1-3	13	29	90	35	23	400	10	5	45	0.7	0.8	9	0.9	100	2.0	800	800	150	15	10	70
4-6	20	44	112	44	30	500	10	6	45	0.9	1.0	11	1.3	200	2.5	800	800	200	10	10	90
7–10	28	62	132	52	34	700	10	7	45	1.2	1.4	16	1.6	300	3.0	800	800	250	10	10	120
Males																					
11–14	45	99	157	62	45	1,000	10	8	50	1.4	1.6	18	1.8	400	3.0	1,200	1,200	350	18	15	150
15–18	66	145	176	69	56	1,000	10	10	60	1.4	1.7	18	2.0	400	3.0	1,200	1,200	400	18	15	150
19–22	70	154	177	70	56	1,000	7.5	10	60	1.5	1.7	19	2.2	400	3.0	800	800	350	10	15	150
23-50	70	154	178	70	56	1,000	5	10	60	1.4	1.6	18	2.2	400	3.0	800	800	350	10	15	150
51+	70	154	178	70	56	1,000	5	10	60	1.2	1.4	16	2.2	400	3.0	800	800	350	10	15	150
Females																					
11–14	46	101	157	62	46	800	10	8	50	1.1	1.3	15	1.8	400	3.0	1,200	1,200	300	18	15	150
15–18	55	120	163	64	46	800	10	8	60	1.1	1.3	14	2.0	400	3.0	1,200	1,200	300	18	15	150
19–22	55	120	163	64	44	800	7.5	8	60	1.1	1.3	14	2.0	400	3.0	800	800	300	18	15	150
23-50	55	120	163	64	44	800	5	8	60	1.0	1.2	13	2.0	400	3.0	800	800	300	18	15	150
51+	55	120	163	64	44	800	5	8	60	1.0	1.2	13	2.0	400	3.0	800	800	300	10	15	150
Pregnant					+30	+200	+5	+2	+20	+0.4	+0.3	+2	+0.6	+400	+1.0	+400	+400	+150	†	+5	+25
Lactating					+20	+400	+5	+3	+40	+0.5	+0.5	+5	+0.5	+100	+1.0	+400	+400	+150	†	+10	+50

From *Recommended Daily Allowances,* 9th ed. Copyright © 1980 National Academy of Sciences, Washington, DC.

that are deficient in one or more essential amino acids. Gelatin (erroneously believed by many people to be a good source of proteins for maintaining healthy nails) is a low quality protein, because it lacks the essential amino acid tryptophan. Eggs have the highest quality protein. If the protein value of eggs is given a score of 100, the score for milk would be 85 and that for meat and soybeans would be about 75.

Many people mistakenly believe that athletes have a much higher requirement for protein than do other people. Simple logic, and a basic understanding of metabolism, will reveal that this cannot be true. Eating more protein cannot produce more muscle tissue (if it could, why lift weights?). Muscles that are increasing in size through weight-training exercises may require slightly more protein intake per day, but muscles certainly don't grow in proportion to the high protein diet that some people ingest. Through deamination of amino acids (chapter 18), the excess protein ingested is simply used for energy or converted into fat.

RDA and U.S. RDA

Early in this century, it was discovered that experimental animals fed a diet consisting only of carbohydrates, lipids, and protein (together with adequate minerals) would die. This indicated that additional substances were required. Some of these substances could be added from milk, thus helping to prolong the animal's life. Since the first of these substances to be isolated was an amine-containing molecule, the term "vitamine" was coined in 1912. As new members of this class of compounds were discovered that did not contain an amine group, the name was shortened to *vitamin.*[3] Vitamins can be defined as organic molecules (exclusive of carbohydrates, lipids, and proteins) that an animal requires in very small amounts in the diet in order to maintain health, grow, and reproduce.

The **recommended daily allowance (RDA)** of vitamins and minerals is provided in table 19.4. These values are set sufficiently high so that variations in the requirements of different

[3]vitamin: L. *vita,* life; plus *amine,* an NH_2 group

Table 19.5 Complementary protein sources that can be used by vegetarians for mutual supplementation

Lentils	+	Wheat, rice
Legumes	+	Cereals (enriched or whole grain)
Sesame or sunflower seeds	+	Leafy vegetables, whole grains
Brewer's yeast	+	Peanuts, legumes, seeds, nuts, whole grains, corn, leafy vegetables
Leafy vegetables	+	Seeds, whole grains, yeast
Traditional Protein Combinations:		
Soybeans	+	Rice (Indochina)
Peas	+	Wheat (Fertile Crescent)
Beans	+	Corn (Central and South America)

healthy people are taken into account. This means that you could need a little less than the RDA or perhaps a little more. If your vitamin and mineral intake is much less than the RDA you may develop a deficiency disease. If you eat much more than the RDA, however, you will not become any healthier; the excess will simply be eliminated from the body (for the water-soluble vitamins) or stored in the body (for the fat-soluble vitamins). Further, vitamins in large doses can be toxic to the body.

People with particular diseases may have different requirements than those in the RDA for particular nutrients, and thus may require special diets devised by clinical nutritionists. Healthy people, however, can safely use the RDA as a guide to their nutritional requirements.

Containers of food are labeled with the **U.S. RDA.** These were derived using the higher values within each range of nutrient recommendations from the RDA tables in 1968. The RDA tables are revised every five years, but the U.S. RDA tables remain the same. This is because of practical considerations (it would be too costly to keep changing the containers when the RDA changes), and because the U.S. RDA is generally slightly higher than the RDA while remaining in a range that is safe.

People can meet their RDA for nutrients by eating a balanced diet. Those who are also interested in limiting their caloric intake can get the nutrients they require by choosing those foods that have a high ratio of nutrients to calories—that is, have a high **nutrient density.** Choosing to eat a potato instead of potato chips, for example, will provide ten times the amount of vitamin C per given number of calories.

Balanced Diet

A person cannot remain healthy by eating just one or two types of food, regardless of how nutritious these foods are by themselves. Neither an exclusive meat and potatoes diet, nor an exclusive vegetable and fruit diet, will provide all of the required nutrients in sufficient amounts. Different types of foods, which each contain different nutrients that complement each other, should be eaten daily for optimum nutrition. This is called a **balanced diet.** A balanced diet contains the four food groups:

1. *Fruits or fruit juices and vegetables.* A wide variety is recommended, since they contain different nutrients.
2. *Grains and cereal products.* This includes bread, breakfast cereals, and rice.
3. *Meat and meat products, or their alternates.* These include not only beef and other red meats, but also eggs, poultry, fish, legumes (beans and peas), and nuts.
4. *Milk and milk products.* Including primarily milk and cheese, the requirements for these are higher during the growing years than during adulthood.

Vegetarians must be particularly careful to obtain the proper amounts of nutrients. In order to obtain enough of the essential amino acids from nonanimal sources, vegetarians should eat foods at the same meal that complement each other in terms of the amino acids they contain. This is called *mutual supplementation,* and is indicated in table 19.5. Since nonanimal foods are completely lacking in vitamin B_{12}, and may be deficient in vitamin D, the addition of milk and eggs to the diet will provide all of the vitamins and essential amino acids required. If the vegetarian will not eat milk and eggs, then soy milk fortified with vitamin B_{12} should be added to the diet.

Many people frequently eat at fast-food restaurants. Although fast-food restaurants can provide a balanced diet, their food has a low nutrient density; that is, they supply an abundance of calories but not enough vitamins and minerals (table 19.6). These foods are generally rich in protein and fat, but low in fiber. Approximately 40% to 58% of the calories in fast food comes from fat (the American Heart Association recommends that fat should not contribute more than 30% of the calories in food). People who eat at fast-food restaurants could improve their nutrition by ordering salad and baked potato, with low-fat milk, in place of french fries and milkshakes.

Labeling Information

The labels on food packages contain a variety of useful information. In addition to the percent of the U.S. RDA of specific nutrients that a serving of the food contains, the ingredients are listed in descending order by weight. Thus, if sugar is the first item on the list of ingredients, sugar is the major ingredient by weight in the product.

Foods that are labeled **enriched** have nutrients added to them that were eliminated during the processing of the food and then added back. These commonly include the B vitamins. When any

Table 19.6 Nutritional value of sample fast-food meals

Meal No.	Sample Meal	Calories	Total Fat	% of Calories from Fat	Approx. Tsp of Fat†	Choles-terol	Sodium	Vitamin A	Vitamin C	Calcium	Precentage of USRDA Vitamin A	Vitamin C	Calcium
			g				*mg*	*IU*		*mg*			
1	Double burger with sauce	625	40	58		105	880	550	7	255			
	Milkshake	410	10	22		35	190	425	3	375			
	French fries (regular size)	240	15	56		15	120	15	8	10			
	Total	*1275*	*65*	*46*	14½	*155*	*1190*	*990*	*18*	*640*	10	30	80
2	Chicken nuggets (6)	310	20	58		70	700	100	2	15			
	Apple pie	280	15	48		5	400	15	10	15			
	Coffee with cream	65	5	69		20	15	55	—	40			
	Total	*655*	*40*	*55*	9	*95*	*1115*	*170*	*12*	*70*	2	20	9
3	Fish sandwich with cheese and tartar sauce	495	25	45		60	676	145	4	140			
	Soda (12 oz)	150	0	0		—	15	—	—	—			
	French fries	240	15	56		13	120	15	8	10			
	Total	*885*	*40*	*53*	9	*73*	*811*	*160*	*12*	*150*	2	20	19
4	Beef tacos (2)	390	20	46		50	565	915	—	190			
	Low-fat milk (8 oz)	105	2	17		10	125	500	2	300			
	Total	*495*	*22*	*40*	5	*60*	*690*	*1415*	*2*	*490*	18	3	61
5	Single burger	290	13	40		45	435	140	3	60			
	Tossed salad with low-calorie dressing	50	1	18		—	445	1590	40	40			
	Low-fat milk	105	2	17		10	125	500	2	300			
	Total	*445*	*16*	*32*	3½	*55*	*1005*	*2230*	*45*	*400*	28	75	50
6	Baked potato (plain)	150	Tr‡	0		0	5	—	30	20			
	Margarine (1 pat)	35	4	100		0	45	155	0	—			
	Tossed salad with low-calorie dressing	50	1	18		0	445	1590	40	40			
	Low-fat milk	105	2	17	10	125	500	2	300				
	Total	*340*	*7*	*18*	1½	*10*	*620*	*2245*	*72*	*360*	28	120	45
7	Cheese pizza (1 slice)	155	5	29		20	455	410	5	145			
	Tossed salad with low-calorie dressing	50	1	18		0	445	1590	40	40			
	Orange juice (8 oz)	110	0	0		0	0	195	95	20			
	Total	*315*	*6*	*17*	1½	*20*	*900*	*2195*	*140*	*205*	27	233	26

*Figures shown represent the average nutrient values for similar items of three or more chains. The nutrient analyses are those of Young et al. The average values as listed may deviate slightly from values published as chain-specific. However, the average values as calculated appear to be fairly representative of what all chains provide.

†One teaspoon of fat is equivalent to approximately 4.5 g of fat.

‡Tr denotes trace.

Reprinted, by permission of The New England Journal of Medicine, September 14, p. 754, 1989.

nutrient is added to the product, even if it was not present in the unprocessed food, it is said to be **fortified.** Examples include the addition of iodine to salt, vitamins A and D to milk, and vitamin C to drinks. When nutrients are added in amounts greater than 50% of their RDA, the label must indicate that the food is **supplemented** with these nutrients.

The word **imitation** can be used only if the product is nutritionally inferior to the natural food it emulates. Note that the term *synthetic,* when applied to vitamins, does not mean "imitation." A synthetic vitamin is exactly the same as the vitamin obtained from plant or animal material; it is not an imitation. The term **low sodium** is used on a label if the product contains less than 140 mg of sodium per serving. **Very low sodium** indicates that the product contains less than 35 mg of sodium. This information is useful for people with high blood pressure who must restrict their sodium intake.

The term **additives** includes ingredients that serve a variety of functions. These include nutrients, agents that thicken or emulsify the product, coloring agents, and preservatives. Most additives are products naturally found in other foods, and all have been approved for safety by the Food and Drug Administration (FDA).

Although most additives are unequivocally safe, a few are somewhat controversial. Artificial sweeteners fall into this category. Cyclamate, for example, is illegal in the United States but legal in Canada, while the reverse is true for saccharine. Although there have been claims to the contrary, it appears that the artificial sweetener known as aspartame is safe for most people. Aspartame is a dipeptide consisting of the amino acids phenylalanine and aspartic acid. Since people with PKU (chapter 11) must avoid foods with phenylalanine in their diet, it is obvious that they should not eat foods sweetened with aspartame.

Nutritional Misinformation

A multimillion dollar industry has been built around, and is dependent upon, nutritional misinformation. This industry grew because of consumer demand; people want to be able to take positive steps to improve their health. What could be easier than popping pills, or choosing certain foods over others, to achieve that goal? While most people who dispense nutritional misinformation are sincere but ignorant, some are simply charlatans who deliberately do this to make money. The rest are easily misled, because many people believe what they read without considering the source. They don't realize that printed matter can contain misstatements of facts and actual lies unless a lawsuit is filed or the government intervenes on a case-by-case basis. Only the labels on foods must by law tell the truth, because this is carefully regulated by the FDA.

How can one know what is fact and what is fiction? These general guidelines have been suggested by Dr. Victor Herbert, a leading expert on nutrition:

> To protect against nutritional cultism and quackery, you should realize that what is true about nutrition is not sensational and what is sensational isn't true. Nutrition is a science and not black magic. You should know that the basis of good nutrition is *moderation in all things,* and that moderate amounts of carbohydrates, fats, proteins, vitamins, and minerals are good for you, but that large amounts are bad for you.
>
> *Nutrition Cultism,* by Victor Herbert, M.D., J. D.

One common example of misleading information is that associated with the term *organic foods.* There is no legal definition of this term. "Organic" simply means that the molecules contain carbon. All food, by this definition, is organic. As used by the health food industry, the term "organic" often indicates that the food was grown with manure or compost, as opposed to chemical fertilizers. There are ecological benefits to be derived from this practice, but plants can't tell the difference between manure and chemical fertilizers. They use the chemical nutrients present in the manure, which are identical to those used in chemical fertilizers. If the chemical fertilizers were less adequate than the manure, the plants wouldn't grow. Tests have demonstrated that, in fact, there is no nutritional difference between a food labeled "organic" and one that is not.

Another misconception is that processed foods, and those with additives and preservatives, are nutritionally inferior to "organic" foods. While some vitamins may be partially inactivated or removed by food processing, they can be added back in foods that are fortified. Processed foods, when used together with fresh vegetables and fruits in the diet, can provide good nutrition. Additives and preservatives help to accomplish this goal by enriching or fortifying the food with vitamins and protecting against decomposition. Minerals, of course, are not damaged if the food is heated during processing.

The idea, promoted by the health food industry, that "natural vitamins" such as vitamin C from rosehips are superior to synthetic vitamins is completely incorrect. The body uses the ascorbic acid (vitamin C), not the rosehip, and one ascorbic acid molecule is the same as another regardless of the source. In fact, synthetic vitamin C is often added to the rosehip preparation because the natural vitamin C content in rosehips is low.

Many people attempt to improve their health by taking amounts of vitamins and minerals that are many times higher than the RDA. This is often called *megavitamin therapy.* There is no scientific justification for this practice. Vitamins do specific jobs in the cells, and can perform their jobs only at certain maximum speeds. When this limit is reached, there is no benefit that can be derived from the presence of excess unemployed vitamins. Contrary to what is often claimed, megavitamin therapy is very unnatural. A person would have to eat fifteen pounds of wheat germ per day, for example, to supply the amount of vitamin E that some people take in capsules. A person would have to eat ten pounds of oranges per day to get as much vitamin C as many people take in pill form. These examples demonstrate that megavitamin therapy is unnatural, unnecessary, and financially wasteful. Since abnormally high amounts of vitamins and minerals are toxic to the body, this practice can also be dangerous.

There are many other nutritional misconceptions. The idea that some foods should not be mixed together during a meal is not only wrong, it can be dangerous. A balanced diet, in fact, requires that different food groups be eaten at the same meal. One food group, contrary to misleading information, does not interfere with the assimilation of another. Honey is not more nutritious than plain table sugar; indeed, the two are nutritionally identical. A person under stress does not need to pop vitamin pills; proper nutrition under stressful conditions can be assured by eating a well-balanced diet.

The health food industry is aware of the above information, and attempts to defend itself by telling people that "science doesn't know everything." Of course it doesn't. But simple testimonials given by people who believe that they were helped is poor evidence. How can you know if these testimonials are sincere? If the testimonials are sincere, how can you know if they represent a real or a placebo effect? If the testimonials are sincere and accurate, scientific studies should support them. If a particular theory doesn't stand up to scientific scrutiny, the theory should not be promoted as fact. Those who choose to ignore this logic may imperil their health.

1. Distinguish between "high quality" and "low quality" proteins. Explain how proteins are used in the body when more amino acids are supplied than are required per day.
2. What are the food groups required for a balanced diet? To what does the term *mutual supplementation* refer?
3. Define the terms *enriched, fortified,* and *supplemented,* as they are used on food package labels.

Vitamins and Minerals

Vitamins are divided into two major categories: water-soluble and fat-soluble. The water-soluble vitamins include the B complex and vitamin C. These function primarily as coenzymes. The fat-soluble vitamins include vitamin A, which is required to prevent night blindness, vitamin D (needed to prevent

Table 19.7 Summary descriptions of the characteristics of vitamins

Vitamin	Action	Deficiency Symptoms	Sources
A	Constituent of visual pigment; strengthens epithelial membranes	Night blindness; dry skin	Yellow vegetables and fruit
B_1 (Thiamine)	Cofactor for enzymes that catalyze decarboxylation	Beriberi; neuritis	Liver, unrefined cereal grains
B_2 (Riboflavin)	Part of flavoproteins (such as FAD)	Glossitis; cheilosis	Liver, milk
B_6 (Pyridoxine)	Coenzyme for decarboxylase and transaminase enzymes	Convulsions	Liver, corn, wheat, and yeast
B_{12} (Cyanocobalamin)	Coenzyme for amino acid metabolism; needed for erythropoiesis	Pernicious anemia	Liver, meat, eggs, milk
Biotin	Needed for fatty acid synthesis	Dermatitis; enteritis	Egg yolk, liver, tomatoes
C	Needed for collagen synthesis in connective tissues	Scurvy	Citrus fruits, green leafy vegetables
D	Needed for intestinal absorption of calcium and phosphate	Rickets; osteomalacia	Fish, liver
E	Antioxidant	Muscular dystrophy	Milk, eggs, meat, leafy vegetables
Folates	Needed for reactions that transfer one carbon	Sprue; anemia	Green leafy vegetables
K	Promotes reactions needed for function of clotting factors	Hemorrhage; inability to form clot	Green leafy vegetables
Niacin	Part of NAD and NADP	Pellagra	Liver, meat, yeast
Pantothenic acid	Part of coenzyme A	Dermatitis; enteritis, adrenal insufficiency	Liver, eggs, yeast

rickets), vitamin K (required to prevent pernicious anemia), and vitamin E. Vitamin D is unique in that it functions as a hormone as well as a vitamin.

As previously described, scientists in the early part of the twentieth century discovered that the addition of milk to a diet could prevent death in animals who were fed only purified carbohydrates, lipids, and proteins. The active agents in the milk, later called vitamins, were then separated into a fat-soluble component of the milk (called "A") and a water-soluble component (called "B"). After this, it was learned that there was more than one vitamin in the component labeled "B"; the term **vitamin B complex** was thus derived.

We now know that there are nine water-soluble vitamins and four fat-soluble vitamins. The water-soluble vitamins include eight in the B complex (thiamine, riboflavin, niacin, B_6, pantothenic acid, biotin, folic acid, and B_{12}) and vitamin C. The fat-soluble vitamins include vitamins A, D, E, and K. A summary of the vitamins and their sources is provided in table 19.7.

The water-soluble vitamins generally act as *coenzymes* (molecules needed by enzymes for their activity). These vitamins therefore affect metabolism in very fundamental ways in a wide variety of organs. As a result, a deficiency in a water-soluble vitamin has multiple effects on the body. Since the B vitamins aid enzymes involved in cell respiration, they are required in order for the cell to obtain energy in the form of ATP. It should be clear, however, that excessive amounts of these vitamins cannot provide "extra energy," as some people claim. The vitamins themselves have little caloric value, and the coenzymes that are derived from these vitamins simply allow enzymes to do their work; they can't goad them into working harder.

The fat-soluble vitamins generally have more specific functions in particular organs. The functions of each vitamin will be considered in the following sections.

The B Vitamins

Thiamine (vitamin B_1) was the first vitamin in the B complex to be purified. This vitamin is needed to prevent the deficiency disease known as *beriberi*. Interestingly, the relationship between nutrition and beriberi was first noted in chickens, which develop a disease much like beriberi when fed a diet of polished rice (rice with the outer husk removed). When the formerly discarded husks were added back to the chicken food the animals recovered from the disease. Thiamine provides a coenzyme needed for cell respiration and for the synthesis of a number of molecules in the cell.

Riboflavin (vitamin B_2) contains a five-carbon sugar (ribose) and is yellow in color (*flav* is Latin for yellow). It is used to produce FAD, a coenzyme that transfers hydrogen atoms in cell respiration (chapter 18). Riboflavin is also a part of other proteins, known as flavoproteins, which are involved in cell respiration and other aspects of metabolism.

Niacin (vitamin B_3) prevents the vitamin-deficiency disease known as *pellagra*[4] (fig. 19.4). This disease, which is characterized by a red rash that turns skin dark and rough, was once quite prevalent in Europe and America. It was found that pellagra could be prevented by meat, yeast, liver, and other foods that provide the B vitamins. The active factor preventing pellagra in these foods was found to be *nicotinic acid*. The name

[4]pellagra: It. *pelle,* skin; *agra,* rough

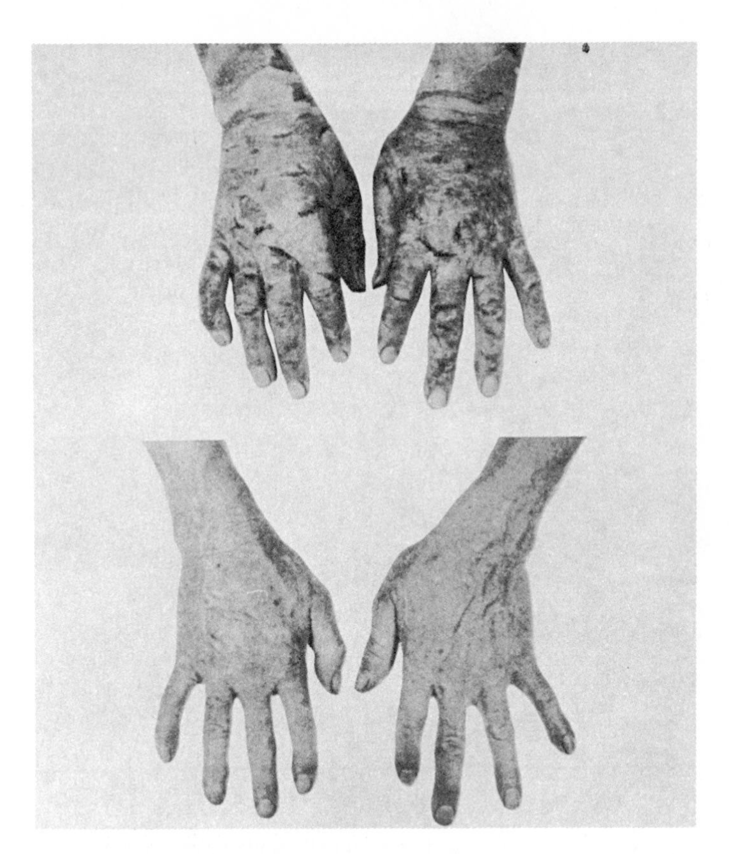

Figure 19.4. *The hands of a person with pellagra before* (a) *and after* (b) *an adequate diet with sufficient amounts of niacin.*
From Spies: "Rehabilitation Through Better Nutrition," Philadelphia, W. B. Saunders © 1974. Reprinted with permission.

is derived from the fact that this vitamin was first isolated from tobacco plants (nicotine in cigarettes has no vitamin value). Niacin is the parent molecule of NAD, which is of vital importance in the production of ATP (chapter 18).

Vitamin B_6, also known as *pyridoxine,* is an important coenzyme in the metabolism of amino acids, as well as for other reactions. The deficiency symptoms for this vitamin are similar to those for niacin and riboflavin.

Biotin is a B vitamin that is produced by bacteria that are normally present within the human intestine. As a result, people do not have to eat biotin in the food and deficiency states are very uncommon. A deficiency in biotin can be produced, however, by extended use of antibiotics which kill the intestinal bacteria. Biotin serves as a coenzyme in many reactions in which carbon dioxide units are transferred from one molecule to another.

Pantothenic acid is another B vitamin that is not associated with a deficiency state. This is because pantothenic acid is so readily available in a wide variety of foods. Indeed, the name pantothenic acid is derived from the Greek "from everywhere."

Folic acid (folacin), as the name implies, is derived from foliage: that is, from the leaves of green, leafy vegetables. A deficiency in folic acid is associated with gastrointestinal problems and with a characteristic form of anemia. Folic acid serves as a cofactor for enzymes that participate in the synthesis of a variety of molecules.

Vitamin B_{12} was the last of the B vitamins to be discovered. This is the only B vitamin that cannot be made by plants; all vitamin B_{12} is produced by microorganisms, such as bacteria. Meat and meat products are the best dietary sources of vitamin B_{12}, and strict vegetarians who do not use milk and eggs may develop a deficiency in this vitamin. In order for vitamin B_{12} to be absorbed in the intestine, a polypeptide called *intrinsic factor,* which is produced by the stomach (chapter 17), is required. A type of anemia known as *pernicious anemia* is caused by a deficiency in vitamin B_{12}.

Vitamin C

Scurvy is the disease caused by a deficiency in vitamin C. Symptoms of scurvy include degeneration of gum tissue, loss of teeth, pain, and degeneration leading to death. This disease was widespread throughout much of human history. Sailors on long journeys were particularly susceptible; it was not uncommon for one-half to two-thirds of a ship's crew to be lost from scurvy.

In 1753, a Scottish ship's surgeon named James Lind published the results of his study demonstrating that eating lemons and oranges could prevent scurvy. The great explorer Captain James Cook therefore took citrus fruits with him on his famous voyages of exploration (1772–1775). Captain Cook also provided his crew with sauerkraut, which was found to protect against scurvy. The English navy was afterwards required to take limes along on its voyages; this is the origin of the nickname "limey" to describe British seamen. Scurvy was also common on land in Europe until the potato (which contains vitamin C) was imported from the New World.

Vitamin C was isolated and identified as *ascorbic acid* by Albert Szent-Györgyi[5] in 1928. The mechanisms by which vitamin C exerts its protective effects in the body are incompletely understood. One of its known effects is to promote the oxidation-reduction reaction whereby the amino acid proline is converted to hydroxyproline. Since hydroxyproline is of major importance in the structure of collagen proteins, which provide structural support for most connective tissues, a deficiency of vitamin C would weaken connective tissues. This weakening could then lead to loss of teeth and other symptoms of scurvy.

The weakening of connective tissues during vitamin C deficiency explains the observation that people with scurvy are susceptible to infections. But does this mean that people who are not deficient in vitamin C, and thus do not have scurvy, can be made more resistant to infections by taking megadoses of vitamin C? One would not predict this to be the case, based on our knowledge of how vitamins work. Nevertheless, there is widespread belief that the taking of vitamin C in doses many times higher than the RDA can protect against the common cold and other infections, and perhaps even against cancer.

These beliefs have been promulgated by Linus Pauling,[6] one of the greatest chemists of the century. As a result of his enor-

[5]Szent-Györgyi, Albert: Hungarian biochemist in America, 1893–
[6]Linus Pauling: American chemist, 1901–

mous prestige, there have been numerous and exhaustive scientific tests of these ideas. The conclusion reached by these investigations is that vitamin C in megadoses has no protective effect, or perhaps only an insignificant effect, against the common cold. The alleged protective effect against cancer has also not been supported by the scientific evidence (chapter 14). Further, since vitamin C is water-soluble, greater amounts than that which is used by the body (estimated at about 30 mg/day) are simply eliminated in the urine. Very high doses over a long period of time may have toxic effects, particularly on the kidneys.

Fat-Soluble Vitamins

Vitamins that are fat-soluble, unlike the water-soluble vitamins, can be stored in the body. As a result, deficiency states take longer to develop than with water-soluble vitamins. There is a negative side to this, however; fat-soluble vitamins can build to toxic levels more easily when taken in excessive amounts than can water-soluble vitamins.

Vitamin A

Vitamin A deficiency is very common in the world, particularly in those areas where children are severely malnourished, with low amounts of body fat to store this vitamin. Vitamin A deficiency causes night blindness, retarded growth, poorly developed tooth enamel, and epithelial damage in the lungs and digestive tract. Cod liver oil, chicken liver, egg yolk, and butterfat are rich in the vitamin A molecule known as *retinol.* Green and yellow vegetables contain the molecule *carotene,* which can be converted into vitamin A (retinol) by animals. Margarine is fortified with carotene and vitamin A to provide the same nutritional value as butter and to give it a similar color.

The ability to cure night blindness by eating liver has been known since antiquity. The reasons for this are now well understood. Liver contains vitamin A, which is a parent molecule for the molecule *retinene.* Retinene, together with protein, forms the visual pigment known as *rhodopsin,* which is found in the photoreceptor cells of the eyes (chapter 20). A deficiency in vitamin A results in a deficiency in rhodopsin, so that images are seen less clearly when the light intensity is low.

Vitamin D

Vitamin D can be obtained from foods and can also be produced by the body. Animal foods rich in the type of vitamin D known as *cholecalciferol* include cod liver oil, liver, egg yolk, and butterfat. Plants make a different form of this vitamin (*ergocalciferol*), but not in large amounts. Milk is commonly fortified with vitamin D.

The body produces cholecalciferol from a derivative (7-dehydrocholesterol) in the skin in response to sunlight. The cholecalciferol, also known as *vitamin* D_3, is secreted by the skin into the blood. The circulation next carries this molecule to the liver, which adds a hydroxyl group (OH) to it to produce 25-hydroxyvitamin D_3. After being secreted by the liver, this molecule is carried to the kidneys, which add another OH group to produce 1,25-dihdyroxyvitamin D_3 (fig. 19.5). This latter molecule is the active form of the vitamin, which, when produced in the body, can properly also be considered a hormone.

The active form of vitamin D stimulates the absorption of calcium from the intestine. Adequate vitamin D during the growing years is therefore essential for the calcification, and thus hardening, of bones. When the intake of calcium is inadequate, a lowering of blood calcium stimulates the secretion of *parathyroid hormone* from the parathyroid glands. This hormone promotes the dissolution of calcium phosphate crystals in bone, and thus helps to raise the blood calcium level back to normal. This is essential for health, but results in a softening of the bones. Inadequate amounts of vitamin D, acting through this mechanism, thus causes demineralization of bones. This disease in children is called **rickets,** and produces a characteristic bow-legged appearance (fig. 19.6). When bone softening occurs in adults, it is called *osteomalacia.*

Vitamin E

Vitamin E was discovered in 1922, in experiments that demonstrated that this vitamin was required for fertility in male rats. When the vitamin was purified in 1936, the molecule was therefore named *tocopherol* (from the Greek "to bear offspring"). This was an unfortunate association, because it has been learned that vitamin E does not specifically improve reproductive ability in humans. Deficiencies in vitamin E are, in fact, rare; when they occur, they are associated with anemia and kidney dysfunction. It has sometimes been stated that eating oysters improves male sexual ability because oysters are rich in vitamin E. This is incorrect on both counts. Foods rich in vitamin E include whole grains, beans, fruits, vegetables, and liver.

Vitamin E functions as an antioxidant, preventing damage to molecules by the oxygen released during certain chemical reactions inside of cells. Acting through this mechanism, vitamin E helps to protect vitamin A from oxidation, and thus to prevent the deficiency states associated with inadequate vitamin A.

Vitamin K

Vitamin K is required for the production of certain blood-clotting factors in the liver (the "K" is derived from the Danish word *koagulation*). Most important among these factors is *prothrombin,* an enzyme needed to convert the soluble protein known as *fibrinogen* into insoluble protein threads called *fibrin.* It is the fibrin threads that provide structural support to a blood clot. Clinically, drugs that interfere with the action of vitamin K (the coumarins) are often used to retard the formation of blood clots. Vitamin K can be obtained from green, leafy vegetables, soybeans, egg yolk, and liver.

Elements

In addition to the organic molecules previously discussed, the body needs a variety of elements (often called minerals) for almost all aspects of body function. These elements are required

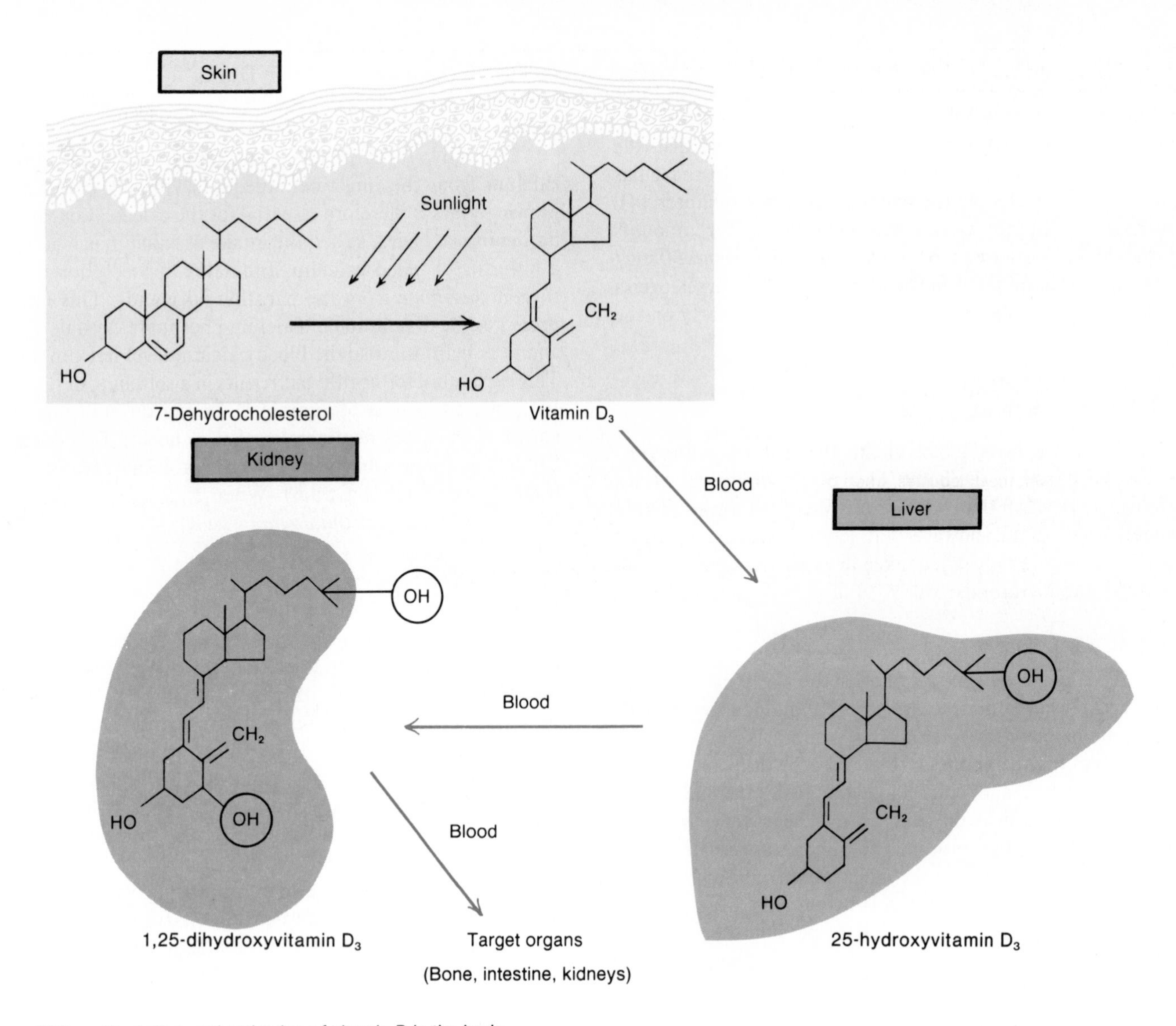

Figure 19.5. *Formation and activation of vitamin D in the body.*

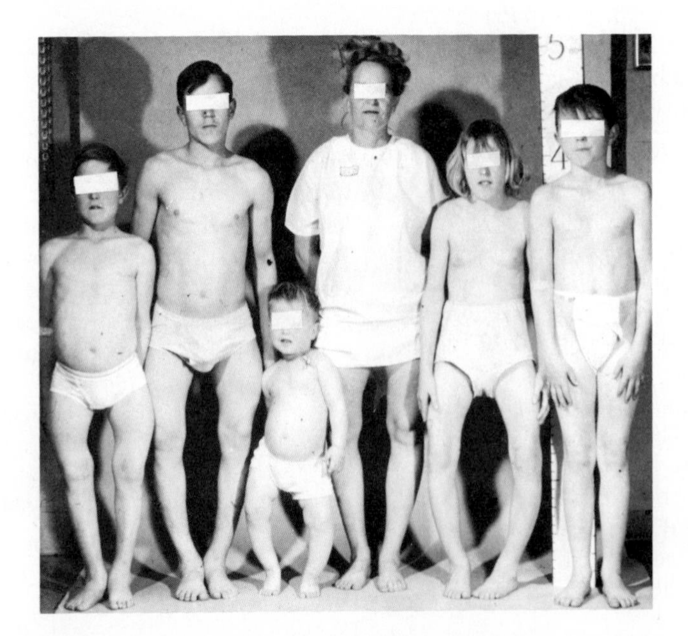

Figure 19.6. *Mother and children with ricketts. The bow legs result from bone softening due to a vitamin D deficiency.*

Table 19.8 Summary of the elements required for human nutrition

Element	Rich Sources	Dietary Allowance for Adults	Function in the Body	Elimination
Calcium	Milk, cheese, some green vegetables	0.8 g daily, 1.2 g in pregnancy and lactation	Bone and tooth formation; coagulation of blood. Regulates muscle contractibility including heartbeat; activates enzymes.	Urine and feces; some in sweat
Phosphorus	Milk, poultry, fish, meats, cheese, nuts, cereals, legumes	0.8 g daily, 1.2 g in pregnancy and lactation	Bone and tooth formation; forms high-energy phosphate compounds for muscular and tissue cell activity; constituent of DNA, RNA, phospholipids, and buffer systems.	Urine and feces
Magnesium	Nuts, cereals, legumes, green vegetables, milk, meat	Women, 300 mg; men, 350 mg; 450 in pregnancy and lactation	Constituent of bone; enzyme activator for energy producing systems; regulates muscles and nerves.	Feces and urine; some in sweat
Iron	Liver, meat, legumes, whole or enriched grains, potatoes, egg yolk, green vegetables, dried fruits	Women, 18 mg; men, 10 mg; 30–60 mg supplement during pregnancy and for 2–3 mo after parturition recommended	Constituent of hemoglobin, myoglobin, and cellular enzymes.	Feces, small amounts in urine and sweat; menstruation blood loss
Iodine	Seafoods, water and plant life in nongoitrous regions; sodium iodide in iodized salt	150 mcg daily; 175 mcg in pregnancy, 200 mcg in lactation	Necessary for formation of thyroid hormones.	Urine
Zinc	Seafoods, meat, liver, eggs, milk	15 mg daily; 20 mg in pregnancy; 25 mg during lactation	Constituent of carbonic anhydrase and other metalloenzymes; growth; sexual maturation; wound healing; taste acuity.	Urine and feces; some sweat and dermal losses
Copper	Liver, nuts, legumes	Estimated safe and adequate intake 2.0–3.0 mg	Aids in utilization of iron in hemoglobin synthesis; constituent of many enzymes; electron transfer; connective tissue metabolism; phospholipid synthesis.	Chiefly in feces—bile
Manganese	Nuts, whole grains, legumes, tea, cloves	Estimated safe and adequate intake 2.5–5.0 mg	Synthesis of mucopolysaccharides, glucose utilization; constituent or activator of several enzymes.	Chiefly in feces—bile
Fluorine	Fluoridated water	Estimated safe and adequate intake 1.5–4.0 mg	Resistance to dental caries.	Urine, feces, and sweat
Chromium	Brewer's yeast, some animal products, whole grains	Estimated safe and adequate intake 0.05–0.2 mg	Glucose metabolism; cofactor for insulin.	Feces and urine
Selenium	Seafoods, kidney, liver	Estimated safe and adequate intake 0.05–0.2 mg	Cellular antioxidant as constituent of enzyme glutathione peroxidase.	Urine and feces; some in breath
Molybdenum	Whole grains and legumes	Estimated safe and adequate intake 0.15–0.5 mg	Constituent of several enzymes (purine and sulfur metabolism).	Urine and feces

From L. Anderson, et al., *Nutrition in Health and Disease,* 17th ed. Copyright © 1982 J. B. Lippincott Company, Philadelphia, PA.

in amounts that range from 50 micrograms (mcg; a millionth of a gram) to 18 milligrams (mg; a thousandth of a gram) per day. Those elements that are required in especially small amounts are called *trace elements*; these include iron, zinc, manganese, fluorine, copper, molybdenum, chromium, and selenium. Elements needed in larger amounts include sodium, potassium, chlorine, calcium, magnesium, and phosphorus. The functions of these elements are so diverse that only their major actions are summarized in table 19.8.

1. Explain why a diet of polished rice can produce beriberi, while a diet of whole rice does not produce this disease.
2. Explain why a deficiency in biotin is rare, and why it might be produced in a person who is taking antibiotics over a long period of time.
3. Trace the origin of the nickname ''limey'' for British seamen.
4. Explain how eating carrots can be good for eyesight.
5. Describe how vitamin D helps to prevent rickets, and explain why this vitamin is also considered to be a hormone.

Summary

Energy Value of Food p. 422

I. The metabolic rate has basal and activity components.
   A. The basal metabolic rate (BMR) is the minimum rate of metabolism in a completely resting state.
   B. Physical activity greatly increases the metabolic rate, resulting in a higher rate of energy consumption.

II. A person's ideal body weight can be estimated from tables.
   A. Obesity is indicated by a body weight that is greater than 15%–25% higher than the ideal weight.
   B. Fat will be lost from the body when the caloric intake over time is less than the caloric expenditure.

Nutritional Values of Foods p. 426

I. In order to maintain tissue structure, eight essential amino acids and one essential fatty acid must be obtained in the diet.

II. Simple carbohydrates are monosaccharide and disaccharide sugars; complex carbohydrates include starch and fiber.
   A. Incomplete proteins are those that are poor in one or more essential amino acid.
   B. Vegetarians must eat a combination of foods in which the deficiencies in one food are complemented by another.

III. The recommended daily allowance (RDA) for nutrients indicates the amounts of vitamins and minerals required for the health of most people.

IV. Foods that are said to be enriched have nutrients added to them that may have been removed during processing.

Vitamins and Minerals p. 430

I. There are eight vitamins within the B complex.
   A. Thiamine, riboflavin, and niacin are particularly important for cellular respiration.
   B. Thiamine prevents beriberi, niacin prevents pellagra, and vitamin B_{12} prevents pernicious anemia.

II. Vitamin C is required for the proper formation of collagen in connective tissue; a deficiency in this vitamin causes scurvy.

III. The fat-soluble vitamins include A, D, E, and K.
   A. Vitamin A is needed for the formation of the visual pigment, and thus to prevent night blindness.
   B. Vitamin D is obtained in the diet and produced in the body; it is needed to prevent rickets.
   C. Vitamin E is an antioxidant, and vitamin K is required to prevent pernicious anemia.

Review Activities

Objective Questions

1. If a person is gaining weight, which of the following statements must be *true?*
   (a) He is eating a diet high in carbohydrates.
   (b) He is eating a diet high in fat.
   (c) He is not eating enough proteins.
   (d) His caloric intake is higher than his caloric expenditure.
   (e) All of these.
2. Suppose a person's caloric intake is 100 calories less than his caloric expenditure per day. How many days will it take to lose five pounds of fat?
   (a) 10 days
   (b) 20 days
   (c) 100 days
   (d) 175 days
   (e) 365 days
3. Which of the following statements is *true?*
   (a) Mutual supplementation is needed when eating low quality proteins.
   (b) Gelatin is a high quality protein.
   (c) Athletes require a far higher protein intake than others.
   (d) Fiber is digested into glucose subunits.
   (e) All of these.
4. Which of the following terms are used to describe a food product in which nutrients are added in amounts greater than 50% of their RDA?
   (a) enriched
   (b) supplemented
   (c) fortified
   (d) imitation
5. Foods that are not labeled "organic"
   (a) are inorganic
   (b) are less nutritious
   (c) are more dangerous
   (d) all of these
   (e) none of these

Match the vitamin to its deficiency disease:

6. Thiamine	(a) vitamin B_{12}
7. Niacin	(b) vitamin C
8. Pernicious anemia	(c) vitamin D
	(d) beriberi
9. Scurvy	(e) pellagra
10. Rickets	

11. Which of the following vitamins is produced by bacteria within the human intestine?
   (a) vitamin C
   (b) pantothenic acid
   (c) biotin
   (d) riboflavin
12. Which of the following vitamins is produced in the human skin in response to sunlight?
   (a) vitamin D
   (b) vitamin A
   (c) vitamin K
   (d) folic acid
   (e) niacin

Essay Questions

1. Explain the benefits of a balanced diet, and compare such a diet with examples of diets that may not be balanced.
2. Suppose a bodybuilder regularly eats half a dozen eggs, a high-protein supplement, and a large steak per day. Will this diet help him build muscles faster than if he ate just his RDA of protein? Explain.
3. Compare the value of "natural vitamins" versus synthetic vitamins, and of "organic" food versus the same items that are not labeled in this way.
4. Can taking B vitamins in amounts above their RDA provide more energy for the body than the RDA amounts? Explain.
5. Evaluate this statement: "Table sugar is a poison; better sweeten your foods with honey instead."

20

Human Perception of the Environment: The Sensory System

Outline

Objectives

By studying this chapter, you should be able to

1. explain the meaning and significance of sensory adaptation
2. describe the law of specific nerve energies
3. describe the structure of taste buds, and enumerate the different modalities of taste
4. describe the structure of olfactory receptors, and explain how this sense is transmitted to the brain
5. identify the structures that comprise the outer and middle ears
6. trace the course of events by which sound waves in air cause movements of the oval window of the cochlea
7. identify the structures that comprise the membranous labyrinth
8. explain how vibrations of the oval window produce vibrations of the basilar membrane, and how this is transduced into nerve impulses
9. explain how pitch discrimination is accomplished
10. describe the structure of the eye, and trace the path of light through the eye
11. define the term *accommodation,* and explain how it is accomplished
12. explain the basis for visual defects such as myopia, astigmatism, and others
13. list the layers of the retina, and describe which way light passes through these layers
14. explain how light activates rods and cones
15. distinguish between the functions of rods and cones

Keys to Pronunciation

chemoreceptor: *ke″mo-re-sep′tor*
eustachian: *u-sta′ke-an*
labyrinth: *lab′i-rinth*
malleus: *mal′e-us*
meatus: *mea′tus*
myopia: *mi-o′pe-ah*
olfaction: *ol-fak′shun*
presbycusis: *pres″be-ku′sis*
presbyopia: *pres″be-o′pe-ah*
proprioceptor: *pro″pre-o-sep′tor*
retina: *ret′i-nah*
retinene: *ret′i-nen*
sclera: *skle′rah*
tympanic: *tim-pan′ik*
tinnitus: *ti-ni′tus*
vitreous: *vit′re-us*

Photo: This child is enjoying the sensory experiences provided by her environment.

Introduction to the Sensory System

Sensory receptors stimulate the production of nerve impulses when they are stimulated by environmental energy. Each type of sensory receptor is most sensitive to a particular type of stimulus. The perceived nature of the stimulus is determined by the regions of the brain that receive the impulses from the receptor.

Our perceptions of the world—its textures, colors, and sounds; its warmth, smells, and tastes—are created by the brain from nerve impulses delivered to it from sensory receptors. These receptors **transduce** (change) different forms of energy in the "real world" into the energy of nerve impulses, which are conducted into the central nervous system by sensory neurons. Different sensory *modalities*—or qualities of sensation, such as sound, light, pressure, and so on—result from differences in neural pathways and synaptic connections. The brain thus interprets impulses in the auditory nerve as sound and in the optic nerve as sight, even though the impulses themselves are identical in the two nerves.

Receptors can be grouped according to the type of sensory information they deliver to the brain. *Proprioceptors*[1] include the receptors in the skeletal muscles, tendons, and joints. These provide a sense of body position and allow fine control of skeletal movements. *Cutaneous receptors* include (1) touch and pressure receptors; (2) warmth and cold receptors; and (3) pain receptors. The receptors that mediate sight, hearing, and equilibrium are grouped together as the *special senses*.

Sensory Adaptation

Some receptors respond with a burst of activity when a stimulus is first applied, but then quickly decrease their firing rate—adapt to the stimulus—when the stimulus is maintained. Receptors with this response pattern are called *phasic receptors*. Receptors that produce a relatively constant rate of firing as long as the stimulus is maintained are known as *tonic receptors* (fig. 20.1).

Phasic receptors alert us to changes in sensory stimuli and are in part responsible for the fact that we can cease paying attention to constant stimuli. This ability is called **sensory adaptation.** Odor, touch, and temperature, for example, adapt rapidly; bathwater feels hotter when we first enter it. Sensations of pain, in contrast, adapt little if at all.

Law of Specific Nerve Energies

Stimulation of a sensory nerve fiber produces only one sensation—touch, cold, pain, and so on. According to the **law of specific nerve energies,** the sensation characteristic of each sensory neuron is that produced by its normal, or *adequate stimulus*

(a)

Stimulus applied

Stimulus withdrawn

Tonic receptor — slow adapting

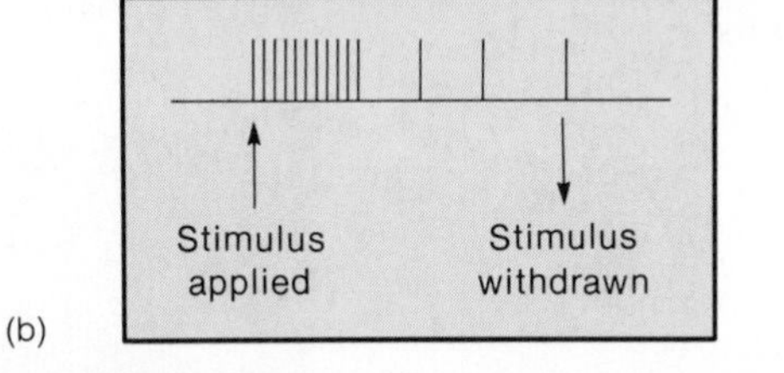

Figure 20.1. *Tonic receptors (a) continue to fire at a relatively constant rate as long as the stimulus is maintained. These produce slowly adapting sensations. Phasic receptors (b) respond with a burst of action potentials when the stimulus is first applied, but then quickly reduce their rate of firing while the stimulus is maintained. This produces rapidly adapting sensations.*

(table 20.1). The adequate stimulus for the photoreceptors of the eye, for example, is light. If these receptors are stimulated by some other means—such as by pressure produced by a punch to the eye—a flash of light (the adequate stimulus) may be perceived.

Paradoxical cold provides another example of the law of specific nerve energies. First, a receptor for cold is located by touching the tip of a cold metal rod to the skin. Sensation then gradually disappears as the rod warms to body temperature. Applying the tip of a warm rod to the same spot, however, causes the sensation of cold to reappear. This paradoxical cold is produced because the heat slightly damages receptor endings, and by this means produces an "injury current" that stimulates the receptor.

Regardless of how a sensory neuron is stimulated, therefore, only one sensory modality will be perceived. This specificity is due to the synaptic pathways within the brain that are activated by the sensory neuron. The ability of receptors to function as sensory filters and be stimulated by only one type of stimulus (the adequate stimulus) allows the brain to perceive the stimulus accurately under normal conditions.

1. "Our perceptions are products of our brains; they are only indirectly related to physical reality." Explain this statement, using examples of vision and the perception of cold.
2. Define the law of specific nerve energies and the adequate stimulus, and explain their significance.
3. Compare sensory adaptation in olfactory and pain receptors.

[1]proprioceptor: L. *proprius,* one's own; *ceptus,* taken

Table 20.1 Classification of receptors based on their normal (or "adequate") stimulus			
Receptor	**Normal Stimulus**	**Mechanisms**	**Examples**
Mechanoreceptors	Mechanical force	Deforms cell membrane of sensory dendrites; or deforms hair cells that activate sensory nerve endings	Cutaneous touch and pressure receptors; vestibular apparatus and cochlea
Pain receptors	Tissue damage	Damaged tissues release chemicals that excite sensory endings	Cutaneous pain receptors
Chemoreceptors	Dissolved chemicals	Chemical interaction affects ionic permeability of sensory cells	Smell and taste (exteroreceptors); osmoreceptors and carotid body chemoreceptors (interoreceptors)
Photoreceptors	Light	Photochemical reaction affects ionic permeability of receptor cell	Rods and cones in retina of eyes

Taste and Olfaction

The receptors for taste and olfaction respond to molecules that are dissolved in fluid, and are thus classified as chemoreceptors. Although there are only four basic modalities of taste, the combination of these, and of the taste sensations with those of olfaction, provide a wide variety of different sensory experiences.

Chemoreceptors that respond to chemical changes in the internal environment are called **interoreceptors**; those that respond to chemical changes in the external environment are **exteroreceptors.** Included in the latter category are *taste receptors,* which respond to chemicals dissolved in food or drink, and *olfactory receptors,* which respond to gaseous molecules in the air. This distinction is somewhat arbitrary, however, because odorant molecules in air must first dissolve in fluid within the olfactory mucosa before the sense of smell can be stimulated. Also, the sense of olfaction has a great effect on the sense of taste, as can easily be verified by eating an onion with the nostrils pinched together.

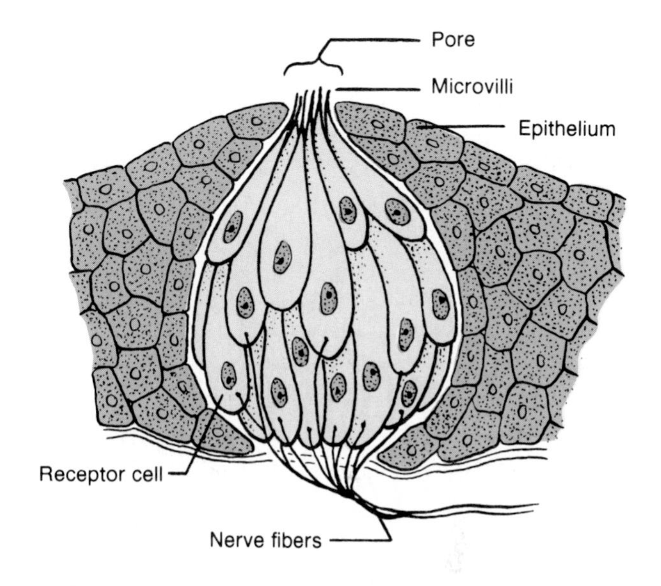

Figure 20.2. *A taste bud.*

Taste

Taste receptors are specialized epithelial cells that are grouped together into barrel-shaped arrangements called *taste buds,* located in the epithelium of the tongue (fig. 20.2). The cells of the taste buds have microvilli at their apical (top) surface, which is exposed to the external environment through a pore in the surface of the taste bud.

Molecules dissolved in saliva at the surface of the tongue interact with receptor molecules in the microvilli of the taste buds. This interaction stimulates the release of a neurotransmitter from the receptor cells, which in turn stimulates sensory nerve endings. Nerve impulses are then transmitted along the ninth and tenth cranial nerves to the brain for interpretation.

There are only four basic modalities of taste, which are sensed most acutely in particular regions of the tongue. These are *sweet* (tip of the tongue), *sour* (sides of the tongue), *bitter* (back of the tongue), and *salty* (over most of the tongue). This distribution is illustrated in figure 20.3.

Sour taste is produced by hydrogen ions (H^+); all acids therefore taste sour. Most organic molecules, particularly sugars, taste sweet to varying degrees. Only table salt (NaCl) has a pure salty taste—other salts, such as KCl (commonly used in place of NaCl by people with hypertension) taste salty but have bitter overtones. Bitter taste is evoked by quinine and seemingly unrelated molecules.

Olfaction

The olfactory receptors are the dendrites of the *olfactory (first cranial) nerve,* in association with epithelial supporting cells. Unlike other sensory modalities, which are relayed to the cerebrum from the thalamus (chapter 6), the sense of olfaction is transmitted directly to the olfactory bulb of the cerebral cortex

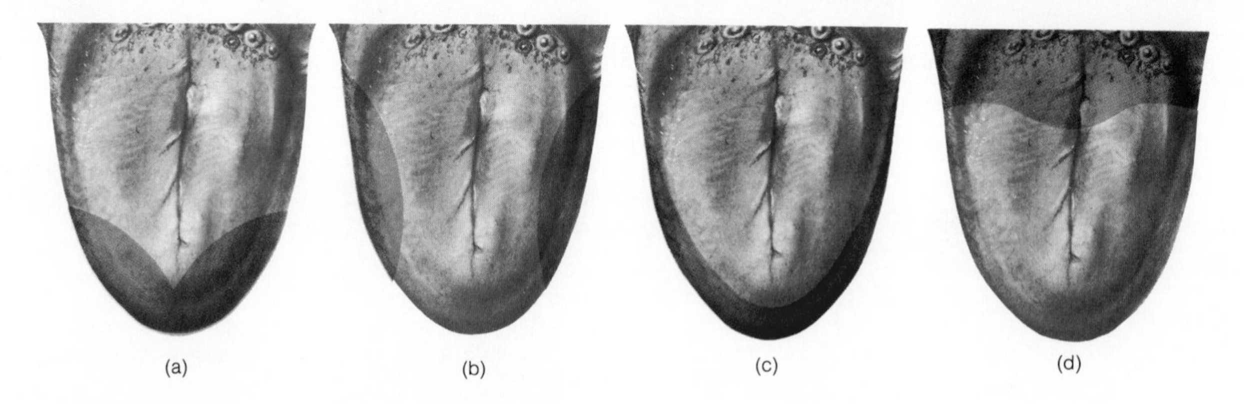

Figure 20.3. *Patterns of taste receptor distribution on the dorsum of the tongue. (a) Sweet receptors; (b) sour receptors; (c) salt receptors; (d) bitter receptors.*

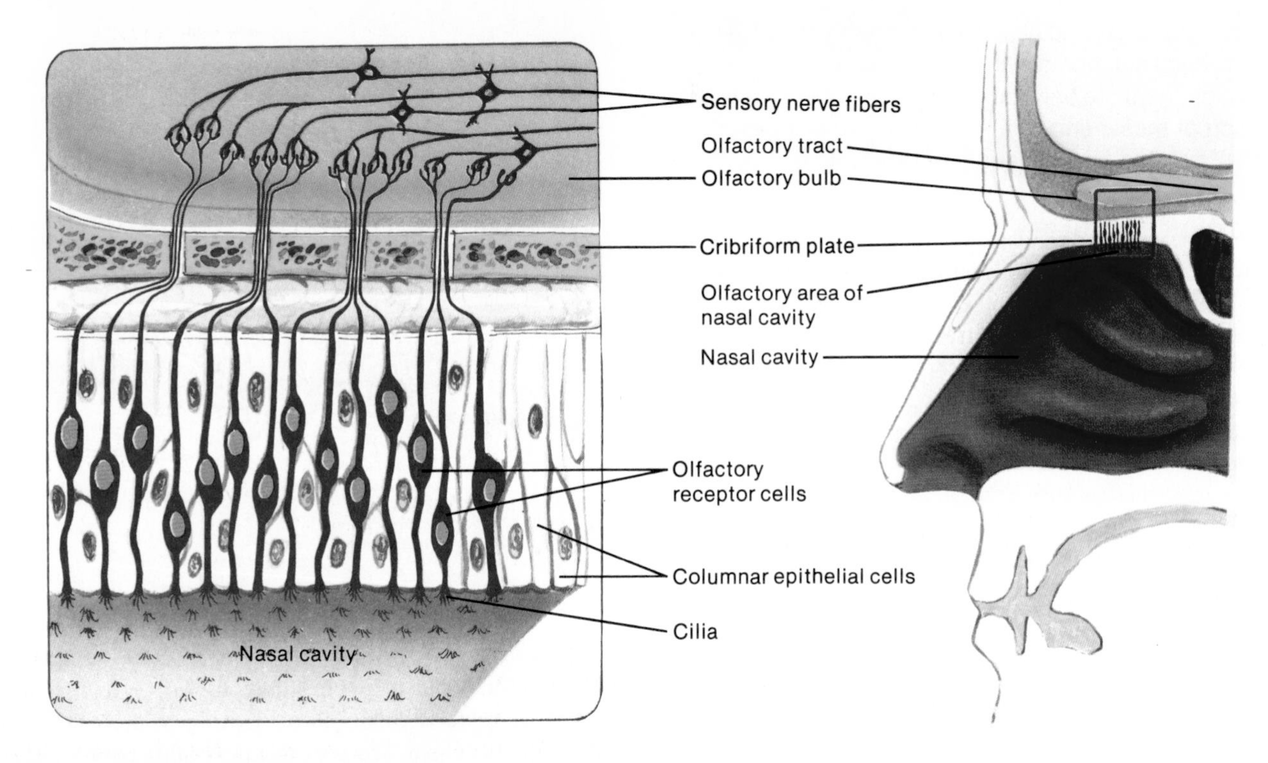

Figure 20.4. *The olfactory epithelium contains receptor neurons that synapse with neurons in the olfactory bulb of the brain.*

(fig. 20.4). This area of the brain is part of the limbic system, which was discussed in chapter 6 as having an important role in the generation of emotions and in memory. Perhaps this explains why the smell of a particular odor, more powerfully than other sensations, can evoke emotionally charged memories.

Unlike taste, which is divisible into only four modalities, many thousands of different odors can be distinguished by people who are trained in this capacity (as in the perfume and wine industries). The molecular basis of olfaction is not understood; although various theories have attempted to explain families of odors on the basis of similarities in molecular shape and/or charges, such attempts have been only partially successful. The extreme sensitivity of olfaction is possibly as amazing as its diversity—at maximum sensitivity, only one odorant molecule is needed to excite an olfactory receptor.

1. Describe the distribution of taste receptors in the tongue.
2. To what types of stimuli do taste buds and olfactory receptors respond?

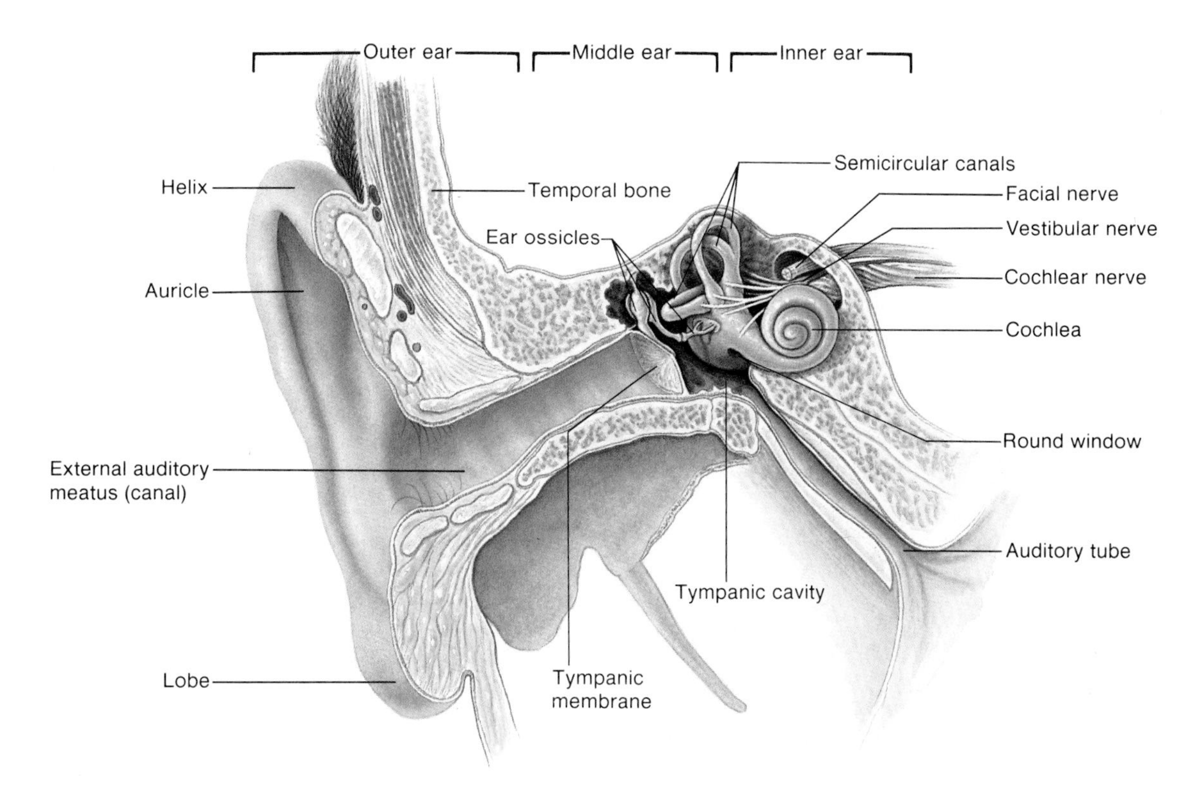

Figure 20.5. *The ear. Note the external, middle, and inner regions of the ear.*

The Ears and Hearing

Sound energy causes vibrations of the tympanic membrane. This, in turn, produces movements of the middle ear ossicles, which press against a membrane called the oval window in the cochlea. The movements of the oval window produce pressure waves within the fluid of the cochlea, which cause movements of a membrane called the basilar membrane. Sensory hair cells are located on the basilar membrane, and the movements of this membrane in response to sound cause the hair cell processes to be bent. This produces nerve impulses which are conducted to the brain where they are interpreted as sound.

Sound waves travel in all directions from their source, like ripples in a pond after a stone is dropped. These waves are characterized by their frequency and their intensity. The **frequency,** or distances between crests of the sound waves, is measured in *hertz (Hz),* which is the modern designation for *cycles per second (cps).* The *pitch* of a sound is directly related to its frequency—the greater the frequency of a sound, the higher its pitch.

The **intensity,** or loudness of a sound, is directly related to the amplitude of the sound waves. This is measured in units known as *decibels (db).* A sound that is barely audible—at the threshold of hearing—has an intensity of zero decibels. Every ten decibels indicates a tenfold increase in sound intensity; a sound is ten times louder than threshold at 10 db, one hundred times louder at 20 db, a million times louder at 60 db, and ten billion times louder at 100 db.

The Outer Ear

Sound waves are funneled by the *pinna,* or *auricle* (flap), into the *external auditory meatus* (fig. 20.5). These two structures comprise the *outer ear.* The external auditory meatus channels the sound waves (while increasing their intensity) to the eardrum, or *tympanic membrane.* Sound waves in the external auditory meatus produce extremely small vibrations of the tympanic membrane; movements of the eardrum during speech (with an average sound intensity of 60 db) are estimated to be about equal to the diameter of a molecule of hydrogen!

The Middle Ear

The middle ear is the cavity between the tympanic membrane on the outer side and the cochlea on the inner side (fig. 20.6). Within this cavity are three **middle ear ossicles** (little bones)—the *malleus* (hammer), *incus* (anvil), and *stapes* (stirrup). The malleus is attached to the tympanic membrane, so that vibrations of this membrane are transmitted, via the malleus and

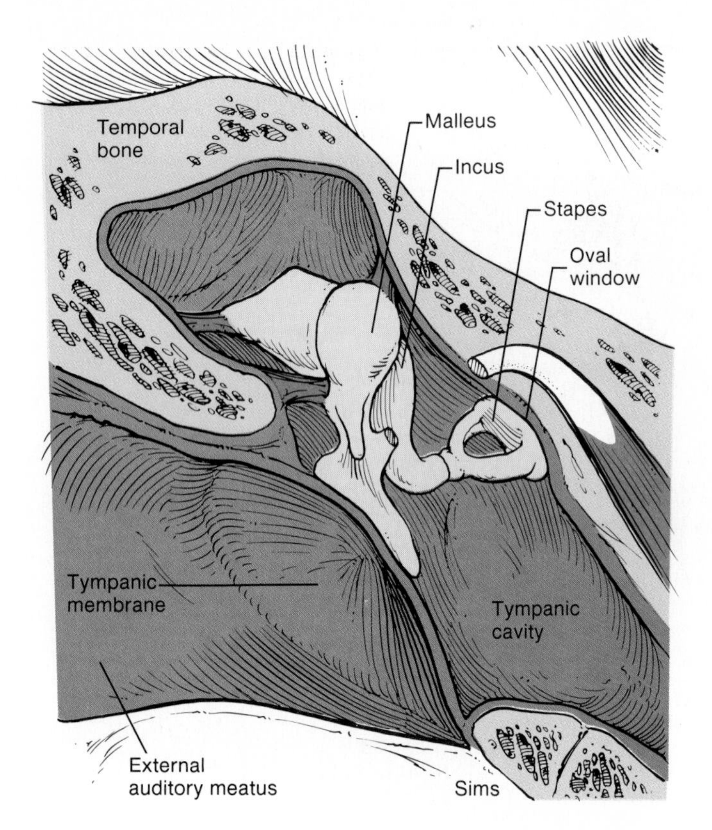

Figure 20.6. *The position of the ear ossicles within the middle ear.*

incus, to the stapes. The stapes, in turn, is attached to a membrane known as the oval window (described later), which thus vibrates in response to vibrations of the tympanic membrane.

> Damage to the tympanic membrane or middle ear ossicles produces **conduction deafness.** This can result from a variety of causes, including *otitis media* and *otosclerosis.* In otitis media, inflammation produces excessive fluid accumulation within the middle ear, which can in turn result in the excessive growth of epithelial tissue and damage to the eardrum. This can occur following allergic reactions or respiratory disease. In otosclerosis, bone is resorbed and replaced by "sclerotic bone" that grows over the oval window and immobilizes the footplate of the stapes (fig. 20.7). In conduction deafness these pathological changes hinder the transmission of sound waves from the air to the cochlea of the inner ear.

The Inner Ear

The sensory structures of the inner ear are located within the **membranous labyrinth,**[2] which is a fluid-filled tube (fig. 20.8) within a bony cavity. The membranous labyrinth of the inner ear is divided into two functional parts: (1) the **vestibular apparatus**; and (2) the **cochlea.**[3] The vestibular apparatus includes the three *semicircular canals,* and is required for the sense of equilibrium. The cochlea is the organ of hearing, and contains a part of the membranous labyrinth known as the **cochlear duct.**

> *Nystagmus*—an uncontrolled oscillation of the eyes—is one of the symptoms of an inner-ear disease called **Meniere's disease.**[4] The early symptom of this disease is often "ringing in the ears," or *tinnitus.* Since the fluid of the cochlea and that of the vestibular apparatus are continuous through a tiny canal, the vestibular symptoms of *vertigo* (loss of equilibrium) and nystagmus often accompany hearing problems in this disease.

The snail-shaped cochlea is divided into three chambers. The middle chamber is the fluid-filled cochlear duct. The upper chamber is the *scala vestibuli,* and the lower chamber is the *scala tympani* (fig. 20.9). When sound causes vibrations of the tympanic membrane (eardrum) and middle ear ossicles, the stapes presses against a membrane called the *oval window,* which is located at the upper scala vestibuli of the cochlea. This produces waves in the coclear duct, which causes its membranes to be pressed down into the lower scala tympani (fig. 20.10). The lower membrane of the cochlear duct is known as the *basilar membrane,* and is particularly significant in the perception of sound.

Located on top of the basilar membrane are numerous cells with "hairs" (actually, cilia) projecting upwards into the duct. These hairs are embedded within a gelatinous *tectorial membrane* within the cochlear duct (fig. 20.11). When the basilar membrane is displaced in response to sound, the hairs are bent. The hair cells are sensory structures that are innervated by dendrites of the auditory (eighth cranial) nerve; bending of their hairs triggers the production of nerve impulses that travel in the nerve fibers to the brain. The association of hair cells, tectorial membrane, and sensory nerve fibers form the functioning units of hearing, known as the **organ of Corti.**[5]

As previously described, sounds of different frequencies are perceived as having different pitches. This is due to the fact that different frequencies of sound cause different parts of the basilar membrane to move more than other parts. Sounds of low frequency cause maximum movement toward the end of the basilar membrane, whereas sounds of higher frequency cause maximum movement earlier in the basilar membrane (in the parts closer to the oval window). This is illustrated in figure 20.12. Different frequencies of sound thus stimulate different organs of Corti, and thus different sensory nerve fibers, to different degrees. This information is perceived by the brain as sounds of different pitches.

Since fluid cannot be compressed, the movements of fluid passing through the cochlear duct to the lower scala tympani cause a flexible membrane, the *round window* (fig. 20.12), to bulge outward into the middle ear cavity. This outward bulging of the round window compensates for the inward displacement of the oval window produced by vibrations of the stapes.

[2]labyrinth: Gk. *labyrinthos,* a system of interconnecting canals
[3]cochlea: L. *cochlea,* snail shell

[4]Meniere's disease: from Prosper Meniere, French physician, 1799–1862
[5]organ of Corti: from Alfonso Corti, Italian anatomist, 1822–1888

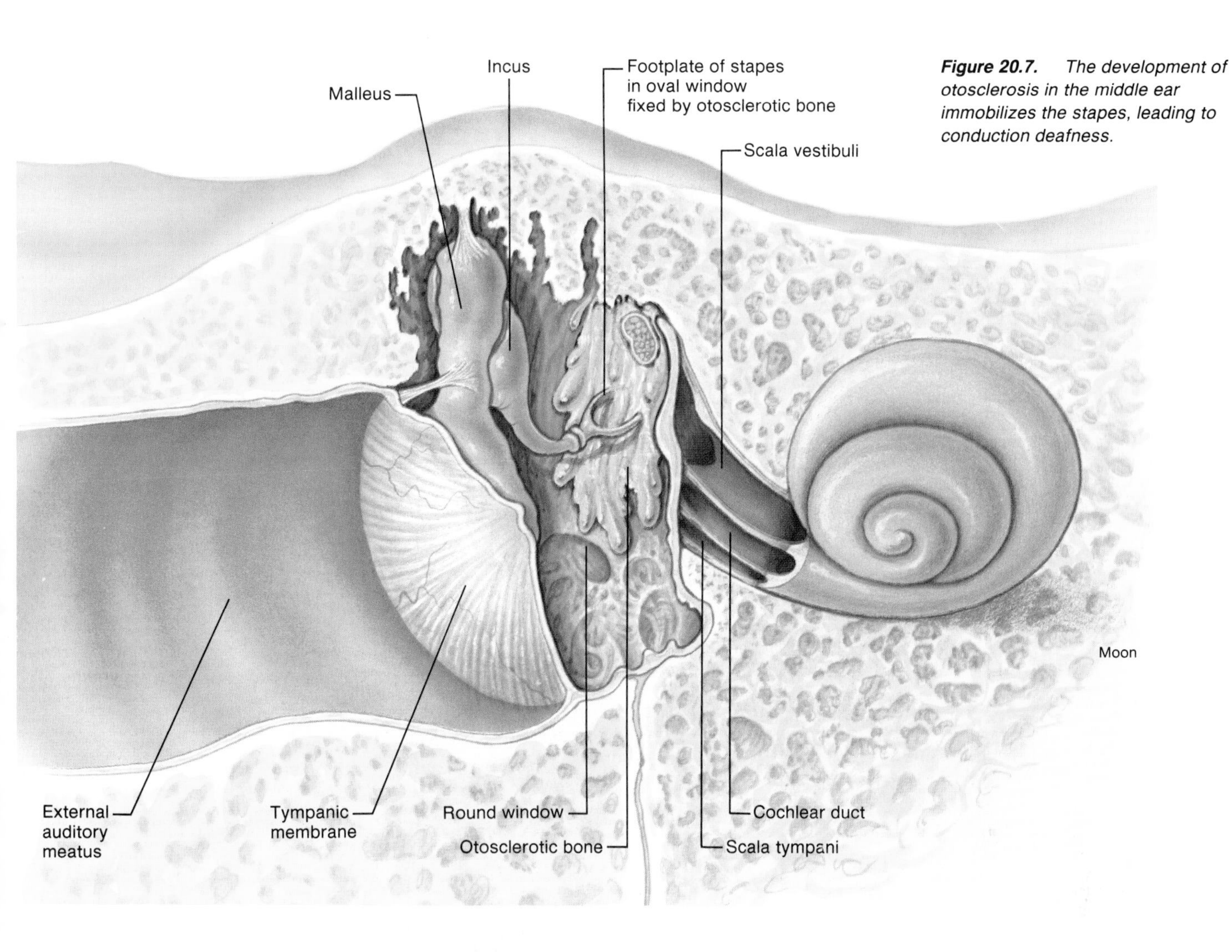

Figure 20.7. *The development of otosclerosis in the middle ear immobilizes the stapes, leading to conduction deafness.*

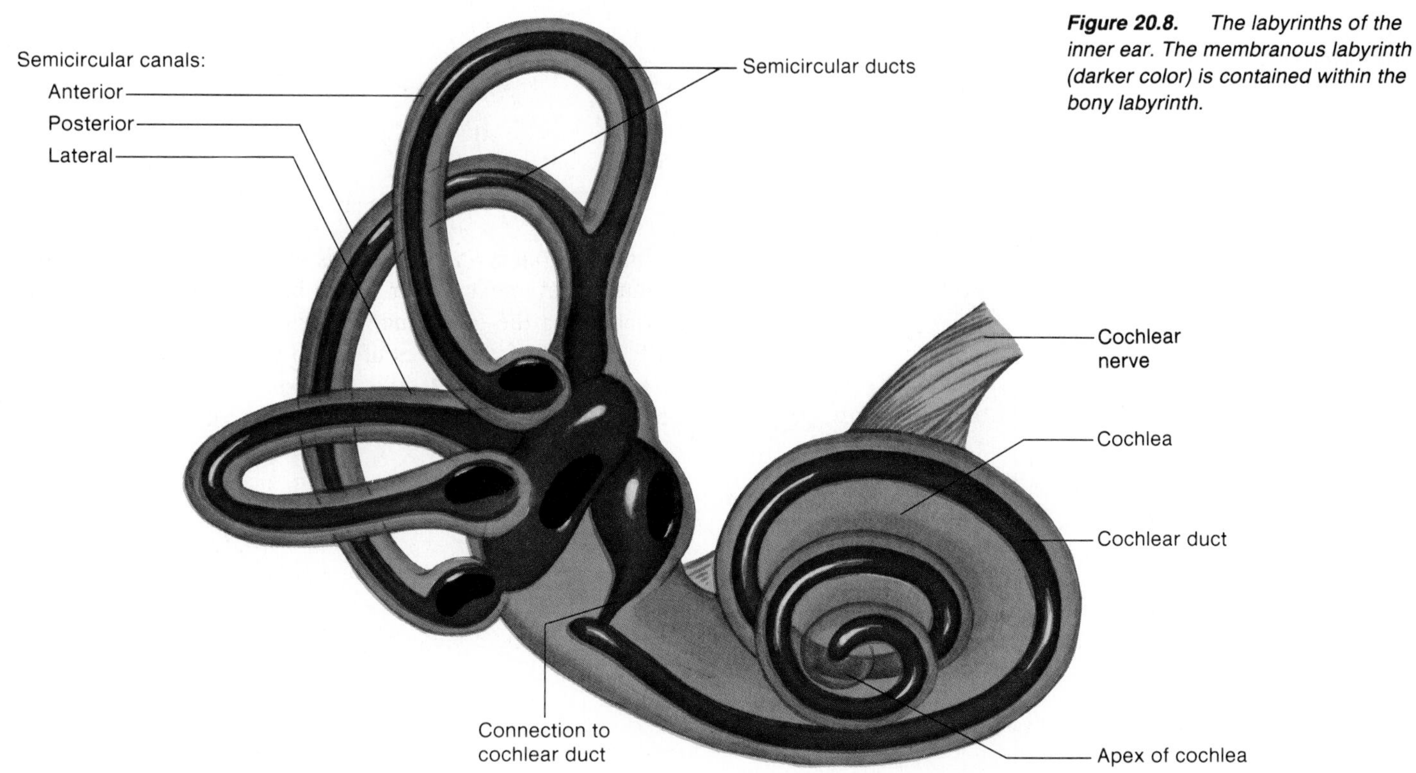

Figure 20.8. *The labyrinths of the inner ear. The membranous labyrinth (darker color) is contained within the bony labyrinth.*

Figure 20.9. *A cross section of the cochlea showing its three turns and its three compartments—scala vestibuli, cochlear duct, and scala tympani.*

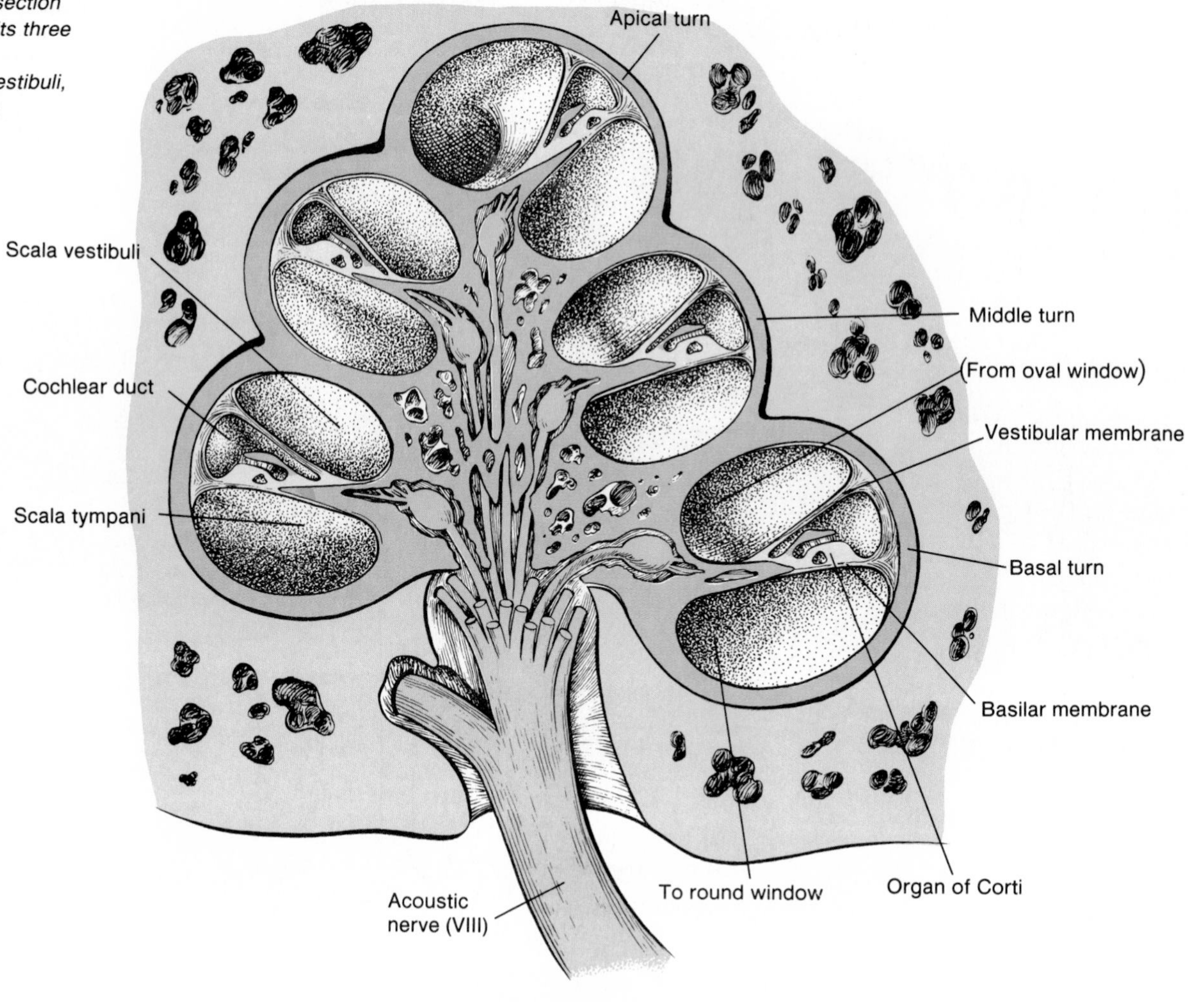

Figure 20.10. *Pressure waves in the scala vestibuli cause deflections of the vestibular and basilar membranes. In this way the pressure is transmitted to the scala tympani.*

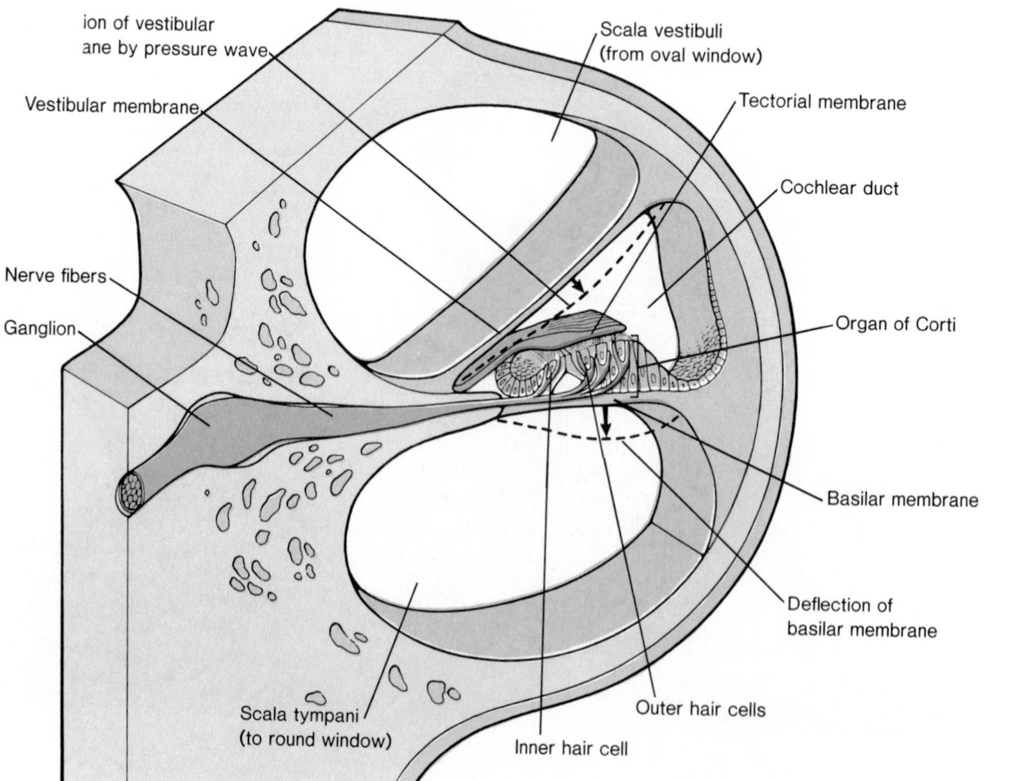

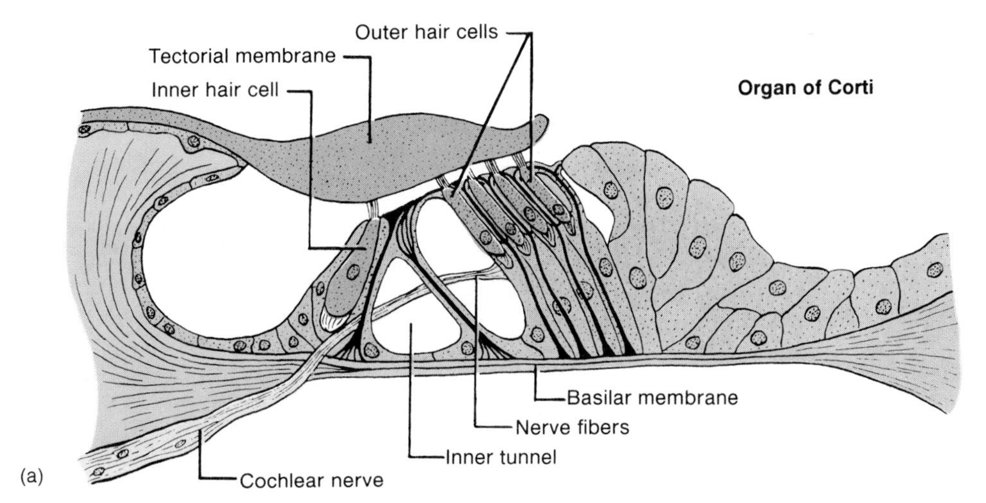

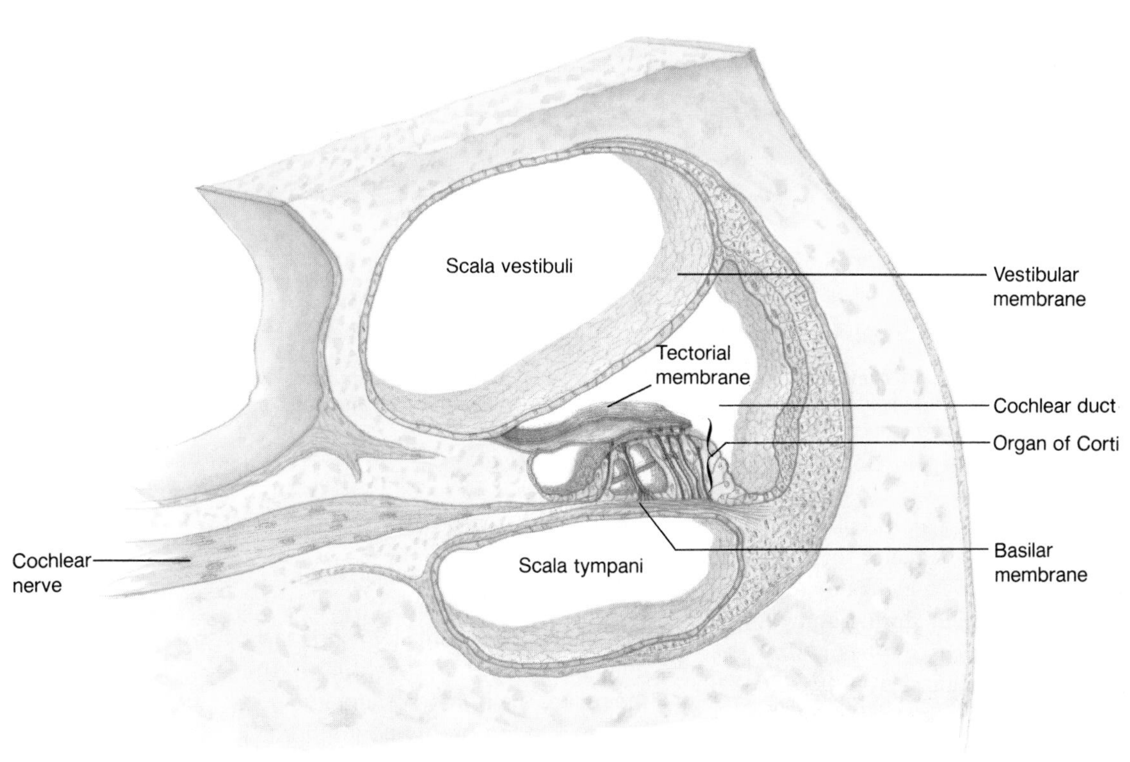

Figure 20.11. *The organ of Corti (a) is located within the cochlear duct of the cochlea (b).*

(a) From J. R. McClintic, Physiology of the Human Body, *2d ed. Copyright © 1978 John Wiley & Sons, Inc.*

The *auditory (eustachian) tube* is a passageway leading from the middle ear to the back of the oral cavity (the pharynx), and serves to equalize the pressure between the two cavities. The auditory tube is usually collapsed, so that debris and infectious agents are prevented from traveling from the oral cavity to the middle ear. The auditory canal opens, however, when you swallow, yawn, or sneeze. People thus sense a "popping" sensation in the ears as they swallow when driving up a mountain, since the opening of the auditory canal permits air to move from the region of higher pressure in the middle ear to the lower pressure in the pharynx.

Hearing Impairments

There are two major causes of hearing loss: (1) **conductive deafness,** in which the transmission of sound waves from air through the middle ear to the oval window is impaired; and (2) **nerve** or **sensory deafness,** in which the transmission of nerve impulses anywhere from the cochlea to the auditory cortex is impaired. Conductive deafness can be caused by middle ear damage from otitis media or otosclerosis, as previously discussed. Nerve deafness may result from a wide variety of pathological processes.

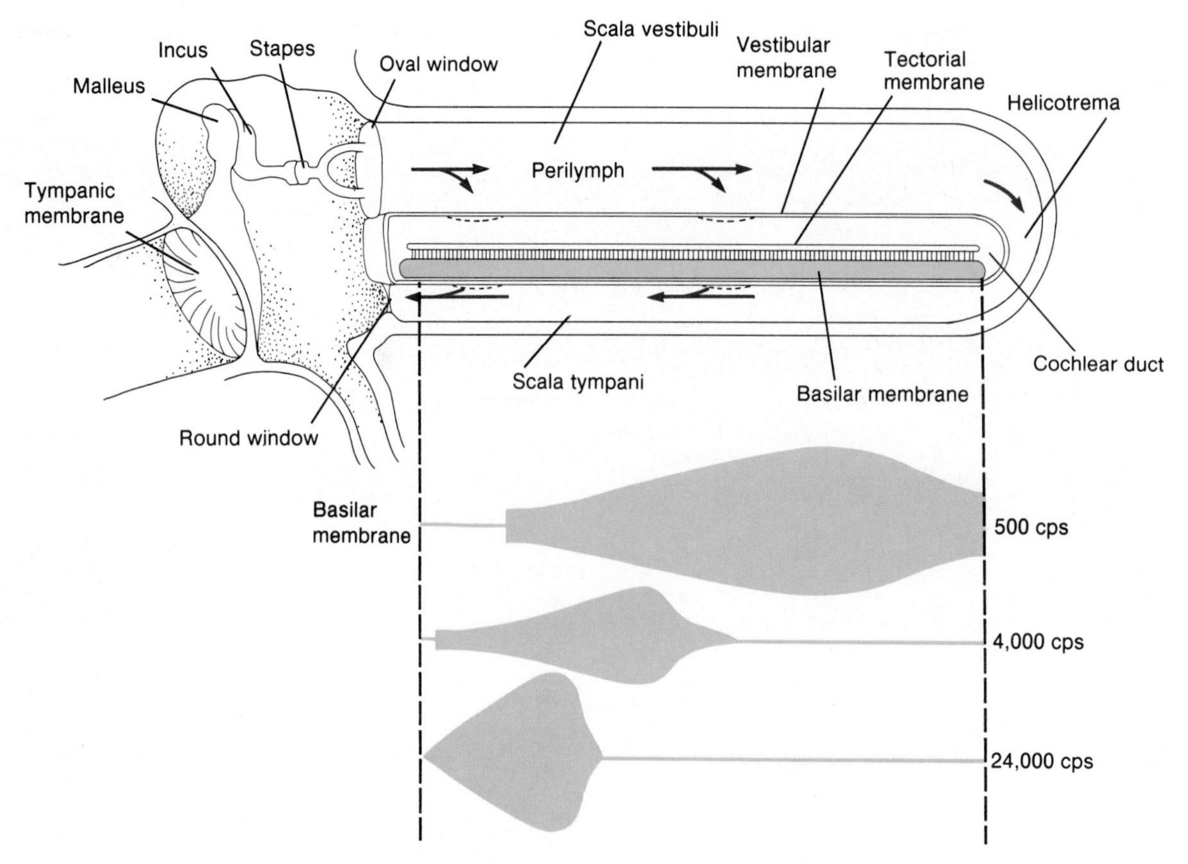

Figure 20.12. Sounds of low frequency cause pressure waves to pass through the helicotrema. Sounds of higher frequency cause pressure waves to "short cut" through the cochlear duct. This causes displacement of the basilar membrane, which is central to the transduction of sound waves into nerve impulses. (cps = cycles per second.)

Conduction deafness impairs hearing at all sound frequencies. Nerve deafness, in contrast, often impairs the ability to hear some pitches more than others. This may be due to pathological processes or to changes that occur during aging. Age-related hearing deficits—called *presbycusis*[6]—begin after age twenty when the ability to hear high frequencies (18,000–20,000 Hz) diminishes. Although the progression is variable, and occurs in men to a greater degree than in women, these deficits may gradually extend into the 4,000–8,000 Hz range. The ability to hear speech is particularly affected by hearing loss in the higher frequencies. People with these types of impairments can be helped by *hearing aids,* which amplify sounds and conduct the sound waves through bone to the inner ear.

1. Use a flow chart to describe how sound waves in air within the external auditory meatus are transduced into movements of the basilar membrane.
2. Explain how movements of the basilar membrane of the organ of Corti can code for different sound frequencies (pitches).

[6]presbycusis: Gk. *presbys,* old; *akousis,* hearing

The Eyes and Vision

Light from an observed object is focused by the cornea and lens onto the photoreceptive layer called the retina at the back of the eye. The focus is maintained by variations in the thickness and degree of curvature of the lens. When light strikes the retina, the photopigment known as rhodopsin dissociates, resulting in the production of nerve impulses.

The eyes transduce energy in the *electromagnetic spectrum* into nerve impulses. Only a limited part of this spectrum can excite the photoreceptors—electromagnetic energy with wavelengths between 400 and 700 nanometers (nm) comprise *visible light.* Light of longer wavelengths, which are in the infrared regions of the spectrum, do not have sufficient energy to excite the receptors but are felt as heat. Ultraviolet light, which has shorter wavelengths and more energy than visible light, is filtered out by the yellow color of the eye's lens.

The outermost layer of the eye is a tough coat of connective tissue called the *sclera.*[7] This can be seen externally as the white of the eyes. The tissue of the sclera is continuous with the transparent *cornea.*[8] Light passes through the cornea to enter the *anterior chamber* of the eye. Light then passes through an

[7]sclera: Gk. *skleros,* hard
[8]cornea: L. *cornu,* horn

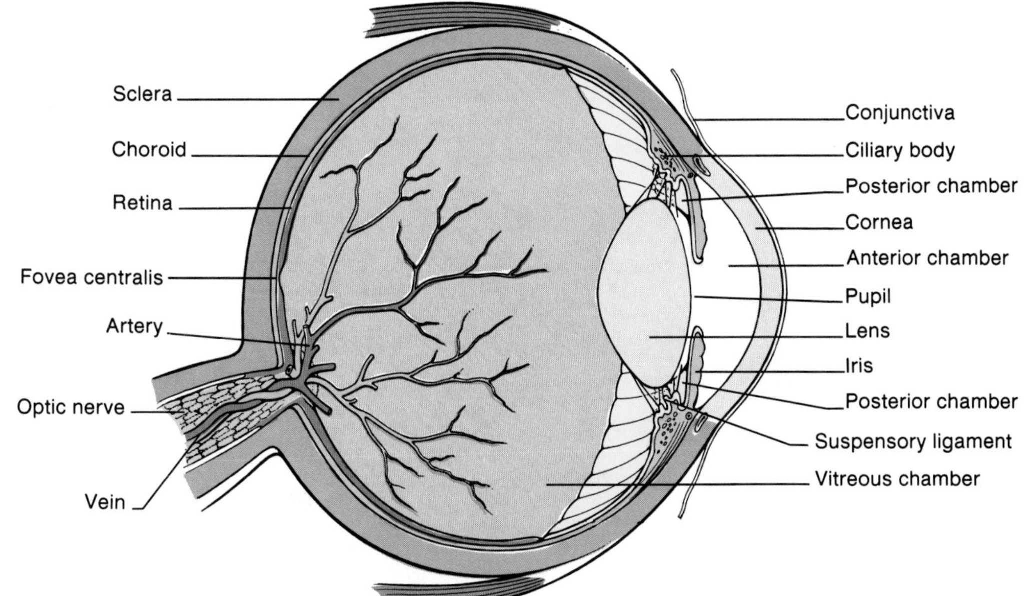

Figure 20.13. *The internal anatomy of the eyeball.*

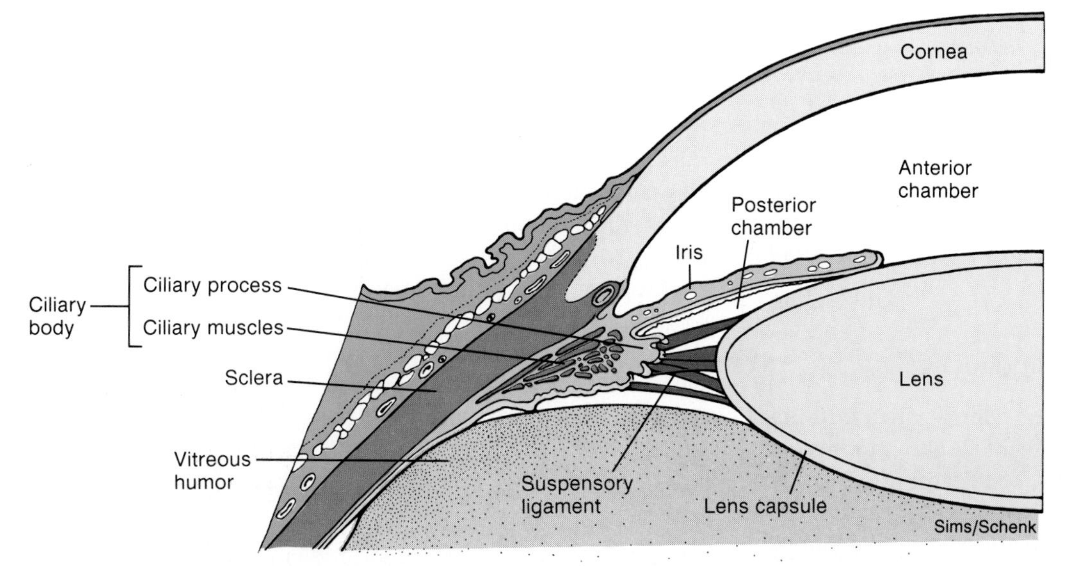

Figure 20.14. *The detailed structure of the anterior portion of the eyeball.*

opening, called the *pupil,* within a pigmented (colored) muscle known as the *iris.*[9] After passing through the pupil, light enters the *lens* (fig. 20.13).

The iris is like the diaphragm of a camera, which can increase or decrease the diameter of its aperture (the pupil) to admit more or less light. Constriction of the pupils is produced by contraction of circular muscles within the iris; dilation is produced by contraction of radial muscles. Variations in the diameter of the pupil are similar in effect to variations in the "f-stop" of a camera.

The iris contains a pigmented epithelium that gives the eye its color. The color of the eye is determined by the amount of pigment—blue eyes have the least pigment, brown eyes have more, and black eyes have the greatest amount of pigment. Albinos, who have a congenital defect in the ability to produce melanin pigment, have eyes that appear pink because the absence of pigment allows blood vessels to be seen.

The lens is suspended from a muscular process called the **ciliary body,** which is connected to the sclera and encircles the lens. *Zonular*[10] *fibers* suspend the lens from the ciliary body, forming a *suspensory ligament* that supports the lens (fig. 20.14).

The portion of the eye located behind the lens is filled with a thick, viscous substance known as the **vitreous**[11] **body.** Light from the lens that passes through the vitreous body enters the neural layer, which contains photoreceptors, at the back of the eye. This neural layer is called the **retina.** While passing through the retina, some of this light stimulates photoreceptors, which in turn activate other neurons. Neurons in the retina contribute

[9]iris: Gk. *irid,* rainbow

[10]zonular: L. *zona,* a girdle

[11]vitreous: L. *vitreus,* glassy

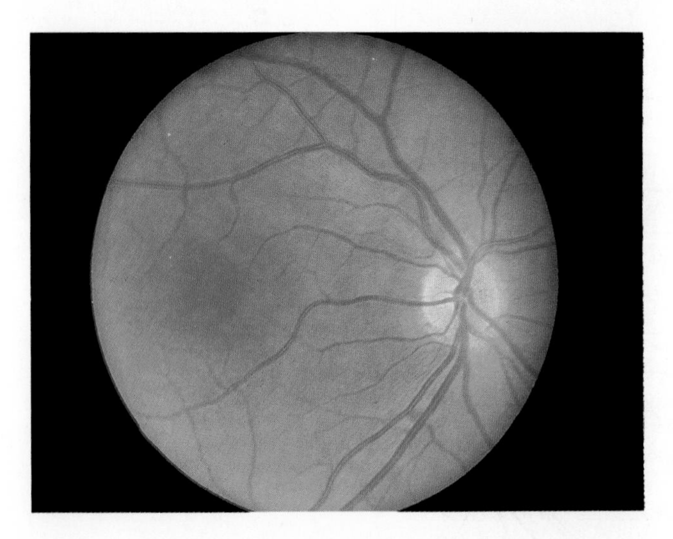

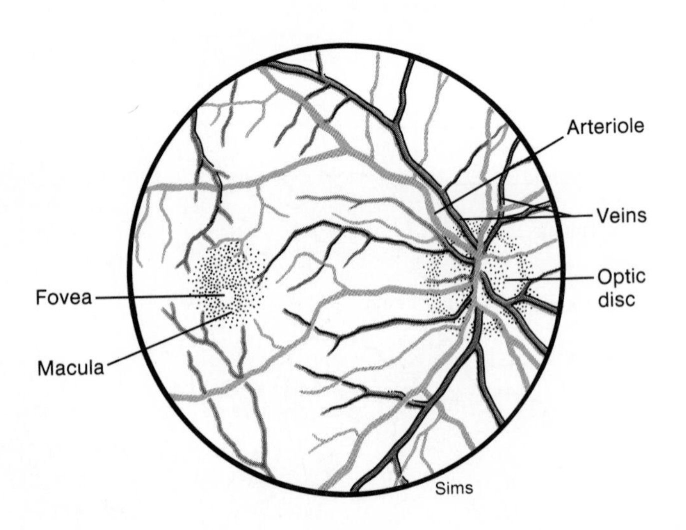

Figure 20.15. *A view of the retina as seen with an ophthalmoscope. Optic nerve fibers leave the eyeball at the optic disc to form the optic nerve. Note the blood vessels that can be seen entering the eyeball at the optic disc.*

fibers that are gathered together at a region called the *optic disc* (fig. 20.15) to exit the retina as the optic nerve. The optic disc is also the site of entry and exit of blood vessels. Since there are no photoreceptors in the optic disc, an image that falls here cannot be seen. The optic disc is thus also known as the *blind spot.*

Refraction of Light and Visual Acuity

Light that passes from a medium of one density into a medium of a different density is *refracted,* or bent. Light entering the eye is refracted by the cornea and lens. The degree of refraction also depends on the curvature of these structures. The curvature of the cornea is constant, while the curvature of the lens can be varied. The refractive properties of the lens can thus provide fine control for focusing light on the retina. As a result of light refraction, the image formed on the retina is upside down and right to left (fig. 20.16).

Accommodation

When a normal eye views an object, parallel rays of light are refracted to a point, or *focus,* on the retina (see fig. 20.19). If the degree of refraction were to remain constant, movement of the object closer to or farther from the eye would cause corresponding movement of the focal point, so that the focus would either be behind or in front of the retina.

The ability of the eyes to keep the image focused on the retina as the distance between the eyes and object is changed is called **accommodation.** Accommodation results from contraction of the ciliary muscle, which is like a sphincter muscle that can vary its aperture (fig. 20.17). When the ciliary muscle is relaxed, its aperture is wide. Relaxation of the ciliary muscle thus places tension on its attached zonular fibers and pulls the lens taut. These are the conditions that prevail when viewing an object which is twenty feet or more from a normal eye; the image is focused on the retina and the lens is in its most flat, least convex form. As the object moves closer to the eyes the muscles of the ciliary body contract. This muscular contraction narrows the aperture of the ciliary body and thus reduces the tension on the zonular fibers which suspend the lens. When the tension is reduced, the lens becomes more round and convex (fig. 20.18).

The ability of a person's eyes to accommodate can be measured by the *near point of vision* test, which is the minimum distance from the eyes that an object can be maintained in focus. This distance increases with age, and indeed accommodation in almost everyone over the age of forty-five is significantly impaired. Loss of accommodating ability with age is known as **presbyopia** (*presby* = old). This loss appears to have a number of causes, including thickening of the lens and a forward movement of the attachments of the zonular fibers to the lens. As a result of these changes, the zonular fibers and lens are pulled taut even when the ciliary muscle contracts. The lens is thus not able to thicken and increase its refraction when, for example, a printed page is brought close to the eyes.

Visual Acuity

Visual acuity[12] refers to the sharpness of vision. The sharpness of an image depends on the *resolving power* of the visual system—that is, on the ability of the visual system to distinguish (resolve) two closely spaced dots. The better the resolving

[12]acuity: L. *acuo,* sharpen

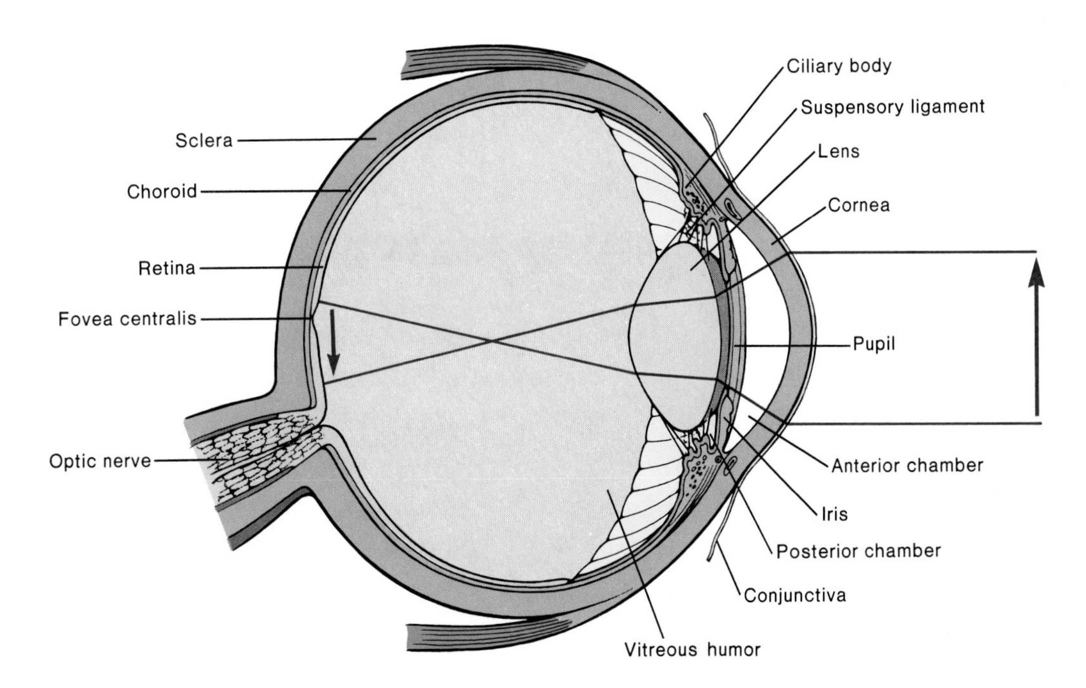

Figure 20.16. *The refraction of light waves within the eyeball causes the image of an object to be inverted on the retina.*

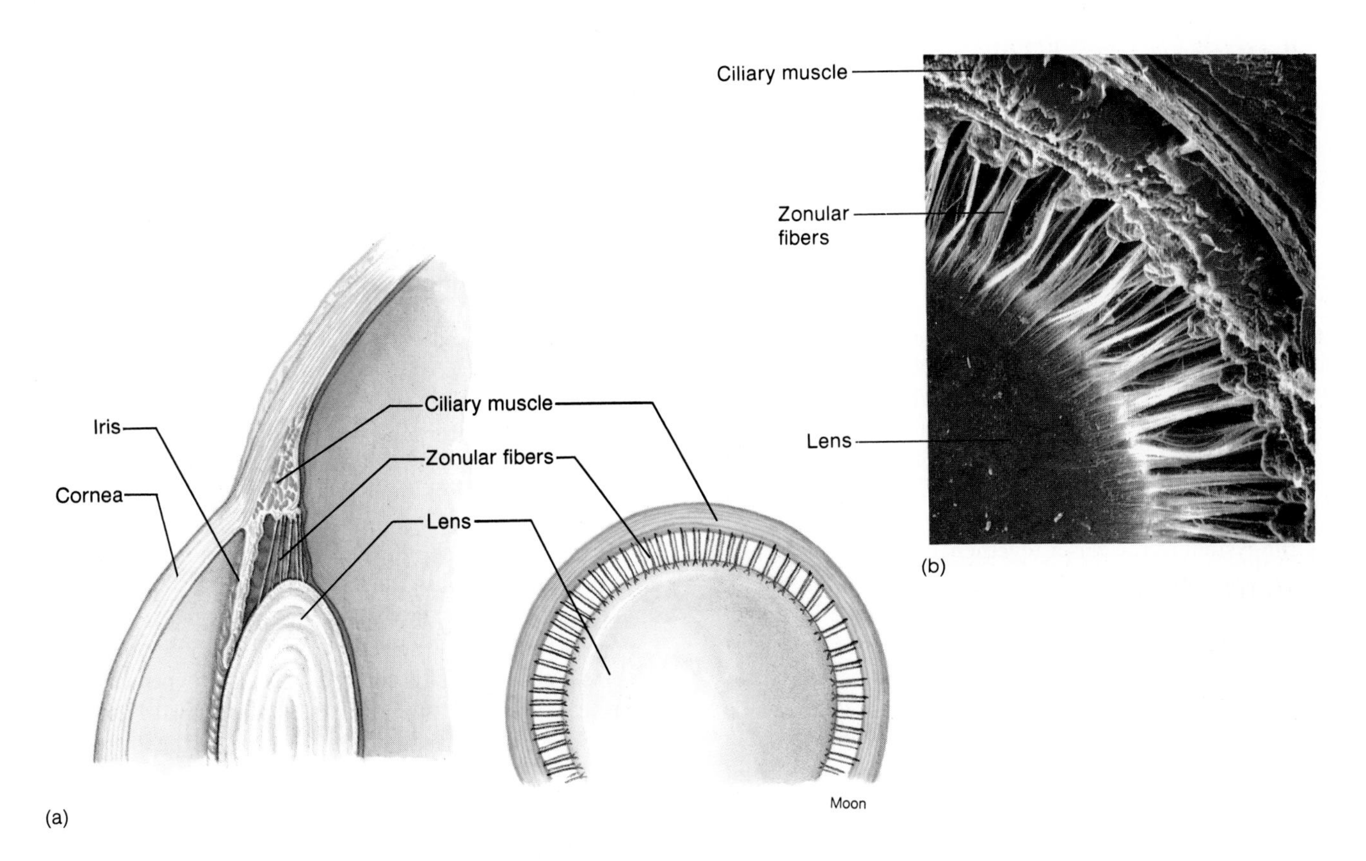

Figure 20.17. *(a) Diagram, and (b) scanning electron micrograph (from the eye of a seventeen-year-old boy), showing the relationship between the lens, zonular fibers, and ciliary muscle of the eye.*

From "How the Eye Focuses," by James F. Koretz and George Handleman.

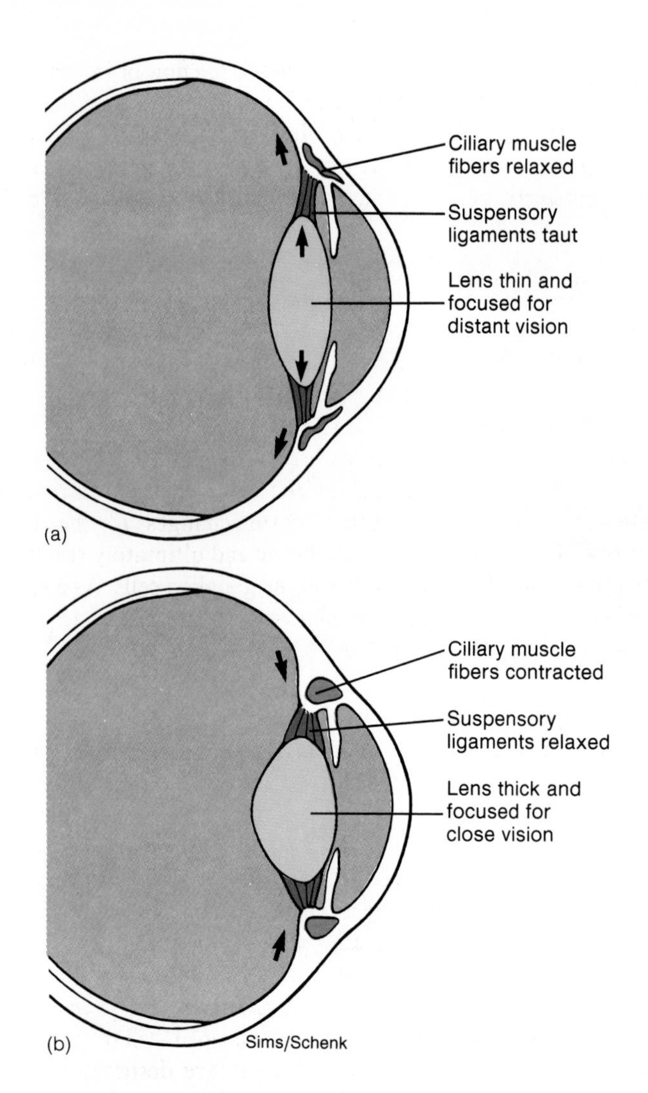

Figure 20.18. *Changes in the shape of the lens during accommodation. (a) The lens is flattened for distant vision when the ciliary muscle fibers are relaxed and the suspensory ligaments are taut. (b) The lens is more spherical for closeup vision when the ciliary muscle fibers are contracted and the suspensory ligaments are relaxed.*

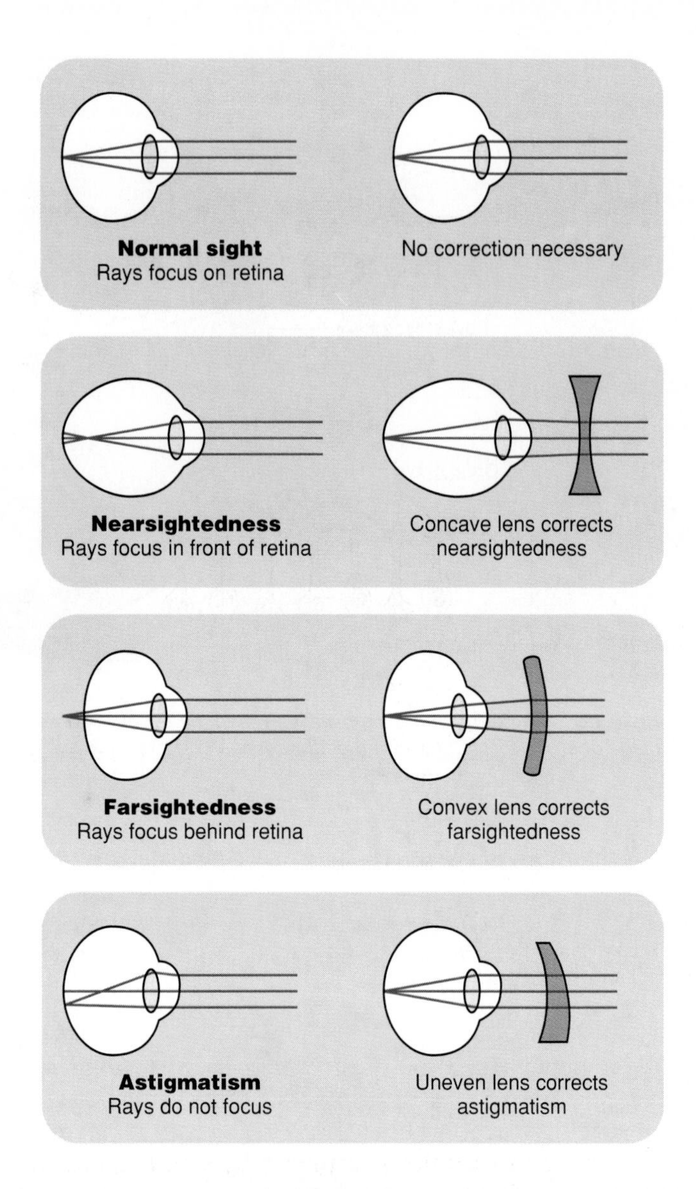

Figure 20.19. *Comparison of the focus of light by the normal eye with eyes that have different problems with light refraction.*

power of the system is, the closer together these dots can be and still be seen as separate. When the resolving power of the system is exceeded, the dots are blurred together as a single image.

When a person with normal visual acuity stands twenty feet from a *Snellen eye chart* (so that accommodation is not a factor influencing acuity), the line of letters marked "20/20" can be read. If a person has **myopia**[13] (nearsightedness), this line will appear blurred because the focus of this image will be in front of the retina. This is usually caused by the fact that the eyeball is too long. Myopia is corrected by glasses with concave lenses that cause the light rays to diverge, so that the point of focus is further from the lens and is thus pushed back to the retina (fig. 20.19).

If the eyeballs are too short, the line marked "20/20" will appear blurred because the distance from the lens to the focused image is longer than the distance to the retina. In order to bring the focus forward to the retina, the object must be placed farther from the eyes. This condition is called **hyperopia** (farsightedness). Hyperopia is corrected by glasses with convex lenses that increase the convergence of light so that the focus is brought closer to the lens and falls on the retina.

The curvature of the cornea and lens is not perfectly symmetrical, so that light passing through some parts of these structures may be refracted to a different degree than light passing through other parts. When the asymmetry of the cornea and/or lens is significant, the person is said to have **astigmatism.**[14] If a person with astigmatism views a circle of lines ra-

[13]myopia: Gk. *myein,* to shut; *ops,* eye

[14]astigmatism: Gk. *a,* without; *stigma,* point

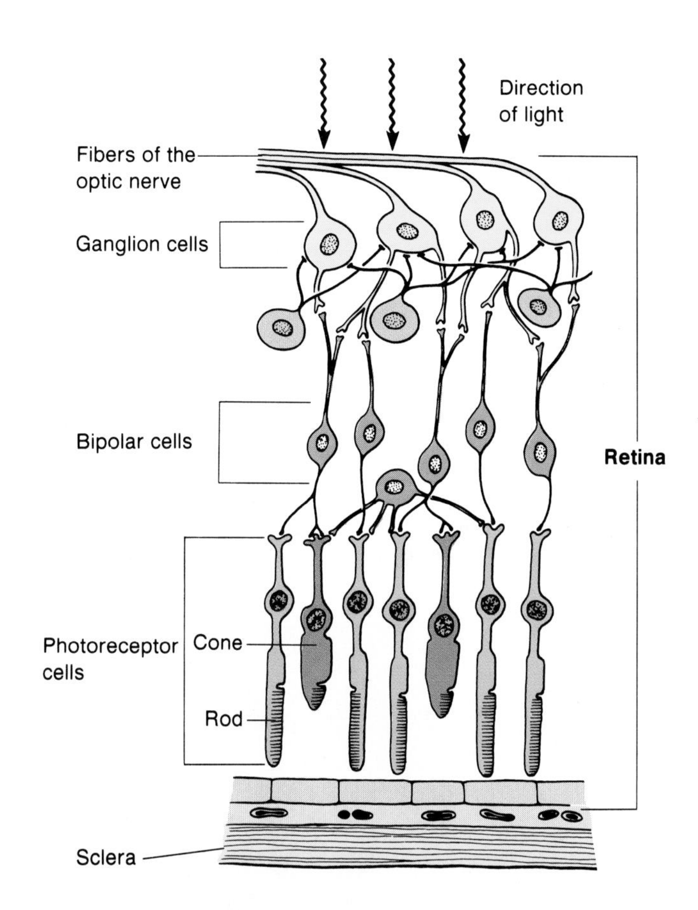

Figure 20.20. *The layers of the retina. The retina is inverted, so that light must pass through various layers of nerve cells before reaching the photoreceptors (rods and cones).*

diating from the center, like the spokes of a wheel, the image of these lines will not appear clear in all 360 degrees; the parts of the circle that appear blurred can thus be used to map the astigmatism. This condition is corrected by cylindrical lenses that compensate for the asymmetry in the cornea or lens.

The Retina

The retina consists of a pigment epithelium, photoreceptor neurons called *rods* and *cones,* and layers of other neurons. The neural layers of the retina are actually a forward extension of the brain. Since the retina is an extension of the brain, the neural layers face outwards, toward the incoming light. Light, therefore, must pass through several neural layers before striking the photoreceptors (fig. 20.20). The photoreceptors then synapse with other neurons, so that nerve impulses are conducted outward in the retina.

The outer layers of neurons that contribute axons to the optic nerve are called *ganglion cells.* This layer receives synaptic input from *bipolar cells* underneath, which in turn receive input from rods and cones. In addition to the flow of information from photoreceptors to bipolar cells to ganglion cells, there are neurons called *horizontal cells,* which synapse with several photoreceptors (and possibly also with bipolar cells), and neurons called *amacrine cells,* which synapse with several ganglion cells.

Effect of Light on the Rod

The photoreceptors—rods and cones (fig. 20.21)—are activated when light produces a chemical change in molecules of pigment. Rods contain a purple pigment known as **rhodopsin.**[15] In response to absorbed light, rhodopsin dissociates into its two components: a pigment called **retinene,** derived from vitamin A (chapter 19), and a protein called **opsin** (fig. 20.22). This reaction is known as the *bleaching reaction.* The dissociation reaction in response to light initiates changes in the ionic permeability of the rod cell membrane and ultimately results in the production of nerve impulses in the ganglion cells. As a result of these effects, rods provide black-and-white vision at night, when the light intensity is low.

Cones and Color Vision

Cones are less sensitive than rods to light, but provide color vision and greater visual acuity. During the day, the high light intensity bleaches out the rods, and color vision with high acuity is provided by the cones. According to the **trichromatic theory** of color vision, our perception of a multitude of colors is due to stimulation of only three types of cones. Each type of cone contains retinene, as in rhodopsin, but this molecule is associated with a different protein than opsin. The protein is different for each of the three cone pigments, and as a result, each of the pigments absorbs light of a given wavelength (color) to a different degree. The three types of cones are designated blue, green, and red, according to the region of the visible spectrum in which each cone pigment absorbs light maximally (fig. 20.23). Our perception of any given color is produced by the relative degree to which each cone is stimulated by any given wavelength of visible light.

While reading or similarly focusing visual attention on objects in daylight, each eye is oriented so that the image falls within a tiny area of the retina called the **fovea**[16] **centralis.** The fovea is a pinhead-sized pit within a yellow area of the retina called the *macula lutea*[17] (see fig. 20.15). There are approximately 120 million rods and 6 million cones in each retina. The photoreceptors are distributed in such a way that the fovea contains only cones, whereas more peripheral regions of the retina contain a mixture of rods and cones.

Cones not only provide color vision, they also produce a sharper image than that produced by rods. This is because each cone in the fovea can excite a ganglion cell and produce nerve

[15]rhodopsin: Gk. *rhodon,* rose; *ops,* eye
[16]fovea: L. *fovea,* small pit
[17]macula lutea: L. *macula,* spot; *luteus,* yellow

Figure 20.21. (a) Structure of a rod and cone. (b) A scanning electron micrograph of rods and cones.

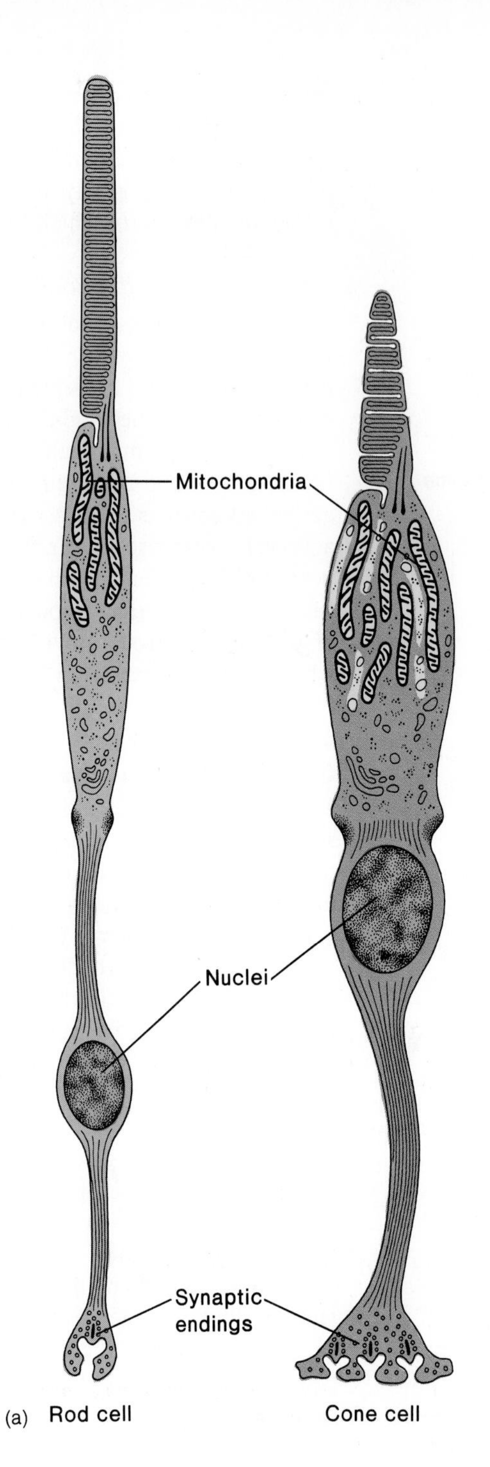

impulses, whereas many rods must be activated to stimulate a ganglion cell. During the day, when we focus images on the fovea, we thus see the images sharply and in color. At night, however, the images are less distinct and in black-and-white because rods are used. Under conditions of low illumination, objects may best be seen from the "corners of the eye" where the rods predominate. People looking at stars through a telescope are thus often instructed not to look directly at the faint celestial object they wish to observe.

---

1. List the structures of the eye, in the order through which light passes in order to reach the photoreceptors.
2. Explain how the ciliary muscle and lens function during accommodation.
3. Describe the visual defects involved in presbyopia, myopia, hyperopia, and astigmatism.
4. Describe the structure of the retina, and distinguish between the location and function of the rods and cones.

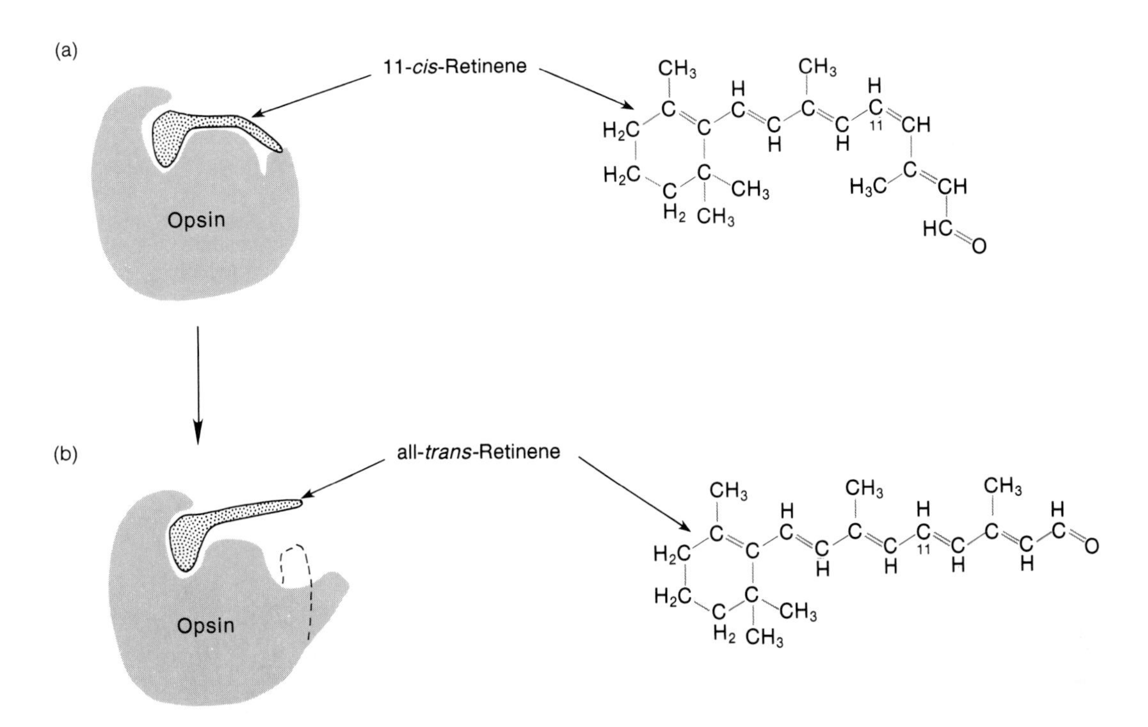

Figure 20.22. *A molecule of rhodopsin before (a) and after (b) the photodissociation reaction in response to light.*

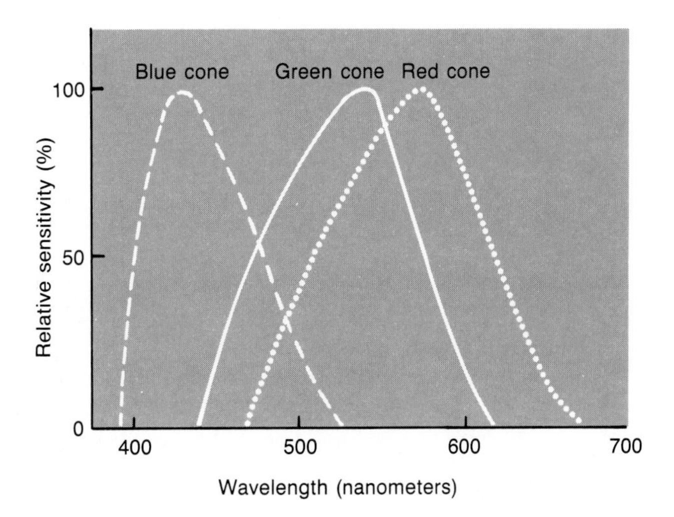

Figure 20.23. *There are three types of cones. Each type of cone contains retinene combined with a different type of protein, producing a pigment that absorbs light maximally at a different wavelength. Color vision is produced by the activity of these blue cones, green cones, and red cones.*

Summary

Introduction to the Sensory System p. 438

I. Sensory adaptation allows us to cease paying attention to constant stimuli.
   A. Receptors respond in a phasic rather than constant fashion.
   B. Olfaction adapts quickly, while pain hardly adapts at all.

II. According to the law of specific nerve energies, the sensation that is perceived when a given receptor is stimulated is that which is evoked by its normal, or adequate, stimulus.

Taste and Olfaction p. 349

I. Taste receptors are taste buds located in the tongue; there are only four modalities of taste: sweet, salty, bitter, and sour.

II. Olfactory receptors are the dendrites of the first cranial nerve; impulses from this nerve are transmitted directly to the brain.

The Ears and Hearing p. 441

I. The tympanic membrane vibrates in response to sound waves channeled to it by the external auditory meatus.

II. Vibrations of the tympanic membrane cause the malleus, incus, and stapes of the middle ear to vibrate.

III. Movements of the stapes against the oval window of the cochlea sets up pressure waves in the fluid of the scala vestibuli.
   - A. These pressure waves push against the cochlear duct, a fluid-filled tube that is part of the membranous labyrinth.
   - B. The basilar membrane, at the base of the cochlear duct, vibrates in response to these pressure waves.
   - C. The organ of Corti, which contains sensory hair cells on the basilar membrane, is stimulated by these vibrations.
   - D. Sounds of different pitch selectively vibrate different parts of the basilar membrane.

The Eyes and Vision p. 446

I. Light enters the eye by passing through the cornea and then the pupil.
   - A. The cornea refracts the light; the pupil is an opening in the middle of the iris.
   - B. The light then passes through the lens and the vitreous body to reach the retina at the back of the eye.

II. The thickness and curvature of the lens is adjusted to maintain a focus on the retina.
   - A. The ability to maintain a focus as an object is brought closer to the eyes is called accommodation.
   - B. If the eyeball is too long, the person has myopia; if it is too short, the person has hyperopia.
   - C. Astigmatism occurs when there is unequal refraction by different regions of the cornea or lens.

III. Photoreceptors are rods and cones.
   - A. When light strikes the rods, rhodopsin dissociates; this results in electrical excitation.
   - B. Photoreceptors synapse with bipolar cells, which in turn synapse with ganglion cells; the axons of these come together at the optic disc (blind spot) and leave the eye as the optic nerve.

IV. Cones are most concentrated at the fovea centralis of the retina.
   - A. There are three types of cones that respond best to three different colors of light: blue, green, and red.
   - B. Cones provide color vision and the most acute vision; rods provide black-and-white vision under conditions of low light levels.

Review Activities

Objective Questions

1. Which of the following is found in the middle ear?
   (a) semicircular canals
   (b) cochlea
   (c) external auditory meatus
   (d) malleus, incus, and stapes
2. Which of the following statements about rhodopsin is *false?*
   (a) It contains retinene.
   (b) Part of the molecule is a derivative of vitamin A.
   (c) It is found in cones.
   (d) It dissociates in response to light.
3. The receptors for taste are
   (a) naked dendrites
   (b) sensitive to four taste modalities
   (c) proprioceptors
   (d) all of these
4. When a person with normal vision views an object from a distance of at least twenty feet
   (a) the ciliary muscles are relaxed
   (b) the suspensory ligament is tight
   (c) the lens is in its most flat, least convex shape
   (d) all of these

Match the following visual defect with its cause:

5. Myopia
6. Presbyopia
7. Hyperopia
8. Astigmatism

   (a) unequal curvature of lens or cornea
   (b) eyeball that is too short
   (c) inability of lens to round out
   (d) eyeball that is too long

9. The ability of the lens to increase its curvature and maintain a focus at close distances is called
   (a) convergence
   (b) accommodation
   (c) astigmatism
   (d) amblyopia
10. Which of the following sensory modalities is transmitted directly to the cerebral cortex without being relayed through the thalamus?
   (a) taste
   (b) sight
   (c) smell
   (d) hearing
   (e) touch

Essay Questions

1. Define accommodation, and explain how it is accomplished. Why is it more of a strain on the eyes to look at a small object close to the eyes than large objects far away?
2. Explain why images that fall on the fovea centralis are seen more clearly than images that fall on the periphery of the retina. Why are the "corners of the eyes" more sensitive to light than the fovea?
3. Explain how the brain is made aware of different pitches of a sound.
4. Some sensory modalities exhibit more sensory adaptation than others. Speculate on the survival value of these differences.
5. "We see with our brain, not our eyes." Defend this statement, using the nature of the retinal image and nerve impulses to support your argument.

21

Human Movement through the Environment: The Musculoskeletal System

Outline

Objectives

By studying this chapter, you should be able to

1. distinguish between the axial and appendicular skeleton
2. describe the microscopic appearance of bone
3. identify the parts of a long bone
4. explain how bone is formed during development, and how a bone can grow in length
5. list the different functions of the skeletal system
6. define the terms *origin* and *insertion* of a muscle
7. describe the different actions of muscles, and explain the meaning of the terms *synergistic* and *antagonistic*
8. distinguish between an isotonic and an isometric muscle contraction
9. explain how muscle fibers are organized within a muscle, and how myofibrils are arranged within a muscle fiber
10. explain how the filaments are arranged to form sarcomeres, and how the filaments and sarcomeres behave during muscle contraction
11. describe how cross-bridges function during the sliding of the filaments, and how the banding pattern of the myofibrils changes during contraction
12. explain the sequence of events by which a nerve stimulates a muscle to contract, and the role of calcium in excitation-contraction coupling
13. distinguish between different muscle fiber types, and describe why one type is more resistant to fatigue than the other type

Keys to Pronunciation

articulation: *ar-tik'u-la'shun*
comminuted: *kom'i-nut''ed*
diaphysis: *di-af'i-sis*
endosteum: *en-dos'te-um*
epiphyseal: *ep'i-fiz'e-al*
epiphysis: *e-pif'i-sis*
gastrocnemius: *gas''trok-ne'me-us*
haversian: *ha-ver'shan*
hemopoiesis: *he''mo-poi-e'sis*
myofibrils: *mi''o-fi'brils*
myoglobin: *mi''o-glo'bin*
myosin: *mi'o-sin*
osteoblast: *os'te-o-blast*
osteoclast: *os'te-o-klast*
periosteum: *per''e-os'te-um*
sarcomeres: *sar'ko-mers*
sarcoplasmic reticulum: *sar''ko-plaz'mik re-tik'u-lum*
synergistic: *sin''er-jis'tik*

Photo: Olympic runners demonstrate the power and beauty of the human musculoskeletal system in action.

The Skeletal System

Bones are living organs that perform a variety of functions. Compact bone is composed of units known as osteons; these units allow osteocytes to remain alive although they are encased within a hard matrix of calcium phosphate. Bones can grow in length through thickening of the growth plates of cartilage located near each end of the bone.

There are a total of 206 bones in the skeletal system (table 21.1). These are often categorized as part of either the **axial skeleton** or the **appendicular skeleton.** The bones of the axial skeleton include the (1) *skull*; (2) *middle ear ossicles* (chapter 20); (3) *hyoid bone* (located beneath the tongue); and (4) *vertebral column.*

The appendicular skeleton is composed of the bones of the (1) *upper extremities* (arm, forearm, wrist, hand, and fingers); (2) *pectoral girdle* (the scapula, or shoulder blade, and the clavicle, or collarbone); (3) *lower extremities* (thigh, lower leg, ankle, foot, and toes); and (4) *pelvic girdle* (the bones of the hip and associated structures).

Bones are usually joined together by specific types of *articulations* (joints). Some of these articulations permit a wide range of movements (such as the ball-and-socket joint that the thigh makes with the hip), some permit limited movement (such as the hinge joint at the elbow), and some allow no movement at all (such as the suture joints between skull bones). The position of some of the major bones within an articulated skeleton is shown in figure 21.1.

Table 21.1 Classification of the bones of the adult skeleton

I. Axial skeleton	
a. Skull	22 bones
8 cranial bones:	
frontal 1	
parietal 2	
occipital 1	
temporal 2	
sphenoid 1	
ethmoid 1	
13 facial bones:	
maxilla 2	
palatine 2	
zygomatic 2	
lacrimal 2	
nasal 2	
vomer 1	
inferior nasal concha 2	
1 mandible	
b. Middle ear ossicles	6 bones
malleus 2	
incus 2	
stapes 2	
c. Hyoid	1 bone
hyoid bone 1	
d. Vertebral column	26 bones
cervical vertebra 7	
thoracic vertebra 12	
lumbar vertebra 5	
sacrum 1 (5 fused bones)	
coccyx 1 (3–5 fused bones)	
e. Rib cage	25 bones
rib 24	
sternum 1	
II. Appendicular skeleton	
a. Pectoral girdle	4 bones
scapula 2	
clavicle 2	
b. Upper extremities	60 bones
humerus 2	
radius 2	
ulna 2	
carpal 16	
metacarpal 10	
phalanx 28	
c. Pelvic girdle	2 bones
os coxa 2 (each os coxa contains 3 fused bones)	
d. Lower extremities	60 bones
femur 2	
tibia 2	
fibula 2	
patella 2	
tarsal 14	
metatarsal 10	
phalanx 28	
Total	206 bones

Structure of Bone

There are two kinds of bone based on their porosity, and most bones have both types (fig. 21.2). **Compact bone** is the hard, outer layer, whereas **spongy,** or **cancellous, bone** is the porous inner portion. The spaces in spongy bone contains bone marrow where blood cells are produced.

In compact bone, the bone cells (*osteocytes*) are arranged in concentric layers around a *central canal,* which contains blood vessels and nerves. Each osteocyte occupies a space called a *lacuna* within the hard extracellular material, or *matrix.* The matrix of bone is arranged in layers, or *lamellae,* of connective tissue hardened by calcium phosphate deposits. This material would cut off the nutrient supply from the central canal to the osteocytes if it were not for the presence of tiny canals, or *canaliculi,* that radiate from the lacunae to the central canal. A central canal with its surrounding osteocytes and lamellae is known as an **osteon,** or **haversian[1] system.**

[1]haversian: from Clopton Havers, English physician and anatomist, 1650–1702

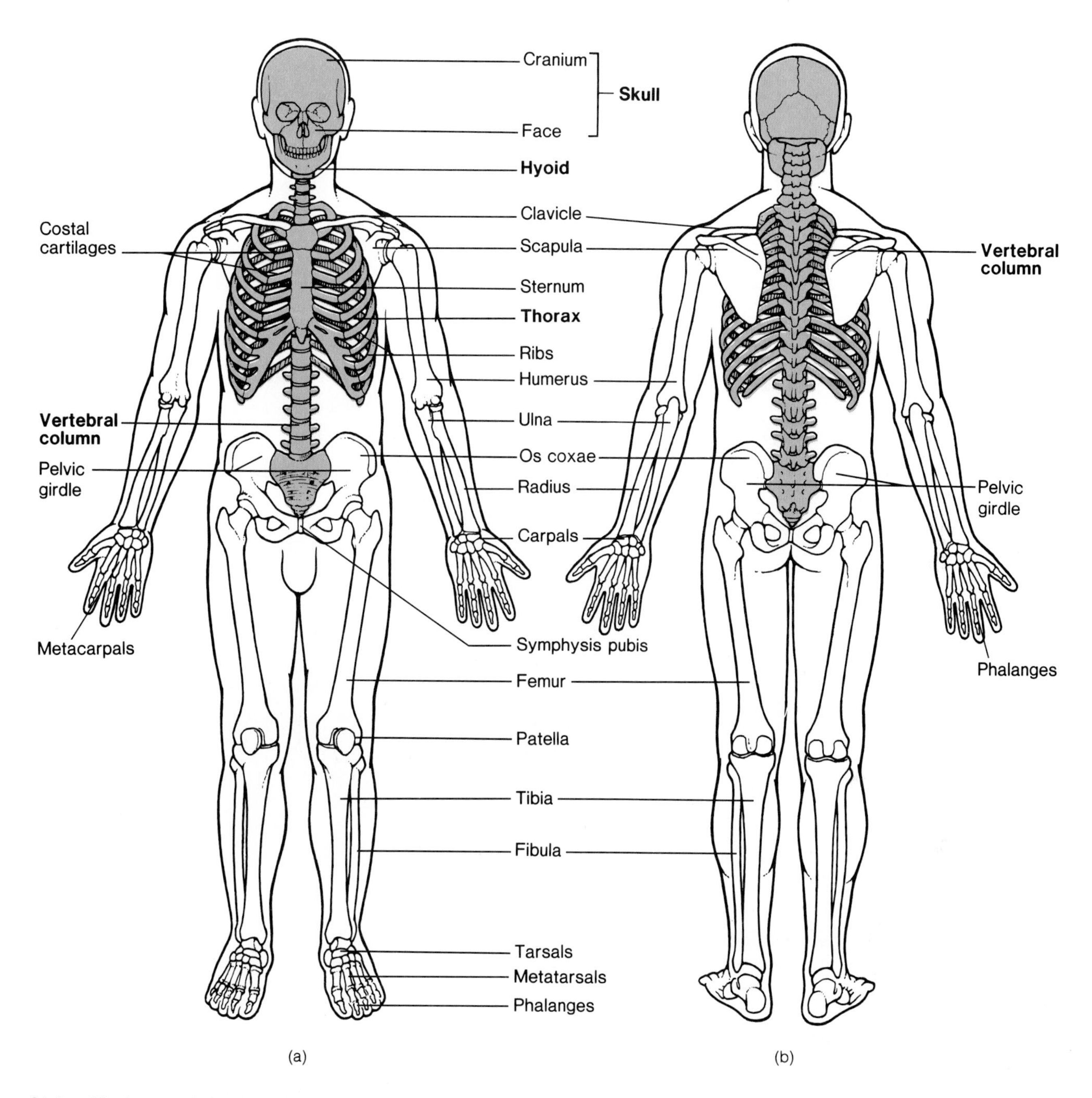

Figure 21.1. *The human skeleton. (a) An anterior view; (b) a posterior view. The axial and appendicular portions are distinguished with color.*

A long bone, of the type found in the arm, for example, is illustrated in figure 21.3. The shaft of the long bone is called the *diaphysis,*[2] and each head of the bone is called the *epiphysis.*[3] Notice that the bone has openings through which blood vessels and nerves can enter; bone is a living organ. This is indeed fortunate; if bone were the dead, dried material that most people envision, it would not be able to repair itself when fractured. (Interestingly, the word "skeleton" comes from a Greek word meaning "dried up.") Each bone is covered by a dense connective tissue covering called a *periosteum,*[4] which delivers blood vessels and nerves to the bone.

[2]diaphysis: Gk. *dia,* throughout; *physis,* growth
[3]epiphysis: Gk. *epi,* upon; *physis,* growth

[4]periosteum: Gk. *peri,* around; *osteon,* bone

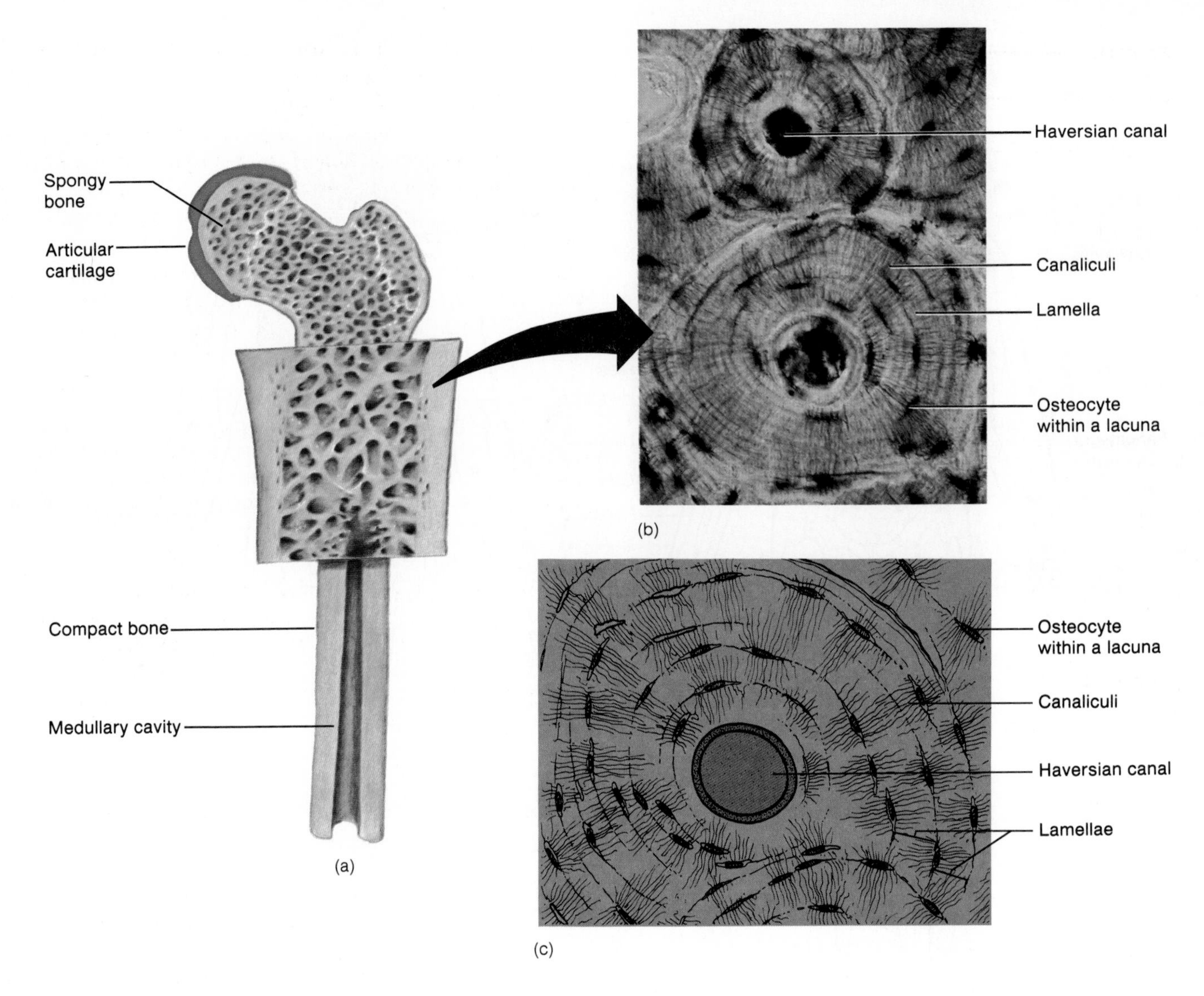

Figure 21.2. *Tissue within bone is arranged into osteons. (b) A photomicrograph and (c) an illustration of the osteons.*

Functions of the Skeletal System

Living bone is dynamic and able to perform many body functions, including support, movement, protection, and others. These are described below.

1. **Support.** The skeleton forms a rigid framework to which are attached the softer tissues and organs of the body.
2. **Protection.** The skull and vertebral column enclose and protect the central nervous system; the rib cage protects the heart, lungs, and other thoracic structures; and the pelvic cavity protects the organs within it.
3. **Body movement.** Bones serve as anchoring attachments for most skeletal muscles. In this capacity, the bones act as levers with the joints functioning as pivots when muscles contract to cause body movements.
4. **Hemopoiesis.** The red bone marrow produces red blood cells, white blood cells, and platelets.
5. **Mineral storage.** The bones are hardened by calcium phosphate crystals. If the blood calcium concentration were to decrease, parathyroid hormone would stimulate bone *resorption.* This term refers to the dissolution of the calcium phosphate crystals by special cells called *osteoclasts.*[5] By this means, additional calcium ions can be added to the blood. If blood calcium remains low for an extended period, as occurs in vitamin D deficiency, the bones can become abnormally softened. This is how rickets is produced (chapter 19).

[5]osteoclast: Gk. *osteon,* bone; *klastos,* broken

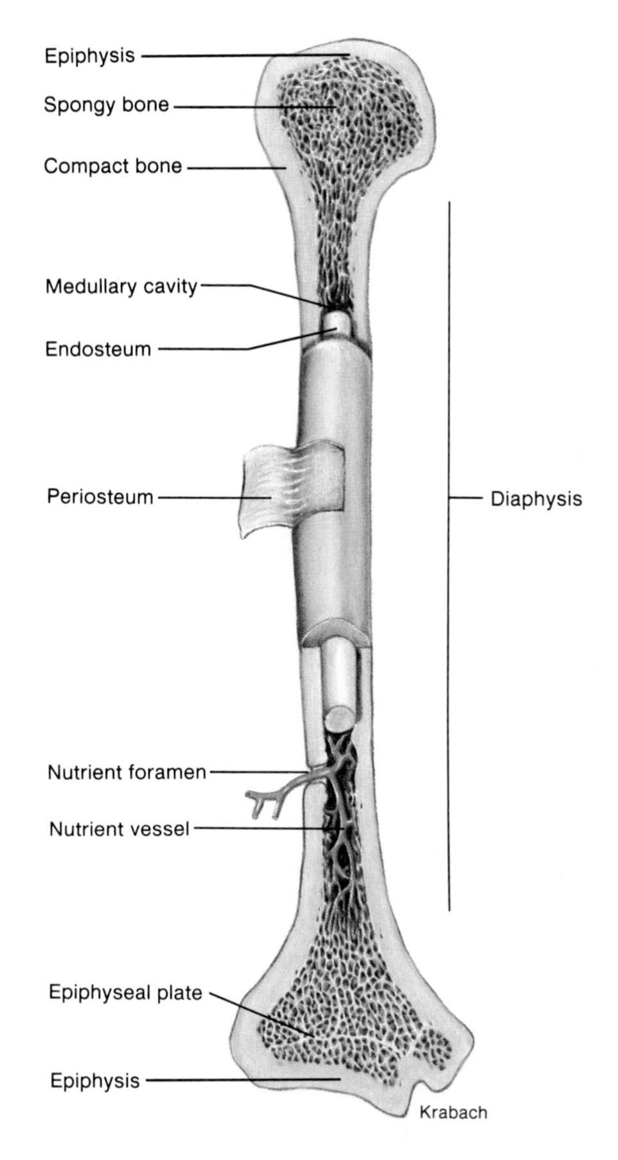

Figure 21.3. *A diagram of a long bone shown in longitudinal section.*

Bone Growth

During embryonic and fetal development, most of what will later become bones are first formed as structures composed of cartilage. These are known as the *cartilage models* of the bones. Later in development, the cartilage models undergo calcification. This causes degeneration of the cartilage, which is replaced by bone tissue in a process called *ossification.* Ossification of a long bone occurs in two locations, the diaphysis and the epiphysis (fig. 21.4).

Bone is formed as a result of this process, but cartilage remains at specific sites. Cartilage forms caps around both epiphyses (fig. 21.4*e, f*) at the regions where the bone will articulate with other bones. This cartilage is thus called **articular cartilage,** and serves to protect the bone surface within the joint. In addition, there is a plate of cartilage located between the epiphysis and the diaphysis in a growing bone. This cartilage is called the **epiphyseal plate** (fig. 21.4*e*).

Long bones can grow in length only at the epiphyseal plates. This is because, unlike bone, cartilage is not hardened by calcium phosphate crystals. Each cartilage cell can thus undergo mitosis to produce daughter cells, and these cells can secrete new cartilage matrix, which pushes the cells apart. As the epiphyseal plates get thicker, the part of the plate located toward the diaphysis of the bone undergoes calcification and ossification. The bone is thus longer than it was, and the remaining cartilage at the plates can repeat the process.

Fracture of a long bone in a growing person may be especially serious if it results in damage to an epiphyseal plate. If such an injury is untreated, or treated improperly, longitudinal growth of the bone may be arrested or retarded, resulting in permanent shortening of the limb.

Growth of bones at the epiphyseal plates is stimulated by growth hormone (chapter 18) and continues until, at a certain age, the entire cartilage is converted to bone (fig. 21.4*f*). Once this occurs, the bone can no longer grow in length. The age at which this occurs is predictable but different for different bones. For this reason, radiologists can determine the ages of people who are still growing by examining X-ray pictures of their bones (fig. 21.5).

After growth in length has ceased, bones can still be remodeled, becoming thicker or thinner in particular places. This occurs in response to pressure or stress on the bones. New bone can be added to the surface of preexisting bone by cells called **osteoblasts,**[6] and bone can be resorbed by osteoclasts as previously described. This is how, for example, orthodontic braces can cause movement of a tooth within its bony socket, or how a fractured bone can be repaired.

There are a variety of different types of bone fractures (fig. 21.6). A *simple,* or *closed, fracture* is one in which the fractured bone does not break through the skin. A *compound,* or *open, fracture* is one in which the fractured bone is exposed to the outside through the skin. A *greenstick fracture* is a break on just one side of the bone, while the other side is bowed.

1. Distinguish between the axial and appendicular skeleton.
2. Describe the structure of a long bone, and list the function of the bones of the skeletal system.
3. Explain how bones are formed from cartilage models, and how a long bone grows in length.

[6]osteoblast: Gk. *osteon,* bone; *blastos,* offspring

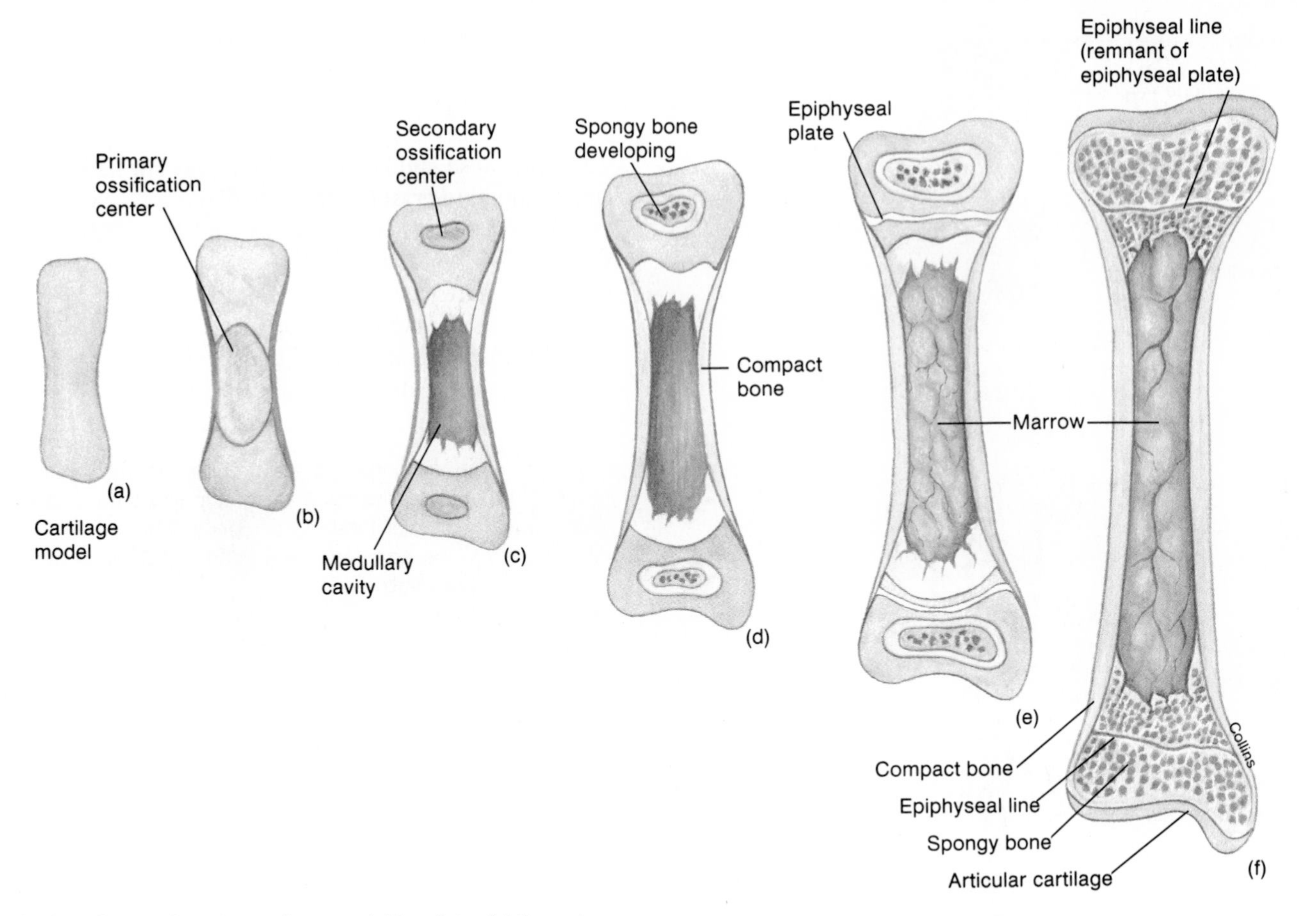

Figure 21.4. *Conversion of a cartilage model in a fetus* (a) *through different stages* (b–e) *to the bone of an adult* (f).

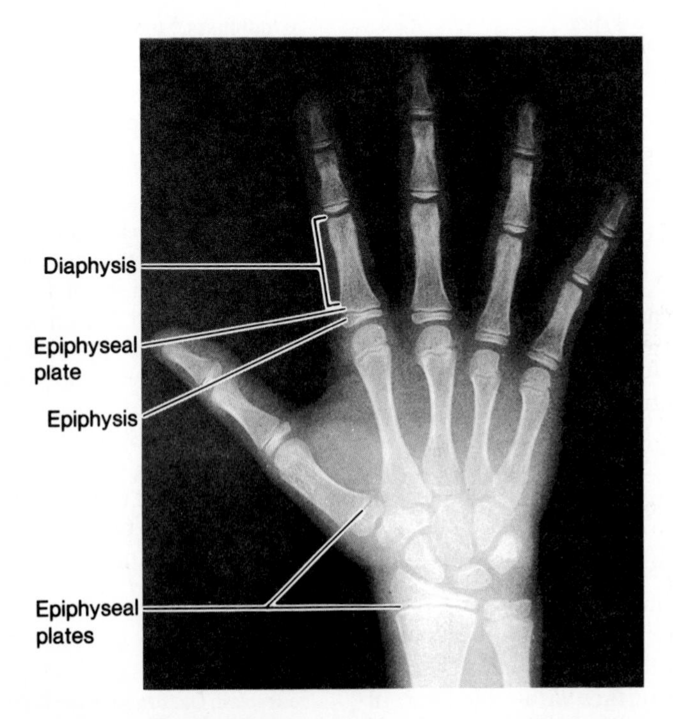

Figure 21.5. *The presence of epiphyseal plates, as seen in an X ray of a child's hand, indicates that bones are still growing in length.*

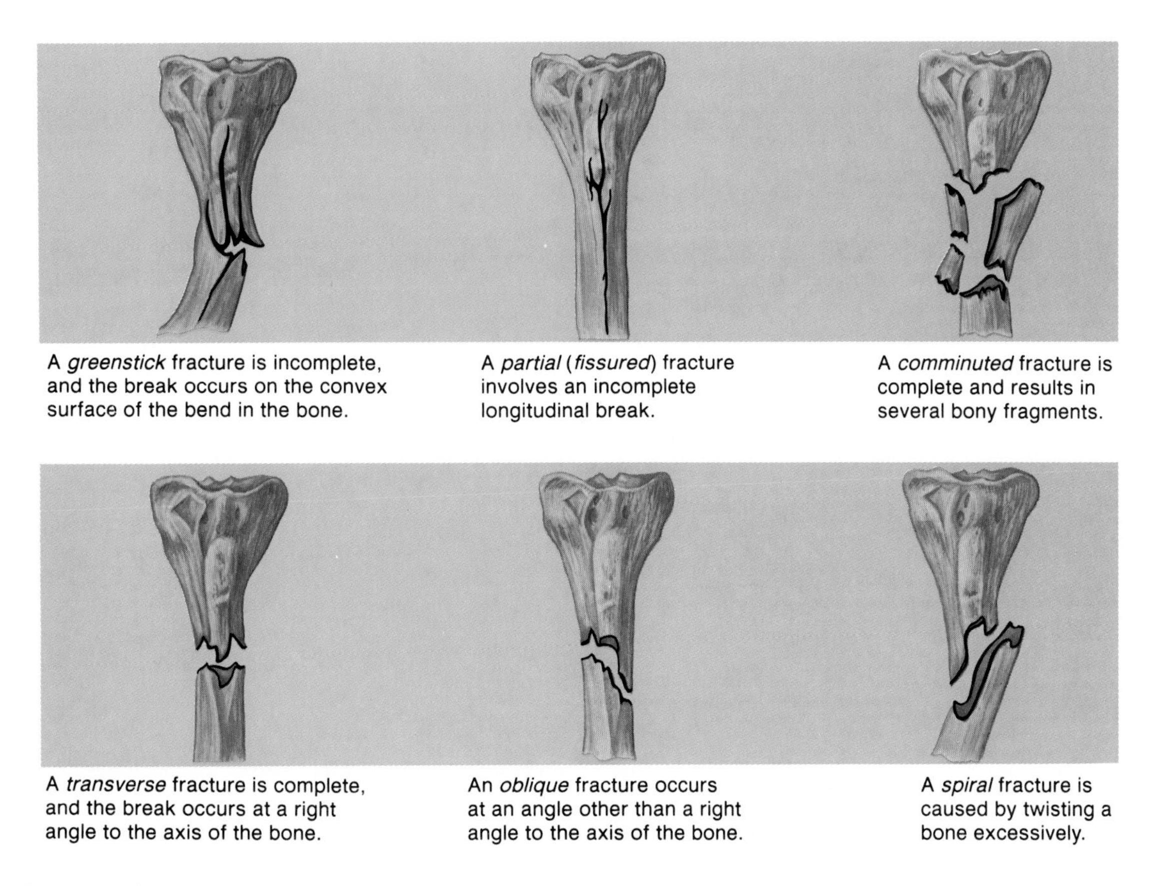

Figure 21.6. *Examples of types of bone fractures.*

Skeletal Muscles

Muscles produce movements of the body by shortening when they contract. Different skeletal muscles produce different categories of movements, such as flexion and extension, as a result of their different attachments to bones. The cells of a muscle are called muscle fibers, and each contains smaller subunits.

Skeletal muscles are under voluntary control, in contrast to cardiac and smooth muscles. Together with cardiac muscle, skeletal muscle is *striated*—it has a cross-banded appearance when viewed under the microscope (as described in chapter 5). The reasons for the striations, and their function in muscle contraction, will be discussed in a later section.

As their name implies, skeletal muscles are usually attached at each end to bone. These attachments are achieved by means of dense, tough bundles of connective tissue known as *tendons*. When skeletal muscles contract, they shorten in length. Both ends of a muscle, however, usually do not move toward the middle. Instead, one end moves toward the other, bringing one bone closer to the other through movement at a joint. The more movable attachment of a muscle is known as its **insertion,** and the less movable attachment is its **origin** (fig. 21.7). When a muscle contracts, therefore, its insertion is moved toward its origin.

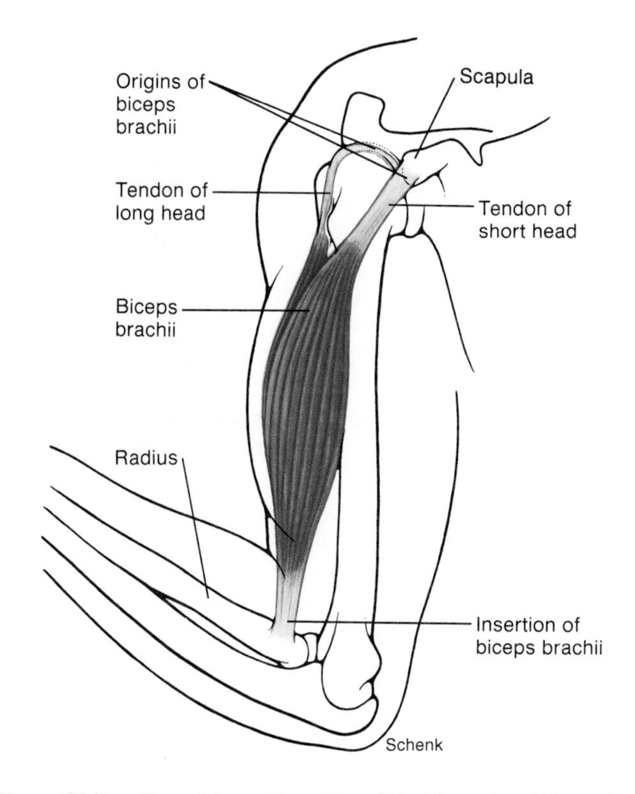

Figure 21.7. *The origin and insertion of the biceps brachii muscle.*

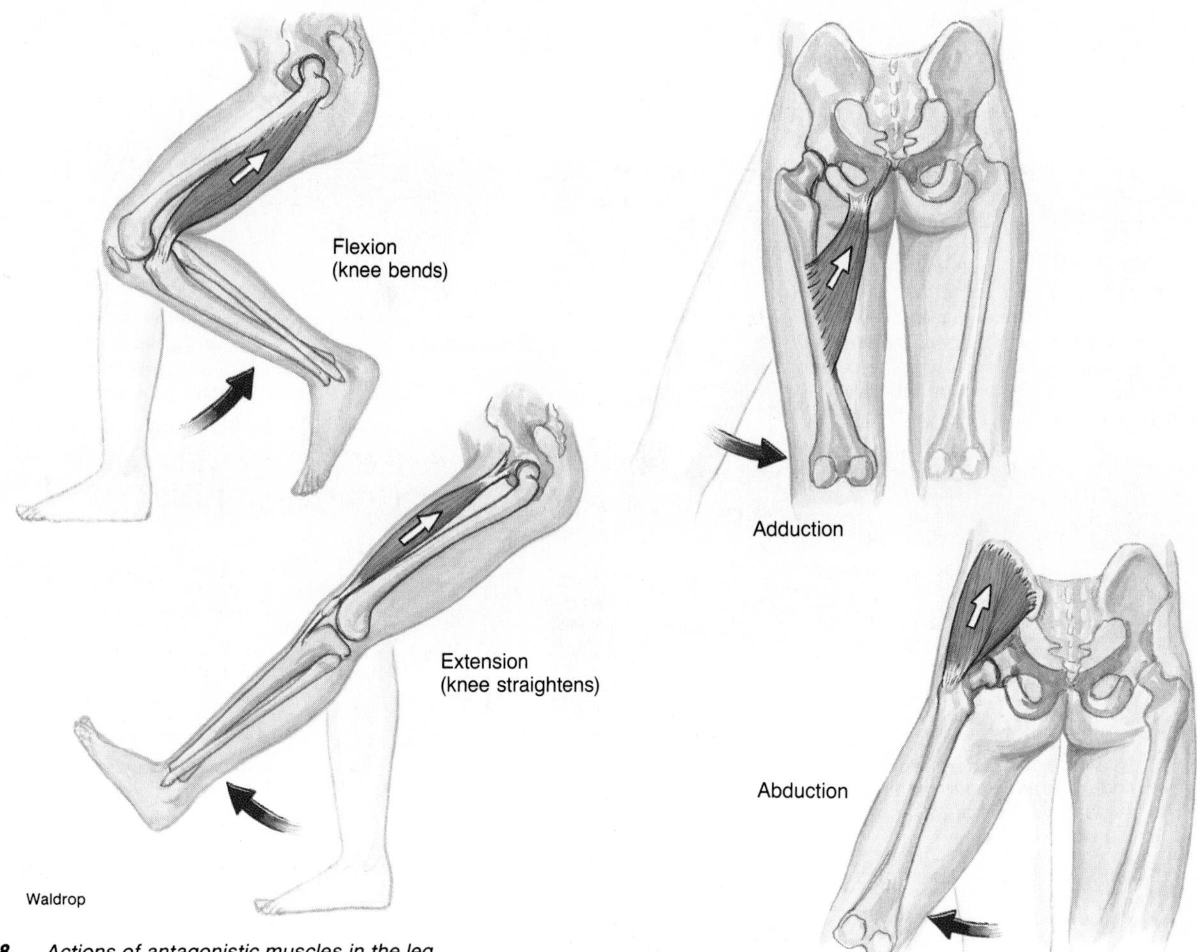

Figure 21.8. *Actions of antagonistic muscles in the leg.*

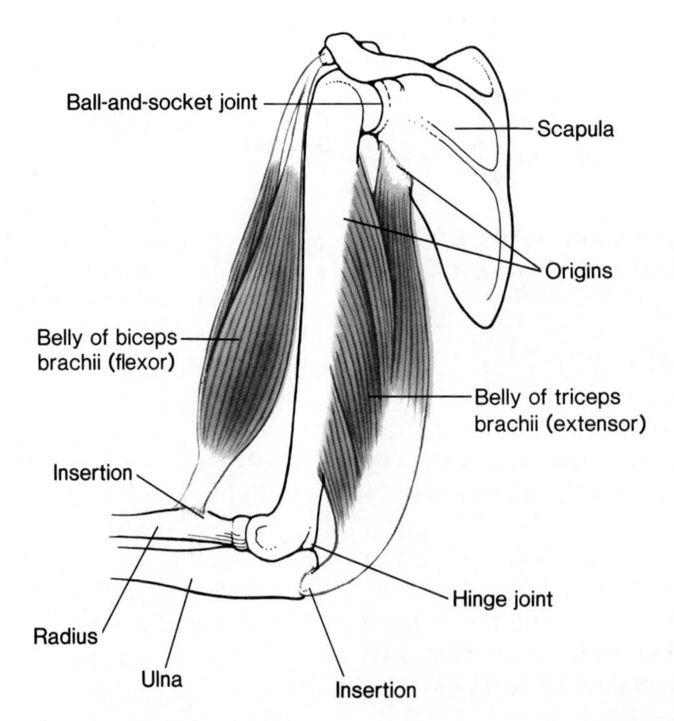

Figure 21.9. *An example of antagonistic muscles. The biceps brachii flexes the arm, whereas the triceps brachii extends the arm.*

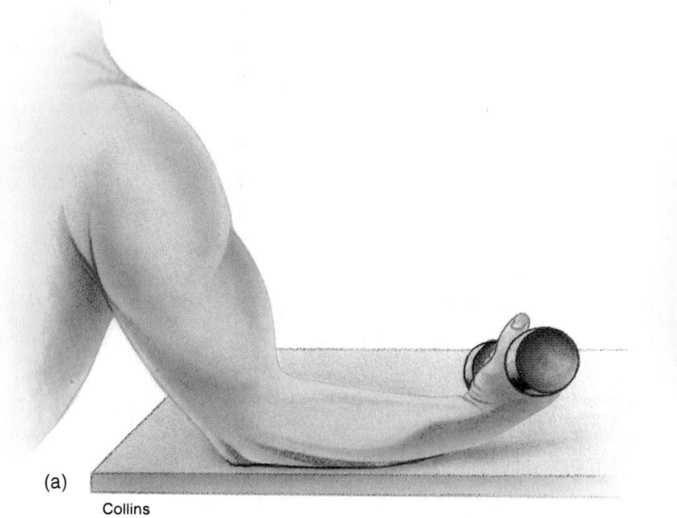

Figure 21.10. *(a) Isometric and (b) isotonic contraction.*

Actions of Muscles

The **action** of a muscle refers to the type of skeletal movement it produces when it contracts. **Flexors** decrease the angle of a joint, while **extensors** increase the angle (fig. 21.8). The *biceps brachii* of the arm, for example, is a flexor that bends the elbow joint, whereas the *triceps brachii* is an extensor of this joint (acting to straighten the joint). Other muscles act as *adductors* (moving an extremity toward the body, as in bringing the feet together) or *abductors* (moving an extremity away from the body, as in coming to a "parade rest" from a stand of "attention"). Still other muscles produce rotation of bones in particular directions. Depending upon their skeletal attachments, some muscles can have a combination of these actions.

In some cases, more than one muscle may produce the same action at the same joint. These muscles are said to have **synergistic[7] actions.** Flexion of the elbow joint, for example, is produced by the biceps brachii, the brachialis, and the brachioradialis. Other muscles have **antagonistic actions.** The actions of the biceps brachii and the triceps brachii, for example are antagonistic (fig. 21.9). It should be understood that both actions are produced by contraction and shortening of a muscle. The only way a muscle can lengthen is through relaxation.

Skeletal movements are produced when a muscle shortens as it contracts. This is the usual type of muscle contraction and is called an **isotonic[8] contraction. Isometric[9] contraction** of muscles occurs when they contract but are not allowed to shorten. This type of muscle contraction occurs, for example, when attempting without success to lift a very heavy object (fig. 21.10). Simultaneous contraction of antagonistic muscles can also cause isometric contractions, as each muscle exerts force that is equal and opposite to the other. This occurs when "making a muscle"; the biceps bulges during this procedure because it is contracting and exerting tension on its bony attachments, even though the forearm does not continue to move.

[7]synergistic: Gk. *synergein,* cooperate
[8]isotonic: Gk. *isos,* equal; *tonos,* tension
[9]isometric: Gk. *isos,* equal; *metron,* measure

Organization of a Muscle

A whole muscle (such as the biceps brachii) is covered by a connective tissue sheath known as the *epimysium.*[10] If the muscle is cut in half and viewed in cross section, it will be seen that connective tissue within the muscle divides the muscle tissue into parallel columns, or **muscle fascicles.**[11] These are the "strings" in stringy meat. The connective tissue that surrounds each muscle fascicle is called the *perimysium.*[12] If you take a portion of stringy meat in your hands, and carefully pull each string (fascicle) apart, you can see the perimysium tear. Often, the muscle fascicles get caught between the teeth when eating stringy meat.

Using a low-power microscope, each muscle fascicle can be dissected further into a large number of parallel **muscle fibers.** Each fiber is a muscle cell. Contraction of the whole muscle is produced by contraction of the muscle cells, or fibers. The muscle fibers are surrounded by a connective tissue covering called the *endomysium*[13] (fig. 21.11). The connective tissue of the endomysium, perimysium, and epimysium is continuous with that of the tendons on each end of the muscle and with the periosteum covering the bones to which the tendons attach. Because of this arrangement, muscles do not usually pull away from their bony attachments when they contract.

Under high power of an ordinary light microscope, or low power of an electron microscope, each muscle cell is seen to be composed of still smaller threadlike structures. These subcellular structures, arranged in parallel within each muscle cell,

[10]epimysium: Gk. *epi,* upon; *myos,* muscle
[11]fascicle: L. *fascis,* bundle
[12]perimysium: Gk. *peri,* around; *myos,* muscle
[13]endomysium: Gk. *endon,* within; *myos,* muscle

Figure 21.11. The relationship between muscle fibers and the connective tissues of the tendon, epimysium, perimysium, and endomysium. Closeup shows an expanded view of a single muscle fiber.

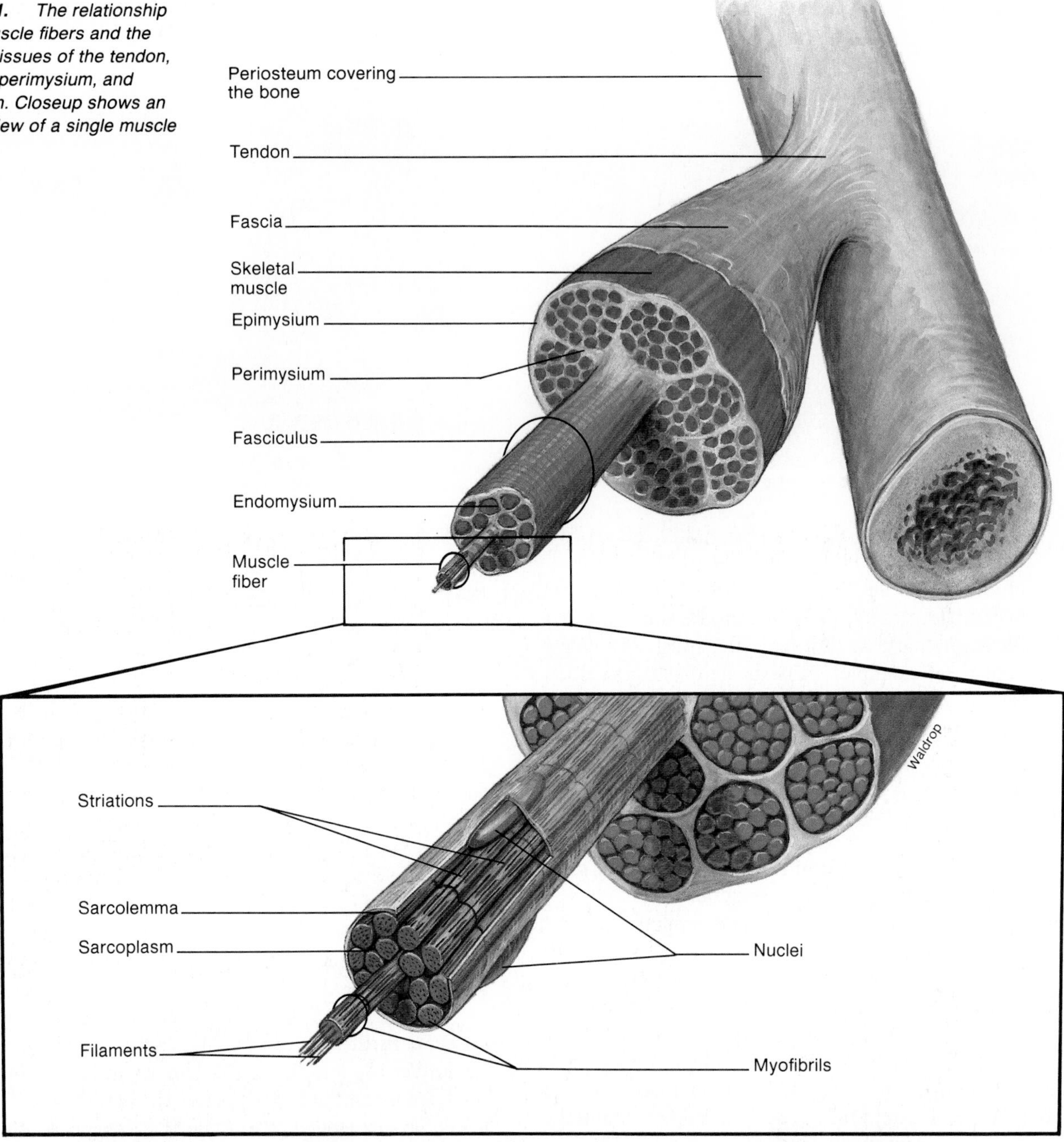

are known as **myofibrils** (fig. 21.11). Contraction of the muscle cell is produced by contraction of its myofibrils. This, in turn, is the result of contraction of still smaller subunits, known as the **muscle filaments,** located within each myofibril. The mechanism by which this is accomplished is described in the next section.

---

1. Describe some of the actions produced by skeletal muscles, citing examples.
2. Distinguish between isotonic and isometric contractions.
3. Going from the smallest structure to the largest, describe the organization of a muscle. Include its connective tissue components.

Mechanism of Muscle Contraction

Each muscle fiber contains smaller structures called myofibrils, and each of these is composed of thick and thin filaments arranged into functional units called sarcomeres. When a muscle contracts, the thick and thin filaments slide over each other. This sliding is produced by the action of cross-bridges that extend from the thick to the thin filaments. Contraction and relaxation is regulated by the concentration of Ca^{++} within the muscle cell.

The appearance of the muscle cells (fibers) under lower magnification of the electron microscope is shown in figure 21.12; striations can clearly be seen. When one muscle fiber is viewed

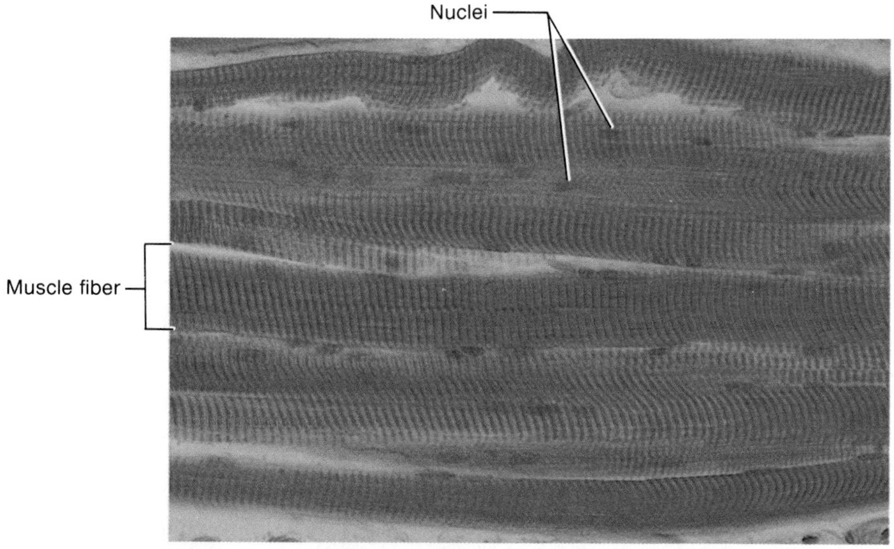

Nuclei

Muscle fiber

Figure 21.12. Appearance of skeletal muscle fibers (cells) under the microscope.

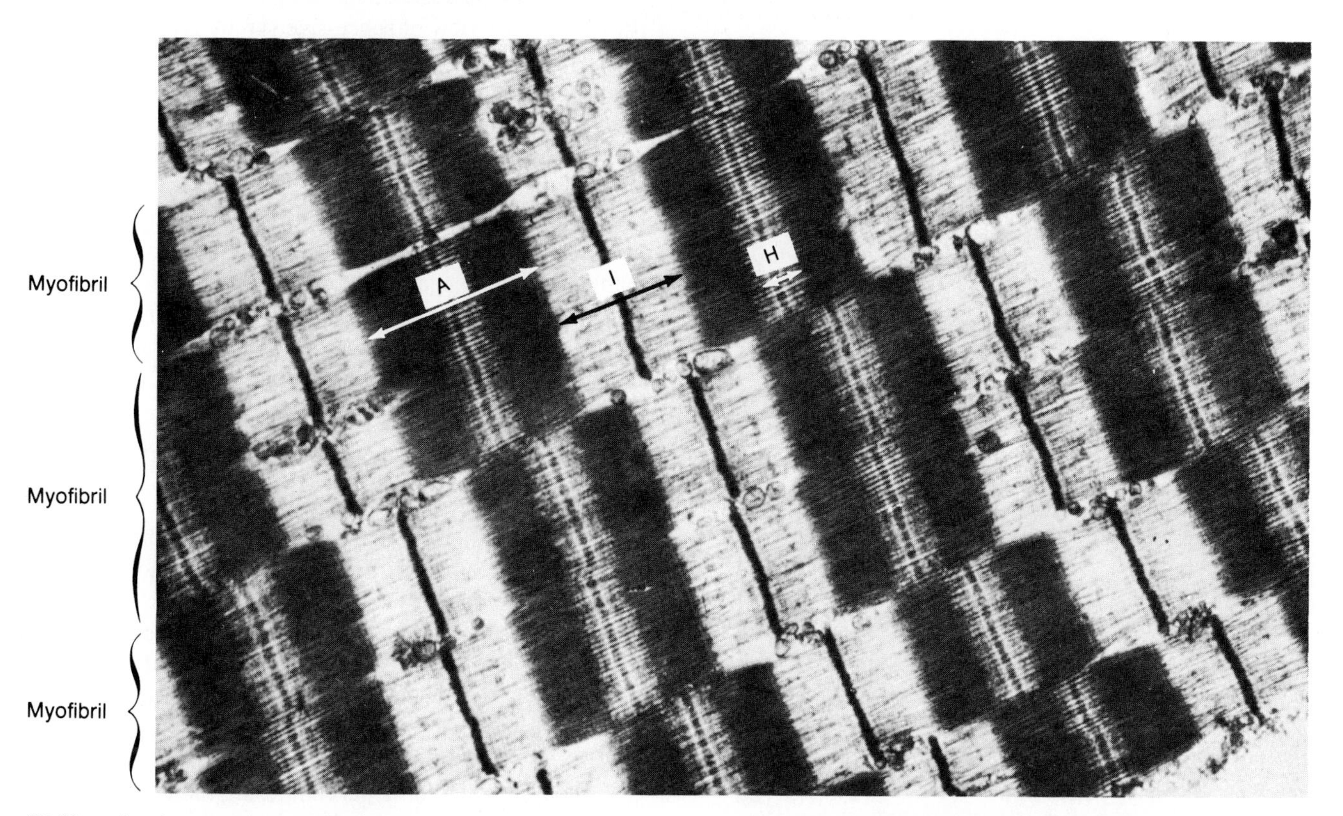

Figure 21.13. An electron micrograph of a longitudinal section of myofibrils, showing A, H, and I bands. Notice how the dark and light bands of each myofibril are stacked in register.

at higher magnification, the myofibrils within it are visible (fig. 21.13). These myofibrils extend in parallel from one end of the muscle fiber to the other. The myofibrils are so densely packed that other organelles—such as mitochondria and intracellular membranes—are restricted to the narrow cytoplasmic spaces that remain between adjacent myofibrils. Using the electron microscope, it can be seen that the myofibrils are striated with dark and light bands. The dark bands are called the **A bands,** and the lighter areas between them are the **I bands** (fig. 21.13).

Thick and Thin Filaments

When a myofibril is observed at high magnification in longitudinal section (side view), the A bands are seen to contain **thick filaments** that are stacked in register. It is these thick filaments that give the A band its dark appearance. The lighter I band, in contrast, contains **thin filaments.** The thick filaments are composed of the protein **myosin,** whereas the thin filaments are composed primarily of the protein **actin.**

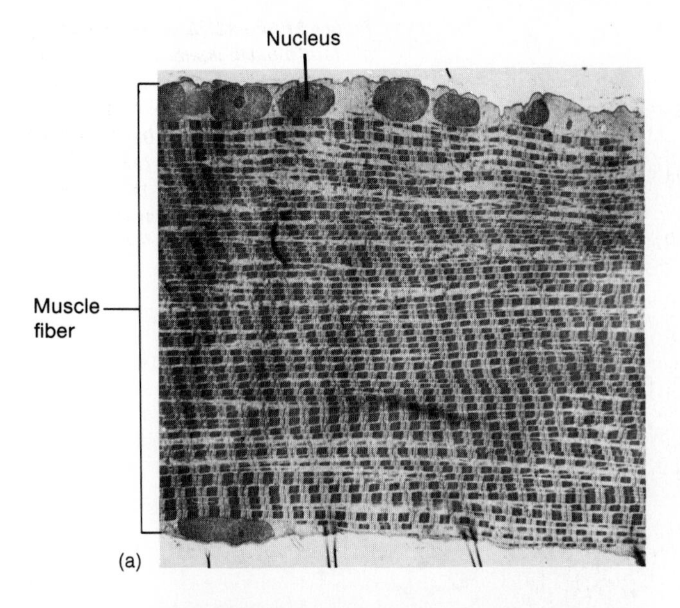

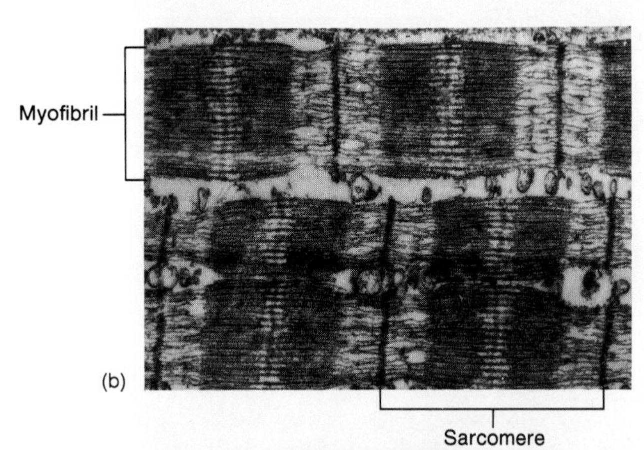

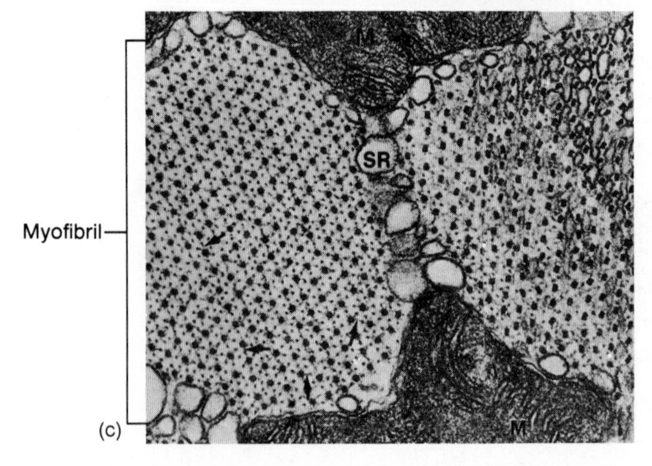

Figure 21.14. *Electron micrographs of myofibrils of a muscle fiber. (a) At low power (1,600×), a single muscle fiber containing numerous myofibrils. (b) At high power (53,000×), myofibrils in longitudinal section. Notice the sarcomeres and overlapping thick and thin filaments. (c) The hexagonal arrangement of thick and thin filaments as seen in cross section (arrows point to cross-bridges; SR = sarcoplasmic reticulum; M = mitochondria).*

(c) From Tissues and Organs: A Text-Atlas of Scanning Electron Microscopy *by Richard G. Kessel and Randy H. Kardon. Copyright © 1979 by W. H. Freeman and Company. Reprinted by permission.*

The I bands within a myofibril are the lighter areas that extend from the edge of one stack of thick myosin filaments to the edge of the next stack of thick filaments. They are light in appearance because they contain only thin filaments. The thin filaments, however, do not end at the edges of the I bands. Instead, each thin filament continues partway into the A bands on each side (between the stack of thick filaments on each side of an I band). Since thick and thin filaments overlap at the edges of each A band, the edges of the A band are darker in appearance than the central region. These central, lighter regions of the A bands are called the *H bands* (for *helle,* a German word for "bright"). The central H bands thus contain only thick filaments that are not overlapped with thin filaments.

In the center of each I band is a thin dark Z line. The arrangement of thick and thin filaments between a pair of Z lines forms a repeating pattern that serves as the basic subunit of striated muscle contraction. These subunits, from Z to Z, are known as **sarcomeres.** A longitudinal section of a myofibril thus presents a side view of successive sarcomeres.

This side view is, in a sense, misleading; there are numerous sarcomeres within each myofibril that are out of the plane of the section (and out of the picture). A better appreciation of the three-dimensional structure of a myofibril can be obtained by viewing the myofibril in cross section. In this view, it can be seen that the Z lines are in reality disc shaped and that the thin filaments that penetrate these Z discs surround the thick filaments in a hexagonal arrangement (fig. 21.14*c*). If one concentrates on a single row of dark thick filaments in this cross section, the alternating pattern of thick and thin filaments seen in longitudinal section becomes apparent.

Sliding Filament Theory of Contraction

When a muscle contracts isotonically, it decreases in length as a result of the shortening of its individual fibers. Shortening of the muscle fibers, in turn, is produced by shortening of their myofibrils, which occurs as a result of the shortening of the distance from Z line to Z line. As the sarcomeres shorten in length, however, the A bands do *not* shorten but instead move closer together. The I bands—which represent the distance between A bands of successive sarcomeres—decrease in length (table 21.2).

The thin actin filaments composing the I band, however, do *not* shorten. Close examination reveals that the thick and thin filaments remain the same length during muscle contraction. Shortening of the sarcomeres is produced, not by shortening of the filaments, but rather by the *sliding* of thin filaments over and between the thick filaments. In the process of contraction, the thin filaments on either side of each A band slide deeper and deeper toward the center, producing increasing amounts of overlap with the thick filaments. The I bands (containing only thin filaments) and H bands (containing only thick filaments) thus become narrower during contraction (fig. 21.15).

Sliding of the filaments is produced by the action of numerous **cross-bridges** that extend out from the myosin toward the actin. These cross-bridges are part of the myosin proteins

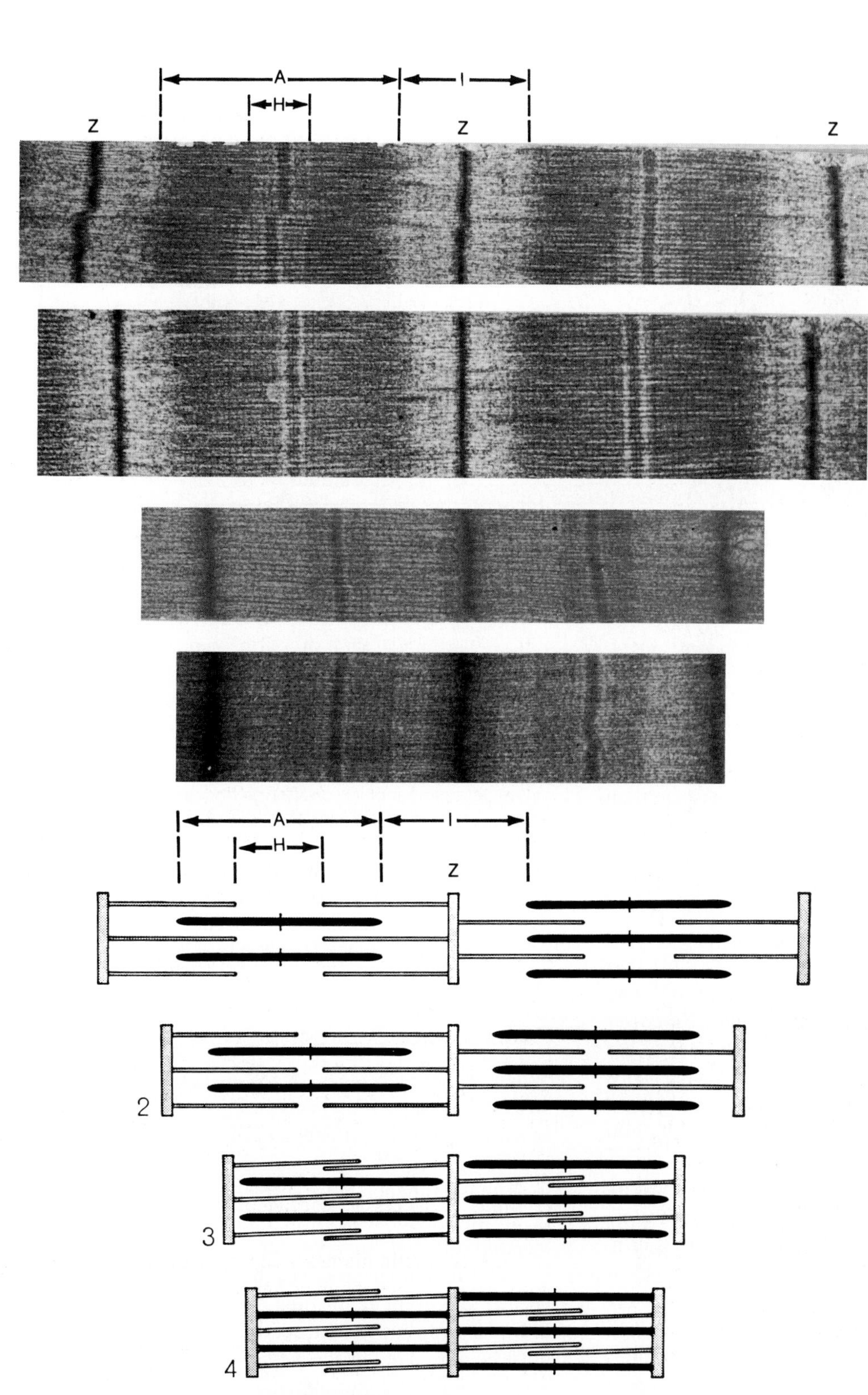

Figure 21.15. *The sliding filament model of contraction. As the filaments slide, the Z lines are brought closer together. The A bands remain the same length during contraction, but the I and H bands get progressively shorter and may eventually become obliterated.*

Table 21.2 Summary of the sliding filament theory of contraction

1. A myofiber, together with all its myofibrils, shortens by movement of the insertion toward the origin of the muscle.
2. Shortening of the myofibrils is caused by shortening of the sarcomeres—the distance between Z lines (or discs) is reduced.
3. Shortening of the sarcomeres is accomplished by sliding of the myofilaments—each filament remains the same length during contraction.
4. Sliding of the filaments is produced by asynchronous power strokes of myosin cross-bridges, which pull the thin filaments (actin) over the thick filaments (myosin).
5. The A bands remain the same length during contraction, but are pulled toward the origin of the muscle.
6. Adjacent A bands are pulled closer together as the I bands between them shorten.
7. The H bands shorten during contraction as the thin filaments from each end of the sarcomeres are pulled toward the middle.

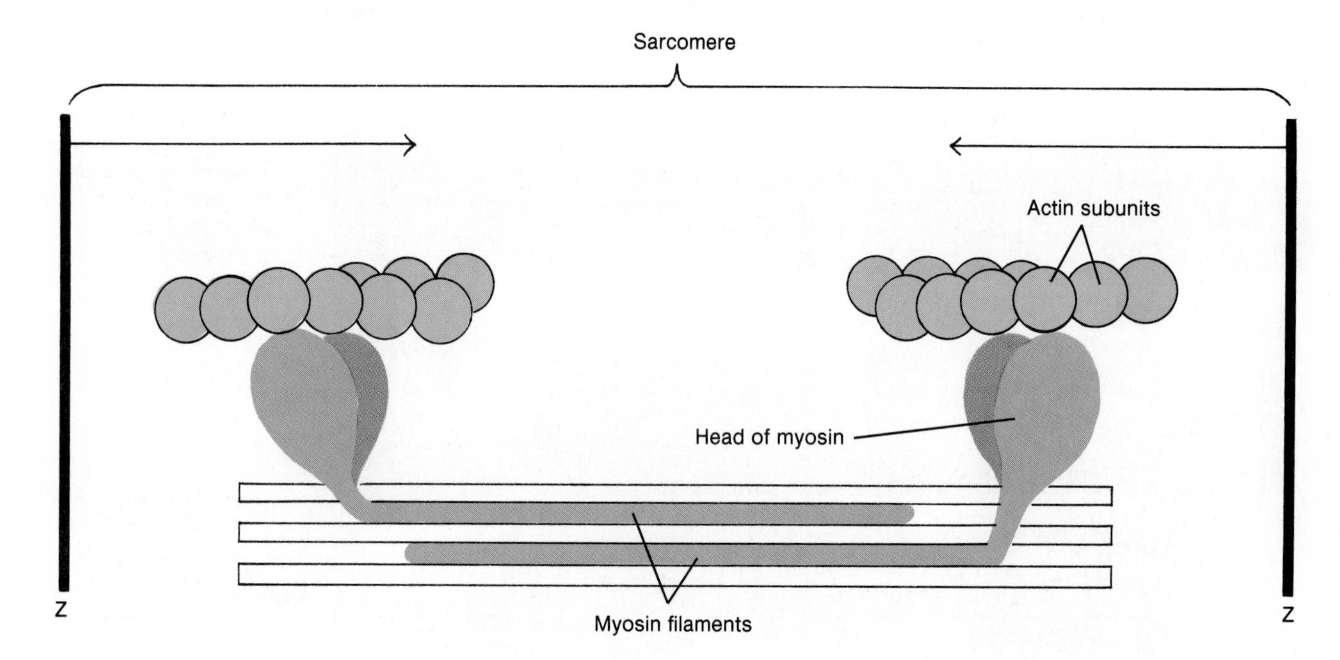

Figure 21.16. Myosin cross-bridges are oriented in opposite directions on either side of a sarcomere.

that extend from the axis of the thick filaments to form "arms" that terminate in globular "heads" (fig. 21.16). The orientation of cross-bridges on one side of a sarcomere is opposite to that on the other side, so that when they attach to actin on each side of the sarcomere they can pull the actin from each side toward the center.

The myosin heads are able to bond to specific attachment sites in the actin subunits. In order to do this, they must first become activated by splitting a molecule of ATP into ADP and P_i. When the cross-bridges bond to actin they produce a *power stroke,* which pulls the thin filaments toward the center of the A bands. At the end of the power stroke each cross-bridge bonds to a fresh ATP molecule. This bonding of the cross-bridge to a new ATP causes the cross-bridge to break its bond with actin and resume its resting orientation.

Because the cross-bridges are quite short, a single contraction cycle and power stroke of all the cross-bridges in a muscle would shorten the muscle by only about 1% of its resting length. Since muscles can shorten up to 60% of their resting lengths, it is obvious that the contraction cycles must be repeated many times. In order for this to occur, the cross-bridges must detach from the actin at the end of a power stroke, reassume their resting orientation, and then reattach to the actin and repeat the cycle.

The detachment of a cross-bridge from actin at the end of a power stroke requires that a new ATP molecule binds to the cross-bridge. The importance of this process is illustrated by the muscular contracture called **rigor mortis,** which occurs due to lack of ATP when the muscle dies. This results in the formation of "rigor complexes" between myosin and actin that cannot detach. In rigor mortis, all of the cross-bridges are attached to actin at the same time.

Regulation of Contraction

When the cross-bridges attach to actin they undergo power strokes and cause muscle contraction. In order for a muscle to relax, therefore, the attachment of myosin cross-bridges to actin must be prevented. The regulation of cross-bridge attachment to actin is a function of two proteins found associated with actin in the thin filaments. These inhibitory proteins are known as *tropomyosin* and *troponin* (fig. 21.17).

The inhibition produced by tropomyosin and troponin is needed to prevent an unstimulated muscle from contracting. In order for a muscle to contract, therefore, the inhibitory proteins must be moved out of the way of the cross-bridges. This occurs when a muscle is stimulated to contract by a somatic motor neuron.

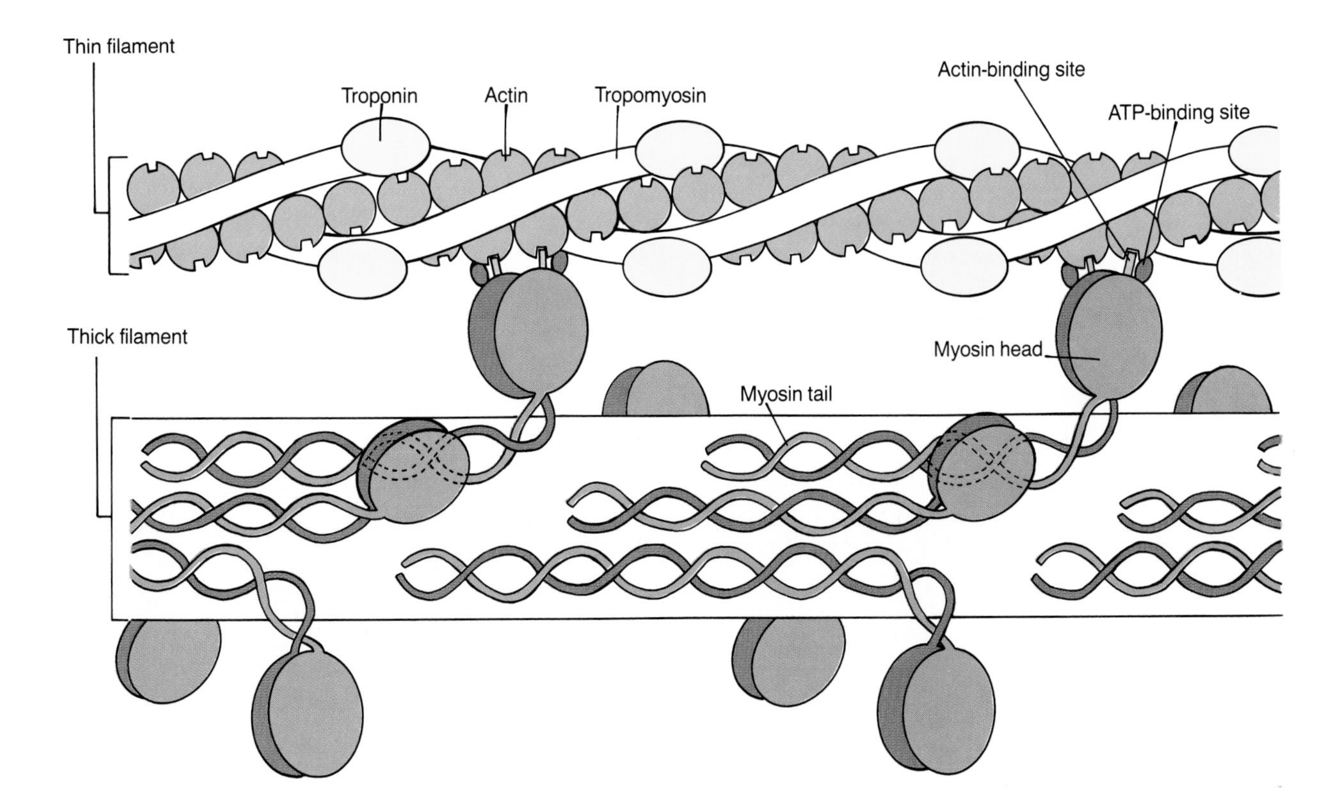

Figure 21.17. *The structure of myosin, showing its binding sites for ATP and for actin.*

When you are going to contract a particular muscle, the somatic motor neurons that go to that muscle are stimulated to produce action potentials (nerve impulses). These action potentials are conducted by the axon of each neuron to its ending, at the synapse that the axon makes with the muscle fibers (fig. 21.18). At this point, acetylcholine (ACh) is released at the synapse (chapter 6), and diffuses to the muscle cell membrane. Through effects at the membrane, the ACh stimulates the production of action potentials in the muscle fiber.

Electrical Excitation of Muscle Fibers

The muscle cell membrane conducts action potentials in the same manner as the axon of a neuron. As described in chapter 6, action potentials are produced by movements of ions across a membrane that separates the extracellular fluid from the cytoplasm of the cell. These electrical impulses are thus conducted along the muscle cell membrane. This membrane, however, is not flat; it tunnels into the cytoplasm of the muscle cell at particular places. The extensions of the cell membrane into the cell form *transverse tubules.* Impulses are conducted into the muscle cell along the transverse tubules.

The transverse tubules come into close proximity with a system of membranous sacs within the muscle cell. These sacs form the *sarcoplasmic reticulum* within the muscle cell (fig. 21.19). When the muscle is at rest, the sarcoplasmic reticulum accumulates calcium ions (Ca^{++}) by active transport, so that the concentration of Ca^{++} within the sacs is much higher than it is in the cytoplasm of the muscle cell. The stage is now set for stimulation of muscle contraction.

Excitation-Contraction Coupling

The conduction of action potentials into the muscle fiber along the transverse tubules stimulates the sarcoplasmic reticulum to release its Ca^{++}. This Ca^{++} then diffuses into the cytoplasm of the muscle cell and attaches to troponin. The bonding of Ca^{++} to troponin causes troponin and the tropomyosin to which it is attached to shift position on the thin filaments. This is like removing the safety catch on a gun. Now, the myosin cross-bridges

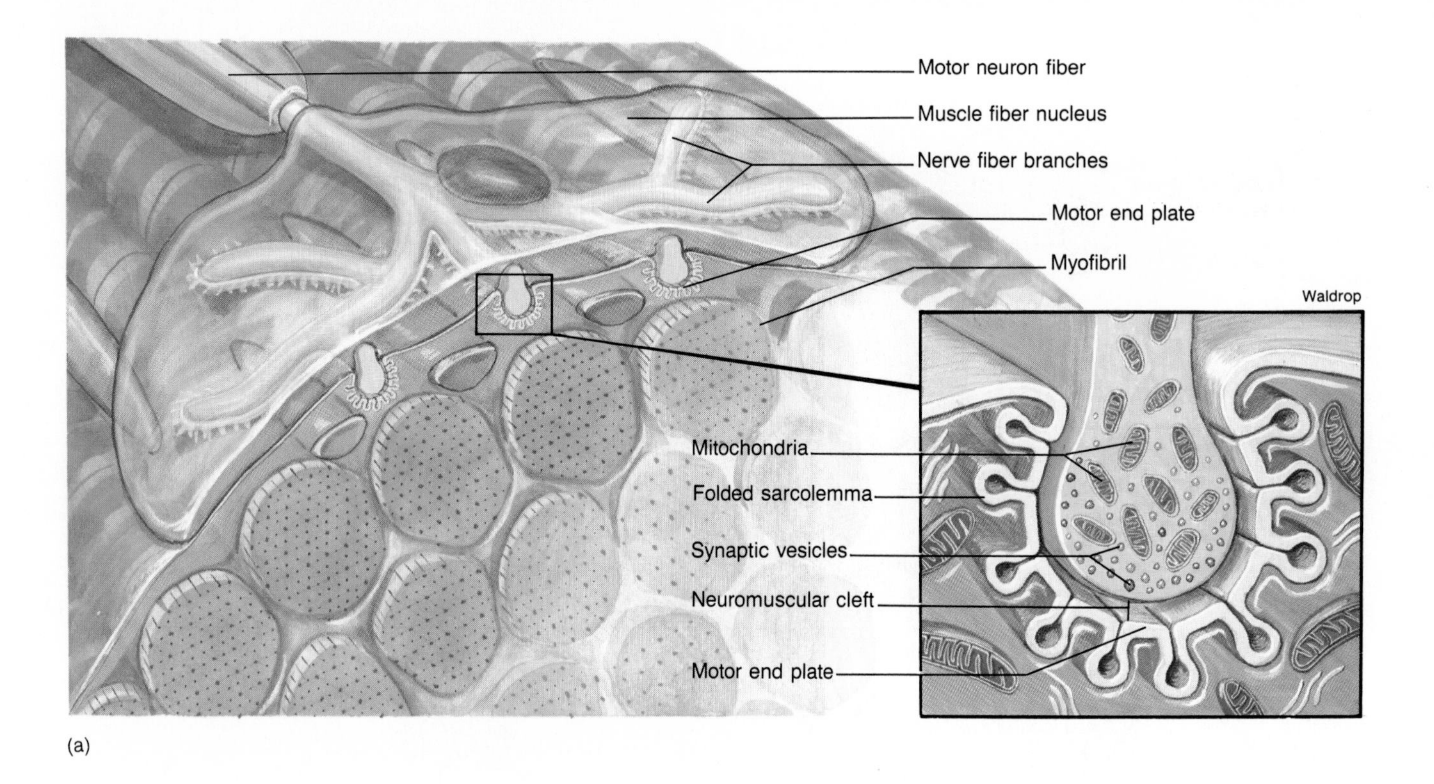

(a)

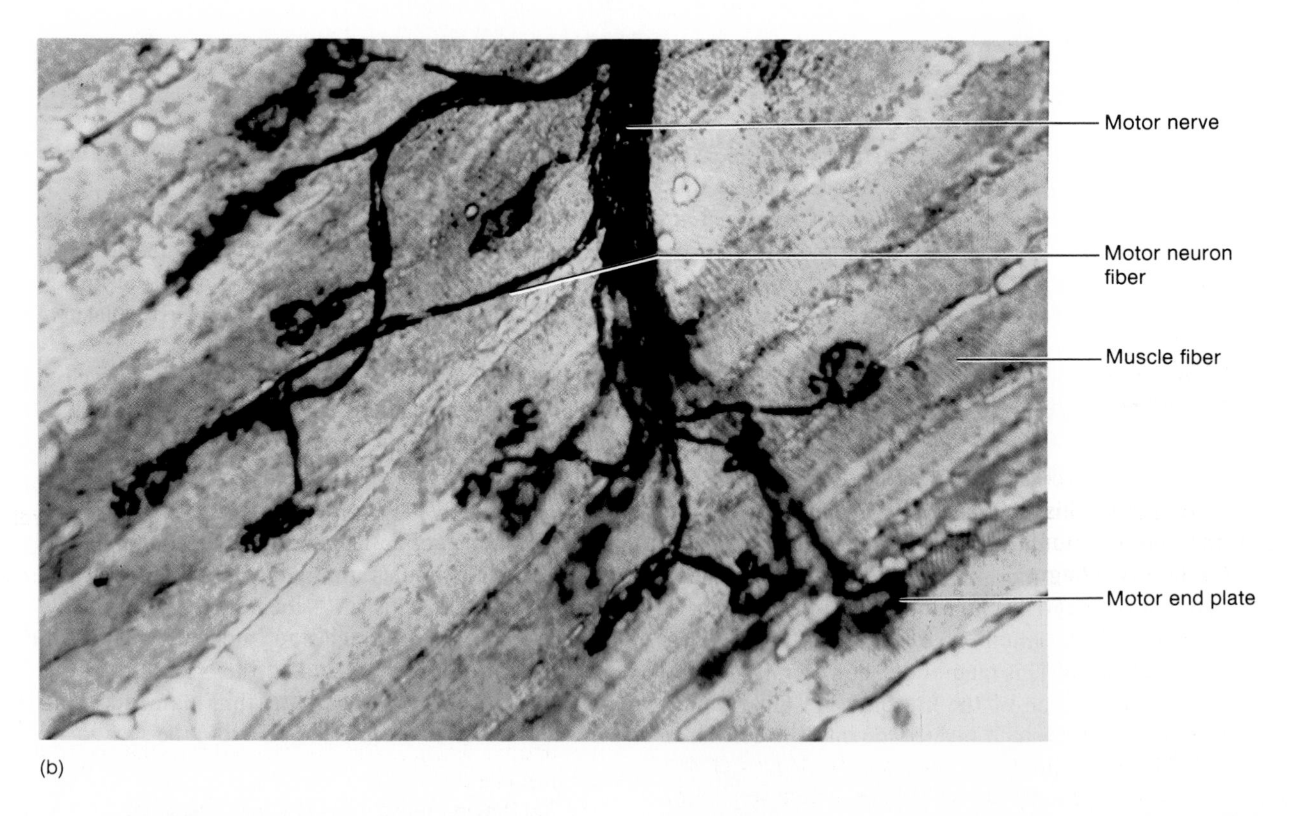

(b)

Figure 21.18. *A motor end plate at the neuromuscular junction.* (a) *A neuromuscular junction is the site where the nerve fiber and muscle fiber meet. The motor end plate is the specialized portion of the sarcolemma of a muscle fiber surrounding the terminal end of the axon. Note the slight gap between the membrane of the axon and that of the muscle fiber.* (b) *A photomicrograph of muscle fibers and motor end plates. A motor neuron and the muscle fibers it innervates constitute a motor unit.*

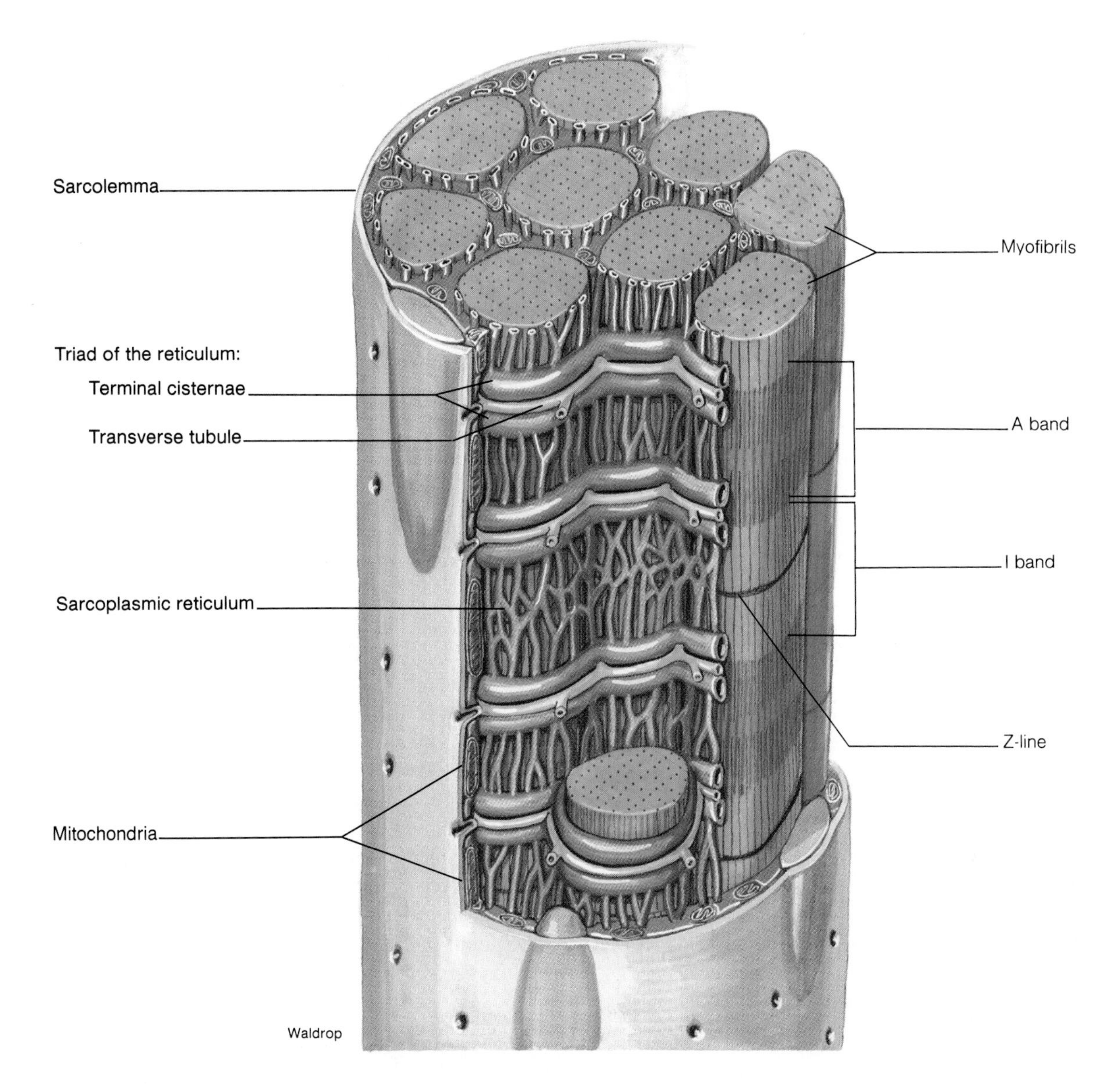

Figure 21.19. *A drawing of myofibrils, showing their relationship to the sarcoplasmic reticulum, transverse tubules, and sarcolemma.*

can attach to actin and produce power strokes (fig. 21.20). The muscle thus contracts. This mechanism by which electrical excitation of the muscle stimulates contraction is known as **excitation-contraction coupling.**

As long as action potentials continue to be produced—which is as long as the neural stimulation of the muscle is continued—Ca^{++} will remain attached to troponin and cross-bridges will be able to undergo contraction cycles. When neural activity and action potentials in the muscle fiber cease, the sarcoplasmic reticulum quickly removes the Ca^{++} from the troponin and cytoplasm. Troponin and tropomyosin move back to their inhibitory position, and muscle relaxation occurs.

1. Describe how the lengths of the A, I, and H bands change during contraction, and explain these changes according to the sliding filament theory.
2. Describe a cycle of cross-bridge activity during contraction, and explain the role of ATP in this cycle.
3. Draw a sarcomere in a relaxed muscle and a sarcomere in a contracted muscle and label the bands in each.
4. Draw a flowchart (using arrows) of the events that occur between the time that ACh is released from a nerve ending and the time that Ca^{++} is released from the sarcoplasmic reticulum.
5. Explain how muscle relaxation is produced.

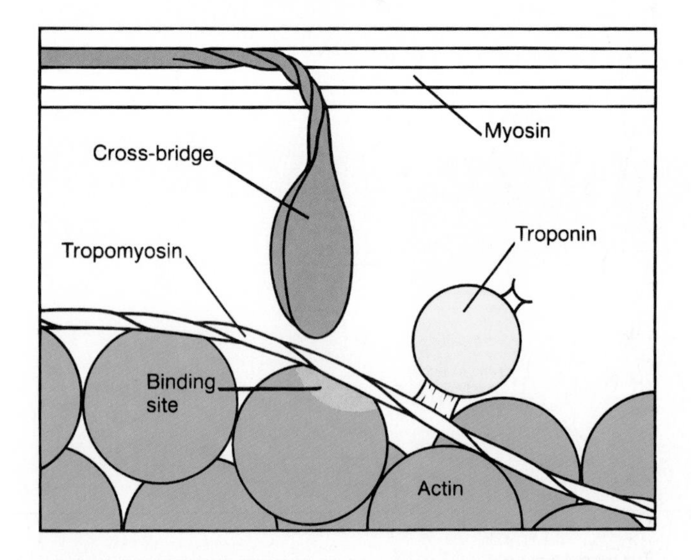

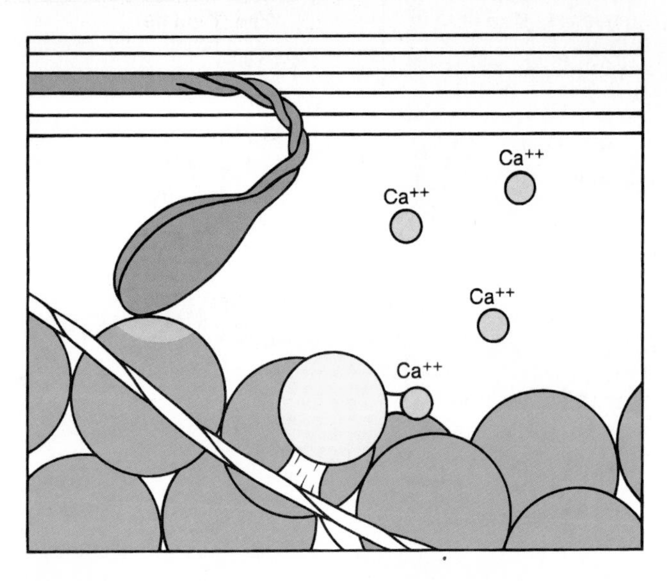

Figure 21.20. The attachment of Ca^{++} to troponin causes movement of the troponin-tropomyosin complex, which exposes binding sites on the actin. The myosin cross-bridges can then attach to actin and undergo a power stroke.

Endurance, Strength, and Muscle Fatigue

Skeletal muscles generate ATP through aerobic and anaerobic respiration. The aerobic and anaerobic abilities of skeletal muscle fibers differ, depending on the type of muscle fiber. Fast-twitch (type I) fibers are adapted for anaerobic respiration and slow-twitch (type II) fibers are adapted for aerobic respiration. The former fibers can produce more power, but the latter have a greater resistance to fatigue.

Skeletal muscles at rest obtain most of their energy from the aerobic respiration of fatty acids (chapter 18). During exercise, muscle glycogen and blood glucose are also used as energy sources. Energy obtained by cell respiration is used to make ATP, which serves as the immediate source of energy for (1) the movement of the cross-bridges for muscle contraction and (2) the pumping of Ca^{++} into the sarcoplasmic reticulum for muscle relaxation.

Slow- and Fast-Twitch Fibers

Skeletal muscle fibers can be divided on the basis of their contraction speed (time required to reach maximum tension) into **slow-twitch,** or **type I fibers,** and **fast-twitch,** or **type II fibers.** The extraocular muscles that move the eyes, for example, have a high proportion of fast-twitch fibers and reach maximum tension in about 7.3 msec (milliseconds—thousandths of a second); the soleus muscle in the leg, in contrast, has a high proportion of slow-twitch fibers and requires about 100 msec to reach maximum tension (fig. 21.21). A muscle such as the gastrocnemius (calf muscle) contains both fast- and slow-twitch fibers, although fast-twitch fibers predominate.

Muscles like the soleus are *postural muscles* that must be able to sustain a contraction for a long period of time without fatigue. This is aided by other characteristics of slow-twitch (type I) fibers that endow them with a high capacity for aerobic respiration. Slow-twitch fibers are served by large numbers of capillaries, have numerous mitochondria and aerobic respiratory enzymes, and have a high concentration of *myoglobin* pigment. Myoglobin is a red pigment—hence the alternate name of *red fibers* for these muscle cells—which is related to the hemoglobin pigment of blood and serves to improve the delivery of oxygen to the slow-twitch fibers.

The thicker, fast-twitch (type II) fibers have a lower capillary supply, fewer mitochondria, and less myoglobin; hence, these fibers are also called *white fibers.* Fast-twitch fibers are adapted to respire anaerobically by a large store of glycogen and a high concentration of glycolytic enzymes. In addition to the type I (slow-twitch) and type II (fast-twitch) fibers, human muscles may also have an intermediate form of fibers. These intermediate fibers are fast-twitch but also have a high aerobic capability. These are sometimes called type IIA, to distinguish them from the anaerobically adapted fast-twitch fibers (which are then labeled IIB). A comparison of these fiber types is summarized in table 21.3.

Muscle Fatigue

Muscle fatigue may be defined as the inability to maintain a particular muscle tension when the contraction is sustained, or to reproduce a particular tension during rhythmic contraction

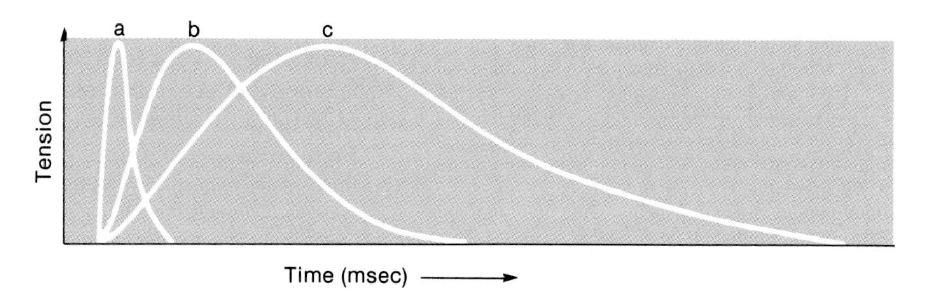

Figure 21.21. *A comparison of the rates with which maximum tension is developed in three muscles. These include the relatively fast-twitch extraocular (a) and the gastrocnemius (b) muscles and the slow-twitch soleus muscle. (c).*

Table 21.3 Characteristics of red, intermediate, and white muscle fibers

	Red (Type I)	Intermediate (Type IIA)	White (Type IIB)
Diameter	Small	Intermediate	Large
Z-line thickness	Wide	Intermediate	Narrow
Glycogen content	Low	Intermediate	High
Resistance to fatigue	High	Intermediate	Low
Capillaries	Many	Many	Few
Myoglobin content	High	High	Low
Respiration type	Aerobic	Aerobic	Anaerobic
Twitch rate	Slow	Fast	Fast
Myosin ATPase content	Low	High	High

over time. Fatigue during a sustained maximal contraction—as when lifting an extremely heavy weight—appears to be due to an accumulation of extracellular K^+. This reduces the membrane potential of muscle fibers and interferes with their ability to produce action potentials. Fatigue under these circumstances lasts only a short time, and maximal tension can again be produced after less than a minute rest.

Fatigue during moderate exercise involving rhythmic contractions occurs as the slow-twitch fibers deplete their reserve glycogen and fast-twitch fibers are increasingly recruited. Fast-twitch fibers obtain their energy through anaerobic respiration, converting glucose to lactic acid. The accumulation of lactic acid results in a rise in intracellular H^+ and a fall in pH. This indirectly interferes with the ability of muscles to contract in response to electrical excitation (the low pH interferes with excitation-contraction coupling). Since the accumulation of lactic acid contributes to muscle pain, however, most people stop exercising before this stage of muscle fatigue occurs.

Adaptations to Exercise

The **maximal oxygen uptake** is the maximum rate that the body can consume oxygen during very strenuous exercise. It averages 50 ml O_2 per minute per kilogram body weight in young males. Females average 25% lower. Top endurance athletes, however, can have maximal oxygen uptakes as high as 86 ml O_2 per minute per kilogram. This is a measure of their greater aerobic respiratory ability.

When a person performs exercise at a low level of effort, such that the oxygen consumption is less than 50% of its maximum, the energy for muscle contraction is obtained almost entirely from aerobic cell respiration. Anaerobic cell respiration, with its consequent production of lactic acid, contributes to the energy requirements as the exercise level rises and more than 60% of the maximal oxygen uptake is required. Top endurance-trained athletes, however, can continue to respire aerobically, with little lactic acid production, at up to 80% of their maximal

Table 21.4 Summary of the effects of endurance training (long-distance running, swimming, bicycling, etc.) on skeletal muscles

1. Improved ability to obtain ATP from oxidative phosphorylation
2. Increased size and number of mitochondria
3. Less lactic acid produced per given amount of exercise
4. Increased myoglobin content
5. Increased intramuscular triglyceride content
6. Increased lipoprotein lipase (enzyme needed to utilize lipids from blood)
7. Increased proportion of energy derived from fat, less from carbohydrates
8. Lower rate of glycogen depletion during exercise
9. Improved efficiency in extracting oxygen from blood

oxygen uptake. These athletes thus produce less lactic acid at a given level of exercise than the average person and, therefore, are less subject to fatigue than the average person.

Endurance training cannot change fast-twitch (type II) fibers to slow-twitch (type I) fibers. All fiber types, however, adapt to endurance training by an increase in myoglobin and aerobic respiratory enzymes, so that the maximal oxygen uptake can be increased by up to 20% through this training. In addition to changes in aerobic capacity, fibers show an increase in their content of triglycerides, which serves as an alternate energy source helping to spare their stores of glycogen. A summary of the changes that occur as a result of endurance training is presented in table 21.4.

Endurance training does not increase the size of muscles. Muscle enlargement is produced only by frequent bouts of muscle contraction against a high resistance—as in weight lifting. As a result of this latter type of training, type II muscle fibers become thicker, and the muscle therefore grows by hypertrophy (increase in cell size, rather than number). This happens as the myofibrils within a muscle fiber get thicker, due to the addition of new sarcomeres and myofibrils. After a myofibril attains a certain thickness it may split into two myofibrils, which may then each become thicker due to the addition of sarcomeres. Muscle hypertrophy, in summary, is associated with an increase in the number and size of myofibrils within the muscle fibers.

1. Describe the characteristics of slow- and fast-twitch fibers.
2. Explain the different causes of muscle fatigue, and relate the causes of fatigue to the fiber types.
3. Describe the effects of endurance training and strength training on the fiber characteristics of the muscles.

Summary

The Skeletal System p. 456

I. The skeletal system is divided into an axial and an appendicular skeleton.

II. Bone consists of osteocytes within a calcified matrix.
  A. The osteocytes are nourished through canaliculi by blood vessels within a central canal.
  B. The units of bone structure are called osteons or haversian systems.
  C. A long bone consists of a shaft, or diaphysis, and the two ends, or epiphyses.

III. Bones serve a variety of functions, including support, protection, body movements, hemopoiesis, and mineral storage.

IV. During fetal development, most bones are first formed as cartilage models.
  A. The cartilage models calcify and are replaced by bone.
  B. Long bones grow in length by thickening of plates of cartilage called the epiphyseal plates.

Skeletal Muscles p. 461

I. When muscles contract, they move their insertion toward their origin.
  A. Muscles can act as flexors, extensors, abductors, adductors, or have other actions.
  B. When a muscle shortens as it contracts, contraction is said to be isotonic; when it is prevented from shortening, the contraction is isometric.

II. The whole muscle is covered by an epimysium.
  A. The perimysium divides the muscle into columns, or fascicles.
  B. Each fascicle consists of many muscle fibers, each surrounded by an endomysium.
  C. Each muscle fiber is a muscle cell, and contains myofibrils.

Mechanism of Muscle Contraction p. 464

I. The dark bands in a muscle fiber are called A bands, and the light bands are the I bands.
  A. The striations of the myofibrils are produced by thick and thin filaments.
  B. Thick filaments, composed of myosin, are stacked to produce the dark A bands.
  C. The light I bands contain thin filaments, composed primarily of the protein actin.
  D. The units that extend from one Z line to the other are known as sarcomeres.

II. When a muscle is at rest, the cross-bridges are prevented from attaching to the actin by two inhibitory proteins called troponin and tropomyosin.
  A. Action potentials are conducted into the muscle fiber by transverse tubules.
  B. This stimulates the sarcoplasmic reticulum to release Ca^{++}, which bonds to troponin.
  C. As a result, the troponin and tropomyosin move away from their inhibitory position, so that the cross-bridges can attach to actin.
  D. When a muscle is to relax, the sarcoplasmic reticulum again accumulates Ca^{++}.

Endurance, Strength, and Muscle Fatigue p. 472

I. There are two major types of skeletal muscle fibers.
  A. Slow-twitch (type I) fibers are adapted for aerobic respiration by an extensive supply of blood capillaries, and abundant amounts of myoglobin and enzymes for aerobic respiration.
  B. Fast-twitch (type II) fibers are adapted for anaerobic respiration by a supply of glycogen and glycolytic enzymes.

II. Type II fibers fatigue much more quickly than type I fibers.
  A. Fatigue during very strenuous work can be produced by an increase in extracellular K^+.
  B. Fatigue during moderate exercise can be produced by the increased production of lactic acid.

III. All muscle fibers can increase their aerobic capacity by endurance training; only type II fibers can become thicker by weight training.

Review Activities

Objective Questions

1. A bone is considered to be a(n)
   (a) tissue
   (b) cell
   (c) organ
   (d) system
2. Which of the following statements is *false?*
   (a) Bones are involved in the production of vitamin D.
   (b) Bones serve to store calcium.
   (c) Red bone marrow is involved in hemopoiesis.
   (d) Most bones develop from cartilage models.
3. Specialized bone cells that resorb bone are
   (a) osteocytes
   (b) osteoclasts
   (c) osteoblasts
   (d) osteons
4. The layman's term "growth plate" when referring to bones is known as the ________ plate and is composed of ________.
   (a) diaphyseal; bone
   (b) diaphyseal; cartilage
   (c) epiphyseal; bone
   (d) epiphyseal; cartilage
5. Adductors and abductors
   (a) decrease and increase the angle of a joint, respectively
   (b) increase and decrease the angle of a joint, respectively
   (c) move a limb toward or away from the body, respectively
   (d) move a limb away from or toward the body, respectively
6. Which of the following is an example of an isometric muscle contraction?
   (a) holding the legs in an "L" position on the rings
   (b) doing arm curls with weights
   (c) jumping
   (d) bicycling
7. Which of the following is the smallest unit within a muscle?
   (a) myofibril
   (b) sarcomere
   (c) fiber
   (d) cable
8. When a muscle shortens during muscle contraction, which of the following statements is *false?*
   (a) The A bands shorten.
   (b) The H bands shorten.
   (c) The I bands shorten.
   (d) The sarcomeres shorten.
9. Electrical excitation of a muscle fiber most directly causes
   (a) movement of troponin and tropomyosin
   (b) attachment of the cross-bridges to actin
   (c) release of Ca^{++} from the sarcoplasmic reticulum
   (d) the splitting of ATP
10. Which of the following statements is *true?*
    (a) Type I muscle fibers have a slower rate of contraction than type II fibers.
    (b) Type I fibers can grow by an increase in the number of muscle fibers they contain.
    (c) Type I muscle fibers can grow by an increase in the size of their muscle fibers.
    (d) Type II fibers are more resistant to fatigue than type I fibers.

Essay Questions

1. Describe how the haversian systems function to keep osteocytes alive within the bone.
2. Explain how a long bone grows in length.
3. Explain how X-ray pictures can be used as an indication of the age and development of a person.
4. Explain how muscles are able to exert tension and shorten during contraction.
5. Trace the course of events that occur when a nerve stimulates a muscle to contract, and explain how muscle relaxation is achieved.

22

Human Interaction with the Environment: Ecology

Outline

Objectives

By studying this chapter, you should be able to

1. define the terms *biosphere, ecosystem, community,* and *population*
2. describe the roles of producers, consumers, and decomposers in an ecosystem, and distinguish between different levels of consumers
3. explain why the biomass or the numbers of organisms at each higher level of a food chain decreases
4. describe the cycles of water, carbon, and nitrogen within the biosphere
5. list the different biomes and describe their characteristics
6. describe the nature and dangers of air pollution
7. explain the nature and dangers of water pollution
8. explain how acid rain is produced and the dangers it presents
9. describe the role of ozone in the stratosphere, and the dangers of ozone depletion
10. describe the greenhouse effect, and indicate the problems that excessive greenhouse warming might engender
11. explain the characteristics of population growth, and how the size of populations is limited by the carrying capacity of the environment
12. describe the growth of the human population, indicate where overpopulation is an imminent threat, and explain the consequences of overpopulation
13. describe the importance of the tropical rain forest biome and the consequences of its destruction

Keys to Pronunciation

biome: *bi'om*
biosphere: *bi'o-sfer*
biota: *bi-o'tah*
chaparral: *chap'-a-ral*
chlorofluorocarbons: *klo''ro-flor''o-car'bonz*
deciduous: *de-sid'u-us*
ecology: *e-kol'o-je*
ecosystem: *ek''o-sis'tem*
eutrophication: *u''tro-fi-ka'shun*
ozone: *o'zon*
taiga: *ti'ga*
trophic: *trof'ik*

Photo: Smoke rising from a paper mill in North Carolina after an inversion layer has lifted.

Basic Concepts in Ecology

Organisms interact with each other and with their physical environment within an ecosystem. Each ecosystem contains plants, animals, and decomposer organisms, which are needed to cycle nutrients within the ecosystem. Large, global ecosystems are known as biomes.

Ecology is the branch of biology that is concerned with the interactions between organisms and their environment. The environment of a given organism includes other members of the organism's species, members of other species with which the organism may interact, and nonliving components of the environment. The word "ecology" is derived from the Greek *oikos,* which can be translated as "home" or "place to live." The study of ecology often involves other disciplines of the natural sciences, including physical geography, geology, climatology, physics, and physiology.

Starting with publication in 1962 of Rachel Carson's book *Silent Spring,* which brought the dangers of pollution to public consciousness, the science of ecology has impacted increasingly with politics and economics. In the 1960s and 1970s, people who were concerned about the environmental consequences of human activity were often described by such derogatory terms as "ecofreaks" and "environmental extremists." The attitude of the general public toward ecological concerns, however, experienced a dramatic reversal in the 1980s.

In 1988, *Time* magazine named earth the *Planet of the Year,* in place of its tradition of naming a "man of the year." The greenhouse warming effect, depletion of atmospheric ozone, the dangers of pollution, and the mass extinction of species (topics covered later in this chapter) have become such serious and widely recognized threats that ecological issues have come to the forefront of global politics. In this era when the cold war appears to be winding down, the dangers of ecological catastrophe are increasing.

Let us hope that the following lines are not prophetic:

> This is the way the world ends
> Not with a bang but a whimper
>
> T. S. Eliot, British poet (1888–1965)

Characteristics of Ecosystems

The term **ecosystem** is used to describe all of the *biotic* (living) and *abiotic* (nonliving) components of the environment with which the organisms interact in a given area. All organisms of an ecosystem considered together form the *biota*[1] of that ecosystem. The earth, seas, lakes, streams, rivers, and atmosphere in which life is found comprise the *biosphere.*

Members of a given species within an ecosystem are referred to as *populations.* Within an ecosystem, different populations of organisms that interact with each other are called *communities.* In the ecosystem of the seashore, for example, there may be a tidepool community, a sandy beach community, a mud flats community, and a rocky shore community.

Roles of Organisms in Ecosystems

There are three major roles that organisms play in every ecosystem: producers, consumers, and decomposers. The **producers** are plants; they are called producers because only they can produce organic molecules from inorganic ones using the process of photosynthesis. The **consumers** are animals. Animals must eat plants or other animals in order to survive. The **decomposers** are bacteria and fungi, which return nutrients from dead plants and animals to the soil. In this way, these nutrients can be reused by plants and thus recycled in the ecosystem.

The consumers are subdivided into different categories. *Herbivores*—animals that eat only plants—are the **primary consumers.** *Carnivores* (flesh-eating animals) that eat the herbivores are known as **secondary consumers.** In some ecosystems, there are **tertiary consumers** that eat the secondary consumers and even **quaternary consumers** that eat the tertiary consumers (fig. 22.1). These different levels of feeding are known as *trophic*[2] levels. When arrows are drawn from the organisms being eaten to the organism doing the eating, as in figure 22.1, a *food chain* is depicted. In reality, the feeding interrelationships between different organisms in an ecosystem can be much more complex. (Most people, for example, will eat grain, crustaceans, herbivorous fish, carnivorous fish, and other organisms.) When these complex interrelationships are indicated, a *food web* is described.

Energy Flow in an Ecosystem

According to the second law of thermodynamics, energy is lost as heat whenever an energy transformation occurs. Whenever one organism eats another, or when a dead organism is decomposed by bacteria and fungi, this law applies. The chemical bond energy present in the organic molecules being eaten or decomposed cannot be transferred wholly to the other organism. In fact, only an average of 10% of the energy from one trophic level can be obtained by the next trophic level. It takes 1,000 pounds of plants (their *biomass*), for example, to produce 100 pounds of herbivore, and it takes 100 pounds of herbivore to produce 10 pounds of secondary consumer. If the biomass, or the numbers of organisms at each trophic level, is represented by a block of a proportionate size, a figure of a *food pyramid* is obtained (fig. 22.2).

Within a meadow ecosystem, for example, there are more plants than animals and more mice than hawks. Put another way, a particular ecosystem can support more primary consumers than secondary consumers, more secondary consumers than tertiary consumers, and so on. This is represented in figure 22.1 by the lengths of the bars. Examination of this figure reveals that a larger human population can be supported if people

[1]biota: Gk. *bios,* life

[2]trophic: Gk. *trophikos,* pertaining to nutrition

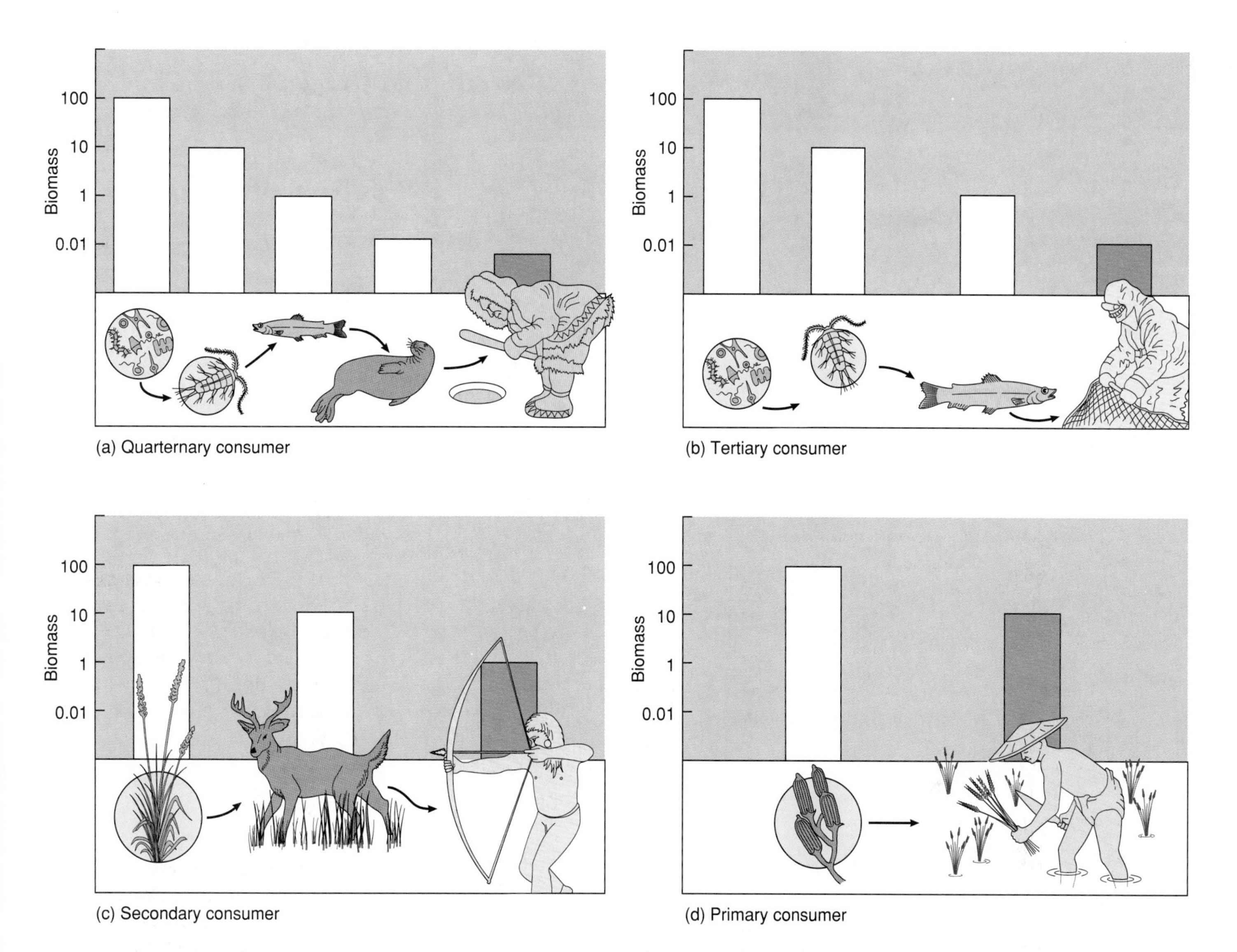

Figure 22.1. *Humans as (a) quaternary; (b) tertiary; (c) secondary; and (d) primary consumers. The lengths of the bars are in proportion to the relative amount of biomass at each level of the food chain.*

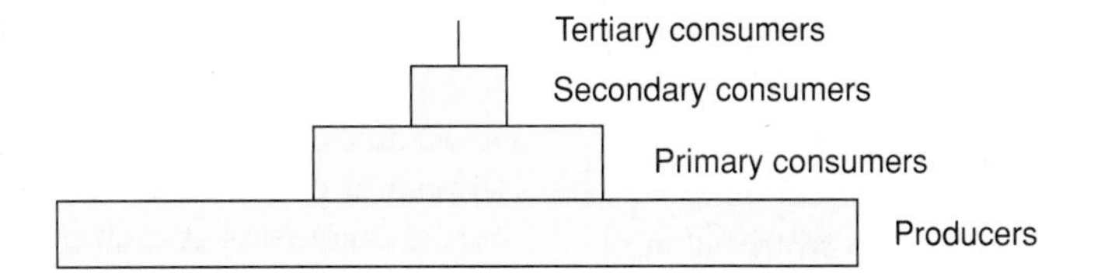

Figure 22.2. *Blocks represent the relative biomass of organisms at each trophic level. This forms a "food pyramid."*

ate an exclusively vegetarian diet instead of the mixed (omnivorous) diet most people prefer. The combined effects of overpopulation and poverty indeed force much of humanity to subsist at the primary consumer level.

Cycles of Matter

All ecosystems would eventually collapse, and life would come to an end, were it not for the continuous input of energy from the sun. A portion of the energy in sunlight is trapped by plants

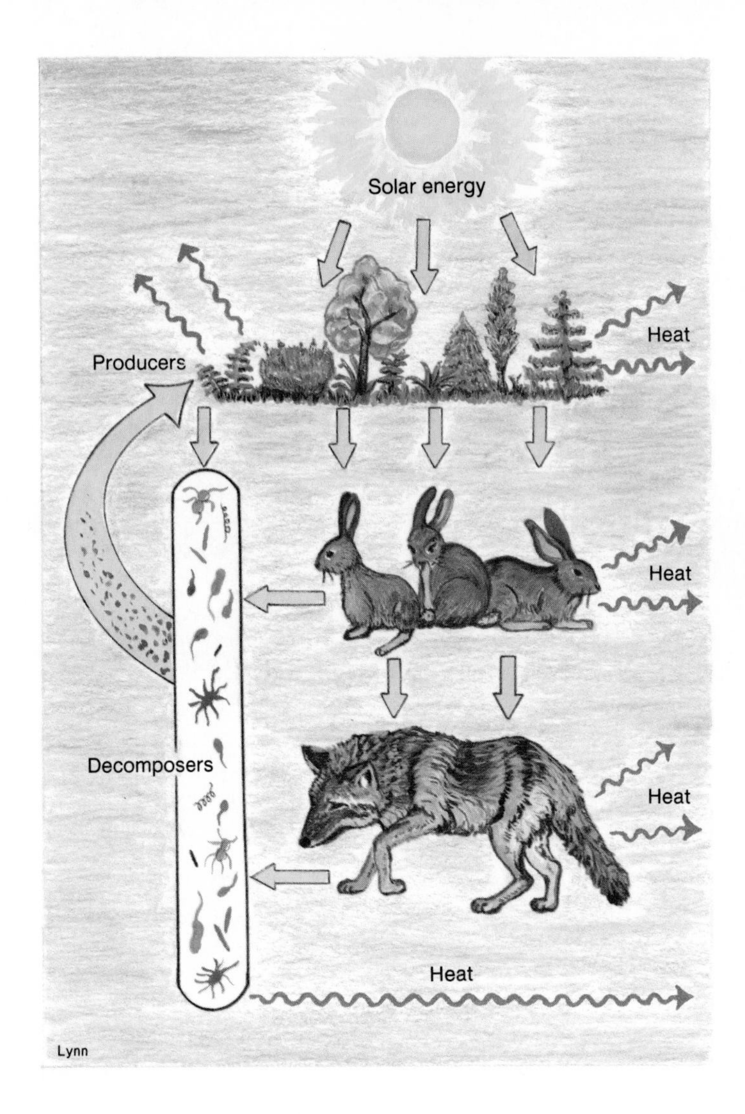

Figure 22.3. Solar energy enters the biota of an ecosystem, and heat energy escapes. Matter is recycled through the action of decomposers.

in the form of chemical bond energy when they produce organic molecules. The components of these molecules—water, carbon, nitrogen, oxygen, sulfur, phosphorus, and others—are recycled within the biosphere. The heat energy lost at every energy transformation ultimately radiates into space. In the biosphere, therefore, *energy flows and matter cycles.* This principle is illustrated in figure 22.3.

The **water cycle** is probably the most obvious of the cycles of matter within an ecosystem. Everyone is aware of the precipitation of water as rain and snow from the atmosphere to the earth and the return of water to the atmosphere by evaporation. The water cycle is illustrated in figure 22.4.

The **carbon cycle** (fig. 22.5) is somewhat more complex than the water cycle. Plants remove carbon dioxide from the air; together with water, the carbon dioxide is used to make glucose and other organic molecules through photosynthesis. This conversion of inorganic carbon in the form of CO_2 to the carbon in organic molecules is known as *carbon fixation.* The carbon atoms in the organic molecules of plants then pass into consumers and decomposers. At each trophic level, the process of cell respiration produces CO_2, which can return to the atmosphere. (Recall that the overall equation for aerobic respiration is fuel $+ O_2 \rightarrow CO_2 + H_2O$.) The same process occurs when wood, coal, and other fuels are burned by human industry. This has

Figure 22.4. *The water cycle.*

From "Threats to the World's Water" by J. W. Maurits la Riviere. Copyright © 1989 SCIENTIFIC AMERICAN, Inc. All rights reserved.

added dangerously high amounts of CO_2 to the atmosphere, contributing to the greenhouse warming of the planet (as discussed in a later section).

The **nitrogen cycle** (fig. 22.6) is still more complex. Some of the nitrogen gas (N_2) in the atmosphere is brought to earth as *nitrite* (NO_2^-) by the effects of lightning. The major source of nitrite in the soil, however, is derived from dead and decaying organisms. The proteins of these organisms is converted to *ammonia* (NH_3) and *ammonium ion* (NH_4^+) by bacteria. Other bacteria convert these compounds to nitrite. Still other bacteria convert nitrite to *nitrate* (NO_3^-). Nitrate is taken into the plant roots, and used by plants to produce amino acids and proteins. This conversion of inorganic nitrogen into organic molecules is called *nitrogen fixation.*

Some of the nitrite and nitrate in soil is converted by bacteria into nitrogen gas, which returns to the atmosphere. This is counterbalanced by the fact that blue-green algae and certain bacteria can produce nitrate from the nitrogen gas of the air. Some of these bacteria are found in the root nodules of *legumes*—plants such as clover, peas, beans, and alfalfa. These plants and bacteria have a symbiotic relationship; that is, one in which each benefits. The plant, in particular, benefits by being able to grow in soil that is not otherwise rich in inorganic nitrogen. In fact, leguminous plants contribute nitrate to the soil, thus enabling plants that are not legumes to grow. This principle is used when crops are rotated in modern farming.

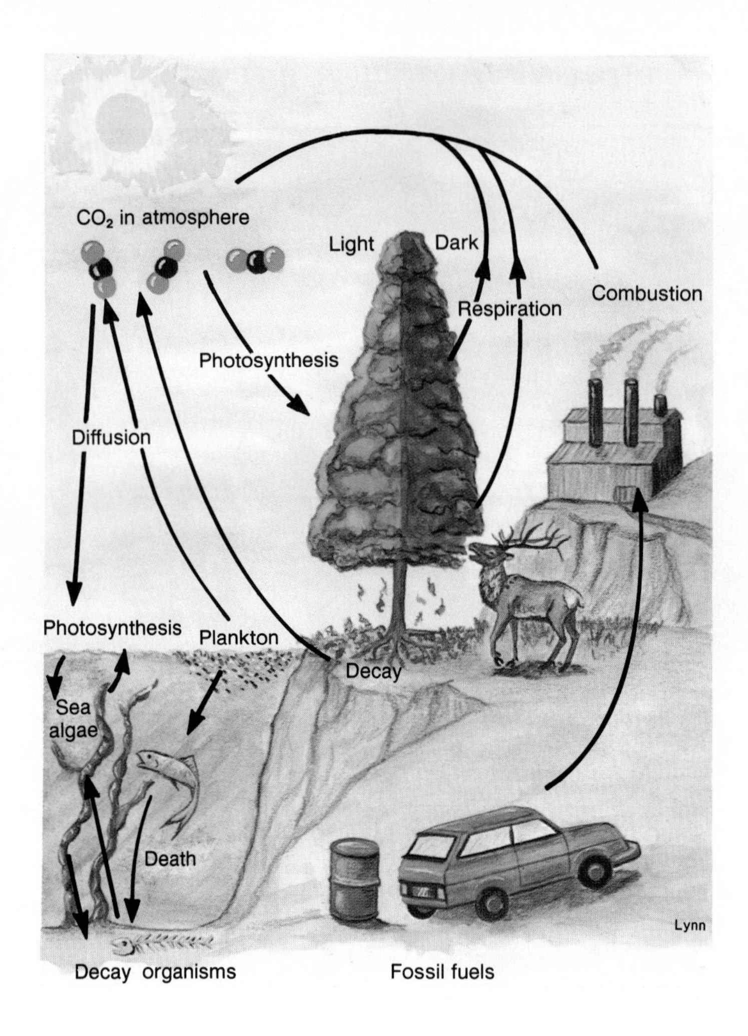

Figure 22.5. *The carbon cycle.*

Biomes

Biomes[3] are large, worldwide ecosystems characterized by particular dominant plants. The different biomes are largely determined by differences in climate. Since climate is influenced by latitude, the global distribution of biomes tends to follow lines of latitude (fig. 22.7).

The **desert biome** is characterized by a scant rainfall of only 25 cm (10 inches) per year. The plants and animals that live in the desert are adapted to its conditions, but the desert as a whole represents a fragile ecosystem that can easily be disrupted by human activity. Human activity that promotes erosion of topsoil has also contributed to the creation of new deserts, particularly in the regions below the Sahara.

The **grassland biome** is, of course, characterized by wide expanses of grasses. Grasslands get more rain than deserts, but not enough rain to promote the extensive growth of trees. Fire is common in grasslands, and also helps to prevent the establishment of trees. The grasslands of North America are known as the prairie and the Great Plains. Elsewhere in the world, grassland biomes are known as the veldt and savanna in Africa, the pampas in South America, and the steppes in Russia. The grasslands support large herds of herbivores, and the soil is rich for agriculture.

The **taiga**[4] biome is the northern coniferous (cone-bearing) forests of pine, fir, spruce, and hemlock. This biome receives long, cold winters and a short summer. The **tundra** biome is located between the taiga and the ice of the Arctic. Tall trees and shrubs are absent, and only the top few inches of soil can thaw each summer. Below this, the soil is permanently frozen and is known as *permafrost*.

The **temperate deciduous forest biome** is common in the eastern United States. The dominant plants are deciduous[5] trees—those that shed their leaves in the fall. This biome exhibits striking and beautiful changes with the four seasons. It receives rainfall throughout most of the year, for a total yearly average of 100 cm (39 inches). Because the hardwood from the deciduous trees is much valued in construction, and because the

[3]biome: Gk. *bios,* life; *-ome,* mass

[4]taiga: Russ. *taiga,* northern coniferous forest

[5]deciduous: L. *decidere,* to fall off

Figure 22.6. *A simplified illustration of the nitrogen cycle.*

area occupied by the forests was needed for the expanding population, humans have eliminated much of the original temperate deciduous forests. The soil that remains is rich in nutrients derived from the decomposition of fallen leaves, though it is not as rich as the soil in grassland biomes.

The **chaparral[6] biome** occurs along the coast of Southern California, in the Mediterranean, and along the coasts of Chile and southern Africa. The temperatures in this biome are mild, with a rainy season in the winter followed by drought in the summer. The plants comprise what is known as an *elfin forest,* because the growth of its evergreen trees is stunted by the low rainfall. Summer droughts make the chaparral very fire-prone. Indeed, the plants of this biome are adapted for fire; they sprout quickly after a fire, and many of their seeds require scorching by fire in order to germinate. Unfortunately, some of the most expensive real estate in the world (along the coasts of Southern California and the Mediterranean) is located in the chaparral, so that human desires and the ecological requirements of the chaparral have often been in conflict.

[6]chaparral: Sp. *chaparro,* an evergreen oak

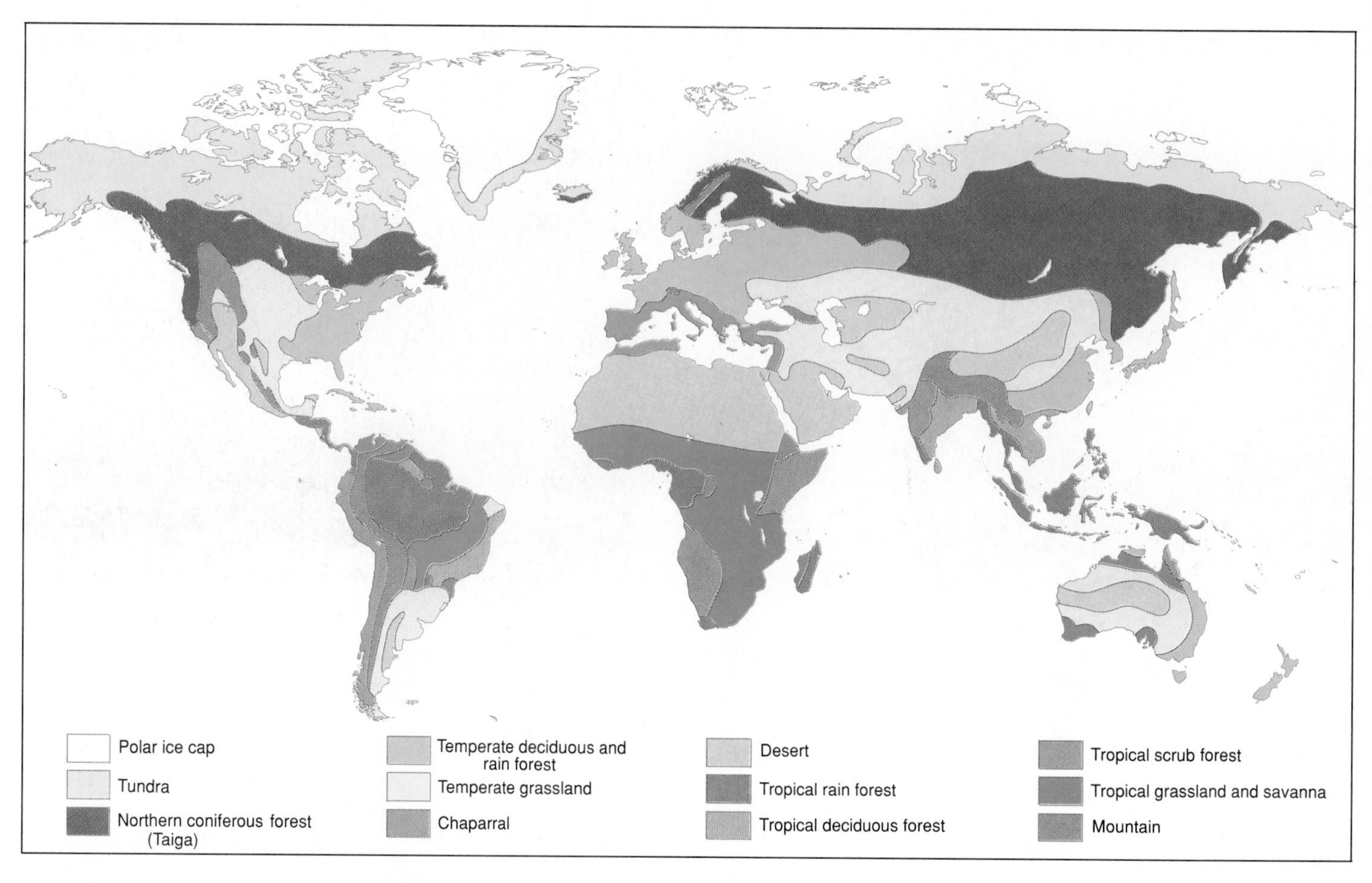

Figure 22.7. *The terrestrial biomes.*

Figure 22.8. *The tropical rain forest.*

The **tropical rain forest biome** (fig. 22.8) receives an average of 2 m (78 inches) of rain, which is evenly distributed throughout the year. The largest tropical rain forest is found in the Amazon Basin of South America; others are located in Indonesia, the Philippines, Central America, the Congo Basin of Africa, and parts of India. Unlike the "jungles" pictured in many movies, the true tropical rain forests are easy to walk through. This is because decomposition of dead organisms occurs quickly, and their nutrients are rapidly taken into the existing vegetation. The predominant plants of this biome are very tall trees, with an average height of 150 feet, that form a continuous leafy canopy far above the forest floor. Of all the biomes, the tropical rain forests have by far the greatest abundance of different plant and animal species. Also, of all the biomes, the tropical rain forests are the most endangered by human activity. This has ominous implications that will be explained in later sections.

---

1. Describe the three different roles of organisms within an ecosystem.
2. Explain, in terms of energy considerations, why an ecosystem can support fewer hawks than mice.
3. Explain the phrase "energy flows and matter cycles," and provide examples.
4. List the different biomes and describe their characteristics.

Figure 22.9. *The town of Times Beach, Missouri, in 1983. Dioxin-contaminated oil was sprayed on the streets ten years earlier to control dust.*

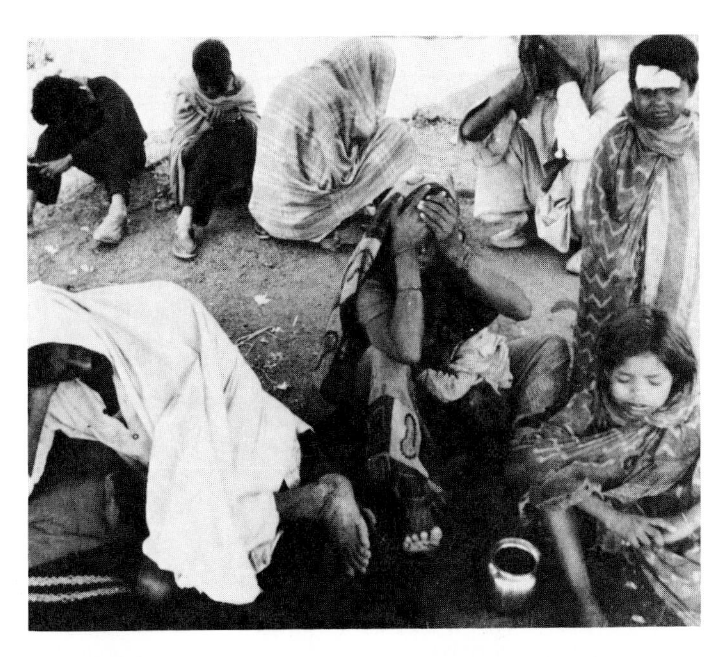

Figure 22.10. *Citizens of Bhopal, India, seeking first aid after a gas leak on December 3, 1984, which killed 1,000 people.*

Pollution of the Environment

Pollution of the air and seas, as well as of lakes, streams, and groundwater, is an increasing threat to human health and welfare. Acid rain has damaged forests, lakes, and streams in different parts of the world. Pollution of the ocean near the coasts and of groundwater is particularly serious.

Humans have dumped their wastes into the earth's land, waters, and air throughout history. This practice was encouraged by the fact that the area receiving our wastes appeared so huge that the small amount of wastes added by one person or group did not appear to be significant. This was particularly true for the wastes put into the air and seas—their effects seemed to be quickly diluted. In an apparently infinite world, the effects of pollution could be disregarded. The error of this worldview was dramatically exposed by the photographs taken of the earth from the moon by *Apollo 14* in 1971. The earth was seen to be a closed, finite, and precious environment. The term *spaceship earth* aptly describes this fragile vessel of all known life.

Starting in September of 1990, a 2.5 acre structure named *Biosphere II*—located near Tuscon, Arizona—will be tested. Four men and four women will enter this structure and be sealed in for two years. Food will be raised by the occupants, and air, water, and wastes will be recycled. Biosphere II will have small areas that mimic seven different biomes, containing over 4,000 species of plants and animals. The purpose of this project is to attempt the creation of a mini-earth (the Biosphere I), which might serve as the prototype for a human colony on the moon or Mars. The problems that must be overcome in this project highlight those faced by the entire human population within our own biosphere.

Even the vast volume of the air and seas has become significantly polluted due to the greatly increased size of the human population and its growing industries. This has produced ecological crises on a global scale. The global problems, however, are often less shocking than local catastrophes (fig. 22.9). One of the most infamous of these incidents occurred in Bhopal, India, on December 3, 1984, in which gas leakage from a Union Carbide pesticide plant caused the immediate deaths of 1,000 people (fig. 22.10). The delayed effects of the poison have killed many more people, so that the official death toll now exceeds 3,000. This tragedy has been aggravated by controversy surrounding the cause of the accident, and regarding the legal and financial responsibility of Union Carbide to the victims.

The air and open seas are *commons*; they are everyone's property and under no one's authority. Without the legal recourse to prevent pollution of these commons by particular groups and nations, all people will suffer. International cooperation is thus essential. In recognition of this fact, the leaders of several developed countries, including the United States, West Germany, Japan, Britain, France, Italy, and Canada met in July of 1989 and resolved to make environmental protection a top international priority.

Air Pollution

Air pollution, forming **smog** (a contraction of "smoke" and "fog"), is a condition that must be endured by the inhabitants of most large urban areas. The intensity of this pollution, and its chemical constituents, vary in different locations. When the city is in a geographic basin, as is Los Angeles, the smog tends to be trapped and concentrated by an *inversion layer* of warm

Oil Spills

The accidental rupture of oil tankers, and the resultant spilling of massive amounts of oil into the ocean, has wrought ecological damage on many occasions. Two tankers crashed, for example, near Trinidad and Tobago in 1979, spilling 300,000 tons of oil. A tanker accident off the coast of South Africa spilled 250,000 tons of oil, and one off the coast of France spilled 223,000 tons. Similar accidents have occurred in the United States: 20,000 tons of oil were spilled in 1976 off the coast of Massachusetts, and 12,000 to 14,500 tons were spilled off the coast of Santa Barbara in 1969.

The largest oil spill in U.S. history occurred on March 24, 1989, when the *Exxon Valdez* spilled 240,000 tons of oil into Prince William Sound near Port Valdez, Alaska (fig. 22.11). A week after the spill, the oil slick covered an area of 900 square miles of ocean, and hundreds of miles of shoreline were covered with oil that was up to six inches deep in some places. Thousands of birds (fig. 22.12) and sea otters have already been killed, and the fishing industry that is so important to the local economy is severely threatened. Despite the fact that many oil spills have occurred in the past, their long-term effects are still uncertain.

This disaster was predicted in 1972 by the Department of the Interior in their environmental impact statement regarding the use of Valdez as an oil shipping port. Considering the economic benefits, and the dangers of relying on foreign oil, Congress decided to ignore the risk. An alternate plan was discarded that would have avoided sea transport by diverting the Alaskan pipeline through Canada to Chicago. Again, economics outweighed ecology in this decision. Considering that the cleanup of oil from this spill has already cost in excess of $100 million, and will probably end up costing much more, this was a bad decision economically as well as ecologically.

The possibility of a major oil spill had been repeatedly predicted. The U.S. oil tanker fleet has aged at a rapid rate, and oil leaks had occurred on a number of previous occasions. Attempts by Alaskan officials to tighten controls on the shipment of oil were blocked by oil companies. The political clout of the oil companies is considerable; Alaska receives royalties, which are divided up among its residents, on the oil it exports. In the wake of the Valdez oil spill, however, the people of Alaska may reevaluate their relationship to the oil companies, and long-term interests may at last come to outweigh short-term economic benefits. The Valdez oil spill demonstrates that protection of the environment is sound economic policy when long-term effects are considered.

Figure 22.11. *Contamination of Eleanor Island with oil after the oil spill from the* Exxon Valdez.

© Tony and Kathy Dawson

Figure 22.12. *Dead bird is examined at the Valdez examination station (left). An eagle on Kachemak Bay is smudged with oil (right).*

Both © Tony and Kathy Dawson

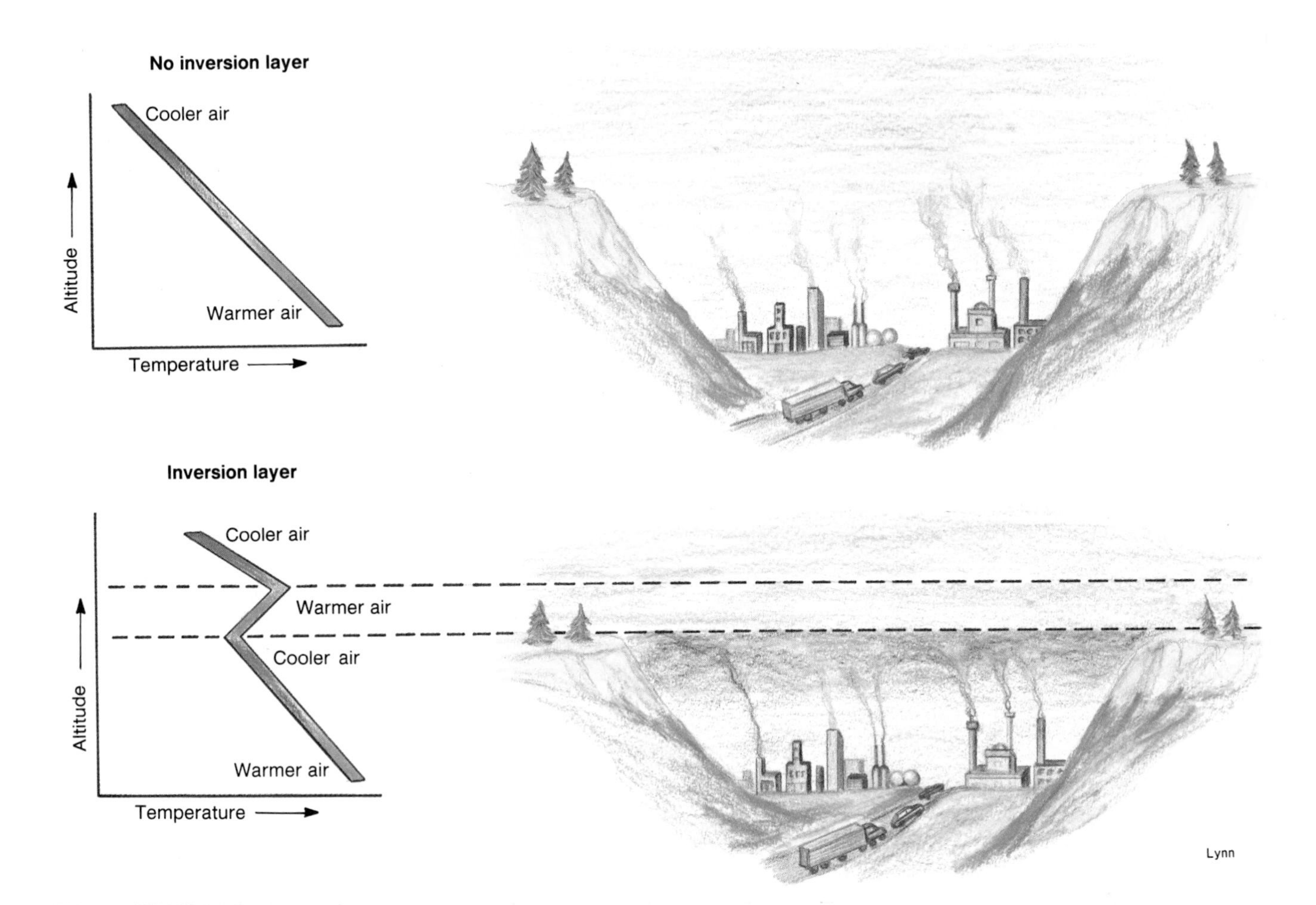

Figure 22.13. *When an inversion layer is not present* (a), *the temperature of the air decreases with altitude in a linear fashion* (left) *and pollutants are free to diffuse in the atmosphere* (right). *When an inversion layer is present* (b), *the warmer air in this layer* (left) *traps pollutants within the basin* (right).

air (fig. 22.13). Air pollution constitutes a serious health hazard; in the Los Angeles basin, for example, it is responsible for an estimated $9 billion per year in health costs. In recognition of this problem, Congress in 1988 reauthorized the Clean Air Act as part of its continuing effort to reduce smog nationwide.

> Automobile emissions constitute a major source of air pollutants. The more miles per gallon a car can travel, however, the less pollutants it will emit. Congress has thus set standards for the automobile industry to produce cars that have a higher fuel efficiency. (This also serves to limit the consumption of oil—a nonrenewable resource—and to decrease the amount of CO_2 emitted, which is important in limiting the greenhouse warming effect.) Gasoline that is reformulated to produce lower emissions and the installation of electrically-heated catalytic converters, or the use of substitute fuels (methanol and compressed natural gas), can also help reduce pollution. The Air Resources Board in California estimates that these cleaner fuels will replace two-thirds of the gasoline used in that state by the year 2010.

Ozone and Carbon Monoxide

Hydrocarbons emitted by cars and various industrial processes react with nitrogen oxides in the air to produce **ozone**[7] (O_3). The formation of this ozone is aided by heat and sunlight, and thus is limited to the daylight hours and is produced in greatest amounts during the summer. It is important to distinguish the ozone in the air we breathe, which is within the *troposphere,* from the ozone in the upper layers of the atmosphere, or *stratosphere.* Whereas the ozone in the stratosphere is required for the maintenance of life on earth (as will be described later), ozone in the troposphere constitutes a health hazard.

The acute, or immediate, dangers to health of breathing ozone are well known (fig 22.14). The extent of these effects varies in different people; about 10% to 15% of the population (including athletes) are very sensitive to ozone. The Environmental Protection Agency has shown that these people lose 40% to 50% of their breathing capacity when they alternate fifteen minutes of exercise with fifteen minutes of rest for two hours when ozone levels are 0.12 parts per million (ppm). (It is not

[7]ozone: Gk. *oze,* stench

Figure 22.14. This bicyclist in Los Angeles is not taking any chances with smog.

R. L. Oliver/Los Angeles Times Photo

uncommon for ozone levels to reach 0.20 ppm on many days in smoggy cities.) The people most affected by ozone are children (who play outdoors), outdoor workers, and the elderly (due to diminished lung function).

In addition to a decreased breathing capacity, other symptoms of ozone poisoning include eye irritation, headaches, chest pains, sore throat, and coughing. The long-term health hazards of ozone in smog are currently under investigation, but are not presently documented.

Carbon monoxide (CO) is another common constituent of smog that is produced in automobile emissions. As discussed in chapter 15, carbon monoxide combines with hemoglobin. This prevents the hemoglobin from bonding to oxygen, and thus decreases the ability of the blood to carry oxygen to the tissues. The lowered ability of the blood to carry oxygen may not be enough to cause noticeable effects in young, healthy people, but it can cause damage in those with pulmonary or heart disease.

Particulates

Small particles suspended in the air are the **particulates** present in smog. These are largely responsible for the decreased visibility that is such a prominent feature of smoggy days (fig. 22.15). When particulates are inhaled, they can work together with ozone to aggravate asthma and bronchitis. Particulates

Figure 22.15. The Los Angeles skyline on a smoggy day.

Los Angeles Times Photo

containing sulfur dioxide, which can cause significant pulmonary problems in children, are most common in the southeastern United States due to the burning of coal and oil.

Nitrogen and Sulfur Compounds

Nitrogen-containing components of automobile exhaust react with oxygen in the air to produce *nitrogen oxides,* particularly NO_2. This is a brownish gas that becomes particularly concentrated around highways. Sunlight converts some of the NO_2 into ammonium nitrate and sodium nitrate, which form particles that decrease visibility. *Nitric acid*—a contributor to eye irritation—is also produced by these chemical reactions.

In addition to nitrogen oxides and their products, the air pollution in some cities contains *sulfur dioxide.* This compound can be damaging by itself, but it can become even more so by being converted into *sulfuric acid.* In addition to posing a health hazard as a component of air pollution, sulfuric and nitric acids in rain produce serious pollution of lakes and streams.

Water Pollution

The fresh water in lakes and streams is a precious resource, accounting for only 0.01% of the total water on earth. This resource can be contaminated by polluted rain and by human wastes. When the human population density is low, bacteria can remove organic wastes from the water in a process called *natural self-purification.* As the human population density and its agricultural needs increase, however, this self-purification can become inadequate. Groundwater is a nonrenewable resource that is particularly susceptible to contamination, because its natural self-purification ability is limited by the lack of oxygen available for the bacteria that promote this process.

In some parts of the world, people dump their excreta in a local river. This transmits the microorganisms that cause typhoid, cholera, and dysentery. Pesticides and heavy metals (such as lead and mercury), which are not degraded by bacteria, can also accumulate in local waters. Toxic wastes stored in metal

Figure 22.16. *Toxic chemicals can leak from rusted drums and contaminate the groundwater.*

drums (fig. 22.16) are time bombs for the pollution of groundwater. When the container rusts away, the toxic chemicals can be released and render the groundwater unhealthful for human consumption.

Pollution of the Great Lakes ecosystem (including Lakes Superior, Michigan, Erie, and Ontario, as well as connecting channels) poses a major health hazard. These lakes contain 20% of the earth's fresh water, and provide drinking water for 10% of the American and 33% of the Canadian populations. Toxic chemicals have entered all levels of the food chain in these ecosystems and impacted the health of thirty-five million Americans and Canadians. A recent scientific study demonstrated that women who regularly eat fish from these lakes give birth to babies who suffer a number of ill effects; women are thus urged to refrain from eating these fish during their reproductive years. Scientists estimate that it will cost in the tens or hundreds of billions of dollars to remove the toxic chemicals from this large polluted ecosystem.

Eutrophication is a different form of water pollution. In this process, nutrients (mainly phosphate) from agriculture, industry, and detergents enter a lake and promote an abnormally large growth of algae. When these large numbers of algae die, they are decomposed by bacteria, which also proliferate as a result. The cell respiration of these bacteria uses up the oxygen in the lake, killing plant and animal life.

Pollution of Coastal Waters

The waters of the coastal areas are the richest in nutrients and thus the areas of greatest biological productivity in the oceans. Indeed, 90% of marine fish that are caught come from within 320 km of the shore. Yet thirteen billion tons of silt per year are dumped into the coastal areas from the mouths of rivers, and the amount of this silt is increasing because of erosion caused by deforestation (destruction of forests).

Pesticides—including DDT and polychlorinated biphenyls (PCBs)—and heavy metals may enter the food chain within the coastal community. These compounds can accumulate within the fatty tissues of organisms because they are not degraded and are fat-soluble. Since a primary consumer eats many producers, and a secondary consumer eats many primary consumers, the concentration of these pesticides and heavy metals in an organism is greater the higher the position it occupies in the food chain. Fish and shellfish within some polluted water, as a result, are unfit for human consumption.

Acid Rain

Rain in unpolluted areas has a pH that is just slightly acidic (pH 5.5–6.5), due to the formation of carbonic acid from carbon dioxide and water. In regions subject to air pollution, nitrogen oxides form nitric acid, and sulfur dioxide forms sulfuric acid; these acids enter the raindrops and lower their pH. (Remember that acidity is inversely related to the pH numbers; a drop of one pH number indicates a tenfold increase in acidity.) The pH of the rain over the northeastern United States in the 1970s was less than 4 (more acidic than lemon juice), and sometimes it had a pH of 3.0 to 3.5. In one documented case, the rain had a pH of 2—the same pH as gastric juice!

Acid rain leaches nutrients from the soil, acidifies streams and lakes (thus killing fish), kills nitrogen-fixing bacteria in the soil, causes destruction of trees and forests, and damages machinery and buildings that are exposed to the rain. These odious effects may not always be caused by local pollution. Since clouds carrying acidified moisture can move from one geographic area to another, acid rain in one country may be caused by air pollution in a different country. Destruction of forests in Ontario, Canada, for example, is largely due to pollution over the northeastern United States, and the destruction of trout and salmon fisheries caused by acid rain in Sweden is primarily the result of pollution in the United Kingdom. International cooperation is thus required for preventing the many problems brought by acid rain.

1. Explain the significance of the phrase "spaceship earth."
2. Describe the constituents of smog, and the health hazards of these substances.
3. Explain how acid rain is produced and how it can damage ecosystems. Also explain why international cooperation is required to eliminate this problem.
4. Explain how eutrophication can kill a lake.

Ozone Depletion and the Greenhouse Effect

Chlorofluorocarbons (CFCs) are destroying the ozone in the stratosphere. With less ozone in the stratosphere, more ultraviolet light will reach the earth and cause skin cancers and other problems. Carbon dioxide is accumulating in the atmosphere due to extensive burning of fossil fuels and of tropical rain forests. The increase in atmospheric carbon dioxide threatens to raise the temperature of the earth through the greenhouse effect.

There are two approaching ecological catastrophes that involve the upper atmosphere. One of these is the depletion of ozone from the stratosphere; the other is the warming of the earth due to accumulation of "greenhouse gases" in the atmosphere. One of these greenhouse gases is the same culprit responsible for ozone depletion. This is a class of compound called **chlorofluorocarbons (CFCs).**

The trade names for CFCs are *Freon* (DuPont) and *Genetron* (Allied-Signal). These compounds are used as aerosol propellants, coolants in refrigerators and air conditioners, solvents for cleaning electronic components, and foaming agents for plastics. One of the properties of CFCs that makes them useful for industry is their lack of chemical reactivity. This property, however, also prevents CFCs from breaking down in the atmosphere. The concentration of CFCs in the atmosphere thus continuously increases as more are added.

Ozone Depletion

The stratosphere contains a layer of ozone (O_3), which is formed from oxygen gas (O_2) under the influence of short-wavelength radiation. This layer of ozone is critically important for life on earth because it absorbs high-energy ultraviolet light from the sun. Without this layer of ozone, the high-energy ultraviolet light would irradiate living organisms, causing damage to their DNA.

Chlorine atoms released by CFCs in the stratosphere destroy ozone through the following chemical reactions:

$$Cl + O_3 \rightarrow ClO + O_2$$

$$ClO + O \rightarrow Cl + O_2$$

Notice that, after destroying a molecule of ozone in the first reaction, the chlorine atom is regenerated in the second reaction. This process can thus be repeated many times; it has been estimated that one chlorine can cause the destruction of 100,000 molecules of ozone. The yearly emissions of CFCs are presently about one million tons; the amount of ozone destroyed is thus enormous.

The significance of this phenomenon was proven in 1985, when the amount of ozone over the Antarctic in its spring (October in the Southern Hemisphere) was found to be substantially decreased (fig. 22.17). This has come to be known as the **ozone hole.** Analysis of records over this area for the spring of

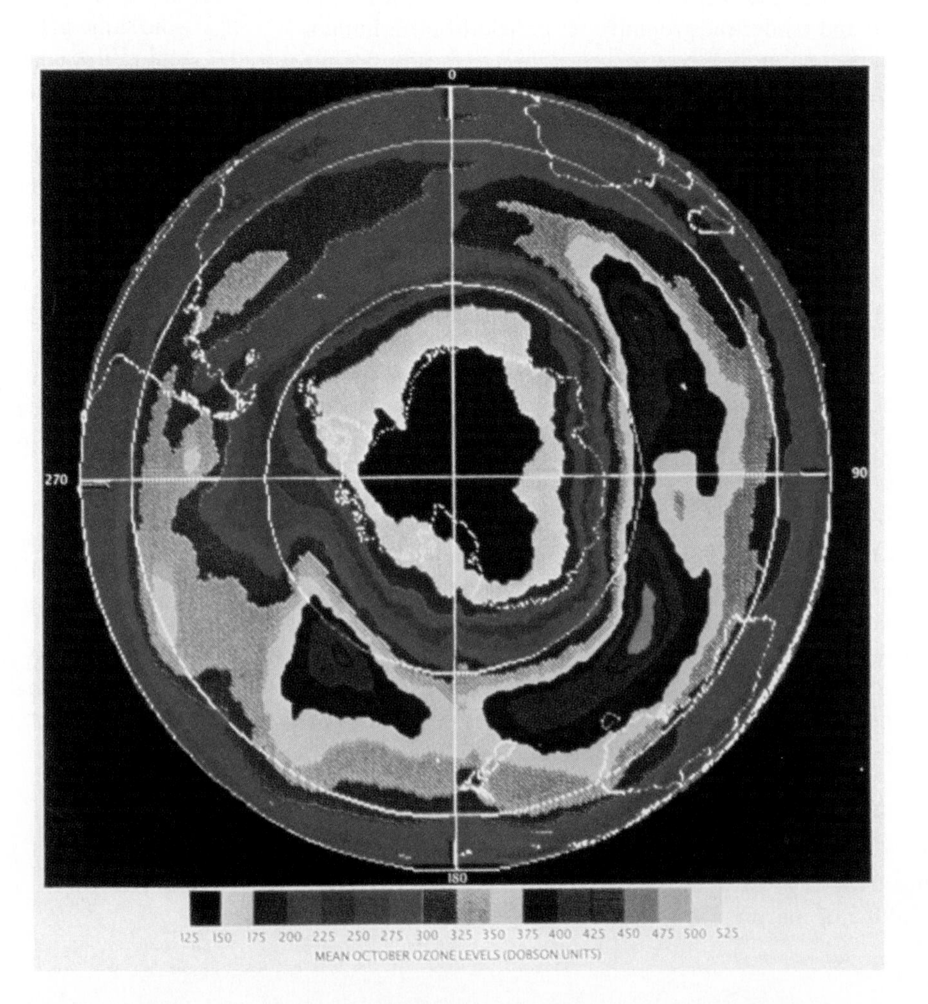

Figure 22.17. *Ozone levels in the Southern Hemisphere on October 5, 1987. The ozone level over Antarctica is about half of what it was a decade ago.*

1955 through 1985 revealed a continual decline in ozone (fig. 22.18), which corresponded to the increasing amounts of CFCs in the atmosphere.

Though first discovered over the Antarctic, there is now evidence that ozone depletion is also occurring in the Northern Hemisphere. The amount of CFCs in the atmosphere is currently increasing at a rate of about 5% per year. Even if CFC production were to halt today, the amount still in the atmosphere would continue to deplete ozone into the next century.

The Environmental Protection Agency has predicted that the increase in ultraviolet radiation reaching earth as a result of ozone depletion will cause millions of new cases of skin cancer and associated deaths. In addition to skin cancer, the depletion of the ozone layer may cause increased incidences of cataracts (clouding of the corneas of the eyes) and immune deficiency diseases. Scientists fear that the increased ultraviolet radiation may also cause significant harm to crops and aquatic ecosystems.

A ban on the use of CFCs in aerosol sprays has been in effect in the United States, Canada, and Scandinavia since the 1970s. In response to the ominous implications of the ozone hole, a United Nations convention met in Montreal in 1987 to set guidelines for reduction in the use of CFCs. The "Montreal protocol" developed from this meeting specified goals of reducing CFC emissions 20% by 1994 and an additional 30% by 1999. This resolve was strengthened in 1990 by a meeting of ninety-three nations, who agreed to completely eliminate CFCs by the year 2000. In addition, they established an international fund to help Third World nations (including China and India) switch to substitutes for CFCs as their economies develop. Such substitutes do exist, and Du Pont (which currently produces 25% of the world's CFCs) is building four new plants that will produce substitutes for their Freon.

Greenhouse Effect

Carbon dioxide, and other gases (CFCs, methane, and nitrogen oxides) to a lesser degree, allow sunlight to penetrate the atmosphere but reflect infrared radiation (heat) from the earth's surface back to the earth (fig. 22.19). This is called the **greenhouse effect,** because the gases serve the same function as the glass panes in a greenhouse. Without a greenhouse effect, the temperature of the earth would be an average of 50°F cooler than it is and would be incompatible with life.

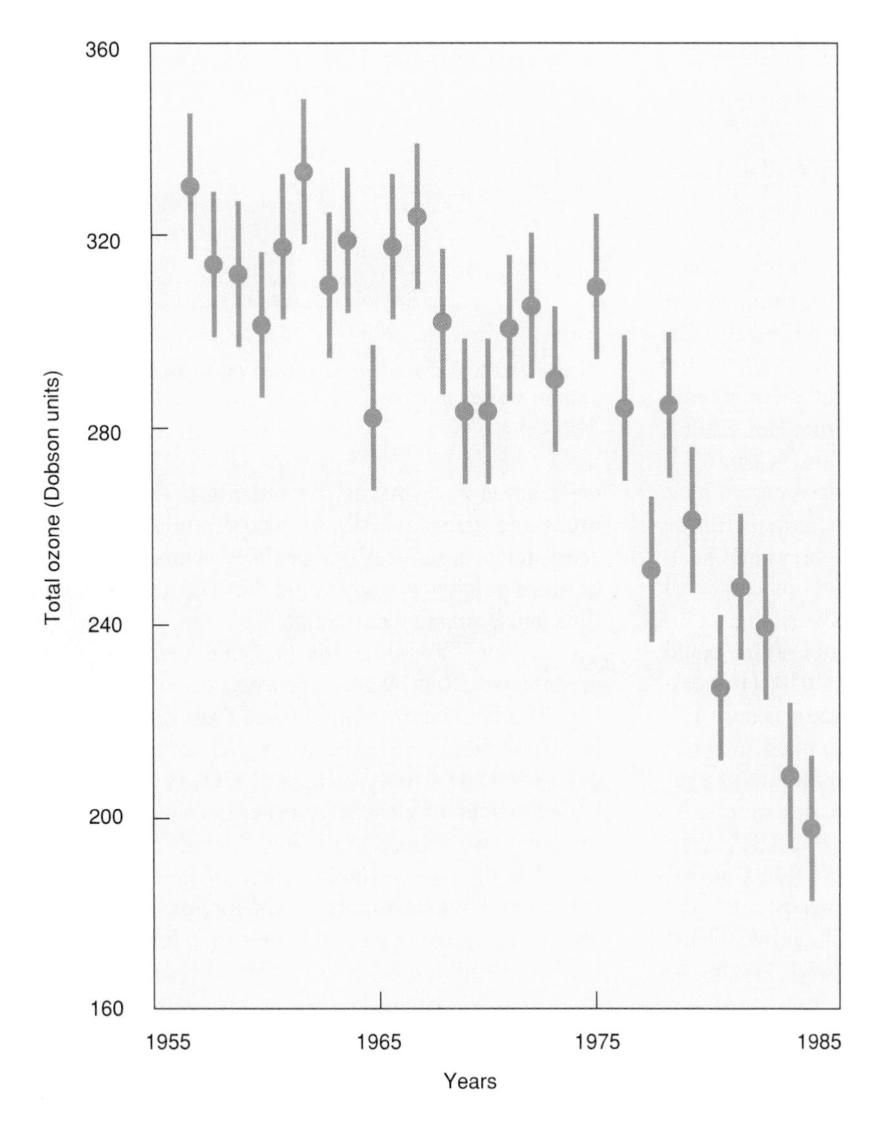

Figure 22.18. The decline in ozone at the British Antarctic Survey base over a thirty-year period.

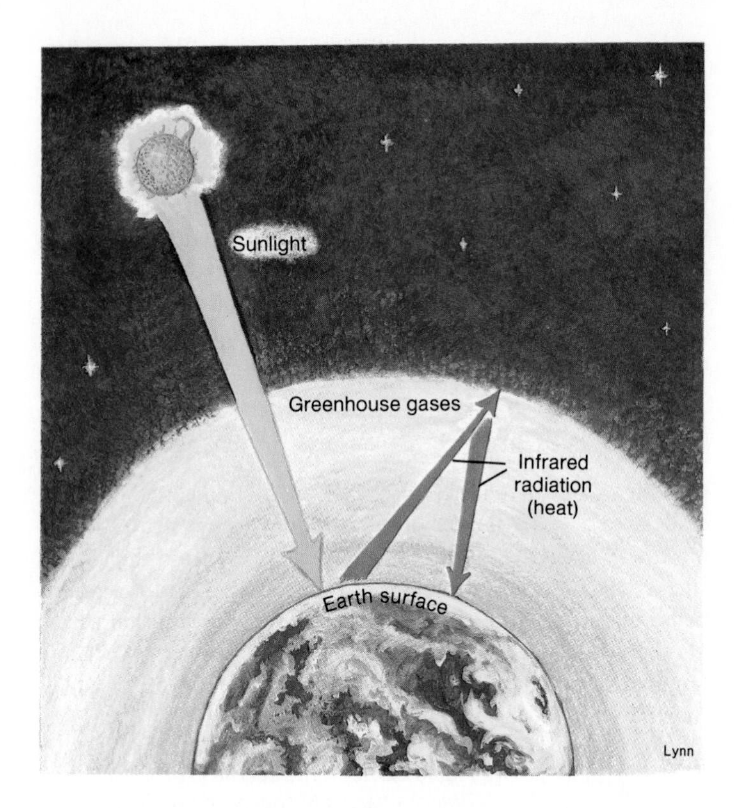

Figure 22.19. *The greenhouse effect. Carbon dioxide and other gases allow sunlight to pass through the atmosphere but provide a barrier to the passage of infrared radiation.*

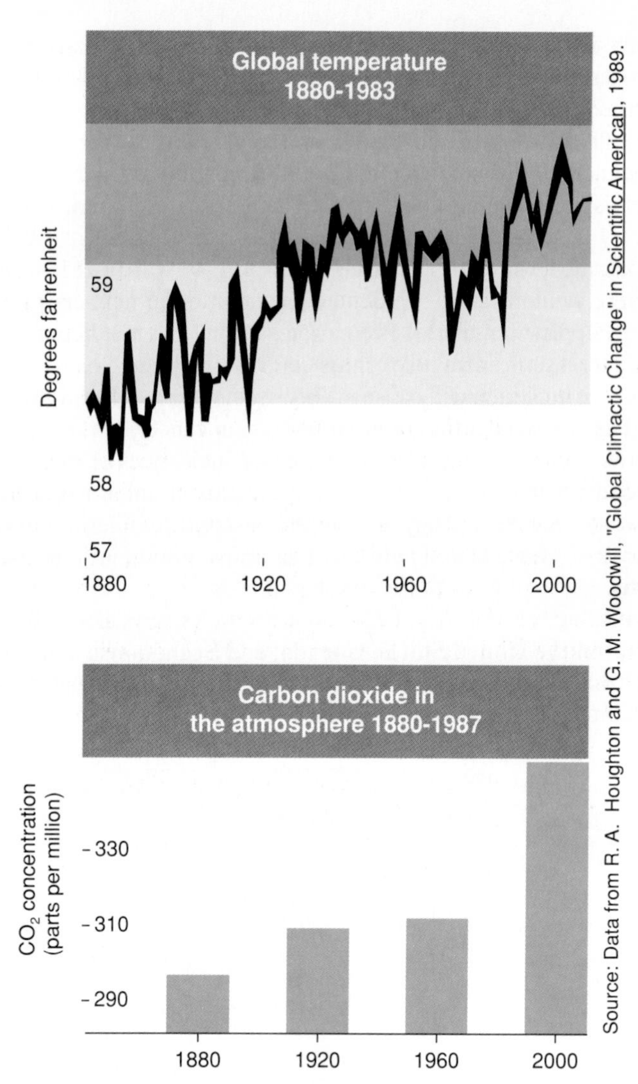

Figure 22.20. *Correlation between the increase in atmospheric carbon dioxide (bottom) and the increase in average global temperature (top) since 1880.*

According to theory, an increase in the amounts of greenhouse gases in the atmosphere should cause a warming of the earth. Evidence suggests that this has, in fact, occurred. The concentration of carbon dioxide has increased from about 290 parts per million (ppm) in the mid-1800s to about 350 ppm now, accompanied by an increase in average temperature (fig. 22.20). Scientists are as yet uncertain if this greenhouse warming is responsible for the fact that the six hottest years on record have all occurred in the last decade. Though its association with the greenhouse effect is controversial, the record-breaking heat, drought, and crop failures of the summer of 1988 have alerted the general public to the reality of greenhouse warming.

Climatologists predict that the global temperature could become an average of 9°F hotter by the year 2050. (In comparison, the global temperature of the last Ice Age was only 4°F cooler than now.) The extent and rate of the temperature rise caused by present greenhouse warming is greater than ever before in earth's history. It could wreck massive destruction by interference with weather patterns, production of drought conditions in the American Midwest and other agricultural areas, production of powerful hurricanes and tornadoes, and the melting of polar ice caps and alpine glaciers. The latter effects would also contribute to a raising of the sea level, leading to destruction of costal regions and cities. The changes in temperature and water availability would produce drastic changes in ecosystems; organisms (particularly plants) that could not adapt or move quickly enough would die. The greenhouse effect could thus cause massive extinctions.

Carbon dioxide is the gas that contributes most to the greenhouse effect. There are a variety of sources of this CO_2 (fig. 22.21). The burning of fossil fuels releases five billion tons of carbon dioxide into the atmosphere each year; each gallon of gas used adds twenty pounds of CO_2 to the atmosphere. One kilowatt-hour of electricity generated in a coal-powered plant produces two pounds of carbon dioxide. The industrialized nations are thus the primary source of greenhouse gases. Countries that have substantial rain forests, however, are also to blame. They are destroying their rain forests at the appalling rate of 50 million acres per year! This aggravates greenhouse warming in two ways: (1) trees remove CO_2 from the air in the process of photosynthesis, so with fewer trees the CO_2 will accumulate in the atmosphere faster; and (2) when the trees are cut down they are burned, releasing CO_2 into the air. This accounts for approximately 30% of the total CO_2 emissions worldwide.

The removal of CFCs from production will help decrease the greenhouse effect, as well as aid the recovery of the ozone layer. An increased emphasis on alternate energy sources—solar power, geothermal power, wind power, and nuclear power—which do not produce greenhouse gases can also help combat greenhouse warming. Realistically, the goal of significantly reducing the production of greenhouse gases can only be met by increased energy efficiency at all levels. This includes increased fuel efficiency of cars, improved mass transit, effective insulation so that less energy is required to heat homes, and so on. Some of these steps should be taken on an individual and local level, since all of us in the developed countries contribute more than our share to the greenhouse effect. Changes on a national and international level, however, must be instituted to reduce the global suffering that will be caused by greenhouse warming.

1. Describe the importance of ozone in the stratosphere, and explain how CFCs are causing destruction of this ozone.
2. Explain what is meant by the "ozone hole."
3. Describe the importance of the greenhouse effect, and the dangers of excessive greenhouse warming.
4. Explain how the burning of fossil fuels and of tropical rain forests contribute to the greenhouse effect.

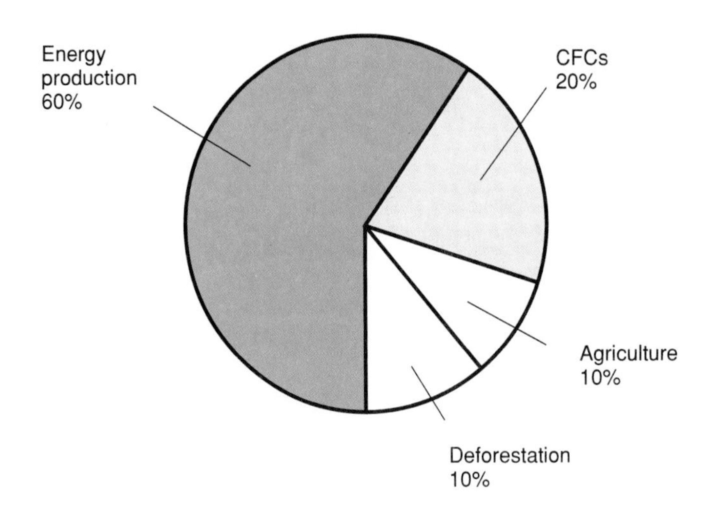

Figure 22.21. *The sources of greenhouse gases.*

Table 22.1 Hypothetical population growth of the housefly

Generation	Numbers If All Survive
1	120
2	7,200
3	432,000
4	25,920,000
5	1,555,200,000
6	93,312,000,000
7	5,598,720,000,000

*This assumes that each female lays 120 eggs and all survive 1 generation.
From Wallace, et al., *Biology: The Science of Life.* © 1981 Goodyear Publishing Co., Santa Monica, CA.

Growth of the Human Population

Populations grow exponentially, but growth is eventually halted when the carrying capacity of the environment is reached. At this point the death rate is equal to or greater than the birthrate. Human populations are rapidly reaching their carrying capacity in the underdeveloped countries, where only extensive use of family planning methods can avert disaster.

A mating pair of organisms can produce more offspring than is required to replace the parents in the population. If the total population consisted of one mating pair, for example, and this pair produced four offspring that survived to maturity, the total population would double in one generation (after the parents died). This doubling adds only two individuals to the population. If the offspring now interbred and repeated this process, a population doubling would occur in which four additional individuals were added. If this is continued, the population would grow by eight in the next generation, sixteen in the next, thirty-two in the next, and so on. This is known as exponential growth, and it is an axiom of biology that *populations grow exponentially.* Notice that, in exponential growth, the number of individuals added increases enormously with each doubling. This can produce population growth at an incredibly rapid rate (table 22.1).

In the real world, populations cannot continue to grow indefinitely in this manner. Eventually, they must reach a point where the death rate counterbalances the birthrate. This occurs when the population has reached the **carrying capacity** of the environment—that is, the maximum size of the population that the environment can support. As the carrying capacity is approached, death occurs rapidly due to inadequate environmental resources (including food and water) and to pollution of the environment from the organism's wastes. At the carrying capacity, the population size may remain constant (fig. 22.22), or the population may crash if the death rate is much greater than the birthrate.

The growth of the human population illustrates some of these principles. There were only about five million people on the entire earth prior to the agricultural revolution ten thousand years ago. This population grew slowly to five hundred million by A.D. 1650. It then took only two hundred years, until 1850, for the population to double to one billion. The next population

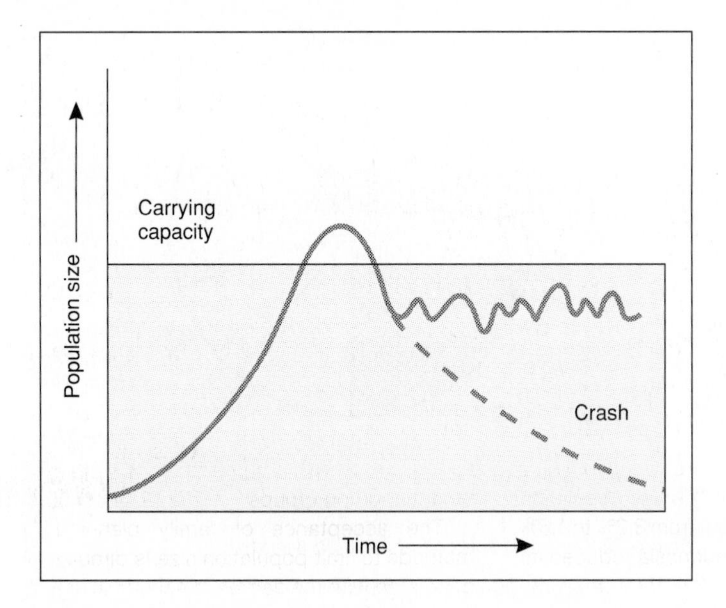

Figure 22.22. *When the population reaches the carrying capacity of the environment, death rates increase to either stabilize the population size or cause it to crash.*

From R. Wallace, et al., Biology: The Science of Life. *Copyright © 1981 Goodyear Publishing Co., Santa Monica, CA.*

doubling, to two billion, occurred only eighty years later in 1930. The population doubled again, to four billion, in 1975 after a span of only forty-five years. Because of the exponential nature of population growth, less time is required for each doubling in size of the population.

The Population Bomb

The human population went past the 5 billion mark in 1987. It is expected to reach 8.5 billion by the year 2025, at which point the carrying capacity of the earth may be reached. Advances in food production technology and genetic engineering cannot keep pace with population growth, because such advances expand the resources of the earth only in an additive fashion, whereas population continues to increase exponentially. The serious and inevitable consequences of overpopulation gave birth to the phrase *population bomb*.

Some commentators, observing the population growth of the developed countries, have written that the population bomb is a dud. This is because the population growth rate of the developed countries is actually declining. The population of the United States will grow from 248 million to 300 million because of the momentum of the baby boom, but should decline in the next century if the birthrates remain the same as now. The growth rate of seven European nations is now zero or negative; the populations of West Germany, Switzerland, and Sweden are expected to decline by 10%, 8%, and 6%, respectively, over the next thirty-five years.

What these commentators have overlooked is the fact that 95% of the additional three billion people that will be added to the world's population will be born in the underdeveloped countries, or "third world." The third world has already added more people since 1960 than the combined populations of North America, Europe, the Soviet Union, and Japan! This population growth occurs because their birthrate greatly exceeds their death rate. The *birthrate* (number of births per woman) in Kenya, Nigeria, and Mexico, for example, is currently 8.12, 6.6, and 4.0, respectively. In comparison, the birthrates are 1.8, 1.7, and 1.4, respectively, in the United States, Canada, and Denmark (table 22.2).

Table 22.2 Birthrates (number of births per woman) of representative countries in 1988

Country	Birth Rate
Kenya	8.12
Saudi Arabia	7.1
Nigeria	6.6
United Arab Emirates	5.9
Mexico	4.0
Indonesia	3.4
China	2.4
Taiwan	2.1
Japan	1.8
United States	1.8
Singapore	1.7
Canada	1.7
Denmark	1.4
West Germany	1.3

Compiled from various sources.

An exercise in arithmetic reveals the catastrophic nature of this population growth. Nigeria, for example, has a population growth rate of 3.4% per year, which yields a doubling time of 22 years. If it continued to grow at that rate indefinitely, the population of Nigeria in 140 years would be equal to the total population of the world today! This is clearly impossible; no country could ever support that many people.

Even at their present population sizes, most of the people in third-world countries barely subsist, and a high proportion (estimated at 37% in India) are at a starvation level. The decline in population of the developed countries, and the continued population growth of underdeveloped countries, will assure that "the rich get richer and the poor get poorer." Those who are fortunate enough to live in the developed countries should not get too smug about this, however. Massive starvations in the third world will inevitably lead to political unrest and war. In addition to purely humanitarian reasons, it is in the practical interests of the developed countries to aid the underdeveloped ones in their battles with overpopulation.

Mass Extinctions

As stated by Harvard biologist E. O. Wilson, "The extinctions ongoing worldwide promise to be at least as great as the mass extinctions that occurred at the end of the Age of Dinosaurs." It seems ironic that, as the human population is threatening to

Family Planning in Underdeveloped Nations

Some people have argued that overpopulation is not a problem, because starvation would disappear if the world's food were more equitably distributed. Although it may indeed be possible to feed the present global population by redistributing food, it will not be possible to keep doing this as the population grows exponentially. Food redistribution is, at best, only a temporary cure. In the meantime, overpopulation causes troubles other than lack of food: increased spread of disease, inadequate medical care, inadequate shelter, decreased educational opportunities, political turmoil, and so on.

The increasingly high population densities of underdeveloped countries are directly responsible for extensive air and water pollution, and for destruction of ecosystems. This has led to deforestation with consequent soil erosion, which further weakens the ability of the environment to support the human population. Population pressure in Nigeria, for example, has resulted in the expansion of the Sahara Desert. Destruction of the tropical rain forest biome, with its attendant calamities, is also hastened by the pressure of overpopulation.

When the birthrate greatly exceeds the death rate, a larger proportion of the population is young. In Africa, for example, one-half of the population is under fifteen years old. This is a time bomb for population growth as all of these young people enter their reproductive years. The population of Kenya, for example, will triple in the next thirty-five years even if extensive family planning were practiced immediately.

The governments of third-world countries almost uniformly promote family planning practices. In Thailand, 70% of couples practice birth control; their population growth rate was thus cut from 3.2% to 1.6% in only fifteen years. Indonesia reduced its birthrate from 5.6 to 3.4 through widespread use of contraceptives. The birthrate in China is now down to 2.4, yielding a population growth rate of only 1.4%, due to government inducements and coercion. Worldwide, sterilization is the most commonly used family planning method. The technique of minilaparotomy, in which a tubal ligation (chapter 8) is performed as a ten-minute outpatient procedure, has gained in popularity in Kenya.

If birth control is not sufficiently practiced, the only other way that population size can be maintained below the carrying capacity of the environment (at which point growth is prevented by an increased death rate) is through abortions. Currently, twenty-eight million abortions are performed per year in the underdeveloped countries. The World Health Organization estimates that approximately 200,000 women die each year from improperly performed abortions. A new abortion pill, designated RU 486, is currently available in France and might be used in the future by third-world countries. This pill is administered in a physician's office to terminate pregnancies prior to nine weeks of gestation. Use of this pill is currently under attack by antiabortion groups.

The acceptance of family planning methods to limit population size is directly correlated with the degree of economic development of a country and with the status of women in the country. In countries that have experienced great improvement in economic conditions, such as Japan, Singapore, and Taiwan, birthrates have declined (to 1.8, 1.7, and 2.1, respectively). If women have low social status and educational opportunities, however, birthrates remain high regardless of economic conditions. Thus, two of the wealthiest countries on earth—Saudi Arabia and the United Arab Emirates—have birthrates of 7.1 and 5.9, respectively. If the developed countries want to defuse the population bomb, they should therefore encourage women's liberation and education as they promote economic development and give technological aid to developing nations.

grow to the point of self-strangulation, countless other organisms with which people share this planet are threatened with extinction. Yet the latter fact is a direct result of the former.

As human populations grow, destruction of habitats threaten the existence of species in all ecosystems. Most alarming, however, is the massive destruction of the tropical rain forests. Although this biome occupies only 7% of the earth's surface, it contains 50% to 80% of all species on earth. The destruction of the tropical rain forests is currently causing extinctions at the phenomenal rate of 4,000 to 6,000 species each year! In the words of biologist Norman Meyer, this is "causing the death of birth."

The massive extinction of species that accompanies the destruction of rain forests is at least as serious as its effects on greenhouse warming. Out of an estimated 5 to 30 million different species that live on earth, only about 1.7 million species have been scientifically described. Most of the presently undescribed species live in tropical rain forests. At their present rates of extinction, most of these species will have vanished before they are even discovered. These vanished species could have provided people with new sources of foods and medicines, new sources for construction, and new energy sources. These are very real possibilities—at least 25% of the medicines currently in use were originally derived from wild plants and other organisms, for example.

In addition to esthetic considerations, therefore, destruction of the tropical rain forests represents a great loss from a practical standpoint. This is even more true in the modern age of genetic engineering, when genes from different species might be cut and spliced together to obtain recombinant organisms that could provide as yet unimagined benefits. As more species become extinct, the size and quality of our genetic library shrinks.

The tropical rain forests are being destroyed for poor economic reasons. After some of this land has been cleared and fertilized by the ash of the burned trees, it can only serve for farming and cattle grazing for two to three years. After this, the infertile soil becomes a rock-hard substance called *laterite*.[8] In order to survive, the people who practice these slash-and-burn agricultural techniques must clear another part of the rain forest and repeat this process. Soon there will be no more rain forests to burn; when this occurs, the people will be more economically deprived than they are now and their rain forests will be gone forever.

It is thus in the long-term interests of the nations which possess tropical rain forests to halt their decimation. Yet, to do this while supporting their population, they need economic and technological aid from the developed nations. In the meantime, biologists are attempting to catalogue and rescue as many of the organisms as they can. The National Cancer Institute is currently testing 10,000 substances from these plants per year for effectiveness in the treatment of cancer and AIDS. Perhaps, with the heightened awareness of the value of rain forests, the destruction can be stopped before it is too late.

Concluding Remarks

The spirit of scientific inquiry, coupled with the desire to understand ourselves and our relation to the natural world, has brought us the knowledge we need to extricate ourselves from the ecological mess that we, in our ignorance, created. In this last decade of the twentieth century, we have a better understanding than ever before of what we are and where we live. We have ceased being just passengers; we are now the crew of spaceship earth. Its survival depends upon our following the course that our scientific explorations have charted.

As expressed so well by T. S. Eliot:

We shall not cease from exploration
And the end of all our exploring
Will be to arrive where we started
And know the place for the first time

---

1. Define the term *carrying capacity*. Explain why populations cannot grow indefinitely.
2. Compare the birthrates of a number of developed and underdeveloped countries.
3. Explain why food redistribution cannot solve the problem of world hunger on a sustained basis.
4. Describe the importance of the tropical rain forests, and explain the problems associated with their destruction.

---

[8]laterite: L. *lateritius*, brick

Summary

Basic Concepts in Ecology p. 478

I. Ecosystems contain producers (plants), consumers (animals), and decomposers (bacteria and fungi).
   A. Primary consumers eat the plants; secondary consumers eat the primary consumers.
   B. Only about 10% of the energy present in the organisms being eaten is transferred to the organisms at the next trophic level of the food chain.

II. Matter, which provides nutrients to organisms, cycles in the biosphere; decomposers are needed to return nutrients such as carbon and nitrogen to the soil so they can be recycled.

III. There are several different biomes with worldwide distribution.

Pollution of the Environment p. 485

I. The biosphere is a closed, finite environment; air and the water of the oceans are commons that receive pollutants from many nations.

II. Smog contains a number of different constituents which pose health hazards.

III. Lakes can be polluted by nutrients that cause an excessive growth of algae; this is called eutrophication.
   A. Many coastal waters are polluted by excess sludge and by pesticides and heavy metals.
   B. Nitric acid and sulfuring acid in polluted air can produce acid rain, which damages soil and aquatic ecosystems.

Ozone Depletion and the Greenhouse Effect p. 490

I. The ozone layer in the stratosphere protects life on earth by absorbing high-energy ultraviolet light from the sun.
   A. Ozone is destroyed by chlorine atoms derived from chlorofluorocarbons (CFCs).
   B. CFCs have caused an ozone hole over the Antarctic.

II. Carbon dioxide, and other gases to a lesser degree, help to prevent too much heat from escaping the earth.
   A. This is called the greenhouse effect, and is needed for life to exist on earth.
   B. Excessive production of carbon dioxide from the burning of fossil fuels and tropical rain forests may cause an excessive warming of the earth, resulting in massive destruction of ecosystems.

Growth of the Human Population p. 493

I. Populations grow exponentially until they are limited by the carrying capacity of the environment.
   A. When population size has reached the carrying capacity, death rates equal or exceed birthrates.
   B. The population of the underdeveloped countries is rapidly growing because the birthrate is much higher than the death rate.

II. Tropical rain forests are being destroyed at an alarming rate as people in these areas practice slash-and-burn agricultural techniques.
   A. This contributes to the greenhouse effect, and is causing the massive extinction of species that are found only in this biome.
   B. The soil of the rain forest is poor, and forms laterite soon after the forest is clear.
   C. Extinction of species decreases the genetic library of life on earth.

Review Activities

Objective Questions

Match the biome with its characteristic:

1. Forests of pine and fir	(a) grasslands
2. Trees display changes with the seasons	(b) chaparral
3. Has the most fertile soil	(c) tropical rain forest
4. Has the greatest diversity of life	(d) taiga
5. Known as the elfin forest	(e) temperate deciduous forest

6. Which of the following statements about an ecosystem is *true?*
   (a) matter flows, energy cycles
   (b) energy flows, matter cycles
   (c) matter and energy cycle
   (d) matter and energy flow
7. Which of the following consumers can an ecosystem support in greatest numbers?
   (a) primary consumers
   (b) secondary consumers
   (c) tertiary consumers
   (d) quaternary consumers
8. Which of the following is *not* required for life on earth?
   (a) greenhouse gases in the atmosphere
   (b) ozone in the stratosphere
   (c) ozone in the troposphere
   (d) bacteria and fungi
9. Chlorofluorocarbons are primarily responsible for
   (a) depletion of ozone
   (b) greenhouse warming
   (c) pollution of lakes
   (d) smog
10. Excessive production of carbon dioxide is primarily responsible for
   (a) depletion of ozone
   (b) greenhouse warming
   (c) pollution of lakes
   (d) smog
11. Which of the following is *true* about population growth?
   (a) Populations can continue to grow indefinitely.
   (b) The human population on earth is experiencing a decline.
   (c) The human population will continue to grow for at least the next five centuries.
   (d) Populations can grow only until the carrying capacity of the environment is reached.
12. When a portion of a tropical rain forest is cleared
   (a) the soil is rich for agriculture
   (b) the soil becomes useless for agriculture after two to three years
   (c) the greenhouse effect is reduced
   (d) the ozone hole is enlarged

Essay Questions

1. Evaluate the following statement: "When the carrying capacity of a country is reached, most people will have to become vegetarians."
2. Explain why the solutions to such problems as acid rain, ozone depletion, and greenhouse warming require international cooperation.
3. Discuss the ecological advantages of reducing oil consumption. What are the ways that this can be done?
4. Why are the oceans and atmosphere called a "commons"? What is the significance of a commons in terms of pollution?
5. Explain why the tropical rain forests are being destroyed, and why this is shortsighted on economic as well as ecological grounds.

Appendix

Answers to Objective Questions

Chapter 1

1. c	5. b	8. e
2. d	6. d	9. c
3. e	7. b	10. e
4. a		

Chapter 2

1. d	7. c	13. a
2. b	8. a	14. c
3. a	9. d	15. a
4. c	10. e	16. e
5. b	11. b	17. b
6. d	12. d	

Chapter 3

1. c	6. c	11. d
2. b	7. d	12. a
3. d	8. d	13. e
4. c	9. b	14. e
5. b	10. b	

Chapter 4

1. d	4. c	7. b
2. b	5. a	8. a
3. d	6. c	9. c

Chapter 5

1. d	4. b	7. b
2. d	5. c	8. a
3. b	6. b	9. c

Chapter 6

1. c	7. d	12. b
2. d	8. a	13. d
3. a	9. c	14. b
4. a	10. c	15. c
5. c	11. d	16. d
6. d		

Chapter 7

1. d	5. d	9. d
2. d	6. a	10. a
3. e	7. b	11. d
4. e	8. e	12. c

Chapter 8

1. c	5. d	9. a
2. d	6. a	10. c
3. b	7. c	11. c
4. b	8. d	12. b

Chapter 9

1. a	5. c	9. c
2. c	6. d	10. a
3. a	7. b	11. a
4. b	8. d	12. c

Chapter 10

1. b	5. d	9. a
2. a	6. e	10. d
3. c	7. b	11. c
4. a	8. a	12. e

Chapter 11

1. c	5. b	9. d
2. d	6. b	10. b
3. c	7. a	11. d
4. d	8. b	12. d

Chapter 12

1. b	6. b	11. c
2. e	7. d	12. c
3. a	8. d	13. c
4. c	9. e	14. b
5. a	10. b	

Chapter 13

1. c	5. c	9. d
2. b	6. b	10. a
3. d	7. d	11. d
4. a	8. a	

Chapter 14

1. c	5. a	9. b
2. a	6. a	10. d
3. d	7. d	11. d
4. a	8. e	12. c

Chapter 15

1. d	6. e	10. b
2. b	7. a	11. c
3. c	8. c	12. d
4. d	9. d	13. a
5. c		

Chapter 16

1. b	5. e	8. a
2. a	6. d	9. e
3. c	7. d	10. c
4. b		

Chapter 17

1. e	5. d	8. d
2. d	6. b	9. e
3. b	7. a	10. e
4. c		

Chapter 18

1. b	6. a	11. b
2. b	7. a	12. a
3. c	8. d	13. b
4. d	9. b	14. d
5. c	10. d	15. d

Chapter 19

1. d	5. e	9. b
2. d	6. d	10. c
3. a	7. e	11. c
4. b	8. a	12. a

Chapter 20

1. d	5. d	8. a
2. c	6. c	9. b
3. b	7. b	10. c
4. d		

Chapter 21

1. c	5. c	8. a
2. a	6. a	9. c
3. b	7. b	10. a
4. d		

Chapter 22

1. d	5. b	9. a
2. e	6. b	10. b
3. a	7. a	11. d
4. c	8. c	12. b

Glossary

A

a-, an- (Gk.) Not, without, lacking.

ab- (L.) Off, away from.

abdomen A body cavity between the thorax and pelvis.

abductor A muscle that moves the skeleton away from the median plane of the body.

ABO system The most common system of classification for red blood cell antigens. On the basis of antigens on the red blood cell surface, individuals can be type A, type B, type AB, or type O.

absorption The transport of molecules across epithelial membranes into the body fluids.

accommodation The ability of the eyes to adjust their curvature so that an image of an object is focused on the retina at different distances.

acetyl CoA Acetyl coenzyme A; an intermediate molecule in aerobic cell respiration which, together with oxaloacetic acid, begins the Krebs cycle. Acetyl CoA is also an intermediate in the synthesis of fatty acids.

acetylcholine (ACh) A molecule that functions as a neurotransmitter chemical in somatic motor nerve and parasympathetic nerve fibers.

acidosis An abnormal increase in the H^+ concentration of the blood that lowers arterial pH below 7.35.

acromegaly A condition caused by the hypersecretion of growth hormone from the pituitary after maturity and characterized by enlargement of the extremities, such as the nose, jaws, fingers, and toes.

ACTH Abbreviation for adrenocorticotropic hormone. A hormone secreted by the anterior pituitary gland which stimulates the adrenal cortex.

actin A structural protein of muscle that, along with myosin, is responsible for muscle contraction.

action potential An all-or-none electrical event in an axon or muscle fiber, in which the polarity of the membrane potential is rapidly reversed and reestablished.

active immunity Immunity involving sensitization, in which antibody production is stimulated by prior exposure to an antigen.

active transport The movement of molecules or ions across the cell membranes of epithelial cells by membrane carriers; the expenditure of cellular energy (ATP) is required.

ad- (L.) Toward, next to.

adductor A muscle that moves the skeleton toward the midline of the body.

ADH Antidiuretic hormone, also known as *vasopressin*. A hormone produced by the hypothalamus and secreted by the posterior pituitary gland; it acts on the kidneys to promote water reabsorption, thus decreasing the urine volume.

adipose tissue Fatty tissue; a type of connective tissue consisting of fat cells in a loose connective tissue matrix.

ADP Adenosine diphosphate; a molecule that, together with inorganic phosphate, is used to make ATP (adenosine triphosphate).

adrenal cortex The outer part of the adrenal gland. The adrenal cortex secretes corticosteroid hormones (such as aldosterone and hydrocortisone).

adrenal medulla The inner part of the adrenal gland. The adrenal medulla secretes epinephrine and (to a lesser degree) norepinephrine.

adrenergic An adjective describing the actions of epinephrine, norepinephrine, or other molecules with similar activity (as in *adrenergic receptor* and *adrenergic stimulation*).

aerobic capacity The ability of an organ to utilize oxygen and respire aerobically to meet its energy needs.

afferent The carrying of something toward a center. Afferent neurons, for example, conduct impulses toward the central nervous system; afferent arterioles carry blood toward the glomerulus.

agglutinate A clumping of cells (usually erythrocytes) due to specific chemical interaction between surface antigens and antibodies.

agranular leukocytes White blood cells (leukocytes) that do not contain cytoplasmic granules; specifically, lymphocytes and monocytes.

albumin A water-soluble protein, produced in the liver, that is the major component of the plasma proteins.

aldosterone The principal corticosteroid hormone involved in the regulation of electrolyte balance (mineralocorticoid).

alleles The pairs of genes located on homologous chromosomes which code for a particular trait.

allergy A state of hypersensitivity caused by exposure to allergens; it results in the liberation of histamine and other molecules with histaminelike effects.

all-or-none law A given response will be produced to its maximum extent in response to any stimulus equal to or greater than a threshold value. Action potentials obey an all-or-none law.

alveoli Plural of *alveolus*; an anatomical term for small, saclike dilations (as in *lung alveoli*).

amniocentesis A procedure to obtain amniotic fluid and fetal cells in this fluid through transabdominal perforation of the uterus.

amnion The inner fetal membrane that contains the fetus in amniotic fluid; this is commonly called the "bag of waters."

an- (Gk.) Without, not.

anabolic steroids Steroids with androgenlike stimulatory effects on protein synthesis.

anabolism Chemical reactions within cells that result in the production of larger molecules from smaller ones; specifically, the synthesis of protein, glycogen, and fat.

anaerobic respiration A form of cell respiration, involving the conversion of glucose to lactic acid, in which energy is obtained without the use of molecular oxygen.

anaerobic threshold The maximum exercise that can be performed before a significant amount of lactic acid is produced by the exercising skeletal muscles through anaerobic respiration. This generally occurs when about 60% of the total aerobic capacity of the person has been reached.

androgens Steroids that have masculinizing effects; primarily those hormones (such as testosterone) secreted by the testes, although weaker androgens are also secreted by the adrenal cortex.

anemia An abnormal reduction in the red blood cell count, hemoglobin concentration, or hematocrit, or any combination of these measurements. This condition is associated with a decreased ability of the blood to carry oxygen.

angina pectoris A thoracic pain, often referred to the left pectoral and arm area, caused by inadequate blood flow (ischemia) to the heart muscle.

antagonistic effects Actions of regulators such as hormones or nerves that counteract the effects of other regulators. The actions of sympathetic and parasympathetic neurons on the heart, for example, are antagonistic.

anterior An anatomical term referring to a forward position of a structure.

anterior pituitary Also called the *adenohypophysis*; this part of the pituitary secretes FSH (follicle-stimulating hormone), LH (luteinizing hormone), ACTH (adrenocorticotropic hormone), TSH (thyroid-stimulating hormone), GH (growth hormone), and prolactin. Secretions of the anterior pituitary are controlled by hormones secreted by the hypothalamus.

antibodies Immunoglobulin proteins secreted by B lymphocytes that have transformed into plasma cells. Antibodies are responsible for humoral immunity. Their synthesis is induced by specific antigens, and they combine with these specific antigens but not with unrelated antigens.

anticoagulant A substance that inhibits blood clotting.

anticodon A base triplet provided by three nucleotides within a loop of transfer RNA, which is complementary in its base pairing properties to a triplet (the codon) in mRNA; the matching of codon to anticodon provides the mechanism for translation of the genetic code into a specific sequence of amino acids.

antigen A molecule able to induce the production of antibodies and to react in a specific manner with antibodies.

antiserum A serum that contains specific antibodies.

aphasia Speech and language disorders caused by damage to the brain. This can result from damage to Broca's area, Wernicke's area, the arcuate fasciculus, or the angular gyrus.

apnea The cessation of breathing.

aqueous humor A fluid produced by the ciliary body that fills the anterior and posterior chambers of the eye.

arteriosclerosis A group of diseases characterized by thickening and hardening of the artery wall and narrowing of its lumen.

artery A vessel that carries blood away from the heart.

astigmatism Unequal curvature of the refractive surfaces of the eye (cornea and/or lens), so that light that enters the eye along certain meridians does not focus on the retina.

atherosclerosis A common type of arteriosclerosis found in medium and large arteries, in which raised areas, or plaques, are formed from smooth muscle cells, cholesterol, and other lipids. These plaques occlude arteries and serve as sites for the formation of thrombi.

atomic number A whole number representing the number of positively charged protons in the nucleus of an atom.

ATP Adenosine triphosphate; the universal energy carrier of the cell.

atrioventricular node The atrioventricular, or AV, node is the specialized mass of conducting tissue located in the right atrium near the junction of the interventricular septum. It transmits the impulse into the bundle of His.

atrioventricular valves The atrioventricular, or AV, valves are one-way valves located between the atria and ventricles. The AV valve on the right side of the heart is the tricuspid, and the AV valve on the left side is the bicuspid or mitral valve.

atrophy The decrease in mass and size of an organ; the opposite of hypertrophy.

auto- (Gk.) Self, same.

autoantibodies Antibodies that are formed in response to and that react with molecules that are part of one's own body.

autocrine A type of regulation in which one part of an organ releases chemicals that help regulate another part of the same organ.

autonomic nervous system The part of the nervous system that involves control of smooth muscle, cardiac muscle, and glands. The autonomic nervous system is subdivided into the sympathetic and parasympathetic divisions.

autosomal chromosomes The paired chromosomes; those other than the sex chromosomes.

axon The process of a nerve cell that conducts impulses away from the cell body.

B

baroreceptors Receptors for arterial blood pressure located in the aortic arch and the carotid sinuses.

basal ganglia Gray matter, or nuclei, within the cerebral hemispheres, which are involved in the control of skeletal movements.

basal metabolic rate (BMR) The rate of metabolism (expressed as oxygen consumption or heat production) under resting or basal conditions fourteen to eighteen hours after eating.

B cell lymphocytes Lymphocytes that can be transformed by antigens into plasma cells that secrete antibodies (and are thus responsible for humoral immunity).

benign Not malignant or life threatening.

bi- (L.) Two, twice.

bile Fluid, produced by the liver and stored in the gallbladder, that contains bile salts, bile pigments, cholesterol, and other molecules. The bile is secreted into the small intestine.

bile salts Salts of derivatives of cholesterol in bile that are polar on one end and nonpolar on the other end of the molecule. Bile salts have detergent or surfactant effects and act to emulsify fat in the lumen of the small intestine.

bilirubin Bile pigment derived from the breakdown of the heme portion of hemoglobin.

blood-brain barrier The structures and cells that selectively prevent particular molecules in the plasma from entering the central nervous system.

bradycardia A slow cardiac rate; less than sixty beats per minute.

bronchioles The smallest air passages in the lungs, which deliver air to alveoli.

brush border enzymes Digestive enzymes in the small intestine that are located in the cell membrane of the microvilli of intestinal epithelial cells.

buffer A molecule that serves to prevent large changes in pH by either combining with H^+ or by releasing H^+ into solution.

bundle of His A band of rapidly conducting cardiac fibers that originate in the AV node and extend down the atrioventricular septum to the apex of the heart. This tissue conducts action potentials from the atria into the ventricles.

C

calorie A unit of heat equal to the amount of heat needed to raise the temperature of one gram of water by 1°C.

cAMP Cyclic adenosine monophosphate; a second messenger in the action of many hormones, such as epinephrine. It serves to mediate the effects of these hormones on their target cells.

cancer A tumor characterized by abnormally rapid cell division and the loss of specialized tissue characteristics. This term usually refers to malignant tumors.

capacitation Changes that occur within spermatozoa in the female reproductive tract that enable the sperm to fertilize ova; sperm that have not been capacitated in the female tract cannot fertilize ova.

capillary The smallest vessel in the vascular system. Capillary walls are only one cell thick, and all exchanges of molecules between the blood and tissue fluid occur across the capillary wall.

carbohydrate Organic molecules containing carbon, hydrogen, and oxygen in a ratio of 1:2:1. This class of molecules is subdivided into monosaccharides, disaccharides, and polysaccharides.

carboxyhemoglobin An abnormal form of hemoglobin in which the heme is bonded to carbon monoxide.

carcinogen A substance that causes cancer.

cardiac muscle Muscle of the heart, consisting of striated muscle cells. These cells are interconnected into a mass called the myocardium.

cardiac output The volume of blood pumped by either the right or the left ventricle per minute.
cardiogenic shock Shock that results from low cardiac output in heart disease.
carrier-mediated transport The transport of molecules or ions across a cell membrane by means of specific protein carriers. It includes both facilitated diffusion and active transport.
casts An accumulation of proteins that produce molds of kidney tubules and that appear in urine sediment.
catabolism Chemical reactions in a cell whereby larger, more complex molecules are converted into smaller molecules.
catalyst A catalyst is a substance that increases the rate of a chemical reaction without changing the nature of the reaction or being changed by the reaction.
catecholamines A group of molecules, including epinephrine, norepinephrine, L-dopa, and related molecules, that have effects that are similar to those produced by activation of the sympathetic nervous system.
cell-mediated immunity Immunological defense provided by T cell lymphocytes, which come into close proximity to their victim cells (as opposed to humoral immunity provided by the secretion of antibodies by plasma cells).
cellular respiration The energy-releasing metabolic pathways in a cell that oxidize organic molecules such as glucose, fatty acids, and others.
centri- (L.) Center.
centrioles Cell organelles that form the spindle apparatus during cell division.
centromere The central region of a chromosome to which the chromosomal arms are attached.
cerebellum A part of the brain which serves as a major center of motor control.
cerebral lateralization The specialization of function of each cerebral hemisphere. Language ability, for example, is lateralized to the left hemisphere in most people.
chemoreceptors Neural receptors sensitive to chemical changes in blood and other body fluids.
chemotaxis The movement of an organism or a cell, such as a leukocyte, toward a chemical stimulus.
cholesterol A twenty-seven-carbon steroid that serves as the precursor of steroid hormones.
cholinergic Denoting nerve endings that liberate acetylcholine as a neurotransmitter, such as those of the parasympathetic system.
chondrocytes Cartilage-forming cells.
chorea The occurrence of a wide variety of rapid, complex, jerky movements that appear to be well coordinated but are performed involuntarily.
chromatids Duplicated chromosomes that are joined together at the centromere and separate during cell division.
chromatin Threadlike structures in the cell nucleus consisting primarily of DNA and protein. This represents the extended form of chromosomes during interphase.
chromosomes Structures in the cell nucleus, containing DNA and associated proteins. The chromosomes are in a compact form during cell division and thus become visible as discrete structures in the light microscope during this time.
chylomicrons Particles of lipids and protein that are secreted by the intestinal epithelial cells into the lymph and are transported by the lymphatic system to the blood.
chyme A mixture of partially digested food and digestive juices within the stomach and small intestine.
cilia Plural of *cilium*; tiny hairlike processes that extend from the cell surface and beat in a coordinated fashion.
cirrhosis Liver disease characterized by the loss of normal microscopic structure, which is replaced by fibrosis and nodular regeneration.
clonal selection theory The theory in immunology that active immunity is produced by the development of clones of lymphocytes able to respond to a particular antigen.
clone A group of cells derived from a single parent cell by mitotic cell division; since reproduction is asexual, the descendants of the parent cell are genetically identical. Term used when cells are separate individuals (as in white blood cells) rather than part of a growing organ.
CNS Central nervous system; that part of the nervous system consisting of the brain and spinal cord.
cochlea The organ of hearing in the inner ear where nerve impulses are generated in response to sound waves.
codon The sequence of three nucleotide bases in mRNA that specifies a given amino acid and determines the position of that amino acid in a polypeptide chain through complementary base pairing with an anticodon in transfer RNA.
coenzyme An organic molecule, usually derived from a water-soluble vitamin, that combines with and activates specific enzyme proteins.
cofactor A substance needed for the catalytic action of an enzyme; it usually refers to inorganic ions such as Ca^{++} and Mg^{++}.
colloid osmotic pressure Osmotic pressure exerted by plasma proteins, which are present as a colloidal suspension. Also called oncotic pressure.
com-, con- (L.) With, together.
cones Photoreceptors in the retina of the eye that provide color vision and high visual acuity.
congestive heart failure The inability of the heart to deliver an adequate blood flow, due to heart disease or hypertension. It is associated with breathlessness, salt and water retention, and edema.
connective tissue One of the four primary tissues, characterized by an abundance of extracellular material.
consumers In terms of ecology, consumers are members of the animal kingom.
contralateral Affecting the opposite side of the body.
cornea The transparent structure forming the anterior part of the connective tissue covering of the eye.
corpus callosum A large, transverse tract of nerve fibers connecting the cerebral hemispheres.
cortex The outer covering or layer of an organ.
corticosteroids Steroid hormones of the adrenal cortex, consisting of glucocorticoids (such as hydrocortisone) and mineralocorticoids (such as aldosterone).
countercurrent multiplier system The interaction that occurs between the descending limb and the ascending limb of the loop of Henle in the kidney. This interaction results in the multiplication of the solute concentration in the interstitial fluid of the renal medulla.
cretinism A condition caused by insufficient thyroid secretion during prenatal development or the years of early childhood. It results in stunted growth and inadequate mental development.
crypt- (Gk.) Hidden, concealed.
cryptorchidism A developmental defect in which the testes fail to descend into the scrotum and, instead, remain in the body cavity.
curare A chemical derived from plant sources that causes flaccid paralysis by blocking ACh receptor proteins in muscle cell membranes.
cyanosis A blue color given to the skin or mucous membranes by deoxyhemoglobin; it indicates inadequate oxygen concentration in the blood.
cyto- (Gk.) Cell.
cytochrome A pigment in mitochondria that transports electrons in the process of aerobic respiration.
cytoplasm The semifluid part of the cell between the cell membrane and the nucleus, exclusive of membrane-bound organelles. It contains many enzymes and structural proteins.
cytoskeleton A latticework of structural proteins in the cytoplasm arranged in the form of microfilaments and microtubules.

D

dark adaptation The ability of the eyes to increase their sensitivity to low light levels over a period of time. Part of this adaptation involves increased amounts of visual pigment in the photoreceptors.
decomposers Decomposers are bacteria and fungi that help recycle molecules within an ecosystem.
delayed hypersensitivity An allergic response in which the onset of symptoms takes as long as two to three days after exposure to an antigen. Produced by T cells, it is a type of cell-mediated immunity.
dendrite A relatively short, highly branched neural process that carries electrical activity to the cell body.
deoxyhemoglobin The form of hemoglobin in which the heme groups are in the normal reduced form but are not bonded to a gas. Deoxyhemoglobin is produced when oxyhemoglobin releases oxygen.

depolarization The loss of membrane polarity in which the inside of the cell membrane becomes less negative in comparison to the outside of the membrane. The term is also used to indicate the reversal of membrane polarity that occurs during the production of action potentials in nerve and muscle cells.

deposition, bone The formation of the extracellular matrix of bone by osteoblasts. This process includes secretion of collagen and precipitation of calcium phosphate in the form of hydroxyapatite crystals.

diabetes insipidus A condition in which inadequate amounts of antidiuretic hormone (ADH) are secreted by the posterior pituitary. It results in inadequate reabsorption of water by the kidney tubules and, thus, in the excretion of a large volume of dilute urine.

diabetes mellitus The appearance of glucose in the urine due to the presence of high plasma glucose concentrations, even in the fasting state. This disease is caused by either a lack of sufficient insulin secretion or by inadequate responsiveness of the target tissues to the effects of insulin.

diastole The phase of relaxation in which the heart fills with blood. Unless accompanied by the modifier term *atrial,* diastole usually refers to the resting phase of the ventricles.

diastolic blood pressure The minimum pressure in the arteries that is produced during the phase of diastole of the heart. It is indicated by the last sound of Korotkoff when taking a blood pressure measurement.

diffusion The net movement of molecules or ions from regions of higher to regions of lower concentration.

digestion The process of converting food into molecules that can be absorbed through the intestine into the blood.

1,25-dihydroxyvitamin D_3 The active form of vitamin D produced within the body by hydroxylation reactions in the liver and kidneys of vitamin D formed by the skin. This is a hormone that promotes the intestinal absorption of Ca^{++}.

diploid Denoting cells having two of each chromosome or twice the number of chromosomes that are present in sperm or ova.

disaccharide The class of double sugars; carbohydrates that yield two simple sugars, or monosaccharides, upon hydrolysis.

diuretic A substance that increases the rate of urine production, and thus decreases the blood volume.

DNA Deoxyribonucleic acid; composed of nucleotide bases and deoxyribose sugar, it contains the genetic code.

dopamine Serves as a neurotransmitter in the central nervous system; also is the precursor of norepinephrine, another neurotransmitter molecule.

ductus arteriosus A fetal blood vessel connecting the pulmonary artery directly to the aorta.

dwarfism A condition in which a person is undersized due to inadequate secretion of growth hormone.

dyspnea Subjective difficulty in breathing.

E

ECG Electrocardiogram (also abbreviated EKG); a recording of electrical currents produced by the heart.

ecology The scientific study of the interactions among groups of organisms and the relationships between organisms and their environment.

ecto- (Gk.) Outside, outer.

-ectomy (Gk.) Surgical removal of a structure.

ectopic Foreign, out of place.

ectopic focus An area of the heart other than the SA node that assumes pacemaker activity.

ectopic pregnancy Embryonic development that occurs anywhere other than in the uterus (as in the fallopian tubes or body cavity).

edema Swelling due to an increase in tissue fluid.

EEG Electroencephalogram; a recording of the electrical activity of the brain from electrodes placed on the scalp.

effector organs A collective term for muscles and glands that are activated by motor neurons.

efferent The transport of something away from a central location. Efferent nerve fibers conduct impulses away from the central nervous system, for example, and efferent arterioles transport blood away from the glomerulus.

elasticity The tendency of a structure to recoil to its initial dimensions after being distended (stretched).

electrolytes Ions and molecules that are able to ionize and thus carry an electric current. The major electrolytes in the plasma are Na^+, HCO_3^-, and K^+.

element, chemical A substance that cannot be broken down by chemical means into simpler compounds. An element is composed of atoms that all have the same atomic number. An element can, however, contain different isotopes of that atom which have different numbers of neutrons and which thus have different atomic weights.

elephantiasis A disease caused by infection with a nematode worm, in which the larvae block lymphatic drainage and produce edema; the lower areas of the body can become enormously swollen as a result.

emphysema A lung disease in which alveoli are destroyed and the remaining alveoli become larger. It results in decreased vital capacity and increased airways resistance.

emulsification The process of producing an emulsion or fine suspension; in the intestine, fat globules are emulsified by the detergent action of bile.

endergonic A chemical reaction that requires the input of energy from an external source in order to proceed.

endo- (Gk.) Within, inner.

endocrine glands Glands that secrete hormones into the circulation rather than into a duct; also called ductless glands.

endocytosis The cellular uptake of particles that are too large to cross the cell membrane; this occurs by invagination of the cell membrane until a membrane-enclosed vesicle is pinched off within the cytoplasm.

endoderm The innermost of the three primary germ layers of an embryo; it gives rise to the digestive tract and associated structures, respiratory tract, bladder, and urethra.

endogenous A product or process arising from within the body; as opposed to exogenous products or influences, which arise from external sources.

endometrium The mucous membrane of the uterus, the thickness and structure of which vary with the phase of the menstrual cycle.

endoplasmic reticulum An extensive system of membrane-enclosed cavities within the cytoplasm of the cell; those with ribosomes on their surface are called rough endoplasmic reticulum and participate in protein synthesis.

endorphins A group of endogenous opioid molecules that may act as a natural analgesic.

endothelium The simple squamous epithelium that lines blood vessels and the heart.

enteric A term referring to the intestine.

entropy The energy of a system that is not available to perform work; a measure of the degree of disorder in a system, entropy increases whenever energy is transformed.

enzyme A protein catalyst that increases the rate of specific chemical reactions.

epi- (Gk.) Upon, over, outer.

epidermis The stratified squamous epithelium of the skin, the outer layer of which is dead and filled with keratin.

epididymis A tubelike structure outside the testes; sperm pass from the seminiferous tubules into the head of the epididymis and then pass from the tail of the epididymis to the vas deferens. The sperm mature, becoming motile, as they pass through the epididymis.

epinephrine Also known as adrenaline; a catecholamine hormone secreted by the adrenal medulla in response to sympathetic nerve stimulation; it acts together with norepinephrine released from sympathetic nerve endings to prepare the organism for "fight or flight."

epithelium One of the four primary tissues, forming membranes that cover and line the body surfaces and forming exocrine and endocrine glands.

EPSP Excitatory postsynaptic potential; a graded depolarization in response to stimulation by a neurotransmitter chemical. EPSPs can be summated, and can stimulate the production of action potentials.

erythroblastosis fetalis Hemolytic anemia in a newborn Rh-positive baby caused by maternal antibodies against the Rh factor that have crossed the placenta.

erythrocytes Red blood cells; the formed elements of blood that contain hemoglobin and transport oxygen.

erythropoietin A hormone secreted by the kidneys which stimulates the bone marrow to produce red blood cells.

essential amino acids Those eight amino acids in adults or nine amino acids in children that cannot be made by the human body and therefore must be obtained in the diet.

estradiol The major estrogen (female sex steroid hormone) secreted by the ovaries.

eutrophication Pollution of a lake or stream with nutrients (primarily phosphates) that cause the excessive growth of algae.

evolution Change in species over a span of many generations, due to random mutation and natural selection.

ex- (L.) Out, off, from.

excitation-contraction coupling The means by which electrical excitation of a muscle results in muscle contraction. This coupling is acheived by Ca^{++}, which enters the muscle cell cytoplasm in response to electrical excitation and which stimulates the events culminating in contraction.

exergonic Denoting chemical reactions that liberate energy.

exo- (Gk.) Outside or outward.

exocrine gland A gland that discharges its secretion through a duct to the outside of an epithelial membrane.

exocytosis The process of cellular secretion in which the secretory products are contained within a membrane-enclosed vesicle; the vesicle fuses with the cell membrane so that the lumen of the vesicle is open to the extracellular environment.

exons Exons are the regions of bases in DNA that code for the production of messenger RNA.

extensor A muscle that, upon contraction, increases the angle of a joint.

extra- (L.) Outside, beyond.

F

facilitated diffusion The carrier-mediated transport of molecules through the cell membrane along the direction of their concentration gradients; it does not require the expenditure of metabolic energy.

FAD Flavin adenine dinucleotide; a coenzyme derived from riboflavin that participates in electron transport within the mitochondria.

fertilization Fusion of an ovum and sperm.

fiber, muscle A skeletal muscle cell.

fiber, nerve An axon of a motor neuron or the dendrite of a pseudounipolar sensory neuron in the PNS.

fibrillation A condition of cardiac muscle characterized electrically by random and continuously changing patterns of electrical activity and resulting in the inability of the myocardium to contract as a unit and pump blood. It can be fatal if it occurs in the ventricles.

fibrin The insoluble protein formed from fibrinogen by the enzymatic action of thrombin during the process of blood clot formation.

flaccid paralysis The inability to contract muscles, resulting in a loss of muscle tone. This may be due to damage to lower motor neurons or to factors that block neuromuscular transmission.

flagellum A whiplike structure that provides motility for sperm.

flexor A muscle that decreases the angle of a joint when it contracts.

follicle A microscopic, hollow structure within an organ. Follicles are the functional units of the thyroid gland and of the ovary.

foramen ovale An opening that is normally present in the atrial septum of a fetal heart and allows direct communication between the right and left atria.

fovea centralis A tiny pit in the macula lutea of the retina that contains slim, elongated cones and provides the highest visual acuity (clearest vision).

Frank-Starling Law of the Heart The Frank-Starling law describes the relationship between end-diastolic volume and stroke volume of the heart. A greater amount of blood in a ventricle prior to contraction results in greater stretch of the myocardium, and by this means produces a contraction of greater strength.

FSH Follicle-stimulating hormone; one of the two gonadotropic hormones secreted from the anterior pituitary. In females FSH stimulates the development of the ovarian follicles, whereas in males it stimulates the production of sperm in the seminiferous tubules.

G

GABA Gamma-aminobutyric acid; it is believed to function as an inhibitory neurotransmitter in the central nervous system.

gametes A collective term for haploid germ cells: sperm and ova.

ganglion A grouping of nerve cell bodies located outside the brain and spinal cord.

gas exchange The diffusion of oxygen and carbon dioxide down their concentration gradients that occurs between pulmonary capillaries and alveoli, and between systemic capillaries and the surrounding tissue cells.

gastric intrinsic factor A glycoprotein secreted by the stomach and needed for the absorption of vitamin B_{12}.

gastric juice The secretions of the gastric mucosa. Gastric juice contains water, hydrochloric acid, and pepsinogen as the major components.

gastrin A hormone secreted by the stomach that stimulates the gastric secretion of hydrochloric acid and pepsin.

gates A term used to describe structures within the cell membrane that regulate the passage of ions through membrane channels. Such gates may be chemically regulated (by neurotransmitters) or voltage regulated (in which case they open in response to depolarization).

gen- (Gk.) Producing.

genetic recombination The formation of new combinations of genes, as by crossing-over between homologous chromosomes.

genetic transcription The process by which RNA is produced with a sequence of nucleotide bases that is complementary to a region of DNA.

genetic translation The process by which proteins are produced with amino acid sequences specified by the sequence of codons in messenger RNA.

gigantism Abnormal body growth due to the excessive secretion of growth hormone.

glomerular filtrate Fluid filtered through the glomerular capillaries into Bowman's capsule of the kidney tubules.

glomerular filtration rate Abbreviated GFR, this is the volume of blood plasma filtered out of the glomeruli of both kidneys per minute.

glomeruli A general term for a tuft or cluster; it is most often used to describe the tufts of capillaries in the kidneys that filter fluid into the kidney tubules.

glomerulonephritis Inflammation of the renal glomeruli, associated with fluid retention, edema, hypertension, and the appearance of protein in the urine.

glucagon A polypeptide hormone that is secreted by the alpha cells of the islets of Langerhans in the pancreas and acts to promote glycogenolysis and raise the blood glucose levels.

glucocorticoids Steroid hormones secreted by the adrenal cortex (corticosteroids) that affect the metabolism of glucose, protein, and fat. These hormones also have anti-inflammatory and immunosuppressive effects; the major glucocorticoid in humans is hydrocortisone (cortisol).

gluconeogenesis The formation of glucose from noncarbohydrate molecules such as amino acids and lactic acid.

glycogen A polysaccharide of glucose—also called *animal starch*—produced primarily in the liver and skeletal muscles. Similar to plant starch in composition, glycogen contains more highly branched chains of glucose subunits than does plant starch.

glycolysis The metabolic pathway that converts glucose to pyruvic acid and yields a net production of two ATP molecules and two molecules of $NADH_2$.

glycosuria The excretion of an abnormal amount of glucose in the urine (urine normally contains only trace amounts of glucose).

Golgi apparatus Stacks of flattened membranous sacs within the cytoplasm of cells, which are believed to bud off vesicles containing secretory proteins.

gonadotropic hormones Hormones of the anterior pituitary gland that stimulate gonadal function—the formation of gametes and secretion of sex steroids. The two gonadotropins are FSH (follicle-stimulating hormone) and LH (luteinizing hormone), which are essentially the same in males and females.

gonads A collective term for testes and ovaries.

graafian follicle A mature ovarian follicle, containing a single fluid-filled cavity, with the ovum located toward one side of the follicle, and perched on top of a hill of granulosa cells.

granular leukocytes Leukocytes with granules in the cytoplasm; on the basis of the staining properties of the granules, these cells are of three types: neutrophils, eosinophils, and basophils.

Graves' disease A hyperthyroid condition believed to be caused by excessive stimulation of the thyroid gland by autoantibodies; it is associated with exophthalmos (bulging eyes), high pulse rate, high metabolic rate, and other symptoms of hyperthyroidism.

gray matter The part of the central nervous system that contains neuron cell bodies and dendrites, but few myelinated axons. This forms the cortex of the cerebrum, cerebral nuclei, and the central region of the spinal cord.

greenhouse effect The ability of carbon dioxide and other gases in the atmosphere to trap some of the heat radiated from the earth.

growth hormone A hormone secreted by the anterior pituitary that stimulates growth of the skeleton and soft tissues during the growing years and that influences the metabolism of protein, carbohydrate, and fat throughout life.

gyrus A fold or convolution in the cerebrum.

H

haploid A cell that has one of each chromosome type and therefore half the number of chromosomes present in most other body cells; only the gametes (sperm and ova) are haploid.

haversian system A haversian canal and its concentrically arranged layers, or lamellae, of bone; it constitutes the basic structural unit of compact bone.

hay fever A seasonal type of allergic rhinitis caused by pollen; it is characterized by itching and tearing of the eyes, swelling of the nasal mucosa, attacks of sneezing, and often by asthma.

hCG Abbreviation for human chorionic gonadotropin. This is a hormone secreted by the embryo which has LH-like actions, and is required for maintenance of the mother's corpus luteum for the first ten weeks of pregnancy.

heart murmur Abnormal heart sounds caused by an abnormal flow of blood in the heart due to structural defects, usually of the valves or septum.

heart sounds The sounds produced by closing of the AV valves of the heart during systole (the first sound) and by closing of the semilunar valves of the aorta and pulmonary trunk during diastole (the second sound).

helper T cells A subpopulation of T cells (lymphocytes) that help stimulate antibody production of B lymphocytes by antigens.

hematocrit The ratio of packed red blood cells to total blood volume in a centrifuged sample of blood, expressed as a percentage.

heme The iron-containing red pigment that, together with the protein globin, forms hemoglobin.

hemoglobin The combination of heme pigment and protein within red blood cells that acts to transport oxygen and (to a lesser degree) carbon dioxide. Hemoglobin also serves as a weak buffer within red blood cells.

hepatic Pertaining to the liver.

hepatitis Inflammation of the liver.

hermaphrodite An organism with both testes and ovaries.

hetero- (Gk.) Different, other.

heterochromatin A condensed, inactive form of chromatin.

heterozygous The condition in which an individual has two different forms of a specific gene. One of these forms (alleles) may be dominant and the other recessive.

high-density lipoproteins Combinations of lipids and proteins that migrate rapidly to the bottom of a test tube during centrifugation; carrier proteins that are believed to transport cholesterol from blood vessels to the liver, and thus to offer some protection from atherosclerosis.

histamine A compound secreted by tissue mast cells and other connective tissue cells that stimulates vasodilation and increases capillary permeability; it is responsible for many of the symptoms of inflammation and allergy.

homeo- (Gk.) Same.

homeostasis The dynamic constancy of the internal environment, which serves as the principal function of physiological regulatory mechanisms. The concept of homeostasis provides a framework for the understanding of most physiological processes.

homologous chromosomes The matching pairs of chromosomes in a diploid cell.

homozygous A condition in which both forms (alleles) of a gene are identical.

hormones Regulatory chemicals secreted into the blood by endocrine cells and carried by the blood to target cells that respond to the hormones by an alteration in their metabolism.

humoral immunity The form of acquired immunity in which antibody molecules are secreted in response to antigenic stimulation (as opposed to cell-mediated immunity).

hyaline membrane disease A disease of some premature infants who lack pulmonary surfactant, it is characterized by collapse of the alveoli (atelectasis) and pulmonary edema. Also called respiratory distress syndrome.

hydrocortisone Also called cortisol; the principal corticosteroid hormone secreted by the adrenal cortex with glucocorticoid action.

hydrophilic A substance that readily absorbs water; literally, "water loving."

hydrophobic A substance that repels, and that is repelled by, water; literally, "water fearing."

hyper- (Gk.) Over, above, excessive.

hyperglycemia Abnormally increased concentration of glucose in the blood.

hyperkalemia Abnormally high concentration of potassium in the blood.

hyperopia Also called farsightedness; a refractive disorder in which rays of light are brought to a focus behind the retina as a result of the eyeball being too short.

hyperplasia An increase in organ size due to an increase in cell numbers as a result of mitotic cell division.

hyperpnea Increased breathing during exercise. Unlike hyperventilation, the arterial blood carbon dioxide values are not changed during hyperpnea, because the increased ventilation is matched to an increased metabolic rate.

hyperpolarization An increase in the negativity of the inside of a cell membrane with respect to the resting membrane potential.

hypersensitivity Another name for allergy; an abnormal immune response that may be immediate (due to antibodies of the IgE class) or delayed (due to cell-mediated immunity).

hypertension High blood pressure. Divided into primary, or essential, hypertension of unknown cause and secondary hypertension that develops as a result of other, known disease processes.

hypertonic A solution with a greater solute concentration and thus a greater osmotic pressure than plasma.

hypertrophy Growth of an organ due to an increase in the size of its cells.

hyperventilation A high rate and depth of breathing that results in a decrease in the blood carbon dioxide concentration below normal.

hypo- (Gk.) Under, below, less.

hypodermis A layer of fat beneath the dermis of the skin.

hypotension Abnormally low blood pressure.

hypothalamic hormones Hormones produced by the hypothalamus; these include antidiuretic hormone and oxytocin, which are secreted by the posterior pituitary gland, and both releasing and inhibiting hormones that regulate the secretion of the anterior pituitary.

hypothalamo-hypophyseal portal system A vascular system that transports releasing and inhibiting hormones from the hypothalamus to the anterior pituitary.

hypothalamus An area of the brain below the thalamus and above the pituitary gland. The hypothalamus regulates the pituitary gland and contributes to the regulation of the autonomic nervous system, among many other functions.

hypoxemia A low oxygen concentration of the arterial blood.

I

immediate hypersensitivity Hypersensitivity (allergy) that is mediated by antibodies of the IgE class and that results in the release of histamine and related compounds from tissue cells.

immunization The process of increasing one's resistance to pathogens. In active immunity a person is injected with antigens that stimulate the development of clones of specific B or T lymphocytes; in passive immunity a person is injected with antibodies made by another organism.

immunosurveillance The function of the immune system to recognize and attack malignant cells that produce antigens not recognized as "self." This function is believed to be cell mediated rather than humoral.

implantation The process by which a blastocyst attaches itself to and penetrates into the endometrium of the uterus.

inhibin Believed to be a water-soluble hormone secreted by the seminiferous tubules of the testes that specifically exerts negative feedback inhibition of FSH secretion from the anterior pituitary gland.
insulin A polypeptide hormone, secreted by the beta cells of the islets of Langerhans in the pancreas, that promotes the anabolism of carbohydrates, fat, and protein. Insulin acts to promote the cellular uptake of blood glucose and, therefore, to lower the blood glucose concentration; insulin deficiency produces hyperglycemia and diabetes mellitus.
inter- (L.) Between, among.
interferons A group of small proteins that inhibit the multiplication of viruses inside host cells and also have antitumor properties.
interleukin-2 A lymphokine secreted by T lymphocytes that stimulates the proliferation of both B and T lymphocytes.
interneurons Also called association neurons; those neurons within the central nervous system that do not extend into the peripheral nervous system; they are interposed between sensory (afferent) and motor (efferent) neurons.
interphase The interval between successive cell divisions; during this time the chromosomes are in an extended state and are active in directing RNA synthesis.
intra- (L.) Within, inside.
introns Noncoding regions of DNA bases that interrupt the coding regions (exons) for mRNA.
in vitro Occurring outside the body, in a test tube or other artificial environment.
in vivo Occurring within the body.
ion An atom or a group of atoms that has either lost or gained electrons and thus has a net positive or a net negative charge.
ionization The dissociation of a solute to form ions.
ipsilateral On the same side (as opposed to contralateral).
IPSP Inhibitory postsynaptic potential. Inhibition of a neuron due to hyperpolarization of the postsynaptic membrane in response to a particular neurotransmitter chemical.
ischemia A rate of blood flow to an organ that is inadequate to supply sufficient oxygen and maintain aerobic respiration in that organ.
islets of Langerhans Encapsulated groupings of endocrine cells within the exocrine tissue of the pancreas. The islets contain alpha cells that secrete glucagon and beta cells that secrete insulin.
iso- (Gk.) Equal, same.
isometric contraction Muscle contraction in which there is no appreciable shortening of the muscle.
isotonic contraction Muscle contraction in which the muscle shortens in length and maintains approximately the same amount of tension throughout the shortening process.
isotonic solution A solution having the same total solute concentration, osmolality, and osmotic pressure as the solution with which it is compared; a solution with the same solute concentration and osmotic pressure as plasma.

J

jaundice A condition characterized by high blood bilirubin levels and staining of the tissues with bilirubin, which gives skin and mucous membranes a yellow color.
junctional complexes The structures that join adjacent epithelial cells together.
juxta- (L.) Near to, next to.

K

keratin A protein that forms the principal component of the outer layer of the epidermis and of hair and nails.
ketoacidosis A type of metabolic acidosis produced by the excessive production of ketone bodies, as in diabetes mellitus.
ketone bodies These include acetone, acetoacetic acid, and β-hydroxybutyric acid, which are derived from fatty acids in the liver. Ketone bodies are oxidized by skeletal muscles for energy.
ketosis An abnormal elevation in the blood concentration of ketone bodies that does not necessarily produce acidosis.
kilocalorie A unit of measurement equal to 1,000 calories, which are units of heat (a kilocalorie is the amount of heat required to raise the temperature of 1 kilogram of water by 1°C). In nutrition the kilocalorie is called a big calorie (Calorie).
Klinefelter's syndrome The syndrome produced in a male by the presence of an extra X chromosome (genotype XXY).
Krebs cycle A cyclic metabolic pathway in the matrix of mitochondria by which the acetic acid part of acetyl CoA is oxidized and substrates are provided for reactions that are coupled to the formation of ATP.

L

lactose Milk sugar; a disaccharide of glucose and galactose.
lactose intolerance The inability of many adults to digest lactose due to loss of the ability of the intestine to produce lactase enzyme.
larynx A structure, consisting of epithelial tissue, muscle, and cartilage, that serves as a sphincter guarding the entrance of the trachea and as the organ responsible for voice production.
lesion A wounded or damaged area.
leukocytes White blood cells.
Leydig cells The interstitial cells of the testes that serve an endocrine function by secreting testosterone and other androgenic hormones.
ligament Dense regular connective tissue, containing many parallel arrangements of collagen fibers, that connects bones or cartilages and serves to strengthen joints.
limbic system A group of brain structures, including the hippocampus, cingulate gyrus, dentate gyrus, and amygdala. The limbic system appears to be important in memory, the control of autonomic function, and some aspects of emotion and behavior.
lipids Organic molecules that are nonpolar and thus insoluble in water. This class includes triglycerides, steroids, and phospholipids.
lipolysis The hydrolysis of triglycerides into free fatty acids and glycerol.
low-density lipoproteins Plasma proteins that transport triglycerides and cholesterol and are believed to contribute to arteriosclerosis.
lumen The cavity of a tube or hollow organ.
lung surfactant A mixture of lipoproteins (containing phospholipids) secreted by type II alveolar cells into the alveoli of the lungs; it lowers surface tension and prevents collapse of the lungs as occurs in hyaline membrane disease, in which surfactant is absent.
luteinizing hormone (LH) A gonadotropic hormone secreted by the anterior pituitary that, in a female, stimulates ovulation and the development of a corpus luteum. In a male, LH stimulates the Leydig cells to secrete androgens.
lymph The fluid in lymphatic vessels that is derived from tissue fluid.
lymphatic system The lymphatic vessels and lymph nodes.
lymphocytes A type of mononuclear leukocyte; the cells responsible for humoral and cell-mediated immunity.
lymphokines A group of chemicals released from T cells that contribute to cell-mediated immunity.
-lysis (Gk.) Breakage, disintegration.
lysosome Organelle containing digestive enzymes and responsible for intracellular digestion.

M

macro- (Gk.) Large.
macromolecules Large molecules; a term that usually refers to protein, RNA, and DNA.
macrophage A large phagocytic cell in connective tissue that contributes to both specific and nonspecific immunity.
malignant A structure or process that is life threatening.
mast cells A type of connective tissue cells that produce and secrete histamine and heparin.
medulla oblongata A part of the brain stem that contains neural centers for the control of breathing and for regulation of the cardiovascular system via autonomic nerves.
mega- (Gk.) Large, great.
meiosis A type of cell division in which a diploid parent cell gives rise to haploid daughter cells; it occurs in the process of gamete production in the gonads.
melanin A dark pigment found in the skin, hair, choroid layer of the eye, and substantia nigra of the brain; it may also be present in certain tumors (melanomas).
melatonin A hormone secreted by the pineal gland that may contribute to the regulation of gonadal function in mammals.
membrane potential The potential difference or voltage that exists between the inner and outer sides of a cell membrane; it exists in all cells, but is capable of being changed by excitable cells (neurons and muscle cells).
membranous labyrinth A system of communicating sacs and ducts within the bony labyrinth of the inner ear.
menarche The age at which menstruation begins.

Meniere's disease Deafness, tinnitus, and vertigo resulting from disease of the labyrinth of the inner ear.

menopause The cessation of menstruation, usually occurring at about age forty-eight to fifty.

menstrual cycle The cyclic changes in the ovaries and endometrium of the uterus, which lasts about a month and is accompanied by shedding of the endometrium, with bleeding. This occurs only in humans and the higher primates.

menstruation Shedding of the outer two-thirds of the endometrium with accompanying bleeding, due to lowering of estrogen secretion by the ovaries at the end of the monthly cycle. The first day of menstruation is taken as day one of the menstrual cycle.

meso- (Gk.) Middle.

mesoderm The middle embryonic tissue layer that gives rise to connective tissue (including blood, bone, and cartilage), blood vessels, muscles, the adrenal cortex, and other organs.

messenger RNA (mRNA) A type of RNA that contains a base sequence complementary to a part of the DNA that specifies the synthesis of a particular protein.

meta- (Gk.) Change.

metabolism All of the chemical reactions in the body; it includes those that result in energy storage (anabolism) and those that result in the liberation of energy (catabolism).

metastasis A process whereby cells of a malignant tumor can separate from the tumor, travel to a different site, and divide to produce a new tumor.

micro- (L.) Small; also, one-millionth.

microvilli Tiny fingerlike projections of a cell membrane; they occur on the apical (lumenal) surface of the cells of the small intestine and in the renal tubules.

micturition Urination.

mitosis Cell division in which the two daughter cells receive the same number of chromosomes as the parent cell (both daughters and parent are diploid).

mono- (Gk.) One, single.

monoclonal antibodies Identical antibodies derived from a clone of genetically identical plasma cells.

monocyte A mononuclear, nongranular leukocyte that is phagocytic and able to be transformed into a macrophage.

monomers A single molecular unit of a longer, more complex molecule; monomers are joined together to form dimers, trimers, and polymers; the hydrolysis of polymers eventually yields separate monomers.

monosaccharides Also called simple sugars; the monomers of more complex carbohydrates. Examples include glucose, fructose, and galactose.

-morph, morpho- (Gk.) Form, shape.

motile Capable of self-propelled movement.

motor cortex The precentral gyrus of the frontal lobe of the cerebrum. Axons from this area form the descending pyramidal motor tracts.

motor neuron An efferent neuron that conducts action potentials away from the central nervous system and innervates effector organs (muscles and glands). It forms the ventral roots of spinal nerves.

motor unit A somatic motor neuron and all of the skeletal muscle fibers stimulated by branches of its axon.

mucous membrane The layers of visceral organs that include the lining epithelium, submucosal connective tissue, and (in some cases) a thin layer of smooth muscle (the muscularis mucosa).

myelin sheath A sheath surrounding axons, which is formed by successive wrappings of a neuroglial cell membrane. Myelin sheaths are formed by Schwann cells in the peripheral nervous system and by oligodendrocytes within the central nervous system.

myocardial infarction An area of necrotic tissue in the myocardium that is filled in by scar (connective) tissue.

myofibrils Subunits of striated muscle fibers that consist of successive sarcomeres; myofibrils run parallel to the long axis of the muscle fiber, and the pattern of their filaments provides the striations characteristic of striated muscle cells.

myoglobin A molecule composed of globin protein and heme pigment, related to hemoglobin, but containing only one subunit (instead of the four in hemoglobin), and found in striated muscles. Myoglobin serves to store oxygen in skeletal and cardiac muscle cells.

myoneural junction Also called the neuromuscular junction; a synapse between a motor neuron and the muscle cell that it innervates.

myopia A condition of the eyes, also called nearsightedness, in which light is brought to a focus in front of the retina due to the eye being too long.

myosin The protein that forms the A bands of striated muscle cells; together with the protein actin, myosin provides the basis for muscle contraction.

myxedema A type of edema associated with hypothyroidism; it is characterized by accumulation of mucoproteins in tissue fluid.

N

NAD Nicotinamide adenine dinucleotide; a coenzyme derived from niacin that functions to transport electrons in oxidation-reduction reactions; it helps to transport electrons to the electron transport chain within mitochondria.

natural selection The environmental influences that cause certain organisms with particular inheritable traits to be more reproductively successful than other members of the population.

necrosis Cellular death within tissues and organs.

negative feedback Mechanisms in the body that act to maintain a state of internal constancy, or homeostasis; effectors are activated by changes in the internal environment, and the actions of the effectors serve to counteract these changes and maintain a state of balance.

neoplasm A new, abnormal growth of tissue, as in a tumor.

nephron The functional unit of the kidneys, consisting of a system of renal tubules and a vascular component that includes capillaries of the glomerulus and the peritubular capillaries.

nerve A collection of motor axons and sensory dendrites in the peripheral nervous system.

neuroglia The supporting tissue of the nervous system, consisting of neuroglial, or glial, cells. In addition to providing support, the neuroglial cells participate in the metabolic and bioelectrical processes of the nervous system.

neurons Nerve cells, consisting of a cell body that contains the nucleus, short branching processes, called dendrites, that carry electrical impulses to the cell body, and a single fiber, or axon, that conducts nerve impulses away from the cell body.

neurotransmitter A chemical contained in synaptic vesicles in nerve endings, which is released into the synaptic cleft and stimulates the production of either excitatory or inhibitory postsynaptic potentials.

neutrons Electrically neutral particles that exist together with positively charged protons in the nucleus of atoms.

niche In terms of ecology, the niche occupied by a population of organisms is the specific role that those organisms play in their ecosystem, and the total environmental requirements of those organisms in their ecosystem.

nodes of Ranvier Gaps in the myelin sheath of myelinated axons, located approximately 1 mm apart. Action potentials are produced only at the nodes of Ranvier in myelinated axons.

norepinephrine A catecholamine released as a neurotransmitter from postganglionic sympathetic nerve endings and as a hormone (together with epinephrine) by the adrenal medulla.

nucleolus A dark-staining area within a cell nucleus; the site where ribosomal RNA is produced.

nucleotide The subunit of DNA and RNA macromolecules; each nucleotide is composed of a nitrogenous base (adenine, guanine, cytosine, and thymine or uracil), a sugar (deoxyribose or ribose), and a phosphate group.

nucleus, cell The organelle, surrounded by a double saclike membrane, called the nuclear envelope, that contains the DNA and genetic information of the cell.

nucleus, brain Aggregation of neuron cell bodies within the brain. Nuclei within the brain are surrounded by white matter and are deep to the cerebral cortex.

nutrient density The ratio of nutrients (vitamins and minerals) to calories in a particular food.

nystagmus Involuntary, oscillatory movements of the eye.

O

obese A person who is excessively fat.

oligo- (Gk.) Few, small.

oligodendrocytes A type of neuroglial cell; it forms myelin sheaths around axons in the central nervous system.

oncogene A gene that causes or contributes to cancer.

oncology The study of tumors.

ontogeny The development of an organism from conception to birth.

oo- (Gk.) Pertaining to an egg.
oogenesis The formation of ova in the ovaries.
optic disc The area of the retina where axons from ganglion cells gather to form the optic nerve and where blood vessels enter and leave the eye; it corresponds to the blind spot in the visual field due to the absence of photoreceptors.
organ A structure in the body composed of a number of primary tissues that perform particular functions.
organelle A membrane-enclosed structure within cells that performs specialized tasks. The term includes mitochondria, Golgi apparatus, endoplasmic reticulum, nuclei, and lysosomes; it is also used for some structures not enclosed by a membrane, such as ribosomes and centrioles.
organ of Corti The structure within the cochlea responsible for hearing. It consists of hair cells and supporting cells on the basilar membrane that help transduce sound waves into nerve impulses.
osmoreceptors Sensory neurons in the hypothalamus that respond to changes in the osmotic pressure of the surrounding fluid.
osmosis The passage of solvent (water) from a more dilute to a more concentrated solution through a membrane that is more permeable to water than to the solute.
osmotic pressure A measure of the tendency for a solution to gain water by osmosis when separated by a membrane from pure water; directly related to the osmolality of the solution, it is the pressure required to just prevent osmosis.
osteo- (Gk.) Pertaining to bone.
osteoblasts Cells that produce bone.
osteocytes Bone-forming cells that have become entrapped within a matrix of bone; these cells remain alive due to nourishment supplied by canaliculi within the extracellular material of bone.
ovaries The gonads of a female that produce ova and secrete female sex steroids.
ovi- (L.) Pertaining to egg.
oviduct The part of the female reproductive tract that transports ova from the ovaries to the uterus. Also called the uterine or fallopian tube.
ovulation The extrusion of a secondary oocyte out of the ovary.
oxidative phosphorylation The formation of ATP by using energy derived from electron transport to oxygen; this occurs in the mitochondria.
oxidizing agent An atom that accepts electrons in an oxidation-reduction reaction.
oxygen debt The extra amount of oxygen required by the body after exercise to metabolize lactic acid and to supply the higher metabolic rate of muscles warmed during exercise.
oxyhemoglobin A compound formed by the bonding of molecular oxygen to hemoglobin.
oxyhemoglobin saturation The ratio, expressed as a percentage, of the amount of oxyhemoglobin compared to the total amount of hemoglobin in blood.
oxytocin One of the two hormones produced in the hypothalamus and secreted by the posterior pituitary (the other hormone is vasopressin); oxytocin stimulates the contraction of uterine smooth muscles and promotes milk ejection in females.
ozone Ozone (O_3) is normally produced in the stratosphere and is needed to shield the earth from ultraviolet light. This layer of ozone is becoming depleted by the effects of man-made chlorofluorocarbons. Ozone in the troposphere is produced by industrial processes and automobile exhausts, and is detrimental to lung function.

P

pacemaker A group of cells that has the fastest spontaneous rate of depolarization and contraction in a mass of electrically coupled cells; in the heart, this is the sinoatrial, or SA, node.
pancreatic juice The secretions of the pancreas that are transported by the pancreatic duct to the duodenum. Pancreatic juice contains bicarbonate and such digestive enzymes as trypsin, lipase, and amylase.
parathyroid hormone (PTH) A polypeptide hormone secreted by the parathyroid glands, PTH acts to raise the blood Ca^{++} levels primarily by stimulating resorption of bone.
Parkinson's disease A tremor of the resting muscles and other symptoms caused by inadequate dopamine-producing neurons in the basal ganglia of the cerebrum. Also called paralysis agitans.
parturition Birth.
passive immunity Specific immunity granted by the administration of antibodies made by another organism.
pathogen Any disease-producing microorganism or substance.
pepsin The protein-digesting enzyme secreted in gastric juice.
peptic ulcer An injury to the mucosa of the esophagus, stomach, or small intestine caused by acidic gastric juice.
perfusion The flow of blood through an organ.
peri- (Gk.) Around, surrounding.
perimysium The connective tissue surrounding a fascicle of skeletal muscle fibers.
periosteum Connective tissue covering bones; it contains osteoblasts and is therefore capable of forming new bone.
peripheral resistance The resistance to blood flow through the arterial system. Peripheral resistance is inversely related to the radius of small arteries and arterioles.
peristalsis Waves of smooth muscle contraction in smooth muscles of the tubular digestive tract, involving circular and longitudinal muscle fibers at successive locations along the tract; it serves to propel the contents of the tract in one direction.
permissive effect A category of hormonal interaction in which one hormone acts to increase the effectiveness of another hormone.
pH The pH of a solution is equal to the logarithm of 1 over the hydrogen ion concentration. The pH scale goes from zero to 14; a pH of 7.0 is neutral, whereas solutions with lower pH are acidic and solutions with higher pH are basic.
phagocytosis Cellular eating; the ability of some cells (such as white blood cells) to engulf large particles (such as bacteria) and digest these particles by merging the food vacuole containing these particles with a lysosome containing digestive enzymes.
phospholipids Lipids containing a phosphate group. These molecules (such as lecithin) are polar on one end and nonpolar on the other end. Phospholipids make up a large part of the cell membrane and function in the lung alveoli as surfactants.
photoreceptors Sensory cells (rods and cones) that respond electrically to light; they are located in the retinas of the eyes.
phylogeny The evolutionary history of a species.
pineal gland A gland within the brain that secretes the hormone melatonin and is affected by sensory input from the photoreceptors of the eyes.
pinocytosis Cell drinking; invagination of the cell membrane to form narrow channels that pinch off into vacuoles; it provides cellular intake of extracellular fluid and dissolved molecules.
plasma The fluid portion of the blood. Unlike serum (which lacks fibrinogen), plasma is capable of forming insoluble fibrin threads when in contact with test tubes.
plasma cells Cells derived from B lymphocytes that produce and secrete large amounts of antibodies; they are responsible for humoral immunity.
plasma membrane Another term for the cell membrane.
platelets Disc-shaped structures, 2 to 4 micrometers in diameter, that are derived from bone marrow cells called megakaryocytes. Platelets circulate in the blood and participate (together with fibrin) in forming blood clots.
pneumothorax An abnormal condition in which air enters the intrapleural space, either through an open chest wound or from a tear in the lungs. This can lead to collapse of one lobe of the lungs.
PNS The peripheral nervous system, including nerves and ganglia.
-pod, -podium (Gk.) Foot, leg, extension.
polar body A small daughter cell formed by meiosis that degenerates in the process of oocyte production.
polar molecule A molecule in which the shared electrons are not evenly distributed, so that one side of the molecule is relatively negatively (or positively) charged in comparison with the other side; polar molecules are soluble in polar solvents such as water.
poly- (Gk.) Many.
polycythemia An abnormally high red blood cell count.
polymer A large molecule formed by the combination of smaller subunits, or monomers.
polymorphonuclear leukocyte A granular leukocyte containing a nucleus with a number of lobes connected by thin, cytoplasmic strands; this term includes neutrophils, eosinophils, and basophils.

polypeptide A chain of amino acids connected by covalent bonds called peptide bonds. A very large polypeptide is called a protein.
polysaccharide A carbohydrate formed by covalent bonding of numerous monosaccharides; examples include glycogen and starch.
portal system Two capillary beds in series, where blood from the first is drained by veins into a second capillary bed, which in turn is drained by veins that return blood to the heart. The two major portal systems in the body are the hepatic portal system and the hypothalamo-hypophyseal portal system.
positive feedback Cause-and-effect relationships that result in the amplification of changes. Positive feedback results in avalanchelike effects, as occurs in the formation of a blood clot or in the production of the LH surge by the stimulatory effect of estrogen.
posterior Anatomical term denoting a backside position.
posterior pituitary The part of the pituitary gland that is derived from the brain; it secretes vasopressin (ADH) and oxytocin, produced in the hypothalamus. Also called the neurohypophysis.
prehormone An inactive form of a hormone that is secreted by an endocrine gland. The prehormone is converted within its target cells to the active form of the hormone.
pro- (Gk.) Before, in front of, forward.
process, cell Any thin cytoplasmic extension of a cell, such as the dendrites and axon of a neuron.
producers In terms of ecology, the producers of an ecosystem are members of the plant kingdom. They are called producers because they can produce organic molecules from inorganic ones.
progesterone A steroid hormone secreted by the corpus luteum of the ovaries and by the placenta. Secretion of progesterone during the luteal phase of the menstrual cycle promotes the final maturation of the endometrium.
prolactin A hormone secreted by the anterior pituitary that stimulates milk production by the female mammary glands.
prophylaxis Prevention or protection.
proprioceptor A sensory receptor that provides information about body position and movement; examples include receptors in muscles, tendons, and joints as well as the sense of equilibrium provided by the semicircular canals of the inner ear.
prostaglandins A family of fatty acids containing a cyclic ring which serves numerous autocrine regulatory functions.
protein The class of organic molecules composed of large polpeptides, in which over a hundred amino acids are bonded together by peptide bonds.
proto- (Gk.) First, original.
proton A unit of positive charge in the nucleus of atoms.
protoplasm A general term that includes cytoplasm and nucleoplasm.
pseudo- (Gk.) False.
pseudohermaphrodite An individual who has some of the physical characteristics of both sexes, but lacks functioning gonads of both sexes; a true hermaphrodite has both testes and ovaries.
pseudopods Footlike extensions of the cytoplasm that enable some cells (with amoeboid motion) to move across a substrate; pseudopods also are used to surround food particles in the process of phagocytosis.
puberty The period of time in an individual's life span when secondary sexual characteristics and fertility develop.
pulmonary circulation The part of the vascular system that includes the pulmonary artery and pulmonary veins; it transports blood from the right ventricle of the heart, through the lungs, and back to the left atrium of the heart.
pupil The opening at the center of the iris of the eye.
purkinje fibers Specialized conducting tissue in the ventricles of the heart that carry the impulse from the bundle of His to the myocardium of the ventricles.
pyrogen A fever-producing substance.

Q

QRS complex The part of an electrocardiogram that is produced by depolarization of the ventricles.

R

reabsorption The transport of a substance from the lumen of the renal nephron to the blood.
reducing agent An electron donor in a coupled oxidation-reduction reaction.
refraction The bending of light rays when light passes from a medium of one density to a medium of another density. Refraction of light by the cornea and lens acts to focus the image on the retina of the eye.
refractory period The period of time during which a region of axon or muscle cell membrane cannot be stimulated to produce an action potential.
releasing hormones Polypeptide hormones secreted by neurons in the hypothalamus that travel in the hypothalamo-hypophyseal portal system to the anterior pituitary gland and stimulate the anterior pituitary to secrete specific hormones.
renal Pertaining to the kidneys.
repolarization The reestablishment of the resting membrane potential after depolarization has occurred.
resorption, bone The dissolution of the calcium phosphate crystals of bone by the action of osteoclasts.
respiratory acidosis A lowering of the blood pH below 7.35 due to the accumulation of CO_2 as a result of hypoventilation.
respiratory alkalosis A rise in blood pH above 7.45 due to the excessive elimination of blood CO_2 as a result of hyperventilation.
respiratory distress syndrome Also called hyaline membrane disease; most frequently occurring in premature infants, this syndrome is caused by abnormally high alveolar surface tension as a result of a deficiency in lung surfactant.
resting potential The potential difference across a cell membrane when the cell is in an unstimulated state. The resting potential is always negatively charged on the inside of the membrane compared to the outside.
retina The layer of the eye that contains neurons and photoreceptors (rods and cones).
rhodopsin Visual purple; a pigment in rod cells that undergoes a photochemical dissociation in response to light and, in so doing, stimulates electrical activity in the photoreceptors.
ribosomes Particles of protein and ribosomal RNA that form the organelles responsible for the translation of messenger RNA and protein synthesis.
rickets A condition caused by a deficiency of vitamin D and associated with interference of the normal ossification of bone.
rigor mortis The stiffening of a dead body, due to the depletion of ATP and the production of rigor complexes between actin and myosin in muscles.
RNA Ribonucleic acid; a nucleic acid consisting of the nitrogenous bases adenine, guanine, cytosine, and uracil; the sugar ribose; and phosphate groups. There are three types of RNA found in cytoplasm: messenger RNA (mRNA), transfer RNA (tRNA), and ribosomal RNA (rRNA).
rods One of the two categories of photoreceptors (along with cones) in the retina of the eye; rods are responsible for black-and-white vision under low illumination.

S

saltatory conduction The rapid passage of action potentials from one node of Ranvier to another in myelinated axons.
sarcomere The structural subunit of a myofibril in a striated muscle, equal to the distance between two successive Z lines.
sarcoplasm The cytoplasm of striated muscle cells.
sarcoplasmic reticulum The smooth or agranular endoplasmic reticulum of skeletal muscle cells; it surrounds each myofibril and serves to store Ca^{++} when the muscle is at rest.
Schwann cell A neuroglial cell of the peripheral nervous system that forms sheaths around peripheral nerve fibers. Schwann cells also direct regeneration of peripheral nerve fibers to their target cells.
sclera The tough white outer coat of the eyeball that is continuous anteriorly with the clear cornea.
second messenger A molecule or ion whose concentration within a target cell is increased by the action of a regulator compound (e.g., hormone or neurotransmitter) and which mediates the intracellular effects of that regulatory compound.

secretion, renal The transport of a substance from the blood through the wall of the nephron tubule into the urine.
semicircular canals Three canals of the inner ear involved with the sense of equilibrium.
semilunar valves The valve flaps of the aorta and pulmonary artery at their juncture with the ventricles.
seminal vesicles The paired organs located on the posterior border of the urinary bladder that empty their contents into the vas deferens and thus contribute to the semen.
seminiferous tubules The tubules within the testes that produce spermatozoa by meiotic division of their germinal epithelium.
semipermeable membrane A membrane with pores of a size that permits the passage of solvent and some solute molecules but restricts the passage of other solute molecules.
sensory neuron An afferent neuron that conducts impulses from peripheral sensory organs into the central nervous system.
Sertoli cells Nongerminal, supporting cells in the seminiferous tubules. Sertoli cells envelop spermatids and appear to participate in the transformation of spermatids into spermatozoa.
serum The fluid squeezed out of a clot as it retracts; serum is plasma without fibrinogen (which has been converted to fibrin in clot formation).
sex chromosomes The X and Y chromosomes; the unequal pairs of chromosomes involved in sex determination (which is due to the presence or absence of a Y chromosome). Females lack a Y chromosome and normally have the genotype XX; males have a Y chromosome and normally have the genotype XY.
shock As it relates to the cardiovascular system, this refers to a rapid, uncontrolled fall in blood pressure, which in some cases becomes irreversible and leads to death.
sickle-cell anemia A hereditary, autosomal recessive trait that occurs primarily in people of African ancestry, in which it evolved apparently as a protection (in the carrier state) against malaria. In the homozygous state, hemoglobin S is made instead of hemoglobin A; this leads to the characteristic sickling of red blood cells, hemolytic anemia, and organ damage.
sinoatrial node The sinoatrial, or SA, node is the normal pacemaker region of the heart and is located in the right atrium near the junction of the vena cavae.
sinus A cavity.
skeletal muscle pump The skeletal muscle pump is the effect skeletal muscle contraction has on the flow of blood in veins. As the muscles contract, they squeeze the veins and in this way help move the blood toward the heart.
sliding filament theory The theory that the thick and thin filaments of a myofibril slide past each other, while maintaining their initial length, during muscle contraction.
smooth muscle Nonstriated, spindle-shaped muscle cells with a single nucleus in the center; involuntary muscle in visceral organs that is innervated by autonomic nerve fibers.
sodium/potassium pump An active transport carrier, with ATPase enzymatic activity, that acts to accumulate K^+ within cells and extrude Na^+ from cells, thus maintaining gradients for these ions across the cell membrane.
soma-, somato-, -some (Gk.) Body, unit.
somatic motor neurons Motor neurons in the spinal cord that innervate skeletal muscles. Somatic motor neurons are divided into alpha and gamma motoneurons.
sounds of Korotkoff The sounds heard when blood pressure measurements are taken. These sounds are produced by the turbulent flow of blood through an artery that has been partially constricted by a pressure cuff.
spastic paralysis Paralysis in which the muscles have such a high tone that they remain in a state of contracture. This may be caused by inability to degrade ACh released at the neuromuscular junction (as caused by certain drugs) or by damage to the spinal cord.
spermatogenesis The formation of spermatozoa, including meiosis and maturational processes in the seminiferous tubules.
sphygmo- (Gk.) The pulse.
sphygmomanometer A manometer (pressure transducer) used to measure the blood pressure.
spindle fibers Filaments that extend from the poles of a cell to its equator and attach to chromosomes during the metaphase stage of cell division. Contraction of the spindle fibers pulls the chromosomes to opposite poles of the cell.
steroid A lipid, derived from cholesterol, that has three six-sided carbon rings and one five-sided carbon ring. These form the steroid hormones of the adrenal cortex and gonads.
stretch reflex The monosynaptic reflex whereby stretching a muscle results in a reflex contraction. The knee-jerk reflex is an example of a stretch reflex.
striated muscle Skeletal and cardiac muscle, the cells of which exhibit cross-banding, or striations, due to arrangement of thin and thick filaments into sarcomeres.
stroke volume The amount of blood ejected from each ventricle at each heartbeat.
sub- (L.) Under, below.
substrate In enzymatic reactions, the molecules that combine with the active sites of an enzyme and are converted to products by catalysis of the enzyme.
sulcus A groove or furrow; a depression in the cerebrum that separates folds, or gyri, of the cerebral cortex.
summation In neural physiology, summation refers to the additive effects of graded synaptic potentials. In muscle physiology, summation refers to the additive effects of contractions of different muscle fibers.
super-, supra- (L.) Above, over.
suppressor T cells A subpopulation of T lymphocytes that acts to inhibit the production of antibodies against specific antigens by B lymphocytes.
surfactant In the lungs, a mixture of phospholipids and proteins produced by alveolar cells that reduces the surface tension of the alveoli and prevents them from collapsing.
sym-, syn- (Gk.) With, together.
synapse A region where a nerve fiber comes into close or actual contact with another cell and across which nerve impulses are transmitted either directly or indirectly (via the release of chemical neurotransmitters).
synergistic Pertaining to regulatory processes or molecules (such as hormones) that have complementary or additive effects.
systemic circulation The circulation that carries oxygenated blood from the left ventricle in arteries to the tissue cells and that carries blood depleted in oxygen via veins to the right atrium; the general circulation, as compared to the pulmonary circulation.
systole The phase of contraction in the cardiac cycle. When unmodified, this term refers to contraction of the ventricles; the term *atrial systole* refers to contraction of the atria.

T

tachycardia Excessively rapid heart rate, usually applied to rates in excess of 100 beats per minute. In contrast to an excessively slow heart rate (below 60 beats per minute), which is termed *bradycardia.*
target organ The organ that is specifically affected by the action of a hormone or other regulatory process.
T cell A type of lymphocyte that provides cell-mediated immunity, in contrast to B lymphocytes that provide humoral immunity through the secretion of antibodies. There are three subpopulations of T cells: killer, helper, and suppressor.
telo- (Gk.) An end; complete; final.
telophase The last phase of mitosis.
tendon The dense regular connective tissue that attaches a muscle to the bones of its origin and insertion.
testosterone The major androgenic steroid secreted by the Leydig cells of the testes after puberty.
tetanus A term used to mean either a smooth contraction of a muscle (as opposed to muscle twitching) or a state of maintained contracture of high tension.
thalassemia A group of hemolytic anemias caused by the hereditary inability to produce either the alpha or beta chain of hemoglobin. It is found primarily among Mediterranean people.
thorax The part of the body cavity above the diaphragm; the chest.
threshold The minimum stimulus that just produces a response.
thrombin A protein formed in blood plasma during clotting which enzymatically converts the soluble protein fibrinogen into insoluble fibrin.
thrombocytes Blood platelets; disc-shaped structures in blood that participate in clot formation.
thrombus A blood clot, produced by the formation of fibrin threads around a platelet plug.
thyroxine Also called tetraiodothyronine, or T_4. The major hormone secreted by the thyroid gland, which regulates the basal metabolic rate and stimulates protein synthesis in many organs; a deficiency of this hormone in early childhood produces cretinism.

tinnitus A ringing sound or other noise that is heard but is not related to external sounds.
tolerance, immunological The ability of the immune system to distinguish self from nonself, and to not attack those antigens that are part of one's own tissues.
total minute volume The product of tidal volume (ml per breath) and ventilation rate (breaths per minute).
toxin A poison.
tracts A collection of axons within the central nervous system, forming the white matter of the CNS.
trans- (L.) Across, through.
transamination The production of a new amino acid by transferring the amine group from one amino acid to either pyruvic acid or one of the Krebs cycle acids.
triiodothyronine Abbreviated T_3; a hormone secreted in small amounts by the thyroid; the active hormone in target cells formed from thyroxine.
trophic level The trophic level of an organism in an ecosystem refers to its position in a food chain.
tropomyosin A filamentous protein that attaches to actin in the thin filaments and that acts, together with another protein called troponin, to inhibit and regulate the attachment of myosin cross-bridges to actin.
troponin A protein found in the thin filaments of the sarcomeres of skeletal muscle. A subunit of troponin binds to Ca^{++}, and as a result, causes tropomyosin to change position in the thin filament.
trypsin A protein-digesting enzyme in pancreatic juice that is released into the small intestine.
TSH Abbreviation for thyroid stimulating hormone, also called thyrotropin. This hormone is secreted by the anterior pituitary and stimulates the thyroid gland.
twitch A rapid contraction and relaxation of a muscle fiber or a group of muscle fibers.
tympanic membrane The eardrum; a membrane separating the external from the middle ear that transduces sound waves into movements of the middle ear ossicles.

U

universal donor A person with blood type O, who is able to donate blood to people with other blood types in emergency blood transfusions.
universal recipient A person with blood type AB, who can receive blood of any type in emergency transfusions.
urea The chief nitrogenous waste product of protein catabolism in the urine, formed in the liver from amino acids.
uremia The retention of urea and other products of protein catabolism due to inadequate kidney function.

V

vaccination The clinical induction of active immunity by introducing antigens to the body, so that the immune system becomes sensitized to those antigens. The immune system will mount a secondary response to those antigens upon subsequent exposures.
vagus nerve The tenth cranial nerve, composed of sensory dendrites from visceral organs and preganglionic parasympathetic nerve fibers. The vagus is the major parasympathetic nerve in the body.
vasa-, vaso- (L.) Pertaining to blood vessels.
vasectomy Surgical removal of a portion of the vas (ductus) deferens to induce infertility.
vasoconstriction Narrowing of the lumen of blood vessels due to contraction of the smooth muscles in their walls.
vasodilation Widening of the lumen of blood vessels due to relaxation of the smooth muscles in their walls.
vein A blood vessel that returns blood to the heart.
ventilation Breathing; the process of moving air into and out of the lungs.
vertigo A feeling of movement or loss of equilibrium.
vestibular apparatus The parts of the inner ear, including the semicircular canals, utricle, and saccule, which function to provide a sense of equilibrium.
villi Fingerlike folds of the mucosa of the small intestine.
virulent Pathogenic, or able to cause disease.
vital capacity The maximum amount of air that can be forcibly expired after a maximal inspiration.
vitamin A term for unrelated organic molecules present in foods which are required in small amounts by the body for normal metabolic function. Vitamins are classified as water-soluble and fat-soluble.

W

white matter The portion of the central nervous system composed primarily of myelinated fiber tracts. This forms the region deep to the cerebral cortex in the brain and the outer portion of the spinal cord.

Z

zygote A fertilized ovum.

Credits

Illustrations

Chapter 1

Fig. 1.4: From Kent M. Van De Graaff, *Human Anatomy*, 2d ed. Copyright © 1988 Wm. C. Brown Publishers, Dubuque, Iowa. All Rights Reserved. Reprinted by permission.

Chapter 2

Fig. 2.7: From *Biology: The Science of Life*, 2/e by Robert A. Wallace, et al. Copyright © 1986, 1981 by Scott, Foresman and Company. Reprinted by permission of HarperCollins Publishers. **fig. 2.16:** From Wilson, et al., *Life on Earth*. Copyright © Sinauer Associates, Inc., Sunderland, MA. Reprinted by permission. **fig. 2.19:** From Norman Wessells and Janet Hopson, *Biology*. Copyright © Random House, New York, NY.

Chapter 3

Fig. 3.5: From John W. Hole, Jr., *Human Anatomy and Physiology*, 5th ed. Copyright © 1990 Wm. C. Brown Publishers, Dubuque, Iowa. All Rights Reserved. Reprinted by permission.

Chapter 4

Fig. 4.10*b*: From Leland G. Johnson, *Biology*, 2d ed. Copyright © 1987 Wm. C. Brown Publishers, Dubuque, Iowa. All Rights Reserved. Reprinted by permission.

Chapter 5

Fig. 5.11: From John W. Hole, Jr., *Human Anatomy and Physiology*, 5th ed. Copyright © 1990 Wm. C. Brown Publishers, Dubuque, Iowa. All Rights Reserved. Reprinted by permission.

Chapter 6

Fig. 6.1: From John W. Hole, Jr., *Human Anatomy and Physiology*, 3d ed. Copyright © 1984 Wm. C. Brown Publishers, Dubuque, Iowa. All Rights Reserved. Reprinted by permission. **fig. 6.3:** From Kent M. Van De Graaff, *Human Anatomy*, 2d ed. Copyright © 1988 Wm. C. Brown Publishers, Dubuque, Iowa. All Rights Reserved. Reprinted by permission. **fig. 6.8:** From John W. Hole, Jr., *Human Anatomy and Physiology*, 5th ed. Copyright © 1990 Wm. C. Brown Publishers, Dubuque, Iowa. All Rights Reserved. Reprinted by permission. **fig. 6.22*b*:** From Kent M. Van De Graaff, *Human Anatomy*, 2d ed. Copyright © 1988 Wm. C. Brown Publishers, Dubuque, Iowa. All Rights Reserved. Reprinted by permission. **fig. 6.23:** From John W. Hole, Jr., *Human Anatomy and Physiology*, 5th ed. Copyright © 1990 Wm. C. Brown Publishers, Dubuque, Iowa. All Rights Reserved. Reprinted by permission. **fig. 6.28:** From John W. Hole, Jr., *Human Anatomy and Physiology*, 3d ed. Copyright © 1984 Wm. C. Brown Publishers, Dubuque, Iowa. All Rights Reserved. Reprinted by permission. **fig. 6.33:** From Kent M. Van De Graaff, *Human Anatomy*. Copyright © 1984 Wm. C. Brown Publishers, Dubuque, Iowa. All Rights Reserved. Reprinted by permission.

Chapter 7

Fig. 7.1*a*: From Kent M. Van De Graaff, *Human Anatomy*, 2d ed. Copyright © 1988 Wm. C. Brown Publishers, Dubuque, Iowa. All Rights Reserved. Reprinted by permission. **fig. 7.7:** From Kent M. Van De Graaff, *Human Anatomy*, 2d ed. Copyright © 1988 Wm. C. Brown Publishers, Dubuque, Iowa. All Rights Reserved. Reprinted by permission. **fig. 7.14:** From John W. Hole, Jr., *Human Anatomy and Physiology*, 5th ed. Copyright © 1990 Wm. C. Brown Publishers, Dubuque, Iowa. All Rights Reserved. Reprinted by permission. **fig. 7.21:** From John W. Hole, Jr., *Human Anatomy and Physiology*, 5th ed. Copyright © 1990 Wm. C. Brown Publishers, Dubuque, Iowa. All Rights Reserved. Reprinted by permission. **fig. 7.22:** From John W. Hole, Jr., *Essentials of Human Anatomy and Physiology*, 3d ed. Copyright © 1989 Wm. C. Brown Publishers, Dubuque, Iowa. All Rights Reserved. Reprinted by permission. **fig. 7.25:** From John W. Hole, Jr., *Human Anatomy and Physiology*, 4th ed. Copyright © 1987 Wm. C. Brown Publishers, Dubuque, Iowa. All Rights Reserved. Reprinted by permission.

Chapter 8

Fig. 8.7: From Kent M. Van De Graaff, *Human Anatomy*, 2d ed. Copyright © 1988 Wm. C. Brown Publishers, Dubuque, Iowa. All Rights Reserved. Reprinted by permission. **fig. 8.12:** From D. W. Fawcett, "Endocrinology: Male Reproductive System" in *Handbook of Physiology*. Copyright © 1975 Williams & Wilkins Company, Baltimore, MD. **fig. 8.14*a*:** From John W. Hole, Jr., *Human Anatomy and Physiology*, 5th ed. Copyright © 1990 Wm. C. Brown Publishers, Dubuque, Iowa. All Rights Reserved. Reprinted by permission. **fig. 8.15:** From John W. Hole, Jr., *Human Anatomy and Physiology*, 5th ed. Copyright © 1990 Wm. C. Brown Publishers, Dubuque, Iowa. All Rights Reserved. Reprinted by permission. **fig. 8.19:** From John W. Hole, Jr., *Human Anatomy and Physiology*, 5th ed. Copyright © 1990 Wm. C. Brown Publishers, Dubuque, Iowa. All Rights Reserved. Reprinted by permission. **fig. 8.20:** From Kent M. Van De Graaff, *Human Anatomy*, 2d ed. Copyright © 1988 Wm. C. Brown Publishers, Dubuque, Iowa. All Rights Reserved. Reprinted by permission. **fig. 8.26:** From Kent M. Van De Graaff, *Human Anatomy*, 2d ed. Copyright © 1988 Wm. C. Brown Publishers, Dubuque, Iowa. All Rights Reserved. Reprinted by permission. **fig. 8.27:** From Ian H. Thorneycroft, et al., *American Journal of Obstetrics and Gynecology*, Vol. III:147. Copyright © 1971 Mosby-Year Book, Inc. Reprinted by permission.

Chapter 9

Fig. 9.1: From Kent M. Van De Graaff, *Human Anatomy*, 2d ed. Copyright © 1988 Wm. C. Brown Publishers, Dubuque, Iowa. All Rights Reserved. Reprinted by permission. **fig. 9.5:** From John W. Hole, Jr., *Human Anatomy and Physiology*, 5th ed. Copyright © 1990 Wm. C. Brown Publishers, Dubuque, Iowa. All Rights Reserved. Reprinted by permission. **fig. 9.7:** From Kent M. Van De Graaff, *Human Anatomy*, 2d ed. Copyright © 1988 Wm. C. Brown Publishers, Dubuque, Iowa. All Rights Reserved. Reprinted by permission. **fig. 9.8:** From John W. Hole, Jr., *Human Anatomy and Physiology*, 5th ed. Copyright © 1990 Wm. C. Brown Publishers, Dubuque, Iowa. All Rights Reserved. Reprinted by permission. **fig. 9.11:** From Kent M. Van De Graaff, *Human Anatomy*, 2d ed. Copyright © 1988 Wm. C. Brown Publishers, Dubuque, Iowa. All Rights Reserved. Reprinted by permission. **fig. 9.12:** From Kent M. Van De Graaff, *Human Anatomy*, 2d ed. Copyright © 1988 Wm. C. Brown Publishers, Dubuque, Iowa. All Rights Reserved. Reprinted by permission. **fig. 9.21:** From John W. Hole, Jr., *Human Anatomy and Physiology*, 5th ed. Copyright © 1990 Wm. C. Brown Publishers, Dubuque, Iowa. All Rights Reserved. Reprinted by permission. **fig. 9.23:** From Kent M. Van De Graaff, *Human Anatomy*, 2d ed. Copyright © 1988 Wm. C. Brown Publishers, Dubuque, Iowa. All Rights Reserved. Reprinted by permission. **fig. 9.24:** From Kent M. Van De Graaff, *Human Anatomy*, 2d ed. Copyright © 1988 Wm. C. Brown Publishers, Dubuque, Iowa. All Rights Reserved. Reprinted by permission. **fig. 9.30:** From Kent M. Van De Graaff, *Human Anatomy*, 2d ed. Copyright © 1988 Wm. C. Brown Publishers, Dubuque, Iowa. All Rights Reserved. Reprinted by permission. **fig. 9.31:** From F. K. Shuttleworth, *Monograph of the Society for Research in Child Development*, Vol. 4, No. 3. Copyright © 1939 University of Chicago Press, Chicago, IL.

Chapter 11

Fig. 11.10: From Ayers, *Cell Biology*, 2d ed. Copyright © Willard Grant Press, Boston, MA. **fig. 11.15:** From G. A. Harrison, et al., *Human Biology*, 3d ed. Copyright © Oxford University Press, Oxford, England.

Chapter 12

Fig. 12.2: From John W. Hole, Jr., *Human Anatomy and Physiology*, 5th ed. Copyright © 1990 Wm. C. Brown Publishers, Dubuque, Iowa. All Rights Reserved. Reprinted by permission. **fig. 12.4:** From John W. Hole, Jr., *Human Anatomy and Physiology*, 5th ed. Copyright © 1990 Wm. C. Brown Publishers, Dubuque, Iowa. All Rights Reserved. Reprinted by permission. **fig. 12.5*a*:** From Kent M. Van De Graaff, *Human Anatomy*, 2d ed. Copyright © 1988 Wm. C. Brown Publishers, Dubuque, Iowa. All Rights Reserved. Reprinted by permission. **fig. 12.6:** From Kent M. Van De Graaff, *Human Anatomy*, 2d ed. Copyright © 1988 Wm. C. Brown Publishers, Dubuque, Iowa. All Rights Reserved. Reprinted by permission. **fig. 12.8:** From John W. Hole, Jr., *Human Anatomy and Physiology*, 5th ed. Copyright © 1990 Wm. C. Brown Publishers, Dubuque, Iowa. All Rights Reserved. Reprinted by permission. **fig. 12.16:** From *Cardiology*. Copyright © Med Publishing, Inc., Plainsboro, NJ. **fig. 12.19:** From E. O. Feigal, *Physiology and Biophysics*, Vol. II. Copyright © 1974 W.B. Saunders, Orlando, FL. **fig. 12.26:** From Kent M. Van De Graaff, *Human Anatomy*, 2d ed. Copyright © 1988 Wm. C. Brown Publishers, Dubuque, Iowa. All Rights Reserved. Reprinted by permission. **fig. 12.27:** From John W. Hole, Jr., *Human Anatomy and Physiology*, 5th ed. Copyright © 1990 Wm. C. Brown Publishers, Dubuque, Iowa. All Rights Reserved. Reprinted by permission.

Chapter 13

Fig. 13.8: From Ivan Roitt, *Essential Immunology.* Copyright © 1974 Blackwell Scientific Publications Limited, Oxford, England.

Chapter 14

Fig. 14.2: From Ian F. Tannock, "Biology of Tumor Growth" in *Hospital Practice,* April 1983. Copyright © 1983 H P Publishing Co., subsidiary of Maclean Hunter Ltd., New York, NY. **fig. 14.4:** From C. Everette Koop, "Smoking and Cancer" in *Hospital Practice,* June 1984. Copyright © 1984 H P Publishing Co., subsidiary of Maclean Hunter Ltd., New York, NY. **fig. 14.11:** From Ronald Levy, "Biologicals for Cancer Treatment Monoclonal Antibodies" in *Hospital Practice,* November 15, 1985. Copyright © 1985 H P Publishing Co., subsidiary of Maclean Hunter Ltd., New York, NY.

Chapter 15

Fig. 15.2: From E. R. Weibel, *Morphometry of the Human Lung.* Copyright © 1963 Springer-Verlag, New York, NY. **fig. 15.3a:** From John W. Hole, Jr., *Human Anatomy and Physiology,* 5th ed. Copyright © 1990 Wm. C. Brown Publishers, Dubuque, Iowa. All Rights Reserved. Reprinted by permission. **fig. 15.10:** From John W. Hole, Jr., *Human Anatomy and Physiology,* 4th ed. Copyright © 1987 Wm. C. Brown Publishers, Dubuque, Iowa. All Rights Reserved. Reprinted by permission. **fig. 15.11:** From Kent M. Van De Graaff, *Human Anatomy,* 2d ed. Copyright © 1988 Wm. C. Brown Publishers, Dubuque, Iowa. All Rights Reserved. Reprinted by permission. **fig. 15.16:** From Kent M. Van De Graaff, *Human Anatomy,* 2d ed. Copyright © 1988 Wm. C. Brown Publishers, Dubuque, Iowa. All Rights Reserved. Reprinted by permission.

Chapter 16

Fig. 16.2: From John W. Hole, Jr., *Human Anatomy and Physiology,* 5th ed. Copyright © 1990 Wm. C. Brown Publishers, Dubuque, Iowa. All Rights Reserved. Reprinted by permission. **fig. 16.6:** From John W. Hole, Jr., *Human Anatomy and Physiology,* 5th ed. Copyright © 1990 Wm. C. Brown Publishers, Dubuque, Iowa. All Rights Reserved. Reprinted by permission. **fig. 16.8:** From John W. Hole, Jr., *Human Anatomy and Physiology,* 5th ed. Copyright © 1990 Wm. C. Brown Publishers, Dubuque, Iowa. All Rights Reserved. Reprinted by permission. **fig. 16.17:** From Kent M. Van De Graaff, *Human Anatomy,* 2d ed. Copyright © 1988 Wm. C. Brown Publishers, Dubuque, Iowa. All Rights Reserved. Reprinted by permission.

Chapter 17

Fig. 17.1: From John W. Hole, Jr., *Human Anatomy and Physiology,* 5th ed. Copyright © 1990 Wm. C. Brown Publishers, Dubuque, Iowa. All Rights Reserved. Reprinted by permission. **fig. 17.2:** From Kent M. Van De Graaff, *Human Anatomy,* 2d ed. Copyright © 1988 Wm. C. Brown Publishers, Dubuque, Iowa. All Rights Reserved. Reprinted by permission. **fig. 17.4*a*:** From John W. Hole, Jr., *Human Anatomy and Physiology,* 5th ed. Copyright © 1990 Wm. C. Brown Publishers, Dubuque, Iowa. All Rights Reserved. Reprinted by permission. **fig. 17.5:** From Kent M. Van De Graaff, *Human Anatomy,* 2d ed. Copyright © 1988 Wm. C. Brown Publishers, Dubuque, Iowa. All Rights Reserved. Reprinted by permission. **fig. 17.8:** From John W. Hole, Jr., *Human Anatomy and Physiology,* 5th ed. Copyright © 1990 Wm. C. Brown Publishers, Dubuque, Iowa. All Rights Reserved. Reprinted by permission. **fig. 17.10:** From Kent M. Van De Graaff, *Human Anatomy,* 2d ed. Copyright © 1988 Wm. C. Brown Publishers, Dubuque, Iowa. All Rights Reserved. Reprinted by permission.

Chapter 18

Fig. 18.17: From *Diabetes,* 17:27–32. Copyright © 1968 American Diabetes Association, Inc. Reproduced with permission of the American Diabetes Association, Inc. **fig. 18.19:** From *Diabetes,* 17:27–32. Copyright © 1968 American Diabetes Association, Inc. Reproduced with permission of the American Diabetes Association, Inc. **fig. 18.21:** From E. J. Barrett and R. A. De Fronao, *Diabetic Ketoacidosis: Diagnosis and Management,* Vol. 19, Issue 4, April 1984. Copyright © 1984 H P Publishing Co., subsidiary of Maclean Hunter Ltd., New York, NY.

Chapter 19

Fig. 19.1: From L. Anderson, et al., *Nutrition in Health and Disease,* 17th ed. Copyright © 1982 J. B. Lippincott Company, Philadelphia, PA.

Chapter 20

Fig. 20.3: From John W. Hole, Jr., *Human Anatomy and Physiology,* 5th ed. Copyright © 1990 Wm. C. Brown Publishers, Dubuque, Iowa. All Rights Reserved. Reprinted by permission. **fig. 20.4:** From John W. Hole, Jr., *Human Anatomy and Physiology,* 5th ed. Copyright © 1990 Wm. C. Brown Publishers, Dubuque, Iowa. All Rights Reserved. Reprinted by permission. **fig. 20.5:** From John W. Hole, Jr., *Human Anatomy and Physiology,* 5th ed. Copyright © 1990 Wm. C. Brown Publishers, Dubuque, Iowa. All Rights Reserved. Reprinted by permission. **fig. 20.8:** From John W. Hole, Jr., *Human Anatomy and Physiology,* 5th ed. Copyright © 1990 Wm. C. Brown Publishers, Dubuque, Iowa. All Rights Reserved. Reprinted by permission. **fig. 20.13:** From Kent M. Van De Graaff, *Human Anatomy,* 2d ed. Copyright © 1988 Wm. C. Brown Publishers, Dubuque, Iowa. All Rights Reserved. Reprinted by permission. **fig. 20.20:** From Kent M. Van De Graaff, *Human Anatomy,* 2d ed. Copyright © 1988 Wm. C. Brown Publishers, Dubuque, Iowa. All Rights Reserved. Reprinted by permission. **fig. 20.21a (*left*):** From John W. Hole, Jr., *Human Anatomy and Physiology,* 5th ed. Copyright © 1990 Wm. C. Brown Publishers, Dubuque, Iowa. All Rights Reserved. Reprinted by permission.

Chapter 21

Fig. 21.1: From Kent M. Van De Graaff, *Human Anatomy,* 2d ed. Copyright © 1988 Wm. C. Brown Publishers, Dubuque, Iowa. All Rights Reserved. Reprinted by permission. **fig. 21.2*a*:** From John W. Hole, Jr., *Human Anatomy and Physiology,* 5th ed. Copyright © 1990 Wm. C. Brown Publishers, Dubuque, Iowa. All Rights Reserved. Reprinted by permission. **fig. 21.2*c*:** From Kent M. Van De Graaff, *Human Anatomy,* 2d ed. Copyright © 1988 Wm. C. Brown Publishers, Dubuque, Iowa. All Rights Reserved. Reprinted by permission. **fig. 21.6:** From John W. Hole, Jr., *Human Anatomy and Physiology,* 5th ed. Copyright © 1990 Wm. C. Brown Publishers, Dubuque, Iowa. All Rights Reserved. Reprinted by permission. **fig. 21.9:** From Kent M. Van De Graaff, *Human Anatomy,* 2d ed. Copyright © 1988 Wm. C. Brown Publishers, Dubuque, Iowa. All Rights Reserved. Reprinted by permission. **fig. 21.11:** From John W. Hole, Jr., *Human Anatomy and Physiology,* 5th ed. Copyright © 1990 Wm. C. Brown Publishers, Dubuque, Iowa. All Rights Reserved. Reprinted by permission. **fig. 21.17:** From Morkin, *Hospital Practice,* 18:6. Copyright © H P Publishing Co., subsidiary of Maclean Hunter Ltd., New York, NY.

Chapter 22

p. 478: From T. S. Eliot. Copyright © Harcourt, Brace & World, Inc., Orlando, FL. **fig. 22.5:** Figure from *Ecology and Field Biology,* 3/ed. by Robert Leo Smith. Copyright © 1980 by Robert Leo Smith. Reprinted by permission of HarperCollins Publishers. **fig. 22.7:** From Curtis: *Biology,* 2/e, 1975. Worth Publishers, New York. Reprinted by permission. **fig. 22.13:** From *Ecoscience.* By P. R. Ehrlich. Copyright © 1977 W. H. Freeman and Company. **fig. 22.18:** Reprinted by permission of *American Scientist,* journal of Sigma Xi, The Scientific Research Society. **fig. 22.21:** Copyright © Union of Concerned Scientists. Reprinted by permission. **p. 496:** From T. S. Eliot. Copyright © Harcourt, Brace & World, Inc., Orlando, FL.

Photos

Table of Contents

Page ix *left:* © Thomas Porett/Photo Researchers, Inc.; **p. ix *middle:*** © Manfred Kage/Peter Arnold, Inc.; **p. ix *right:*** © David Scharf/Peter Arnold, Inc.; **p. x *left:*** © J. & L. Weber/Peter Arnold, Inc.; **p. x *middle:*** © Leonard Lessin/Peter Arnold, Inc.; **p. x *right:*** © Francis Leroy, Biocosmos/Science Photo Library/Photo Researchers, Inc.; **p. xi *left:*** © Junebug Clark/Photo Researchers, Inc.; **p. xi *middle:*** © Jean-Luc Tabuteau/The Image Works; **p. xi *right:*** © Alan Carey/The Image Works; **p. xii *left:*** © Tom McCarthy/Unicorn Stock Photos; **p. xii *middle:*** © Cecil Fox/Photo Researchers, Inc.; **p. xii *right:*** © Steve Allen/Peter Arnold, Inc.; **p. xiii *left:*** © Walker/Photo Researchers, Inc.; **p. xiii *middle:*** © Cecil Fox/Science Source/Photo Researchers, Inc.; **p. xiii *right:*** © Mike Greenlar/The Image Works; **p. xiv *left:*** © Jack Spratt/The Image Works; **p. xiv *middle:*** © Bob Daemmrich/The Image Works; **p. xiv *right:*** © Tom Hollyman/Photo Researchers, Inc.

Chapter 1

Opener: © Thomas Porett/Photo Researchers, Inc.; **1.2:** © Historical Pictures Service, Inc.; **1.5:** © Dr. Sheril D. Burton; **1.6*a*:** © M. P. L. Fogden/Bruce Coleman, Inc.; **1.6*b*:** © Rod Williams/Bruce Coleman, Inc.; **1.6*c*:** © Bob Burch/Bruce Coleman, Inc.; **1.6*d*:** © C. B. & D. W. Frith/Bruce Coleman, Inc.; **1.6*e*:** © Jen & Des Bartlett/Bruce Coleman, Inc.

Chapter 2

Opener: © Tom McHugh/Photo Researchers, Inc.; **2.1:** © Historical Pictures Service, Inc.; **2.2:** © John Moss/Black Star; **2.10:** © K. W. Fink/Bruce Coleman, Inc.; **2.10** (inset photo): © Ted Schiffman/Peter Arnold, Inc.; **2.12 both:** © M. W. F. Tweedie/Bruce Coleman, Inc.; **2.14:** By John Amos © National Geographic Society; **2.20:** © 1985 David Brill; **2.21:** *Human Origins,* SCIENCE 245, 1989, p. 1343 by E. L. Simon. Photo courtesy of Peter Schmidt, PhD.; **2.22:** © John Reader/Photo Researchers, Inc.; **2.24:** Courtesy of the French Government Tourist Office

Chapter 3

Opener: © Manfred Kage/Peter Arnold, Inc.; **3.22:** © Edwin Reschke

Chapter 4

Opener: © David Scharf/Peter Arnold, Inc.; **4.3 *bottom:*** © Edwin Reschke; **4.4*a, b*:** © Kwang W. Jeon; **4.5 all:** © R. H. Albertin, M. A.; **4.7:** © Sandra L. Wolin; **4.9:** © Richard Chao; **4.10*a*:** © K. R. Porter; **4.11*a*:** © K. R. Porter; **4.12:** © E. G. Pollack; **4.13, 4.20:** © Richard Chao

Chapter 5

Opener: © Petit Format/Quigoz-Edelmann/Science Source/Photo Researchers, Inc.; **5.1; 5.2; 5.3; 5.4*a*; 5.6; 5.9; 5.10:** © Edwin Reschke

Chapter 6

Opener: © J. & L. Weber/Peter Arnold, Inc.; **6.2:** © Dr. John D. Cunningham/Visuals Unlimited; **6.6:** H. Webster, from Hubbard, John *The Vertebrate Peripheral Nervous System* Plenum Press, 1974; **6.9:** © Andreas Karschin, Heinz Wassle, and Jutta Schnitzer/Scientific American; **6.13:** Bell et al: *Textbook of Physiology and Biochemistry* 10th edition, © Churchill Livingston, Inc.; **6.16:** © John Heuser, Washington University, School of Medicine, St. Louis, MO

Chapter 7

Opener: © Leonard Lessin/Peter Arnold, Inc.; **7.1*b*:** © Edwin Reschke; **7.8*a–d*:** *American Journal of Medicine* 20 (1956) 133; **7.16*b*:** © Journalism Services; **7.17:** © Martin Rotker/Taurus Photos; **7.18; 7.20:** © Lester Bergman and Associates; **7.24:** © Edwin Reschke

Chapter 8

Opener: © Francis Leroy, Biocosmos/Science Source/Photo Researchers Inc.; **8.2:** Wisniewski, L. P. and Hirshhorn, K. *Guide to Human Chromosome Defects* 2/e, White Plains: March of Dimes Birth Defects Foundation, 1980; **8.3*a*:** © David M. Phillips; **8.4*a*2; 8.4*b*2; 8.4*c*1; 8.4*d*1; 8.4*e*; 8.21*a*:** © Edwin Reschke; **8.22*a*:** Blandau, R. J. *A Textbook of Histology* 10th ed., © W. B. Saunders Co., 1975; **8.11*a*:** Biophoto Assoc./Photo Researchers, Inc. **8.22*b*:** © Landrum Shettles; **8.23:** Bloom, W. and Fawcett, D. W.; *A Textbook of Histology* 10th ed., © W. B. Saunders Co., 1975; **8.24:** © Landrum Shettles; **8.29:** © Martin Rotker/Taurus Photos; **8.31*a*:** © SIU/Peter Arnold, Inc.; **8.31*b–f*:** © Bob Coyle

Chapter 9

Opener: © Junebug Clark/Photo Researchers, Inc.; **9.1*c*:** © David Phillips/Visuals Unlimited; **9.2:** Lucian Zamboni from Greep, Roy and Weiss, Leon; *Histology* 3/ed, © McGraw Hill Book Co. 1973; **9.6*a*; 9.6*b*; 9.6*c*; 9.6*d*:** © R. G. Edwards; **9.7*a*; 9.7*b*:** © Landrum Shettles; **9.13*a*:** © Donald Yaeger/Camera MD Studios; **9.13*b*:** © Landrum Shettles; **9.13*c*; 9.19*a*; 9.19*b*:** © Donald Yaeger/Camera MD Studios; **9.23 *right*; 9.23 *left*:** © Dr. Landrum Shettles; **9.25:** © Gregory Dellore, M.D. and Steven L. Clark, M.D.; **9.33:** © Carolina Biological Supply Co., Burlington, NC

Chapter 10

Opener: © Jean-Luc Tabuteau/The Image Works, Inc.; **10.1:** © Richard Chao; **10.7*a*:** O. L. Miller, Jr., *Journal of Cell Physiology* 74 (1969); **10.10:** © Alexander Rich; **10.15*a*:** © Daniel S. Friend; **10.20:** SCIENCE, Vol. 240, pp. 1616–1618, 17 June 1988, *DNA Fingerprinting Takes the Witness Stand* Marx, J. Photo by Lifecodes Corporation;

10.22: © Andrew Hiatt. Reprinted by permission from *NATURE* Vol. 342, p. 76, 1989, Macmillan Magazines, Ltd.

Chapter 11

Opener: © Alan Carey/The Image Works, Inc.; **11.4 all:** From *Sickle Cell Disease,* 2nd ed., by Solanki, Phillip L., M.D. and McCurdy, Paul R., M.D., by Medcom, Inc. Reprinted by permission of Medcom, Inc.; **11.5:** © Moravske Museum-Mendelianum; **11.6:** © Historical Pictures Service, Inc.; **11.8:** © Springer/The Bettmann Archive; **11.9:** © The Bettmann Archive; **11.19:** © Dr. Thomas G. Brewster; **11.20:** © M. Coleman/ Visuals Unlimited

Chapter 12

Opener: © Bill Longcore/Science Source/ Photo Researchers, Inc.; **12.12:** © Richard Menard; **12.18*a*; 12.18*b*:** © American Heart Association; **12.28:** © Edwin Reschke/Peter Arnold, Inc.

Chapter 13

Opener: © Tom McCarthy/Unicorn Stock Photos; **13.4:** Computer graphic by Arthur J. Olson, PhD/1985, all rights reserved; Scripps Clinic & Research Foundation, LaJolla, CA; **13.12:** © Dr. Hans Gelderblom; **13.14:** © Noel R. Rose; **13.16 both:** © Custom Medical Stock Photo; **13.17*b*:** © Acarology Lab, Ohio State University

Chapter 14

Opener: © Cecil Fox/Photo Researchers, Inc.; **14.1 both:** © Dr. G. Steven Martin; **14.9 both:** © Andrejs Liepins, Dept. of Biology, Memorial University, St. John's, Newfoundland; **14.10 both:** © Dr. Steven A. Rosenberg et al, The New England Journal of Medicine, Dec. 22, 1988 edition, p. 1676, *A Special Report.* Reprinted by permission

Chapter 15

Opener: © Steve Allen/Peter Arnold, Inc.; **15.1; 15.3*b*; 15.4:** American Lung Association; **15.5 both:** Edward C. Vasquez, RT, CRT, Dept. of Radiological Technology, Los Angeles City College; **15.9 both:** Comroe, J. H.: *Physiology of Respiration* © 1974 Yearbook Medical Publishers, Inc.; **15.12 both:** © Martin Rotker/Peter Arnold, Inc.

Chapter 16

Opener: © Walker/Photo Researchers, Inc.; **16.3:** © CNRI/Science Photo Library/Photo Researchers, Inc.; **16.7:** © Daniel Friend from Bloom, William, and Fawcett, Donald, *Textbook of Histology* 10/e, © W. B. Saunders, Co.; **16.18 all:** © Abbott Laboratories

Chapter 17

Opener: © Cecil Fox/Science Source/Photo Researchers, Inc.; **17.4*b*; 17.6:** Courtesy of Utah Valley Hospital, Dept. of Radiology; **17.7:** © Edwin Reschke; **17.11:** © Manfred Kage/Peter Arnold, Inc.; **17.14 *a & b*:** © Keith Porter; **17.14*c*:** Alpers, D. H. and Seetharan, B.: *New England Journal of Medicine* 296 (1977) 1047; **17.17:** Sodeman, W. A. and Watson, T. M.: *Pathologic Physiology,* 6/e, W. B. Saunders, Co. © 1969; **17.18:** After Sabota, from Bloom, Ward and Fawcett, Donald: *A Textbook of Histology* 10/e, © W. B. Saunders, Co. 1975; **17.22*a*:** © Carroll Weiss/RBD; **17.22*b*:** © Dr. Sheril Burton

Chapter 18

Opener: © John Griffin/The Image Works, Inc.; **18.23:** © Lester V. Bergman and Associates; **18.24:** *American Journal of Medicine* 20, (1956), 133

Chapter 19

Opener: © Mike Greenlar/The Image Works, Inc.; **19.2 both; 19.3:** World Health Organization; **19.6:** © E. W. Lovrien

Chapter 20

Opener: © Jack Spratt/The Image Works, Inc.; **20.15:** © Thomas Sims; **20.17*b*:** © P. N. Farnsworth, University of Medicine and Dentistry, New Jersey Medical School; **20.21:** © F. Werblin

Chapter 21

Opener: © Bob Daemmrich/The Image Works, Inc.; **21.2*b*:** © Edwin Reschke; **21.5:** Courtesy Utah Valley Hospital, Dept. of Radiology; **21.12:** © John D. Cunningham/ Visuals Unlimited; **21.13:** H. E. Huxley: *The Structure and Function of Muscle* Vol. 1, 2/e, 1972, © Academic Press, Inc.; **21.14 *a & b*; 21.15:** © H. E. Huxley; **21.18*b*:** © John D. Cunningham/Visuals Unlimited

Chapter 22

Opener: © Tom Hollyman/Photo Researchers, Inc.; **22.8:** © G. J. James/BPS; **22.9:** © Scott Dine; **22.10:** © AP/Wide World Photos, Inc.; **22.16:** © Ray Fortner/Peter Arnold, Inc.; **22.17:** NASA

Illustrators

Bowring Cartographics

11.15, 22.7

Sam Collins

14.3

Illustrious, Inc.

2.9, 2.13, 2.19, 4.6, 10.17, 10.18, 10.19, 11.18, 13.6, 13.10, 13.11, 13.13, 14.2, 14.6, 14.7, 14.8, 14.9, 14.12, 15.19, 19.1, 22.1

Teryl Lynn

2.3, 2.6, 2.8, 2.11, 2.23, 22.3, 22.4, 22.5, 22.6, 22.13, 22.19

Felecia J. Paras

2.15, 2.17, 2.18, 12.5*b*

Index

A

D

E

F

Q

R

S